MW01493208

Earth from Space

This image from NASA's Suomi-NPP "Marble" series is centered on the Eastern Hemisphere. The swirling cloud patterns and storm systems reflect not only the distribution of moisture in the atmosphere, but also the transfer of vast amounts of energy among regions of varying temperature and pressure.

Views of our planet from space can help us grasp how the atmosphere, oceans, land, and life itself are all interconnected.

This image shows a region including Africa, the Middle East, and Europe. It is a composite of nighttime images assembled from data acquired by the same Suomi NPP satellite as was used for daylight image, but this view is based on instrumentation that observes light emanating from the ground. Note how strongly major cities show up in the image, as well as flares in the Middle East from the burning of natural gas byproducts from petroleum mining activities.

forecasting
student success

with

Understanding
Weather and Climate

real world applications

The seventh edition of *Understanding Weather and Climate* combines student-friendly writing, relevant applications, stunning visualizations, and integrated mobile media for the most comprehensive and dynamic introduction to meteorology.

12-3 FOCUS ON THE ENVIRONMENT AND SOCIETAL IMPACTS

Superstorm Sandy, 2012

In October 2012 Hurricane Sandy—also called Superstorm Sandy—made its mark as one of the most devastating North American weather events in recent years (Figure 12–3–1). First reaching tropical storm status on October 22, it hit Jamaica as a Category 1 hurricane on the 22nd, strengthened further as it migrated northward, and hit eastern Cuba as a Category 3 hurricane on the early morning of October 25.

After departing Cuba, Sandy temporarily lost hurricane status, but it reintensified to a hurricane while passing east of southern Florida. The hurricane then continued along a northward track roughly parallel to the U.S. East Coast. On October 29 it hit the mainland south of New York City at Brigantine, New Jersey, with 130 km/hr (81 mph) maximum sustained winds.

▲ **FIGURE 12–3–1 Damage from Superstorm Sandy.** Six months after the storm, the remains of destroyed houses in Mantoloking, New Jersey were still evident.

▲ **FIGURE 15–2 Sea Ice in Antarctica.** Seasonally varying sea ice extent is one of the important factors in the way oceans influence the climate.

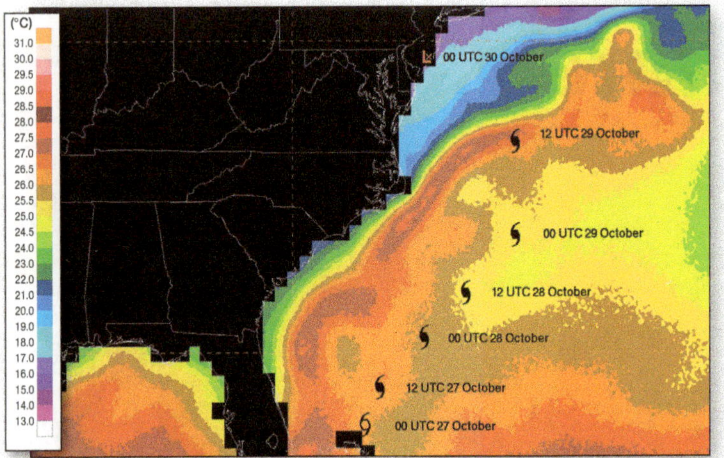

SST's (°C) on 28 Oct 2012 with the best track of Sandy plotted at 12 hour intervals

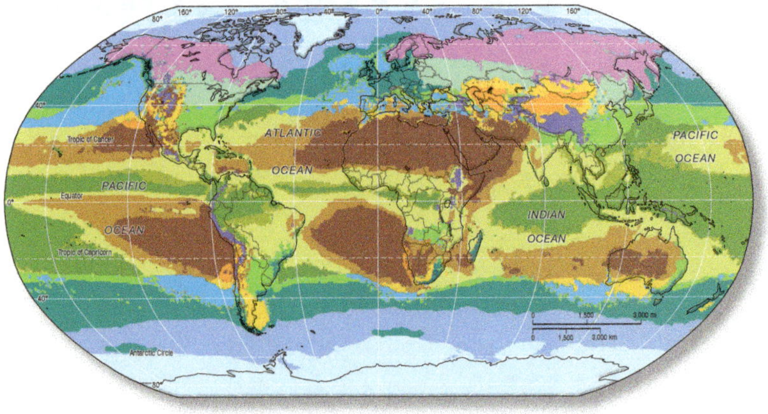

▲ **NEW! *Focus on the Environment and Societal Impacts*** features explore real-world impacts of weather hazards on people and society, illustrating the broad impacts on people and the decision-making that goes into coping with weather events.

▲ **NEW! Coverage of Oceans & Climate** in Chapters 8, 15, and 16 that emphasize how the atmosphere and oceans are interconnected, and how the role of the oceans is key to understanding precipitation patterns, the formation of tropical cyclones, and the impacts of climate change.

the latest science

▼ **UPDATED!** ***Focus on Aviation*** features explore the impacts of various atmospheric phenomena on aviation. Examples include: discussion of winter storms and air travel (Chapter 1); using altimeters to measure altitude (Chapter 4); impacts of icing on aircraft (Chapter 5); recommended pilot responses to icing in different types of clouds (Chapter 6); density altitude and aircraft performance (Chapter 9); lightning and aircraft (Chapter 11); and airport and airline responses to hurricanes (Chapter 12).

▼ **UPDATED!** ***Focus on Severe Weather*** features discuss dramatic and dangerous hazardous weather phenomena, including coverage of many recent events like the deadly 2011 and 2013 tornado seasons and recent deadly cyclones.

9-3 FOCUS ON SEVERE WEATHER

A Cold Front Chills Louisiana

Cold fronts can lead to major changes in weather over short time periods. A cold front that moved across the United States in January 2013 provides an excellent example. Figure 9–3–1 shows the position of an advancing cold front on the morning of January 11 and corresponding observed minimum temperatures. A cold front extends from northwest Mexico to the Dakotas (Figure 9–3–1a) and continues northeastward as a stationary front. Behind the front, daily minimum temperatures were low, especially across Montana and northern Wyoming (Figure 9–3–1b).

By the next morning the cold front had advanced well to the southeast and extended from North Texas to the western Great Lakes (Figure 9–3–2a). As expected, the central part of the country experienced a substantial drop in temperatures and extreme cold covered much of the western Great Plains (Figure 9–3–2b). Twenty-four hours later, the cold front continued its eastward migration but at a slower rate than it had been moving previously (Figure 9–3–3a), and the map of minimum temperatures (Figure 9–3–3b) reflects this movement. Figure 9–3–4 shows the position of the front and the corresponding minimum temperatures on January 14.

Figure 9–3–5 shows three 24-hour plots of temperature illustrating how temperatures and dew points can change with the

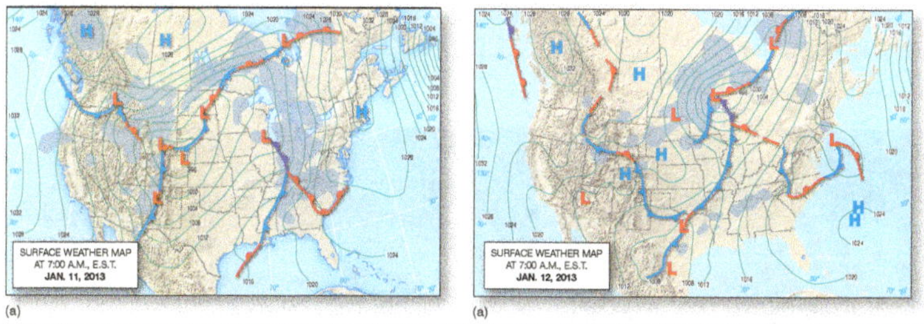

11-5 FOCUS ON AVIATION

Microbursts

Microbursts can pose a serious threat to aircraft, especially during takeoffs and landings (Figure 11–5–1). The horizontal spreading of a microburst creates strong wind shear when it reaches the surface. For example, air may flow westward on one side of the microburst while spreading eastward on the opposite side. Imagine what this might do to an aircraft attempting to land in a microburst. As the plane enters the microburst, a headwind provides lift, to which the pilot might respond by turning the aircraft downward. As soon as the plane passes the core of the downdraft, however, the headwind not only disappears, it is replaced by a tailwind, decreasing lift. Coming after the pilot's earlier downward adjustment, this causes the plane to abruptly drop in altitude. Because the plane is not far above the ground when these events occur, the pilot may not have time to compensate before a deadly crash occurs.

Fortunately, such disasters are rare. They are also becoming less likely because the installation of Doppler radar at about 40 U.S. airports has proven highly effective at detecting microbursts, with a detection rate of about 95 percent.

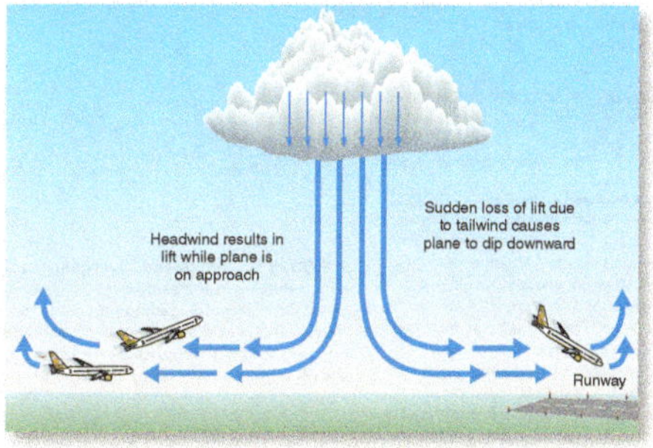

▲ **FIGURE 11–5–1** Microbursts can make aircraft landing and takeoff perilous. A plane flying into the headwinds of a microburst gets a sudden increase in lift. This lift suddenly disappears and is replaced by a tailwind as it exits the downdraft, thereby reducing the lift. If the pilot overcompensates and guides the plane downward while entering the downdraft, a dangerous drop in altitude may occur. Notice the curl at the ends of the downdrafts, which mark the outer limit of the microburst at the ground.

11–5–1 Explain how microbursts create a threat to aircraft.

11–5–2 Describe the efforts that have been undertaken to reduce aircraft vulnerability to microbursts.

▶ **UPDATED!** Coverage of **climate change** impacts and projections integrated throughout, including the latest findings of the IPCC's 5th Assessment Report.

▲ **FIGURE 16–5 Evidence from Tree Rings.** The blue dots on the middle panel correspond to each tenth-year.

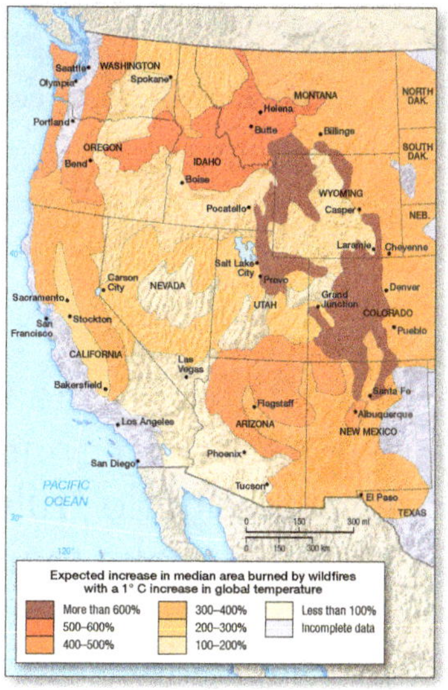

▲ **FIGURE 16–26 Projected Changes in Wildfires.** The map shows the expected percentage increase in median area burned by wildfires with a 1 °C increase in global temperature.

principles & tools of meteorology

11–4 FORECASTING

Doppler Radar

Just as we are able to distinguish different colors of light by their wavelengths, so can we differentiate sounds by the length of their sound waves. If an object making a sound is moving away from a listener, the sound waves are stretched out and assume a lower pitch. Sound waves are compressed when an object moves toward the listener, making them higher pitched. Unconsciously, we use this principle, called the **Doppler effect**, to determine whether an ambulance siren is coming closer or moving away. If the pitch of the siren seems to become higher, we know the ambulance is getting nearer (of course, the siren would also sound louder). A similar process occurs when electromagnetic waves are reflected by a moving object: The light shifts to shorter wavelengths when reflected by an object moving toward the receiver and to longer wavelengths as it bounces off an object moving away from the receiver.

Applying the Doppler Effect

Doppler radar is a type of radar system that takes advantage of this principle. It allows the user to observe the movement of raindrops and ice particles (and thus determine wind speed and direction) from the shift in wavelength of the radar waves, as well as the intensity of precipitation. Like any other type of radar, Doppler radar has a transmitter that emits pulses of electromagnetic energy with wavelengths on the order of several centimeters. Depending on the wavelength used, water droplets and snow crystals above certain critical sizes reflect a portion of the radar's electromagnetic energy back to the transmitter/receiver. In the case of particularly violent tornadoes that pick up large objects from the ground, the radar will observe this airborne material and display it as a *debris ball*.

Doppler radar is special in its ability to observe the motion of the cloud constituents. If a cloud droplet is moving away from the radar unit, the wavelength of the beam is slightly elongated as it bounces off the reflector. Such reflections are normally indicated on the display monitor as reddish to yellow. Likewise, a droplet moving toward the radar unit undergoes a shortening of the wavelength. Echoes from these constituents are displayed as blue or green on the radar screen.

Radar Scans

A radar unit must rotate 360 degrees to get a complete picture of the weather situation surrounding the transmitter/receiver unit. When the transmitter makes one complete rotation at a fixed angle, it is said to have completed a **sweep**. The angle can then be increased as a second sweep is taken that depicts a higher cloud level. This can be repeated several times so that the radar can peer into multiple levels of the cloud. The compilation of all the individual sweeps takes approximately 5 to 10 minutes and produces a **volume sweep**.

Figure 11–4–1 shows a pair of Doppler radar images of a major storm near Dallas–Fort Worth, Texas, on March 29, 2000. Figure 11–4–1a shows the reflectivity of the storm, with redder regions indicating intense precipitation and green areas representing less intense precipitation. The white arrows point toward a hook echo (described in the main text of this chapter). Figure 11–4–1b displays the *storm radial velocity (SRV) pattern*, which describes the motions taking place within the cloud. SRV displays use redder colors to represent winds blowing away

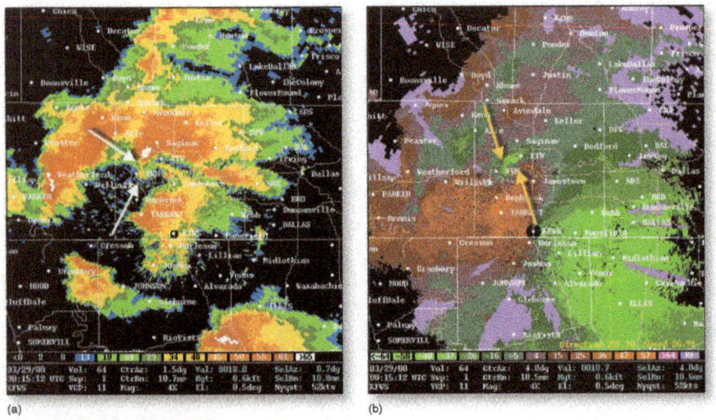

Video Identifying Tornadic Thunderstorms Using Radar Velocity Data
http://goo.gl/szdHJQ

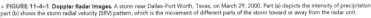

▲ **FIGURE 11–4–1 Doppler Radar Images.** A storm near Dallas–Fort Worth, Texas, on March 29, 2000. Part (a) depicts the intensity of precipitation; part (b) shows the storm radial velocity (SRV) pattern, which is the movement of different parts of the storm toward or away from the radar unit.

10–3 PHYSICAL PRINCIPLES

A Closer Look at Divergence and Convergence

Upper-level divergence and convergence are changes in the horizontal area occupied by an air parcel and result from changes in vorticity as air flows. This relationship between divergence and vorticity can be summarized in the simple equation

$$-\frac{1}{\zeta}\frac{\Delta\zeta}{\Delta t} = div \quad \text{vorticity and divergence}$$

where $-\frac{1}{\zeta}\frac{\Delta\zeta}{\Delta t}$ is the standardized change (here, the decrease) in absolute vorticity (ξ) with respect to time (t), and *div* = divergence. If absolute vorticity increases, convergence must result. If absolute vorticity decreases, divergence must result.

Divergence and convergence can occur in two ways. The first is by an increase or a decrease in the speed of air as it flows. The second is by a stretching out or pinching inward of the air, in a direction perpendicular to the direction in which it is moving. The divergence and convergence described earlier in this chapter can take either form.

Speed Divergence and Speed Convergence

Speed divergence and **speed convergence** occur when air moving in a constant direction either speeds up or slows down. Consider the two parcels of air, *A* and *B*, in Figure 10–3–1a. Both parcels are moving in the same direction, but parcel *B* moves faster, as indicated by the length of the arrows. Because the leading parcel has greater speed than the one behind it, the distance between the two increases with time Figure 10–3–1b. This is an example of speed divergence.

This form of divergence is analogous to what might happen in a race with many entrants at the starting line. Initially, the runners cluster together, with little space between them. When the starting gun goes off, the people at the front of the

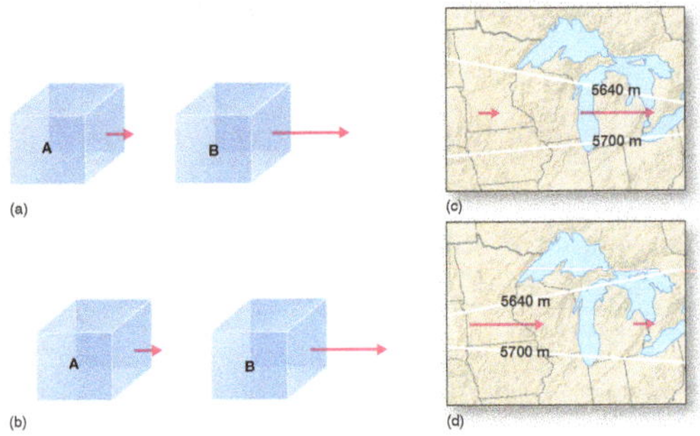

▲ **FIGURE 10–3–1 Speed Divergence and Convergence.** (a) Two hypothetical parcels of air are moving in the same direction, with the one in front moving faster. (b) At some interval of time later, the leading parcel has moved even farther ahead, creating speed divergence. This is also illustrated in (c), with the tighter spacing of height contours to the east creating speed divergence. Speed convergence is occurring in (d). Note that the values shown on the lines in (c) and (d) represent the height of the 500 mb level in meters.

▲ UPDATED! ***Forecasting*** features apply the topics of the chapter to forecasting principles and often include simple "rules of thumb" that help students make their own forecasts. This text contains numerous examples of how physical principles are employed in weather forecasting.

▲ UPDATED! **Physical Principles** complement the main narrative by delving deeper into qualitative topics. More mathematical in nature than the rest of the text, these boxes accommodate students who have a more quantitative interest in the topic.

structured learning

Understanding Weather and Climate provides an active structured learning path to help guide students towards mastery of key meteorological concepts.

◀ **UPDATED!** ***Learning Outcomes*** listed at the beginning of the chapter and now integrated within chapter sections help students prioritize key concepts and skills.

◀ **NEW!** ***Word Clouds*** at the start of each chapter emphasize the key topics and concepts of the chapter.

Visual Analysis

This satellite image shows a large, powerful storm over the North Atlantic Ocean on March 26, 2014.

4.1. The wispy faint clouds blowing from west to east in the top part of the image are far above the friction layer. Assuming gradient flow, draw lines showing the orientation of height contours.

4.2. At lower levels do clouds appear to spiral into or out of the storm's center? Is this a cyclone or anticyclone?

▲ **Superstorm of March 2014.** After crossing the continental U.S., the storm intensified over the Atlantic Ocean.

▲ **NEW!** ***Visual Analysis Activities*** at the end of chapters draw on visualizations of real-world meteorology phenomena and data, asking students to make observations and predictions, and to demonstrate critical thinking, image interpretation, and data analysis skills.

CHECKPOINT

15.1 What is climate?

15.2 What considerations enter the decision regarding the length of period to use in a climatic average?

15.3 What are some of the factors that determine the climate of a location?

15.4 What are some of the problems encountered when trying to divide the Earth into climatic zones?

▲ **UPDATED!** ***Checkpoint Questions*** are integrated throughout the chapters after major sections, giving students a chance to stop, practice, and apply their understanding of key chapter content.

BEFORE-CLASS LEARNING

Mobile Media & Reading Assignments Ensure Students Come to Class Prepared

▼ NEW! *Mobile-Enabled Quick Response (QR)* codes integrated throughout each chapter empower students to use their mobile devices to learn as they read, providing instant access to over 120 Videos and Animations of real-world atmospheric phenomena and visualizations of key physical processes.

Video
Hot Towers and Hurricane Intensification

http://goo.gl/jJmpo

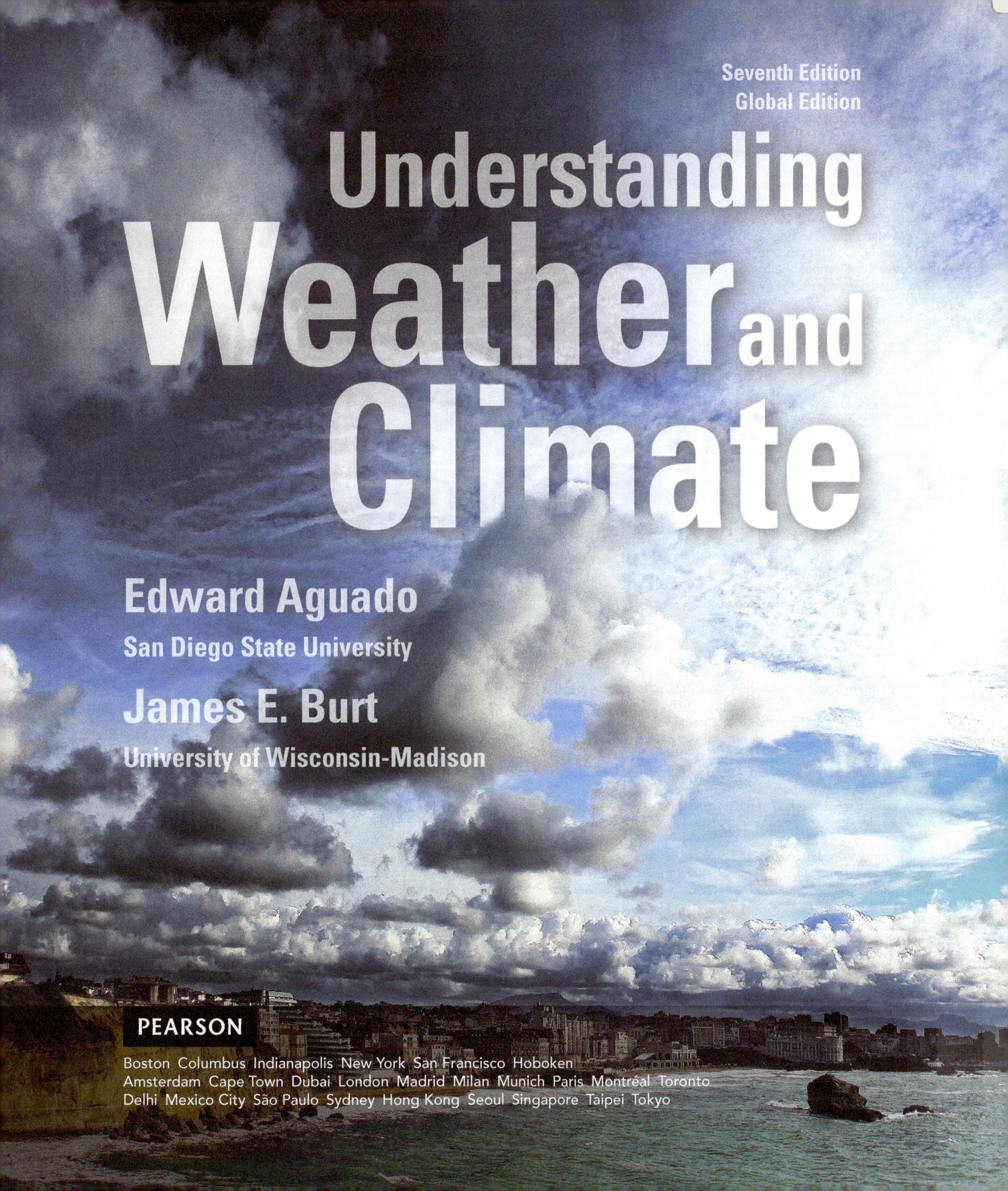

Seventh Edition
Global Edition

Understanding Weather and Climate

Edward Aguado
San Diego State University

James E. Burt
University of Wisconsin-Madison

PEARSON

Boston Columbus Indianapolis New York San Francisco Hoboken
Amsterdam Cape Town Dubai London Madrid Milan Munich Paris Montréal Toronto
Delhi Mexico City São Paulo Sydney Hong Kong Seoul Singapore Taipei Tokyo

Senior Geography & Meteorology Editor: Christian Botting
Head of Learning Asset Acquisition, Global Edition: Laura Dent
Executive Marketing Manager: Neena Bali
Program Manager: Anton Yakovlev
Project Manager: David Zielonka
Project Manager: Sean Hale
Director of Development: Jennifer Hart
Executive Development Editor: Jonathan Cheney
Team Lead, Geosciences, Environmental Science, Chemistry: David Zielonka
Media Producer: Ziki Dekel
Editorial Assistant: Amy De Genaro
Assistant Acquisitions Editor, Global Edition: Murchana Borthakur
Associate Project Editor, Global Edition: Binita Roy
Marketing Assistant: Ami Sampat
Senior Manufacturing Controller, Production, Global Edition: Trudy Kimber
Illustrations: International Mapping Associates
Art Studio: Precision Graphics
Image Lead: Maya Melenchuk
Photo Researcher: Lauren McFalls
Rights and Permissions Manager: Rachel Youdelman
Associate Project Manager, Rights and Permissions: William Opaluch
Design Director: Mark Ong
Cover Image: © Francois Roux/Shutterstock
Cover Designer: Lumina Datamatics
Interior Designer: Wanda Espanda
Operations Specialist: Maura Zaldivar

Credits and acknowledgments borrowed from other sources and reproduced, with permission, in this textbook appear on pages 581–583.

Pearson Education Limited
Edinburgh Gate
Harlow
Essex CM20 2JE
England

and Associated Companies throughout the world

Visit us on the World Wide Web at:
www.pearsonglobaleditions.com

© Pearson Education Limited 2015

The rights of Edward Aguado and James E. Burt to be identified as the authors of this work have been asserted by them in accordance with the Copyright, Designs and Patents Act 1988.

Authorized adaptation from the United States edition, entitled Understanding Weather and Climate, 7th edition, ISBN 978-0-321-98730-3, by Edward Aguado and James E. Burt, published by Pearson Education © 2015.

All rights reserved. No part of this publication may be reproduced, stored in a retrieval system, or transmitted in any form or by any means, electronic, mechanical, photocopying, recording or otherwise, without either the prior written permission of the publisher or a license permitting restricted copying in the United Kingdom issued by the Copyright Licensing Agency Ltd, Saffron House, 6–10 Kirby Street, London EC1N 8TS.

All trademarks used herein are the property of their respective owners. The use of any trademark in this text does not vest in the author or publisher any trademark ownership rights in such trademarks, nor does the use of such trademarks imply any affiliation with or endorsement of this book by such owners.

ISBN 10: 1-292-08780-3
ISBN 13: 978-1-292-08780-1

British Library Cataloguing-in-Publication Data
A catalogue record for this book is available from the British Library

10 9 8 7 6 5 4 3 2 1

Typeset in 10 ITC Garamond Std by S4Carlisle Publishing Services.

Printed by Ashford Colour Press Ltd, Gosport.

Brief Contents

Contents

PART THREE

Distribution and Movement of Air 240

8 Atmospheric Circulation and Pressure Distributions 242

9 Air Masses and Fronts 288

PART FOUR

Disturbances 312

Preface

The atmosphere is the most dynamic of all Earth's spheres. In no other realm do events routinely unfold so quickly, with so great a potential impact on humans. Some of the most striking atmospheric disturbances (such as tornadoes) can take place over time scales on the order of minutes—but nevertheless have permanent consequences. Events such as the California drought, which began in 2011 and showed no signs of abatement by mid-2014, take longer, but can have much more widespread effects. Water levels have dropped precipitously in reservoirs, the state's huge agricultural industry has been severely impacted, water allocations have been reduced, and many areas have been threatened by unusually dangerous wildfires. While catastrophes such as this are momentous, even the most mundane of atmospheric phenomena influence our lives on a daily basis (for instance, the beauty of blue skies or red sunsets, rain, or the daily cycle of temperature).

Atmospheric processes, despite their immediacy on a personal level and their importance in human affairs on a larger level, are not readily understood by most people. This is probably not surprising, given that the atmosphere consists primarily of invisible gases, along with suspended, frequently microscopic particles, water droplets, and ice crystals.

Understanding Weather and Climate is a college-level text intended for both science majors and nonmajors taking their first course in atmospheric science. We have attempted to write a text that is informative, timely, engaging to students, and easily used by professors. In this book, our overriding goal is to bridge the gap between abstract explanatory processes and the expression of those processes in everyday events. We have written the book so that students with little or no science background will be able to build a nonmathematical understanding of the atmosphere.

That said, we do not propose to abandon the foundations of physical science. We know from our own teaching experience that physical laws and principles can be mastered by students of widely varying backgrounds. In addition, we believe one of meteorology's great advantages is that reasoning from fundamental principles explains so much of the field. Compared to some other disciplines, this is one in which there is an enormous payoff for mastering a relatively small number of basic ideas.

Finally, our experience is that students are always excited to learn the "why" of things, and to do so gives real meaning to "what" and "where." For us, therefore, the idea of forsaking explanation in favor of a purely descriptive approach has no appeal whatsoever. Rather, we propose merely to replace mathematical proof (corroboration by formal argument) with qualitative reasoning and appeal to everyday occurrences. As the title implies, the goal remains understanding atmospheric behavior.

New to the Seventh Edition

- **NEW Quick Response (QR) codes** integrated throughout each chapter empower students to use their mobile devices for learning as they read, providing instant access to over 120 videos and animations of real-world atmospheric phenomena and visualizations of key physical processes.
- **NEW *Focus on the Environment and Societal Impacts*** features explore the impact of weather hazards on people and society, not just by looking at the physical principles, but also by looking at the broader human issues, as well as mitigation policies and strategies. These case studies are grounded in real-world examples that illustrate the broad effects of weather on people and society and the decision making that goes into coping with weather events.
- **NEW Emphasis on oceans and their role in regulating weather and climate**, including unique dedicated sections on oceans and our climate in Chapters 8, 15, and 16. These sections emphasize how the atmosphere and oceans are interconnected parts of the larger Earth system, and how the role of the oceans is key to understanding such important phenomena as precipitation patterns, the formation of tropical cyclones, and the impacts of climate change.
- **NEW and UPDATED *Focus on Aviation*** features explore the meteorological impacts of weather on aviation. New topics include *Using Altimeters to Monitor Altitude* (Chapter 4), *Density Altitude and Aircraft Performance* (Chapter 9), and *Airports' and Airlines' Response to Hurricanes* (Chapter 12).
- **UPDATED coverage of climate change** integrated throughout, including the findings of the IPCC's Fifth Assessment Report.
- **NEW *Word Clouds*** presented at the start of each chapter emphasize the key topics and concepts of the chapter.
- **NEW *Visual Analysis Activities*** at the end of each chapter draw on visualizations of real-world meteorology phenomena and data, asking students to make observations and predictions and to demonstrate higher-level critical thinking and data analysis skills.
- **NEW Equations** are highlighted and defined throughout the text for easy reference.
- **NEW** sections and features on the **scientific method** and its role in atmospheric science (Chapter 1), the Coriolis force (Chapter 4), and maintaining the general circulation (Chapter 8).
- **NEW full-color reference globes and maps** at the beginning and the end of the book provide students with dynamic satellite and cartographic reference imagery.

- The **latest data, case studies, applications,** and current **examples from meteorology today** are integrated to make the seventh edition the most current and relevant introduction to meteorology. For example, Chapter 3 includes the most up-to-date values of the global energy balance; Chapter 8 discusses El Niño types and their differing effects on U.S. temperatures; Chapter 10 offers a scientifically current description of midlatitude cyclones against the background of the original polar front theory; Chapter 11 presents the most recent instructions on tornado safety that reflect outcomes from the deadly tornado outbreaks of 2013 and new material on floods and flash floods; and Chapter 12 presents updated statistics on hurricane incidence. Chapter 13 focuses on the extreme winter of 2013 to illustrate the methods and pitfalls of short and long-term forecasting. Chapter 15 has new maps of global climates based on state-of-the art ocean and land data sets. Chapter 16 has new material on climate change, including synopses of the Fifth IPCC Assessment report and the 2014 Third National Climate Assessment (NCA).

Distinguishing Features

Scientific Literacy and Currency We emphasize scientific literacy throughout the book. This emphasis gives students an opportunity to develop a deeper understanding about the building blocks of atmospheric science and serves as tacit instruction regarding the workings of all the sciences. For instance, in Chapter 2 we cover the molecular changes that occur when radiation is absorbed or emitted, items that are often considered a "given" in introductory texts. In Chapter 3 these basic ideas are used to help build student understanding of why individual gases radiate and absorb particular wavelengths of radiation and to illustrate how processes operating at a subatomic level can manifest themselves at global scales. Similarly, our discussion of anthropogenic warming in Chapter 16 includes cloud, water vapor, and lapse rate feedbacks in order to provide a more complete account of the uncertainties surrounding this critical environmental topic.

An emphasis on scientific literacy is effectively implemented only if it is accompanied by careful attention to currency. We believe that two kinds of currency are required in a text: an integration of current *events* as they relate to the topic at hand, and an integration of current *scientific thinking.* For instance, the reader will find discussion of both recent hurricane activity and the most recent theories regarding the mechanisms that generate severe storms. Scientific literacy also calls for attention to language—after all, precision of language is an important distinguishing characteristic of science, one that sets it apart from other intellectual activities.

Instructor Flexibility During the writing process, we have enjoyed interacting with many of our colleagues who teach courses in weather and climate on a regular basis. It was especially interesting to see how little consensus exists regarding topic order (truth be told, the authors of this book don't agree on the optimal sequence). With this in mind, we tried to minimize the degree to which individual chapters depend on material presented earlier. Thus, instructors who prefer a chapter order different from the one we ultimately chose will not be disadvantaged.

Emphasis on Climate Change In 2013 and 2014, the Intergovernmental Panel on Climate Change (IPCC) released its latest report on the current knowledge of climate change and human impacts. The seventh edition of *Understanding Weather and Climate* makes heavy reference to that work, and has updated climate statistics through 2014, and post-IPCC developments are included throughout. These sections present physically based explanations behind the changes that have occurred and are likely to occur in the future.

Emphasis on Forecasting In addition to a comprehensive chapter on the topic, this text contains numerous examples of how physical principles are employed in weather forecasting. We have included several discussions of the use of thermodynamic diagrams in weather forecasting and analysis. These charts are extremely valuable but not immediately comprehensible to most students. To alleviate this problem, we introduce thermodynamic diagrams in a sequential fashion. That is, their use for plotting vertical temperature profiles is presented in the chapter on temperature. We expand on this in the chapter on atmospheric moisture to show how various measures of humidity can also be determined with the aid of the charts. Thus, instructors can teach their students how to use these diagrams without inundating them with excessive detail all at once.

Current Applications of Meteorology This edition presents a greater number of weather maps and images to illustrate how atmospheric phenomena occur in everyday settings. The new examples have been selected for currency and illustrative value. Special attention has been given to some of the most notable hurricanes and typhoons of recent years, along with 2013 tornado outbreaks, and the brutal winter of 2013–2014 in eastern North America.

Readability In contrast to the more formal scientific style used in many science textbooks, we have chosen to adopt more casual prose. Our goal is to present the material in language that is clear, readable, and friendly to the student reader. We employ frequent headings and subheadings to help students follow discussions and identify the most important ideas in each chapter. As a rule, we keep technical language to a minimum.

Dynamic Media

A fundamental feature of this book is the integration of the classic print textbook model with instructional technology. Many online videos and animations are available for students to access directly from the print textbook pages with QR codes that can be scanned with mobile devices.

Focus on Learning

The chapters offer a number of study aids:

- *Learning Outcomes* are outlined at the beginning of each chapter and within chapter sections, helping students prioritize key concepts and skills.
- *Checkpoint Questions* are integrated throughout the chapters after major sections, giving students a chance to stop, practice, and apply their understanding of key chapter content.
- *Did You Know* features highlight interesting meteorological facts in every chapter.
- *Focus on the Environment and Societal Impacts* features highlight environmental and human impact issues as they relate to the study of the atmosphere.
- *Focus on Severe Weather* features focus on dramatic and dangerous severe and hazardous weather phenomena, including coverage of many recent events like the deadly 2011 and 2013 tornado seasons.
- *Focus on Aviation* features explore the impacts of various atmospheric phenomena on aviation. Examples include discussion of winter storms and air travel (Chapter 1), impacts of icing on aircraft (Chapter 5), recommended pilot responses to icing in different types of clouds (Chapter 6), and lightning and aircraft (Chapter 11).
- *Physical Principles* features are more mathematical in nature and accommodate students who have a more quantitative interest in the topic.
- *Forecasting* features apply the principles discussed in the chapter to forecasting and often include simple "rules of thumb" that help students make their own forecasts.
- *Summary.* Each chapter concludes with a chapter summary highlighting the main points in the chapter.
- *Review Questions.* These Review Questions test reading comprehension and can be answered from information presented in the chapter.
- *Critical Thinking.* These questions require students to use material presented in the chapter to work out answers relevant to real-world questions.
- *Problems & Exercises* encourage students to work out solutions to numerical questions to gain a better understanding of chapter material.
- *Visual Analysis Activities.* These activities ask students to make observations and predictions and to demonstrate higher-level critical thinking and data analysis skills.
- *Key Terms* are printed in boldface when first introduced. Most are also listed at the end of each chapter, along with the page number on which each first appears. All key terms are defined in the glossary at the end of the book.

Acknowledgments

The authors would like to extend their thanks to the editorial and production team at Pearson Education: our Senior Editor Christian Botting, who oversaw the project; Executive Development Editor Jonathan Cheney, who provided expert advice throughout the revision process; Project Manager Kristen Sanchez; Program Manager Anton Yakovlev; and Media Producer Ziki Dekel. Among many other things, we greatly appreciate their management of a complicated schedule. The book is much improved for their efforts. A number of other people were extremely helpful on the production side of things, including Mary Tindle at S4Carlisle Publishing Services.

We also benefited greatly from the advice of many professional educators and meteorologists. Mark Moede of the San Diego office of the National Weather Service provided valuable information on the day-to-day activities of weather forecasters, and Rick Smith, Warning Coordination Meteorologist at the National Weather Service Office in Norman, Oklahoma, provided first-hand knowledge of the deadly Oklahoma tornadoes of 2013 and what they taught us about tornado safety.

We must also offer special thanks to the many colleagues who spent valuable time and energy preparing in-depth reviews of our early efforts, many of whom have continued in this role through multiple revisions. We are particularly grateful to the accuracy reviewers Lou McNally, Redina Herman, and Jonathan D. W. Kahl, who read over the current edition with exceptional care and made many excellent suggestions. Additionally, we thank the following educators and researchers who provided reviews of the current and previous editions:

Rafique Ahmed, *University of Wisconsin–LaCrosse*
Al Armendariz, *Southern Methodist University*
David Barclay, *SUNY Cortland*
Greg Bierly, *Indiana State University*
Mark Binkley, *Mississippi State University*
David Brommer, *University of North Alabama*
Gerald Brothen, *El Camino College*
David P. Brown, *University of Arizona*
Adam W. Burnett, *Colgate University*
Gregory Carbone, *University of South Carolina*
R. E. Carlson, *Iowa State University*
Donna J. Charlevoix, *University of Illinois, Urbana–Champaign*
Christopher R. Church, *Miami University of Ohio*
John H. E. Clark, *Penn State University*
Andrew Comrie, *University of Arizona*
William T. Corcoran, *Missouri State University*
Eugene Cordero, *San Jose State University*
Mario Daoust, *Southwest Missouri State University*
Michael Davis, *Kutztown University*
Arthur (Tim) Doggett, *Texas Tech University*

Dennis M. Driscoll, *Texas A&M University*
Ted Eckmann, *Bowling Green State University*
Neil I. Fox, *University of Missouri–Columbia*
James Gammack-Clark, *Florida Atlantic University*
Victor A. Gensini, *College of DuPage*
Mario A. Giraldo, *California State University of Northridge*
Christopher M. Godfrey, *University of Oklahoma*
Thomas Guinn, *Embry-Riddle Aeronautical University*
Redina Herman, *Western Illinois University*
Rex J. Hess, *University of Utah*
Jay S. Hobgood, *Ohio State University*
Edward J. Hopkins, *University of Wisconsin–Madison*
Scott A. Isard, *University of Illinois*
Eric Johnson, *Illinois State University*
Jonathan D. W. Kahl, *University of Wisconsin–Milwaukee*
Scott Kirsch, *University of Memphis*
Thomas Kovacs, *Eastern Michigan University*
Daniel James Leathers, *University of Delaware*
Gong-Yuh Lin, *California State University–Northridge*
Anthony Lupo, *University of Missouri*
Jason Martinelli, *Creighton University*
Lou McNally, *Embry-Riddle Aeronautical University*
Deborah Metzel, *University of Massachusetts–Boston*
Thomas L. Mote, *University of Georgia*
Gregory D. Nastrom, *St. Cloud State University*
Gerald R. North, *Texas A&M University*
Jim Norwine, *Texas A&M University–Kingsville*
John E. Oliver, *Indiana State University*
Stephen Podewell, *Western Michigan University*
David Privette, *Central Piedmont Community College*
Sarah Pryor, *Indiana University–Bloomington*
Azizur Rahman, *University of Minnesota, Crookston*
Peter S. Ray, *Florida State University*
Robert V. Rohli, *Louisiana State University*
Steven A. Rutledge, *Colorado State University*
Erinanne Saffell, *Arizona State University*
Arthur N. Samel, *Bowling Green State University*
Hans Peter Schmid, *Indiana University*
Justin Schoof, *Southern Illinois University–Carbondale*
Marshall Shepherd, *University of Georgia*
Brent Skeeter, *Salisbury University*
Stephen Stadler, *Oklahoma State University*
S. Elwynn Taylor, *Iowa State University*
Mingfang Ting, *University of Illinois*
Graham Tobin, *University of South Florida*
Paul E. Todunter, *University o f North Dakota*
Liem Tran, *Florida Atlantic University*
Donna Tucker, *University of Kansas*
Audrey Wagner, *Southern Illinois University*
Barry Warmerdam, *Kings River Community College*
Thompson Webb III, *Brown University*
Jack Williams, *University of Wisconsin–Madison*
Thomas B. Williams, *Western Illinois University*
Morton Wurtele, *University of California–Los Angeles*
Douglas Yarger, *Iowa State University*
Charlie Zender, *University of California–Irvine*
Suzanne Zurn-Birkhimer, *Purdue University*

Digital Resources for Instructors

- ***Instructor Resource Manual* (download only) by Doug Gamble, University of North Carolina, Wilmington** The *Instructor Resource Manual* is intended as a resource for both new and experienced instructors. It includes a variety of lecture outlines, additional source materials, teaching tips, advice about how to integrate visual supplements (including the Web-based resources), and various other ideas for the classroom.
  See **www.pearsonglobaleditions.com/Aguado**.
- **TestGen® Computerized *Test Bank* (download only) by Jonathan D. W. Kahl, University of Wisconsin–Milwaukee [0321992539]** TestGen® is a computerized test generator that lets instructors view and edit *Test Bank* questions, transfer questions to tests, and print tests in a variety of customized formats. This *Test Bank* includes more than 2000 multiple-choice, fill-in-the-blank, and short-answer/essay questions. Questions are correlated to the text's Learning Outcomes, Pearson's Global Science Outcomes, the section of each chapter, the revised U.S. National Geography Standards, and Bloom's taxonomy to help instructors better map the assessments against both broad and specific teaching and learning objectives. The *Test Bank* is also available in Microsoft Word and is importable into systems such as Blackboard.
  See **www.pearsonglobaleditions.com/Aguado**.
- ***Other Instructor Resources*** This is a collection of resources to help teachers make efficient and effective use of their time. The following Instructor Resource content is available online via the Instructor Resources section of www.pearsonglobaleditions.com/Aguado.
  - All textbook images as JPEGs, PDFs, and PowerPoint™ presentations
  - Preauthored Lecture Outline PowerPoint™ presentations, which outline the concepts of each chapter with embedded art and can be customized to fit teachers' lecture requirements
  - CRS "Clicker" questions in PowerPoint™, which correlate to the text's Learning Outcomes, U.S. National Geography Standards, and Bloom's taxonomy

About the Authors

Edward Aguado is Professor of Geography in the Department of Geography of San Diego State University. He received his Ph.D. from the University of Wisconsin–Madison, and his M.A. and B.A. from UCLA. His research interests are in the precipitation and hydrology of western U.S. mountains. He regularly teaches introductory and advanced meteorology, climatology, and physical geography, and often serves as a consultant and expert witness on climatology and weather.

Jim Burt is Professor of Geography in the Department of Geography of the University of Wisconsin–Madison. He received his Ph.D. from UCLA. His research interests are in physical geography, climatology, and spatial analysis. His most recent projects involve data-driven geospatial modeling and knowledge discovery from historical maps. He regularly teaches courses in physical geography, climatology, quantitative methods and geocomputing.

To Lauren, William, and Babsie June

—EA

To my parents, Martha F. Burt
and Robert L. Burt

—JEB

PART

1 Energy and Mass

CHAPTER 1

Composition and Structure of the Atmosphere

CHAPTER 2

Solar Radiation and the Seasons

CHAPTER 3

Energy Balance and Temperature

CHAPTER

4

Atmospheric Pressure and Wind

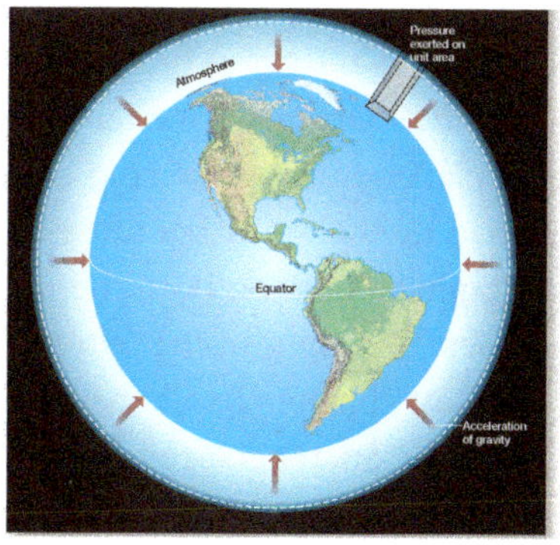

◀ *Dense fog at dawn in Dubai, United Arab Emirates. Differences in pressure cause winds to flow onto the land from the surrounding gulf during the day. At night emitted radiation causes water vapor carried by those winds to cool and reach the point of condensation.*

The atmosphere is remarkably variable. Its characteristics are quite different from place to place and from the surface to its upper reaches. It is also subject to subtle movements (such as a gentle breeze) or violent motions (such as tornadoes). After an introduction to the scientific methods used in the study of weather and climate, these chapters look at the composition of the atmosphere and how it is distributed around the planet, how the Sun heats the air, and how pressure and wind patterns are created.

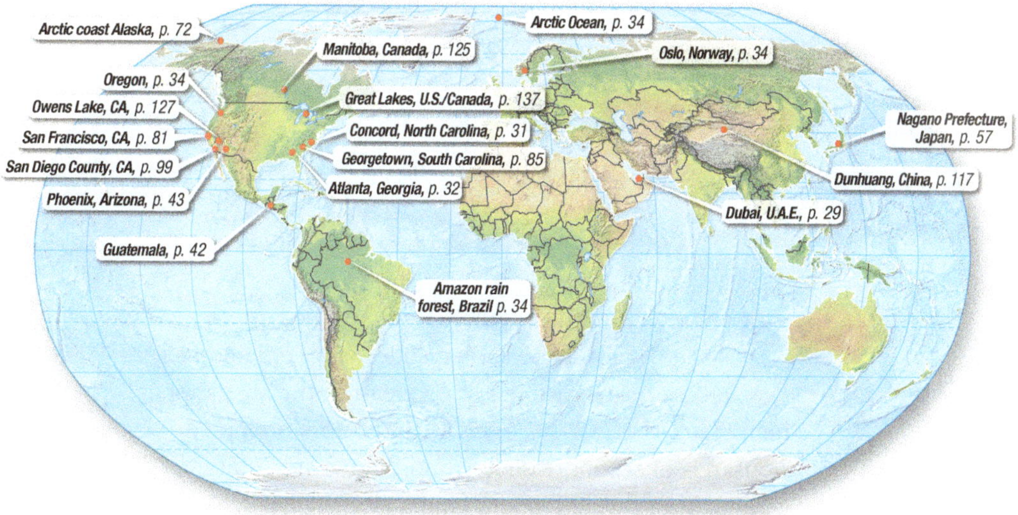

CHAPTER

1 Composition and Structure of the Atmosphere

January 2014 was a brutal month for the eastern United States. Repeated episodes of record and near-record cold struck every state east of the Mississippi. Temperatures of −30 °C (−22 °F) and below prompted thousands of school closings in northern states, and interstate highways were rendered impassible by heavy snows. Late in the month a mix of snow and ice literally paralyzed commuters in Atlanta, and grounded thousands of flights at America's busiest airport. Mainstream media seized on "polar vortex" as the explanation for these cold snaps, deploying a term that had rarely if ever been used before in popular accounts. Commentators on both the political right and left did their best to represent the extreme cold as confirming their views of climate change generally and global warming in particular. Such occurrences are but one example of the atmosphere's importance in human affairs.

◂ *A single storm in mid-February 2014 brought up to 40 cm of snow (16 inches) to parts of North Carolina accustomed to only a few inches in a typical year. Students celebrated school closings (as seen in this photo), but over a million of homes lost power, airports all along the seaboard were paralyzed, and 22 people died as result of the storm.*

Learning Outcomes

After reading this chapter, you should be able to:

- **1.1** Distinguish between weather and climate.
- **1.2** Explain the scientific method.
- **1.3** Describe the thickness of the atmosphere and the vertical distribution of gases within it.
- **1.4** Describe the behavior of gas molecules in the atmosphere, including residence times and the roles of vertical mixing and gravitational settling.
- **1.5** Describe the composition of the atmosphere.
- **1.6** Explain how air pressure arises and describe the vertical variation of pressure and density.
- **1.7** Identify and describe the layers of Earth's atmosphere.
- **1.8** Explain the evolution of the atmosphere during Earth's history.
- **1.9** Identify the basic types of data found on weather maps.
- **1.10** List major events in the history of meteorology.

The list of examples includes many other types of disasters that have resulted in huge human and financial costs (Table 1–1). On a personal level none of us is immune to even routine events, whether that means adjusting plans for a picnic or reveling in the beauty of an exceptional sunset.

1–1 FOCUS ON AVIATION

Winter Storms and Air Travel

Video
The Benefits of Doppler Radar
http://goo.gl/wNyeKp

Winter weather can play havoc with air travel in the United States. Naturally, some years are worse than others, and the winter of 2010–2011 was especially difficult. Between November and early February, snow and ice conditions associated with four major storms caused U.S. airlines to cancel some 86,000 flights, with thousands of other flights encountering major delays. These events amounted to a significant portion of all scheduled flights for an industry recovering from some economic difficulties. In December 2010, 3.7 percent of all U.S. flights were cancelled because of winter storms, in contrast to the 2.9 percent cancelled the previous December. Even airports not normally subject to crippling winter storms, such as Hartsfield-Jackson Atlanta International Airport, were subject to the cancellation of thousands of flights when a single January storm left behind 6 inches of snow in a city that usually receives half that much in an entire season (Figure 1–1–1).

One of the problems associated with winter storms is that aircraft loaded to capacity with passengers can be forced to wait on taxiways between terminal gates and runways for extended periods. For example, the time aircraft must wait before being de-iced contributes to delays. An outcry of consumer criticism led to a policy taking effect in April 2010 that allows the U.S. Department of Transportation to levy fines for air carriers of up to $27,500 per passenger when flights are forced to remain on the tarmac for more than three hours. Some believe this will motivate airlines to cancel flights more readily than in the past, causing greater inconvenience for passengers forced to wait hours or even days for another flight to get them to their destination. On the other hand, passengers are now much less likely to spend half a day on a crowded aircraft just waiting to get to the runway.

Often the disruption of air travel is due to what happens outside the airport during winter weather. Sometimes it is easier for airlines to fly crews in from other areas than it is to wait for scheduled personnel who are delayed by impassable highways on their way to the airport. And of course, flights scheduled to depart San Diego on a warm, sunny day may be unable to do so because the aircraft needed for the flight is stranded at an East Coast airport.

Weather in other seasons also poses different hazards to commercial aviation, as will be discussed in later chapters of this book.

▲ **FIGURE 1–1–1 Snowbound Airplanes in Atlanta, January 2011.**

1–1–1 In the depths of winter, what are the approximate chances of your plane being grounded because of weather in the United States? Much less than 1%? A few percent? Between 5% and 10%? More?

1–1–2 Beyond the immediate conditions at an originating airport, what weather-related factors might cause a flight to be cancelled?

Oddly enough, although we are continually surrounded and affected by the atmosphere, most of us know relatively little about how and why the atmosphere behaves as it does. In the chapters that follow, we hope to provide an account of both the how and the why, in ways that will lead you to understand the underlying physical processes. This chapter introduces the most basic elements of meteorology, laying the foundation for much of the rest of the book.

The Atmosphere, Weather, and Climate

1.1 Distinguish between weather and climate.

The **atmosphere** is a mixture of gas molecules, small suspended particles of solid and liquid, and falling precipitation. **Meteorology** is the study of the atmosphere and the processes

TABLE 1–1

Three Decades of Billion-Dollar U.S. Weather Disasters

Dollar Amounts Are Adjusted to 2013 Values											
Year	No. of Events	No. of Deaths	Total Cost	Year	No. of Events	No. of Deaths	Total Cost	Year	No. of Events	No. of Deaths	Total Cost
2013	7	109	23	2003	5	138	17.0	1993	4	338	45.9
2012	11	377	115.6	2002	3	28	17.7	1992	6	87	54.4
2011	14	764	51.3	2001	2	46	9.1	1991	3	43	8
2010	4	46	9.4	2000	2	140	8.1	1990	3	13	9.3
2009	6	26	11.6	1999	5	676	15.2	1989	4	207	21.1
2008	9	296	61.4	1998	7	419	32.4	1988	1	7500	78.8
2007	5	37	12.2	1997	5	114	11.3	1987	0	0	0
2006	6	95	13.5	1996	1	233	20.7	1986	1	21	1.3
2005	5	2002	90.4	1995	4	99	20.8	1985	5	228	11.9
2004	5	172	56.6	1994	6	133	12.6	1984	1	80	1.1

(such as cloud formation, lightning, and wind movement) that cause what we refer to as the "weather." **Weather** is distinct from **climate** in that the former deals with short-term phenomena and the latter with characteristic long-term patterns. A rough analogy can help with the distinction. Most of us have an image of New York's Brooklyn Bridge during the morning rush hour. If our mental picture of slow-moving congestion is the bridge's "traffic climate," weather would be the particular combination of individual cars, buses, and trucks found there on a single day. Take a look outside your window and what you will see is weather. The current temperature, humidity, wind conditions, amount and type of cloud cover, and the presence or absence of precipitation—these are all elements of weather.

DID YOU KNOW?

Tornadoes are not strictly a U.S. phenomenon. Italy, for example, is ranked sixth in the world for tornado density. Twelve were reported in 2012 in an area about the size of Arizona (which had one that year).

Climatology concerns itself with the same elements of the atmosphere that meteorology does, but on a different time scale. Rather than focusing on a single point in time, climatology relies on averages taken over a number of years in order to gauge typical atmospheric conditions for locations across Earth's surface. When people joke about summer conditions as "Sahara-like," they are implicitly making a climatological reference to average conditions in North Africa. Averages are very important, but climatologists also want to know the variability of the weather elements just as, in addition to the average speed, the bridge commuter wants to know about traffic variability. In the case of the atmosphere, it might be useful to know that Boulder, Colorado, has an average April temperature of 7 °C (45 °F), but this figure becomes more meaningful when one understands just how far the temperature might depart from the value on any given day. Frequencies of occurrence of weather events—such as extreme heat, hail, or lightning—are also aspects of climates. Finally, a particularly important part of climatology is concerned with changes in Earth's climate and the factors responsible for those changes.

This book's focus on weather and climate correctly suggests that the atmosphere is of primary interest, but we cannot understand our atmospheric environment without reference to land and ocean processes. For example, as illustrated in Figure 1–1, moist Pacific climates in western Oregon give way to desert-like conditions a short distance eastward because of mountain topography. Lush forests in the Amazon basin control a number of processes that are key to the region's climate. Similarly, bright, highly reflective snow and ice surfaces in polar locations contribute greatly to extreme cold. Hurricanes that batter coastal locations could not form without the fuel provided by a warm ocean surface. Western Europe would be frigid if not for the heat imported by ocean currents, and on much longer time scales, large shifts in ocean circulation have led to major climatic changes. Furthermore, the composition of the atmosphere and Earth climate cannot be explained without considering the exchange of material between the solid Earth and the atmosphere. Clearly, an integrated approach that considers all components of the Earth system is necessary. The diagram in Figure 1–1 shows one approach to conceptualizing the components and their interactions. Note the presence of external natural processes and human activities as agents of change.

CHECKPOINT

1.1 Define weather and climate in your own words.

1.2 Compare the concerns of the sciences of meteorology and climatology, giving some examples of different phenomena they might investigate.

1.3 List some places you have visited whose climate is affected by proximity to an ocean or by its position deep within a continent.

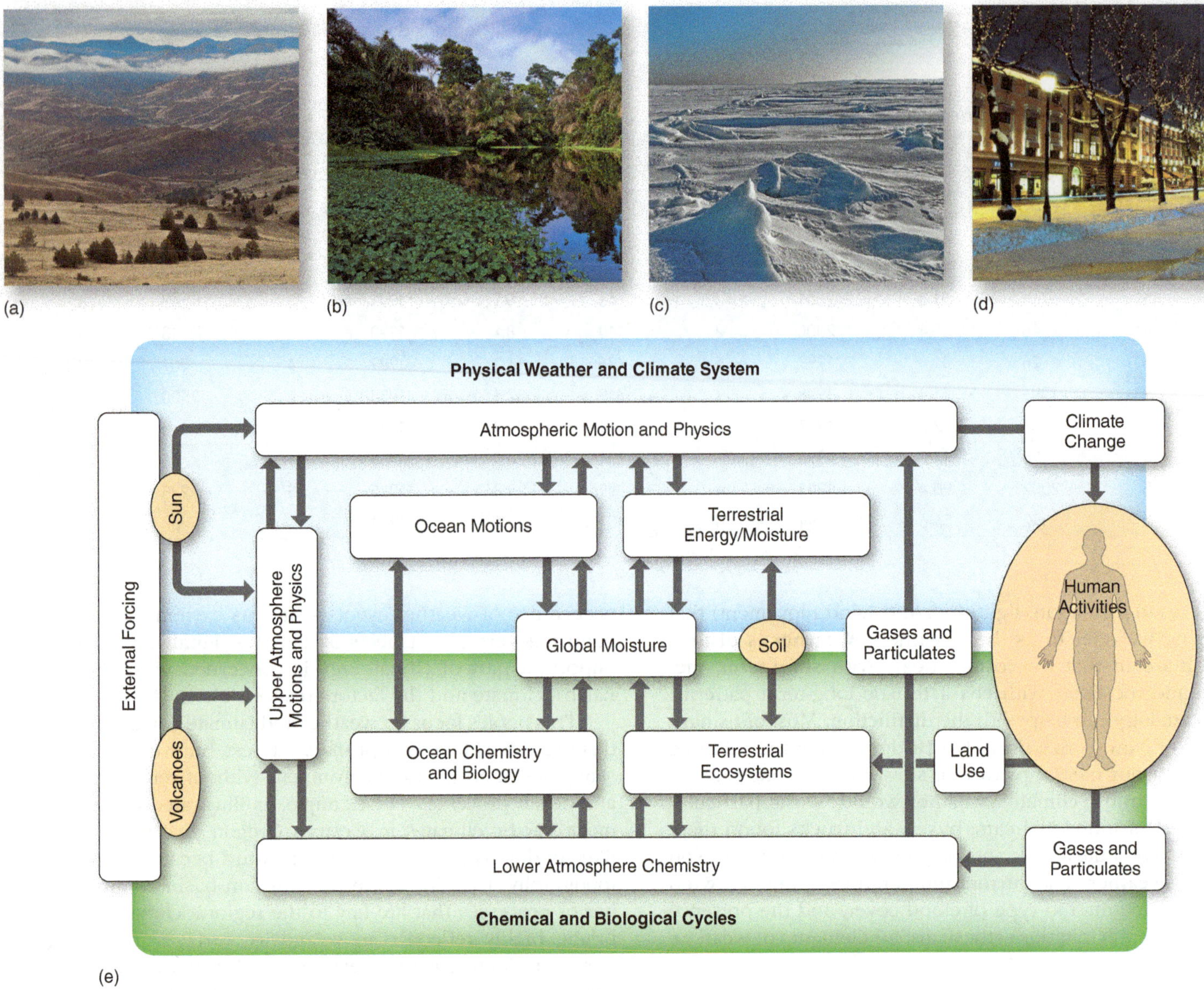

▲ **FIGURE 1–1 Earth as a System.** (a–d) These photos show examples of interactions: (a) dryness in Eastern Oregon caused by mountains to the west, (b) dense Amazon vegetation that recycles water between the surface and atmosphere, (c) bright snow enhancing Arctic cold, and (d) ocean currents that make winter in Oslo, Norway, warmer than places much closer to the equator. (e) This figure is a simplified view of the Earth system. The upper part of the diagram represents purely physical aspects of Earth, such as ocean currents, winds, cloud formations, and temperature distributions. The bottom half depicts the constant exchange of material throughout the system, a process known as *cycling*. These exchanges occur between and among the living and nonliving realms, and they both affect and are themselves affected by the physical components of the Earth system.

The Scientific Method and Atmospheric Science

1.2 Explain the scientific method.

Like other physical sciences, meteorology and climatology rely on the **scientific method**. Although it might sound like a strict process, this is really a framework for answering scientific questions. It begins with an inquiry regarding the physical world. For example, having observed tornadoes, we could ask what is responsible for their spin. Or, knowing from landforms glaciers left behind that glaciers have waxed and waned over eons, we might ask if variations in the Sun's output could be involved in glacial outbreaks. Or perhaps we want to know if El Niños (warm episodes in the Tropical Pacific Ocean) affect climate in some other part of the world. The scientific method provides a way to address such questions. It consists of elements that collectively provide us with a way to learn about natural phenomena; it amounts to a convention regarding what it means to "know" something. As individuals we obviously obtain knowledge in various ways; for example, we don't need the scientific method to know that a glowing ember will burn one's hand. The scientific method, however, is particularly useful when our goals are to obtain a shared understanding and to resolve ambiguity. On a purely practical level, its value has been proven over centuries of use, and everyday life abounds with discoveries and inventions that would have been impossible to achieve otherwise.

Applying the Scientific Method

The scientific method has the following elements: question, hypothesis, prediction, experiment, analysis, and conclusion (Figure 1–2a). To illustrate these elements, we will use research from about a decade ago concerning the climatic effects of jet aircraft (Figure 1–2b).[1] In particular, researchers wanted to investigate the possibility that clouds produced by aircraft (contrails) could influence temperature near the ground. Thus for this example, the question is simply "Can contrails affect surface temperature?" The idea that clouds produced by aircraft contrails could influence temperature near the surface had been debated for decades, but the grounding of aircraft in the United States following the U.S. terrorist attacks in 2001 provided a chance to study the question in a new way. In particular, the sudden absence of jet contrails offered a "before and after" test covering the entire country.

Hypotheses follow from the question and play a particularly important role in the scientific method. They can be proposed explanations for previously observed measurements, or claims about the role of some process in a phenomenon of interest. By definition, scientific hypotheses must be testable. If there is no conceivable way of evaluating a hypothesis, it has no role in the scientific method. As a practical matter this means the hypothesis must lead to predictions. In the contrails example the hypothesis specifies how contrails would affect surface temperatures. Based on prior knowledge about radiation transfer in the atmosphere, it amounts to an educated guess about how contrails might exert influence on air temperature near the ground.

Predictions can be a forecast of what will happen in the future, but more commonly are statements about what one should observe if particular data are analyzed. The important thing is that predictions are logical consequences that follow from hypotheses. The *experiment* can be a procedure performed in a laboratory, or a computer simulation, or anything else that produces data bearing on the prediction. In the contrail example, the "experiment" was to compile air temperature data from weather stations throughout the continental United States. In this case the experiment was almost trivial, because it merely required retrieval of air temperature data routinely collected for other purposes. But easy or not, the experimental methods must be clearly described. In this case it meant documenting the data (how many weather stations, where, etc.). In other cases it could mean describing the computer model used to generate new data or describing exactly what measurement procedures were used.

The *analysis* step evaluates the predictions and thereby renders information about the hypotheses, which is reflected in the *conclusion*. In this example, analysis showed that the grounding period was highly unusual: The average day–night temperature range increased dramatically during the grounding period. In fact, the increase was larger than in any of the prior 30 years. As explained in the figure, a larger range is expected based on the hypothesis. Thus, the analysis supports the hypothesized connection between jet contrails and air temperature. Here again the methods used must be clearly specified. The reason is that reproducibility is a critical aspect of the scientific method. Other researchers employing the same methods must achieve the same outcomes. If some aspect of the process were to depend on the particular individuals involved it would not be scientific. In other words, things such as a researcher's experience, reputation, "common sense," and ability to provide deep insights do not carry any weight when it comes to confirming or refuting hypotheses.

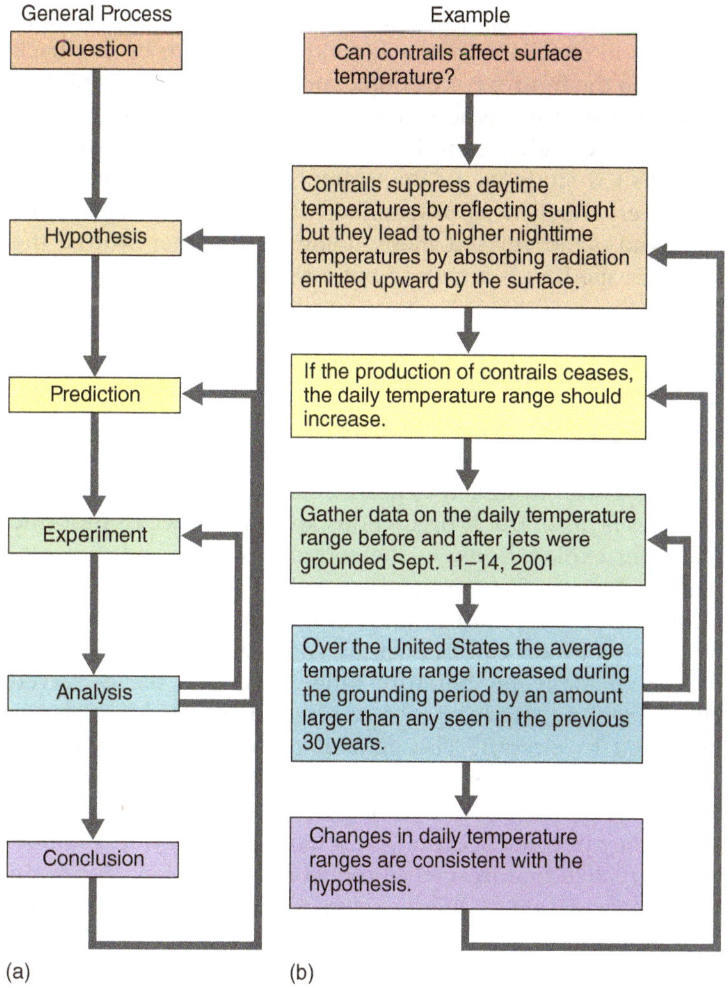

▲ **FIGURE 1–2 Elements of the Scientific Method in (a) General Terms and Illustrated by an (b) Example.**

Variations on the Scientific Method

The scientific method is often presented as a sequence of steps, and it is true that the outcomes of a scientific study can usually be placed in the categories shown in Figure 1–2.

[1]Travis, David J., Andrew M. Carleton, and Ryan G. Lauritsen, "Contrails Reduce Daily Temperature Range," *Nature*, 418 (2002), p. 601.

However, the actual practice by an individual or research group is likely to involve backtracking as shown by feedback arrows in the figure. For example, preliminary analysis of data may suggest other hypotheses and predictions calling for different or expanded experiments. What appears ultimately in a published study as a tidy sequence of logically connected parts very likely was assembled after multiple false starts, dead-ends, and revised methodologies. In other words, the scientific method is anything but a rigid process.

The Nature of Hypotheses and Scientific Theories

It is important to realize that the scientific method can only disprove hypotheses. That is, should a prediction not materialize in a well-designed experiment, we can be sure the hypothesis is false. But if predictions are correct, we cannot rule out other explanations. In the contrail example, other explanations have in fact emerged to explain the observed reduction in daily temperature range. Subsequent studies showed that high-altitude clouds such as contrails have only a small impact on daily temperature range, and that the observed change was mostly likely the result of changes in low clouds unrelated to aircraft. The general point is that hypotheses are never proven; they merely withstand attempts at disproof. When hypotheses hold up long enough under repeated assaults they become accepted as **scientific theory**. It is important to realize that in science the word *theory* does not mean vague speculation. A theory refers to ideas that are accepted as a description or explanation of how things are, but allowing for the possibility that new information might emerge that will modify those ideas. In science if something is said to be so, what the statement really means is that the idea has been subject to extreme scrutiny and there is no contradictory evidence or competing theory that can explain what has been observed.

Not surprisingly, much of science involves critically analyzing the work of others, searching for weaknesses in procedures and alternative explanations. As you read this text, we urge you to follow the logic behind the claims that are made rather than simply memorize "facts." As you'll see, there is much uncertainty when it comes to the atmosphere and many of the ideas presented in the following chapters will change in the coming years. More importantly, a number of pressing environmental issues have scientific questions at their core. Appreciating the basis and limits of scientific understanding regarding the atmosphere is essential in order to be informed on questions of climate change, air pollution, water availability, food production, and many other timely topics.

CHECKPOINT

1.4 List the elements of the scientific method.

1.5 Suppose you are inclined to try a new weight-loss pill. Devise a scientific approach to evaluating its effectiveness.

Thickness of the Atmosphere

1.3 Describe the thickness of the atmosphere and the vertical distribution of gases within it.

Every child has wondered "How high is the sky?" There is no definitive answer to that question, however, because Earth's atmosphere becomes thinner at higher altitudes. A person in a rising hot-air balloon would be surrounded by an atmosphere that gradually becomes less dense. At some height, the air becomes so thin that the balloonist would pass out from a shortage of oxygen—but there would still be an atmosphere. At an altitude of 16 km, or 10 mi, the density of the air is only about 10 percent of that at sea level, and at 50 km (30 mi) it is only about 1 percent of what it is at sea level. But even at heights of several hundred kilometers above sea level, there is some air and, hence, an atmosphere. We have no way to establish its upper boundary, however, because no universally accepted definition exists of how much air in a given volume constitutes the presence of an atmosphere. You would probably not say only one molecule of air per cubic kilometer constitutes an atmosphere, but how much is enough?

Viewed from Earth's surface, the atmosphere appears to be extremely deep. In reality, however, most of the atmosphere is contained within a relatively shallow envelope surrounding the planet. Let's assume for the sake of discussion that the upper limit of the atmosphere occurs at 100 km (60 mi) above sea level (in fact, 99.99997 percent of the atmosphere is below this height). By comparing this 100 km thickness with the 6400 km (4000 mi) radius of Earth, we see that the depth of the atmosphere adds less than 2 percent to Earth's cross-sectional size. This is evident in Figure 1–3, an image of Earth and its atmosphere taken from space. The top of the thunderstorm cloud probably has an altitude of about 12 km (7.5 mi), but when viewed from space, it appears to hug the ground. Though impressive when we look up at them, those clouds are no thicker than the skin of an apple, comparatively speaking.

Given the shallowness of the atmosphere, its motion over large areas must be primarily horizontal. Indeed, with some very notable exceptions, horizontal wind speeds are typically hundreds to thousands of times greater than vertical wind speeds. However, we cannot overlook the wind's vertical motions. As we shall see, even small vertical displacements of air have an impact on the state of the atmosphere. Paradoxically, the least impressive motions—vertical—despite being hardest to detect and forecast, turn out to determine much of atmospheric behavior.

CHECKPOINT

1.6 Nearly all of the atmosphere lies below an altitude of 100 km (60 mi). As a percentage of Earth size, about how thick is such a layer?

1.7 Explain why horizontal motions are so much stronger than vertical motions within the atmosphere.

▲ **FIGURE 1–3 Earth's Atmosphere.** Despite its appearance from the surface, the atmosphere is extremely thin relative to the rest of Earth.

DID YOU KNOW?

The total mass of the atmosphere, 5.4×10^{15} kilograms (kg),[1] is equivalent to 5.95 million million tons—the amount of water that would fill a lake the size of California to a depth of 13 km (7.7 mi).

[1] 5,140,000,000,000,000 kg

Factors Affecting the Composition and Vertical Structure of the Atmosphere

1.4 Describe the behavior of gas molecules in the atmosphere, including residence times and the roles of vertical mixing and gravitational settling.

The atmosphere is composed of a mixture of invisible gases and a large number of suspended microscopic solid particles and water droplets. Molecules of the gases can be exchanged between the atmosphere and Earth's surface by physical processes, such as volcanic eruptions, or by biological processes, such as plant and animal respiration. Molecules can also be produced and destroyed by purely internal processes, such as chemical reactions between the gases.

Gas Exchange, Equilibrium Conditions, and Residence Times

Consider a gas that is constantly exchanged between the atmosphere and Earth's surface. If we think of the atmosphere as a reservoir for this gas, the gas concentration in the reservoir will remain constant so long as the input rate from the surface is equal to the output rate. Under such conditions, we say that the concentration of the gas exists in a *steady state* or *equilibrium* condition. The total number of molecules in the atmosphere is constant, but individual molecules move in and out. The average length of time that individual molecules of a given substance remain in the atmosphere is called the *residence time*. The residence time is found by dividing the mass of the substance in the atmosphere (in kilograms) by the rate at which the substance enters and exits the atmosphere (in kilograms per year). Figure 1–4 illustrates both the concepts of steady state and residence time. Parts (a) and (b) show two gases entering and leaving the atmosphere at the same rate. In both cases input equals output, so both gases are in steady state. However, the gas in (b) is much less abundant, thus its molecules spend relatively little time in the atmosphere before being removed and the residence time is short. The same transfer rate in (a) is small compared to that atmosphere's store of molecules, therefore the residence time is larger.

DID YOU KNOW?

The atmosphere teems with insects flying and riding on air currents at altitudes up to several thousand meters (12,000 ft) in search of better locations for food, nesting places, and mating partners. The amount varies greatly by location and season, but a recent study in England in the summer found an average of 3 billion insects crossing 1 km^2 (0.39 mi^2) each day.

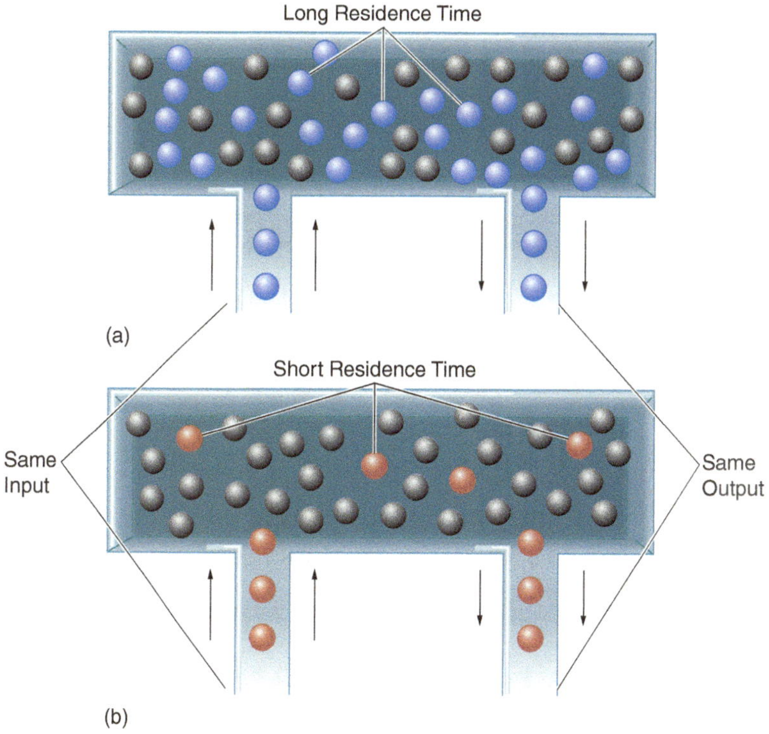

▲ **FIGURE 1–4 Residence Times of Atmospheric Gases.** (a) Gas that is abundant relative to its input and output has a long residence time (see all molecules shown in blue). (b) If input and output are rapid compared to the atmosphere's store, residence time is short (see all molecules shown in red).

The Homosphere and Heterosphere

The lowest 80 km (50 mi) of the atmosphere is sufficiently stirred by mixing that differences in molecular weight have no role in the distribution of gases. In other words, in this region vertical motions are more important than gravitational settling, and thus processes other than settling under gravity must explain any variations present. This layer is known as the **homosphere**, because it reflects the homogenizing role of wind and other motions. Above that lies the **heterosphere**, where gases segregate according to molecular weight. Heavier gases are more abundant in the lower heterosphere, whereas lighter gases such as hydrogen and helium become relatively more abundant with increasing altitude. Considering that only a tiny fraction of atmosphere remains above 80 km, we will not pursue these details of the heterosphere but will instead deal almost exclusively with the homosphere.

Within the homosphere atmospheric gases are often categorized as being permanent or variable, depending on whether or not their concentration is uniform. **Permanent gases** are found everywhere in nearly the same proportion, whereas **variable gases** are those whose distribution in the atmosphere is uneven in both time and space. Notice that describing the amount of a gas is not completely straightforward. It is natural to use a percentage, but how do we compute the percentage? Do we find the percent of the atmosphere's mass represented by a gas, or do we find the percent of the atmosphere's volume occupied by that gas? The answer is that both are used! Percent by volume is particularly attractive because—at least in the homosphere—percent by volume also represents percent by number of molecules. For example, if a gas is 20 percent by volume, 20 percent of the molecules in the atmosphere are that gas. The presence of variable gases, however, complicates the problem of calculating these percentages. If the quantity of some variable gas increases, the percentage of other gases must decrease even though no molecules are removed. In the face of this complication, atmospheric scientists calculate permanent gas percentages after excluding variable gases from the calculation. Fortunately, variable gases account for so little of the atmosphere that global average values are hardly affected. Both of these types of gases are discussed in the next section.

CHECKPOINT

1.8 What is atmospheric residence time?

1.9 What is the difference between permanent and variable gases?

Composition of the Atmosphere

1.5 Describe the composition of the atmosphere.

Earth's atmosphere is mostly permanent gases, and so we describe them first in the next section. Variable gases, which include water vapor, carbon dioxide, ozone, and methane, account for a much smaller proportion of atmospheric gases and are considered after that.

The Permanent Gases

Video
Global Changes in Carbon Dioxide Concentrations
http://goo.gl/i7XIcL

Whether considered by mass or volume, the atmosphere is more than 99 percent permanent gases. The most abundant by far are nitrogen and oxygen, followed by the inert (unreactive) gases argon and neon in small amounts and several other gases in even smaller amounts (Table 1–2). Atmospheric **nitrogen** occurs primarily as paired nitrogen atoms bonded together to form single molecules denoted N_2. Nitrogen gas has a molecular weight of 28.01, meaning that, on average, the mass of one N_2 molecule is slightly in excess of that of a combined total of 28 protons and neutrons. As with other elements, some versions of nitrogen have a different number of neutrons, giving somewhat different molecular weights. The importance of various elemental forms, or *isotopes* will be seen in later chapters. For now we simply note this is the explanation for fractional parts in molecular weights.

Nitrogen is a largely unreactive gas that accounts for 78 percent of the volume of all the permanent gases, or 75.5 percent of their mass. The processes that add and remove nitrogen from the atmosphere occur very slowly, so nitrogen has a very long residence time, measured in millions of years.

The second most abundant gas, **oxygen** (O_2), constitutes 21 percent of the volume of the atmosphere and 23 percent of its mass. Oxygen is crucial to the existence of virtually all forms of life. Like nitrogen, the oxygen molecules of the atmosphere consist mostly of paired atoms, called *diatomic oxygen*. Their residence time is about 6000 years. Together, nitrogen and oxygen account for 99 percent of all the permanent gases, with **argon** making up most of the remainder. Removal processes are so slow for argon that its residence time is extremely long. Note that the abundance of a gas gives no clue about its importance in weather. Indeed, despite the fact that the permanent gases in Table 1–2 make up nearly the entire atmosphere, none can explain storms, clouds, the deep cold that often follows a clear night, or other events we associate with weather.

TABLE 1–2

Permanent Gases of the Atmosphere

Constituent	Formula	Percent by Volume	Molecular Weight
Nitrogen	N_2	78.08	28.01
Oxygen	O_2	20.95	32.00
Argon	Ar	0.93	39.95
Neon	Ne	0.002	20.18
Helium	He	0.0005	4.00
Krypton	Kr	0.0001	83.80
Xenon	Xe	0.00009	131.30
Hydrogen	H_2	0.00005	2.02

TABLE 1–3
Variable Gases of the Atmosphere

Constituent	Formula	Percent by Volume	Molecular Weight
Water vapor	H_2O	0.25	18.01
Carbon dioxide	CO_2	0.0396	44.01
Ozone	O_3	0.01	48.00
Methane	CH_4	0.00018	16.04

Variable Gases

The variable gases account for only a small percentage of the total mass of the atmosphere (Table 1–3). Despite their relative scarcity, some of these gases profoundly affect the behavior of the atmosphere—and even your own physical comfort.

WATER VAPOR The most abundant of the variable gases, **water vapor**, accounts for about one-quarter of 1 percent of the total volume of the atmosphere on average. Because the source of water vapor in the atmosphere is evaporation from Earth's surface, its concentration normally decreases rapidly with altitude, and most atmospheric water vapor is found in the lowest 5 km (3 mi) of the atmosphere. As explained later in Chapter 5, water vapor condenses to a liquid at relatively low levels in the homosphere. Thus, although vertical mixing extends to great heights, water vapor is anything but uniformly distributed with altitude.

Water is constantly exchanged between the planet and the atmosphere in what is called the **hydrologic cycle**, which is described in Chapter 5. Water continuously evaporates from both open water and foliage into the atmosphere, where it eventually condenses to form liquid droplets and ice crystals. These liquid and solid particles are removed from the atmosphere by precipitation as rain, snow, sleet, or hail. Because of the rapidity of global evaporation, condensation, and precipitation, water vapor has a very short residence time of only 10 days.

We all know how damp and muggy the air feels when the water vapor content is high and how parched our skin can get when the air is dry (not to mention "bad hair days" when the moisture content is extremely high or low). Despite the wide range of physical comfort levels we experience in response to variations in water vapor content, the actual range of water vapor content is really quite limited. Near Earth's surface, the water vapor content ranges from just a fraction of 1 percent of the total atmosphere over deserts and polar regions to about 4 percent in the tropics. This means that at most we would find that 4 out of every 100 air molecules are water vapor. (Outside of the tropics, water vapor content does not usually exceed 2 percent.) At higher altitudes, water vapor is even rarer.

Despite being a relatively small portion of the atmosphere, water vapor is extremely important. Not only is it the source of the moisture needed to form clouds, it is a very effective absorber of energy emitted by Earth's surface. (We describe radiant energy in the next chapter.) Its ability to absorb Earth's thermal energy makes water vapor one of the "greenhouse gases" we discuss in Chapter 3.

Keep in mind that water vapor is not the same as small droplets of liquid water. Water vapor exists as individual gas molecules. Unlike the molecules of liquids and solids, water vapor molecules are not bound together. In that regard, water vapor is similar to N_2, O_2, and other atmospheric gases. Unlike other gases, however, it readily changes phase into liquid and solid forms both at Earth's surface and in the atmosphere. As will be seen in later chapters, such changes in phase involve large amounts of energy that are very important in weather and climate.

Although water vapor is an invisible gas, some satellite systems can indirectly measure the amount of water vapor in the air and display its varying content on images (Figure 1–5). These images typically show that water vapor exhibits large changes over very short distances, more so than other variable

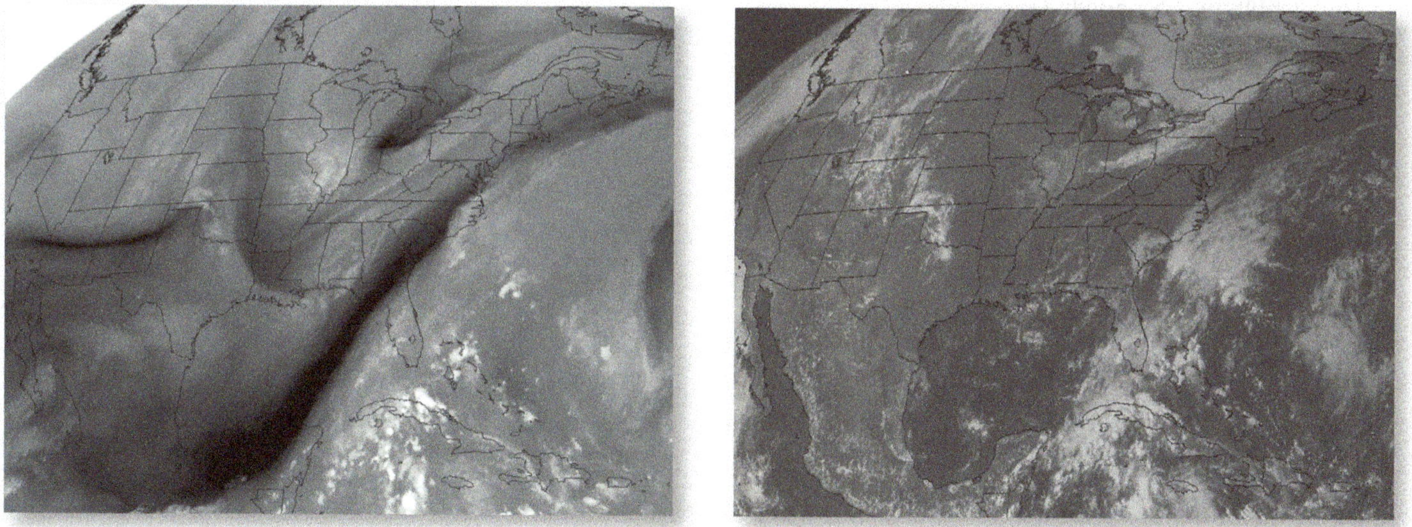

(a) (b)

▲ **FIGURE 1–5 Water Vapor in the Atmosphere.** (a) Satellite measurements of upwelling radiation can be used to estimate water vapor amount. (b) A visible rendition for the same day shows that the water vapor is more broadly distributed than might be guessed from clouds alone.

gases. Displays of water vapor content can be quite valuable to forecasters, who use the images to determine broad-scale wind patterns in the middle and upper atmosphere. The images also help meteorologists identify the boundaries between adjoining bodies of air.

CARBON DIOXIDE Another variable gas is **carbon dioxide**, CO_2. We do not physically sense variations in the amount of carbon dioxide present in the atmosphere, as we do with water vapor, and yet, as you will see shortly, we cannot ignore these variations. Increases in the carbon dioxide content of the atmosphere may have some important climatic consequences that could greatly affect human societies.

Carbon dioxide currently accounts for about 0.0396 percent of the atmosphere's volume. When a gas occupies such a small proportion of the atmosphere, we often express its content as parts per million (ppm) rather than percent (parts per hundred). Thus, the current atmospheric concentration of CO_2 is about 396 ppm. It is supplied to the atmosphere by plant and animal respiration, the decay of organic material, volcanic eruptions, and natural and *anthropogenic* (human-produced) combustion. Carbon dioxide is removed from the atmosphere by **photosynthesis**, the process by which green plants convert light energy to chemical energy. Photosynthesis uses energy from sunlight to convert water absorbed by the roots of plants and carbon dioxide taken in from the air into the chemical compounds known as carbohydrates, which both support plant processes and ultimately nourish the animal kingdom. In this process, oxygen molecules are released into the atmosphere as a by-product (see *Box 1–2, Physical Principles: Photosynthesis, Respiration, and the Global Carbon Cycle*).

For many decades, the rate of carbon dioxide input to the atmosphere has exceeded the rate of removal, leading to a global increase in concentration. Figure 1–6 plots the record of carbon dioxide content obtained from the Mauna Loa Observatory. (Taken at an elevation of 3400 m, or 11,150 ft, on one of the Hawaiian Islands, these observations are considered representative for the Northern Hemisphere.) Since the 1950s, the concentration of CO_2 has increased at a rate of about 2.15 ppm per year. The increase has occurred mainly because of an increase in anthropogenic combustion and, to a lesser extent, deforestation of large tracts of woodland (Figure 1–7). The increase has received considerable scientific and media attention because CO_2 (like water vapor) effectively absorbs radiation emitted by Earth's surface. The CO_2 increases projected to the year 2100 would double existing levels of carbon dioxide and could lead to a continued warming of the lower atmosphere that began more than a century ago.

Figure 1–6 shows not only the overall increase in CO_2 levels in the Northern Hemisphere but also seasonal oscillations. Specifically, the amount of carbon dioxide in the atmosphere is greatest in the early spring and lowest in late summer. The springtime maximum occurs because during the winter plant growth is slow, so plants take less CO_2 from the air. In addition, leaf litter has been decomposing all winter, which means carbon in the litter has been combined with oxygen (oxidized) to form CO_2 and has entered the air. During the period of summer regrowth, carbon is removed from the atmosphere, and so carbon dioxide levels fall.

For every CO_2 molecule removed from the atmosphere by photosynthesis, plants produce one O_2 molecule. Atmospheric O_2 does not show any seasonal variation, however, because the photosynthetic contribution is so small relative to the huge atmospheric O_2 reservoir. Carbon dioxide has a residence time of about 150 years.

OZONE The form of oxygen in which three O atoms are joined to form a single molecule is called **ozone**, O_3. This

▶ **FIGURE 1–6 Atmospheric Carbon Dioxide.** The carbon dioxide content of the atmosphere has been increasing for the past 150 years due to human activities. Only recent decades are shown in this figure. The data were obtained from the Mauna Loa Observatory on a high mountain in Hawaii and are representative of a broad band of Northern Hemisphere latitudes. The zigzag line reveals the seasonal cycle in the growth and decay of plants. Comparable measurements taken in the Southern Hemisphere are similar, except the seasonal cycle is of course reversed.

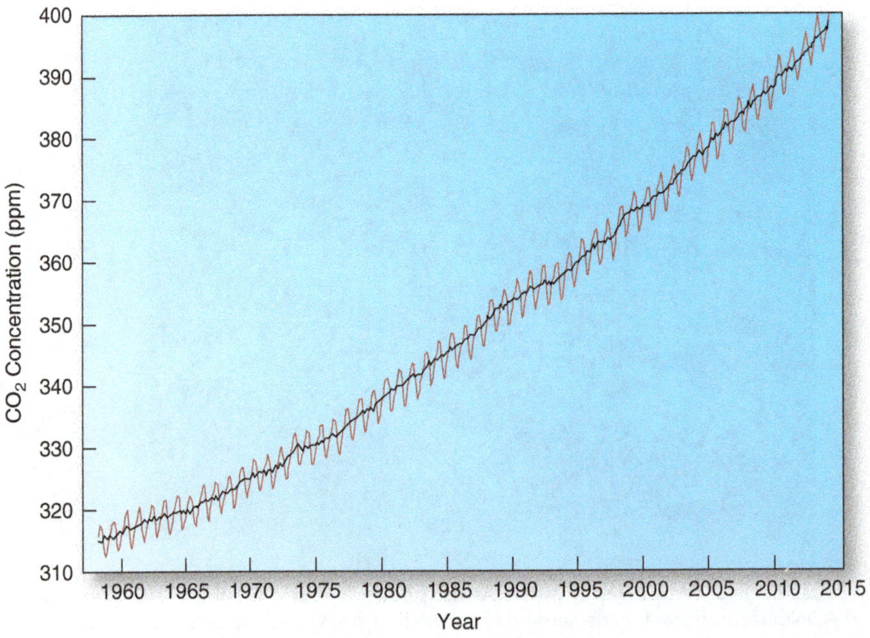

1–2 PHYSICAL PRINCIPLES

Photosynthesis, Respiration, and the Global Carbon Cycle

Without the process of photosynthesis, Earth would have an entirely different atmosphere—and would probably be without life as we know it. Through photosynthesis, plants utilize light energy from the Sun to make food. Because it requires sunlight, photosynthesis occurs only during the day. It also requires chlorophyll, an organic substance found in green plants and some single-celled organisms. Photosynthesis converts solar energy, water, and carbon dioxide into simple carbohydrates. These can then be converted to complex carbohydrates, starches, and proteins, all of which supply plants with the material for their own growth. Plants in turn provide the basic nutrients for grazing and browsing animals.

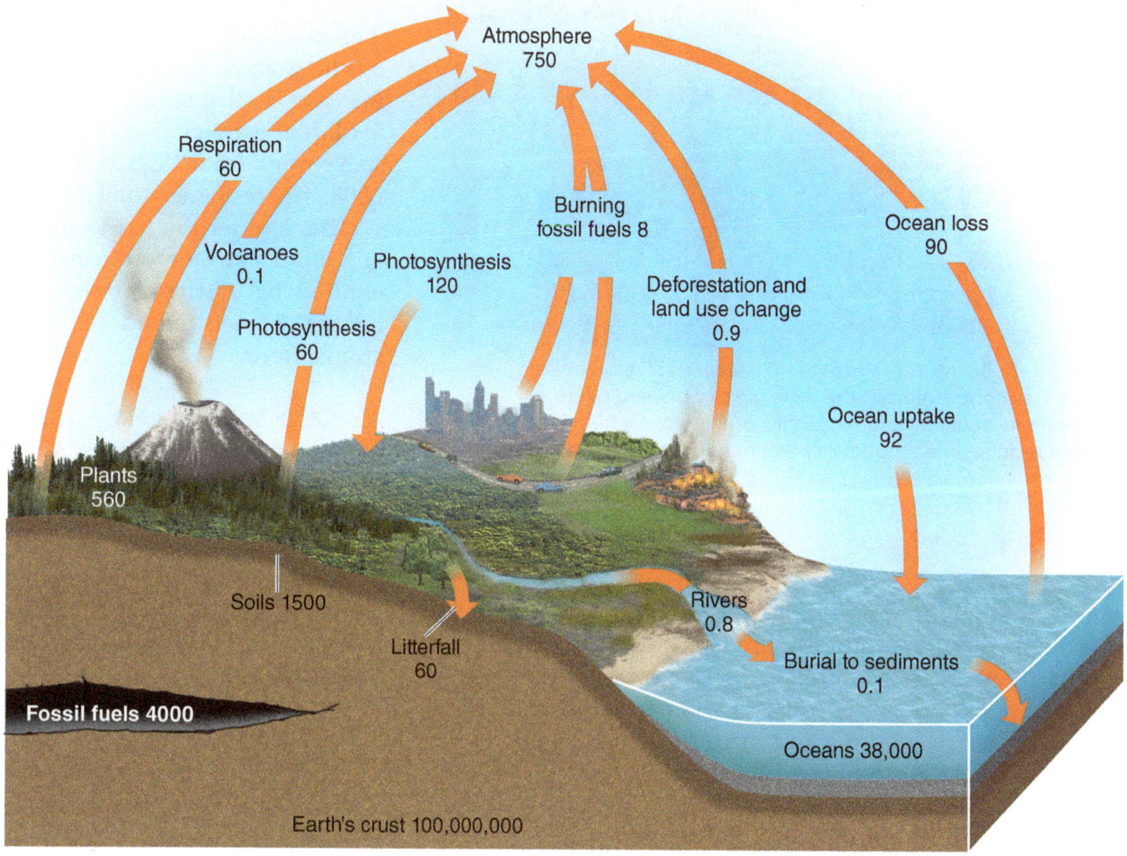

▲ **FIGURE 1–2–1 The Global Carbon Cycle.** Transfers and stores of carbon are shown in billions of metric tons (about 1.1 billion U.S. tons).

In addition to photosynthesis, an exchange of gases takes place through the leaves of plants—**respiration**. Respiration provides the mechanism by which plants obtain the oxygen they need to perform their metabolic processes. (For animals, respiration is synonymous with breathing. After a plant dies, however, it no longer takes in CO_2 from the air. Instead, its stored carbon is oxidized and released back to the atmosphere as CO_2 as the plant decomposes. If no major changes occur in the amount and distribution of vegetation, CO_2 intake (for photosynthesis) balances CO_2 output (from respiration and decay) over the course of a year, and the total store of atmospheric CO_2 is unaffected by plant growth.

There is, however, another factor governing the CO_2 balance. Under some circumstances, dead plant material is quickly buried beneath the surface, does not decompose, and therefore does not release its stored carbon back into the atmosphere. Instead, over millions of years the material is transformed into fossil fuels, such as petroleum or coal, which we extract from the ground and burn for heat and energy, yielding CO_2 as a combustion product. In doing so, we release carbon that otherwise would have remained underground for eons. Although we speak of "adding" to the atmosphere, it is probably more accurate to think of "moving" carbon back to the atmosphere. Regardless, the result is certainly an increase in atmospheric CO_2, with potentially global consequences.

Figure 1–2–1 shows these exchanges and others comprising the global carbon cycle. The values shown are billions of metric tons (a metric ton is a little more than a U.S. ton). Although no carbon disappears, some branches of the cycle are obviously very long lived, such as the return of carbon buried in rocks. In addition, the figure is accurate only for the present time. Most importantly, the fossil fuel branch has been growing for the past 150 years and will continue to do so in the coming century. We should mention that the transfers shown in the figure are the best that science can produce, yet they do not balance—they imply a loss of 2 to 3 billion metric tons per year from the atmosphere that is not accounted for in transfers to other reservoirs. Evidence increasingly suggests that the balance is taken up by the biosphere, but it is not clear where or how. As will be seen later, this uncertainty has implications for understanding both past and future changes in CO_2.

Video
Global Carbon Uptake by Plants
http://goo.gl/hDHbF

1–2–1 What is the largest carbon reservoir on Earth?

1–2–2 What process is most important in moving carbon from the surface to the atmosphere?

▲ **FIGURE 1–7 Burnt Rain Forest in Guatemala.** Deforestation by humans has contributed to the increase of carbon dioxide in the atmosphere because it releases carbon previously held in plant tissues and reduces the amount of plant material that can carry out photosynthesis.

substance is something of a paradox. The small amount of it that exists in the upper atmosphere is absolutely essential to life on Earth; near Earth's surface, however, it is a major component of air pollution, causing irritation to lungs and eyes and damage to vegetation. Fortunately, ozone occurs only in minute amounts in the lower atmosphere, so that even in highly polluted urban air its concentration may be only about 0.15 ppm (that is, 15 out of every 100 million molecules are ozone). Aloft, at altitudes of 25 km (15 mi), its concentration might be 50 to 100 times higher. Yet with a concentration of only 15 ppm at that altitude, ozone is just a tiny part of the atmosphere. As we've said, the homosphere is essentially nothing but permanent gases.

Ozone in the part of the upper atmosphere called the *stratosphere* is vital to life on Earth because it absorbs lethal ultraviolet radiation from the Sun. Why is ozone mostly found in the stratosphere rather than at Earth's surface? The answer is complicated, as literally hundreds of chemical reactions govern its abundance. To simplify, it can be said that ozone forms when atomic oxygen (O) collides with molecular oxygen (O_2). Atomic oxygen is produced in the very upper reaches of the atmosphere, but little ozone forms there because of low air density at those altitudes. Nearer Earth's surface (but still high up in the atmosphere), the chances of O atoms colliding with O_2 molecules are greater, so the highest ozone values are found there.

When it absorbs ultraviolet radiation, ozone splits into its constituent parts ($O + O_2$) which can then recombine to form another ozone molecule. Through these reactions, ozone is continually being broken down and re-formed to yield a relatively constant concentration in the ozone layer (see *Box 1–3, Focus on the Environment and Societal Impacts: Depletion of the Ozone Layer*).

METHANE Another variable gas is **methane**, CH_4. For hundreds of thousands of years, the concentration of methane has varied cyclically, rising and falling on roughly a 23,000-year basis. The pattern was broken about 5000 years ago, when methane began increasing in a departure from the expected downward trend. With industrialization the rate of increase accelerated, so that values have more than doubled during the past 200 years, reaching about 1.8 ppm at present (Figure 1–8). Most of this increase is attributed to rice cultivation, biomass burning, and fossil fuel extraction (coal and petroleum mining). The roughly stable trend from 1998 to 2007 gave way to increasing values in 2008. The residence time for methane is about 10 years.

Despite its low concentration in the atmosphere, methane is an extremely effective absorber of thermal radiation emitted by earth's surface. Thus, increases in atmospheric methane levels could play a role in the warming of the atmosphere.

AEROSOLS Small solid particles and liquid droplets in the air (excluding cloud droplets and precipitation) are collectively known as **aerosols**.[2] Although we associate them with polluted urban atmospheres, aerosols are formed by both human and natural processes. Thus, they are ever present, even in areas far from human activity. They normally occur at concentrations of about 10,000 particles per cubic centimeter (cm^3) over land surfaces, which is equivalent to roughly 17,000 particles per cubic inch ($in.^3$).

DID YOU KNOW?

On average, each breath a person takes brings into the lungs about 1000 cm^3 (1 liter, or 61 in^3.) of air. Given the average size and concentration of aerosols, each of us draws about 1 trillion aerosols into our lungs several times each minute, or about two tablespoons of solids each day.

The smallest of these particles have radii on the order of 0.1 micrometer (μm; one-millionth of a meter) and are believed to form primarily from the chemical conversion of sulfate gases to solids or liquids. Larger particles are introduced into the air directly as wind-generated dust, volcanic ejections, sea spray, and combustion by-products. Because these particles are extremely small, most fall so slowly that even the smallest of vertical motions keeps them suspended in the atmosphere. (The most effective mechanism for removing aerosols is their capture by falling precipitation.) Aerosols typically have life spans of a few days to several weeks.

Aerosols have some noticeable effects on the atmosphere. Urban smog, which includes aerosols, severely reduces visibility, and dust storms can reduce visibility to near zero when large

[2]The term **particulate** is often used interchangeably with *aerosol*. This terminology may be confusing for some people, who understand *particulate* to mean solid particles only. However, at these microscopically small sizes, it is difficult to make a distinction between the liquid and solid states; thus, there is little reason to prefer one term over the other.

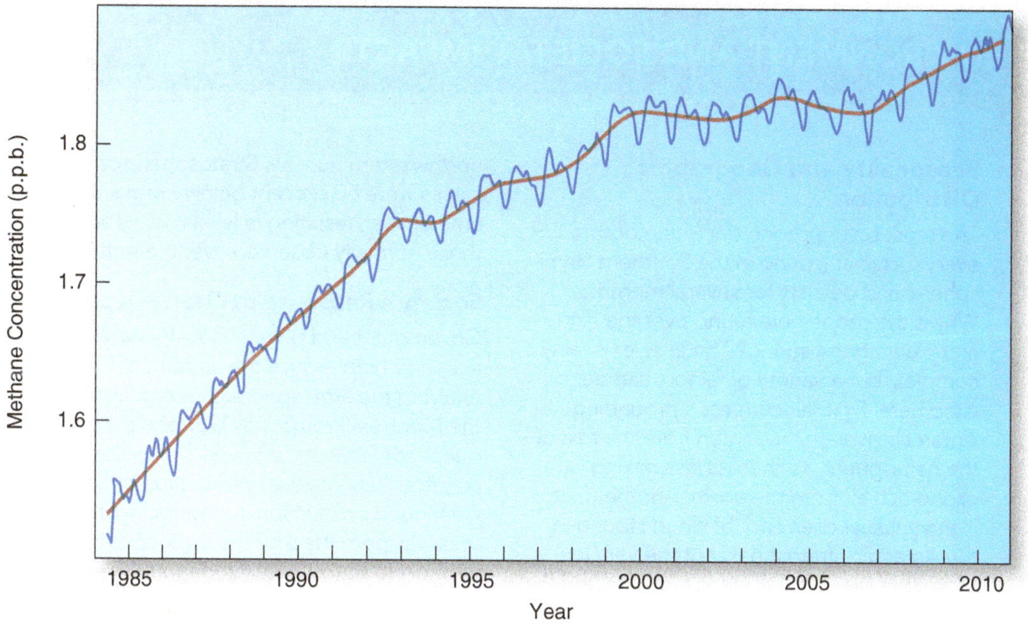

◀ **FIGURE 1–8 Methane Levels in the Atmosphere.** Methane is a gas that has varied greatly over long time scales. The most recent increase began about 5000 years ago from a value of 600 ppb. Methane amounts roughly doubled in the past 200 years, but the rate of increase has slowed in the past two decades.

volumes of soil are dislodged from the surface (Figure 1–9). Aerosols also play a major role in the formation of cloud droplets because virtually all cloud droplets that form in nature do so on suspended aerosols called **condensation nuclei**. Condensation nuclei are discussed in more detail in Chapter 5.

CHECKPOINT

1.10 What are permanent gases? Variable gases?

1.11 Classify the gases in the atmosphere according to whether they are permanent or variable.

1.12 Only a few of the gases in Tables 1–2 and 1–3 are important in weather and climate. Which ones?

▼ **FIGURE 1–9 Dust Storm Hitting Phoenix, Arizona, July 21, 2012.**

Atmospheric Density and Pressure

1.6 Explain how air pressure arises and describe the vertical variation of pressure and density.

As we have seen, the atmosphere has no distinct upper boundary; the air simply becomes less and less dense with increasing altitude. In this section, we look first at the changing density of the atmosphere, then pressure (a closely related variable).

Density

The **density** of any substance is the amount of mass of the substance (expressed in kilograms)[3] contained in a unit of volume (1 cubic meter, m^3). In gases, individual molecules have no attachment to one another and move about randomly. One characteristic of gases is that no definite limit exists to the amount of mass that can fill a given volume—molecules can always be added to or removed from the volume, resulting in a density change. Alternatively, the volume containing a fixed mass of gas can decrease (as happens in a bicycle pump as the piston moves inward and compresses the air), resulting in a density increase as the constant mass is squeezed into a smaller and smaller volume.

Like any other assemblage of gases, therefore, the atmosphere is compressible. When we feel the weight of something, we are being subjected to the downward gravitational force

[3]Although a kilogram corresponds to about 2.2 pounds in the English system of measurement used in the United States, there is a difference between the two. A kilogram is a unit of mass. A pound, however, is a unit of weight equal to mass times the acceleration of gravity. On Earth 1 kilogram of a substance has a weight of 2.2 pounds. If the same kilogram goes to the Moon, its mass will not change but its weight will be only one-sixth of what it was on Earth. This is true because the acceleration of gravity on the Moon is one-sixth as strong as the acceleration of gravity on Earth.

1–3 FOCUS ON THE ENVIRONMENT AND SOCIETAL IMPACTS

Depletion of the Ozone Layer

In 1972 Sherwood Rowland and Mario Molina, atmospheric chemists then working at the University of California at Irvine, proposed that certain human-produced chemicals called chlorofluorocarbons (CFCs) could be carried naturally into the stratosphere and damage the ozone layer. At that time CFCs were widely used in refrigeration and air conditioning, in the manufacture of plastic foams, and as solvents in the electronics industry.

Solving the Ozone Puzzle

Knowing that CFCs do not easily react with other molecules in the lower atmosphere, Rowland and Molina needed to explain how CFCs could affect ozone. They proposed that these molecules could reach the stratosphere intact, where they would break down and release free atoms of chlorine (Cl). Under certain circumstances, chlorine atoms can effectively destroy ozone molecules. In the first step of this process, a chlorine atom reacts with an ozone molecule to produce O_2 and chlorine monoxide (ClO). Next, an oxygen atom (O) reacts with ClO, creating another O_2 molecule while freeing the chlorine atom (Cl). Note that the chlorine atom that first reacted with the ozone molecule is still present and capable of reacting again with another ozone molecule. This fact makes chlorine atoms able to repeatedly break down ozone molecules. In fact, as many as 100,000 ozone molecules can be removed from the atmosphere for every chlorine molecule present. Rowland and Molina's theory is now accepted, and the two scientists were rewarded for their work with a Nobel Prize in 1995.

Seasonality and Geographic Distribution

The most severe ozone depletion occurs every October (spring in the Southern Hemisphere) and persists for several months. Why is the ozone hole found over the Antarctic during the spring? The answer is fairly complex, but a variety of factors can be cited here. First, air currents surrounding Antarctica isolate the region from the rest of the hemisphere, so there is less mixing with ozone-rich air from the north. Another factor is the unusual chemistry of clouds found in the Antarctic stratosphere. At the very low temperatures found there, clouds are largely made up of nitric acid and water, rather than ordinary water ice. Processes involving these clouds allow certain chlorine compounds to accumulate. When the dark Antarctic winter ends, the burst of ultraviolet radiation breaks apart these compounds to create free chlorine atoms (Cl). These Cl free radicals readily destroy ozone, as described previously. A very recent discovery shows that depletion begins at the rim of Antarctica, where sunlight arrives first, and works its way poleward. Thus, depletion begins in June at 65° S but not until late August at 75° S. We should emphasize that chlorine is not usually abundant in the Antarctic stratosphere. Rather, what chlorine exists is in a form able to destroy ozone, thanks to the unique conditions that take place in the spring (see Figure 1–3–1).

Significant stratospheric ozone depletion has also been detected over much of Europe and North America, including a less pronounced hole over the Arctic. Nobody knows how long the ozone will continue to decrease, or how depleted it may become, but in November 1999 satellite and land-based data revealed abnormally low ozone levels over northwestern Europe. Stratospheric ozone values were 30 percent below normal for that time of year, resulting in levels nearly as low as those normally observed over the Antarctic.

Society's Response to Ozone Depletion

Government and private industry have taken action to help reduce the amount of CFCs entering the atmosphere. In accordance with the Montreal Protocol of 1987 and subsequent conferences, the world's developed countries have ceased production of CFCs. Continued production in developing countries was permitted by the protocol and many have done so (by 2007 China was the world's largest producer). CFCs have lifetimes in the atmosphere of about 100 years, so an immediate reduction in the ozone hole will not occur. But progress has been made; since 1997 there has been a decline in the amount of chlorine in the stratosphere, and the size of the ozone hole appears to have stabilized, although significant year-to-year variations still occur due to weather conditions. A 2006 United Nations study predicted that Antarctic ozone levels may return to pre-1980 levels sometime between 2060 and 2075—a slow but significant improvement. Thus, curbing CFC emissions illustrates how solid science, combined with international cooperation, can have a major impact on the protection of the environment.

1–3–1 Where and when does the ozone hole occur?

1–3–2 Think about the production of CFCs and the release of carbon dioxide from fossil fuel burning. If society wanted to limit future carbon dioxide levels, would it be easy to achieve success like that seen with CFCs?

exerted by the overlying mass. At lower altitudes, there is more overlying atmospheric mass than at higher altitudes. Because air is compressible and subjected to greater compression at lower elevations, the density of the air at lower levels is greater than that aloft. Figure 1–10 illustrates this principle. Despite its simplicity, this concept is a key to understanding concepts presented later.

DID YOU KNOW?

Olympic speed skaters account for variations in air density. When racing at high elevations, they glide farther in the comparatively thin air and take 14 fewer strokes per lap.

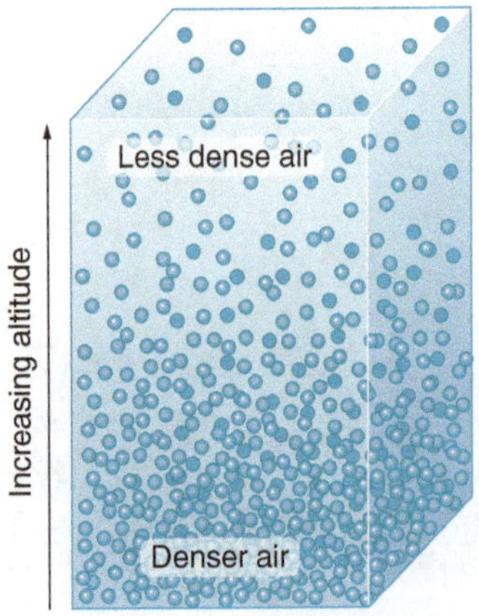

◀ **FIGURE 1–10 Density in Relation to Altitude.** Because of compression, the atmosphere is more dense near its base and progressively "thins out" with altitude.

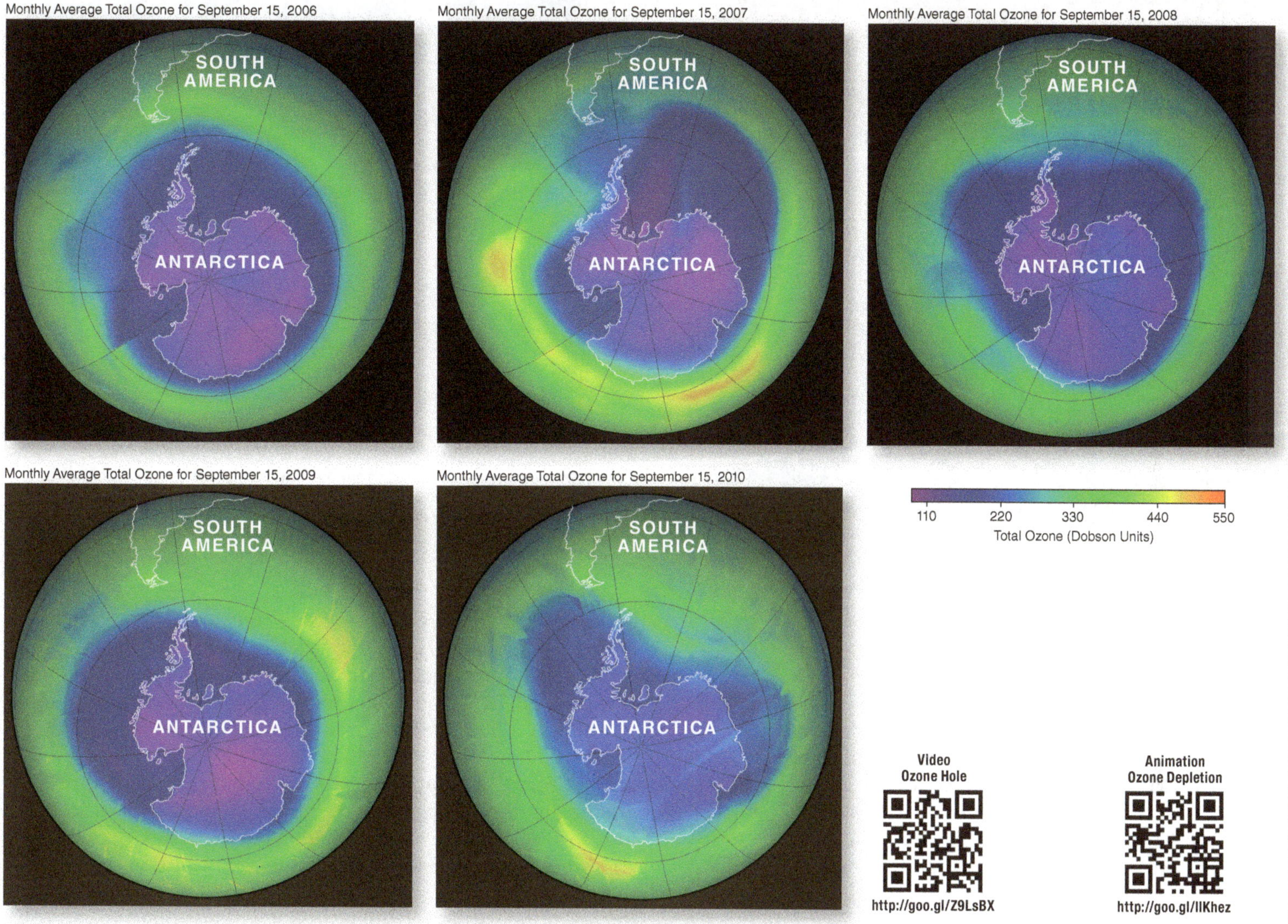

Video
Ozone Hole
http://goo.gl/Z9LsBX

Animation
Ozone Depletion
http://goo.gl/IlKhez

▲ **FIGURE 1–3–1 Changes in the Ozone Layer.** This series of satellite images shows a reduction of the stratospheric ozone over Antarctica. Mapped values are Dobson Units (DU), a measure of total ozone amount in the atmospheric column, nearly all of which is found in the stratosphere. The global average is about 340 DU. The boundary of the "ozone hole" is considered to be 220 DU, shown in dark blue. Values that low were not observed prior to 1979.

Air density decreases gradually with increasing altitude. At sea level, the air density is normally about 1.2 kg/m^3. By comparison, at Denver, Colorado, the "Mile High City," the air density is only about 85 percent of that. As a result, punted footballs and batted baseballs experience a corresponding decrease in air resistance and travel farther than they would at sea level.

One way to think about density is in terms of the average distance a molecule travels before colliding with another. Near Earth's surface, molecules move a mere 0.0001 millimeter (mm) before collision. In contrast, at 150 km (93 mi) above sea level, a molecule travels about 10 m (33 ft) before colliding with another; at 250 km (155 mi), a molecule is likely to travel a full kilometer (0.62 mi) before meeting another.

Pressure

We don't feel it, but the atmosphere's great mass is far from weightless as the crushed can in Figure 1–11 demonstrates. At sea level the weight is about 14.7 pounds per square inch,

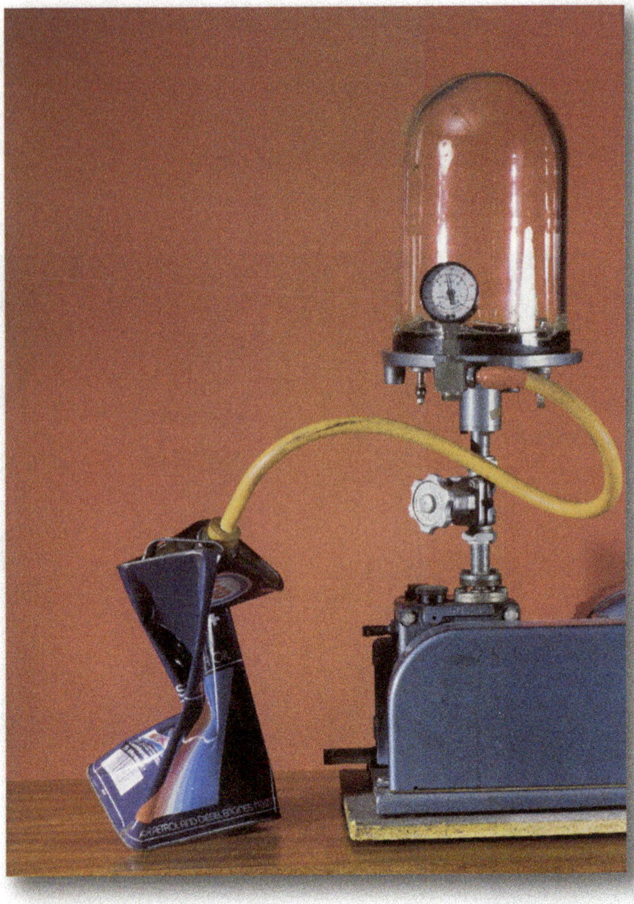

▲ **FIGURE 1–11 Demonstrating Atmospheric Pressure.** A can is crushed by atmospheric pressure after air is removed by a vacuum pump.

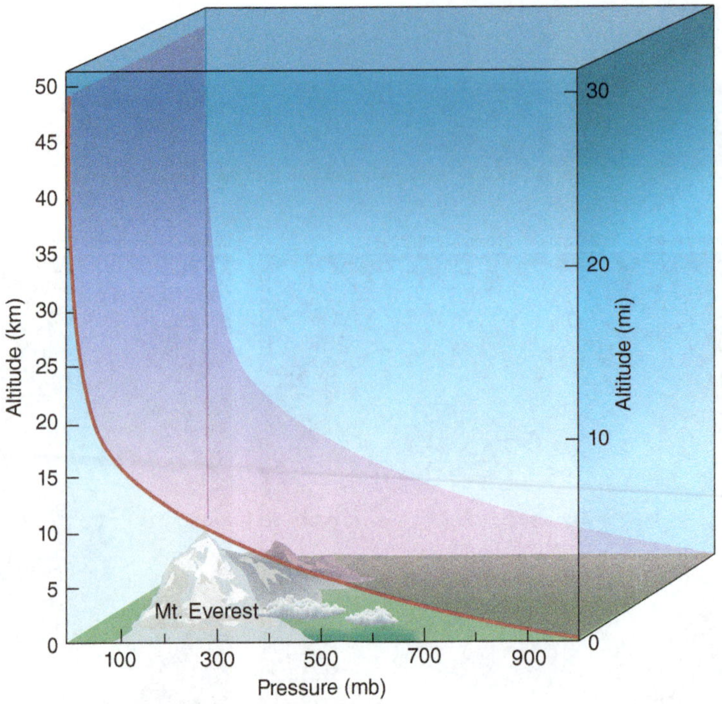

▲ **FIGURE 1–12 Pressure in Relation to Altitude.** Pressure decreases with increasing altitude, falling by about 50 percent for every 5 km (3 mi). Atmospheric density behaves similarly. The nonlinear trend in both is a consequence of the compressibility of air.

which in everyday language is known as sea-level air pressure. The corresponding metric unit of pressure used in the United States is the **millibar (mb)** and in Canada the **kilopascal (kPa)**. Specified in these terms sea-level pressure is 1013 mb and 101.3 kPa. Chapter 4 will discuss this more thoroughly. The main points for now are that pressure arises because of the atmosphere's mass, and its value reflects the mass of atmosphere above a given point. We don't feel the atmosphere's weight because our body cavities are full of air matching the surrounding air pressure. The closest we get to a sensation of pressure is the "pop" in our ears as airplanes ascend and descend, reflecting a temporary pressure imbalance between the cabin air and the air inside the eardrum. It follows that as one goes upward through the atmosphere, the amount of overlying air is less and the pressure is necessarily lower. You can see this by inspecting Figure 1–10 and imagining a count of the molecules above any height. The count would decrease with height, and with that the corresponding pressure.

Pressure must decrease vertically, but the rate of decrease is not constant. Again looking at Figure 1–10, it seems clear that in the lower part of the column one passes through many molecules in a short vertical distance and pressure falls rapidly, but at higher altitudes lower density means one must go farther to find the same pressure decrease. Rather than a uniform decrease with height, the pressure falls rapidly near the surface and more slowly aloft as seen in Figure 1–12. Instead of a straight line corresponding to a constant absolute change with height, the curve suggests change by a constant percentage per unit distance. In fact, as a rough rule of thumb, pressure falls by about 50 percent for every 5 km (3 mi) of altitude. The graph shows a pressure of about 500 mb at 5 km (1000 divided by 2), 250 mb at 10 km (500 divided by 2), and so forth. This nonlinear relationship between pressure and height arises because of the atmosphere's compressibility, just as the density variation with altitude arises because of compressibility. Moreover, vertical changes in density follow the same rough rule of thumb (ignoring small effects produced by temperature differences and assuming no vertical accelerations are present[4]). Pressure and density are clearly related, and for understanding the structure of the atmosphere there is no reason to prefer one over the other. That said, pressure does have an advantage in being much easier to measure, and it is much more commonly used in most aspects of meteorology.

CHECKPOINT

1.13 How does air pressure arise? How does it vary vertically?

1.14 Why do air pressure and density show much the same vertical trend?

[4]Chapter 4 will discuss these assumptions in more detail and will explain the remarkable reliability of this approximation. It will also make the connection between density and temperature explicit.

Layers of the Atmosphere

1.7 Identify and describe the layers of Earth's atmosphere.

We know that the atmosphere's composition remains nearly constant up to a height of about 80 km (50 mi)—that is, within the homosphere. Yet despite this gradual change in density and nearly constant chemical composition with height, meteorologists still find it convenient to divide the atmosphere vertically into several distinct layers. Some layers are distinguished by electrical characteristics, some by chemical composition (this is the homosphere/heterosphere distinction discussed earlier), and some by temperature characteristics. Together with the change in density with height, this layering gives the atmosphere its structure.

Layering Based on Temperature Profiles

Compared to horizontal motions in the atmosphere (i.e., wind), vertical motions are usually slow and limited in range because of the shallowness of the atmosphere. Nonetheless, such motions greatly influence the likelihood of cloud development, precipitation, and thunderstorm activity. Air whose temperature decreases rapidly with height rises readily, while air whose temperature either decreases slowly or increases with height resists such motion. Scientists therefore divide the atmosphere into four layers based not on chemical composition (which is relatively constant throughout most of the atmosphere), but rather on how mean temperature varies with altitude. The average temperature profile shown in Figure 1–13, called the **standard atmosphere**, shows the four layers: troposphere, stratosphere, mesosphere, and thermosphere.

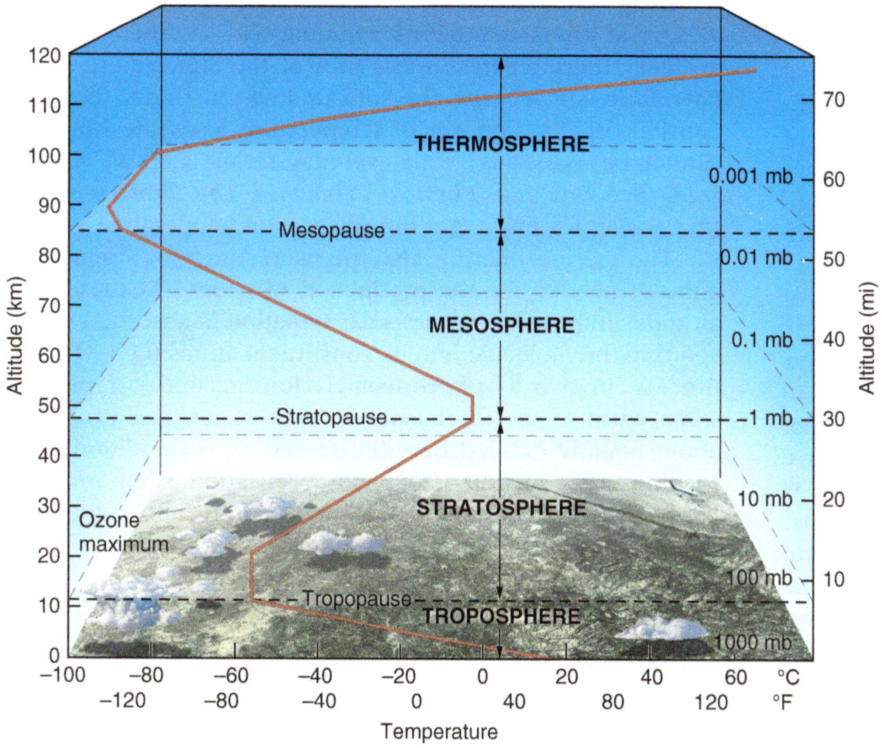

▲ **FIGURE 1–13 Layers of the Atmosphere.** The temperature profile of the atmosphere results in four layers based on thermal characteristics.

THE TROPOSPHERE Unless you get a chance to fly in a military jet, you will spend your entire life in the **troposphere**, the lowest of the four temperature layers. The name is derived from the Greek word *tropos* ("turn") and implies an "overturning" of air resulting from the vertical mixing and turbulence characteristic of this layer. The troposphere is where the vast majority of weather events occur and is marked by a general pattern in which temperature decreases with height. Despite being the shallowest of the atmosphere's four layers, the troposphere contains about 85 percent of its mass. This is possible, of course, because air is compressible.

The depth of the troposphere varies considerably, ranging from 8 to 20 km (3.6 to 12 mi), with a mean of about 15 km (9 mi). The altitude at which the troposphere ends depends largely on its average temperature, being highest where the air is warm and lowest in cold regions. The troposphere is therefore thicker over the tropics than over the polar regions and thicker during the summer than during the winter.

Temperatures vary greatly from bottom to top in the troposphere. The average global temperature is about 15 °C (59 °F) near the ground but only about −59 °C (−71 °F) at the top of the troposphere, an average decrease of about 6.5 °C/km (3.6 °F/1000 ft). Thus, you might feel perfectly comfortable inside an airplane at an altitude of 10 km (33,000 ft), but the temperature outside would likely be in the neighborhood of −50 °C (−58 °F). At the top of the troposphere, a transition zone called the **tropopause** marks the level at which temperature ceases to decrease with height.

An apparent paradox is associated with the trend toward decreasing temperature with height in the troposphere. We know that Earth is heated almost entirely by the Sun, which might make you think temperature should *increase* with height during the day (because as we move upward from the surface, we get closer to the Sun). The explanation for this paradox is that the atmosphere is relatively transparent to most types of radiant energy emitted by the Sun. In other words, a large portion of the sunlight passing through the atmosphere is not absorbed and therefore does not contribute greatly to its warming. As we shall see in Chapter 3, the most important direct source of energy for the atmosphere is not downward-moving solar radiation but rather energy emitted upward from Earth's surface.

Despite the strong tendency for temperature to decrease with altitude in the troposphere, it is not uncommon for the reverse situation to occur. Such situations, where temperature increases with height, are known as **inversions**. An inversion is significant because it inhibits upward motion and thereby allows high concentrations of pollutants to be confined to the lowest parts of the atmosphere.

THE STRATOSPHERE Above the tropopause is the **stratosphere**, a name derived from the Latin word for *layer*. Except for the penetration of some strong thunderstorms into the lower stratosphere, little

▲ **FIGURE 1–14 Thunderstorm Clouds.** Most clouds exist in the troposphere, but the tops of strong thunderstorm clouds can extend kilometers into the stratosphere when violent updrafts occur. The flattened area at the top of this cumulonimbus cloud is in the stratosphere.

weather occurs in this region (Figure 1–14). In the lowest part of the stratosphere, the temperature remains relatively constant at about –57 °C (–71 °F) up to a height of about 20 km (12 mi). From there to the top of the stratosphere (called the **stratopause**), about 50 km (30 mi) above sea level, the temperature increases with altitude until it reaches a mean value of –2 °C (28 °F).

In the upper stratosphere, heating is almost exclusively the result of ultraviolet radiation being absorbed by ozone. Therefore, as solar energy penetrates downward through the stratosphere, there is less and less ultraviolet radiation available and a resultant decrease in temperature. In the part of the stratosphere where temperature does not vary with height, heating is the result of both absorption of solar ultraviolet radiation and absorption of thermal radiation from below. Thus, as we move up or down in this region, the reduction in solar heat is offset by the increase in heat given off by Earth. The net result is the straight vertical line of no temperature change shown in Figure 1–13.

The stratosphere contains about 19.9 percent of the total mass of the atmosphere. Thus, the troposphere and stratosphere together account for 99.9 percent of the total mass of the atmosphere. The fact that gases are so compressible allows the major portion of the mass of the atmosphere to be contained in the two lowest layers. Note that despite being more than triple the depth of the troposphere, the stratosphere contains only one-quarter the mass: troposphere = 80 percent of total atmospheric mass; stratosphere = 20 percent.

Within the stratosphere is the **ozone layer**, a zone of increased ozone concentration at altitudes between 20 and 30 km (12 and 18 mi). Despite its name, the ozone layer is not composed primarily of ozone. In fact, at 25 km (15 mi) above sea level, where the percentage of ozone in this region is greatest, its concentration might be only about 10 ppm. Despite its scarcity, ozone is an extremely important constituent of the stratosphere. It is largely responsible for absorbing the solar energy that warms the stratosphere, and it also protects life on Earth from the lethal effects of ultraviolet radiation.

Being well removed from Earth's surface, where evaporation supplies water vapor to the atmosphere, the stratosphere has a very low moisture content. Moreover, as seen in Figure 1–13, the temperature characteristics of the stratosphere inhibit vertical motions that favor the formation of clouds (to be discussed in Chapter 6). These conditions also inhibit precipitation, and consequently particulates from volcanic eruptions can remain in the stratosphere for many months. Furthermore, the strong winds of the stratosphere can cause its aerosol content to be distributed across the globe, creating a veil of material that can affect the penetration of sunlight to the surface. For a couple of years after the 1991 eruption of Mount Pinatubo in the Philippines, for example, the Northern Hemisphere experienced redder-than-normal sunrises and sunsets as a result of aerosols in the stratosphere accompanied by a measurable decrease in global temperature.

THE MESOSPHERE AND THERMOSPHERE Of the 0.1 percent of the atmosphere not contained in the troposphere and stratosphere, 99.9 percent exists in the **mesosphere**, which extends to a height of about 80 km (50 mi) above sea level. As in the troposphere, temperature in the mesosphere decreases with altitude. The only significant source of heat is absorption of solar radiation near the base of the mesosphere. This is dispersed upward only weakly by vertical air motions, thus temperatures fall rapidly with increasing altitude.

Above the mesosphere is the **thermosphere**, where temperature increases vertically to values in excess of 1500 °C (2732 °F). These high temperatures can be misleading, however, if we overlook the distinction between high temperature and high heat content. The temperature of the air is an expression of its internal energy, which is related to the speed at which its molecules move. The amount of heat contained in a volume of air reflects not only its temperature but also its mass. Because there are so few gas molecules in this layer, the air cannot have a high heat content no matter what its temperature is. In fact, density is so low in the upper reaches of the thermosphere that a gas molecule will normally move as much as several kilometers before colliding with another. Thus, an ordinary thermometer in this part of the atmosphere would have little contact with the surrounding air. Under these circumstances, the concept of temperature loses meaning and cannot be associated with everyday terms such as *hot* and *cold*. (See *Box 1–4, Physical Principles: The Three Temperature Scales,* for important information on temperature scales.)

The thermosphere is also the main source of atomic oxygen (O) needed for the production of ozone. At these high altitudes intense ultraviolet solar radiation is absorbed by O_2, splitting the molecule into its constituent atoms (O + O). This process, known as **photodissociation**, occurs so rapidly in some parts of the thermosphere that the oxygen present is about equally divided between O and O_2. In addition to its role in the production of O, photodissociation is important for heating the thermosphere and—most significantly—shielding the surface from lethal doses of ultraviolet radiation.

DID YOU KNOW?

Although Venus is approximately the same size as Earth, its atmosphere is vastly different. The surface density of Venus's atmosphere is some 50 times greater than Earth's and it is 96.5 percent carbon dioxide. Our planetary neighbor has a vastly different temperature as well—averaging about 475 °C, or 890 °F!

1–4 PHYSICAL PRINCIPLES

The Three Temperature Scales

Fahrenheit and Celsius

At one time, the temperature scale used all over the world was the **Fahrenheit scale**. Invented in the early 1700s by Gabriel Fahrenheit, it assigns values of 32° and 212° to the freezing and boiling points of water.* Thus, there are 180 Fahrenheit degrees between freezing and boiling. Fahrenheit developed his scale by assigning 0° and 100° to temperatures he could produce in his laboratory. For the lower temperature, he used a mixture of water, ice, and a salt, and (according to one account) his wife's armpit served to establish the other as the temperature of a human body. Although the Fahrenheit scale has been replaced in Canada and nearly every other country around the world, in the United States it is still the scale used by the general public.

The other familiar scale for measuring temperature is the **Celsius scale**, named for Anders Celsius, who formulated it in 1742. The Celsius scale assigns values of 0° and 100° to the freezing and boiling points of water, so there are only 100 Celsius degrees between the two points. This means that a Celsius degree is larger than a Fahrenheit degree. Thus, for example, a 2 °C change is larger than a 2 °F change. However, this is *not* to say a temperature expressed in °C is always higher than the same temperature expressed in °F; it can be higher or lower. To convert from Celsius to Fahrenheit, we use the following formula:

$$°F = 9/5°C + 32$$

To convert from Fahrenheit to Celsius, we use

$$°C = 5/9\,(°F - 32)$$

You can use these formulas to verify that −40 °F = −40 °C.

Kelvin

Though useful in everyday applications, both the Fahrenheit and Celsius scales have a significant shortcoming—they allow for negative values. This is not the case for other units of measurement. For example, buildings do not have negative heights and weights, cars do not travel at negative speeds, and children do not have negative ages. But negative temperature values give the impression that substances can have negative heat contents—a situation that is physically impossible. To overcome this problem, scientists use a different scale for the measurement of temperature, called the **Kelvin scale**. In this system, the temperature 0 K is the lowest possible temperature that can exist in the universe. (Notice that we omit the degree notation with this scale and just refer to the number of kelvins,† K.) A temperature of 0 K implies no heat, and it is therefore impossible for subzero temperatures to exist with this scale.

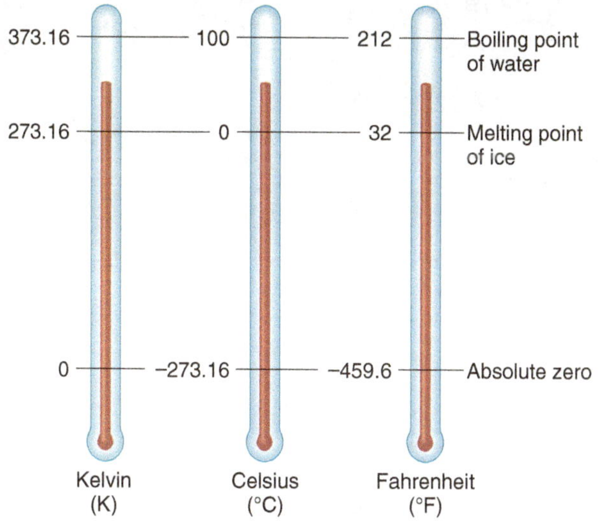

▲ **FIGURE 1–4–1 The Kelvin, Celsius, and Fahrenheit Scales.**

The Kelvin scale is really a modified form of the Celsius scale insofar as the increments of the two are equal. Thus, if the temperature increases 1 degree Celsius, it also increases 1 kelvin. The only difference between the two is the starting point; 0 K corresponds to −273.15 °C. Therefore, conversion from Celsius to Kelvin is simply

$$K = °C + 273.15$$

To convert from Kelvin to Celsius, we use

$$°C = K - 273.15$$

Figure 1–4–1 shows the Kelvin, Celsius, and Fahrenheit scales.

1–4–1 Convert 10 °C to °F.

1–4–2 If the temperature of a location were to increase by 2 °C, would the corresponding change in °F be larger, smaller, or the same?

*To be precise, these values apply to pure water at sea level. In addition, water doesn't freeze spontaneously at 32 °F, so "melting point" is a better term than "freezing point."

†The Kelvin scale is abbreviated with a capital K; the unit of measurement is spelled with a lowercase k.

A Layer Based on Electrical Properties: The Ionosphere

The four layers of the atmosphere described previously are delineated on the basis of temperature profiles. An additional layer, called the **ionosphere**, can be defined based on its electrical properties. This layer, which extends from the upper mesosphere into the thermosphere, contains large numbers of electrically charged particles called *ions*. **Ions** are formed when electrically neutral atoms or molecules lose one or more electrons and become positively charged ions or gain one or more electrons and become negatively charged ions. In the ionosphere, atoms and molecules lose electrons as they are bombarded by solar energy, thus creating positively charged ions and free electrons. The resulting ion-rich layers wax and wane on a daily basis with the rise and fall of the sun's ionizing radiation. (Like dissociation, ionization also warms the thermosphere, contributing further to the high temperatures found there.)

The ionosphere affects radio communication with satellites, including those of the Global Positioning System. Without accounting for ionospheric disturbances, GPS receivers would have much larger positional errors. The ionosphere is also responsible for the **aurora borealis**, or northern lights, and **aurora australis**, the southern lights (Figure 1–15). In the ionosphere, subatomic particles from the Sun are captured by Earth's magnetic field (the same field that makes compass

▲ **FIGURE 1–15 Aurora Australis.** Subatomic particles from the Sun are captured by Earth's magnetic field, causing an agitation of molecules and the emission of light with different colors.

needles point to the north). Collisions with atmospheric gases ionize some atoms and *excite* others (meaning the electrons of the atoms jump to greater orbital distances from their nuclei). Radiation is emitted when electrons fall back to lower orbits, or when ionized atoms regain a free electron. Thus the aurora does not reflect radiant energy as do clouds, but rather emits light that is much like a neon lamp.

CHECKPOINT

1.15 List the four temperature layers of the atmosphere and describe their characteristics.

1.16 Temperature decreases with altitude in the troposphere but increases with altitude in the stratosphere. What is the explanation for this?

Origin and Evolution of the Atmosphere

1.8 Explain the evolution of the atmosphere during Earth's history.

It is generally believed that Earth was formed 4.6 billion years ago, but our atmosphere is a comparatively recent development. Its origin is traced to gases locked up in the interior of Earth and released through volcanic **outgassing**. Outgassing was possibly augmented, maybe even dwarfed, by material brought to Earth in small comets on the order of 15 m (49 ft) in diameter. Recent satellite observations suggest that there is a steady rain of these water-bearing "cosmic snowballs," with from 5 to 30 striking Earth at any one time. If they have been falling at this rate since the formation of Earth, they might well be the source of most of the world's water vapor.

Evolution of the Atmosphere

The atmosphere has been greatly transformed since its early days rich in water vapor and carbon dioxide. The removal of the water vapor is easily explained: Upon cooling in the prehistoric atmosphere, it readily condensed and precipitated to the surface to form the oceans. Precipitation also contributed to the removal of carbon dioxide from the atmosphere. As raindrops fell through the CO_2-rich atmosphere, some CO_2 was dissolved in every drop. (Soft drinks provide an everyday example of how CO_2 is dissolved in water). As they fell, the CO_2 drops transferred carbon from the atmosphere to the oceans, where it combined with material eroding from the continents and was buried in seafloor sediments.

The transformation to an atmosphere high in oxygen depended on the advent of primitive, anaerobic bacteria (those that survive in the absence of oxygen) about 3.5 billion years ago. These primitive life-forms were the first in a long line of organisms that removed carbon dioxide from the air and replaced it with oxygen (these gases are exchanged freely between the oceans and the atmosphere). All of these processes gradually led to an increase in atmospheric oxygen at the expense of carbon dioxide. But one other transformation had to take place to create an atmosphere that could support life at the surface. Recall that without an ozone layer in the upper atmosphere, sunlight reaching the surface would contain lethal levels of ultraviolet radiation. Thus, a protective ozone layer had to develop before life could exist outside the oceans. Fortunately, this occurs naturally when the ultraviolet radiation breaks down diatomic oxygen molecules into individual oxygen atoms. The oxygen atoms could then recombine with O_2 in the upper atmosphere to form the ozone layer. Once this happened, the amount of ultraviolet radiation reaching the surface was reduced sufficiently to allow plants to survive on land. This increase in plant cover in turn accelerated the rate at which photosynthesis replaced atmospheric carbon dioxide with oxygen.

None of this accounts for the high concentration of nitrogen in the atmosphere. Although it constitutes a small portion of the material released by outgassing, nitrogen is removed from the atmosphere very slowly. As a result, its concentration has gradually increased to the point that it is now the main constituent of the atmosphere. Finally, argon, the third most abundant gas, is explained by slow seepage from the solid Earth of argon-40 with its 18 protons and 22 neutrons. Argon-40 is produced by the radioactive decay of a rare form of potassium. Another isotope, argon-36, is abundant in the universe but nearly absent on Earth. This is additional evidence that any primordial atmosphere was lost. Argon is an inert gas, so if the current atmosphere had formed at the same time as Earth, argon-36 should be present in much higher amounts.

Video
The Influence of Volcanic Ash
http://goo.gl/oYEzbE

CHECKPOINT

1.17 What two gases were most abundant in the early atmosphere of Earth?

1.18 Give two examples of how biological processes (life) have affected the composition of the atmosphere. Are they important today?

Some Weather Basics

1.9 Identify the basic types of data found on weather maps.

All of us are familiar with daily weather forecasts, from which we routinely receive information on the present and predicted state of the atmosphere. In recent years, the amount of weather information available to the public has exploded, most notably via the Internet. Detailed maps and weather reports that were once available only to professional meteorologists can now be accessed by anybody with a computer, tablet, or smartphone connected to the Internet. Such access makes learning the fundamentals of weather and climate far more enjoyable, because we can now look at map, satellite, radar, and other resources and see how the principles of meteorology play out on a daily basis. We now present an overview of the fundamentals of weather, along with an introduction to weather maps.

Temperature

Temperature is one of the most obvious and important weather components. Everyday experience indicates that temperature usually varies gradually from one place to another. In other words, as you take an 80 km (50 mi) car trip over flat terrain you are not likely to experience temperatures that vary greatly. Instead, you will probably observe only gradual changes in temperature as you drive along. On the other hand, there are times when substantial temperature differences appear over short distances, or when major shifts in temperature occur over short time periods at a particular location. These major changes in temperature often occur due to the presence of **fronts**, fairly narrow boundary zones separating relatively warm and cold air. As you drive across the frontal zones, you will experience notable temperature shifts. Likewise, if a front passes over your location, the temperature will change substantially.

Four types of fronts exist, which are discussed later in this text. For now, let's concern ourselves with two particular types: *cold fronts* and *warm fronts*. A cold front is a boundary separating cold and warm air. It is shown on a map as a heavy blue line with triangles pointing in the direction the cold air is moving. A warm front is formed when warm air flows toward cold air and is indicated by a red line with semicircles showing the direction of warm air movement. In Figure 1–16a, a cold front extends southwestward from the low-pressure region over Ontario to northwest Kansas, and then westward to eastern Nevada. A warm front extends southeastward from the Ontario low to the northern Great Lakes.

Humidity

You have undoubtedly heard the term **relative humidity**. Relative humidity is just one of several ways of expressing the amount of water vapor in the air. (Remember, water vapor is a gas!) It indicates the amount of water vapor present relative to the maximum possible; thus, it is usually reported as a percentage. Though a commonly used indicator of water vapor content, relative humidity has some serious shortcomings. For this reason, another index called the **dew point temperature** (or simply the dew point) is often preferred. For now let's just say that the higher the dew point, the greater the amount of water vapor in the air. Dew points above about 15 °C (59 °F) or so indicate humid air, and dew points above about 20 °C (68 °F) are very uncomfortable. Dew points less than about 5 °C (41 °F) are relatively dry. In Figure 1–16a, much of the southeastern and south-central United States has humid air, while the interior western and north central United States are drier. We will see how to read temperature and dew from maps later in the chapter.

Atmospheric Pressure and Wind

Atmospheric **pressure** is one of the most fundamental weather characteristics because of its role in producing wind. Three generalizations about surface pressure are particularly important. First, air tends to blow away from regions of high pressure toward areas of lower pressure. In other words, it is the horizontal variation in air pressure that generates **winds**. If the pressure were uniform from place to place, the air would be continually calm. Second, the greater the change in pressure over a given distance, the greater the resulting wind speed. The third generalization is that air tends to rise in areas of low surface pressure and sink in zones of high pressure. This is important because rising motions favor the formation of clouds, while sinking motions promote clear skies.

Mapping Air Pressure

For all these reasons maps of air pressure are an essential tool in understanding weather. However, if one were to map pressure as measured at the surface, the result would be almost useless! The decline with altitude is so strong that the resulting map would essentially be a map of elevation. Thus, standard practice is to convert measured values to what would be expected at sea level. That is, for locations high above sea level, the measured value is increased substantially. Elevations near sea level get little correction. By using a common altitude (0 km) for all locations, mapped differences in pressure reflect horizontal gradients that can set the air in motion.

Sea-level pressure is routinely plotted on maps with lines called **isobars**. Each isobar connects points having equal air pressure. Figure 1–16a illustrates how isobars depict the distribution of pressure. Notice that an isobar labeled 1032 encircles extreme northern Washington, northern Idaho, western Montana, and southern British Columbia. Any point on that line has a pressure that is equal to 1032 mb (which corresponds to 103.2 kPa).

There are three major low-pressure areas over North America in Figure 1–16a: one northwest and offshore of the high pressure centered over northern Washington, another northwest of the Great Lakes in Ontario, and one off the coast of the southeastern United States (Hurricane Noel). We know from the first generalization about high- and low-pressure systems given above that wind must be flowing into each of these areas from the surrounding regions. The spacing of the isobars tells us where the wind is blowing most strongly. Notice that the isobars are closer together near the low-pressure system over central Canada and relatively far apart over the south-central and southwestern United States. It therefore follows that at this point in time the wind speeds are generally greater over central Canada than farther to the south.

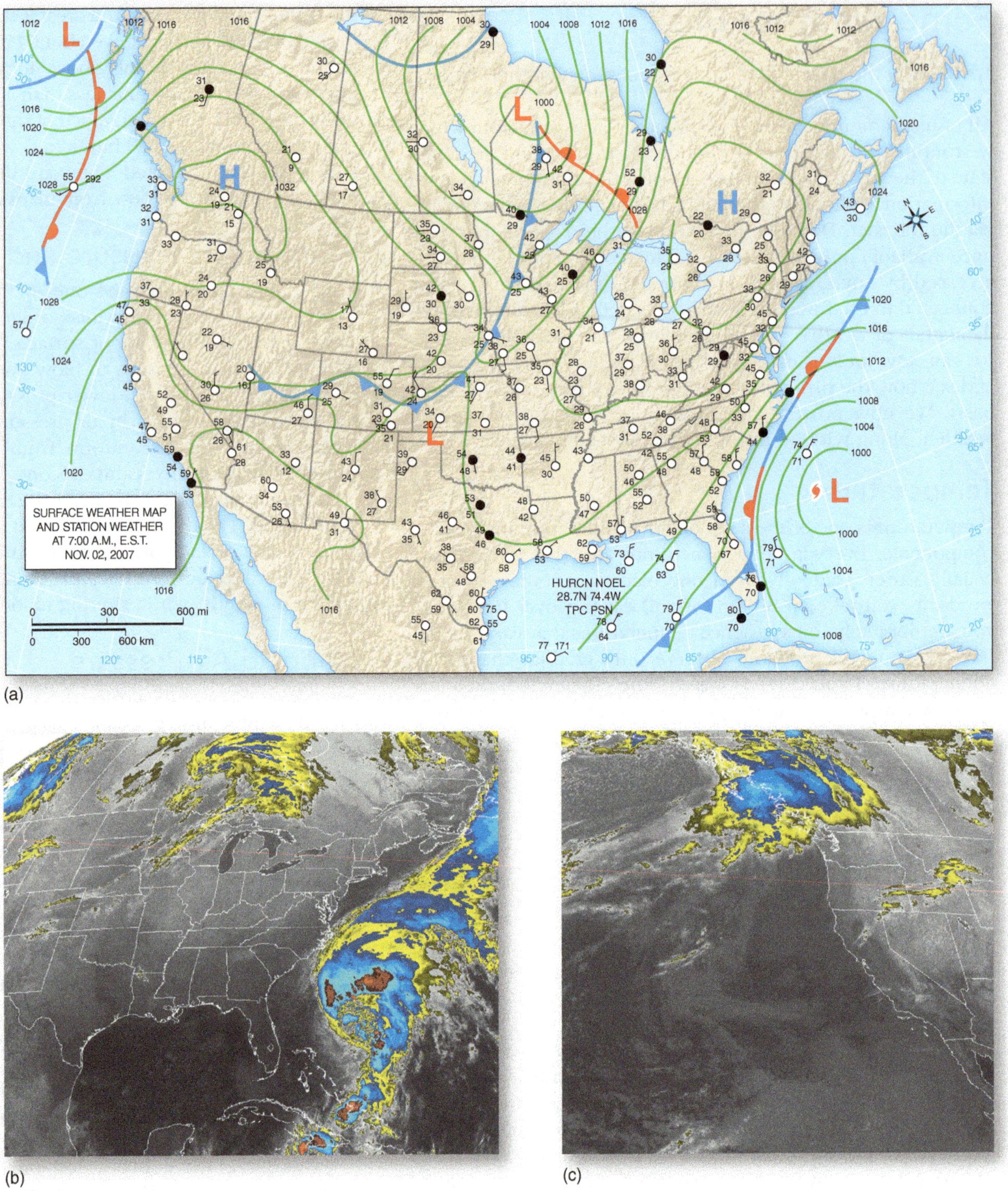

▸ **FIGURE 1–16 Weather Maps and Satellite Images.** (a) A typical surface weather map. (b) Satellite image of the eastern United States and Canada. Areas in bright red depict heavy precipitation. (c) As in part (b) but for the western United States and Canada.

Station Models

More detailed information regarding wind speed and direction can be obtained on weather maps by looking at the **station models**, which contain symbols and numbers giving detailed weather information for particular locations. In Figure 1–16a, the station model for Cheyenne in southeastern Wyoming contains an open circle with a line pointing toward the northwest, which indicates the direction that the wind is blowing *from*. At the end of the line is a single tick mark. The number and type of tick marks at the end of the line give an approximate wind speed, using the conventions shown in Figure 1–17a. In this example, Cheyenne has winds coming from the northwest at 4 to 13 km/hr (3 to 8 mph).

Station models also give cloud cover information. Open circles, such as that shown for Cheyenne in Figure 1–16a,

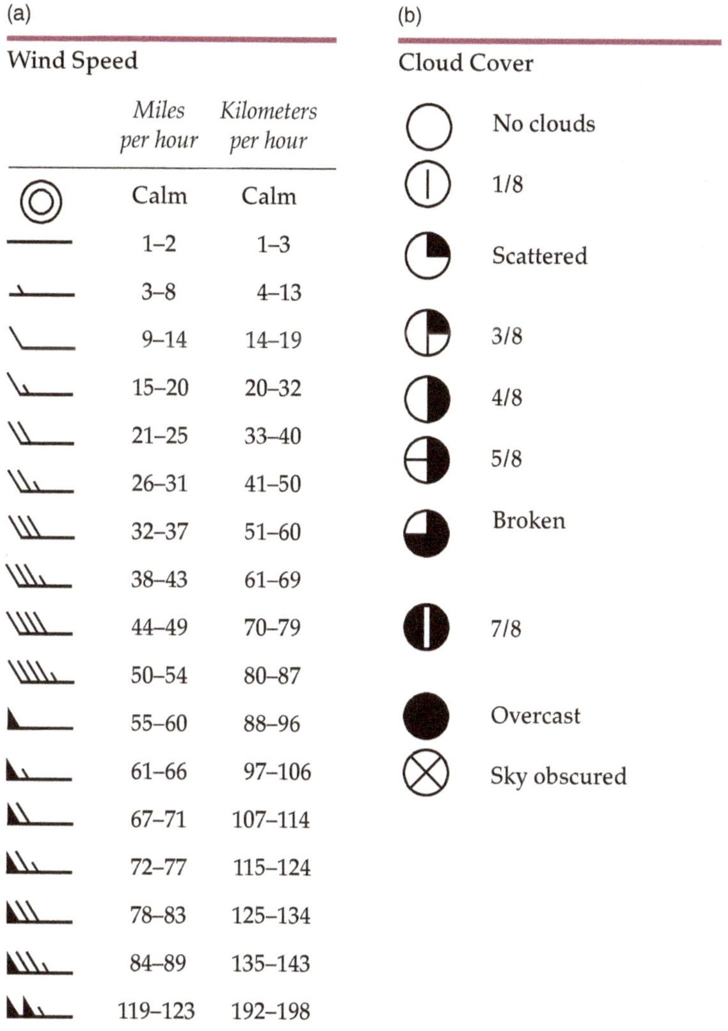
(a)
Wind Speed
Miles per hour
Kilometers per hour
Calm Calm
1–2 1–3
3–8 4–13
9–14 14–19
15–20 20–32
21–25 33–40
26–31 41–50
32–37 51–60
38–43 61–69
44–49 70–79
50–54 80–87
55–60 88–96
61–66 97–106
67–71 107–114
72–77 115–124
78–83 125–134
84–89 135–143
119–123 192–198
(b)
Cloud Cover
No clouds
1/8
Scattered
3/8
4/8
5/8
Broken
7/8
Overcast
Sky obscured

▲ **FIGURE 1–17 Station Model Symbols.** (a) Wind speed is shown by tics. (b) Circle patterns indicate cloud amount.
Note: You might wonder why the wind ticks have such odd values. The ticks represent changes of 5 and 10 nautical miles per hour, or knots. Thus one small tick mark corresponds to 5 knot winds, a large tick mark 10 knots, and so on. However, this scale is not widely used outside of some aviation and maritime situations. When converted to miles or kilometers per hour, the numbers above are the result.

indicate clear skies, while fully shaded circles indicate overcast conditions. Intermediate amounts of cloud cover are indicated by the patterns shown in Figure 1–17b. Temperature and dew point in degrees Fahrenheit are plotted to the left of the cloud circle. The upper figure is temperature, whereas the lower is dew point. Thus, Cheyenne has an air temperature of 17 °F and a dew point of 13 °F.

Look again at Figure 1–16a. Notice how the isobar pattern and station model information verify the generalizations about pressure distributions, wind movement, and cloud cover. The wind barbs show that the air does indeed move outward from the region of high pressure, while air is flowing into the low-pressure area north of the Great Lakes. Furthermore, the satellite images in Figures 1–16b and c show that much of the area around the high-pressure systems has clear skies, in contrast to the general overcast near the three low-pressure systems.

CHECKPOINT

1.19 How are wind and sea-level pressure related?

1.20 Using the map in Figure 1–16a and referring to Figure 1–17, describe the weather along the coast of North Carolina.

A Brief History of Meteorology

1.10 List major events in the history of meteorology.

The amount of information available to even casual weather observers is enormous compared to just a couple of decades ago. Computers and mobile devices almost instantly deliver better information than a professional weather forecaster had available in the late 1900s. But the information explosion did not happen overnight; for centuries the advances in meteorological knowledge came about in piecemeal increments.

Early Atmospheric Scientists

For atmospheric science to become quantitative, basic instrumentation was required. Galileo led the way in 1593 with a prototype version of the basic thermometer, which ultimately led to the development of the Fahrenheit and Celsius scales in 1714 and 1736, respectively. In 1643, Evangelista Torricelli invented the mercury barometer, the instrument still used as a standard for measuring atmospheric pressure, but it was not until the late 1700s that an instrument became available for measuring water vapor.

Nineteenth-Century Progress

The balloon ascents taken by J. L. Gay-Lussac and Jean Biot in 1804 were landmark events in the scientific understanding of the atmosphere and in human courage. The first ride, taken by both scientists, rose to a height of 3962 m (13,000 ft) (Figure 1–18). A second ascent taken solo by Gay-Lussac went up to 7000 m (23,000 ft). At that height, the average temperature is some 30 degrees below zero Celsius (−22 °F) and the air density is only 40 percent of its value at sea level. Gay-Lussac continued to make temperature and pressure observations until he passed out from oxygen deficiency.

Although most weather observers have not been as daring as Gay-Lussac, a network of observers must exist to operate weather instruments and maintain permanent records of weather elements. Accurate data are needed to discern atmospheric patterns and make forecasts. The first network of this type in North America was authorized in 1847 when the Smithsonian Institution allocated the considerable sum of $1000 to purchase instruments for a network of volunteer weather observers across the United States. These observers submitted their subjective observations and instrument data to the Institution each month, providing useful climatological information. Still, the inability to collect, map, and rapidly disseminate the data made operational forecasting impossible. At about the same time, however, a new invention—the telegraph—enabled the rapid transmission of data from sites across the country to a central collecting station, thus making

(a) (b)

▲ **FIGURE 1–18 The Long History of Balloons in Meteorology.** (a) In 1804 Joseph Guy-Lussac and Jean Biot ascended to an altitude of about 4 km (2.5 mi) measuring temperature, pressure, and humidity. (b) Today thousands of weather balloons are launched worldwide.

possible the forecasting of weather by simply noting the location and movement of current weather conditions.

The birth of U.S. weather forecasting occurred in 1870, when President Ulysses S. Grant authorized the establishment of the Army Signal Service. In 1891 the agency was renamed the Weather Bureau and transferred to the Department of Agriculture, where it remained until 1940. The agency was then transferred to the Department of Commerce and in 1970 was renamed the **National Weather Service** as part of the newly established National Oceanic and Atmospheric Administration.

The Development of Modern Weather Forecasting

Over time, an ever-growing network of surface stations became important to understanding and forecasting weather, but the greatest amount of the atmosphere lies well above the surface. Unmanned weather balloons became a valuable source of data in the 1920s, and since the 1940s the tracking of balloon-carried instrument packages called *radiosondes* has become a regular source of meteorological and forecasting data.

The era following World War II brought with it enormous advances in meteorological knowledge and technology. Weather radars, first developed during the war, have evolved into units that can peer into clouds and observe their internal characteristics and motions. Weather satellites have likewise become essential to meteorologists since the launching of Tiros I (Television and Infrared Observation Satellite) in 1960. And technological advances in computer hardware and software that have revolutionized most other aspects of society have had an incredible impact on atmospheric understanding and forecasting since they were introduced to meteorology in the 1950s. Indeed, weather prediction was one of the first uses for digital computers and remains a driving force behind their continued development.

CHECKPOINT

1.21 Is it possible a librarian will one day discover a temperature record going back to the Middle Ages? If not, what is the longest possible temperature record?

1.22 When did computer-based weather forecasting begin?

Summary

1.1 Distinguish between weather and climate.

- Weather is the state of the atmosphere at a particular point in time, whereas climate has to do with long-term averages and other statistical properties.

1.2 Explain the scientific method.

- The scientific method provides a framework for constructing knowledge about the natural world.
- The main elements are formulation of a question and testable hypotheses, predictions that follow from the hypotheses, analysis, and conclusions. Conclusions reached by the scientific method are repeatable and open to scrutiny by others.

1.3 Describe the thickness of the atmosphere and the vertical distribution of gases within it.

- Most of Earth's atmosphere is contained within a relatively shallow envelope surrounding the planet. The depth of the atmosphere adds less than 2 percent to Earth's cross-sectional size.
- Given the shallowness of the atmosphere, its winds are mainly horizontal. Note, however, that the atmosphere's smaller vertical motions determine much of atmospheric behavior.

1.4 Describe the behavior of gas molecules in the atmosphere, including residence times and the roles of vertical mixing and gravitational settling.

- Gases are constantly exchanged between the atmosphere and Earth's surface.
- The average length of time that molecules of a given substance remain in the atmosphere is their residence time.
- Winds and other motions thoroughly mix the gases in the homosphere.
- Gases in the heterosphere segregate according to their molecular weight.

1.5 Describe the composition of the atmosphere.

- The atmosphere is a mixture of gases plus liquid and solid particles.
- The main gases are nitrogen and oxygen, which are present in nearly uniform amounts (78% and 21% by volume, respectively).
- Other important gases, such as water vapor, ozone, and carbon dioxide, are highly variable over space and time.

1.6 Explain how air pressure arises and describe the vertical variation of pressure and density.

- Air pressure reflects the mass of overlying air and therefore decreases with increasing altitude. Because the atmosphere is compressible, density declines vertically as well. Both decrease rapidly near the surface and more slowly as altitude increases.

1.7 Identify and describe the layers of Earth's atmosphere.

- Troposphere—the lowest 15 km (9 mi) of the atmosphere where nearly all weather events occur and most water vapor is found. Temperature decreases with increasing altitude.
- Stratosphere—15 to 50 km (9 to 31 mi). Initially steady temperatures with altitude, then increasing toward the upper stratosphere. The ozone layer is within the stratosphere.
- Mesosphere—50 to 80 km (31 to 50 mi). Decreasing temperature with altitude. This layer has little meteorological significance.
- Thermosphere—above 80 km (50 mi), no definite top. Increasing temperature with altitude but very low density.

1.8 Explain the evolution of the atmosphere during Earth's history.

- Over time, the amounts of carbon dioxide and water vapor decreased in abundance, while amounts of nitrogen and oxygen increased.
- Both physical and biological processes were important. Most oxygen is the result of photosynthesis.

1.9 Identify the basic types of data found on weather maps.

- Surface maps show the distributions of temperature, dew point, winds, clouds, and sea-level air pressure. Station models are used to show values at individual locations.

1.10 List major events in the history of meteorology.

- The thermometer was invented in 1593 and the barometer in 1643.
- The Army Signal Service, established in 1870, later become the U.S. National Weather Service.
- Radiosonde use began in the 1940s, and the first weather satellite was launched in 1960.
- Since the 1950s, weather forecasting has relied on the use of computers and their ever-increasing computational power.

Key Terms

aerosols *p. 42*
argon *p. 38*
atmosphere *p. 32*
aurora australis *p. 49*
aurora borealis *p. 49*
carbon dioxide *p. 40*
Celsius scale *p. 49*
climate *p. 33*
climatology *p. 33*
condensation nuclei *p. 43*
density *p. 43*
dew point temperature *p. 51*
Fahrenheit scale *p. 49*
front *p. 51*
heterosphere *p. 38*
homosphere *p. 38*
hydrologic cycle *p. 39*
inversion *p. 47*
ionosphere *p. 49*
ions *p. 49*
isobar *p. 51*
Kelvin scale *p. 49*
kilopascal (kPa) *p. 46*
mesosphere *p. 48*
meteorology *p. 32*
methane *p. 42*
millibar (mb) *p. 46*
National Weather Service *p. 54*
nitrogen *p. 38*
outgassing *p. 50*
oxygen *p. 38*
ozone *p. 40*
ozone layer *p. 48*
particulate *p. 42*
permanent gases *p. 38*
photodissociation *p. 48*
photosynthesis *p. 40*
pressure *p. 51*
relative humidity *p. 51*
respiration *p. 41*
scientific method *p. 34*
scientific theory *p. 36*
standard atmosphere *p. 47*
station model *p. 52*
stratopause *p. 48*
stratosphere *p. 47*
thermosphere *p. 48*
tropopause *p. 47*
troposphere *p. 47*
variable gases *p. 38*
water vapor *p. 39*
weather *p. 33*
wind *p. 51*

Review Questions

1. Explain why television newscasts have weather segments but not climate segments.
2. "Prediction" is an important element of the scientific method. Does this mean forecasting the future? Please explain.
3. Why is it difficult to define an absolute top of the atmosphere?
4. What are the homosphere and the heterosphere?
5. What is the difference between the permanent and variable gases of the atmosphere? Which gases contribute most to the total mass of the atmosphere?
6. Given that variable gases are so rare, why consider them at all?
7. What is ozone, and why is it both beneficial and harmful to life on Earth?
8. Why has the concentration of carbon dioxide in the atmosphere been increasing during the past century?

9. What are aerosols? Are they formed only by human activities or do they occur naturally?
10. Convert the following Celsius temperatures to Fahrenheit: −20 °C, 10 °C, 40 °C.
11. Convert the following Fahrenheit temperatures to Celsius: −22 °F, 50 °F, 113 °F.
12. How do photosynthesis, respiration, and decay affect the carbon dioxide balance of the atmosphere?
13. In what way does the density of the atmosphere vary with altitude?
14. What are the distinguishing characteristics of the troposphere, stratosphere, mesosphere, and thermosphere?
15. What is the tropopause?
16. In which thermal layer of the atmosphere is the ozone layer found? Why is the term "ozone layer" somewhat misleading?
17. The troposphere is less than half as thick as the stratosphere but contains 10 times as much mass. Please explain.
18. What percentage of the total mass of the atmosphere is contained in the troposphere and the stratosphere combined?
19. A station at elevation 900 m (3000 ft) has a surface pressure of 900 mb. An airplane is flying 5 km above the surface. Use the rule of thumb in the section on pressure to guess the air pressure surrounding the plane.
20. How is the ionosphere distinct from the layers of the atmosphere defined by their temperature profiles?
21. What is outgassing and why was it important?
22. Why were anaerobic bacteria important to the evolution of the atmosphere?
23. What are isobars?
24. Briefly describe the effect that variations in pressure exert on other weather elements.
25. What are station models and what useful information do they depict?

Critical Thinking

1. Volcanic eruptions continue to occur and outgas water vapor, carbon dioxide, and other gases. Do you think that this will be a significant factor in increasing the concentration of these gases during the next century? Why or why not?
2. Temperatures usually decrease with height in the troposphere but increase with height in the stratosphere. Why do the two layers have such different profiles?
3. The thermosphere has extremely high temperatures, but a person exposed to the thermosphere would rapidly freeze. Explain the apparent contradiction in terms of what you know about heat and temperature.
4. The United Nations has sponsored an agreement (the Kyoto Protocol) that limits carbon dioxide emissions from many developed and developing countries. It has not been ratified by all countries that originally signed the protocol. What are the pros and cons of doing so?

Problems & Exercises

1. Keep a record of daily weather in your area and download current weather maps from one of the available websites. Do you notice any patterns on the maps that tend to be associated with particular weather conditions?
2. Look at today's weather map and observe the contrasting weather conditions across the United States and Canada. Do any areas exhibit significant changes in weather from adjacent regions? How well defined are the boundary zones? Repeat this exercise for several days and see if these transition regions show movement.
3. The National Climatic Data Center (NCDC) has a website at **http://www.ncdc.noaa.gov/oa/climate/severeweather/extremes.html** that provides a wealth of information on extreme weather and climate events. Examine the site to see how much information is available. Do any of the topics relate to weather events in your hometown?
4. Examine the map of National Weather Service Offices at **www.nws.noaa.gov/organization.php#maps**. Is there a weather service office near you? If so, consider visiting the office and seeing firsthand what goes into producing a forecast.

Visual Analysis

Consider this surface weather map for March 19, 2014. There is a winter storm centered on Chicago, Illinois, with a cold front running through Louisiana, Missouri, and Illinois.

1.1. Compare temperatures on either side of the cold front where it passes through Arkansas. Which side is colder? About how large is the temperature difference?

1.2. The temperature in Indianapolis, Indiana, is 46 °F. In 24 hours the cold front will have moved eastward almost to the East Coast. What temperature would you forecast for Indianapolis that day? Get the map for March 20 from the following NOAA website: **http://www.hpc.ncep.noaa.gov/dailywxmap/dwm_stnplot_20140320.html**. Compare your forecast to the actual value.

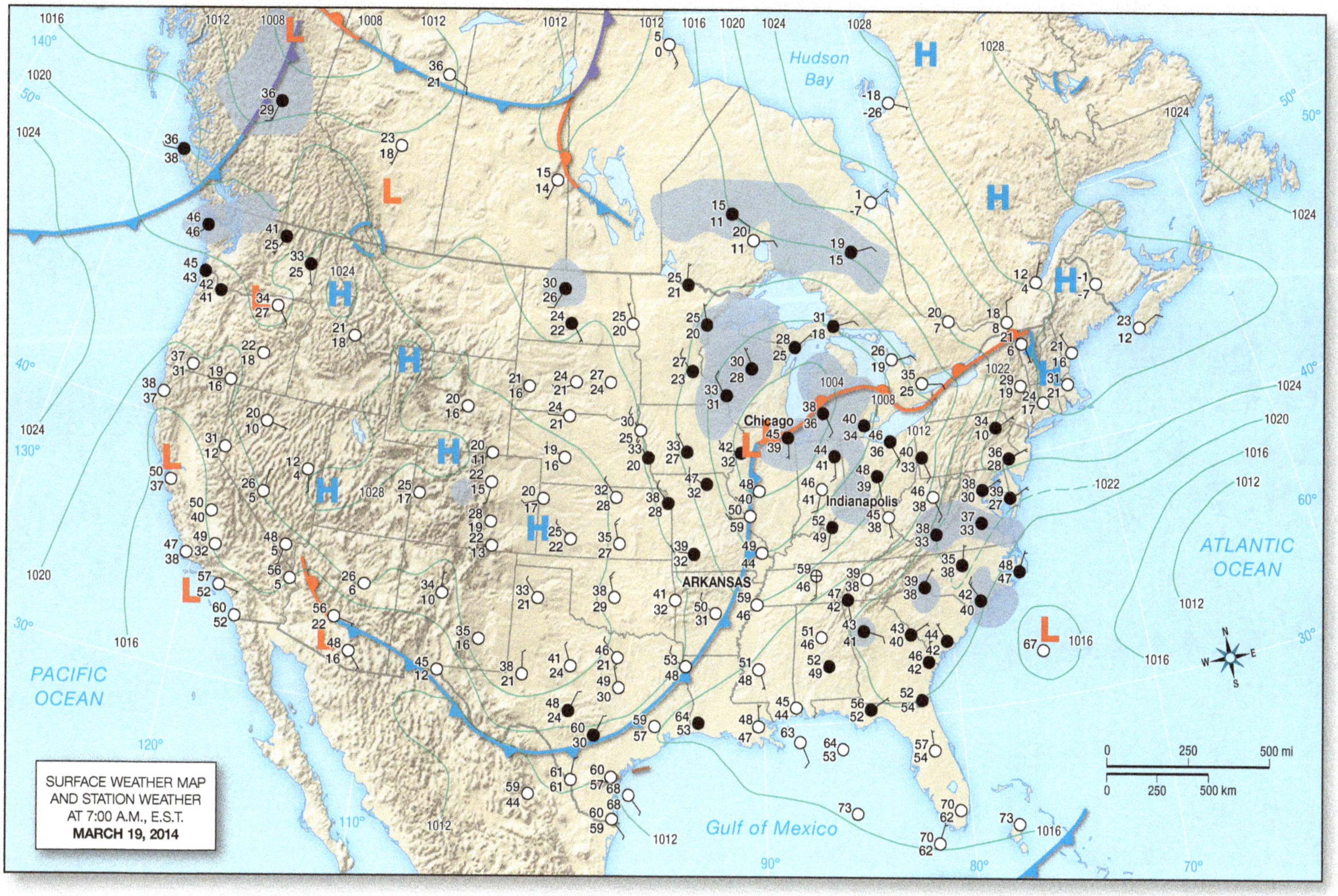

Map
Indianapolis
Weather Map

http://goo.gl/mmoHU5

CHAPTER

2 Solar Radiation and the Seasons

We tend to associate certain events to occur with the change of seasons. Some places are renown for certain seasonal changes, such as Japan with its spring cherry blossoms and fall colors. In some areas the first snowfall of the year marks the beginning of ski season. In Southern California many people welcome the late fall months, as that is when rain storms normally come in and break up the summer drought.

While seasonal changes like these are familiar to most of us, many people don't know the reasons why such changes occur. In fact, many people have misconceptions about what causes these differences in seasons and climates. For example, some people incorrectly believe that variations in the distance between Earth and the Sun cause the seasons, with summer occurring when Earth and the Sun are closest together. But Earth and the Sun are closest to each other on or about January 4—in the midst of the Northern Hemisphere winter! Likewise, when the Sun is farthest from Earth on July 4, it's summer in the Northern Hemisphere. The purpose of this chapter is to provide an explanation of how Earth's orientation toward the Sun creates the seasons and how the amount of solar energy Earth receives varies with latitude.

We all know northern regions of North America are vulnerable to such events, and primarily during winter months. But there are reasons *why* these events are limited to certain regions and times of year. We know that Honolulu, Hawaii,

◀ *Fall colors mark the coming of autumn in Nagano Prefecture, Japan.*

Learning Outcomes

After reading this chapter, you should be able to:

2.1 Identify the types of energy and how they can be transmitted.

2.2 Explain the laws governing the amount and type of radiation emitted by objects.

2.3 Identify the solar constant and explain how much energy Earth receives from the Sun.

2.4 Recognize the characteristics of Earth's orbit around the Sun and how they create the seasons.

2.5 Explain three ways in which Earth's changing orientation with respect to the Sun affects the receipt of solar radiation.

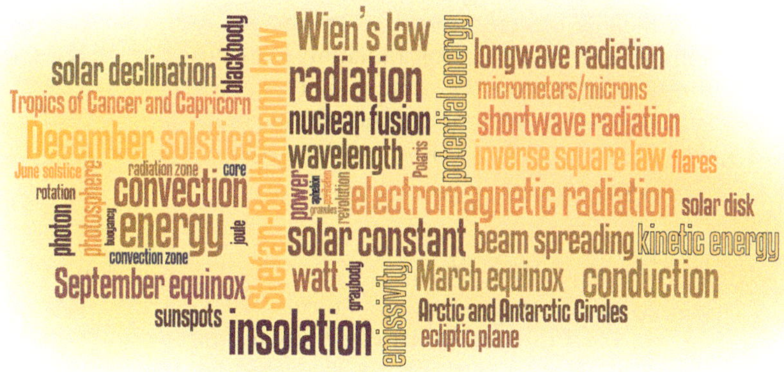

is not subject to this type of weather—but exactly why may be a little unclear to many people. In fact, many people have misconceptions about what causes these differences in seasons and climates. For example, some people incorrectly believe that variations in the distance between Earth and the Sun cause the seasons, with summer occurring when Earth and the Sun are closest together. But Earth and the Sun are closest to each other on or about January 4—in the midst of the Northern Hemisphere winter! Likewise, when the Sun is farthest from Earth on July 4, it's summer in the Northern Hemisphere. The purpose of this chapter is to provide an explanation of how Earth's orientation toward the Sun creates the seasons and how the amount of solar energy Earth receives varies with latitude.

Energy

2.1 Identify the types of energy and how they can be transmitted.

Energy is traditionally defined as "the ability to do work." This definition isn't entirely accurate and raises its own questions (What is work?), but it is impossible to do better in just a few words. Rather than travel far afield in search of a precise definition, we'll assume that everyone has at least a vague idea of energy as an agent capable of setting an object in motion, like warming a teapot, or otherwise manifesting itself in everyday events. The standard unit of energy in the International System (SI) used in scientific applications is the **joule** (abbreviated as J). Although students may be more familiar with the calorie as the unit for energy, the joule is preferred in this context (1 joule = 0.239 calorie). A related term, **power**, is the rate at which energy is released, transferred, or received. The unit of power is the **watt** (W), which corresponds to 1 joule per second.

Even the simplest activity requires a transfer of energy. In fact, while you read these words, an energy transfer is occurring as the chemical energy from food you have eaten is converted into the kinetic energy (energy of motion) needed to move your eyes across this line of type. But your body, like any other machine, is not perfectly efficient; it loses some thermal energy as chemical energy is converted into kinetic energy. Thus, your eye muscles give off heat as they contract and relax.

The same concept applies to our atmosphere. About half of one billionth of the energy emitted by the Sun is transferred to Earth as **electromagnetic radiation**, some of which is directly absorbed by the atmosphere and surface. This radiation provides the energy for the movement of the atmosphere, the growth of plants, the evaporation of water, and an infinite variety of other activities.

Kinds of Energy

Energy can occur in a variety of forms. We often speak of radiant, electrical, nuclear, and chemical energy; but, strictly speaking, all forms of energy fall into the general categories of **kinetic energy** and **potential energy**. These are illustrated in Figure 2–1.

KINETIC ENERGY Kinetic energy can be viewed as energy in use and is often described as the energy of motion. Motion can occur on a large scale, as in the movement of an object from one place to another. Examples that occur in nature include falling raindrops (Figure 2–2a), water flowing through a river channel, and grains of dust transported by the wind. The motion of kinetic energy can also occur at a microscale, as in the case of molecular vibration or rotation (illustrated for water in Figure 2–2b).

A solid object may seem to be standing still, but its atoms or molecules undergo a certain amount of vibration. Gas and liquid molecules, in contrast, are not fixed in space but move about randomly (Figure 2–3). In solids, liquids, or gases, the rate of vibration or random movement determines the temperature of the object.

POTENTIAL ENERGY If kinetic energy is energy in use, potential energy is energy that hasn't yet been used. Potential energy can assume many forms. For example, a plant's carbohydrates have potential energy that can be consumed by animals or by the plant itself and then metabolized to yield the energy needed for all of its biological activity. When our own bodies metabolize food, we are using this potential energy, converting it to kinetic energy, and releasing heat as a by-product.

Another form of potential energy results from an object's position. Consider, for example, a cloud droplet that occupies

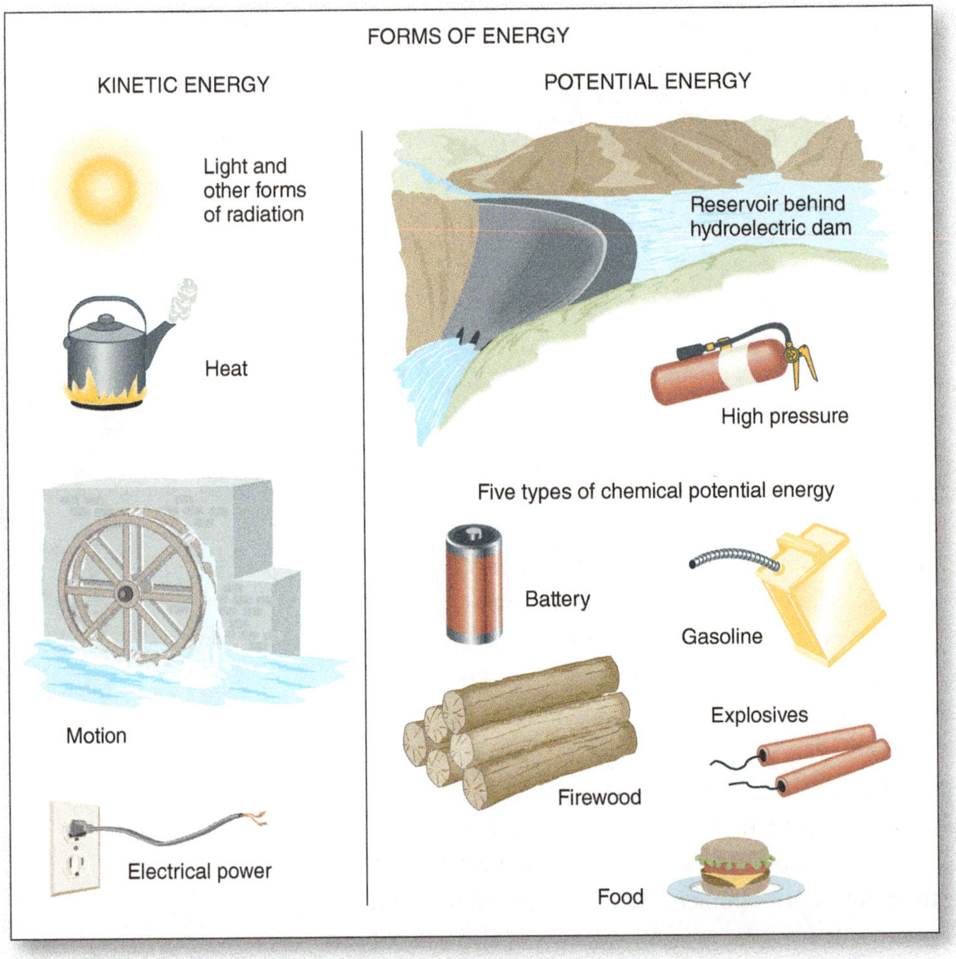

▲ **FIGURE 2–1 Energy.** There are several different forms of energy, but each of these is a form of either kinetic energy (the energy of motion) or potential energy.

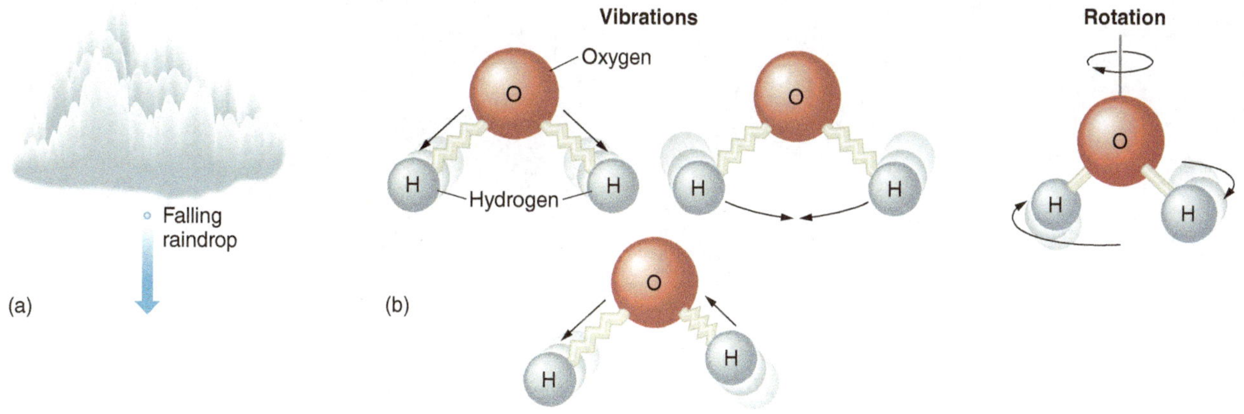

▲ **FIGURE 2–2 Kinetic Energy.** The motion associated with moving objects, kinetic energy can occur as in the falling raindrop in (a), or as molecular vibration or rotation, as depicted for water molecules in (b). The greater the rate of vibration or rotation, the higher the temperature of the substance.

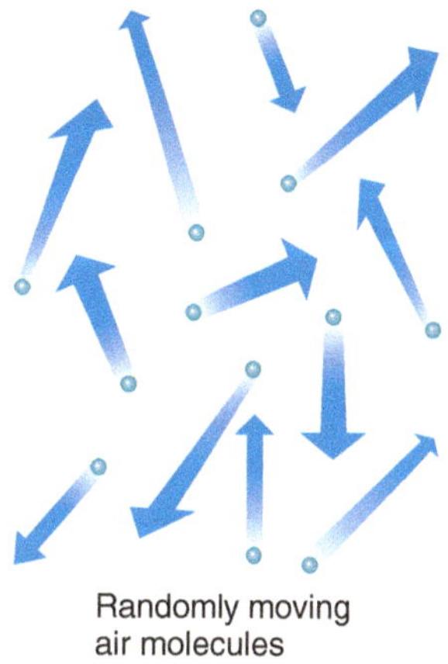

▲ **FIGURE 2–3 Air Molecules.** Random motion characterizes the movement of air molecules.

▲ **FIGURE 2–4 Heat Transfer.** The fire below the pan transfers heat due to the radiation of energy, along with the convection of air circulating between the pot and the fire. The boiling water inside the pot circulates heat via convection. The person's hand holding the pot handle will feel it warm as conduction occurs.

some position above Earth's surface. Like all other objects, the droplet is subject to the effect of gravity. As it falls toward Earth's surface, the object's potential energy is converted to kinetic energy. Obviously, the higher the droplet's altitude, the greater the distance it is capable of falling and the greater its potential energy. It is important to recognize that the droplet did not attain its height by magic, because energy was used to elevate its mass in the first place.

CHECKPOINT

2.1 What is potential energy?

2.2 Explain what happens in terms of kinetic and potential energy as a raindrop falls from a cloud.

Heat Transfer Mechanisms

To understand heat transfer it is useful to invoke two fundamental laws of thermodynamics: The *first law of thermodynamics* dictates that energy is a conserved property that is neither created nor destroyed, but may change form and travel from place to place. The *second law of thermodynamics* informs us that heat is always transferred from regions of high temperature to low temperature. Heat can be transferred from one place to another by three processes: conduction, convection, and radiation (Figure 2–4).

CONDUCTION **Conduction** is the movement of heat through a substance without appreciable movement of molecules. A simple example is a metal rod, one end of which is placed over a campfire. The part of the rod above the flame is warmed, and molecules there gain energy. Some of this energy is passed to neighboring molecules, which, in turn, heat adjacent molecules. (The exact mechanism of molecular "passing" depends on the substance—in metals, it is mainly accomplished by electrons.) This process occurs throughout the length of the rod so that after a few minutes the entire piece of metal becomes too hot to handle. The transfer of heat from the warmer to the colder part of the rod is conduction. Note

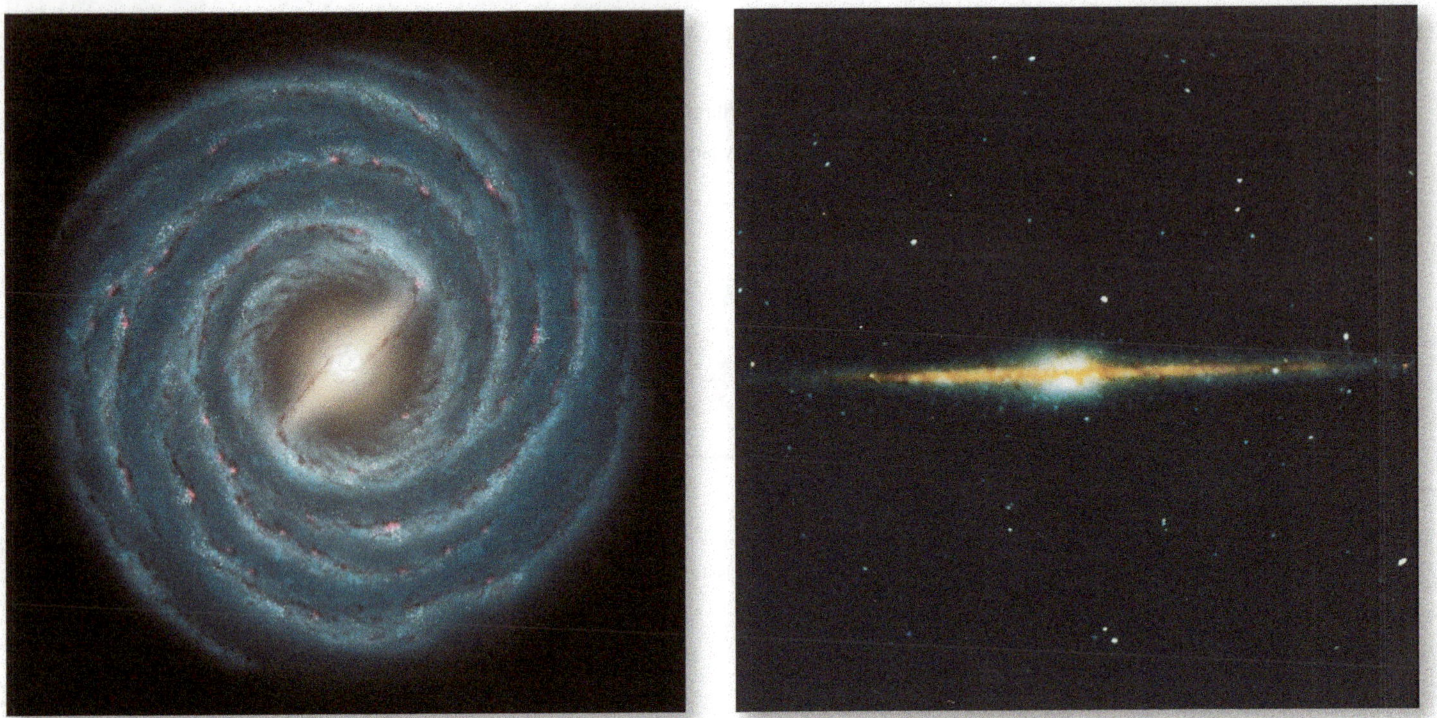

(a) (b)

▲ **FIGURE 2–5 Milky Way Galaxy.** Our solar system is part of the Milky Way galaxy, which contains more than 100 billion stars. (a) An artist's rendition. (b) A wide-angle infrared image of the plane and bulge of our Milky Way.

that although heat travels through the rod, the molecules that make up the rod do not move. Conduction is most effective in solid materials, but as we will see in Chapter 3, it also is an important process in a very thin layer of air near Earth's surface.

CONVECTION The transfer of heat by the mixing of a fluid is called **convection**. Unlike conduction, convection is accomplished by movement of the liquid or gas in which the process occurs. You can observe this process by watching a pot of water boil on a kitchen stove. The water at the bottom of the pot is closest to the source of energy and warms most rapidly. In warming, the water expands ever so slightly, becomes less dense, and rises to the surface. The rising water must, of course, be replaced from above, so water formerly at the surface sinks to the base of the pot. These rising and sinking motions cause a rapid movement not only of mass but also of the thermal energy within the circulating water.

Convection in the atmosphere is not much different from that within a pot of boiling water. During the daytime, heating of Earth's surface warms a very thin layer of air (on the order of 1 mm thick) in contact with the surface. Above this thin *laminar layer*, in which convection does not occur, air heated from below expands and rises upward because of the inherent **buoyancy** of warm air. Buoyancy is the tendency for a light fluid (a liquid or gas) to float upward when surrounded by a heavier fluid. Unlike water in a pot, the atmosphere can undergo convection even in the absence of buoyancy through a process called *forced convection*, the vertical mixing that happens as the wind blows. We discuss these processes in more detail in Chapter 3.

RADIATION Of the three energy transfer mechanisms, **radiation** is the only one that can be propagated without a transfer medium. In other words, unlike conduction or convection, the transfer of energy by radiation can occur through empty space. Virtually all the energy available on Earth originates from the nearby (in astronomical terms) star we call the Sun, a member of the Milky Way galaxy (Figure 2–5). The atmosphere also has other sources of energy: Minute amounts of radiation are received from the billions of other stars in the universe, and some energy reaches the surface from Earth's interior. However, the contribution of these sources is minuscule compared to the energy from the Sun.

We will now examine the characteristics of radiation and the way Earth's orientation affects the radiation received. The spatial and seasonal variations in the receipt of solar energy are not mere abstractions; they are, in fact, the driving force for virtually all the processes discussed in the rest of this book.

CHECKPOINT

2.3 How is energy transferred in conduction, convection, and radiation?

2.4 Describe the types of energy transfer involved in heating soup in an iron cooking pot suspended above the glowing coals of a campfire.

Characteristics of Radiation

2.2 Explain the laws governing the amount and type of radiation emitted by objects.

Radiation is emitted by all matter. Thus, *everything*—including the stars, Earth, ourselves, and this book—is constantly emitting electromagnetic energy. We are all familiar with electromagnetic energy in many of its forms. We see the environment around us because a type of radiation we

Video
The Sun in Ultraviolet
http://goo.gl/ZdFhEG

call *visible light* impinges on our eyes, which then send signals to our brains to produce visual images. A different type of electromagnetic energy is used when we warm a meal in a microwave oven; the radiation agitates the molecules of the food and thereby increases its temperature. Other types of radiation may be less beneficial or even harmful, such as ultraviolet radiation, which can lead to sunburns, malignancies, or even death. Although different types of radiation have different effects, they are all very similar in that they are transmitted as a sequence of waves.

Think of a wave created by a rock tossed into a pond. The wave is revealed by an oscillation in the water surface with alternating crests (high points in the ripples) and troughs (low points). When you observe the regular rise and fall of the surface as the wave passes, you know energy is being transferred.

In the case of radiation, the waves are electrical and magnetic oscillations. That is, radiation consists of both an electrical and a magnetic wave. With the proper instruments, we would detect these electrical and magnetic variations—hence the term *electromagnetic radiation*. The electric and magnetic waves are perpendicular to one another and rise and fall in unison, as shown in Figure 2–6.

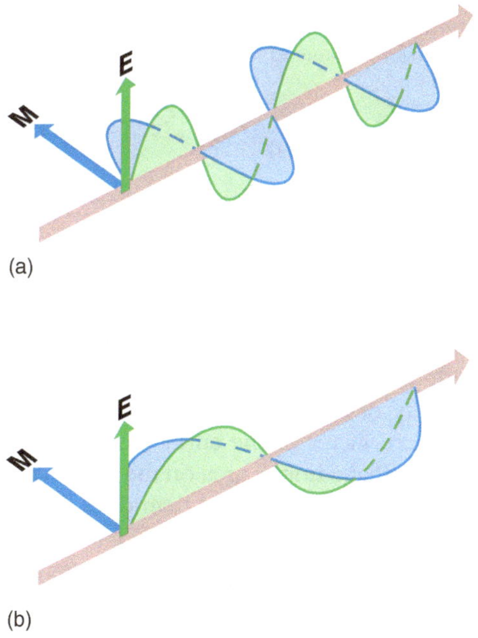

▲ **FIGURE 2–6 Electromagnetic Radiation.** The main components of electromagnetic radiation are an electric wave (*E*) and a magnetic wave (*M*). As radiation travels, the waves migrate in the direction shown by the pink arrow. The waves in (a) and (b) have the same amplitude, so the radiation intensity is the same. However, (a) has a shorter wavelength, so it is qualitatively different from (b). Depending on the exact wavelengths involved, the radiation in (a) might pass through the atmosphere, whereas that in (b) might be absorbed.

Radiation Quantity and Quality

To describe electromagnetic radiation completely, we need to provide information about the amount of energy transferred (quantity) and the type (quality), of the energy. This is similar to describing someone's weight, where we might state quantity in pounds and indicate quality using words such as "mostly flab." In the case of radiation, quantity is associated with the height of the wave, or its *amplitude*. Everything else being equal, the amount of energy carried is directly proportional to wave amplitude.

The quality, or "type," of radiation is related to another property of the wave, the distance between wave crests. Figure 2–6 shows waves of electromagnetic radiation moving in the same direction. All have the same amplitude, but the distance between the individual wave crests is smaller for the waves depicted at the top. The upper waves therefore have a shorter **wavelength**, which is the distance between any two corresponding points along the wave (crest-to-crest, trough-to-trough, etc.). Because of their shorter wavelengths, the waves in Figure 2–6a are qualitatively different, and might produce different effects, from the waves in Figure 2–6b. For example, X-ray radiation has an extremely short wavelength and is able to penetrate soft tissues. On the other hand, ordinary light, having a somewhat longer wavelength, can be absorbed or reflected by the skin. Compared to everyday objects, the radiation of interest here has very small wavelengths. It is therefore convenient to specify wavelengths using small units called **micrometers** (or **microns**). One micrometer—signified by µm—equals one-millionth of a meter, or one-thousandth of a millimeter (0.00004 in.).

All forms of electromagnetic radiation, regardless of wavelength, travel through space at the speed of light, which is about 300,000 km (186,000 mi) per second. At that speed, it takes 8 minutes for energy from the Sun to reach Earth. The energy received from the other, more distant stars takes even longer to arrive at Earth. For instance, radiation from the next nearest star, Proxima Centauri, must travel through space for 4.3 years before reaching us. Though this may seem like a long time, it is minuscule compared to the *billions* of years needed for light from a distant star to arrive at Earth.

Electromagnetic energy comes in an infinite number of wavelengths, but we can simplify things by categorizing wavelengths into just a few individual "bands," as indicated in Figure 2–7 and Table 2–1. The band with the shortest wavelengths consists of gamma rays, with a maximum wavelength of 0.0001 µm. Successively longer wavelength bands include X-rays, ultraviolet (UV), visible, near-infrared (NIR), thermal infrared (IR), microwave, and radio waves. Note that there is nothing unique or special about the visible portion of this electromagnetic spectrum other than the fact that our eyes and nervous systems have evolved to be able to sense this type of energy. Except for their wavelengths, visible rays are just like any other form of electromagnetic energy.

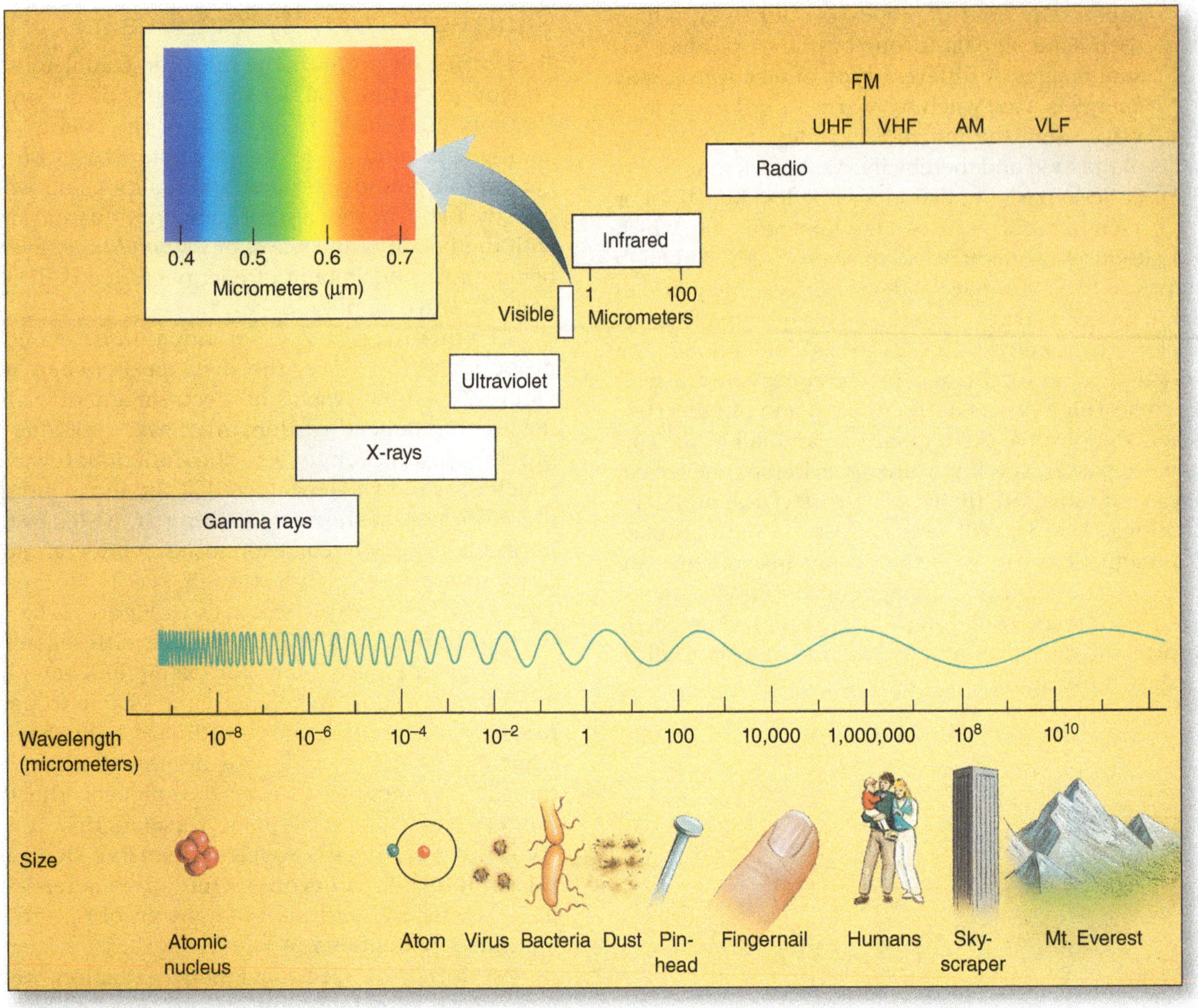

▲ **FIGURE 2–7 Electromagnetic Energy.** Electromagnetic energy can be classified according to its wavelength.

TABLE 2–1
Wavelength Categorizations

Type of Energy	Wavelength (Micrometers)
Gamma	<0.0001
X-ray	0.0001 to 0.01
Ultraviolet	0.01 to 0.4
Visible	0.4 to 0.7
Near infrared (NIR)	0.7 to 4.0
Thermal infrared (IR)	4.0 to 100
Microwave	100 to 1,000,000 (1 meter)
Radio	>1,000,000 (1 meter)

CHECKPOINT

2.5 What is electromagnetic radiation?

2.6 Explain how the electromagnetic spectrum is categorized into bands.

Intensity and Wavelengths of Emitted Radiation

All objects radiate energy, not merely at one single wavelength, but over a wide range of wavelengths. The curves in Figure 2-8 show the intensity of radiation emitted at all wavelengths every second by a square meter of the surface of the Sun and Earth. We can readily see that a unit of area on the Sun emits much more radiation (about 174,000 times more) than does the same amount of surface area on Earth. The shape of the curve showing the intensity of energy emitted by Earth at different wavelengths is similar to that of the Sun, but the total energy released is much less, and the peak of the curve corresponds to a longer wavelength.

Of course, the amount of radiation emitted and its wavelengths are not the result of mere chance; they obey some fundamental physical laws. Strictly speaking, these laws apply only to perfect emitters of radiation, so-called **blackbodies**. Blackbodies are purely hypothetical bodies—they do not exist in nature—that emit the maximum possible radiation at every wavelength. Earth and the Sun are close to blackbodies and therefore nearly follow the laws described shortly. Other

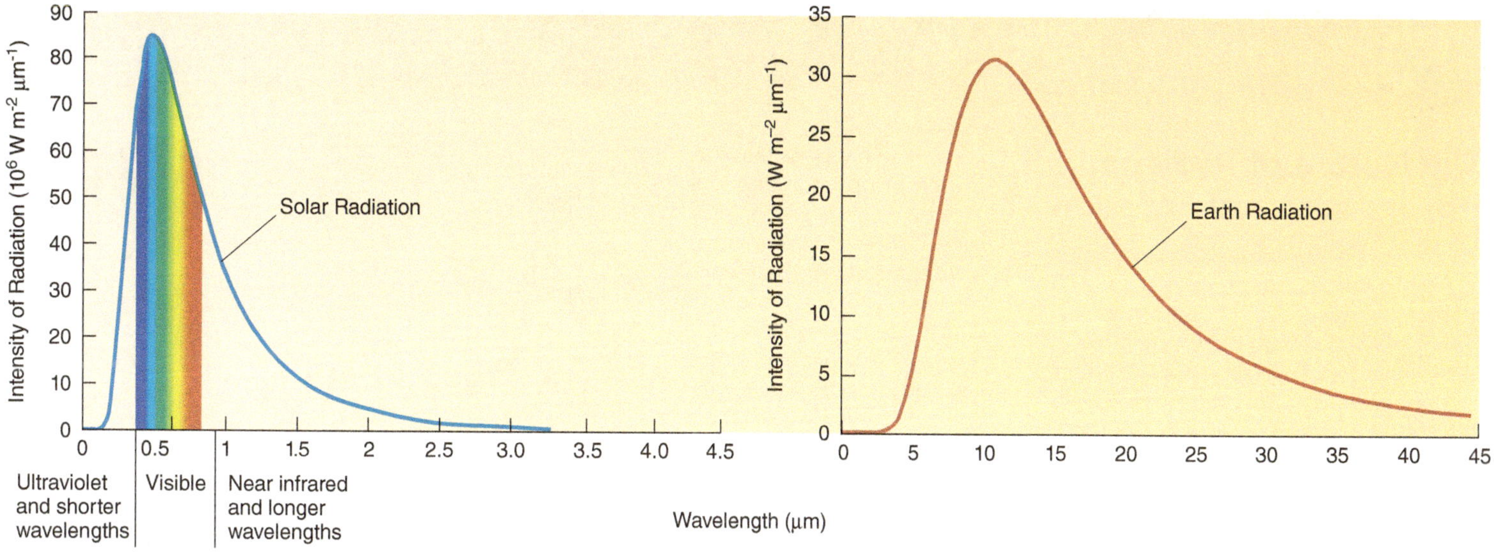

▲ **FIGURE 2–8 Radiation.** Substances emit radiation over a wide range of wavelengths. Because of its higher temperature emission from the Sun is vastly greater than that of Earth. [Notice the scale factor of 1,000,000 used for the solar curve.] The Sun also emits at shorter wavelengths than does Earth. Essentially all Solar radiation is shorter than 4 micrometers, whereas essentially all Earth radiation is found at longer wavelengths.

materials may or may not approximate blackbodies. In particular, the atmosphere, composed mainly of gases, is especially far from a blackbody, so we will not treat it as one.

STEFAN-BOLTZMANN LAW The single factor that determines how much energy a blackbody radiates is its temperature. Hotter bodies emit more energy than do cooler ones; thus, not surprisingly, a glowing piece of hot iron radiates more energy than an ice cube. Interestingly, though, the amount of radiation emitted by an object is not linearly proportional to its temperature. In other words, a doubling of temperature produces *more* than a doubling of the amount of radiation emitted. Specifically, the intensity of energy radiated by a blackbody increases according to the fourth power of its absolute temperature. This relationship, the blackbody version of the **Stefan-Boltzmann law**, is expressed as

Stefan-Boltzmann law $I = \sigma T^4$

where I denotes the intensity of radiation in watts per square meter, σ (Greek lowercase sigma) is the Stefan-Boltzmann constant (5.67×10^{-8} watts per square meter per K^4), and T is the temperature of the body in kelvins (see *Box 1–5, Physical Principles: The Three Temperature Scales*).

Because the intensity of radiation depends on the temperature raised to the fourth power, a doubling of temperature leads to a 16-fold increase in emission. Solving the Stefan-Boltzmann equation using the mean temperature of Earth's surface (about 288 K, 15 °C, or 59 °F) reveals that a square meter emits about 390 watts of power. In contrast, the surface of the Sun, with its temperature of about 5800 K (5500 °C, or 9900 °F), emits about 64 million watts per square meter.

Although true blackbodies do not exist in nature, they provide a useful model for understanding the maximum amount of radiation that can be emitted. Most liquids and solids can be treated as **graybodies**, meaning that they emit some percentage of the maximum amount of radiation possible at a given temperature. Whereas some substances (for example, water) are highly efficient at radiating energy, others (for example, aluminum) are less efficient. The percentage of energy radiated by a substance relative to that of a blackbody is referred to as its **emissivity**. Emissivities range from just above zero to just below 100 percent and are denoted by the Greek letter epsilon (ε). By incorporating the emissivity of any body, we derive the graybody version of the Stefan-Boltzmann law:

$$I = \varepsilon \sigma T^4$$

Including the emissivity factor means that the electromagnetic energy emitted by any graybody will be some fraction of what would be emitted by a blackbody. Note that even though the graybody form of the Stefan-Boltzmann law shows radiation intensity to be a function of both emissivity and temperature, most natural surfaces have emissivities above 0.9 (that is, above 90 percent of blackbody emission). In most cases, therefore, differences in emission are governed by temperature differences. The atmosphere is an exception to this because emission depends on a number of factors, such as the amount of water vapor and other gases in the air, each of which has its own radiative properties (see *Box 2–1, Physical Principles: The Nature of Radiation, Absorption, and Emission*). Still, we can say that the atmosphere is not a perfect emitter of radiation and therefore emits less radiation at any particular temperature than would a blackbody.

CHECKPOINT

2.7 What does the Stefan-Boltzmann law tell us about the relationship between an object's temperature and the amount of radiation it emits?

2.8 How would an increase in Earth's temperature influence the amount of radiation emitted by Earth's surface?

2–1 PHYSICAL PRINCIPLES

The Nature of Radiation, Absorption, and Emission

We commonly describe electromagnetic energy traveling through space as a sequence of waves, but in some contexts it behaves as a stream of particles. The particle nature of radiation applies at the smallest scale of observation, as when visible light is emitted by a single atom or molecule. When light is emitted, there is a change in the orbital characteristics of the electrons in the emitter. As the orbit changes, a small bundle of energy, called a **photon**, is released.

Emission of Photons

Using the simplest example, imagine a hydrogen atom (with one proton and a single orbiting electron). The electron is not free to assume just any orbital distance from the nucleus. Instead, it is confined to fixed orbital distances, called *shells*. Each shell is associated with a given energy level; the greater the distance from the nucleus, the greater the energy level. As shown in Figure 2–1–1, when sufficient energy is absorbed by the hydrogen atom, it can become "excited," and its electron jumps from its "ground state" to a higher shell. Similarly, if the electron jumps back to its previous energy level, it gives off energy in the form of a photon. Because only a few discrete shells exist, only certain energy changes are possible. This means that a photon emitted by an atom can contain only certain discrete amounts of energy, corresponding to the atom's decrease in energy.

Absorption of Photons

Similarly, an atom is restricted in what photons it can absorb, namely those with energies that push the atom into an allowable state. It so happens that the energy of a photon depends only on its wavelength: Photons at shorter wavelengths have more energy than photons at longer wavelengths. Consequently, if you know the wavelength of a photon, you can know its energy, and you can know whether or not the atom can emit or absorb that photon. This is a long-winded way of explaining that an atom will absorb and emit radiation only at certain wavelengths.

Behavior of Radiation in the Atmosphere

Of course, our atmosphere does not consist of single hydrogen atoms; it is composed mainly of molecules of gases. For these gases, changes in energy level are more complex than for simple hydrogen atoms, but the fact remains that emission and absorption involve a decrease or increase in energy level as a photon is released or absorbed. Furthermore, emission and absorption are again confined to just those wavelengths that cause the molecule to move into an allowable energy state. That is, atmospheric gases, just like hydrogen, are selective absorbers and emitters. This is not true for liquids and solids, which tend to emit and absorb a wide range of wavelengths.

One very important consequence of all this is that the atmosphere and surface respond differently to radiation. In addition, the atmosphere responds quite differently to radiation of various wavelengths. As will be seen later, these basic principles go a long way toward explaining Earth's climate and climate change.

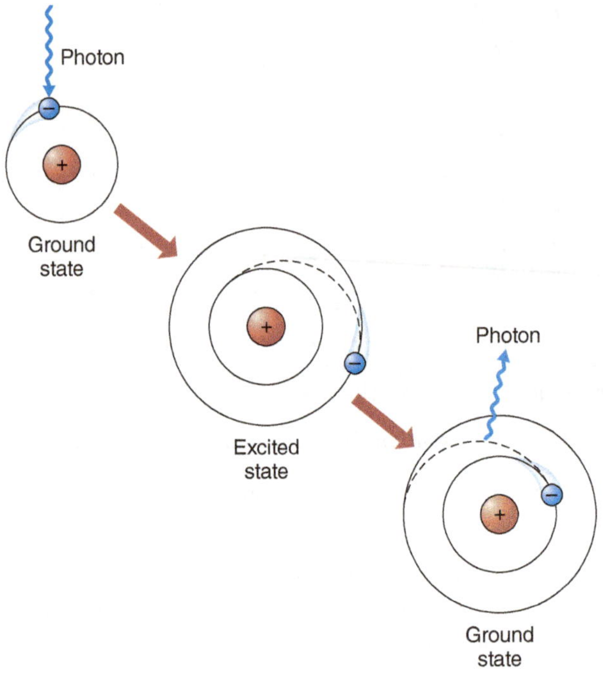

▲ **FIGURE 2–1–1 Shells.** Electrons orbit the nucleus of atoms in prescribed zones called *shells*. This figure depicts a single electron orbiting the nucleus of a hydrogen atom. Upon receiving energy in the form of a photon, the electron is in an excited state and jumps to its next shell. When the electron returns to its ground state, it releases energy in the form of a photon. Note that the energy emitted by such atoms must occur in discrete packets; at the atomic scale, units of energy are divided into individual parcels.

2–1–1 What are "shells" in regard to electrons?

2–1–2 What is the relationship between a photon's energy and its wavelength?

DID YOU KNOW?

To be comfortable, humans need to maintain a skin temperature of about 306 K (33 °C or 91 °F). At this temperature, an average-sized person with surface area of 1.8 square meters emits 895 watts of power—about the same as that used by nine household light bulbs. Of course, the wavelengths emitted are in the thermal infrared range, so we don't light up a dark room.

WIEN'S LAW As we saw in Figure 2–8, the radiation emitted by the Sun and Earth (or any other body) is not of a single wavelength, nor are all wavelengths emitted in equal amounts. In the case of the Sun, the wavelength emitted more than any other is 0.5 μm, while energy radiated by Earth peaks at about 10 μm. For any radiating body, the wavelength of peak emission (in micrometers) is given by **Wien's** (pronounced "weens") **law**:

Wien's law $$\lambda_{max} = \text{constant}/T$$

where λ_{max} refers to the wavelength of energy radiated with greatest intensity. More specifically, the constant in the preceding equation rounds off to the value 2898 for T in kelvins and λ_{max} in micrometers. Thus, we can determine the peak wavelength of radiation emitted as:

$$\lambda_{max} = 2898/T$$

Wien's law tells us that hotter objects radiate energy at shorter wavelengths than do cooler bodies. This is not surprising given that shorter wavelengths correspond to higher energies. Hot objects, possessing more thermal energy, necessarily radiate a higher proportion of energy at those shorter, more energetic wavelengths. Thus, for example, solar radiation is most intense in the visible portion of the spectrum, though it emits over a wide range of wavelengths. Most of the radiation has wavelengths of less than 4 μm, which we generically refer to as **shortwave radiation**. Of the radiation emitted by the Sun, about 46.5 percent is near and thermal infrared, 46.8 percent is visible light, and 6.7 percent is ultraviolet.

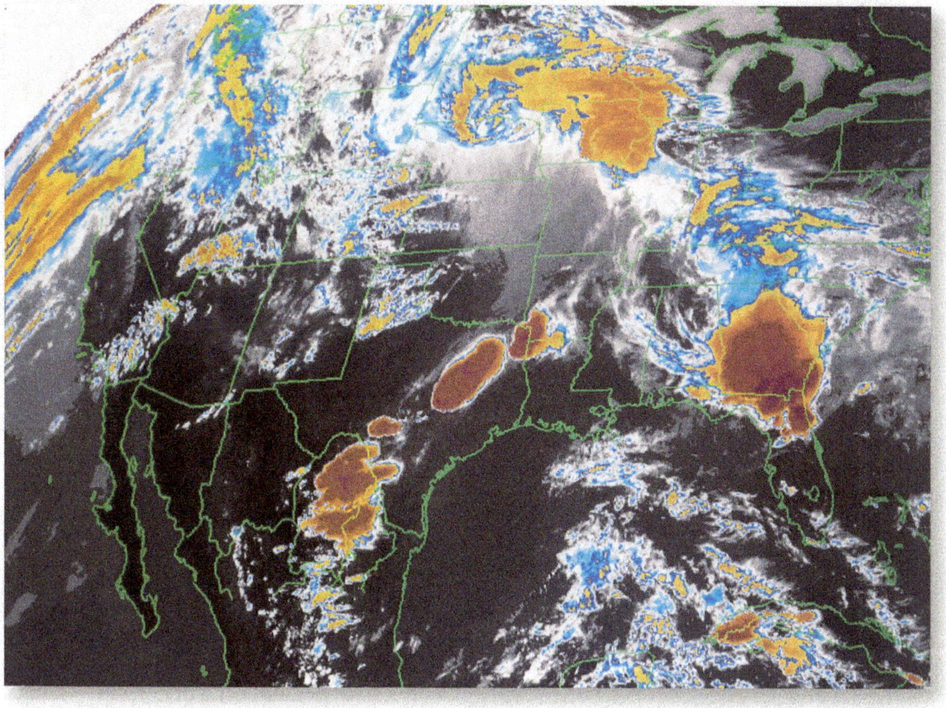

▲ **FIGURE 2–9 Satellite Imagery.** A color-enhanced satellite image of North America. Cloud-top heights are determined by applying radiation laws; redder colors denote higher cloud tops.

Radiation emanating from Earth's surface and atmosphere consists mainly of that having wavelengths longer than 4 μm. This type of electromagnetic energy is called **longwave radiation**.

It is important to note that hotter bodies radiate more energy than do cooler bodies *at all wavelengths*. For example, the Sun radiates energy with wavelengths of about 0.5 μm most effectively and puts out far less energy at λ = 10 μm. Nevertheless, the Sun emits more radiation at those wavelengths than does Earth, despite the fact that Earth emits virtually nothing but longwave radiation (see *Box 2–2, Physical Principles: The Sun*).

The Stefan-Boltzmann law and Wien's law have some very useful and interesting applications. You have undoubtedly seen color-enhanced satellite images showing the distribution of clouds across North America, such as the one in Figure 2–9. This type of image depicts the height of cloud tops, which can be used as an indicator of the intensity of precipitation occurring below. The images are obtained by measuring the intensity of infrared radiation emitted by the cloud tops. Colder surfaces radiate less intense energy than do warm bodies. Weather satellites measure radiation intensity to determine the cloud-top temperatures across a target region. Because higher clouds tend to be colder than lower level clouds (remember that in the troposphere temperature tends to decrease with altitude), temperatures can be used to infer the height of the clouds and thus their relative thickness. Thicker clouds, in turn, usually yield more intense precipitation. Infrared imagery can be obtained at night, as well as during the daytime, because it relies on energy radiated from the cloud tops rather than reflected light.

CHECKPOINT

2.9 What does Wien's law tell us about the relationship between the temperature of an object and the wavelength at which it radiates?

2.10 How has Wien's law been applied in the understanding of satellite images of clouds?

The Solar Constant

2.3 Identify the solar constant and explain how much energy Earth receives from the Sun.

We all know that the Sun is extremely hot and we are protected from its great energy by our distance from the solar surface. But the electromagnetic energy moving through space is not depleted as it moves toward Earth. Radiation traveling through space carries the same amount of energy and has the same wavelength as when it left the solar surface. However, at greater distances from the Sun, it is distributed over a greater area, which reduces its intensity.

Consider a sphere completely surrounding the Sun whose radius is equal to the mean distance between Earth and the Sun, or 1.5×10^{11} m (Figure 2–10). As the distance from the Sun increases, the intensity of the radiation diminishes in proportion to the distance squared. This relationship is known as the **inverse square law**. By dividing total solar emission (3.865×10^{26} W) by the area of our imaginary sphere surrounding the Sun (the area of any sphere is

2-2 PHYSICAL PRINCIPLES

The Sun

The Sun may seem special to us, but compared to the 100 billion or more other stars in our galaxy, it is not particularly unique. Although stars vary considerably in size, temperature, brightness, and density, the Sun is about average in terms of these characteristics. The Sun's general regions are shown in Figure 2–2–1.

The Interior

The innermost portion of the Sun consists of its core, the radiation zone, and the convection zone. The **core** is characterized by extremely high temperatures (about 15 million °C, or 27 million °F) and high densities that lead to the energy-generating process of **nuclear fusion**. In this reaction, hydrogen atoms combine under tremendous heat and pressure to form a smaller number of heavier helium atoms. A certain amount of mass is lost in the process, and radiant energy is released—the same energy that works its way to the solar surface, travels through space, and ultimately warms Earth. The amount of this energy is staggering. Try to imagine the explosion of 100 billion hydrogen bombs—that is equivalent to the amount of energy released in the core *every second!*

Energy initially travels outward from the core as electromagnetic energy through the **radiation zone** and into the base of the **convection zone**, where upwelling of the solar gases transfers the energy to the relatively thin solar surface.

The Photosphere

The layer of the Sun that radiates most of the energy away from the Sun is called the **photosphere**. It is the layer of the Sun we actually see as the **solar disk**. Although radiation travels from the photosphere to Earth in only about 8 minutes, the transfer of radiant energy within the Sun is incredibly slow. In fact, it takes about a million years for the energy unleashed in the core to travel to the base of the photosphere; thus, the energy reaching Earth is ancient.

The photosphere is marked by a number of features of varying sizes and lifetimes. **Granules** are the ever-present tops of convection cells that transport energy from the base of the photosphere to its surface. These features, analogous to bubbles in a pot of boiling water, are about 1000 km (600 mi) in diameter with life spans on the order of 5 to 10 minutes. At any given time, there are millions of these on the surface of the photosphere.

Sunspots (each lasting a few weeks or months) are dark regions on the photosphere with diameters of about 10,000 km (6000 mi) and temperatures about 1500 °C cooler than the surrounding surface. They form in response to locally strong magnetic fields, a thousand times more intense than those of the surrounding photosphere, which block the upwelling of heat from below.

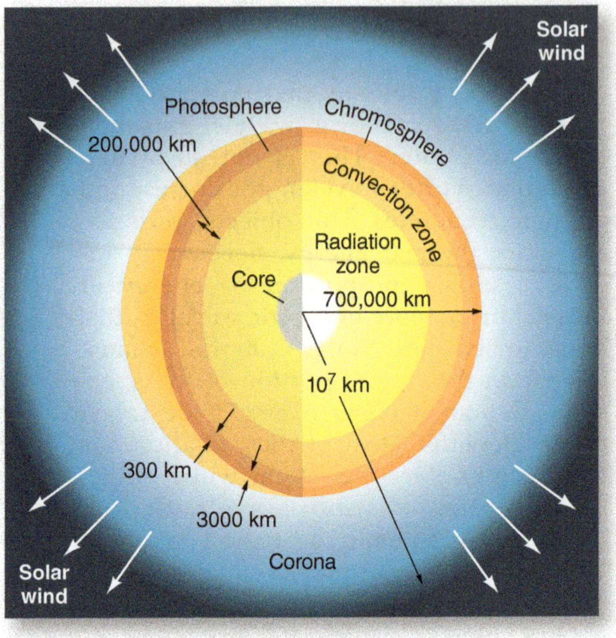

▲ **FIGURE 2–2–1 Layers of the Sun.** Energy is produced in the core of the Sun by nuclear fusion. Within the Sun, the energy is radiated to the base of the convection zone, where mixing transfers the energy upward to the base of the photosphere (the layer of the Sun visible from Earth).

Video Solar Power http://goo.gl/9GvFmT

Video Tour of the Electromagnetic Spectrum http://goo.gl/q1q6Gq

given as $4\pi r^2$), we can determine the amount of solar energy received by a surface perpendicular to the incoming rays at the mean Earth–Sun distance. This incoming radiation is equal to

$$\frac{3.849 \times 10^{26}\ \text{W}}{4\pi(1.5 \times 10^{11}\ \text{m})^2} = 1361\ \text{W/m}^2$$

We refer to the value 1361 W/m^2 as the **solar constant** (although minor variations in solar output and other factors allow for some minor departures from this "constant"). For the sake of comparison, using the same procedure, we can determine that the solar constant for Mars (2.25×10^{11} m from the Sun) is 445 watts per square meter.

Because the Sun is so much larger than Earth, its rays approach our planet as essentially parallel rays, as if the solar surface were a giant wall emitting the radiation.

CHECKPOINT

2.11 How does the intensity of solar radiation vary with distance from the Sun?

2.12 What is the solar constant and how is it derived using the inverse square law?

The Causes of Earth's Seasons

2.4 Recognize the characteristics of Earth's orbit around the Sun and how they create the seasons.

Although the Sun emits a nearly constant amount of radiation, on Earth we experience significant changes in the amount of radiation received during the course of a year. These variations in energy manifest themselves as the seasons. We also know that the low latitudes (for example, the

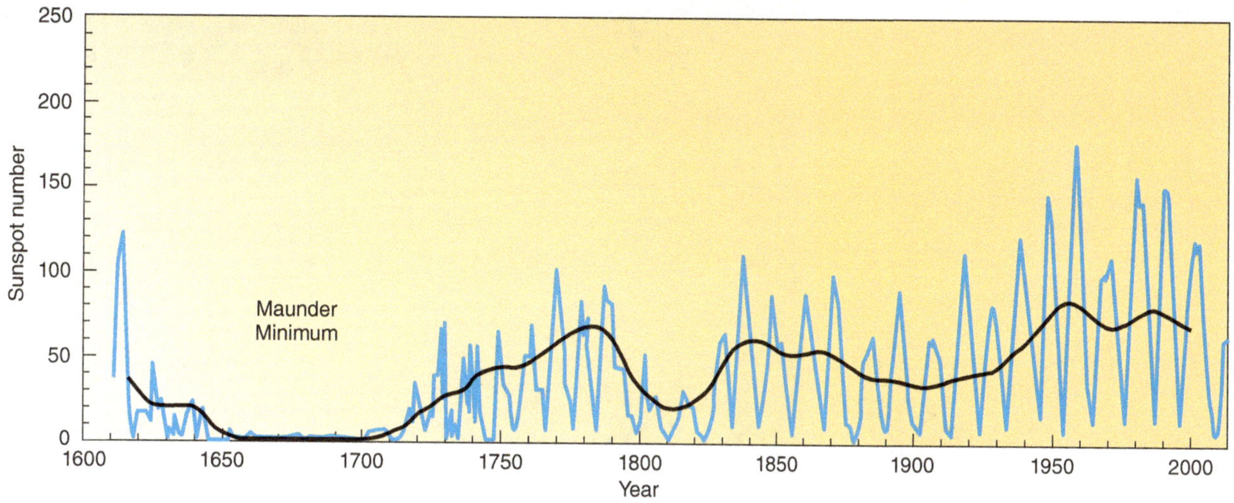

▲ **FIGURE 2–2–2 Sunspots.** With peak occurrences normally observed about every 11 years, sunspots appear with some regularity. As seen in the smoothed curve (black), there are also long-term changes in average sunspot number. Much of the 17th century, for example, was marked by minimal sunspot activity. This period is known as the Maunder minimum.

Sunspots remain fixed in place and appear to move because of the rotation of the Sun (which takes about 24 days and 16 hours to complete one turn of its axis). The number of sunspots tends to peak every 11 years (Figure 2–2–2). Although the cycle is usually well defined, long episodes of minimal or unusually high sunspot activity have appeared during historic times. Figure 2–2–2 shows a long period of reduced sunspot activity during the 17th cenury. Associated with these changes in sunspot number are very small changes in solar radiation on the order of about 0.1 percent. In light of such small radiation changes, it is not surprising that no definitive connections have been established between sunspots and climate.

Perhaps the most spectacular of solar disturbances are **flares**, intensely hot flashes (perhaps 100 million °C) across the photosphere surface due to magnetic instabilities. Temperatures within flares can achieve a staggering 100,000,000 K. They exist for only a matter of minutes but they emit a huge amount of energy, particularly in the form of X-ray and ultraviolet radiation.

The Chromosphere

The chromosphere is a very thin layer about 2,000 km (1200 mi) in thickness. The chromosphere has an extremely low density and is almost always invisible from Earth. However, the chromosphere can be evident during total solar eclipses.

2–2–1 What are the three layers of the Sun?

2–2–2 Describe the types of features that episodically appear on the photosphere.

tropics and subtropics) receive more solar radiation per year at the top of the atmosphere than do regions at higher latitudes (for example, the Arctic and Antarctic). In this section, we will discuss how Earth's orbit around the Sun and its orientation with respect to incoming radiation influence the seasonal and latitudinal receipt of incoming solar radiation (often called **insolation**[1]).

Earth's Revolution and Rotation

As we know, and as Figure 2–11 shows, Earth orbits the Sun once every 365¼ days as if it were riding along a flat plane. We refer to this imaginary surface as the **ecliptic plane** and to Earth's annual trip about the plane as its **revolution**.

The orbit is not quite circular but instead sweeps out an elliptical path, so that the distance between Earth and the Sun varies over the course of the year. Earth is nearest the Sun—at a point called **perihelion**—on or about January 4, when Earth–Sun distance is about 147,000,000 km (91,000,000 mi). Earth is farthest from the Sun—at a point called **aphelion**—on or about July 4, when Earth–Sun distance is about 152,000,000 km (94,000,000 mi). Thus, on perihelion Earth is 3 percent closer to the Sun than on aphelion. But because the intensity of incoming radiation varies inversely with the square of Earth–Sun distance (recall the inverse square law), the radiation is almost 7 percent more intense. (As mentioned at the outset of this chapter, however, this variation in intensity is not what causes the change of seasons.)

In addition to its revolution, Earth also undergoes a spinning motion called **rotation**. Rotation occurs every 24 hours (23 hours and 56 minutes, to be exact) around an imaginary line,

[1]You may see the term *insolation* described as an acronym for "incoming solar radiation." Such is not the case; it is from a Latin word and was first used in the 1600s.

▸ **FIGURE 2–10 Reduction of Energy Received Due to Distance.** The intensity of a beam of solar radiation does not weaken as it travels away from the Sun. However, as radiation travels farther from the Sun it is distributed over a greater area, thereby making it less effective at warming or illuminating any surface it hits. Imagine two spheres encompassing the Sun (such as the one with a radius equal to the mean Earth–Sun distance). All the radiation emitted from the Sun would be captured by this surrounding sphere. Now imagine that the surrounding sphere has a radius equal to the mean distance between the Sun and Mars. This sphere is larger than the previous one, so the energy emitted must be distributed over a greater area.

Video
January
Global Movie
http://goo.gl/zIDgr2

Video
Seasonal
Changes in Global
Snow Cover
http://goo.gl/A8D8Ay

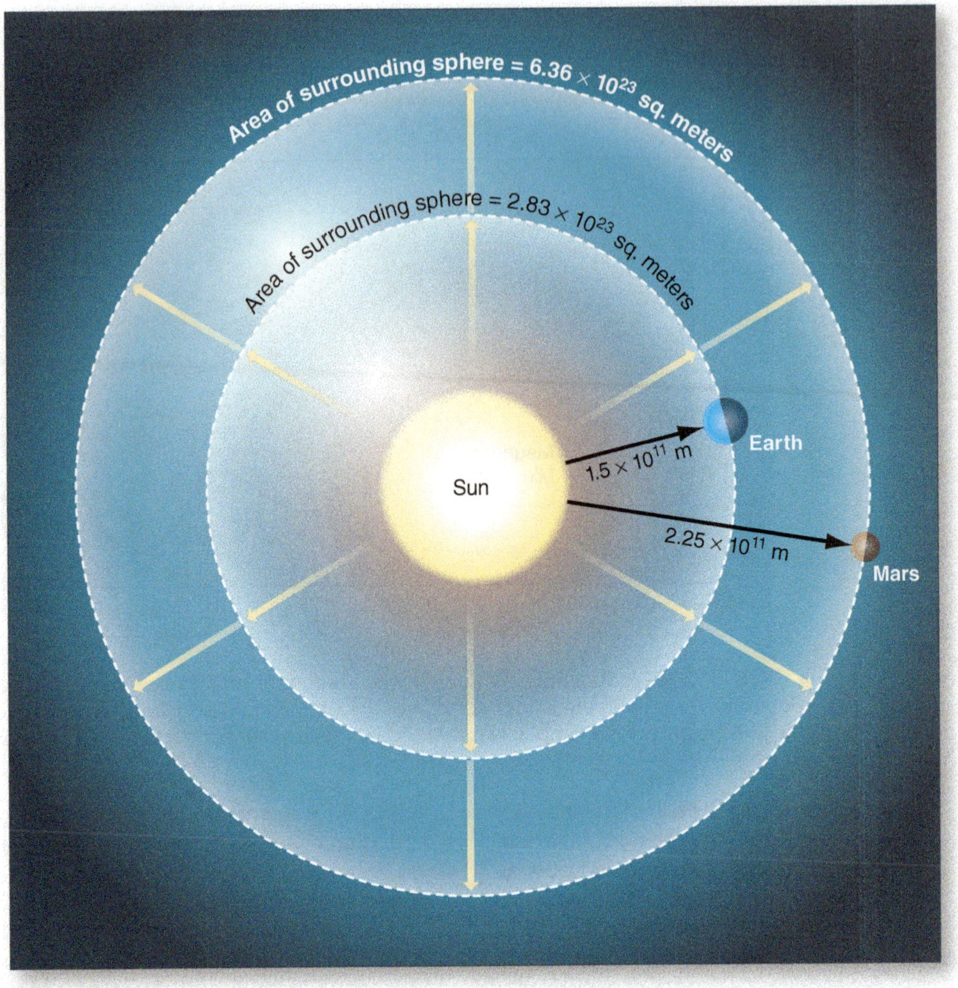

▸ **FIGURE 2–11 Elliptical Orbit of Earth.** If you could look down on Earth's orbit from a point in space, the orbit would appear to be nearly circular, although it is in fact slightly elliptical. Earth is nearest to the Sun (perihelion) on about January 4 and farthest away (aphelion) around July 4.

Video
July Global
Movie
http://goo.gl/wjjusN

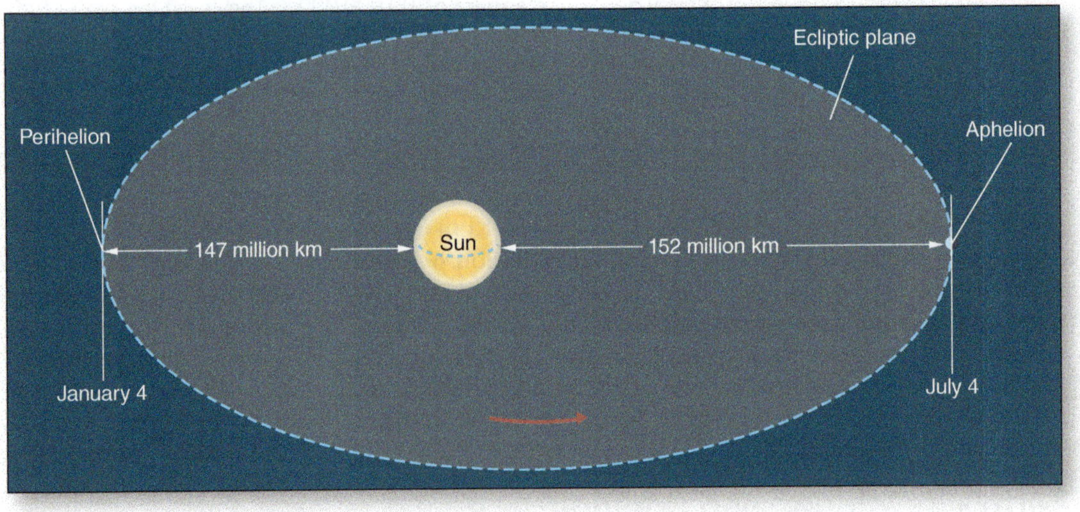

called *Earth's axis*, connecting the North and South Poles. The axis is not perpendicular to the plane of the orbit of Earth around the Sun but is tilted 23.5° from it, as shown in Figure 2–12. Moreover, no matter what time of year it is, the axis is always tilted in the same direction and always points to a distant star called **Polaris** (the North Star). The constant direction of the tilt means that for half the year the Northern Hemisphere is oriented somewhat toward the Sun, and for half the year it is directed away from the Sun. The changing orientation of the hemispheres with regard to the Sun is the actual cause of the seasons—not the varying distance between Earth and the Sun.

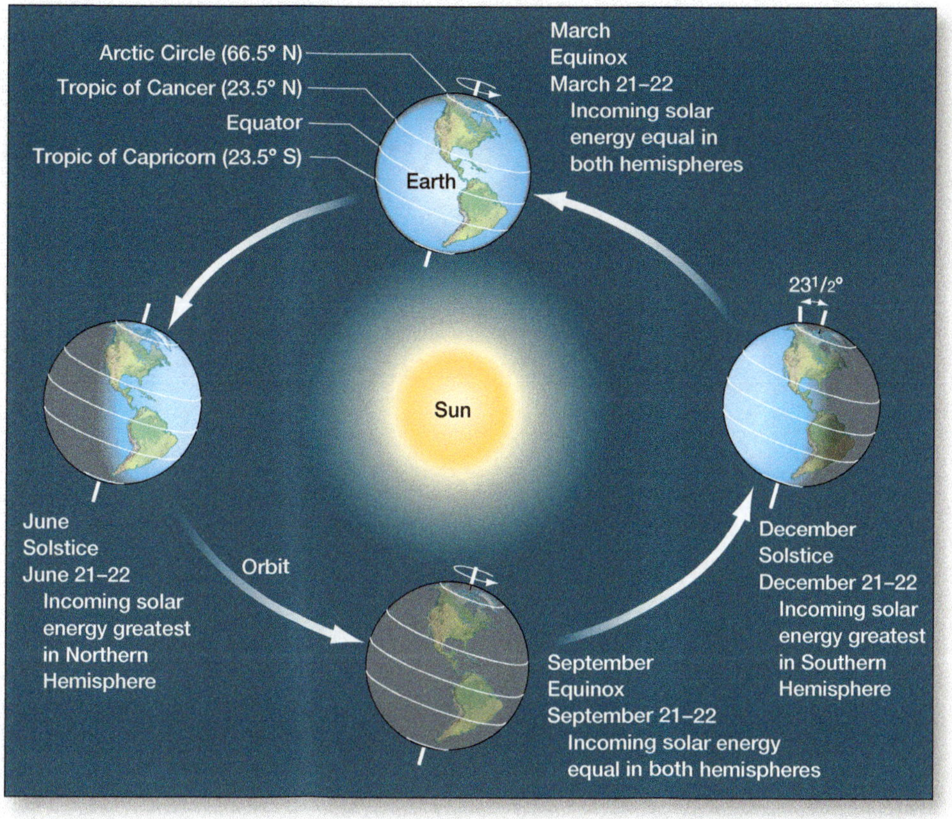

◂ **FIGURE 2–12 Orientation of Earth's Axis.** When Earth is at the far left position, the Northern Hemisphere has its maximum tilt toward the Sun (the summer solstice for that hemisphere). The Northern Hemisphere will then have a stronger orientation toward the Sun than will the Southern Hemisphere. The opposite situation exists when Earth is in the far right position. The situation at the bottom and top of the diagram represents the equinoxes, in which neither hemisphere has a stronger orientation toward the Sun.

Video
Global Variations in Insolation Through the Year
http://goo.gl/2VT9ka

HYPOTHETICAL EXAMPLE: EFFECT OF 90° TILT It is easier to visualize how the tilt of the axis influences the seasons if we consider a hypothetical situation in which the axis is tilted not 23.5°, but rather a full 90°, as depicted in Figure 2–13. Actually, this is the case for Uranus, so what we describe is not entirely hypothetical. Examine the situation when Earth is in position #1. The Northern Hemisphere is oriented directly toward the Sun so that it is fully illuminated over the entire 24-hour period of rotation. Meanwhile, the Southern Hemisphere undergoes 24 hours of continual darkness. This situation favors greater warmth in the Northern than in the Southern Hemisphere. Furthermore, a person standing at the North Pole would observe the Sun as being directly overhead during the entire day. Moving away from the North Pole toward the equator, there is a gradual reduction in the angle of the Sun above the horizon, until at the equator the Sun appears to be right on the horizon. South of that line, the Sun is below the horizon and nighttime covers the Southern Hemisphere.

Now refer to position #3, which occurs 6 months after position #1. In this situation, the Southern Hemisphere is in continual sunlight, while the Northern Hemisphere is subjected to 24 hours of darkness. Furthermore, someone standing at the South Pole would see the Sun directly overhead, and the apparent position of the Sun would shift toward the horizon for viewers located closer to the equator.

Finally, observe the intermediate positions, #2 and #4. In these two situations, the 90° tilt of the axis is neither toward nor away from the Sun, and the tilt becomes irrelevant to the receipt of insolation. Moreover, in positions #2 and #4 every place on Earth receives 12 hours of daylight and 12 hours of darkness because every latitude is half sunlit and half dark. Finally, note that at noon (when the longitude of any place in question is aligned directly toward the Sun), a person standing at the equator would observe the Sun to be directly overhead. Thus, Earth's revolution causes seasonal changes in the amount of heating of the surface. When either the Northern or Southern Hemisphere is oriented toward the Sun, that hemisphere receives a greater amount of insolation and therefore warms more effectively. Whichever hemisphere is oriented away from the Sun receives less radiation.

Solstices and Equinoxes

Refer back to Figure 2–12, which shows the true seasonal change in the orientation of Earth with respect to the Sun based on the actual 23.5° tilt of the axis. Although the axis is tilted only 23.5°, and not 90°, the principle just described still applies. During 6 months of the year, the Northern Hemisphere receives more sunlight than does the Southern Hemisphere; during the other 6 months, the Southern Hemisphere receives a greater amount of insolation. The four positions shown in the diagram represent 4 days that have particular significance.

In the farthest left position in Figure 2–12, the Northern Hemisphere has its maximum tilt toward the Sun. This occurs on or about June 21, which we refer to as the **June solstice** (this is also called the *summer solstice* according to the corresponding Northern Hemisphere season). Although we designate this as the first day of summer, it actually represents the day on which the Northern Hemisphere has its greatest availability of insolation. Six months later (on or about December 21), the Northern Hemisphere has its minimum availability of solar radiation on the **December solstice** (*winter solstice* in the Northern Hemisphere), which is called the first day of winter in the Northern Hemisphere and the first day of summer in the Southern Hemisphere. Intermediate between the

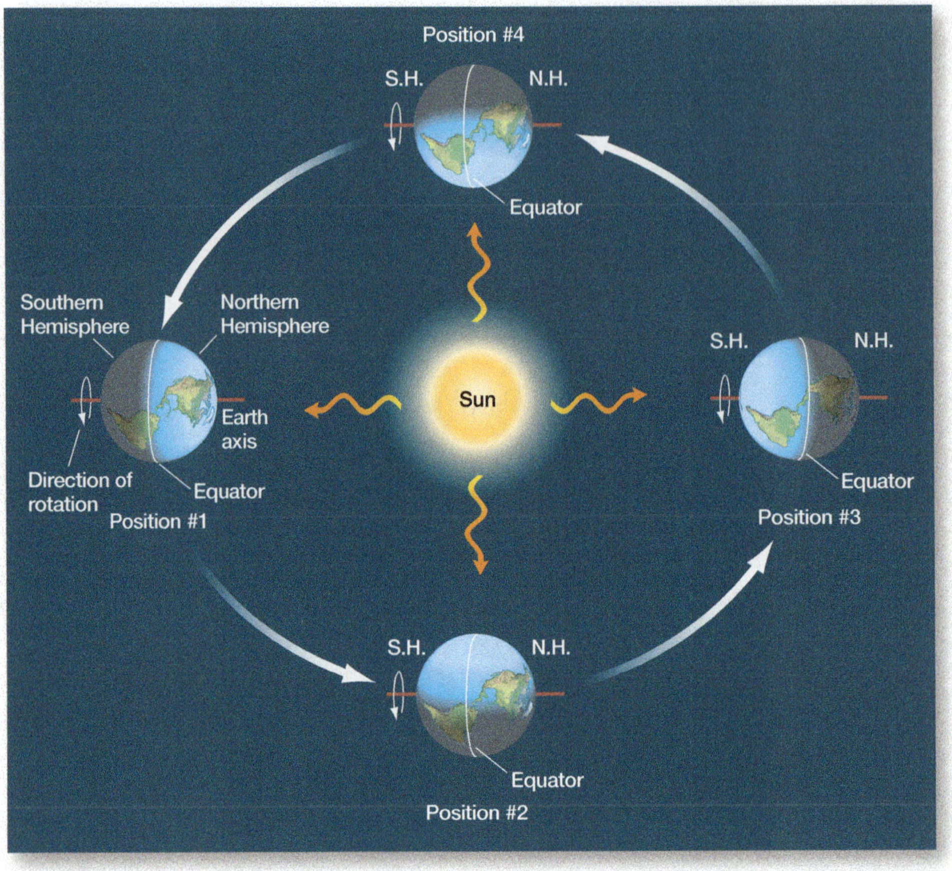

▸ **FIGURE 2–13 Hypothetical Tilt of Earth's Axis.** A hypothetical situation wherein Earth's axis is aligned along the ecliptic plane. In position #1, the Northern Hemisphere receives much energy from the Sun while the Southern Hemisphere is in constant darkness. The situation reverses six months later (position #3). In positions #2 and #4, both hemispheres receive equal amounts of solar energy.

Animation
Earth-Sun Relations

http://goo.gl/HXefff

two solstices are the **March equinox** (often called the *vernal* or *spring equinox* for the Northern Hemisphere) on or about March 21 and the **September equinox** (called the *autumnal equinox* in the Northern Hemisphere) on or about September 21. On the equinoxes, every place on Earth has 12 hours of day and night (the word equinox refers to "equal night"), and both hemispheres receive equal amounts of energy.

Of course, the transitions between the four positions shown in Figure 2–12 do not occur in sudden leaps; instead, a steady progression occurs from one position to the next. As shown in Figure 2–14, the 23.5° tilt of the Northern Hemisphere toward the Sun on the June solstice causes the *subsolar point* (the point on Earth where the Sun's rays meet the surface at a right angle—and where the Sun appears directly overhead) to be located at 23.5° N. This is the most northward latitude at which the subsolar point is located. The fact that the Sun never appears directly overhead poleward of 23.5° N gives that latitude special significance, and we call it the **Tropic of Cancer**.

Likewise, on the December solstice, the sun is directly overhead at 23.5° S, the **Tropic of Capricorn**. On the two equinoxes, the subsolar point is on the equator. Thus, the subsolar point migrates 47° (that is, between 23.5° N and 23.5° S) over a 6-month period, and on any particular day it is located somewhere between the Tropics of Cancer and Capricorn. This seasonal movement of the subsolar point is similar to what would happen if instead of Earth orbiting the Sun, its axis were to slowly rock back and forth, toward and away from the Sun.

The latitudinal position of the subsolar point is the **solar declination**, which can be visualized as the latitude at which the noontime Sun appears directly overhead. Table 2–2 provides values of the solar declination for several days of the year.

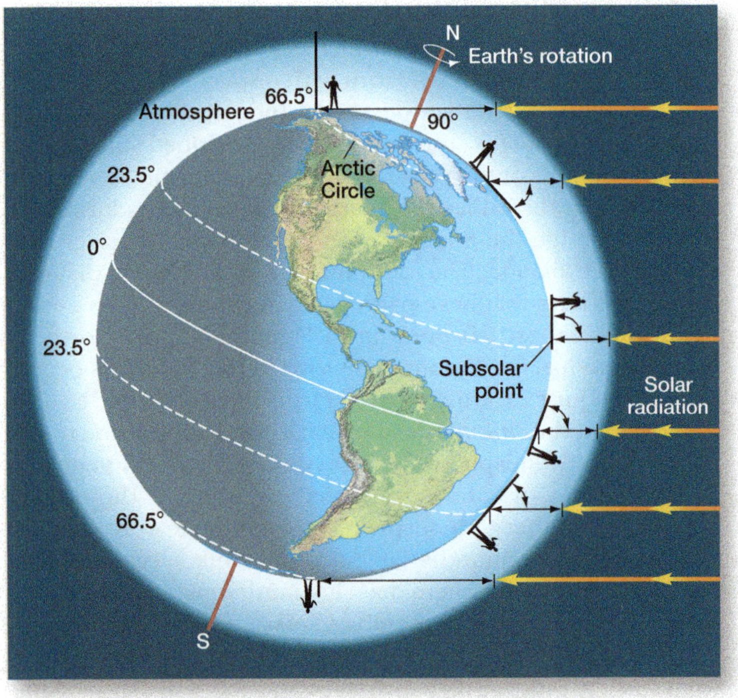

▴ **FIGURE 2–14 Subsolar Point.** Because Earth's axis is tilted 23.5°, the subsolar point is at 23.5° N during the June solstice.

CHECKPOINT

2.13 Describe the orientation of Earth's axis at the winter solstice, summer solstice, and the equinoxes in the Northern Hemisphere.

2.14 Explain how the position of the subsolar point varies from one solstice to the next.

TABLE 2–2

Solar Declination on Selected Dates

Date	Jan 21	Feb 21	Mar 21	Apr 21	May 21	Jun 21	Jul 21	Aug 21	Sep 21	Oct 21	Nov 21	Dec 21
Declination	20° S	11° S	0°	12° N	20° N	23.5° N	21° N	12° N	0°	10° S	20° S	23.5° S

Effects of Earth's Changing Orientation

2.5 Explain three ways in which Earth's changing orientation with respect to the Sun affects the receipt of solar radiation.

The changing orientation of Earth with respect to the Sun directly affects the receipt of insolation through three mechanisms: (1) the length of the period of daylight during each 24-hour period, (2) the angle at which sunlight hits the surface, and (3) the amount of atmosphere that insolation must penetrate before it can reach Earth's surface.

Period of Daylight

One way the tilt of the axis influences energy receipt on Earth is by its effect on the lengths of day and night. We have already seen that if the axis were tilted 90° from the plane of Earth's orbit there would be 1 day when the entire Northern Hemisphere underwent 24 hours of daylight and an equivalent period six months later of continuous darkness. But the axis is only tilted 23.5° from the plane of the orbit, so only the latitudes poleward of 66.5° (that is, 90° minus 23.5°) experience a 24-hour period of continuous daylight or night. These lines of latitude are the **Arctic Circle** (in the Northern Hemisphere) and the **Antarctic Circle** (in the Southern Hemisphere). This is illustrated in Figure 2–15. On the June solstice, any place north of the Arctic Circle has 24 hours of daylight (Figure 2–16). Just a short distance south of the Arctic Circle there is almost (but not quite) 24 hours of daylight. Moving toward the equator, the period of daylight decreases until reaching the equator, where the day and night are both 12 hours long. Moving into the Southern Hemisphere, the day's length shrinks until 66.5° S, where night is 24 hours long. The opposite pattern, of course, holds for the December solstice.

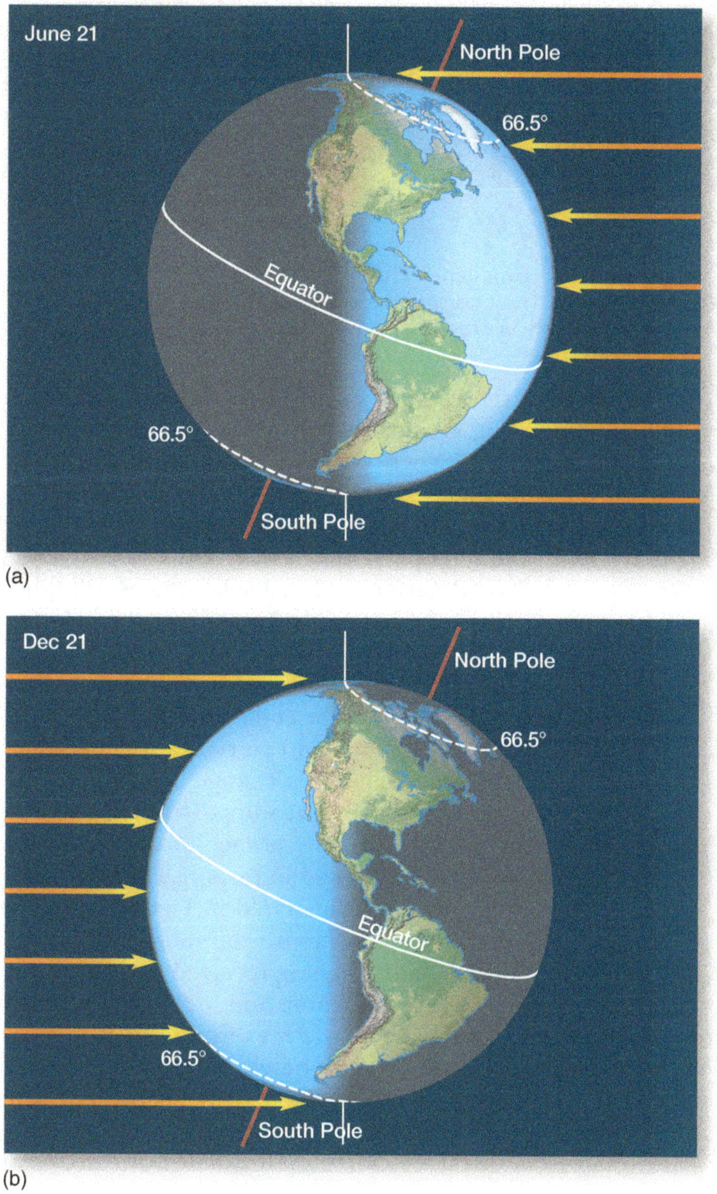

▲ **FIGURE 2–15 Varying Period of Daylight.** On the June solstice (a), every point north of 66.5° N has 24 hours of daylight and every point south of 66.5° S has continual night. During the December solstice (b), the situation is reversed. These latitudes are called the Arctic Circle and Antarctic Circle, respectively.

Solar Angle

Solar angles (the angle of the sun above the horizon) vary with the time of year (the solar declination), the latitude of a given location, and the time of day. Though a universal formula exists for calculating solar angles at any time or place, it is not as simple as the one used to determine the angle at noontime.

NOONTIME SUN ANGLES Noontime sun angles for any given latitude can be easily determined if the solar declination is known. To do this, you simply go through the following procedure:

1. Subtract the latitude of a given location from 90°.
2. Then add the solar declination if it is in the same hemisphere as the location in question (subtract the solar declination if it is in the opposite hemisphere).

Let's use Toronto, Ontario (latitude 44° N), on the June solstice (solar declination = 23.5°) as an example. The noontime solar angle is found as:

$$(90^\circ - 44^\circ) + 23.5^\circ = 69.5^\circ$$

Six months later on the December solstice the noontime angle is

$$(90^\circ - 44^\circ) - 23.5^\circ = 22.5^\circ.$$

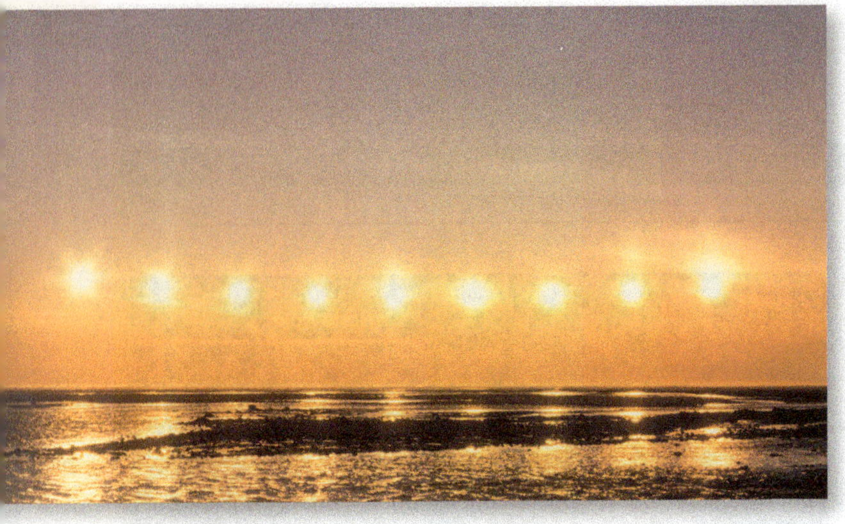

▲ **FIGURE 2–16 Time-lapse Photo of the Sun in June on the Arctic Coast of Alaska.** This photo was taken at night, which meant the camera faced north in order to image the Sun.

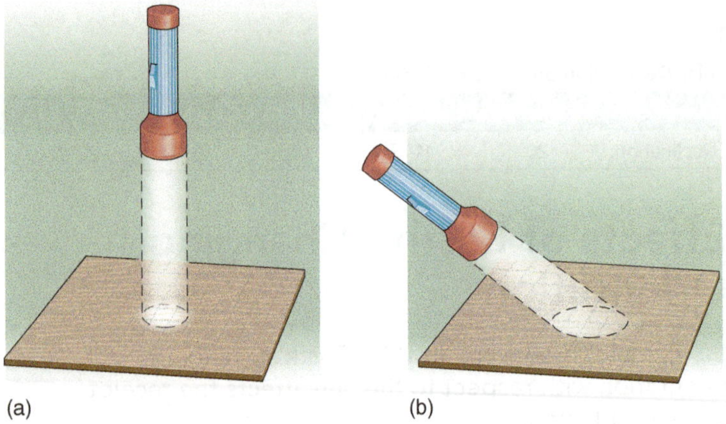

▲ **FIGURE 2–17 Beam Spreading.** The intensity of radiant energy hitting a surface is affected by its angle of incidence. (a) More direct illumination causes a more intense amount of heating than does (b) a less direct angle.

The 47° difference between the two dates is the result of the corresponding change in declination. This is a considerable difference, and when coupled with the much shorter period of daylight there in the winter, it is easy to see why Toronto winters aren't known for being mild.

If we had used Memphis, Tennessee (latitude 35° N), as an example, we would have calculated solar angles 9° higher for both dates, reflecting its latitude 9° closer to the equator.

Video
Net Radiation at the Top of the Atmosphere
http://goo.gl/E9I2G0

EFFECT OF SOLAR ANGLE: BEAM SPREADING

We all know that as the sun rises higher above the horizon, the ground and air typically get warmer into the afternoon. This change is largely due to a decrease in beam spreading. **Beam spreading** is the increase in the surface area over which radiation is distributed in response to a decrease of solar angle, as illustrated in Figure 2–17. The greater the spreading, the less intense the radiation. In part (a) of the figure, the incoming light is received at a 90° angle, which concentrates it on a small area and increases its ability to heat the surface. In part (b), the rays hit the surface more obliquely and the energy is distributed over a greater area, leading to a less intense illumination (less energy per unit area). Thus, a beam of light is more effective at illuminating or warming a surface if it has a high angle of incidence (that is, the angle at which it hits the surface).

The change in noontime solar angle over the course of a year can cause significant differences in the intensity of sunlight hitting the surface due to beam spreading. For example, beam spreading of the noontime sun at Toronto on the December solstice is almost two and a half times greater than on the June solstice. Thus, in the absence of other effects, the midday sun will be less than half as intense in the winter as in the summer.

Differences in the amount of beam spreading have a major role in causing the seasons. During the 6-month period between the March and September equinoxes, any latitude in the Northern Hemisphere has a more direct angle of incidence than does its Southern Hemisphere counterpart. Thus, insolation available to the Northern Hemisphere is subjected to less beam spreading, which promotes greater warming of the surface. During the following six months, the situation is reversed and the Southern Hemisphere has, on the whole, higher Sun angles.

Figure 2–18 shows the effects of noontime solar angle and length of daylight period for the solstices and equinoxes. On the June solstice, both factors work together to enhance warming in the Northern Hemisphere; on the December solstice, they combine for less effective heating in the north.

EFFECT OF SOLAR ANGLE: ATMOSPHERIC BEAM DEPLETION

Think back to the last time you spent a few hours lying out in the Sun. If you went outside early in the day when the Sun was low above the horizon, you probably did not feel a great deal of warming from its rays, even if you positioned yourself so that the sun would hit your skin at nearly a 90° angle. But as the Sun got higher in the sky, it became more effective at warming your body. This reduction in the sun's ability to warm your skin is the result of the greater amount of atmosphere it must pass through when the sun is low above the horizon. You can see in Figure 2–19a, insolation approaching the surface at a 90° angle passes through the atmosphere as directly as possible. Compare this to Figure 2–19b, in which sunlight approaches the surface at a low angle (as it does around sunrise or sunset). In this situation, a beam of sunlight must pass through a greater amount of atmosphere. Although the atmosphere is mostly transparent to incoming sunlight, some radiation is absorbed and even more is reflected back to space. The greater the thickness traveled, the more the beam is depleted. Because the solar altitude is lowest in the Northern Hemisphere near the December solstice, at that time more energy is lost due to atmospheric effects than at any other time of the year.

Overall Effects of Period of Daylight, Solar Angle, and Beam Depletion

We can now summarize some of the important points regarding controls on energy reaching Earth's surface. We must emphasize that we have not considered differences in atmospheric transparency or cloudiness at all; thus, the discussion assumes uniform optical properties. In other words, the

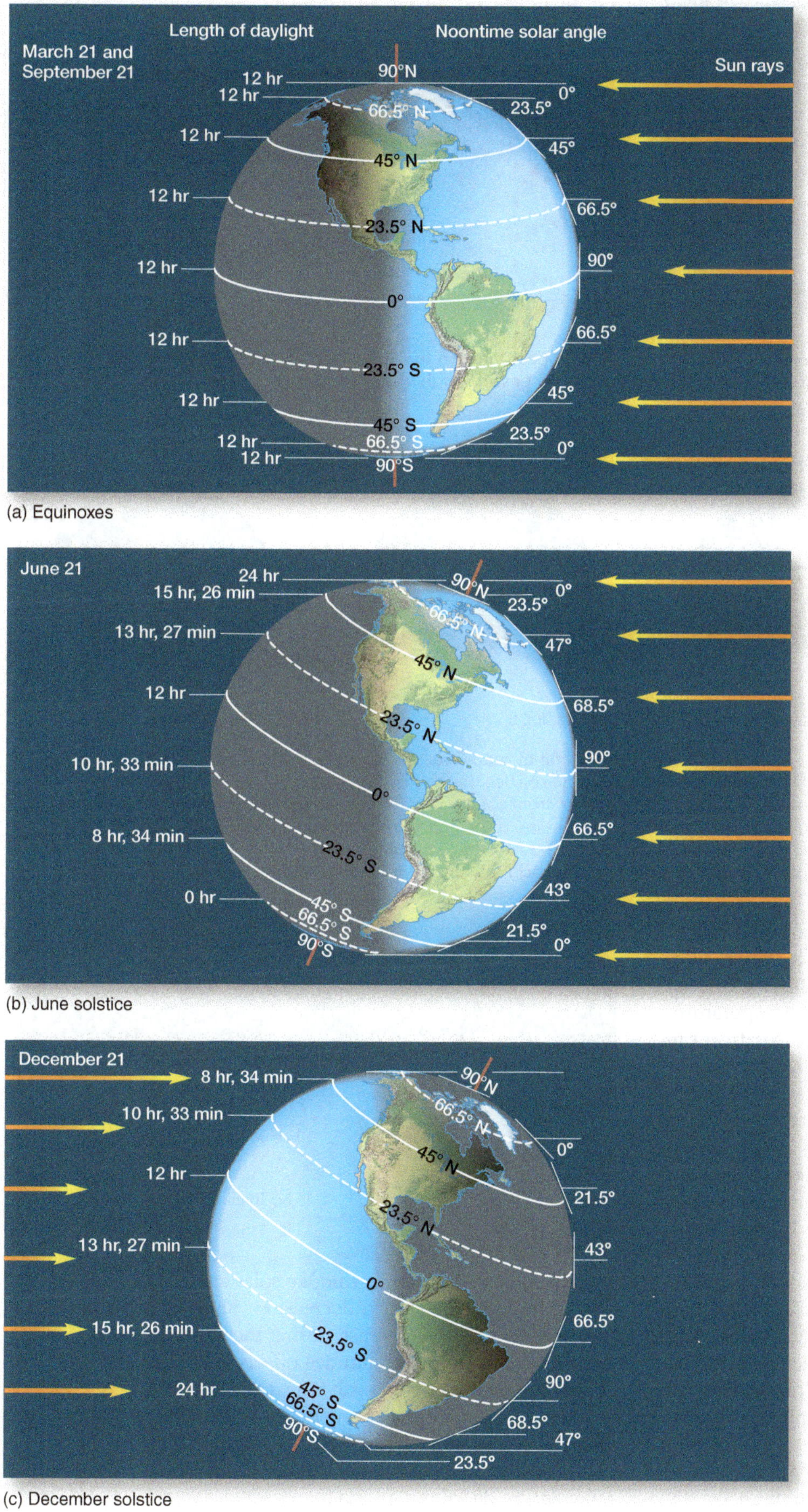

▲ **FIGURE 2–18 Length of Daylight.** The length of day (left) and the noon solar angle (right) are shown for the (a) equinoxes, (b) June solstice, and (c) December solstice.

patterns described below arise solely from geometrical considerations.

1. **June solstice:**
   - **a.** Solar declination = 23.5° N.
   - **b.** Every latitude in the Northern Hemisphere receives more energy than the corresponding latitude in the Southern Hemisphere.
   - **c.** Every place north of the Arctic Circle receives 24 hours of daylight, and every place south of the Antarctic Circle has 24 hours of night.
   - **d.** The equator receives 12 hours of daylight.
   - **e.** For the Northern Hemisphere as a whole, sunlight travels through less atmosphere than it does in the Southern Hemisphere.
2. **December solstice:**
   - **a.** Solar declination = 23.5° N.
   - **b.** Every latitude in the Southern Hemisphere receives more energy than the corresponding latitude in the Northern Hemisphere.
   - **c.** Every place south of the Antarctic Circle receives 24 hours of daylight, and every place north of the Arctic Circle has 24 hours of night.
   - **d.** The equator receives 12 hours of daylight.
   - **e.** For the Southern Hemisphere as a whole, sunlight travels through less atmosphere than it does in the Northern Hemisphere.
3. **Equinoxes:**
   - **a.** Solar declination is 0°.
   - **b.** Every place on Earth has 12 hours of daylight.
   - **c.** Both hemispheres receive equal amounts of insolation.

Figure 2–20 shows the paths traversed by the direct rays of the sun at several latitudes at different times of year. Notice how the paths are more curved at higher latitudes and that the Sun sweeps a straight-line path across the sky at all times over the equator.

Changes in Energy Receipt with Latitude

We have seen that seasonal changes in the orientation of Earth with respect to the Sun

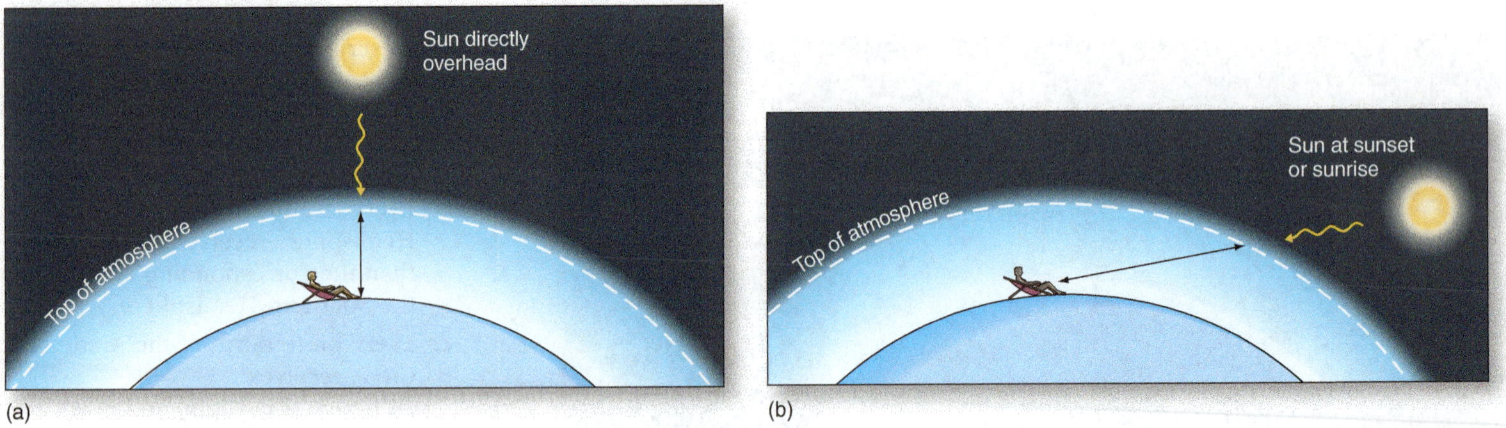

▲ **FIGURE 2–19 Atmospheric Path Lengths.** A high solar angle (a) allows sunlight to pass through the atmosphere with a relatively short path. Lower sun angles such as those near sunrise and sunset (b) require that the energy pass through more of the atmosphere. This increase in atmospheric mass results in a greater depletion of energy than in (a).

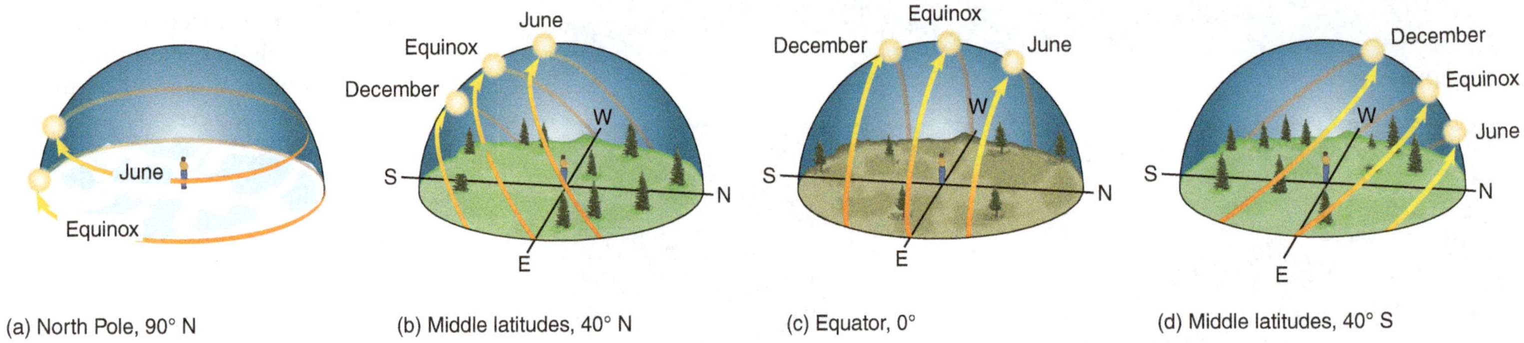

▲ **FIGURE 2–20 Solar Paths.** Apparent path of the Sun at different locations and time of year. Over the North Pole (a) the Sun appears to sweep a circle across the horizon on the equinoxes and at 23.5° above the horizon on the June solstice. In the middle latitudes of the Northern Hemisphere (b) the Sun sweeps an arcuate path from the southeast to the southwest in the winter, from east to west on the equinoxes, and from northeast to northwest on the June solstice. At the equator (c) the Sun appears to sweep a straight-line path regardless of the time of year, but with the path displaced north or south, depending on the time of year. In the middle latitudes of the Southern Hemisphere (d) the Sun sweeps a path similar to that observed in the Northern Hemisphere, except that the Sun swings to the north in the middle of the day rather than to the south.

determine the availability of solar radiation across the globe. Obviously, a large supply of energy is favorable for warm temperatures and a lack of energy leads to cold conditions. But the impact of radiation availability does not end with temperature. As we will see in later chapters, energy receipts also affect the distribution of pressure, which, in turn, affects winds, cloudiness, precipitation, and other aspects of weather and climate.

Having examined the effects of latitude and time of year separately, we can now look at how their combined effects set up differences in available insolation from one latitude to another. Table 2–3 presents the values of noontime solar angle and length of day for Winnipeg, Manitoba, Canada (50° N), and Austin, Texas (30° N), for the December and June solstices. (Remember that the values shown in the table represent the amount of energy available at the top of the atmosphere. Atmospheric conditions—especially the amount and type of cloud coverage—will reduce the amount of radiation that actually reaches the surface. The processes that affect the receipt of the energy at the surface are described in Chapter 3.) During the winter, Winnipeg has only 62 percent as much available solar radiation as does Austin because of its lower solar angle and the shorter period of daylight. During the June solstice, however, Winnipeg has a slightly greater amount of radiation because its greater period of daylight more than offsets its lower midday solar angle. Thus, during the winter, both factors lead to decreasing values of available insolation with latitude. In contrast, during the summer, the increasing day's length at the high latitudes offsets the effect of lower Sun angles. The result is a weak latitudinal gradient in summer and strong north–south differences in winter. As will be seen in coming chapters, this has profound implications for seasonal temperature distributions and for seasonal changes in the vigor of large-scale atmospheric motion.

TABLE 2–3

Variations in Solar Angle and Daylength

	Solar Angle at Noon	Length of Day	Total Radiation for Day (megajoules/m^2)
December 21			
Winnipeg (50° N)	16.5°	7 hr, 50 min	7.1
Austin (30° N)	36.5°	10 hr, 04 min	18.6
June 21			
Winnipeg (50° N)	63.5°	16 hr, 10 min	44.5
Austin (30° N)	83.5°	13 hr, 56 min	43.9

Video
Solar Eclipse
http://goo.gl/BfxU6Q

CHECKPOINT

2.15 What are the two general ways in which solar declination and latitude combine to affect the heating of the surface?

2.16 Why are seasonal differences in temperature generally greater at higher latitudes than lower latitudes?

DID YOU KNOW?

On the June solstice, more solar radiation is available at the top of the atmosphere at the Arctic Circle than at either the Tropic of Cancer or the equator. In this particular instance the effect of the 24-hour period of sunlight at the Arctic Circle outweighs the lower Arctic midday Sun angles.

Summary

2.1 Identify the types of energy and how they can be transmitted.

- All matter emits electromagnetic energy in a continuum of different wavelengths, but not all substances or objects radiate the same amount of energy, nor do they emit it at the same dominant wavelengths.

2.2 Explain the laws governing the amount and type of radiation emitted by objects.

- The Stefan-Boltzmann law tells us that hot objects emit radiation more intensely than do cooler objects.
- Wien's law dictates that hotter objects emit radiation at shorter wavelengths.
- Together, these laws mandate that the unit of surface area for the Sun puts out hundreds of thousands of times more energy than does the same area of Earth's surface, and that the energy from the Sun is emitted at shorter wavelengths.

2.3 Identify the solar constant and explain how much energy Earth receives from the Sun.

- The solar constant is the average value of the amount of solar radiation received by the entire Earth. This is a misnomer to some extent because the radiation received does undergo some variability.
- Earth receives on average 1361 W/m^2 from the Sun, the result of the temperature of the photosphere and the Earth–Sun distance.

2.4 Recognize the characteristics of Earth's orbit around the Sun and how they create the seasons.

- The change in Earth–Sun distance due to the planet's elliptical orbit has a relatively minor influence on seasonal energy receipts.
- A more important factor in creating the seasons is the orientation of the northern and southern hemispheres with respect to the Sun.

2.5 Explain three ways in which Earth's changing orientation with respect to the Sun affects the receipt of solar radiation.

- Changes in solar angle result in varying amounts of beam spreading and atmospheric path length through which radiation must travel, as well as the length of the daytime period.

Key Terms

aphelion *p. 69*
Arctic and **Antarctic Circles** *p. 73*
beam spreading *p. 74*
blackbody *p. 64*
buoyancy *p. 62*
conduction *p. 61*
convection *p. 62*
convection zone *p. 68*
core *p. 68*
December solstice *p. 71*
ecliptic plane *p. 69*
electromagnetic radiation *p. 60*
emissivity *p. 65*
energy *p. 60*
flares *p. 69*
granules *p. 68*
graybody *p. 65*
insolation *p. 69*
inverse square law *p. 67*
joule *p. 60*
June solstice *p. 71*
kinetic energy *p. 60*
longwave radiation *p. 67*
March equinox *p. 72*
micrometers or **microns** *p. 63*
nuclear fusion *p. 68*
perihelion *p. 69*
photon *p. 66*
photosphere *p. 68*
Polaris *p. 70*
potential energy *p. 60*
power *p. 60*
radiation *p. 62*
radiation zone *p. 68*
revolution *p. 69*
rotation *p. 69*
September equinox *p. 72*
shortwave radiation *p. 67*
solar constant *p. 68*
solar declination *p. 72*
solar disk *p. 68*
Stefan-Boltzmann law *p. 65*
sunspots *p. 68*
Tropics of Cancer and **Capricorn** *p. 72*
watt *p. 60*
wavelength *p. 63*
Wien's law *p. 66*

Review Questions

1. Give several examples of kinetic and potential energy as they exist on Earth.
2. Conduction and convection are alike in that both transfer heat within a substance. What is the critical difference between them?
3. We have discussed sunlight and X-rays as two examples of electromagnetic radiation. Describe radiation as a wave phenomenon, and explain what is meant by the term *electromagnetic*.
4. Why is wavelength important in radiation transfer? That is, when discussing radiation, why isn't it enough to specify the amount or rate of energy transfer?
5. Place the following wavelength bands in correct order of wavelength: visible, X-rays, ultraviolet, microwave, infrared.
6. Describe how the wavelengths and total energy emitted change as the temperature of an object increases.
7. Why is the Kelvin scale superior to the Fahrenheit and Celsius scales in many scientific applications?
8. The solar constant is about 1361 W/m^2. If the distance between Earth and Sun were to double, what would be the new value?
9. What is the most important factor responsible for seasons on Earth?
10. Describe the annual march of solar declination.
11. What is the significance of the Arctic and Antarctic Circles?
12. Pick a day in the Northern Hemisphere winter. Describe the changes in daylength and solar position you would encounter if you were to travel from the North Pole to the South Pole. Do the same for a day in the Northern Hemisphere summer.
13. If the tilt of Earth's axis were 10°, where would we find the Arctic and Antarctic Circles? Would this cause a change in the dates of the solstices, equinoxes, and perihelion and aphelion?
14. Explain how changes in solar position influence the intensity of radiation on a horizontal surface.
15. Explain why the equator always has 12 hours of sunlight.
16. If you were to travel from the equator to the North Pole, on what day would variations in solar radiation be smallest? Why? Explain how daylength and solar angle change as you move poleward.
17. Burlington, Vermont, is located at 44.5° N. What is the angle of the noontime Sun on either of the equinoxes and on the solstices?

Critical Thinking

1. Some old buildings are warmed by radiators. And the latest trend in residential heating is "radiant heat" supplied by hot water running through tubing in floors. Is this a truly accurate descriptor of how rooms in those buildings are actually warmed?
2. Goose down is composed of a large number of very small filaments. These separate the air within a parka or sleeping bag into many small pockets that do not readily circulate. How does this feature make down such a good insulating material? How does this feature of goose down relate to the material in this chapter?
3. Why is it not completely accurate to describe the energy coming from the Sun as visible radiation?
4. If Earth's speed of rotation were to change, would there be a corresponding change in the amount of energy the planet as a whole receives?
5. At noon the solar angle is always greater at Tucson, Arizona, than at Laramie, Wyoming. Is the same also true for 6 P.M.?
6. Locations near the equator typically have less seasonality than do locations farther away from the tropics. Explain why this is so.
7. How might the temperature change in the course of a day differ on east-facing vs. west-facing slopes?
8. Why is it that the solar angle cannot be considered the sole influence on the amount of radiation reaching Earth's surface? Is the situation different for the Moon?
9. At noon at 45° N latitude, the solar angle is 45° above the southern horizon. What would the angle of incidence be on a north-facing slope of 45°? Would the slope of the surface affect both beam spreading and atmospheric path length?
10. Describe the apparent path of the Sun to a person standing at the North Pole on June 22.
11. On the equinoxes a person at the equator would see the sun rise exactly to the east, pass directly overhead at noon, and set exactly in the west—all over a 12-hour period. How will this change on the solstices?

Problems & Exercises

1. An instrument measures the radiation emitted from an ocean surface as 365 watts per square meter. What law would you apply to determine the ocean surface temperature? (More advanced question: What would the ocean surface temperature actually be? *Hint*: You will need to rearrange one of the equations given in this chapter.)
2. Assume that a body has an emissivity of 0.9 and a temperature of 300 K. Which would have a greater impact on the intensity of radiation emitted: a 50 percent reduction in the emissivity or a 5 percent reduction in the absolute temperature?
3. One factor that influences the amount of insolation available is the varying Earth–Sun distance. Using the distances for perihelion and aphelion of 1.47×10^{11} m and 1.52×10^{11} m, respectively, determine the intensity of solar radiation at the top of Earth's atmosphere on those two days.
4. Saturn is about 1.42×10^{12} m from the Sun, or about 9.5 times as far from the Sun as Earth is. Calculate the solar constant for Saturn. Do you suppose the distribution of wavelengths of the sunlight received at that distance is different from the distribution Earth receives?
5. What is the difference in the noontime solar angle between the two solstices at a latitude of 10° N? How does this compare to the range of noontime solar angles at 30° N? Can you think of any significant outcomes of this difference?
6. Go to **http://aa.usno.navy.mil/data**. Check the first of the two hypertext options under the category "Positions of the Sun and Moon." This site will allow you to determine the solar angle throughout the course of any date you select. Plot the solar angle of the Sun throughout the daylight period for the solstices and the equinoxes and notice how the pattern changes throughout the year. Next, for the solstices and equinoxes, plot and compare the differing solar elevations over the course of the day for East Lansing, Michigan; Knoxville, Tennessee; and Gainesville, Florida. What generalizations can you make?

Visual Analysis

Note the colors of the logs in the fireplace shown here and answer the following questions:

2.1. What type of energy is being emitted when you stand by a fire such as this one and feel the heat?

2.2. How would the color of the logs change if you were able to get the fire hotter?

2.3. Which law discussed in the chapter must be invoked to answer the preceding two questions?

CHAPTER

3 Energy Balance and Temperature

The winter of 2013–2014 was a particularly dry one over California, with hardly any precipitation over most of the state during what is normally its wet season. Then in the middle of February a long, wide plume of moisture—often referred to as an atmospheric river—streamed northeastward over the Hawaiian Islands and into central California, bringing some badly needed rain to the lower elevations and snowpack into the mountains. Though it didn't end the state's drought, it did reduce the water supply problem to a degree. But the event also brought up an interesting question: Exactly where did the energy come from to transport all this moisture thousands of kilometers? Our cars don't run without fuel, nor do our bodies perform without stored energy from the food we eat. The atmosphere must have some means of obtaining this heat. In this chapter we examine how that happens.

◄ *In the midst of the severe drought California received a welcome drenching in early February 2014. Though the precipitation caused local flooding the region soon returned to the ongoing drought conditions. Here a pedestrian makes his way through the downpour near the Bay Bridge in San Francisco.*

Learning Outcomes

After reading this chapter, you should be able to:

3.1 Explain how incoming solar radiation is affected by the atmosphere.

3.2 State the proportions of incoming solar energy that are reflected, scattered, or absorbed.

3.3 Explain how energy is transferred between the surface and the atmosphere.

3.4 Explain the role of the greenhouse effect in warming the atmosphere.

3.5 Describe the global distribution of temperatures on Earth's surface.

3.6 List the factors that influence temperatures on Earth's surface.

3.7 Describe daily and annual temperature patterns.

3.8 Explain how temperatures are measured.

3.9 Define the concept of temperature means and ranges.

3.10 Define some useful indexes of temperature.

3.11 Make use of thermodynamic diagrams and understand vertical temperature profiles.

3.12 Summarize observations of climate change/global warming.

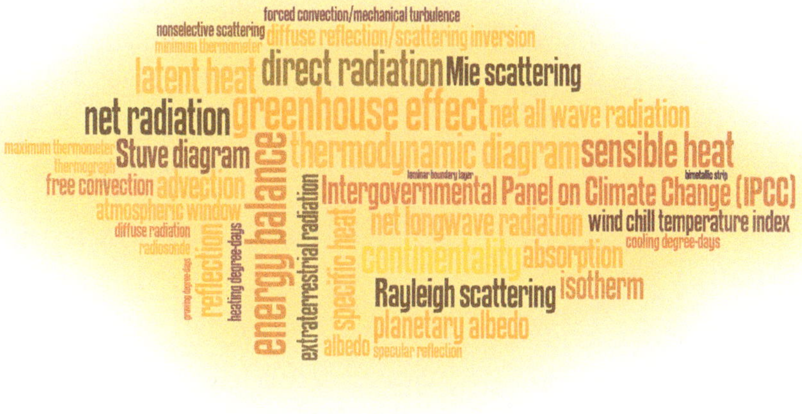

As we saw in Chapter 2, solar radiation provides virtually all the energy that drives the atmosphere. But most of the energy contained in the atmosphere does not accrue by the *direct* absorption of solar radiation. Instead, the majority of the energy comes *indirectly* from the Sun after first having been absorbed by Earth's surface. From there, several processes combine to transfer this absorbed energy to the atmosphere. But in the end, as much energy goes into the Earth surface–atmosphere system as goes out of it. We refer to this input and output as Earth's **energy balance**. In this chapter, we examine the energy transfer process, which provides the fuel for all things we think of as weather.

Atmospheric Influences on Insolation

3.1 Explain how incoming solar radiation is affected by the atmosphere.

Three processes—the absorption, scattering, and transmission of solar radiation—directly affect the distribution of temperature throughout the atmosphere. They also explain a number of atmospheric phenomena of everyday interest, such as the blue sky on a clear day or the redness of a sunset.

Solar radiation does not pass unimpeded through the atmosphere, but rather is attenuated by a variety of processes. The atmosphere absorbs some radiation directly and thereby gains heat. Another portion of radiation disperses as weaker rays going out in many different directions through a process we call *scattering*. Some of the scattered radiation is directed back to space; the remainder is scattered forward as the light we see from the portion of the sky away from the solar disk. In either case, the energy that is scattered is not absorbed by the atmosphere and therefore does not contribute to its heating. The end result of all this is what we see when we look upward, as in Figure 3–1.

The remaining insolation is neither absorbed nor scattered—it passes through the atmosphere without modification, reaching the surface as direct radiation. But not all the energy reaching the surface is absorbed. Instead, a fraction is scattered back to space and, like the radiation scattered by the atmosphere, it does not contribute to the heating of the planet.

In this section, we explore the processes affecting incoming radiation.

Absorption

Atmospheric gases, particulates, and droplets all reduce the intensity of insolation by **absorption**. Absorption represents an energy transfer to the absorber. This transfer has two effects: The absorber gains energy and warms, while the amount of energy delivered to Earth's surface is reduced.

The gases of the atmosphere are not equally effective at absorbing sunlight, and different wavelengths of radiation are

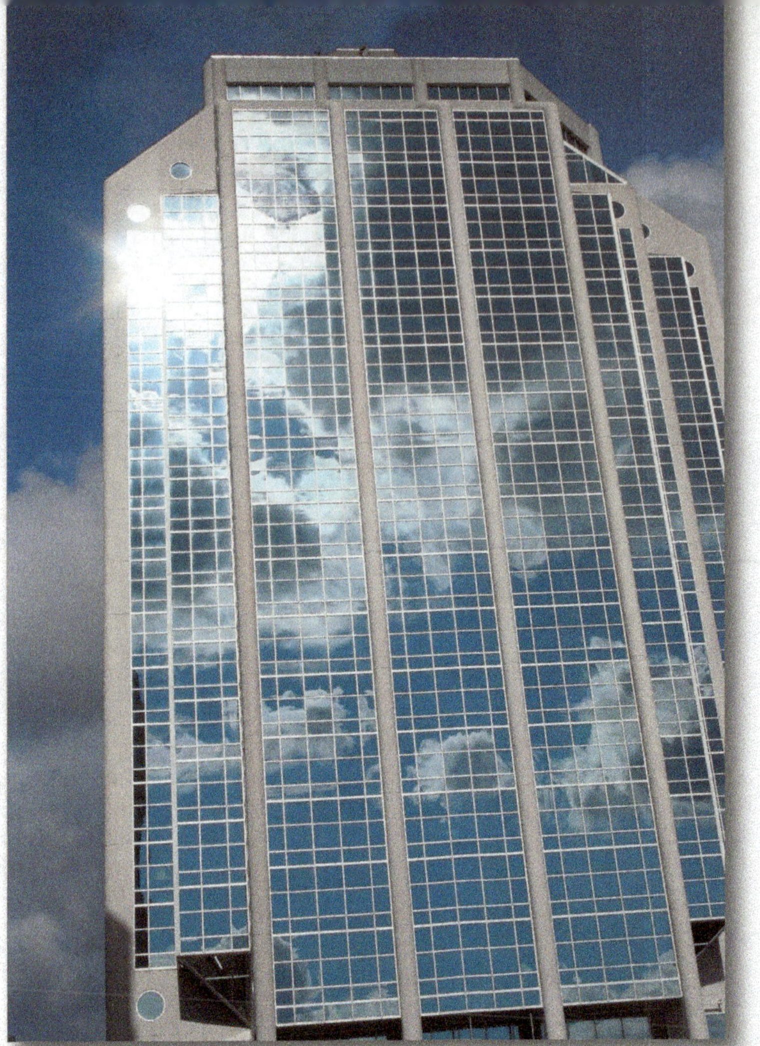

▲ FIGURE 3–1 Reflection of Sun and Cloud Off a Building.

not equally subject to absorption. Ultraviolet radiation, for example, is almost totally absorbed by ozone in the stratosphere. Visible radiation, in contrast, passes through the atmosphere with only a minimal amount of absorption. This is of no minor consequence, because if the atmosphere *were* able to absorb all the incoming solar energy, the sky would appear completely dark. Artificial lights would be useless, because their radiation would likewise be absorbed. The very fact that we can see great distances suggests that the atmosphere is not particularly good at absorbing visible radiation, an impression that turns out to be correct.

Near-infrared radiation, which represents nearly half the radiation emitted by the Sun, is absorbed mainly by two gases in the atmosphere—water vapor and (to a lesser extent) carbon dioxide. So when the air contains a lot of water vapor, the amount of shortwave radiation reaching Earth's surface is reduced substantially, even though it may be uncomfortably hot and humid. This is why direct sunlight in the desert feels so hot and shade is so welcome. In humid regions the apparent temperature difference between standing in direct sunlight and standing in shade is relatively small. When the humidity is high, water vapor absorbs a significant portion of near-infrared radiation, thereby reducing the amount of energy available to warm your skin. On dry days, the lack of water vapor allows a greater amount of near-infrared radiation to penetrate the atmosphere and raise your skin temperature.

Reflection and Scattering

The **reflection** of energy is a process whereby radiation making contact with some material is simply redirected away from the surface without being absorbed. The reason we are able to see is that the human eye can detect the receipt of visible radiation. Visible energy travels in all directions as it is reflected off objects in our field of view. Some of the reflected light comes into contact with our eyes, which in turn send signals to optical centers in our brains. All substances reflect visible light, but with vastly differing effectiveness. Objects do not reflect all wavelengths equally. A shirt, for example, will appear green if it most effectively reflects wavelengths in the green portion of the spectrum. A fresh patch of snow very effectively reflects visible light, while a piece of coal reflects only a small portion of the visible radiation hitting its surface. The percentage of insolation reflected by an object or substance is called its **albedo**. Some typical albedo values are presented in Table 3–1.

Light can be reflected off a surface in a couple of different ways. When light strikes a mirror, it is reflected back as a beam of equal intensity in a manner known as **specular reflection**. In contrast, when a beam is reflected from an object as a larger number of weaker rays traveling in many different directions, it is called **diffuse reflection**, or **scattering**. When scattering occurs, you cannot see an image of yourself on the reflecting surface as you can in a mirror. Consequently, although a surface of fresh snow might reflect back most of the visible light incident on it, you would not be able to check out your appearance by looking at it. The vast majority of natural surfaces are diffuse rather than specular reflectors.

In addition to large solid surfaces, gas molecules, particulates, and small droplets scatter radiation. Furthermore, although much radiation is scattered back to space, much is also redirected forward to the surface. The scattered energy reaching Earth's surface is thus **diffuse radiation**, which is in contrast to unscattered **direct radiation**. Figure 3–2 illustrates the process of scattering and the transformation of direct radiation to diffuse radiation. You can think of it this way: The blocking of direct radiation is what creates shadows, but a surface in the shadow of the direct radiation is not

TABLE 3–1

Typical Surface Albedos

Surface	Albedo
Cities	0.14–0.18
Dense forest	0.05–0.10
Forest	0.14–0.20
Grassland	0.16–0.20
Green crops	0.15–0.25
Sand	0.18–0.28
Snow (fresh)	0.75–0.95
Snow (old)	0.40–0.60
Soils	0.05–0.40

Source: Peixoto and Oort, *Physics of Climate*, 1992.

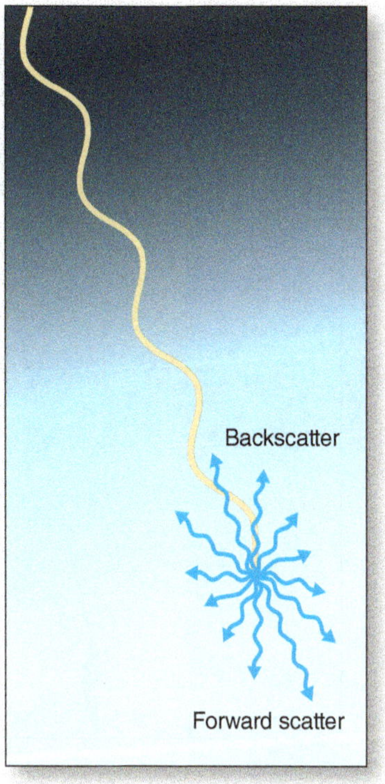

◄ **FIGURE 3–2 Scattering.** In the process of scattering, a beam of radiation is broken down into many weaker rays redirected in other directions.

completely dark because it is illuminated by diffuse radiation. Notice that whether accomplished by a gas molecule, particulate, or droplet, this result is still a scattering process in which the radiation is redirected but not absorbed. The radiation can be scattered forward or backward (backscattering).

The characteristics of radiation scattering by the atmosphere depend on the size of the scattering agents (the air molecules or suspended particles) relative to the wavelength of the incident electromagnetic energy. Three very general categories of scattering exist: Rayleigh scattering, Mie scattering, and nonselective scattering.

RAYLEIGH SCATTERING Scattering agents smaller than about one-tenth the wavelength of incoming radiation disperse radiation in a manner known as **Rayleigh scattering**. Rayleigh scattering is performed by individual gas molecules in the atmosphere. It primarily affects shorter wavelengths. Rayleigh scattering is particularly effective for visible light, especially those colors with the shortest wavelengths, so blue light is more effectively scattered by air molecules than is longer-wavelength red light. Furthermore, Rayleigh scattering disperses radiation both forward and backward. Combined with its greater effectiveness in scattering shorter wavelengths than longer wavelengths, this characteristic leads to three interesting phenomena: the blue sky on a clear day, the blue tint of the atmosphere when viewed from space, and the redness of sunsets and sunrises.

Figure 3–3 illustrates how Rayleigh scattering produces a blue sky. As parallel beams of radiation enter the atmosphere, a portion of the light is redirected away from its original direction. A person looking upward, away from the direction of the Sun, can see some of the scattered light that has been redirected toward the viewer. Because blue light is among

▸ **FIGURE 3–3 Why the Sky Is Blue.** The sky appears blue because the gases and particles in the atmosphere scatter some of the incoming solar radiation in all directions. Air molecules scatter shorter wavelengths most effectively. Someone at the surface looking skyward perceives blue light, the shortest wavelength of the visible portion of the spectrum.

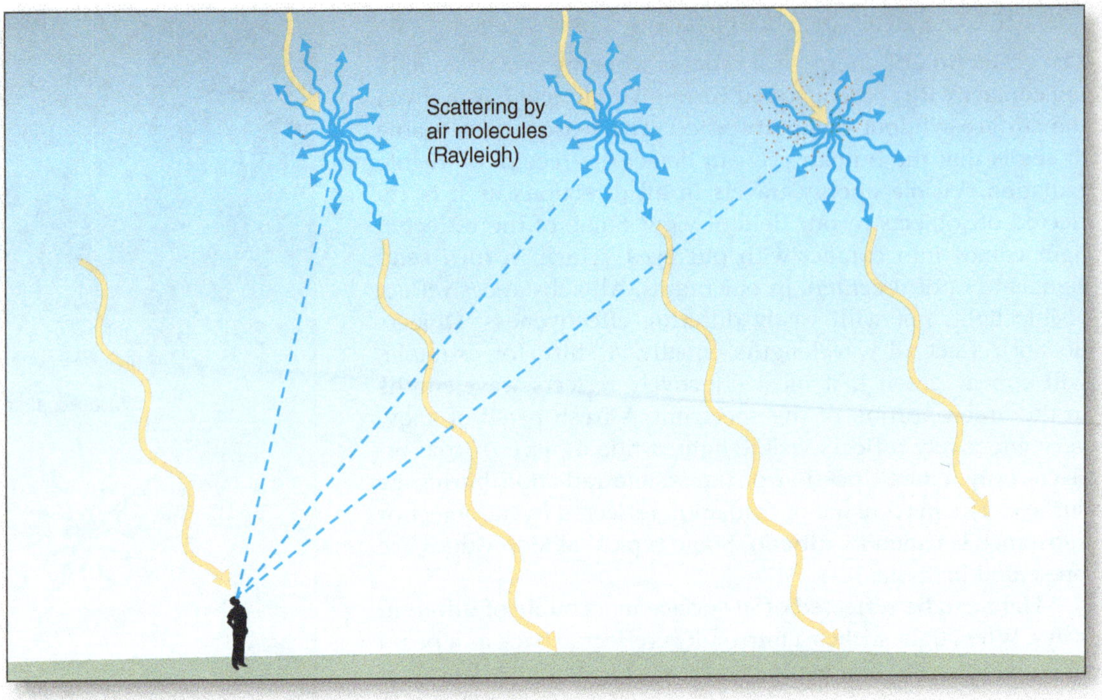

the shortest (and therefore most readily scattered) of the visible wavelengths, the scattered radiation contains a higher proportion of blue light than yellow, green, or other longer-wavelength light. Rayleigh scattering occurs at every point in a clear atmosphere and diverts energy toward a viewer from all directions, so no matter where you look on a cloudless day, the sky is blue. Of course, not all the incoming radiation is scattered on a clear day. In fact, the amount of diffuse radiation received at the surface under cloudless skies is normally about one-tenth that of the direct radiation.

On the Moon, which has no atmosphere, the "sky" appears black (Figure 3–4). As a viewer looks toward the horizon on the Moon, there is no downward scattered light because of the absence of an atmosphere, and the sky appears little different from the way it does at night. All that can be seen is the energy reflected off the lunar surface and Earth.

The same process that leads to the blue sky as seen from the surface also produces the bluish tint of the atmosphere as viewed from space. Like forward scattering, backscattering is biased toward blue wavelengths, so diffuse radiation directed back to space appears blue.

Rayleigh scattering is also largely responsible for the redness of sunrises and sunsets. Figure 3–5 shows how this happens. When the Sun is barely over the horizon, sunlight must travel a greater distance through the atmosphere than it does during the middle of the day, and the longer path increases the amount of Rayleigh scattering. As the direct beam travels its long path, the shortest wavelengths of radiation are depleted, so the longer wavelengths constitute an increasing percentage of the direct sunlight. The sky in the general vicinity of the Sun thereby takes on a reddish tint due to the depletion of the green and blue (shorter-wavelength) light.

DID YOU KNOW?

Because ultraviolet radiation (especially that having the shortest wavelengths in the UV band, known as UV-B) has very short wavelengths, Rayleigh scattering breaks down much of the energy into diffuse radiation. The fact that so much of the UV is diffuse and coming from different parts of the sky means that standing in the shade of a tree might not offer a great deal of protection against sunburn or skin cancer. Though you may be shielded from direct radiation by the tree, much of the diffuse radiation can still reach your skin.

▾ **FIGURE 3–4 "Earthrise" from the Moon.** Although this photo was taken during the day, the Moon has no blue sky. This is due to the absence of an atmosphere to scatter incoming solar radiation. Notice the blue tint of Earth, the result of Rayleigh scattering.

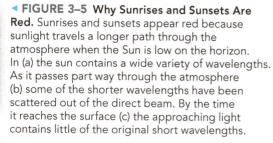

◀ **FIGURE 3–5 Why Sunrises and Sunsets Are Red.** Sunrises and sunsets appear red because sunlight travels a longer path through the atmosphere when the Sun is low on the horizon. In (a) the sun contains a wide variety of wavelengths. As it passes part way through the atmosphere (b) some of the shorter wavelengths have been scattered out of the direct beam. By the time it reaches the surface (c) the approaching light contains little of the original short wavelengths.

Animation Earth-Sun Relations

http://goo.gl/HXefff

MIE SCATTERING Vertical motions in the atmosphere are sufficiently strong that the atmosphere always contains suspended aerosols. This is true not only in cities, which tend to have higher air pollution concentrations, but also in rural areas far removed from urban activities. The microscopic aerosol particles are considerably larger than air molecules and scatter sunlight by a process known as **Mie** (pronounced "mee") **scattering**. Unlike Rayleigh scattering, Mie scattering is predominantly forward, diverting relatively little energy backward to space. Furthermore, Mie scattering does not have nearly the tendency to scatter shorter-wavelength radiation that Rayleigh scattering does. Thus, on hazy or polluted days (when there are high concentrations of aerosols) the sky appears gray, because the whole range of the visible part of the spectrum is effectively scattered toward the surface.

Mie scattering causes sunrises and sunsets to be redder than they would be due to Rayleigh scattering alone, so episodes of heavy air pollution often result in spectacular sunsets (Figure 3–6). Fires can also trigger enhanced Mie scattering. Residents of the western United States observed this phenomenon firsthand when major fires burned across the region during the summer of 2002. If a fire is large enough, Mie scattering can be increased great distances downwind. In 1988, for example, fires in Yosemite National Park reddened the sky as far away as Minneapolis, Minnesota. Volcanic eruptions, such as the major eruption of Mount Pinatubo in 1991, can even enhance the color of sunrises and sunsets across an entire hemisphere, as stratospheric winds transport aerosols far from their source.

NONSELECTIVE SCATTERING The water droplets in clouds are considerably larger than suspended particulates; therefore, they scatter sunlight in yet another way. In the aggregate, clouds reflect all wavelengths of incoming radiation about equally, which is why they appear white or gray. Because of the absence of preference for any particular wavelength, scattering by clouds is sometimes called **nonselective scattering**.

Clouds are by far the most important agent in nonselective scattering, and they exert a tremendous impact on the global receipt of solar radiation by reflecting large amounts of energy back to space.

▼ **FIGURE 3–6 Effect of Particulates on Scattering.** The scattering of shorter wavelengths enhances the redness of sunrises and sunsets during episodes of heavy particulate concentrations. A paper mill at Georgetown, South Carolina.

Transmission

When solar radiation travels through the vacuum of outer space, there is no modification of its intensity, direction, or wavelength. However, when it enters the atmosphere, only some of the radiation can pass unobstructed to the surface. The amount varies greatly, depending on atmospheric conditions. A clear, dry atmosphere might transmit as much as 80 percent of the incoming solar radiation as direct beam radiation without scattering or absorption. This is what you experience on a sunny, unpolluted day with sharp, distinct shadows. In contrast, when it is cloudy or hazy, only a small fraction of solar radiation will reach the surface as direct radiation. Under these conditions, there is both a reduction in the amount of radiation reaching the surface and a shift from direct radiation to diffuse, or scattered, radiation.

Spatial Distribution of Solar Radiation

The amount of solar radiation reaching the surface depends on two factors: the amount of insolation available at the top of the atmosphere (**extraterrestrial radiation**) and the reduction in that amount due to absorption and backscattering by the atmosphere. Figure 3–7 shows the annual average solar radiation received by a horizontal surface over the United States for January (a), July (b), and annually (c). The pattern for January solar radiation receipt is strongly affected by latitude, with the northern tier states showing lower values than those to the south. This reflects the combined influence of shorter periods of daylight and lower solar angles at higher latitudes during the winter months.

During the summer months, as shown in (b), the reduction in midday solar angles with increasing latitude is partially offset by the longer period of daylight. Thus, atmospheric conditions, especially cloud cover, dominate the pattern of solar radiation receipt. The highest insolation values occur over the western United States (except for the extreme northwest), where cloud cover is typically less than that found in the humid east. On an annual basis (c), the southwestern United States receives the greatest amount of insolation.

CHECKPOINT

3.1 What is the difference between specular reflection and scattering?

3.2 Compare and contrast Rayleigh, Mie, and nonselective scattering.

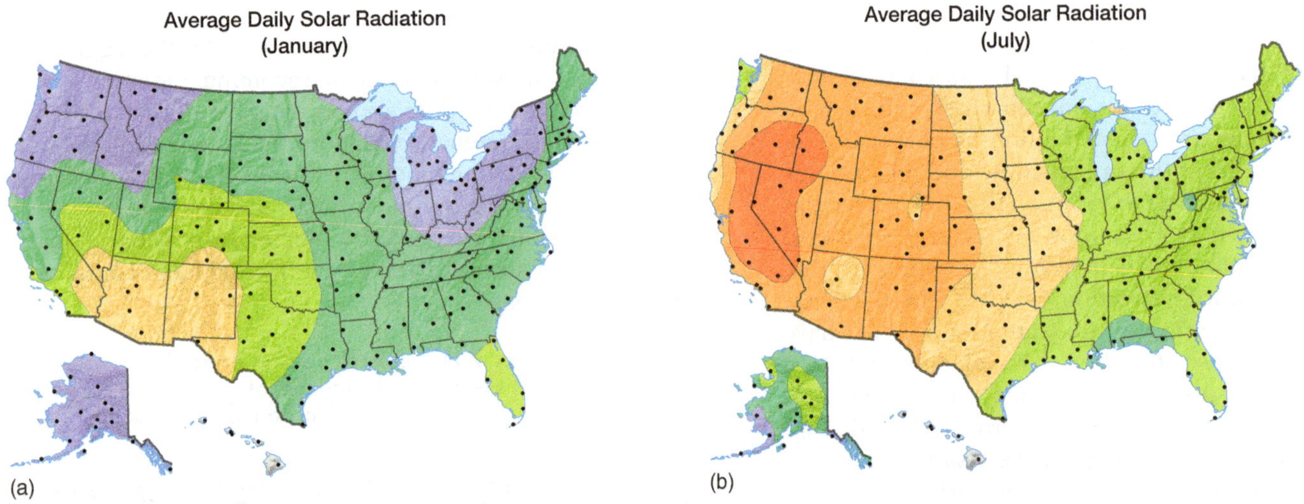

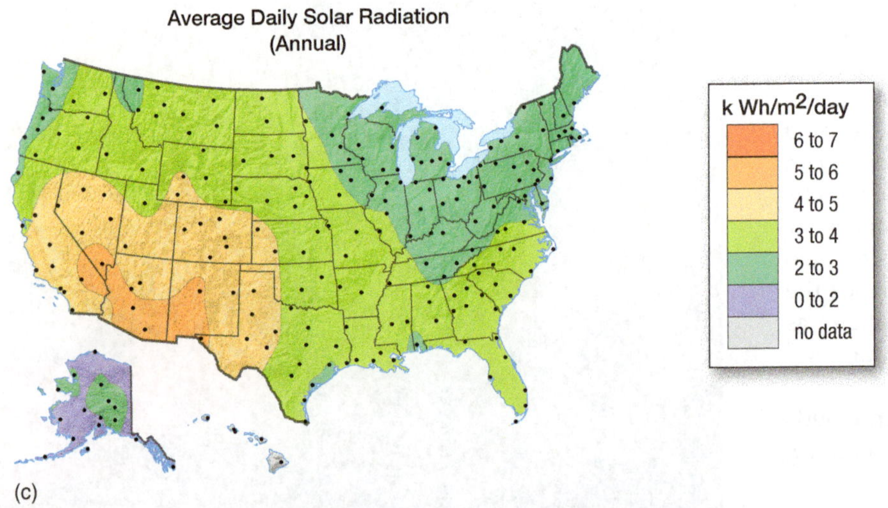

◂ **FIGURE 3–7 Solar Radiation.** Average receipt of solar radiation at the surface in (a) January, (b) July, and (c) annually. Values are in kilowatt hours per square meter per day. The black dots show the locations of the measuring stations.

The Fate of Solar Radiation

3.2 State the proportions of incoming solar energy that are reflected, scattered, or absorbed.

Because Earth's orbit around the Sun is not perfectly circular, there are slight seasonal variations in the availability of insolation, with almost 7 percent more solar energy available on perihelion than on aphelion (Chapter 2). Despite this variation, it is useful to think of a constant supply of radiation at the top of the atmosphere and to examine what happens, on average, to this energy. In other words, we need to account for the relative amount of radiation that is transmitted through the atmosphere, absorbed by the atmosphere and surface, and scattered back to space.

Such an exercise is more than a mere bookkeeping activity, because the amount of radiation absorbed by the atmosphere and surface will greatly influence their temperatures. For the sake of simplicity, we will assume that 100 units of insolation are available at the top of the atmosphere and then compare the amount of energy scattered back to space and absorbed by the atmosphere and surface to these 100 units. Bear in mind that the values presented in this discussion are annual and global averages; they do not necessarily apply to any particular place or time (see Figure 3–8). And although there is some degree of uncertainty, the numbers shown here represent the most recent values published in 2009.

Atmospheric Absorption

As a global average, the atmosphere absorbs 23 of the 100 units available at the top of the atmosphere. Six of the 23 units are ultraviolet radiation absorbed in the stratosphere by ozone, with most of the remainder being near-infrared radiation absorbed in the troposphere by gases (mostly water vapor). Thus, most of the radiation absorbed by the atmosphere is not visible radiation—a situation that suits us well, because if visible radiation were strongly absorbed by the atmosphere, it would be harder for us to see.

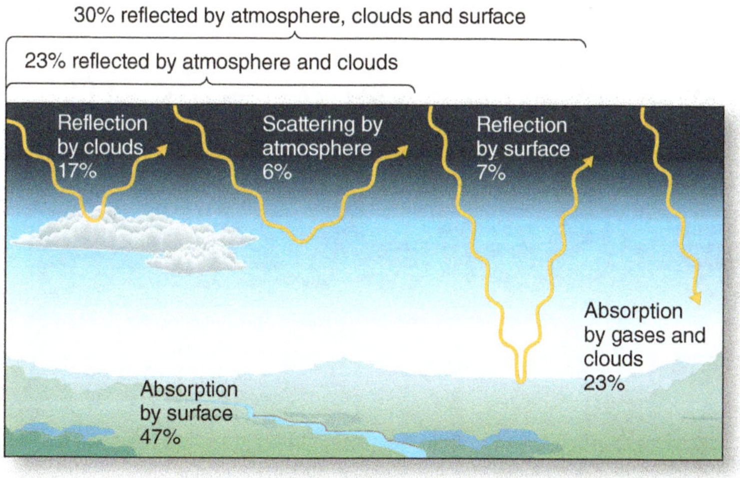

▲ **FIGURE 3–8 Processes Affecting Incoming Solar Radiation.** A number of processes affect solar radiation as it passes through the atmosphere. The clouds and gases of the atmosphere reflect 17 and 6 units, respectively, of insolation back to space. The atmosphere absorbs another 23 units. Only 54 percent of the insolation available at the top of the atmosphere actually reaches the surface, where another 7 units are reflected back to space. The net solar radiation absorbed by the surface is 47 units.

Atmospheric Backscattering

Though comparatively ineffective at absorbing shortwave radiation, cloud droplets scatter backward a large percentage of the incident energy. On average, the global cloud cover reflects 17 units of the incoming solar radiation back to space. But clouds are not the only agent of backscatter. The backscattering of radiation by atmospheric gases and aerosols accounts for another 6 of the 100 units of insolation at the top of the atmosphere. Together, scattering by clouds and gases reflects back to space 23 units (17 from clouds and 6 from the clear atmosphere).

Absorption and Reflection at the Surface

After atmospheric absorption and backscattering, 54 units of insolation are able to reach the surface. But not all the radiation reaching the surface is absorbed because Earth's surface is not completely black. Of the 54 units, 7 are reflected back to space from the surface. Overall, a total of 30 units of solar radiation (23 from the atmosphere and 7 from the surface) are scattered back to space, resulting in a **planetary albedo** of 30 percent. Note that the amount of insolation reflected from the surface is slightly smaller than that backscattered by atmospheric gases. In other words, when viewed from space, the planet shines more from atmospheric reflection than from surface reflection.

The end result of these processes is that the atmosphere absorbs 23 units of energy, while the surface takes in 47 units. If this were the end of the story, we would all be in big trouble, because the constant supply of heat would cause continued warming of the planet. In fact, if this energy were stored in the upper few centimeters of Earth's surface, the surface would be heated at a rate of several hundred degrees Celsius each day!

Obviously, we do not observe the oceans boiling away, nor the land surface melting; thus, the surface must be continually losing energy. The same is true for the atmosphere, and for the Earth–atmosphere system as a whole. In other words, in the absence of rapid climate change, the surface, the atmosphere, and the planetary system must give up as much energy as they obtain. We now discuss the mechanisms involved in the maintenance of Earth's energy balance.

CHECKPOINT

3.3 What happens to incoming solar radiation as it moves through the atmosphere?

3.4 Describe the geographic and seasonal variability of solar radiation reaching the surface over the United States.

Energy Transfer Processes

3.3 Explain how energy is transferred between the surface and the atmosphere.

Earth's atmosphere and surface continually exchange energy with each other. Much of this energy exchange is accomplished by the emission and absorption of radiation, but other mechanisms are also at work. This section describes the processes by which the energy transfer occurs.

Surface–Atmosphere Radiation Exchange

Like all other objects at terrestrial temperatures, Earth's surface and atmosphere radiate energy almost completely in the longwave (primarily thermal infrared) portion of the spectrum. Any discussion of longwave energy transfer is somewhat more complex than that of solar radiation because longwave energy has no obvious beginning or end point.

LONGWAVE RADIATION The atmosphere absorbs most of the longwave radiation emitted by Earth's surface. This increases the temperature of the atmosphere, which causes it to radiate more energy outward. The energy radiated by the atmosphere is transferred in all directions, including downward, and so the surface receives a considerable portion of this radiation. This causes further surface heating, which leads to an increase in longwave radiation emission from the surface, which again warms the atmosphere, and so on and so on. In other words, there is an infinite cycle of exchange, with energy constantly transferring back and forth.[1]

As the longwave radiation is cycled back and forth between the atmosphere and the surface, 70 units of longwave energy escape to space—12 of those units escape to space from the surface and 58 units radiate upward from the atmosphere. Note that the 70 units lost by the surface and atmosphere combined is exactly equal to the 70 units of shortwave radiation absorbed.

The middle panels of Figure 3–9 illustrate the flow of longwave radiation (the panels on the left incorporate information from Figure 3–8, and we discuss the right-hand panels later in this section). We see that while the surface emits 116 units of longwave energy on average, it also gets back 98 units radiated downward by the atmosphere. The end result is a **net longwave radiation** of minus 18 units (98 – 116 units) for Earth's surface. Meanwhile the atmosphere radiates 156 units of longwave radiation while absorbing 104 units radiated by the surface, yielding a net radiation of minus 52 units (104 – 156 units) for the atmosphere. These numbers might appear to be counterintuitive, given that they exceed the amount of radiation available from the Sun! But these seemingly large numbers are possible because of the constant cycling of longwave energy and because the atmosphere and surface have acquired energy in the form of absorbed shortwave energy.

While the values above are long-term averages, we see considerable variation on a given day. The clear atmosphere absorbs longwave radiation far better than solar radiation, mainly due

[1]Although tempting to do so, we should not think of longwave radiation being cycled back and forth, or "bouncing" between the surface and atmosphere. First, natural surfaces do not reflect much longwave radiation. Second, when radiation is absorbed, the energy no longer exists as radiation, so it can hardly bounce anywhere.

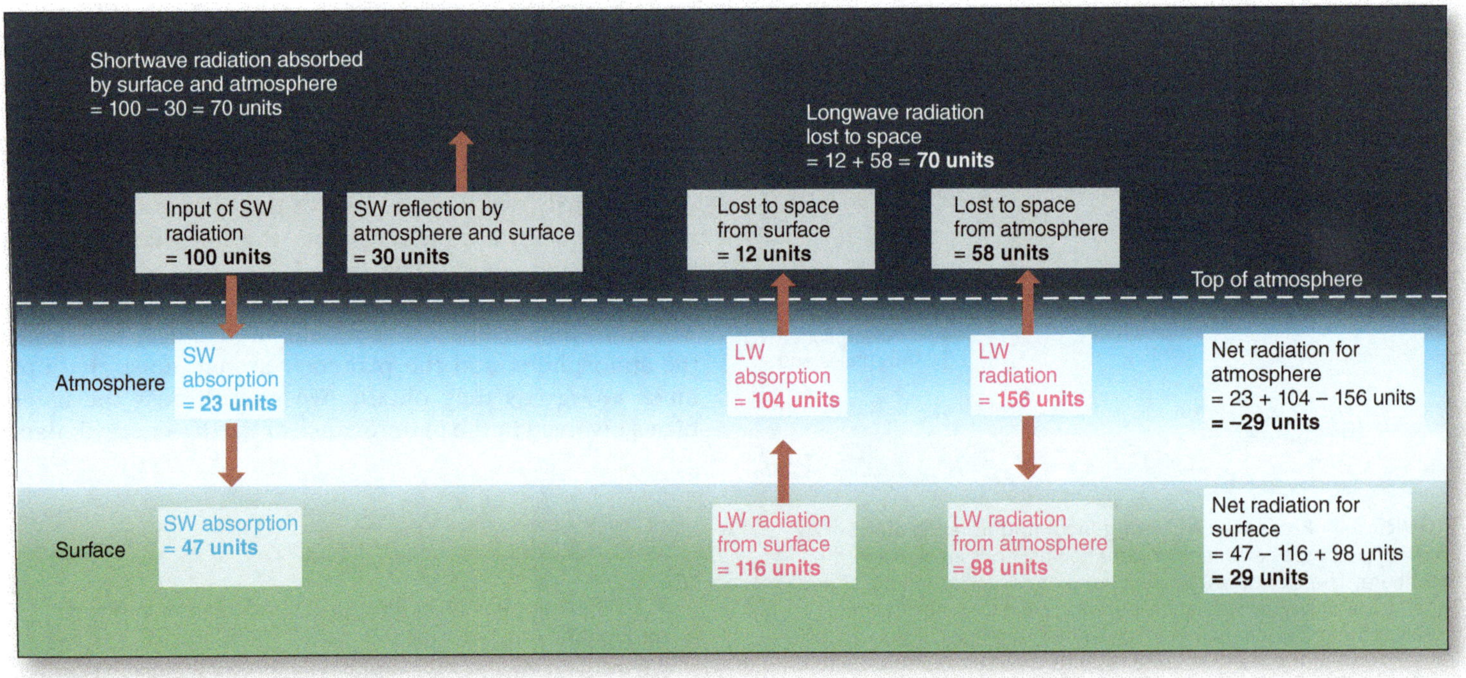

▲ **FIGURE 3–9 Earth's Energy Balance.** Net radiation is the end result of the absorption of shortwave radiation and the absorption and radiation of longwave radiation. The surface has a net radiation surplus of 29 units, while the atmosphere has a deficit of 29 units.

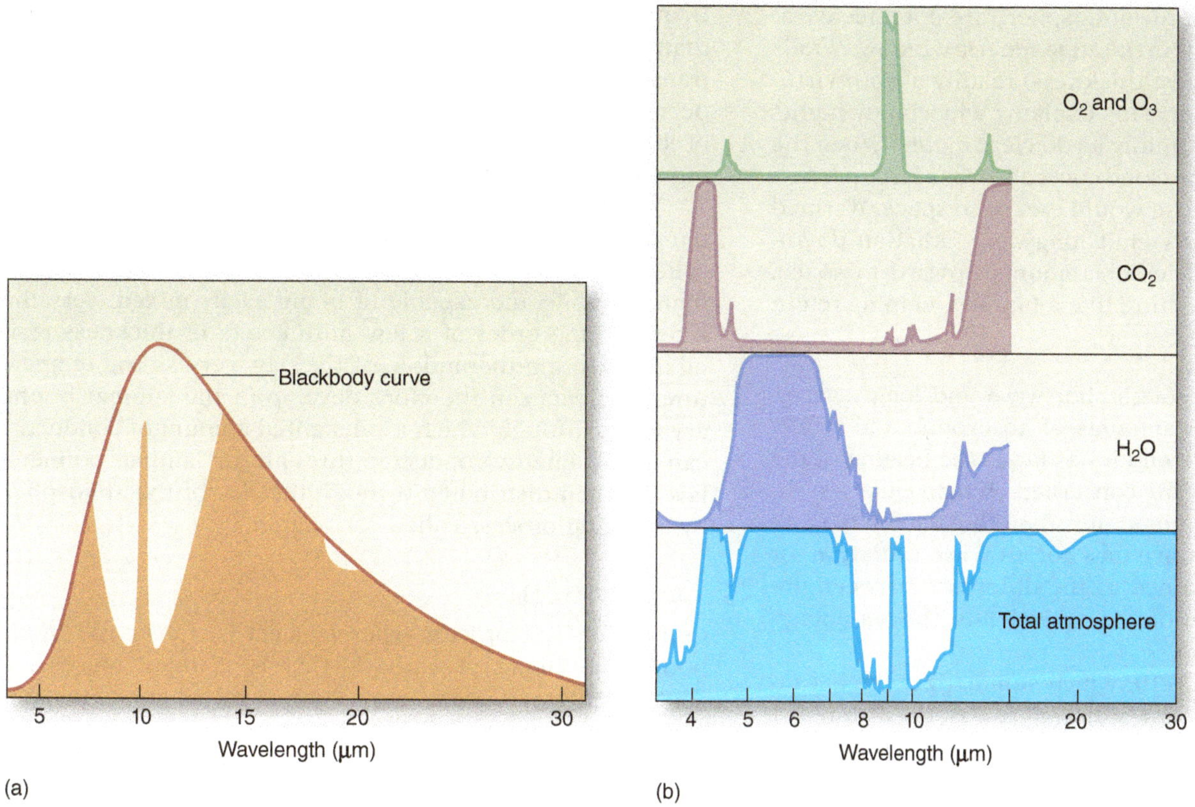

◂ **FIGURE 3–10 Earth as an Emitter of Radiation.** (a) Earth's surface acts nearly as a blackbody in its emission of radiation, but the gases of the atmosphere absorb most of the energy with wavelengths outside of the range of 8 to 12 μm. The shaded area in (a) indicates the energy absorbed by the atmosphere. (b) This figure shows the effectiveness of individual gases in absorbing the energy. The percentage of area shaded indicates the percentage of longwave energy absorbed.

to the presence of water vapor and carbon dioxide. As shown in Figure 3–10, both of these gases are good absorbers of longwave radiation, with strong absorption bands in the longwave part of the spectrum (see *Box 3–1, Physical Principles: Selective Absorption by Water Vapor and Carbon Dioxide*).

Although water vapor, carbon dioxide, and other greenhouse gases are good at absorbing most wavelengths of longwave radiation, a portion of the longwave spectrum can pass through the atmosphere relatively unimpeded. Interestingly, wavelengths in this band, 8 to 12 μm, happen to match those radiated with greatest intensity by Earth's surface. This range of wavelengths, not readily absorbed by atmospheric gases, is called the **atmospheric window**. The atmospheric window must not be thought of as a place in the atmosphere or the absence of some gas; it represents just a certain range of wavelengths of special importance to the radiation balance.

3–1 PHYSICAL PRINCIPLES

Selective Absorption by Water Vapor and Carbon Dioxide

Water vapor and carbon dioxide are the two gases that absorb the most longwave radiation emitted by the surface (see Figure 3–10). Why are these gases so selective, being nearly transparent for shortwave radiation but nearly opaque for longwave? Recall from Chapter 2 that isolated atoms have discrete energy states, with only certain energy states possible. As energy is absorbed and emitted by a gas molecule, its energy state rises and falls by discrete amounts from one allowable state to another. We have also seen that the energy associated with a photon of radiation is discrete and depends on its wavelength. Knowing the wavelength, we know the energy level of the photon.

It must be, therefore, that gas molecules absorb only certain photons, namely, those that push the molecule into allowable energy states. Photons with higher or lower energy values will not be absorbed but will instead pass through the gas. Because a unique wavelength is associated with every energy level, this is equivalent to saying that only certain wavelengths can be absorbed by any particular gas. (The same is not true of liquids and solids, whose molecules interact to give much more continuous absorption.) Whether or not a particular wavelength can be absorbed depends on the molecular structure of the absorber (the configuration of electrons, etc.). As it happens, the gases in the atmosphere do not have strong absorption bands in the visible part of the spectrum. But some of them, including water vapor and carbon dioxide, do have molecular structures that permit absorption of longwave radiation. Combined, the various gases absorb most of the longwave energy passing through the atmosphere.

3–1–1 Which two gases are most effective at absorbing longwave radiation?

3–1–2 Why is it that gases do not absorb all wavelengths of energy?

Although the gases of the atmosphere are not effective at absorbing the wavelengths in the atmospheric window, clouds (even those of fairly modest thickness) readily absorb virtually all longwave radiation. This explains why cloudy nights do not cool off nearly as rapidly as do clear nights. When the evening sky is overcast, the cloud cover absorbs a large portion of the energy that otherwise would escape to space. Warmed by this energy, the clouds emit longwave radiation downward to the surface and lesser amounts upward to space. The clouds thus act something like a blanket, helping retain heat.

NET RADIATION Even though shortwave and longwave radiation undergo different amounts of absorption and reflection, they are not separate entities as far as the heating of the atmosphere and surface are concerned. When either is absorbed, the absorber is warmed. It is therefore natural to combine longwave and shortwave into **net all wave radiation**, or simply **net radiation**, defined as the difference between absorbed and emitted radiation, or equivalently, the net energy gained or lost by radiation.

Refer back to Figure 3–10, which summarizes the net radiation balance for Earth. The atmosphere absorbs 23 units of solar radiation but undergoes a net loss of 52 units of thermal radiation, for a net deficit of 29 units. The surface absorbs 47 units of solar radiation but has a longwave deficit of 18, resulting in a net radiative surplus of 29 units. In other words, the atmosphere has a net deficit of radiative energy exactly equal to the surplus attained by the surface.

If radiation were the only means of exchanging energy, the surplus of radiative energy obtained by the surface would result in perpetual warming, while the deficit of the atmosphere would lead to continual cooling. Eventually our feet would be scorched by a terrifically hot ground while the rest of our bodies would freeze, surrounded by a bitterly cold atmosphere. This, of course, is not about to happen, because energy is transferred from the surface to the atmosphere and within the atmosphere by two other forms of heat transfer: conduction and convection. The net transfer of energy by these two processes allows the radiation surplus at the surface to be eliminated while at the same time offsetting the radiation deficit of the atmosphere.

CHECKPOINT

3.5 What is the atmospheric window?

3.6 Explain how the atmosphere and surface end up with net losses of longwave radiation.

Conduction

Conduction, described in general terms in Chapter 2, helps transfer energy near the surface. As radiant energy is absorbed by a solid Earth surface during the middle of the day, a temperature gradient (a rate of change of temperature over distance) develops in the upper few centimeters of the ground. In other words, temperatures near the surface become greater than those a few centimeters below. As a result, conduction transfers energy downward. At night the reverse process occurs; the topmost portion of the ground cools by the loss of longwave radiation and becomes colder than the underlying volume. Energy is then conducted upward.

Warming of the ground during the day also sets up a temperature gradient within a very thin, adjacent sliver of air called the **laminar boundary layer**. Although air is usually highly mobile and capable of being easily mixed, very thin layers on the order of a few millimeters in thickness resist mixing. During the middle of the day, very strong temperature gradients can therefore develop in the laminar boundary layer, through which a substantial amount of conduction can occur. Energy conducted through the laminar boundary layer is then distributed through the rest of the atmosphere by a mixing process called *convection*.

Convection

Convection is a process whereby heat is transferred by the bodily movement of a fluid—that is, a liquid or gas. In contrast to conduction, convection involves the actual displacement of molecules. Unlike conduction, which transfers energy from the surface to the atmosphere, convection circulates this heat between the very lowest and the remaining portions of the atmosphere. The direction of heat transfer is upward when the surface temperature exceeds the air temperature (the normal situation in the middle of the day). At night the surface typically cools more rapidly than the air, and energy is transferred downward. Convection can be generated by two processes in fluids: local heating (free convection) and mechanical stirring (forced convection).

FREE CONVECTION **Free convection** is the mixing process related to buoyancy, the tendency for a lighter fluid to float upward when surrounded by a denser fluid. Recall your days as a child, when you could annoy your parents by blowing bubbles through a straw in a glass of milk. Once the air was injected into the milk, it would immediately rise upward because of its lesser density and cause turbulent mixing. This was free convection at work.

Free convection (shown in Figure 3–11a) often occurs when a localized parcel of air is heated more than the nearby air. Because warm air is less dense than cold, it is relatively buoyant and rises. On a warm summer day, we can see the effect of free convection by observing a circling hawk (Figure 3–11b) that stays airborne without flapping its wings. This flight is possible because the hawk's wings are designed to catch the rising parcels of buoyant air that carry it upward. Convection can have far more important impacts than helping to keep hawks in the air—it can lead to intense precipitation.

FORCED CONVECTION **Forced convection** (also called **mechanical turbulence**) occurs when a fluid breaks into disorganized swirling motions as it undergoes a large-scale flow. When water flows through a river channel, for example, it

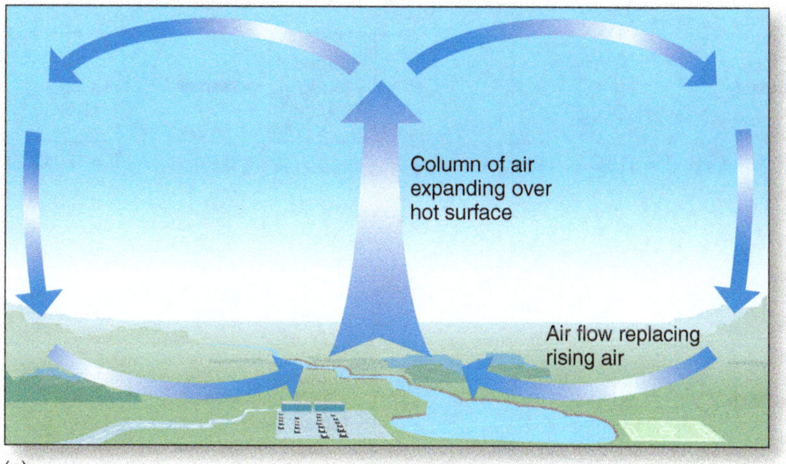

(a)

(b)

▲ **FIGURE 3–11 Free Convection.** (a) Convection is a heat transfer mechanism involving the mixing of a fluid. In free convection, local heating can cause a parcel of air to rise and be replaced by adjacent air. (b) Free convection can create updrafts able to keep a hawk airborne without it having to flap its wings.

▲ **FIGURE 3–12 Forced Convection.** Air is forced to mix vertically because of its low viscosity (ability to be held together) and the deflection of wind by surface features.

does not flow uniformly, as would a very thick syrup. Instead, the flow breaks down into numerous *eddies*. Forced convection in the atmosphere is shown in Figure 3–12. Horizontally moving air undergoes the same type of turbulence as flowing water. Instead of moving as a uniform mass, the air breaks up into numerous small parcels, each with its own speed and direction that are superimposed on the larger-scale flow. Because there is a strong vertical component to the eddy motions, the forced convection helps transport energy from the top of the laminar boundary layer upward during the day.

Generally speaking, higher wind speeds generate greater forced convection. Mechanical turbulence is also enhanced when air flows across rough surfaces (for example, forests and cities) rather than smooth ones such as glaciers. Both free and forced convection transfer two types of energy: sensible heat and latent heat.

CHECKPOINT

3.7 What is the role of convection in the atmosphere?

3.8 Compare and contrast free convection and forced convection.

Sensible Heat

The transfer of energy as **sensible heat** is simple. When energy is added to a substance, an increase in temperature can occur that we physically sense (hence the term *sensible*). This is what you experience when you rest outside on a warm, sunny day; the increase in your skin temperature results from a gain in sensible heat. The magnitude of temperature increase is related to two factors, the first of which is **specific heat**, defined as the amount of energy needed to produce a given temperature change per unit mass of the substance. In SI units, specific heat is expressed in joules per kilogram per kelvin. Everything else being equal, a substance with high specific heat warms slowly because much energy is required to produce a given temperature change. Likewise, it also takes longer for a substance with a high specific heat to cool off, assuming the same rate of energy loss.

The temperature increase resulting from a surplus of energy receipt also depends on the *mass* of a substance. Not surprisingly, a given input of heat results in a greater rise in temperature if it is applied to only a small amount of mass. For example, compare the amount of energy needed to boil water for a cup of tea to that needed to take a warm bath. In just a few minutes, a single burner on a kitchen stove can have the water ready for the tea, but your hot water heater must supply considerably more energy for the large amount of bath water. These relationships are shown in Figure 3–13.

Sensible heat travels by conduction through the laminar boundary layer and is then dispersed upward by convection. Through these mechanisms, 5 of the 29 units of net radiation surplus for the surface (Figure 3–9) are transferred to the atmosphere, where they help offset the net radiation deficit. The remaining 24 units are transferred to the atmosphere by the convection of heat in another form.

Latent Heat

Latent heat is the energy required to change the phase of a substance (that is, its state as a solid, liquid, or gas). In

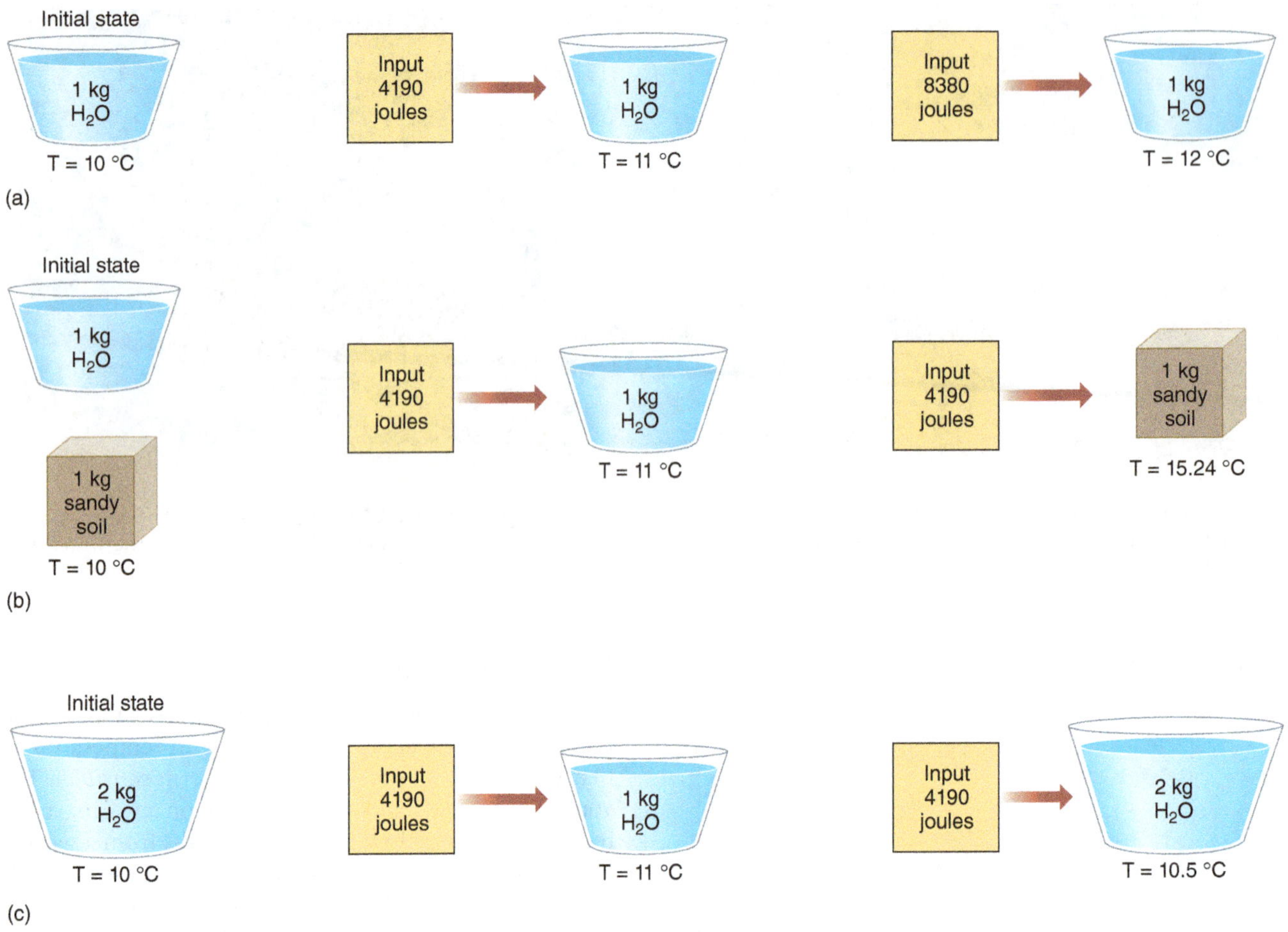

▲ **FIGURE 3–13 Heat Content.** Several factors determine the heat content of a substance. (a) T input of 4190 J of energy to a kilogram of water increases its temperature 1 °C, while a doubling of the energy input causes twice as much heating. The specific heat of a substance also influences the amount of temperature change resulting from an input of energy. (b) T application of 4190 J to 1 kg of sandy soil produces more than five times the increase in temperature than it would for a kilogram of water. The amount of mass also affects the temperature change accruing for a given energy input. (c) Note that the temperature increase for 2 kg of water is half as much as that for 1 kg.

meteorology we are concerned almost exclusively with the heat involved in the phase changes of water.

Recall that all physical processes require energy. The evaporation of water and the melting of ice are no exceptions to this rule—for either process, energy must be supplied. In the case of melting ice, the energy is called the *latent heat of fusion*. For the change of phase from liquid to gas, the energy is called the *latent heat of evaporation*. It takes seven and a half times more energy (2,500,000 joules) to evaporate a kilogram of liquid water than it does to melt the same amount of ice (335,000 joules). Although both forms of latent heat can be important locally, on a global scale the latent heat of evaporation is far more influential.

When radiation is received at the surface, it can raise the temperature of the land or the water. If water happens to exist at the surface (or can be brought up from below the surface through the root systems of plants), some of the energy that might have been used to increase the surface temperature is instead used to evaporate some of the water. This results in a smaller temperature increase than would occur for a dry surface. You have probably experienced this while walking barefoot on a hot pavement. If the pavement is watered down, the surface cools as energy is taken from the ground and used for evaporation. The amount of energy consumed can be quite large, as much as 90 percent of absorbed solar radiation for a completely wet surface.

Let's use another common example to illustrate the concept of latent heat. We all know that perspiration is a mechanism that lowers our body temperatures and keeps us from overheating. But how does it work? Clearly, it is not a matter of sweat being cold—it is as warm as the body producing it. The reason sweat cools a person is latent heat. As you exercise, the heat produced as a by-product causes your body temperature to rise. However, if your skin is covered with water and that water is free to evaporate, some of the energy produced by your body is used to evaporate the moisture rather than increase your body temperature.

DID YOU KNOW?

When you drop an ice cube in a glass of water, the cooling that results is not primarily due to the lower temperature of the ice cube. It is mostly due to the latent heat of fusion involved in melting the ice. The energy used to melt the ice is taken from the water, thereby lowering its temperature.

The energy needed to evaporate water or melt ice is said to be latent because it does not disappear. It is held, "latent" in the atmosphere, to be released later when the reverse process occurs—the condensation of water vapor into liquid cloud or fog droplets. In effect, then, the evaporation of water makes energy available to the atmosphere that otherwise would have warmed the surface. It thus acts as an energy transfer mechanism, taking heat from the surface to the atmosphere. On a global average basis, the amount of energy transferred to the atmosphere as latent heat amounts to 24 units, which makes latent heat considerably more important as a mode of heat transfer than sensible heat (5 units). Perhaps this is not surprising, given that the planet is mostly covered by ocean. The mean annual values of the net radiation components, latent and sensible heat, are depicted in Figure 3–14.

Net Radiation and Global Temperature

The balance between incoming and outgoing radiation is not merely fortuitous; physical laws dictate that it must be the case. To understand why, consider what would happen if the Sun were suddenly to increase its radiative output and raise the temperature of Earth. Because warm bodies radiate more energy than cooler ones, the planet would respond by emitting more longwave energy back to space. As long as the incoming energy from the Sun were to exceed that emitted by Earth, the global temperature would continue to rise, and emission to space would increase accordingly. Eventually the planetary temperature would increase to the point where outgoing energy equaled incoming energy, and a new equilibrium temperature would be established. The individual values in Figure 3–14 would all be changed, but the global inputs and outputs of radiation would still sum to zero.

Latitudinal Variations

The overall balance between incoming and outgoing radiation discussed in previous sections applies to the planet as a whole but not to any particular place. Figure 3–15 shows that the balance between incoming and outgoing radiation varies with latitude. On an annual basis, the Earth surface–atmosphere system gains more radiation than it loses between latitudes at about 38° north and south, and it loses more than it gains poleward of these two parallels. The boundary between zones of radiant energy surpluses and deficits migrates seasonally. During the Northern Hemisphere summer, most of the area north of about 15° S gains more radiant energy than it loses. During the Northern Hemisphere winter, most areas south of about 15° N take in more radiation than they emit.

If no other processes were involved, the net gain of radiation between 38° north and south latitudes and the deficits outside this zone would cause the tropics and subtropics to undergo continual heating and the extratropical regions to constantly cool. This does not occur, however, because the energy surplus at low latitudes is offset by the horizontal movement, or **advection**, of heat poleward. This transfer is accomplished primarily by the global wind systems (75 percent), and secondarily by the oceanic currents (25 percent), as illustrated in Figure 3–16. As the wind and water currents move, they carry with them their internal heat, which is redistributed across the globe.

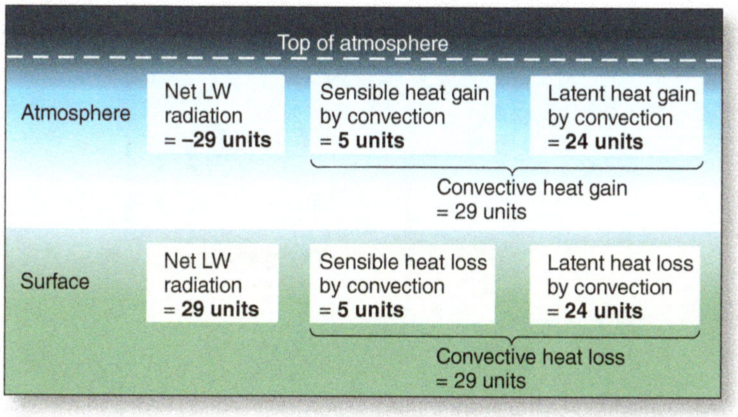

▲ **FIGURE 3–14 Surface–Atmosphere Energy Balance.** Both the surface and atmosphere lose exactly as much energy as they gain. The surface has a surplus of 29 units of net radiation, which is offset by the transfer of sensible and latent heat to the atmosphere. The atmosphere offsets its 29 units of radiation deficit by the receipt of sensible and latent heat from the surface.

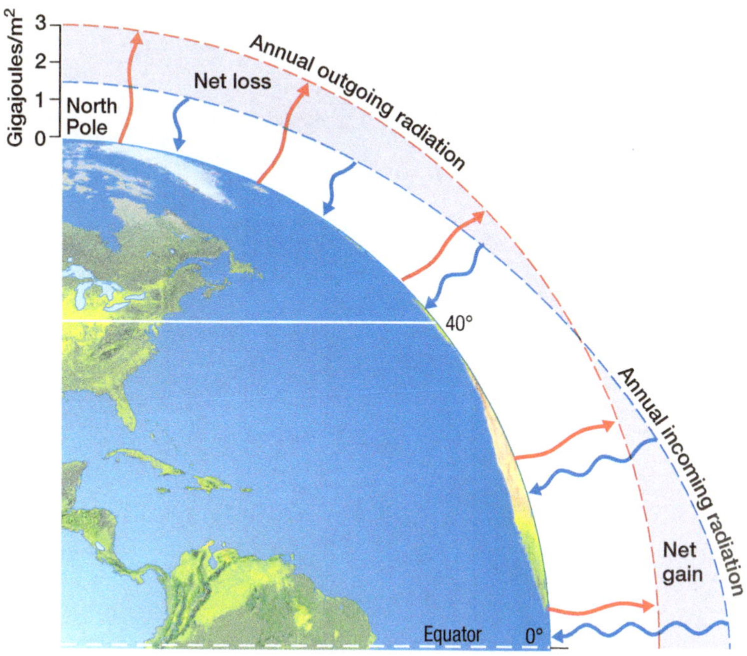

▲ **FIGURE 3–15 Annual Average Net Radiation Values for the Atmosphere and Surface Combined.** At latitudes between about 38° north and south, a radiant energy surplus exists. Poleward of these latitudes, the atmosphere and surface lose more radiant energy than is gained.

CHECKPOINT

3.9 What are the two forms of heat transferred between the surface and atmosphere by convection?

3.10 How do the relative amounts of incoming and outgoing net radiation vary across the globe?

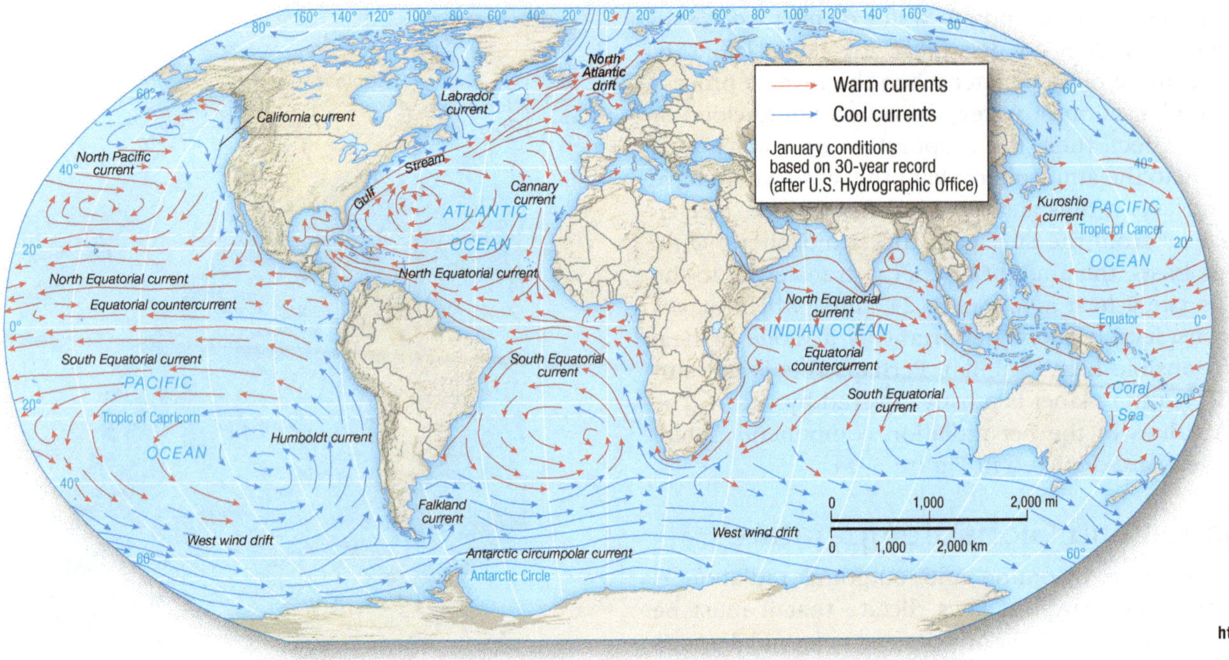

Video
Global Sea-Surface Temperatures—Climatology

http://goo.gl/VF3k7T

▲ **FIGURE 3–16 Circulation of Ocean Currents.** Currents that are moving warm water are depicted by red arrows, those moving cold water by blue arrows.

The Greenhouse Effect

3.4 Explain the role of the greenhouse effect in warming the atmosphere.

The interactions that warm the atmosphere are often collectively referred to as the **greenhouse effect**, but the analogy to a greenhouse is not strictly accurate. Greenhouses, such as the one shown in Figure 3–17, are made primarily of glass, which is transparent to incoming shortwave radiation but opaque to outgoing longwave radiation. The glass therefore allows in more radiation than is allowed to escape, causing the temperature inside the structure to be warmer than outside. In that regard, a greenhouse is similar to the atmosphere, which also transmits most of the incoming solar energy but absorbs the vast majority of longwave radiation emitted upward by the surface.

The analogy breaks down, however, when we incorporate the effect of convection. A greenhouse not only reduces the loss of energy by longwave radiation but also prevents the loss of sensible and latent heat by convection. In contrast, the *greenhouse gases* (those that absorb longwave radiation) of the atmosphere do not impede the transfer of latent and sensible heat. Thus, it would be more accurate if the term "greenhouse effect" were replaced by "atmospheric effect."

If the atmosphere had none of the "greenhouse gases" that absorb outgoing longwave radiation, Earth would be considerably colder, on average, and the temperature would oscillate wildly from day to night. In fact, without greenhouse gases and clouds, Earth's surface would have an average temperature of −18.6 °C (−1.4 °F), rather than the observed mean temperature of 15 °C (59 °F). The greenhouse

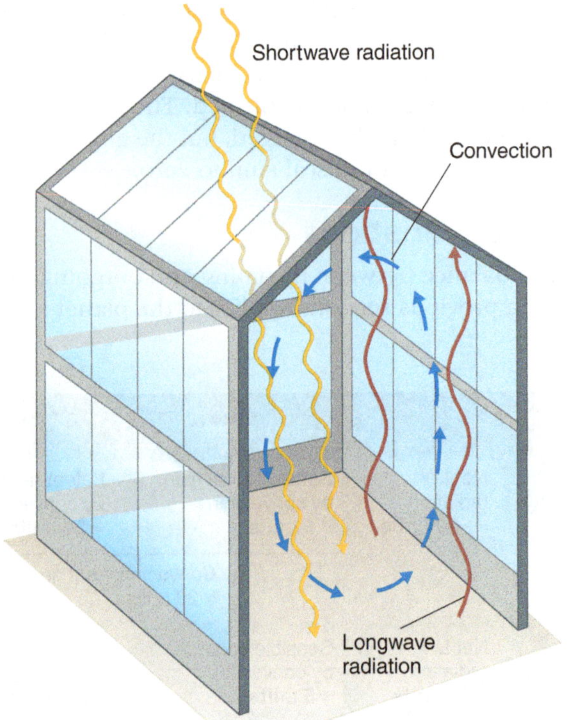

▲ **FIGURE 3–17 The Greenhouse Effect.** The air inside a greenhouse is warmer than that outside because glass allows solar radiation to enter but is opaque to outgoing longwave radiation. It also precludes the movement of heat away from the surface by convection. This latter effect makes the action of a greenhouse different from that of the atmosphere. Therefore, the term "greenhouse effect" is not completely appropriate when applied to the atmosphere.

3–2 PHYSICAL PRINCIPLES

Earth's Equilibrium Temperature

If Earth had no atmosphere, and therefore no greenhouse effect, the mean temperature of the planet would be much colder. Using the principles discussed thus far, we can easily estimate the magnitude of the greenhouse effect. To do so, we will compute the equilibrium temperature for a planet having no atmosphere. By comparing the computed and observed temperatures, we will see how the atmosphere influences Earth's temperature.

First, assume the planet acts as a blackbody with regard to longwave radiation, that the planetary albedo is 30 percent, and that the solar constant is 1361 watts per square meter (W/m^2). If Earth were a flat disk perpendicular to incoming radiation, each square meter would receive 1361 joules per second (J/sec). But Earth is not a flat disk; it is a sphere, the surface area of which is four times larger than that of a disk of the same radius. Thus, the intensity of radiation averaged over the sphere is one-fourth as large as for the imaginary disk. Because of this, each square meter of Earth receives 1361/4, or 340 W/m^2. (These 340 watts per square meter are the 100 units of solar radiation discussed earlier in this chapter in the section titled *The Fate of Solar Radiation*.) Given the planetary albedo of 30 percent, it must be that 70 percent of this incoming radiation is absorbed. In other words, total absorbed radiation is

$$1361\ W/m^2 \times 0.25 \times 0.7 = 238.2\ W/m^2$$

The planet must lose exactly as much energy as it gains, and the intensity of radiation for a blackbody is determined by rearranging and applying the Stefan-Boltzmann law. Recall that the Stefan-Boltzmann law for a blackbody states that

$$I = \sigma T^4 \quad \text{Stefan-Boltzmann}$$

We know, however, that the intensity of radiation for the planet without an atmosphere must be 238.2 W/m^2. We then rearrange the equation to solve for T, rather than I, to get

$$T^4 = I/\sigma$$

which can be reduced to

$$T = (I/\sigma)^{0.25}$$

Using the values $\sigma = 5.67 \times 10^{-8}\ W/(m^2\ K^4)$ and $I = 238.2\ W/m^2$, the equilibrium temperature works out to 254.6 K (−18.6 °C, −1.4 °F). Thus, the mean temperature of Earth would be far colder without an atmosphere.

Note that our calculation is highly simplified and somewhat questionable. For example, we used 30 percent for the planetary albedo, but that value arises in part from the albedo of the atmosphere, which our imaginary planet lacks. Should we therefore have used present-day surface albedo in the computation? Perhaps, but surface albedo on the real Earth is in part the result of temperature, the very thing we are trying to compute! Ideally, we would treat albedo as a variable, allowing it to respond to changes in temperature.

We see that even a beginning question about the global effect of greenhouse gases raises complications that are not easy to address with a simple model. Given this, it is not surprising that realistic computer models of atmospheric behavior are enormously complex, requiring huge computer resources. Nonetheless, computer models have become indispensable for daily weather forecasting (see Chapter 13) and tell us much of what we know about potential climatic changes due to human activity (see Chapter 16).

3–2–1 Outline the general process of determining Earth's equilibrium temperature.

3–2–2 How does Earth's equilibrium temperature compare to its actual temperature near the surface? Why is there a discrepancy between the two temperatures?

effect keeps Earth warmer by absorbing most of the longwave radiation emitted by the surface, thereby warming the lower atmosphere, which in turn emits radiation downward. (A more quantitative discussion of this topic is presented in *Box 3–2, Physical Principles: Earth's Equilibrium Temperature*.)

These days you find considerable discussion in the media about the possibility of climatic warming due to the anthropogenic increase of greenhouse gases such as carbon dioxide and methane. We briefly discuss how human activities have contributed to a warming atmosphere later in this chapter. A more general discussion of climatic change appears in Chapter 16.

CHECKPOINT

3.11 What is one drawback of using a greenhouse as an analogy for how energy behaves in the atmosphere?

3.12 What role does the greenhouse effect play in keeping Earth a habitable planet?

Global Temperature Distributions

3.5 Describe the global distribution of temperatures on Earth's surface.

One of the most immediate and obvious outcomes of radiation gain or loss is a change in the air temperature. Figure 3–18 shows the mean air temperature distributions for January and July and the differences between the two.

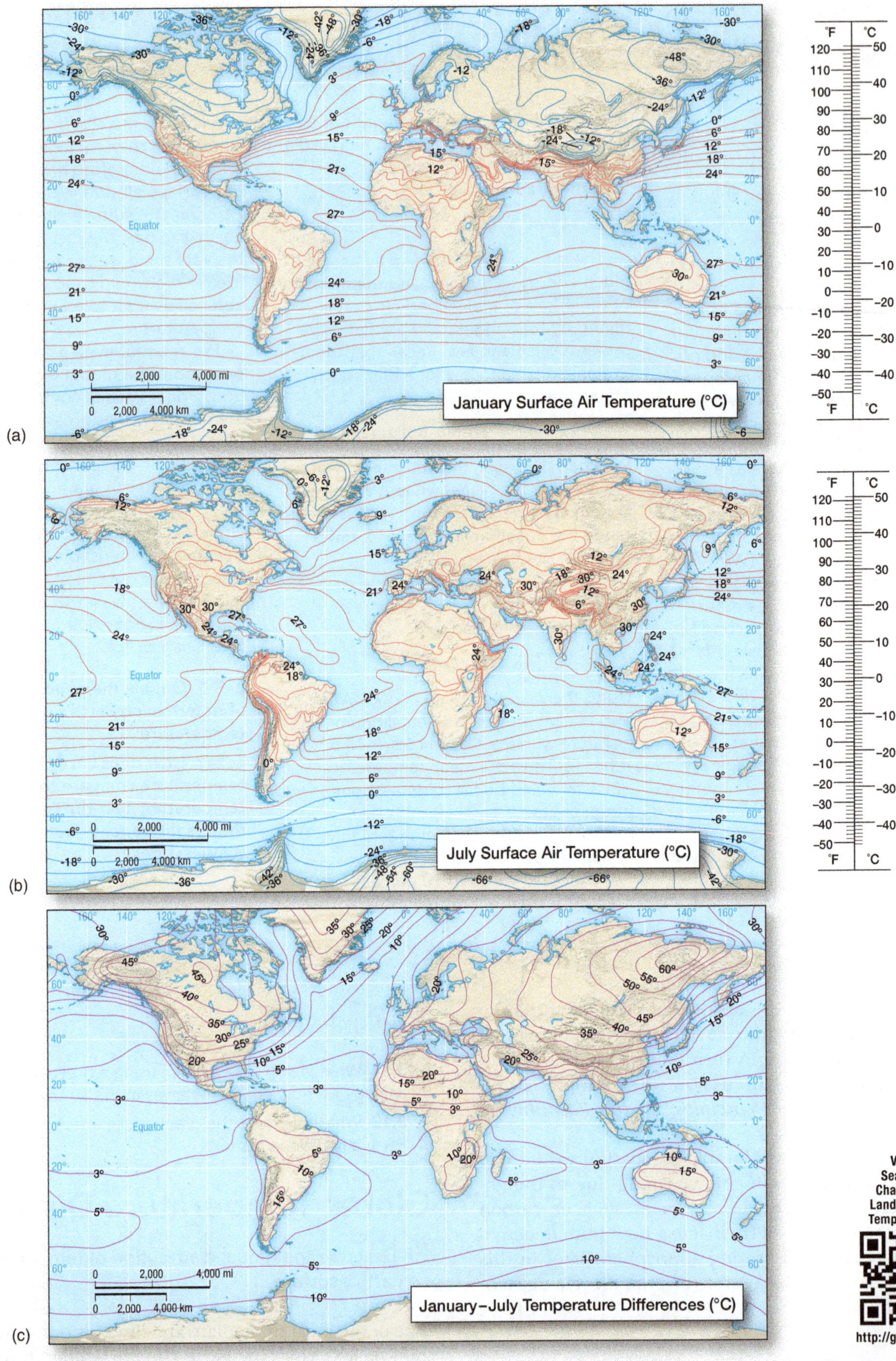

Video
Seasonal Changes in Land Surface Temperatures
http://goo.gl/Lv1cl

▲ **FIGURE 3–18 Surface Air Temperatures.** Distribution of mean January (a) and July (b) surface air temperatures. The difference in temperature between the two months is shown in (c).

Each line on the maps, called an **isotherm**, connects points of equal temperature. Several patterns of large-scale temperature are apparent in (a) and (b). First, as expected, temperatures tend to decrease poleward in both hemispheres. Second, the latitudinal temperature gradient is greatest in the hemisphere experiencing winter (that is, the Northern Hemisphere in January and the Southern Hemisphere in July). The strong gradients in the winter hemisphere occur because the midday sun angles and the length of day both decrease with latitude. During summer, the lower midday sun angles at the higher latitudes are offset by longer days, so the temperature gradients are relatively weak.

DID YOU KNOW?

The winter of 2013–2014 was a brutal one for eastern North America, as wave after wave of intense cold and snow hit the continent. Despite the severity of the winter, January 2014 was the fourth warmest on record for Earth as a whole. It was such a warm January out west that residents of California might think of it as "the winter that wasn't." Other parts of the world also experienced an unusually warm January, including much of China, Europe, and Australia.

The third feature of the maps in Figure 3–18 is that the isotherms shift poleward over land in the hemisphere experiencing summer and shift equatorward over land during the winter. In other words, temperatures over land tend to be warmer in the summer than over adjacent water bodies, and colder during winter. Finally, the Northern Hemisphere has a steeper temperature gradient during its winter (a) than does the Southern Hemisphere during its winter (b). This is because the Southern Hemisphere consists of a much greater proportion of ocean than land. As you can see from (c), landmasses have much greater annual ranges in temperature than do ocean bodies, for reasons discussed later in this chapter.

CHECKPOINT

3.13 Describe the way isotherms bend over continents in the Northern Hemisphere for summer and for winter.

3.14 Why does the Northern Hemisphere have steeper temperature gradients with latitude than does the Southern Hemisphere?

Influences on Temperature

3.6 List the factors that influence temperatures on Earth's surface.

Certain geographical factors combine to influence temperature patterns across the globe. These factors include latitude, altitude and elevation, atmospheric circulation patterns, continentality, ocean current characteristics along coastal locations, and local conditions.

Latitude

Most people know that outside the tropics, the annual mean temperature decreases with latitude—Santa Claus, living at the North Pole, must have a wardrobe suitable for extremely cold conditions. Not only does latitude influence the average temperature, it also affects seasonal patterns. As described in Chapter 2, the tilt of Earth's axis influences the amount of solar radiation available at any latitude on any particular day. Within the tropics, there is relatively little annual variation in the length of day and the midday solar angle, so energy receipts exhibit little change through the course of the year. Outside the tropics, the noontime solar angles exhibit a range of 47°, with the lowest solar angles coinciding with the periods of shorter days. As a result, the availability of incoming radiation (and therefore the temperature) is more variable as distance from the equator increases.

Altitude and Elevation

Any point within the atmosphere has some particular *altitude* (its height above mean sea level). Altitude is not synonymous with *elevation*, the distance above sea level for a land surface. For example, some particular city may have an elevation of 1000 m above sea level, while air that is 1000 m above the city would have an altitude of 2000 m. Altitude and elevation both deal with position relative to sea level, but altitude is related to the atmosphere and elevation refers to land.

As was shown in Figure 1–12, temperatures in the troposphere typically decrease with altitude above sea level. This occurs because the surface is the primary source of direct heating for the troposphere, and increased altitude implies a greater distance from the energy source. Figure 3–19 contrasts day and night temperatures at three locations. Position A is located a couple of meters (about 6 ft) above the surface at sea level. Position B is 3000 m (about 10,000 ft) directly above A, and position C is 3000 m above sea level but just a couple of meters above the mountain surface.

During the middle of the day, position A responds to the absorption of solar radiation at the surface and warms as the surface transfers energy upward by convection and the emission of longwave energy. Position B, a considerable distance from the surface, undergoes virtually no warming because the energy emanating from the surface has largely been absorbed by the air below. Position C, although at the same altitude as B, is nearer to the primary source of warming, and its daytime temperature rises appreciably. On the other hand, the thinner atmosphere at C does not allow as much energy to be absorbed as at A, so its daytime temperatures are somewhat lower.

At night, the surface below position A cools by the emission of longwave radiation. The air 2 m above the surface also undergoes a lowering of temperature in response to the cooling of the underlying surface. At position B, the air undergoes little cooling because of its distance from the surface. Position C chills rapidly because the sparse atmosphere above does not effectively absorb the outgoing radiation from the surface. Cooling is often enhanced by rapid evaporation into the drier high-altitude air. Perhaps while on a mountain camping trip you've noticed how

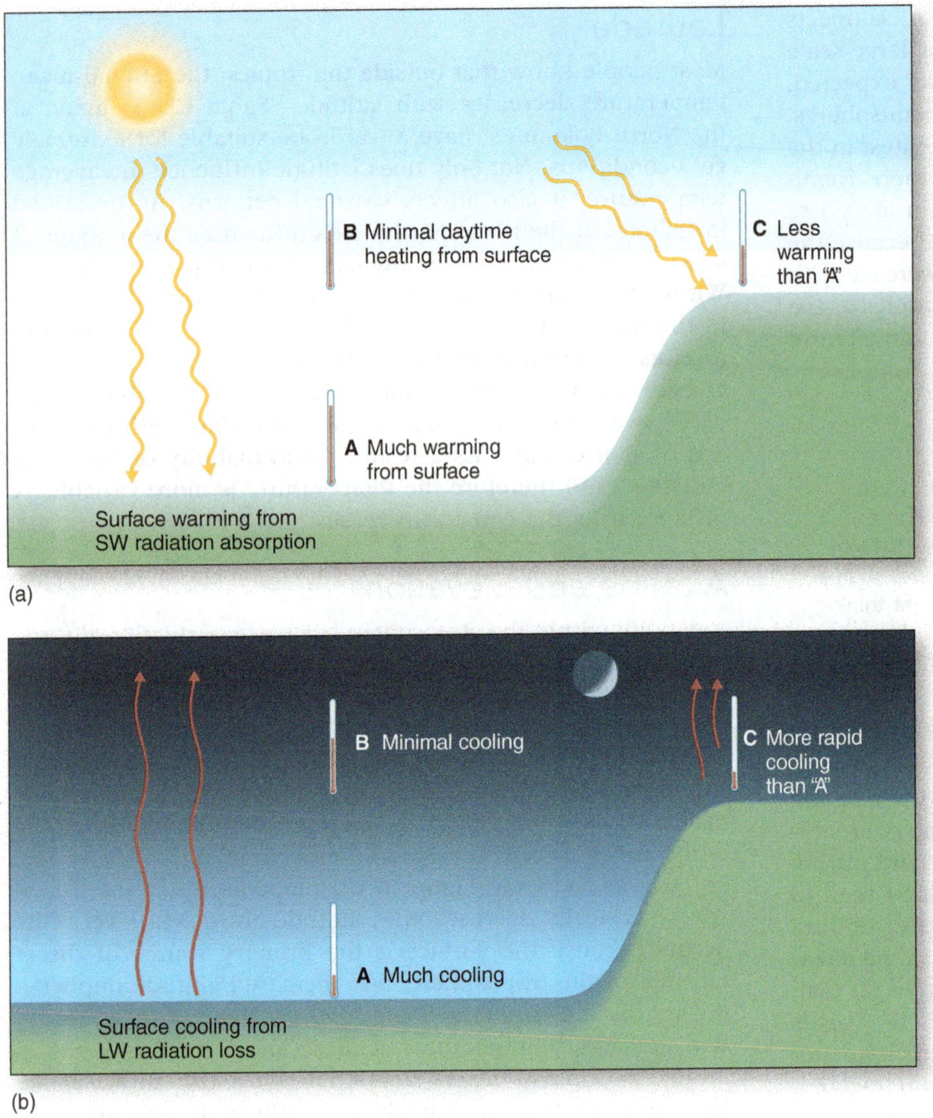

▲ **FIGURE 3–19 Effects of Elevation and Altitude on Daily Temperature Patterns.** During the day (a) lower elevation sites (a) undergo more rapid warming than area farther from the surface (B) or at higher surface elevations (c). At night (b) higher elevation locations (C) have more rapid cooling than do lower elevations (A) or the air at the same altitude above sea level but farther from the surface (B).

rapidly a hot meal becomes stone cold. This is a consequence of strong evaporation and weak atmospheric counterradiation, compared to lower elevations. The overall effect of elevation above sea level and altitude above the ground, therefore, is that cooling and warming cycles are minimal high above the surface, and the air just above a high-elevation surface will undergo greater cycles of cooling and warming than will air at the same altitude but farther above the surface. It is not uncommon for the nighttime loss of longwave radiation at the surface to lead to lower temperatures near the surface than aloft. This reversal of the normal pattern in the troposphere (lower temperatures with increasing distance from the surface) is known as an **inversion**. The characteristics of inversions are described in Chapter 6.

Atmospheric Circulation Patterns

As we will discuss in Chapter 8, an organized pattern of mean atmospheric pressure and airflow across the globe strongly influences the movement of warm and cold air, with a direct effect on temperature. These large-scale circulation patterns also influence the development of cloud cover, which has an indirect effect on temperature. Subtropical areas (latitudes 20° to 30° in both hemispheres), for example, tend to be regions of minimal cloud cover, and insolation passing through the atmosphere undergoes less loss due to backscatter and absorption on its way to the surface. In contrast, equatorial regions are often cloudy in the afternoon and experience a greater reduction of incoming solar radiation. The result is that the highest temperatures on Earth tend to occur not at the equator but in the subtropics. Many other patterns of atmospheric circulation affect regional temperatures.

Contrasts Between Land and Water

Because the atmosphere is heated primarily from below, it should be no surprise that the type of surface influences air temperature. The greatest influence arises because of contrasts between land and water. Water bodies are far more conservative than land surfaces with regard to their temperature, meaning that they take longer to warm and cool when subjected to comparable energy gains and losses.

San Francisco, California, and Omaha, Nebraska, together provide an interesting example of the effect of **continentality**—the effect of an inland location that favors greater temperature extremes (Figure 3–20). They are at similar latitudes and not significantly different elevations, and both are subject to a predominantly west-to-east airflow. Yet San Francisco, situated along the Pacific Coast, has a much smaller temperature range than does its inland counterpart.

Four reasons cause water bodies to be more conservative than landmasses with regard to temperature:

1. The specific heat of water is about five times as great as that of land.
2. Radiation received at the surface of a water body can penetrate to several tens of meters deep and distribute its energy throughout a very large mass. In contrast, the insolation absorbed by land heats only a very thin, opaque surface layer.
3. The warming of a water surface can be reduced considerably because of the vast supply of water available for evaporation. Because much energy is used in the evaporative process, less warming occurs.
4. Unlike solid land surfaces, water can be easily mixed both vertically and horizontally, allowing energy surpluses from one area to flow to regions of lower temperature.

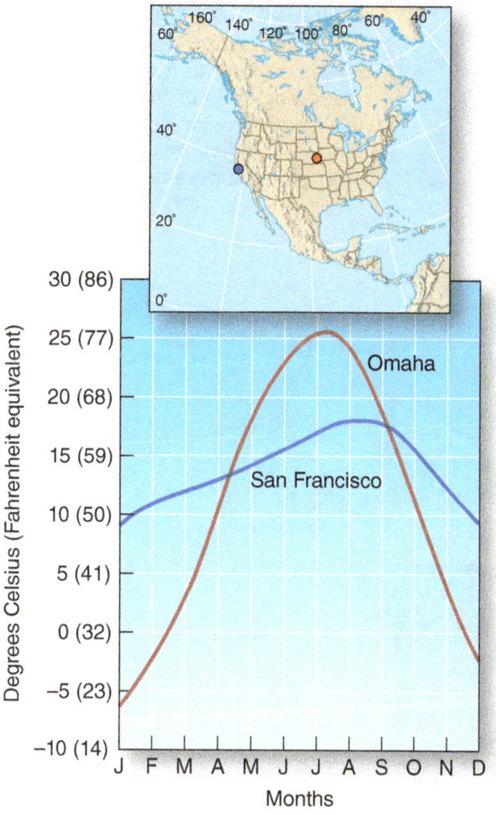

▲ **FIGURE 3–20 Continentality.** Areas near the oceans (like San Francisco) typically have smaller temperature ranges over the course of a year than locations well within continental interiors (Omaha).

Warm and Cold Ocean Currents

Figure 3–16 shows warm and cold ocean currents. The warm currents typically move poleward in the western portion of the ocean basins near the east coasts of continents in the middle latitudes, carrying large amounts of energy with them. Similarly, along the eastern margins of oceans, cold ocean currents dominate in the middle latitudes. Where the water temperatures are high, heat is transferred to the atmosphere and promotes higher air temperatures. Thus, the existence of a warm ocean current offshore can cause a location along the east coast of a continent to have higher temperatures than would a cold current offshore along a west coast.

Compare, for example, Los Angeles, California, and Charleston, South Carolina, two coastal locations at similar latitudes and elevations. Summers are considerably warmer at Charleston than at Los Angeles, largely (but not entirely) because the temperature of the ocean off Charleston is higher than the temperature of the ocean off Los Angeles. The high water temperatures of the Gulf of Mexico and the western margin of the Atlantic Ocean allow the transfer of an enormous amount of heat to the atmosphere, and the average July temperature in Charleston of 31 °C (88 °F) is substantially warmer than that of Los Angeles (23 °C; 73 °F). In winter, temperatures are lower at Charleston than at Los Angeles because the westerly winds blowing toward Los Angeles are subject to the moderating effects of the Pacific, whereas Charleston's winter temperatures are affected by colder prevailing winds passing over the continental interior. Thus, the influence of atmospheric circulation, which we noted previously, can interact with the position along the edge of a continent in influencing temperature patterns.

Local Conditions

A number of site-specific factors, such as slope orientation and steepness, amount and type of vegetation cover, and surface albedo can influence the temperature characteristics of an area. In the Northern Hemisphere, south-facing slopes receive midday sunlight at a more direct angle than do those oriented in other directions, thereby promoting greater energy receipt and higher surface temperatures. The heating of south-facing slopes often results in a greater amount of drying than on the opposite, north-facing slopes. Vegetation patterns often respond to the change in microclimate, with plants intolerant of dry conditions occupying north-facing slopes. Such a pattern is shown in Figure 3–21.

Densely wooded areas also have different temperature regimes than areas devoid of vegetation cover. In a region like that shown in Figure 3–22, a tall, dense vegetation cover reduces the amount of sunlight hitting the surface during the day, and considerable evaporation of water from leaf surfaces occurs. At night the plant canopy reduces the net longwave radiation losses. These factors lead to lower daytime temperatures but warmer evenings.

The effects of vegetation on local climate can often be put to use in a manner that increases human comfort. For example, it is often a good idea to plant deciduous trees on the equatorward side of houses. During the warm summer months, trees can cast shade on houses, thereby keeping interior temperatures down or reducing air conditioning costs. When winter arrives, the loss of the leaves from the deciduous trees minimizes the reduction in sunlight, and this can help keep the building warm.

▼ **FIGURE 3–21 Effect of Slope Aspect on Temperatures.** One of the local factors affecting temperature is slope aspect. In the Northern Hemisphere, north-facing slopes typically receive less intense daytime heating and therefore exhibit lower temperatures. This retards the rate of surface evaporation, making more water available for plant life. Looking east in this photo taken in San Diego County, California, you can see denser vegetation cover on the north-facing slopes (right) than on the south-facing slopes (left).

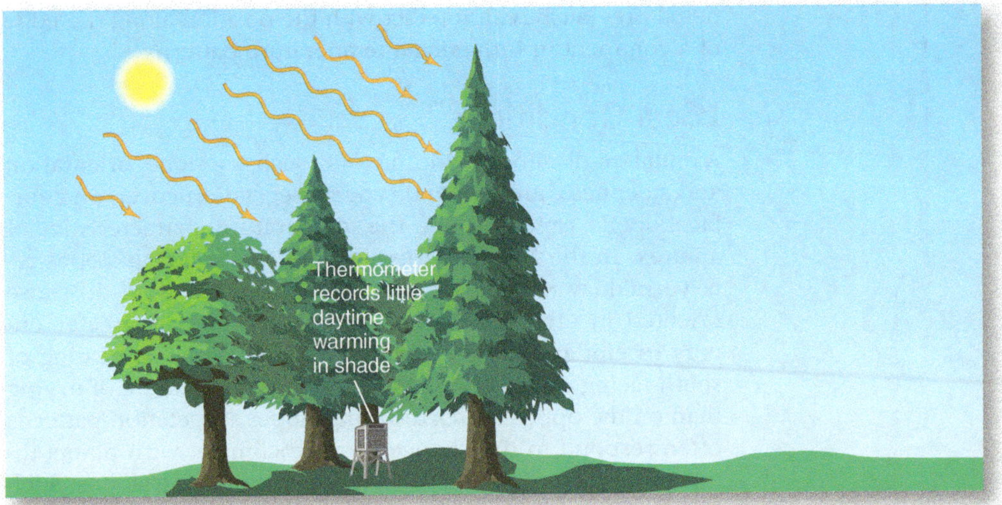

(a)

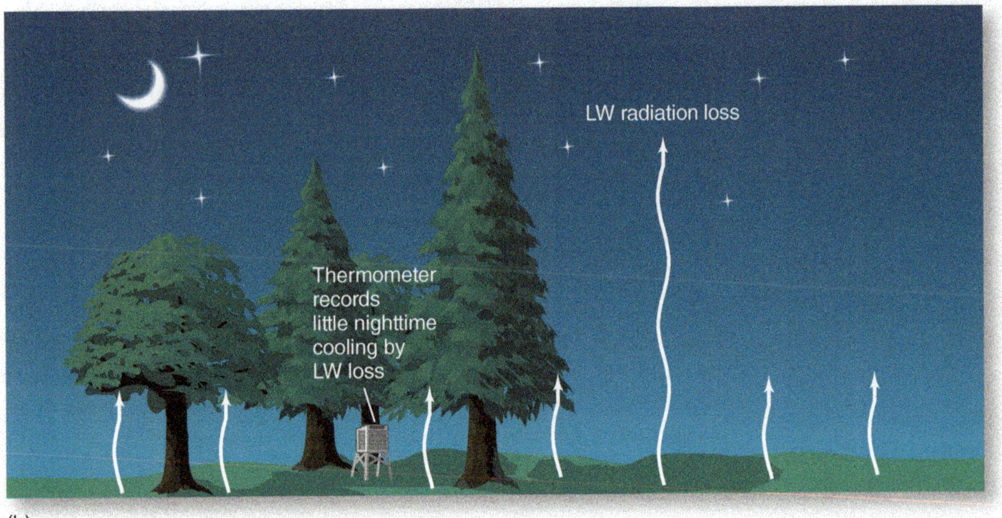

(b)

◄ **FIGURE 3–22 Effect of Vegetation Cover on Temperatures.** (a) Dense vegetation cover lowers daytime temperatures because of its shadowing effect on incoming solar radiation. (b) At night the forest canopy retards the loss of longwave radiation to space, resulting in higher nighttime temperatures than in the open.

Albedo plays an important role as well. Black asphalt surfaces are notoriously hard on bare feet on hot sunny days, in part because such a high percentage of solar radiation is absorbed at the surface. In contrast, a lighter concrete will reflect a larger percentage of incoming shortwave radiation.

CHECKPOINT

3.15 How does an ocean current affect the temperature of nearby land?

3.16 Seattle, Washington, and Fargo, North Dakota, are both at similar latitudes. Which would you expect to have the lower temperatures in January? Why?

Daily and Annual Temperature Patterns

3.7 Describe daily and annual temperature patterns.

The principles of heat transfer discussed in Chapter 2 and earlier in this chapter directly affect the daily (often called *diurnal*) and seasonal temperature changes that occur at any location. Let's first consider typical daily patterns.

Daytime Heating and Nighttime Cooling

First consider what happens over a 24-hour period on a cloudless day, beginning at sunrise (Figure 3–23a). As the Sun emerges above the horizon, the incoming energy does not effectively heat the surface due to the beam spreading and atmospheric depletion associated with the low solar angle. At sunrise and for some time period afterward, the receipt of solar radiation does not offset the loss of longwave radiation at the ground, leading to steady or slowly decreasing temperatures. (Though some people may assume that temperatures begin to rise as soon as the Sun comes up, such is not the case.) As the morning proceeds, the Sun sweeps a path from where it rose over the eastern sky toward the south (or north in the Southern Hemisphere), while at the same time rising upward from the horizon. Thus, as the outgoing longwave radiation remains roughly constant, the incoming solar radiation increases. At some point the Sun rises far enough above the horizon for its energy to offset the loss of longwave radiation, and warming occurs.

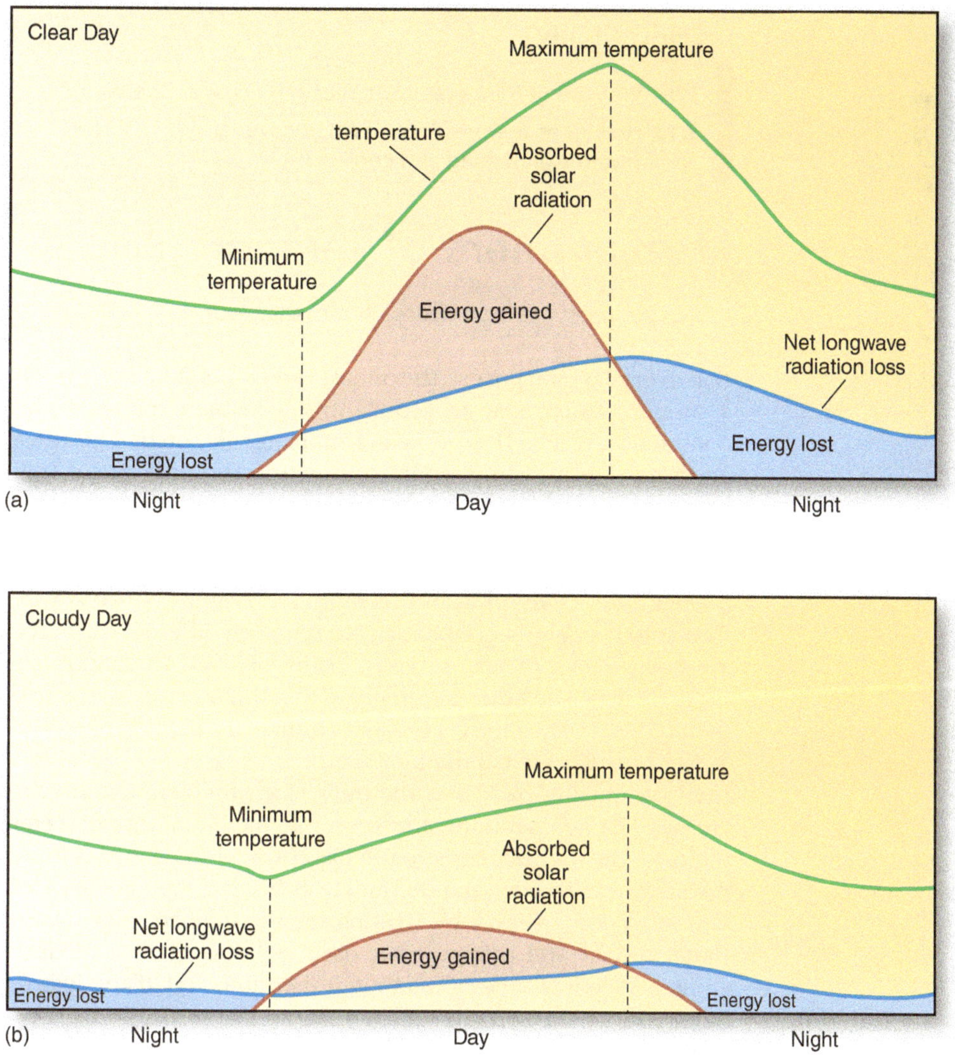

▲ **FIGURE 3–23 Diurnal Temperature Changes in Relation to Energy Gains/Losses.** Surface temperature increases whenever the energy gains exceed energy losses. (a) On clear days the availability of solar radiation in the middle of the day produces a large surplus that persists into the afternoon, but at night the longwave radiation loss results in substantial cooling. (b) Overcast conditions suppress the diurnal change in temperature mainly by reducing the incoming solar radiation gain during the day and by supplying more downward longwave radiation from the atmosphere at night.

Video
Heavy Convection Over Florida
http://goo.gl/PHJb7i

How long after sunrise does the ground begin to warm? That depends on several factors, including the rate at which the solar angle increases over the course of the morning, which in turn varies with latitude and time of year. Other factors include air temperature, wind speed, surface wetness, and the amount of heat conducted through the ground, to name a few.

At solar noon the Sun achieves its greatest angle above the horizon and the surface receives its greatest input of solar radiation. But experience tells us that noon is not the warmest time of the day and that maximum air temperatures (as measured at the usual 5 ft above the surface) are more likely to occur at least a couple hours afterward. Recall that surface temperature increases as long as energy gained by the surface is greater than energy lost. Although solar radiation begins to decrease after noon, the net energy gain by all processes often exceeds the loss for some time afterward, and warming continues. Eventually the energy balance becomes negative, and the surface cools. Typically the cooling rate is much slower than the previous warming, largely because as the surface becomes colder than the ground beneath, heat is conducted upward. This is not enough to prevent surface temperatures from falling, but it does slow the decrease.

Ground heat flux to the surface continues through the night for as long as the surface temperature continues to fall. These surface temperature trends are largely mirrored by air temperature near the ground, except that the amplitude of daily change is less, and the time of maximum air temperature lags farther behind solar noon.

Effects of Cloud Cover and Wind

The amplitude of the daily temperature pattern is reduced under overcast conditions. During the day the cloud cover can greatly reduce the daytime input of solar radiation and likewise reduce the magnitude of the net longwave radiation loss over the entire 24-hour period (Figure 3–23b). The overall effect is to lower the daytime maximum temperature while keeping nighttime temperatures somewhat higher than they would be on a clear night. Overall, the temperature difference over the 24-hour period is much less than in the previous example.

Strong winds also moderate daily temperature ranges. Recall that higher wind speeds promote greater forced turbulence. When turbulence increases, the enhanced vertical movement causes any small parcel of air immediately in contact with the ground to be quickly displaced upward and replaced by another parcel. The result is that the sensible heat transferred from the ground is distributed to a greater mass of air, which reduces the increase in temperature near the surface. (Not only do strong winds reduce the rate of warming, they also make the air feel colder at any particular temperature, as discussed later in this chapter.) The same process works at night to inhibit cooling (Figure 3–24). If a lot of forced convection occurs, no parcel of air will remain in contact with the chilled surface long enough to undergo substantial cooling. In the absence of forced convection, a shallow layer of air is subject to cooling for an extended time period, and temperatures can drop rapidly.

Annual Cycle

The relative amounts of incoming and outgoing radiation also affect energy budgets and temperatures on an annual basis. The Northern Hemisphere experiences its greatest amount of

Time (CST)	Temperature °F (°C)	Wind (mph)
9 P.M.	41.0 (5.0)	W 16
10 P.M.	39.0 (3.9)	WNW 26
11 P.M.	39.0 (3.9)	W 26
12 A.M.	37.9 (3.3)	W 24
1 A.M.	37.9 (3.3)	W 21
2 A.M.	37.9 (3.3)	W 17
3 A.M.	37.9 (3.3)	W 20
4 A.M.	37.9 (3.3)	W 21
5 A.M.	37.9 (3.3)	W 20
6 A.M.	37.9 (3.3)	W 23
7 A.M.	37.9 (3.3)	W 20
8 A.M.	37.9 (3.3)	W 18
9 A.M.	37.9 (3.3)	WNW 23

▲ **FIGURE 3–24 Effect of Clouds and Wind on Temperature.** Cloudy, windy conditions can lead to very little cooling at night. This was illustrated at Chicago's Midway Airport on October 25–26, 2001, a day with total overcast. The combined effects of cloud cover and vigorous winds allowed only a 3.1 °F (1.7 °C) drop in temperature between 9 P.M. and 9 A.M.

solar radiation, but not its highest temperatures, on the June solstice. For about 4 to 6 weeks following the solstice, many places continue to have a positive energy balance, which allows warming to continue until sometime in July or August when maximum temperatures are achieved.

CHECKPOINT

3.17 Why aren't temperatures typically highest at noontime?

3.18 How does the presence of cloud cover and wind affect daily patterns of heating and cooling?

Measurement of Temperature

3.8 Explain how temperatures are measured.

For everyday purposes, the measurement of temperature is a simple and routine procedure in which the expansion and contraction of the fluid in a thermometer is noted. The most accurate thermometers contain mercury, the only metal that exists as a liquid at normal Earth temperatures. Less expensive models are available that contain dyed alcohol.

It is often useful to know the daily maximum and minimum temperatures. A **maximum thermometer** (Figure 3–25a) is very similar to a regular thermometer, but with two differences. Unlike a regular thermometer, it must contain mercury—it cannot use dyed alcohol. The other difference is that in the tube just beyond the bulb is a very narrow constriction that allows the mercury to expand outward when the temperature increases but prevents it from contracting back into the bulb when the temperature decreases. The temperature shown on the maximum thermometer indicates the highest temperature experienced since the last time it was reset. Resetting the thermometer is easy; the mercury can be forced down simply by shaking the thermometer downward (note that an oral thermometer used at home or in a medical clinic is merely a type of maximum thermometer). If the instrument is mounted on a pivoting base, it can be reset by being spun so that centripetal force propels the mercury back into the bulb.

A **minimum thermometer** (see Figure 3–25b) is also similar to a regular thermometer, except that it can only contain dyed alcohol and has within it a small index shaped like a

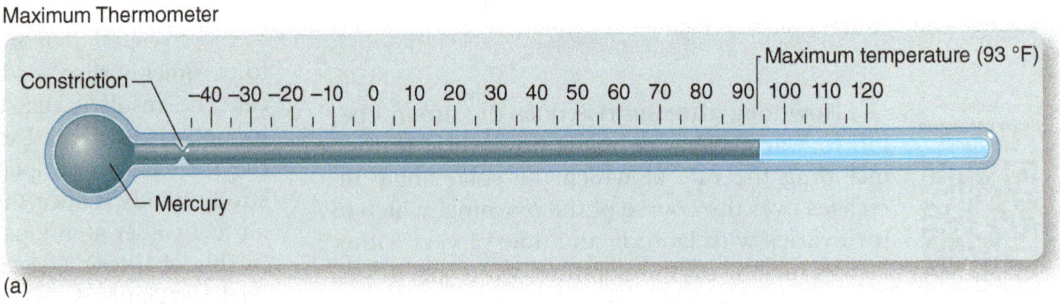

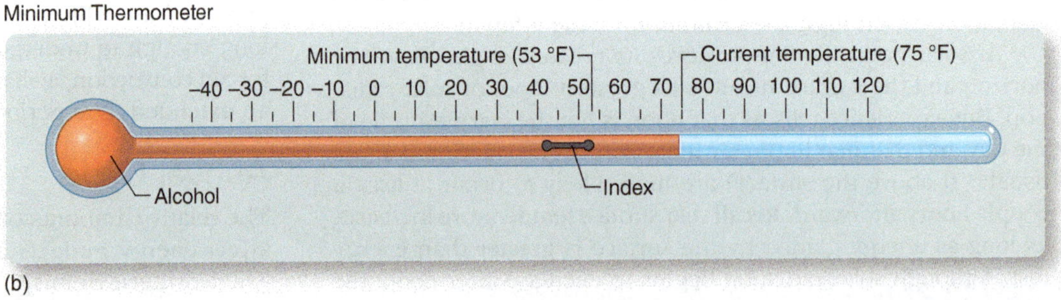

▶ **FIGURE 3–25 Maximum and Minimum Thermometers.** (a) Maximum thermometers have a small constriction that only allows the mercury to move outward from the bulb. (b) Minimum thermometers have a sliding index indicating the lowest temperature since the instrument was last reset.

▲ **FIGURE 3–26 Thermographs.** Thermographs rely on bimetallic strips of different materials bonded together. Because all metals have different rates of expansion and contraction with temperature, one undergoes a greater change in length than the other, causing the strip to bend. An inked pointer and scale are attached to the bimetallic strip, whose bending is amplified by a lever.

weightlifter's dumbbell. If the index is at the end of the alcohol and the temperature is decreasing, surface tension (the force that holds water molecules together) pulls the index toward the bulb. When the temperature increases, the index remains at its present position as the alcohol expands away from the bulb. Minimum thermometers are mounted horizontally with a latch that can be released to allow the instrument to be inverted. The instrument is reset by turning it upside down, allowing the index to slowly slide down to the end of the alcohol.

Another instrument for the measurement of temperature is the **bimetallic strip**. The two different metals in the strip have different rates of thermal expansion and contraction, thus bending the strip an amount that depends on temperature. When this mechanism is coupled with a rotating drum and a pen, the resulting **thermograph** (Figure 3–26) gives a continuous record of temperature. Thermostats for the heating and air conditioning units of many homes use bimetallic strips to determine the room temperature.

The temperature sensors described above are no longer used at airports and other sites having electronic sensors, but they are still used widely at thousands of weather-observing sites worldwide that report their data to national meteorological centers (discussed in Chapter 13). Considerably more sophisticated instruments for measuring temperature are used at airports and other primary sites. Among these are **resistance thermometers**, instruments that send an electrical current through a very thin filament made of conductor or semiconductor material exposed to the air. The temperature of the filament is the same as that of the surrounding air, and its temperature influences its resistance to the electrical current. The instrument registers the amount of resistance and uses the reading to determine the air temperature.

Resistance thermometers are fast-response instruments, meaning that they rapidly register changes in the ambient temperature. Consider what might happen if you were to put a thermometer (a slow-response device) in a refrigerator for an hour or so and then remove it. Although your kitchen is probably much warmer than the inside of your refrigerator, it would take several minutes before the instrument accurately measured the temperature of the room. Fast-response instruments must be used in situations where rapid temperature changes may be encountered, such as at experimental research stations measuring sensible and latent heat exchanges.

Instrument Shelters

You have no doubt heard statements such as this one: "It was 100 degrees in the shade." This is a common expression but one that might have originated from the Department of Redundancy Department because, to be meaningful, temperature must *always* be measured in the shade. A thermometer exposed to direct sunlight will be warmed by the absorption of insolation and assume a temperature greater than that of the air. But when we refer to "air" temperature, we are really concerned with the temperature of the air—not that of a thermometer. As a result, temperature-measuring instruments should always be kept in an instrument shelter like the one shown in Figure 3–27.

An instrument shelter is designed to reduce the influence of incoming radiation on the instruments. For example,

▼ **FIGURE 3–27 Instrument Shelter.** Temperature instruments must be kept in shelters that protect them from the absorption of solar radiation. Note that the box is painted white and has slats to allow the movement of air. It also has a door that opens on the north side (in the Northern Hemisphere).

shelters are painted white to reduce the absorption of sunlight and paneled with slats to permit the free flow of air and the removal of any heat that might otherwise accumulate. Finally, instrument shelters used at official weather-observing stations must conform to a standardized height, such that that the thermometers are mounted at 1.52 m (5 ft) above the ground.

In the United States, temperature is observed hourly by automated systems at National Weather Service offices and Federal Aviation Agency (FAA) facilities at airports across the country. The automated systems (described in Chapter 13) use resistance thermometers for temperature readings. This network is supplemented by observations made at a large number of cooperative agencies (such as U.S. Forest Service stations) and by individual volunteers, often using maximum and minimum thermometers.

Temperature Means and Ranges

3.9 Define the concept of temperature means and ranges.

In just about all aspects of daily life, we use descriptive statistics to talk about the things around us. Although the concept of an average, or mean, value of a property is fairly straightforward, applying the concept involves occasional complications. For example, trying to determine a daily mean temperature poses a dilemma—exactly how many times during the day must we measure it to obtain a true mean? We could make observations every hour, every minute, or even every second, with each method giving us a separate value.

MM **MapMaster** North America Physical Environment: Avg. January Temp, Avg. July Temp

DID YOU KNOW?

Hawaii and Alaska share the highest temperature ever recorded in each of those states: 37.8 °C (100 °F). The situation is a little different with regard to the minimum temperature, however. While the lowest reading ever taken in Hawaii was −11 °C (12 °F), the record low for Alaska was −62 °C (−80 °F).

The standard procedure is simple—the *daily mean* is defined as the average of the maximum and minimum temperature for a day. The advantage of this method is that the daily mean can be obtained even at weather stations having just the most basic instrumentation; a minimum and maximum thermometer are all we need to compute daily mean temperatures. The disadvantage of using just the maximum and minimum temperature is that it introduces a bias. Observe the daily temperature pattern for a particular day, shown in Figure 3–28. Notice that the nighttime temperatures remain nearly equal to the minimum throughout most of the night, while afternoon temperatures are near the maximum for only a few hours. If we were to obtain a daily mean by taking 24 hourly spaced observations and dividing by 24, the mean value would be lower than that obtained by using just the maximum and minimum temperatures. Nonetheless, averaging the maximum and minimum temperatures is the accepted method for obtaining daily mean temperatures even though it introduces a bias.

The *daily temperature range* is obtained simply by subtracting the minimum temperature from the maximum.

Having obtained daily mean temperatures for an entire month, we calculate the *monthly mean temperature* simply by summing the daily means and dividing by the number of days in the month. Similarly, an *annual mean temperature* is obtained by summing the monthly means for a year and dividing by 12. The *annual range* is the difference between the highest and lowest monthly mean temperatures.

Global Extremes

More impressive than mean temperatures are the maximum and minimum values ever recorded. Not surprisingly, they tend to occur at continental locations. The highest temperature ever recorded in North America was at Death Valley, California. Death Valley is located only a couple of hundred kilometers from the Pacific Ocean, but the Sierra Nevada mountain range presents a barrier that eliminates any moderating influence of the water. In addition, Death Valley's position below sea level and its sparse vegetation cover further promote high temperatures. The world's highest measured temperature was the 56.7 °C (134 °F) reading recorded on July 10, 1913, at Greenland Ranch in Death Valley.

DID YOU KNOW?

Of the 50 U.S. states, Montana holds the record for the greatest spread between its all-time maximum and minimum temperatures. That state recorded a high temperature of 47 °C (117 °F) at Medicine Lake in 1937 and low temperature of −57 °C (−70 °F) at Rogers Pass in 1954, for a range of 104 °C (187 °F).

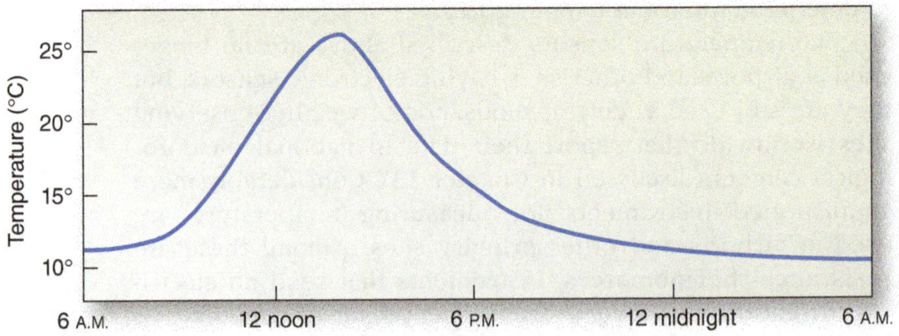

▸ **FIGURE 3–28 Daily Temperature Range.** This graph shows a continuous plot of temperature over a 24-hour period with clear skies. Note that the temperature is near that of the maximum for a relatively short time period. In contrast, the air temperature is near the daily minimum throughout most of the night.

3–3 FOCUS ON THE ENVIRONMENT AND SOCIETAL IMPACTS

Recent Severe Heat Waves

Summer heat waves are certainly no rarity in the United States and Canada. Sooner or later everybody endures an episode of unpleasantly high temperatures. But heat waves can cause much more than a few days of discomfort—they can kill. One of the most notable heat waves of the past few decades was the relatively brief but severe event in mid-July 1995 in the north-central United States.

Although extremely high temperatures occurred from the Great Plains to the Atlantic Coast, nowhere was the problem more acute than in Chicago, Illinois, where 525 people died from the heat. The heavy mortality resulted from a combination of high temperatures (Midway Airport recorded an all-time high temperature of 41.1 °C, or 106 °F) and unusually high humidities. The heat and humidity combined to make the "apparent temperature" equivalent to 47 °C (117 °F). Though the searing daytime heat created plenty of misery on its own, it is believed that the major factor leading to the many deaths is the fact that the extreme heat went uninterrupted, with the apparent temperature exceeding 31.5 °C (89 °F) for nearly 48 consecutive hours. (Recent research suggests that such conditions pose a greater danger than do brief periods of more extreme heat.)

Four years later, in 1999, another major July heat wave occurred in the eastern two-thirds of the United States. Once again, Illinois was in the center of the action, with more than half of the 232 fatalities across the Midwest occurring in Chicago. Missouri was the second hardest hit state, with 61 fatalities. All across the region, power outages occurred from excessive demand, roads buckled, and crops wilted in the fields.

As July gave way to August, the heat moved eastward toward the Atlantic states, where it broke numerous weather records. Charleston, South Carolina, had an all-time high temperature of 40.5 °C (105 °F). The high temperature in Augusta, Georgia, of 39.4 °C (103 °F) came on the sixth consecutive day in which the record for the daily maximum temperature was tied or exceeded. And on August 8, Raleigh-Durham, North Carolina, broke the 100 °F mark (37.8 °C) for the 11th time that summer.

Severe heat has also impacted other parts of the world. Some 35,000 people died from heat-related causes during a heat wave across Europe in July and August of 2003. England recorded its all-time highest temperature on August 10, with a reading of 38.1 °C (100.6 °F) at Gravesend-Broadness. France, suffering through its hottest summer since World War II, was especially hard hit, with over 11,000 heat-related fatalities. In January 2014 Australia experienced a heat wave severe enough to force postponement of the Australian Open tennis tournament, as a ball boy collapsed under temperatures reaching 43.4 °C (110 °F). Other recent Australian heat waves have led to record-breaking wildfires.

The decade of the 1990s and the first decade and a half of the 2000s were remarkably warm relative to other periods in recorded history. This is particularly noteworthy because the topic of human-induced climatic warming has been a major issue for scientists and policy makers. This matter will be discussed further in later chapters of this book.

3–3–1 What aspect of the 1995 heat wave made it particularly deadly?

3–3–2 How do the global temperatures of recent decades compare to the historical record?

The lowest temperature ever recorded occurred at the Vostok Research Station in Antarctica in 1983, with a reading of −89 °C (−129 °F). The research station is located atop thousands of meters of glacial ice, thereby combining the effects of high latitude, a continental locale, and a sparse atmosphere. In North America, the record low temperature of −63 °C (−81 °F) was observed at Snag, Yukon, in February 1947.

Undoubtedly, more extreme temperatures have occurred across the globe and North America at locations without temperature observation stations. *(See Box 3–3, Focus on the Environment and Societal Impacts Recent Severe Heat Waves.)*

CHECKPOINT

3.19 How would you determine the daily temperature range for a location?

3.20 How are daily mean temperatures determined?

Some Useful Temperature Indices

3.10 Define some useful indexes of temperature.

There are several situations in which basic temperature readings can be combined with other variables to provide a better understanding of the effects of temperature. Three of these indices are described in this section. An additional temperature index that incorporates the effect of humidity will be discussed in Chapter 4, which addresses atmospheric moisture.

Wind Chill Temperatures

Temperature by itself exerts a major impact on human comfort, but the discomfort caused by high or low temperatures can be compounded by other weather factors. If low temperatures are accompanied by windy conditions, a person's body loses heat much more rapidly than it would under calm conditions, due to an increase in sensible heat loss. Thus, a windy day with a temperature of –10 °C (14 °F) might feel colder than a calm day at –20 °C (–4 °F). As a result, when temperatures are low, it is common for weather reports to state both the actual temperature and how cold that temperature actually feels, the **wind chill temperature index** (or simply the *wind chill temperature*) (see Tables 3–2 and 3–3).

While useful, the wind chill index has certain shortcomings. For example, its calculations do not take into account the potential warming effect of sunlight on a person's body. Nonetheless, the index provides people with guidance in the way they should prepare for cold and windy conditions, which is

TABLE 3–2
Wind Chill Temperature (° C)

Wind (km/hr)	Temperature (°C) 5	0	–5	–10	–15	–20	–25	–30	–35	–40
5	4	–2	–7	–13	–19	–24	–30	–36	–41	–47
10	3	–3	–9	–15	–21	–27	–33	–39	–45	–51
15	2	–4	–11	–17	–23	–29	–35	–41	–48	–54
20	1	–5	–12	–18	–24	–31	–37	–43	–49	–56
25	1	–6	–12	–19	–25	–32	–38	–45	–51	–57
30	0	–7	–13	–20	–26	–33	–39	–46	–52	–59
35	0	–7	–14	–20	–27	–33	–40	–47	–53	–60
40	–1	–7	–14	–21	–27	–34	–41	–48	–54	–61
45	–1	–8	–15	–21	–28	–35	–42	–48	–55	–62
50	–1	–8	–15	–22	–29	–35	–42	–49	–56	–63
55	–2	–9	–15	–22	–29	–36	–43	–50	–57	–63
60	–2	–9	–16	–23	–30	–37	–43	–50	–57	–64
65	–2	–9	–16	–23	–30	–37	–44	–51	–58	–65
70	–2	–9	–16	–23	–30	–37	–44	–51	–59	–66
75	–3	–10	–17	–24	–31	–38	–45	–52	–59	–66
80	–3	–10	–17	–24	–31	–38	–45	–52	–60	–67

TABLE 3–3
Wind Chill Temperature (° F)

Wind (mph)	Temperature (°F) 40	35	30	25	20	15	10	5	0	–5	–10	–15	–20	–25	–30	–35	–40
5	36	31	25	19	13	7	1	–5	–11	–16	–22	–28	–34	–40	–46	–52	–57
10	34	27	21	15	9	3	–4	–10	–16	–22	–28	–35	–41	–47	–53	–59	–66
15	32	25	19	13	6	0	–7	–13	–19	–26	–32	–39	–45	–51	–58	–64	–71
20	30	24	17	11	4	–2	–9	–15	–22	–29	–35	–42	–48	–55	–61	–68	–74
25	29	23	16	9	3	–4	–11	–17	–24	–31	–37	–44	–51	–58	–64	–71	–78
30	28	22	15	8	1	–5	–12	–19	–26	–33	–39	–46	–53	–60	–67	–73	–80
35	28	21	14	7	0	–7	–14	–21	–27	–34	–41	–48	–55	–62	–69	–76	–82
40	27	20	13	6	–1	–8	–15	–22	–29	–36	–43	–50	–57	–64	–71	–78	–84
45	26	19	12	5	–2	–9	–16	–23	–30	–37	–44	–51	–58	–65	–72	–79	–86
50	26	19	12	4	–3	–10	–17	–24	–31	–38	–45	–52	–60	–67	–74	–81	–88
55	25	18	11	4	–3	–11	–18	–25	–32	–39	–46	–54	–61	–68	–75	–82	–89
60	25	17	10	3	–4	–11	–19	–26	–33	–40	–48	–55	–62	–69	–76	–84	–91

especially important when one considers the fact that extreme cold is a major cause of fatalities directly attributable to weather.

Just as windy conditions can make low temperatures feel even colder, high humidities can cause warm days to feel oppressively hot. As a result, a heat index has been calculated that incorporates the effect of high atmospheric moisture at high temperatures. This index is discussed in Chapter 5.

Heating and Cooling Degree-Days

Because climate exerts a major impact on the energy demands of people across the world, it is useful to create an index that can help anticipate energy needs. One such index, called **heating degree-days**, is based on the notion that buildings generally need artificial heating to bring their interior temperatures up to the desired temperature of 21 °C (70 °F) when the mean temperature for a day falls below 18 °C (65 °F)[2]. Each degree of temperature for which the daily mean is lower than 18 °C (65 °F) constitutes one degree-day. Hence, a day with an average temperature of 13 °C (56 °F), for example, would account for 5 Celsius or 9 Fahrenheit heating degree-days. These values are then summed for each day of the year in which the temperature falls below the critical value, yielding the total number of heating degree-days for a location. Figures 3–29a and b map the heating degree-days for the United States and Canada, respectively.

Cooling degree-days are the warm season analogy to heating degree-days. They are obtained by subtracting a base temperature—often but not always 18 °C (65 °F)—from the daily average temperature and summing those values. Figure 3–30 plots United States cooling degree-days using the base temperature of 65 °F. (Environment Canada has not produced an equivalent map for that country.)

Growing Degree-Days

Growing seasons for agricultural products are to a large extent determined by the length of time in which the temperature remains above some *base temperature* that varies by crop type. If a particular crop requires a daily average temperature of 10 °C (50 °F) to grow, the growing season will extend between the first day of spring and the last day of fall with mean temperatures above that base value. **Growing degree-days** are calculated by subtracting a crop's base temperature from the daily mean temperature and summing those numbers over the growing season.

Video
Temperatures and Agriculture
http://goo.gl/HOy00

CHECKPOINT

3.21 Why is the wind chill index a useful measure?

3.22 What do heating and cooling degree-days indicate?

[2]Note that the daily mean temperature is the average of the day's maximum and minimum temperatures. Since indoor temperatures do not fluctuate as rapidly as outdoor temperatures, the 18 °C (65 °F) works out to be a better base value than 21 °C (70 °F).

Thermodynamic Diagrams and Vertical Temperature Profiles

3.11 Make use of thermodynamic diagrams and understand vertical temperature profiles.

The distribution of temperatures across Earth's surface is a basic component of daily weather. But temperature variations do not occur only horizontally; temperatures are just as likely to vary vertically, and those variations can greatly affect atmospheric behavior. Rapid decreases with height, for example, increase the potential for cloud development and even severe thunderstorms. Temperature inversions (increasing temperature with altitude) have their own implications for the atmosphere, including the suppression of atmospheric mixing that can concentrate air pollutants near the surface.

Thermodynamic diagrams, which depict the vertical profiles of temperature and humidity with height above the surface, provide extremely important information to weather forecasters. These charts enable forecasters to determine the height and thickness of existing clouds and the ease with which the air can be mixed vertically (a consideration in the near-term development of clouds and precipitation). The data on the charts are obtained from instrument packages called **radiosondes** that are carried aloft by weather balloons twice a day at weather stations across the globe.

Though extremely valuable, thermodynamic diagrams are not altogether simple to understand, and some of the information one can obtain from them is beyond the scope of this text. Consequently, rather than trying to present a complete guide to the use of these charts all at once, we will introduce various elements of thermodynamic diagrams as they pertain to the individual discussions in this text. As we progress, you will see just how crucial such charts can be to day-to-day forecasting. For now we discuss only the plotting of vertical temperature profiles on these charts. This will provide a foundation for further understanding in later chapters.

Thermodynamic diagrams come in several varieties. The *skew-T graph* is commonly used by meteorologists but can be conceptually difficult for beginning students. The easiest to use is the **Stuve** (pronounced "STU-vay") **diagram**, a highly simplified version of which is shown in Figure 3–31. (We will expand greatly on the information contained in this diagram in subsequent chapters.) The Stuve diagram starts with a rectangular grid, with temperature plotted on the horizontal axis and pressure on the vertical axis. Three things must be noted at the onset: First, the chart plots temperature as a function of the *pressure level*—not the height above the surface. The altitude of a particular pressure level—such as the 500 mb level—is not a constant, but varies over time and from one location to another. This convention allows a more direct application of meteorological laws than would plotting temperature against altitude.[3]

[3]For reference, the height of the average altitude of several pressure levels is often indicated on the diagram.

▸ **FIGURE 3–29 Heating Degree-Days.** The map shows the distribution of heating degree-days in the United States in degrees Fahrenheit (a) and Canada in degrees Celsius (b).

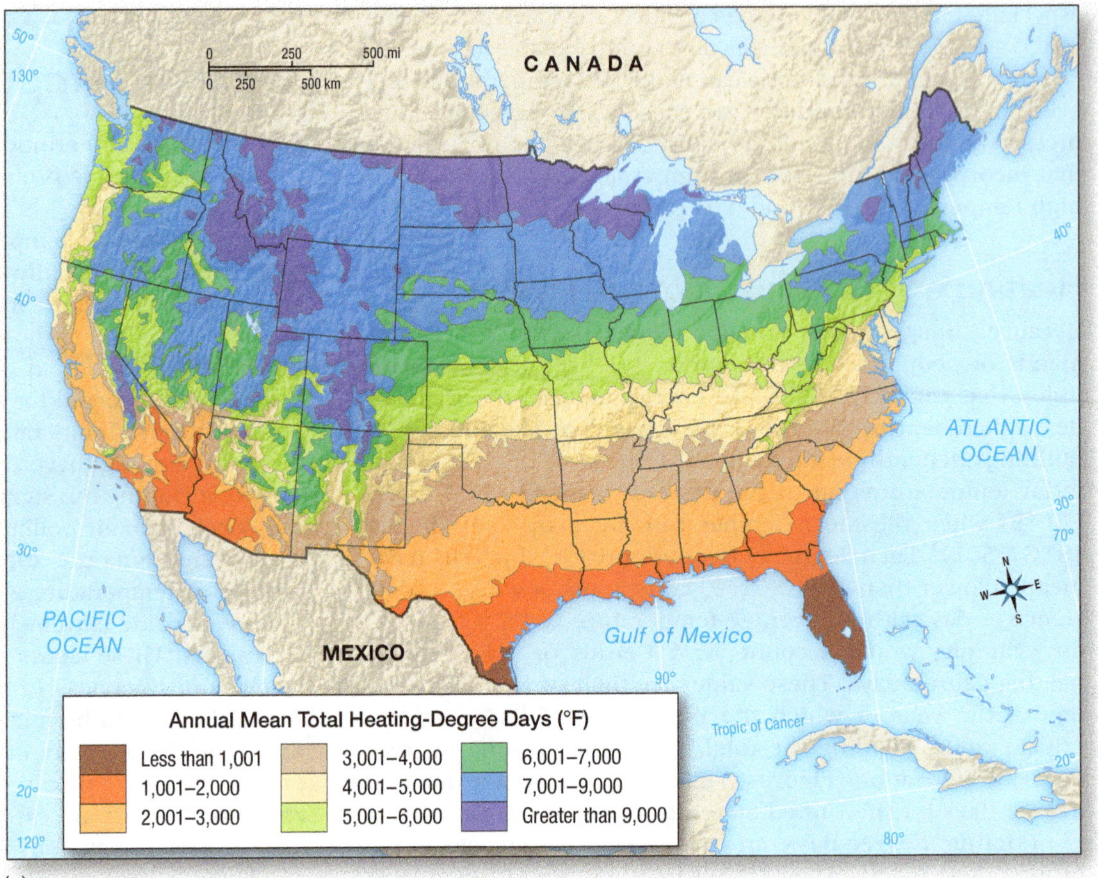

(a)

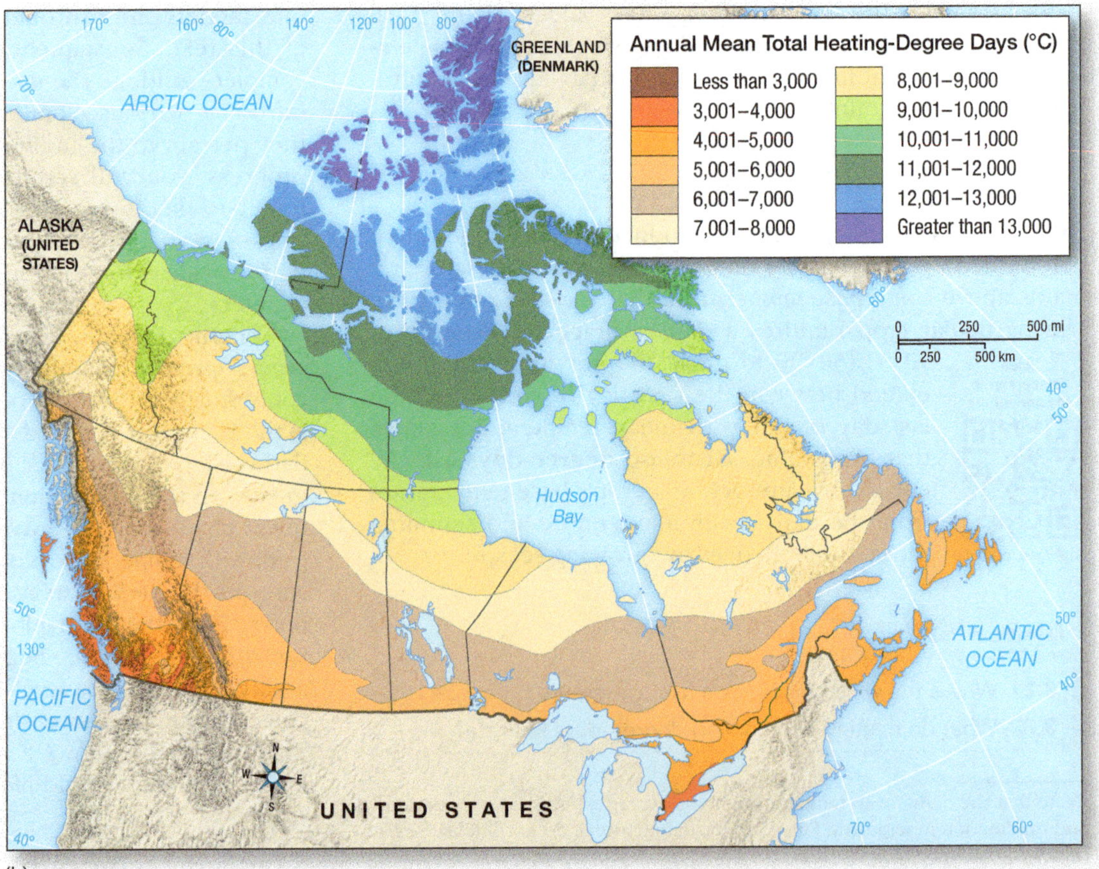

(b)

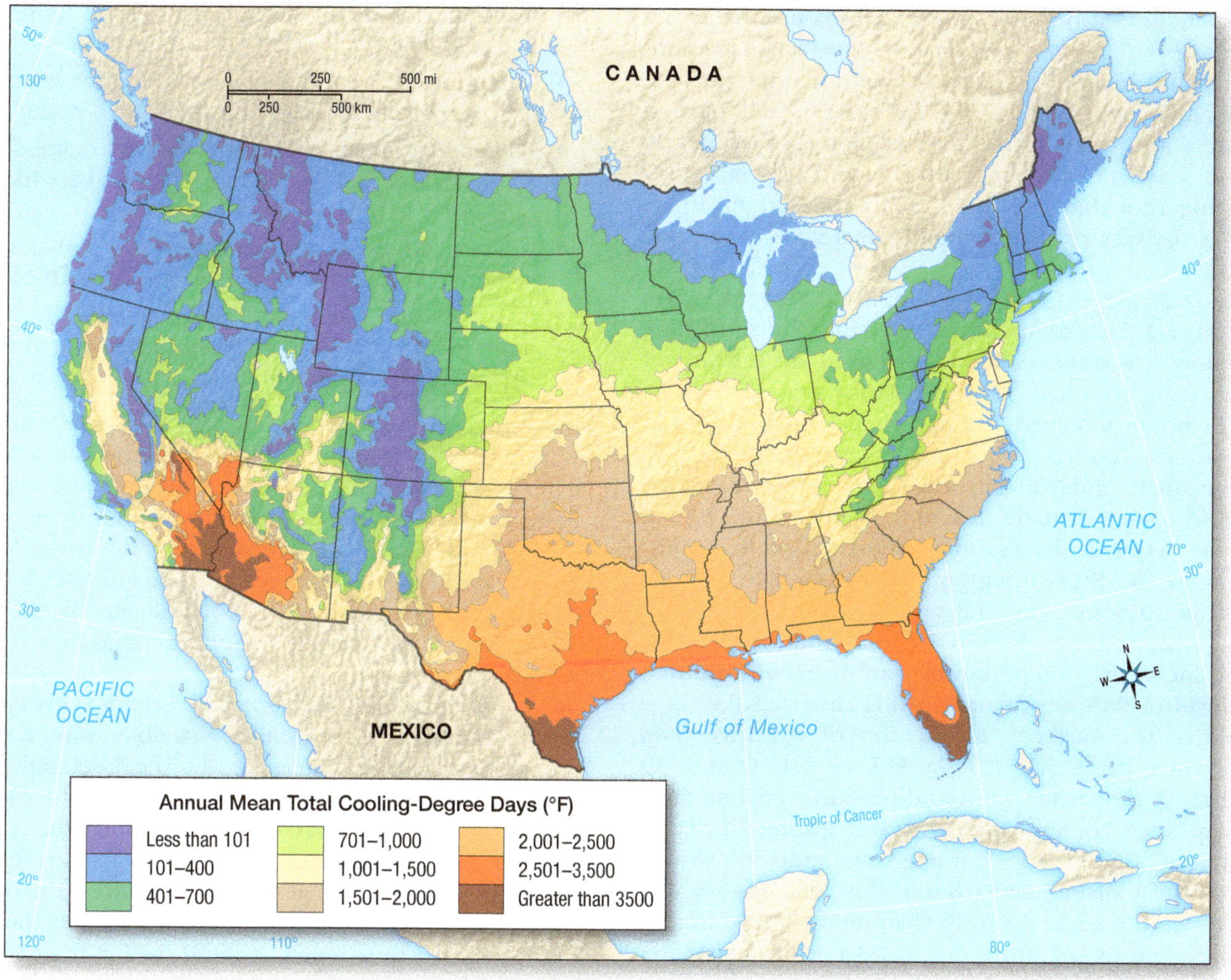

▲ **FIGURE 3–30 Cooling Degree-Days.** Distribution of cooling degree-days in the United States (°F).

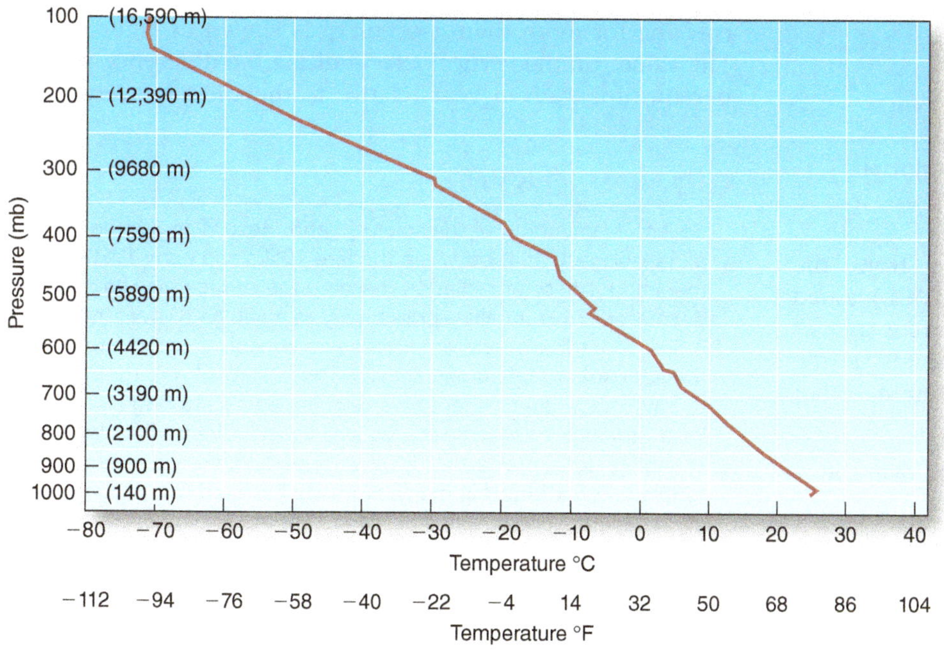

◀ **FIGURE 3–31 Simplified Stuve Diagram.** This diagram shows the temperature profile for Slidell, Louisiana, on the morning of June 9, 2002. The values in parentheses are the heights of the pressure pressure level, in meters.

Second, note that the pressure decreases upward along the vertical axis. This is a response to the fact that atmospheric pressure invariably decreases with height from the surface. Finally, observe that the pressure is not plotted on a linear scale. Instead, the axis is plotted on a nearly logarithmic scale so that the 1000 mb and 900 mb lines are closer together than are the 300 mb and 200 mb lines. This mimics the way pressure actually decreases in the real atmosphere.

Let's look at the example of a simplified Stuve diagram in Figure 3–31. The temperature plotting (or *sounding*, as these plots are often called) was based on the ascent of a radiosonde launched at Slidell, Louisiana, on June 9, 2002, at 12 noon, Greenwich mean time (7 A.M. CDT). In all, observations were recorded at 52 levels from the surface all the way up to the 100 mb level (well into the stratosphere). The data are made available in text form, which is useful for detailed observations, and then plotted automatically onto the Stuve diagram. At the surface, the air pressure was 1015 mb and the temperature was 22.8 °C (73.0 °F). As the radiosonde initially rose from the surface, it recorded increasing temperatures—an inversion—up to a pressure level of 988 mb (which in this case was 238 m—781 ft—above the surface). At the top of the inversion, the temperature was 25.0 °C (77.0 °F), or 2.2 °C (4.0 °F) warmer than at the surface. Ground-based inversions such as these are very common at night and during the early morning hours due to the loss of longwave radiation (inversions were briefly discussed earlier in this chapter—we will return to them in greater detail in Chapter 6). Above the inversion layer, the temperature decreases fairly consistently with altitude until about the 140 mb level (located at about 14.7 km—9.1 mi—above sea level), where the air temperature becomes nearly constant with height at about −70 °C (−90 °F), the point at which the air temperature ceases to decrease with height, which marks the tropopause (as described in Chapter 1).

Our Warming Planet

3.12 Summarize observations of climate change/global warming.

In 2013 the international group of experts from the **Intergovernmental Panel on Climate Change (IPCC)** issued their fifth assessment report summarizing the current state of knowledge about climate change. This report builds on the previous report issued in 2007 and concludes, among other things, the following:

- The atmosphere has warmed by about 0.85 °C over the 1880–2012 period.
- Almost all portions of the world with reliable data have experienced warming during the past century.
- Changes in many extreme temperature events have been observed since about 1950, with the frequency of heat waves having increased in large parts of Europe, Asia, and Australia.
- The upper 75 m of the ocean surface has warmed by 0.11 °C per decade over the 1971–2010 period.
- The annual mean Arctic sea ice extent decreased over the 1979–2012 period with a rate that was very likely in the range of 3.5 to 4.1 percent per decade.
- It is very likely that the mean rate of global average sea level rise was about 1.7 mm/yr between 1901 and 2010.
- It is extremely likely that human activities have caused more than half of the observed increase in global average surface temperature since the 1950s.

In some regards all this was an old story—scientists have been warning about changing climate and the role of human activities for some time. But the news media and scientific community placed enormous importance on this report—and for good reason. The IPCC was created by the World Meteorological Organization to periodically summarize what the science community currently knows about climate change and its impacts (an overview of the organization and how it goes about its mission is presented in *Box 16–2, Focus on the Environment: Intergovernmental Panel on Climate Change*). It is the most authoritative organization assembled on the subject, and its findings are widely accepted by scientists and policy makers around the world. In this chapter we briefly summarize its findings specifically related to temperature. Other chapters will refer to the IPCC where appropriate. Chapter 16 presents a detailed analysis of climate change, including the processes by which human activities have altered the climate, the techniques by which climate scientists have arrived at their findings, and related impacts.

Recent Observed Changes in Temperature

There is no question that global temperatures have been increasing for more than a century, as shown in Figure 3–32. The values on the vertical axes indicate the difference in temperature for a given year relative to the climatic average for

DID YOU KNOW?

Recent examination of the solar radiation record across the globe suggests that between the late 1950s and early 1990s the amount of solar radiation reaching the surface may have decreased substantially—possibly by as much as 10 percent. A likely contributor was an increase in aerosols emitted into the atmosphere by human activity. These aerosols are thought to have had a direct effect by absorbing and scattering solar radiation back to space, as well as an indirect effect by changing cloud properties. It now appears that environmental regulations instituted in many countries have led to a reduction in human-introduced aerosols, which has led to a recovery in the amount of solar radiation reaching the surface. While having many positive effects, the increase in incoming radiation could also contribute to future global warming.

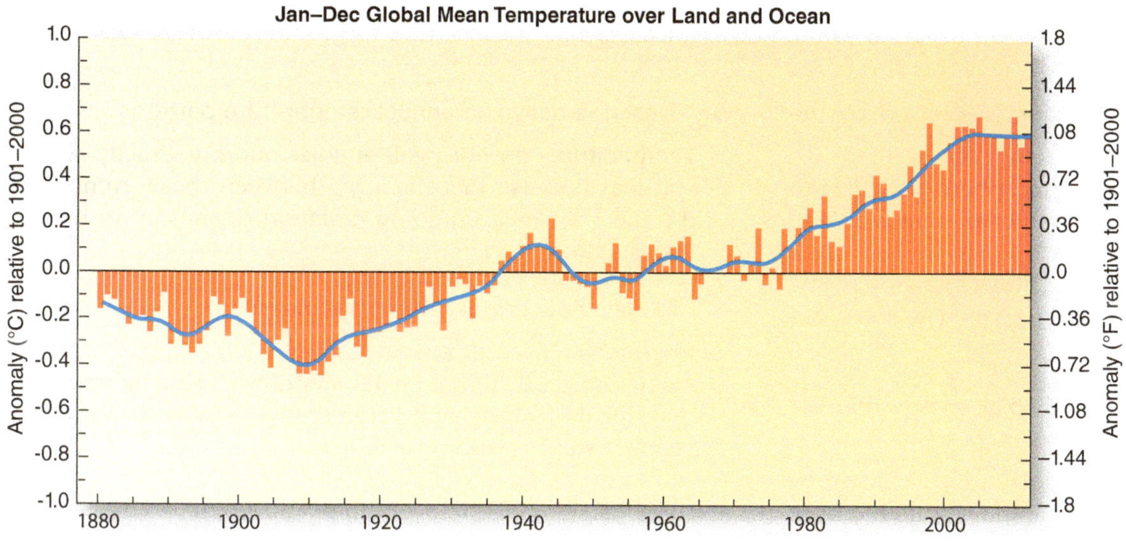

Animation
Global Warming
http://goo.gl/E1JEYu

Video
Global Warming Predictions
http://goo.gl/Z7toq9

▲ **FIGURE 3–32 Global Temperatures, 1800–2012.** The graph plots temperatures as departures from the 1901–2000 average.

the period. Every year from 1880 through the late 1930s experienced below-normal temperatures. Following the late 1930s, most years had above-normal temperatures, with marked warming beginning in 1976. Over the entire period global surface temperatures increased by about 0.7 °C (1.26 °F) per century, but the rate of temperature increase since about 1976 has been about three times as great as that during the entire 20th century.

The last decade of the 20th century and the early years of the 21st century were remarkably warm, with 10 of the 11 warmest years on record as of 2013 having occurred between 2002 and 2013.

Changes in the Occurrences of Extreme Heat and Cold

An important issue related to temperature change involves the frequency, intensity, and duration of extremely hot or cold events. The evidence suggests that extreme events may play a significant part in the observed warming. In other words, episodes of uncommonly high temperatures may account for a greater part of the recent warming than increases in more frequent day-to-day temperatures.

MM **MapMaster** North America Physical Environment: Temperature Anomalies El Niño/La Niña Years

But because rare events are by definition infrequent, such a trend in unusual events may be hard to verify statistically. For example, what if a heat wave of a particular intensity occurs five times over a given 25-year period but then occurs seven times over the next 25-year period? Would that signify a trend or might it just be the result of random variation? Despite this difficulty, numerous studies have looked at the change in the frequency of extreme heat and cold events in different parts of the world. These studies have shown no consistent pattern that applies for all regions, though, generally speaking, during the past half century there has been a greater decrease in extremely low temperatures than there has been an increase in occurrence of severe heat. Since 1979, however, the increase in extreme maximum temperatures has outpaced the increase in anomalous minimum temperatures.

MM **MapMaster** World Physical Envionment: Global Surface Warming, Worst Case Projections

An increase in the severity and frequency of extreme heat can have a serious impact on human mortality. (Recall the discussion of recent severe heat outbreaks and the fact that severe heat annually kills more people in the United States than does lightning, tornadoes, or hurricanes.) To make matters worse, the increase in nighttime temperature has been greater than the increase in daytime temperatures, and these high temperatures have been coupled with increasing humidity levels that create even higher "apparent temperatures" (discussed in Chapter 5). This is important because experts believe that the most dangerous health threat occurs not from very high daytime temperatures but from the persistence of high apparent temperatures for several days without intervening cool nights. On the other hand, the reduced frequency and intensity of major outbreaks of cold conditions can reduce loss of life due to that type of weather. Studies suggest that with continued global warming, some countries will see a decrease in cold-related deaths that will more than offset warm-event deaths, whereas in other countries the reverse will be true.

CHECKPOINT

3.23 Describe some of the important temperature trends that have been observed globally.

3.24 To what extent, if any, does the IPCC believe that humans are responsible for observed warming?

Summary

3.1 Explain how incoming solar radiation is affected by the atmosphere.

- Solar energy passing through the atmosphere is subject to absorption and scattering, while a certain amount is able to be transmitted to the surface.
- Scattering reduces the intensity of the direct radiation reaching the surface but at the same time creates blue skies, red sunsets, and white clouds.

3.2 State the proportions of incoming solar energy that are reflected, scattered, or absorbed.

- About a quarter of incoming solar radiation is absorbed by the atmosphere with about an equal amount scattered back to space. Thus, about half the energy reaches the surface, with some of that likewise reflected back to space.

3.3 Explain how energy is transferred between the surface and the atmosphere.

- Having absorbed solar radiation, the surface and the atmosphere radiate longwave energy back and forth with a substantial amount escaping back to space.
- The end result of the transfer of solar and longwave radiation is that the surface has a net radiation surplus while the atmosphere has a deficit of equal value. The radiation deficit of the atmosphere and the surplus at the surface are offset by the convection of sensible and latent heat.

3.4 Explain the role of the greenhouse effect in warming the atmosphere.

- The greenhouse effect describes the manner in which shortwave radiation enters the Earth atmosphere–surface system, but is retained by the absorption of longwave energy by greenhouse gases.

3.5 Describe the global distribution of temperatures on Earth's surface.

- In the most general sense, temperatures tend to decrease with latitude outside of the tropics.
- Land regions tend to be warmer in the summer than adjacent oceanic areas, and colder in the winter.
- The hemisphere experiencing winter exhibits stronger latitudinal temperature gradients than that having summer.

3.6 List the factors that influence temperatures on Earth's surface.

- Latitude, altitude/elevation, atmospheric circulation patterns, continentality, the effect of warm versus cold ocean currents, and local factors all influence temperatures.

3.7 Describe daily and annual temperature patterns.

- Temperatures do not peak at solar noon or exactly on the solstices because there is a lag between those events and the time at which outgoing radiation catches up to incoming radiation.

3.8 Explain how temperatures are measured.

- Thermometers can provide simple readings of current, maximum, and minimum temperatures using mercury or dyed alcohol. More sophisticated devices are used for major weather stations and research sites.

3.9 Define the concept of temperature means and ranges.

- Daily temperature averages are determined by adding the maximum and minimum temperatures and dividing by two. Monthly and annual values are obtained by the traditional method of calculating mean values.
- Daily ranges are obtained by finding the differences between maximum and minimum temperatures, while annual ranges are the differences between the warmest and coldest monthly averages.

3.10 Define some useful indexes of temperature.

- Wind chill temperatures, heating and cooling degree-days, and growing degree-days are useful to people in planning for outdoor activities, seasonal energy use, and agriculture.

3.11 Make use of thermodynamic diagrams and understand vertical temperature profiles.

- Thermodynamic diagrams provide valuable information for vertical transects of the atmosphere. This chapter introduced the simplest of their applications: plotting temperature.

3.12 Summarize observations of climate change/global warming.

- Earth's climate has substantially warmed during the past century or so, especially since the middle 20th century. This increase is almost certainly due to the activities of humans, such as burning fossil fuels, which increase atmospheric carbon dioxide, a greenhouse gas.

Key Terms

absorption *p. 82*
advection *p. 93*
albedo *p. 83*
atmospheric window *p. 89*
bimetallic strip *p. 103*
continentality *p. 98*
cooling degree-days *p. 107*
diffuse radiation *p. 83*
diffuse reflection or **scattering** *p. 83*
direct radiation *p. 83*
energy balance *p. 82*
extraterrestrial radiation *p. 86*
forced convection or **mechanical turbulence** *p. 90*
free convection *p. 90*
greenhouse effect *p. 94*
growing degree-days *p. 107*
heating degree-days *p. 107*
Intergovernmental Panel on Climate Change (IPCC) *p. 110*
inversion *p. 98*
isotherm *p. 97*
laminar boundary layer *p. 90*
latent heat *p. 91*
maximum thermometer *p. 102*
Mie scattering *p. 85*
minimum thermometer *p. 102*
net all wave radiation *p. 90*
net longwave radiation *p. 88*
net radiation *p. 90*
nonselective scattering *p. 85*
planetary albedo *p. 87*
radiosonde *p. 107*
Rayleigh scattering *p. 83*
reflection *p. 83*
resistance thermometer *p. 103*
sensible heat *p. 91*
specific heat *p. 91*
specular reflection *p. 83*
Stuve diagram *p. 107*
thermodynamic diagram *p. 107*
thermograph *p. 103*
wind chill temperature index *p. 106*

Review Questions

1. Explain how the absorption and scattering of radiation in the atmosphere affect the receipt of solar radiation at the surface.
2. How do specular reflection and diffuse reflection differ?
3. Which two gases are the most effective at absorbing longwave radiation?
4. What does the term *albedo* mean?
5. What characteristics of Rayleigh scattering cause it to create a blue sky?
6. Why are overcast days typically gray?
7. What properties of Mie scattering distinguish it from Rayleigh scattering?
8. What is the numerical value of Earth's planetary albedo?
9. Which type of scattering accounts for the majority of Earth's planetary albedo?
10. Describe quantitatively how much solar radiation is absorbed and reflected by Earth's atmosphere and surface.
11. What is the atmospheric window?
12. Why is it incorrect to state that longwave radiation bounces back and forth between clouds and the surface?
13. Explain why the incoming and outgoing radiation for the Earth system (radiation entering and leaving the top of the atmosphere) must equal each other.
14. What is the difference between free convection and forced convection?
15. How do conduction and convection work together to transfer heat upward?
16. Describe sensible and latent heat.
17. How do the net input and output of radiation vary with latitude?
18. Which two processes transport energy from zones of radiation surplus to zones of radiation deficit?
19. Why does the term *greenhouse effect* inaccurately describe how the atmosphere is heated?
20. Discuss how geographic factors such as latitude and altitude influence the distribution of temperature across Earth's surface.
21. How do the various instruments that are used to observe temperature work?
22. Explain how daily, monthly, and annual mean temperatures are computed from observed temperatures. Discuss some of the factors that can bias resulting values.
23. Describe the horizontal and vertical scales on Stuve diagrams.

Critical Thinking

1. Shorter wavelength radiation is more subject to Rayleigh scattering than is longer wave radiation. Explain how this might affect the value of facing directly toward the Sun to pursue an even suntan.
2. Even on cloudy days, excessive exposure can lead to a danger of sunburn. What does this imply about the effect of clouds on ultraviolet radiation?
3. Desert areas are often photographed with spectacular sunsets. Can you think of any reasons why they may be more inclined to have particularly flashy skies at dusk?
4. Our eyes are sensitive only to wavelengths between 0.4 and 0.7 micrometers. Would the sky appear any different on clear days if our eyes could also perceive wavelengths between 0.2 and 0.4 micrometers? How would the ratio of perceived diffuse to direct radiation change?
5. Would you expect both the Northern and Southern Hemispheres to have the same average albedo? What factors might cause the two hemispheres to reflect different percentages of insolation back to space?
6. Net radiation values in the summer may be higher in forested areas than desert areas, despite the higher temperatures in the desert. How can this be?
7. Snow often melts more rapidly in wooded areas immediately adjacent to trees than in nearby openings. What type of energy transfer processes could lead to this effect?
8. The ratio of sensible to latent heat loss from a surface is called the *Bowen ratio*. How do you suppose Bowen ratios might differ among desert, wooded, and urban landscapes?
9. Clouds can reduce the amount of insolation reaching Earth's surface, but they can also reduce the amount of longwave radiation from the surface that escapes to space. How might this affect maximum and minimum temperatures? Do you think all types of clouds would produce similar effects?
10. Figure 3–16 shows the mean distribution of ocean currents. It is believed that climate change, through a variety of mechanisms, could cause a shift in the position of some currents. Can you identify any land regions whose climate could be vulnerable to shifts in nearby currents? Are there any localities whose climates could cool even if the average global temperature were to warm?
11. An orchard farmer hears a weather forecast for overnight low temperatures to hover just above the freezing point of 0 °C (32 °F), but with wind chill temperatures expected to drop significantly lower. Will the wind chill increase the possibility of frost damage? Why or why not?
12. Instrument shelters protect thermometers from the heating effect of absorbed sunlight. Is it also true that shelters protect the thermometers from the chilling effect of wind?

Problems & Exercises

1. Go to **http://rredc.nrel.gov/solar/old_data/nsrdb/1961-1990/redbook/atlas/**. Select mean solar radiation as data type and horizontal flat plate for instrument orientation. Examine the maps of average solar radiation for January, April, July, and October. Describe each of the patterns. How much do you think Earth–Sun relationships affect the distribution relative to the effect of cloud cover and other weather elements?
2. The latent heat for water is 2,500,000 joules per kilogram (J/kg), and the specific heat of water is 4190 J/kg per degree Celsius of temperature change. Assume that a kilogram of water begins with a temperature of 20 °C (68 °F). Compare the amount of energy needed to bring the water to the boiling point to the amount of energy needed to evaporate the same amount of water.
3. Go to **http://earthobservatory.nasa.gov/Observatory/Datasets/netflux.erbe.html** and observe the seasonal shift in net radiation for the surface and atmosphere. What is the most obvious pattern? Are there significant differences between land bodies and adjacent oceans?
4. View the maps of mean minimum temperatures in January and mean maximum temperatures in July at **www.climatesource.com/map_gallery.html**. Assess the relative importance of latitude, altitude/elevation, continentality, ocean current, and local conditions to these distributi ons.

Visual Analysis

This map from the IPCC depicts the observed change in temperature across the globe between 1901 and 2012.

3.1. Which regions have generally experienced the greatest warming?

3.2. How does the warming over land compare to the warming over the oceans? Explain your answer.

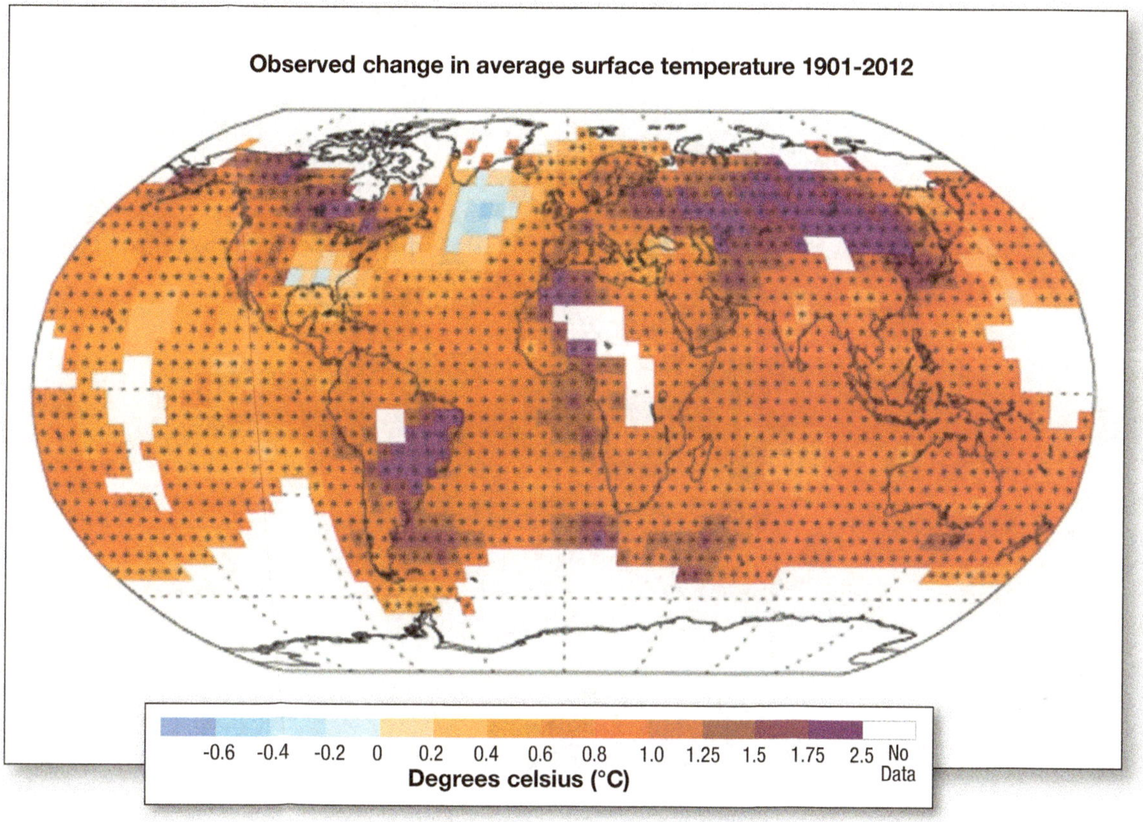

CHAPTER

4 Atmospheric Pressure and Wind

Dust storms form when fine particles are scoured from the surface and lifted high into the atmosphere. The storm pictured at left was particularly strong, forming a vast pool of dust visible on satellite images (inset). Storms like this provide evidence of the atmosphere's motion on huge scales. On much smaller scales, winds are part of our everyday experience. For example, we take for granted that light winds are far more likely near sunrise than in the midafternoon, and we plan activities accordingly. Winds might be familiar and obvious, but their origin lies in a less tangible aspect of the atmosphere: pressure. Indeed, as concerned as we sometimes are with wind conditions, few of us pay much attention to atmospheric pressure. After all, how many times have you canceled a picnic because the pressure was too low? Or how many people do you know who have special clothes they wear only on days of high pressure?

Building on Chapter 1, this chapter provides in-depth coverage of atmospheric pressure and its vertical and horizontal distributions. We discuss the relationship between pressure and other atmospheric variables and the processes that create horizontal and vertical variations in pressure. With this foundation, we can go on to discuss storm patterns in later chapters.

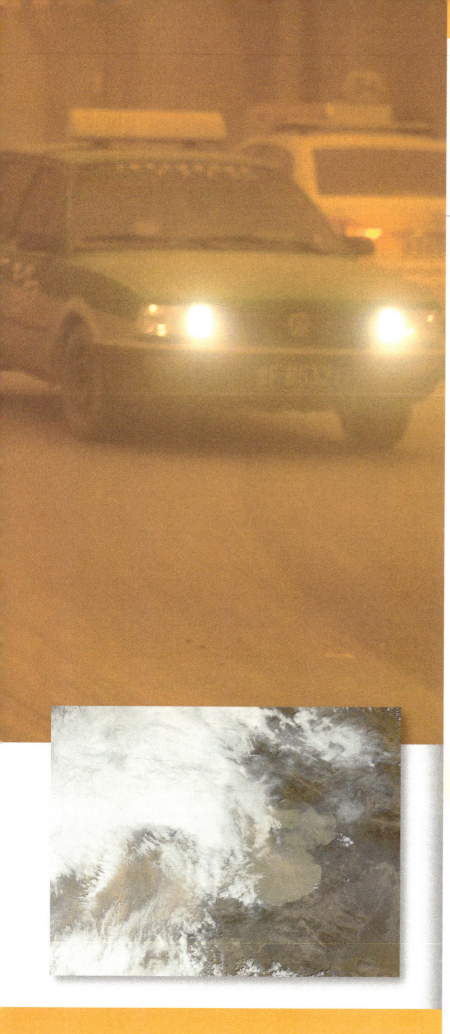

◀ *Dunhuang, China Dust Storm, April 23, 2014.*
Dust storms form frequently in northern China in the spring powerful fronts move through desert landscapes, and make life miserable for inhabitants. This one was massive in both area and strength. The resulting wall of dust is clearly visible on the inset satellite image as a sharp line dividing relatively clear air to the east from choking conditions within the storm.

Learning Outcomes

After reading this chapter, you should be able to:

4.1 Explain the concept of air pressure.

4.2 Describe how pressure changes vertically and horizontally in the atmosphere.

4.3 Apply the equation of state to calculate air density.

4.4 Identify devices used to measure air pressure.

4.5 Describe the distribution of air pressure across the globe, at sea level, and in the upper atmosphere.

4.6 Describe how pressure gradients in the upper atmosphere arise and how they are mapped.

4.7 Explain the factors that affect wind speed and direction.

4.8 Describe winds in relation to pressure gradients in the upper atmosphere and near the surface.

4.9 Explain the forces that produce anticyclones, cyclones, troughs, and ridges.

4.10 Describe how meteorologists measure wind.

The Concept of Pressure

4.1 Explain the concept of air pressure.

The atmosphere contains a tremendous number of gas molecules being pulled toward Earth by the force of gravity. These molecules exert a force on all surfaces with which they come in contact, and the amount of that force exerted per unit of surface area is **pressure** (see *Box 4–1, Physical Principles: Velocity, Acceleration, Force, and Pressure*). Of course, the concept of pressure is not confined to meteorology—it is fundamental to all the physical sciences. In most physical science applications, the standard unit of pressure is the **pascal (Pa)**, but in the United States meteorologists use the **millibar (mb)**, which equals 100 Pa. Canadian meteorologists use yet another unit, the **kilopascal (kPa)**, equal to 1000 Pa, or 10 mb. For purposes of comparison, air pressure at sea level is typically roughly 1000 mb (100 kPa)—or more precisely, the global average is 1013.2 mb.

Video
Growth of Wind Power in the US
http://goo.gl/FhzBgs

To understand the characteristics of pressure, refer to Figure 4–1, which depicts a sealed container of air. The enclosed air molecules move about continually and exert a pressure on the interior walls of the container (a). The pressure of the air is proportional to the rate of collisions between the molecules and walls. We can increase the pressure in two ways. The first way is by increasing the density of the air either by pumping more air into the container or by decreasing the volume of the container (b). The second is by increasing the air temperature, in which case the molecules exert higher

4–1 PHYSICAL PRINCIPLES

Velocity, Acceleration, Force, and Pressure

It is quite common in everyday conversation to hear the terms *force* and *pressure* used interchangeably, just as *velocity* and *speed* are often considered synonymous. In the language of science, however, intermixing these terms can lead to great confusion. Let us look briefly at how they differ and their connection to a related concept, acceleration.

Velocity, Speed, and Acceleration

Any object that moves has a particular **speed**, defined as the distance traveled per unit of time. Speed is related to, but not the same as, velocity. **Velocity** incorporates direction as well as speed. Think, for example, of two cars traveling at 20 meters per second (44 mph) but moving in opposite directions. Though they have the same speed, their velocities are not equal because of their different directions. This distinction is crucial for understanding our next quantity, **acceleration**, the change in velocity (not speed) with respect to time.

Because velocity includes both speed and direction, a change in either speed or direction is an acceleration. Consider a car that at one moment in time travels at 20 m/sec; one second later, the same car has a speed of 19 m/sec; one second later, the speed is 18 m/sec, and so on. As each second goes by, the car's speed decreases 1 m/sec (note that acceleration can be either positive or—as in this example—negative). An acceleration can also occur as a change in direction with respect to time, even for an object whose speed does not change. A car traveling at a constant speed but gradually turning undergoes an acceleration, just like a car whose speed is changing.

In meteorology there is one particular acceleration of utmost importance—**gravity** (*g*). This acceleration, 9.8 m/sec^2 (32.1 ft/sec^2), is nearly constant across the globe. It describes the acceleration an object would experience if gravity were the only force affecting its movement. There is a slight decrease in *g* from equator to pole, and also a very small difference in *g* from the surface to the upper atmosphere. For most applications, however, these variations in *g* are so slight they can be ignored.

Force and Pressure

One of the most important tenets of physical science is Newton's Second Law, which relates the concept of **force** (denoted *F*) to mass (*m*) and acceleration (*a*). Specifically, Newton's Second Law tells us that the acceleration of an object is proportional to the force acting on it and inversely proportional to its mass. Symbolically, this is expressed as

$$a = \frac{F}{m} \text{ or}$$

$$F = ma \qquad \text{Newton's Second Law}$$

Imagine that a fully loaded 18-wheeler truck is stopped at a traffic light next to a bicycle. Suppose that when the light turns green, both begin to accelerate at the same rate and therefore remain next to each other. For them to accelerate at the same rate, it is easy to see that the much more massive truck requires a larger force (and more powerful "engine"). Likewise, if two bodies with equal mass are subjected to different forces, the one subjected to the greater force will undergo a greater acceleration.

Keep in mind that *F* in the equation above is the net force acting on the object. If various forces are acting simultaneously, they must all be considered together to determine the acceleration; we must account for both the magnitude and direction of each. As we will see with regard to falling raindrops (Chapter 7), forces acting in opposite directions reduce the net force and resulting acceleration, sometimes to zero.

Let's apply Newton's Second Law to our atmosphere. The mass of the atmosphere is 5.14×10^{18} kg. (To get an idea of what 5.14×10^{18} kg weighs, picture a million boxcars, each containing a billion elephants.) Multiplying the mass of the atmosphere by the acceleration of gravity, we determine that the force exerted on the atmosphere is about 5.0×10^{19} newtons (a newton [N] is the unit of force it takes to accelerate 1 kg one meter per second every second).

Force divided by the area on which it is exerted equals pressure. So dividing this force of 5.0×10^{19} newtons by the surface area of Earth gives us the average force per unit area, or average surface pressure of about 10.132 newtons per square centimeter.

pressure because they are moving more rapidly (c). Thus, pressure reflects both the density and temperature of a gas.

If the air in the container is a mixture of gases (as it is in the atmosphere), each gas exerts its own specific amount of pressure, referred to as its *partial pressure*. The total pressure exerted is equal to the sum of the partial pressures. This relationship is known as **Dalton's law**.

On Earth the container in Figure 4–1 is surrounded by the atmosphere, which exerts pressure on the exterior walls. Consider what would happen if we were to remove the lid of the container or puncture a hole in its side. If the pressure *outside* the container were greater than that within, the outside air would be forced inward until the pressure equalized. (The force of this equalization is what causes the "whoosh" when you open a vacuum-packed container.) On the other hand, if the pressure were greater *inside* the container, air would be forced outward until the internal pressure decreased to match the surrounding air. In either case, within moments the air pressure exerted on the outside of the container would become exactly equal to that on the inside. This example introduces us to another characteristic of air: It constantly moves to establish an equilibrium between areas of high and low pressure.

DID YOU KNOW?

Many people know that air molecules are constantly moving about randomly, but people are surprised to learn just how fast they travel. On a day with a comfortable temperature, molecules near Earth's surface move at a rate of several hundred kilometers per second—comparable to the speed of a rifle bullet.

This is equivalent to 1013.2 mb, or about 14.7 pounds per square inch.

Having made the distinction between force and pressure, we now should address the question of how this distinction applies to the atmosphere. The answer is that despite the nearly constant total force of the atmosphere, its gases are not uniformly distributed across the planet. Consider a particular area at Earth's surface with an imaginary column extending upward to the top of the atmosphere, as shown in the top right of Figure 4–1–1. Greater surface pressures exist at the bases of atmospheric columns that contain a greater number of molecules, and lower surface pressures are found where less air occupies the column. Just how these differences in pressure arise is considered later in this chapter; for the time being, the important point is that surface pressure reflects the mass of atmosphere within the column, as shown in Figure 4–1–1.

4–1–1 What is the difference between force and pressure?

4–1–2 If the acceleration of gravity were to double with no change in the atmosphere's mass, what would happen to air pressure?

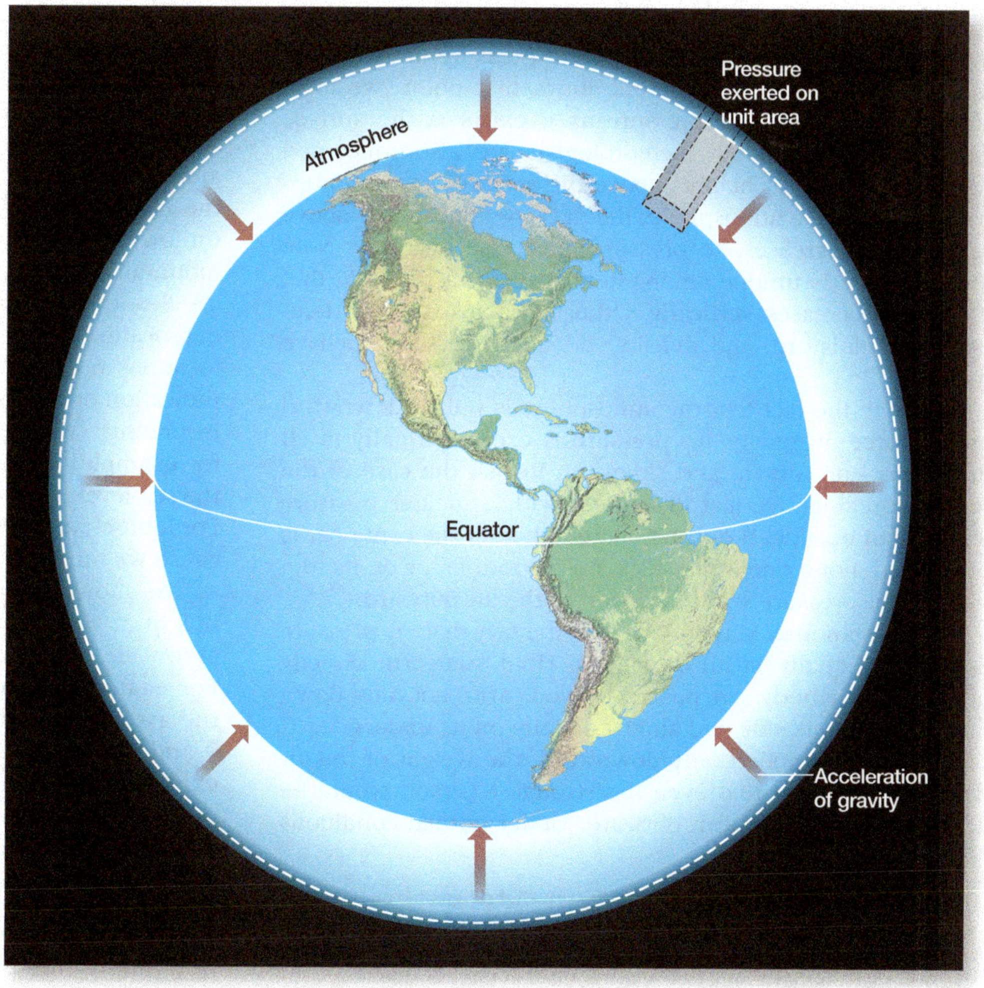

▸ **FIGURE 4–1–1 Gravity, Mass, and Pressure.** The downward force of the atmosphere is equal to the mass of the entire atmosphere times the acceleration of gravity. Because the amount of mass and the acceleration of gravity are constant through time, the total force of the entire atmosphere does not change. Pressure is defined as the amount of force exerted per unit of area. Thus, the shaded area in the figure experiences a certain amount of pressure. Pressure varies because the mass of overlying air varies from place to place and time to time.

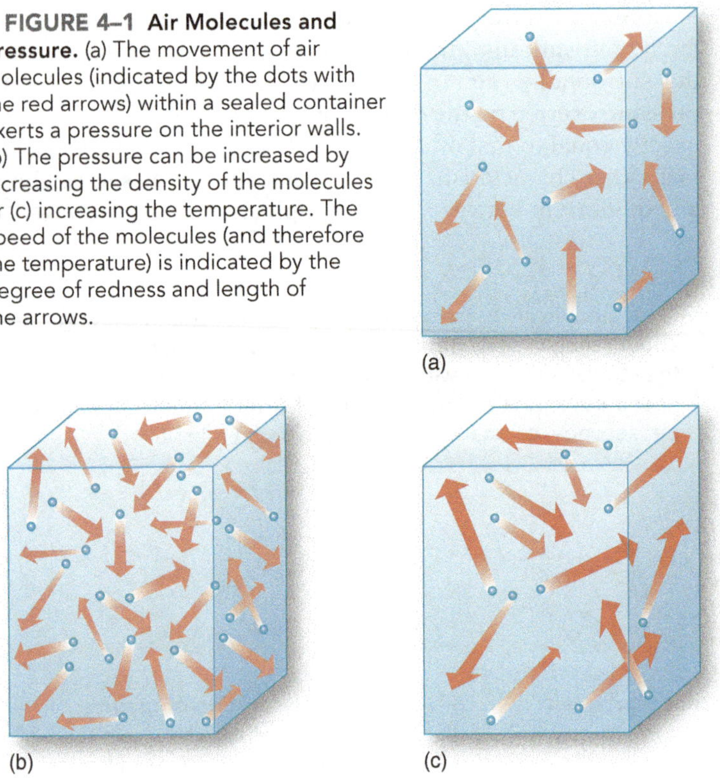

▸ **FIGURE 4–1 Air Molecules and Pressure.** (a) The movement of air molecules (indicated by the dots with the red arrows) within a sealed container exerts a pressure on the interior walls. (b) The pressure can be increased by increasing the density of the molecules or (c) increasing the temperature. The speed of the molecules (and therefore the temperature) is indicated by the degree of redness and length of the arrows.

As described in Chapter 1, what we experience as atmospheric pressure is in fact the mass of the air above us being pulled downward by gravity. The pressure at any point reflects the mass of atmosphere above that point. We most commonly measure air pressure as it exists at the surface (**surface pressure**), but meteorologists are also concerned with air pressure at heights above the surface. As we go upward through the atmosphere, the mass of atmosphere above necessarily decreases, so pressure must also decrease. Note that pressure is unique among atmospheric variables in that it always decreases vertically. Other variables (such as temperature, moisture, and density) do not necessarily behave this way.

Despite the fact that the atmosphere is pulled downward by the force of gravity, air pressure is exerted equally in all directions—up, down, and sideways. Revisit the case of the sealed container we just described, with a greater pressure inside than outside. It doesn't matter whether the container is punctured along one of its sides, on its bottom, or on its top; the greater pressure within still forces the air outward.

Here is another way to visualize the fact that air pressure is exerted equally in all directions. Hold your arm directly out from your body. Air pushes on your arm, not only down but also along its side and almost equally on its underside.[1] If pressure were applied only downward, the weight of the air would be so great that even the strongest person would be unable to extend her arm outward. Under normal conditions at sea level, the force would equal about 14.7 pounds on every square inch—quite a load for even a strong arm!

CHECKPOINT

4.1 In what way is pressure related to the movement of molecules in a gas?

4.2 What units are used in measuring pressure?

Vertical and Horizontal Changes in Pressure

4.2 Describe how pressure changes vertically and horizontally in the atmosphere.

As mentioned in Chapter 1, we need to measure and compare the differences in pressure that arise in different locations, because these differences produce horizontal movements of air. The job is complicated by the fact that elevation varies from place to place. (Recall that high elevations have lower pressure simply because there is less overlying air.) If we used just surface measurements for comparisons, it would be impossible to separate the effects of elevation from the true pressure differences that lead to wind. To overcome this problem, meteorologists use the concept of sea-level pressure.

Surface pressure is the pressure actually observed at a particular location, whereas **sea-level pressure** is the pressure that would exist if the observation point were at sea level. Because most land surfaces are above sea level, surface pressure is almost always the lower of the two. Compare, for example, the surface pressures at the tall mountain peak and the nearby valley in Figure 4–2. Although the atmosphere is uniformly distributed over the area, surface pressure at the mountain location is considerably less than at the valley.

Sea-level pressure allows us to compare pressure at different locations, taking into account differences in elevation. For locations not very much above sea level, we can get a good indication of sea-level pressure by assuming a uniform change in pressure with elevation. At an elevation of 150 m (500 ft), for example, we add 14 mb to the surface pressure to obtain the sea-level pressure (roughly a 1 mb increase for every 10 m). For high-elevation sites, however, this method is unreliable because we must account for compressibility of the atmosphere.

As shown in Figure 4–3, pressure does not decrease with height at a uniform rate. Instead, it decreases most rapidly at low elevations and gradually tapers off at greater heights. For example, from sea level to 1 km (0.6 mi) in elevation, the average pressure decreases about 100 mb, but between 9 and 10 km it drops only half as much. This nonlinearity exists because, as we know, air is compressible. With atmospheric mass packed more densely at low levels, a small elevation change at low levels takes you through a large number of molecules, resulting in a large pressure drop.

Though surface pressure also varies from place to place, horizontal pressure differences are very small compared to vertical differences. For example, a sea-level pressure of 1050 mb

[1]The pressure on the bottom of your arm is very slightly greater than at the top. This is because pressure always decreases with altitude, and the top of your arm is a few centimeters higher than the bottom. This difference is extremely small, however, and can be disregarded in this example.

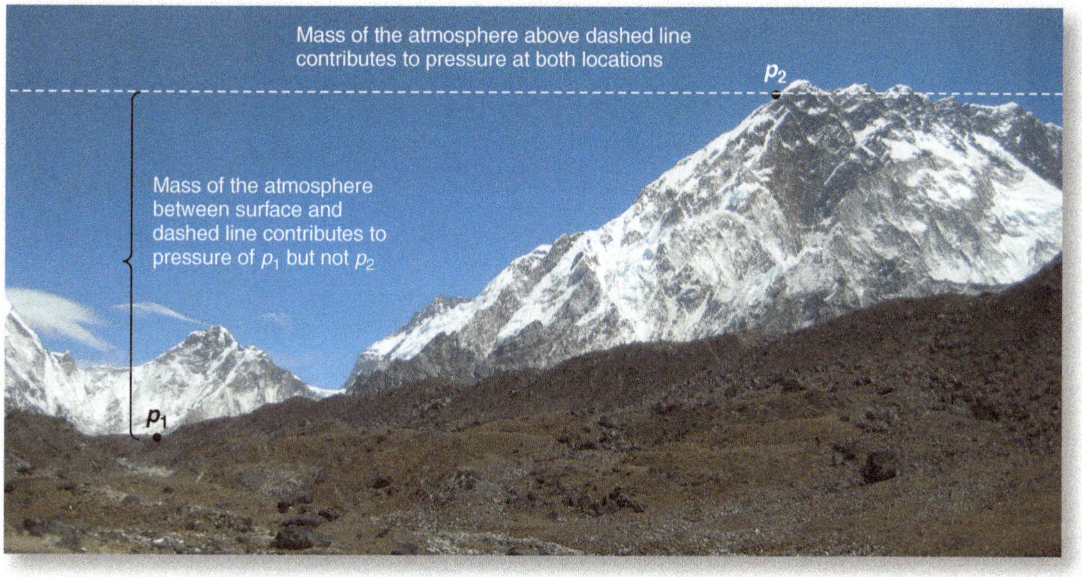

◀ **FIGURE 4–2 Atmospheric Mass and Pressure.** Because atmospheric pressure is a response to the weight of the overlying atmosphere, it always decreases with elevation. The pressure at the top of the mountain, p_1, is less than that at the base of the mountain, p_2, because of the greater amount of overlying air.

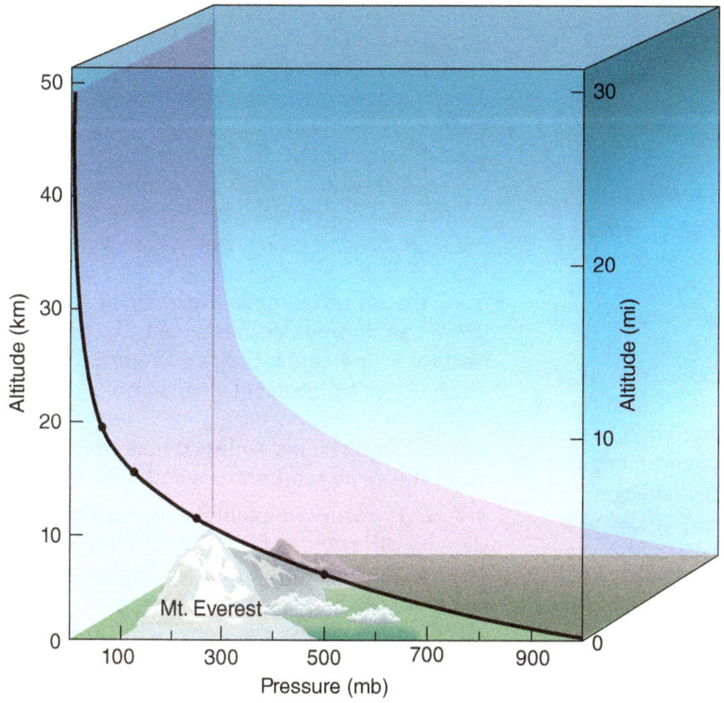

▲ **FIGURE 4–3 Vertical Variation in Pressure.** Pressure decreases with altitude by about half for each 5.5 km (3.3 mi).

is considered very high, yet it is only 4 percent greater than the global average sea-level pressure. Moreover, the difference between the highest and lowest sea-level pressures over North America on a given day might amount to only about 25 mb—less than a 2.5 percent difference—and even this small percentage difference in pressure would normally be realized over a distance of many hundreds of kilometers. In contrast, we need only go to the top of a modest hill or tall building to find an equivalent pressure change.

CHECKPOINT

4.3 Why does air pressure decrease as you go upward through the atmosphere?

4.4 How is surface pressure converted to sea-level pressure?

The Equation of State

4.3 Apply the equation of state to calculate air density.

Everyday experience indicates that gases tend to expand when heated and become denser when cooled. This suggests that temperature, density, and pressure are related to one another. As a matter of fact, their relationship is quite simple. It is described by the **equation of state** (also called the **ideal gas law**),

$$p = \rho RT \quad \text{Ideal gas law}$$

in which p is pressure expressed in pascals, ρ (the Greek letter rho) is density in kilograms per cubic meter, R is a constant equal to 287 joules per kilogram per kelvin, and T is temperature (in kelvins). To put the equation in words, it tells us that if the air density increases while *temperature is held constant*, the pressure will increase. Similarly, at constant density, an increase in temperature leads to an increase in pressure. (*Box 4–2, Physical Principles: Variations in Density*, presents more information on the equation of state.)

CHECKPOINT

4.5 What variables appear in the equation of state?

4.6 "Whenever the pressure is high at a location, the temperature is also high." Do you agree with this statement? Why or why not?

DID YOU KNOW?

Throughout the troposphere the atmosphere consists of an extremely high concentration of molecules. At sea level, there are some 1 million billion molecules of air occupying each cubic millimeter of volume. But despite this large number of molecules, the atmosphere consists primarily of empty space. The air molecules are in fact spaced far apart relative to their size so that about only one-tenth of 1 percent of the volume is occupied by atoms. In contrast, liquids, which are much denser than gases, have much smaller intermolecular distances; about 70 percent of their volume is occupied by matter.

4–2 PHYSICAL PRINCIPLES

Variations in Density

Perhaps you have wondered how much air weighs. The air around you has a particular density, and any volume of air contains a certain amount of mass. Changes in the density of air affect many everyday phenomena. For example, the density of the atmosphere influences how much lift a plane gets as it accelerates down a runway in preparation for takeoff. Likewise, automobile fuel injectors must account for variations in density to deliver the right mixture of gasoline and air into the car's engine. The density of air can even affect the amount of resistance the air exerts on a batted baseball, thereby influencing its distance traveled.

Effect of Temperature on Density

But are the variations in density really substantial? We can use the equation of state to see exactly how much variations in temperature affect the air's density. To do this, let's first rearrange the equation to the following:

$$\rho = p/RT \quad \text{Equation of state}$$

Let's now also compare the air density for two situations: a warm day with a temperature of 308 K (35 °C or 95 °F) and a cold one with a temperature of 278 K (5 °C or 41 °F). For consistency, we will assume that the pressure is 100,000 Pa (1000 mb; 100 kilopascals) in both instances.

Applying the equation of state for the warmer day, we find that the density of the air is

$$\rho = \frac{100{,}000 \text{ Pa}}{287 \text{ J kg}^{-1} \text{ K}^{-1} \times 308 \text{ K}} = 1.13 \text{ kg/m}^{3*}$$

When we lower the air temperature to 278 K (5 °C or 41 °F), the equation yields an air density of

$$\rho = \frac{100{,}000 \text{ Pa}}{287 \text{ J kg}^{-1} \text{ K}^{-1} \times 278 \text{ K}} = 1.25 \text{ kg/m}^3$$

This is nearly 11 percent greater than the density the air had on the warmer day—a nontrivial amount.

Effect of Humidity on Density

In addition to temperature and pressure, the humidity of the air exerts an influence (although only a very minor one) on density. Let's see how. Molecular oxygen (O_2) and nitrogen (N_2) make up most of the mass of the atmosphere and exist in a constant proportion. Other, lesser constituents of the atmosphere are present in different amounts at different places and times, and because each has its own unique molecular weight (an expression of the relative amount of mass for molecules), their relative abundance can slightly affect the density of the atmosphere. Among these gases, water vapor usually accounts for about 1 percent of the atmospheric mass. Intuitively, we might assume that a greater humidity would favor a denser atmosphere. Actually, just the opposite is true.

Compare the amount of mass contained in individual molecules of water vapor and of the most abundant atmospheric gases. The molecular weights of nitrogen and oxygen are 28.01 and 32.00, respectively, and the mean molecular weight of the dry atmosphere is 28.5. Water vapor, on the other hand, has a molecular weight of only 18.01. Thus, as the proportion of the air occupied by water vapor increases, an accompanying reduction in the mean molecular weight of the atmosphere must occur. All other things being equal, humid air is less dense than dry air.

Incorporating the effect of varying moisture content requires only a small modification to the equation of state. Calculations using the revised formula show that at 15 °C (59 °F), air density declines by only 0.6 percent for a 1 percent increase in water vapor (from dry air to 1 percent water vapor).

*To get the units to balance, you must reduce the units of pascals (Pa) and joules (J) to their fundamental dimensions. Thus, Pa = kg m^{-1} sec^{-2} and J = kg m^2 sec^{-2}.

4–2–1 Why is humid air less dense than dry air at the same temperature and pressure?

4–2–2 The worked example showed a density difference of 0.13 kg/m^3 over a temperature change of 20 °C. Suppose the example had compared temperatures 40 °C apart. Would the density difference be twice as large?

Measuring Pressure

4.4 Identify devices used to measure air pressure.

Any instrument that measures pressure is called a *barometer*. Two types of barometers are most common for routine observations: one consisting of a tube partially filled with mercury and another that uses collapsible chambers.

Mercury Barometers

The standard instrument for the measurement of pressure is the **mercury barometer** (Figure 4–4), invented by Evangelista Torricelli in 1643. It is a simple device made by filling a long tube with mercury and then inverting the tube so that the mercury spills into a reservoir. Although the tube is turned over completely, it does not empty. Instead, the air pushes downward on the pool of mercury and forces some of it up into the tube. The greater the air pressure, the higher the column of mercury.

Barometric pressure is often expressed as the height of the column of mercury in a barometer, which at sea level averages 76 cm (29.92 in.). This measure is inconsistent with the concept of pressure, however, because pressure does not have units of length. In other words, expressing barometric pressure in centimeters or inches is as incongruous as stating somebody's age as "30 miles per hour" or weight as "$1.99." The length measurements obtained from a barometer are a response to the atmospheric pressure but are not direct observations of pressure itself. Meteorologists

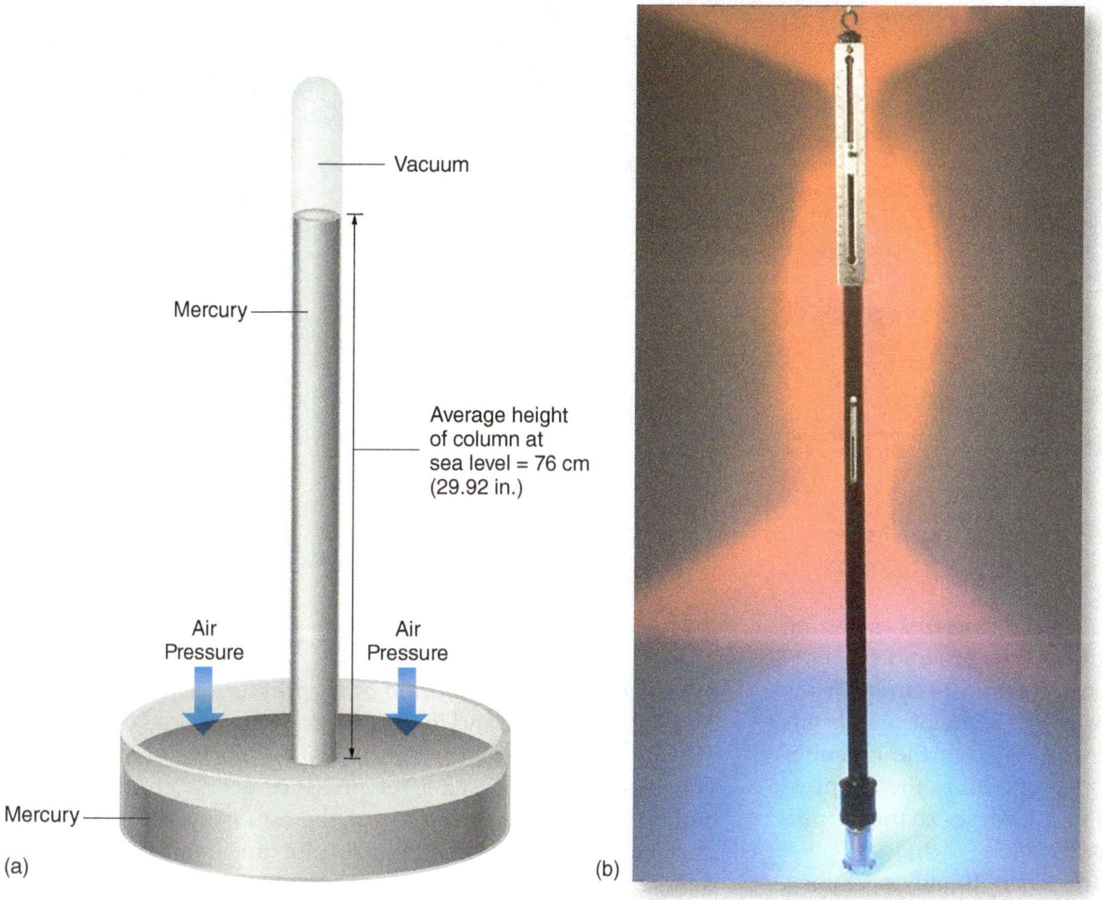

▲ **FIGURE 4–4 Mercury Barometers.** (a) A schematic showing how a mercury barometer works. (b) An actual mercury barometer.

prefer a unit that measures force per unit area, such as pounds per square inch or millibars. The simple conversion formulas for converting barometric heights to millibars are

$$1 \text{ centimeter} = 13.32 \text{ mb and}$$
$$1 \text{ inch} = 33.865 \text{ mb}$$

Mercury is an excellent fluid for use in a barometer because it is extremely heavy, with a density 13.6 times greater than that of water. This feature allows the instrument to be of a manageable size. Consider that if water were used instead of mercury, the column of water would need to be about 10 m (33 ft) tall to counterbalance the pressure of the atmosphere. On the other hand, although a three-story, water-filled barometer would be less than portable, it would make for very precise measurements because even small changes in pressure would translate into large height changes.

CORRECTIONS TO MERCURY BAROMETER READINGS

One of the most important tools of a meteorologist is the weather map, which among other things plots the distribution of air pressure across the surface. Before barometer data can be used on the map, however, three corrections must be made to compensate for local factors that affect the readings.

The first correction compensates for the influence of elevation (described earlier in this chapter). If surface pressure values were plotted on weather maps, they would give a false representation of the distribution of the atmosphere. This happens because high elevations have lower surface pressures than do low elevations, even if the sea-level pressures are the same. To standardize the observations, we must convert surface pressure readings to sea-level values. For a station situated 100 m (328 ft) above sea level, about a centimeter (0.4 in.) is added, corresponding to about 13 mb. At higher elevations, a much greater adjustment might be needed. At Denver, Colorado (the "Mile High City"), for example, the correction is about 16 cm (6.24 in.), or 213 mb.

The second correction deals with the similarity between a mercury barometer and a thermometer. Just as the mercury in a thermometer expands with increasing temperature, so does the mercury in a barometer. The expansion reduces the density of the fluid and requires that it attain a greater height to offset the pressure of the atmosphere. In other words, on a hot day the height of the mercury column is greater than on a cold day, even if the atmospheric pressure is the same. For this reason, mercury barometers always have a thermometer attached to determine the temperature of the instrument, and a correction table tells us what height the mercury column would be if the temperature were at the standard value of 0 °C (32 °F). At normal room temperature, this correction is small, requiring the subtraction of about 0.25 mm (0.01 in.).

The third correction accounts for the slightly greater acceleration of gravity at higher latitudes. To standardize the readings from all latitudes, we convert them to what they would be if the local gravity were equal to that at 45° north or south, or midway between the equator and poles. The latitudinal changes in gravity are small, however, and corrections are usually on the order of 0.25 mm (0.01 in.).

Aneroid Barometers

Mercury barometers are precise instruments, but they are also expensive and inconvenient to relocate. An alternative

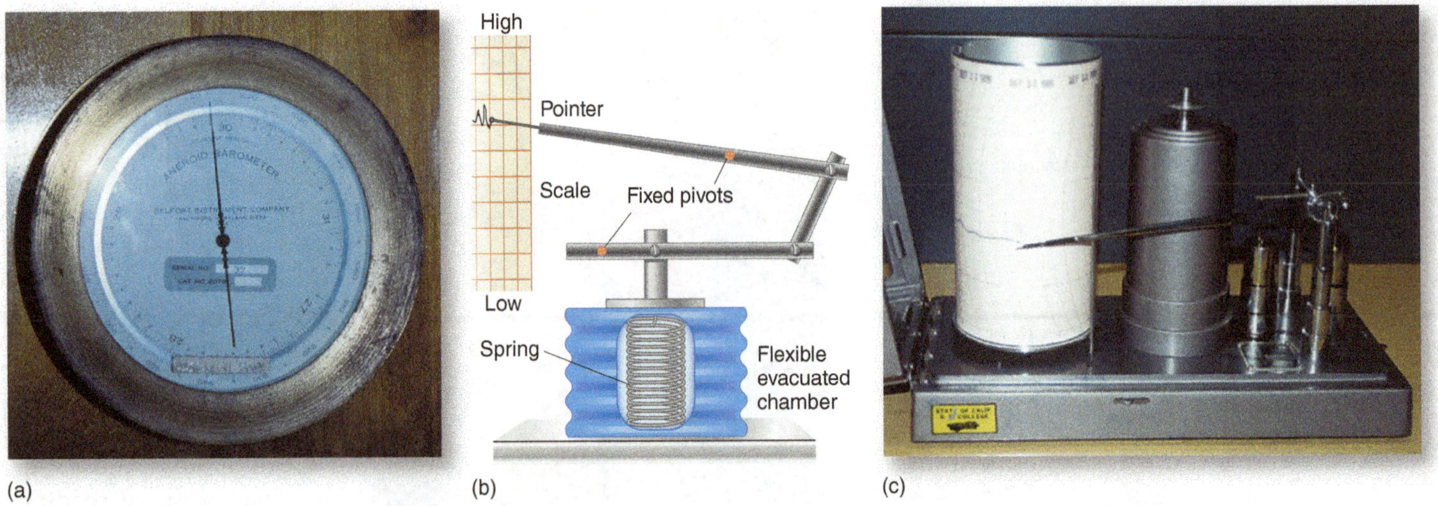

▲ **FIGURE 4–5 Aneroid Barometers.** An (a) aneroid barometer and (b) its workings. (c) A barograph.

instrument for measuring pressure is the **aneroid** (meaning "without liquid") **barometer** (Figure 4–5a). Aneroid barometers are relatively inexpensive and can be quite accurate. They contain a collapsible chamber from which some of the air has been removed (b). The atmosphere presses on the chamber and compresses it by an amount proportional to the air pressure. A pointing device connected to a lever mechanism indicates the air pressure on a graduated scale.

Aneroid barometers, which are often found in homes, must be calibrated when first installed. The user simply finds out the current sea-level pressure and sets the instrument by turning a small screw on the back of the casing. Because there is no expandable fluid in an aneroid barometer, the instrument requires no temperature correction. Furthermore, the effects of altitude and latitude are already accounted for when the instrument is first calibrated. Thus, once calibrated, an aneroid barometer gives the sea-level pressure without corrections or adjustments.

Sometimes it is useful to have a continual record of pressure through time. Aneroid devices that plot continuous values of pressure are called **barographs** (Figure 4–5c). A rotating drum (usually set to one rotation per week) turns a chart so that a pen traces a permanent record of the changing pressure.

CHECKPOINT

4.7 How does a mercury barometer work?

4.8 What corrections need to be made to readings from a mercury barometer and why are they necessary?

The Distribution of Pressure

4.5 Describe the distribution of air pressure across the globe, at sea level, and in the upper atmosphere.

The distribution of sea-level pressure across the globe is a highly variable characteristic of the atmosphere. To visualize this distribution, meteorologists plot lines called *isobars* on weather maps.

Each **isobar** is drawn so that it connects points having exactly the same sea-level pressure, and locations between any two isobars have pressures between those represented by the two lines. Isobars are drawn at intervals of 4 mb on U.S. surface weather maps, so the pressure difference between adjacent isobars is the same everywhere on the map. The advantage of this is that the distance between adjacent isobars provides information about how rapidly pressure changes from one place to another. In other words, the spacing of the isobars indicates the strength of the **pressure gradient**, or rate of change in pressure, in the same way that spacing of isotherms reveals temperature gradients. A dense clustering of isobars indicates a steep pressure gradient (a rapid change in pressure with distance), while widely spaced isobars indicate a weak gradient.

By way of example, Figure 4–6 maps the sea-level pressure distribution as it existed on May 13, 2013. On that day there was high pressure over much of the southeastern United States, and low-pressure cells in eastern and central Canada. Pressure gradients were strongest on the east side of the low in the province of Manitoba. Most other places, such as Illinois, had much weaker pressure gradients.

Pressure Gradients

Pressure gradients provide the impetus for the movement of air we call *wind*. Imagine two people pushing against each other. The person who exerts the greater force pushes the other one back, and the greater the difference in force applied, the faster the pushed person will move. The same

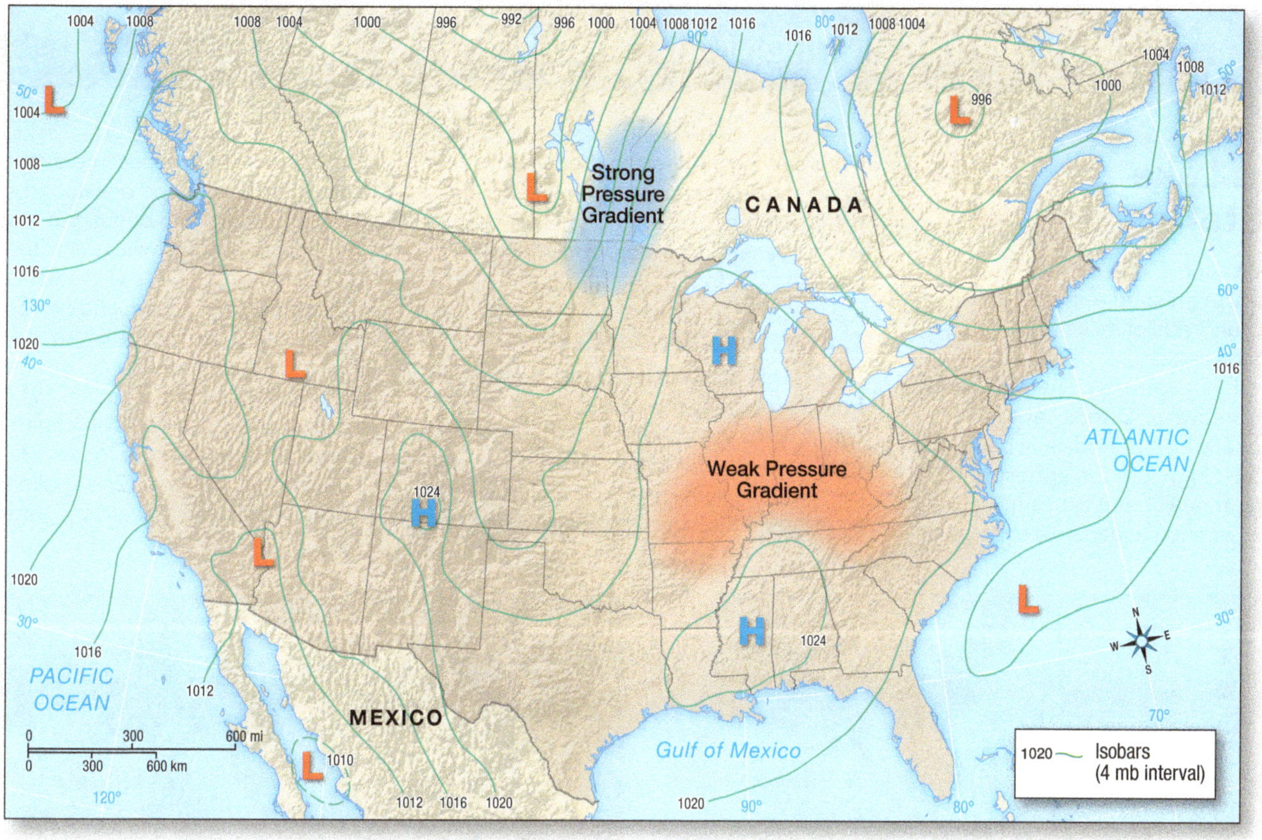

▲ **FIGURE 4–6 Map of Sea-Level Pressure.** A weather map showing the distribution of sea level air pressure on May 13, 2013. Shaded areas show places where isobars are close together (strong pressure gradients) and far apart (weak pressure gradients). The resulting wind speeds were respectively high and low in those locations.

concept applies to air. If the air over one region exerts a greater pressure than the air over an adjacent region, the higher-pressure air will spread out toward the zone of lower pressure as wind. The pressure gradient gives rise to a force called the **pressure gradient force** that sets the air in motion. For pressure gradients measured at constant altitude, we use the term *horizontal* pressure gradient force and call the resulting motion *wind.* Everything else being equal, the greater the pressure gradient force, the greater the wind speed. When we observe the work of wind in waving tree branches, creating ocean waves, or moving blocks of ice as in Figure 4–7, the origin lies in the pressure gradient force.

(a)

(b)

▲ **FIGURE 4–7 Lake Ice Pileups in Manitoba, Canada, May 13, 2013.** The pressure gradient force seen in Figure 4–6 provided enough energy to create walls of ice, destroying homes and cottages in the process.

4–3 FOCUS ON THE ENVIRONMENT AND SOCIETAL IMPACTS

Owens Lake Dust Storms

Strange as it seems, one of the largest single air pollution sources in the United States lies hundreds of miles from heavy industry, smokestacks, or roadways clogged with automobiles. The polluter is the nearly dry bed of Owens Lake in eastern California, which periodically belches clouds of dust high into the atmosphere that endanger health and cause millions of dollars of economic damage each year (Figure 4–3–1). Although they might seem to be natural occurrences, these storms are the direct result of actions taken a century ago to benefit growing populations in Southern California.

The Drying of Owens Lake

Owens Lake is an ancient body of water fed from the north by the Owens River. During glacial times thousands of years ago, the lake was so full of water that it spilled over high passes to the south, with some of its water reaching what is now Death Valley. As the climate warmed and became drier, the lake became shallower and developed into a terminal lake with no outlet. But it was still a substantial body of water until the early 20th century, complete with steamships transporting silver, lead, and zinc ore taken from mines in the mountains to the east.

All this changed with construction of the Los Angeles Aqueduct in 1913. This massive water project is an engineering marvel. Every second it moves about 10 cubic meters of water (2642 gallons) more than 200 miles southward by gravity alone,

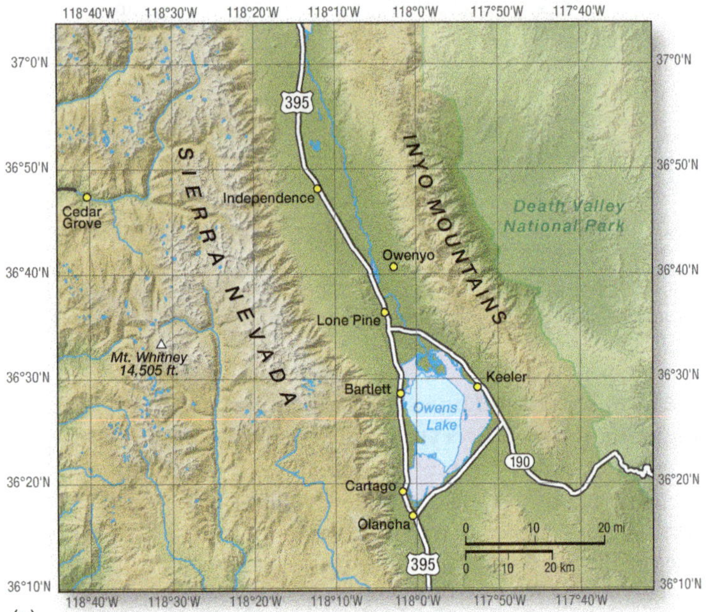

(a)

(b)

▲ **FIGURE 4–3–1 Owens Lake, California.** As seen in the (a) map and (b) satellite image Owens Lake lies between the Sierra Nevada and Inyo Mountains.

HORIZONTAL PRESSURE GRADIENTS Meteorologists are very concerned with the distribution of pressure when examining weather maps. The map of sea-level pressure shown in Figure 4–6 is fairly typical, having low- and high-pressure areas of average magnitude. Notice that the changes in pressure across the map are small. The lowest pressure observed is about 990 mb, while the highest is about 1025 mb. This 35 mb difference represents a mere 3 percent or so of the average pressure. Note also that the physical distance separating the areas of highest and lowest pressure is about 2500 km (1586 mi). In the most general sense, then, the pressure gradients across the map are on the order of 35 mb per 2500 km, or about 1 mb per 70 km. Clearly, on a continental scale at least, pressure gradients are usually small.

On a smaller scale, horizontal pressure gradients can be much greater. Strong gradients are needed to create dust storms such as those described in *Box 4–3, Focus on the Environment and Societal Impacts: Owens Lake Dust Storms.* Hurricanes require even stronger gradients to produce their violent and destructive winds. Yet even a hurricane may have a pressure in its interior that is only about 50 mb less than that just outside the storm, some 300 km (180 mi) away. Such a hurricane would have a pressure change of 1 mb per 6 km, yielding only a 5 percent difference in pressure over a considerable distance. This is in marked contrast to vertical pressure gradients, wherein a drop of 50 mb can occur within a vertical distance of only half a kilometer (0.3 mi).

▲ **FIGURE 4–3–2 Owens Lake Today.** The mostly dry lake viewed from the south.

and it generates electricity as well. But its effect on Owens Lake was anything but benign. Within 17 years the lake was essentially dry, appearing much as it does today (Figure 4–3–2). With this drying came huge dust storms caused by winds picking up fine-grained material from the former lake bed.

Dust Storms

The dry lake is about 280 km^2 (108 mi^2), about 40 percent of which is covered by minerals left behind by evaporation of briny water. Compared to much older natural lake beds in the West that have had time to equilibrate with their climate, the floor of Owens Lake is very active. When winds are strong enough, sand-sized rock fragments bounce along the ground, breaking the surface deposits into small particles that can be lofted high into the atmosphere. Particularly worrisome are particles smaller than 10 μm in diameter, so-called PM_{10} particles. These travel great distances and are a serious health hazard for animals and people unlucky enough to be exposed to them (see Chapter 14 for more information about PM_{10} particles).

The geography of the Owens Valley means strong winds and associated dust storms are common. It was originally believed that the dust storms were caused by winds blowing across the Sierra Nevada Mountains and down into the valley from the west. But more recent research has revealed that strong westerly winds are rare, and most storms have winds that blow along the valley axis, from either the south or north. High winds arise from several mechanisms, including the passage of cold fronts, downward momentum transfer from fast-moving air high in the atmosphere, and channeling of air through valley sidewalls. Dust storms can occur any time of year, but are most common in the spring.

The net effect is that the Owens Lake bed can produce very large quantities of dust, by one estimate about half a million metric tons of dust every year (about 550 million U.S. tons). This amounts to 6 percent of the total U.S. production from one very small area. Communities in the surrounding area within the valley exceed federal air quality standards for particulates 48 times per year on average. Dust reaches high elevations in surrounding mountains, where it has the potential to damage sensitive tree species. In some cases dust storms affect visibility as much as 250 km (150 mi) away. Dust is also blown into nearby protected wilderness areas that have minimum visibility requirements set by federal law.

Remediation

There has never been any doubt about the origin of this dust or the responsibility of the city of Los Angeles in its production. Following an historic agreement reached in the late 1990s, the city has been charged with making efforts to control Owens Lake dust. It has employed several methods, the most successful of which is to simply pump water back into the lake and thereby reduce the amount of dry sediment exposed to scouring winds. There is no doubt these efforts have had great effect, with measurements typically 10 times lower than before. However, dust levels remain 10 times higher than federal standards and are higher than similar dry locations elsewhere. At present the city of Los Angeles claims it has done what was promised, while other stakeholders say more must be done. As so often happens with issues related to water, the matter is now in federal court.

4–3–1 What atmospheric factors contribute to dust storms in the Owens Valley?

4–3–2 If you were the federal judge hearing the case, what factors would you consider in deciding if Los Angeles has done enough to reduce dust?

DID YOU KNOW?

The difference in atmospheric pressure between the top and bottom of the Empire State Building is about the same as the pressure difference between the center and the outside of a strong hurricane.

VERTICAL PRESSURE GRADIENTS Though we don't often think about vertical changes in pressure and usually aren't exposed to strong vertical air motions, both are important in the atmosphere. We've already seen that atmospheric pressure always decreases with altitude. Notice, for example, in Figure 4–3 that the mean sea-level pressure of 1013.2 mb decreases to 500 mb at an altitude of 5640 m (about 18,000 ft). Thus, the average vertical pressure gradient in the lower half of the atmosphere is about 500 mb per 5640 m, or just less than 1 mb per 10 m (33 ft). Compare that to the horizontal pressure gradient of an average hurricane, which we saw to be about 1 mb per 6000 m. The *average* vertical pressure gradient in this example is 600 times greater than the *extreme* horizontal pressure gradient associated with a hurricane! In sum, vertical pressure gradients are very much greater than changes in horizontal pressure and, as we will see in Chapter 8, strongly affect general atmospheric motions.

Hydrostatic Equilibrium

You already know that a pressure gradient force causes wind to flow from high to low pressure and that air pressure rapidly decreases with altitude. Given these two facts, you might infer that the wind must always blow upward. If this were the case, it would have troublesome implications for humans on the surface, who would suffocate as all the air around us literally exploded out to space in response to the vertical pressure gradient force.

Before you panic, however, consider a second relevant fact: Gravity pulls all mass, including the atmosphere, downward. Then why doesn't the atmosphere collapse all the way down to the point where we would be able to breathe only by getting on our hands and knees and sucking up the air that has fallen to the surface? That doesn't happen because the vertical pressure gradient force and the force of gravity are normally of nearly equal value and operate in opposite directions, a situation called **hydrostatic equilibrium**.

When the gravitational force exactly equals the vertical pressure gradient force in magnitude, no vertical acceleration occurs. When the gravitational force slightly exceeds the vertical pressure gradient force, downward motions result. Such downward motions are always very slow. On the other hand, the upward-directed pressure gradient force sometimes greatly exceeds the gravitational force, and updrafts in excess of 160 km/hr (100 mph) can develop. Such updrafts are associated with powerful thunderstorms. Although the gravitational and vertical pressure gradient forces are normally almost in balance, the exact value of each varies from place to place and time to time. The downward gravitational force on a volume of air is proportional to its mass (remember that force = mass × acceleration), so a dense atmosphere experiences a greater gravitational force than does a sparse atmosphere. Thus, if a dense atmosphere is to remain in hydrostatic equilibrium, it must have a greater vertical pressure gradient force to offset the gravitational force.

Examine the two identical columns of air in Figure 4–8a. Both have a surface pressure of 1000 mb that decreases to 500 mb at 5640 m above the surface. If the column on the right is heated, as shown in (b), it still contains as much mass as it did before, but the 500 mb level now lies at 5700 m above the surface. Thus, a greater distance is required for the pressure to decrease 500 mb; in other words, there is a lessened vertical pressure gradient. As a result of more slowly decreasing pressure in the warm column, pressures aloft are higher than those in the cold column. At the same time, the density of the air decreases because the same amount of mass occupies a larger volume. So the weaker vertical pressure gradient and the decreased density are interrelated. This relationship has a major effect on horizontal motions in the upper atmosphere. In particular, because the pressure aloft is higher in the warm column, the pressure gradient force would tend to push the air toward the cold column. This is discussed at length in the next section. For now, before moving on, be sure to understand the reason for higher pressure aloft in the warm column. (For mathematical details on the relationship between density and the vertical pressure gradient, see *Box 4–4, Physical Principles: The Hydrostatic Equation*.)

CHECKPOINT

4.9 What is the pressure gradient force?

4.10 Use the concept of hydrostatic equilibrium to explain why a cool, dense column of air has a greater vertical pressure gradient than a warmer, less dense column of air.

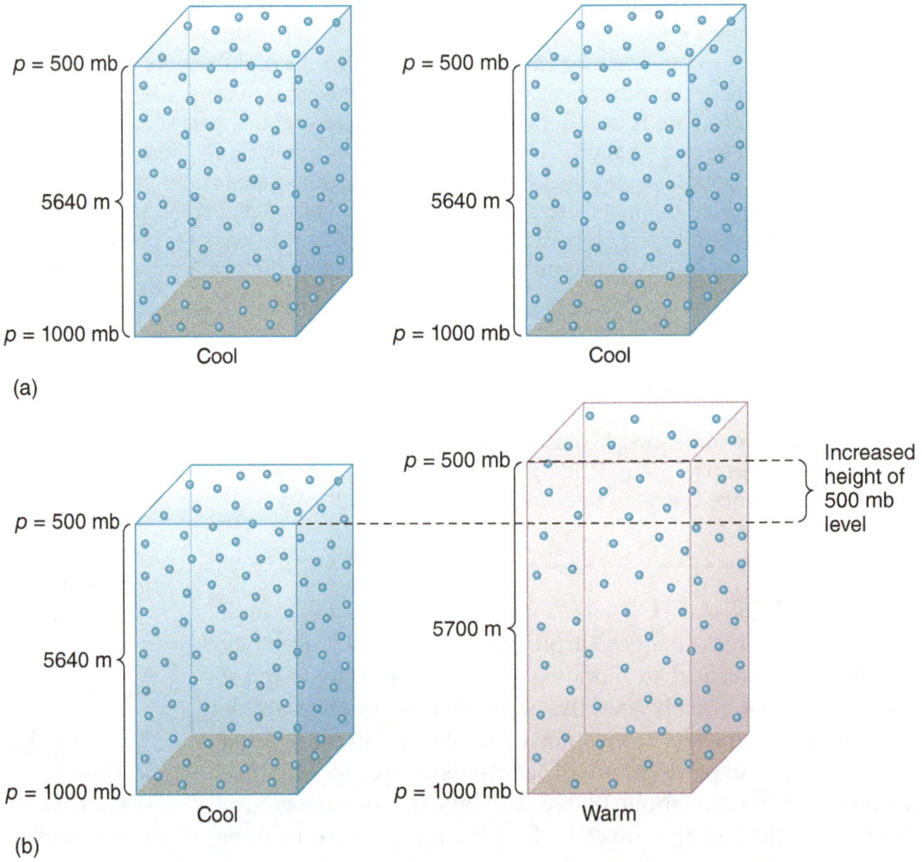

▸ **FIGURE 4–8 Temperature and Pressure.** (a) Two columns of air with equal temperatures, pressures, and densities. (b) Heating the column on the right causes it to expand upward. It still contains the same amount of mass, but it has a lower density to compensate for its greater height. The pressure drops 500 mb over 5700 m within the warm air; it only takes 5640 m of ascent for the pressure to drop the same amount in the cool air. Thus, the cool air has the greater vertical pressure gradient.

4-4 PHYSICAL PRINCIPLES

The Hydrostatic Equation

The concept of hydrostatic equilibrium (in which the vertical pressure gradient force is equal and opposite to the gravitational force) can be succinctly summarized by the **hydrostatic equation**:

$$\frac{\Delta p}{\Delta z} = -\rho g \quad \text{Hydrostatic equation}$$

By convention, the Greek letter delta (Δ) stands for "change in." In this case, Δp refers to a change in pressure, while Δz refers to a change in altitude. Thus, $\Delta p/\Delta z$ on the left side of the equation refers to the change in pressure for a unit of increase in altitude.

We have met the symbols ρ and g before as density and the acceleration of gravity, respectively. The negative sign on the right-hand side accounts for the fact that pressure decreases with height; that is, the left-hand side is always negative. For the two sides to balance, the right-hand side must also be negative.

Thus, the hydrostatic equation states that the rate at which pressure decreases with height equals the product of the air density and the acceleration of gravity. But because the acceleration of gravity is virtually constant, the rate at which pressure declines with altitude is determined almost completely by the density of the atmosphere. In particular, higher-density air has a greater vertical pressure gradient.

As an example, let us compare the two columns of air in Figure 4–8b, supposing that their temperatures are 0 °C and 40 °C. Using the surface pressure of 1000 mb, the equation of state gives the density of the warm air as 1.1 kg per cubic meter. At the same pressure, the cool air must have higher density, in this case 1.3 kg per cubic meter. Assuming hydrostatic equilibrium, the corresponding vertical pressure gradients at the surface are as follows:

Warm Air Column

$$\frac{\Delta p}{\Delta z} = -1.1(9.8) = -10.8 \text{ Pa/m}$$

Cool Air Column

$$\frac{\Delta p}{\Delta z} = -1.3(9.8) = -12.8 \text{ Pa/m}$$

This confirms our earlier reasoning, where we concluded that pressure declines more rapidly in a cool, dense air column than in a warm air column. As we discuss in the body of the text, this sets up an upper-level horizontal pressure gradient between warm and cool air.

4-4-1 Why does a negative sign appear in the hydrostatic equation?

4-4-2 On a clear night with no wind, an air column cools. Does the vertical pressure gradient increase, decrease, or remain unchanged?

Horizontal Pressure Gradients in the Upper Atmosphere

4.6 Describe how pressure gradients in the upper atmosphere arise and how they are mapped.

Just as horizontal pressure gradients occur near the surface, they likewise occur in the upper atmosphere. As we have already seen, atmospheric pressure decreases more rapidly with height (that is, vertically) in a cold, dense air column than in a warm, less dense air column. Looking at Figure 4–8b, you can see that for the cool column at 5640 m above the surface, the air pressure is 500 mb; for the warm column the air pressure at 5640 m above the surface is greater than 500 mb. Thus, in the midtroposphere around the 5640 m height there is a horizontal pressure gradient, with lower pressure over the cool column. Equivalently, the height of the 500 mb level is lower in the cool column.

If we look at the height of some given pressure level (such as the 500 mb level) on a map, its height will often vary from one place to another. More specifically, the surface will slope downward in the direction of colder air, because cold dense air has stronger vertical pressure gradients and will attain a particular air pressure at a lower height (assuming for simplicity's sake that sea-level pressure is uniform). Where a horizontal pressure gradient exists, there must also be a slope in the height of a particular pressure level, with heights decreasing toward colder air. It so happens that the horizontal pressure gradient force is proportional to the slope of the height of that pressure level. If we know the slope, we know the pressure gradient force. Figure 4–9 shows the distribution of the 500 mb level in an idealized atmosphere, with a gradually decreasing temperature toward the North Pole. Notice that the decrease in temperature toward the pole causes a decrease in the height of the 500 mb level, with the surface sloping downward toward colder air. On the "ground" of the diagram are contour lines showing the height of the 500 mb surface. The contour labels tell how high you must go to find a pressure of 500 mb. For example, if you stand on a line labeled 5500 m, the 500 mb surface is 5.5 km above you. The heights decrease toward the north; thus, the contour values also decrease northward.

Figure 4–10 shows a real 500 mb map for May 3, 1995. The height contours are labeled in meters; thus, heights range from 5880 m in the south to 5220 m in the extreme northwest.

▸ **FIGURE 4–9 Idealized Latitudinal Gradient in Upper Atmosphere Pressure.** The gradual poleward decrease in mean temperature results in denser air occurring at high latitudes. As indicated by the hydrostatic equation, pressure drops more rapidly with height at high latitudes and lowers the height of the 500 mb level. The dashed lines depict the heights of the 500 mb level as they would be drawn on a 500 mb weather map.

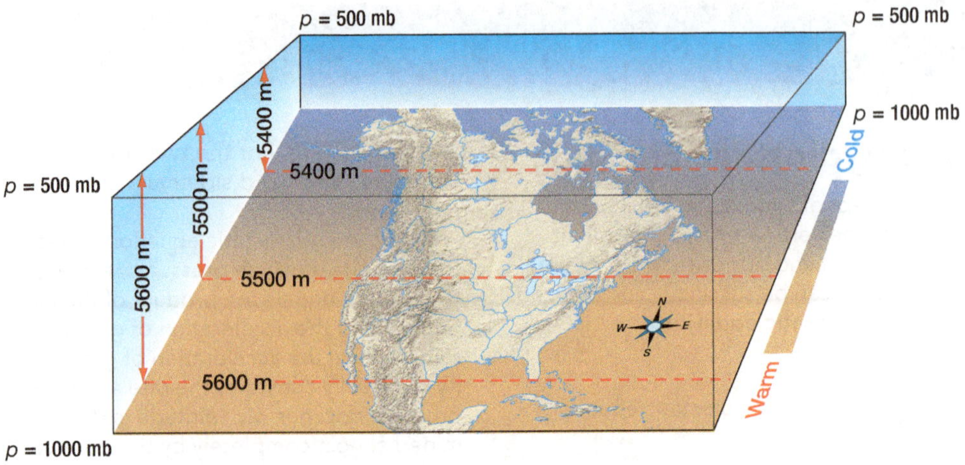

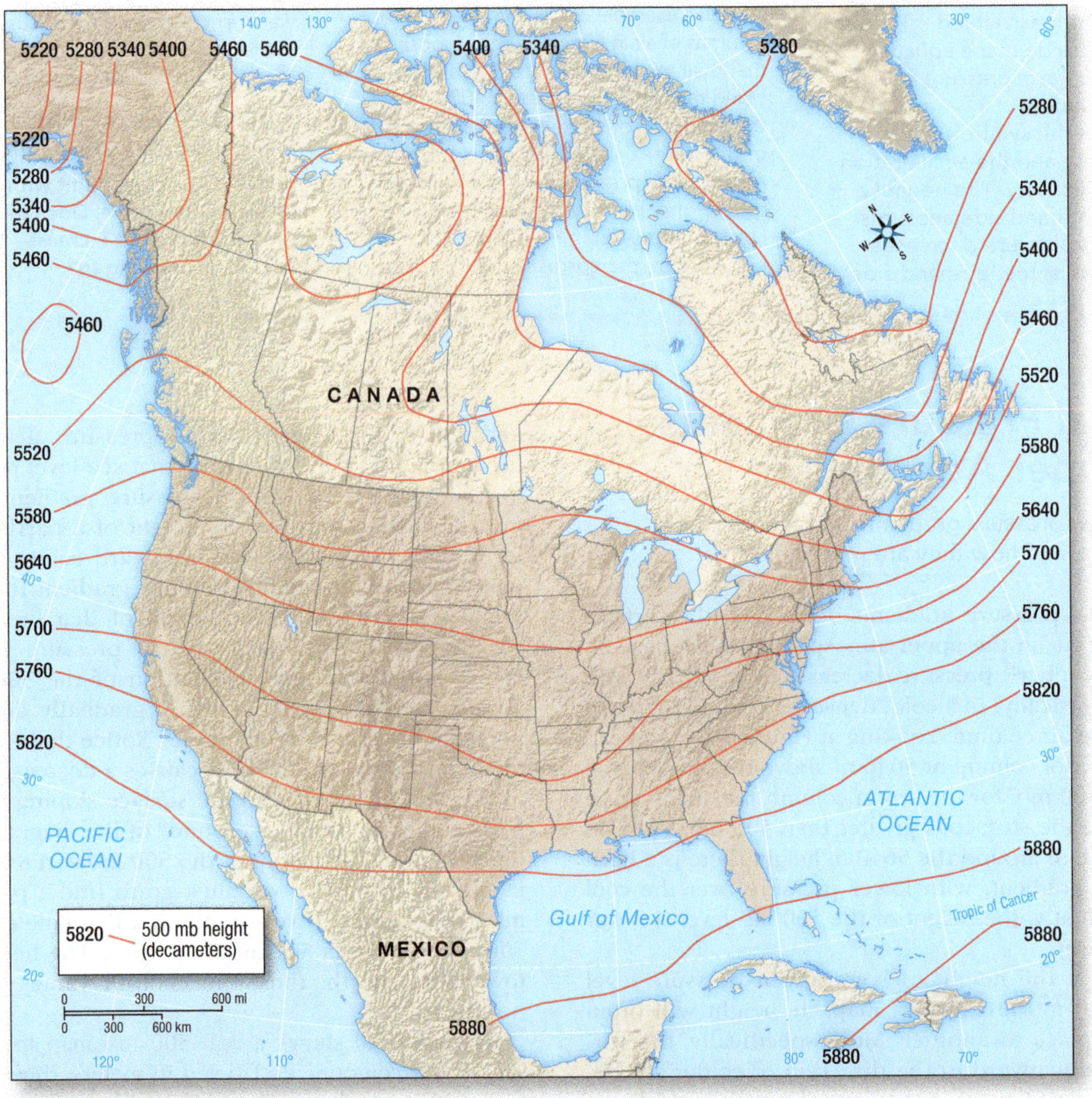

▴ **FIGURE 4–10 Map of 500 mb Heights.** The distribution of the height of the 500 mb level on May 3, 1995. The height contours are labeled in meters.

Contours for 500 mb maps are drawn at 60 m intervals. Overall, the pattern is consistent with a decrease in height from south to north, following the temperature gradient from south to north.

Where the height contours are close together, the pressure gradient force is large. Thus, on that day, upper-level winds were strongest over Newfoundland, with speeds above 160 km/hr (100 mph). Over Nebraska, the pressure gradient force was much weaker, and winds were only 35 km/hr (22 mph).

Notice that the range in heights is only about 660 m from highest to lowest, a change of about 10 percent across a huge distance. Like horizontal pressure gradients at the surface, these upper-atmosphere gradients are small and the pressure surfaces are nearly horizontal. Nevertheless, these weak gradients can produce significant accelerations and high winds, especially in the upper atmosphere where friction is nearly absent and density is low.

Upper-level maps are produced twice each day by the National Weather Service and similar agencies in other countries. They are extremely important for weather forecasting, as we will discuss in Chapter 13. In addition to the maps of the 500 mb level, similar maps are produced for the 850, 700, and 300 mb levels. These correspond to conditions at about 1500, 3000, 10,000, and 13,000 m above sea level (5000, 10,000, 33,000, and 43,000 ft), respectively. Keep in mind that these maps depict the varying heights of these individual pressure surfaces.

CHECKPOINT

4.11 How do horizontal changes in temperature in the lower atmosphere affect the height of the 500 mb level?

4.12 For the map in Figure 4–10, describe how the 500 mb level and the pressure gradient change along the Pacific Coast from Alaska to California.

Forces Affecting the Speed and Direction of the Wind

4.7 Explain the factors that affect wind speed and direction.

The unequal distribution of air across the globe establishes the horizontal pressure gradients that initiate the movement of air as wind. If no other forces were involved, the wind would always flow in the direction of the pressure gradient force. However, the situation is complicated somewhat by the effect of two other forces. The first arises from planetary rotation and alters the direction of the wind. The second force, friction, slows the wind.

The Coriolis Force

To appreciate the Coriolis force, first imagine that Earth does not rotate and arrows are launched in the four cardinal directions as seen in Figure 4–11a. Newton's law dictates that objects move in a straight line unless acted on by another force, and therefore the arrows would follow the paths depicted.[2] In contrast, on our actual rotating planet the arrows would appear to follow curving arcs as seen in Figure 4–11b. Notice

[2]Notice that the easterly and westerly arrows do not follow a line of latitude. Except for the equator, latitude lines are not straight. The arrows would need to turn in order to follow a parallel, and that would require one or more additional forces.

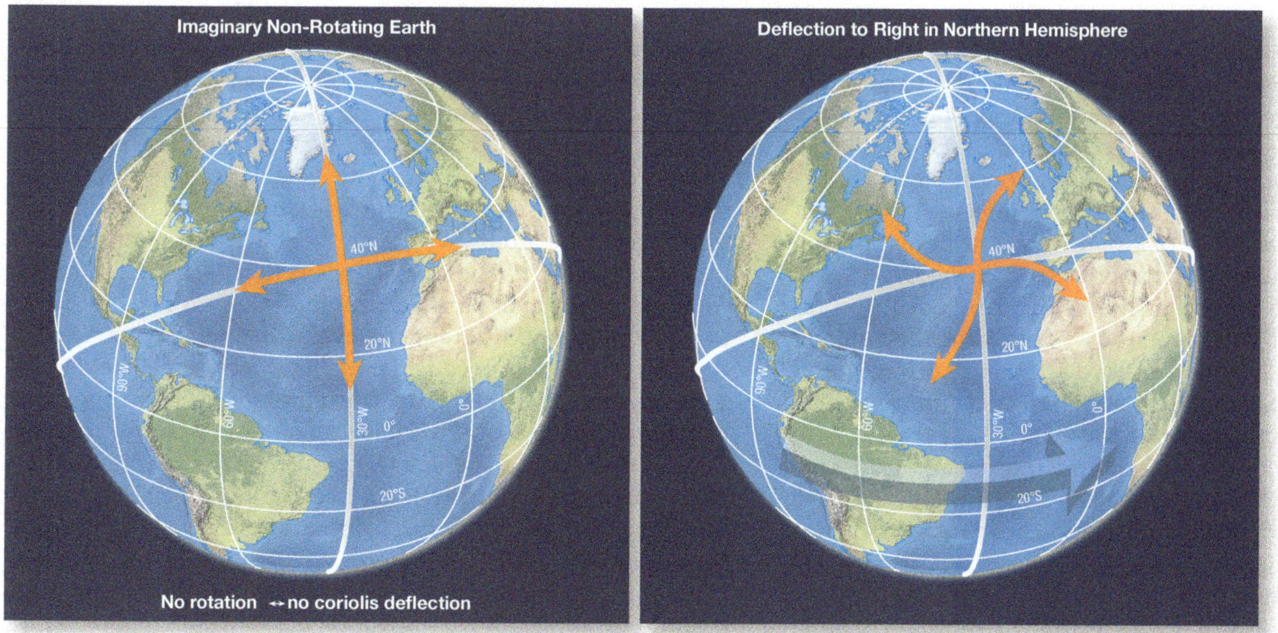

(a) (b)

Video
Coriolis Effect
on a Merry-Go-Round
http://goo.gl/WujYp

Animation
Coriolis Effect
http://goo.gl/slLc3a

▲ **FIGURE 4–11 Visualizing Coriolis Deflection.** Four arrows are launched in the cardinal directions from latitude 40°N, longitude 30°W and move in straight paths. (a) Their paths on a nonrotating planet. (b) On a rotating planet the same objects follow curving paths relative to the surface.

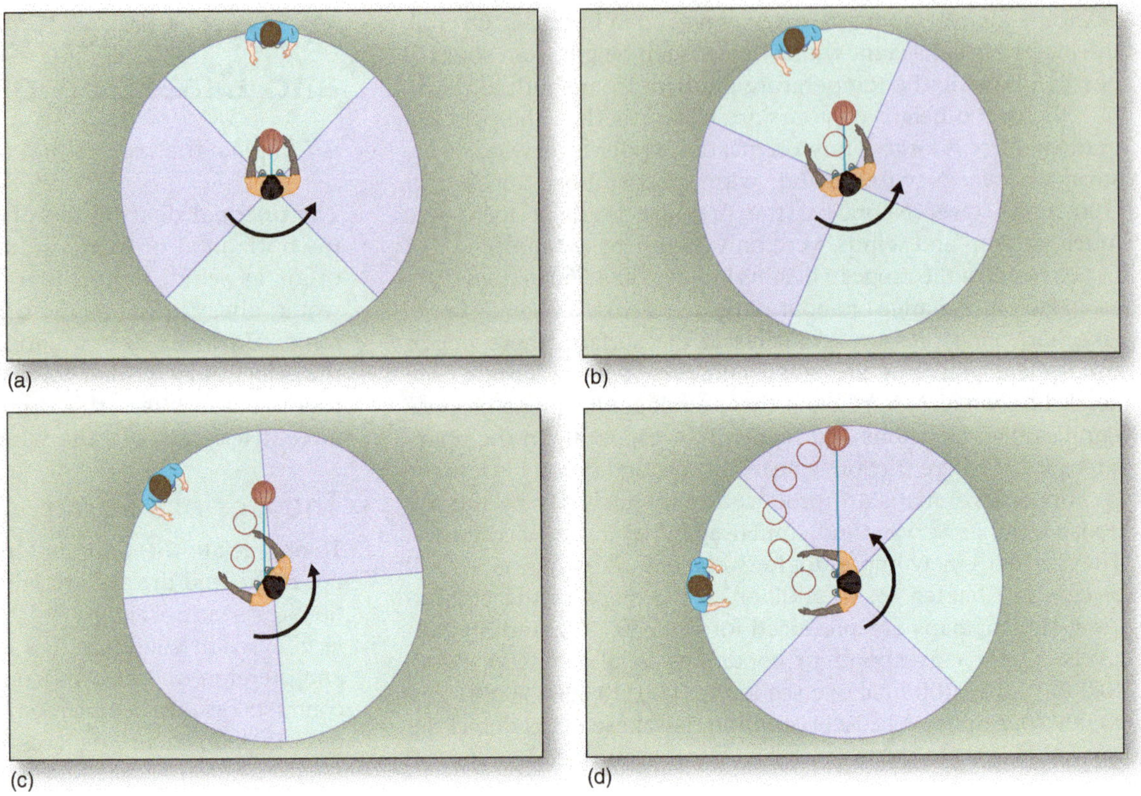

▸ **FIGURE 4–12 Movement Across a Rotating Platform.** As an object moves along a rotating surface, its motion appears to curve away from the target. The ball in the center of the counterclockwise-rotating platform in (a) is about to move toward the target at the top of the figure. As the ball moves toward the target, the rotation of the platform causes the target to move away from its original position. This continues as the ball moves away from the center (b, c), so that by the time it nears the edge of the platform the ball appears to have curved to its right (d). Because all points on Earth (except along the equator) undergo some rotation, all moving objects experience this apparent displacement, which is ascribed to a force called the *Coriolis force.* This force not only acts on projectiles, it acts on the movement of the atmosphere as well.

that all four arrows turn to the right along the direction of motion. All Northern Hemisphere tracks would behave similarly. In the Southern Hemisphere the arrows would turn to the left. These deflections are illusionary, because when viewed from a stationary perspective the paths are straight. This phenomenon, called the **Coriolis** (pronounced "core-ee-OH-liss") **force**,[3] also causes a deflection relative to Earth's surface in the flight of cannonballs, migrating birds, and jet aircraft. In fact, it has an impact on anything that moves in any direction, including the wind. In all cases, however, the deflections are *apparent*, meaning they emerge only because we track motions across the rotating Earth. Therefore, this "force" is in one sense fictional. For that reason some texts prefer the term *Coriolis effect.*

To understand why Earth's rotation affects the path of moving objects, consider a rotating carousel with one person standing at the center and another person at the edge, as in Figure 4–12a. If the person at the center lobs a ball toward the person at the edge of the carousel's platform, the ball will travel in a straight direction toward the original target. But at the same time the rotating platform moves the intended receiver away from the path of the ball. Figures 4–12b and c show the position of the ball and the movement of the platform after successive time increments. By the time the ball reaches the edge of the platform (d), the person who was supposed to have caught it has moved far enough away that he or she has no chance of catching it. To the individual on the platform, the ball would appear to be turning to its right, even though it is in fact moving in a straight line. As it moves in a straight direction, it sweeps a curved path relative to the turning platform beneath it.

The motion in the carousel example is much like an object originating at the North Pole. After leaving the Pole the object would appear to follow a path that curves to the right. Thus if your room were located at the North Pole, a ball rolling across the floor in a straight line would appear to turn to the right. The deflection appears to be rightward because, when looking down at the floor, the room is rotating counterclockwise. A room at the South Pole has clockwise rotation when viewed from above, thus a similar ball would appear to turn leftward. In both cases the reference frame (the floor) rotates about the vertical direction. Although hard to visualize, locations away from the Poles also rotate around the local vertical. The vertical direction is a line running from the center of Earth through the surface. This direction is obviously different for every location, as we saw in Chapter 2 when we considered solar position (see Figure 2–14). Locations other than the Poles don't turn once per day around their local vertical, but there is some rotation rate between 0 and 360 degrees/day. In fact, the rotation rate decreases away from the Poles, becoming zero at the equator. Thus this imaginary force operates everywhere except at the equator. In addition, it increases with increasing distance from the equator because the rotation rate increases with latitude.

The magnitude of the Coriolis force also *increases* with wind speed (though this may seem counterintuitive at first). It is critically important to understand that the *Coriolis force*

[3]For G. G. de Coriolis (1792–1843), who gave the first mathematical explanation for this phenomenon.

involves the deflection that occurs over a given increment of time (say, 1 second)—not the amount of deflection that occurs when an object moves from one particular location to another. Consider this: If an object has the minimum speed possible (zero), it would undergo no deflection. A very slow-moving object might be deflected slightly, but over any brief increment of time the deflection would be minimal. A fast-moving object would cover a considerable distance over the same time interval and therefore be subject to a greater deflection. (See *Box 4–5, Physical Principles: The Coriolis Force*, for further information.)

You should have a solid grasp of the following four fundamental characteristics of the Coriolis force:

1. The Coriolis force produces an apparent deflection in all moving objects, regardless of their direction of motion. The deflection is to the right in the Northern Hemisphere and to the left in the Southern Hemisphere.
2. The Coriolis force is zero at the equator and increases with increasing latitude, reaching a maximum level at the poles.
3. The Coriolis force acting on any moving object increases with the object's speed.
4. The Coriolis force changes only the direction of a moving object, never its speed.

The Coriolis force is not large relative to other commonly encountered forces. To produce detectable effects, the Coriolis force must act over relatively long periods of time. It is important mainly for the motion of objects traveling long distances, such as the air circulating around a hurricane. For motions that occur over short distances, its effect is negligible. Thus, although a basketball shot undergoes a minor deflection because of Earth's rotation, the deflection is so small that it cannot be used as an excuse for a miss. For the same reason, the Coriolis force does not materially influence the motion of water spiraling into a bathtub or kitchen drain, contrary to what people commonly believe. Whether the water spirals clockwise or counterclockwise is usually determined by an asymmetry in the shape of the basin, or by an initial direction of spin imparted by the water supply. The Coriolis acceleration is present, but it cannot produce significant changes in direction because it has so little time to act before the water reaches the drain.

Friction

The other factor that influences the movement of air is **friction**, the force resisting the movement of a fluid or object as it passes along a surface or an adjacent gas or liquid. Air in contact with the surface experiences frictional drag, which decreases wind speed. Air just above, in contact with the slower-moving surface layer, likewise experiences frictional drag, but from the underlying air rather than from the surface. As this layer slows down, air at higher levels is similarly affected. The effects of friction therefore originate from the surface but are found throughout the lower atmosphere. As discussed below, friction affects the relationship between the direction of the pressure gradient force and the resulting wind. It follows that the effect of friction on the wind direction varies greatly with altitude.

Generally speaking, friction is important within the lowest 1.5 km (1 mi) of the atmosphere—often called the **planetary boundary layer** or just the **boundary layer**. Friction lowers the wind speed for a given pressure gradient, which reduces the Coriolis force as well. In contrast, air in the **free atmosphere**, above about 1.5 km, experiences negligible friction. In the absence of friction, winds in the upper atmosphere are fundamentally simpler. We begin our discussion, therefore, with a description of upper-level wind patterns.

CHECKPOINT

4.13 What is the Coriolis force?

4.14 In what ways do latitude, wind speed, and direction affect the intensity of the Coriolis force?

Winds Aloft and Near the Surface

4.8 Describe winds in relation to pressure gradients in the upper atmosphere and near the surface.

The ground exerts a frictional force on the air moving along it. This frictional force gradually diminishes with distance from the surface, and at some height (often around 2 km—or a little more than a mile) the frictional force becomes negligible. This decrease in frictional force with altitude helps to account for differences between winds in the upper atmosphere and winds near the surface.

Winds in the Upper Atmosphere

Given that friction aloft is unimportant, winds in the upper atmosphere depend on the interaction of only two forces: the pressure gradient and Coriolis forces. In thinking about air movements, it is helpful to imagine a small volume of air, perhaps something like a sphere a meter or so in diameter. Meteorologists call these **air parcels** as an analogy to packages that might be mailed from one place to another. In deciding how air will move, we can examine the forces acting on these imaginary volumes.

4–5 PHYSICAL PRINCIPLES

The Coriolis Force

In describing wind and other motions, convention says we take the surface of Earth as a reference frame. For example, when we say there is a 10-meter-per-second wind, we mean the air is moving past the surface at 10 m/sec. Because the surface rotates, we are describing motions relative to a rotating reference frame. The result is that an object moving in a straight line with respect to the stars appears to follow a curved path on Earth's surface, as shown in Figure 4–12b. To maintain our Earth-bound reference frame, we need to introduce a force that produces the apparent deflection. The Coriolis force is just that force. It can be thought of as an accounting device that lets us describe winds on a rotating Earth without violating Newton's law. Although the Coriolis force technically acts in three dimensions, we will considerate it to be a horizontal force. This is an acceptable approximation given that motions on Earth are essentially horizontal thanks to how thin the atmosphere is compared to the size of the planet.

Rotation Around the Local Vertical

The key to understanding the Coriolis force is to realize that what matters is rotation around the **local vertical direction**. In other words, no matter where we are we must ask: What direction is "up", and how much rotation is there around that direction? Clearly, the vertical direction is a line running from the center of Earth through the location in question. In other words, it is a line perpendicular to the surface. At the poles the vertical direction corresponds to Earth's axis of rotation, and thus the local reference frame rotates 360 degrees per day. At 89° latitude, the local vertical is not perfectly aligned with Earth's axis, and so the effective rotation rate is somewhat less. At still lower latitudes the rotation rate is even less because the misalignment is greater (Figure 4–5–1). At the equator the local vertical is at a right angle to and not aligned at all with Earth's axis. Thus, there is no rotation around the local vertical. As seen in Figure 4–5–1, the local vertical sweeps around the axis of rotation, but remains perpendicular throughout the day.

To find the rotation at any latitude, we need to determine the angle between the axis and the local vertical. This amounts to projecting the axis onto the local vertical, and is given by the sine of latitude. In other words,

$$\text{Local rotation} = (\text{Earth rotation}) \times \text{sine (latitude)}$$

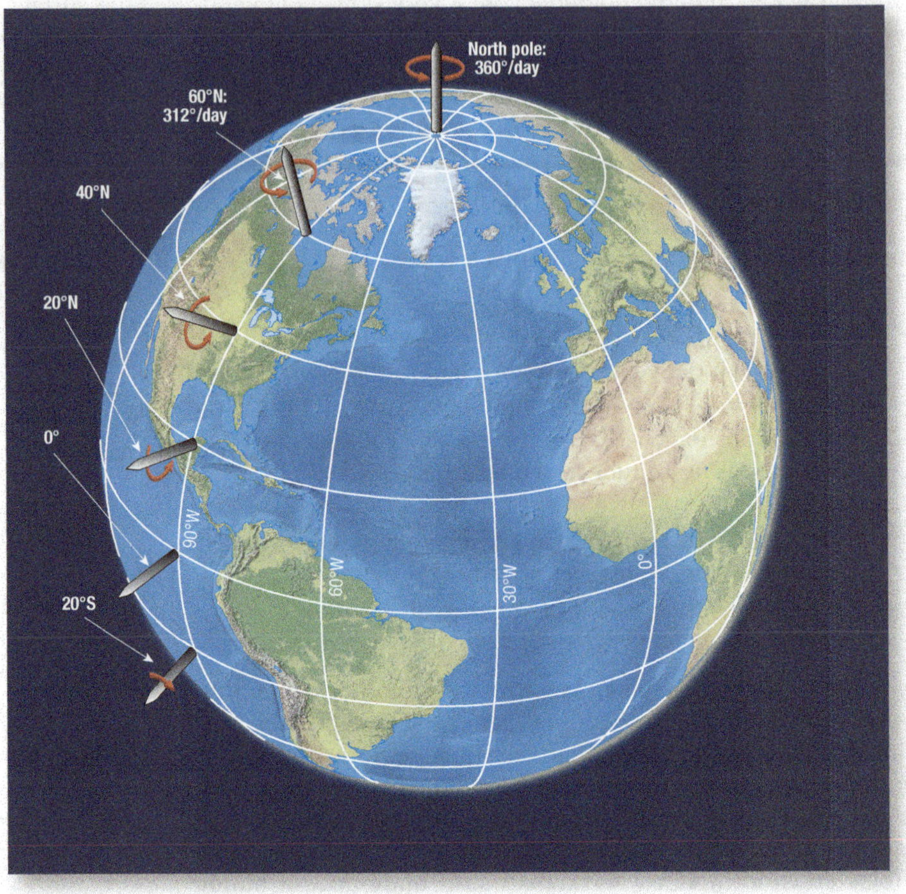

▲ **FIGURE 4–5–1 Rotation Around the Vertical Direction.** Straight arrows show the direction of the local vertical. Rotation around the local varies from zero at the equator to once per day at the poles.

Figure 4–5–1 shows some representative values. Notice that Southern Hemisphere rotations are given negative values, which reflects their opposite sense of rotation. Figure 4–5–2 shows the Coriolis effect on four projectiles moving at 500 km/hr (311 mph). In three hours Earth rotates by 45 degrees and all projectiles move a total of 1500 km. However, their paths curve to the right, with more curvature at higher latitudes than lower latitudes. (If launched at the same latitude in the Southern Hemisphere, these projectiles would curve leftward by the same amounts.)

Wind Speed and the Coriolis Force

The other factor affecting the Coriolis force is the object's speed. Look at Figure 4–5–3 and notice the paths for objects moving half as fast, 250 km/hr (155 mph), as those in Figure 4–5–2. In three hours they have traveled half as far as the objects traveling at 500 km/hr and have experienced much less total deflection. If the deflection over the same time period is less, the corresponding deflective "force" must be smaller. In fact the Coriolis force is directly proportional to wind speed. Putting the two factors together, the Coriolis force (F_C) is given by

$$F_C = 2\,\Omega \sin(\varphi)\, v \quad \text{Coriolis acceleration}$$

where Ω is Earth's rotation rate, φ is latitude, and v is wind speed.

As written here, the equation gives the Coriolis force per unit mass; that is, the force per kilogram of moving air.

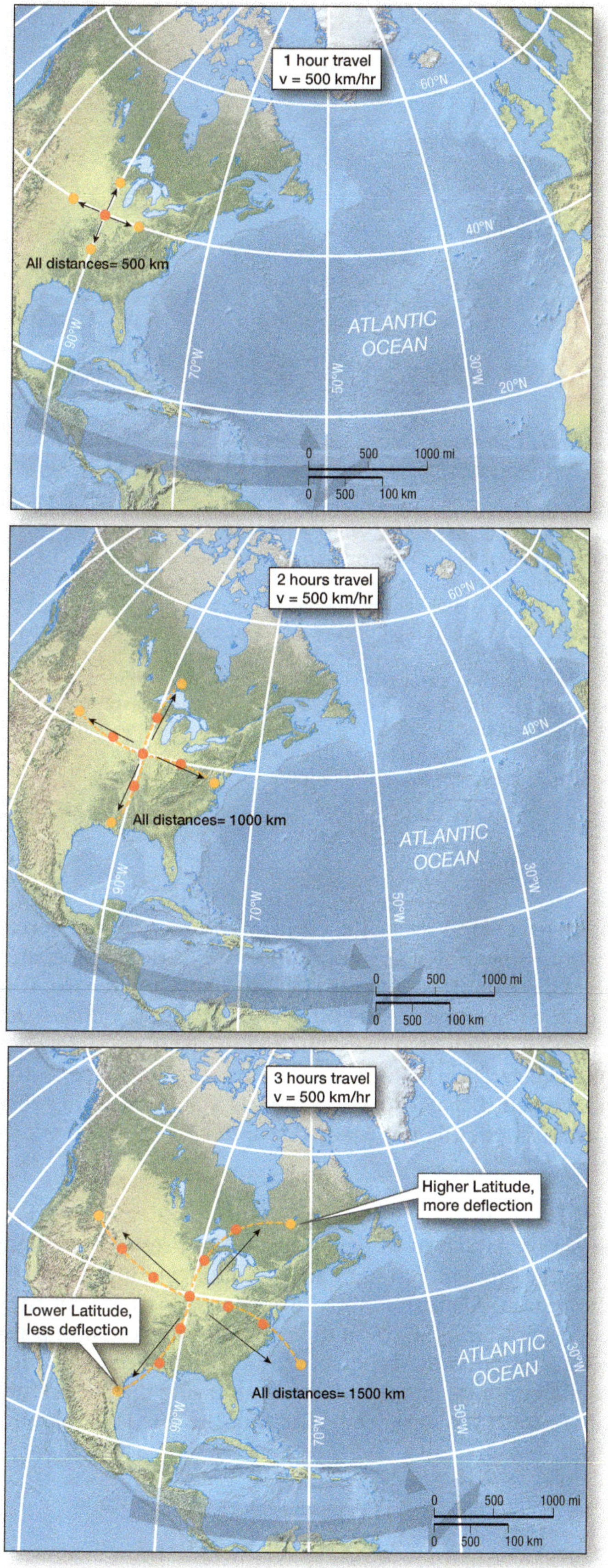

◄ **FIGURE 4–5–2 Travel Paths of Fast-Moving Objects.** Four projectiles are launched north, south, east, and west from a common point in the Northern Hemisphere. The figures show their positions after one, two, and three hours of travel. All cover the same distance and follow paths that curve to the right. Paths extending farther north have more deflection and a stronger than average Coriolis force.

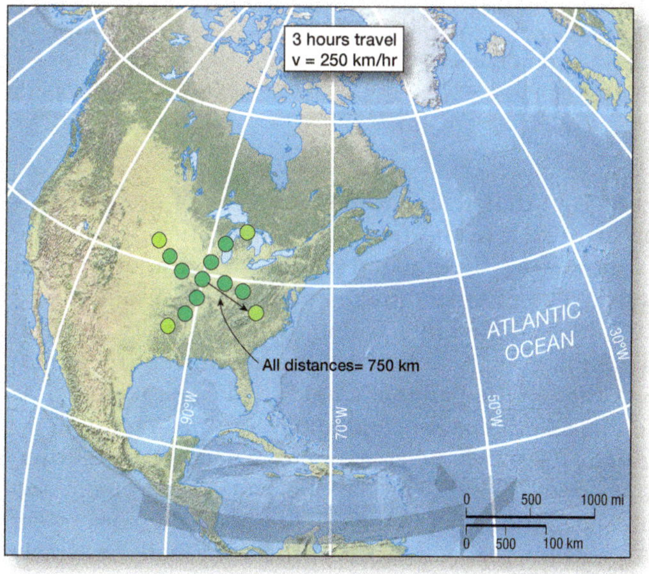

▲ **FIGURE 4–5–3 Travel Paths of Slower Objects.** Four projectiles moving half as fast as in Figure 4–5–2. With less time to travel there is less deflection. For each unit of the travel time the Coriolis Force is half as strong as in Figure 4–5–2.

Combining the information with Newton's Second Law, which tells us that acceleration is force per unit mass, we see that F_C is an acceleration—specifically, the Coriolis acceleration. That said, from here forward we won't draw a distinction between the Coriolis acceleration and the Coriolis force, but will instead use the two interchangeably.

4–5–1 Calculate the local rotation rate for latitude 45°. Is it half of the value at the poles?

4–5–2 Suppose you wanted to calculate the Coriolis force on Jupiter. What factors would you have to consider?

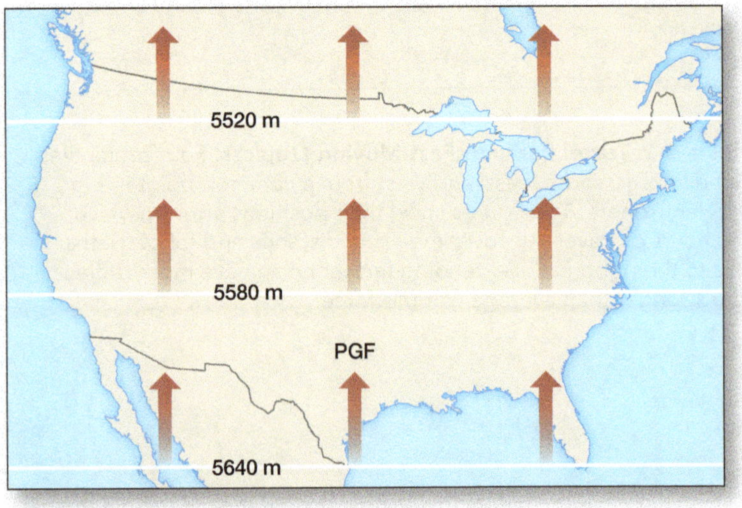

▲ **FIGURE 4–13 Idealized Upper Air Map and Pressure Gradient Force.** The pressure gradient force is directed from south to north in this hypothetical distribution of the height of the 500 mb level. On a nonrotating planet, this would cause air to move from south to north.

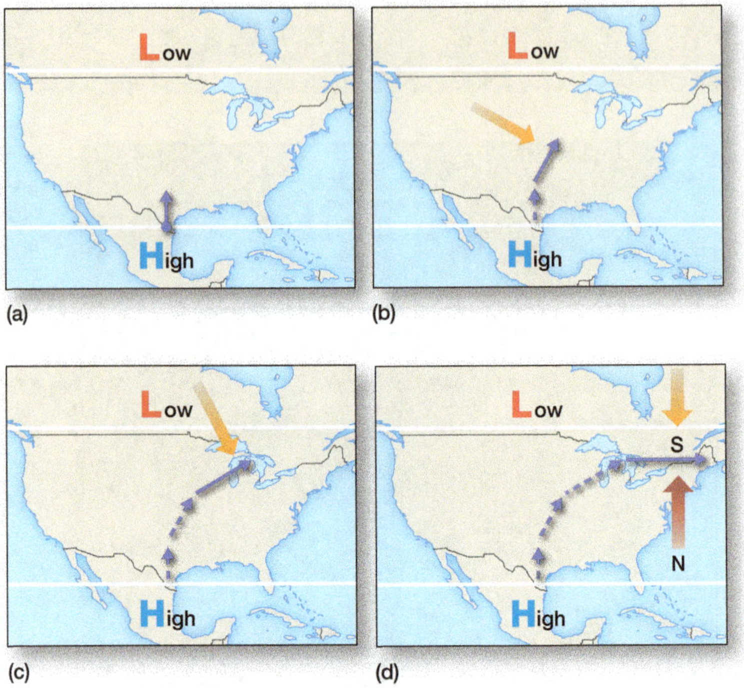

▲ **FIGURE 4–14 Geostrophic Wind.** (a) Assume that there is a stationary parcel of air in the upper atmosphere subjected to a south-to-north pressure gradient force. If the parcel is tethered to an imaginary pole, no movement of the parcel can take place. (b) Once the imaginary cord is cut, the horizontal pressure gradient accelerates the parcel northward. Initially, when the wind speed is low (as indicated by a short arrow), the Coriolis force is small. (c) As the parcel speeds up (longer arrow), the strength of the Coriolis force increases and causes greater displacement to the right. (d) Eventually, the wind speed increases the Coriolis force sufficiently to cause the air to flow perpendicular to the pressure gradient force.

Figure 4–13 illustrates the simplest type of pressure pattern that can exist in the free atmosphere. In this case the height contours are straight and parallel to one another, with a pressure gradient force directed northward. Assume that a parcel of air contained in a balloon with a density equal to the surrounding air is tethered to a pole in the free atmosphere. Although the balloon is held in place, it nonetheless is susceptible to the pressure gradient force acting on it.

Figure 4–14 shows what happens when the connecting cord is severed. Initially, the balloon had no movement and therefore no Coriolis acceleration. But after the cord is cut (a), the pressure gradient force propels the balloon slowly toward the low-pressure area. As the pressure gradient force causes the balloon to move faster, there is an accompanying increase in the Coriolis force, and the balloon accelerates farther to its right (in the Northern Hemisphere) (b, c). Eventually, the air flows parallel to the height contour lines (d). At this point, the Coriolis force and pressure gradient balance one another, so no net force is acting on the balloon.

From here on, the balloon moves parallel to the height contours, which in this case is a straight line with a constant speed. In other words, the airflow becomes unaccelerated, moving with unchanging speed and direction. Such nonaccelerating flow is called **geostrophic flow** (or **geostrophic wind**), and it occurs when the pressure gradient force equals the Coriolis force. Geostrophic flow occurs only in the upper atmosphere where friction is absent and only the Coriolis and pressure gradient forces apply.

You may wonder why the airflow in Figure 4–14d does not turn all the way back toward the high-pressure area. The simplest answer is that if it were to move in that direction, it would have to flow against the pressure gradient force. This would slow down the air, reduce the Coriolis force, and thereby cause the flow to turn back to its left. Likewise, if it were to turn northward, the air would receive a boost from the pressure gradient force. This would lead to a stronger Coriolis acceleration, which would turn the air back to the south. We see that geostrophic flow is stable, meaning that once established, it is not easily disrupted.

Gradient Flow

In the simple situation in Figure 4–14, the pressure gradient force is uniform, with straight and parallel contours throughout the region. Such situations do occur in nature, but they are the exception rather than the rule. A more common pressure distribution is the type shown in Figure 4–15, in which the height contours curve and assume varying distances from one another. In the absence of friction, the air flows parallel

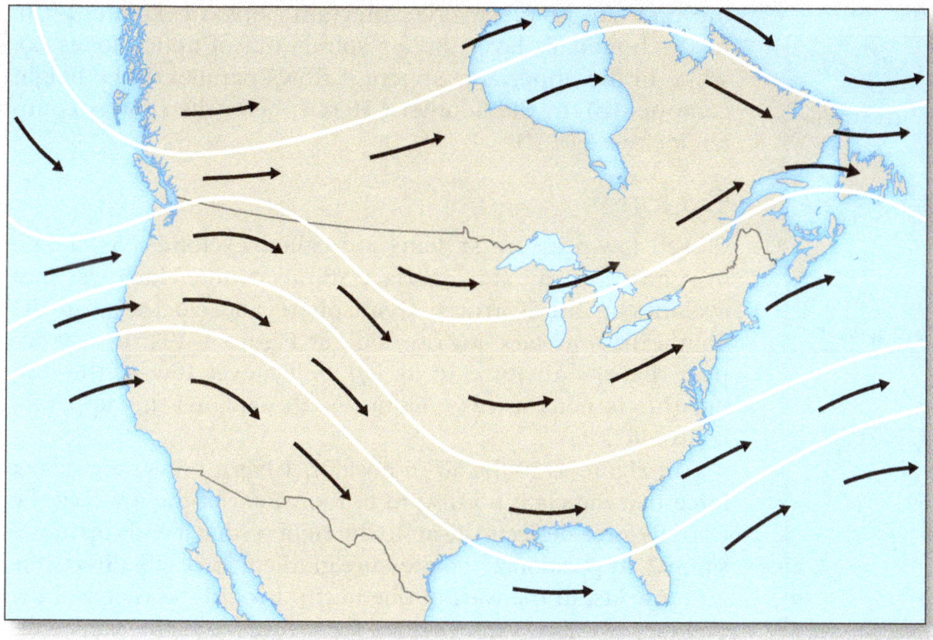

◀ **FIGURE 4–15 Gradient Wind.** As seen in this 500mb map, gradient flow occurs in the upper atmosphere where the flow is unaffected by friction from the surface. Geostrophic flow is a special case of gradient flow where there is no acceleration.

to the contours, for the same reasons as in geostrophic flow. But this type of flow is not truly geostrophic because it is constantly changing direction and therefore undergoing an acceleration. In order for the air to follow the contours, there must be a continual mismatch between the pressure gradient and Coriolis forces. In some places the air is flowing faster than it would in geostrophic flow, and some places more slowly. (Later it will be seen that around high-pressure areas the flow exceeds geostrophic speeds and the converse is true around lows.) Meteorologists refer to this movement as **gradient flow** (or as **gradient wind**). Like geostrophic flow, gradient flow develops only in the absence of friction, and the wind flows perpendicular to the pressure gradient. In fact, geostrophic flow is simply a special case of gradient flow, arising if the contours happen to be straight and parallel. The cloud movements seen in Figure 4–16 are an example of gradient and geostrophic movements. The straight segments are geostrophic, whereas the curved portions are gradient flow.

▼ **FIGURE 4–16 Cloud Streaks over the Great Lakes, December 13, 2013.** At high altitudes friction is negligible, and the motions seen are parallel to isobars.

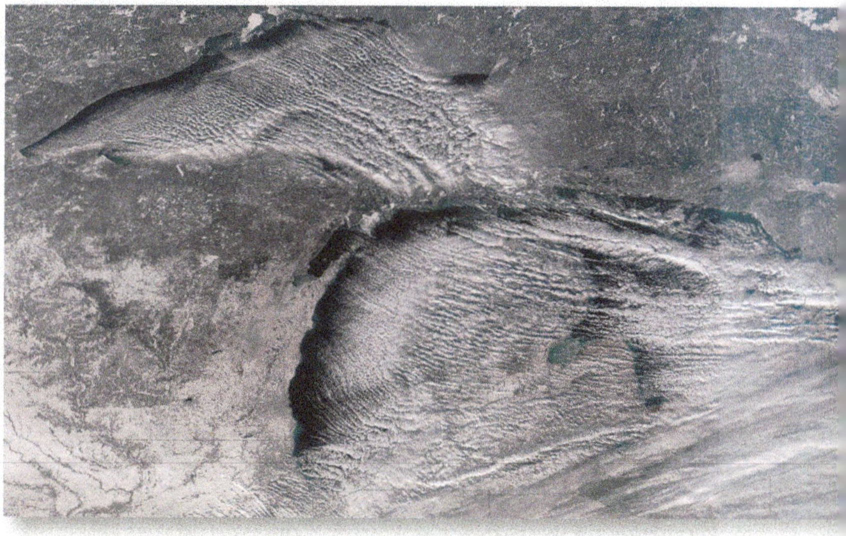

Near-Surface Winds

Friction makes winds near the surface slower than those in the middle and upper atmosphere, given equal pressure gradients. The lower wind speeds reduce the Coriolis force and thereby prevent the flow from becoming gradient or geostrophic. Thus, the winds in the boundary layer do not flow parallel to the isobars; rather, they cross the isobars at an angle as they blow from high to low pressure. As always, there is Coriolis deflection to the right in the Northern Hemisphere and to the left in the Southern Hemisphere (Figure 4–17). The angle of airflow relative to the pressure gradient is not constant, being greater at higher latitudes (because of the stronger Coriolis force) and over smooth surfaces where friction is minimized (such as oceans and large lakes).

DID YOU KNOW?

Clusters of wind turbines have been considered as a way to reduce the destructive power of hurricane winds. One study shows that 78,000 turbines placed around New Orleans would have turned 2005's Hurricane Katrina into a relatively modest tropical storm.

CHECKPOINT

4.15 How are geostrophic flow and gradient flow similar? How are they different?

4.16 In the absence of friction, why does the air flow parallel to the isobars whether or not they are curved? In other words, why doesn't the air cross the isobars?

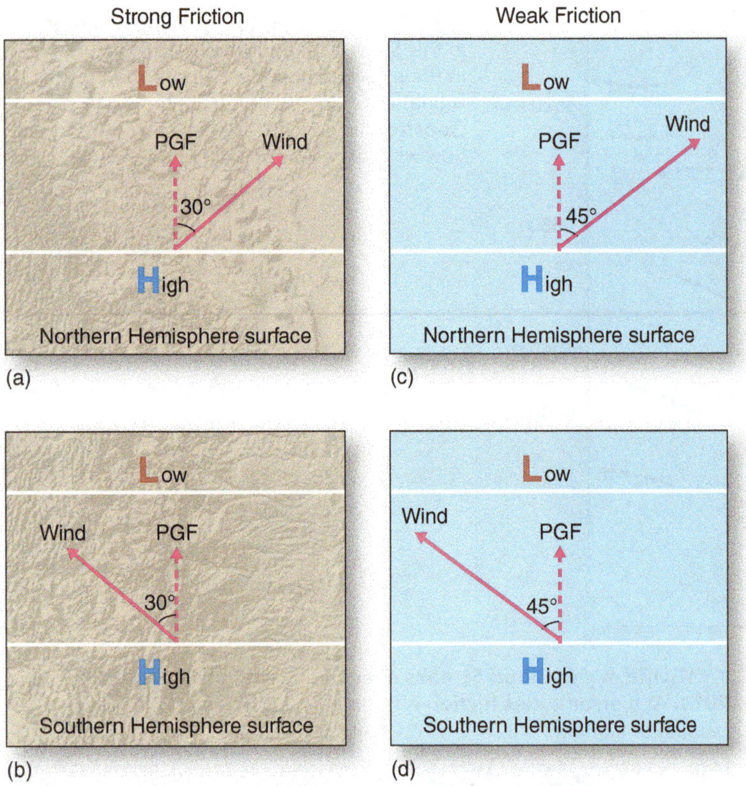

▲ **FIGURE 4–17 Low-Level Forces and Wind.** Geostrophic flow cannot exist near the surface. Friction slows the wind, so that the Coriolis force is less than the pressure gradient force. Thus, the air flows at an angle to the right of the pressure gradient force in the Northern Hemisphere (a, c) and to the left in the Southern Hemisphere (b, d). When friction is high, wind is closer to perpendicular to the isobars (c, d).

Anticyclones, Cyclones, Troughs, and Ridges

4.9 Explain the forces that produce anticyclones, cyclones, troughs, and ridges.

Experience tells us that the sea-level pressure across the globe is not haphazardly distributed into disorganized, widely scattered zones of high and low pressure. Instead, it is usually organized into a small number of large high- and low-pressure regions. Thus, on a given day North America might have four or five major pressure centers at one time.

Anticyclones

Enclosed areas of high pressure marked by roughly circular isobars or height contours are called **anticyclones**. The wind rotates clockwise around anticyclones in the Northern Hemisphere, as the Coriolis force deflects the air to the right and the pressure gradient force directs it outward (Figure 4–18). In the boundary layer the air spirals out of anticyclones (a), while in the upper atmosphere it flows parallel to the height contours (b). In the Southern Hemisphere, the flow is counterclockwise (c, d).

Cyclones

Closed low-pressure systems are called **cyclones**. As shown in Figure 4–19, air at the surface spirals counterclockwise into cyclones in the Northern Hemisphere and clockwise in the Southern. Now take a close look at Figure 4–19a. It may appear that the air turns to its *left* as it moves toward the low, but this is not the case. Figure 4–20 explains this apparent contradiction.

In Figure 4–20 the air in position 1 has a pressure gradient force that directs it northward to the center of the low, but the Coriolis force deflects the air to the right so that it ends up in position 2. At position 2, the pressure gradient force still directs the air (now just to the west of due north) toward the center of the cyclone, and again the Coriolis force turns it to the right. This shift in the direction of the pressure gradient gives the trajectory

Animation
Cyclones and Anticyclones
http://goo.gl/0HarBu

▼ **FIGURE 4–18 Anticyclonic Flow.** Air spirals clockwise out of anticyclones in the Northern Hemisphere (a) and rotates clockwise around the high in the upper atmosphere (b). The flow is reversed in the Southern Hemisphere (c, d).

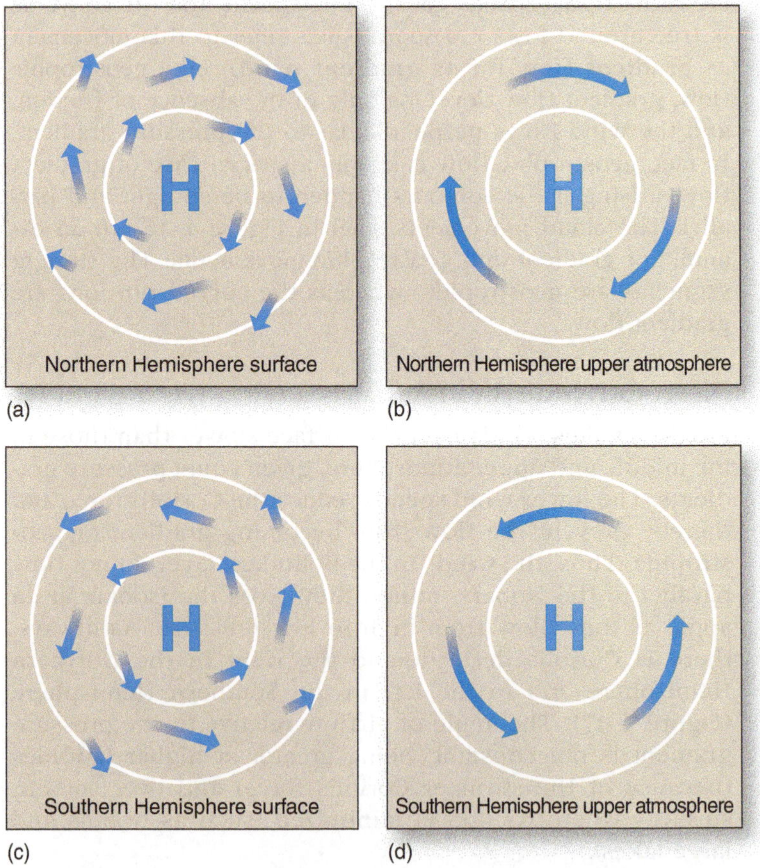

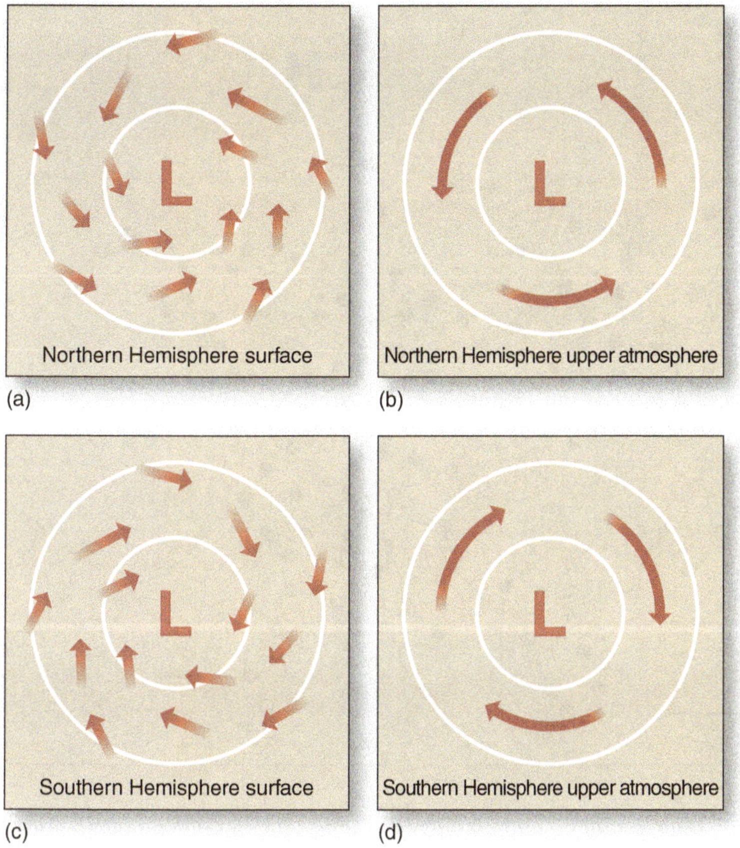

▲ **FIGURE 4–19 Cyclonic Flow.** (a) Air spirals counterclockwise into surface cyclones in the Northern Hemisphere and (c, d) rotates counterclockwise around an upper-level low. The flow is reversed in the Southern Hemisphere.

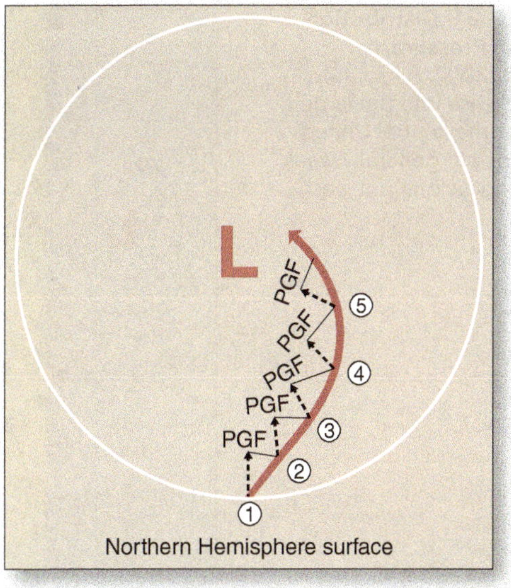

▲ **FIGURE 4–20 Flow into a Low.** The counterclockwise flow of air into surface cyclones in the Northern Hemisphere turns to the left but is nevertheless deflected to the right by the Coriolis force. If there were no Coriolis force, it would cross the isobars at a right angle; instead, it is everywhere deflected to the right.

the appearance of being deflected to the left. But at any moment in time, it is accelerated to the right of the direction of motion.

The term *cyclone* sometimes causes confusion because people associate it with major tropical storms near southern Asia. In fact, any closed low-pressure system (even one that produces nothing other than gentle breezes and a few clouds) is a cyclone. Although the violent tropical storm goes by the same name, in this discussion we use the term in its most generic sense. We will discuss the more specific type of cyclone in Chapter 12.

In addition to their characteristic horizontal winds, particular vertical motions are associated with cyclones and anticyclones. As it approaches the center of the cyclone, air moving into a low-pressure system at the surface has nowhere to go but up. The rising motions, as we will see in the next chapter, promote cloudy or stormy weather. In contrast, the air in anticyclones moving out from the center is replaced by sinking air. Anticyclones therefore typically have clear skies and fair weather. As a general rule, anticyclones are larger than cyclones and have weaker horizontal pressure gradients and weaker winds.

Figure 4–21 shows the distribution of sea-level pressure across much of North America on October 18, 2007, with a large cyclone occupying much of the north-central United States. Over the Dakotas the wind generally approaches the low pressure from the northwest. The wind barbs show the direction that the wind is blowing *from* and the number of tick marks indicate the approximate wind speed (see the left side of Figure 1–17 for further information on interpreting the wind barbs). Airflow approaches the low out of the southeast over Indiana and generally comes out of the east over the Great Lakes region. A high-pressure area dominates much of the western United States. From western Montana to central California there are few isobars, indicating a weak pressure gradient. Not surprisingly, wind speeds are very low over most of the west, with the exception being over the extreme Pacific Northwest where a well-developed cyclone approaches from the west.

Troughs and Ridges

Pressure systems often occur as closed cells, which are fully enclosed in a somewhat circular fashion. However, at high altitude, most low- and high-pressure systems occur instead as elongated areas called **troughs** (low pressure) and **ridges** (high pressure), as depicted in Figure 4–22. As air approaches the ridge in (a), it begins turning to the right in order to remain parallel to the pressure gradient. For this to happen the Coriolis force must increase, which means the speed must increase. In other words, the wind speed must exceed what is given by geostrophic balance. Conversely, air flowing around the trough must slow down so that it can turn to the left. The Coriolis force is still present and pushing to the right of the wind direction, but it is weaker than the pressure gradient force and thus the wind can divert from a straight geostrophic path.

There is a tendency for pressure to be distributed as cyclones and anticyclones at the surface and gradually give way to ridges and troughs in the upper atmosphere. This is depicted in Figure 4–23, which shows simultaneously weather maps obtained for the 850, 700, 500, and 300 mb levels.

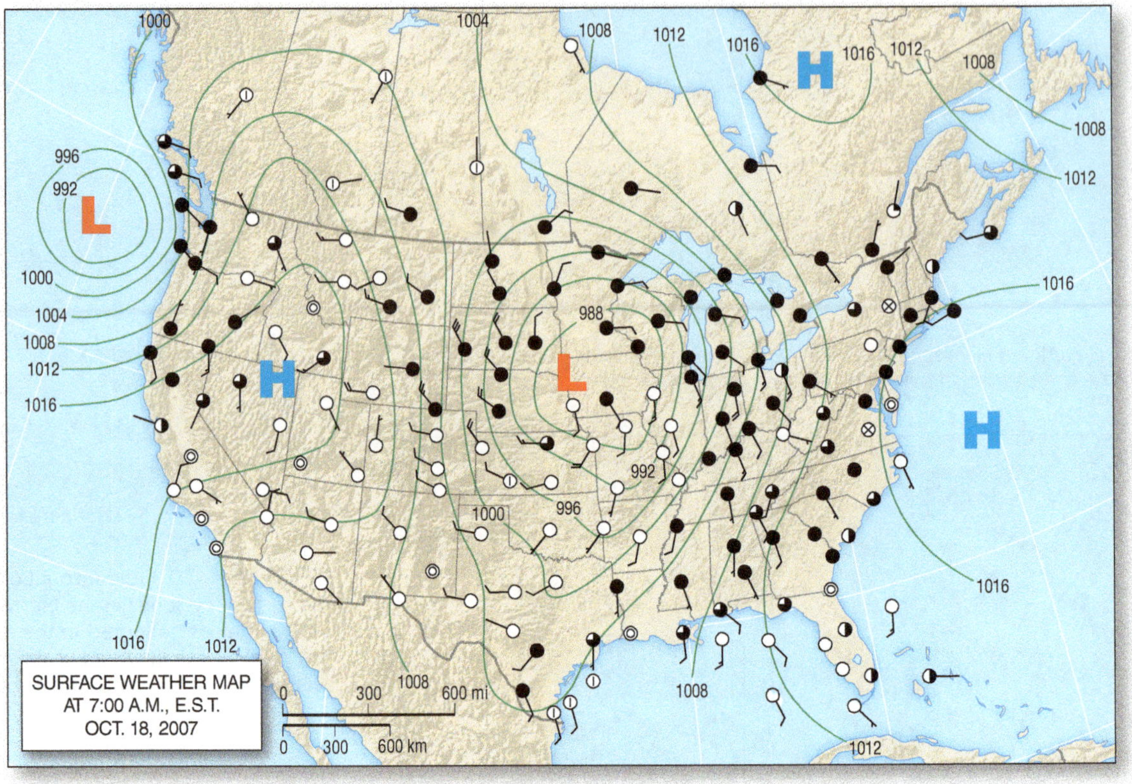

▸ **FIGURE 4–21 Distribution of Sea-Level Pressure, October 18, 2007.** Air rotates counterclockwise into the large cyclone over the central United States. Winds are generally weak out of the anticyclone out west.

Video
Hurricane Winds

http://goo.gl/Pp6GnB

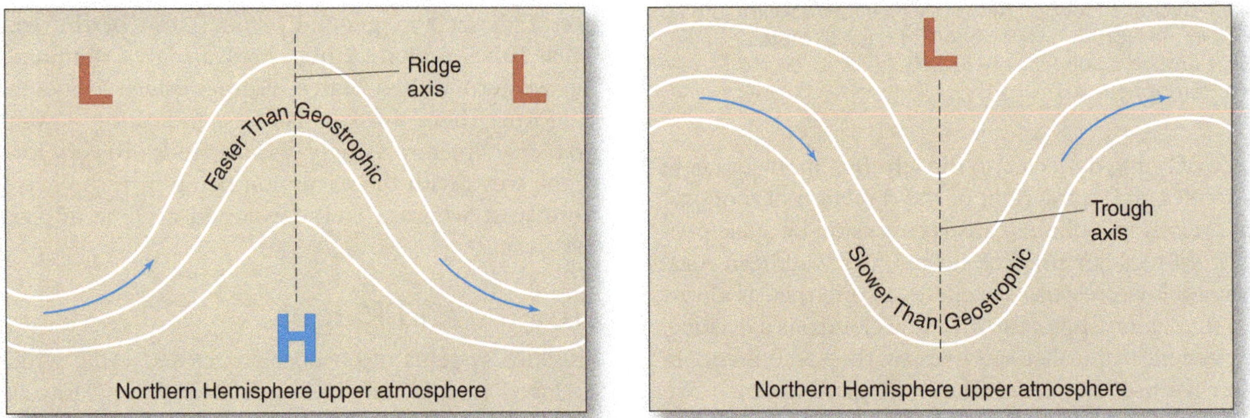

(a) (b)

▲ **FIGURE 4–22 Ridges and Troughs.** Elongated zones of high and low pressure are called ridges (a) and troughs (b), respectively. These are Northern Hemisphere examples, and so the Coriolis force is everywhere pushing to the right of the wind direction.

Troughs and ridges also appear on surface maps. Figure 4–24 shows the distribution of sea-level pressure on October 21, 2007, three days after the situation shown in Figure 4-21. In this map, a new anticyclone has developed over the West, but in this case the high-pressure system has a much stronger pressure gradient than was found in Figure 4-21. Notice the northwestward bending of the isobars along the California coast (highlighted by the dashed line) extending from the region of low pressure off the southern California coast. This intrusion of lower pressure is a trough. This example is a noteworthy one because the strong pressure gradient over much of California fostered very strong winds, high temperatures, and dry air, which led to one of the worst outbreaks of wildfires in California history, with over 1500 homes destroyed.[4]

CHECKPOINT

4.17 What is the difference between a cyclone and a trough?

4.18 In Figure 4–24, what resulted from the trough represented by the dashed line along the California coast? Explain.

[4]This was a classic example of a Santa Ana wind, which is explained in Chapter 8.

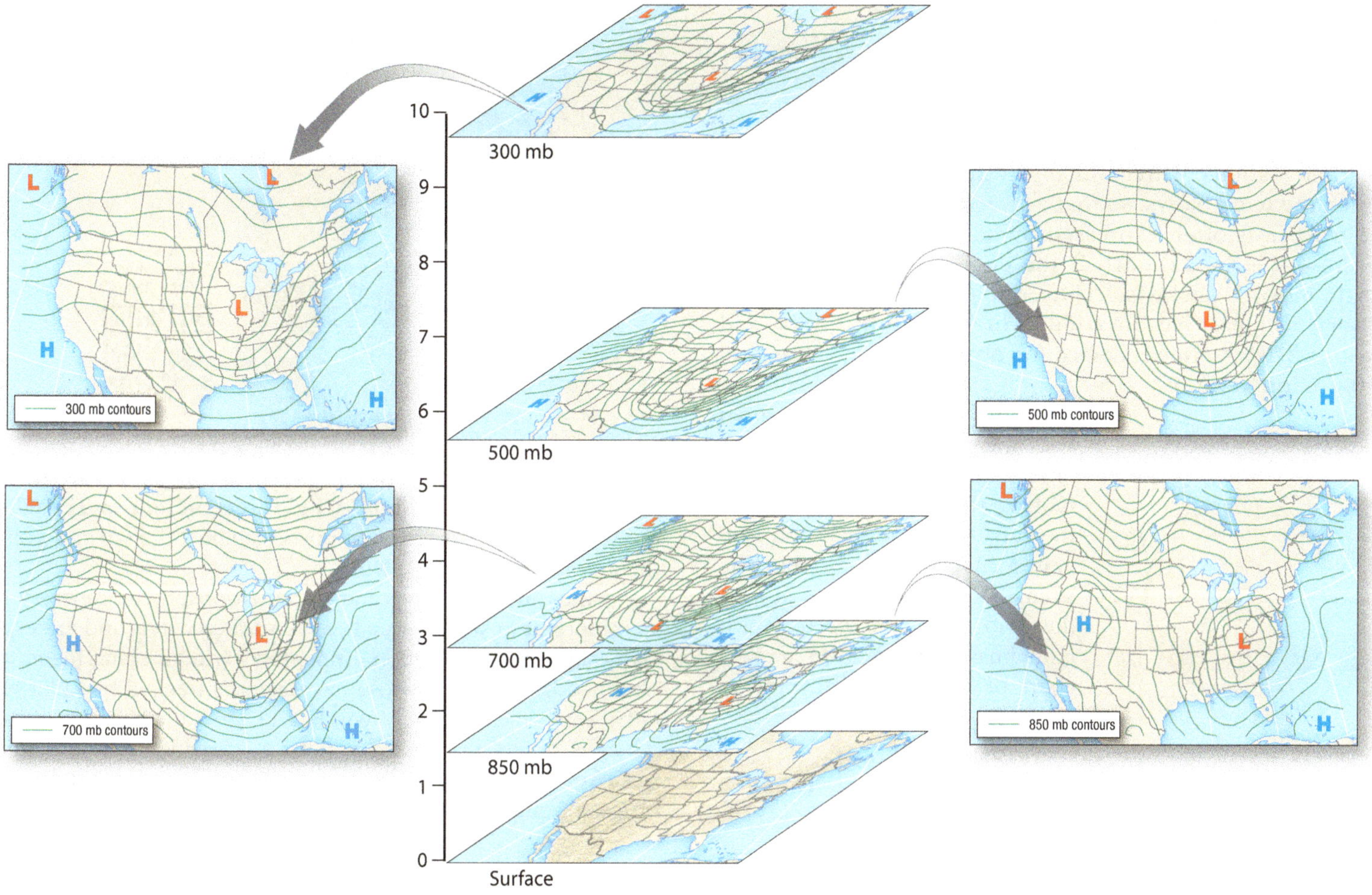

▲ **FIGURE 4–23 Maps of Standard Pressure Levels.** Maps of the 850, 700, 500, and 300 mb levels for the same date and time. Observe how the pattern of cyclones and anticyclones at the 850 mb level gradually gives way to one of troughs and ridges.

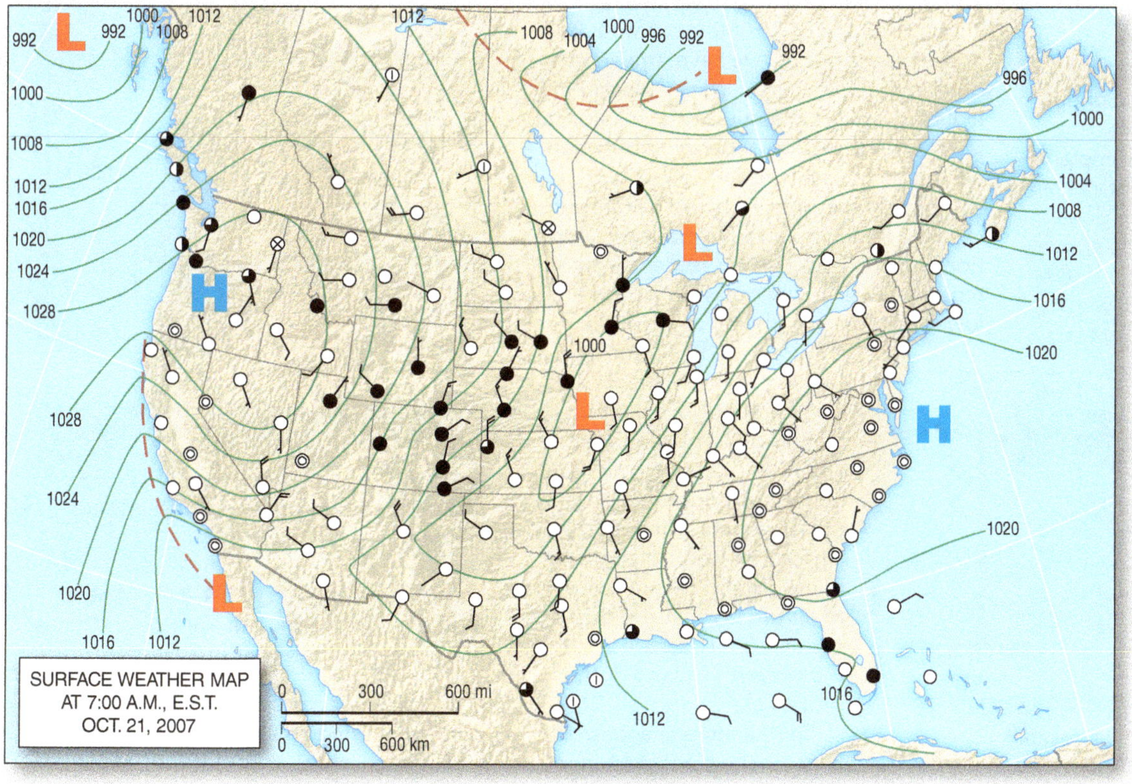

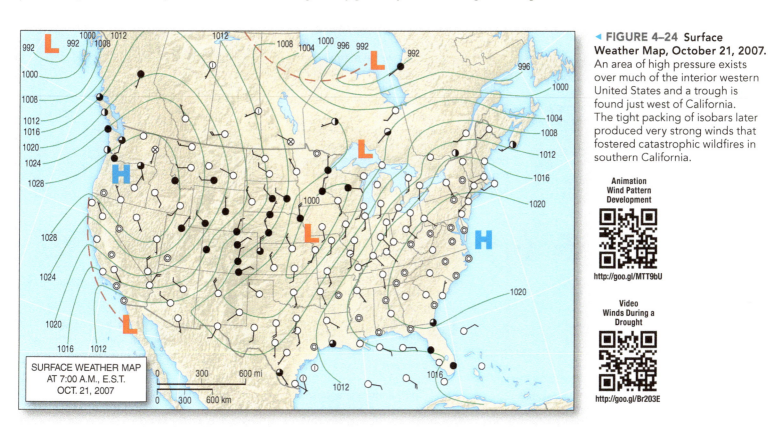

◄ **FIGURE 4–24 Surface Weather Map, October 21, 2007.** An area of high pressure exists over much of the interior western United States and a trough is found just west of California. The tight packing of isobars later produced very strong winds that fostered catastrophic wildfires in southern California.

Animation
Wind Pattern Development
http://goo.gl/MTT9bU

Video
Winds During a Drought
http://goo.gl/Br203E

Measuring Wind

4.10 Describe how meteorologists measure wind.

Wind direction and velocity are measured at all major weather stations. Direction is always given as that from which the wind blows, so that a "westerly" wind is one from the west. It is often expressed by its azimuth. As shown in Figure 4–25, the **azimuth** is the degree of angle from due north, moving clockwise. Thus, due north is represented as 0° (or 360°), east has an azimuth of 90°, south of 180°, and west of 270°. A simple device for observing wind direction is the **wind vane**.

Wind speeds are measured with anemometers. **Anemometers** have rotating cups mounted on a moving shaft. Wind blowing into the cups turns the shaft and generates an electrical current. The strength of the current is proportional to the wind speed, which is ultimately plotted on a strip chart or entered digitally into a computer.

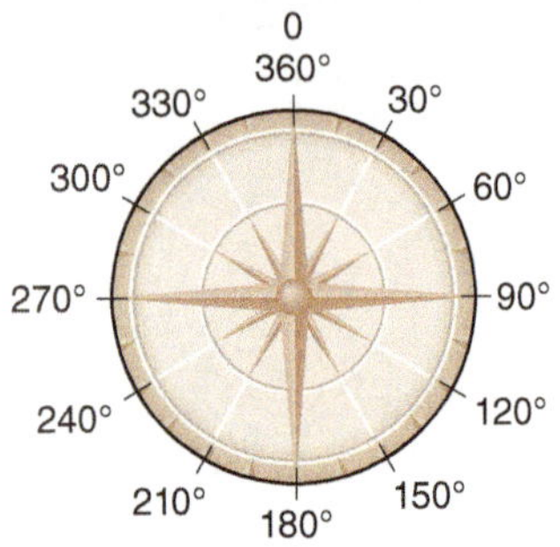

▲ **FIGURE 4–25 Wind Azimuth.** The azimuth expresses direction based on the angle away from north (0°), moving clockwise. Thus, for example, due east is represented as 90°, south 180°, and west 270°.

◄ **FIGURE 4–26 An Aerovane.** The propeller's rotation rate increases with wind speed. The aerovane has a propeller to generate an electroc.

Video
Forecasting
Wind Patterns

http://goo.gl/pliWTF

Looking like an airplane without wings, an **aerovane** (Figure 4–26) indicates both wind direction and speed. When the wind changes direction, it pushes against the tail and points the aerovane toward the wind.

Winds in the middle and upper troposphere are every bit as important to weather analysis as surface winds and are especially significant to aviation. Upper-level wind measurements are obtained by balloons whose movement is tracked by radar and GPS. Some measure only winds and are called **rawinsondes**, whereas others measure a variety of variables and are called **radiosondes**. These are launched twice a day at meteorological offices throughout the world.

4-6 FOCUS ON AVIATION

Using Altimeters to Monitor Altitude

To maintain the correct altitude throughout a flight most airplanes use a *pressure altimeter*—essentially an aneroid barometer with a scale reading in feet instead of inches of mercury, mb, or hPa. An altimeter does not directly measure "true" altitude above the ground or mean sea level. Instead, the altimeter reflects *pressure altitude*, which is the height of a given pressure surface.

When the pilot says "we have attained our cruising altitude of 33,000 feet," what he or she means is that "we have arrived at our assigned pressure surface." (Assigning flights to a given pressure surface helps to maintain safe vertical separation between aircraft on the same route).

Altimeter settings must be calibrated with sea level pressure. Without recalibration, a plane flying across isobars from high pressure to low pressure will be at a lower altitude than the altimeter indicates (Figure 4–6–1). To ensure that the plane will fly at a safe height above a mountain range, for example, the altimeter's settings must be updated frequently using air-pressure data from nearby airports or weather stations. Recalibrating the altimeter to air pressure at the destination airport assures that the aircraft will land at the proper elevation.

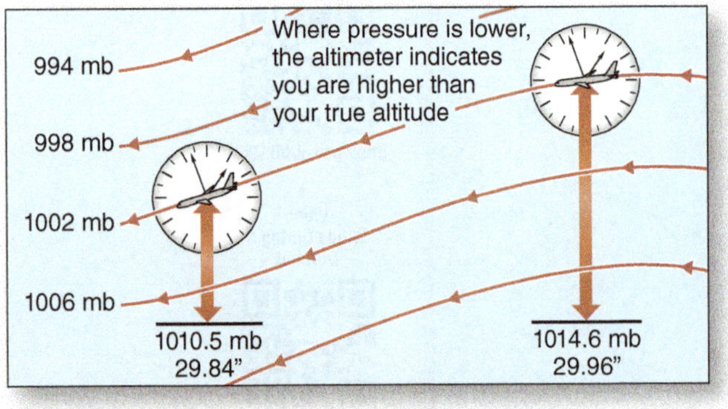

▲ **FIGURE 4–6–1** Effect of pressure on reliability of altimeter readings for an uncorrected altimeter.

Temperature changes beneath the plane also affect altimeter readings (Figure 4–6–2). Where temperature is lower, the altimeter indicates you are higher than your true altitude.

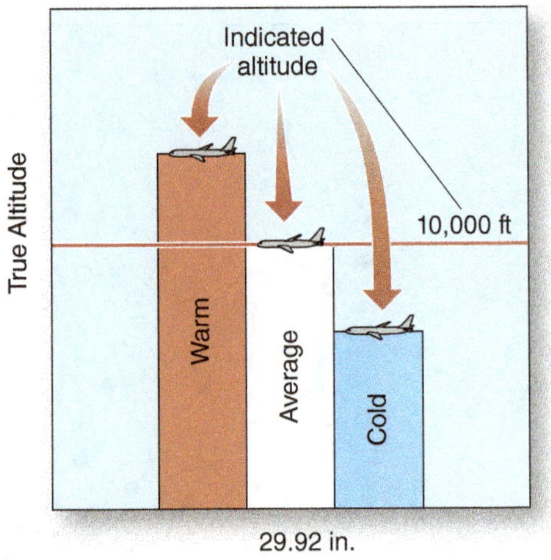

▲ **FIGURE 4–6–2** Relationship between temperature, indicated altitude, and true altitude.

4–6–1 What factors must a pilot take into account in using an altimeter to keep an aircraft at a constant, true altitude? Explain.

Summary

4.1 Explain the concept of air pressure.

- Air pressure arises from the weight of overlying atmosphere.
- Higher pressure corresponds to more air molecules in the air column.

4.2 Describe how pressure changes vertically and horizontally in the atmosphere.

- Pressure always decreases vertically (moving upward from the surface).
- Pressure varies horizontally by only a small percentage over great distances, but this is enough to set the atmosphere in motion.

4.3 Apply the equation of state to calculate air density.

- At constant density, pressure increases with temperature. At constant pressure, density increases with temperature. In general, all three are variable, so there is no universal relationship between any two variables.

4.4 Identify devices used to measure air pressure.

- Mercury barometers rely on the weight of the atmosphere to lift a column of mercury to a height that reflects prevailing pressure.
- Aneroid barometers use compression and expansion of a metal chamber to sense variations in pressure.

4.5 Describe the distribution of air pressure across the globe, at sea level, and in the upper atmosphere.

- Surface pressure largely follows elevation, with high elevations having lower pressure than sea-level locations.
- In the upper atmosphere, pressure decreases moving from warm areas toward cold. Equivalently, isobaric heights decrease moving toward colder locations.

4.6 Describe how pressure gradients in the upper atmosphere arise and how they are mapped.

- Because cold air is denser than warm air, in a cold air column pressures aloft are usually lower than at the same altitude in a warm column.
- Isobaric surfaces slope downward from warm air toward cold air.
- Pressure gradients aloft are mapped using the height of standard pressure surfaces.

4.7 Explain the factors that affect wind speed and direction.

- The pressure gradient force gives rise to winds.
- The Coriolis force deflects wind to the right in the Northern Hemisphere and to the left in the Southern Hemisphere.
- Friction opposes the direction of motion and slows the wind. It is important in the lowest few kilometers of the atmosphere but not at higher altitudes.

4.8 Describe winds in relation to pressure gradients in the upper atmosphere and near the surface.

- Near the surface, winds cross the isobars. Above the friction layer, wind is parallel to the pressure gradient (geostrophic or gradient flow).

4.9 Explain the forces that produce anticyclones, cyclones, troughs, and ridges.

- Differences in heating and cooling cause expansion and contraction of air columns, leading to pressure gradients.
- These are often expressed as low-pressure features in the form of cyclones at low levels and troughs aloft. The corresponding high-pressure features are called anticyclones and ridges.

4.10 Describe how meteorologists measure wind.

- Aerovanes are used near the surface to capture both wind direction and speed.
- Balloons (rawinsondes and radiosondes) are used to track winds aloft. Thousands of them are launched twice a day by weather agencies throughout the world.

Key Terms

acceleration *p. 118*
aerovane *p. 142*
air parcels *p. 133*
anemometer *p. 142*
aneroid barometer *p. 124*
anticyclones *p. 138*
azimuth *p. 142*
barograph *p. 124*
barometric pressure *p. 122*
Coriolis force *p. 132*
cyclones *p. 138*
Dalton's law *p. 119*
equation of state or **ideal gas law** *p. 121*
force *p. 118*
free atmosphere *p. 133*
friction *p. 133*
geostrophic flow or **geostrophic wind** *p. 136*
gradient flow or **gradient wind** *p. 137*
gravity *p. 118*
hydrostatic equation *p. 129*
hydrostatic equilibrium *p. 128*
isobar *p. 124*
kilopascal (kPa) *p. 118*
local vertical direction *p. 134*
mercury barometer *p. 122*
millibar (mb) *p. 118*
pascal (Pa) *p. 118*
planetary boundary layer or **boundary layer** *p. 133*
pressure *p. 118*
pressure gradient *p. 124*
pressure gradient force *p. 125*
radiosondes *p. 142*
rawinsondes *p. 142*
ridges *p. 139*
sea-level pressure *p. 120*
speed *p. 118*
surface pressure *p. 120*
troughs *p. 139*
velocity *p. 118*
wind vane *p. 142*

Review Questions

1. What is a partial pressure?
2. What is Dalton's law?
3. How do speed and velocity differ? How do force and pressure differ?
4. Why does pressure always decrease with altitude?
5. What is the difference between surface pressure and sea-level pressure?
6. What are the equation of state and the hydrostatic equation, and what do they tell us?
7. What two variables determine air pressure?
8. Describe how mercury and aneroid barometers measure air pressure, and explain why corrections need to be used for the observations made from one but not the other.
9. In what way is it misleading to express pressure in inches of mercury?
10. Explain the concept of hydrostatic equilibrium.
11. Explain how air temperature affects the vertical pressure gradient.
12. Describe the roles (if any) that wind speed, latitude, and direction of motion have in determining the magnitude of the Coriolis force.
13. What is meant by "rotation around the local vertical"? How does it vary with latitude?
14. What are geostrophic and gradient flows? Why don't they occur near the surface?
15. Explain how the pressure gradient force, the Coriolis force, and friction determine the movement of air in the free atmosphere and in the planetary boundary layer.
16. What do anemometers and aerovanes measure?
17. Define the terms *cyclone, anticyclone, trough,* and *ridge*.
18. Briefly describe the movement of air around cyclones and anticyclones in the Northern and Southern Hemispheres.

Critical Thinking

1. Would a particular pressure gradient produce the same exact wind speed over an Arizona desert that it would over a dense forest of tall trees? Why or why not?
2. On a particular day, the vertical pressure gradient at the surface is −11 pascals per meter. What is the vertical pressure gradient in units of millibars per kilometer? Would you be able to use this gradient to exactly determine the pressure at the top of a building 200 m tall?
3. If a low-pressure region were to instantaneously replace a high-pressure system (assuming normally encountered values of high and low pressure), do you think you would be able to notice the difference by the pressure in your ears? Why or why not?
4. Pressurized cans of shaving cream advise users not to expose the product to excessive heat. What might happen if that advice is not followed? Will this potential problem remain throughout the life of the product?
5. The pilot of a small plane wants to fly at a constant height above the surface. Can the pilot fly at a constant pressure level (such as 500 mb) to ensure the constant height above the ground? Why or why not?
6. Consider a 90-story skyscraper with high-speed elevators. Would a person ascending from the 46th to the 90th floor undergo the same degree of ear popping as a person ascending from the 1st floor to the 45th? Why or why not?
7. A rule of thumb is that the 850 mb level often marks the boundary between the free atmosphere and the boundary layer. Are there parts of North America where this relationship is likely not to be valid? If so, where?

Problems & Exercises

1. Refer to any website below that produces surface and 500 mb maps. Examine the current surface map and identify the major cyclones, anticyclones, troughs, and ridges at the surface. Then look at the 500 mb map. Does the same general pattern emerge? Do the troughs and ridges at the 500 mb level occur directly over the corresponding features on the surface map? (Relationships between the surface and 500 mb levels will be discussed further in Chapter 10.)
2. Examine today's 500 mb weather map. You will probably find a trend toward decreasing 500 mb heights with increasing latitude. Are there any exceptions on the map to that general pattern? If so, observe the surface temperatures across North America. Do the temperature patterns

have any association with the 500 mb pattern? If so, describe them.

3. On a daily basis, go to the Weather Channel's website at www.weather.com. Read the narrative describing the general weather pattern across the United States and identify the most notable weather events occurring across the country. Then look at the surface and 500 mb weather maps. This process will help you to become more familiar with normal pressure distributions and the type of weather often associated with them. As you proceed through this text, the pressure patterns and their association with daily weather will become more meaningful to you.

4. Observe a surface weather map that plots isobars and station models A good resource is www.atmo.arizona.edu. At this website, search for maps and charts under the weather tab. Do the airflow patterns around cyclones and anticyclones shown by the station models completely correspond with the generalizations made in this chapter? If not, why not?

Visual Analysis

This satellite image shows a large, powerful storm over the North Atlantic Ocean on March 26, 2014.

4.1. The wispy faint clouds blowing from west to east in the top part of the image are far above the friction layer. Assuming gradient flow, draw lines showing the orientation of height contours.

4.2. At lower levels do clouds appear to spiral into or out of the storm's center? Is this a cyclone or anticyclone?

▲ **Superstorm of March 2014.** After crossing the continental U.S., the storm intensified over the Atlantic Ocean.

PART

2 Water in the Atmosphere

CHAPTER 5

Atmospheric Moisture

CHAPTER 6

Cloud Development and Forms

CHAPTER 7

Precipitation Processes

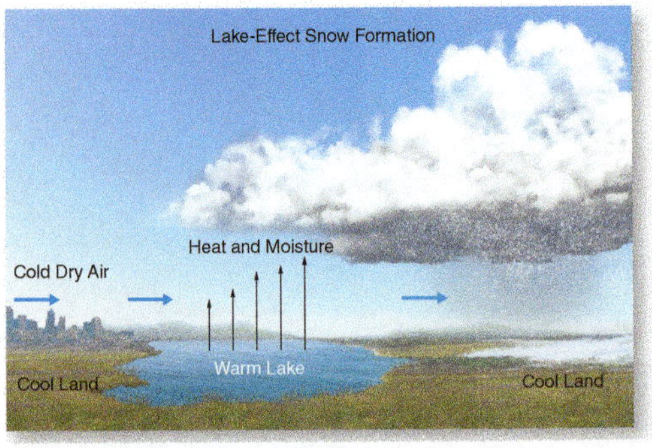

◀ *Snow, ice, water and fog in Patagonia, in South America.*

Water exists in the atmosphere in three distinct phases: vapor, liquid, and ice. This part of the book describes the conditions and processes involved in the transformation of water through these phases that lead to the creation of clouds. The discussion then turns to the processes by which liquid water and ice crystals in clouds grow sufficiently to fall as precipitation, and the processes that determine the type of precipitation to occur.

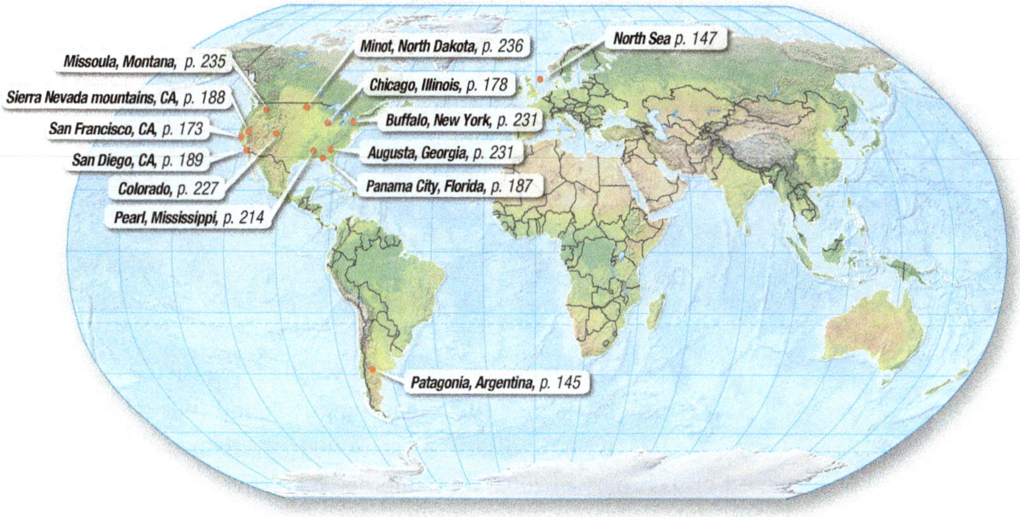

CHAPTER

5 Atmospheric Moisture

When one thinks of dangerous weather, fog isn't usually the first thing that comes to mind. But fog can have disastrous consequences, as happened on February 5, 2002, on California State Highway 99, south of Fresno. As a widespread fog began to set in that reduced visibility to as little as 15 m (50 ft), California Highway Patrol "pace cars" were dispatched to lead traffic at safe speeds. But, as has happened many times in the past, the effort was unsuccessful. Eighty-seven cars were involved in two chain reaction pileups that left two persons dead and many others injured. Just one month earlier, a similar set of accidents caused two other fatalities along Interstate 5 near the Sacramento airport. Just how thick are these fogs? There are restaurant billboards along Highway 99 touting pea soup as thick as the fog.

Local authorities have been proactive since 2002. They have installed weather stations, cameras, and sensors that detect the fog, along with 39 electronic message signs along Highway 99, to reduce the number of crash events and the number of vehicles involved in such occurrences.

◀ *Clouds can be produced by several processes and take on many forms. Here in the North Sea a very low fog called sea smoke forms just above water that is warmer than the air. Slightly higher, turbulence in the wake of wind turbines promotes vigorous mixing of cold and warm air and leads to thicker clouds. At high levels wispy clouds arise from more common processes described in this chapter.*

Learning Outcomes

After reading this chapter, you should be able to:

- **5.1** Describe the hydrologic cycle.
- **5.2** Explain the concept of saturation.
- **5.3** Identify the indices used in measuring the atmosphere's water vapor content.
- **5.4** Describe how humidity is measured.
- **5.5** Describe how water vapor is distributed in the atmosphere.
- **5.6** Explain the processes that lead to saturation.
- **5.7** Explain the factors that affect condensation.
- **5.8** Describe how diabatic and adiabatic processes produce cooling and condensation.
- **5.9** List different forms of condensation.
- **5.10** Describe the distribution of fog.
- **5.11** Describe the formation and dissipation of cloud droplets.
- **5.12** Explain how the effects of humidity on human discomfort are rated.
- **5.13** Describe possible effects of global warming, including its effects on evaporation rates and atmospheric water vapor content.

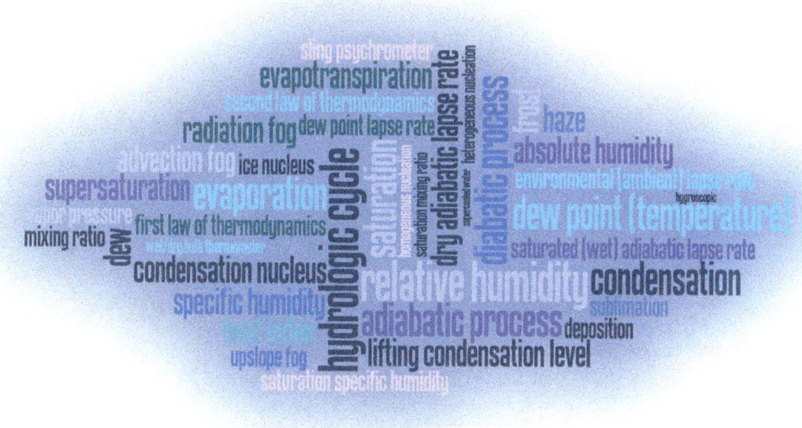

The Hydrologic Cycle

5.1 Describe the hydrologic cycle.

Precipitation in all its forms—rain, snow, hail, and so on—is for many people the most notable feature of the atmosphere. Though the amount and timing of precipitation vary markedly from one region of Earth to another, the total amount of precipitation for the entire globe is relatively constant from one year to the next—at about 104 cm (41 in.) per year. And yet the atmosphere doesn't run out of water! Clearly, there is continual replenishment of water lost through precipitation. In other words, just as we have described for other gases, water vapor is constantly added to and removed from the atmosphere. As long as these exchanges balance, the atmospheric store will remain constant. The movement of water between and within the atmosphere and Earth is referred to as the **hydrologic cycle**. As it happens, the hydrologic cycle, depicted in Figure 5–1, is among the fastest of all geochemical cycles; atmospheric residence time for water vapor is only 10 days or so. The hydrologic cycle is a continuous series of processes that occur simultaneously. As with any cycle, the entire process has no real end or beginning.

Let's begin this discussion with the evaporation of water from Earth's surface into the atmosphere. Evaporation can occur directly from the oceans or from water bodies on land surfaces (such as lakes and rivers). It can also occur indirectly through plants via a process called **transpiration**. The combined effects of transpiration and evaporation are referred to as **evapotranspiration**. The water vapor that goes into the air eventually becomes water droplets or ice crystals in the form of clouds or fog. Many fog and cloud droplets or crystals will evaporate back into the air, but others precipitate down to the surface.

The shortest route the cycle can take is from the ocean to the atmosphere and back to the ocean. The situation is a bit more involved if water precipitates onto land. There, some precipitation might not reach the surface directly and may instead fall onto vegetation, accumulating as a coating of water or ice. The water that undergoes this process, called *interception*, might then drip or trickle down (after melting, if the precipitation fell as snow) the plant to the surface or evaporate back into the air. In some environments a sizable percentage—as much as 40 percent in some situations—of the precipitation that has fallen can be evaporated back into the atmosphere after interception.

Rainfall that does reach the land surface, either directly or after interception, might then flow above the surface into rivers, which then transport the water into a lake or ocean, or it can evaporate directly back into the atmosphere. If the precipitation falls as ice (the most obvious example being snow), it might temporarily remain on the ground before melting, or it might be locked away for eons as part of a glacier.

Liquid water at the ground does not always flow along the land surface but instead penetrates into the ground in a process called *infiltration*. Such water is pulled downward by gravity and can collect in the pores of underlying soil or rock as *groundwater*. Much of this groundwater eventually seeps into rivers for eventual transport toward the ocean, where the cycle continues. But much is almost immediately withdrawn by plants and transpired to the atmosphere. Still smaller quantities enter the animal kingdom as plants are consumed by browsers of all sizes, from microorganisms to elephants. And an ever-increasing fraction is drawn off by humans for agriculture (from which most is transpired), industry, and residential uses.

Video
Hydrologic
Cycle
http://goo.gl/h7pTt6

Animation
Earth's Water
and the
Hydrologic Cycle
http://goo.gl/a10BWC

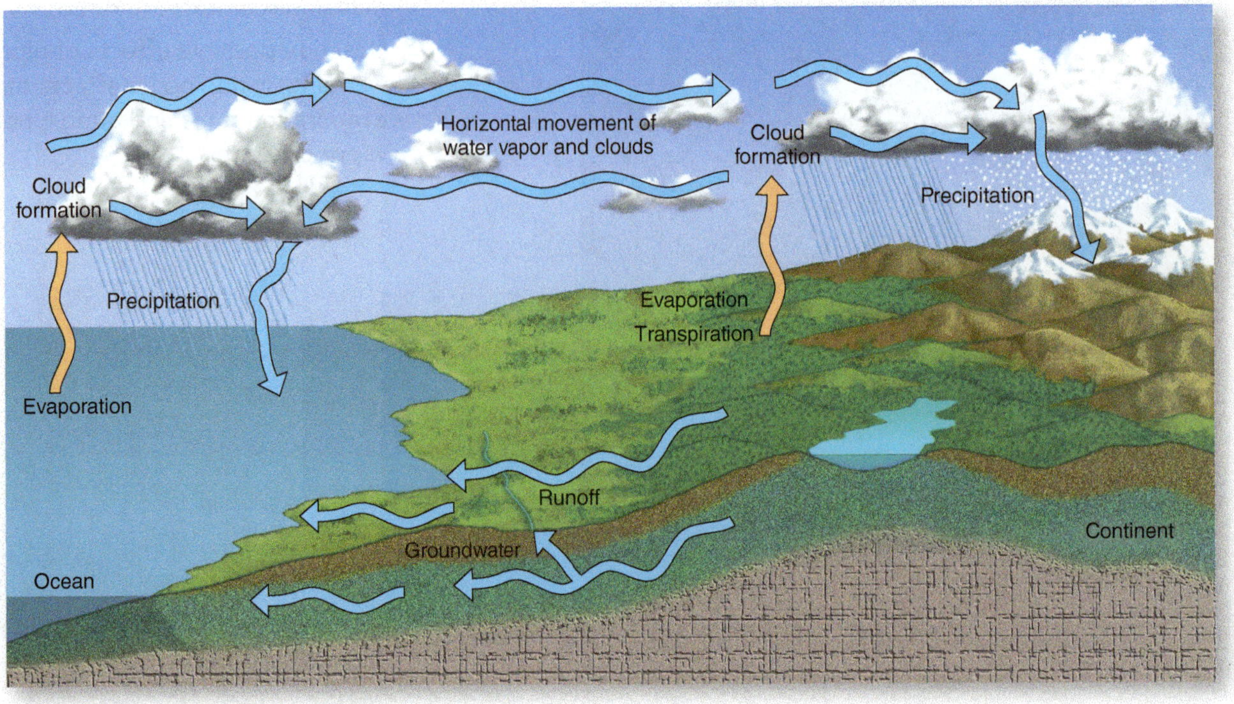

▲ **FIGURE 5–1** The Hydrologic Cycle.

DID YOU KNOW?

Just like the planet on which you live, you have your own hydrologic cycle. The human body is 50 to 60 percent water, on average, and none of the water molecules in your body reside there permanently. In Chapter 1 we saw that the residence time of water vapor in the atmosphere can be calculated by dividing its mass by the rate at which it enters and leaves the atmosphere. The same calculation can be made for the water in your body, and that calculation yields a residence time of about 14 days—not all that different from the residence time of water in the atmosphere (10 days). Also, most of the water molecules currently in your body are ancient, because water is a very stable molecule. Thus, those molecules in your body are well traveled and have probably been a part of the lives of many a famous person—at least temporarily.

This chapter deals with water in its various states and its transformation to and from the liquid and solid phases in the atmosphere. The two chapters following will examine clouds, the processes that form them, and precipitation. In reading those sections, keep in mind that the processes discussed represent individual components of the hydrologic cycle.

CHECKPOINT

5.1 Explain two processes in the hydrologic cycle by which water vapor enters the atmosphere.

5.2 Describe the steps and processes by which a molecule of water vapor in the atmosphere could return to the ocean via the land.

Water Vapor and Liquid Water

5.2 Explain the concept of saturation.

Although matter in the gaseous phase is highly compressible, the density of a gas cannot be increased to an arbitrarily high level. At some point a limit is reached, forcing a change to a liquid or solid state. For one atmospheric gas, water vapor, that limit is routinely achieved at temperatures and pressures found on Earth. (Other gases, such as nitrogen and oxygen, can be liquefied only at very low temperatures.) Air that contains as much water as possible is said to be *saturated*,[1] and the introduction of additional water vapor results in formation of water droplets or ice crystals. The concept of saturation is fundamental to understanding the processes that form clouds and fog. We begin our discussion with a hypothetical laboratory experiment that describes the general principles of evaporation and condensation. We then apply those principles to the processes that take place in the real atmosphere.

Evaporation and Condensation

Figure 5–2 depicts a hypothetical experiment, in which a tightly sealed container is partially filled with pure water (H_2O). Although it may seem obvious at this juncture, let's stipulate that the water in the jar has a perfectly flat surface. Furthermore, assume that at the onset of the experiment the water surface is covered by an impermeable coating, so no water vapor exists in the volume of the container above the water surface. Whether the volume above the water surface contains any air is entirely irrelevant to this experiment. The volume can contain normal air, pure hydrogen, methane, or fumes from French perfume—it can even be a complete

[1] As we will see later in this chapter, the concept of saturated air is a little more complex than is given here. Nonetheless, this description works well in the early stages of this discussion.

Animation
Water Phase Changes

http://goo.gl/ZChN5n

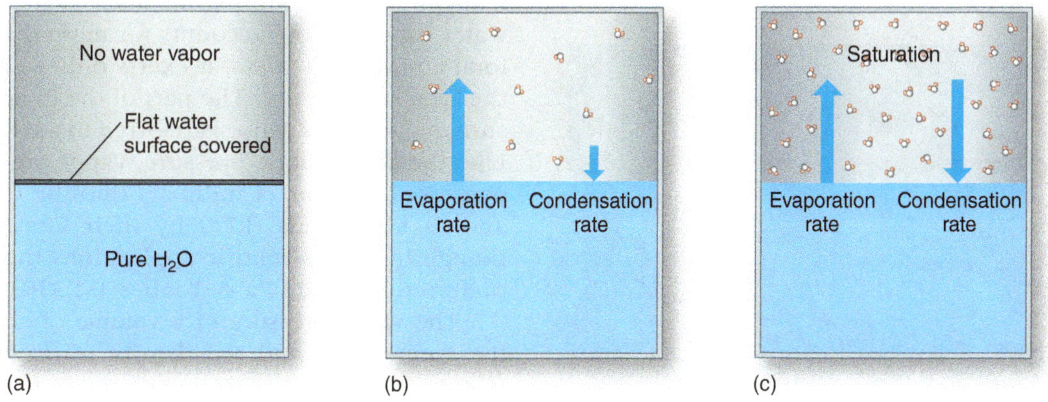

▲ **FIGURE 5–2 Evaporation, Condensation, and Saturation.** (a) The diagrams show a hypothetical jar containing pure water with a flat surface and an overlying volume that initially contains no water vapor. (b) When evaporation begins, water vapor accumulates in the volume above the liquid water. Initially, no condensation can occur because of the absence of water vapor above the liquid. But as evaporation contributes moisture to the overlying volume, some condensation can occur. (c) Evaporation exceeds condensation for a while and thereby increases the water vapor content. Eventually enough water vapor is above the liquid for condensation to equal evaporation. At this point, saturation occurs.

vacuum. All that matters with respect to the evaporation/condensation process is that no water vapor be present initially.

Figure 5–2b shows what happens when we remove the covering on the liquid water surface. Without the covering, some of the molecules at the surface can escape into the overlying volume as water vapor. The process whereby molecules break free of the liquid volume is known as **evaporation**. The opposite process is **condensation**, wherein water vapor molecules randomly collide with the water surface and bond with adjacent molecules. At the beginning of our hypothetical experiment, no condensation could occur because no water vapor was present. As evaporation begins, however, water vapor starts to accumulate above the surface of the liquid.

At the early stages of evaporation, the low water vapor content prevents much condensation from occurring, and the rate of evaporation exceeds that of condensation. This leads to an increase in the amount of water vapor present. With increasing water vapor content, however, the condensation rate likewise increases. Eventually, the amount of water vapor above the surface is enough for the rates of condensation and evaporation to become equal, as shown in Figure 5–2c. A constant amount of water vapor now exists in the volume above the water surface due to offsetting gains and losses by evaporation and condensation. The resulting equilibrium state is called **saturation**. When this equilibrium exists in the atmosphere, the air is said to be saturated.

The state of saturation described here can occur whether or not air (or other gases, for that matter) exists in the container. In other words, the water vapor is not "held" by the air (although this erroneous statement is frequently made). Water vapor is a gas, just like the other components of the air. Thus, it does not need to be "held" by air any more than the oxygen, nitrogen, argon, and other gases of the atmosphere need to be held by water vapor! When the air is saturated, there is simply an equilibrium between evaporation and condensation; the dry air plays no role in achieving this state. It is also important to realize that the exchange of water vapor and liquid described here applies as well to the change of phase between water vapor and ice. The change of phase directly from ice to water vapor, without passing into the liquid phase, is called **sublimation**. The reverse process (from water vapor to ice) is called **deposition** (Figure 5–3).

▼ **FIGURE 5–3 Deposition.** This process refers to the direct transfer of vapor to ice.

CHECKPOINT

5.3 What are evaporation and condensation?

5.4 In the hypothetical experiment described, how do rates of evaporation and condensation change until saturation is achieved?

Indices of Water Vapor Content

5.3 Identify the indices used in measuring the atmosphere's water vapor content.

Humidity refers to the amount of water vapor in the air. Humidity can be expressed in a number of ways—in terms of the density of water vapor, the pressure exerted by the water vapor, the percentage of the amount of water vapor that can actually exist, or several other methods. There is no single "correct" measure, but, rather, each has its own advantages and disadvantages, depending on the intended use. All measures of humidity have one thing in common, however—they apply exclusively to water vapor, and not to liquid droplets or ice crystals suspended in or falling through the air. Let's now take a look at these measures.

Vapor Pressure

In Chapters 1 and 4, we saw that the air exerts pressure on all surfaces. Each gas that makes up the atmosphere contributes to the total air pressure, with the most abundant permanent gases accounting for most of the pressure. Because water vapor seldom accounts for more than 4 percent of the total atmospheric mass, it exerts only a small percentage of the total air pressure. The part of the total atmospheric pressure due to water vapor is referred to as the **vapor pressure**. Like the atmospheric pressure, vapor pressure is commonly expressed in units of millibars (mb) by U.S. meteorologists, and as kilopascals (kPa) by their Canadian counterparts, though in most scientific applications the pascal (Pa) is the preferred unit (100 Pa = 1 mb = 0.1 kPa).

The vapor pressure of a volume of air depends on both the temperature and the density of water vapor molecules (Figure 5–4). If the air temperature is high, water vapor

▸ **FIGURE 5–4 Vapor Pressure.** The movement of water vapor molecules exerts a pressure on surfaces, called *vapor pressure*. The vapor pressure increases with concentration and temperature.

molecules (along with all the other gaseous constituents of the atmosphere) move more rapidly and exert a greater pressure. Similarly, a greater concentration of water vapor molecules means that a greater amount of mass is available to exert pressure. In practice, temperature influences are small compared to density changes, so vapor pressure closely follows changes in the density or abundance of water molecules.

Because there is a maximum amount of water vapor that can exist, there is a corresponding maximum vapor pressure, called the **saturation vapor pressure**. The saturation vapor pressure does not represent the current amount of moisture in the air; rather, it is an expression of the maximum that *can* exist. The saturation vapor pressure depends on only one variable—temperature. Figure 5–5 shows the relationship between saturation vapor pressure and temperature, with higher temperatures having higher saturation vapor pressures. For example, at 40 °C the saturation vapor pressure is 73.8 mb, while at 0 °C it is only 6.1 mb, less than one-tenth as much.

The increase in saturation vapor pressure with temperature is not linear. At low temperatures only a modest increase in saturation vapor pressure is seen, but at high temperatures saturation vapor pressure grows rapidly. For example, a 2 °C increase in temperature, from 0 °C to 2 °C, increases the saturation vapor pressure from 6.1 to 7.1 mb, only a 1 mb difference. Raising the temperature the same amount from a higher starting point, from 40 °C to 42 °C, raises the saturation vapor pressure by 7.7 mb, from 73.8 to 81.5 mb. This nonlinear behavior is captured in a simple statement: At temperatures normally encountered near Earth's surface the saturation vapor pressure approximately doubles for every 10 °C (18 °F) increase in temperature.

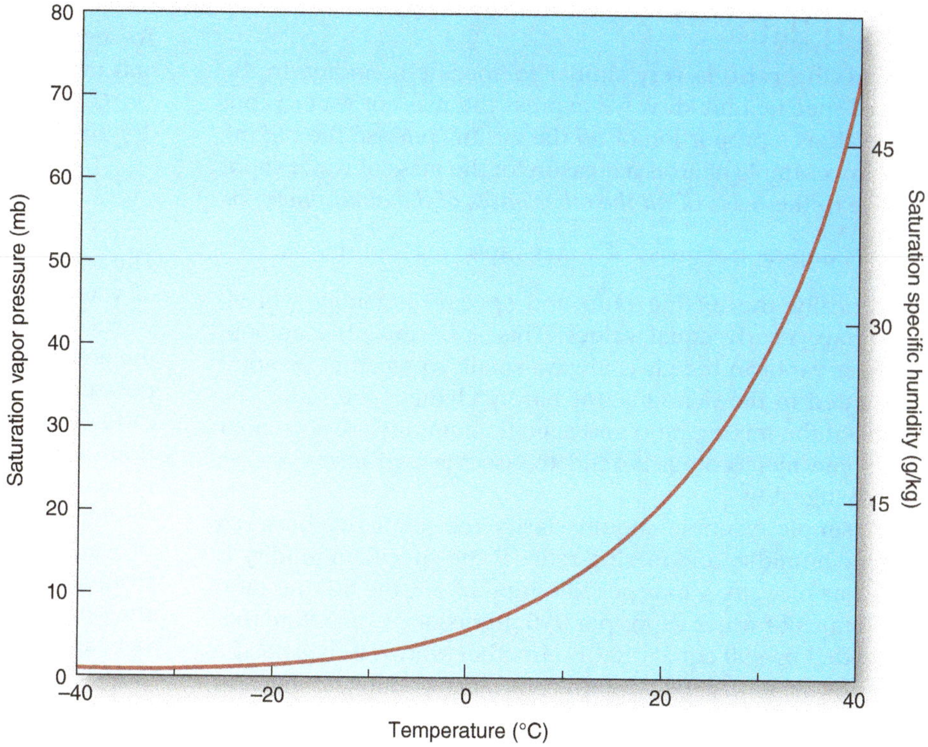

▴ **FIGURE 5–5 Saturation Vapor Pressure, Saturation Specific Humidity, and Temperature.** The curve is steeper at higher temperatures, meaning that saturation vapor pressure is more sensitive to temperature changes when the air is warm.

Absolute Humidity

Another measure of water vapor content is the **absolute humidity**, which is simply the density of water vapor, expressed as the number of grams of water vapor contained in a cubic meter of air. Because absolute humidity represents the amount of moisture contained in a volume of air, its value changes whenever air expands or contracts. Thus, for example, if an air parcel expands (as it does when it is heated or lifted upward), its absolute humidity will fall, even though no water vapor is removed from the parcel. Because absolute humidity suffers from this drawback and has no strong advantage over any other index, it is not widely used.

Specific Humidity

Although not normally encountered outside scientific applications, **specific humidity** is a useful index for representing atmospheric moisture. Specific humidity expresses the mass of water vapor existing in a given mass of air. Consider, for example, a volume containing exactly 1 kg of air (at sea level such a volume would be about 0.8 cubic meter, or about 27 cu ft). Of that kilogram, some number of grams would be water vapor. The proportion of the atmospheric mass accounted for by water vapor is the specific humidity. Most often, specific humidity is expressed as the number of grams of water vapor per kilogram of air. Because the water vapor outside the tropics usually is less than 2 percent of the mass of the air near the surface, specific humidities are normally less than 20 grams of water vapor per kilogram of air. Specific humidity for a sample of air is formally expressed as

$$\text{Specific humidity} = (\text{grams of water vapor/kilogram of air})$$

Unlike vapor pressure, specific humidity is affected to a small degree by atmospheric pressure, because it depends in part on the total mass of the atmosphere.

Unlike absolute humidity, specific humidity has the advantage of not changing as air expands or contracts. When a kilogram of air expands, its mass is unchanged (it is still 1 kg), and the proportion that is water vapor is unchanged. As a result, the specific humidity is unaffected. Likewise, specific humidity is not temperature dependent. If a kilogram of air contains 1 g of water vapor, it still contains 1 g after heating. For this reason, specific humidity is a good indicator for comparing water vapor in the air at different locations whose air temperatures might be different from each other.

For example, if Toronto, Ontario, has a specific humidity of 10 grams of water vapor per kilogram of air on a given day and Albuquerque, New Mexico, has 5 g/kg, we can infer that Toronto has twice as much water vapor in the air as does Albuquerque, no matter what their temperatures are. This may not seem very profound, but the direct correspondence between specific humidity and water vapor content does not hold for the more frequently used index of moisture, relative humidity. Thus, specific humidity is a useful measure of water vapor whose only real drawback is the general public's unfamiliarity with the term.

Because there is a maximum amount of water vapor that can exist at a particular temperature, there is likewise a maximum specific humidity. This maximum is called the **saturation specific humidity**. This property is directly analogous to the saturation vapor pressure and increases in the nonlinear manner shown in Figure 5–5.

Mixing Ratio

The **mixing ratio** is very similar to specific humidity. In the case of specific humidity, we express the mass of water vapor in the air as a proportion of *all* the air. In contrast, the mixing ratio for a sample of air is a measure of the mass of water vapor relative to the mass of *all the other gases of the atmosphere*, or

$$\text{Mixing ratio} = \text{mass of water vapor}/\text{mass of dry air}$$

Numerically, the mixing ratio and specific humidity will always have nearly equal values. This is because the amount of water vapor in the air is always small, so whether or not it is counted in the denominator hardly changes the ratio. The values of the mixing ratio and specific humidity are so similar that some meteorologists tend to use the two terms almost interchangeably.

A simple example should clarify the similarity between specific humidity and mixing ratio. If the specific humidity is 10 grams of water vapor per kilogram of air, the mixing ratio is 10 grams of water vapor per 990 grams of dry air. Note that 10 divided by 990 equals 10.011. In other words, if the specific humidity is 10.0 g/kg, the mixing ratio is only 1.1 percent higher, or 10.011 g/kg.

Using the mixing ratio as an index of moisture content offers the same advantages as using specific humidity. The maximum possible mixing ratio is called the **saturation mixing ratio**. Values of the saturation mixing ratio are presented in Table 5–1.

TABLE 5–1
Saturation Mixing Ratios

Temperature in °C (°F)	Saturation Mixing Ratio (g/kg)
−40 (−40)	0.1
−30 (−22)	0.3
−20 (−4)	0.8
−10 (14)	1.8
0 (32)	3.8
5 (41)	5.4
10 (50)	7.6
15 (59)	10.6
20 (68)	14.7
25 (77)	20.1
30 (86)	27.2
35 (95)	36.6
40 (104)	49.0

Relative Humidity

The most familiar measure of water vapor content is **relative humidity (RH)**, which relates the amount of water vapor in the air to the maximum possible at the current temperature. The World Meteorological Organization defines the relative humidity[2] as

Relative humidity $$RH = (\text{mixing ratio}/\text{saturation mixing ratio}) \times 100\%$$

To see how this works, refer to Figure 5–6a, in which the actual mixing ratio is 6 g/kg of dry air, and the temperature of 14 °C (57.2 °F) yields a saturation mixing ratio of 10 g/kg of dry air. The relative humidity would thus be

$$RH = \frac{6}{10} \times 100\% = 0.6 \times 100\% = 60\%$$

The relative humidity is not uniquely determined by the amount of water vapor present. Because more water vapor can exist in warm air than in cold air, the relative humidity depends on both the actual moisture content and the air temperature. If the temperature of the air increases, more water vapor can exist and the ratio of the amount of water vapor in the air relative to saturation decreases. Thus, the relative humidity declines even if the moisture content is unchanged. To see a specific example, refer to Figure 5–6b and consider what would happen if the amount of water vapor remains constant, but the temperature increases from its original 14 °C to 25 °C (77 °F). At the new temperature, the saturation mixing ratio increases to 20 grams of water vapor per kilogram of dry air, and the relative humidity becomes

$$RH = \frac{6}{20} \times 100\% = 0.3 \times 100\% = 30\%$$

[2]The American Meteorological Society has a slightly different definition of relative humidity, wherein the vapor pressure is divided by the saturation vapor pressure. The two methods yield almost identical results.

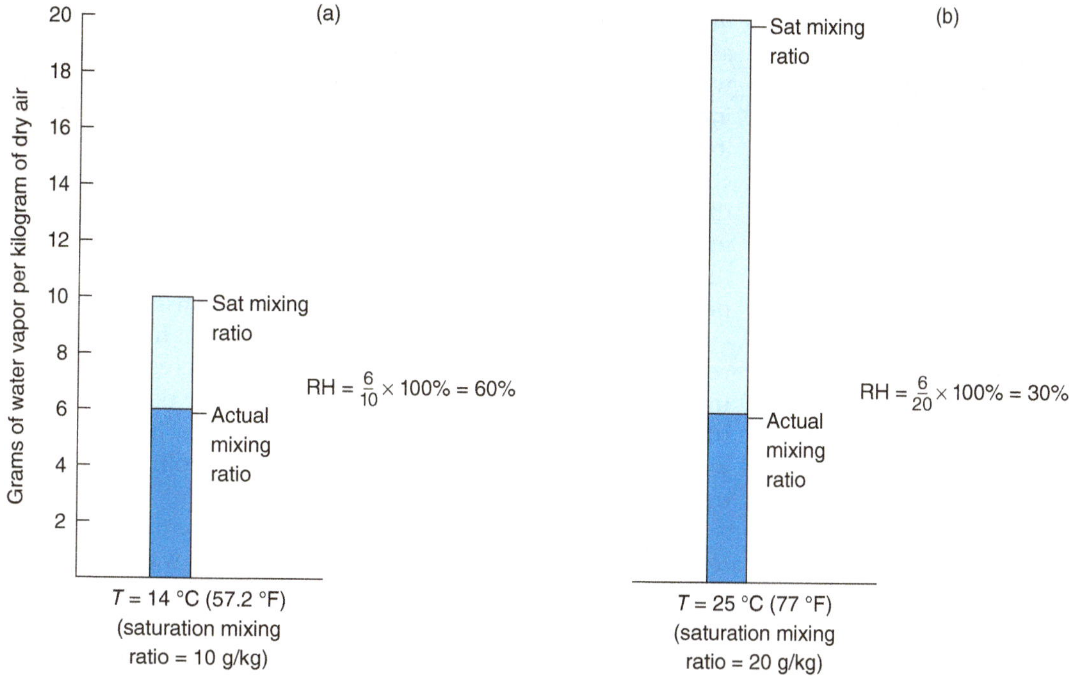

◀ **FIGURE 5–6 Relative Humidity.** This measure of moisture is dependent not only on the amount of water vapor in the air but also on the air temperature (which determines the saturation mixing ratio). Assume for both (a) and (b) that the mixing ratio is 6 grams of water vapor per kilogram of dry air. In example (a), we assume that the air temperature is 14 °C (57.2 °F), which (based on Figure 5–5) has a saturation mixing ratio of 10 grams of water vapor per kilogram of dry air; hence, the air contains 60 percent of what it could contain and the relative humidity is 60 percent. The same 6 grams of water vapor per kilogram of dry air is plotted in (b), but the temperature is now 25 °C (77 °F). At this temperature the saturation mixing ratio is 20 grams of water vapor per kilogram of dry air, so the relative humidity is only 30 percent—half of what it was in (a) even though the same amount of water vapor exists.

The relative humidity decreased even though the amount of water vapor remained constant! This is a significant drawback to any index that is supposed to be a measure of humidity.

Because of its dependence on temperature, the relative humidity will change throughout the course of the day even if the amount of moisture in the air is unchanged. Relative humidity is usually highest in the early morning—not because of abundant water vapor, but simply because the temperature is lower. As the day warms up, the relative humidity typically declines because the saturation mixing ratio increases. Figure 5–7 shows a typical pattern of daily temperature and relative humidity values. Notice how relative humidity varies by a factor of 3 over the course of the day. Almost all of that variation arises from the change in air temperature.

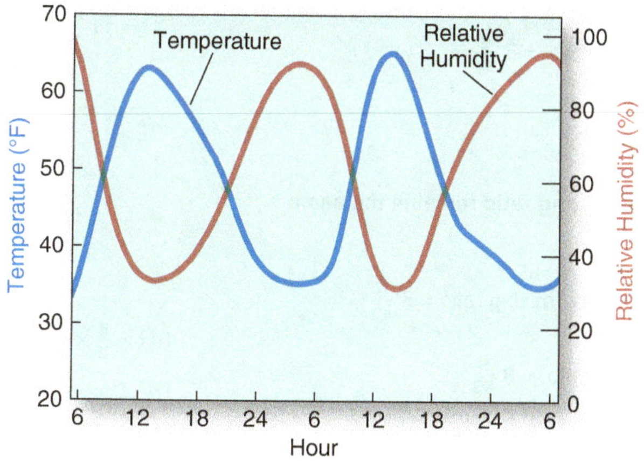

▲ **FIGURE 5–7 Temperature and Relative Humidity.** The graph shows a typical plot of temperature and relative humidity over a 48-hour period. Observe that as the temperature increases, the relative humidity decreases, and vice versa. In this example there is a substantial change in the relative humidity even though the actual water vapor content over the course of the day underwent only minimal changes. This shows the strong dependence of relative humidity on air temperature (and thus one of its serious limitations as an indicator of moisture content).

The influence of temperature on relative humidity creates another problem—it confounds direct comparisons of moisture contents at different places with unequal temperatures. Consider, for example, a cold morning in Montreal, Quebec, where the temperature is −20 °C (−4 °F) and the mixing ratio is 0.7 g/kg. At −20 °C, the saturation mixing ratio is 0.78 gram of water vapor per kilogram of dry air, and the resultant relative humidity is [(0.70 ÷ 0.78) × 100%], or 89.7%. Now compare that to the warmer situation at Atlanta, Georgia, where the temperature is 10 °C (50 °F) and the mixing ratio is 6.2 g/kg (nearly nine times greater than at Montreal!). At 10 °C, the saturation mixing ratio is 7.7 grams of water vapor per kilogram of dry air, so the relative humidity is [(6.2 ÷ 7.7) × 100%], or 80.5%. Notice that the relative humidity is lower at Atlanta than at Montreal despite the fact that it contains much more water vapor. This illustrates why relative humidity is a poor choice for comparing the amount of water vapor in the air at one place to that at another.

Some people are confused about the true meaning of relative humidity. Some believe the term represents the percentage of the air that is water vapor. This is not correct. To see why, consider an instance in which the relative humidity is 100 percent. If the air were 100 percent water vapor, it would include no nitrogen or oxygen, and we would have a difficult time breathing, let alone discussing water vapor content! Another common misperception is about how high the relative humidity can be on a hot, humid day. Many people would estimate that on such a day the relative humidity would be about 99 percent. But in reality, very hot days never have relative humidities approaching that value. That is because at high temperatures the saturation mixing ratio is very much higher than the actual mixing ratios likely to be encountered. For example, if the temperature is 35 °C (95 °F), the saturation mixing ratio is 36.8 grams of water vapor per kilogram of dry air. But we have seen that outside the tropics it is unusual for the mixing ratio to exceed 20 g/kg—even when the air is humid. Thus, a 99 percent relative humidity is not a realistic possibility at that temperature. Indeed, warm days can be extremely uncomfortable even with relative humidities of only about 50 percent.

DID YOU KNOW?

Humid air is not heavier than dry air. It is easy to imagine why people might have this misconception, given that water is much heavier than air. But that only applies for water in its liquid form, and when we discuss humidity we are dealing with *water vapor*. Water vapor is in fact less dense than either nitrogen or oxygen, the two most prevalent gases of the atmosphere. So when the humidity is high, those gases are replaced by the lighter water vapor. But this effect is very small, and the difference in atmospheric density with humidity variations is very small.

Dew Point

A useful moisture index that is free of the temperature relationship just described is the **dew point temperature** (or simply the **dew point**), the temperature at which saturation occurs. This quantity may seem confusing at first because it is expressed as a temperature, but it is a simple index to use and easy to interpret. And it is dependent almost exclusively on the amount of water vapor present.

Consider the parcel of unsaturated air in Figure 5–8. Initially the air temperature was 14 °C (57.2 °F), yielding a saturation mixing ratio of 10 grams of water vapor per kilogram of dry air. The initial mixing ratio was 8 g/kg. The relative humidity was therefore 80 percent. As the air cools, its relative humidity increases, and if the air is cooled sufficiently its relative humidity reaches 100 percent and it becomes saturated. Any further cooling leads to the removal of water vapor by condensation. In this example, the dew point is 10 °C (50 °F) because that is the temperature at which the saturation mixing ratio is 8 g/kg. Notice that even though the relative humidity increased as the temperature decreased, the dew point remained constant at 10 °C.

What would have happened if the specific humidity had remained constant and the temperature had increased from its initial 14 °C? The relative humidity would have decreased, yet the dew point would have remained constant. The dew point would not have changed because eventual cooling of the air to 8 °C still would have led to saturation.

The dew point is a valuable indicator of the moisture content; when the dew point is high, abundant water vapor is in the air. Moreover, when combined with air temperature, it is an indicator of the relative humidity. When the dew point is much lower than the air temperature, the relative humidity is very low. When the dew point is nearly equal to the air temperature, the relative humidity is high. Furthermore, when the air temperature and the dew point are equal, the air is saturated and the relative humidity is 100 percent.

EFFECT OF TEMPERATURE CHANGES ON DEW POINTS Unlike relative humidity, the dew point does not change simply because air temperature changes. Moreover, if one location has a higher dew point than another, it will also have a greater amount

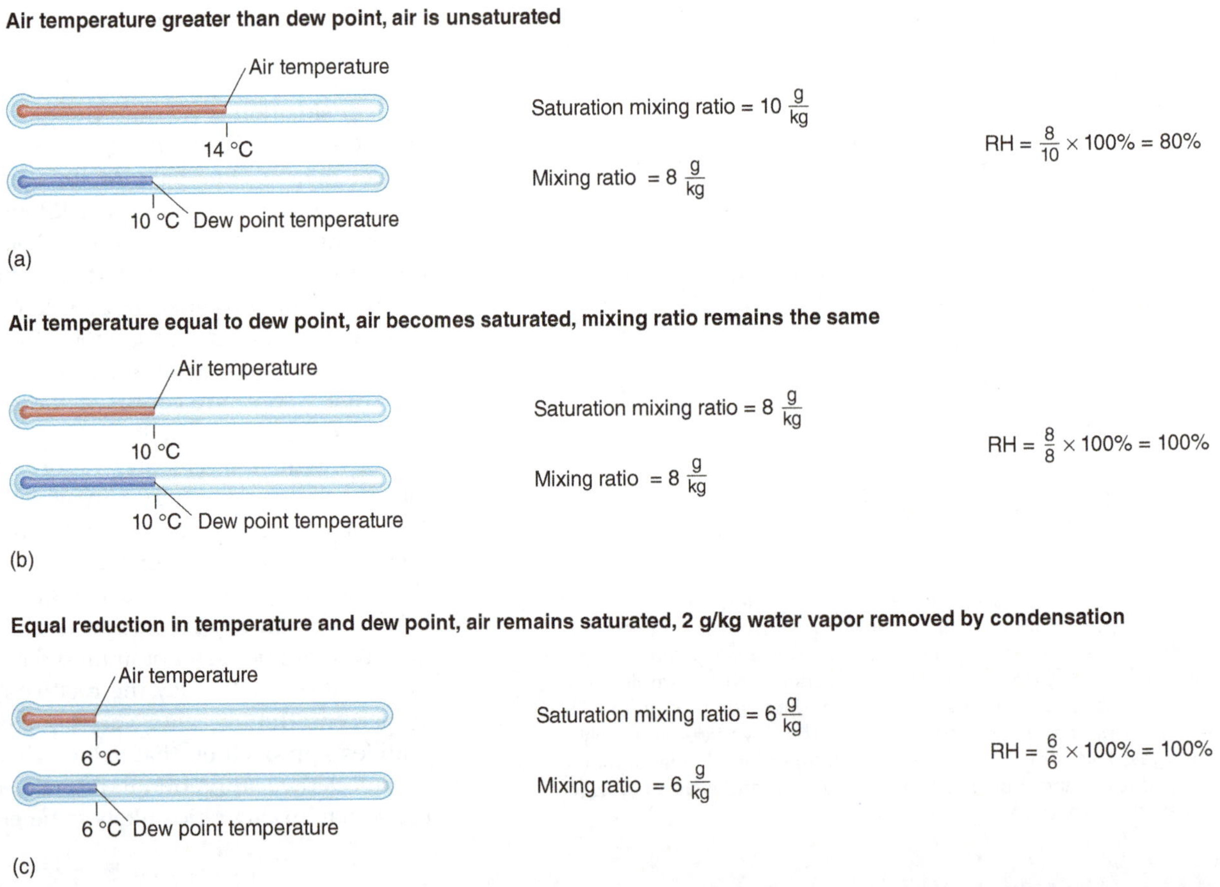

▸ **FIGURE 5–8 Dew Point and Water Vapor Content.** The dew point is an expression of water vapor content, although it is expressed as a temperature. (a) The temperature exceeds the dew point and the air is unsaturated. (b) When the air temperature is lowered so that the saturation mixing ratio is the same as the actual mixing ratio, the air temperature and dew point are equal. (c) Further cooling leads to an equal reduction in the air temperature and dew point so that they remain equal to each other.

of water vapor in the air, assuming the same air pressure. Once you are familiar with dew point, it is probably the most effective index of water vapor content. Dew points on very humid, hot days are typically in the low 20s on the Celsius scale (low 70s Fahrenheit). (When you see a dew point of 70 °F or higher, you can plan on a sleepless night unless you have air conditioning.) On comfortable days that are neither humid nor dry, dew points may be in the low teens Celsius (mid-50s Fahrenheit); very dry days can have dew points in the minus 20s or lower on the Celsius scale (0 °F). The dew point temperature can sometimes serve as a predictor of overnight cooling, as explained in *Box 5–1, Forecasting: Dew Point and Nighttime Minimum Temperatures.*

The dew point is always equal to or less than the air temperature; under no circumstances does it ever exceed the temperature. So what happens if the air temperature is lowered to the dew point and then cooled further? In that case, the amount of water vapor exceeds the amount that can now exist, and the surplus is removed from the air. This happens by condensation of the water vapor to form a liquid or by the formation of ice crystals. In either case, the dew point decreases at the same rate as the air temperature, and the two remain equal to each other. This is illustrated in Figure 5–8b and c. When the temperature was lowered to 10 °C in Figure 5–8b, the air became saturated with 8 grams of water vapor per kilogram of dry air. As the air cooled further to 6 °C (Figure 5–8c), the saturation mixing ratio decreased to 6 g/kg. Because the mixing ratio, by definition, cannot exceed the saturation mixing ratio, 2 grams of water vapor (8 grams minus 6 grams) had to be removed from each kilogram of dry air by condensation. The removal of water vapor kept the specific humidity equal to that of the saturation specific humidity and also lowered the dew point. Note that when the temperature at which saturation would occur is below 0 °C (32 °F), we use the term **frost point** instead of dew point.

MM MapMaster North America Physical Environment
Average Distribution of Dew Points January and July (U.S.)

CHECKPOINT

5.5 What is the dew point?

5.6 What happens if air temperature drops below the dew point?

Measuring Humidity

5.4 Describe how humidity is measured.

Considering the fact that water vapor is an invisible gas mixed with all the other gases of the atmosphere, you might suspect that its measurement would entail some highly sophisticated instrumentation. Such is not the case. The simplest and most widely used instrument for measuring humidity, the **sling psychrometer** (Figure 5–9a), consists of a pair of thermometers, one of which has a cotton wick around the bulb that is saturated with water. The other thermometer has no such covering and simply

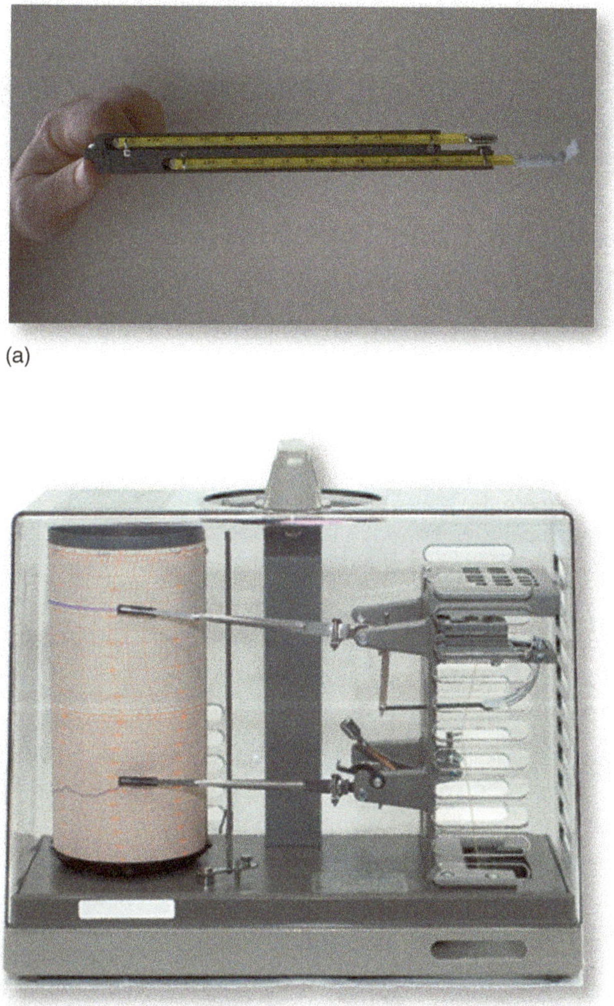

(a)

(b)

▲ **FIGURE 5–9 Measuring Humidity.** (a) Sling psychrometer and (b) hygrothermograph.

measures the air temperature. The two thermometers, called a **wet bulb** and **dry bulb thermometer**, respectively, are mounted to a pivoting device that allows them to be circulated ("slung") through the surrounding air. If the air is unsaturated, water evaporates from the wet bulb, whose temperature falls as latent heat is consumed. After about a minute or so of circulating, the amount of heat lost by evaporation is offset by the input of sensible heat from the surrounding, warmer air, and the cooling ceases. Thereafter, the wet bulb maintains a constant temperature no matter how long the instrument is swung around.

The difference between the dry and wet bulb temperatures, called the **wet bulb depression**, depends on the moisture content of the air. If the air is completely saturated, no net evaporation occurs from the wet bulb thermometer, no latent heat is lost, and the wet bulb temperature equals the dry bulb temperature. On the other hand, if the humidity is low, plenty of evaporation will take place from the wet bulb, and its temperature will drop considerably before reaching an equilibrium value. To determine the moisture content, first you note the difference between the dry and wet bulb temperatures. Then, using tables such as Tables 5–2 and 5–3, you obtain the dew point, relative humidity, or any other humidity measure by finding the value corresponding to the row for the air temperature and the column for the wet bulb depression.

5–1 FORECASTING

Dew Point and Nighttime Minimum Temperatures

Knowledge of the current dew point temperature is a useful tool to the forecaster for the prediction of the following morning's low temperature. If no major wind shifts or other weather changes are anticipated, the minimum temperature will often approximate the dew point. Consider a hypothetical evening with an air temperature of 15 °C (59 °F) and a dew point of 5 °C (41 °F). The spread between the air temperature and the dew point temperature is not very large, and a 10 °C (18 °F) lowering of the air temperature is feasible. If the air temperature does indeed drop to the dew point and there is little or no wind, a radiation fog has a good chance of forming (Figure 5–1–1). The fog would then inhibit further cooling, partly because latent heat is released and partly because water droplets are extremely effective at absorbing longwave radiation from the surface. Without the loss of radiation, the surface temperature would remain almost constant, and the overnight low would equal the dew point temperature.

The relationship between dew point and minimum temperature will not hold under certain conditions. The first has to do with the changes in the big weather picture. Imagine, for example, that a mass of warmer air is moving into the forecast region. This large body of air can replace the one present at the time of the forecast and bring with it higher nighttime temperatures. Similarly, the passage of an advancing cold front (briefly described in Chapter 1 and discussed in more detail in Chapter 9) can lead to significant drops in temperature below the current dew point.

Both heavy cloud cover and strong winds inhibit a drop in air temperature, and their presence may keep minimum air temperatures above the dew point. Cloud cover achieves this effect because of its absorption and downward reradiation of longwave energy. Strong winds prevent large temperature decreases at the surface by vertical mixing. A shallow layer of cold air that would otherwise develop is easily disrupted, leading to higher surface temperatures and more uniform temperatures with height.

Minimum temperatures won't go down to the dew point if the difference between the air temperature and the dew point temperature is very large. One can readily see how this might occur if a location has a high temperature of 45 °C (113 °F) and a dew point of 0 °C (32 °F). Even with calm winds and no cloud cover, a cooling of 45 °C is unlikely over the course of a short summer night. Though the temperature won't always drop down as low as the dew point, it is always certain that unless a front passes through or the wind direction changes significantly, the minimum temperature will not fall below the evening dew point.

5–1–1 Why are dew points often good predictors of nighttime minimum temperatures?

5–1–2 Under what conditions are dew points not useful predictors of nighttime low temperatures?

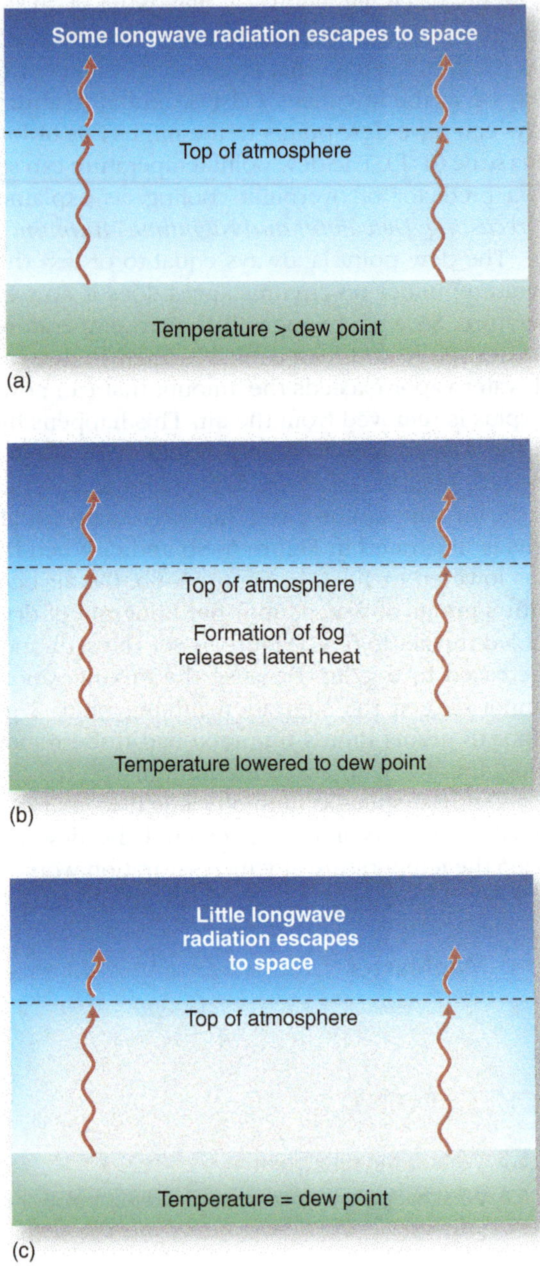

▲ **FIGURE 5–1–1 Dew Points and Minimum Temperatures.** Dew points are often good predictors of overnight minimum temperatures. (a) Initially the absence of cloud or fog droplets allows much longwave radiation to escape to space. (b) When the temperature drops to the dew point, fog can form, which releases latent heat into the air. (c) The presence of fog then inhibits the escape of longwave radiation to space, inhibiting further cooling.

TABLE 5–2

Dew Points

Dew point temperatures

Dry Bulb (Air) Temperature: Dry Bulb (°C)	Wet Bulb Depression, °C (Dry Bulb Temperature Minus Wet Bulb Temperature = Wet Bulb Depression): 1	2	3	4	5	6	7	8	9	10	11	12	13	14	15	16	17	18	19	20	21	22
−20	−33																					
−18	−28																					
−16	−24																					
−14	−21	−36																				
−12	−18	−28																				
−10	−14	−22																				
−8	−12	−18	−29																			
−6	−10	−14	−22																			
−4	−7	−22	−17	−29																		
−2	−5	−8	−13	−20																		
0	−3	−6	−9	−15	−24																	
2	−1	−3	−6	−11	−17																	
4	1	−1	−4	−7	−11	−19																
6	4	1	−1	−4	−7	−13	−21															
8	6	3	1	−2	−5	−9	−14															
10	8	6	4	1	−2	−5	−9	−14	−28													
12	10	8	6	4	1	−2	−5	−9	−16													
14	12	11	9	6	4	1	−2	−5	−10	−17												
16	14	13	11	9	7	4	1	−1	−6	−10	−17											
18	16	15	13	11	9	7	4	2	−2	−5	−10	−19										
20	19	17	15	14	12	10	7	4	2	−2	−5	−10	−19									
22	21	19	17	16	14	12	10	8	5	3	−1	−5	−10	−19								
24	23	21	20	18	16	14	12	10	8	6	2	−1	−5	−10	−18							
26	25	23	22	20	18	17	15	13	11	9	6	3	0	−4	−9	−18						
28	27	25	24	22	21	19	17	16	14	11	9	7	4	1	−3	−9	−16					
30	29	27	26	24	23	21	19	18	16	14	12	10	8	5	1	−2	−8	−15				
32	31	29	28	27	25	24	22	21	19	17	15	13	11	8	5	2	−2	−7	−14			
34	33	31	30	29	27	26	24	23	21	20	18	16	14	12	9	6	3	−1	−5	−12	−29	
36	35	33	32	31	29	28	27	25	24	22	20	19	17	15	13	10	7	4	0	−4	−10	
38	37	35	34	33	32	30	29	28	26	25	23	21	19	17	15	13	11	8	5	1	−3	−9
40	39	37	36	35	34	32	31	30	28	27	25	24	22	20	18	16	14	12	9	6	2	−2

(a) Dew Points, Celsius

Dew point temperatures

Dry Bulb (°F)	Wet Bulb Depression (°F) (Dry Bulb Temperature Minus Wet Bulb Temperature = Wet Bulb Depression): 1	2	3	4	5	6	7	8	9	10	11	12	13	14	15	16	17	18	19	20	21	22	23	24	25	26	27	28	29	30	31	32	33	34	35
0	−7	−20																																	
5	−1	−9	−24																																
10	5	−2	−10	−27																															
15	11	6	0	−9	−26																														
20	16	12	8	2	−7	−21																													
25	22	19	15	10	5	−3	−15	−51																											
30	27	25	21	18	14	8	2	−7	−25																										
35	33	30	28	25	21	17	13	7	0	−11	−41																								
40	38	35	33	30	28	25	21	18	13	7	−1	−14																							
45	43	41	38	36	34	31	28	25	22	18	13	7	−1	−14																					
50	48	46	44	42	40	37	34	32	29	26	22	18	13	8	0	−13																			
55	53	51	50	48	45	43	41	38	36	33	30	27	24	20	15	9	1	−12	−59																
60	58	57	55	53	51	49	47	45	43	40	38	35	32	29	25	21	17	11	4	−8	−36														
65	63	62	60	59	57	55	53	51	49	47	45	42	40	37	34	31	27	24	19	14	7	−3	−22												
70	69	67	65	64	62	61	59	57	55	53	51	49	47	44	42	39	36	33	30	26	22	17	11	2	−11										
75	74	72	71	69	68	66	64	63	61	59	57	55	54	51	49	47	44	42	39	36	32	29	25	21	15	8	−2	−23							
80	79	77	76	74	73	72	70	68	67	65	63	62	60	58	56	54	52	50	47	44	42	39	36	32	28	24	20	13	6	−7	−53				
85	84	82	81	80	78	77	75	74	72	71	69	68	66	64	62	61	59	57	54	52	50	48	45	42	39	36	32	28	24	19	12	3	−12		
90	89	87	86	85	83	82	81	79	78	76	75	73	72	70	69	67	65	63	61	59	57	55	53	51	48	45	43	39	36	32	28	24	19	11	1
95	94	93	91	90	89	87	86	85	83	82	80	79	78	76	74	73	71	70	68	66	64	62	60	58	56	54	52	49	46	43	40	37	33	29	24
100	99	98	96	95	94	93	91	90	89	87	86	85	83	82	80	79	77	76	74	72	71	69	67	65	63	61	59	57	55	52	50	47	44	41	37
105	104	103	101	100	99	98	96	95	94	93	91	90	89	87	86	84	83	82	80	78	77	75	74	72	70	68	67	65	63	61	58	56	54	51	48
110	109	108	106	105	104	103	102	100	99	98	97	95	94	93	91	90	89	87	86	84	83	81	80	78	77	75	73	72	70	68	66	64	62	60	57
115	114	113	112	110	109	108	107	106	104	103	102	101	99	98	97	96	94	93	92	90	89	87	86	84	83	81	80	78	76	75	73	71	69	67	65
120	119	118	117	115	114	113	112	111	110	108	107	106	105	104	102	101	100	98	97	96	94	93	92	90	89	87	86	84	83	81	80	78	76	75	73
125	124	123	122	121	119	118	117	116	115	114	112	111	110	109	108	106	105	104	103	101	100	99	97	96	95	93	92	90	89	88	86	84	83	81	80
130	129	128	127	126	124	123	122	121	120	119	118	116	115	114	113	112	110	109	108	107	106	104	103	102	100	99	98	96	95	94	92	91	89	88	86

(b) Dew Points, Fahrenheit

TABLE 5–3

Relative Humidities

Dry Bulb (°C)	Wet Bulb Depression, °C (Dry Bulb Temperature Minus Wet Bulb Temperature = Wet Bulb Depression)																				
Dry Bulb (Air) Temperature	1	2	3	4	5	6	7	8	9	10	11	12	13	14	15	16	17	18	19	20	21	22
−20	28																					
−18	40																					
−16	48	0																				
−14	55	11																				
−12	61	23																				
−10	66	33	0																			
−8	71	41	13																			
−6	73	48	20	0																		
−4	77	54	32	11																		
−2	79	58	37	20	1																	
0	81	63	45	28	11																	
2	83	67	51	36	20	6																
4	85	70	56	42	27	14																
6	86	72	59	46	35	22	10	0														
8	87	74	62	51	39	28	17	6														
10	88	76	65	54	43	33	24	13	4													
12	88	78	67	57	48	38	28	19	10	2												
14	89	79	69	60	50	41	33	25	16	8	1											
16	90	80	71	62	54	45	37	29	21	14	7	1										
18	91	81	72	64	56	48	40	33	26	19	12	6	0									
20	91	82	74	66	58	51	44	36	30	23	17	11	5									
22	92	83	75	68	60	53	46	40	33	27	21	15	10	4	0							
24	92	84	76	69	62	55	49	42	36	30	25	20	14	9	4	0						
26	92	85	77	70	64	57	51	45	39	34	28	23	18	13	9	5						
28	93	86	78	71	65	59	53	45	42	36	31	26	21	17	12	8	4					
30	93	86	79	72	66	61	55	49	44	39	34	29	25	20	16	12	8	4				
32	93	86	80	73	68	62	56	51	46	41	36	32	27	22	19	14	11	8	4			
34	93	86	81	74	69	63	58	52	48	43	38	34	30	26	22	18	14	11	8	5		
36	94	87	81	75	69	64	59	54	50	44	40	36	32	28	24	21	17	13	10	7	4	
38	94	87	82	76	70	66	60	55	51	46	42	38	34	30	26	23	20	16	13	10	7	5
40	94	89	82	76	71	67	61	57	52	48	44	40	36	33	29	25	22	19	16	13	10	7

Relative humidity values

(a) Relative Humidities, Celsius

Dry Bulb (°F)	Wet Bulb Depression (°F) (Dry Bulb Temperature Minus Wet Bulb Temperature = Wet Bulb Depression)																																		
	1	2	3	4	5	6	7	8	9	10	11	12	13	14	15	16	17	18	19	20	21	22	23	24	25	26	27	28	29	30	31	32	33	34	35
0	67	33	1																																
5	73	46	20																																
10	78	56	34	13																															
15	82	64	46	29	11																														
20	85	70	55	40	26	12																													
25	87	74	62	49	37	25	13	1																											
30	89	78	67	56	46	36	26	16	6																										
35	91	81	72	63	54	45	36	27	19	10	2																								
40	92	83	75	68	60	52	45	37	29	22	15	7																							
45	93	86	78	71	64	57	51	44	38	31	25	18	12	6																					
50	93	87	80	74	67	61	55	49	43	38	32	27	21	16	10	5																			
55	94	88	82	76	70	65	59	54	49	43	38	33	28	23	19	11	9	5																	
60	94	89	83	78	73	68	63	58	53	48	43	39	34	30	26	21	17	13	9	5	1														
65	95	90	85	80	75	70	66	61	56	52	48	44	39	35	31	27	24	20	16	12	9	5	2												
70	95	90	86	81	77	72	68	64	59	55	51	48	44	40	36	33	29	25	22	19	15	12	9	6	3										
75	96	91	86	82	78	74	70	66	62	58	54	51	47	44	40	37	34	30	27	24	21	18	15	12	9	7	4	1							
80	96	91	87	83	79	75	72	68	64	61	57	54	50	47	44	41	38	35	32	29	26	23	20	18	15	12	10	7	5	3					
85	96	92	88	84	81	77	73	70	66	63	59	57	53	50	47	44	41	38	36	33	30	27	25	22	20	17	15	13	10	8	6	4	2		
90	96	92	89	85	81	78	74	71	68	65	61	58	55	52	49	47	44	41	39	36	34	31	29	26	24	22	19	17	15	13	11	9	7	5	3
95	96	93	89	86	82	79	76	73	69	66	63	61	58	55	52	50	47	44	42	39	37	34	32	30	28	25	23	21	19	17	15	13	11	10	8
100	96	93	89	86	83	80	77	73	70	68	65	62	59	56	54	51	49	46	44	41	39	37	35	33	30	28	26	24	22	21	19	17	15	13	12
105	97	93	90	87	84	81	78	75	72	69	66	64	61	58	56	53	51	49	46	44	42	40	38	36	34	32	30	28	26	24	22	21	19	17	15
110	97	93	90	87	84	81	78	75	73	70	67	65	62	60	57	55	52	50	48	46	44	42	40	38	36	34	32	30	28	26	25	23	21	20	18
115	97	94	91	88	85	82	79	76	74	71	69	66	64	61	59	57	54	52	50	48	46	44	42	40	38	36	34	33	31	29	28	26	25	23	21
120	97	94	91	88	85	82	80	77	74	72	69	67	65	62	60	58	55	53	51	49	47	45	43	41	40	38	36	34	33	31	29	28	26	25	23
125	97	94	91	88	86	83	80	78	75	73	70	68	66	64	61	59	57	55	53	51	49	47	45	44	42	40	38	37	35	33	32	30	29	27	26
130	97	94	91	89	86	83	81	78	76	73	71	69	67	64	62	60	58	56	54	52	50	48	47	45	43	41	40	38	37	35	33	32	30	29	28

Relative humidity values

(b) Relative Humidities, Fahrenheit

Video Forecasting Relative Humidity

http://goo.gl/0zww9F

Some psychrometers are equipped with fans that circulate air across the bulbs of the two thermometers. These **aspirated psychrometers** save the user the effort needed to sling the thermometers through the air (as well as the aggravation of cleaning up the mess after accidentally striking nearby objects). Another alternative to the sling psychrometer is the **hair hygrometer**, whose basic part is a band of human hair. Hair expands and contracts in response to the relative humidity. By connecting strands of hair to a lever mechanism, we can easily determine the water vapor content. Often, the hygrometer is coupled with a bimetallic strip and rotating drum to give a continuous record of temperature and humidity. Such a **hygrothermograph** is shown in Figure 5–9b. Though more sophisticated instruments are used for keeping track of humidity at major weather platforms, such as those at airports as described in Chapter 13, hygrothermographs are still often housed in instrument shelters (Chapter 3) at cooperative agencies. (The distribution of water vapor across the globe is routinely observed by water vapor imagery. This is described in *Box 5–2, Forecasting: Water Vapor Satellite Imagery*.)

CHECKPOINT

5.7 Why does a sling psychrometer include two thermometers?

5.8 Why is a wet bulb reading lower than a dry bulb reading?

5–2 FORECASTING

Water Vapor Satellite Imagery

We know that water vapor is an invisible gas, so it may seem strange that meteorologists look at satellite images showing the distribution of water vapor. These images are obtained by sensors on board weather satellites that observe the amount of radiation received at two particular wavelengths of electromagnetic energy, 6.7 and 7.3 μm (Figure 5–2–1). Water vapor is effective at emitting and absorbing this portion of the infrared band of wavelengths, so the energy emitted by the surface at these wavelengths is absorbed by water vapor before it can exit the atmosphere. At the same time, energy at 6.7 and 7.3 μm emitted by water vapor in the upper part of the troposphere does escape to space and is observed by the satellite. Thus, the water vapor imagery actually depicts the distribution of water vapor (and clouds) in the upper troposphere making it extremely useful, in particular for discerning high-level moisture distributions. When a sequence of images taken at regular time intervals is combined to make a movie loop (such as the Weather in Motion loop), the resultant movie is extremely effective in showing upper atmospheric motions. These motions, which were introduced in Chapter 4, will be discussed more fully in later chapters.

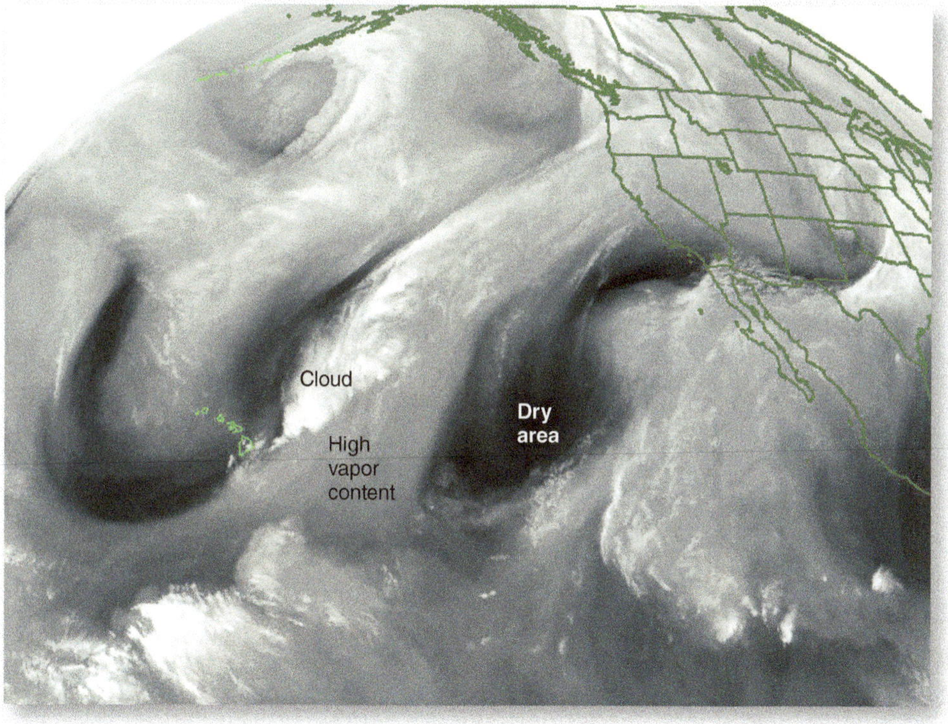

▲ **FIGURE 5–2–1 Water Vapor Image of North America and the East Pacific Ocean.** These images are obtained from weather satellites with sensors that observe radiation emitted by water vapor in the upper troposphere. Regions shown in white have high water vapor contents; those that are dark are relatively dry.

5–2–1 Explain how water vapor imagery can sense the presence of vapor despite the fact that it is an invisible gas.

5–2–2 What characteristic of water vapor imagery makes it especially useful for detecting atmospheric motions?

Distribution of Water Vapor

5.5 Describe how water vapor is distributed in the atmosphere.

Water vapor at a point in the atmosphere is the result of either local evaporation or horizontal or vertical transport of moisture from other locations or altitudes. The effect of horizontal transport (advection) on the distribution of water vapor is clearly evident in Figure 5–10, which shows the spatial distribution of mean dew points (in °C and in °F) across the United States in January (top) and July (bottom). Looking at the eastern two-thirds of the country first, it is clear that for both months the amount of water vapor generally decreases with distance from the Gulf of Mexico and Atlantic Ocean. Because of the Gulf's high water temperatures, moisture is readily evaporated into the atmosphere year-round, and this moisture can be transported northward. The decline in water vapor content is seen in a north–south direction and also moving westward from about the Mississippi River toward the Rocky Mountains during the summer. During the winter months, the amount of moisture extending into the Great Plains is low and only a minimal amount of east–west variation exists.

The effect of distance from the source of moisture is also evident in the West, with water vapor generally decreasing from the Pacific Coast to the Rocky Mountains. The most rapid drop occurs very near the coast because local mountains block off substantial amounts of moisture from inland areas.

Comparing the two maps, you will note a substantial increase in the amount of water vapor in the air in July over that in January. This should not be surprising, because lower January temperatures preclude the existence of high water vapor contents. Thus, along the Ohio River Valley, for example, the average dew point increases from about −7 °C (20 °F) in January to perhaps 17 °C (63 °F) in July. This is why residents of much of the country, especially those east of the Rockies, are subject to uncomfortable dry skin in the winter—only to find themselves sweating profusely during the summer.

The general patterns described for the United States also apply to most of Canada. In this section we looked at the horizontal distribution of water vapor, but water vapor content also varies with distance from the surface, generally decreasing upward. *Box 5–3, Forecasting: Vertical Profiles of Moisture,* provides information on the analysis of vertical water vapor distributions.

Video
January Water Vapor
http://goo.gl/QzzGWg

Video
July Water Vapor
http://goo.gl/gUHFi0

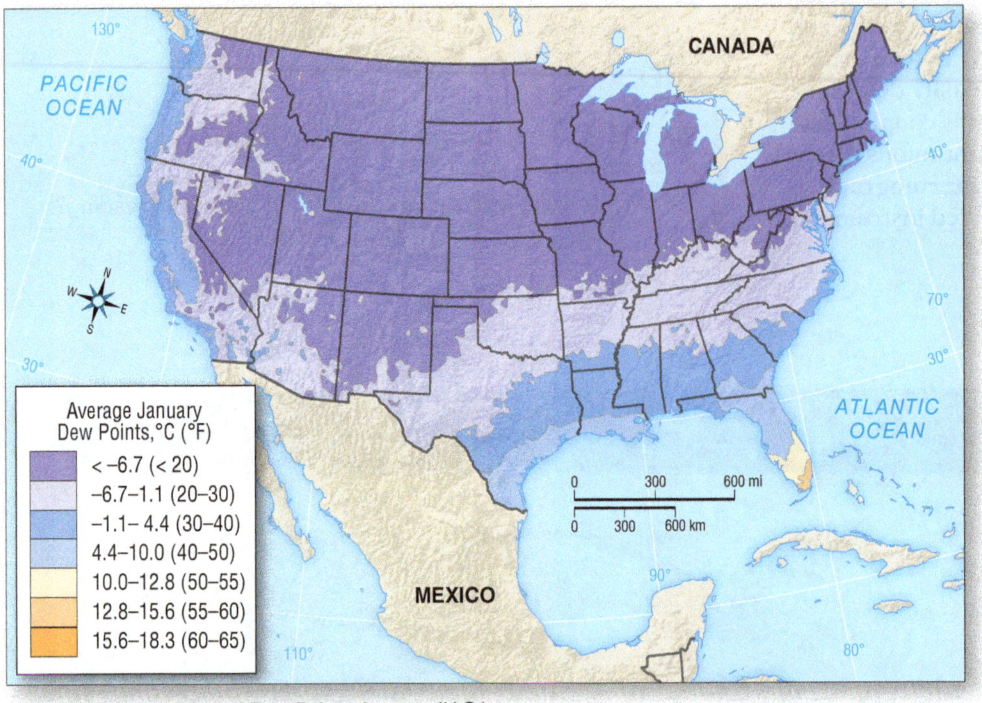

(a) Average Distribution of Dew Points January (U.S.)

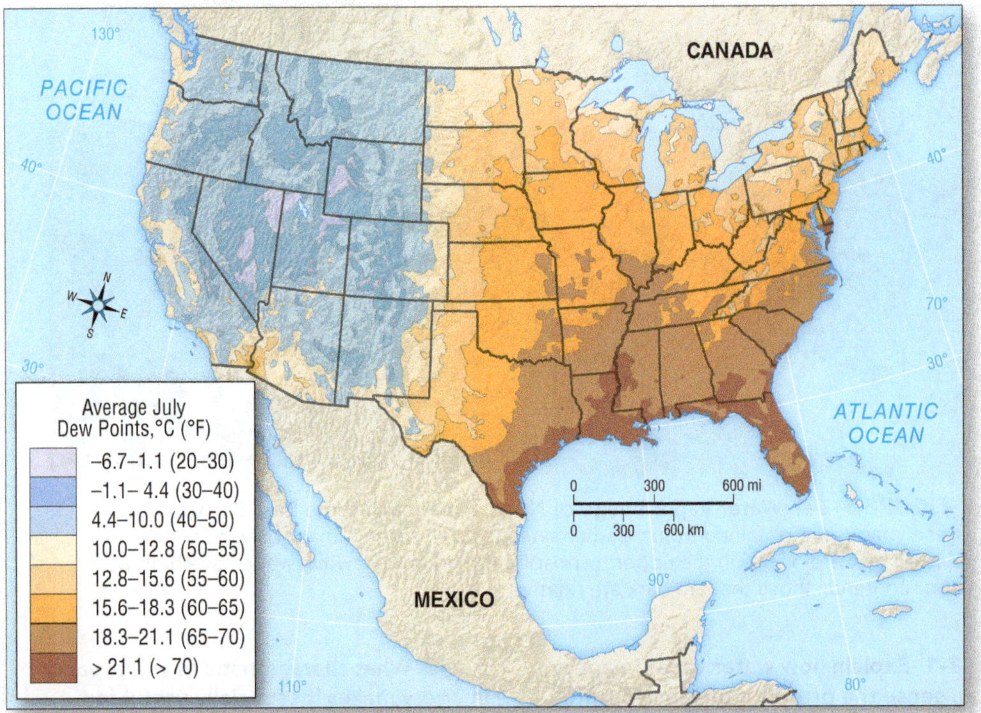

(b) Average Distribution of Dew Points July (U.S.)

▲ **FIGURE 5–10 Average Distribution of Dew Points.** Across the United States, dew point varies greatly from (a) January to (b) July.

MM **MapMaster** North America
Physical Environment
Average Distribution of Dew Points January and July (U.S.)

5-3 FORECASTING

Vertical Profiles of Moisture

In Chapter 3, you saw how simplified thermodynamic diagrams can be used to plot the vertical profile of temperature. Because dew point values are likewise expressed as temperatures, they, too, can be plotted on thermodynamic charts. In fact, by plotting temperatures and dew points simultaneously, you can obtain considerable information about cloud conditions. Refer to the profiles (also called soundings) of temperature and dew point shown in the Stuve diagram (explained in Chapter 3) in Figure 5–3–1, taken from Stapleton Airport, near Denver, Colorado, at midnight, Greenwich mean time, on April 12, 2002.

Reading a Thermodynamic Diagram

The example in Figure 5–3–1 can first be used to illustrate how plots of temperature (the curve on the right) and dew point (left) can give us information about cloud cover. The temperature at the surface is 17 °C (63 °F), and the dew point is −4 °C (25 °F). Given the disparity between the temperature and dew point, we know that the surface relative humidity is low. But this changes at about the 560 mb level (at a height of about 4850 m, or 15,900 ft, above the surface), where the two values become nearly equal to each other. At this level the air is saturated and we have the base of a cloud deck (a slight measurement error accounts for the plotted temperature and dew point values not being exactly equal). The air remains saturated up to about the 460 mb level, where the cloud tops out. Because the air is saturated in the layer of air between 4850 and 6500 m above the surface, we can infer that the cloud is about 1650 (5400 ft) m thick.

Mixing Ratios on a Thermodynamic Diagram

The thermodynamic diagram shown here is slightly more complex than the one in Chapter 3, because it includes an additional set of lines that provides one more type of moisture information. The dashed red lines that slope gently to the left as they extend upward can be used to estimate the mixing ratio and the saturation mixing ratio at any level. To determine the mixing ratio at any level, you observe the position of the dew point plot and interpolate between the dashed red lines. Let's first see how the plot of the air temperature profile can be used to determine the saturation mixing ratio at a given pressure level, in this case the 700 mb level. At that level the dew point is barely greater than −10 °C (14 °F). Notice that the point on the graph where the dew point profile crosses the 700 mb level occurs right between the two sloping lines labeled 2 and 4. That indicates that actual mixing ratio at the 700 mb level is right between 2 (g/kg) air and 4 (g/kg)—with 3 g/kg a very good approximation.

We can follow a similar procedure to find out what the saturation mixing ratio is at the 700 mb level. At the 700 mb level the air temperature is 2 °C (35.6 °F). The point of intersection for the temperature profile falls right on the mixing ratio line labeled as 7. Thus, the air at the 700 mb level has a saturation mixing ratio of just under 7 grams of water vapor per kilogram of dry air.

To determine the relative humidity at the 700 mb level we can use the values just obtained to get,

$$RH = (\text{mixing ratio}/\text{saturation mixing ratio}) \times 100\% = (3/7) \times 100\% = 43\%$$

This procedure can be performed at the surface or any other level of the atmosphere.

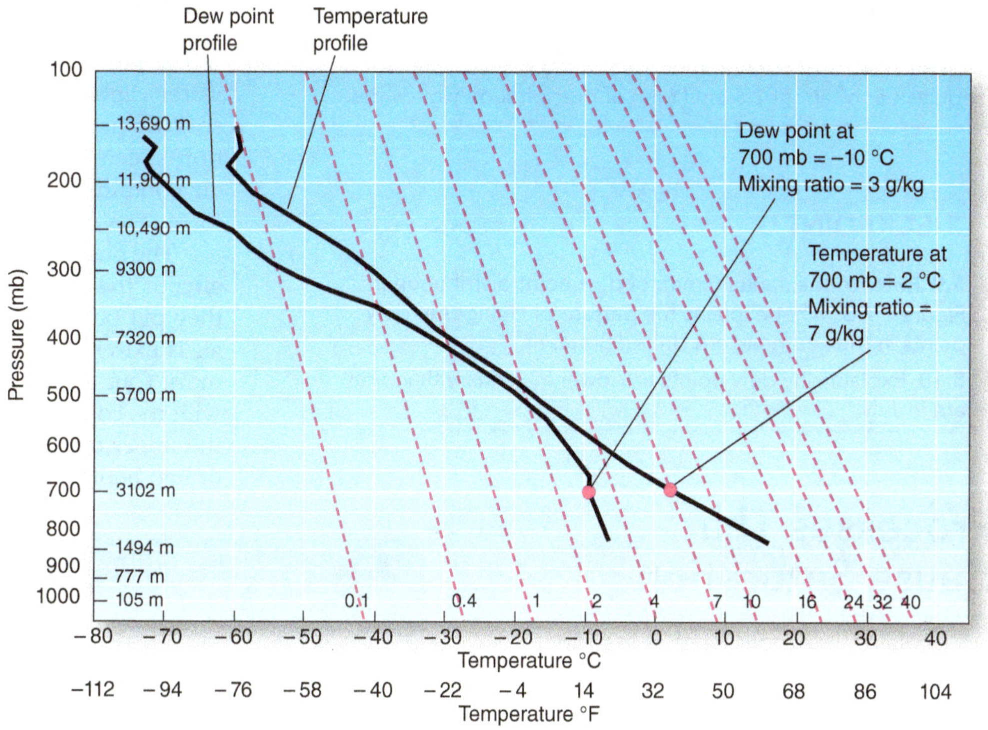

▲ **FIGURE 5–3–1 A Sounding of Temperature and Dew Point.** This chart plots temperature and dew points throughout the troposphere and much of the stratosphere. The slightly sloping red lines depict values of mixing ratio in grams of water vapor per kilogram of dry air (labeled just above the x-axis). Meteorologists use these lines along with the plots of temperature to determine the saturation mixing ratio; these lines and the dew point profiles are used to obtain the actual mixing ratio.

Thermodynamic Diagrams and Forecasting

Later in this chapter, we will see how thermodynamic diagrams can give us information in forecasting the likelihood of cloud development.

5–3–1 What index of moisture content is directly plotted on thermodynamic diagrams?

5–3–2 How can we estimate the mixing ratio from a thermodynamic diagram?

DID YOU KNOW?

Averaged over the entire planet, Earth's atmosphere contains 13×10^{15} kilograms of water vapor (the equivalent of 3.4×10^{15} gallons when condensed to liquid). While this may sound like a lot, it represents a mere one-thousandth of 1 percent of all the free H_2O on the planet. If all of this water vapor were to condense and collect at Earth's surface, it would be the equivalent of about 2.5 cm (1 in.) of precipitation worldwide.

CHECKPOINT

5.9 Refer to the January map of dew point distribution in Figure 5–10. What explains the pattern of changes in dew points in the southeastern United States?

5.10 Explain why dew points are lower in January than they are in July.

Processes That Cause Saturation

5.6 Explain the processes that lead to saturation.

Air can become saturated by any one of three general processes: adding water vapor to the air; mixing cold air with warm, moist air; and lowering the temperature to the dew point. The first of these processes can be seen in your bathroom when you take a warm shower. The warm water from a showerhead evaporates moisture into the air in the room and brings it to the saturation point. Condensation first forms on your mirrors and other surfaces, and then a general fog develops. In the natural environment the evaporation of water from falling raindrops can raise the dew point in the air beneath the cloud from which the rain falls. If enough vapor is added to the air to saturate it, a **precipitation fog** forms beneath the cloud.

Condensation from the second process, the mixing of cold with warm moist air, is illustrated in Figure 5–11. Consider parcels A and B of unsaturated air. Parcel A has a temperature of 0 °C (32 °F) and a specific humidity of 3 grams of water vapor per kilogram of dry air; parcel B has a temperature of 30 °C (86 °F) and a specific humidity of 20 g/kg. If equal amounts of the two parcels are mixed together, the new parcel has a temperature of 15 °C (59 °F), exactly midway between the temperatures of the original parcels. However, such is not the case for the specific humidity.

Although the total H_2O content in the mixed parcel is exactly between the amounts of the original parcels, some of the moisture occurs in the form of a liquid rather than as water vapor. Here is why: The midpoint between the original specific humidity values is 11.5 g/kg, but at 15 °C the saturation specific humidity is only 10.8 g/kg. In other words, the air contains 0.7 g/kg more water than can exist in the vapor form. The surplus therefore condenses to form fog droplets.

The preceding process is what causes contrails to form behind aircraft traveling at high altitudes. As the jet engines burn fuel, they put out a large amount of heat as well as water vapor. The air is extremely turbulent in the wake of the aircraft, so the hot, moist (but unsaturated) exhaust from the engines rapidly mixes with the cold surrounding air. At the subfreezing temperatures of the upper troposphere, the vapor directly forms into ice crystals or into liquid droplets that eventually freeze to form the contrail.

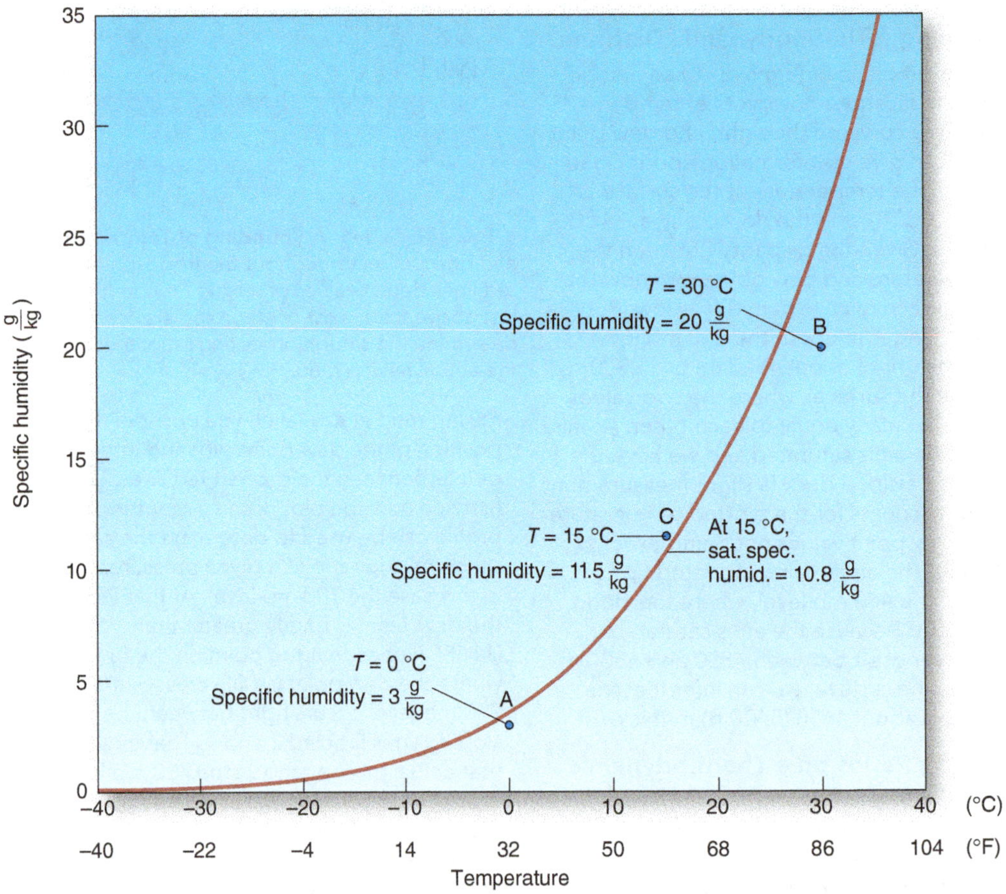

▲ **FIGURE 5–11 Saturation by the Mixing of Warm, Moist Air with Cold Air.** All parcels that fall below the curve in this graph have specific humidities less than the saturation specific humidity for a given temperature, so the parcels are unsaturated. Parcels above the curve have specific humidities that exceed the saturation specific humidity, so the excess moisture in the parcel condenses. In this example parcel A has a low temperature and can therefore contain only a small amount of water vapor. The warm parcel (parcel B) has a higher moisture content. Parcel C results when parcels A and B are mixed together; it has a temperature in between those of the original parcels. The amount of moisture in the air is greater than that which can exist at the new temperature, so the excess water vapor condenses.

▲ **FIGURE 5–12 Steam Fog.** In Baniff National Park, Canada.

A similar but naturally occurring phenomenon is known as **steam fog**. As we learned in Chapter 3, water bodies are rather slow to change temperature. As a result, lakes can remain relatively warm well into the fall or early winter, even as air temperatures fall. Because of evaporation and the upward transfer of sensible heat, a thin, transitional layer of air exists just above the water surface that is warmer and moister than the air above. If a mass of cold air abruptly passes over the warm lake, the warm, moist transitional air mixes with the overlying layer of cold air to form a layer of fog a meter or two thick (Figure 5–12).

Although we can cite several examples of how clouds form by the increase of moisture content or by mixing warm, moist air with cold, dry air, experience shows that most clouds form when the air temperature is lowered to the dew point. There are several ways this cooling can occur, requiring considerable explanation, as we will see later in this chapter. For now we simply note that this third mechanism, atmospheric cooling, is by far the most common process for cloud formation.

CHECKPOINT

5.11 What process that leads to saturated air causes the "fog" that forms in the bathroom when you take a shower?

5.12 How does steam fog form?

Factors Affecting Saturation and Condensation

5.7 Explain the factors that affect condensation.

This chapter opened with a hypothetical experiment in which water in a jar attained an equilibrium between condensation and evaporation. The experiment, assuming pure water with a flat surface, provided a foundation for understanding saturation. But meteorology studies the atmosphere—not what goes on in hypothetical jars. In the real atmosphere, we are concerned with the rates of evaporation and condensation across the surfaces of suspended cloud and fog droplets. Such droplets are neither flat nor made up of pure H_2O. We therefore expand on our discussion of evaporation, condensation, and equilibrium to take into account the effects of curvature of cloud and fog droplets and the fact that they are not made of pure H_2O.

Effect of Curvature

Water droplets exist in nature not as tiny cubes with flat sides, but rather as microscopically small spheres with considerable curvature. Compare the two droplets shown in Figure 5–13. The one on the left is much larger than the one on the right and therefore has less pronounced curvature. We could even consider a more extreme example—Earth itself. Most of us are pretty certain by now that the planet is not flat but spherical. However, because of Earth's large size, its curvature is inconspicuous to anybody standing on it, and a centimeter

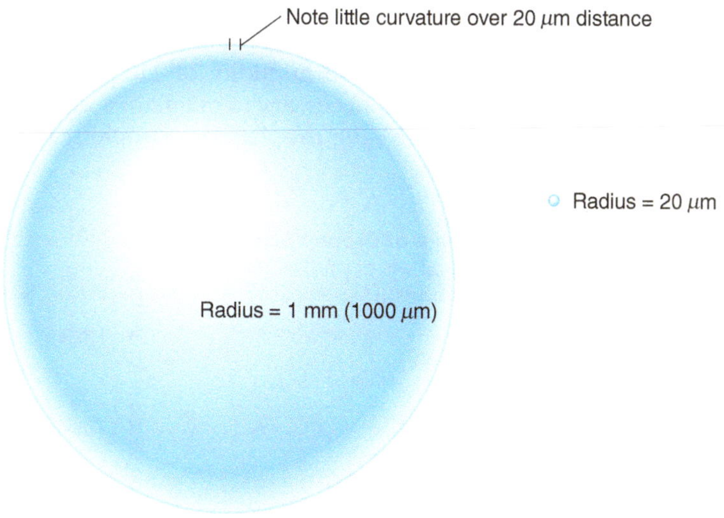

▲ **FIGURE 5–13 Relationship of Droplet Size and Curvature.** Large droplets are less curved than small droplets. The larger droplet on the left has virtually no apparent curvature along a 20-μm distance across its surface, unlike the much smaller droplet on the right. Droplets with highly curved surfaces evaporate more rapidly than droplets with less curved surfaces, and therefore require greater surrounding vapor pressures to remain in equilibrium.

of distance across its surface essentially forms a straight line. Moreover, a straight-edge ruler can lie flat against its surface. But hold a ruler over a tennis ball and only a small part of it is in contact with the ball's surface; the ball is more strongly curved than Earth.

Cloud droplets are much smaller than tennis balls, of course, and thus have even less semblance to a flat surface and curve markedly within a very short distance. But what does curvature have to do with saturation? The answer is that curvature has an effect on evaporation from cloud droplet surfaces and therefore on the vapor pressure necessary for saturation. Effects arising from surface tension lead to differences in the saturation point, as described shortly.

The graph of saturation vapor pressure versus temperature, shown in Figure 5–5, applies only to flat surfaces of pure H_2O. For curved water surfaces, the evaporation rate is greater, which is significant because water droplets suspended in the air are spherical and therefore do not have flat surfaces. The enhanced rate of evaporation requires that condensation also be increased for the two to remain in balance. Thus, a highly curved droplet of pure water at any given temperature has a higher saturation vapor pressure than indicated in Figure 5–5. Stated another way, highly curved droplets of pure water require relative humidities in excess of 100 percent to keep them from evaporating away. Air that has a relative humidity in excess of 100 percent is said to be **supersaturated**, a situation that can exist because the highly curved nature of suspended water droplets would cause them to rapidly evaporate in an atmosphere at 100 percent relative humidity with respect to flat water surfaces.

Figure 5–14 illustrates the effect of droplet size on the relative humidity needed to maintain an existing droplet of pure H_2O. For very small droplets (those with radii of about 0.1 μm), relative humidities in excess of 110 percent are needed to achieve an equilibrium between evaporation and condensation. In other words, those droplets require a supersaturation of 10 percent (i.e., a relative humidity of 110 percent). The degree of supersaturation necessary to maintain a droplet rapidly decreases with increasing droplet size. Droplets with radii larger than 10 μm require supersaturations of less than 1 percent.

The Role of Condensation Nuclei

If the atmosphere were devoid of any aerosols, condensation would occur only by **homogeneous nucleation**, in which droplets form by the chance collision and bonding of water vapor molecules under supersaturated conditions. Such droplets would necessarily have only a small number of molecules and a high degree of curvature, and therefore would only exist at high levels of supersaturation. This process seldom if ever occurs, because certain **hygroscopic** (water-attracting) aerosols in the atmosphere assist the formation of droplets at relative humidities far below those necessary for homogeneous nucleation. The formation of water droplets onto hygroscopic particles is called **heterogeneous nucleation**, and the particles onto which the droplets form are called **condensation nuclei**. These particles enhance condensation through two processes, as described below.

EFFECT OF SOLUTION Some condensation nuclei readily dissolve in water to form a solution. For any solution a certain number of the molecules at the surface are those of the *solute* (the material dissolved in the water) rather than H_2O molecules. With fewer H_2O molecules at the surface, the rate of evaporation is lower than that for pure water. As a result, solutions require less water vapor above the surface to maintain an equilibrium between evaporation and condensation. We see that cloud droplets formed from soluble condensation nuclei will have a lower saturation point than droplets of pure water. Thus, after formation, they will be less likely to evaporate than any formed by homogenous nucleation. The solute effect therefore promotes condensation, which is counter to the effect of curvature, and condensation normally occurs at relative humidities near or slightly below 100 percent.

Although the proportion of aerosols in the air that are hygroscopic is small, the atmosphere contains so many

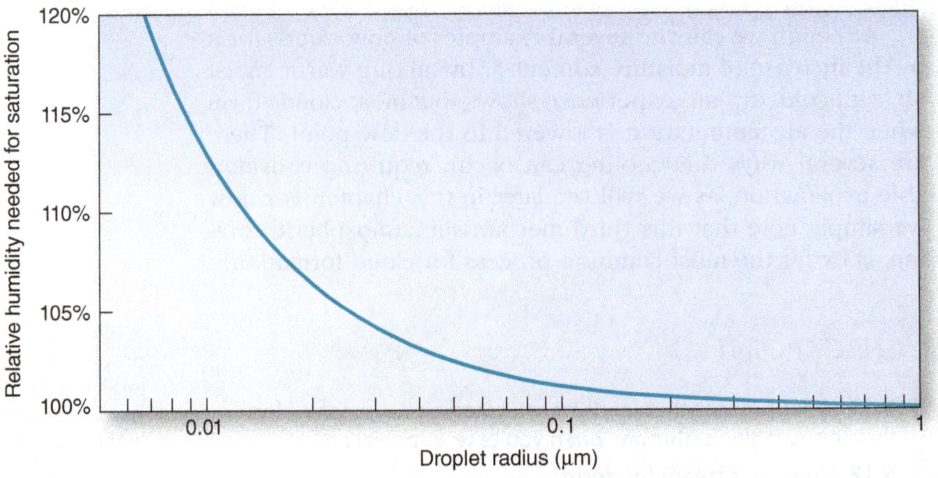

▸ **FIGURE 5–14 Relationship of Droplet Size and Relative Humidity.** Small droplets of pure water require relative humidities above 100 percent to remain in equilibrium.

▲ **FIGURE 5–15 Haze over Victoria Harbour, Hong Kong.**

suspended particles that condensation nuclei are always abundant. Some materials are more hygroscopic than others, and large aerosols are generally more effective than smaller ones. Some are even capable of attracting water at relative humidities below 90 percent and forming extremely small droplets. We observe such droplets as **haze** (Figure 5–15).

Condensation nuclei originate from natural and human processes. Scientists once believed that most condensation nuclei in the atmosphere were natural, consisting mostly of continental dust, sea salt, and aerosols derived from volcanic eruptions, natural fires, and gases given off by marine phytoplankton. Research undertaken during the 1990s indicated that the role of human activity was previously understated and that anthropogenic sources may account for the majority of condensation nuclei over industrialized areas in the Northern Hemisphere. Still, there is considerable uncertainty about which materials are the most active cloud condensation nuclei.

ICE NUCLEI So far we have examined the formation of liquid water droplets when air becomes saturated. But saturation can occur at very low temperatures, which suggests that rather than liquid water droplets, ice crystals may form. On the other hand, many of us have walked through fog composed of liquid droplets even though the temperature was below 0 °C (32 °F). So what really happens when saturation occurs at temperatures below the freezing point? Does this lead to the condensation of liquid droplets or to the deposition of ice crystals? The answer is that it depends. Strangely, although ice always melts at 0 °C, water in the atmosphere does not normally freeze at 0 °C.

If saturation occurs at temperatures between 0 °C and −4 °C (32 °F and 29 °F), the surplus water vapor invariably condenses to form **supercooled water** (water having a temperature below the melting point of ice but nonetheless existing in a liquid state). Ice does not form within this range of temperatures. Just as the formation of liquid droplets at relative humidities near 100 percent requires the presence of condensation nuclei, the formation of ice crystals at temperatures near 0 °C requires **ice nuclei**. Unlike condensation nuclei, which are always abundant, ice nuclei are rare in the atmosphere. This is because an ice nucleus must have a six-sided structure that mimics the alignment of molecules in an ice crystal (although exceptions to this rule exist).

A material's ability to act as an ice nucleus is temperature dependent, and no materials are effective ice nuclei at temperatures above −4 °C. Though ice can exist at temperatures between −4 °C and 0 °C, it does not form spontaneously in the atmosphere in this range. (In fact, there is little ice in the atmosphere at temperatures above −10 °C [14 °F].) Thus, between −4 °C and 0 °C, the removal of water vapor occurs only by the condensation of supercooled water.

As temperature decreases, the likelihood of ice formation increases, and at temperatures between about −10 °C and −40 °C (14 °F and −40 °F) saturation can lead to the nucleation of ice crystals, supercooled droplets, or both. In clouds having temperatures within this range, liquid droplets and solid crystals will usually coexist, but with a greater proportion of ice at lower temperatures. At temperatures below −30 °C (−22 °F), the cloud will be mostly ice crystals. For temperatures below −40 °C, saturation leads to the formation of ice crystals only, with or without the presence of ice nuclei. As we will see in Chapter 7, the coexistence of ice crystals and liquid droplets in clouds is extremely important for the development of precipitation outside the tropics. (*Box 5–4, Focus on Aviation: Icing,* describes how the formation of ice aloft creates a hazard for aircraft.)

Among the materials that serve as ice nuclei are components of natural soils called *clays*. Clay materials have a platy structure, microscopic sizes, and strong electrical attractions. These characteristics make them very difficult to dislodge from the surface and incorporate into the atmosphere, which largely explains their scarcity. Clays occur in a variety of compositions, of which the most effective seems to be ***kaolinite*** (aluminum silicate). Other types of clay serve as ice nuclei but are active only at lower temperatures.

CHECKPOINT

5.13 How does droplet size affect rates of evaporation and condensation?

5.14 Why are ice nuclei often required for the formation of ice crystals?

5–4 FOCUS ON AVIATION

Icing

The formation of ice onto aircraft, *icing*, can create a serious hazard to aircraft. In the period from 1997 to 2006, icing was responsible for 10 percent of all weather-related accidents and 22 percent of those involving fatalities. Icing most commonly forms on the body of an aircraft (*structural icing*), but ice can also develop within the air intake system (*induction icing*) and cause the engine to shut down.

Effects of Icing

Even a relatively thin film of ice can have a drastic effect on the aerodynamics of a plane, increasing drag (wind resistance), reducing the amount of lift provided by the wings, lowering the stall speed, and hindering aircraft maneuverability. Ice also adds additional weight to the aircraft, although this effect is secondary relative to that of the aerodynamic impacts.

Forms of Structural Icing

Structural icing usually occurs in one of the following three forms:

Clear ice. Sometimes called *glazed ice*, this type of ice is translucent and has a smooth appearance. It is caused when a plane flies through a cloud containing supercooled drops that slowly freeze onto the wings and fuselage. This type of ice occurs at temperatures not far below the freezing point of water, so that the water spreads from the leading edge of the plane and migrates slowly backward, gradually forming a continuous film of ice. Although it generally maintains a continuous coating, the ice can also form lumps and undulations, causing even greater impacts on the plane's aerodynamics.

Rime. Rime ice usually forms at lower temperatures than does clear ice, generally within the range of −10 °C to −20 °C (14 °F to −4 °F). At these lower temperatures water can freeze rapidly and entrap small amounts of air, giving it a rough, milky appearance. Because the ice forms quickly, the water does not flow across a wing or fuselage prior to freezing. It instead accumulates at the front of the plane.

Mixed ice. This is the most common form of icing, consisting of a combination of clear ice and rime.

Distribution of Icing

Aircraft icing is most threatening when clouds contain a large amount of supercooled liquid water. As shown in Figure 5–4–1, winter icing is more likely to occur near British Columbia–Washington State, the coast of Alaska, over the eastern Great Lakes, and over extreme eastern Canada. Pilots must know what to do if icing occurs, but it is absolutely essential that they anticipate its formation before taking off and while in flight. The National Weather Service Aviation Weather Center provides detailed information on current icing conditions at http://aviationweather.gov/adds/icing (Figure 5–4–2).

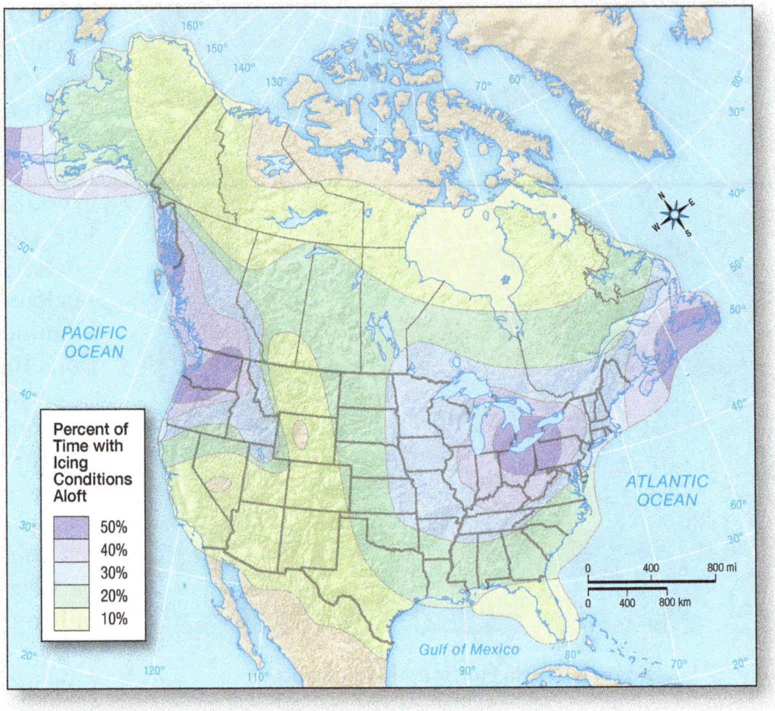

▲ **FIGURE 5–4–1 Occurrence of November–March Icing Across North America.**

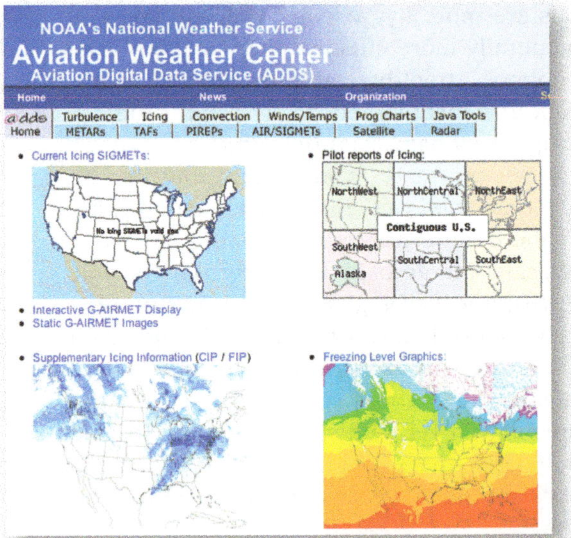

▲ **FIGURE 5–4–2 Icing Maps for Aviation.** The National Weather Service Aviation Weather Center provides up-to-date information on icing conditions.

5–4–1 Explain how icing presents a danger to aircraft.

5–4–2 What are two possible types of icing that can form on an aircraft?

Cooling the Air to the Dew or Frost Point

5.8 Describe how diabatic and adiabatic processes produce cooling and condensation.

Although condensation can occur from an increase in the amount of water vapor or from mixing cold air with warm, moist air, the most common mechanism for cloud formation is the lowering of the air temperature to the dew or frost point. We might expect that air temperature will change only in response to gains or losses of energy, but such is not the case. Air temperature changes can occur from two very general classes of processes: those that involve the removal or input of energy, and those that do not. These are referred to as *diabatic* (DIE-a-bat-ic) and *adiabatic* (A-dee-a-bat-ic), respectively.

Diabatic Processes

A **diabatic process** is one in which energy is added to or removed from a system. A pot of water placed over a stove warms diabatically, as does air that is warmed by conduction when in contact with a warm surface. Likewise, air that passes over a cool surface loses energy by conduction into the surface and therefore cools diabatically. Note that the direction of heat transfer is in accordance with the **second law of thermodynamics**, which dictates that energy moves from regions of higher to lower temperatures. Diabatic processes are frequently responsible for the formation of fog but are secondary to adiabatic processes for the development of clouds.

Adiabatic Processes

Processes in which the temperature changes but no heat is added to or removed from a substance are said to be **adiabatic** (which literally means "not diabatic"). Although changes in temperature without the exchange of heat may at first seem counterintuitive, adiabatic processes are common in the atmosphere and provide the most important mechanism for the formation of clouds (see *Box 5–5, Physical Principles: Adiabatic Lapse Rates and the First Law of Thermodynamics*).

Using a hand pump to inflate a bicycle tire is an excellent example of an adiabatic process. Even though no heat is added, compressing the air causes its temperature to increase, as you can readily observe by touching the base of the pump. It is the compression of the air alone that causes its temperature to increase. When air is allowed to escape from the tire valve, its temperature decreases because of expansion, and the jet of outgoing air feels cold to a hand placed near the stream of outgoing air. It doesn't feel cold because the flow of air cools your skin, but because it is actually chilled due the expansion that takes place when the air escapes from the tire.

Figure 5–16 applies this concept to a parcel of unsaturated air that is displaced upward. As the air rises, it encounters lower surrounding pressure, expands, and cools. The rate at which a rising parcel of unsaturated air cools, called the **dry adiabatic lapse rate (DALR)**, is very nearly 1.0 °C/100 m (5.5 °F/1000 ft). Thus, a parcel of unsaturated air cools 1 °C for every 100 meters of ascent, despite the fact that no heat is removed. Likewise, the downward movement of unsaturated air leads to compression and warming at the same rate. It is important to note that the term *lapse* refers to a decrease in temperature with altitude. We therefore leave out the minus sign preceding the 1.0 °C/100 m.

The process shown in Figure 5–16 also applies in the opposite way when unsaturated air sinks. In this event the air warms at 1 °C/100 meters of descent.

If a parcel of air rises high enough and cools sufficiently, expansion lowers its temperature to the dew or frost point, and condensation or deposition commences (Figure 5–17). The altitude at which this occurs is known as

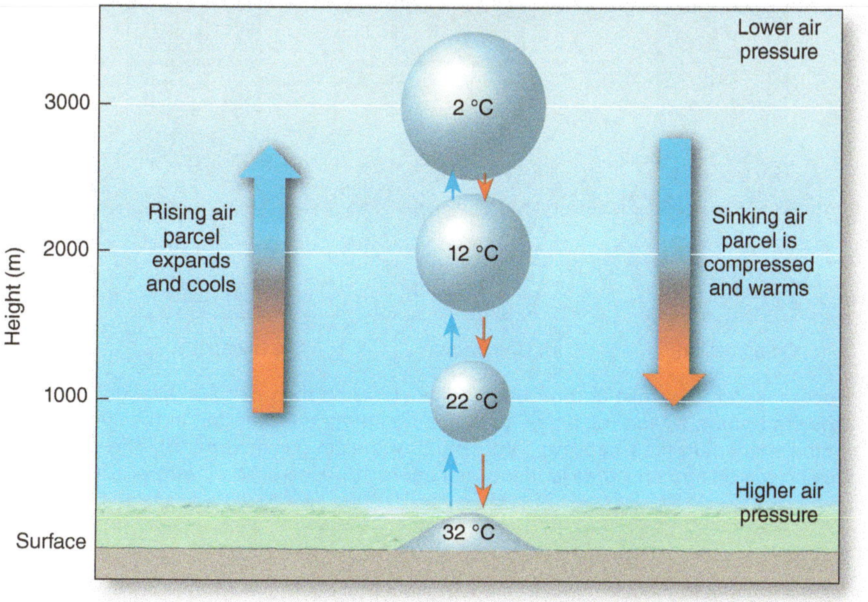

◀ **FIGURE 5–16 Dry Adiabatic Lapse Rate (DALR).** As a parcel of unsaturated air rises, its temperature decreases at the dry adiabatic lapse rate of 1 °C per 100 meters. In this example, unsaturated air at the surface has an initial temperature of 32 °C, which decreases by 1 °C for every hundred meters of ascent. Descending unsaturated air warms at the same rate.

5-5 PHYSICAL PRINCIPLES

Adiabatic Lapse Rates and the First Law of Thermodynamics

To fully understand diabatic and adiabatic processes, we refer to a version of the **first law of thermodynamics**, which states what happens when heat is added to or removed from gases. Specifically, if heat is added, there will be some combination of an expansion of the gas and an increase in its temperature. The law is given in numerical form as

First Law of Thermodynamics
$$\Delta H = p \bullet \Delta\alpha + c_v \bullet \Delta T$$

where ΔH = heat added to the system, p is the air pressure, $\Delta\alpha$ is the change in volume (positive for expansion and negative for contraction), c_v is the specific heat for air (assuming a constant volume), and ΔT is the change in temperature. (Note that the Greek letter delta, Δ, preceding a symbol represents a change in the value of the quantity.) The first term on the right-hand side of the equation, $p \bullet \Delta\alpha$, is the work performed by the gas as expansion occurs. The second term, $c_v \bullet \Delta T$, refers to the change in internal energy. The important thing for us is that heat added to the air does not simply disappear but rather is apportioned between temperature and volume changes.

The first law of thermodynamics describes the underlying principle of what occurs in the cylinder of an internal combustion engine in an automobile, as shown in Figure 5–5–1. As the air–fuel mixture burns, it expands and pushes down on the piston (this is the work performed) and ultimately propels the car. In addition, there is an increase in the internal energy of the gas, which we observe as an increase in temperature. (Just ask anyone who has ever burned a hand on an exhaust manifold!) The energy unleashed with the combustion of the fuel is therefore manifested as work performed and an increase in temperature. Of course, good automotive design calls for the engine to convert most of the chemical energy to work performed, with little going toward an increase in internal energy. In other words, engine heat represents wasted energy, and a cold exhaust manifold would be the mark of good engineering. The same relationship among heat, temperature, and volume applies to our atmosphere.

An adiabatic process represents a special case of the first law of thermodynamics in which the left-hand side of the equation equals 0 (no heat is added or removed). Substituting 0 for ΔH yields

$$0 = p \bullet \Delta\alpha + c_v \bullet \Delta T$$

which can be rearranged as

$$p \bullet \Delta\alpha = -c_v \bullet \Delta T$$

or

$$-p \bullet \Delta\alpha = c_v \bullet \Delta T$$

Stated in words, the adiabatic form of the first law indicates that if no heat is added or removed from the system, work performed *by* the air (the expansion of the gas) causes a decrease in internal energy (a decrease in temperature) and work performed *on* the gas (compression) leads to warming. Stated even more succinctly, expanding air cools and air undergoing compression warms.

5-5-1 What is the first law of thermodynamics?

5-5-2 What changes in work performed and internal energy occur when a parcel of air is lifted?

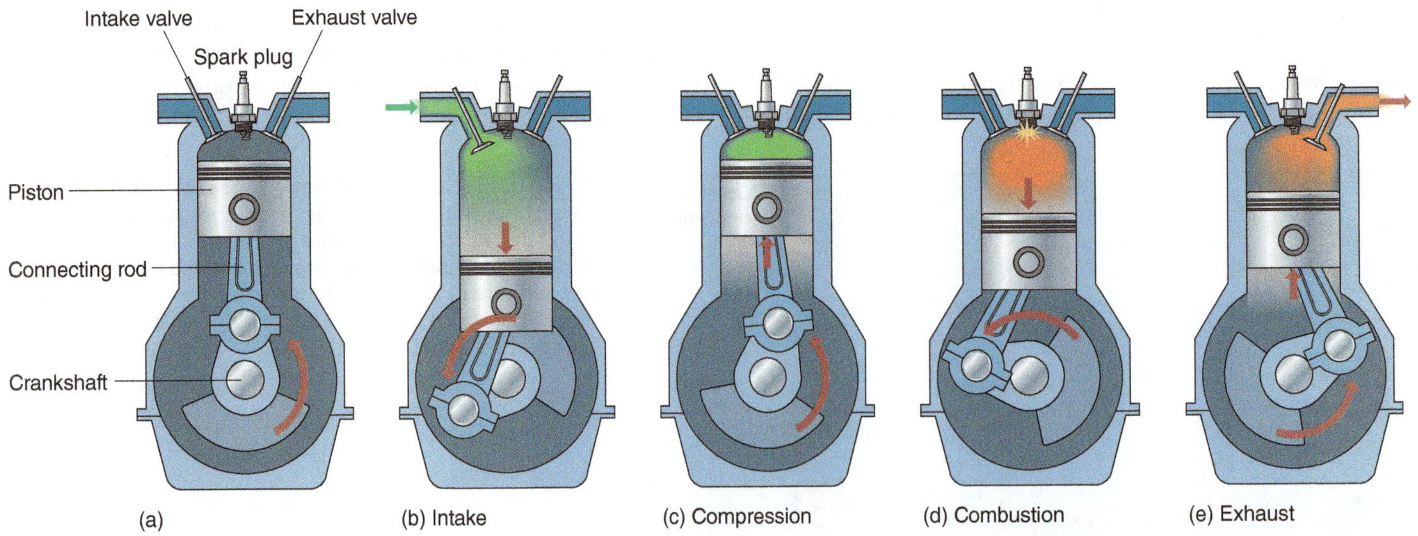

▲ **FIGURE 5-5-1 Car Engine as an Illustration of Adiabatic Processes.** A four-stroke automobile engine works on the principle invoked by the first law of thermodynamics. As the piston is pulled down by the crankshaft, (a) and (b), an air–fuel mixture enters the cylinder. In the second stroke the mixture is compressed (c). The third stroke occurs when the spark plug fires, causing combustion of the air–gas mixture (d). The energy released by the burning fuel is manifested as work done by the moving piston and an increase in internal energy (the increase in temperature). The burned fuel is expelled during the fourth stroke (e).

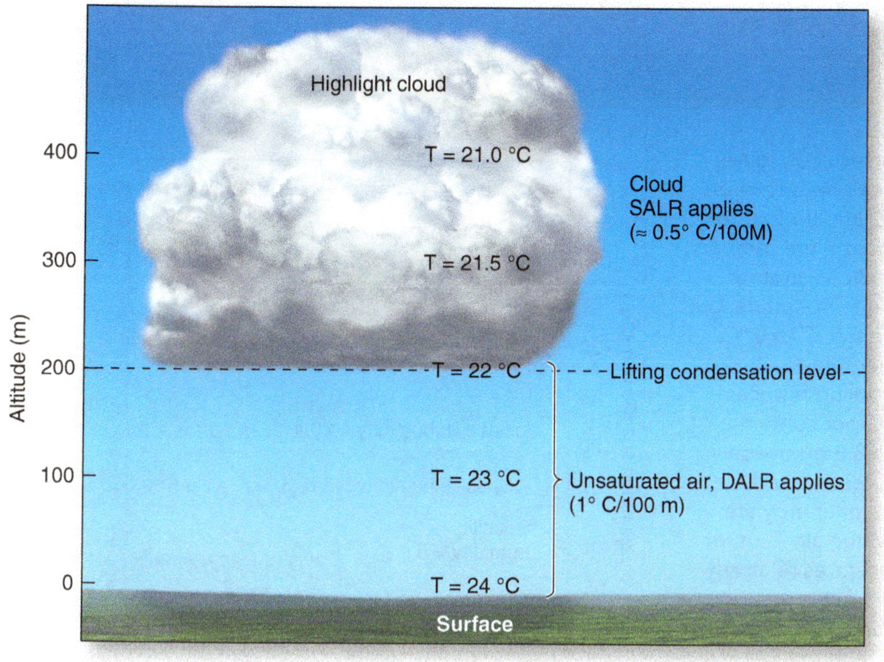

◄ **FIGURE 5–17 Adiabatic Lapse Rates.** The dry adiabatic lapse rate applies to rising or sinking air when it is unsaturated, as in the air from the surface to 200 meters in this example. If lifted sufficiently, an unsaturated parcel can have its temperature lowered to the dew point at some particular height called the lifting condensation level (LCL). Further lifting results in a lowering of the temperature according to the saturated adiabatic lapse rate (SALR) of ≈ 0.5 °C/100 m.

the **lifting condensation level (LCL)**. As the saturated air rises beyond the LCL, expansion continues to lower its temperature, but the cooling is partially offset by the release of latent heat from condensation (or the deposition to ice). Thus, the lifting of saturated air results in a less rapid cooling than the lifting of unsaturated air. The rate at which saturated air cools is the **saturated adiabatic lapse rate (SALR)**, which is about 0.5 °C/100 m (3.3 °F/1000 ft). The terms *wet adiabatic lapse rate and moist adiabatic lapse rate* are often used interchangeably with saturated adiabatic lapse rate. Unlike the dry adiabatic rate, the SALR is not a constant value but instead varies with temperature, as we explain in *Box 5–6, Physical Principles: The Varying Value of the Saturated Adiabatic Lapse Rate.*

DID YOU KNOW?

Water vapor is not the only familiar gas that can condense as a result of cooling due to expansion. Carbon dioxide in a bottle of beer or soda exists as a gas dissolved in the drink and is also found in the neck of the bottle between the top of the drink and the bottle cap. When you open the bottle, the carbon dioxide above the surface of the liquid expands out of the bottle neck and into the surrounding atmosphere. In doing so, its temperature drops adiabatically, and some condenses to form a visible miniature CO_2 cloud wafting out of the bottle top.

The Environmental Lapse Rate

The adiabatic lapse rates are not to be confused with the **environmental** (or **ambient**) **lapse rate (ELR)**, which applies to the vertical change in temperature through still air. A large mass of air is not likely to have a constant temperature; rather, its temperature usually decreases with altitude in the troposphere. The rate at which the ambient temperature decreases with height is the ELR. The ELR for the troposphere is highly variable. It changes from day to day, from place to place, and even from one altitude to another within the atmospheric column. An atmosphere in which temperature decreases rapidly with elevation is said to have a "steep" environmental lapse rate.

As an analogy, we can contrast the changes in temperature that would occur inside a rising balloon with that of the surrounding air. In Figure 5–18, a balloon rises through the

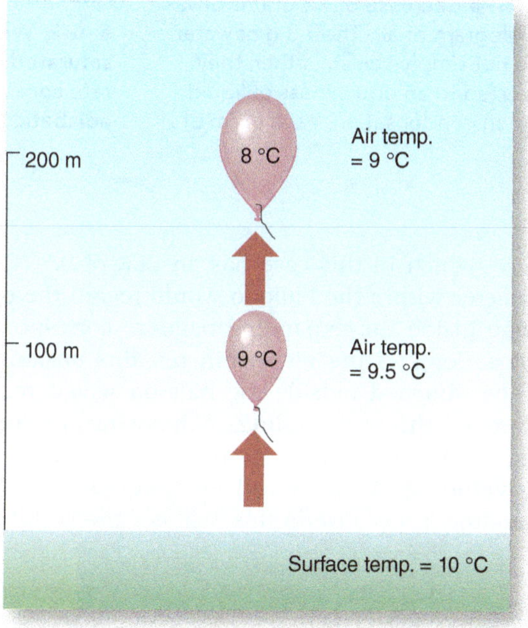

▲ **FIGURE 5–18 Environmental Lapse Rate (ELR).** Do not confuse the adiabatic lapse rates with the environmental (or ambient) lapse rate. A balloon rising through air with an ELR of 0.5 °C/100 m passes through air whose temperature decreases from 10 °C at the surface, to 9.5 °C at 100 m, and 9.0 °C at 200 m. The air within the balloon cools at the dry adiabatic lapse rate of 1.0 °C/100 m, faster in this example than the ELR, and therefore attains a temperature of 8 °C at the 200 m level.

5–6 PHYSICAL PRINCIPLES

The Varying Value of the Saturated Adiabatic Lapse Rate

The dry adiabatic lapse rate (DALR) has a constant value of 1 °C/100 m. The saturated adiabatic lapse rate (SALR) is usually about half that value, because the release of latent heat as saturated air rises partially offsets the cooling by expansion. But unlike the value of the dry adiabatic lapse rate, the SALR is not constant. Rather, it depends on the temperature of the saturated air parcel, with higher temperatures causing lower lapse rates. Figure 5–6–1 can help us see why this is the case. Recall that if air is saturated, its actual specific humidity is equal to the saturation specific humidity. As a rising parcel of saturated air is cooled adiabatically, the amount of water vapor that can exist decreases and the surplus water vapor is removed by condensation (or deposition).

Now observe what happens when the temperature of warm, saturated air decreases 5 °C (9 °F), from 30 °C to 25 °C. As the air cools, its specific humidity changes from 27.7 grams of vapor per kilogram of air to 20.4—a decrease of 7.3 grams of vapor per kilogram of air. The 7.3 g of water vapor do not simply vanish; rather, they are converted to an equal mass of liquid water. Upon condensation, each gram of water releases 2500 joules of latent heat, for a total of 18,250 J. Compare that to what happens if the temperature undergoes another 5 °C drop in temperature, but this time from 5 °C to 0 °C. At this lower temperature, the 5° of cooling reduces the water vapor content from 5.5 to 3.8 g/kg—only a 1.7 g decrease—and only 4250 J of energy are released to the air. Thus, at low temperatures relatively little latent heat is released to offset the cooling due to expansion, and the SALR is nearly equal to the DALR. When warm, saturated air is lifted, a greater amount of latent heat is available to offset the cooling by expansion, and the SALR assumes a lower value.

5–6–1 How does temperature affect the value of the saturated adiabatic lapse rate?

5–6–2 Why isn't the saturated adiabatic lapse rate constant, like the dry adiabatic lapse rate?

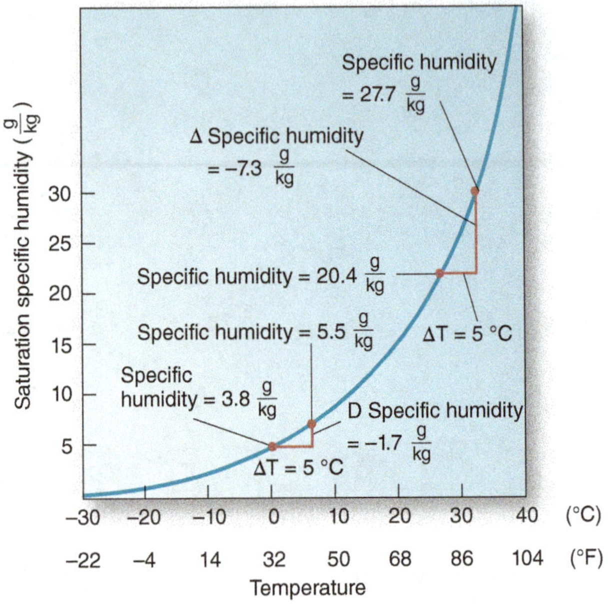

▲ **FIGURE 5–6–1 Saturated Adiabatic Lapse Rate (SLR).** Unlike the DALR, the SALR is not a constant value. When warm, saturated air cools, it causes more condensation (and hence more latent heat release) than does cold, saturated air. For example, if saturated air cools from 30 °C to 25 °C (a 5° decrease), the specific humidity decreases from 27.7 grams of water vapor per kilogram of air to 20.4. A 5 °C drop in temperature from 5 °C to 0 °C lowers the specific humidity only 1.7 grams for each kilogram of air. This brings about less warming to offset the cooling by expansion, as well as a greater saturated adiabatic lapse rate.

atmosphere, which in this case has an ELR of 0.5 °C/100 m. A thermometer within the balloon would record the temperature change inside the expanding balloon, corresponding to the DALR as long as the air within remains unsaturated. A thermometer attached outside the balloon would record the temperature of the surrounding, nonmoving air, reflecting the environmental lapse rate. As the balloon rises, the temperature within the balloon will be lower than that of the air surrounding it, because in this instance the DALR exceeds the ELR.

CHECKPOINT

5.15 What happens as a parcel of unsaturated dry air rises? Explain.

5.16 What is the difference between the environmental lapse rate and the adiabatic lapse rate?

Forms of Condensation and Deposition

5.9 List different forms of condensation and deposition.

Saturation can lead to the formation of liquid water or ice crystals. The condensation or deposition can occur in the air as cloud or fog, or onto a surface as dew or frost. In this section we discuss the processes that give rise to these various types of condensation. Table 5–4 summarizes the general types of condensation and the processes that form them.

Dew

Dew (Figure 5–19a) is liquid condensation on a surface, often occurring during the early morning hours. At night the loss of longwave radiation can cause the surface to cool diabatically. Air immediately in contact with the cold surface cools

TABLE 5–4

General Types of Condensation and Deposition

Form	Predominant Processes	Characteristics
Dew	Lowering of temperature to the dew point near the surface. Favored under clear skies and no wind. Diabatic process.	Appears as coating of liquid water on surfaces.
Frost	Lowering of air temperature to saturation point, when the saturation point is below 0 °C (32 °F). Diabatic process.	Appears as large number of small, white crystals on surfaces.
Frozen dew	Formation of dew at temperatures above 0 °C, followed by cooling to temperatures below 0 °C. Diabatic process.	Continuous layer of solid ice on surface.
Fog	Usually by cooling of layer of air with light winds. Sometimes by evaporating water from falling precipitation or by mixing warm, moist air with cold air. Diabatic or adiabatic process.	Large concentration of suspended droplets in layer of air near ground. Under extreme cold, can consist of suspended ice crystals.
Radiation fog	Cooling of air to dew point by longwave radiation loss. Diabatic process.	Same as above.
Advection fog	Cooling of air to dew point as it passes over cool surface. Diabatic process.	Same as above.
Upslope fog	Cooling of air as it flows upslope. Adiabatic process.	Same as above.
Precipitation fog	Increasing the water vapor content of the air by evaporation from falling droplets. Adiabatic process.	Same as above.
Steam fog	Mixing warm, moist air with cold air. Adiabatic process.	Same as above.
Clouds	Usually by lifting of air and adiabatic cooling.	Concentration of suspended droplets and/or ice crystals in air well above the surface.

by conduction, and if the temperature decreases all the way to the dew point, condensation forms. Dew is most likely to form on clear, windless nights, when the absence of clouds allows much longwave radiation to escape to space and the lack of wind precludes the mixing of warmer air from above. Together these conditions promote rapid cooling within a shallow layer of air immediately adjacent to the surface.

Frost

The formation of **frost** is similar to that of dew, except that saturation occurs when the temperature is below 0 °C. When the air temperature is lowered to the frost point, very small ice crystals are deposited onto solid surfaces, giving them a bright white appearance, as shown in Figure 5–19b. This type of deposition, sometimes referred to as *white frost* or *hoar frost*, occurs by the transformation of water vapor directly into ice, without going through the liquid phase.

Because it consists of a huge number of separate ice crystals rather than a solid, continuous coating of ice, frost on the windshield of a car is often easy to remove by a swift brushing with a credit card or window scraper. This is in contrast to a more troublesome type of condensation called *frozen dew.*

Frozen Dew

Frozen dew differs from frost in both its structure and its manner of formation. It begins when saturation forms liquid dew at temperatures slightly above 0 °C. When further cooling brings its temperature below the freezing point, the liquid solidifies into a thin, continuous layer of ice. In contrast to frost, frozen dew is neither milky white nor easy to remove but instead bonds tightly to any surface on which it forms, as shown in Figure 5–19c.

Because it is a continuous coating of solid ice, a mere brushing does not come close to removing frozen dew. In addition to coating your car's windshield, the ice can make it difficult or impossible to get your key into the lock mechanism of your door. And even if you are lucky enough to be able to turn the key, the door can become frozen to the car frame as if it were welded shut. On the other hand, it may be just as well for you that you are not able to get into your car because of *black ice*, the smooth coating of frozen dew that forms on road surfaces. This is especially likely to form over bridges, causing dangerous, slippery driving conditions.

Fog

Fog is essentially a cloud whose base is at or near ground level (Figure 5–20). It can be extremely shallow, on the order of a meter or so in depth, or it can extend for tens of meters above the surface. Like any other form of condensation, fog can form by the lowering of the air temperature to the dew point, an increase in the water vapor content, or the mixing of cold air with warm, moist air.

PRECIPITATION AND STEAM FOGS We have already described one example of condensation resulting from the addition of water vapor to the air, *precipitation fog*, which results from the evaporation of falling raindrops. Another, although very localized, type of fog occurs from adding water vapor to the air—the type you see right in front of you when you exhale on a cold day. We have also discussed the formation of *steam fogs*, which occur when cold, dry air mixes with warm, moist air above a water surface. All other types of fog result from a cooling of the air to the dew point. They include radiation fogs, advection fogs, and upslope fogs.

▲ FIGURE 5–19 **Forms of Condensation and deposition.** (a) Dew, (b) frost, and (c) frozen dew.

▲ FIGURE 5–20 **Fog at Morton's Overlook Great Smoky Mountains National Park.**

RADIATION FOG **Radiation fogs** (sometimes called *ground fogs*) develop when the nighttime loss of longwave radiation causes cooling to the dew point. Like dew, a radiation fog is most likely to form on cloudless nights when longwave radiation from the surface easily escapes to space. Unlike dew, it is most likely to form with light winds of about 5 km/hr (3 mph) rather than in perfectly still air. Light breezes promote a gentle stirring of the lower atmosphere, which permits condensation to form throughout a layer of air. When the wind speed is much greater than 5 km/hr, the excess turbulence brings warm air down toward the surface and thereby inhibits cooling to the dew point.

Most radiation fogs begin to dissipate within a few hours of sunrise. Although we sometimes talk about a fog "lifting," that is not what really happens. It is probably better (although still somewhat imprecise) to describe it as "burning off." When sunlight penetrates the fog, it warms the surface, which in turn warms the overlying air. As the air temperature increases, the fog droplets gradually evaporate. Because the evaporation of droplets is most rapid near the surface, the fog appears to lift, although it really undergoes no vertical displacement.

When radiation fogs are especially well developed, they can scatter backward the greater part of incoming solar radiation. Because the amount of energy reaching the surface is now reduced, the fog can persist throughout the day, especially in the winter when the days are short and the sun angles are low. Under the most extreme circumstances, fogs can persist for days on end.

A prime location for a persistent radiation fog is the Central Valley of California (Figure 5–21). To the west of the valley, the Coast Ranges block the moderating effects of the Pacific Ocean, while the Sierra Nevada Mountains isolate the valley from the east. In addition to the clear skies and light

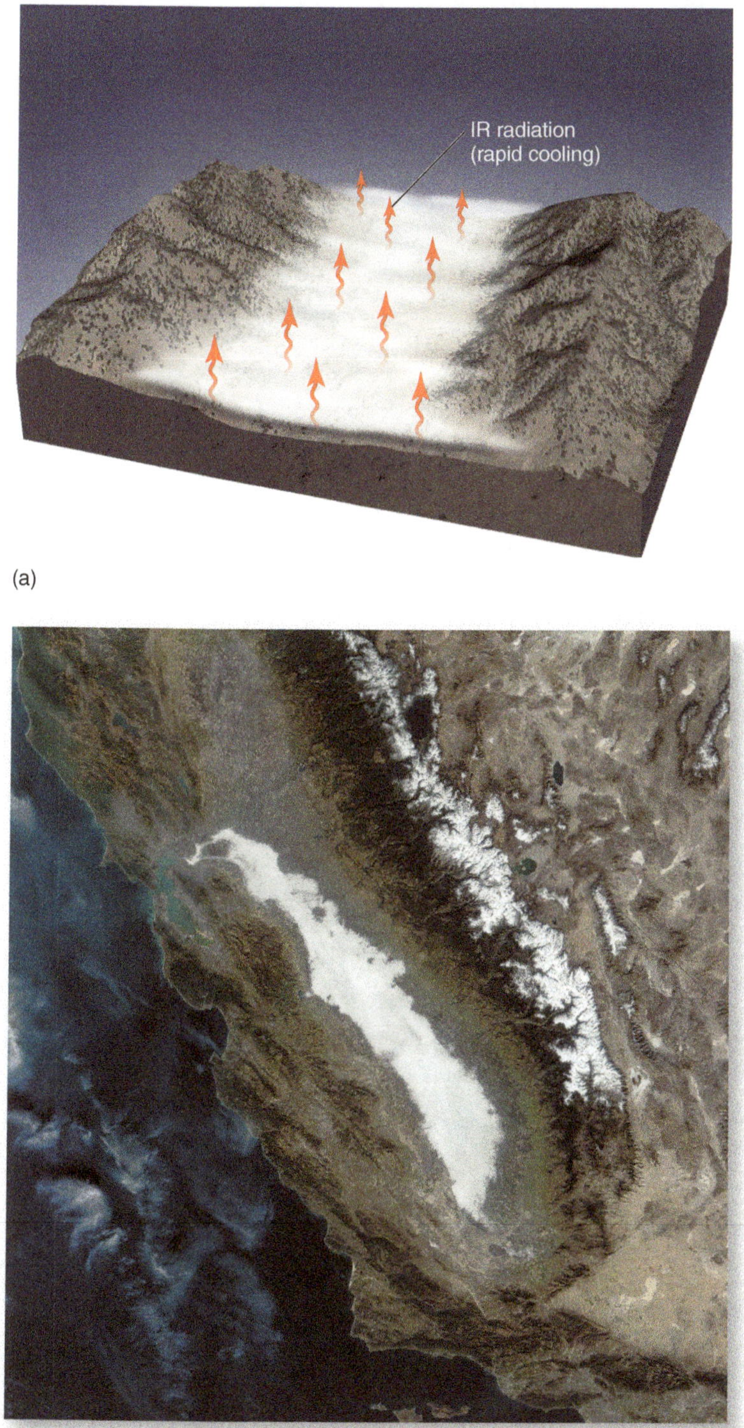

▲ **FIGURE 5–21 Radiation Fog.** (a) The Central Valley of California often gets radiation fogs in winter that last for days on end. (b) These persistent fogs are visible from space.

winds that often dominate during the winter, this heavily agricultural region has copious amounts of moisture evaporated into the air from the irrigated farmland. When the radiation fog (locally referred to as a *tule fog*) covers the valley, visibility along the two major north–south highways can be dangerously reduced to near zero. The opening pages of this chapter describe an incidence of a tule fog and its consequences.

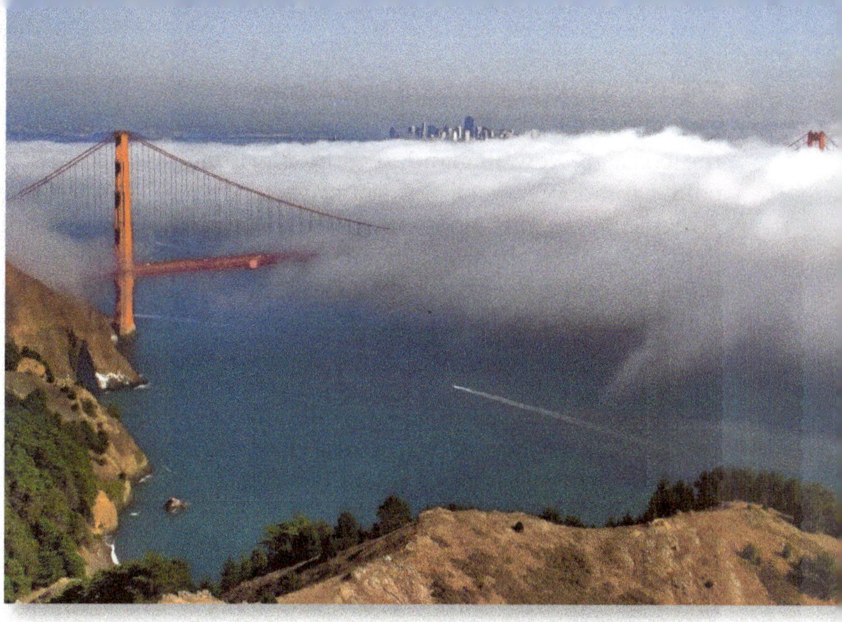

▲ **FIGURE 5–22 Advection Fog.** This type of fog frequently forms when air flows from a region of warmer water over a cold ocean current. This is particularly common over the San Francisco Bay area.

Radiation fogs—and the risks they impose on travelers—are frequent occurrences wherever air has the opportunity to cool at night with gentle stirring. Various state transportation agencies have tried to alleviate the threat of such accidents by painting reflective "fog lines" on at-risk highways and installing automated weather stations to provide real-time weather information, but measures such as these cannot overcome the inherent danger imposed by dense fog.

Radiation fogs are formed by diabatic cooling and are therefore associated with cold air. Because cold air is denser than warm air with otherwise similar characteristics, radiation fogs often settle into local areas of low elevation, where they are called *valley fogs*. It is tempting to attribute the fog's high density to the heavy droplets of liquid water, but they are not the cause. It is the low temperature—not the existence of water droplets—that causes the fog to be dense and settle into valleys. Because all the water droplets replace an equal mass of water vapor from which they formed, their presence does not make the air any heavier.

ADVECTION FOG **Advection fogs** (Figure 5–22) form when relatively warm, moist air moves horizontally over a cooler surface (the term *advection* refers to horizontal movement). As the air passes over the cooler surface, it transfers heat downward; this causes it to cool diabatically. If sufficient cooling occurs, a fog forms. Such fogs can be advected for considerable distances and persist well downwind of the area over which they form. One of the most famous examples of this phenomenon occurs during the summer months over the San Francisco Bay area. As relatively warm Pacific air drifts eastward, it passes over the narrow, cold, southward-flowing California ocean current. The cooling of the air offshore forms the fog, which drifts eastward toward San Francisco. It is quite common for the fog to fully engulf San Francisco while Oakland and Berkeley, across the bay, remain warm and sunny.

Advection fogs can also form over water when warm and cold ocean currents are in proximity to each other. Off the coast of New England and the Maritime Provinces, the cold Labrador current flows just to the north of the Gulf Stream. When the moist air from the Gulf Stream region drifts over the Labrador current, it can be cooled to the dew point to form a persistent, dense fog that can last for weeks. Although most common in summer, it can occur any time during the year.

Another example of an advection fog is the fog that often covers London during the winter. Warm air passing over the warm Gulf Stream is advected over England, where it is chilled by the surface to form a thick fog.

The winds associated with advection fogs are often greater than those of radiation fogs. This promotes greater turbulence and allows the droplets to circulate to greater heights. Advection fogs can have thicknesses up to about a half kilometer (1500 ft).

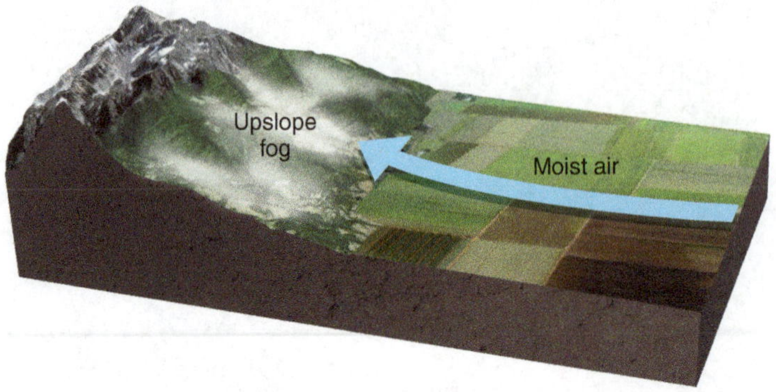

▲ **FIGURE 5–23 Upslope Fog.** Upslope fogs form by adiabatic cooling of air as it flows up a hill or mountainside.

DID YOU KNOW?

The water droplets suspended in foggy air do not cause the air to be heavier or denser than it otherwise would be. When water vapor condenses to form water droplets, there is no net gain in the mass of the air—the liquid droplets weigh as much as the water vapor used to form them. But foggy air is often dense. Why? Because the low temperatures that favor the formation of fog cause the air to be denser (recall the equation of state in Chapter 4). Thus, the fog and the dense air are both the result of the same cause—lowered air temperature—and one does not cause the other.

UPSLOPE FOG Of the three types of fog caused by the cooling of air, only **upslope fog** is formed by adiabatic cooling. When air flows along a gently sloping surface, it expands and cools as it moves upward. The western slope of the Great Plains of the United States provides an excellent setting for this type of condensation (Figure 5–23). Westward from the Mississippi River valley, the elevation gradually increases toward the foothills of the Rocky Mountains. Moist air from the Gulf of Mexico cools adiabatically as it ascends the slope of the plains to create widespread fog. Figure 5–24 presents a highly generalized description of the types of fog most prevalent over regions of the United States and Canada.

▶ **FIGURE 5–24 Fog Distribution.** The map shows where the different types of fog are commonly found in North America.

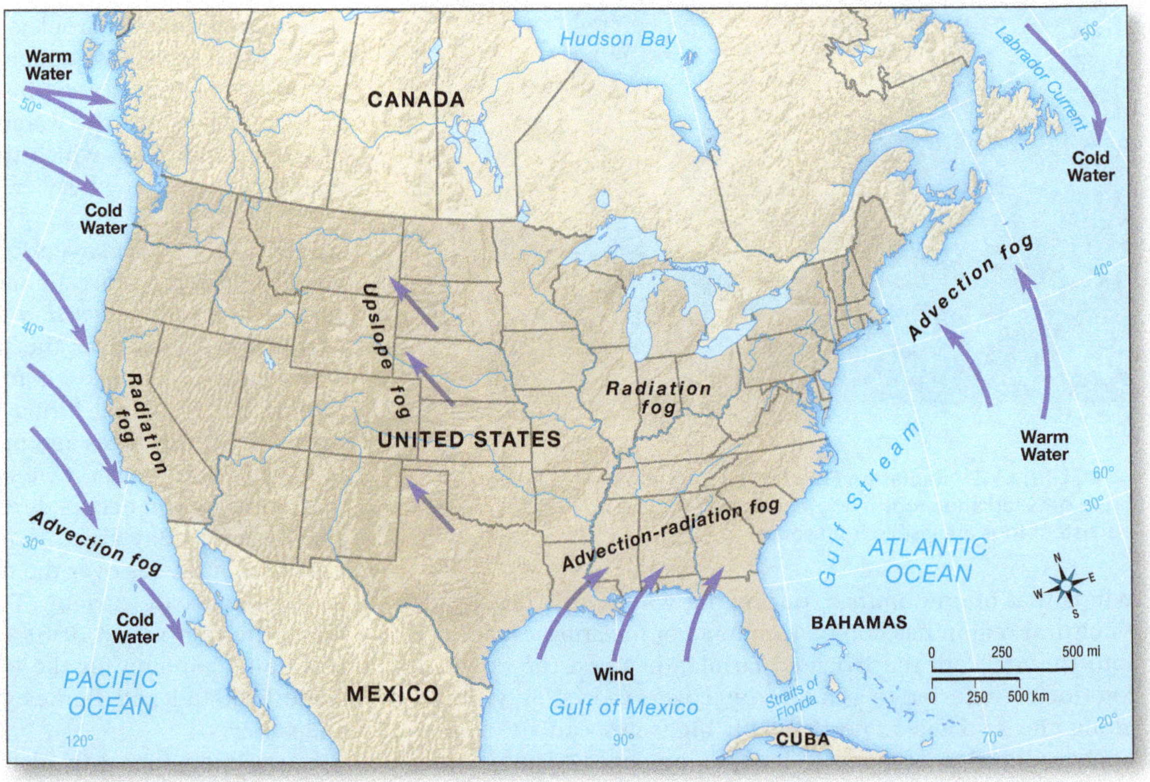

CHECKPOINT

5.17 What are the four general types of condensation that occur on or near Earth's surface?

5.18 Identify the different types of fog and explain how they form.

Distribution of Fog

5.10 Describe the distribution of fog.

Figure 5–25 depicts the number of heavy fog days (defined as limiting visibility to a quarter of a mile or less) across the 48 conterminous United States. The three significant centers of heavy fog are along the Pacific Northwest, New England, and the middle and southern Appalachians. The Pacific Northwest and the coast of British Columbia experience numerous advection fogs, as westerly winds advect moist air over the cold California current. But it is not correct to attribute all of the fog formation to the cooling effect of the cold surface waters. As damp, cool air reaches the shore, fog formation is abetted by the *orographic* effect (the lifting effect caused as air crosses a mountain or similar barrier) as air approaches the steep Coast Ranges. Cape Disappointment, Washington, wins the prize for the foggiest U.S. location, being shrouded in heavy fog nearly one-third of the time. The zone of most persistent fog is confined primarily to the coastal region because the Coast Ranges block the flow of moisture inland.

New England and the Maritime Provinces experience a large number of heavy fog days. Along the coast, advection fogs dominate, with the coast of Maine having the highest incidence of dense cover. The advection fogs are most prevalent in summer. Inland, radiation fogs are very common at some of the higher elevation areas of New Hampshire and Vermont.

The southern Appalachians, the third major focus of fog, undergo a large number of radiation fogs, particularly in late summer and fall. Not surprisingly, fog is at a minimum in the desert Southwest.

CHECKPOINT

5.19 What type of fog is likely to form on a cool, cloudless night with light winds?

5.20 Compare and contrast the processes that form advection fogs and upslope fogs.

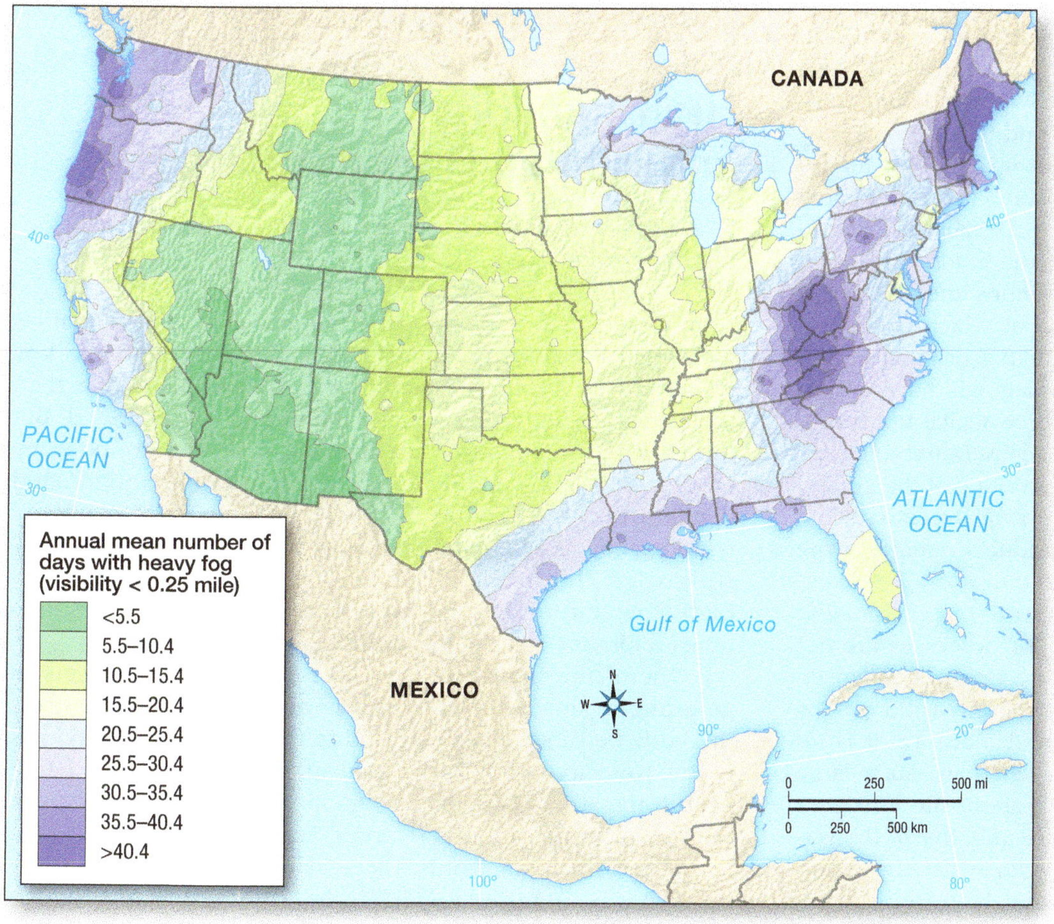

◄ **FIGURE 5–25 Average Annual Number of Days with Heavy Fog in the United States.**

Video
Satellite View of Fog
http://goo.gl/ECLZLM

MM **MapMaster** North America Physical Environment Days with Heavy Fog

Formation and Dissipation of Cloud Droplets

5.11 Describe the formation and dissipation of cloud droplets.

Farther from the surface than fog, dew, or frost, clouds are usually the result of the adiabatic cooling associated with rising air parcels. In this section we take a closer look at such parcels as they rise, become saturated, and continue upward above the level at which condensation first occurs. In Chapter 6 you will see how these processes, together with mechanisms that lift air, lead to the development of clouds.

For reasons we will not discuss here, the dew point decreases as the air rises, at the rate of about 0.2 °C/100 m (1.1 °F/1000 ft). This decrease is called the **dew point lapse rate.** As unsaturated air is lifted, its temperature therefore approaches the dew point by 0.8 °C for every 100 m of ascent (i.e., 1.0 °C minus 0.2 °C). Thus, if the air temperature and dew point start out at 18 °C and 10 °C, respectively, an ascent of 1000 m is necessary to cause the air to be saturated.

Raising an air parcel above the lifting condensation level (LCL) initially leads to the formation of small cloud droplets. But at about 50 m or so above the LCL, all the condensation nuclei in the air will have attracted water, and further uplift leads only to the growth of existing water droplets. In other words, no new droplets form as the air continues to rise; instead, the existing droplets grow larger.

The processes that lead to the formation of a cloud do not continue forever. At some point in time, lifting will cease and no further condensation will occur. When this happens, further cloud development comes to an end.

Now consider what happens if a rising parcel of air reverses its movement and begins to subside. The parcel now warms at the *saturated* adiabatic lapse rate because the warming by compression is partially offset by the gradual evaporation of the droplets. The evaporation continues until the parcel has descended back to the original level at which condensation occurred. If the air then sinks below the original LCL, it warms at the DALR, because all the droplets will have evaporated. If brought down to the initial level at which uplift first began, the air will reassume its original temperature and dew point. The net effect of all this is that the cooling of the air and the condensation of liquid water are *reversible processes.* Note that we assume all the condensation products remain in the parcel and are thus available for evaporation during descent. Of course, the real atmosphere *loses* moisture by precipitation of rain and snow, which means that these processes are not strictly reversible. But a relatively small portion of condensed water falls out, so the concept remains generally valid.

In thinking about vertical motions and moisture, it is interesting that very small displacements can have such large consequences. After all, we hardly think about the effects of horizontal movements covering 100 km (62 mi). But in the vertical, movement of just 1 km (0.62 mi) can make the difference between a fine day and a ruined picnic, because rising air leads to clouds, and clouds give rise to precipitation. Like the formation of clouds, precipitation is no simple matter, as we will see in Chapter 7.

▲ **FIGURE 5–26 Hazardous Heat.** The combination of high temperature and high humidity can cause extreme discomfort and even medical emergencies.

CHECKPOINT

5.21 What is the dew point lapse rate and why is it important?

5.22 Describe what would happen if a parcel of unsaturated air were lifted up above the lifting condensation level and then returned to its original altitude.

High Humidity and Human Discomfort

5.12 Explain how the effects of humidity on human discomfort are rated.

Temperature is one of the most important weather variables with regard to human comfort, and excessively high and low temperatures account for more North American fatalities than do hurricanes, lightning, floods, and tornadoes combined. But the effects of temperature extremes can be compounded by other factors, such as humidity, the intensity of sunlight, and strength of wind (Figure 5–26). In Chapter 3 we saw how wind speeds can be combined with low temperatures to create a wind chill index. Similarly, the effect of humidity and high temperatures can be expressed in a **heat index**, sometimes referred to as the **apparent temperature**.

One of the human body's most effective mechanisms in guarding against excessive heat is perspiration. Sweat cools the body because, when released to the surface of the skin, it is free to evaporate into the atmosphere and consume latent heat. If the atmosphere has a high moisture content, however, the rate of evaporation is retarded and the loss of latent heat is reduced. In other words, the sweat is unable to effectively do what it is supposed to do. The heat index accounts for this effect (Table 5–5).

TABLE 5–5

Heat Index

Temp. (°F) °C	Relative Humidity (%)											
	40	45	50	55	60	65	70	75	80	85	90	95
(110) 47	(136) 58											
(10°) 43	(130) 54	(137) 58										
(106) 41	(124) 51	(130) 54	(137) 58									
(104) 40	(119) 48	(124) 51	(131) 55	(137) 58								
(102) 39	(114) 46	(119) 48	(124) 51	(130) 54	(137) 58							
(100) 38	(109) 43	(114) 46	(118) 48	(124) 51	(129) 54	(136) 58						
(98) 37	(105) 41	(109) 43	(113) 45	(117) 47	(123) 51	(128) 53	(134) 57					
(96) 36	(101) 38	(104) 40	(108) 42	(112) 44	(116) 47	(121) 49	(126) 52	(132) 56				
(94) 34	(97) 36	(100) 38	(103) 39	(106) 41	(110) 43	(114) 46	(119) 48	(124) 51	(129) 54	(135) 57		
(92) 33	(94) 34	(96) 36	(99) 37	(101) 38	(105) 41	(108) 42	(112) 44	(116) 47	(121) 49	(126) 52	(131) 55	
(90) 32	(91) 33	(93) 34	(95) 35	(97) 36	(100) 38	(103) 39	(106) 41	(109) 43	(113) 45	(117) 47	(122) 50	(127) 53
(88) 31	(88) 31	(89) 32	(91) 33	(93) 34	(95) 35	(98) 37	(100) 38	(103) 39	(106) 41	(110) 43	(113) 45	(117) 47
(86) 30	(85) 29	(87) 31	(88) 31	(89) 32	(91) 33	(93) 34	(95) 35	(97) 36	(100) 38	(102) 39	(105) 41	(108) 42
(84) 29	(83) 28	(84) 29	(85) 29	(86) 30	(88) 31	(89) 32	(90) 32	(92) 33	(94) 34	(96) 36	(98) 37	(100) 38
(82) 28	(81) 27	(82) 28	(83) 28	(84) 29	(84) 29	(85) 29	(86) 30	(88) 31	(89) 32	(90) 32	(91) 33	(93) 34
(80) 27	(80) 27	(80) 27	(81) 27	(81) 27	(82) 28	(82) 28)	(83) 28	(84) 29	(84) 29	(85) 29	(86) 30	(86) 30

Category	Heat Index	Possible heat disorders for people in high-risk groups
Extreme Danger	(130 °F or higher) 54 °C or higher	Heat stroke likely.
Danger	(105 – 129 °F) 41 – 54 °C	Muscle cramps and/or heat exhaustion possible with prolonged exposure and/or physical activity.
Extreme Caution	(90 – 105 °F) 32 – 41 °C	Muscle cramps and/or heat exhaustion possible with prolonged exposure and/or physical activity.
Caution	(80 – 90 °F) 27 – 32 °C	Fatigue possible with prolonged exposure and/or physical activity.

Source: National Weather Service Office, Birmingham, AL.

The apparent temperatures caused by the combination of heat and humidity provide useful guidelines for people. At values between 41 °C and 54 °C (105 °F and 129 °F), muscle cramps or heat exhaustion are likely for high-risk people, and even people who are not at high risk face the threat of heat stroke (a potentially fatal increase in the body's internal temperature). If the National Weather Service forecasts that the apparent temperature will exceed 41 °C (105 °F) for more than 3 hours, it issues a ***heat advisory***. Apparent temperatures above 54 °C (129 °F) are considered extremely dangerous, and heat stroke is likely for at-risk people.

The apparent temperatures shown in Table 5–5 should be considered approximate guidelines. Some people react differently to heat, and the index does not account for variables such as exposure to bright sun or wind. And always keep in mind that high apparent temperatures can be fatal, especially if coupled with prolonged activity. In 2007 unseasonable heat at the Chicago Marathon may have been linked to the death of one participant. Forty-nine people ended up hospitalized and hundreds more collapsed during the event (see *Box 5–7, Focus on the Environment and Societal Impacts: High Temperatures and Human Health*).

CHECKPOINT

5.23 What two atmospheric variables are combined to yield the heat index?

5.24 Why is sweat not able to cool a person effectively during periods when the heat index is high?

DID YOU KNOW?

Bangkok, Thailand, one of the hottest cities in the world, has such persistently high temperatures and humidity that it exceeds the U.S. National Weather Service threshold for issuing heat advisories for more than half the days of a typical year.

5–7 FOCUS ON THE ENVIRONMENT AND SOCIETAL IMPACTS

High Temperatures and Human Health

Aside from the discomfort they can cause, high apparent temperatures can have very substantial impacts on human welfare. Some effects are indirect; for example, the combination of high temperatures and humidities increase the growth rates of household mold and dust mites. Others effects, such as *heat stroke* and its precursor, *heat exhaustion*, are obvious and present an overriding danger to human health.

Anybody who has watched or participated in physical outdoor sports on hot, humid days is familiar with the cramping that sometimes occurs. Though painful, the situation is not serious and can be remedied by stretching and ingesting sports drinks that help replenish electrolytes. A more serious condition is heat exhaustion, which occurs after continued exertion in high apparent temperatures. Victims of heat exhaustion attain elevated core body temperatures and exhibit symptoms such as light-headedness, nausea, general weakness, and fatigue. The victim's skin may also feel cold and moist. Fortunately, victims of heat exhaustion usually remain cognizant of the situation and can be relieved of symptoms by getting to a cooler location and resting, taking a cool shower, and drinking cold, nonalcoholic beverages. The situation should have no long-term repercussions.

When the body's temperature rises to 40 °C (104 °F) or above, the individual has heat stroke. Unlike heat exhaustion, this is an emergency condition marked by skin that is flushed and warm to the touch (it may be dry if brought on just by hot conditions or moist if the result of strenuous exercise in addition to exposure to hot conditions), nausea, a racing heart rate, unconsciousness or disorientation, and rapid breathing. These symptoms indicate that emergency care is needed. Even if the condition persists for just a few hours, long-term damage to the heart, brain, or kidneys can develop.

▲ **FIGURE 5–7–1 2007 Chicago Marathon.** The unusual heat at this event led to one fatality and numerous hospitalizations.

Heat stroke is more likely to occur in people exposed to continuously hot conditions for several days than during brief episodes of extreme apparent temperatures. Typically, young children and people over age 65 are more likely to get heat stroke, as are people with certain conditions such as obesity and those taking certain prescription medications. But anybody—even people who are young and fit—should take care when participating in physically demanding activities during hot, humid conditions. Experienced runners in high endurance events like the Chicago Marathon are well aware of the danger of heat stroke (Figure 5–7–1).

5–7–1 Name a potentially harmful, nonphysiological effect of high temperature and humidity.

5–7–2 What is the difference between heat exhaustion and heat stroke?

Atmospheric Moisture and Climate Change

5.13 Describe possible effects of global warming, including its effects on evaporation rates and atmospheric water vapor content.

We have seen in this chapter that the amount of water vapor that can exist in the air increases with temperature. It is also clear that higher water temperatures lead to increased evaporation rates; and, in fact, ocean surface temperatures *have* increased in conjunction with higher air temperatures during recent decades.

Let's take the situation a step further. In Chapter 3 we saw that water vapor, like carbon dioxide, is an effective gas at absorbing longwave radiation emitted from Earth's surface. As such it is a greenhouse gas that keeps the planetary temperature at a value greater than it would otherwise be. We therefore need to consider whether global warming leads to increases in evaporation rates and atmospheric water vapor contents, and whether such increases would in turn contribute to further atmospheric warming—and by extension even greater water vapor contents. Self-perpetuating situations such as these are referred to as *positive feedbacks*.

The Intergovernmental Panel on Climate Change (IPCC) has extensively reviewed the published research on this phenomenon and reiterated its findings in its Fifth Assessment Report, issued in 2013 (see Chapter 3). The panel cited studies around the world that have found that increases in specific humidity near the surface have been associated with increasing temperatures since 1976. These increases in water vapor are not restricted to the surface; increases in water vapor content even into the upper troposphere have likewise occurred.

Observations suggest a 4 percent overall increase between 1970 and the early 21st century.

Despite the increase in moisture content, the average relative humidity across the globe has not increased. This is due to the effect of temperature on the saturation point. As the amount of water vapor in the air has increased, so, too, has the amount of water vapor that *can* exist because of higher temperatures. Over most oceanic areas the relative humidities have remained fairly constant, as increasing water vapor contents have been offset by increases in the saturation specific humidity. Over certain land areas the increases in specific humidity have been more than offset by increases in the saturation levels, leading to locally unchanged or slightly reduced relative humidities.

Carbon dioxide contents will continue to increase at least for some time to come, which will likely lead to further atmospheric warming. With greater warming we can also anticipate further increases in specific humidity and more effective absorption of outgoing longwave radiation. So will this positive feedback ultimately result in cataclysmic temperature increases? Not necessarily, because strong negative feedbacks also operate (Chapter 16). The most important of these is increasing emission of longwave radiation as temperature rises. Recall from Chapter 2 that for blackbodies, emission grows with the fourth power of temperature. Although Earth does not emit to space as a perfect blackbody, increasing longwave emission acts counter to the water vapor feedback.

Another feedback is related to clouds: It might be that increasing water vapor concentrations will lead to increases in global cloud cover. The effects of cloud cover on shortwave and longwave radiation depend on the type of cloud cover, the height of the clouds, and their location. Under some conditions an increase in cloud cover can lead to a substantial reduction in absorbed solar radiation that suppresses further warming. However, clouds are strong absorbers of longwave radiation, thus they also limit radiation emitted to space. This reduction in outgoing radiation can exceed the reduction in solar radiation absorption, thereby promoting further warming. Even on a global basis, it is an open question as to whether clouds will emerge as a negative or positive feedback in the future. This issue will be discussed in Chapter 16.

Video
Global Evaporation Rates

http://goo.gl/bBKAbl

CHECKPOINT

5.25 What kind of change in moisture content has occurred since the mid 1970s?

5.26 How will negative feedbacks affect further changes in moisture content in the atmosphere?

Summary

5.1 Describe the hydrologic cycle.

- Water in its three phases—solid, liquid, and gas—constantly moves across the interface between the atmosphere and Earth's surface through what is known as the *hydrologic cycle*.
- Despite the fact that water accounts for only a small proportion of the mass of the air, in its three phases it is extremely important to the atmosphere—and to all life on Earth.

5.2 Explain the concept of saturation.

- The process of saturation begins at the surface of liquid water as molecules constantly move about. Some randomly break free of the surface to become water vapor.
- As long as the rate of evaporation exceeds condensation, the amount of water vapor increases. When the condensation rate becomes equal to the evaporation rate, and saturation occurs.
- In the atmosphere, clouds and fog persist as long as there is equilibrium between evaporation and condensation rates. When evaporation and condensation are in equilibrium, the air is saturated and small droplets remain in the air without evaporating.

5.3 Identify the indices used in measuring the atmosphere's water vapor content.

- The amount of water vapor, or humidity, is affected by the pressure it exerts (vapor pressure).
- The specific humidity and mixing ratio are ratios that relate the mass of water vapor to the air in which it is contained and to the mass of the other gases, respectively.
- Relative humidity is the amount of water vapor in the air relative to the amount that can exist at the current temperature.
- The dew point is the temperature to which the air temperature must be lowered for saturation to occur.
- Condensation is controlled by two factors having opposite effects: curvature retards condensation, and the abundance of hygroscopic nuclei facilitates it.

5.4 Describe how humidity is measured.

- Humidity can be determined using paired thermometers (a psychrometer) that provide dry and wet bulb temperatures. The difference between the two temperatures is the wet bulb depression, which, when combined with the dry bulb temperature, permits the use of simple tables to determine relative humidity and dew point.

5.5 Describe how water vapor is distributed in the atmosphere.

- The water vapor content changes both seasonally and spatially. In North American summers dew points are typically greater east of the Rockies than out west, and are higher in the deep South, decreasing northward.

- Because of the lower temperatures experienced in winter, dew points are considerably lower on average in winter than in summer.

5.6 Explain the processes that lead to saturation.

- For any kind of condensation (such as dew, fog, or clouds) to form, the dew point must equal the air temperature. This can result from raising the vapor content of the air to the saturation level; mixing warm, moist air with cooler, dry air; or by lowering the air temperature to the dew point. The latter process is most important for cloud formation.

5.7 Explain the factors that affect condensation.

- Water droplets have considerable curvature, which increases the amount of moisture needed for them to be maintained relative to larger masses of water with flat surfaces.
- The curvature effect is largely offset by the fact that droplets do not occur as pure water but instead exist as solutions.

5.8 Describe how diabatic and adiabatic processes produce cooling and condensation.

- Lowering the air temperature does not require that heat be removed from the air. In fact, most clouds form by adiabatic cooling—the lowering of the air temperature without the removal of heat. Adiabatic cooling results from the expansion of air that occurs when it is lifted and is a direct application of the first law of thermodynamics. The most important diabatic process is the cooling that takes place when the surface loses longwave radiation, cools, and chills the air in contact with it.

5.9 List different forms of condensation.

- Noncloud forms of condensation—dew, frost, frozen dew, and fog—are distinguished from clouds by their proximity to or direct contact with Earth's surface.
- Though one type of fog (upslope fog) results from adiabatic processes, the others result from diabatic cooling of the air, involving the loss of energy.

5.10 Describe the distribution of fog.

- Fog in the United States is most prevalent along the Pacific Northwest, New England, and the southern and middle Appalachians.

5.11 Describe the formation and dissipation of cloud droplets.

- Unsaturated air that is lifted cools at 1 °C per every 100 meters of ascent until the air temperature is lowered to the dew point. At that level, the air begins to cool at the saturated adiabatic lapse rate (approximately 0.5 °C/100 m).
- New cloud droplets initially form after saturation occurs, but after a certain amount of lifting, condensation then occurs onto existing droplets. If the air is then lowered, the air and dew point are the same at each height as they were during the ascent of the air parcel.

5.12 Explain how the effects of humidity on human discomfort are rated.

- Humidity affects a human's susceptibility to heat-related dangers. This susceptibility is reflected in the heat index, a combination of the temperature and humidity.
- The apparent temperature reflects the combination of high temperature and high humidity.
- When the air is warm, high amounts of humidity increase the rate of discomfort and susceptibility to heat exhaustion and heat stroke.

5.13 Describe possible effects of global warming, including its effects on evaporation rates and atmospheric water vapor content.

- Because temperatures have risen over much of the globe, the specific humidity has increased near the surface over most regions, although not all regions have experienced a concomitant increase in relative humidity. This is important not just for its immediate effect, but also because it amplifies greenhouse gas warming and because of its potential to impact cloud formation and patterns.

Key Terms

absolute humidity *p. 153*
adiabatic process *p. 169*
advection fogs *p. 175*
apparent temperature *p. 178*
aspirated psychrometer *p. 161*
condensation *p. 152*
condensation nuclei *p. 166*
deposition *p. 152*
dew *p. 172*
dew point (temperature) *p. 156*
dew point lapse rate *p. 178*
diabatic process *p. 169*
dry adiabatic lapse rate (DALR) *p. 169*
dry bulb thermometer *p. 157*
environmental (ambient) lapse rate (ELR) *p. 171*
evaporation *p. 152*
evapotranspiration *p. 150*
first law of thermodynamics *p. 170*
frost *p. 173*
frost point *p. 157*
frozen dew *p. 173*
hair hygrometer *p. 161*
haze *p. 167*
heat index *p. 178*
heterogeneous nucleation *p. 166*
homogeneous nucleation *p. 166*
humidity *p. 152*
hydrologic cycle *p. 150*
hygroscopic *p. 166*
hygrothermograph *p. 161*
ice nuclei *p. 167*
lifting condensation level (LCL) *p. 171*
mixing ratio *p. 154*
precipitation fog *p. 164*
radiation fogs *p. 174*
relative humidity (RH) *p. 154*
saturated (wet or moist) adiabatic lapse rate (SALR) *p. 171*
saturation *p. 152*

saturation mixing ratio *p. 154*
saturation specific humidity *p. 154*
saturation vapor pressure *p. 153*
second law of thermodynamics *p. 169*
sling psychrometer *p. 157*
specific humidity *p. 153*
steam fog *p. 165*
sublimation *p. 152*
supercooled water *p. 167*
supersaturated *p. 166*
transpiration *p. 150*
upslope fog *p. 176*
vapor pressure *p. 152*
wet bulb depression *p. 157*
wet bulb thermometer *p. 157*

Review Questions

1. What is the hydrologic cycle?
2. Why is it incorrect to refer to the air as "holding" water vapor?
3. Explain the concepts of equilibrium and saturation.
4. What is vapor pressure? In what units of measure is it expressed?
5. What are deposition and sublimation?
6. What units of measurement are used to describe mixing ratio and specific humidity? Why are the two values nearly equal?
7. Why is absolute humidity seldom used?
8. Define *relative humidity*.
9. Why is relative humidity a poor indicator of the amount of water vapor in the air?
10. Define *dew point*. What characteristics make this measure superior to relative humidity?
11. Why can't the dew point temperature exceed the air temperature? What happens if the air temperature is lowered to a value below the initial dew point?
12. Describe the distribution of average dew point across the United States in summer and winter.
13. What are the three general methods by which the air can become saturated?
14. Why doesn't homogeneous nucleation form water droplets in the atmosphere?
15. What are the effects of droplet curvature and solution on the amount of water vapor needed for saturation?
16. What are condensation nuclei and ice nuclei? Are they typically made of the same materials? Which is more abundant in the atmosphere?
17. What is supercooled water?
18. What are psychrometers? How do they work?
19. What is the heat index?
20. What is the first law of thermodynamics and how does it apply to cloud development?
21. Define *dry bulb temperature, wet bulb temperature,* and *wet bulb depression.*
22. Explain the difference between diabatic and adiabatic processes.
23. What are the numerical values of the dry and saturated adiabatic lapse rates? Under what circumstances are they applicable?
24. What does the environmental lapse rate refer to?
25. Describe the various processes that can lead to the formation of fog.
26. Describe the various processes that can lead to the formation of dew.
27. What is the difference between frozen dew and frost?

Critical Thinking

1. When rubbing alcohol is applied to a person's skin, it feels colder than the application of water would. Why?
2. A person parks her car in the driveway on a warm afternoon and notices a small puddle of water beneath the car a few minutes later. Explain how using the car's air conditioning can account for the puddle.
3. A crowded classroom is filled with students. In what way might the presence of the students affect the dew point and relative humidity in the room?
4. A person sleeps through the night without waking up, but awakes in the morning weighing slightly less than the night before. What happened?
5. The temperature within a forest is −2 °C (28 °F) and there is frost on the trees but no fog. Outside the woods there is a fog. Why wasn't this fog in the woods?
6. A map of North America shows the average distribution of vapor pressure across the continent. Will the distribution on the map be *only* a function of the amount of water vapor in the air, or will the distribution be affected by another factor as well? Explain.
7. At Wheeling, West Virginia, the evening temperature is 13 °C (55 °F) and the dew point is 9 °C (48 °F). How would you assess the likelihood of fog forming overnight?

8. Is fog more likely to occur downwind or upwind of an oil refinery? Why?

9. All fogs are made of water droplets or ice crystals. Despite the fact that they have the same composition, how would you know if a particular fog is a radiation, advection, or upslope fog?

10. Diesel engines, like four-stroke engines, work because of burning fuel, but they do not require a spark plug or similar device for initiating the burning. Apply your knowledge of the first law of thermodynamics to explain how the fuel can be forced to burn.

Problems & Exercises

1. Assume that a kilogram of air consists of 995 g of dry air and 5 g of water vapor. Show that the specific humidity and mixing ratio are very nearly equal.

2. Assume that a kitchen measures 4 meters by 5 meters by 3 meters. If the air density is 1 kg/m^3 and the specific humidity is 10 g water vapor per kilogram of air, how much water vapor is in the room? If the doors and windows were sealed shut, would boiling 1 kg of water into the air make a substantial change in the humidity of the room?

3. The dry and wet bulb readings in Honolulu are 27 °C and 21 °C (80 °F and 69 °F), respectively. At Charlottesville, Virginia, the readings are 10 °C and 7 °C (50 °F and 45 °F).
   a. Use Tables 5–2 and 5–3 to determine the relative humidity and the dew point for both locations.
   b. Which of the two locations is more humid?

4. Assume a parcel of air starts out at the surface with a temperature of 12 °C (54 °F) and a dew point of 9.6 °C (49 °F). Then it is lifted.
   a. What will the air temperature be at the 100 m level?
   b. What will the dew point be at the 100 m level? (*Hint:* Don't forget the dew point lapse rate.)
   c. At what height will condensation occur?
   d. What will the temperature be when condensation occurs?
   e. What will the dew point be when condensation occurs?
   f. What will the temperature be 500 m above the surface?
   g. If the parcel of air is lowered back to the surface (assuming none of the condensed moisture was removed as rain), what will the temperature and dew point be?

6. The dry and wet bulb temperatures are 21 °C and 12 °C (70 °F and 54 °F). Use Tables 5–2 and 5–3 to answer the following questions:
   a. What are the dew point and the relative humidity?
   b. What will the dew point and relative humidity be if the air temperature increases to 27 °C (80 °F)? (*Hint:* Do not assume the same wet bulb depression.)
   c. What will the dew point and relative humidity be if the air temperature drops to 4 °C (39 °F)? (*Hint:* You don't need the tables for this one.)
   d. What will the dew point and relative humidity be if the air temperature drops further, to 2 °C (35 °F)?

5. The numerical value of the specific heat for air in the first law of thermodynamics, c_v, is strictly valid only for air with no water vapor. Water vapor has a specific heat approximately twice as great as that of c_v. Therefore, will a humid mass of air undergoing expansion undergo more or less cooling than would a dry mass of air?

Visual Analysis

Observe the map plotting dew point temperatures across the United States on June 15, 2014, and answer the following questions:

5.1. Define the region in which the dew points indicate uncomfortably muggy conditions?

5.2. Which areas indicate dry conditions?

5.3. Why might the western United States have substantially different dew point readings from those east of the Rocky Mountains?

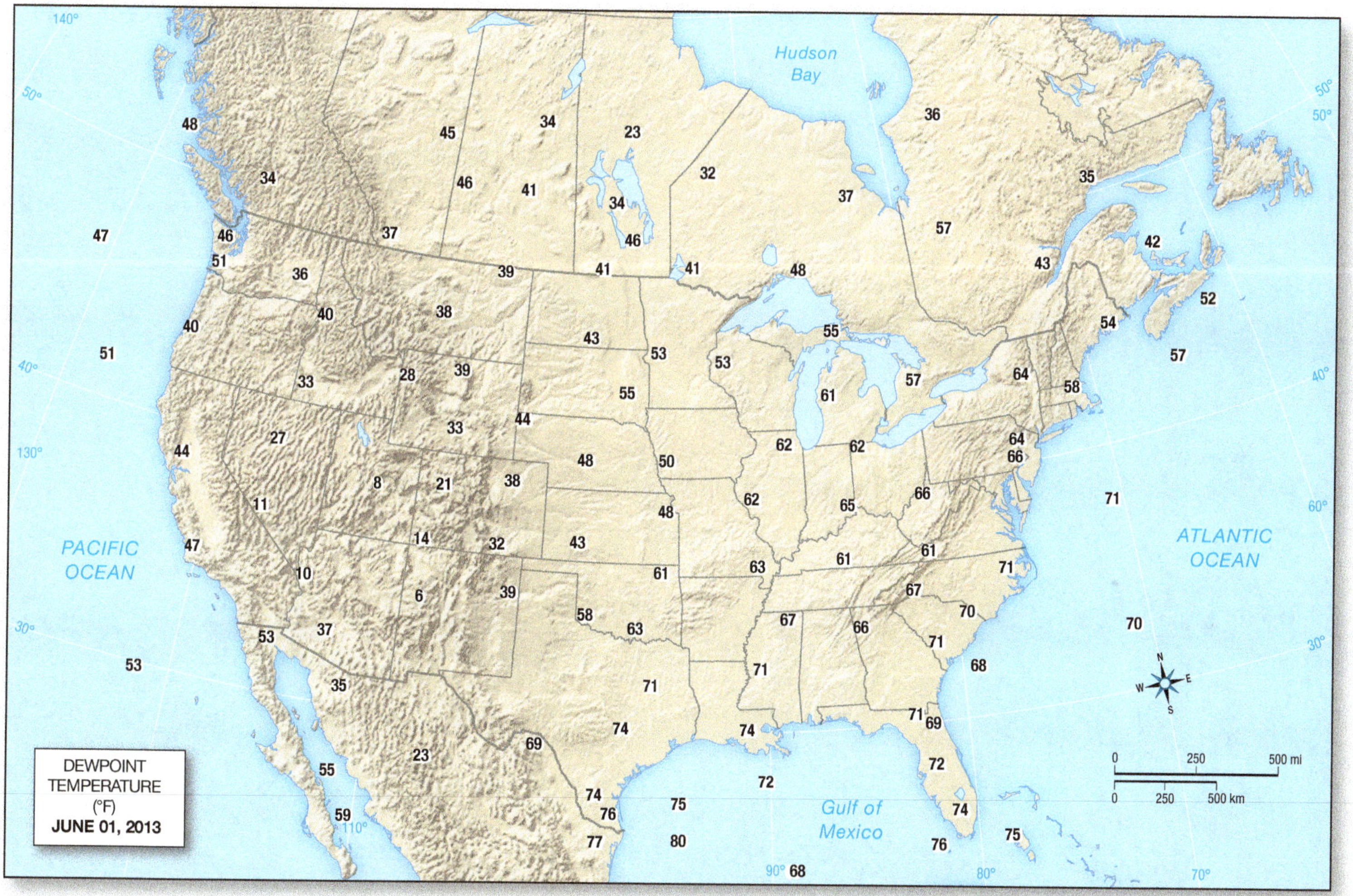

CHAPTER

6 Cloud Development and Forms

It's not too surprising to walk outside and catch sight of some clouds above the ground. They are so commonplace that we often forget to look for them, which is unfortunate because they can present some very beautiful skyscapes. We know that lifted air cools adiabatically and that if lifted sufficiently it can cool to the dew point and become saturated. In this chapter we will review some of the common mechanisms that allow such lifting to occur. But, as this photo shows, such lifting might occur in some unexpected locations. On this particular day in February 2012, nearly saturated air containing a lot of hygroscopic (see Chapter 5) salt particles from the adjacent beach at Panama City, Florida, drifted in over the seaside buildings. Being nearly saturated to begin with, the air required minimal uplift before producing these unusual and striking clouds. Unusual cloud formations are among the many fascinating features the atmosphere can present to us on a nearly daily basis. This chapter discusses the processes and conditions associated with the formation of clouds due to upward motions. It also describes the cloud types that form as a result of those processes.

◂ *Clouds form as very moist air rises over buildings in Panama City, Florida .*

Learning Outcomes

After reading this chapter, you should be able to:

6.1 Describe four mechanisms that lift air.

6.2 Explain the factors that determine the stability of parcels of air.

6.3 Explain the factors that influence the environmental lapse rate.

6.4 Explain what causes unstable air to stop rising.

6.5 Explain what causes stable air and inversions to develop.

6.6 Describe the different types of clouds.

6.7 Describe how meteorologists measure cloud coverage.

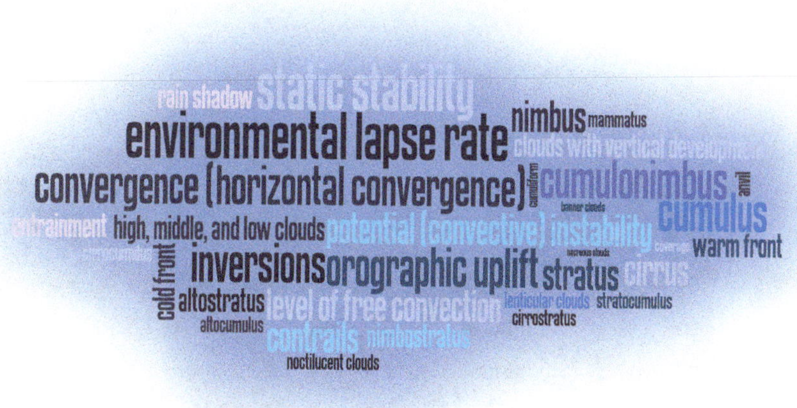

Mechanisms That Lift Air

6.1 Describe four mechanisms that lift air.

In Chapter 5 we saw that condensation or deposition can occur by (1) adding water vapor to the air; (2) mixing warm, moist air with cold air; or (3) lowering the air temperature to the dew point by adiabatic cooling of rising air. Although the first two processes can lead to cloud formation in many situations, lowering of air temperature to the dew point is the most common mechanism of cloud formation, especially for clouds that cause precipitation.

Four mechanisms lift air so that condensation and cloud formation can occur:

1. Orographic lifting, forcing air above a mountain barrier
2. Frontal lifting, displacement of one air mass over another
3. Convergence, when low level winds flow toward a location from multiple directions
4. Localized convective lifting due to buoyancy.

Orographic Uplift

As shown in Figure 6–1a, air flowing toward a hill or mountain will be deflected around and over the barrier. The upward

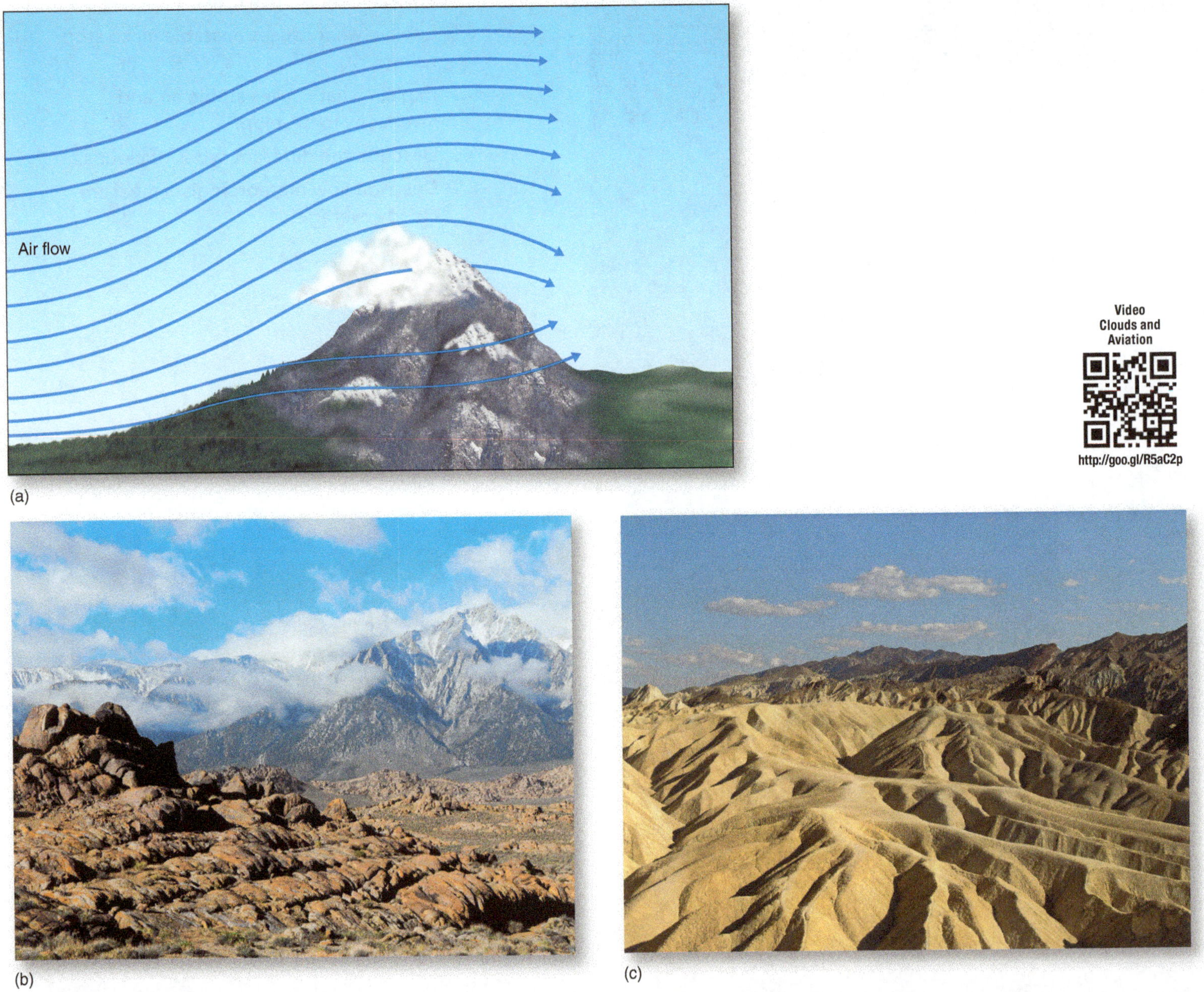

Video
Clouds and
Aviation
http://goo.gl/R5aC2p

▲ **FIGURE 6–1 Orographic Uplift.** (a) When air approaches a topographic barrier, it can be lifted upward or deflected around the barrier. The Sierra Nevada range forms a major barrier to winds that generally blow from the west. (b) This promotes enhanced precipitation on the windward side, and (c) a rain shadow on the leeward side.

displacement of air that leads to adiabatic cooling and possible cloud formation is called **orographic uplift** (or the **orographic effect**). The height to which these *orographic clouds* can rise is not limited to the height of the hill or mountain; their tops can extend many hundreds of meters higher and even into the lower stratosphere. The height of cloud tops is strongly related to characteristics of the air that vary from day to day, an issue we will describe in more detail later in this chapter.

Downwind of a mountain ridge, on its leeward side, air descends the slope and warms by compression to create a **rain shadow**, an area of lower precipitation. The Sierra Nevada mountain range provides a dramatic illustration of this effect. The ridge crest of the Sierra runs north to south and is essentially perpendicular to the predominantly west-to-east airflow. With much of the range being higher than 3500 m (11,500 ft), precipitation on the western, windward side is greatly enhanced because of orographic lifting (refer to Figure 6–1b); in places, the mean annual precipitation exceeds 250 cm (100 in.). The eastern slope of the range is extremely steep and the valley floor is low, sometimes below sea level. Thus, the descending air on the leeward slope creates one of the strongest rain shadows on Earth. Death Valley, one of the driest places in North America, is just east of the range, whose windward slopes accumulate the majority of California's usable water (Figure 6–1c). A comparable rain shadow exists in South America, where the Andes Mountains form an abrupt barrier to the westerly winds. Figure 6–2 illustrates the rapid changes that can take place in orographic cloud development.

Frontal Lifting

Although temperature normally changes from place to place, experience tells us that such changes are usually quite gradual. In other words, if the temperature is 10 °C (50 °F) in Toronto, Ontario, chances are the temperature in Buffalo, New York, about 100 km (60 mi) away, will not be very much different. Sometimes, however, transition zones exist in which great temperature differences occur across relatively short distances. These transition zones, called *fronts*, are not like vertical walls separating warm and cold air but rather slope gently, as we will discuss in Chapter 9.

Airflow along frontal boundaries results in the widespread development of clouds in either of two ways. When cold air advances toward warmer air (a situation called a **cold front**), the denser cold air displaces the lighter warm air ahead of it, as shown in Figure 6–3a. When warm air flows toward a wedge of cold air (a **warm front**), the warm air is forced upward in much the same way that the orographic effect causes air to rise above a mountain barrier (Figure 6–3b).

Convergence

Because the mass of the atmosphere is not uniformly distributed across Earth's surface, large areas of high and low surface pressure exist. These pressure differences set the air in motion in the familiar effect we call the wind. Not surprisingly, the

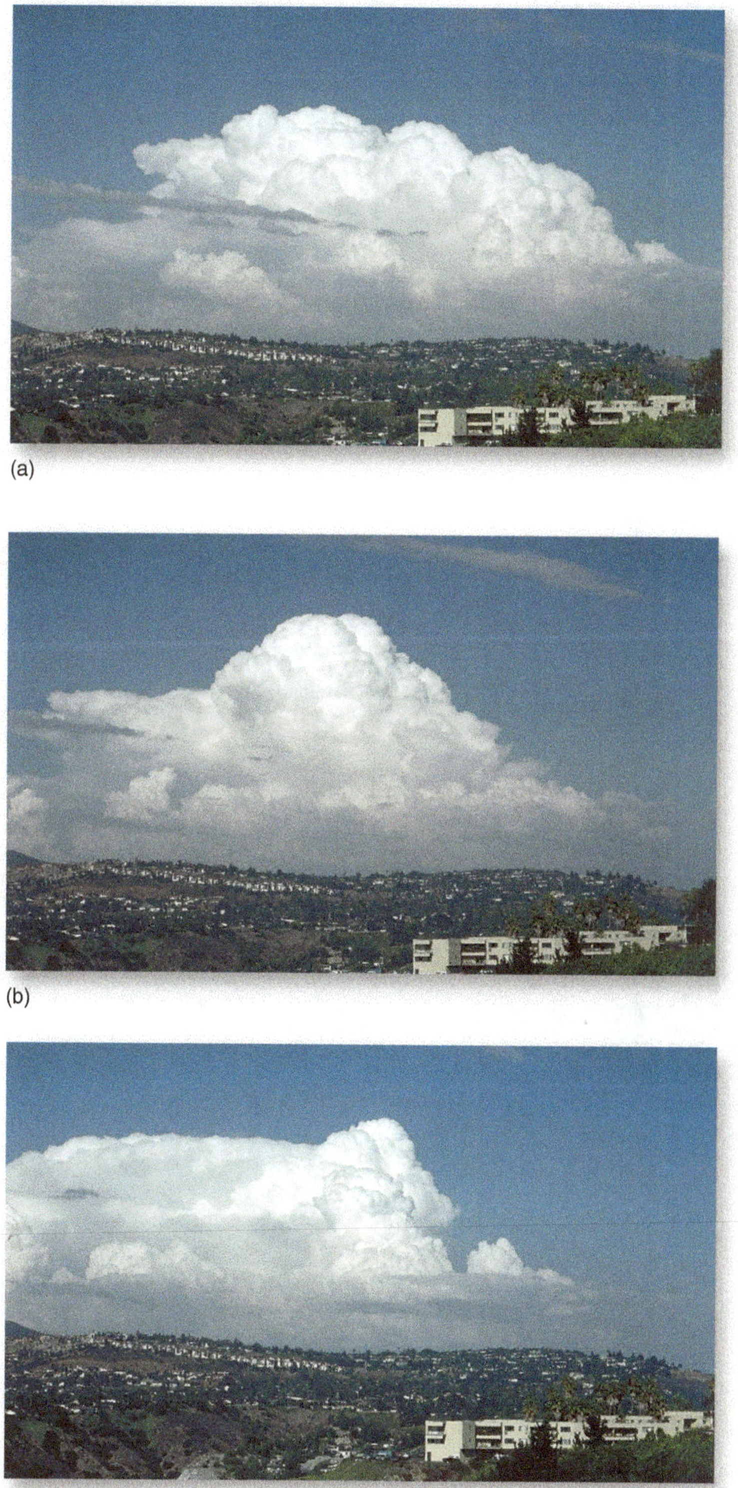

(a) (b) (c)

▲ **FIGURE 6–2 Cumulus Clouds.** The development of cumulus clouds over the mountains east of San Diego, California. Notice the increased vertical development between (a) and (b). In (c), the cloud has decreased in depth due to precipitation. Photos were taken at about 10-minute intervals.

pattern of wind that results is very much related to the pattern of pressure.

In particular, when a low-pressure cell is near the surface, winds in the lower atmosphere tend to converge on the

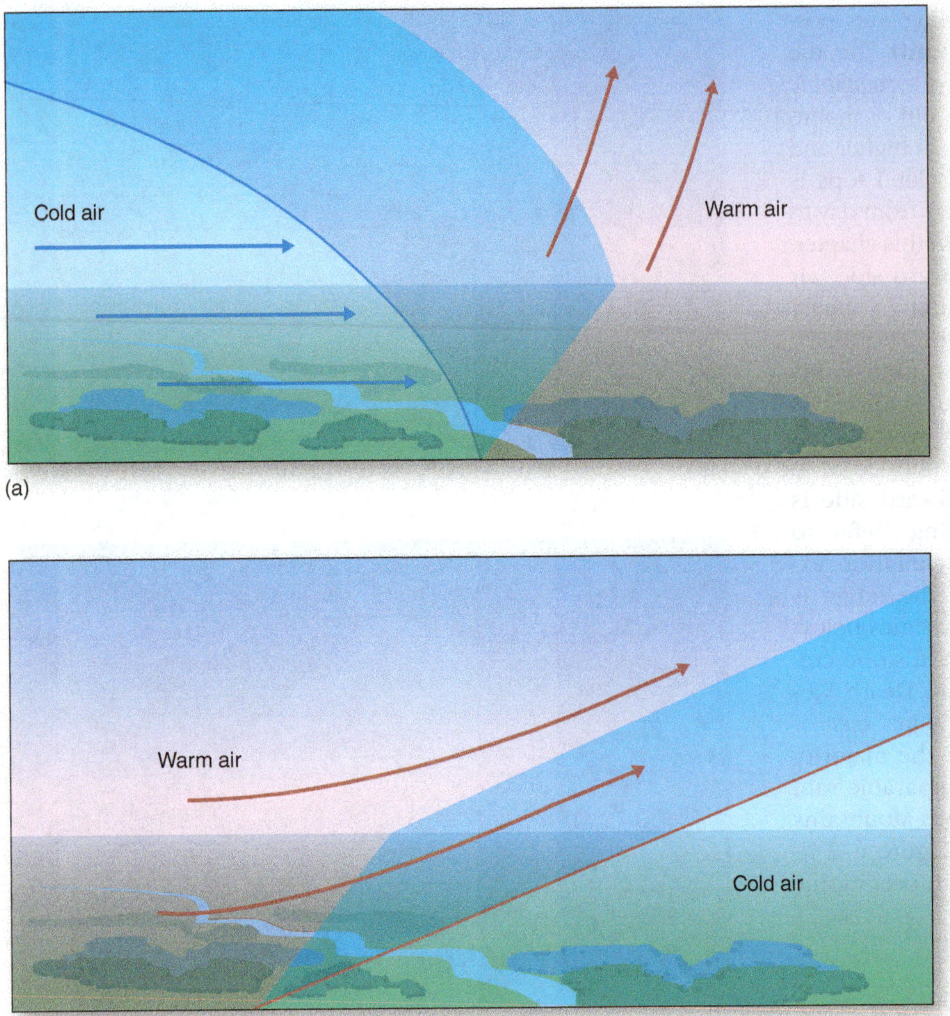

▲ **FIGURE 6–3 Frontal Boundaries.** (a) Cold fronts cause uplift as cold air advances on warmer, less dense air. Horizontal flow in the warm air is often parallel to the front. (b) Uplift occurs along a warm front when warm air overruns the cold wedge of air ahead of it.

center of the low from all directions. Horizontal movement toward a common location is called **horizontal convergence**, or just **convergence** for short (Figure 6–4). Does convergence lead to increasing density near the center of the low, with imported air confined to its original altitude? No—instead, vertical motions near the center of the low carry away about as much mass as is carried in. Air will rise. This will be explained in more detail later, but for now we can just make the connection between low-level convergence and rising air with adiabatic cooling.

Localized Convection

We saw in Chapter 3 that free convection is lifting that results from heating the air near the surface. It is often accompanied by updrafts strong enough to form clouds and precipitation. During the warm season, heating of Earth's surface can produce free convection over a fairly limited area and create the brief afternoon thunderstorms that disrupt summer picnics Figure 6–5 shows the sequence of warming and lifting. In Canada and the United States east of the Rocky Mountains, high moisture content of the air sometimes allows for tall clouds with bases at relatively low altitudes. Such conditions favor vigorous precipitation over small regions (free convection by its nature does not create updrafts more than several tens of meters in diameter). Even in the deserts of the Southwest, which are usually low in water vapor, intense heating can lead to localized convection intense enough to cause thundershowers.

Free convection arises from buoyancy, the tendency for a lighter fluid to float upward through a denser one. By itself, buoyancy can initiate uplift. But buoyancy can also speed or slow the uplift begun by the orographic effect, frontal lifting, and convergence. As we will see next, meteorology uses the concept of static stability to summarize the effect of buoyancy on uplift.

Video
Gravity Wave Clouds
http://goo.gl/ykZrY5

CHECKPOINT

6.1 List four mechanisms that lift air.

6.2 Which lifting mechanism also produces a rain shadow effect?

◄ **FIGURE 6–4** Convergence is a lifting mechanism that occurs when air flows into a given area from different directions.

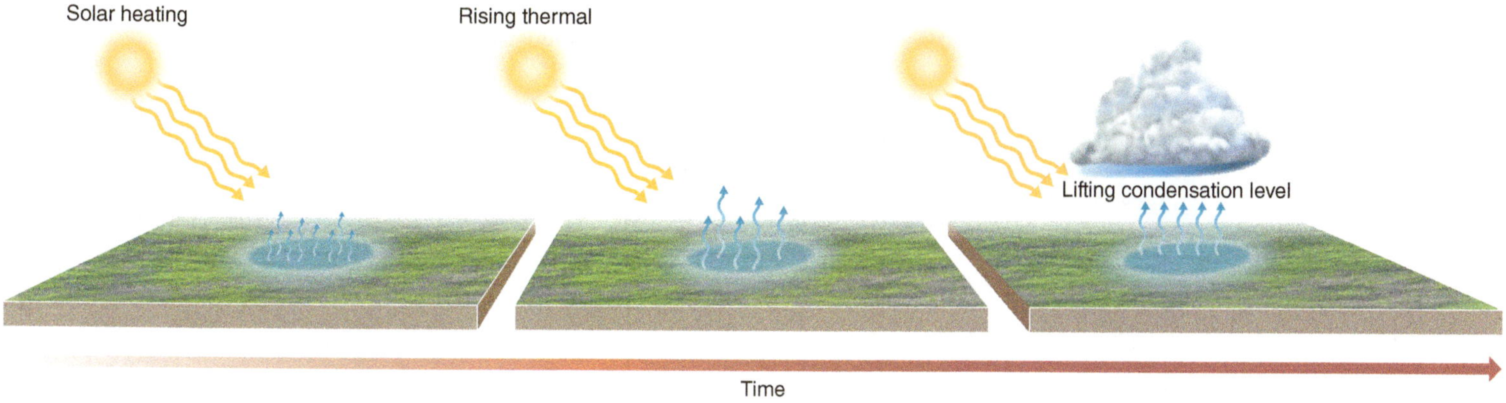

▲ **FIGURE 6–5 Localized Convection.** Surface heating leads to rising air (blue arrows) which may reach saturation.

Static Stability and the Environmental Lapse Rate

6.2 Explain the factors that determine the stability of parcels of air.

Sometimes the atmosphere is easily displaced and an air parcel given an initial boost upward continues to rise, even after the original lifting process ceases. At other times, the atmosphere resists such lifting. The air's susceptibility to uplift is called its **static stability**. *Statically unstable* air becomes buoyant when lifted and continues to rise if given an initial upward push; *statically stable* air resists upward displacement and sinks back to its original level when the lifting mechanism ceases. *Statically neutral air* neither rises on its own following an initial lift nor sinks back to its original level; it simply comes to rest at the height to which it was displaced. The concept of stability is an extremely important one for understanding cloud formation.

Animation
Atmospheric Stability
http://goo.gl/bqxtGo

Buoyancy, Static Stability, and Lapse Rates

Static stability is closely related to buoyancy. The term *static* is used to differentiate this kind of stability from other kinds of instability, including potential instability (described later in this chapter in *Box 6–2, Forecasting: Potential Instability*). When a parcel of air is less dense than the air around it, it has a positive buoyancy and floats upward. (In fact, buoyant parcels of air increase in speed as they move upward—even to the point of creating violent updrafts.) Air that is denser than its surroundings sinks if not subjected to continued lifting forces. Density differences between a parcel and its surroundings are determined by their temperatures. If the parcel is warmer than the surrounding air, it will be less dense and experience a lifting force. If it is colder, it will be more dense and have "negative" buoyancy.

If a rising parcel cools at a rate that makes it colder than the surrounding air, it will become relatively dense. This will tend to suppress uplift. If the lifted air cools more slowly than its surroundings, it will become warm relative to the surroundings and have positive buoyancy. This creates a tendency for a parcel to rise on its own, even in the absence of other lifting mechanisms. Thus, the buoyancy of a rising air parcel depends on its rate of cooling relative to the surrounding air. Temperatures in the parcel are governed by either the dry or saturated adiabatic lapse rate, whereas the surroundings are governed by the **environmental lapse rate (ELR)**. (The adiabatic and environmental lapse rates were explained in Chapter 5.)

Consider a parcel of air near the surface that is lifted through the surrounding air. The lifted air cools at one or the other of the adiabatic lapse rates, and the surrounding air maintains its original temperature profile. The relative density of the rising parcel thus depends on two conditions: (1) whether or not the parcel is saturated (which determines the applicable adiabatic lapse rate) and (2) the ELR. These factors combine to produce different types of air with regard to their static stability. These are *absolutely unstable, absolutely stable*, and *conditionally unstable*. Let's use the following examples to see how static stability can be determined by comparing the environmental lapse rate to the saturated and dry adiabatic lapse rates.

Absolutely Unstable Air

Figure 6–6a illustrates what happens when a parcel of unsaturated air is lifted and the ELR is greater than the dry adiabatic lapse rate (DALR). In other words, in Figure 6–6a, the rising air is cooling more slowly than its surroundings.

Let's suppose the air temperature at the surface is 10 °C and the ELR is 1.5 °C/100 m, which means that the air is cooling at the rate of 1.5 °C for every 100 m of height. As our parcel rises, it cools at the DALR (recall from Chapter 5 that the DALR is 1 °C/100 m). When it is lifted to the 100 m level, the rising parcel cools to 9 °C—half a degree warmer than the surrounding air. If the parcel is lifted to the 200 m level, its temperature becomes 8 °C, or 1 °C warmer than the surrounding air. Thus, the lifted parcel is becoming progressively warmer and more buoyant than the surrounding air.

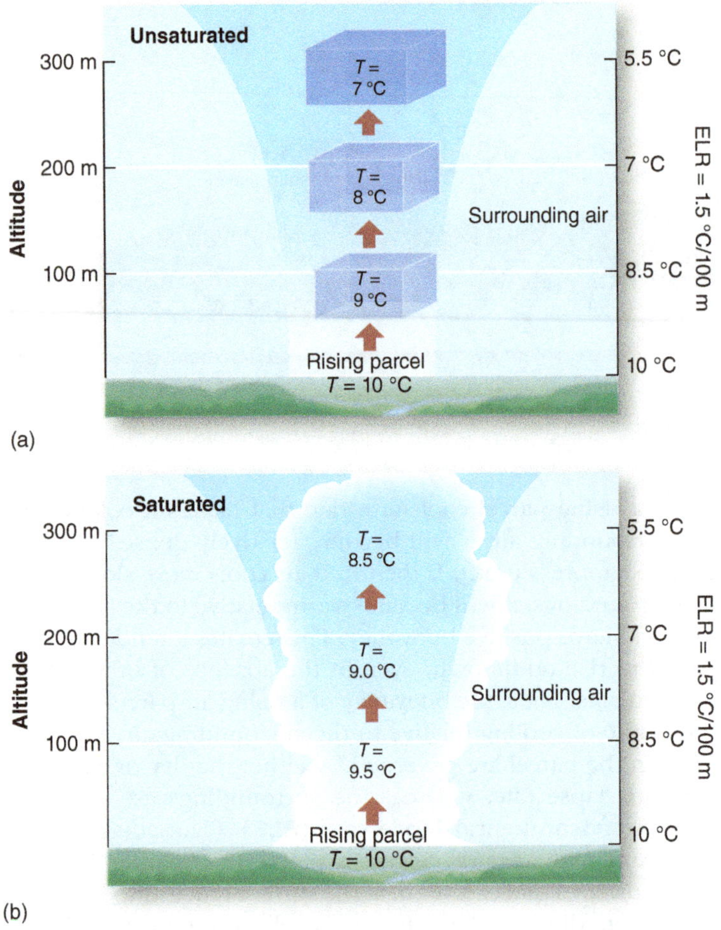

▲ **FIGURE 6–6 Absolutely Unstable Air.** In both examples, the ELR of 1.5 °C/100 m exceeds the DALR. (a) Lifted unsaturated air cools at 1 °C/100 m (the DALR). The rising parcel does not cool as rapidly with height as does the air surrounding it. It therefore becomes warmer and more buoyant than the surrounding air. (b) The lifted saturated air likewise cools less rapidly (0.5 °C/100 m) with height than the surrounding air. Thus, with an ELR greater than the DALR, air forced to rise becomes warmer and more buoyant than the surrounding air, whether it is unsaturated or saturated. Air with such temperature profiles is said to be absolutely unstable.

The air in this instance is said to be *absolutely unstable* because once a parcel within is lifted, it continues to move upward. Not only does the parcel rise, but it does so at an ever-increasing speed. This occurs because the temperature difference between it and the surrounding air continually increases, leading to greater buoyancy, and also because it gathers momentum as it rises.

Figure 6–6b provides a second example of absolutely unstable air. In this case, the ELR is still 1.5 °C/100 m, but the air is now *saturated*. The lifted parcel of air therefore cools more slowly, at the saturated adiabatic lapse rate (SALR), and will be warmer than an unsaturated parcel. The temperature difference between the warm parcel and colder surrounding air is greater, giving rise to a stronger buoyant force. We conclude that the air is again unstable, even more so than in the previous example.

These two examples show that *whenever the environmental lapse rate exceeds the dry adiabatic lapse rate, the air is absolutely unstable and a parcel contained within it will continue to rise once lifted, regardless of whether or not it is saturated.* (Of course, upward motions cannot continue forever. But for now, we will put aside the issue of how far unstable air can rise so we can focus on the main concept associated with what happens to air in a zone of instability.)

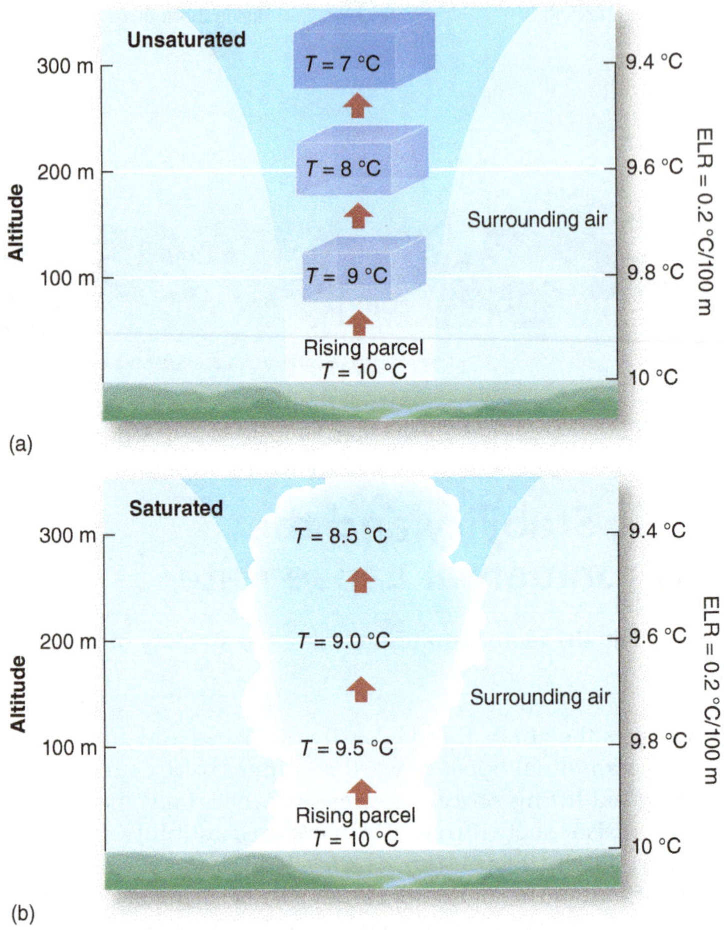

▲ **FIGURE 6–7 Absolutely Stable Air.** Both examples show what would happen if the parcels were forced to rise. In both cases, the ELR of 0.2 °C/100 m is less than the SALR of 0.5 °C. It does not matter whether the air is unsaturated (a) or saturated (b); rising air parcels would become colder and less buoyant than the surrounding air. Regardless of whether the air is unsaturated or saturated, temperature profiles that exhibit ELRs less than the SALR indicate absolutely stable air.

Absolutely Stable Air

Figures 6–7a and b illustrate what happens when the ELR, in this case 0.2 °C/100 m, is less than the saturated adiabatic lapse rate. As we see in (a), if a saturated parcel of air were to somehow rise, its temperature would drop more rapidly than the temperature of the air around it, making the parcel relatively heavier and less buoyant. Because of its negative buoyancy, the lifted air would sink back to its initial level if the lifting mechanism stops. Such air is *absolutely stable.* The same principle applies in part (b) of the figure. The saturated parcel becomes colder than the air around it. Like the unsaturated parcel in (a), it has a tendency to sink back to its original position.

From these two examples we can conclude that *whenever the environmental lapse rate is less than the saturated adiabatic lapse rate, the air will be absolutely stable and will resist lifting, regardless of whether or not it is saturated.* Note that it is possible for the ELR to be such that the temperature does not change at all with height, or even for the temperature to increase with height, as will be discussed later in this chapter. Though we will not provide examples here, the logic described in this section applies to these situations as well. If the ELR = 0 °C/100 m or if the temperature increases with height (a negative ELR), the air will be absolutely stable.

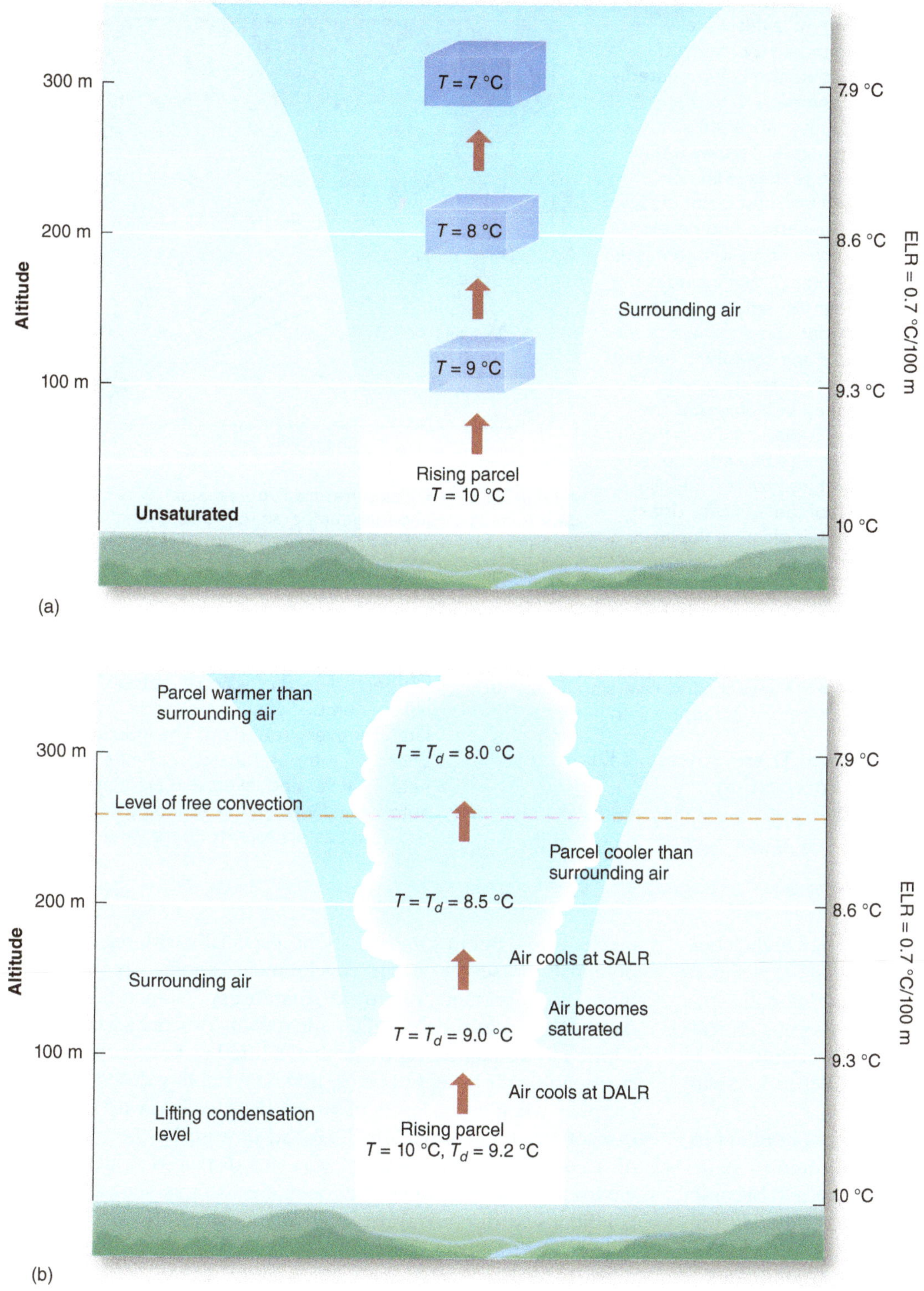

▲ **FIGURE 6–8 Conditionally Unstable Air.** The atmosphere is conditionally unstable when the ELR is between the dry and saturated adiabatic lapse rates. (a) The ELR is 0.7 °C/100 m and the air is unsaturated. As a parcel of air is lifted, its temperature is less than that of the surrounding air, so it has negative buoyancy. (b) A parcel starts off unsaturated but becomes saturated at 100 m, where it is cooler than the surrounding air. Further lifting cools the parcel at the SALR. At the 200 m level, it is still cooler than the surrounding air, but if taken to 300 m, it is warmer and buoyant. Thus, if the air is lifted sufficiently, the parcel continues to rise by virtue of its buoyancy.

Conditionally Unstable Air

The preceding examples describe what happens when the ELR is less than the SALR or greater than the DALR. But what happens when the ELR is between the dry and saturated adiabatic lapse rates? In this environment, the air is said to be *conditionally unstable*, and the tendency for a lifted parcel to sink or continue rising depends on whether or not it becomes saturated and how far it is lifted.

Let's suppose there is an ELR of 0.7 °C/100 m in the atmosphere through which an unsaturated parcel is rising (Figure 6–8a). Because the lifted parcel becomes colder (and therefore denser) than the surrounding air, it resists further uplift. The situation at this point is much the same as was described for absolutely stable air.

Figure 6–8b illustrates how the same environmental lapse rate can lead to a lifted parcel of air becoming buoyant relative to its surroundings if it is lifted to a great enough height. We again apply the ELR of 0.7 °C/100 m to a lifted parcel that eventually becomes saturated and forms a cloud. In this example, the air has an initial temperature of 10 °C and a dew point, T_d, of 9.2 °C. It cools at the DALR until it reaches saturation (the lifting condensation level) at the 100 m level, where $T = T_d = 9.0$ °C. There the lifted parcel is colder and denser than the surrounding air; it is not buoyant and will not rise further unless something forces it to do so. If the parcel is lifted farther, it will cool at the saturated adiabatic rate, which is less than the ELR. At the 200 m level, the lifted parcel is still colder than the surrounding air, but the difference in temperature between it and the surrounding air is less than it was at the 100 m level. If the parcel is lifted to the 300 m level, it then

6–1 FORECASTING

Determining Stability from Thermodynamic Diagrams

The static stability of air can be determined numerically by comparing the environmental lapse rate to the saturated and dry adiabatic lapse rates. Thermodynamic diagrams can also be very useful in this regard. Figure 6–1–1 provides a simple schematic showing how this comparison is done. The figure compares three hypothetical temperature profiles against the SALR and the DALR on a portion of a simplified thermodynamic diagram. The lines labeled as *moist adiabat* and *dry adiabat* plot the change in temperature a saturated or unsaturated parcel of air would experience if it were lifted or lowered. The three hypothetical temperature profiles illustrate examples in which the air is absolutely unstable (profile 1—the temperature decreases more rapidly than the DALR), conditionally unstable (profile 2—ELR is between the DALR and SALR), and absolutely stable (profile 3—ELR is less than the SALR).

Reading a Thermodynamic Diagram

In the real world, the ELR varies from the surface upward. For example, at one level the air might be absolutely stable, whereas at another level it might be conditionally or absolutely unstable. Thermodynamic diagrams allow the forecaster to observe the resultant changes in stability at different levels visually, rather than by having to compute the ELR repeatedly for comparison to the adiabatic lapse rates. Figure 6–1–2 shows a temperature (heavy line on right) and dew point (heavy line on left) sounding plotted on a Stuve diagram (see **Chapter 3**) that includes dry and moist adiabats. The dry adiabats are shown in green and slope steeply to the left as they extend upward. The dashed blue lines are the moist adiabats. From the surface to 850 mb, the temperature profile is parallel to that of the adjacent dry adiabats, indicating the layer has neutral stability. Above that is a shallow layer, which is statically stable. From 800 mb to about 650 mb, the air is conditionally unstable. And just above that, there is a very shallow inversion layer with temperature increasing with height.

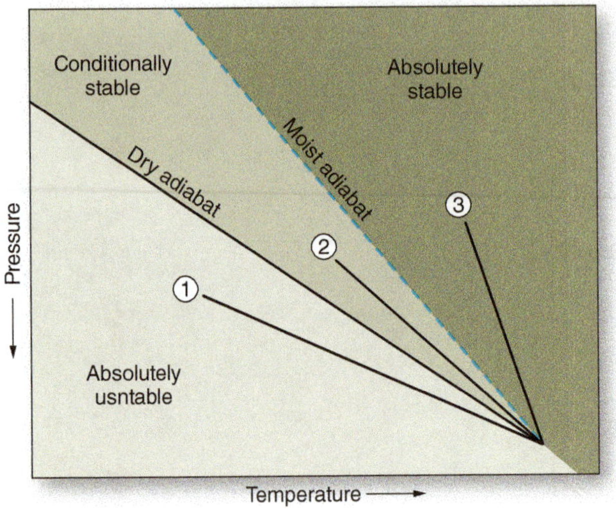

▲ **FIGURE 6–1–1 Temperature Profiles.** Stability can be determined by comparing temperature profiles with the slope of the dry and wet adiabats. Profile 1 is absolutely unstable, profile 2 conditionally unstable, and profile 3 absolutely stable.

Use of Thermodynamic Diagrams in Forecasting

The changes in stability at different levels may appear to make the use of the thermodynamic diagram a daunting task for the forecaster, but the situation has several remedies. Professional meteorologists have a number of numerical indexes calculated for every sounding. The indexes are based on temperature–dew point combinations at varying levels and are computed automatically when the soundings are plotted. Forecasters refer to these values for

becomes *warmer* than the ambient air. The lifted parcel thus becomes buoyant and will now rise on its own, even in the absence of external lifting. Thus, if the atmosphere is conditionally unstable, an air parcel becomes buoyant if lifted above some critical altitude. That altitude, called the **level of free convection**, is the height to which a parcel of air must be lifted for it to become buoyant and to rise on its own.

The condition in the term *conditionally unstable* refers to a parcel's ability to become buoyant only if it is lifted to some particular level. Air not lifted to that level does not become buoyant and will rise only if subjected to some other lifting mechanism (such as the passage of a front). When a parcel of conditionally unstable air rises above that level, it is common for clouds to rapidly increase in depth and yield precipitation. (*Box 6–1, Forecasting: Determining Stability from Thermodynamic Diagrams,* provides more information on the analysis of stability.)

Static and Potential Instability

The types of static stability discussed in this section pertain to a parcel of air's ability to rise when subjected to uplifting mechanisms. But in some situations, especially with regard to the potential for severe storm development, another type of stability is important: **potential instability** (sometimes called **convective instability**). While static stability describes what would happen to a small parcel of air that is lifted or lowered while the surrounding air is kept in place, potential stability describes what happens when entire layers of air are displaced upward (as in the case of a mass of warm air displaced upward by the movement of a cold front). When such layers are displaced upward, their environmental lapse rates may be altered so that statically stable conditions within the layer can change to statically unstable conditions, or vice versa. Potential instability is a feature that is particularly important with the forecasting of severe thunderstorms, as described in *Box 6–2, Forecasting: Potential Instability.*

CHECKPOINT

6.3 What is static stability?

6.4 How does air that is statically stable behave differently from air that is statically unstable?

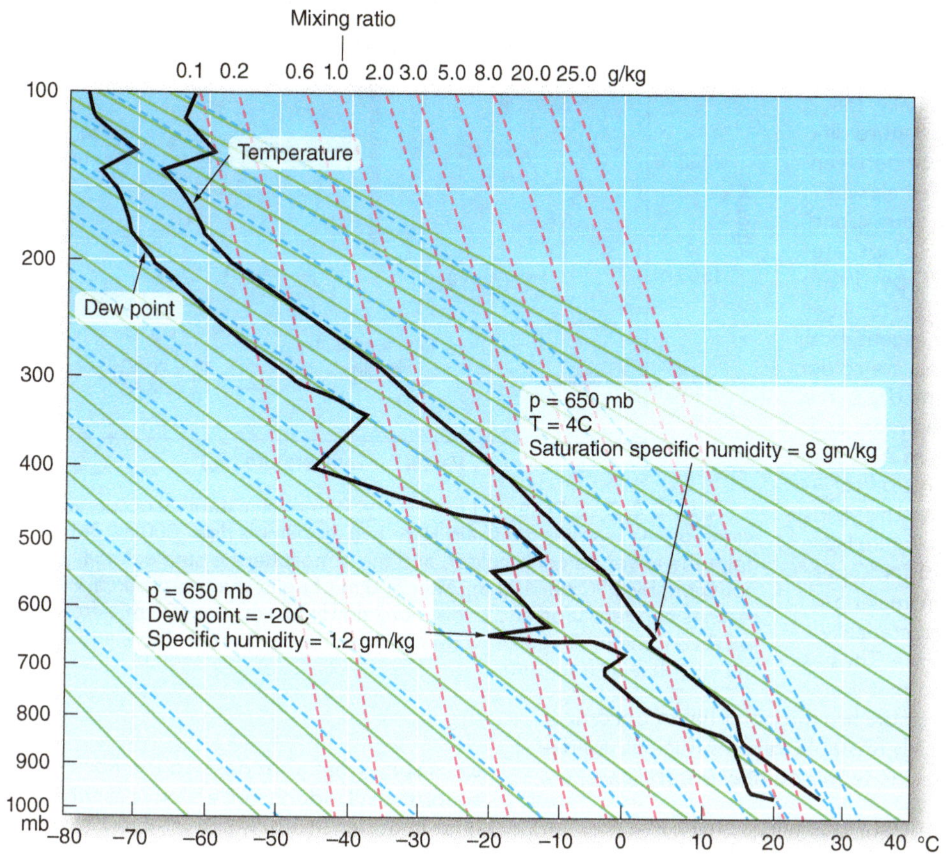

FIGURE 6–1–2 Thermodynamic Diagram. Thermodynamic diagram plotting temperature (heavy line on right) and dew point (left) profiles for Detroit, Michigan, on June 27, 2002. The dry adiabats (showing how a parcel of unsaturated air would change if lowered or lifted) are shown in green and slope steeply to the left as they extend upward. The dashed blue lines are the moist adiabats, showing how the temperature of saturated parcels would change if displaced vertically. The slightly arcing, red dashed lines show mixing ratio values. The profile in heavy black on the right represents the temperature profile. At any pressure level you can see what the saturation specific humidity is by interpolating between the red dashed lines. The profile on the left depicts the dew point. That line can be used to interpolate the actual specific humidity at any pressure level.

initial guidance in their interpretations of how stability conditions will influence the likelihood of cloud cover, precipitation, or violent weather.

6–1–1 How can comparisons between temperature soundings and dry adiabats be used to provide stability information?

6–1–2 How do forecasters deal with the fact that stability conditions change with altitude when trying to predict the formation of clouds, precipitation, or violent weather?

Factors Influencing the Environmental Lapse Rate

6.3 Explain the factors that influence the environmental lapse rate.

ELRs are highly variable in space and time. Just as the surface air temperature at a location is subject to change, so is the vertical temperature profile. The following three factors can bring about changes in the ELR.

Heating or Cooling of the Lower Atmosphere

During the daytime, solar radiation heats Earth's surface, which in turn warms the atmosphere in contact with it. Because it is heated more rapidly than the air aloft, the lower atmosphere typically has a larger ELR during the midday, as shown in Figure 6–9 by lines that slope steeply toward colder termperatures. The initial temperature profile indicated by the solid line changes through the course of the day, and steeper profiles (shown by the dashed lines) can result for several hundred meters above the surface. The effect of solar radiation on the lapse rate is (of course) greatest on clear, sunny days, especially above unvegetated land surfaces, where abundant solar radiation is available and little energy is expended on evaporation.

Cooling of the surface, such as occurs at night, chills the lower atmosphere and decreases the ELR. With sufficient cooling, the air near the surface can become colder than the air above and create a situation in which air temperature *increases* with height. This is known as a **temperature inversion**, a condition of extremely stable air that we discuss later in the chapter.

Advection of Cold and Warm Air at Different Levels

Temperature profiles can be influenced by differences between wind directions at low and high levels. In Figure 6–10a, for example, low- and high-level winds are both from the west,

6–2 FORECASTING

Potential Instability

Air with a large temperature lapse rate is said to be statically unstable. Another type of instability that influences vertical air motions, called potential instability, arises when a layer of dry air rests above moist air below. Lifting the two layers can cause the temperature lapse rate to increase, thus making the air statically unstable.

The Development of Potential Instability

Consider the situation shown in Figure 6–2–1. In parcel 1, located in the moist layer, the temperature (Ta) equals 30 °C and the dew point (Td) is 26 °C. In parcel 2, located in the dry layer, the temperature and dew point are 25 °C and 18 °C, respectively. Now consider what happens if some process lifts the two layers containing the parcels. Both parcels are initially unsaturated, so they cool at the dry adiabatic lapse rate (DALR) of 1 °C per 100 m, and their dew points decrease at 0.2 °C per 100 m. After 500 m of ascent, the temperature of both parcels has fallen by the same amount, so the temperature difference between them is unchanged. However, parcel 1 is now saturated, so further lifting will cause its temperature to decline at the saturated adiabatic lapse rate (SALR). Meanwhile, parcel 2 is still unsaturated, so further lifting will cause it to cool at the DALR as before. Now let's lift the two parcels another 500 m. Assuming an SALR of 0.6 °C/100 m, the lower parcel cools by 3 °C to 22 °C, while the upper parcel cools at the DALR to 15 °C. We can now see how uplift of the column of air containing the two parcels has affected its stability. Initially, the temperature difference between the parcels was 5 °C, corresponding to a laps rate of 1 °C per 100 m (neutral stability). After ascent, both parcels are colder, but the dry parcel cooled more and became 7 °C colder than the parcel below. This yields a temperature lapse rate of 1.4 °C per 100 m, making the air statically unstable. Thus, the air that is statically stable has the potential to become statically unstable, given sufficient uplift—hence the term *potential instability*.

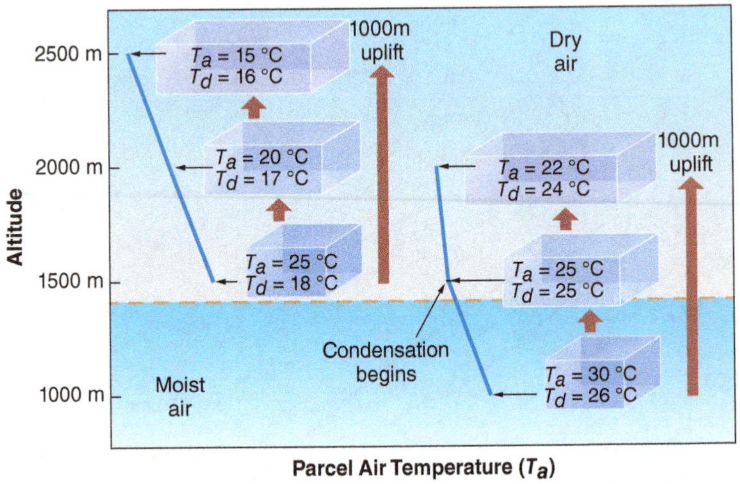

▲ **FIGURE 6–2–1 Potential Instability.** Potential instability occurs when warm, dry air overlies moist air. Here two air parcels rise 1000 m. Before uplift the lower parcel is colder, and the atmosphere is stable. Initially both cool by 1 °C per 100 m. After 500 m of uplift the moist parcel reaches condensation and cools more slowly, 0.4 °C per 100 m. After the full ascent the moist parcel is warmer than the dry parcel above and thus the layer has destabilized.

Potential Instability and Severe Thunderstorms

Both theory and experience show that potential instability is an important factor in the development of severe thunderstorms. During spring and summer, the southern Great Plains region often has warm, humid air near the surface advected from the Gulf of Mexico. In the middle troposphere above the region, westerly winds bring dry air from the southern Rockies. This air in the middle troposphere sinks somewhat after passing the Rockies to form a subsidence inversion (described later in this chapter), which inhibits the development of localized thunderstorms. But given sufficient uplift, the surface layer of air can become statically unstable and severe thunderstorms can develop.

6–2–1 What kind of temperature and moisture pattern causes potential instability?

6–2–2 How does potential instability affect the weather in the southern Great Plains?

where surface and upper-level temperatures are 10 °C and 9.5 °C, respectively. The lapse rate is thus 0.5 °C/100 m. In Figure 6–10b, surface winds are unchanged, but upper-level flow now comes from the colder northeast, so that temperatures aloft are lower, only 9.0 °C. Cold air has been advected (transported horizontally) above the surface, resulting in a steeper lapse rate. Warm air can be similarly advected, if winds happen to blow from warmer toward colder locations.

Of course, the advection of warm or cold air can occur at any level. For example, if cold air is advected at low levels, the lapse rate declines, producing greater stability. Moreover, advection is not confined to a single altitude; we should not think of a moving slab of air at some height unconnected to the rest of the atmosphere. Most commonly, there is a gradual change in wind direction (and speed) with height. Go outside on a cloudy, windy day and you will probably observe differences in the movement of clouds at different levels. Depending on how the winds are oriented relative to the temperature distribution, each altitude can have differing amounts of cold or warm advection. This does not mean advection is haphazard or random. As you will see in a later chapter, definite systematic relations exist between the wind and pressure fields. The point here is that the presence of advection is variable, both from day to day and from altitude to altitude within a column, and the effects on atmospheric stability are likewise variable.

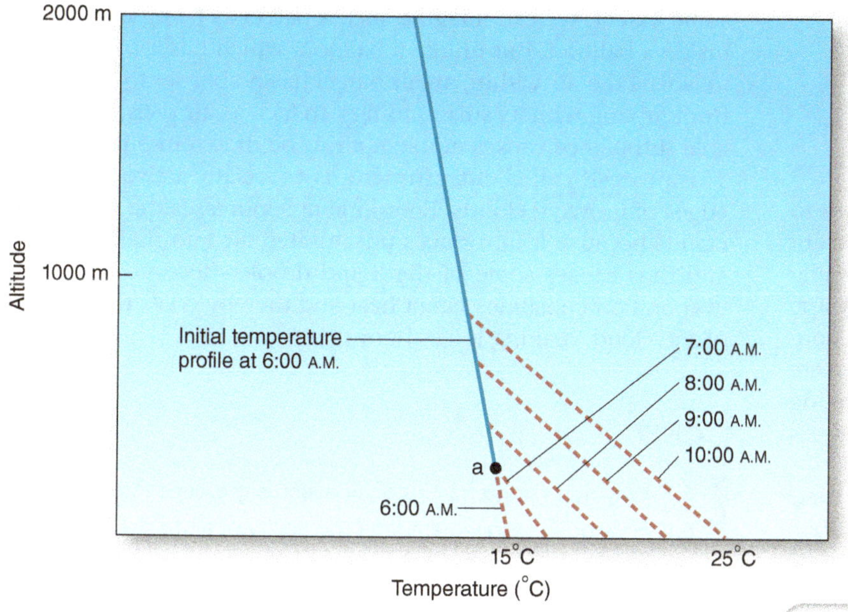

▲ **FIGURE 6–9 Surface Heating.** The ELR can be changed by heating of the surface, as shown by the sequential changes in this temperature profile. At 6 A.M. this hypothetical profile shows a constant change in temperature with height. At 7 A.M. the lower atmosphere has warmed the most right near the surface, with less heating aloft, until there is no warming of the air above the point labeled "a."

Advection of an Air Mass with a Different ELR

The atmosphere has a strong tendency to be arranged into large areas distinguished by small horizontal differences in temperature and humidity. These so-called air masses maintain their temperature and moisture characteristics as they move from one place to another. When an air mass migrates to a particular area and replaces another, the initial ELR at that location gives way to that of the new air mass. In Figure 6–11, for example, the air at position A initially has a low temperature lapse rate, while an air mass approaching from the left of the diagram has a greater lapse rate (a). The new air mass with the steeper lapse rate reaches position A, bringing with it more unstable conditions (b).

CHECKPOINT

6.5 Name three factors that affect the environmental lapse rate.

6.6 How can differences in advection between different levels affect the ELR?

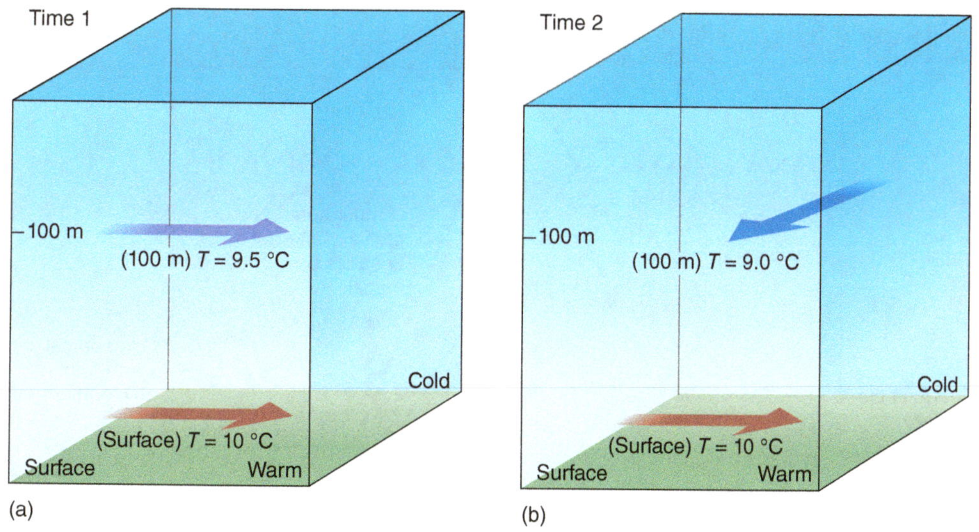

◀ **FIGURE 6–10 Differential Advection.** The ELR can be changed by the advection of air with a different temperature aloft. (a) The winds at the surface and the 100 m level bring in air with temperatures of 10 °C and 9.5 °C, respectively, yielding an ELR of 0.5 °C/100 m. (b) The surface winds still bring in air with a temperature of 10 °C. But the wind direction at the 100 m level has shifted to northeasterly, and the advected air has a temperature of 9.0 °C. This yields a steeper ELR of 1.0 °C/100 m.

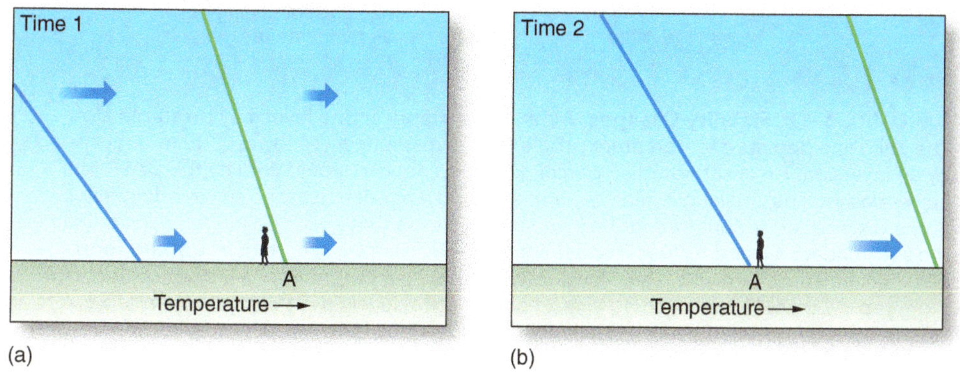

◀ **FIGURE 6–11 ELR Changes.** The ELR changes when a new air mass replaces one that has a different lapse rate. In (a), The green line represents the temperature profile originally encountered at position A. In (b) that profile has been replaced by the profile of the advected air mass (blue line).

Limitations on the Lifting of Unstable Air

6.4 Explain what causes unstable air to stop rising.

We now know that once unstable air is lifted, it continues to rise and even increases in speed. This brings up an important question: What causes unstable air to quit rising? If a rising parcel continued to rise forever, it would eventually escape Earth, never to be seen again. Given enough time, the continued loss of the unstable air parcels would entirely deplete the atmosphere. Of course, our atmosphere is not exploding out to space, so something must occur to eventually suppress uplift.

The primary braking mechanism for rising parcels is their ascent into a layer of stable air. A second process that inhibits upward motions is called *entrainment*.

A Layer of Stable Air

The solid blue line in Figure 6–12 plots one of the infinite number of temperature profiles that can exist. From Earth's surface to the 500 m level, the air is unstable; above 500 m, it is stable. If a parcel of air rises from the surface (for simplicity, we assume it is unsaturated), it becomes buoyant throughout the lowest 500 m. Above 500 m, however, the rising parcel cools more rapidly than the ambient air and eventually becomes cooler than its surroundings. The parcel does not come to a screeching halt at that point, however, because it still has considerable upward momentum. Instead, the parcel gradually slows to a stop and then sinks back down because of its greater density relative to the surrounding air. The parcel may then bob up and down before coming to rest at some equilibrium level.

Is a stable layer always present at some altitude to contain uplift? The answer is yes, because, if nothing else, a rising parcel will eventually encounter the stratosphere, which is extremely stable. As a result, even the most rapidly ascending parcels of air must slow down and reach an equilibrium level above the tropopause. Although severe thunderstorms can have updrafts of more than 200 km/hr (120 mph), uplift seldom extends above the lowest kilometer or so of the stratosphere.

Entrainment

When we talk about a rising air parcel, we mean a small mass that undergoes motions distinct from the surrounding atmosphere. To some extent, we can imagine such a parcel as being like the air inside a balloon. But unlike a balloon, which has a rubber film to isolate the air within, an air parcel has no barrier to prevent it from mixing with its surroundings. In fact, as air rises, considerable turbulence causes ambient air to be drawn into the parcel. This process, called **entrainment**, is especially active along the edges of growing clouds. Entrainment suppresses the growth of clouds because it introduces unsaturated air into their margins and thus causes some of the liquid droplets to evaporate. The evaporation consumes latent heat and thereby cools the margin of the cloud, making it less buoyant.

CHECKPOINT

6.7 Why don't rising parcels of air eventually escape into space?

6.8 How does entrainment limit the ability of air to rise?

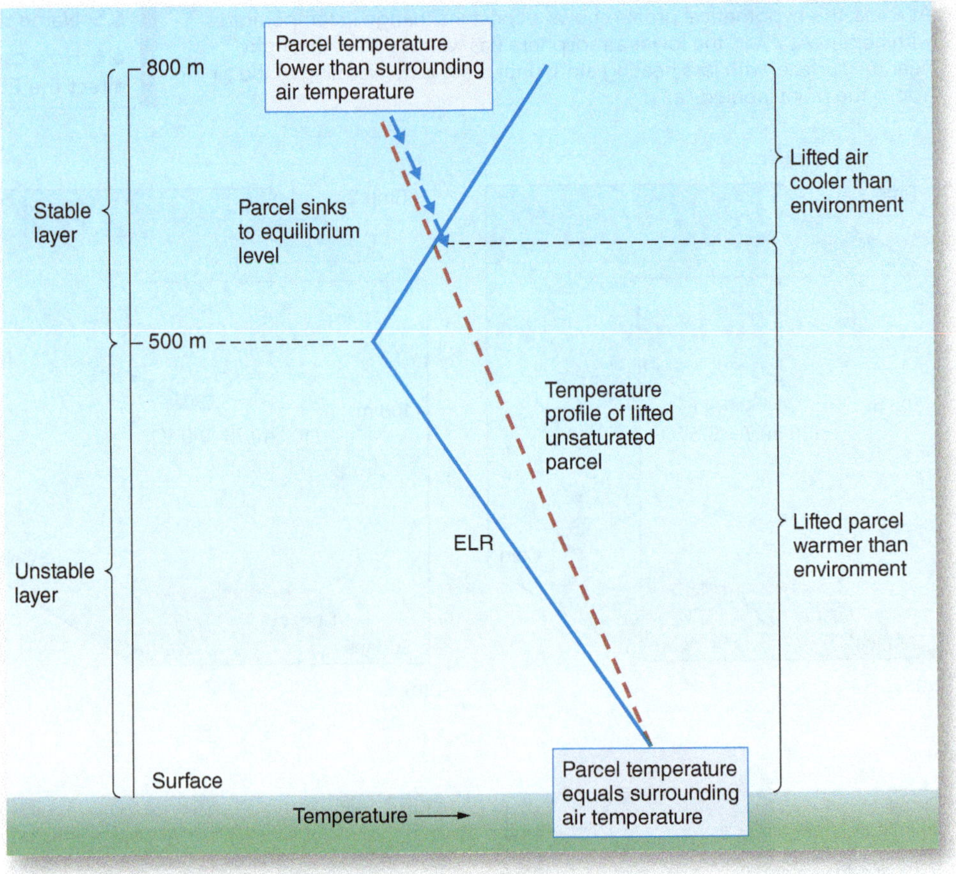

▲ **FIGURE 6–12 Stability Changes.** Air that is unstable at one level may be stable aloft. The solid line depicts a temperature profile that is unstable in the lowest 500 m but capped by an inversion. An unsaturated air parcel displaced upward would cool by the DALR (the heavy dashed line), making it initially warm and buoyant relative to the surrounding level. Some distance after penetrating the inversion layer, the rising air is no longer warmer than the surrounding air, and further lifting is suppressed. The parcel continues upward for some distance, however, because it has considerable upward momentum. As it does so, it cools more rapidly than the surrounding air and becomes relatively dense. After coming to a stop, the heavy parcel of air sinks back down and eventually comes to rest at some equilibrium level.

Extremely Stable Air: Inversions

6.5 Explain what causes stable air and inversions to develop.

So far, we have been concerned with the mechanisms that cause air to rise, and the influence of instability on the effectiveness of those mechanisms. Now it is useful to examine the most extreme forms of stable air, those associated with inversions.

Although on average the temperature in the troposphere decreases with elevation, in some situations temperature *increases* with altitude to form a *temperature inversion*, or inversion for short. Air parcels rising through inversions encounter ever-warmer surrounding air and therefore eventually become colder than their surroundings and have strong negative buoyancy. Inversions are thus extremely stable and resist vertical mixing.

Several different processes can cause different types of inversions to develop. One of the most common is a *radiation inversion*, which results from diabatic cooling of the surface. On cloud-free nights with little or no wind, longwave radiation emitted by the surface easily escapes to space. This lowers the ground temperature, which in turn chills the air immediately in contact with it. Because the lower air chills more rapidly than that farther from the surface, an inversion develops at ground level.

If cooling is sufficient to lower the temperature to the dew point, a radiation fog forms. Inversions are associated with all radiation fogs, but if the cooling does not lower the temperature to the dew point, a radiation inversion can exist without the appearance of fog.

Radiation inversions occur throughout the world. Though they are usually restricted to fairly shallow depths above the surface, they can have important ramifications for agriculture and other activities. We discuss two examples in *Box 6–3, Focus on the Environment and Societal Impacts: Radiation Inversions and Human Activities.*

6–3 FOCUS ON THE ENVIRONMENT AND SOCIETAL IMPACTS

Radiation Inversions and Human Activities

On clear, still nights, air near the ground can cool rapidly. Of course, air higher up will not be subjected to the same cooling as air in contact with the surface, and temperature differences of several degrees Celsius can be observed over just a few meters in height. Thus, subfreezing temperatures can exist near the ground, while just a short distance above, temperatures may be safely above the freezing level. This has some important effects on agriculture in the southern United States.

When temperatures drop to the freezing point, winter crops are vulnerable to frost damage. If the ground is relatively flat, the cold air can develop locally. In hilly regions, cooling of the air can occur at higher levels and gravity may cause the chilled air to flow downward and accumulate at the foot of the hills.

To offset the frost problem, growers often set large "wind machines" atop masts, as shown in Figure 6–3–1a. When temperatures near the ground fall dangerously low, the machines are turned on to force the warm air in the upper part of the inversion down toward the surface. Thus, the potentially damaging air near the surface is replaced as warmer air circulates downward.

At even lower temperatures, growers may also activate *smudge pots* that burn heating oil, as shown in Figure 6–3–1b. Although the emission of longwave radiation by the smudge pots helps somewhat to protect the crops, their greater contribution is to produce free convection. Like the wind machines, the smudge pots produce a continual mixing of air between low and higher levels, so that surface air stays above the freezing point. Yet another tactic is to spray citrus crops with water. As water freezes, latent heat is released, which keeps the fruit within a few degrees of freezing, warm enough to prevent severe damage. Of course, on very cold nights inversions are so deep and cold that none of these remedies is effective. In such a case, frost damage can ruin an entire crop.

Problems associated with radiative cooling and the resultant inversions are not restricted to agriculture. The strong stability of an inversion can suppress the vertical motions that dilute the concentration of pollutants near the surface. It is common for urban dwellers to notice a low-lying layer of sooty haze near the surface on cold, clear mornings when radiation inversions are most likely to form.

(a)

(b)

▼ **FIGURE 6–3–1 Wind Machines.** (a) Agricultural wind machines blow air downward during times of potential frost damage to force warm air in the upper part of a radiation inversion toward the crops. (b) During times of greater stress, growers also may employ smudge pots.

6–3–1 By what mechanism(s) do smudge pots and wind machines mitigate frost damage?

6–3–2 Why does spraying crops with water decrease the likelihood of frost damage?

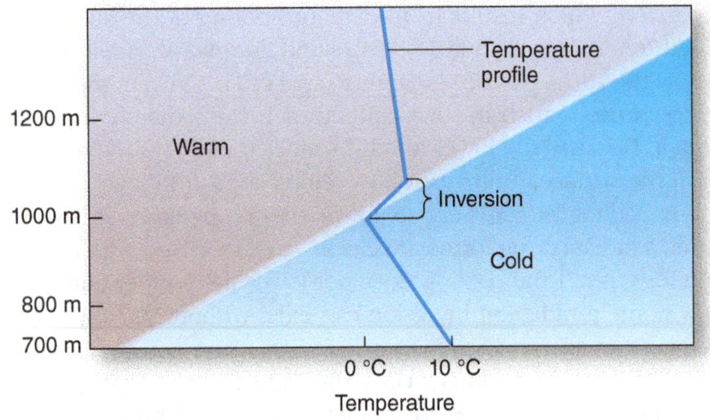

▲ **FIGURE 6–13 Frontal Inversions.** The temperature is plotted with the solid blue line.

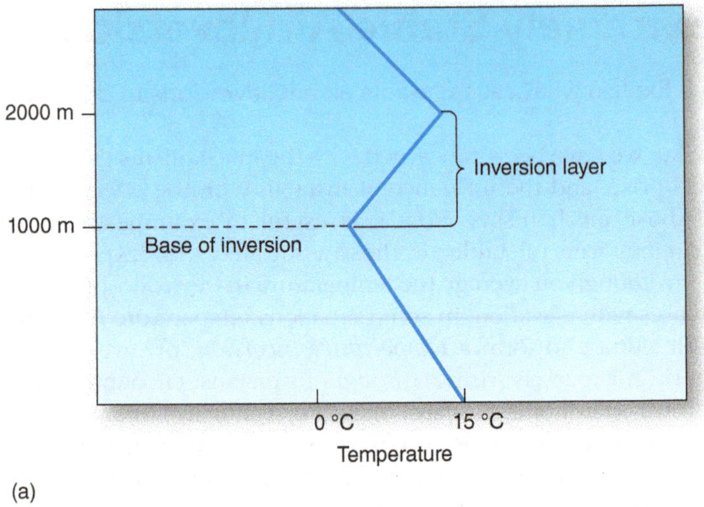

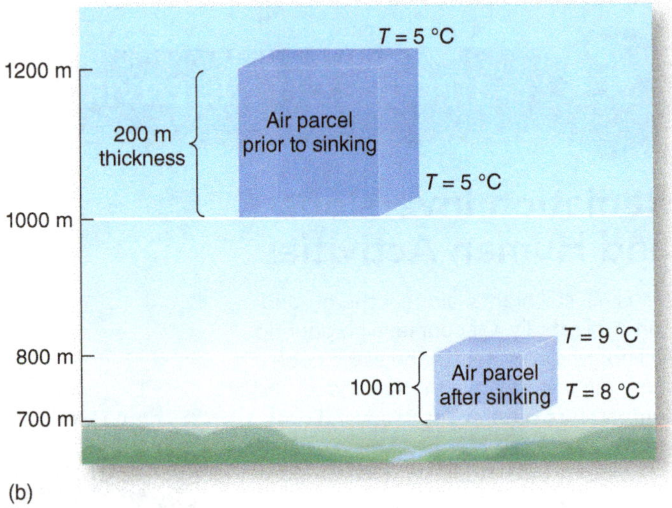

▲ **FIGURE 6–14 Subsidence Inversions.** Subsidence inversions occur when air descends toward but not all the way to the surface. (a) The base of the inversion marks the lowest point to which air has subsided. (b) We see the cause of the inversion: As a column of air sinks, it undergoes compression. In this example the layer of air has a uniform temperature of 5 °C. When it is lowered the upper portion sinks further than the bottom portion, and therefore warms 2 °C, while the bottom of the layer only warms 1 °C.

Factors other than diabatic cooling of the surface can also produce inversions. When a cold or warm front is present, for example, a transition zone separates warm and cold air masses. The boundary is not horizontal but rather forms a wedge of cold air that underlies warmer air, as shown in Figure 6–13. The horizontal extent of these *frontal inversions* can be up to several hundred kilometers, and the height of the inversion increases with distance from surface position of the front. Rain falling into a very cold surface layer can freeze before reaching the ground (resulting in sleet), or on the ground, resulting in the much more dangerous freezing rain.

More extensive and meteorologically important than frontal inversions are the ***subsidence inversions*** that result from sinking (or *subsiding*) air. Recall that a layer of air is compressed and warmed during descent. As it is compressed, its thickness decreases, which means that the top of the layer descends a greater distance than does the bottom of the layer. The longer descent leads to greater temperature increases at the top of the layer than the bottom, and, if enough sinking occurs, an inversion forms.

Subsidence is common along the eastern margins of large areas of high pressure and downwind of major mountain ranges. Because the subsiding air does not descend all the way to the surface, the base of the inversion can be several hundred meters above the surface. As a result, subsidence inversions are clearly distinguishable from radiation inversions, the bases of which are at ground level. Figure 6–14 shows a typical temperature profile for a subsidence inversion. Notice that the top of the inversion layer is more than 10 °C (18 °F) warmer than its base. This is a large difference, but not at all uncommon.

A spectacular example of a subsidence inversion can be found every year between the months of April and September in the United States. A large high-pressure system called the Hawaiian High often forms over the middle latitudes of the North Pacific. As upper-level air rotates out of the high-pressure center, subsidence occurs over coastal southern California and forms an inversion. This provides a "cap" for the vertical dispersal of pollutants that helps to give Los Angeles some of the most polluted air in North America.

CHECKPOINT

6.9 What is an inversion?

6.10 How does a subsidence inversion form?

Cloud Types

6.6 Describe the different types of clouds.

As indicated earlier, clouds form when air becomes saturated at some level above the ground. The saturation most often results from the upward displacement of air to the lifting condensation level. Clouds can assume a variety of shapes and sizes and can occur near the surface or at high altitudes. Most cloud types occur in the troposphere, but some appear in the stratosphere and even in the mesosphere. Clouds can contain liquid droplets, ice crystals, or a mixture of the two. They can be thick or thin and have high or low liquid water or ice contents. It is easy to see why meteorologists would want a classification scheme to distinguish the many types of clouds from one another.

DID YOU KNOW?

Have you ever looked out the window of an airplane and found faint clouds formed right above the plane's wings? This is the result of adiabatic expansion of the air above the wing that can bring the immediate air temperature down to the dew point.

The first widely accepted system for cloud classification was devised by English naturalist Luke Howard in 1803. It divided clouds into four basic categories:

1. **Cirrus**—thin, wispy clouds of ice
2. **Stratus**—layered clouds
3. **Cumulus**—clouds having vertical development
4. **Nimbus**—rain-producing clouds.

Our current classification scheme is a modified version of Howard's typing that retains his four categories and also allows new combinations (for instance, *cirrostratus* clouds have the characteristics of cirrus clouds and stratus clouds). The 10 principal types of clouds that result are then grouped according to their height and form:

1. **High clouds**—cirrus, cirrostratus, and cirrocumulus
2. **Middle clouds**—altostratus and altocumulus
3. **Low clouds**—stratus, stratocumulus, and nimbostratus
4. **Clouds with extensive vertical development**—cumulus and cumulonimbus.

The 10 principal cloud types based on this scheme are outlined in Table 6–1 and Figure 6–15. Several of the 10 cloud types described in Table 6–1 can also be divided into subtypes, called species. These species are described in Table 6–2.

High Clouds

High clouds are generally above 6000 m (19,000 ft). They are almost universally composed of ice crystals instead of liquid droplets. Recall that within the troposphere the average temperature decreases from 15 °C (59 °F) at sea level at a rate of 6.5 °C/1000 m (3.6 °F/1000 ft). As a result, the average temperature for high clouds is usually no higher than –35 °C (–31 °F); cooling to the frost point causes the formation of ice crystals instead of supercooled droplets.

Where surface temperatures are very low, clouds composed exclusively of ice can occur at altitudes as low as 3000 m (about 10,000 ft). Thus, the definition of a high cloud (above 6000 m) is somewhat temperature dependent.

TABLE 6–1
Ten Principal Cloud Types

High Clouds (Heights Greater Than 6000 m, or 19,685 ft)		
Cirrus (Ci)	(Figure 6–16)	Thin, white, wispy clouds resembling mares' tails.
Cirrostratus (Cs)	(Figure 6–19)	Extensive, shallow clouds somewhat transparent to sunlight, producing a halo around the Sun or Moon.
Cirrocumulus (Cc)	(Figure 6–20)	High, layered cloud with billows or parallel rolls.
Middle Clouds (Heights Between 2000 and 6000 m, or 6562 and 19,685 ft)		
Altostratus (As)	(Figure 6–21)	Extensive, watery, layered cloud. Allows some penetration of sunlight but Moon or Sun appears as bright spot within cloud.
Altocumulus (Ac)	(Figure 6–22)	Shallow, midlevel cloud containing patches or rolls.
		Generally more opaque and having less distinct margins than cirrocumulus.
Low Clouds (Below 2000 m, or 6562 ft)		
Stratus (St)	(Figure 6–23)	Uniform layer of low cloud ranging from whitish to gray.
Nimbostratus (Ns)	(Figure 6–24)	Low cloud producing light rain. Produces darker skies than altostratus.
Stratocumulus (Sc)	(Figure 6–25)	Low-level equivalent to altocumulus.
Clouds with Vertical Development (May Extend Through Much of Atmosphere)		
Cumulus (Cu)	(Figures 6–26 and 6–27)	Detached billowy clouds with flat bases and moderate vertical development. Sharply defined boundaries.
Cumulonimbus (Cb)	(Figure 6–28)	Clouds with intense vertical development with characteristic anvil. May be tens of thousands of meters thick. Appear very dark when viewed from below.

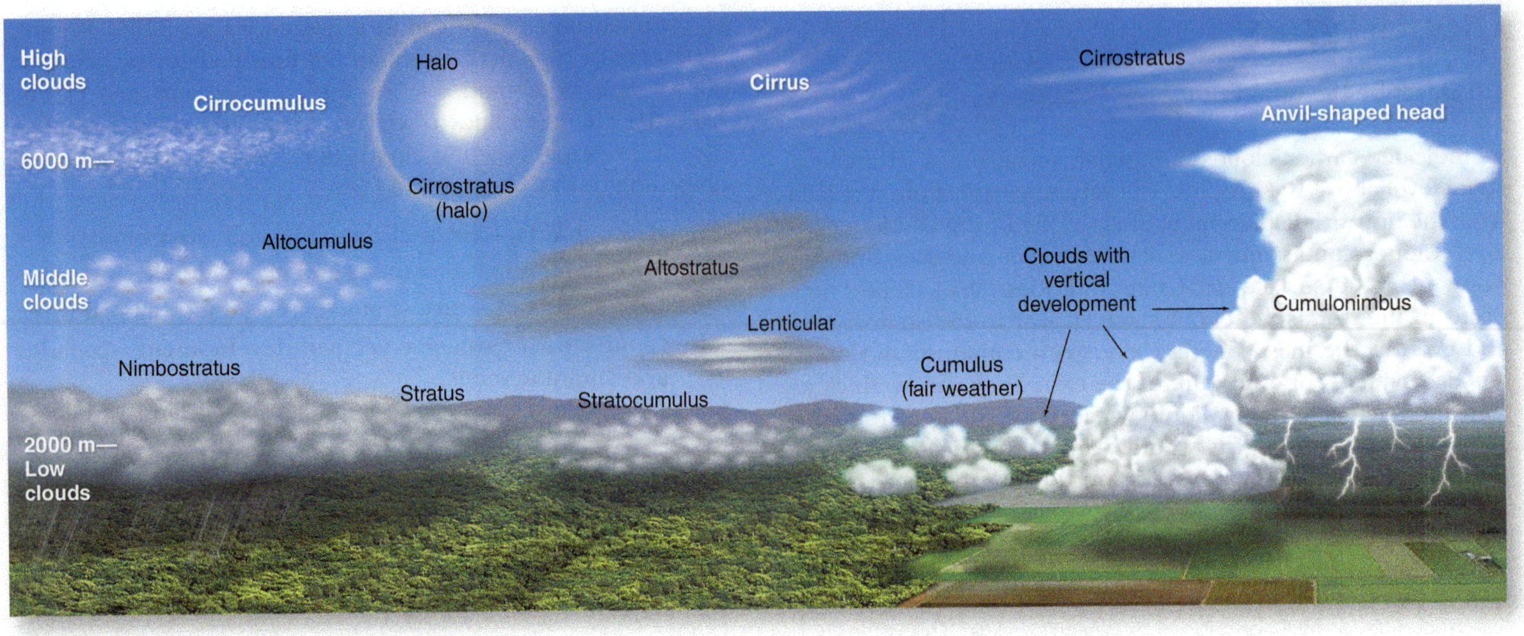

▲ **FIGURE 6–15 Generalized Cloud Chart.**

TABLE 6–2
Cloud Species

Name	Types of Cloud	Description
Calvus	Cumulonimbus	Upper portion of cumulonimbus loses distinct outline as ice crystals form in lieu of water droplets.
Capillatus	Cumulonimbus	Fibrous upper portion consists of ice crystals. Further developed than cumulonimbus calvus.
Castellanus	Cirrus, cirrocumulus, altostratus, and stratocumulus	Tower-like vertical development on portion of cloud.
Congestus	Cumulus	Undertaking rapid and significant vertical development.
Fractus	Stratus and cumulus	Broken or ragged.
Humilis	Cumulus	Having slight vertical extent.
Incus	Cumulonimbus	Having anvil-shaped top.
Lenticularis	Stratocumulus, altocumulus, cirrocumulus	Lens shaped.
Mammatus	Cumulonimbus	Pouch-like.
Pileus	Cumulus, cumulonimbus, stratocumulus	Cap-like cloud immediately above larger cloud.
Translucidus	Altocumulus, altostratus, stratocumulus, and stratus	Somewhat transparent, so the Sun, Moon, or higher clouds may be discernible.

CIRRUS The simplest of the high clouds are *cirrus* (abbreviated Ci), which are wispy aggregations of ice crystals (Figure 6–16). The average thicknesses of cirrus clouds is about 1.5 km (1 mi), but they can be as thick as 8 km (5 mi). Given the very low temperatures at which they exist, they have little water vapor from which to form ice. So although they may be easily visible, the water content of these clouds is extremely low. In fact, the ice content of cirrus clouds is typically only about 0.025 grams per cubic meter, or about one-thousandth of an ounce per cubic yard.

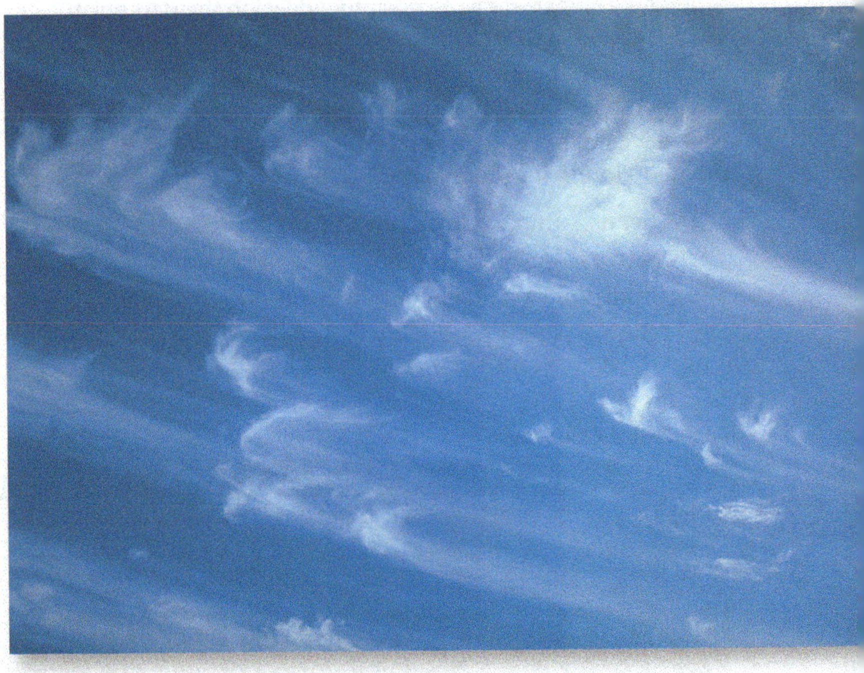

▲ **FIGURE 6–16 Cirrus Clouds.** These types of clouds are wispy clouds of ice crystals.

Although the entire mass of ice contained in a cirrus cloud is small, the individual crystals can be as long as 8 mm (0.3 in.). These crystals fall at a speed of about 0.5 meters per second (about 1 mile per hour), which is sufficient for them to overcome updrafts and descend as *fall streaks* (Figure 6–17).

CONTRAILS A type of ice cloud, known as a **contrail** (Figure 6–18), is frequently caused by jet aircraft. The very hot engine exhaust contains considerable water vapor as a result of fuel combustion, and turbulence in the wake of the aircraft rapidly mixes the exhaust with the cold, ambient air. As explained in Chapter 5, the mixing of warm, moist air with cold air can lead to saturation and, in this case, the rapid formation of ice crystals.

▲ **FIGURE 6–17** Fall streaks result from falling ice crystals.

▲ **FIGURE 6–18 Aircraft Contrails.**

CIRROSTRATUS Like cirrus clouds, **cirrostratus** (Cs) clouds (Figure 6–19) are composed entirely of ice but tend to be more extensive horizontally and have a lower concentration of crystals than cirrus. Though cirrostratus clouds reduce the amount of solar radiation reaching the surface, enough direct sunlight penetrates to allow objects at the surface to cast shadows. Furthermore, they do not fully obscure the Moon or Sun behind them. Instead, when viewed through a layer of cirrostratus, the Moon or Sun has a whitish, milky appearance but a clear outline. Cirrostratus are often found above warm fronts.

A characteristic feature of cirrostratus clouds is the *halo*, a circular arc around the Sun or Moon formed by the refraction (bending) of light as it passes through the ice crystals. Ice crystals bend much of the passing sunlight or moonlight 22° away from its initial direction. So if you face toward the Sun or Moon, refracted sunlight will be directed toward you from a ring that surrounds the Sun at a 22° angle.

▲ **FIGURE 6–19 Cirrostratus Clouds.** Such clouds often create a halo around the Sun or Moon.

CIRROCUMULUS Among the most beautiful of clouds, **cirrocumulus** (Cc) are composed of ice crystals that are arranged into long rows of individual, puffy clouds (Figure 6–20). Cirrocumulus form during episodes of ***wind shear***, a condition in which the wind speed and/or direction changes with height. Wind shear often occurs ahead of advancing storm systems, so cirrocumulus clouds are often a precursor to precipitation. Because of their resemblance to fish scales, cirrocumulus clouds are associated with the term "mackerel sky."

Middle Clouds

Middle clouds occur between 2000 and 6000 m (6500 and 19,000 ft) above the surface and are usually composed of

▼ **FIGURE 6–20 Cirrocumulus Clouds.** These frequently occur in rows of individual, puffy clouds.

▲ **FIGURE 6–21 Altostratus Clouds.** These middle-level, layered clouds are composed of water droplets.

▲ **FIGURE 6–22 Altocumulus Clouds.** Layered, midlevel altocumulus clouds are often arranged in bands.

DID YOU KNOW?

For three days following the tragedy of September 11, 2001, all commercial aircraft were grounded in the United States. Formation of contrails stopped immediately; according to one study, this increased the daily temperature range across the country during that period. The evidence suggests that the temporary disappearance of contrails increased the amount of sunlight reaching the surface during the daytime (thus raising daytime surface temperatures), while increasing longwave radiation losses at night (thereby lowering nighttime temperatures). Researchers believe that the daily temperature ranges across the country averaged about 1.8 °C (3.2 °F) more than they normally would over the period, though this finding has been questioned by some who think the change may have been the result of unusually clear skies. It is likely that over the Midwest, where contrails are most likely to form, the increase in the daily range may have been even greater.

▲ **FIGURE 6–23 Stratus Clouds.** These are low-level, layered clouds.

liquid droplets. The two major categories in this group are both prefixed by *alto*, which means "middle."

ALTOSTRATUS The middle-level counterparts to cirrostratus, **altostratus** (As) clouds (Figure 6–21), differ from cirrostratus in that they are more extensive and composed primarily of liquid water. Altostratus clouds scatter a large proportion of incoming sunlight back to space, thereby reducing the amount of sunlight that reaches the surface. The insolation that does make its way to the surface consists primarily or exclusively of diffuse radiation, so one way to distinguish altostratus from cirrostratus is by the absence of shadows. Furthermore, when viewing the Sun or Moon behind altostratus, one sees a bright spot behind the clouds instead of a halo.

ALTOCUMULUS Layered clouds called **altocumulus** (Ac) (Figure 6–22) form long bands or contain a series of puffy clouds arranged in rows. They are often gray in color, although one part of the cloud may be darker than the rest. Consisting mainly of liquid droplets rather than ice crystals, they usually lack the beauty of cirrocumulus.

Low Clouds

Low clouds have bases below 2000 m (6562 ft). *Stratus* (St) (Figure 6–23) are layered clouds that form when extensive areas of stable air are lifted. They are normally between 0.5 and 1 km (0.3 to 0.6 mi) thick, in marked contrast to their horizontal

▲ **FIGURE 6–24 Nimbostratus Clouds.** This type of cloud produces light precipitation.

▲ **FIGURE 6–25 Stratocumulus Clouds.** These are low, layered clouds with some vertical development.

extent, which can exceed that of several states. Usually the rate of uplift producing a stratus cloud is only a few tens of centimeters per second (less than 1 mph), and its water content is low, perhaps just a few tenths of a gram per cubic meter.

STRATUS A type of cloud that does not necessarily form from the lifting of air, stratus clouds may also result from the turbulence associated with strong winds. Consider a situation in which the air is at rest and there is a moderate decrease in temperature and dew point with height. When the wind begins to blow, it stirs the air by forced convection. This causes a slight decrease in dew point near Earth's surface and an increase in water vapor at the top of the layer, as moisture is redistributed vertically. At the same time, the air temperature decreases more rapidly with altitude after mixing because the uplifted air cools at the DALR rather than the original (more gradual) ELR. The reduction in air temperature and increase in moisture content in the upper portion of the layer causes the air to become saturated.

NIMBOSTRATUS Low, layered clouds that yield precipitation are called **nimbostratus** (Ns) (Figure 6–24). Because nimbostratus clouds have low liquid water contents and weak updrafts to replenish moisture, they yield only light precipitation. Seen from below, these clouds look very much like stratus, except for the presence of precipitation.

STRATOCUMULUS Low, layered clouds with some vertical development are called **stratocumulus** (Sc) (Figure 6–25). Their darkness varies when seen from below because their thickness varies across the cloud. Thicker sections appear dark, and thinner areas appear as bright spots.

CHECKPOINT

6.11 What two criteria form the major basis of cloud classification?

6.12 How do cirrocumulus, altocumulus, and stratocumulus clouds differ from each other?

Clouds with Vertical Development

CUMULIFORM **Cumuliform** clouds have substantial vertical development and occur when the air is absolutely or conditionally unstable. Vertical velocities within these clouds are commonly several meters per second, but they can achieve speeds well in excess of 50 m/sec (112 mph). In other words, updrafts in certain cumuliform clouds can be more rapid than the horizontal winds found in weak hurricanes! Liquid water contents are several times greater than those of stratiform clouds. *Cumulus* (Cu) clouds fall into several subgroups distinguished by the extent of their vertical development.

CUMULUS HUMILIS Fair-weather cumulus (Figure 6–26a), called *cumulus humilis*, have a single plume of rising air that often results from localized heating at the surface. They do not yield precipitation (hence the name, "fair-weather cumulus"), and they can evaporate away soon after their formation. Notice in Figure 6–26b that the clouds and the open areas between them form an invisible circulation system. The clouds mark the zone of rising air, and the cloud-free areas occur where the air sinks.

CUMULUS CONGESTUS More intensely developed clouds are *cumulus congestus* (Figure 6–27). Unlike cumulus humilis, they consist of multiple towers, and each tower has several cells of uplift. This gives them a fortress-like appearance with numerous columns of varying heights. Their strong vertical development implies that these clouds form in unstable air.

(a)

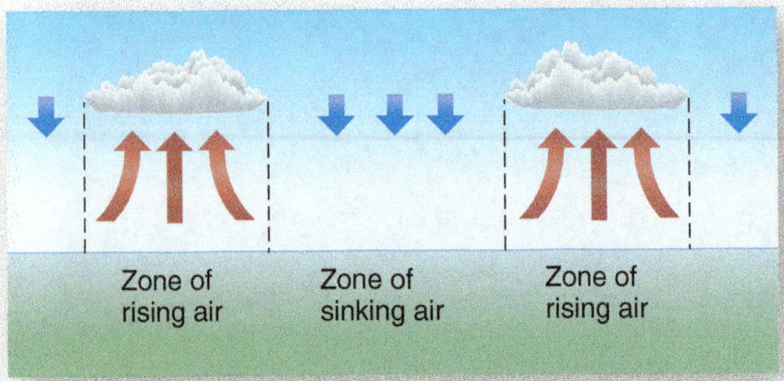

(b)

◂ **FIGURE 6–26 Cumulus Humilis.** (a) Puffy, white cumulus humilis are sometimes called fair-weather cumulus. (b) Scattered fair-weather cumulus clouds form by rising air parcels, but the area between clouds has weak downdrafts.

The individual towers of a cumulus congestus have life spans of only tens of minutes and are constantly being replaced by newly forming ones. Because of their vertical extent, cumulus congestus clouds can have large temperature differences from top to bottom, and even on hot days their upper portions can have subfreezing temperatures. Liquid droplets do not freeze into ice instantaneously as updrafts carry them into the cold portion of the cloud, however; instead, they can remain in their supercooled state for some time. Eventually the supercooled droplets do freeze, and the liquid cloud becomes dominated by ice crystals. We can actually observe the result of this process from the ground. When a portion of the cloud becomes *glaciated* (composed entirely of ice), it does not exhibit sharply defined edges like the portions consisting of water. Instead, it has a washed-out appearance that makes it readily distinguishable from the liquid portion of the cloud. The upper portion of the cloud in Figure 6–27 provides a good example of glaciation.

CUMULONIMBUS Because they produce the most intense thunderstorms, **cumulonimbus** (Cb) clouds (Figure 6–28) are considered the most violent of all clouds. In warm, humid, and unstable air, they can have bases just a few hundred meters above the surface and tops extending into the lower stratosphere. In other words, these clouds can occupy almost the entire depth of the troposphere and more!

A cumulonimbus is frequently distinguished by the presence of an **anvil** (so named for its resemblance to the blacksmith's tool). This feature, composed entirely of ice crystals, is formed by the high winds of the upper troposphere that extend the cloud forward. The anvil appears as a wedge of ice at the top of the cloud that gradually thins out as it gets farther from the main body of the cloud. Strong winds can propel hailstones toward the anvil, where they are ejected and fall from the cloud. This is why airline pilots avoid flying near the anvil.

▾ **FIGURE 6–27 Cumulus Congestus.** Note the substantial vertical development.

▾ **FIGURE 6–28 Cumulonimbus.** Extending into the lower stratosphere, these clouds can create violent weather.

DID YOU KNOW?

The ice crystals that compose the anvil of a cumulonimbus can exist long after the rest of the cloud has completely dissipated. The resultant cirrus clouds can then be transported many hundreds of kilometers by upper-level winds before they sublimate away. Thus, the cirrus observed by residents of the eastern half of the United States and Canada may be the remnants of a major storm from a couple of days earlier and a considerable distance away.

Although cumulonimbus clouds have extremely strong updrafts, the updrafts vary across the cloud. The most rapidly rising air is found in about the upper third or so of the cloud, with weaker motions below. Furthermore, updrafts along the margins of the cloud are generally less intense than those in the interior because of entrainment. Cumulonimbus even have regions where air descends. Although commercial aircraft can fly through such clouds, the rapid and dramatic shifts in vertical winds cause extreme turbulence that would violently jostle the plane and its occupants. Commercial pilots wisely fly around rather than through cumulonimbus clouds. Other aspects of aviation safety are discussed in *Box 6–4, Focus on Aviation: Responding to Icing in Different Types of Clouds.* In *Box 6–5, Forecasting: Why Clouds Have Clearly Defined Boundaries,* we examine an interesting aspect of cumulus clouds.

CHECKPOINT

6.13 Which type of cloud can extend into the lower stratosphere?

6.14 What are two reasons that airline pilots might want to avoid clouds of this type?

Unusual Clouds

Certain cloud types do not fall neatly into the categories mentioned above. **Lenticular clouds** form downwind of mountain barriers and have curved shapes like eyeglass lenses (Figure 6–29). They form when mountain ranges disrupt the flow of air to form a series of waves. As the air rises in each wave, adiabatic cooling leads to condensation; as the air descends, adiabatic warming causes the cloud droplets to gradually evaporate. Although new droplets constantly form on the upwind side of lenticular clouds and old droplets evaporate on the downwind side, the clouds remain in a fixed position. Thus, the flow

6–4 FOCUS ON AVIATION

Responding to Icing in Different Types of Clouds

In Chapter 5 we talked about the threat that icing presents to aviation. The first line of defense when having to fly through icing conditions is to activate the plane's anti-icing equipment. But pilots sometimes can also alter the course of the flight to lessen the threat, with the appropriate action depending on the cloud type.

When icing appears to occur within stratiform clouds, it is a good idea to change elevation upward or downward by about 900 meters—3000 feet (Figure 6–4–1). By definition, stratiform clouds have relatively little vertical development, so changing altitude by that amount will likely get you out of the region of supercooled droplets.

Icing can occur rapidly and severely in cumuliform clouds, especially in updrafts and at temperatures ranging from −2 °C to −20 °C (28 °F to −4 °F). Because cumuliform clouds have very substantial vertical development, changes in altitude may be a less effective maneuver than in stratiform clouds. It is a good idea if possible to simply fly around such clouds. Icing in cumuliform clouds normally occurs at altitudes below about 8000 meters (about 27,000 ft).

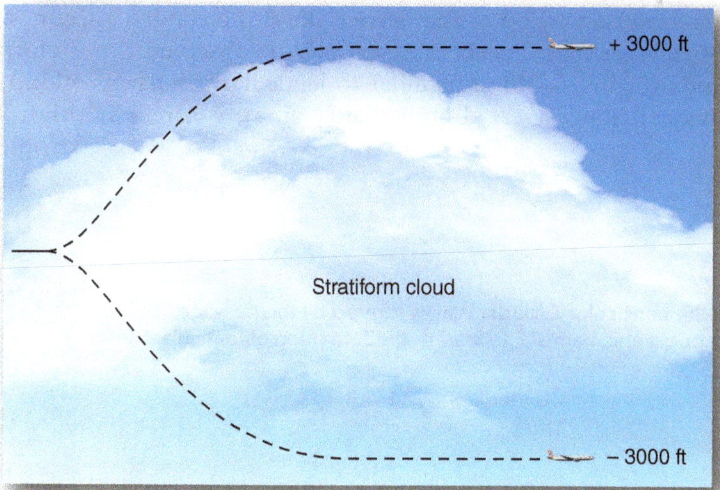

▲ **FIGURE 6–4–1 Icing.** Pilots can avoid icing conditions in stratiform clouds by increasing or decreasing altitude.

6–4–1 What should a pilot do if ice is encountered in a stratiform cloud?

6–4–2 Why might changing altitude not immediately solve the problem of icing in a cumuliform cloud?

6–5 FORECASTING

Why Clouds Have Clearly Defined Boundaries

The next time you see a cumulus cloud, notice its base and edges. You will see that the boundaries of the cloud are marked by flat, sharply defined bottoms and edges that are clearly distinct from the surrounding air. This clear definition of the cloud base is partly due to the rapid growth of droplets as they form just above the lifting condensation level.

Recall that liquid water at the base of a cloud initially forms onto condensation nuclei, of which there are a finite (but very large) number. Within a few tens of meters of the lifting condensation level, all the available condensation nuclei have attracted moisture, and further condensation occurs only onto existing droplets. When droplets first form, they are very small; but they quickly attain diameters of about a micrometer, which makes them effective at scattering visible light—hence the clearly visible base. (If instead the droplets grew slowly, there would be only a gradual increase in the number capable of scattering visible light from the base of the cloud. The cloud would therefore have a faintly visible base that would gradually become more discernible with height.)

Another factor is that cloud droplets evaporate within a very short distance of the cloud base. To see why, consider that if a droplet falls into unsaturated air below the cloud, it shrinks by evaporation. Because they are so small, cloud droplets fall slowly to begin with and fall even more slowly as they evaporate. As a result, the distance a cloud droplet can fall without evaporating is minuscule, on the order of a centimeter. Viewed from a great distance, this will look like a flat surface. (By way of contrast, raindrops have survival distances measured in kilometers because they are so much larger.)

The sharp boundaries along the sides of cumulus clouds are the result of entrainment. When unsaturated air just outside the margins of the cloud is drawn into the cloud by turbulence, water droplets rapidly evaporate, leaving behind a distinct boundary between the unsaturated, ambient air and the large droplets remaining within the cloud.

6–5–1 How does the size of cloud droplets forming at the base of a cloud create the sharp boundary between the cloud and dry air?

6–5–2 What process works to give the sides of clouds well-defined boundaries?

of moisture into and out of lenticular clouds is similar to the movement of objects on a conveyor belt, with as much mass removed as is added. Usually no more than two or three lenticular clouds form downwind of the barrier, but if the conditions are just right, as many as six or seven may be observed. **Banner clouds** (Figure 6–30) are similar to lenticular clouds but are individual clouds located immediately above isolated peaks.

Sometimes portions of cumulonimbus clouds hang downward in sac-like shapes called **mammatus** (Figure 6–31). These features occur where downdrafts force water droplets below the cloud base or sloping edge of the cloud, usually near the anvil. Because of the high liquid water content of these clouds, the droplets contained in the downward-moving

Video Identifying Clouds in Satellite Imagery

http://goo.gl/94G0U4

▼ **FIGURE 6–29 Lenticular Clouds.** Waves formed by the passage of air over a topographic barrier can lead to the formation of lenticular clouds.

▼ **FIGURE 6–30 Banner Clouds.** These form atop isolated mountain peaks.

▲ **FIGURE 6–31 Mammatus.** These clouds are found on the margins of cumulonimbus clouds, are formed by downdrafts, and sometimes are distorted by complex motions.

▲ **FIGURE 6–32 Nacreous Clouds.** These stratospheric clouds are only observed at high latitudes.

▲ **FIGURE 6–33 Noctilucent Clouds.** Located in the mesosphere, noctilucent clouds can illuminate the sky at high latitudes during the twilight hours.

air require substantial descent (and resultant adiabatic warming) before they fully evaporate. Thus, the mammatus extend some distance below the droplets of the surrounding cloud base.

Like most other weather features, the majority of clouds exist in the troposphere. However, two rare cloud types exist at higher levels and can be seen during the twilight hours of winter at high latitudes. **Nacreous clouds** (Figure 6–32) consist of supercooled droplets or ice crystals in the stratosphere at heights of about 30 km (20 mi). They have a soft, whitish appearance and sometimes are called *mother of pearl* clouds. Even higher are the **noctilucent clouds** (Figure 6–33), whose location in the mesosphere allows them to be illuminated after sunset (or before sunrise) when the surface and the lower atmosphere are in Earth's shadow.

DID YOU KNOW?

The sighting of noctilucent clouds in North America has historically been restricted to latitudes near the Canadian–United States border and northward. They have now recently been observed as far south as Colorado and Utah. This could reflect a decrease in the temperature of the stratosphere associated with climate warming in the lower atmosphere. If so, their increased appearance over the United States may have significant ramifications—beyond making the nighttime sky more dramatic.

CHECKPOINT

6.15 What is a lenticular cloud?

6.16 What role do adiabatic warming and cooling play in the formation of lenticular clouds?

Cloud Coverage and Observation

6.7 Describe how meteorologists measure cloud coverage.

In addition to their height and form, another characteristic of clouds is their breadth or **coverage**. Meteorologists use several terms to describe coverage, as shown in Table 6–3.

Clouds on any given day are not restricted to a single height above the surface. They can occur simultaneously at several different levels in the atmosphere, and each level can have different cloud types and a different amount of coverage. For example, a detailed cloud report might describe the sky as having scattered cumulus at 1000 m, broken altostratus at 4000 m, and a layer of overcast cirrostratus at 7000 m.

Surface-Based Observation of Clouds

Cloud heights and coverages up to a height of 3650 m (12,000 feet) are now routinely determined by automated devices called laser *ceilometers* as part of automated sensing units installed at many airports (Figure 6–34). The laser units emit a brief pulse of energy upward that gets reflected downward by cloud droplets or ice crystals. The laser beam travels at a known speed, so the amount of time it takes for the pulse to make its round-trip can easily be translated to the height of the cloud base.

Ceilometers can also reveal the amount of coverage for up to three layers above the surface. This is done by evaluating repeated measurements over the previous 30 minutes and noting the amount of time that clouds were present at each level. To make the estimate more representative of the current time period, observations over the previous 10 minutes are given twice as much weight in the calculation as those of the other 20 minutes. See *Box 6–6, Physical Principles: The Surprising Composition of Clouds,* to get a better idea of what we are really looking at when we observe clouds.

▲ **FIGURE 6–34 Ceilometer.** The height of multiple layers of clouds can be determined using a ceilometer. Laser beams are emitted upward and a portion of the energy is reflected back to the instrument from each cloud layer. Cloud heights are determined by the time required for the laser pulse to return to the ceilometer.

Cloud Observation by Satellite

Virtually every television weather report shows the distribution and movement of cloud cover—both regionally and locally—as observed by satellite. In some cases, the satellites view the cloud cover by sensing reflected visible radiation in much the same way a digital camera would view it if fixed on an orbiting platform. On these *visible images*, areas of deep cloud cover appear as a brighter shade of white (Figure 6–35a). Such images have both advantages and shortcomings: They are useful for observing most clouds, but are not very good for distinguishing between high-, medium-, and low-level clouds; and, of course, they are unable to provide any information at night.

Better information can be obtained when visible imagery is coupled with *infrared images* (Figure 6–35b). Unlike visible imagery, which relies on solar radiation scattered

TABLE 6–3
Cloud Coverage

Amount of Cloud Coverage	Condition
0	Clear
1/8 to 2/8	Few*
3/8 to 4/8	Scattered
5/8 to 7/8	Broken
8/8	Overcast

* Any cloud coverage at all up to 2/8 is classified as "few."

6–6 PHYSICAL PRINCIPLES

The Surprising Composition of Clouds

We think of clouds as being liquid water and/or ice, but by far the greatest amount of the mass contained in a cloud is air. Although clouds contain a very large number of suspended droplets or particles—typically about 1000 per cubic centimeter—even the largest of these droplets are extremely small. Therefore, despite their large numbers they amount to relatively little mass. This might not be too surprising if you recall that water vapor only accounts for a small percentage of the mass of the atmosphere. Because the water vapor is so limited, it stands to reason that only a small amount of ice or liquid water can be deposited or condensed.

We can apply some simple arithmetic to determine the relative mass of air and liquid water in a cloud. The average cloud droplet has a radius of about 0.001 cm, and the volume of a sphere is equal to

$$\frac{4}{3}\pi r^3$$

where r is the radius. Substituting 0.001 cm for r, we find that the average droplet has a volume of about 4×10^{-9} cubic centimeters.

Because the density of water is about 1 g/cm^3, it follows that the mass of each droplet is about 4×10^{-9} g. Multiplying this value by the 1000 droplets per cubic centimeter normally found in a cloud gives us a liquid water content of 0.000004 g/cm^3.

Now we can compare the 0.000004 g/cm^3 of water to the mass of the air. At an altitude of 5.5 km above sea level, for example, the density of the air is about half that at sea level, or roughly 0.0006 g/cm^3. Compare this to the mass of the liquid, and you will find about 150 times more air than water in the cloud.

You can think of this another way: If a cloud has a horizontal area of 1 square kilometer and a height of 1 km, it contains about 4 million kilograms of water. This is far less than the approximately 600 million kilograms of air in the same cloud.

6–6–1 Why don't clouds contain much condensed liquid water? What factor is responsible for the limited amount of water that can exist in clouds?

6–6–2 Overall, how does the amount of water in a cloud compare to the amount of air in it?

back toward space by cloud tops, infrared imagery senses the amount of electromagnetic energy *emitted* by clouds. (Infrared images are similar to the water vapor images discussed in Chapter 5, but they are based on the detection of different wavelengths, making them less sensitive to the presence of water vapor but better for tracking clouds.) As discussed in Chapter 2, all materials radiate energy, with the amount of energy and the wavelengths radiated dependent on temperature. Generally, cloud tops at higher elevations will have lower temperatures than those closer to the surface (recall that temperatures within clouds decrease with altitude), and thus emit less radiation than lower-altitude cloud tops. On the most basic types of infrared imagery, clouds of greater thickness appear in a brighter shade of white. Because thicker clouds are more likely to produce precipitation (or severe weather), such areas on satellite imagery are significant.

Infrared imagery can be color enhanced to provide more detailed information to the user. Areas on *color-enhanced infrared* images (Figure 6–35c) having the deepest convection are modified to appear as bright red or purple, while those of less-intense activity are assigned yellow and green hues. Lowest-level cloud tops are presented in white.

Cloud movie loops and images such as those shown in Figure 6–35 are readily available on the web from many government, commercial, and university sources. More information on using these images will be provided in Chapter 13.

DID YOU KNOW?

Juneau, Alaska, is the cloudiest U.S. city, receiving only 30 percent of its possible sunshine. Prince Rupert, British Columbia, is the cloudiest city in Canada, with overcast conditions occurring 70 percent of the time. In stark contrast, Yuma, Arizona, receives 90 percent of its possible sunshine on average, annually, making it the sunniest city in the United States and Canada.

CHECKPOINT

6.17 What is a ceilometer?

6.18 What are the advantages (and disadvantages, if any) of satellite images of clouds made using visible radiation and infrared radiation?

(a)

(b)

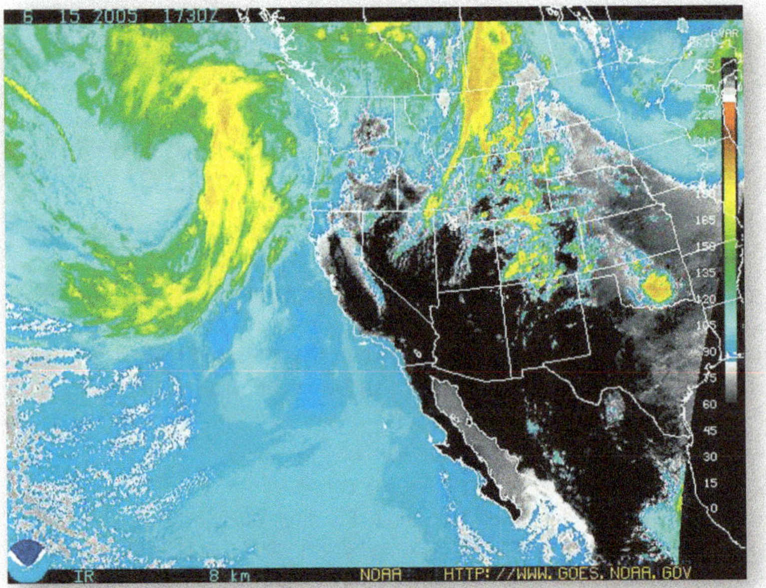

(c)

◂ **FIGURE 6–35 Satellite Imagery.** (a) Visible, (b) infrared, and (c) color-enhanced infrared satellite images of western North America taken on the morning of June 15, 2005. The comma-shaped cloud pattern off the coast of Oregon and Washington is a typical mid-latitude storm system. The whitish area off the coast of southern and Baja California is a large zone of low clouds and fog, typical of the region in the spring. Notice that the storm system appears well in all three images, with the color-enhanced image providing the best detail of the cloud structure. Notice also that the visible image in this instance provides the most distinct image of the low cloud and fog area off the southern coast.

Video
Clouds Developing Over Florida
http://goo.gl/sVW2cl

Video
Is That a Cloud?
http://goo.gl/W80Jyw

Summary

6.1 Describe four mechanisms that lift air.

- Air can be lifted so that it cools adiabatically to the dew point through orographic uplift, frontal uplift, convergence, and convection.

6.2 Explain the factors that determine the stability of parcels of air.

- Lifting of air can be facilitated or hindered by the air's static stability. Stable air resists displacement, whereas unstable air rises on its own if given an initial push upward. Conditionally unstable air rises on its own if lifted sufficiently to what is called the level of free convection.
- The static stability of air can be altered by a condition called potential instability, which involves what happens when changing environmental lapse rates (and therefore static stability) lift whole layers of air.

6.3 Explain the factors that influence the environmental lapse rate.

- The inflow of warm and cold air at different altitudes can change the environmental lapse rate at a location. At other times an entire air mass can be displaced by another one, resulting in entirely different conditions, including the ELR.

- The heating or cooling of the air at the surface changes the ELR by increasing or lowering the temperature of the lowest portion of the atmosphere.

6.4 Explain what causes unstable air to stop rising.

- An air parcel that encounters a stable layer will slow down and quit rising, due to its loss of buoyancy.
- The introduction of nearby air into a rising air parcel can cause entrainment, affecting the temperature and buoyancy of the rising air.

6.5 Explain what causes stable air and inversions to develop.

- Inversions right at the surface (radiation inversions) are caused by cooling the lower atmosphere.
- As fronts pass by there will be a mass of warmer air overlying the colder air, causing a frontal inversion.
- Inversions can form away from the surface by the sinking of air (subsidence inversions). The air undergoes compression as it sinks so that the upper portion of the layer undergoes more adiabatic warming than the lower portion.

6.6 Describe the different types of clouds.

- Most clouds are classified based on combinations of their height (high, medium, low, or having extensive vertical development) and form (stratiform, cirriform, and cumuliform). Some unusual clouds can also be found that do not fit into those standard groupings.

6.7 Describe how meteorologists measure cloud coverage.

- Cloud decks are usually divided into five categories based on the percentage of sky covered.
- Skies are mostly covered by clouds during broken and overcast conditions. The presence of some cloud cover that does not cover more than half of the sky is categorized as few or scattered conditions.

Key Terms

altocumulus *p. 204*
altostratus *p. 204*
anvil *p. 206*
banner clouds *p. 208*
cirrocumulus *p. 203*
cirrostratus *p. 203*
cirrus *p. 201*
clouds with extensive vertical development *p. 201*
cold front *p. 189*
contrails *p. 202*
convergence or **horizontal convergence** *p. 190*
coverage *p. 210*
cumuliform *p. 205*
cumulonimbus *p. 206*
cumulus *p. 201*
entrainment *p. 198*
environmental lapse rate (ELR) *p. 191*
high, middle, and **low clouds** *p. 201*
lenticular clouds *p. 207*
level of free convection *p. 194*
mammatus *p. 208*
nacreous clouds *p. 209*
nimbostratus *p. 205*
nimbus *p. 201*
noctilucent clouds *p. 209*
orographic uplift or **orographic effect** *p. 189*
potential (convective) instability *p. 194*
rain shadow *p. 189*
static stability *p. 191*
stratocumulus *p. 205*
stratus *p. 201*
temperature inversion *p. 195*
warm front *p. 189*

Review Questions

1. Describe the four mechanisms that lift air and promote cloud formation.
2. Explain how buoyancy affects the air's susceptibility to uplift.
3. Describe the situations that can cause air to be absolutely stable, absolutely unstable, or conditionally unstable.
4. What will determine whether air that is conditionally unstable will become buoyant?
5. What two factors can ultimately stop rising parcels of air from continuing upward?
6. What is the level of free convection?
7. Describe the processes that bring about changes in the environmental lapse rates, and thus the stability of the atmosphere.
8. Define the term *inversion*, and describe the mechanisms that can cause the various types of inversions.
9. What is entrainment, and how does it affect the growth of clouds?
10. Describe the classification scheme for clouds based on their height and form.
11. List the major subtypes of high, middle, and low clouds.
12. What type of cloud produces a characteristic halo?
13. Other than height, what significant difference exists between altocumulus and cirrocumulus clouds?
14. How do cumulus humilis and cumulus congestus clouds differ from each other?
15. What conditions generate lenticular and banner clouds?
16. What distinctive feature characterizes a cumulonimbus cloud?
17. What are mammatus?
18. List and describe the types of clouds that exist above the troposphere.

Critical Thinking

1. Is the stability of the air more likely to change rapidly near the surface or aloft? At what time of day are major changes in the ELR most likely?
2. Except for the shallow zone near the surface, it is rare for the atmosphere to be absolutely unstable. Why is it difficult for a very steep ELR to develop in the middle and upper atmosphere?
3. Localized convection might be more vigorous over the desert in the summer than over a wooded region, but precipitation is more likely in the wooded environment. Why?
4. Orographic uplift can cause cloud or fog to form. How might stability be a factor in determining which develops?
5. Some of the higher peaks in Hawaii have "cloud forests" at certain levels. What type of environmental situations would favor ubiquitous cloud cover?
6. Is it possible for radiation and subsidence inversions to occur simultaneously?
7. Unlike the Sierra Nevada range in the west, the Appalachian Mountains do not exhibit very strong rain shadow effects. Why not?
8. What time of year will unstable conditions be most common over the continental United States and Canada?
9. In many regions, the orographic effect causes precipitation to increase with elevation. Can you think of any reason why this might not be true all the way up to the top of Mt. Everest?

Problems & Exercises

1. On a daily basis, examine the current surface weather maps and the satellite images depicting cloud cover from any of the websites offering those products. Do you notice any relationships between cloud cover and surface pressure distributions? How do clouds tend to appear when associated with cold or warm fronts? Are there any mountainous regions that seem to have a higher incidence of cloud cover? Describe the patterns you observe.
2. Assume that a parcel of air starts out with a temperature of 12 °C and a dew point of 10.4 °C and that the ELR = 0.7 °C/100 m. Determine the following:
   - **a.** Is the air stable, unstable, or conditionally unstable?
   - **b.** At what level will the air become saturated?
   - **c.** Where is the level of free convection?
3. Redo Problem 2, but assume that the ELR = 0.4 °C. Explain why there will be no level of free convection.

Visual Analysis

Examine the photo and answer the following questions.

6.1. How many different general cloud forms can you identify?

6.2. At how many levels do clouds appear in this photo?

6.3. What might you infer about the stability of the air, as indicated by the cloud forms?

▲ **Multiple Cloud Types.** The Sun is behind the highest cloud in this photo.

CHAPTER

7 Precipitation Processes

On March 18, 2013, a major storm system moved across the southeastern United States. The National Weather Service had issued numerous advisories to the public across the region, warning about the potential for severe weather, including strong winds, tornadoes, and large hail. The advisories turned out to be justified. Among the major weather conditions accompanying the storm were multiple instances of large hail. Near Jackson, Mississippi, some of the hail achieved baseball-sized proportions. The largest observed hail stone turned out to be softball size, with a diameter of 10 cm (4.25 in.)—making it the seventh largest ever recovered in Mississippi.

As one might expect, the danger from fast-falling projectiles of this size is great, and in this case there were reports of major damage and at least one serious injury. Cars traveling on local highways were seen sporting new dents and broken windshields. One car dealer had 250 cars

◀ *Large hail did widespread damage across Mississippi during a hailstorm on March 18, 2013, including harm to these cars in Pearl, MS.*

seriously damaged while another reported that every vehicle on its 20-acre lot had been damaged. The hail was severe enough to damage many house roofs, and at least three people were injured by the falling hail.

Like other types of precipitation, hail events need not be damaging, and they do provide the environment with critically important water. But the threat always exists that precipitation can pose a risk to humans and the environment. This chapter explains the processes by which precipitation occurs.

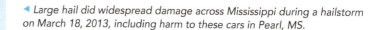

Learning Outcomes

After reading this chapter, you should be able to:

- **7.1** Describe the processes involved in the growth of cloud droplets.
- **7.2** Describe the distribution of precipitation and explain how different types of precipitation form.
- **7.3** Describe how precipitation is measured.
- **7.4** Summarize efforts to induce precipitation through cloud seeding.
- **7.5** Explain the factors that can cause floods.

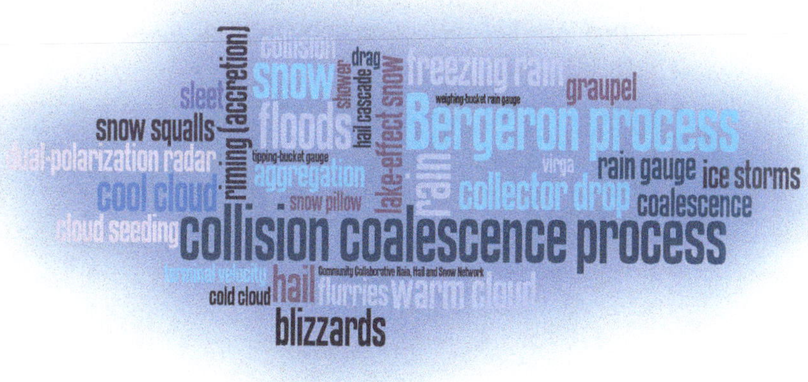

Growth of Cloud Droplets

7.1 Describe the processes involved in the growth of cloud droplets.

Acting alone, gravity would accelerate all objects toward the surface. But gravity is not the only force acting on a falling object; at the same time, the air exerts an opposing resistance or **drag**. As speed increases, so does resistance, until its force equals that of gravity and the acceleration ceases. The object falls, but at a constant speed, its **terminal velocity**. More than anything else, terminal velocity depends on size, with small objects falling much more slowly than large objects. (We examine details of the relationship between size and terminal velocity in *Box 7–1, Physical Principles: Why Cloud Droplets Don't Fall.*) Cloud droplets fall slowly because they are so tiny. Their small size is largely explained by the fact that condensation nuclei are very abundant; thus, cloud water is spread across numerous small droplets rather than being concentrated in fewer large drops. With their small size, cloud droplets initially have extremely low terminal velocities, making it impossible for them to reach the surface.

This effect is apparent in Figure 7–1, which shows terminal velocities for various cloud constituents. The smallest are the condensation nuclei, on which liquid droplets form. (For the sake of simplicity, the figure applies only to clouds consisting of liquid water alone, without ice crystals.) Condensation nuclei are so small that they fall at an imperceptibly slow rate. Larger cloud droplets (but not falling as precipitation) typically range from about 10 μm to about 50 μm in radius (recall that 1 μm is one-millionth of a meter). These have fall speeds ranging from about 1 cm/sec (0.02 mph) to about 25 cm/sec (0.5 mph). By way of contrast, the much larger raindrops shown in the figure fall at 650 cm/sec, about 25 times faster.

Raindrops fall to the surface when they become large enough that gravity overcomes the effect of updrafts. How large is large enough? In terms of radius, raindrops are about 100 times bigger than typical cloud droplets. But in terms of volume or mass of water, raindrops are larger than cloud drops by a factor of a million, rather than just 100. The difference arises because volume for a sphere is proportional to the cube of the radius. If the radius is 100 times larger, the volume is 100 × 100 × 100 (1 million) times larger. Raindrops are not truly spherical, but the principle holds: Precipitation particles are vastly more massive than cloud drops. Although we do not think of clouds yielding massive falling objects, they certainly do, at least from the point of view of a cloud droplet. In the paragraphs that follow, we outline the processes that give rise to these "massive" falling objects.

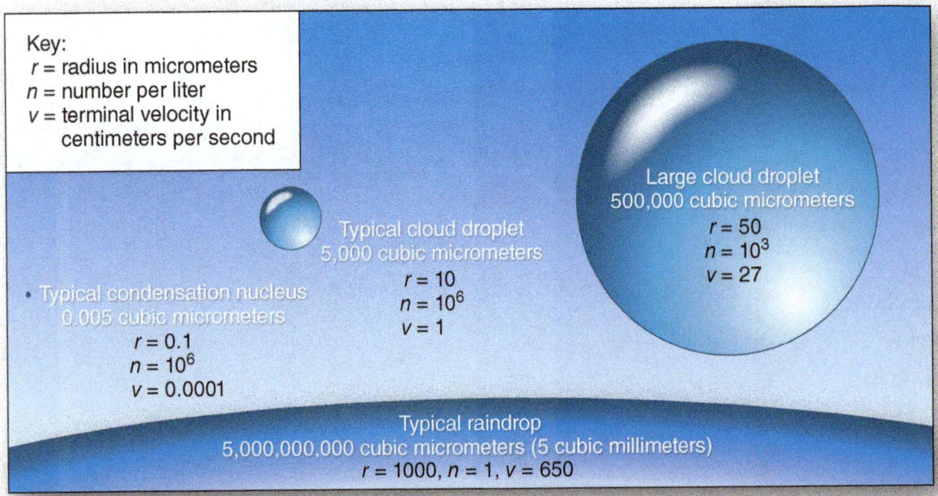

▲ **FIGURE 7–1 Cloud Composition.** This figure shows the average characteristics of cloud constituents.

Growth by Condensation

When cloud droplets begin to form by the adiabatic cooling of ascending air, they do so on condensation nuclei. But within a few dozen meters above the lifting condensation level, all the available condensation nuclei have attracted water, and any further condensation can only occur on existing droplets.

Condensation can lead to rapid growth for very small water droplets, but only until they achieve radii up to about 20 μm—far smaller than necessary to fall as precipitation. Beyond this point, further growth by condensation is minimal. To understand why, recall that relatively little water vapor is available for condensation. With so many droplets competing for a limited amount of water, none can grow very large. It is clear that if growth by condensation were the only process operating, we would experience little, if any, precipitation on Earth. We should therefore think of condensation as only the starting point for rain and snow. Two other processes are responsible for further droplet growth; their relative importance depends on the temperature of clouds.

Growth in Warm Clouds

Most precipitating clouds in the tropics, and some in the middle latitudes, are **warm clouds**, those having temperatures greater than 0°C throughout. In warm clouds, the **collision–coalescence process** causes precipitation. This process depends on the differing fall speeds of different-sized droplets.

Cloud droplets come in different sizes, and therefore attain different terminal velocities. Refer to Figure 7–2 and consider what will happen when the largest droplet (the **collector drop**) falls through a warm cloud. As the collector drop falls, it overtakes some of the smaller droplets in its path because of its greater terminal velocity. This provides the opportunity for collisions and coalescence.

7–1 PHYSICAL PRINCIPLES

Why Cloud Droplets Don't Fall

You are probably familiar with the legendary, late-sixteenth-century experiment of Galileo Galilei, who dropped two objects—a light one and a heavy one—off the Leaning Tower of Pisa. The objects, being subjected to the same gravitational acceleration, hit the ground at nearly the same time. Galileo's demonstration may seem inconsistent with our everyday experience, because an ant would surely take longer than a golf ball to fall from the top of a tall building. It is also at odds with our claim that small droplets fall slowly. The solution must be that a force besides gravity acts on falling objects; that force is called wind resistance, or drag. By examining how these two forces work together, we will gain some insight into why cloud droplets do not fall. To keep the discussion simple, we will assume spherical droplets throughout—using more realistic shapes, however, would not change our conclusions.

Newton's Second Law tells us that if a net force is applied to a mass, it will undergo an acceleration (or change in velocity through time). For a given mass, the acceleration is directly proportional to the net force. In equation form, the law is given as

$$\text{Net force} = \text{mass} \times \text{acceleration} \quad \text{Newton's Second Law}$$

Notice that Newton's Second Law says that we must consider the net force, the result of all the forces acting on the object. As far as a falling droplet is concerned, there is the downward gravitational force, which is opposed by the force of wind resistance (drag). A droplet suddenly released in the atmosphere falls at increasing speed, but not indefinitely. Eventually the force of drag (F_d) balances the force of gravity (F_g), resulting in no net force:

$$\text{Net force} = F_g - F_d = 0$$

With no net force, there is no acceleration, and the droplet falls at its terminal velocity. How fast does it fall? To answer that, we need to know something about the magnitude of the two forces.

Force of Gravity

The force of gravity is equal to mass times the acceleration of gravity, g. Whenever we step on a scale, we measure this force. For a liquid droplet presumed to be spherical, mass is the density of water, ρ, times the droplet volume, $4/3\pi r^3$, where r is the droplet radius. We therefore obtain F_g as follows:

$$F_g = \rho \frac{4}{3}\pi r^3 g \quad \text{Gravitational Force}$$

Force of Drag

Drag between the droplet and surrounding air depends on the rate of fall and on the size of the droplet. Just like an automobile on a highway, a faster-moving droplet experiences greater resistance as it moves through the air. In fact, to a good approximation, the drag force increases with the square of wind speed, v^2. So how does size influence drag?

It can be shown that the drag can be expressed as

$$F_d = 0.5 C_D \rho_a v^2 \pi r^2 \quad \text{Drag}$$

where C_D is a constant referred to as the drag coefficient, and ρ_a is the density of air (about one-thousandth the density of water). The value of C_D is not important here; what matters is that F_d is proportional to the square of both the fall rate (v^2) and the radius (r^2).

Terminal Velocity

For a droplet falling at terminal velocity, we have said that gravity and drag are equal. If we use v_t for terminal velocity, we get

$$F_g = F_d$$

$$g\rho \frac{4}{3}\pi r^3 = 0.5 C_D \rho_a v_t^2 \pi r^2$$

To find the terminal velocity, we rearrange and first solve for v_t^2:

$$v_t^2 = \left(g\rho \frac{4}{3}\pi r^3\right) / \left(0.5 C_D \rho_a \pi r^2\right)$$

By consolidating the numerical values and constants in the above equation into a single constant, k_1, we get

$$v_t^2 = k_1 r$$

or

$$v_t = k_2 \sqrt{r} \quad \text{Terminal Velocity}$$

where $k_2 = \sqrt{k_1}$. From this equation, we see that as droplet radius increases, so does terminal velocity. Large droplets fall faster than small droplets. What happens physically is that both F_g and F_d increase with radius, but the gravitational force increases more than drag, and a higher fall rate is therefore required to cancel F_g. Notice that as far as the droplet is concerned, falling through a still atmosphere at v_t is the same as remaining stationary in an updraft of speed v_t. Thus, the equation says that a strong updraft is needed to hold a large droplet aloft, whereas a small droplet is easily suspended.

Going back to the Leaning Tower of Pisa situation described at the beginning of this box, we can now understand why Galileo's objects fell at nearly the same speed. With large and therefore heavy objects, the gravitational forces far exceeded the drag forces throughout their short fall. With negligible drag, gravity accelerated both at nearly the same rate. If he had used objects of greatly different size, or if the objects had fallen far enough to reach their terminal velocities, differences in v_t would have emerged. The old, familiar story would be about wind resistance, and books like this would have no need for a feature on the topic.

7–1–1 How do the forces of gravity and drag interact to give objects their terminal velocities?

7–1–2 What is the formula for estimating the terminal velocities of falling raindrops? What does that formula tell us about how dramatically a change in radius affects the terminal velocity?

COLLISION As it falls, a collector drop collides with only some of the droplets in its path. The likelihood of a **collision** depends on both the absolute size of the collector and its size relative to the droplets below. If the collector drop is much larger than those below, the percentage of collisions (the *collision efficiency*) will be low. Figure 7–3 illustrates why. As the collector drop falls, it compresses the air in its path. The compressed air creates a small gust of wind that pushes the smaller droplets out of the way. The small gust of wind cannot push aside larger droplets, however, and the collector is able to collide

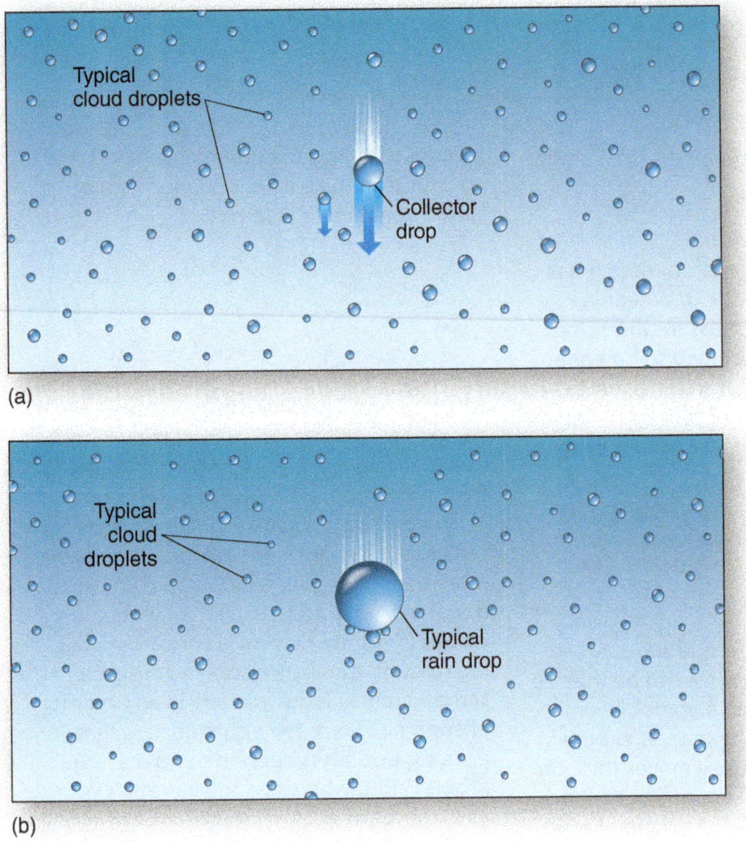

▲ **FIGURE 7–2 Collision–Coalescence.** (a) Because of their greater mass, collector drops have greater terminal velocities (indicated by the length of the downward-pointing arrows) than do the smaller droplets in their path. (b) Collector drops overtake and collide with the smaller ones.

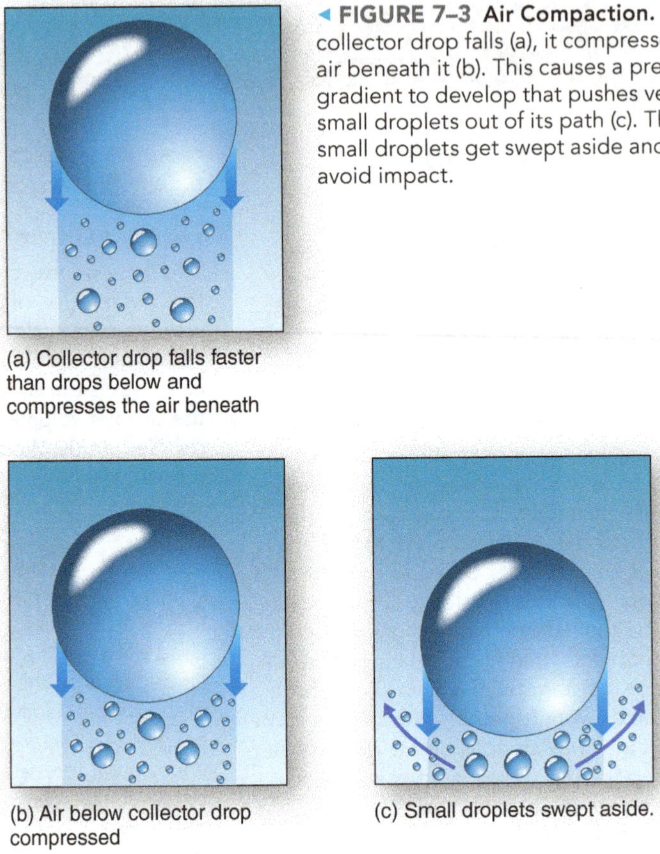

(a) Collector drop falls faster than drops below and compresses the air beneath
(b) Air below collector drop compressed
(c) Small droplets swept aside.

◄ **FIGURE 7–3 Air Compaction.** As a collector drop falls (a), it compresses the air beneath it (b). This causes a pressure gradient to develop that pushes very small droplets out of its path (c). The small droplets get swept aside and avoid impact.

with them. As a result, the collision efficiency is lower for droplets that are very much smaller than the collector drops.

You have probably witnessed a similar phenomenon on a larger scale while driving down a country road in summer, with your windshield turning relatively large flying insects into "bug juice." Too heavy to get swept aside by the compressed air immediately ahead of the windshield, the bugs follow their own paths until the fateful moment of impact. Smaller bugs, in contrast, get blown out of harm's way.

Collision efficiencies are low for droplets nearly equal in size to the collector drop because their terminal velocities are so close to the collector's velocity that it is difficult for the collector to catch up to and collide with them. Continuing with the car analogy, collisions between vehicles are unlikely as long as all move at the same speed and direction.

Under certain situations, collision efficiencies can actually exceed 100 percent, and the collector can collide with more droplets than are in its path. A falling drop creates turbulence that can entrain small droplets outside its path and carry them back toward the top of the collector, where collision occurs.

COALESCENCE When a collector drop and a smaller drop collide, they can either combine to form a single, larger droplet or bounce apart. Most often the colliding droplets stick together. This process is called **coalescence**, and the percentage of colliding droplets that join together is the *coalescence efficiency*. Because most collisions result in coalescence, coalescence efficiencies are often assumed to be near 100 percent.

Collision and coalescence together form the primary mechanism for precipitation in the tropics, where warm clouds predominate. In the middle latitudes, most precipitating clouds have freezing temperatures, at least in their upper portions. This favors the growth of precipitation by another mechanism involving the coexistence of ice crystals and supercooled water droplets, the Bergeron process (also known as the Bergeron-Findeisen or ice crystal process) described in the next section.

Video
The Importance of Wind Resistance
http://goo.gl/WcRqp7

CHECKPOINT

7.1 What are warm clouds and where are they most likely to be found?

7.2 What is the role of the collector drop in the formation of precipitation in a warm cloud?

Growth in Cool and Cold Clouds

Unlike their counterparts in the tropics, at least a portion of most midlatitude clouds have temperatures below the melting point of ice. Some, such as the one in Figure 7–4a, have temperatures below 0 °C throughout and consist entirely of ice crystals, supercooled droplets, or a mixture of the two. These are referred to as **cold clouds**.

Cool clouds (Figure 7–4b), on the other hand, have temperatures above 0 °C in the lower reaches and subfreezing

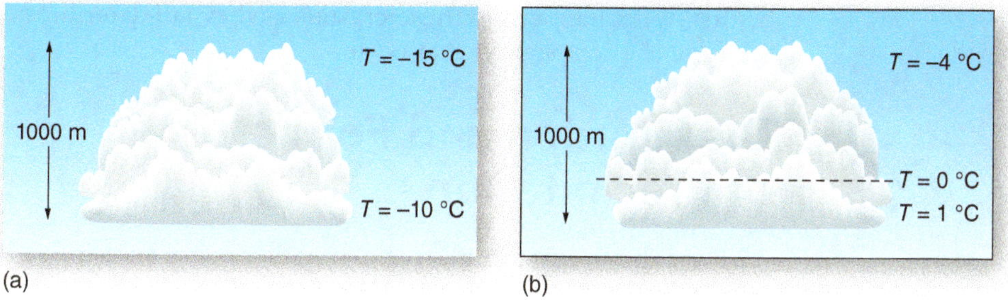

◄ **FIGURE 7–4 Cold and Cool Clouds.** (a) Clouds with temperatures below 0 °C from their base to their top are called cold clouds. (b) Cool clouds have temperatures above 0 °C in their lower portions with subfreezing temperatures above.

conditions above. As we discussed in Chapter 5, saturation at temperatures between about −4 °C (25 °F) and −40 °C (−40 °F) can lead to the formation of ice crystals, if ice nuclei are present, or to the formation of supercooled liquid droplets, if ice nuclei are absent. Thus, a well-developed cumulus cloud might be composed entirely of water droplets in its lower portion, a combination of supercooled droplets and ice crystals in its middle section, and exclusively ice crystals in its upper reaches (Figure 7–5). The processes described in this section operate in cold and cool clouds having a mixture of ice and liquid water.

As we will now see, the coexistence of ice and supercooled water droplets is essential to the development of most precipitation outside the tropics. The process by which droplets and crystals in midlatitude clouds grow to precipitation size was first described by one of the preeminent figures of modern meteorology, Tor Bergeron. This process is therefore often referred to as the Bergeron process.

BERGERON PROCESS The principle underlying the **Bergeron process** is that the saturation vapor pressure over ice (the amount of water vapor needed to keep it in equilibrium) is less than that over supercooled water at the same temperature. This is because molecules in an ice crystal bond to each other more tightly than molecules of liquid water. You should recall that saturation exists when the vapor pressure of the air is at the point where evaporation from a water droplet would be exactly offset by condensation back onto it, or if sublimation from an ice crystal would be offset by deposition. Within a certain range of temperatures below 0 °C, both ice crystals and supercooled droplets can coexist in a cloud. Ice crystals, however, do not sublimate ice to vapor as rapidly as water droplets evaporate liquid to vapor. Thus, ice crystals do not require as high a surrounding vapor pressure as do water droplets to remain in equilibrium, and they are said to have a lower saturation vapor pressure. As a result, if there is just enough water vapor in the air to keep a supercooled droplet from evaporating away, then there is more than enough water vapor to maintain an ice crystal. Let us see how that leads to precipitation.

▲ **FIGURE 7–5 Cool Cloud Composition.** The lower portion of this cumulonimbus cloud consists entirely of liquid droplets, the middle a mixture of ice and liquid, and the upper portion entirely of ice. Note the less sharply defined margins of the glaciated portion composed of ice.

Refer to Figure 7–6 and consider the situation in which ice crystals and supercooled droplets coexist, and the vapor pressure is equal to that needed to keep the droplets in equilibrium. In (a) the rate of condensation onto the liquid droplet equals the rate of evaporation. But while vapor pressure in the cloud equals the saturation vapor pressure for the droplet, it exceeds that for the ice. This causes some of the water vapor in the air to be deposited directly on the ice. The vapor content of the air then falls, which in turn causes the liquid droplet to evaporate as it gives up water to restore equilibrium (b).

The process does not end there, because evaporation from the droplet increases the water vapor content of the air, which causes further deposition onto the ice crystals (c). This leads to a continuous transfer in which the liquid droplets surrender water vapor, which is subsequently deposited onto the ice crystals. In other words, the ice crystals continually grow at the expense of the supercooled droplets. Although Figure 7–6 suggests this process involves distinct steps, evaporation and deposition actually occur simultaneously.

The growth of ice crystals by the deposition of water vapor initiates precipitation. As the ice crystals grow, their increasing mass enables them to fall through the cloud and collide with droplets and other ice crystals. The collisions cause two other important processes to occur that greatly accelerate the growth rate of the ice crystals: riming and aggregation.

RIMING AND AGGREGATION We have seen that the formation of ice crystals in the atmosphere usually requires the presence of ice nuclei, or particles that initiate freezing. It so happens that ice itself is a very effective ice nucleus. Thus, when ice crystals fall through a cloud and collide with supercooled droplets, the liquid water freezes onto them. This process, called **riming** (or **accretion**), causes rapid growth of

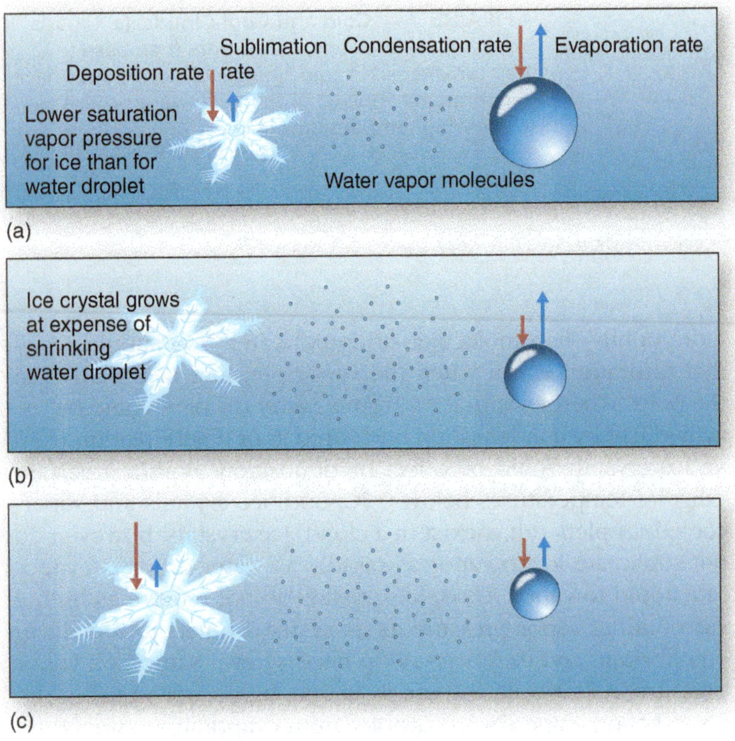

▲ **FIGURE 7–6 Bergeron Process.** Ice crystals in cool clouds grow at the expense of supercooled water droplets. (a) If the vapor pressure is such that water droplets have an equilibrium between evaporation and condensation, then there will be an excess of deposition over sublimation for ice crystals. (b) Ice crystals grow by the deposition of water vapor, thereby removing water vapor and cause the water droplets to become smaller. (c) Eventually the ice crystals become large enough to fall from the cloud.

the ice crystals, which further increases their fall speeds and promotes further riming.

Another process in the development of precipitation is **aggregation**, the joining of two ice crystals to form a single, larger one. Aggregation occurs most easily when the ice crystals have a thin coating of liquid water to make them more "adhesive." Water is more likely to be present when the cloud temperature is not much below 0 °C, so adhesion is more common at the warmer end of cold clouds. (Perhaps you have noticed that very large snowflakes are more common during warm, early season snows, as opposed to those that come in the dead of winter.)

The combination of riming and aggregation allows ice crystals to grow much faster than by the deposition of water vapor to ice alone. In fact, growth rates from the three processes combined allow the formation of precipitation-sized crystals within about half an hour from the initial formation of the ice. When the ice crystals begin to fall, precipitation begins. What happens to these crystals as they fall determines the type of precipitation that occurs.

CHECKPOINT

7.3 What is the difference between a cool cloud and a cold cloud?

7.4 How does the fact that the saturation vapor pressure over ice is less than the saturation vapor pressure over supercooled water lead to precipitation?

Distribution and Forms of Precipitation

7.2 Describe the distribution of precipitation and explain how different types of precipitation form.

The processes described above are ultimately responsible for all the various kinds of precipitation. Figure 7–7 shows the distribution of mean annual precipitation based on worldwide weather records.

In the tropics, precipitation occurs primarily by the collision–coalescence process, and it can therefore occur only as rain. In the middle latitudes, where ice crystal processes dominate, precipitation occurs as a solid or a liquid, depending on the temperature profile of the air through which it falls. If precipitation reaches the surface without ever having melted, we recognize it as snow. If it melts on the way down, it might reach the surface as rain. But raindrops sometimes freeze again before, or immediately after, reaching the surface, and then a different type of precipitation results. We now discuss the various types of precipitation.

Snow

Precipitation as **snow** is initiated by the Bergeron process and the growth of crystals by riming and aggregation. Snow occurs if the falling crystals do not melt prior to reaching the ground. Therefore, cloud temperatures must be less than −4 °C (25 °F) and temperatures from the surface to the cloud base not much more than 0 °C. (Snow might not completely melt if it falls through a shallow layer of air not much warmer than 0 °C.) Figure 7–8 illustrates a typical temperature profile required for snowfall.

Ice crystals in clouds can have a wide variety of shapes, including six-sided plates, columns, solid or hollow needles, and complex dendrites with numerous long, narrow extensions (Figure 7–9). The structure depends on the temperature and moisture conditions that exist when the crystal is formed.

If all of a crystal's growth occurs under similar conditions, its structure can be quite simple. If, on the other hand, the temperature and moisture conditions change during growth, a complex mixture of plate, needle, and dendrite can develop. Consider, for example, a crystal that originates in the cold, upper reaches of a cloud and gradually falls through a warmer environment. Because each combination of moisture and temperature tends to favor a different type of structure, the crystal can have a particular form at its nucleus, with other forms superimposed.

Snowflakes exist in a wide range of shapes and sizes. They can be as small as about 50 μm or as large as 5 mm (0.2 in.). Where riming is the dominant growth process (which is the case in relatively warm clouds), the crystals tend to form a dense "wet" snowpack that is ideal for snowball fights, but is no friend to snowblowers. In contrast, very cold snow typically forms small snowflakes that accumulate on the ground

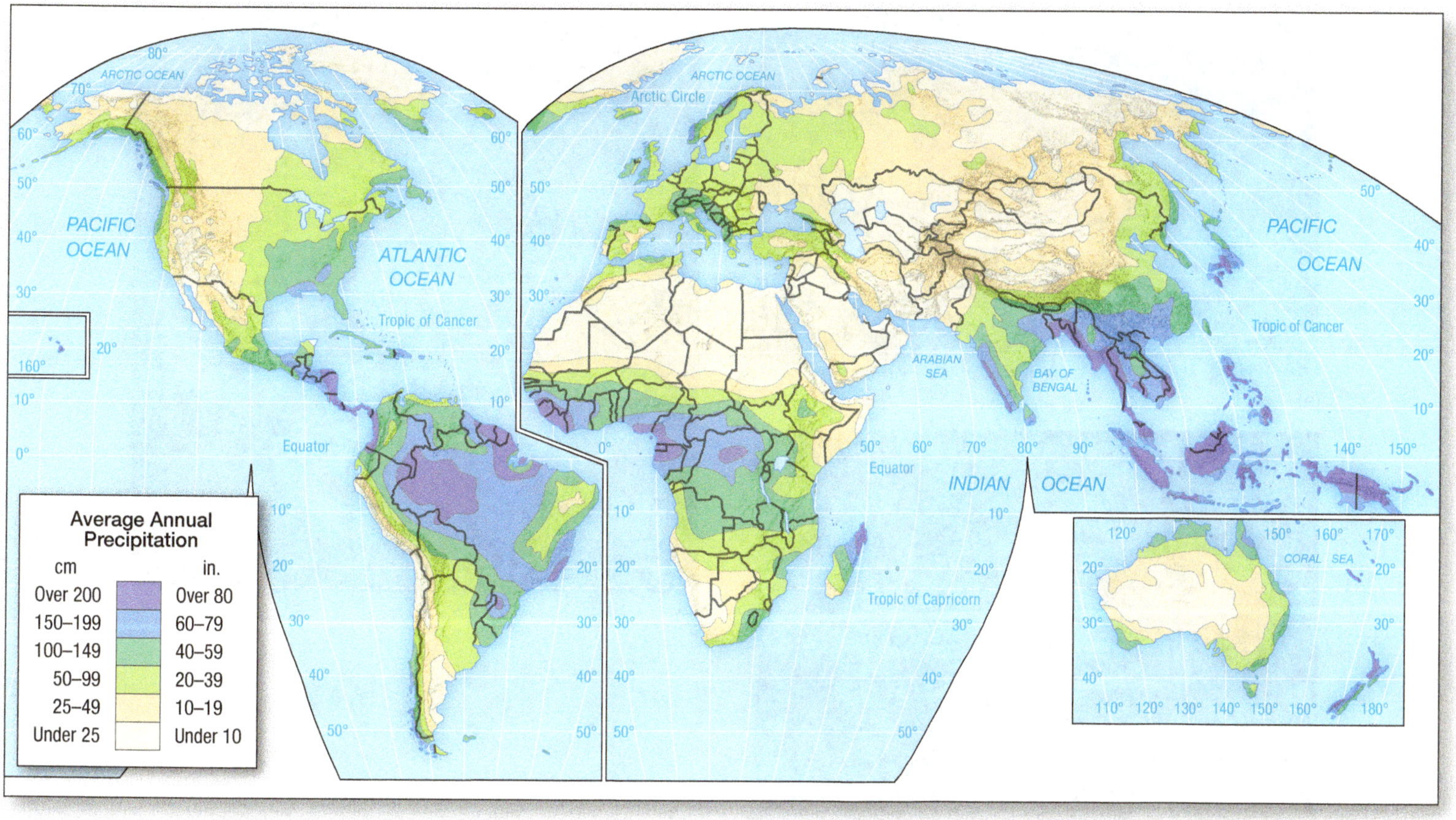

▲ **FIGURE 7–7 Average Annual World Precipitation.**

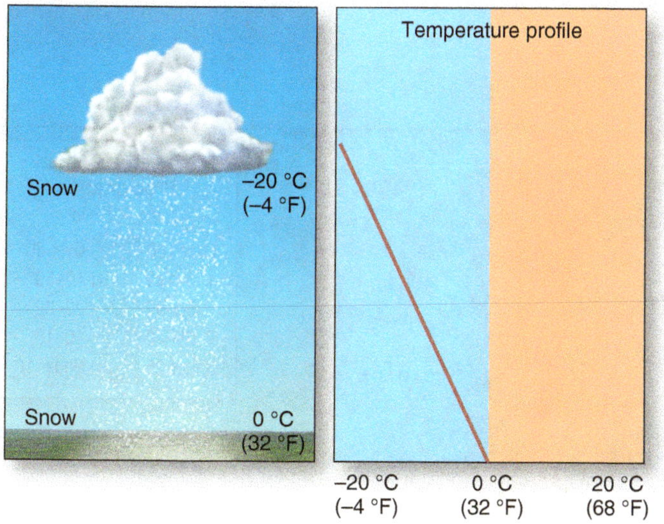

▲ **FIGURE 7–8 Snow.** This form of precipitation is initiated by the Bergeron process and reaches the ground only if the crystals falling to the ground never encounter temperatures above 0 °C.

with a lower density. Because of their low temperature, these crystals have less adhesion and are difficult to pack. Skiers know this type of snow as *powder*. There is a widely held misconception about snow. Some people believe it can be too cold to snow, but this is not the case. It can be too cold to snow *a lot*, but it is never too cold to snow *at all*. Because mass is conserved, any ice crystals that form must do so at the expense of the water vapor content of the air. At very low

MM **MapMaster** World Physical Environment
Average Annual Precipitation

temperatures, only a small amount of water vapor can exist in the air. And without an ample supply of water vapor, cooling of the air can cause the deposition of only a limited supply of ice. It is still possible for some snow to occur, no matter how low the temperature.

The intensity and duration of snowfall is subject to considerable variability. Snowfalls can occur over extensive time periods, but can also occur only lightly and intermittently. Snowfalls of only short duration are referred to as **flurries**, and generally produce only trace accumulations on the ground. At the other end of the intensity spectrum are **snow squalls**, brief but moderate to heavy snow events accompanied by strong winds. Continuous episodes of heavy snow, reduced visibility, and winds exceeding 56 km/hr (35 mph) are called **blizzards**.

NORTH AMERICAN DISTRIBUTION Figure 7–10a maps the distribution of mean annual snowfall across Canada and the United States. In the western portion of North America, the distribution of snowfall is governed largely by the presence of north–south mountain ranges (the Coast Ranges, the Sierra Nevada, the Cascades, and the Rockies) that provide orographic uplift and enhance the precipitation from passing storm systems. At high elevations, these ranges have winter temperatures low enough that most precipitation occurs as snow. Over the eastern two-thirds of the continent, there is an increase in

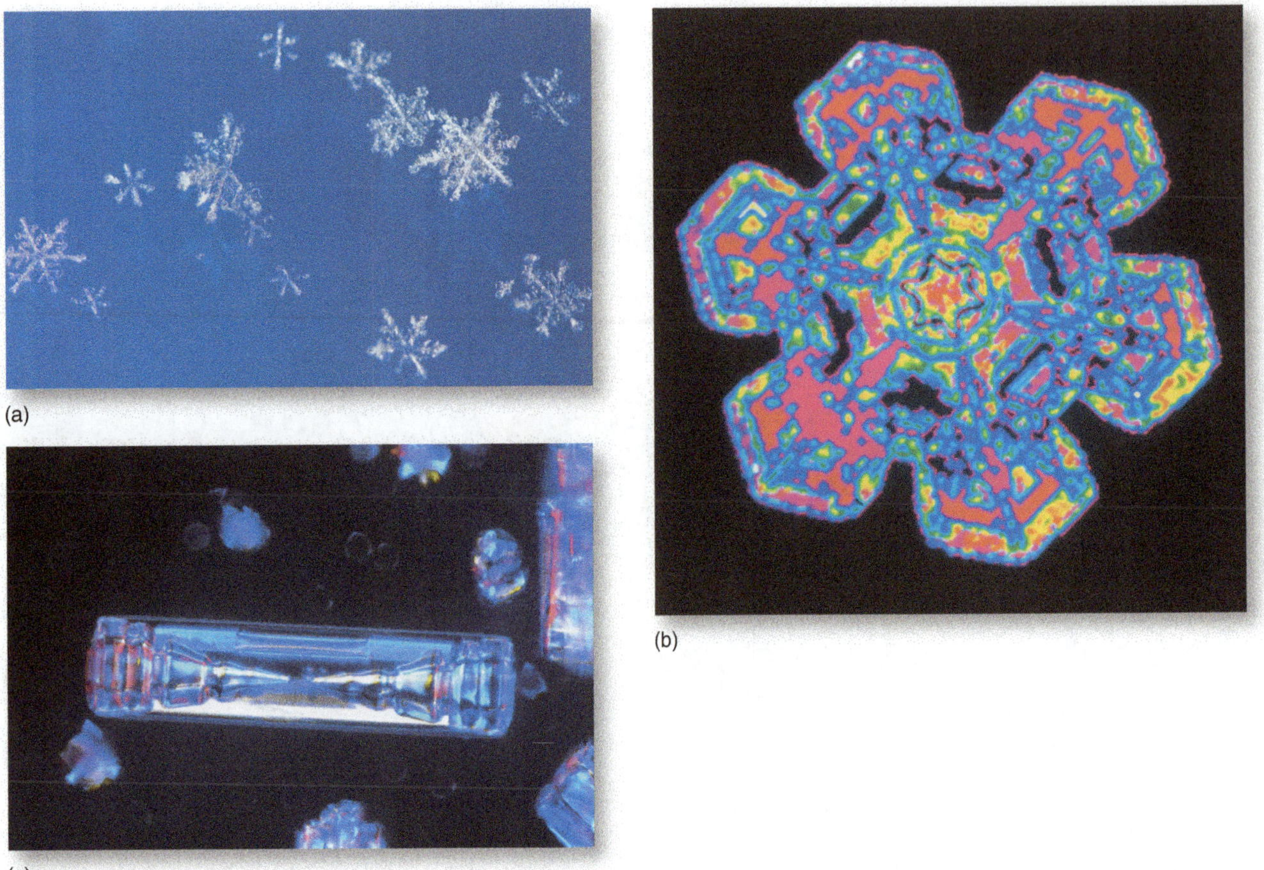

(a) (b) (c)

▲ **FIGURE 7–9 Ice Crystal Shapes.** Many forms of ice crystals can occur, including (a) dendrites (b) plates and (c) columns. Each is favored under certain conditions of moisture content and temperature.

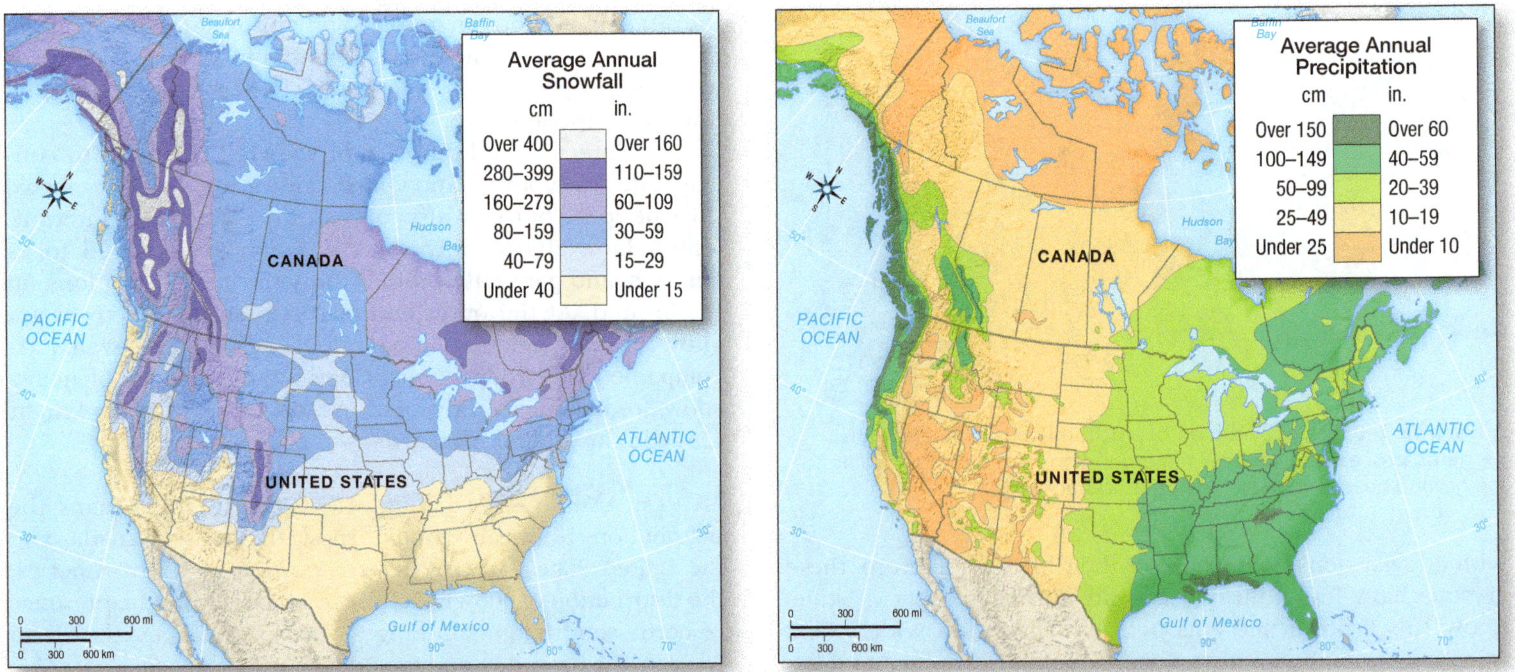

(a) (b)

▲ **FIGURE 7–10 Precipitation.** (a) Average annual snowfall in Canada and the United States and (b) average annual precipitation.

MM® **MapMaster** North America Physical Environment Average Annual Snowfall and Precipitation mean snowfall with latitude, mostly because the lower temperatures at higher latitudes favor snow rather than rain.

The distribution of annual snow is in marked contrast to the distribution of annual precipitation—rain plus the water equivalent of snow (Figure 7–10b). Total annual precipitation over the eastern two-thirds of North America decreases with latitude rather than increases, largely due to the fact that there is less water vapor in the air with increasing distance from the Gulf of Mexico. Furthermore, lower temperatures typically found at higher latitudes reduce the amount of water vapor that can exist in the air. Note, too, the decrease in precipitation westward across the Great Plains, revealing a rain shadow in the lee of the Rockies.

LAKE-EFFECT AND LAKE-ENHANCED SNOW One feature of the snow distribution not shown in Figure 7–10 is the strong creation or local enhancement of snowfall that occurs downwind of the Great Lakes and other large bodies of water, referred to as **lake-effect** or **lake-enhanced snow** (Figure 7–11). This is most prevalent during the late fall and early winter, when lake temperatures are still moderately high but cold air can pass over from the north. As shown in Figure 7–12, lake-effect snow occurs when the lake warms and evaporates moisture into the lower atmosphere. As the lower atmosphere warms, it can become unstable as the temperature lapse rate increases. Thus, air that was originally dry and stable becomes moist and statically unstable. When the air passes over land, the effects of topography, vegetation, and other features of the land surface slow the wind. The decrease in wind speed causes convergence, a mechanism for uplift and adiabatic cooling discussed in Chapter 6. Thus, the passage of cold air over the lakes and subsequent landfall provides the three mechanisms favorable for precipitation: unstable air, sufficient water vapor, and a mechanism for uplift. In the case of lake-enhanced snow, existing snowstorms yield greater accumulations because of moisture added to the storm by the evaporation of warm lake water.

This lake effect often produces snow showers restricted to a strip of land that can be anywhere from 1 to 80 km (0.6 to 50 mi) long and can extend more than 100 km (60 mi) inland. (It can also increase the amount of snowfall from storm systems passing over the lakes.) Lake-effect snow is most common along the eastern and southern shores of the Great Lakes, as shown in Figure 7–13. Figure 7–14 shows how intense the lake effect can be for a particular event, plotting the distribution of snow downwind of Lakes Erie and Ontario on January 27, 2014. Notice how large amounts of snow fell over a small area of northern New York, with relatively little accumulation elsewhere.

▲ **FIGURE 7–11 Heavy Lake-Effect Snow.** Buffalo, New York.

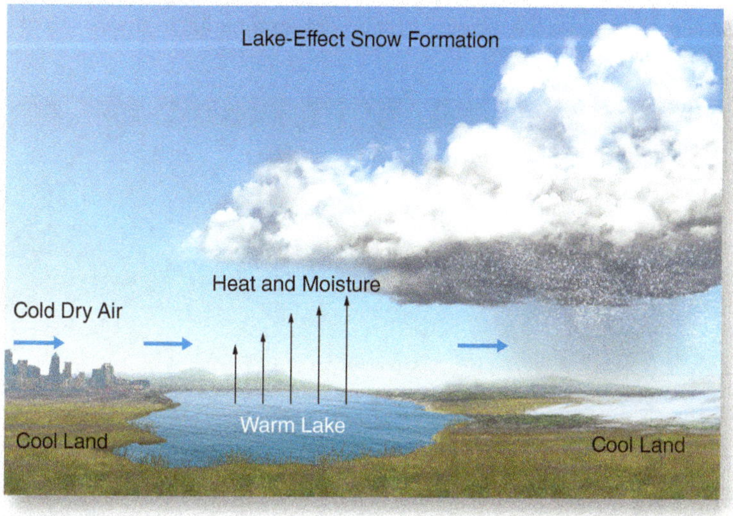

▲ **FIGURE 7–12 Lake-Effect Snow Formation.** As cold air moves over a warm lake, heat and moisture are brought into its base. The moist, unstable air mass is then subject to surface convergence as it slows down upon reaching the downwind shoreline, sometimes resulting in heavy snowfalls.

While lake-effect snow is most likely to occur on the eastern and southern shores of the Great Lakes, the passage of low-pressure systems (Chapter 4) will create a counterclockwise flow that can favor lake-enhanced snow over the north shores. Toronto, Ontario, experienced such an event in January 2014 in which blinding snows led to the rescue of more than 400 people stranded on highways in their vehicles.

Video
Lake Effect Snow
http://goo.gl/3hPLBF

DID YOU KNOW?

The greatest recorded seasonal snowfall occurred at Mt. Baker Lodge, Washington, at 1500 m elevation (5000 ft) over the winter of 1998–99. The total of 2736 cm (1140 in., or 90 ft!) exceeded the previous record observed at Mt. Rainier, Washington, during the winter of 1971–72 (2693 cm, or 1122 in.).

CHECKPOINT

7.5 What is lake-effect snow and where does it typically occur in the United States?

7.6 How does the movement of cold, dry air over a warm lake produce lake-effect snow?

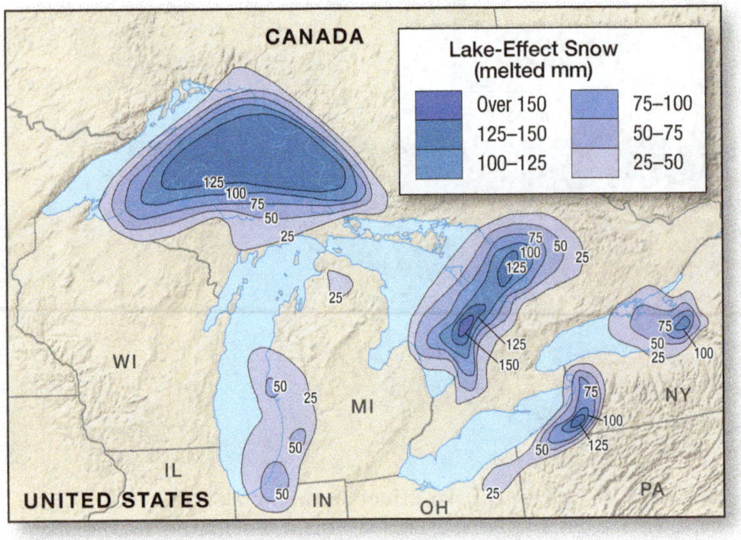

▲ **FIGURE 7–13 Effect of Great Lakes.** The map shows the average increase in snowfall due to lake-effect or lake-enhanced snowfall in millimeters. Note the increase in snow cover immediately downwind of the lakes.

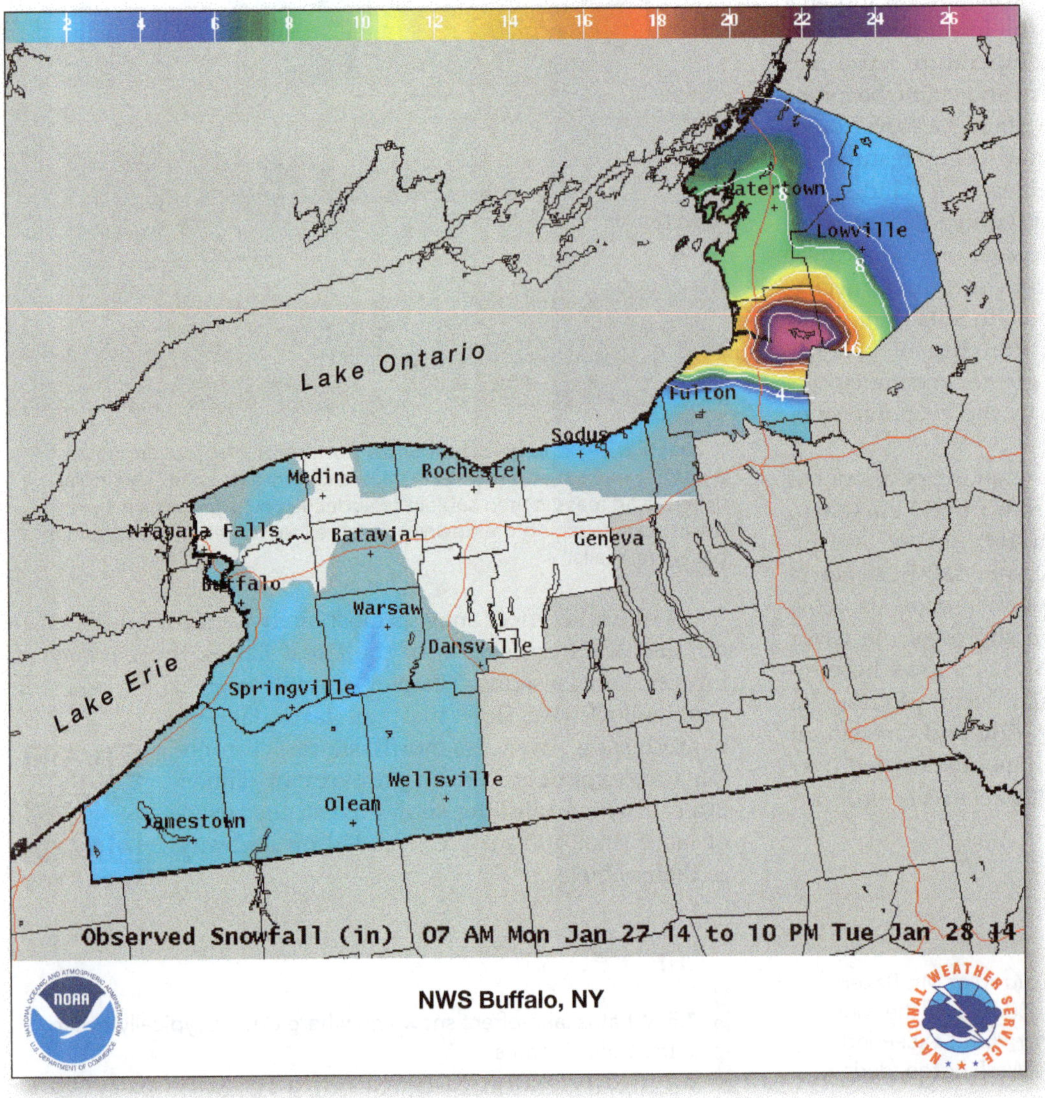

▲ **FIGURE 7–14 January 27, 2014.** Note the localized enhancement of snowfall in the north.

Rain

As we have already seen, most precipitation in the tropics comes from warm clouds whose temperatures are somewhat above the melting point of ice. Furthermore, the air temperature below the clouds is well above freezing, so the rain does not freeze after leaving the base of the cloud. Thus, virtually all precipitation in this part of the world occurs as **rain**, except in high mountains such as Kilimanjaro in Tanzania. Figure 7–15a illustrates a typical temperature profile that could occur with rain of this type. In the middle latitudes, precipitation is usually initiated by the Bergeron process, so most rainfall results from the melting of falling snow (Figure 7–15b).

RAIN SHOWERS Convection can lead to the development of cumuliform clouds and precipitation within a few minutes. The episodic precipitation from these rapidly developing clouds is called **showers**. Showers can occur as either rain or snow, but because convection is usually most vigorous in the warm season, showers are more likely to occur as rain.

During a steady rain, droplets occur in a wide variety of sizes. In a shower, the first droplets all tend to be large and widely spaced, but within a short period of time the large droplets give way to a greater number of smaller ones. What happens is quite simple: Large and small droplets fall from the base of the cloud together, but the larger raindrops have greater terminal velocities and reach the surface while the smaller ones are still falling through the air.

Another factor favors the occurrence of large drops at the beginning of a rain shower. Because they take longer to fall through the unsaturated air, small raindrops are more likely to evaporate before reaching the surface. (Evaporation does decrease after a few minutes, however, when the first drops have sufficiently increased the moisture content of the air.)

VIRGA Sometimes droplets that fall from a cloud fall through a layer of warm dry air and evaporate before ever hitting the ground. When this situation, called **virga**, occurs, a shaft of rain is visible beneath the cloud but it gradually thins out and disappears (Figure 7–16).

RAINDROP SHAPE One prevailing myth about weather is that raindrops are teardrop shaped. In reality, raindrops are initially spherical

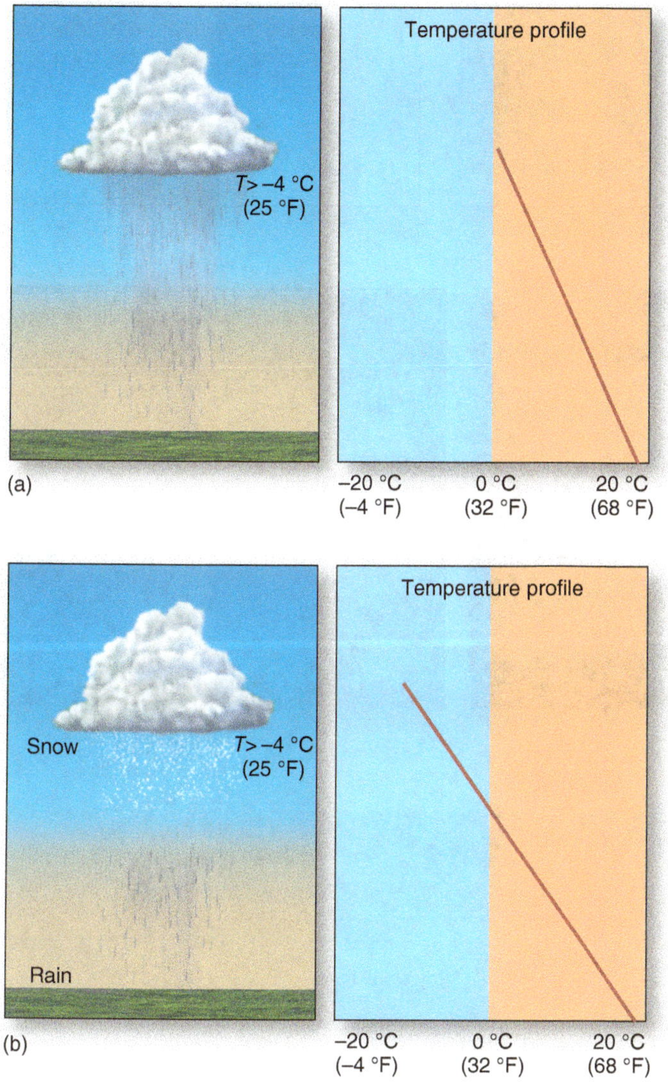

▲ **FIGURE 7–15 Rain.** (a) This type of precipitation can occur when the collision–coalescence process occurs, or (b) when falling ice crystals fall from a cloud and melt on the way to the ground.

DID YOU KNOW?

The average precipitation for the entire world is 97 cm (38.8 in.) per year, and actual global precipitation occurring in any given year seldom departs much from this average. In fact, every year during the period 1900 to 2005 had global precipitation within ±5 cm (2 in.) of the average amount. So drier than normal years in some parts of the world are generally offset by wet conditions elsewhere.

(Figure 7–17a). As they grow by collision and coalescence, their velocities increase, and the greater wind resistance flattens them along the bottom to give them a parachute or mushroom shape (b). As the bottom of the drops flattens out, the greater surface area increases the wind resistance. At this point, the drop begins to deform. The flat bottom becomes concave, somewhat in the shape of a parachute with the bottom surrounded by a relatively thick, doughnut-shaped rim (c). Eventually, wind resistance exceeds the surface tension that

▲ **FIGURE 7–16 Virga.** Rain that completely evaporates prior to hitting the surface is called virga.

holds the droplets together, and the ring at the base of the drop and the bubble-shaped top break into multiple, smaller droplets (d). The resulting small droplets then begin to grow again by collision and coalescence.

The breakup of falling drops explains why collision and coalescence do not produce enormously large droplets. If the drops were to grow continually on their way down, they could conceivably attain the size of basketballs! Under special conditions, droplets can have diameters of up to 5 mm (0.2 in.) or more, but they are seldom larger.

Graupel and Hail

When an ice crystal takes on additional mass by riming, its original six-sided structure becomes obscured and its sharp edges are smoothed out. The new ice may contain very small air bubbles that give it a spongy texture and milky-white appearance. This type of modified ice crystal is called **graupel** (pronounced "GRAU-pull"). Graupel pellets attain diameters up to 5 mm or so, giving them terminal velocities of about 2.5 m/sec (5 mph). Graupel pellets can fall to the ground as precipitation, but under other circumstances they can remain in the cloud and provide the nuclei upon which hailstones form.

Hail consists of multiple layers of ice that are usually no more than a few millimeters in thickness, but under extreme circumstances can achieve sizes comparable to softballs (Figure 7–18). Hailstones are often relatively concentric, though some have irregular shapes. No other type of precipitation is capable of producing drops or particles nearly as large as hail, so it should not be surprising that large amounts of water must be available in the cloud in which hailstones grow and that very strong updrafts are required to allow precipitation this large to remain airborne. Hail therefore forms from cumulonimbus clouds.

Hail development begins with the formation of a small particle called an *embryo*, usually consisting of graupel, in a region of cloud where supercooled droplets coexist with ice crystals (Figure 7–19). The cumulonimbus clouds that produce hail do not have uniformly strong updrafts across their entire

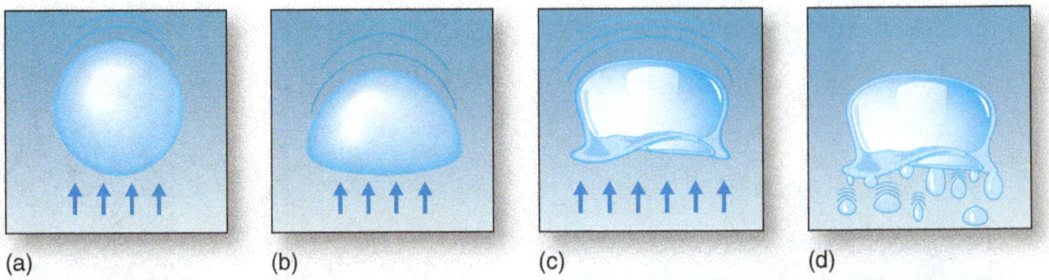

▶ **FIGURE 7–17 Raindrop Shape.** Contrary to popular belief, raindrops are not teardrop shaped. They are initially spherical (a) but flatten out on the bottom as they fall (b). Wind resistance deforms the bottom of the drop, stretching it inward to resemble a parachute with a doughnut-shaped ring at the base (c). Eventually, the droplet breaks apart (d).

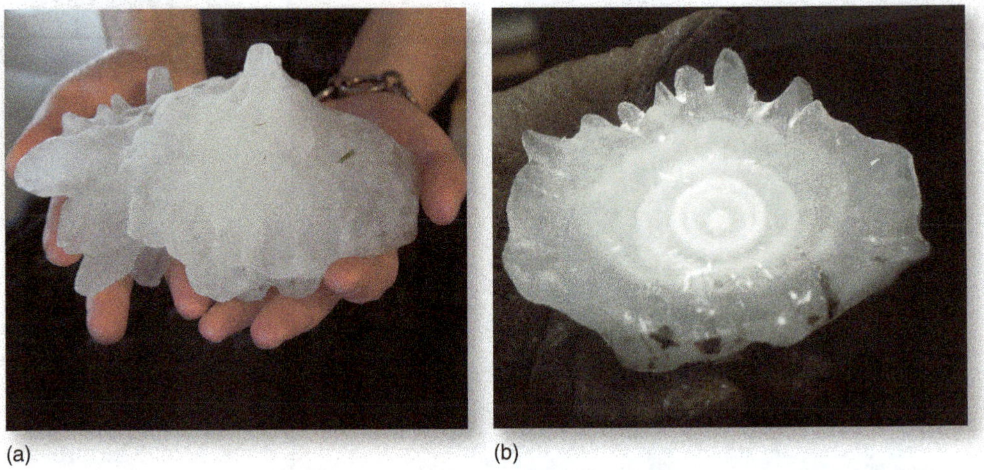

▶ **FIGURE 7–18 Hail Size.** (a) Though usually no larger than pea sized, hail can be remarkably large. (b) Its growth accrues by the repeated addition of new layers of ice.

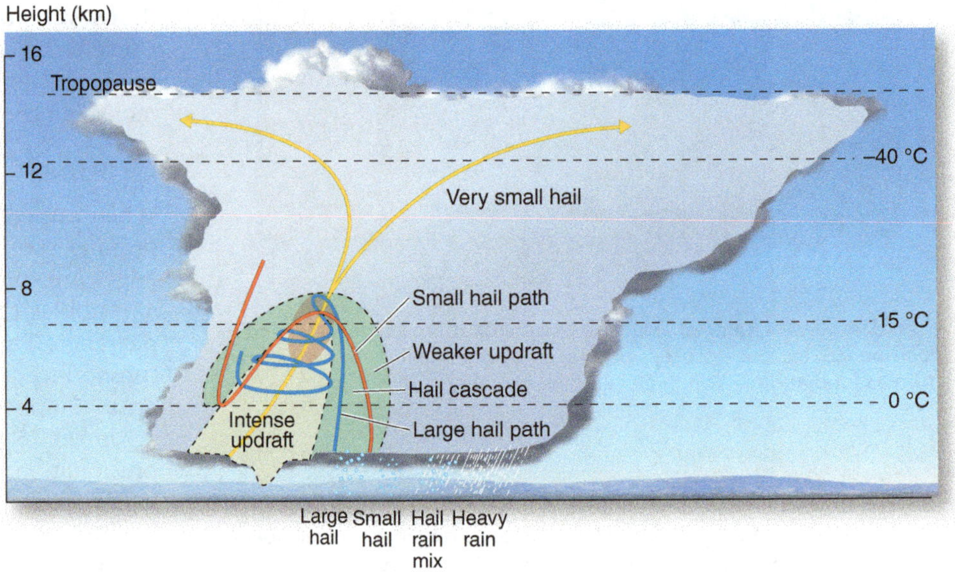

▶ **FIGURE 7–19 Hail Formation.** Embryos near zones of strong updrafts may be blown into the updraft, rapidly lifted upward, and ejected as small hail. Hail that forms as it rotates around these updrafts can accumulate more ice to become large and fall through a region called a hail cascade.

width; they have localized areas of extremely violent vertical motions surrounded by zones of relatively weak updrafts. Hail growth typically begins just outside these intense updrafts.

Some developing hailstones get blown from the edge of the strong updrafts into the core, where they are rapidly taken to the upper reaches of the cloud and ejected. When this happens the stones do not have time to accumulate large amounts of ice, so they fall from the cloud as relatively small hail.

Large hail can develop if the growing particles rise slowly around the more intense core of the updraft. This provides time for the hail to come into contact with supercooled liquid droplets that freeze onto the existing ice. This process can take place in one of two ways: the *dry growth* or *wet growth regimes*.

The dry growth regime occurs at temperatures well below 0 °C (32 °F), wherein supercooled water droplets collide with and freeze onto ice crystals. As the water freezes it releases latent heat that warms the crystal, but not sufficiently to bring its temperature to 0 °C.[1] The ice forms rapidly, allowing it to incorporate very small air bubbles that give it a whitish, opaque appearance.

The wet growth regime from the latent heat released when water freezes onto the ice is sufficient to bring its temperature to 0 °C. The water can remain as a liquid for some time, during which the air bubbles can be released. When the water does freeze, the lack of air bubbles gives it a much clearer, translucent appearance. The existence of supercooled water

[1]This latent heat of fusion is the heat released when liquid water freezes. It is analogous to the latent heat of evaporation released when water vapor condenses to form a liquid.

also allows the ambient wind conditions to deform the layer of water so that when it freezes it takes on a nonconcentric shape, as in Figure 7–18a.

Cumulonimbus clouds are extremely dynamic, and things change readily from one part of the cloud to another and over very short time increments. As a result, growing hailstones can rapidly undergo changes in environment between the wet and dry regimes, causing multiple layers of ice having different appearances from their adjacent ones.

It was once believed that this layering was the result of hailstones making multiple passes upward and downward across the freezing level in a cloud. This is not the case, however; we now know that the creation of multiple ice layers can result from growth during a single passage upward through a cloud, with the hailstone falling when it becomes too heavy to be maintained by the updrafts. Hailstones often fall in a localized area outside the zone of strongest updrafts, called a **hail cascade** (Figure 7–19).

Although most hailstones are pea sized, they can become as large as marbles, golf balls, or, under the most violent of conditions, even softballs!

DID YOU KNOW?

The largest hailstone ever recorded was found in Vivian, South Dakota, on July 23, 2010. It was officially measured at 19 cm (8 in.) in diameter and weighed 0.88 kg (1 pound, 15 ounces). A local ranch hand found the hailstone and immediately stored it in a freezer. Apparently the stone was even larger when initially recovered, but the storm knocked out power and the stone partially melted before it could be measured.

Hailstones consist mainly of ice with only a small amount of air mixed in. Because ice is relatively dense—90 percent as dense as liquid water—hailstones can become fairly heavy. Compare hailstones to snowflakes, whose volume is occupied mostly by air. Snowflakes have relatively little mass and low terminal velocities, so they flutter to the ground and make barely a sound. Hailstones, on the other hand, sound like a barrage of falling marbles as they hit the surface. Table 7–1 lists the terminal velocities of various sizes of hailstones.

Hailstones the size of baseballs (radius = 3.5 cm) contain about 160 g of ice (weighing about a third of a pound) and fall at about 40 m/sec (88 mph)! No wonder they are capable of producing tremendous damage. Hailstorms present a major threat to the Great Plains of the United States (Figure 7–20) and Canada (Alberta and interior British Columbia), where they are known to destroy entire fields of crops in a matter of minutes. Because they are most common in the spring and summer, it is often too late to replant the acreage with new seed.

TABLE 7–1

Terminal Velocities of Hailstones

$V_t = \sqrt{\frac{1}{3}\rho/k}\,\sqrt{r} = 20\sqrt{r}$ (r in cm, V_t in m/sec) See Box 7–1 for explanation.

Radius (cm)	Terminal Velocity (Meters per Second)
0.1	6 (13 mph)
1.0	20 (44 mph)
2.0	28 (62 mph)
3.0	35 (78 mph)

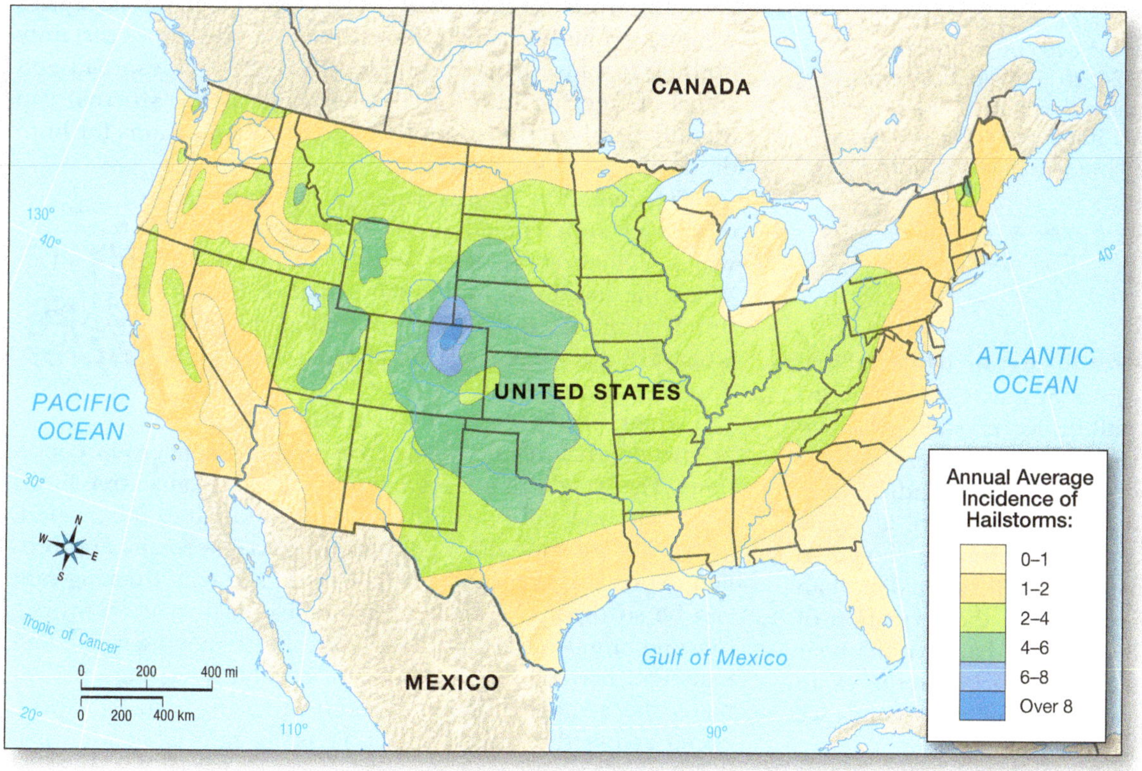

◂ **FIGURE 7–20** Annual Average Number of Hailstorms over the United States.

MM **MapMaster**
North America
Physical Environment
Average Number of Hail Storms (U.S.)

7-2 PHYSICAL PRINCIPLES

The Effect of Hail Size on Damage

You might wonder why large hailstones are more damaging than small ones. After all, isn't the issue how much ice falls to the ground, and aren't many small stones roughly equivalent to a smaller number of large stones? As it happens, this is far from true. Damage done by hail increases very rapidly in a nonlinear fashion with increasing hailstone size. What matters is the amount of kinetic energy hail carries as it falls to the surface.

Kinetic energy (KE) depends on mass (m) and speed (v) according to

$$KE = \frac{1}{2}mv^2 \quad \text{Kinetic energy}$$

The mass of a stone is its density (ρ) times its volume

$$\left(\frac{4}{3}\rho\pi r^3\right)$$

Velocity can be found using the formula in Table 7–1. We can thus write

$$KE = \frac{1}{2}\left(\rho\frac{4}{3}\pi r^3\right)(20^2 r)$$
$$= 838\rho r^4$$

We see that the kinetic energy of a hailstone is proportional to the fourth power of its radius. As a result, a hailstone with a radius of 1 cm (0.4 in.) packs not twice as much punch as a 0.5 cm (0.2 in.) hailstone, but 16 times as much. The threshold for automobile windshield damage due to falling hailstones is about 5 cm (2 in.) in diameter.

7-2-1 What is the relationship between a falling hailstone's radius and the stone's kinetic energy?

7-2-2 What effect does the doubling of a hailstone's radius have on its kinetic energy?

(See *Box 7–2, Physical Principles: The Effect of Hail Size on Damage,* for more information about the destructive potential of hail.)

DID YOU KNOW?

Aviators are well advised to avoid cumulonimbus clouds for a variety of reasons, including the possibility of damaging hail. Hail can even be encountered near the anvil of cumulonimbus clouds as winds transport the stones large distances from the area in which they formed. But hail can even be a problem for aircraft on the ground. On May 24, 2011, major hailstorms caused damage to dozens of commercial aircraft on the ground at airports over the southern United States. American Airlines alone had to cancel some 700 flights so that planes could undergo inspections and repairs.

Sleet

Sleet forms when raindrops freeze in the air while falling. Because most rain outside the tropics originates from the ice crystal process, sleet begins as falling ice crystals or snowflakes. As the ice falls through the air, it encounters warmer air and melts to form a raindrop. If the falling raindrop then encounters a lower layer of air whose temperature is below 0 °C, it can refreeze to form sleet. This process, shown in Figure 7–21, results in semitransparent pellets smaller than about 5 mm (0.2 in.) in diameter. Because the formation of sleet requires that a droplet fall through air that is cooler near the surface than aloft (Figure 7–22), it necessarily requires an inversion, usually one associated with a warm front (which we describe in Chapter 9).

Of course, a raindrop will not freeze instantaneously; sufficient cooling must take place as it falls through the surrounding air. Thus, for sleet to develop, the layer of cold air beneath an inversion must be fairly deep. If it is too shallow, another type of frozen precipitation will occur, freezing rain.

Freezing Rain

Freezing rain (Figure 7–23) is one of the more deceptive weather events. It usually looks like a gentle rain—certainly nothing to cause major problems. In reality, widespread episodes of freezing rain (often referred to as **ice storms**) can literally paralyze transportation and communications for hundreds of square kilometers.

Video Global Precipitation
http://goo.gl/6rvTi

Freezing rain begins when a light rain or drizzle of supercooled drops falls through air with a temperature at or slightly below 0 °C. When the raindrops hit the surface, they form a thin film of water, but only for a moment. Soon afterward the water freezes to form a slick, continuous coating of ice.

When freezing rain hits roadways, the loss of friction that results leads to extremely dangerous conditions. The weight of accumulated ice also can cause tree limbs and telephone and power lines to snap and fall to the ground. When you imagine downed lines, impassable roads, and broken debris scattered about, it is easy to see why freezing rain can be so disruptive to human life. To make matters worse, freezing rain is often associated with slowly moving storm systems and may therefore persist for several days. Figure 7–24 maps the average number of hours freezing rain is experienced across the United States. Freezing rain is clearly most

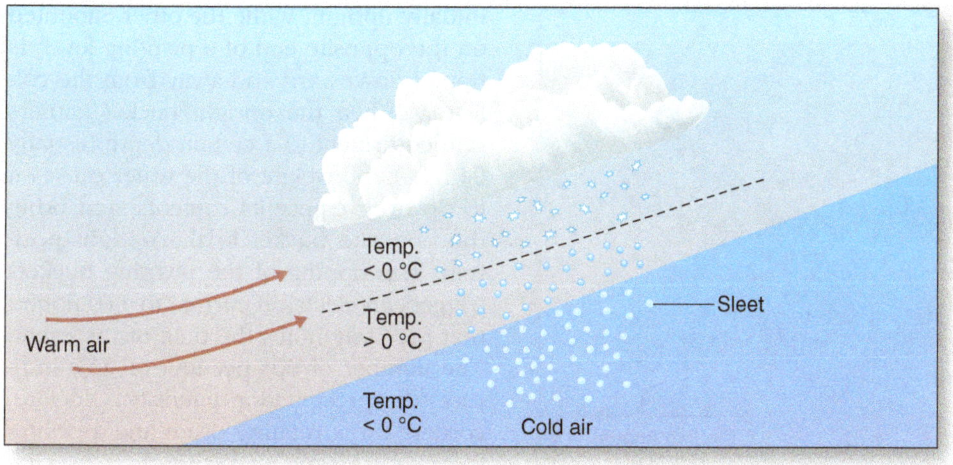

▲ **FIGURE 7–21 Sleet Formation.** This solid form of precipitation falls as rain but then passes through a cold layer and freezes into small pellets. This is most common along warm fronts.

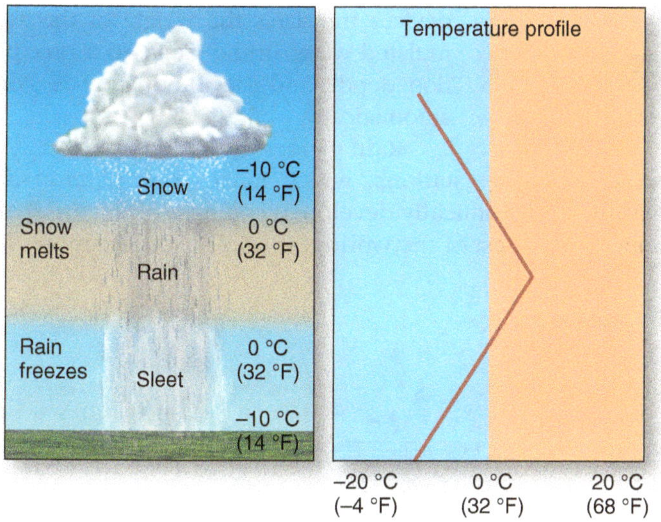

▲ **FIGURE 7–22 Sleet Temperature Profiles.** Sleet requires an inversion that allows freezing of precipitation to occur as it approaches the surface.

▼ **FIGURE 7–23 Freezing Rain in Griffin, Georgia.**

common in the northeastern United States and eastern Canada.

One of the most destructive ice storms in recent decades occurred from the Great Lakes through the southern Great Plains on December 9–11, 2007. Tens of thousands of homes and businesses lost power across Kansas, Missouri, Iowa, and Illinois, but the storm delivered its strongest punch to Oklahoma, where up to 3.5 cm (1.5 in.) of ice accumulated across the state. The entire state was declared a federal disaster area as 640,000 customers lost electrical power, some of them for up to a week. At least 27 people died from the storms, mostly from traffic accidents on the slick roads. If there were one favorable outcome from the event, it was the postponement of final exams across many of the colleges and universities in the Great Plains.

CHECKPOINT

7.7 Briefly define graupel, hail, sleet, and freezing rain.

7.8 How are the conditions under which sleet and freezing rain form similar? How are they different?

Measuring Precipitation

7.3 Describe how precipitation is measured.

Given the effects of precipitation on everyday activities, it is no surprise that we measure precipitation at many locations. Just as precipitation occurs in several forms, different types of instruments exist for measuring it. Each method has its own advantages and disadvantages.

Rain Gauges

Rainfall is usually measured with a **rain gauge** (Figure 7–25a). Standard gauges have collecting surfaces with diameters of 20.3 cm (8 in.). The precipitation funnels into a tube with one-tenth the surface area of the collector, so the depth of accumulated water undergoes a 10-fold increase. This amplification lets us measure the precipitation level precisely by simply inserting a calibrated stick into the water, removing it, and noting the depth of the wet portion, rather like checking oil with a dipstick. Note that the measuring stick has a correspondingly graduated scale so the 1 cm mark is actually 10 cm from the base of the scale.

An automated collector known as a **tipping-bucket gauge** (Figure 7–25b) provides a record of the timing and intensity of precipitation. This instrument funnels precipitation from the top like a standard rain gauge, but as the water accumulates it is stored in one of two pivoting buckets. One of the buckets is

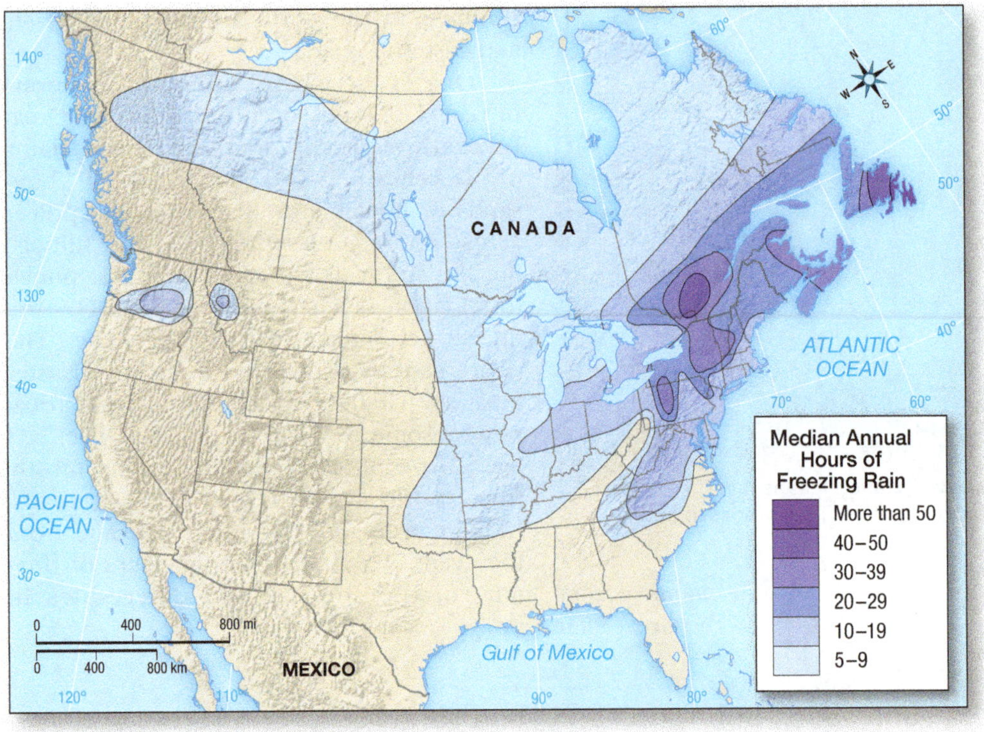

▲ **FIGURE 7–24 Distribution of Freezing Rain.** Number of hours per year on average of freezing rain.

initially upright, while the other, mounted on the opposite end of a pivoting lever, is tipped downward and away from the collector. When the upright bucket gathers rain equivalent to a certain depth (usually 0.01 in.), the weight of the water causes it to tip over, empty its contents, and bring the opposite bucket to the upright position. The tipping of the pivoting buckets triggers an electrical current to a computer that precisely notes the time of the event. The number of tips per unit of time indicates the precipitation intensity. Older recorders use a rotating drum and a printer to provide an analog record of the precipitation rate.

Weighing-bucket rain gauges are similar to tipping-bucket gauges insofar as new accumulations of rain are constantly recorded. A weighing mechanism in these devices translates the weight of the accumulated water in the gauge to a precipitation depth, and the information is stored automatically.

Rain gauges are found at virtually all weather-recording stations, which makes precipitation data plentiful in economically developed countries. Data are scarce in much of the rest of the world, especially over the more than

(a)

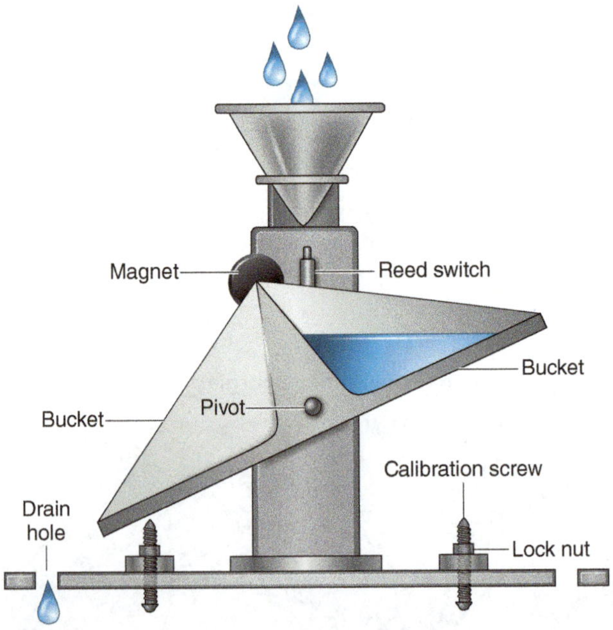

(b)

▲ **FIGURE 7–25 A Standard Rain Gauge with Its Component Parts.** (a) Rain captured by the collector (bottom left) is funneled into the narrow tube (right). The calibrating stick is inserted into the collection tube, and the length of the wetted portion indicates the precipitation accumulation. (b) The interior workings of a tipping-bucket gauge.

70 percent of Earth's surface covered by ocean. Furthermore, measurement accuracy is a concern (even with modern instruments), due to such problems as evaporation from the gauge and winds that can prevent rain from entering the gauge.

RAIN GAUGE MEASUREMENT ERRORS We tend to put unquestioned confidence in the readings we get from precipitation gauges, but unfortunately they have several inherent sources of error. First of all, they are *point measurements*, meaning that they represent the amount of precipitation that has fallen at only a single point or location—not across the street, 100 m away, or down the block. Compared to air temperature, precipitation shows wide variations across different locations, so we face some difficulty in trying to generalize from gauge readings.

Precipitation gauges have other flaws. For example, wind-generated turbulence near the top of a gauge can deflect precipitation away from the collecting surface or some water can splash out on impact with the gauge, leading to an underestimate of the true value. The moral here is that the distribution of precipitation is not as well recorded as we might suppose, nor as well recorded as befits its extreme value to agriculture and human welfare.

PRECIPITATION MEASUREMENT BY WEATHER RADAR During recent years, the network of rain gauges has been augmented by radar measurements. Though radar will be discussed in more detail in Chapter 11, for now we can say that weather radars estimate the intensity of precipitation by emitting microwave radiation with wavelengths of several centimeters. Precipitating droplets, ice crystals, and hailstones scatter some of the emitted radiation back to the radar unit, which records the intensity of the backscattered radiation. In general, the more intense the backscattered radiation, the more intense the precipitation. Meteorologists have developed schemes that relate the intensity of backscattered radiation to the rate of precipitation.

Radiation is not emitted continuously by the transmitter, but just for very brief periods that are interspersed with momentary pauses. Sufficient time is allowed for each pulse of radiation to echo back to the transmitter/receiver unit before the next beam is emitted. The closer the precipitation is to the radar, the quicker the pulse will return to the unit. By measuring the strength of the return radiation and the time taken for it to return to the unit, a profile can be created that shows how much precipitation is occurring and how far from the radar it is.

The transmitter/receiver slowly rotates as it emits and receives the radar beams, thus allowing a two-dimensional depiction of the precipitation for several hundred kilometers from the radar. The information can be continually monitored and stored, so that the total amount of precipitation occurring over a fairly extensive region can be estimated (Figure 7–26). Such measurements

DID YOU KNOW?

On average, at any moment the atmosphere over the 48 conterminous United States contains the equivalent of 175 trillion liters (40 trillion gallons) of liquid water in the form of water vapor. Slightly more than one-tenth of this moisture condenses and falls to the surface daily as precipitation—enough to yield an average of 76 cm (30 in.) of precipitation annually across the country.

have proven to be particularly useful in providing short-term forecasting for potential floods.

DUAL-POLARIZATION RADAR In the middle of 2013 the National Weather Service completed a major overhaul of its network of weather radar stations across the country to improve the network's ability to identify and measure precipitation. The enhancement of the radar network is based on the principle shown in Figure 2–6, which illustrates the nature of electromagnetic radiation. Recall that all radiation (including that emitted by radar) consists of both electric and magnetic waves, oriented perpendicular to each other. Prior to 2013 radar units transmitted their pulses with a single orientation, so that the electric waves have only a single orientation—horizontal. But in **dual-polarization radar** (also called *dual-polarimetric radars*) the energy is transmitted with two orientations, horizontally and vertically, so that the electric waves from the two modes are perpendicular to each other. This is important because cloud droplets, hail, and snow backscatter radar beams

Video
Record-Breaking Hailstorm as Seen by Radar
http://goo.gl/sE6wy9

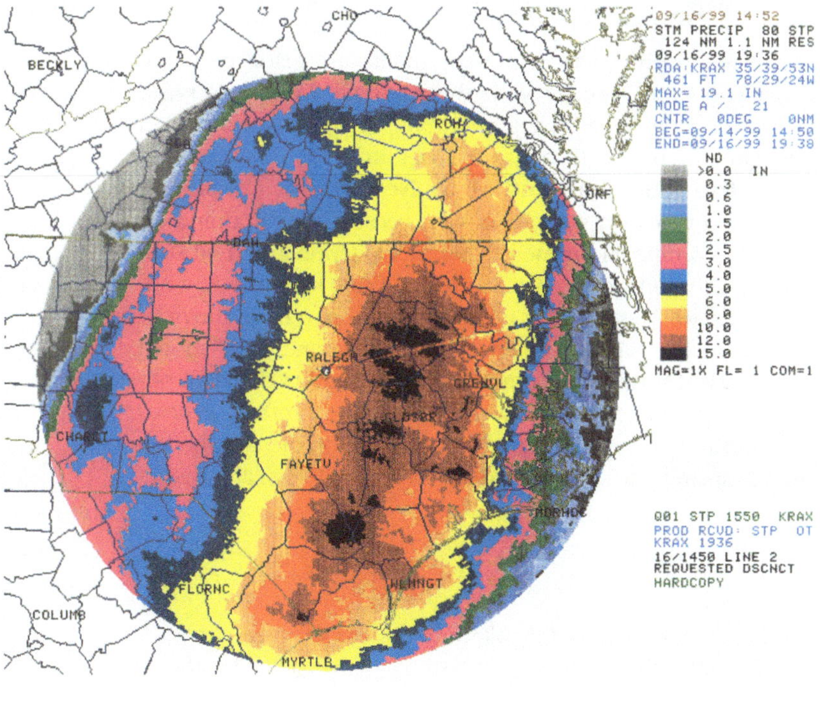

▲ **FIGURE 7–26 Doppler Radar Image.** Precipitation estimates depicted by Doppler radar.

differently for the vertical and horizontal orientations. Thus, meteorologists can better discern precipitation rates and forms, and even differentiate cloud constituents from smoke, debris, and other particles suspended in the air.

Snow Measurement

Rain gauges are particularly unreliable when precipitation occurs as snow, because captured snow can obstruct the inlet to the collecting tube. To estimate the precipitation in these environments, we measure the depth of accumulated snow. The *water equivalent* of the snow, which is the depth of water that would result if all the snow melted, can then be roughly estimated using a conversion ratio of 10:1. In reality, the ratio of snow depth to water equivalent can vary greatly—from about 4:1 to 50:1.

In remote mountainous areas, particularly in the western United States and Canada, observations of snow cover have been made for decades at hundreds of *snow courses*. Usually about 10 observations are made at each snow course by pushing a collection tube into the snow, extracting the tube and its contents, and weighing them on a spring balance. The weight of the snow-filled tube is directly proportional to the water equivalent of the snow cover, and the average of the 10 or so readings is used as the representative value.

Manual snow course measurements are still obtained but are frequently augmented by automated snow pillows. **Snow pillows** are large air mattresses filled with an antifreeze liquid and connected to pressure recorders. As snow accumulates on a pillow, the increased weight is recorded and converted to a water equivalent. These instruments have radio devices that transmit the data to a centralized receiving station.

Community Collaborative Rain, Hail and Snow Network

Although many precipitation measurements are made by agencies that cooperate directly with the National Weather Service, thousands of gauges measuring rain, hail, and snow are provided by the general public through a program called **Community Collaborative Rain, Hail and Snow Network**, or CoCoRaHS. The program relies on volunteers who receive basic training on the use of instruments for measuring precipitation and then submit the observations to a central website. While supported by numerous agencies, the network receives a large part of its financial support from the National Science Foundation and the National Oceanic and Atmospheric Administration (NOAA). The data obtained are used by the National Weather Service and other agencies and organizations that need good information on the distribution of rainfall during particular events or on a climatological basis.

CHECKPOINT

7.9 Explain how each of the following devices works to measure precipitation: tipping-bucket gauge, weighing-bucket gauge, snow pillow.

7.10 What are some sources of error in measuring precipitation?

Cloud Seeding

7.4 Summarize efforts to induce precipitation through cloud seeding.

Since the late 1940s, people have tried to induce precipitation from clouds, most often to alleviate droughts. This process, called **cloud seeding**, involves injecting one of two materials into nonprecipitating clouds. The objective is to convert some of the supercooled droplets in a cool cloud to ice and cause precipitation by the Bergeron process.

One of the materials, *dry ice* (frozen carbon dioxide), promotes freezing because of its very low temperature (below −78 °C or −108 °F). Very small shavings of dry ice can be ejected from a plane flying through the cloud. When introduced into a cloud, dry ice lowers the temperature of the droplets so that freezing can occur by homogeneous nucleation (see Chapter 5). (Recall that at temperatures below about −40 °C water droplets require no ice nuclei to freeze.)

The second agent for cloud seeding, *silver iodide*, initiates the Bergeron process by acting as an ice nucleus at temperatures as high as −5 °C (23 °F). Silver iodide owes its effectiveness as an ice nucleus to its six-sided structure. Like dry ice, silver iodide can be introduced directly into a cloud from aircraft. More often, it is mixed with a material that produces smoke when ignited in ground-based burners. Updrafts then carry the smoke and the silver iodide into the cloud. If a portion of the seeded cloud is cold enough, some of the supercooled droplets will freeze and begin the Bergeron process.

The cost effectiveness of cloud seeding is widely debated. Under ideal circumstances, it can supplement water supplies somewhat. Take, for example, the case of the Sierra Nevada mountain range, which supplies much of the water for California and Nevada. Under the right wind, temperature, and moisture conditions, silver iodide released from the ground can enhance snowfall (which is later released as spring melt) by perhaps 10 percent. The right conditions are not usually met, however. Cloud seeding trials in the mountains of Colorado, Utah, and Montana provided disappointing results.

Strong theoretical reasons provide grounds to doubt the usefulness of cloud seeding, except in regions with a continued uplift of air (such as where an orographic effect exists). Recall that water vapor accounts for only a small portion of the air, and that the liquid water content of a cloud is relatively small compared to that of the mass of the air contained within. Thus, the formation of heavy precipitation requires a constant resupply of moisture into clouds by updrafts. If such updrafts are already occurring, precipitation will probably occur with or without seeding. Consequently, many meteorologists believe that under most circumstances cloud seeding yields little or no additional precipitation.

Another question raises an ethical concern about cloud seeding. Let's assume that seeding a cloud produces rain. Would the cloud have yielded precipitation farther downwind if it had not been seeded? Residents downwind might argue

7–3 FOCUS ON AVIATION

Fog Seeding at Airports

Seeding is sometimes used to clear fogs along airport runways. If a fog exists at temperatures below 0 °C (32 °F), introducing dry ice can instigate the Bergeron process, as it does in clouds. Some of the water droplets freeze into ice crystals, which grow at the expense of water droplets and fall out of the fog as snow, leaving a local area of clear air. Of course, this technique can only work for fogs containing supercooled droplets, which is the exception rather than the norm.

Missoula (Montana) International Airport has had ongoing success seeding fog since 2006. Carbon dioxide is sprayed from a couple of pickup trucks into the air just upwind from the runways to stimulate localized snowfall out of the fog. Flights that would ordinarily have been cancelled are now often allowed to take off and land, sometimes with only minor delays. On average this technique is used about five times each year at Missoula.

Several other methods have been tried for dispersing warm fogs, including flying helicopters over the runways to force down the warm air within a radiation inversion and introducing salt crystals into the fog with the intent of making some droplets larger—thereby accelerating growth by collision and coalescence (Figure 7–3–1).

▲ **FIGURE 7–3–1** Fog over the Missoula Valley, Montana.

7–3–1 Through what process does cloud seeding assist in fog dispersal?

7–3–2 What chemical is used in airport fog dispersal?

they were deprived of precipitation that would have occurred naturally over their own fields and drainage basins. Such matters have in fact been litigated in civil court. In short, a number of open questions still exist about the value of cloud seeding for enhancing precipitation.

Seeding has also been attempted to reduce hail intensity. It was once believed that seeding hail-producing clouds could increase the number of growing ice pellets. Because clouds contain a limited amount of water that can freeze onto growing hailstones, increasing the number of hailstones would theoretically reduce their average size. The decrease in size would reduce the kinetic energy of the falling hailstones and thereby lower the likelihood of damage near the ground. However, a multiyear experiment in northeastern Colorado in the 1970s failed to support the usefulness of seeding as a hail suppression measure, and such attempts have been discontinued in the United States and Canada. Attempts to reduce the intensity of hurricanes by cloud seeding have likewise failed to yield convincing results.

The principles involved in cloud seeding also apply to the removal of fog. See *Box 7–3, Focus on Aviation: Fog Seeding at Airports,* to learn more about this.

DID YOU KNOW?

The viability of using dry ice to promote precipitation formation was discovered serendipitously by Vincent Schaefer in 1946. Working with a home freezer in his lab to test whether certain materials could work as ice nuclei, he introduced dry ice into the freezer with moist air to cool it further. He then saw some of the water droplets immediately turn to ice. Further tests by Schaefer and Irving Langmuir proved that introducing dry ice into real clouds could trigger the formation of ice crystals. A few years later, Bernard Vonnegut (brother of novelist Kurt Vonnegut) discovered that silver iodide could also promote ice crystal formation.

CHECKPOINT

7.11 What are two substances that have been used in cloud seeding?

7.12 Suppose that you are a wheat farmer on the High Plains in Montana. Would cloud seeding intended to save your crop during a drought be cost effective? Explain why or why not.

7–4 FOCUS ON THE ENVIRONMENT AND SOCIETAL IMPACTS

Floods in the Midwestern United States

During the spring of 1993 residents of the Midwestern United States witnessed flooding on an unprecedented scale. The Mississippi River, normally about 800 m (0.5 mi) wide near St. Joseph, Missouri, stretched out to as much as 10 km (6 mi), putting nearly half of St. Charles County under water. At Kansas City, Missouri, the Missouri River rose 6.7 m (22 ft) above its banks. Across the Midwest, tens of thousands of homes were damaged or destroyed by the flooding, as entire neighborhoods and 77 small towns ended up under water.

But what many would have thought would be a once-in-a-lifetime event was repeated 15 years later, when another round of extensive flooding soaked the Midwest in 2008. A series of heavy rains hit the region in early June, including a few exceptionally strong ones that brought more than 22 cm (9 in.) of rain in a 2-day period. Gays Mills, Wisconsin, which had been inundated by flood waters from the Kickapoo River for the second time in 10 months, was forced to consider relocating the town on higher ground farther up the floodplain to avoid similar events in the future.

Indiana, Michigan, Illinois, and Missouri were beset by record-breaking floods from exceptionally heavy rain. But no state was hit harder than Iowa, where 83 of its 99 counties were declared disaster areas, more than 8 percent of the corn and soybean acreage was under water, and damage was estimated at about $1.5 billion. Many towns and cities fought rising water with sandbags, but often unsuccessfully, as in Cedar Rapids . In contrast to the 1993 floods, the 2008 floods were the result of more intense rainfall events that occurred over a shorter time period and had faster falling river levels after the peak of the flooding.

▲ **FIGURE 7–4–1** Flood waters pour into sections of Minot, ND, from the Souris River.

But the story doesn't end there. Major flooding in the Mississippi River basin resumed in April 2011 and carried into the summer of that year. The 2011 floods resulted from a combination of factors. Heavy rains during the previous summer and autumn saturated the soils, which were then unable to absorb melting snow and additional rain that spring. By May 10 the Mississippi River had reached its greatest extent of overbank flooding at Memphis, Tennessee. Soon thereafter some floodwaters were purposely released further downstream in order to prevent widescale flooding of Baton Rouge, Louisiana.

Though the Mississippi River had reached its maximum stage by midspring, much of the north-central United States and the provinces of Saskatchewan and Manitoba were subject to record flooding owing to the same combination of snowmelt and an abundance of rain for multiple months. One of the most dramatic flood events occurred in Minot, North Dakota, when one quarter of the town had to be evacuated as the Souris River topped its banks (Figure 7–4–1). In the end, 4100 homes ended up under water.

7–4–1 In the United States, what years have had major floods since the 1990s?

7–4–2 What were the causes of the three major U.S. floods since the 1990s?

Floods

7.5 Explain the factors that can cause floods.

It is impossible to overstate the importance of precipitation to every aspect of life on Earth. But there are times when nature delivers an excess of water and people find themselves inundated. The result is flooding, a natural hazard that can cause billions of dollars in damage and also lead to fatalities. According to NOAA the average annual cost of **floods** between 2003 and 2012 was more than $10 billion per year, along with an average of 76 U.S. fatalities. And those figures do not even include losses from coastal flooding associated with landfalling tropical storms and hurricanes.

Video
Rain and Flooding
http://goo.gl/KokICh

Several factors can serve to create flood conditions. Among them is the simple occurrence of excessive precipitation over a widespread area. When rain falls over a land cover, a portion of it can soak directly into the soil—but once the soil is saturated all further rainfall remains at the surface. Surface accumulations of rain can then run downslope and eventually be routed into a nearby stream or the rain can pool up at low elevations. Given enough time, streams begin to approach flood stage and are subject to overtopping their banks.

When precipitation falls as snow, it can remain on the ground for months. But if deep snowpacks accrue over the course of a winter, they are subject to rapid melt under periods of warm weather or the influence of warm rainfall. Like falling rain, melted snow can percolate into an unsaturated soil. But if the input of melt water exceeds the rate at which the soil can absorb it, heavy runoff and potential flooding may occur. Such flooding can be exacerbated if large blocks of ice get carried into rivers and cause localized blockage.

Sometimes particularly intense storms can produce rapid flooding. Such *flash floods* are likely to be more localized than other kinds of flooding, but their sudden onset can make them extremely deadly. Flash floods are discussed in Chapter 11.

Sometimes all the factors that can lead to flooding work together to create widespread problems. The central United States has been subject to three particularly large and severe floods since 1993. The events are discussed in *Box 7–4, Focus on the Environment and Societal Impacts: Floods in the Midwestern United States.*

CHECKPOINT

7.13 Describe how the varying moisture contents of soils can impact an area's vulnerability to floods.

7.14 Explain how previous precipitation episodes can lead to flooding months later.

Summary

7.1 Describe the processes involved in the growth of cloud droplets.

- Some cloud droplet growth occurs through condensation onto existing droplets, but further growth depends on other processes.
- In the tropics, the primary mechanism for droplet growth is collision and coalescence. Outside the tropics, the Bergeron process dominates, wherein ice crystals grow at the expense of supercooled droplets and then grow further via riming and aggregation.

7.2 Describe the distribution of precipitation and explain how different types of precipitation form.

- Precipitating ice crystals that do not melt before reaching the surface form snow.
- Graupel and hail form when supercooled water attaches to ice crystals and freezes. In the case of hail, growth occurs when water continually freezes onto existing ice pellets.
- Sleet occurs as raindrops that freeze before hitting the surface.
- Freezing rain occurs when supercooled precipitation freezes on contact with the ground.

7.3 Describe how precipitation is measured.

- The standard instrument for measuring precipitation is the simple rain gauge.
- The timing and intensity of precipitation can be recorded with a modified precipitation gauge called a *tipping-bucket gauge.*
- Radar is a very useful tool for getting information on accumulated precipitation and rainfall rates.

7.4 Summarize efforts to induce precipitation through cloud seeding.

- Cloud seeding involves the introduction of materials into clouds to stimulate precipitation by the Bergeron process.
- Cloud seeding can be accomplished with dry ice, which causes supercooled droplets to freeze by homogeneous nucleation, or with silver iodide, which serves as an ice nucleus.

7.5 Explain the factors that can cause floods.

- Floods can form when heavy rainfall occurs over periods of several months, thereby overwhelming the soil's ability to absorb the excess water. Unusually heavy snowfall can have a similar effect when high temperatures cause rapid melting of accumulated ice.

Key Terms

aggregation *p. 222*
Bergeron process *p. 221*
blizzards *p. 223*
cloud seeding *p. 234*
coalescence *p. 220*
cold cloud *p. 220*
collector drop *p. 218*
collision *p. 219*
collision–coalescence process *p. 218*
Community Collaborative Rain, Hail and Snow Network *p. 234*
cool cloud *p. 220*
drag *p. 218*
dual-polarization radar *p. 233*
floods *p. 236*
flurries *p. 223*
freezing rain *p. 230*
graupel *p. 227*
hail *p. 227*
hail cascade *p. 229*
ice storms *p. 230*
lake-effect or **lake-enhanced snow** *p. 225*
rain *p. 226*
rain gauge *p. 231*
riming or **accretion** *p. 221*
shower *p. 226*
sleet *p. 230*
snow *p. 222*
snow pillow *p. 234*
snow squalls *p. 223*
terminal velocity *p. 218*
tipping-bucket gauge *p. 231*
virga *p. 226*
warm cloud *p. 218*
weighing-bucket rain gauge *p. 232*

Review Questions

1. What determines the terminal velocity of falling droplets and raindrops?
2. Describe the characteristics that distinguish warm, cool, and cold clouds.
3. How do collision and coalescence increase the size of cloud droplets?
4. Why isn't growth by condensation able to create precipitation-sized droplets on its own?
5. How do the growth processes of droplets in warm and cold clouds differ?
6. Explain how variations in the saturation vapor pressure for ice crystals and supercooled water droplets affect the development of precipitation.
7. Why can't the Bergeron process take place in warm clouds?
8. What are riming and aggregation?
9. Why is precipitation greater in Mississippi than in Michigan?
10. How do lakes enhance precipitation downwind?
11. Why do rain showers start with only large drops?
12. It is never too cold for snow to occur. Is that also true of sleet?
13. Explain why the formation of sleet requires an inversion.
14. Why does hail consist of multiple layers of ice?
15. What are some inherent sources of error in rain gauges?
16. How do weighing-bucket and tipping-bucket gauges measure rainfall?
17. What conditions lead to the development of floods?
18. What materials are used in cloud seeding, and how do they stimulate (or inhibit) precipitation?
19. Explain how snow pillows measure snow accumulation.

Critical Thinking

1. How might a warming of the atmosphere change how rainfall forms in the middle latitudes?
2. It is not possible for a cloud to precipitate all of its ice crystals or water droplets. Why not?
3. Industrial activity can increase the number of cloud condensation nuclei. Would an increase in the number of such nuclei tend to promote the formation of rainfall or inhibit it? Why?
4. *Precipitable water* refers to the depth of water that would precipitate if all the water in a column of air above the surface were to rain out. Typically, precipitable water is on the order of about 2.5 cm (1 in.), but precipitation amounts can greatly exceed 2.5 cm. How is this possible?
5. Using weather radar to examine vertical profiles of clouds and precipitation shafts, it is easy to determine the height at which the temperature is 0 °C (32 °F). Why?
6. Aircraft icing is a serious threat to aviation at temperatures of about −10 °C to 0 °C (14 °F to 32 °F). Why is it less of a problem at lower temperatures?
7. Typical raindrops fall at a speed of about 6 m/sec—roughly 25 km/hr (15 mph). Snowflakes obviously fall more slowly. What does this imply about the depth of cloud required to yield precipitation through the collision–coalescence process compared to through the Bergeron process?
8. Hail sometimes shoots out of a cumulonimbus cloud near the anvil. How can large hailstones be found in this relatively shallow portion of the cloud?
9. A shampoo company once advertised that its product was "pure as rainwater." Do you think this is true, and if so, does this speak well of the shampoo?
10. Ice in the upper reaches of a cumulonimbus cloud over Colorado may be observed 2 days later over the eastern United States. How can these ice crystals manage to survive without having been precipitated out of the cloud or sublimated away?

Problems & Exercises

1. A particular collector drop has a fall speed of 0.26 m/sec while a smaller, 10 μm droplet directly below falls at a rate of 0.01 m/sec. How long will it take for the two to collide? How far will each of them have fallen prior to collision?
2. The terminal velocities of spherical falling droplets and raindrops are proportional to the square root of their radii.
   a. If a cloud droplet with a radius of 10 μm falls at 0.01 m/sec, how fast would a droplet with a radius of 100 μm fall?
   b. If both droplets are within a cloud, 100 m above the cloud base, how long would it take the two of them to fall to the bottom of the cloud, assuming no growth or diminution in size?

3. On a regular basis, examine the weather radar map at http://radar.weather.gov/Conus/. Then click on any region for a closer, regional view of the precipitation. Would you describe the precipitation intensity as uniform or spotty? This pattern is likely to vary by season and region. How might season, type of precipitation, and geography influence the spatial variability of precipitation within a storm?

Visual Analysis

This image shows a radar analysis of precipitation falling along the Middle-Atlantic states on March 3, 2014. The radar image uses the new dual-polarization technology to distinguish different types of precipitation.

7.1. What are the different types of precipitation occurring across the region?

7.2. Which regions might be dealing with the most severe disruptions?

7.3. If you were a forecaster, how might the narrowness of the regions of different precipitation types make your job difficult?

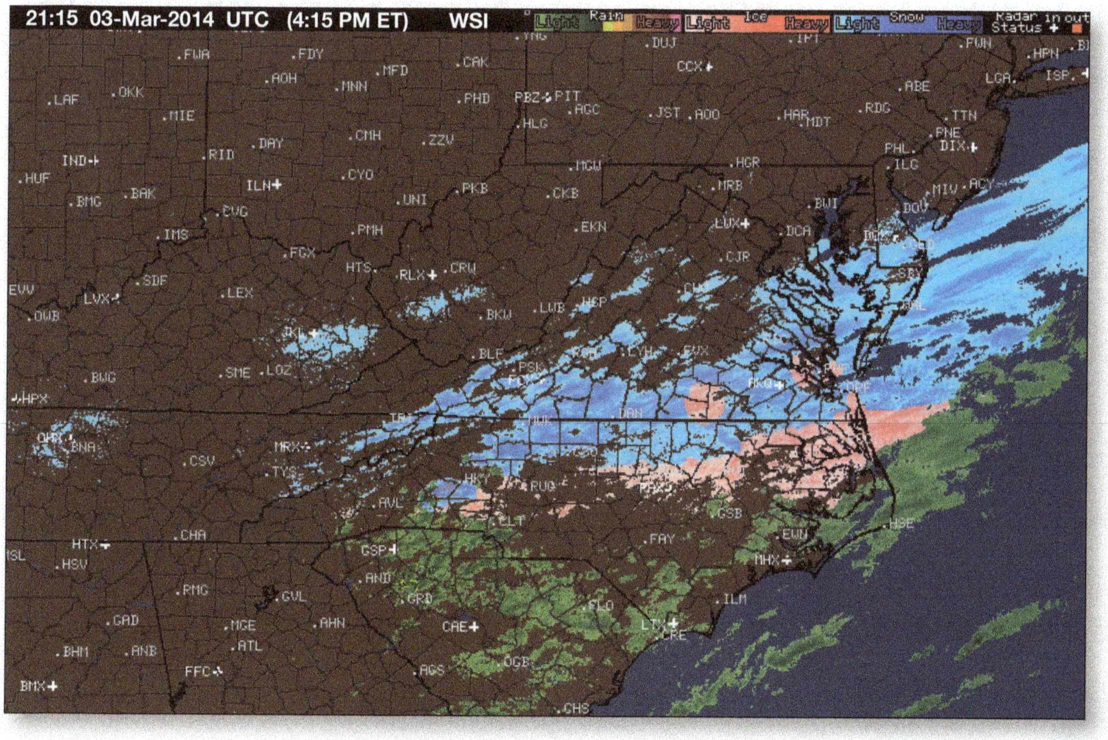

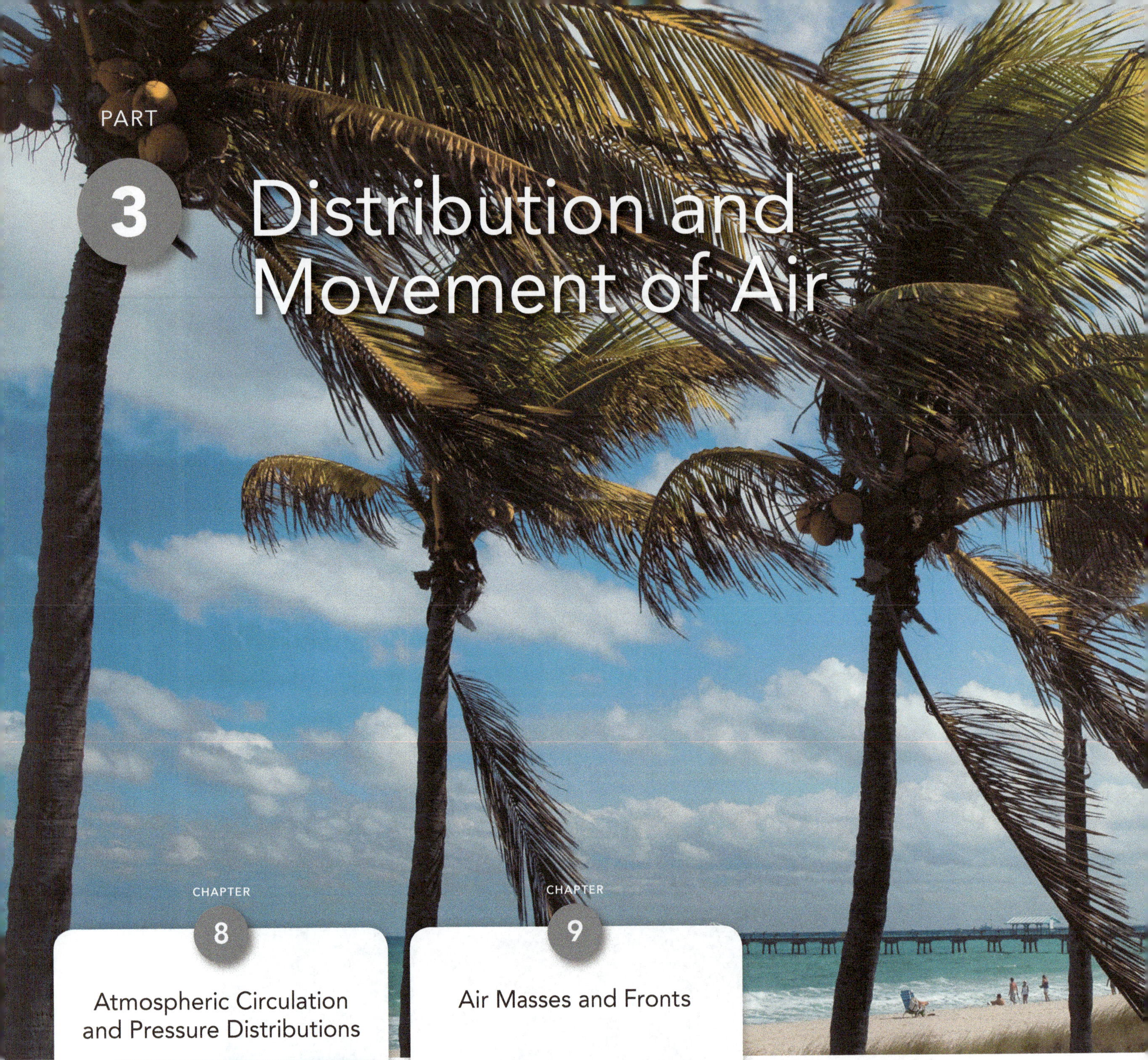

PART

3 Distribution and Movement of Air

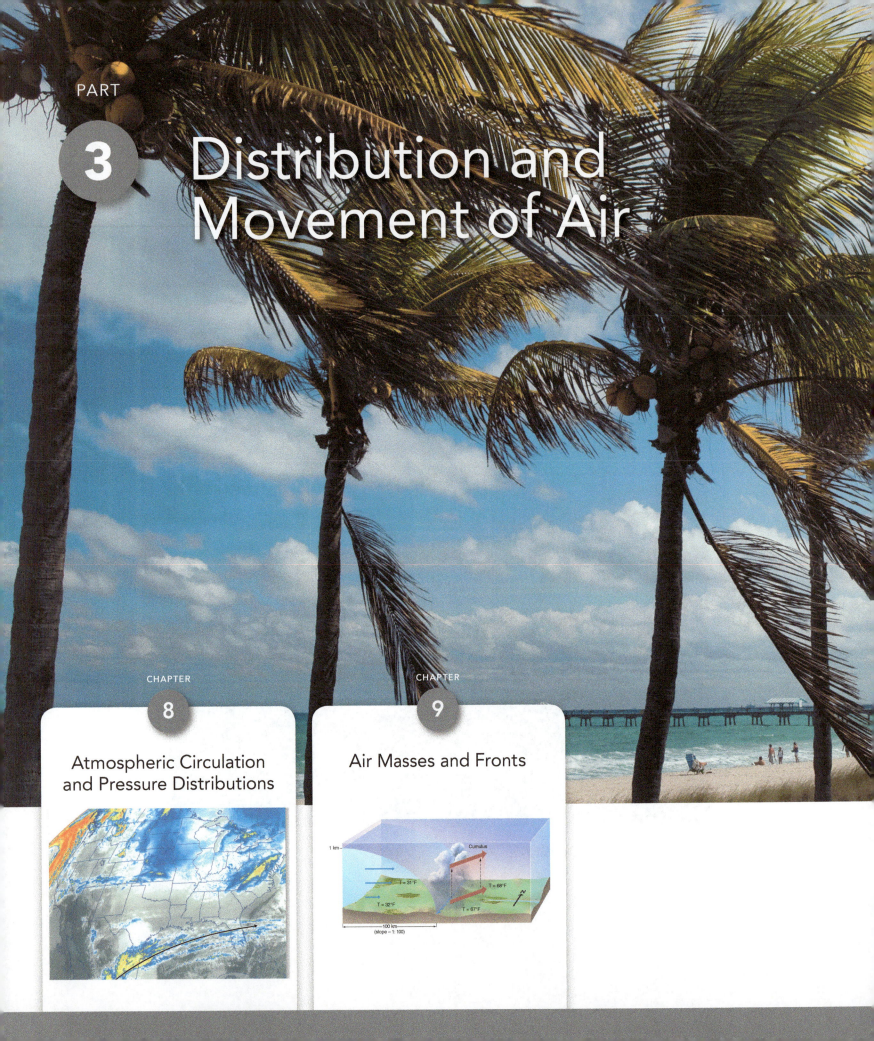

CHAPTER

8

Atmospheric Circulation and Pressure Distributions

CHAPTER

9

Air Masses and Fronts

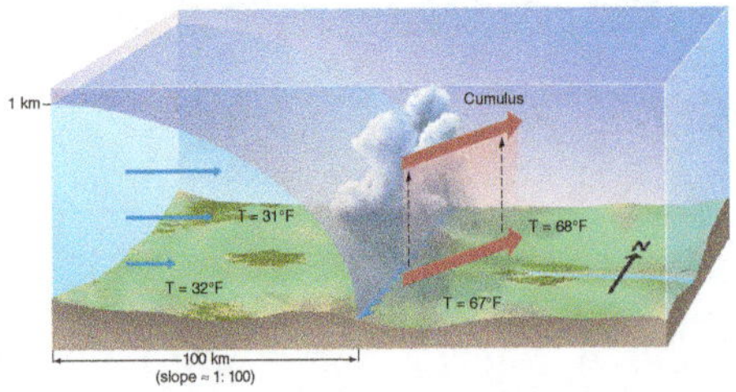

◀ *A summer beach scene looking south on the Atlantic Coast of Florida near Fort Lauderdale. Large-scale prevailing winds and small-scale temperature differences between land and water combine to make onshore winds like those seen here very common.*

Atmospheric pressure, wind, temperature and moisture patterns are not haphazardly distributed across Earth's surface. Certain patterns of pressure and wind occupy preferred positions across the globe at different times of the year and appear at various spatial scales. At the largest scale, interactions between the atmosphere, continents and oceans produce occasional shifts in global patterns of circulation and weather. At smaller scales, it is common to find large areas of fairly uniform temperature and moisture separated from each other by well-defined boundaries called fronts. This section of the book examines the causes and distributions of the resultant pressure, wind, temperature, and humidity patterns that dominate much of our daily weather.

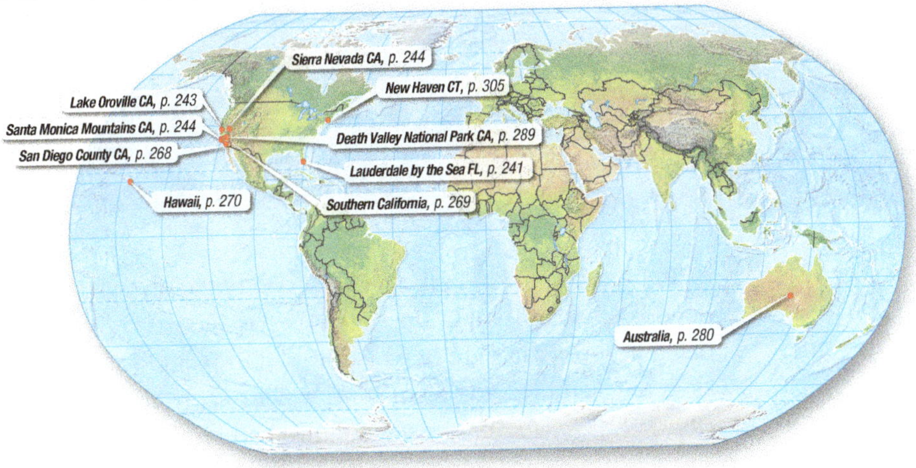

CHAPTER 8

Atmospheric Circulation and Pressure Distributions

In 2014 California was in the midst of one of the driest periods on record. All of the state was classified as being in severe drought or worse (Figure 8–1). Snow packs in the Sierra Nevada Mountains, which supply about one-third of the water for cities and farms, had received little replenishment in the preceding winter (Figure 8–2) and snow depths were only 18 percent of normal on the final day of measurement in May. This was the third year of a drought that saw farmers cutting down almond trees they could no longer afford to irrigate, wine makers abandoning highly profitable vines, and water conservation measures enforced in cities everywhere under a statewide declaration of emergency. Things were not much better in Texas (see Figure 8–1), where a drought that began in 2010 continued.

◂ *Lake Oroville, California as seen in 2014 during a prolonged drought and under normal conditions (inset).*

Learning Outcomes

After reading this chapter, you should be able to:

8.1 List the terms used to describe the spatial scales of weather phenomena.

8.2 Explain the single-cell model of atmospheric general circulation and the origin of east–west (zonal) winds.

8.3 Describe the three-cell model and how accurately it represents observed motions.

8.4 Describe the distribution and effects of semipermanent pressure cells.

8.5 Explain the distribution of wind and pressure in the upper troposphere.

8.6 Explain how the atmosphere affects the movement of ocean waters.

8.7 Identify major wind systems such as monsoons; foehn, chinook, and Santa Ana winds; katabatic winds; and sea and land breezes.

8.8 Explain ocean–atmosphere interactions such as El Niño/La Niña, Walker circulation, and Pacific Decadal, Arctic, and North Atlantic oscillations.

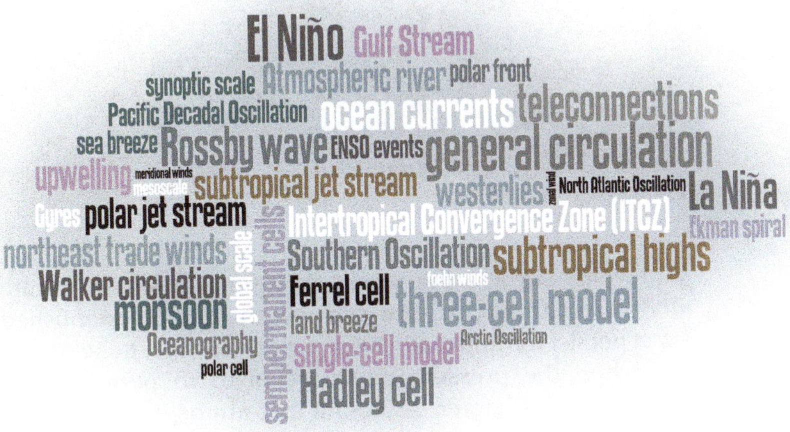

▸ **FIGURE 8–1 Drought in the West.** The map shows drought conditions across the United States in June 2014, with widespread dryness in the far West. Similarly extreme drought conditions were found in Texas, continuing a trend that began in 2010.

MM **MapMaster**
North America
Physical Environment
Recent Peak Drought Record

Video
Floods and Droughts
http://goo.gl/ECRmJ9

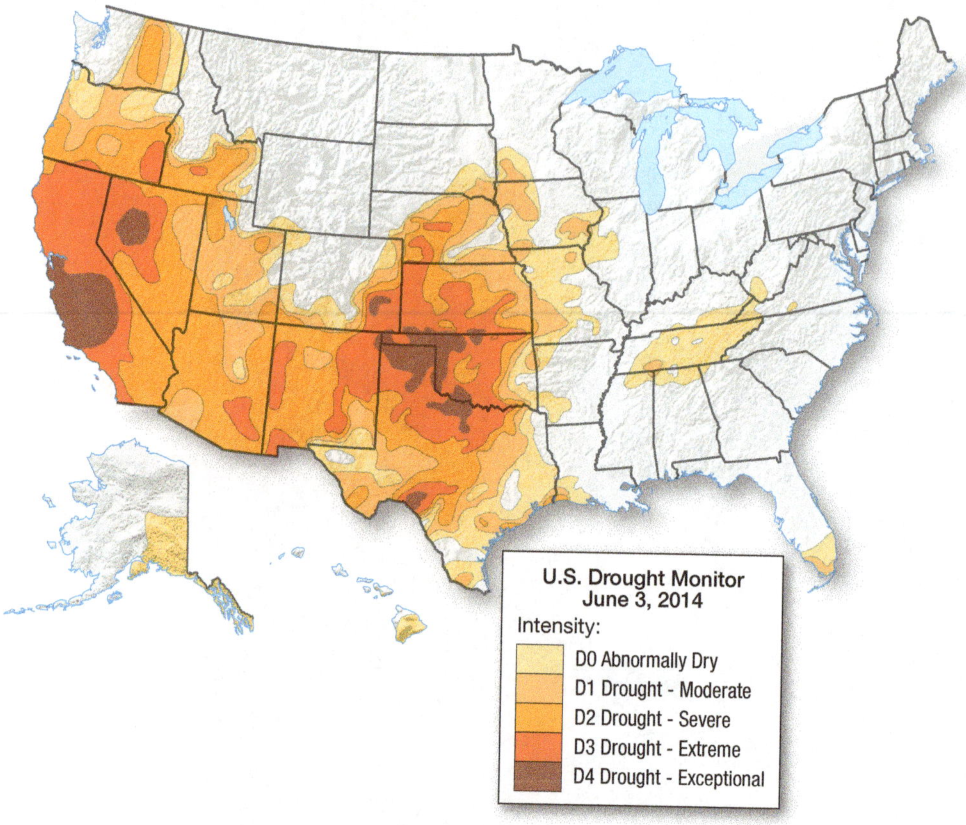

▾ **FIGURE 8–2 January Snow Cover in California. (a) 2013. (b) 2014.** Snowpacks in the Sierra Nevada Mountains in 2014 were a fraction of those in the previous year.

(a) 2013

(b) 2014

Events such as the recent California drought typically result from pressure patterns that, once established, persist for unusually long periods of time. Relief comes only when the pressure pattern evolves to permit wetter conditions. In Chapter 4 we saw that atmospheric pressure varies from one place to another, but its distribution is not haphazard. Instead, well-defined patterns dominate the distribution of pressure and winds across the global surface. The largest-scale patterns, called the **general circulation**, can be considered the background against which unusual events occur, such as the drought described above. Likewise, even mundane daily wind and pressure variations can be thought of as departures from the general circulation.

Our first goal in this chapter is to describe dominant planetary wind motions and look at the processes that generate them. In particular, we examine the interrelationships between the winds of the upper and lower atmosphere and the connections that occur at the boundary between the surface of the oceans and the lower atmosphere. We then consider wind and pressure patterns at sequentially smaller spatial and temporal scales. The chapter concludes with a discussion of ocean–atmosphere interactions.

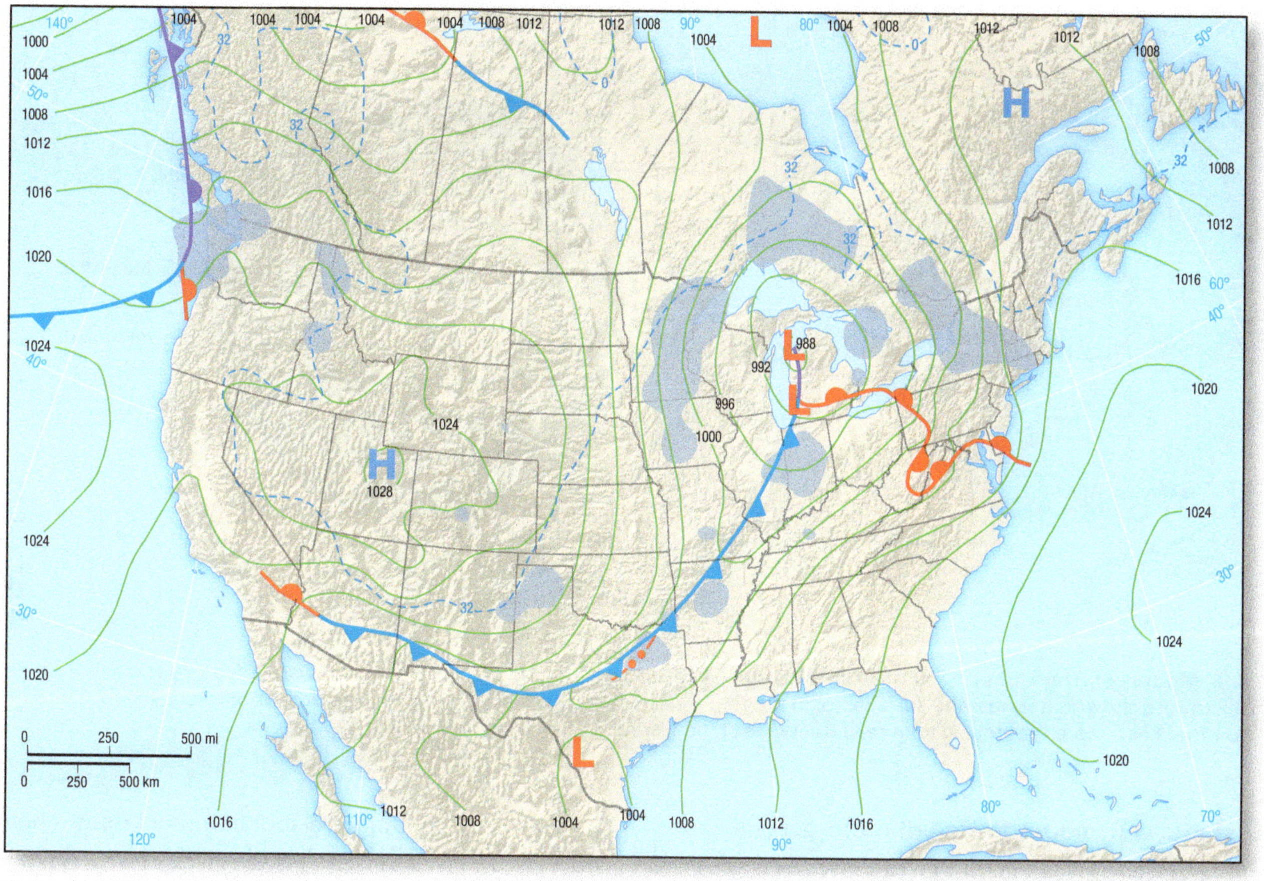

▲ **FIGURE 8–3 Scale of Atmospheric Phenomena.** The sample weather map shows features at the synoptic scale (the high-pressure and low-pressure systems) and mesoscale (shaded areas showing precipitation).

The Concept of Scale

8.1 List the terms used to describe the spatial scales of weather phenomena.

Those who deal with atmospheric phenomena must be concerned with the notion of scale—in terms of both aerial coverage and time span. Some features of the atmosphere cover large portions of Earth and are maintained over an extensive time period (perhaps for weeks). Such features exist at what is called the **global scale**. Given the size and long duration of these features, you may not hear reference to them on local media weather reports. But as we examine large-scale wind and pressure patterns in this chapter, you will see that global scale phenomena provide the background on which all weather and climate events take place.

High-pressure and low-pressure patterns over large parts of continents occur at what is called the **synoptic scale**, meaning that they cover hundreds or thousands of square kilometers. Synoptic scale features persist for periods of days to as much as a couple of weeks. These are the most salient features that one sees on daily weather maps, such as the high- and low-pressure systems shown in Figure 8–3.

Other elements of daily weather operate at the **mesoscale**, which may cover anywhere from just a few square kilometers to hundreds of square kilometers and for periods from as brief as half an hour to perhaps a large part of a day. A localized thunderstorm over western Oklahoma would be an excellent example of a mesoscale event, as would an organized cluster of storms covering several counties. The smallest exchanges of mass and energy operate at the **microscale**, such as those that might cause ripples to form on snow or a sandy beach—or the swirling of smoke emanating from an unattended fry pan. Microscale features are usually covered in more advanced texts.

This chapter examines typical patterns of pressure and wind, beginning with those at the largest scale and progressing to more localized situations. Box 8–1, *Physical Principles: Maintaining General Circulation*, provides details on the physical processes underlying atmospheric general circulation.

CHECKPOINT

8.1 Distinguish global-scale from synoptic-scale atmospheric phenomena.

8.2 Give examples of mesoscale and microscale phenomena.

Single-Cell Model

8.2 Explain the single-cell model of atmospheric general circulation and the origin of east–west (zonal) winds.

Scientists have sought to describe general circulation patterns for centuries. As early as 1735 a British physicist, George Hadley (1685–1768), proposed a simple circulation pattern

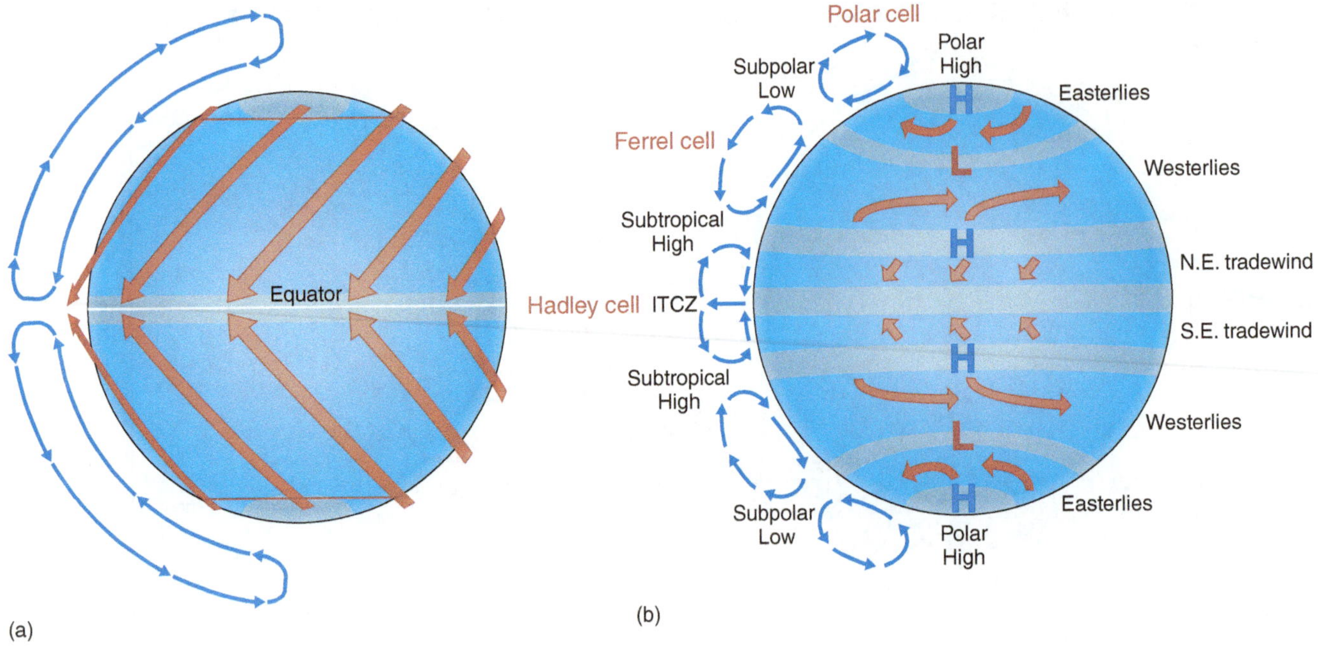

Animation
Global Wind Patterns
http://goo.gl/ffBUHg

▲ **FIGURE 8–4 Models of Atmospheric Pressure and Wind.** (a) In the single-cell model, air expands upward, diverges toward the poles, descends, and flows back toward the equator near the surface. (b) In the three-cell model, thermally driven Hadley circulation is confined only to the lower latitudes. Two other cells (more theoretical than real) exist in each hemisphere, the Ferrel and polar cells.

called the **single-cell model** to describe the general movement of the atmosphere. One of his primary goals was to explain why sailors so often found winds blowing east to west in the lower latitudes. (Winds blowing east to west or west to east are referred to as **zonal winds**; those moving north to south or south to north are called **meridional**[1].) Hadley's idealized scheme, shown in Figure 8–4a, assumed a planet covered by a single ocean and warmed by a fixed Sun that remained overhead at the equator. Hadley suggested that the strong heating at the equator caused a circulation pattern in which air expanded vertically into the upper atmosphere, diverged toward both poles, sank back to the surface, and returned to the equator. Hadley did not think winds would simply move north and south, however. He believed instead that the rotation of Earth would deflect air to the right in the Northern Hemisphere and to the left in the Southern Hemisphere, leading to the east–west surface winds shown in the figure.[2]

Hadley's main contributions were to show that differences in heating give rise to persistent large-scale motions (called *thermally direct circulations*) and that zonal winds can result from deflection of meridional winds. His idea of a single huge cell in each hemisphere was not so helpful, however.

[1]The wind is seldom purely zonal or purely meridional, but instead usually moves in some intermediate direction. In that case we will think of the wind as having both a zonal and a meridional component. It is possible for the two components to be equal (as in a southwesterly wind), but in general one component will be larger than the other.

[2]Hadley lived from 1685 to 1768, before Gaspard de Coriolis (1792–1843) quantitatively described the acceleration due to Earth's rotation. Nonetheless, Hadley had a qualitative knowledge of this force and incorporated it into his model.

A somewhat more elaborate model does a better—though still simplified—job of describing the general circulation. This **three-cell model** (Figure 8–4b) was proposed by U.S. meteorologist William Ferrel (1817–1891) in 1865.

CHECKPOINT

8.3 What are zonal winds? Meridional winds?

8.4 What were two main contributions of Hadley's single-cell model?

The Three-Cell Model

8.3 Describe the three-cell model and how accurately it represents observed motions.

According to the three-cell model, the circulation of each hemisphere is composed of three distinct cells: the heat-driven **Hadley cell** that circulates air between the tropics and subtropics, a **Ferrel cell** in the middle latitudes, and a **polar cell**. Each cell consists of one belt of rising air with low surface air pressure, a zone of sinking air with surface high pressure, a surface wind zone with air flowing generally from the high-pressure belt to the low-pressure belt, and an airflow in the upper atmosphere from the belt of rising air to the belt of sinking air. Though more realistic than the single-cell model, the three-cell model is so general that only fragments of it actually appear in the real world. Nonetheless, the names for many of its wind and pressure belts have become well established in our modern terminology, and it is important that we understand where these hypothesized belts are located.

DID YOU KNOW?

You might wonder why the general circulation of the atmosphere is approximated by three cells, rather than some other number. The wind and pressure belts that make up these cells arise because of interplay between Earth's rotation rate and the energy gradient between the equator and the poles. If Earth rotated faster, we would expect more belts. Thus, Jupiter, which turns on its axis every 11 hours, has many belts, not just three in each hemisphere.

The Hadley Cell

Along the equator, strong solar heating causes air to expand upward and diverge toward the poles. This creates a zone of low pressure at the equator called the **equatorial low**, or the **intertropical convergence zone (ITCZ)**. The upward motions that dominate the region favor the formation of heavy rain showers, particularly in the afternoon. Heavy precipitation associated with the ITCZ is observable on weather maps and satellite images (Figure 8–5). Notice in the figure that the equatorial low exists not as a band of uniform cloud cover but rather as a zone containing many clusters of convectional storms. The ITCZ is the rainiest latitude zone in the entire world, with many locations accumulating more than 200 days of rain each year. Imagine how listless you might feel in such an environment, with hot, humid afternoons giving way to heavy rain showers all year long. The area is also one in which winds can become light or nonexistent for extended time periods. This monotony is the basis for the old nautical term still used today, the *doldrums*.

▲ **FIGURE 8–5 Intertropical Convergence Zone (ITCZ).** The ITCZ is observable on this satellite image as the band of convective clouds and showers extending from northern South America into the Pacific.

Within the Hadley cell, air in the upper troposphere moves poleward to the subtropics, to about 20° to 30° latitude. As it travels, it acquires increasing west-to-east motion. This westerly component is so strong that the air circles Earth a couple of times before reaching its ultimate destination in the subtropics. In other words, the upper-level air follows a great spiraling path out of the tropics, with zonal motion much stronger than the meridional component. Among other things, this explains why material ejected by tropical volcanic eruptions spreads quickly over a wide range of longitudes.

Upon reaching about 20° to 30° latitude, air in the Hadley cell sinks toward the surface to form the **subtropical highs**, large bands of high surface pressure. Because descending air warms adiabatically, cloud formation is greatly suppressed and desert conditions are common in the subtropics. The subtropical highs generally have weak pressure gradients and light winds. Such conditions exert minimal impact on long-distance travel today. But in preindustrial days when oceangoing vessels depended on the wind, its prolonged absence could be catastrophic. Ships crossing the Atlantic from Europe to the New World risked getting stranded in mid-ocean while crossing the subtropics. Often among the cargo were cattle and horses to be brought to the New World, and legend has it that the crews of stalled ships threw their horses overboard. The jettisoned cargo has lent its name to what we colloquially call the **horse latitudes**.

In the Northern Hemisphere, as the pressure gradient force directs surface air from the subtropical highs to the ITCZ, the weak Coriolis force deflects the air slightly to the right to form the **northeast trade winds** (or simply the *northeast trades*). In the Southern Hemisphere, the northward-moving air from the subtropical high is deflected to the left to create the **southeast trade winds**. The Hadley Cell cross-section in Figure 8–4b shows that the trade winds are fairly shallow. Moving upward through the troposphere, the easterly motions weaken and are eventually replaced by westerly motion

8–1 PHYSICAL PRINCIPLES

Maintaining General Circulation

Unceasing motion is undoubtedly the most conspicuous feature of the atmosphere. Across the planet winds ranging from gentle breezes to hurricane strength are the rule. Even when calm conditions arise in some locales, they don't last long. This is obvious from even the most casual observation, but the reasons are not so obvious. Why must the atmosphere continually move? How are the circulation systems described in this chapter maintained? As described in this box, answers to these questions involve considering transformations of energy and momentum within the Earth–atmosphere system.

The Global Atmospheric Heat Engine

At the largest scale, that of the globe, we know from Chapter 3 that energy gained by the planet from the Sun balances longwave radiation emitted by the atmosphere and surface. In particular, about 70 percent of incoming solar radiation is absorbed (238 W/m^2) and an equal amount of radiation is emitted to space. (To keep the discussion simple, we are ignoring the prospect for climate change.) Atmospheric motions represent kinetic energy (KE), all of which is ultimately traceable to the input solar radiation. So, some part of absorbed solar radiation must be converted to the energy of motion. Just as a car's engine converts the heat of burning fuel into motion, the atmospheric heat engine converts solar radiation into motion. However, in the atmosphere transformation of heat energy to KE is far less direct than in a car engine.

Recall from Chapter 2 that potential energy (PE) is energy that could possibly be converted to motion. The child in Figure 8–1–1a generates PE by tossing sand onto a sand pile. Lifting sand against the force of gravity requires energy, and that energy is stored as PE in the sand pile. Frequently, part of the sand pile collapses and some of the sand tumbles down. These minor landslides occur when the difference in potential energy between sand grains exceeds the ability of friction to prevent movement. Clearly, the PE provided by the child is converted to KE. Similar considerations regarding PE and KE apply to the atmosphere. The potential energy of an air column arises partly from its temperature. A warm column has more PE than a cold column. Another part of PE, about 40 percent, arises from the vertical position of molecules in the column. To lift a parcel of air against the force of gravity requires energy, and that energy is stored as PE.

In the atmospheric heat engine, the Sun warms the planet and raises the potential energy of the atmosphere, just as the child raises the potential energy of the sand. If the surface is warmed by the Sun and heats the overlying air column, its PE increases. Not only does the temperature increase, but the column expands vertically and therefore air molecules are on average higher than before. Thus, both parts of the column's PE increase: PE due to temperature and PE due to the molecules' vertical position.

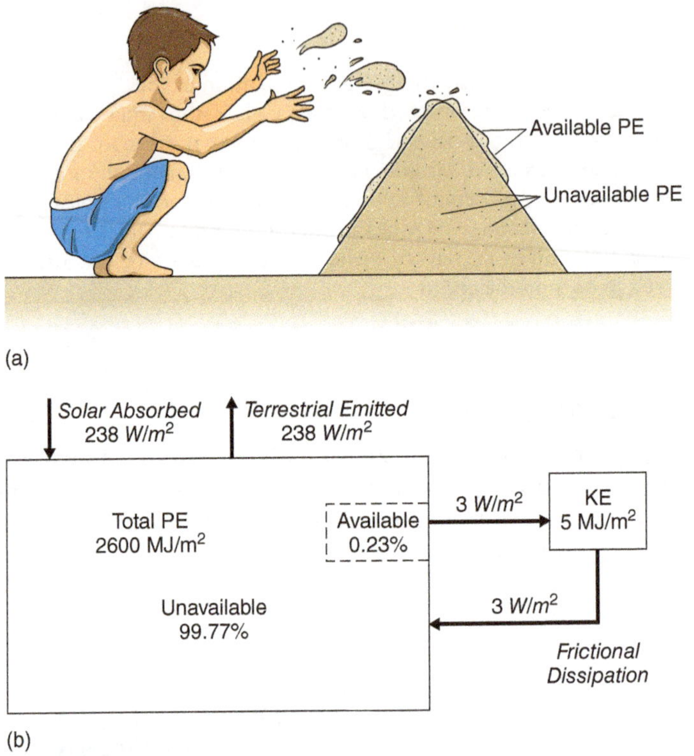

▲ **FIGURE 8–1–1 Potential and Kinetic Energy Conversions.** (a) The child adds potential energy (PE) to the pile by raising sand grains against the force of gravity. As grains tumble down, their PE is converted to the energy of motion, kinetic energy (KE). (b) Atmospheric motion is maintained by continuous conversion of PE to KE within the Earth system. Winds eventually shed their energy through frictional heating, which helps maintain longwave emission by the planet to space.

About half of the atmosphere's mass is below the 500 mb level and so the center of gravity for the atmosphere must be at about 5.5 km. If this leads to you think that vast amounts of PE are held in the atmosphere, you are right. However, most of a column's PE is not available for conversion to KE, because air parcels aloft are supported by parcels below. The same is true of the sand pile; all of the grains in the interior are supported from below and cannot be moved. Only if circumstances are right can conversion to kinetic energy result. In a sand pile, movement happens on the slope, where there is a gradient in PE—and so it is in the atmosphere. Loosely speaking, if the height of the center of gravity in two adjacent columns of air is not the same, we can expect KE (and wind) to result. Or, if dense air is found above less dense air, overturning of the air column will result, again converting PE to KE.

Figure 8–1–1b illustrates this production and conversion of energy in a schematic form. Energy inputs and outputs to Earth balance, maintaining a PE store of about 2600 million joules (MJ) per square meter of Earth. Only a tiny fraction of that huge inventory, 0.23 percent, is available for conversion to KE. As seen in the figure, the rate of conversion is about 3 W/m^2. This is enough to sustain all of the motions we observe, such as the huge belts of prevailing winds and ocean currents seen on the maps. It also includes all manner of "disturbances," ranging from hurricanes and tornadoes down to the smallest puffs of wind. Notice, however, that at 3 W/m^2 the average rate of energy conversion is small, roughly 1.3 percent of incoming solar radiation. Clearly, the atmosphere engine is quite inefficient.

If 3 W/m^2 is extracted from the store of PE, that same amount of heat energy must be returned so that planetary PE doesn't decrease over time and so that Earth can emit the full 233 W/m^2 required for energy balance. In other words, it can't remain as kinetic energy but must be added back to the pool of PE. By the same token, kinetic energy can't simply accumulate indefinitely, because that would lead to infinitely high winds! In other words, the store of KE must be constant, which requires losses that balance the input.

The conversion back to PE is accomplished by friction. Frictional losses occur as air blows past the surface. They also occur within the atmosphere because of turbulence. If you blow hard air rushes out of your lungs, but the puff doesn't cross the room. There is friction as it moves through the surrounding air, and that friction robs the puff of energy. Eventually the room air is as quiet as before, and the energy of motion has been converted to heat. This is called **frictional dissipation** and corresponds to production of PE. As seen in the figure, the average rate of frictional dissipation equals the generation of kinetic energy. Most of this occurs in the lower troposphere, where friction is especially important.

Constant movement implies a pool of KE, just as there is a pool of PE. The KE store is not large compared to PE, only about 5 million joules per square meter (MJ/m^2). It is interesting to think about how long that store would last if depleted at a rate of 3 W/m^2. Remembering that a watt is one joule per second, the calculation is straightforward:

$$\frac{5{,}000{,}000 \text{ J m}^{-2}}{3 \text{ J sec}^{-1}\text{m}^{-2}} = 1{,}666{,}667 \text{ sec}$$

We see that there is enough KE present to maintain motions for about 1.6 million seconds, or just 19 days! Just as the atmosphere's store of water is small compared to evaporation and precipitation, the store of KE is small compared to its production and loss.

Momentum Exchange and Atmospheric Motion

Maintaining atmospheric motion also involves momentum transfers. In this case **angular momentum** is the quantity in question. Imagine a basketball player spinning a ball on his or her finger. Even though the ball doesn't move through space, it possesses momentum called angular momentum. To stop the ball from spinning or to make it go faster, we must remove or add angular momentum. Likewise, the Earth–atmosphere system has angular momentum by virtue of its rotation. Most is earth momentum, but the atmosphere contributes to the total because it spins with the surface below. Ignoring weak tidal forces from other objects in the solar system, nothing is pushing on the planet, and the total angular momentum for the planet is nearly constant.

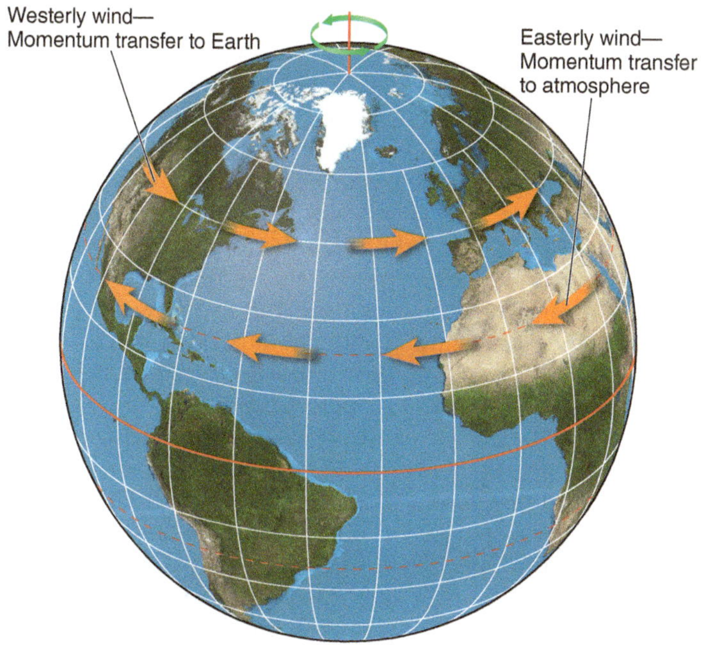

▲ **FIGURE 8–1–2 Angular Momentum Transfer to and from the Surface.** Westerly winds blow in the direction of rotation and transfer momentum to the surface below. By contrast, easterly winds represent momentum transfer to the atmosphere.

Consider locations where the average surface winds are westerly. Blowing from west to east, they are pushing in the direction of rotation, and clearly add angular momentum to the earth (Figure 8–1–2). In other words, westerly winds transfer angular momentum from the atmosphere to Earth. Easterly winds do just the opposite, transferring angular momentum from Earth to the atmosphere. Clearly, it must be that averaged over the whole planet, easterly and westerly momentum exchange must balance in order to maintain atmospheric motion. Just as there is no net energy loss or gain for the planet, total angular momentum is constant. Were that not the case, winds would either cease or increase forever.

As a final point, we should emphasize that the numbers given are global averages. In some places the production of PE by solar radiation is not too different than the rate of conversion to KE, and winds are sustained on a more-or-less steady basis. The Hadley cells are one example. The fairly continuous nature of the trade winds is maintained by roughly continuous conversion of PE at a rate close to what is provided by the Sun in the tropics. This is like a playground carousel, where spinning is maintained by repeated small pushes, whose energy transfer is just large enough to offset frictional costs. On the other hand, oftentimes KE production rates in the atmosphere are thousands of times larger than solar energy input, as seen in tornado winds that move fully loaded train cars. Or, to take another example, how could a thunderstorm develop at night when the Sun isn't even shining? Clearly, storms don't violate energy conservation and their KE doesn't come out of nowhere. Such disturbances are produced when available PE is concentrated in small areas and released in bursts of KE production. Just how that happens is a story of its own covered in later chapters.

8–1–1 Car engines are about 20 percent efficient. How does this compare to the atmospheric heat engine?

8–1–2 What role does frictional dissipation play in the heat engine?

8–1–3 Do easterly winds add or withdraw angular momentum from the solid earth?

aloft. Together, the subtropical highs, the equatorial low, the trade winds, and the upper-level westerly motions form the Hadley cells. Because it is produced thermally, the Hadley circulation is strongest in the winter season, when temperature gradients are strongest.

The Ferrel and Polar Cells

According to the three-cell model, the Hadley cell accounts for the movement and distribution of air over about half of Earth's surface. Immediately flanking the Hadley cell in each hemisphere is the Ferrel cell, which circulates air between the subtropical highs and the **subpolar lows**, or areas of low pressure. On the equatorial side of the Ferrel cell, air flowing poleward away from the Northern Hemisphere subtropical high undergoes a substantial deflection to the right, creating a wind belt called the **westerlies**. In the Southern Hemisphere, the pressure gradient force propels the air southward, but the Coriolis force deflects it to the left—thus producing a zone of westerlies in that hemisphere as well. Unlike the Hadley cell, the Ferrel cell is envisioned as an indirect cell, meaning that it does not arise from differences in heating, but is instead caused by turning of the two adjacent cells. Imagine three logs placed side by side, touching one another. If the two outer logs are turned in the same direction, friction will set the middle one in motion. Thus, the Ferrel cell shows the same kind of overturning as the other cells, but for different reasons.

In the polar cells of the three-cell model, surface air moves from the **polar highs** to the subpolar lows. Like the Hadley cells, the polar cells are considered thermally direct circulations. Compared to the poles, air at subpolar locations is slightly warmer, resulting in low surface pressure and rising air. Very cold conditions at the poles create high surface pressure and low-level motion toward the equator. In both hemispheres, the Coriolis force turns the air to form a zone of **polar easterlies** in the lower atmosphere.

The Three-Cell Model vs. Reality: The Bottom Line

Do the wind and pressure belts of the three-cell model adequately describe real-world patterns? The answer is: sort of. We have already seen that the ITCZ is real enough to be observed from space and that many deserts exist in their predicted locations. Furthermore, the trade winds are the most persistent on Earth. We would have to say that the Hadley circulation provides a good account of low-latitude motions. On the other hand, the Ferrel and polar cells are not quite as well represented in reality, though they do have some manifestation in the actual climate.

With regard to surface winds, much of the middle latitudes experience the strong westerly winds depicted by the model, especially in the Southern Hemisphere. Of course, local conditions often override this tendency (in fact, much of the central United States is dominated by a southerly flow during the summer). It is even more difficult to observe a persistent pattern of polar easterlies in the overall wind regime. They emerge in long-term averages, but are not a prevailing wind belt as the trades are.

With regard to upper-level motions, the three-cell model is not realistic at all. For example, where the Ferrel cell implies easterly motion in the upper troposphere, there is overwhelming westerly wind. Moreover, large overturning cells do not exist outside of the Hadley zones. Thus, the three-cell model mainly provides a starting point for a more detailed account. But perhaps its failures aren't surprising, given that it doesn't consider land–ocean contrasts or the influence of surface topography, two factors that surely ought to influence planetary winds and pressure.

CHECKPOINT

8.5 How do rainfall and wind vary across a Hadley cell, from the equator to about 20° to 30° latitude? Explain.

8.6 What are the shortcomings of the three-cell model?

Semipermanent Pressure Cells

8.4 Describe the distribution and effects of semipermanent pressure cells.

The three-cell model provides a good beginning for describing the general distribution of wind and pressure, but the real world is not covered by a series of belts that completely encircle the globe. Instead, we find a number of alternating **semipermanent cells** of high and low pressure, as shown in Figure 8–6. They are called *semipermanent* because they undergo seasonal changes in position and intensity over the course of the year. Some of these cells result from temperature differences, and others from dynamical processes (meaning that they are related to the motions of the atmosphere). Among the most prominent features in the Northern Hemisphere during winter (Figure 8–6a) are the **Aleutian** and **Icelandic lows**—over the Pacific and Atlantic Oceans, respectively—and the **Siberian high** over central Asia. In summer (Figure 8–6b), the best-developed cells are the **Hawaiian** and **Bermuda–Azores highs** of the Pacific and Atlantic Oceans and the **Tibetan low** of southern Asia.

The size, strength, and locations of the semipermanent cells undergo considerable change from summer to winter. During the winter, a strong Icelandic low occupies a large portion of the North Atlantic, while the Bermuda–Azores high appears as a small, weak anticyclone. During summer, the Icelandic low weakens and diminishes in size, and the Bermuda–Azores high strengthens and expands. Even more dramatically, the Siberian high of the winter in interior Asia gives way to the Tibetan low of summer. As we will see

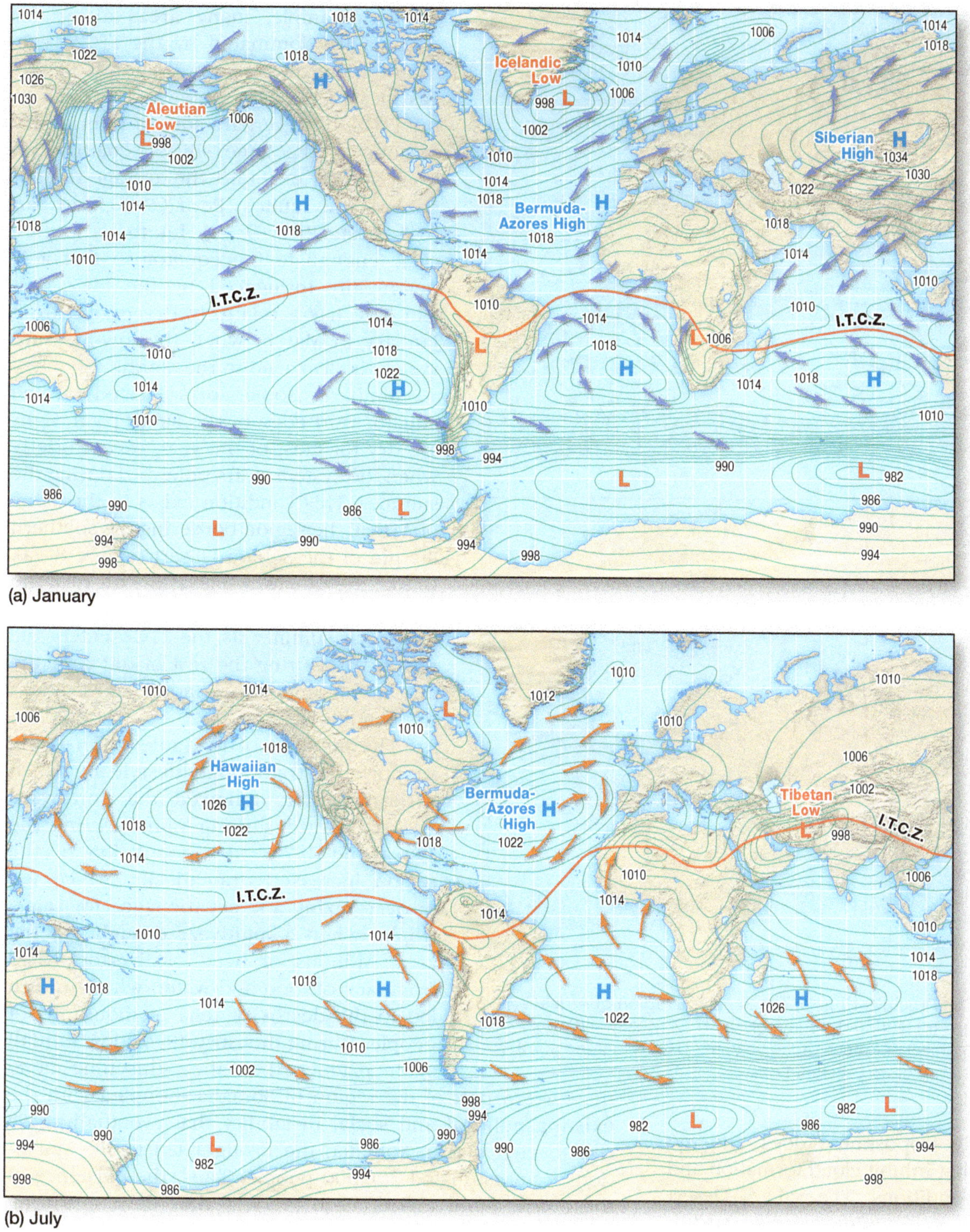

◂ **FIGURE 8–6 Mean Sea Level Pressure.** Sea-level pressure for (a) January and (b) July.

later in this chapter, the seasonal shift in the distribution of semipermanent cells plays a major role in one of Earth's most important circulation patterns—the monsoon of southern and southeastern Asia.

We mentioned earlier that the Hadley cell is fairly easy to see in the real world, with the ITCZ appearing over much of the equatorial regions and the subtropical highs existing over much of the subtropics. But as you can also see in Figure 8–6, the subtropical highs exist primarily over the oceans (as the Hawaiian and Bermuda–Azores highs over the Northern Hemisphere) and not over land. Despite the absence of pronounced high sea-level pressure over the subtropical land masses, the

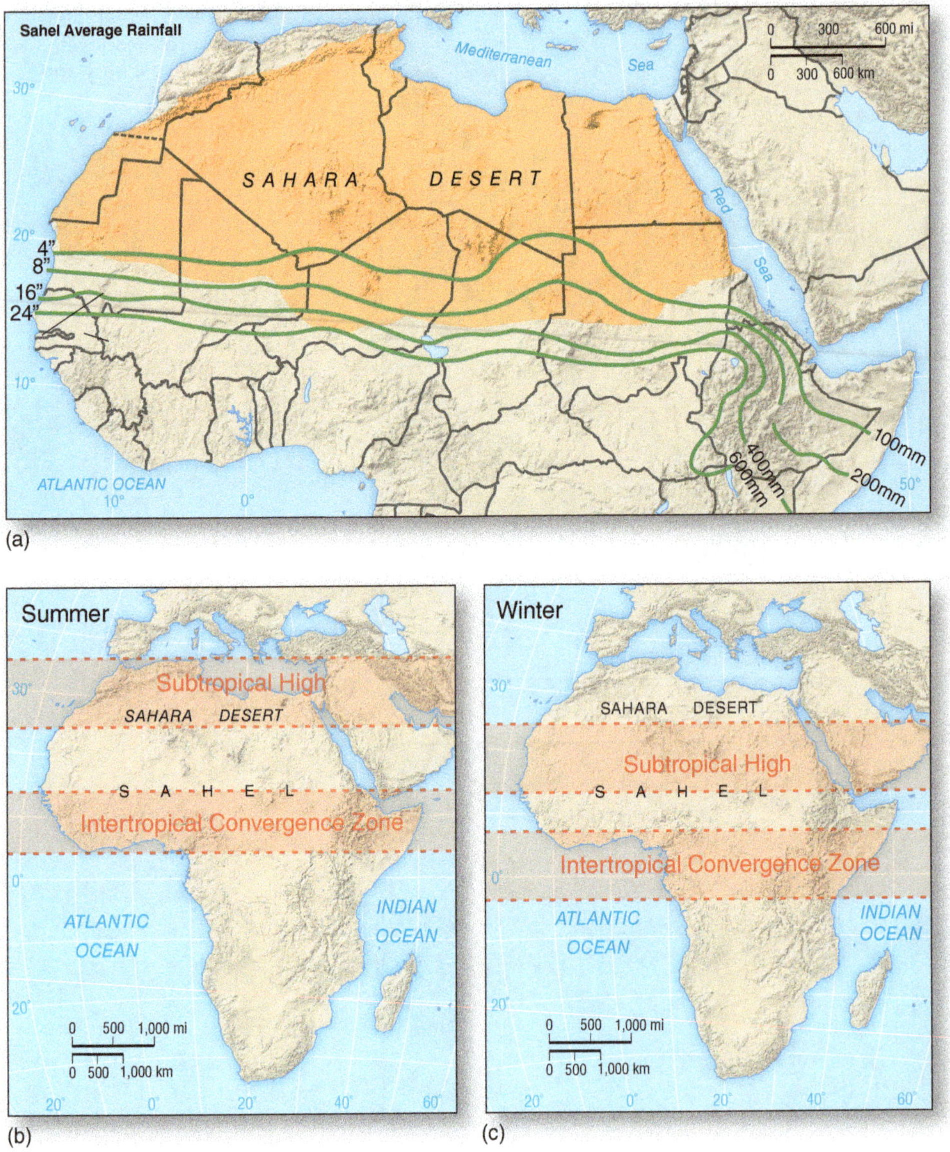

▲ **FIGURE 8–7 Effect of the ITCZ on Precipitation.** (a) The Sahel is a region of Africa bordering the southern Sahara Desert. (b) During the summer, the ITCZ usually shifts northward and brings rain to the region. (c) For much of the year, the ITCZ is located south of the Sahel, and the region receives little or no precipitation.

Video
Global Fire Patterns
http://goo.gl/gne63L

air in the middle troposphere does undergo sinking motions that inhibit cloud formation and promote desert conditions. The Sahara Desert of northern Africa, the interior desert of Australia, and the deserts of the southwestern United States and northwest Mexico clearly reflect this sinking process.

As the solar declination changes seasonally, so does the zone of most intense heating. Knowing that the Hadley cells are thermal, we might expect the associated pressure and wind belts to move seasonally, and indeed they do. Although with a lag of several weeks, the ITCZ, subtropical highs, and trade winds all follow the "migrating Sun." This movement has a major impact on many of the world's climates—and on the people who inhabit them.

For example, many areas along the equator are dominated by the ITCZ year-round and experience no dry season. Areas located near the poleward margins of the ITCZ, however, are subject to brief dry seasons as the zone shifts equatorward. Compare, for example, the average rainfall patterns for Iquitos, Peru (3° S), and San Jose, Costa Rica (9° N). Iquitos is located close enough to the equator so that it is perennially influenced by the ITCZ, but San Jose has a relatively dry period from January to March, when the low-pressure system is displaced to the south.

Similarly, some areas located on the equatorward edge of the subtropical highs are dry for most of the year, except briefly when the system shifts poleward during the summer. This condition exists in the Sahel of Africa, the region bordering the southern margin of the Sahara Desert (Figure 8–7a). Unlike the Sahara, which is dry all year, the Sahel normally experiences a brief rainy period each summer as the ITCZ enters the region (b). During the rest of the year, the descending air of the Hadley cell leads to dry conditions (c).

The migration of the Hadley cell has long supported a traditional lifestyle in which African herders followed the northward and southward shifting rains. During the 1960s and 1970s, the population of the region increased dramatically, which led to overgrazing and set the stage for catastrophe when a multiyear drought hit the area. Millions of head of livestock died from lack of food and water, which in turn led to the deaths of tens of thousands of people. During the early 1990s, low rainfall again plagued the Sahel—this time along the eastern part of the region in Somalia. Coupled with the existing political and social instability that eventually gave rise to civil war in Somalia, the drought led to the starvation of an estimated 300,000 people. Thus, the existence of these cells—and variations in the way they develop during unusual years—are more than mere abstractions. They have real-world ramifications for millions of people.

CHECKPOINT

8.7 What are the best-developed semipermanent pressure cells during winter in the Northern Hemisphere?

8.8 Explain how and why the subtropical highs affect the climates of the continents adjacent to them.

8.9 How does the seasonal shift of the ITCZ affect the lives of people in sub-Saharan Africa?

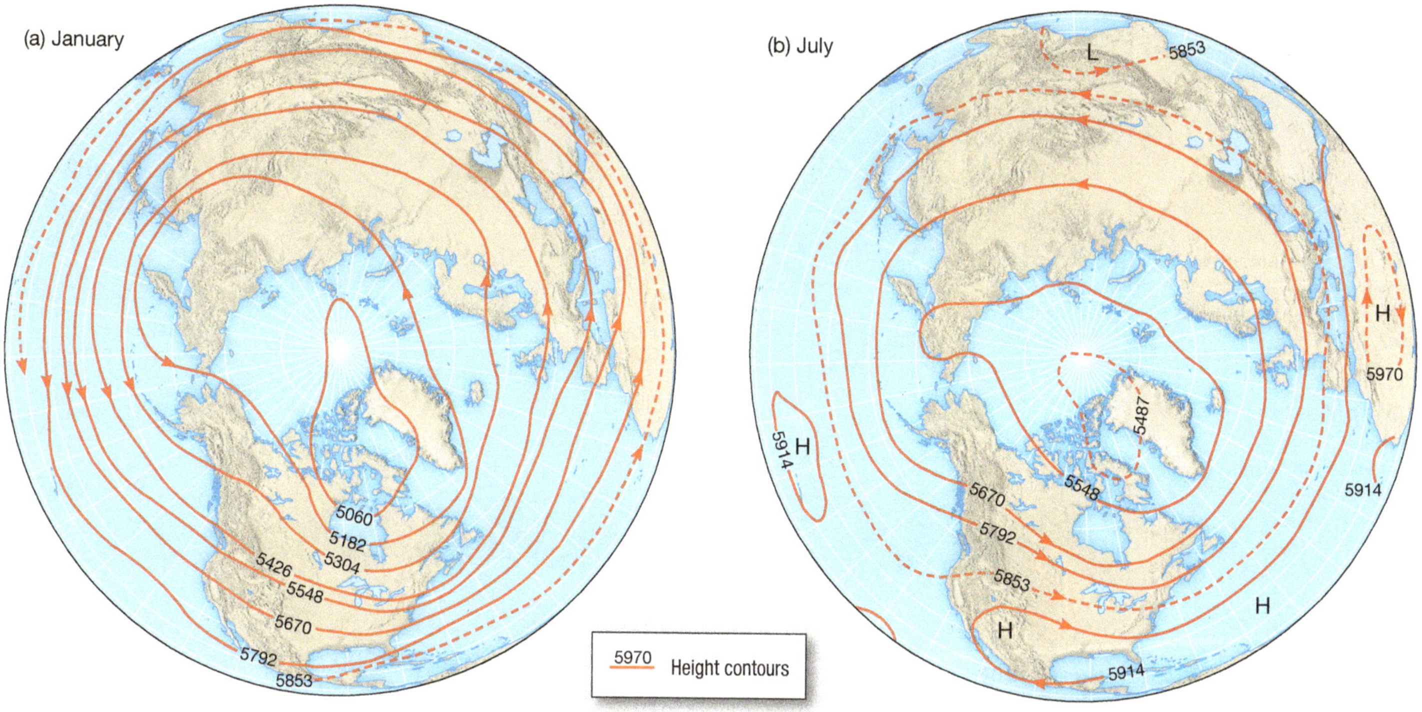

▲ **FIGURE 8–8 Mean Heights of the 500 mb Level.** Mean heights (in meters) for (a) January and (b) July with arrows showing the direction of flow. The pattern is mostly zonal with decreasing heights toward the poles.

The Upper Troposphere

8.5 Explain the distribution of wind and pressure in the upper troposphere.

In Chapter 4 we saw that pressure decreases more rapidly with altitude where the air is cold. We also know that temperature in the lower troposphere generally decreases from the subtropics to the polar regions. These two principles are critical in understanding the distribution of wind and pressure in the upper troposphere.

Figure 8–8 maps the global distributions of the mean height of the 500 mb surface (a convenient level representing conditions in the middle troposphere) for January (a) and July (b). In both months, the height of the 500 mb level exhibits a strong tendency to decrease toward the polar regions due to the lower temperatures found at higher latitudes. In January, the average height of the 500 mb surface over much of the southern United States is about 5670 m (18,600 ft), while over northern Canada it decreases to less than 5300 m (17,400 ft). A similar but less extreme change occurs in July as well.

Three features stand out from the maps in Figure 8–8. First, for both January and July, the 500 mb heights are greatest over the tropics and decrease with latitude. Second, the gradient in height is greater in the winter (the Southern Hemisphere in July and Northern Hemisphere in January). Third, at all latitudes the height of the 500 mb level is greater in the summer than during the winter. All three of these features result from the general distribution of temperature in the lower-middle atmosphere; areas of warm air have greater 500 mb heights.

Westerly Winds in the Upper Atmosphere

Recall from Chapter 4 that height differences correspond to pressure differences and that when the 500 mb surface slopes steeply a strong pressure gradient force exists. We can therefore infer from the 500 mb maps that there is always a pressure gradient force across the middle latitudes trying to push the air toward the poles. Of course, in the absence of friction, the winds do not blow poleward, but rather blow parallel to the height contours, from west to east. The pressure gradient force is strongest in winter (the height contours are closely spaced), so the upper-level westerlies are strongest in winter. What does this mean to you? For one thing, it explains why most midlatitude weather systems migrate from west to east. In other words, it tells us why a storm over Chicago might find its way over the East Coast a day or two later, but seldom if ever does such a storm make the reverse trip.

The predominance of westerly winds in the upper troposphere also affects aviation. For example, a commercial aircraft going from Chicago to London has an estimated flight time of about 7.5 hours, while the return trip normally takes an hour longer because it must overcome headwinds. The difference in flight time would be even greater were it not

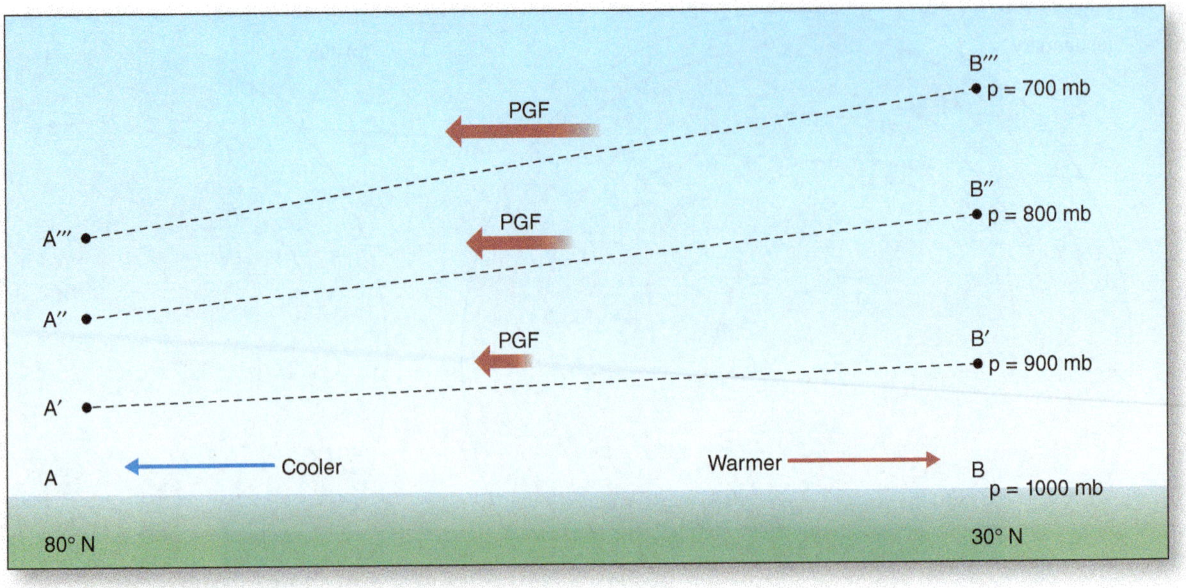

▸ **FIGURE 8–9 Pressure Gradients and Latitude.** Latitudinal temperature gradients cause pressure surfaces to slant poleward. In this example, we assume a constant gradual decrease in temperature with latitude. The intensity of the pressure gradient force remains constant from one latitude to another but increases with altitude.

MapMaster
North America
Physical Environment
Pressure

for the fact that airlines route their planes to take advantage of tailwinds and avoid headwinds.

Wind speeds generally increase with height between the surface and the tropopause. Partly this is because of decreasing friction, but more importantly, the pressure gradient force is typically stronger at high altitudes. As illustrated in Figure 8–9, the surfaces representing the 900, 800, and 700 mb levels all slant downward to the north, but not by the same amount. Higher surfaces slope more steeply, which means that the pressure gradient force is greater. You may recall from Chapter 4 that the pressure gradient force is directly proportional to slope, without regard to density.

But why do those higher surfaces slope more steeply? The air is warmer at point *B*, so the layer of air from 900 mb to 1000 mb is thicker at point *B* than at point *A*. Similarly, the thickness from 800 mb to 900 mb is greater at point *B*. In other words, the height change from *B′* to *B″* is greater than the change from *A′* to *A″* It must be, therefore, that the 800 mb surface slopes downward more than the 900 mb surface (*B″* − *A″* is greater than *B′* − *A′*). The difference in heights between successive surfaces continues to increase upward, leading to stronger winds.

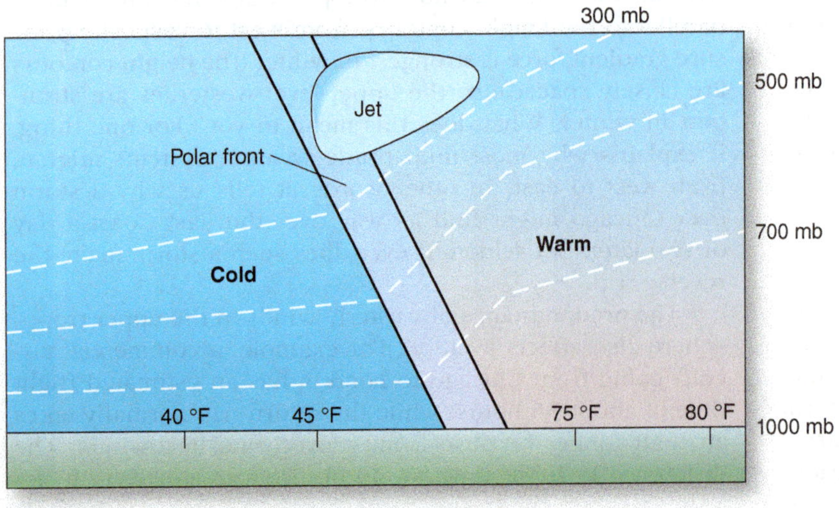

▲ **FIGURE 8–10** The polar jet stream is situated above the polar front near the tropopause, where temperature contrasts are strongest.

The Polar Front and Jet Streams

The gradual change in temperature with latitude depicted in Figure 8–9 does not always occur in reality. Instead, areas of gradual temperature change often give way to narrow, strongly sloping boundaries between warm and cold air. One such boundary, the **polar front**, is shown in Figure 8–10.

Outside of the frontal zone, the changes in temperature with latitude are gradual (as they were in Figure 8–9), and the slopes of the 900, 800, and 700 mb levels respond accordingly. But within the front, the slope of the pressure surfaces increases greatly because of the abrupt horizontal change in temperature. With steeply sloping pressure surfaces, a strong pressure gradient force occurs, resulting in the **polar jet stream**. Thus, we see the jet stream as a consequence of the polar front, arising because of the strong temperature gradient. At the same time, the jet stream reinforces the polar front. In Chapter 10, we will see that a jet stream is necessary to maintain the temperature contrast across the front.

Jet streams can be thought of as meandering “rivers” of air, usually 9 to 12 km (30,000 to 40,000 ft) above sea level (Figure 8–11). Their wind speeds average about 180 km/hr (110 mph) in winter and about half that in summer, though peak winds can exceed twice these values. Like rivers on land, jet streams are highly turbulent, and their speeds vary considerably as they flow. Unlike rivers on land, they have no precisely defined banks. Furthermore, a single jet stream will often diverge at some point and fork off into two distinct jets. Thus, the locations and boundaries of these features on weather maps are often difficult to accurately pinpoint.

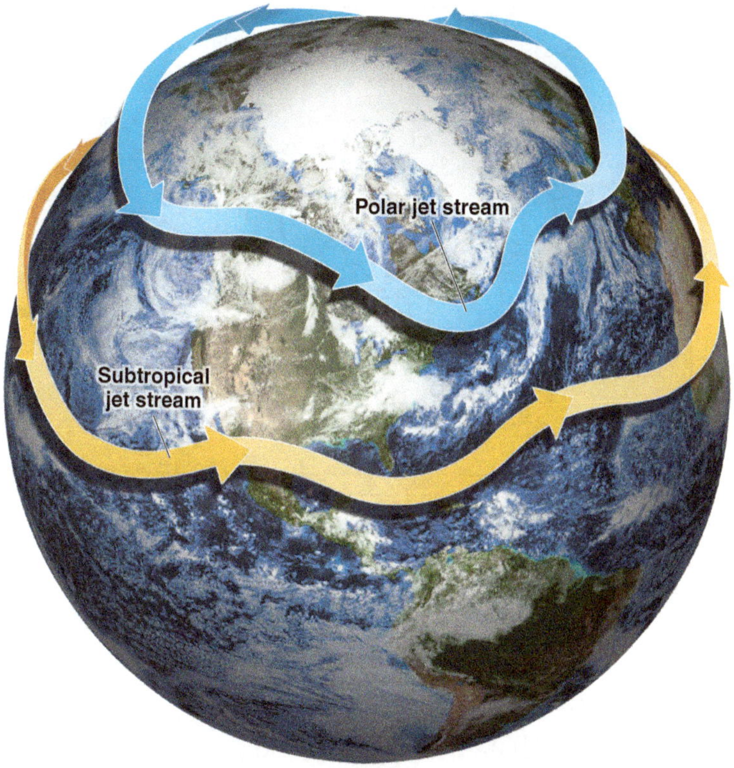

▲ **FIGURE 8–11** Jet streams are localized areas of intense winds in the upper troposphere. We often see both a polar jet stream and subtropical jet stream in each hemisphere.

The polar jet stream greatly affects daily weather in the middle latitudes. Nearer the equator is another prominent jet stream, the **subtropical jet stream**, associated with the Hadley cell. As the upper-level air flows away from the ITCZ, the conservation of angular momentum imparts ever-growing westerly motion. When moving toward the northeast, the subtropical jet stream can bring with it warm, humid conditions. Figure 8–12 shows the flow of moisture associated with a subtropical jet stream.

Troughs and Ridges

Although on average 500 mb heights decrease toward the poles, at any given time significant departures from the general trend will exist. Typically undulations, or waves, are superimposed on the overall decrease in height toward the poles. Figure 8–13 illustrates this. It shows an axis of low height in the middle of the United States, flanked by "mountains" of high heights on either side. The valley of low heights is called a *trough*, and the upward bulges are called *ridges*. Also shown on the diagram are height contours—notice that they, too, have a wavelike character. The air flows parallel to the contours, so there is wavelike motion to the air stream as well.

Figure 8–14 shows simplified contour maps depicting the relationship between 500 mb heights and ridges and troughs. No east–west height changes (no trough, no ridge) occur in (a), and the flow is completely zonal. In (b), on the other hand, a trough is in the midsection of the country. Going from point 1 to 2, there is a height decrease from 5610 m to 5580 m. Going from 2 to 3, heights increase again. In other words, going from 1 to 3 requires us to cross a valley (trough) of low heights. We see that height contours are displaced toward the equator in troughs and toward the pole in ridges. We also see that air winding its way poleward around ridges and equatorward around troughs will have a meridional component as well as a zonal component. In fact, when waves are pronounced, we say the flow is "meridional," as opposed to "zonal" when the flow is nearly all westerly.

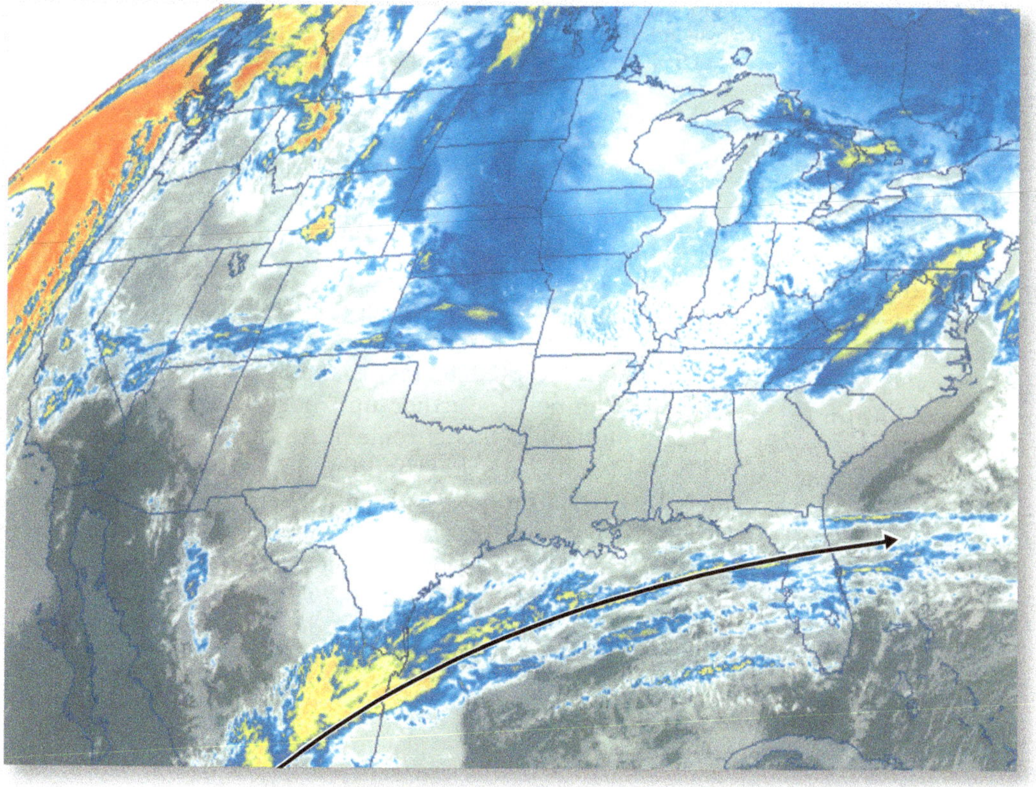

◄ **FIGURE 8–12** The subtropical jet stream appears in this infrared satellite image as the band of cloudiness extending from Mexico through Florida. The flow in this subtropical jet stream is from southwest to northeast as shown by the black arrow. Note that redder colors generally indicate thicker cloud cover.

Animation
Jet Stream and Rossby Waves
http://goo.gl/QUMn5i

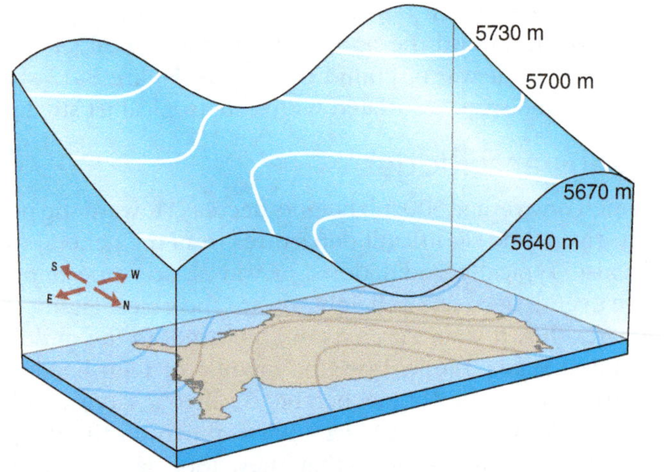

▲ **FIGURE 8–13 Hypothetical View of the 500 mb Surface Looking from Northeast to Southwest.** Heights decrease from south to north but also rise and fall through the ridges and trough. Vertical changes are highly exaggerated in the figure. Actual height changes are very small compared to the size of the continent.

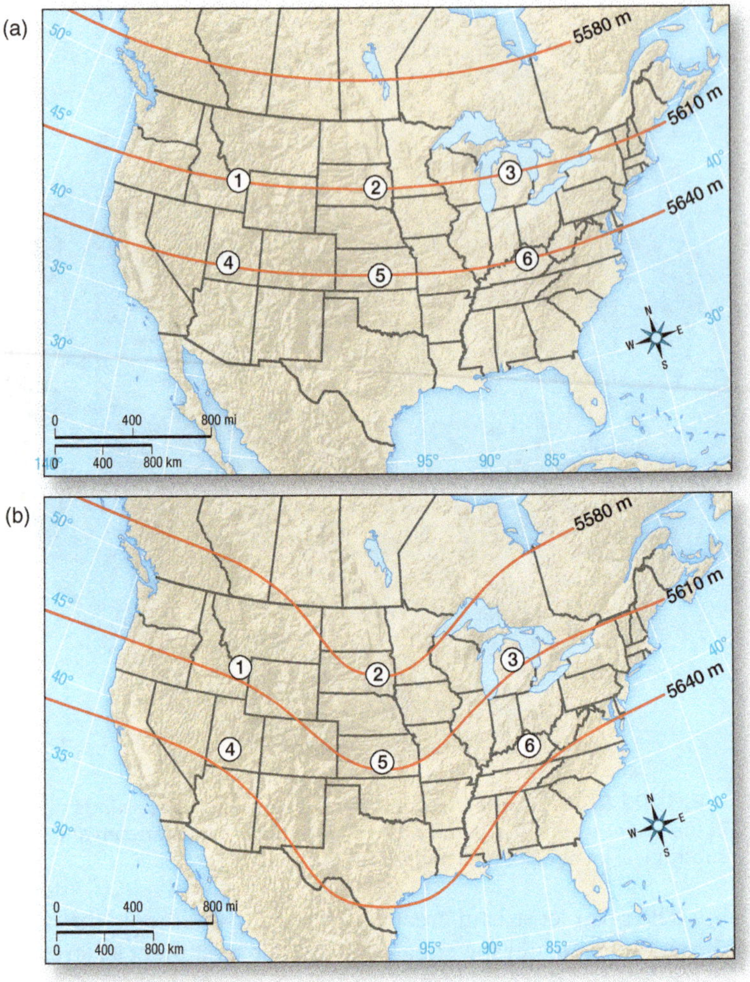

▲ **FIGURE 8–14 Comparing Zonal Flow to a Trough at the 500 mb Level.** Troughs occur in the middle troposphere where the 500 mb height contours dip equatorward. (a) The positions 1–3 have the 500 mb level at 5610 m. Farther to the south, at positions 4–6, the 500 mb level is at 5640 m. (b) The contour lines are in the same position over the East and West Coasts as they were in (a), but they shift equatorward over the central portion of the continent. Thus, positions 2 and 5 have lower pressure than the areas east and west of them. The zone of lower pressure over the central part of the continent is a trough.

Rossby Waves

We have seen that ridges and troughs give rise to wave-like flow in the upper atmosphere of the middle latitudes. The largest of these are called **Rossby waves.**[3] Usually, there are anywhere from three to seven Rossby waves circling the globe. Like other waves, each has a particular *wavelength* (the distance separating successive ridges or troughs) and *amplitude* (its north–south extent). Though Rossby waves can remain in fixed positions, they also migrate west to east (Figure 8–15), or on rare occasions from east to west.

Rossby waves undergo distinct seasonal changes from summer to winter. They tend to be fewer in number, have longer wavelengths and contain their strongest winds during winter. The latter two characteristics—wind speed and wavelength—affect the rate at which Rossby waves migrate downwind (see *Box 8–2, Physical Principles: The Movement of Rossby Waves*).

Rossby waves exert a tremendous impact on day-to-day weather, especially when they have large amplitudes. They are capable of transporting warm air from subtropical regions to high latitudes or cold polar air to low latitudes. Because upper-level air tends to change temperature only slowly in the absence of strong vertical motions, Rossby waves can bring anomalous temperatures to just about any place within the middle to high latitudes. This is illustrated by Figure 8–16, which shows a strong Rossby wave over North America on September 22, 1995. Record-breaking low temperatures for the date were observed over much of the central United States as the wave brought cold air from the far north. Farther upwind, a southwesterly flow brought mild air to the extreme northwest of North America, with Fairbanks, Alaska, basking in temperatures in the mid-20s Celsius (mid-70s Fahrenheit).

In addition to redistributing cold or warm air, Rossby wave patterns have a less obvious impact on local weather. Changes in the flow along the wave lead to *divergence* and *convergence*. When air in the upper atmosphere diverges (or spreads out), it

[3]Named for Carl-Gustaf Rossby, who contributed much pioneering research on upper-level air flow in the early 1900s.

8–2 PHYSICAL PRINCIPLES

The Movement of Rossby Waves

Like all atmospheric phenomena, the movement of Rossby waves results not from mere chance, but from the combined actions of numerous physical forces. Three factors determine the rate at which a Rossby wave propagates: (1) the westerly component of its internal wind speed, (2) its latitudinal position, and (3) its wavelength. In particular, Rossby waves with vigorous winds and shorter wavelengths move most rapidly, as indicated by the formula

$$C = U - bL^2/4\pi^2 \qquad \text{Rossby Formula}$$

where C is the speed at which the wave propagates downwind (m/sec), U is the average westerly component of the wind speed within the wave (m/sec), b is a function of latitude equal to 1.6×10^{-11} m^{-1} sec^{-1} at 45°, and L is the wavelength (m).

The rates of downwind migration for Rossby waves at 45° latitude with various wavelengths and wind speeds are presented in Table 8-2-1. As you can see, a wave having a westerly wind speed of 20 m/sec (45 mph) and a wavelength of 3000 km advances 60 percent faster than does one with the same speed but a wavelength of 5000 km (16 m/sec vs. 10 m/sec).

TABLE 8-2-1

Rates of Downwind Rossby Wave Migration at Latitude 45° N or 45° S

Wavelength	20 m/sec Wind Speed	40 m/sec Wind Speed
3000 km	16 m/sec wave speed	36 m/sec wave speed
5000 km	10 m/sec wave speed	30 m/sec wave speed

For a given wind speed and latitude, there is some particular wavelength, L_{crit}, at which the waves do not migrate at all. Waves longer than this critical value actually migrate from east to west and are said to be *retrograding waves*. By rearranging the equation and setting $C = 0$ (that is, by assuming the waves are stationary), we can determine the critical wavelength:

$$L_{crit} = 2p\sqrt{U/b} \qquad \text{Critical Wavelength}$$

For example, at 45° a Rossby wave with a jet stream speed of 40 m/sec will migrate upwind if its wavelength is more than about 10,000 km. This, however, is a particularly long wavelength. At high wind speeds, retrograding motion occurs only for exceptionally long waves, which are quite rare. For lower wind speeds, the critical wavelength is less (only 5000 km for wind at 10 m/sec). As it happens, shorter waves (with lower wind speeds) are more common. Thus, when westward movement appears, it tends to occur when upper-level winds are weak, not strong.

8–2–1 How is a Rossby wave's wavelength related to wave speed?

8–2–2 What is a *retrograding wave* and under what conditions do they commonly occur?

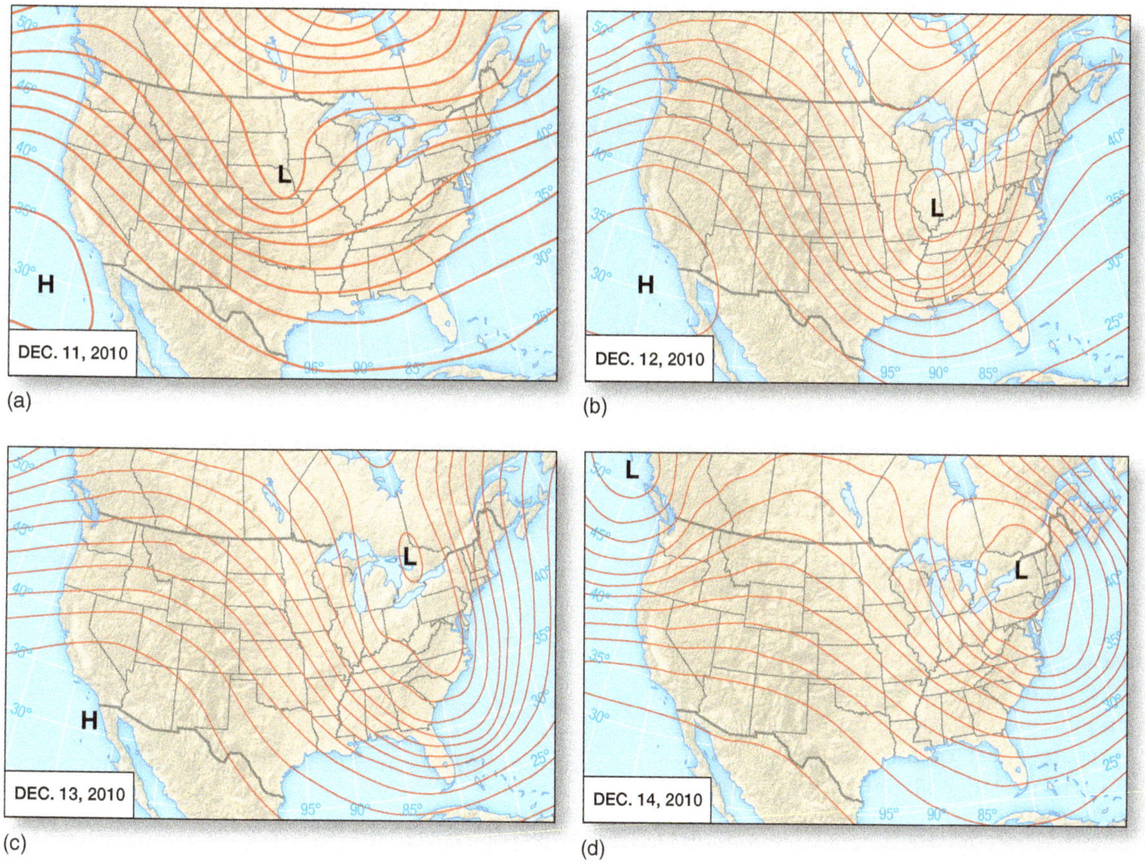

◀ **FIGURE 8–15 Migration of Rossby Waves.** This sequence of 500 mb maps shows the migration of Rossby waves at 24-hour intervals.

▶ **FIGURE 8–16 Rossby Wave Advection.** (a) Rossby waves in the upper atmosphere can advect cold or warm air from one location to another. (b) On September 22, 1995, such a wave brought very mild conditions to interior Alaska, while southward-moving air brought record-breaking low temperatures to much of the central United States.

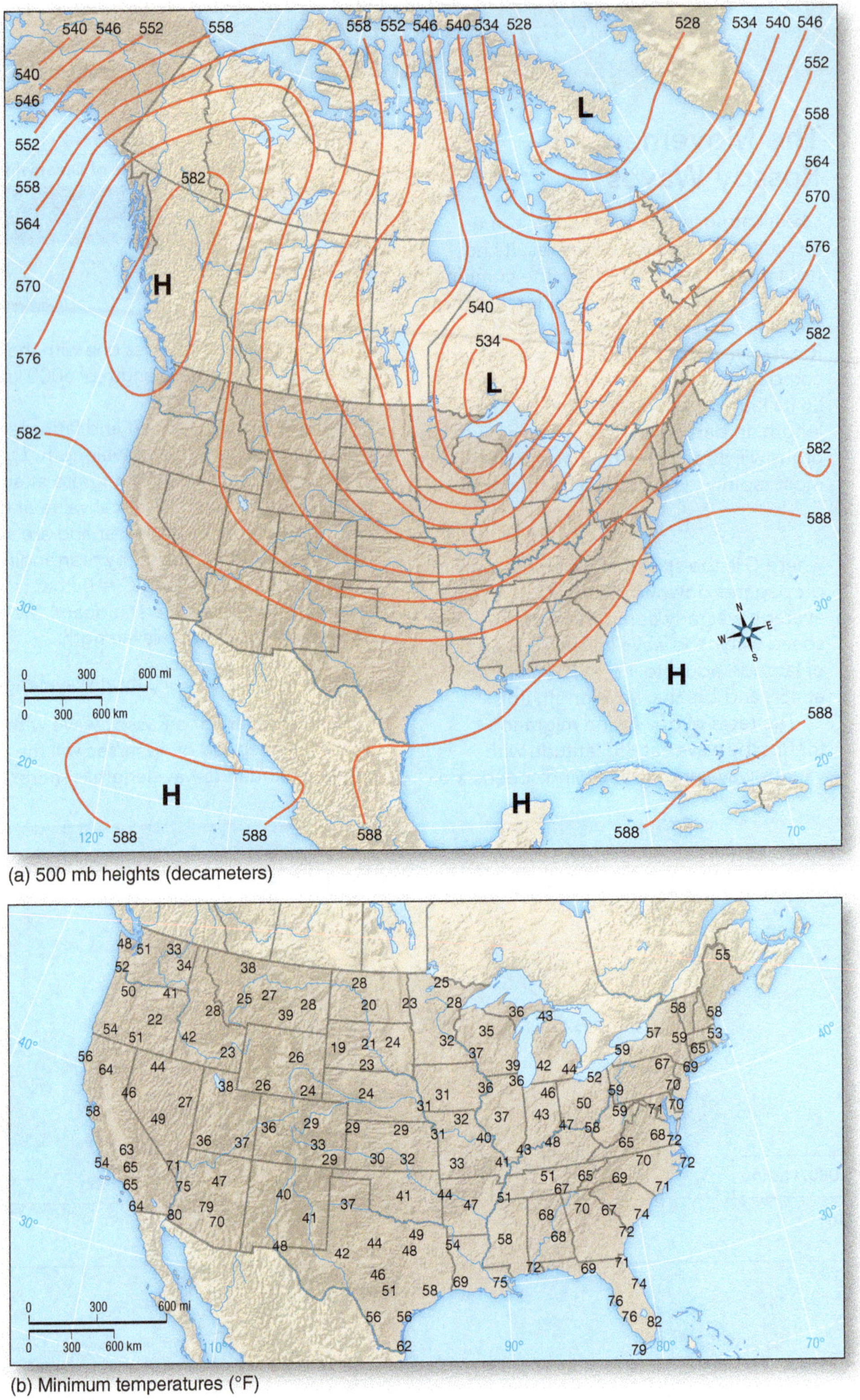

draws air upward from below, causing adiabatic cooling. Thus, divergence in the upper atmosphere can serve as a mechanism for cloud development and precipitation. Convergence in the upper atmosphere has the opposite effect of forcing air downward and inhibiting cloud formation. We will return to this concept in Chapter 10. (*Box 8–3, Physical Principles: The Dishpan Experiment,* elaborates on the nature of Rossby waves.)

CHECKPOINT

8.10 How do winter and summer compare with respect to Rossby waves?

8.11 Why are Rossby waves important in midlatitude weather?

8–3 PHYSICAL PRINCIPLES

The Dishpan Experiment

Although Rossby waves are the largest of the atmospheric waves, other swirling motions of varying sizes likewise exist. Why this complexity? At the simplest level, the behavior of the upper atmosphere is the inevitable result of three factors: (1) the unequal heating of the atmosphere from the equator to the poles, (2) the rotation of the planet, and (3) the inherently turbulent nature of the atmosphere.

To illustrate the interaction of these three, we can reproduce the migrating waves and eddy motions of the upper atmosphere with a relatively simple piece of hardware—a pan of water that rotates at a constant speed with a cooling of the fluid near the center and warming along the edge (Figure 8–3–1). The "dishpan experiment" simulates the rotating Earth with a surplus of net incoming radiation at low latitudes, and a net deficit closer to the poles. Even this very simple exercise yields motions of the fluid that in many ways resemble those of the upper troposphere.

Long waves form in the pan, resembling atmospheric Rossby waves. Superimposed on the long waves are smaller-scale eddies similar to smaller flows on Earth. Changes in the speed of rotation or the differential heating between the edge and center of the pan cause observable changes in the waves and eddies, with more extreme differences in heating and slower rotation rates leading to an increase in the amplitude of large waves at the expense of smaller-scale eddies. This implies that the oscillations in the atmosphere represent an inherent characteristic of any fluid (liquid or gaseous) on a rotating surface with spatially varying inputs of heat. Such observations are not restricted to simple dishpan experiments; elaborate computer models that simulate the motions of the atmosphere reveal similar patterns.

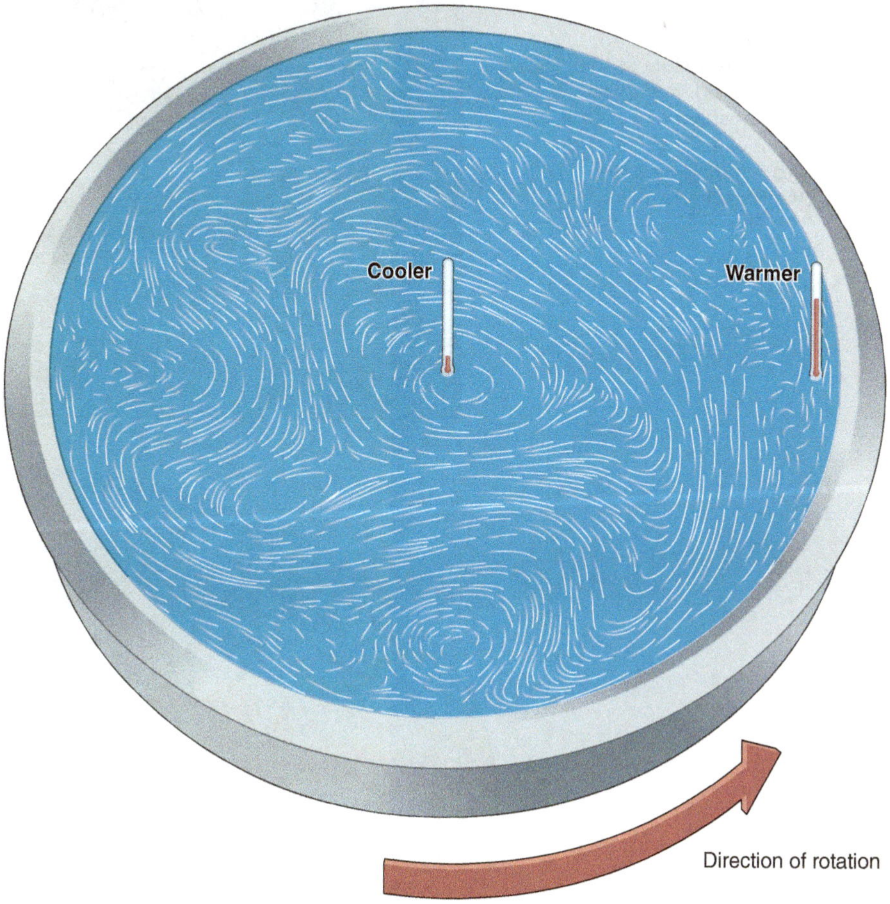

▲ **FIGURE 8–3–1 Eddies.** Pattern of eddies of different size in a "dishpan experiment."

8–3–1 What elements of Earth's system are represented in the dishpan and how?

8–3–2 What are some important factors affecting Earth's general circulation not represented in the dishpan experiment?

Atmospheric Rivers

Water vapor is transported by atmospheric circulation just like every other constituent. But unlike nitrogen, oxygen, and other constant gases, water vapor amounts are highly localized and the movement of water vapor is anything but uniform. Outside of the tropics, most water vapor moves in narrow sinuous corridors about 400 km (250 mi) wide and perhaps 2000 km (1240 mi) long (Figure 8–17a). These are called **atmospheric rivers**, which is appropriate considering their shape and that their flow volume is comparable to the world's largest rivers on land. But it is important to remember that they are streams of vapor, not liquid water. At any given time there are likely to be between three and five atmospheric rivers in various places around the world, each lasting about a day on average. Globally, about 130 occur each year, and they are estimated to be responsible for 90 percent of the meridional transport of water vapor in the middle latitudes.

Atmospheric rivers form in two different ways. Some are rooted in evaporation from warm subtropical oceans, which supplies water vapor transported poleward out of the subtropical highs. Figure 8–17a provides a good example. Other concentrations of water vapor form from convergence of ocean-supplied water vapor along cold fronts in midlatitude cyclones, as will be discussed in Chapter 10. Thus atmospheric rivers are common over oceans in both the subtropics and midlatitudes

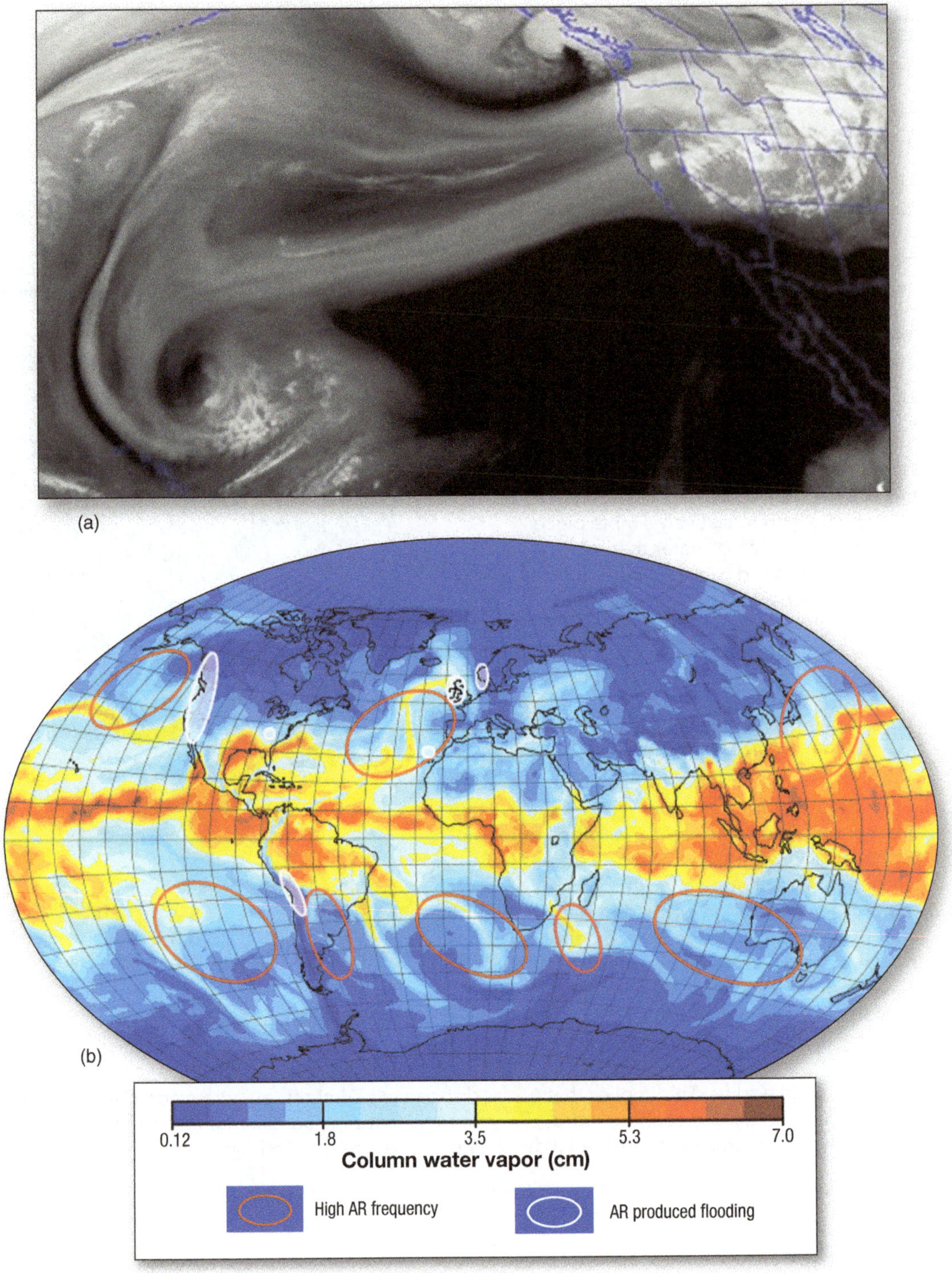

▸ **FIGURE 8–17 Atmospheric Rivers (ARs).** (a) An infrared satellite image for September 30, 2013 shows a strong atmospheric river flowing the Pacific near Hawaii toward the California coast. (b) This map shows total water vapor for November 13, 2009. Several rivers are visible as thin streaks of yellow and red. Red ovals are places were atmospheric rivers are often found. White ovals show locales that are prone to flooding by water transported by atmospheric rivers.

(Figure 8–17b). They can bring very heavy rain and flooding to land areas (white areas in Figure 8–17b) and are given colorful names by forecasters because of their importance. For example, an atmospheric river hitting the West Coast of the United States is called a "Pineapple Express." The "Maya Express" carries moisture northward from the Gulf of Mexico and is linked to flooding in the U.S. Midwest. Considering their importance in extreme rainfall and the statistics cited above, the importance of atmospheric rivers as a moisture supply to the middle latitudes is obvious. We can no more ignore them than we would a river like the Nile bringing water to otherwise dry locales.

CHECKPOINT

8.12 What are jet streams?

8.13 What roles do atmospheric rivers play in the movement of water vapor?

Ocean Circulation

8.6 Explain how the atmosphere affects the movement of ocean waters.

As we have seen, the movement of the atmosphere is strongly influenced by the input of heat from the surface. We have also seen that land and water undergo differential rates of heating and cooling; in part this is because of the vertical and horizontal motions of the water's surface. In this section, we take a close look at how the atmosphere affects the movement of oceanic waters. Later in this chapter, we look at some of the mutual interactions in the ocean–atmosphere systems. The subjects covered in this section are often taught in another field of science known as **oceanography**. Oceanography, also known as marine science, includes the physical processes discussed below and many other aspects of our watery world, including marine biology and ocean ecosystem dynamics.

DID YOU KNOW?

The global sea level has risen by about 106 mm/yr (4 in./yr) during the past 60 years, but because of varying winds, currents, and other factors the increase is by no means uniform across the globe. Increases in parts of the tropical Pacific have been several times larger, to the point that the bodies of World War II soldiers have been washed from their graves on low-lying islands.

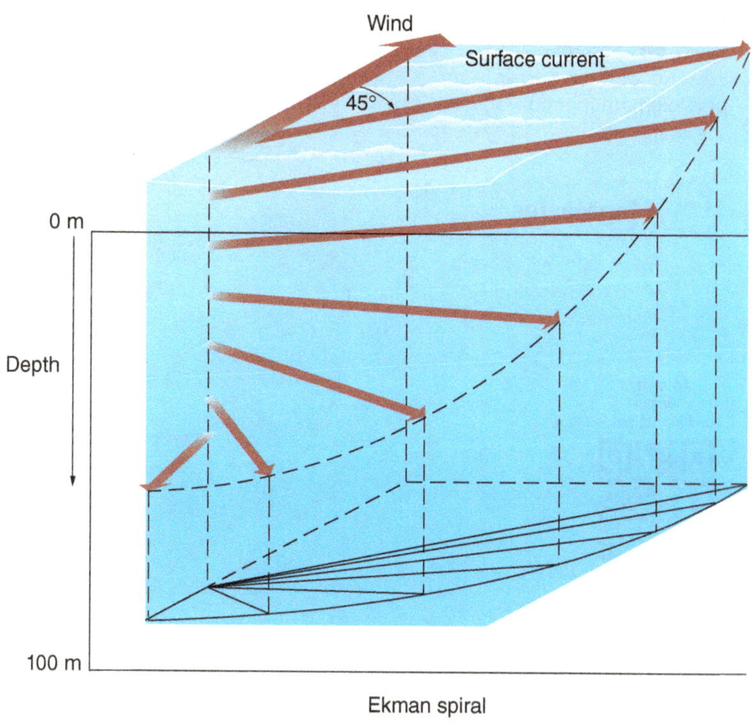

▲ **FIGURE 8–18 The Ekman Spiral.** Surface currents flow at an angle 45° to the right (in the Northern Hemisphere) of the winds that drive them and continue to shift clockwise as their speed decreases. At a depth of about 100 m, the current approaches the opposite direction of the surface current and begins to die out.

Animation
Ekman Spiral
Coastal Upwelling/ Downwelling
http://goo.gl/aJMMST

Causes of Ocean Currents

Ocean currents are horizontal movements of surface water that are often found along the rims of the major basins. These currents, discussed briefly in Chapter 3, have a great impact on the exchange of energy and moisture between the oceans and the lower atmosphere. Ocean currents, along with atmospheric motions, are also important in moving heat poleward from low latitudes. In fact, very near the equator they move much more heat poleward than the atmosphere. Their importance extends to middle latitudes, but in those regions atmospheric motions do most of the heavy lifting with respect to energy transport. But even there the effect of ocean currents on climate is conspicuous, with warm ocean currents, for example, favoring the existence of warm, humid air.

Surface ocean currents are driven by winds in the lower atmosphere that exert a drag on the water. Just like winds, they are affected by the Coriolis force, friction, gravity, and the position and shape of continents. Contrary to what you might expect, the Coriolis force causes the surface water to move not in the same direction as the wind, but at an angle of 45° to the right of the airflow in the Northern Hemisphere (to the left in the Southern Hemisphere). Furthermore, neither the direction nor the speed of the current is uniform with depth. Friction with water at lower levels sets it in motion, and that flow is in turn affected by the Coriolis force. Thus, the current turns increasingly to the right (in the Northern Hemisphere) and decreases in speed at greater depths. At about 100 m below the surface, the direction of the current approaches 180° to the direction of the wind, and the current dies out. Figure 8–18 illustrates this pattern, called the **Ekman spiral**.

The movement of the current at the surface corresponds well to the global wind patterns described earlier in this chapter, as shown in Figure 8–19. Let's have a look at the surface currents of the Atlantic, whose pattern is quite typical of those in the other major oceans. Looking at Figure 8–19, note the clockwise circulation cell corresponding to the oceanic subtropical high discussed earlier. Winds on the southern side of the high just north of the equator (the easterly trade winds) drag the surface water westward as the **North Equatorial Current**. Upon reaching North America, most of the westward moving water turns north, but some is deflected to the south toward the equator. A similar pattern appears in the South Atlantic because of the subtropical high centered there. Trade winds drive water westward as the **South Equatorial Current**. Most is deflected southward when it reaches South America, but some is deflected northward to the equator. The water from the North and South Equatorial Currents converges and piles up in the western equatorial Atlantic to create the eastward-moving **Equatorial Countercurrent**.

In the Northern Hemisphere, most of the North Equatorial Current reaching the South American coast turns northward to form the warm **Gulf Stream** (Figure 8–20). Near 40° N, the westerlies force the current to the east, where it becomes the **North Atlantic Drift**. The current remains warm as it

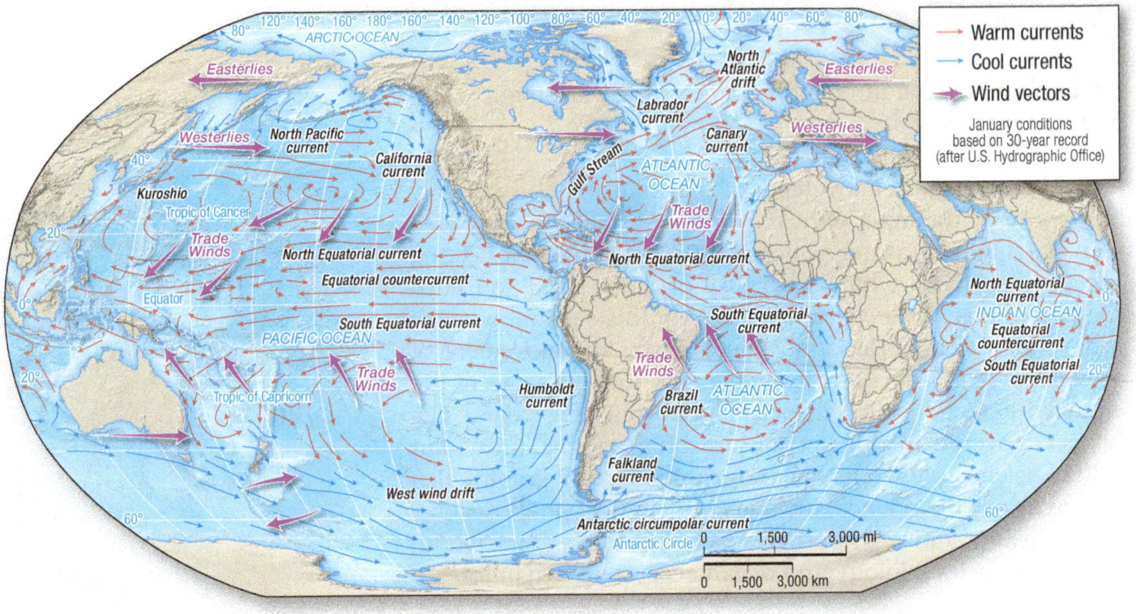

▶ **FIGURE 8–19 Ocean Currents.** The map shows ocean currents of the world with the approximate location of some of the major global wind systems.

MapMaster
World
Physical Environment
Ocean and Wind Currents/
Circulation Patterns

Animation Ocean Circulation
http://goo.gl/QIYk8o

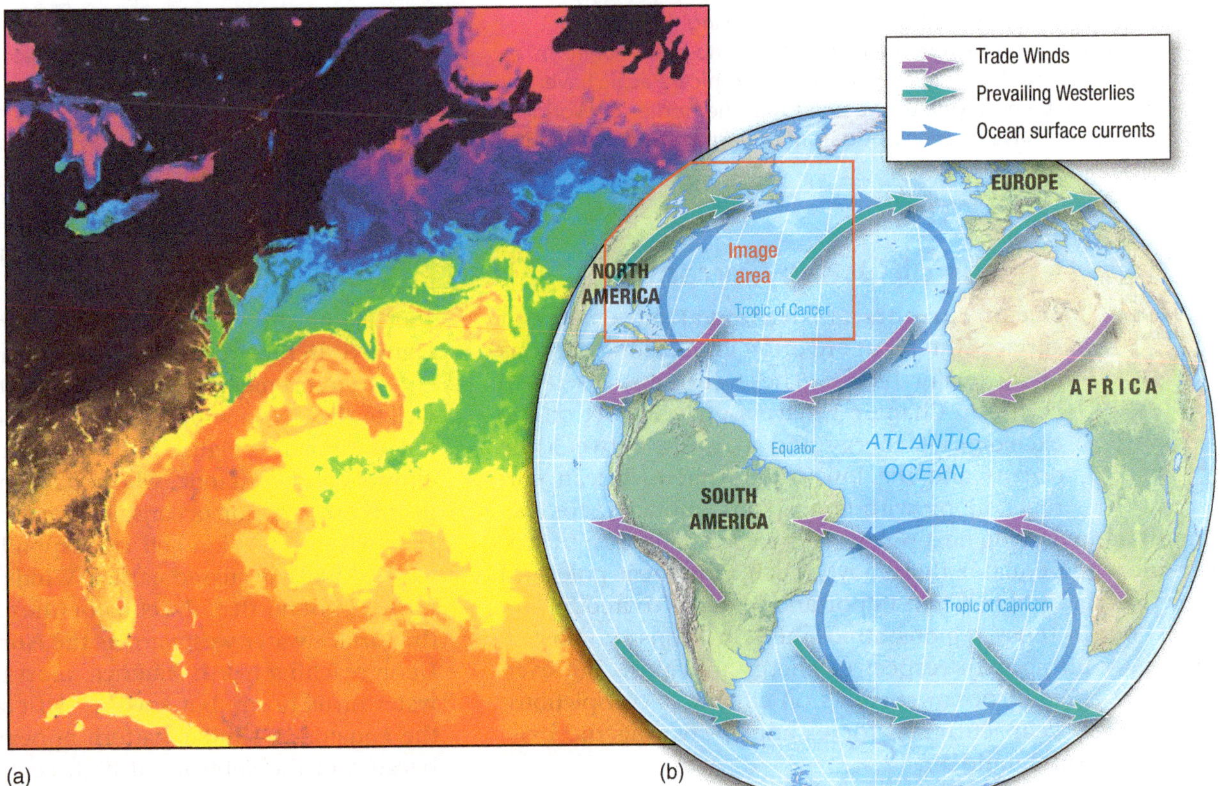

(a) (b)

▶ **FIGURE 8–20** The Gulf Stream flows in a complex pattern with eddies of different size superimposed. (a) The infrared satellite image shows temperatures in colors ranging from blue (cold) to warm (red). (b) A map view shows how the Gulf Stream is part of a large gyre in the North Atlantic, and that a similar gyre exists in the South Atlantic.

Animation The Gulf Stream
http://goo.gl/VxNztu

UNIVERSITY OF MIAMI
ROSENSTIEL SCHOOL of MARINE & ATMOSPHERIC SCIENCE

flows toward northern Europe, which makes winter conditions there unusually mild for those latitudes. Even the Scandinavian countries have surprisingly warm winter temperatures, considering their far northerly position. The southern branch of the North Atlantic Drift gradually cools and becomes the cold **Canary Current** as it turns southward. At corresponding middle latitudes, temperatures are considerably warmer over the western part of the Atlantic than they are over the east. Thus, for example, the average temperature off the New Jersey shore is about 8 °C (14 °F) warmer than off northern Portugal. The effect of a cold ocean current is also seen along the beaches of California, where even during the summer the water temperature seldom rises much above 22 °C (72 °F).

Finally, the cold **Labrador Current** that flows southward along the Maritime Provinces of Canada is fed by the **East** and **West Greenland Drift**.

The South Atlantic has a very similar pattern (see Figures 8–10 and 8–20b). There is the westward moving South Equatorial Current, which feeds a warm current flowing poleward in the western ocean margin (the Brazil Current). In the east a cold equatorward current flows off the coast of southern Africa (the Benguela Current). In both the North Atlantic and South Atlantic, we see the circulation largely organized in great nearly circular **gyres** in accordance with winds circulating around subtropical high pressure. Notice that similar gyres appear in the Pacific Ocean as well as the South Indian Ocean.

Upwelling and Downwelling

In addition to the major ocean currents that circulate large masses of water horizontally, localized vertical motions of ocean water have a bearing on weather and climate. Because solar radiation is mostly absorbed in the uppermost layer of the ocean, the surface is usually warmer than the waters below. But strong *offshore* movement of water along a coastal region sometimes draws up cooler waters from below to take their place. This process, called **upwelling**, greatly influences sea-surface temperatures (and thus weather and climate) over the eastern portions of the major oceans.

Nowhere is upwelling better illustrated than along the western coast of South America. The equatorial coast of Colombia is one of the rainiest places in the world, with average annual precipitation approaching 700 cm (290 in.). But nearby, between the latitudes of about 7° and 30° S, lies the world's driest desert—the Atacama. At the heart of this desert is Arica, Chile (18° S), which does not have a single month in the year that averages as much as 1 mm (0.04 in.) of precipitation—and most years receives no precipitation at all! This dryness is due largely to the upwelling of cold water along the coast, because cold water chills the lower atmosphere and favors the development of stable air.

The same process occurs along the central California coast during the summer (Figure 8–21a). Air circulating out of the Hawaiian high sets up an Ekman spiral, with a net flow of water away from the coast. Upwelling is particularly strong just north of the San Francisco Bay region, making the water there particularly cool. Beachgoers in southern California are also familiar with the occasional effects of upwelling. Episodes of hot, dry Santa Ana winds (discussed later in this chapter) bring huge crowds to the beaches. But although the high air temperatures and clear skies are great for sunbathing, the offshore winds lead to strong upwelling that creates low water temperatures and keeps many people out of the water.

DID YOU KNOW?

The Gulf Stream, the largest of all the ocean currents, transports about 100 times as much water per unit of time as do all the world's rivers put together.

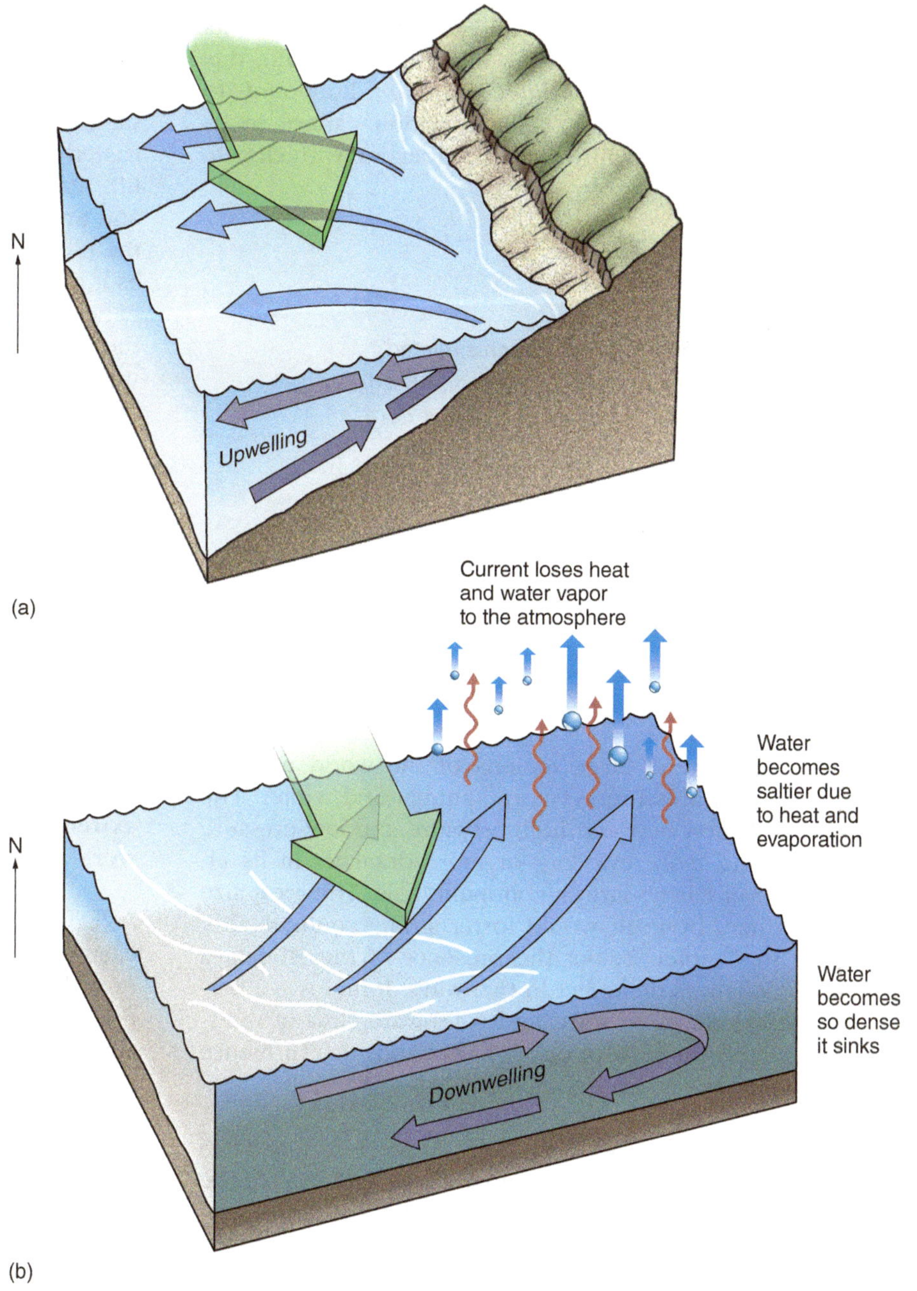
N
Upwelling
(a)
Current loses heat and water vapor to the atmosphere
Water becomes saltier due to heat and evaporation
N
Water becomes so dense it sinks
Downwelling
(b)

▲ **FIGURE 8–21 Upwelling and Downwelling.** (a) Upwelling can occur if a wind blows parallel to a coast. In this example, a wind blowing from north to south in the Northern Hemisphere propels surface water offshore. Water from below rises to replace the displaced surface water. (b) Downwelling occurs if surface water become denser than water below.

If surface waters become very dense, they respond to the stronger gravitational force and sink to lower levels. Such **downwelling** is caused when surface waters cool and also when they lose moisture through evaporation. Salt is left behind, which makes the water denser than the previously fresher water (Figure 8–21b). A good example of this is seen in the North Atlantic, where the warm North Atlantic Drift loses huge quantities of heat and moisture to the atmosphere above. Very cold, dense water near Greenland and Labrador downwells to a depth of 2 to 4 km (1.2 to 2.5 mi) and flows far southward, crossing the equator and eventually joining deep waters flowing around Antarctica and beyond. Because of its connection to temperature and the mineral halite (salt), this circulation is called *Thermohaline Circulation*. It amounts to a global three-dimensional ocean circulation pattern in its own right and is discussed further in connection with climate change in Chapter 16.

CHECKPOINT

8.14 How are winds and ocean currents related?

8.15 What are the Gulf Stream and the North Atlantic Drift, and how do they affect northwestern Europe?

Major Wind Systems

8.7 Identify major wind systems such as monsoons; foehn, chinook, and Santa Ana winds; katabatic winds; and sea and land breezes.

Monsoons

We know that several large regions of the world undergo changes in mean pressure between summer and winter. The seasonal oscillation between high-pressure and low-pressure cells is nowhere more evident—or more dramatic in its effects—than over Earth's largest continent, Asia. The great size of this continent by itself would foster strong continentality (described in Chapter 3), but the presence of the Himalaya Mountains enhances the effect in two ways: It imposes a barrier that blocks the northward and southward flow of moisture, and it alters the flow of upper-level winds that influence surface conditions.

SUMMER AND WINTER MONSOONS Figure 8–22 illustrates the seasonal reversal in surface winds that characterizes the **monsoon** (from the Arabic word meaning "season") of southern Asia. Although the term is often erroneously associated only with the heavy summer rains that occur over southern Asia, *monsoon* refers to the climatic pattern in which heavy precipitation alternates with hot, dry conditions on an annual basis. During January (a), the winds generally flow southwestward toward the Indian Ocean from the southern Himalayas. The descending air is compressed and warmed, leading to dry conditions over most of India and Southeast Asia. But offshore flow does not arise from the Tibetan high, as might be surmised from the sea-level map in Figure 8–6, because the Tibetan high is quite shallow and the air cannot cross the Himalayas into southern India. Air flows down the southern slope of the Himalayan Mountains due to more rapid cooling over land than over the Indian Ocean. This flow is enhanced by a portion of the jet stream flowing along the southern flank of the mountain range. Interactions between the jet stream and mountains cause convergence in the upper troposphere, which leads to sinking air that gets forced southward. As depicted in the inset, waters in the northern Indian Ocean flow westward in the form of the North Equatorial Current.

The situation changes abruptly during late spring or early summer, when heating of the continent contributes to a reversal in the wind direction at both low and high levels. Aloft is an easterly jet stream, with often divergent motion promoting uplift. At low levels, onshore flow occurs, bringing warm, moist, and unstable air from the Indian Ocean to the southern part of the continent, where it rises orographically and by convection as it passes over the hot surface. Cloud formation is further enhanced by a much stronger orographic effect as the air ascends the southern slopes of the Himalayas. A steady flow of warm water is supplied by the clockwise ocean gyre that establishes itself in the summer months (see map inset in Figure 8–22).

MONSOON DEPRESSIONS The combination of moist air and strong uplift produces precipitation in amounts unimaginable to inhabitants of more moderate climates. The heavy rainfall in northeastern India is enhanced even further by the movement of **monsoon depressions** (or monsoon lows), areas of low pressure superimposed in the southeasterly airflow out of the Bay of Bengal. Consider, for example, the average distribution of monthly precipitation at Cherapunji, India (Figure 8–23). Rainfall amounts are generally low during the winter months, but by early summer mean values approach 300 cm (125 in.) *per month!* Keep in mind that this extreme climate exists in a very heavily populated area of the world, affecting the lives of millions of people.

ARE THERE MONSOONS IN NORTH AMERICA? The southeastern part of North America is geographically similar to Asia. Like Asia, it lies to the north of a body of warm water, the Gulf of Mexico, and it undergoes a seasonal reversal of the surface winds due to warming and cooling of the land mass. But despite this similarity, the southern United States does not have a strong monsoon climate. Part of the reason for this is that the smaller size of North America results in weaker oscillations in pressure than those of Asia. More important, however, is the fact that the southern United States does not have a major east–west mountain chain comparable to the Himalayas. In fact, the area is essentially flat, except for the relatively low southern Appalachians that extend southwest to northeast. The Appalachians do not create a strong orographic barrier to the southerly wind flow of summer, and therefore they do not promote the extreme rainfall seen in south Asia. They also do not block the passage of winter storms from the north, as do the Himalayas, nor do they promote persistent

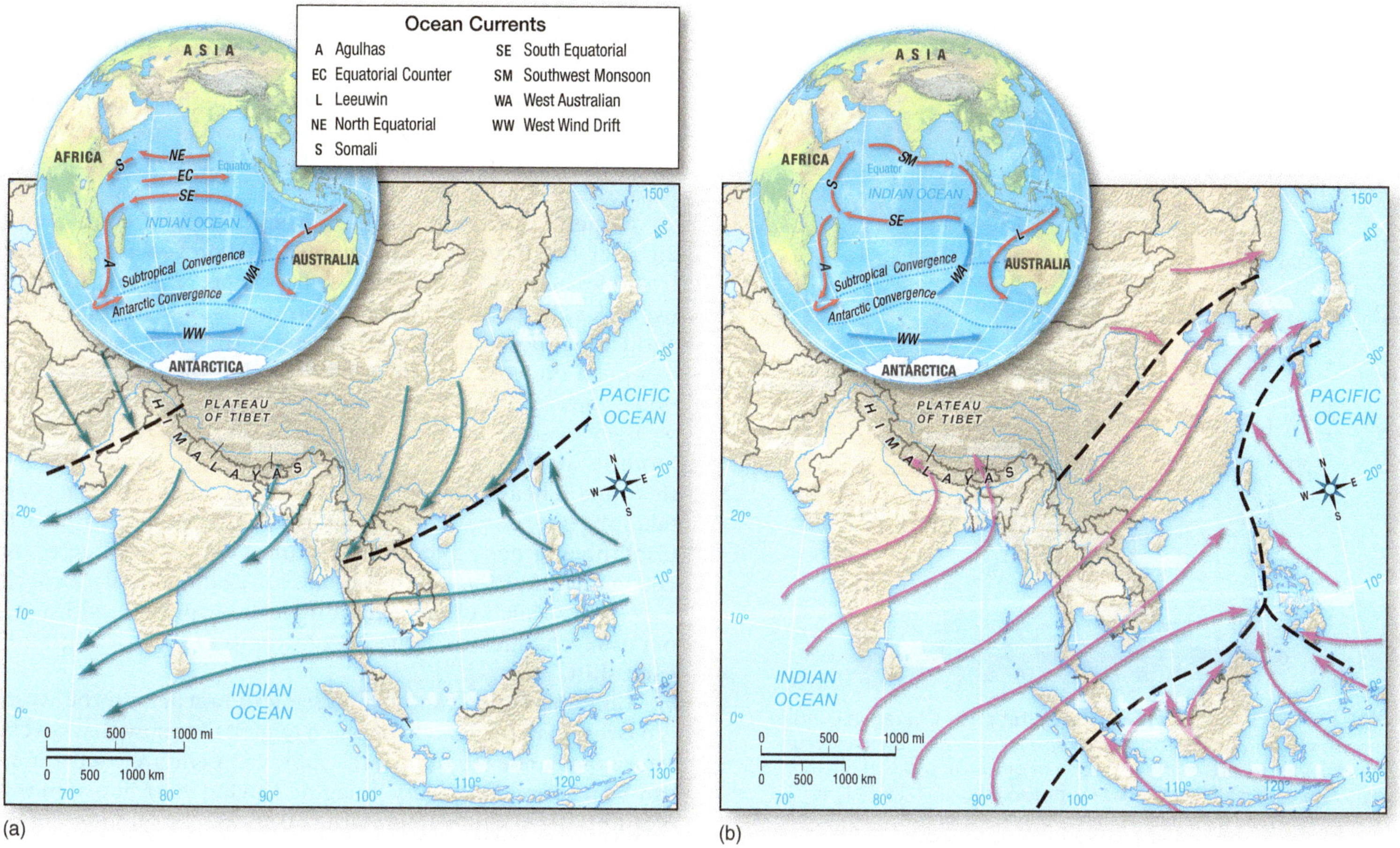

▲ **FIGURE 8–22 Monsoon Winds.** The monsoon of Asia results from a reversal of the winds between the winter and summer. (a) During winter, dry air flows southward from the Himalayas. The North Equatorial Current is strong (inset) as would be expected with strong trade winds. (b) When summer arrives, moist air is drawn northward from the equatorial oceans. Surface heating, convergence, and a strong orographic effect cause heavy rains over the southern part of the continent. The dashed lines represent areas of convergence, where winds from different directions tend to come together. In the ocean clockwise circulation forms a gyre not present in the winter months.

upper-level convergence. With all of that, it is easy to see why winter is hardly a dry season in eastern North America.

The desert of the southwestern United States experiences what residents often call the *Arizona monsoon*.[4] In spite of its name, the Arizona monsoon bears little resemblance to the real thing. It occurs each summer when the desert southwest heats considerably, creating low pressure over Arizona and extreme southeastern California (Figure 8–24). Warm, moist air flowing in from the south and southwest and strong surface heating can lead to heavy convection to trigger scattered thundershowers. Unlike the intense precipitation of the Asian monsoon, that of the Arizona monsoon does little to alleviate the desert conditions of the American Southwest.

CHECKPOINT

8.16 Why does the summer monsoon produce heavy rainfall?

8.17 How do continentality and the Himalayas help cause the monsoon in southern Asia?

[4]Also called the *Southwest monsoon* or the *Mexican monsoon*.

Foehn, Chinook, and Santa Ana Winds

Foehn, Chinook, and Santa Ana winds are those that flow downslope in response to the distribution of high- and low-pressure systems over and near large mountain areas. Compression of the descending air leads to adiabatic warming. We discuss each of these individually.

FOEHN WINDS **Foehn** (pronounced "fern" like the common plant) is the generic name for synoptic-scale winds that flow down mountain slopes, warm by compression, and introduce hot, dry, and clear conditions to the adjacent lowlands. Although the term *foehn* strictly applies to winds coming from the Alps of Europe, we generally use it to describe this type of wind anywhere in the world. In Europe, foehns develop when midlatitude cyclones approach the Alps from the southwest. The air rotates counterclockwise toward the center of low pressure and descends the northern slopes. These winds bring unseasonably warm conditions to much of northern Europe during the winter, when they are most prevalent.

CHINOOK WINDS When winds warmed by compression descend the eastern slopes of the Rocky Mountains in

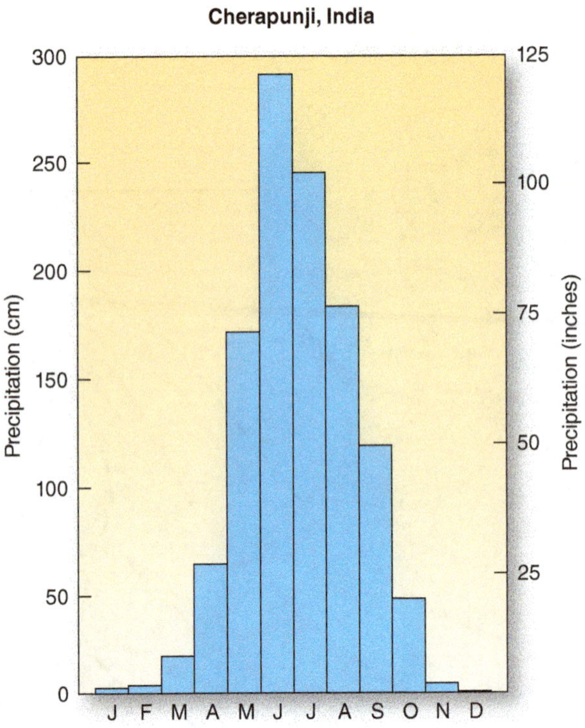

▲ **FIGURE 8–23 Monthly Mean Precipitation at Cherapunji, India.** The graph highlights the sudden increase in precipitation that occurs when the south winds of the monsoon begin. Over much of the monsoon region, there is an abrupt increase in precipitation in May or June.

North America, they are called **chinooks**. Low-pressure systems east of the mountains cause these strong winds to descend the eastern slopes at speeds that can exceed 150 km/hr (90 mph) when funneled through steep canyons. Like their European counterparts, chinooks are most common during the winter when midlatitude cyclones routinely pass over the region.

Sometimes the presence of a large mass of cold, dense air near the base of a mountain range prevents a chinook from flowing all the way down the slope. The hot air then overrides the cold air, and no warming is observed near the surface. If the chinook strengthens sufficiently, however, it can push the cold air out of its path, and the foothill region undergoes a rapid temperature increase. But if the hot winds weaken even momentarily, the cold air can return to the foothills and bring another sudden change in temperature—this time a nearly instantaneous cooling. Such reversals can take place repeatedly over a short period of time, with each bringing another rapid and sometimes extreme temperature change. Rapid City, South Dakota, for example, once experienced three such cycles over a three-hour period, with temperature changes as large as 22 °C (40 °F) occurring with each shift.

Chinook winds can be a blessing to ranchers in the western Great Plains who rely on them to melt the snow that covers their rangelands. To others, the rapid temperature oscillations can be a source of misery. Imagine leaving your house in the morning when the temperature is 0 °C (32 °F), getting out of your car when it is 38 °C (100 °F), going for lunch when the temperature is down to –10 °C (14 °F), and returning from lunch when it is up to 35 °C (95 °F). Such changes have been known to occur from chinooks, and there is anecdotal evidence (although nothing has been proven) suggesting an increase in violent crime, depression, and suicide during such episodes. Other problems not related to human discomfort can also arise from chinooks. During the 1988 Winter Olympics at Calgary in Alberta, Canada, chinook winds melted the snow cover and forced a postponement of the ski competition for several days.

In parts of the western Great Plains, chinooks are frequent enough and strong enough to increase the average winter temperatures, as exemplified by Rapid City and Sioux Falls, South Dakota. Rapid City is situated in the foothills of the Black Hills and commonly experiences chinooks, whereas Sioux Falls is several hundred kilometers to the east and well out of their range. But despite the fact that its

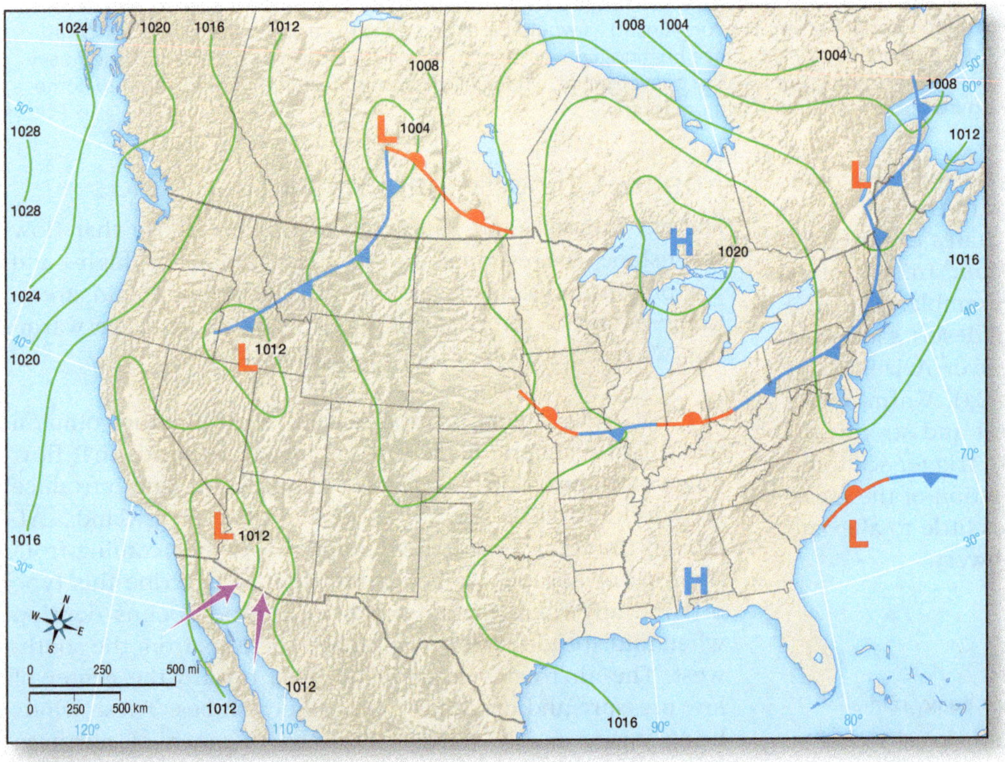

▲ **FIGURE 8–24 North America's Monsoon.** The "Arizona monsoon" occurs each summer as intense heating creates low pressure at the surface. Moist air flows toward the low pressure from the south and southwest, and localized convection can trigger heavy thundershowers.

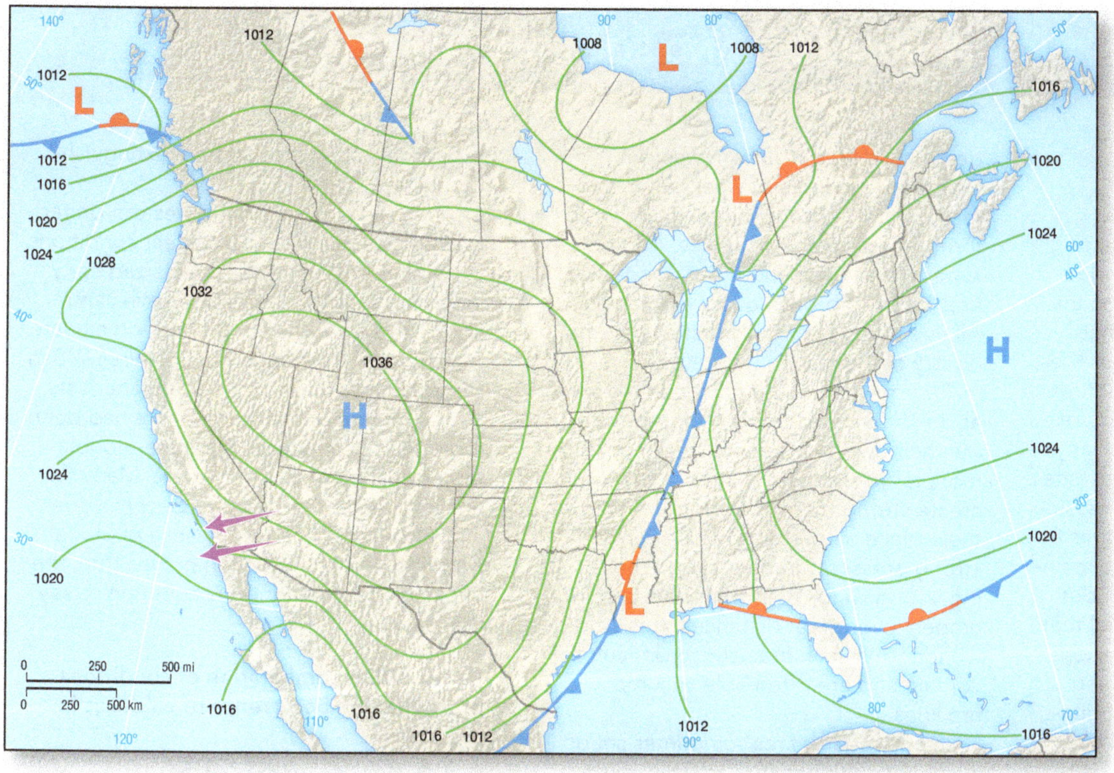

◄ **FIGURE 8–25 Santa Ana Winds.** The weather map shows a high-pressure system over the Rockies, causing a Santa Ana wind over southern California.

Video
Black Carbon Aerosoles Trace Global Winds
http://goo.gl/Pqfv5u

Animation
Seasonal Pressure & Precipitation Patterns
http://goo.gl/GGgL4q

elevation is 500 m (1650 ft) greater, Rapid City has an average January temperature 4 °C (7 °F) higher than that of Sioux Falls, thanks to frequent chinook winds.

SANTA ANA WINDS The **Santa Ana winds** of California are similar to foehns and chinooks, but they arise from a somewhat different synoptic pattern. These winds, common in the fall and to a lesser extent in the spring, occur when high pressure develops over the Great Basin (Figure 8–25). Air flowing away from the high pressure descends to lower elevations and warms by compression, just as air flowing along the eastern slopes does for the chinooks. The difference is that the Santa Ana winds occur in response to a large area of high pressure causing air to flow out of the Rockies, whereas the chinook forms in response to air flowing across the range.

During Santa Ana conditions, the sinking air can warm by 30 °C (54 °F) or more and attain temperatures in excess of 40 °C (104 °F) near the coast. Contrary to what some people believe, Santa Anas are *not* warm because they pass over hot desert surfaces—it is compression, and compression only, that causes their high temperatures. In fact, during a well-developed Santa Ana, the coastal areas of California are usually hotter than interior desert locations such as Las Vegas, Nevada.

Santa Ana winds often contribute to the spread of tremendously destructive fires in California (see *Box 8–4, Focus on the Environment and Societal Impacts: Wildfires*). The natural vegetation of the region is dominated by an assemblage of species collectively referred to as *chaparral* (Figure 8–26), which is dry and highly flammable. When Santa Anas develop, the combination of hot, dry winds, low humidity, and an abundant source of fuel can set the stage for a major conflagration.

▲ **FIGURE 8–26 Drought-Adapted Vegetation.** Much of coastal California is covered by chaparral, a vegetation type adapted to the summer-drought conditions of the region.

In October 2007 three large fires spawned by Santa Ana winds ravaged San Diego County. The fires forced the evacuation of more than half a million people as about 1500 km2 (560 mi^2) of land and nearly 1600 homes burned. Seven people died as a result of the fires. Smoke was so thick in places that the Sun appeared as a subdued red spot in the sky (Figure 8–27). And yet the residents of the area could say they had been through worse—a mere four years earlier in October 2003, the same dry and windy conditions spread fires that ravaged

8–4 FOCUS ON THE ENVIRONMENT AND SOCIETAL IMPACTS

Wildfires

In the summer and fall of 2002, the landscape of the western United States was particularly dry due to an extended drought. That situation, coupled with the accumulation of potential fuel that had built up over the years since previous fires, set the potential for devastating wildfires. Many ecologists believe that fire suppression, though it may save homes and large areas of woodland in the short run, creates the conditions for uncontrollable fires. According to this argument, allowing fires to burn simply allows the ecosystem to maintain its normal conditions and precludes the buildup of highly volatile material that ultimately sets off massive conflagrations.

In 2002 the worst fears were indeed realized. By the first day of summer, hundreds of thousands of acres of land had already burned across the West. Unusually high temperatures and strong winds set the stage for the first of the most destructive fires in June in western Colorado. A couple of weeks later, two major fires in Arizona joined together to form a massive wall of flames. And before summer was over, enormous, long-lived fires had also broken out over Oregon. All three fires were the worst ever in the history of the respective states.

Part of what makes massive fires such as these so devastating is that they create their own weather in a way that fosters more rapid spreading. Intense flames create strong thermal updrafts that lower the surface air pressure. This sets up strong pressure gradients, and the resulting strong winds help transport burning embers over great distances. It is little wonder that such fires are so difficult to contain and are often able to jump fire lines.

In one regard, fires sometimes set up weather conditions that actually assist fire crews. The same convection that can set up strong pressure gradients can also trigger rain showers that help extinguish the flames.

The 2014 drought in the western United States once again set up perfect conditions for wildfires. The fire season started early in California with seven wildfires in January 2014. Santa Ana conditions and a heat wave in May intensified multiple fires in San Diego county that burned simultaneously. By the middle of the month there had been more than 1400 fires in the state, double the normal number for the date. Many scientists predict the West will become drier in the coming decades because of global warming. If that prediction proves true, the implications for wildfires is sobering to say the least.

8–4–1 What combination of conditions promotes the occurrence of wildfires in the West?

8–4–2 How do wildfires create their own weather that can both spread the fires and limit them?

▲ **FIGURE 8–27 California Wildfires, October 2007.** Smoke was so thick during the San Diego County fires that the sun barely appeared in the sky as a dull, red ball.

an even larger area, destroyed more than 2400 homes, and killed 16 people (Figure 8–28). At the same time, another 1100 homes were destroyed and 24 other people perished in San Bernardino County, east of Los Angeles.

DID YOU KNOW?

Several explanations, some rather implausible, have been offered for the origin of the name *Santa Ana*. The most widely accepted explanation is that during the early part of the 20th century a local newspaper in Orange County reported on such a wind blowing out of the nearby Santa Ana Canyon. Eventually the name was used to describe the type of wind rather than its location.

Such fires are not restricted to southern California. In October 1991, Santa Ana–like winds fanned a wildfire through the Oakland Hills, east of the San Francisco Bay area. It, too, destroyed more than 1000 homes and killed 24 people.

A recent paper published by the California Climate Change Center combined observations with numerical models to conclude that the frequency of Santa Ana winds has declined during the last half of the 20th century and will likely continue to decline through the current century due to global climate change. This is associated with complex interactions at both the synoptic and mesoscales.

Katabatic Winds

Like foehn and Santa Ana winds, **katabatic winds** warm by compression as they flow down slopes (Figure 8–29). Unlike foehns and Santa Anas, however, katabatic winds do not

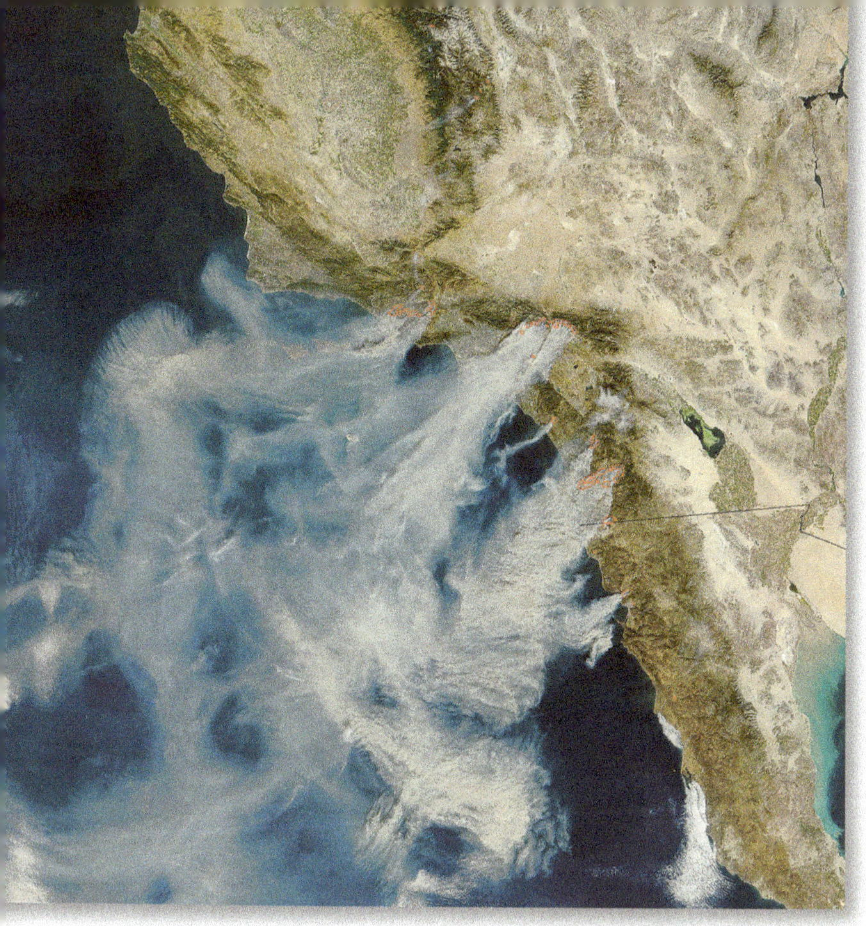

▲ **FIGURE 8–28 Southern California During the October 2003 Fires.** Santa Ana winds caused the fires to spread rapidly westward during the onset of the fires and created the large plumes of smoke that were blown offshore. Few clouds appear in this satellite image except over a couple of bays along Baja California and over the extreme bottom portion.

result from the migration of surface and upper-level weather systems. Rather, they originate when air is locally chilled over a high-elevation plateau. The air becomes dense because of its low temperature and flows downslope. The two best locations for such winds are along the margins of the Antarctic and Greenland ice sheets.

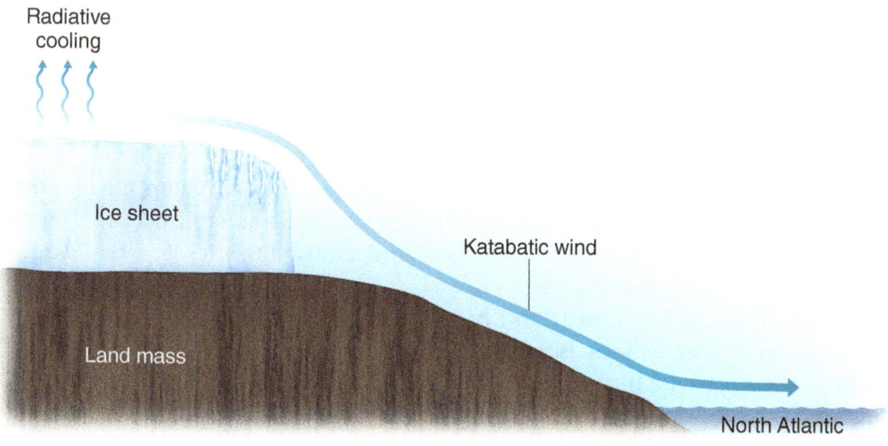

▲ **FIGURE 8–29 Katabatic Winds.** Formed over high plateaus, katabatic winds occur especially over large ice sheets such as over Greenland and Antarctica. Radiative cooling causes air over the ice sheet to become dense and flow downward toward lower elevations.

Katabatic winds usually occur as light breezes, but when funneled through narrow, steep canyons they can attain speeds in excess of 100 km/hr (60 mph). They happen sporadically because it can take some time for the air over the plateau to chill sufficiently. As soon as a mass of cold air forms over a plateau, gravity pulls it downslope and sets the wind in motion. When the cold air has been depleted, the wind ceases.

Much of coastal Antarctica is characterized by these alternating gusts and lulls of wind. And they can be very strong; one site, Cape Denison, occasionally experiences gusts up to 200 km/hr (120 mph) and holds the distinction of having the greatest average recorded wind speed on Earth.

Katabatic winds are not restricted to Greenland and Antarctica. They also flow out of the Balkan Mountains toward the Adriatic coast, where they are called *boras*. In France they are called *mistrals* as they flow out of the Alps and into the Rhone River Valley.

CHECKPOINT

8.18 Under what conditions do foehns, chinooks, and Santa Ana winds form? Why are these winds warm?

8.19 How do katabatic winds differ from foehns, chinooks, and Santa Ana winds? Explain.

Sea and Land Breezes

Near coastal regions or along the shores of large lakes, the differential heating and cooling rates for land and water form a diurnal (daily) pattern of reversing winds. During the daytime (especially in summer), land surfaces warm more rapidly than the adjacent water, which causes the air column overlying the land to expand and rise upward (Figure 8–30). At a height of about 1 km (0.6 mi), the rising air spreads outward, which causes an overall reduction in the surface air pressure. Over the adjacent water less warming takes place, so the air pressure is greater than that over land. The air over the water moves toward the low-pressure area over the land, which sets up the daytime **sea breeze** (remember, winds are always named for the direction *from* which they blow). (See Figure 8–31.)

As a sea breeze encroaches landward, a distinct boundary exists between the cooler maritime air and the continental air it displaces. This boundary, called the **sea breeze front**, usually produces a small but abrupt drop in temperature as it passes. This does not mean that the temperatures will not continue to rise after the sea breeze front moves on—only that there is a temporary lull in the rate of warming. Farther inland more heating can occur before the sea breeze passes, and temperatures become higher than those along the coastal strip.

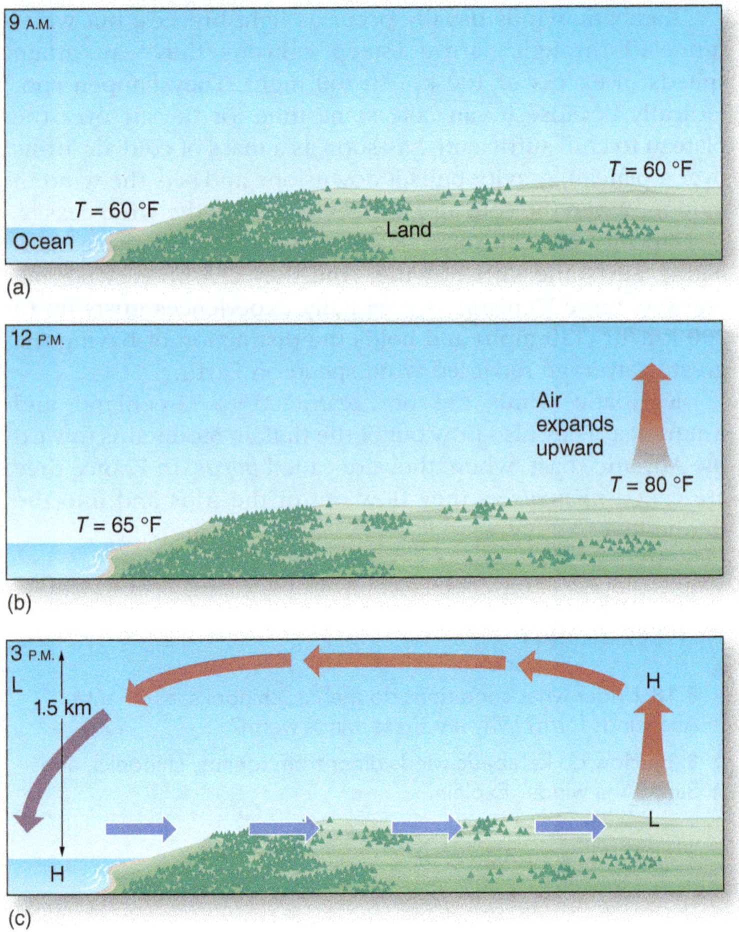

▲ **FIGURE 8–30 Development of a Sea Breeze.** Heating over the inland area causes air to expand upward and diverge at higher altitudes. This creates a surface low-pressure area, and the sea breeze flows inland from the sea.

Note that though the rising air in the heated column creates low pressure at the surface, it also creates higher pressure in the middle atmosphere. But we know that pressure always decreases with height. So how can this be? Remember that the terms *high pressure* and *low pressure* are relative—that is, they mean that the pressure is higher or lower than surrounding air *at the same level.* Thus, in this example the surface pressure over land is less than that over the adjacent ocean because the outflow of air above the 1 km (0.6 mi) level reduces the total amount of air over the surface. In the middle atmosphere, however, the rising of air from below increases the amount of mass above a particular level, and thereby increases the pressure relative to that of the surrounding air (though the pressure is still less than that below).

At night, when the land surface cools more rapidly than the water, the air over the land becomes dense and generates a **land breeze**. That is, lower land temperatures make for higher surface pressure and offshore flow. Compared to land breezes, sea breezes are usually more intense and last for a longer period of time each day. Sea breezes tend to

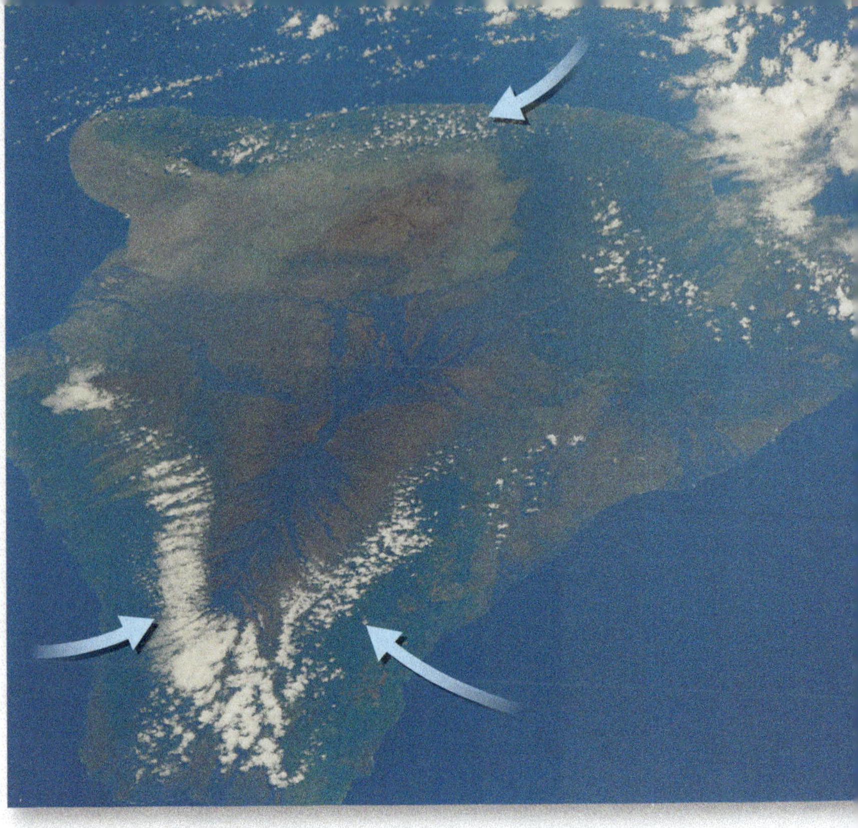

▲ **FIGURE 8–31 Sea Breeze.** This image of Hawaii shows the effect of a sea breeze. Heating of the land causes the air to expand upward. Coastal air flowing toward the interior is lifted as it passes over the mountains, causing orographic cloud cover.

TABLE 8–1

Average Wind Speed and Direction—Los Angeles, CA

	Winter		Spring		Summer		Fall	
	Speed		Speed		Speed		Speed	
Time (PST)	Dir	(m/sec)	Dir	(m/sec)	Dir	(m/sec)	Dir	(m/sec)
4 A.M.	ENE	1.0	E	0.5	WSW	0.4	ENE	0.6
10 A.M.	ENE	1.4	WSW	1.9	WSW	3.2	WSW	1.0
1 P.M.	WSW	2.5	WSW	5.3	WSW	5.5	WSW	4.5
4 P.M.	WSW	3.5	WSW	5.5	WSW	5.8	W	5.1
10 P.M.	NE	0.5	W	1.6	WSW	2.4	W	0.7
1 A.M.	ENE	1.0	–	0	WSW	1.0	NNE	0.2

Source: California Air Resources Board.

be strongest in the spring and summer when the greatest daytime temperature contrasts occur between land and sea. A typical sea/land breeze pattern is described in Table 8–1, which shows average wind characteristics for Los Angeles, California.

The sea/land breeze type of circulation is not confined to coastlines but also occurs along the shores of large lakes, in which case it creates daytime **lake breezes**. Such systems occur along the margins of the Great Lakes in the United States and Canada, but there they occupy a narrower zone than do those along the Pacific or Atlantic coasts.

Valley and Mountain Breezes

A diurnal pattern of reversing winds similar to the land/sea breeze system also exists among mountains and valleys. During the day, mountain slopes oriented toward the sun heat most intensely. The air over these sunny slopes warms, expands upward, and diverges outward at higher altitudes in much the same way as it does over inland areas when a sea breeze develops. The **valley breeze** occurs when air flows up from the valleys to replace it (Figure 8–32a).

At night, the mountains cool more rapidly than do low-lying areas, so the air becomes denser and sinks toward the valleys to produce a **mountain breeze** (Figure 8–32b). Mountain breezes are usually just about as intense as valley breezes but tend to be somewhat gustier. Whereas valley breezes blow fairly continuously at speeds below 15 km/hr (10 mph), the nighttime air may be still for several minutes and then suddenly flow downslope.

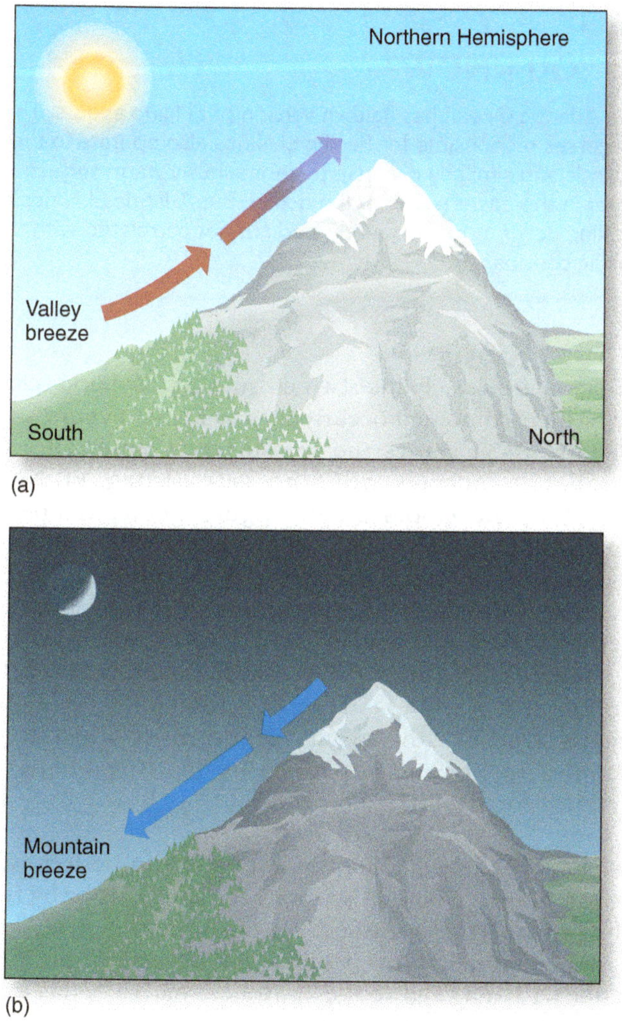

▲ **FIGURE 8–32 Valley and Mountain Breezes.** (a) A valley breeze forms when daytime heating causes the mountain surface to become warmer than nearby air at the same altitude. The air expands upward and the air flows from the valley to replace it. (b) Nocturnal cooling makes the air dense over the mountain and initiates a mountain breeze.

Ocean–Atmosphere Interactions

8.8 Explain ocean–atmosphere interactions such as El Niño/La Niña, Walker circulation, and Pacific Decadal, Arctic, and North Atlantic oscillations.

Earlier in this chapter we described how atmospheric circulations propel ocean currents. In a similar manner, the oceans exert an important influence on the input of heat and moisture to the atmosphere. Warm surface waters heat the overlying atmosphere by the transfer of sensible and latent heat. The addition of this heat, in turn, affects atmospheric pressure. Thus, the atmosphere and the ocean are linked as a complex system.

Compared to those in the atmosphere, oceanic motions and temperature changes are exceedingly slow. Temperatures in the lower troposphere can change tens of degrees in a matter of hours, whereas oceanic temperatures are very stable. Because the ocean surface changes so slowly, information about the current distribution of temperature can be a useful tool for atmospheric scientists when making long-term weather forecasts. In this section we examine some of the important interrelationships between the ocean and the atmosphere, including the famous El Niño and La Niña.

El Niño, La Niña, and the Walker Cell

El Niño and La Niña events are patterns of tropical Pacific sea-surface temperatures that may persist for months to perhaps a year. Though they are oceanographic phenomena, they are closely linked with the atmosphere through what is referred to as the **Walker cell** (also called *Walker circulation*). This is an atmospheric circulation cell in the tropical Pacific that is oriented east–west (zonal) rather than north–south like the meridional Hadley cells. The normal state of the Walker cell is shown in Figure 8–33. The trade winds move warm surface waters near the equator westward, causing higher temperatures and even a difference in sea level—about half a meter greater in the western Pacific compared to the eastern Pacific. Warmer water in the western Pacific leads to higher air temperatures, lower surface air pressure, and more convective precipitation. Thus, the Walker cell has rising air over Indonesia, which is consistent with the very moist climates found there. At the same time, the upwelling of colder water from below replaces the warm water migrating westward in the eastern Pacific. Relatively cold water favors high air pressure at the surface and sinking motions. The Walker cell, like the Hadley cells, is a thermally direct cell with rising air in the warm branch and sinking air in the cold branch.

▸ **FIGURE 8–33 Walker Cell.** The Walker cell involves a westward flow of surface air over the equatorial Pacific. The surface flow drags warm surface waters into the western Pacific. If the westerly flow weakens sufficiently or reverses, the warm waters to the west migrate eastward and cause an El Niño. By contrast, a La Niña forms if the westerly flow intensifies.

Video
El Niño

http://goo.gl/Izs5E

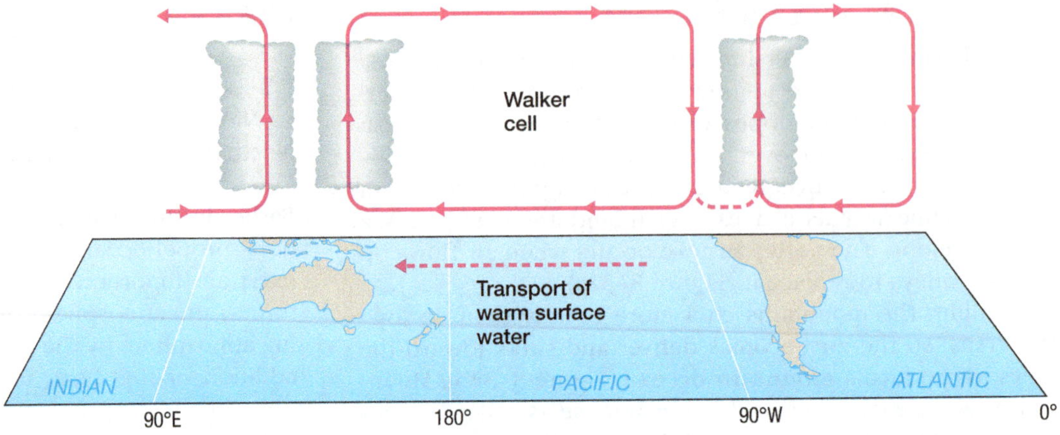

EL NIÑO Before its dramatic reappearance in 1983, the phenomenon known as **El Niño** was largely unknown to the public. But that changed in the early part of the year when the unusually warm waters in the eastern Pacific Ocean that mark an El Niño helped spawn a series of powerful storms in southern California. The storms not only caused severe flooding but also generated heavy surf that caused extensive coastal property damage and completely washed the sand away from many beaches. Another El Niño in 1997–1998 was also unusually strong, again leading to episodes of heavy surf, landslides, and flooding in southern California. In addition, precipitation across the southern tier of states was well above normal and severe storms were more frequent; some storms spawned very damaging tornadoes. Residents of the northern United States and eastern Canada also experienced anomalous weather conditions with unusually mild temperatures during the winter. High water temperatures promote two conditions favorable for major storm activity: increased evaporation into the air and reduced air pressure. Therefore, there is little doubt that the unusual episodes of 1983 and 1997–1998, in which the surface waters were as much as 6 °C (11 °F) warmer than normal, played a role in the formation and passage of the storms. The 1997–1998 event was strong enough to cause a spike in global average temperature that wouldn't be surpassed for a decade.

So what exactly is El Niño and how does it form? At two- to seven-year intervals (every 40 months on average), the surface waters of the eastern Pacific, especially near the coast of Peru, become unusually warm. (The tendency for this warming to occur near the Christmas season led to the name El Niño, a reference to the Christ child.)

An El Niño develops when the trade winds weaken or even reverse and flow eastward. The warm water normally found in the western Pacific gradually "sloshes" eastward as a slowly moving wave, in part due to the higher sea level, and eventually makes its way to the coast of North and South America. The eastern Pacific warms, upwelling weakens, and the warm surface layer deepens both there and throughout the tropical Pacific basin. The change in sea-surface conditions linked with the change in the atmospheric pressure distribution is called the **Southern Oscillation**. The El Niño and Southern Oscillation are closely intertwined, and atmospheric scientists refer to their combined occurrences as **ENSO events**.

DID YOU KNOW?

The Atlantic Ocean has its own version of El Niño events. The processes responsible for Pacific El Niños also operate in the Atlantic and similarly give rise to anomalously warm surface water in the eastern part of the ocean. But Atlantic El Niños are smaller, occur more frequently, and are not correlated with their Pacific cousins.

Figure 8–34 shows the sea-surface temperature (SST) anomalies associated with the very strong El Niño of 1997–1998. The colors on the figure depict temperature anomalies, the differences in temperature from normal conditions. Of course, no two ENSO events look exactly alike. The shape and exact locations of the pools of warm water vary between events, as do the magnitudes of the SST anomalies. Figure 8–35 shows the overall average distribution of November–March SST anomalies for eight average-to-large El Niño events. Figure 8–36 illustrates the variability in El Niño situations by plotting temperature anomalies for six different events.

LA NIÑA When an El Niño dissipates, it can be followed either by a return to normal sea-surface conditions or by further cooling of the tropical eastern Pacific. If the waters cool to below-normal temperatures, we have the reverse of an El Niño, called a **La Niña** (Figure 8–37). Notice that La Niñas amount to strengthening of the Walker cell, whereas El Niños occur when the Walker cells weakens. See *Box 8–5, Physical Principles: What Causes El Niños and La Niñas?,* for a discussion of the factors that lead to changing ENSO conditions.

Variations in the Walker cell and accompanying occurrences of El Niños and La Niñas are common. Figure 8–38

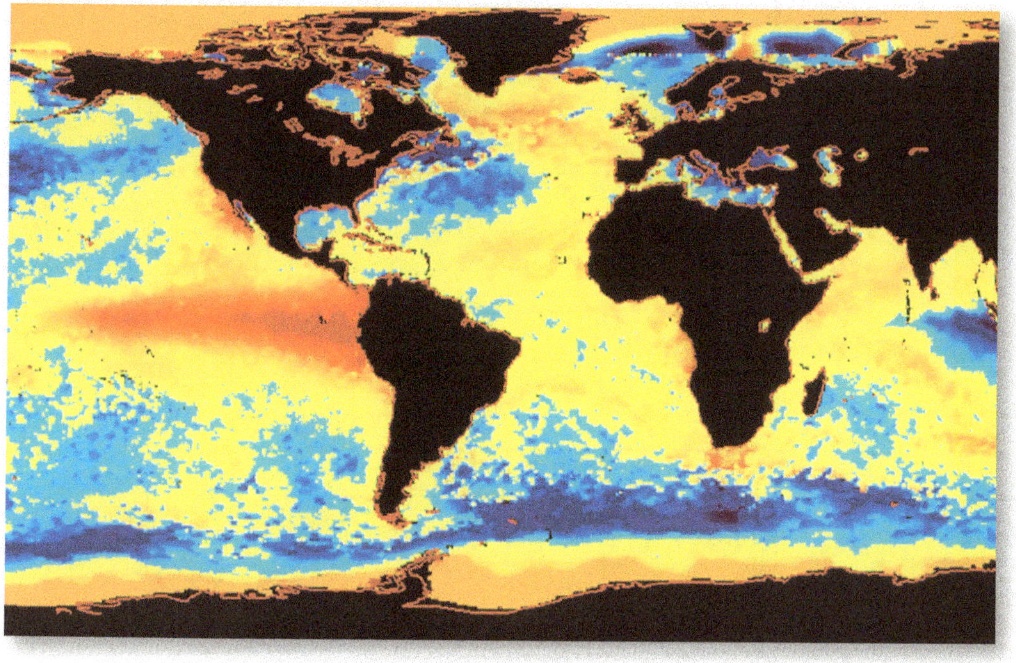

▲ **FIGURE 8–34 The 1997–1998 El Niño Event.** This El Niño contained a very large area of much above-normal sea-surface temperatures over the tropical east Pacific (shown in bright red). The image is based on satellite data obtained on November 3, 1997.

Animation
El Niño and
La Niña
http://goo.gl/pm1Wb6

shows one measure of ENSO activity, the so-called **Oceanic Niño Index (ONI)** from 1950 though mid-2014. The ONI tracks departures from average in sea-surface temperature for a range of longitudes in the central Pacific centered on the equator. The ONI has become the standard measure used by NOAA for identifying both El Niños and La Niñas. Values between –1 and 1 are considered within normal conditions, and values outside are either El Niños (above 1) or La Niñas (below –1). In Figure 8–38 you can see that these events come and go at irregular intervals. There is some tendency for events to alternate between warm and cold, but nothing prevents multiple El Niños or multiple

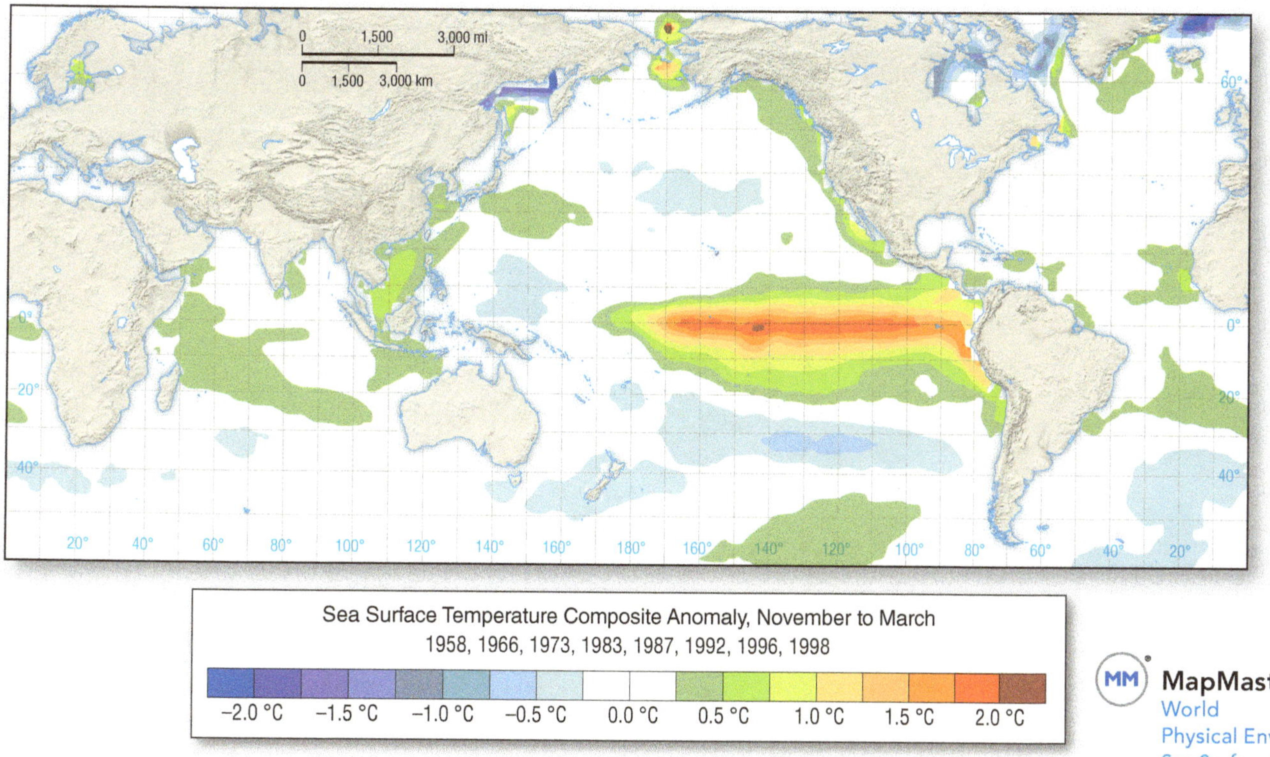

MM MapMaster
World
Physical Environment
Sea Surface Temperature El Niño/La Niña

▲ **FIGURE 8–35 Composite of El Niño Temperature Anomalies.** The maps shows the average sea-surface temperature differences from normal observed during the November through March period for eight El Niño episodes. The most prominent feature is the +2 °C (4 °F) increase in temperatures in the equatorial tropical east Pacific.

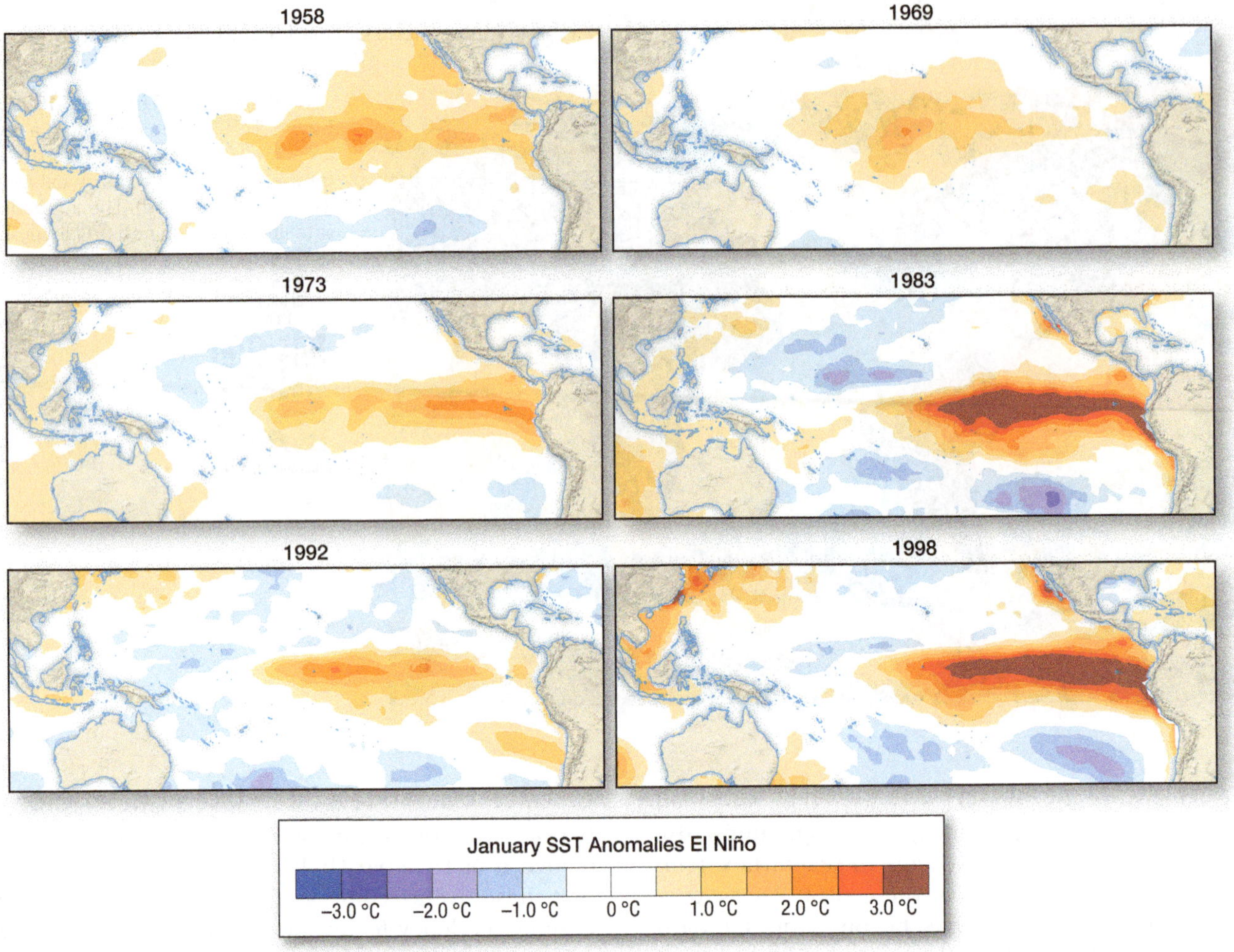

▲ **FIGURE 8–36 Temperature Anomalies for Six El Niño Events.** Every El Niño event looks somewhat different from all others; some are larger than others and some have greater temperature anomalies than others. This figure depicts the observed sea-surface temperature anomalies for six El Niño events.

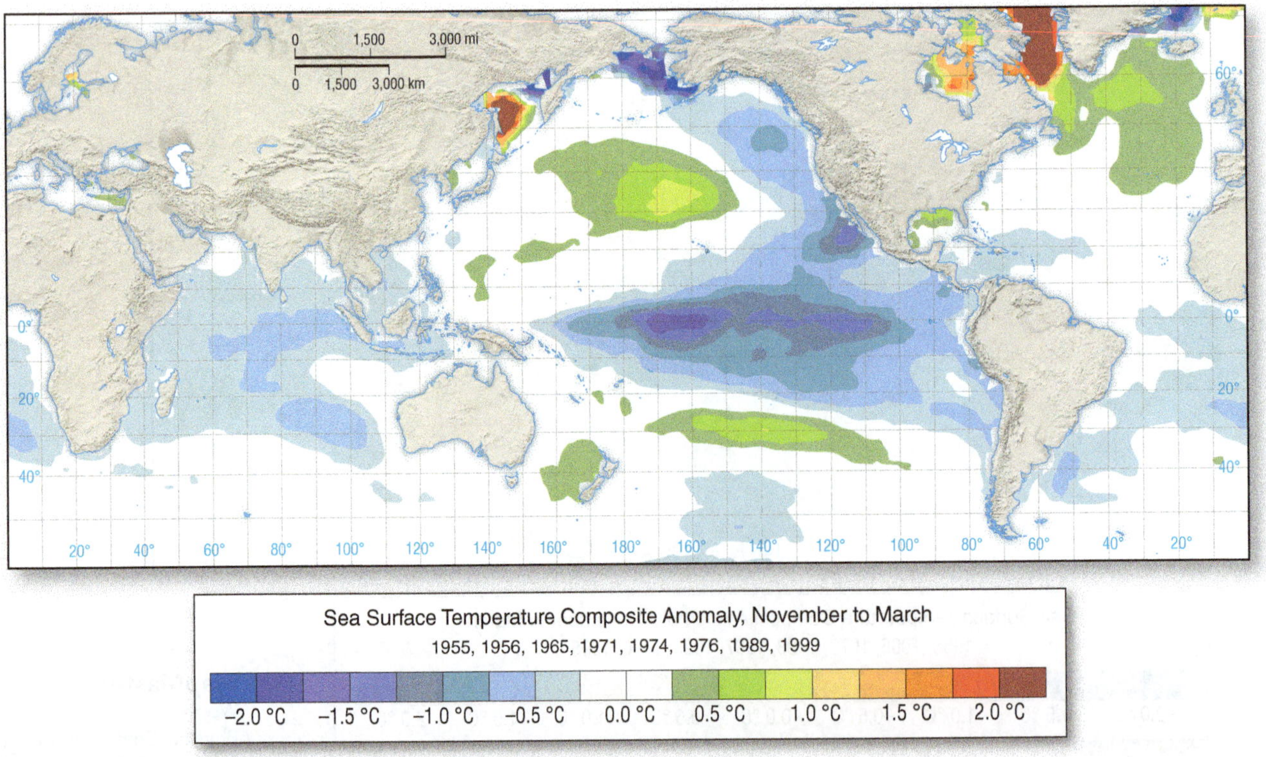

▲ **FIGURE 8–37 Composite of La Niña Temperature Anomalies.** The map shows the average sea-surface temperature differences from normal observed during the November through March period for eight La Niña episodes. Note the large area of cooler-than-normal temperatures over much of the eastern Pacific, extending poleward along the North and South American coasts. Cold waters are also found in the Indian Ocean.

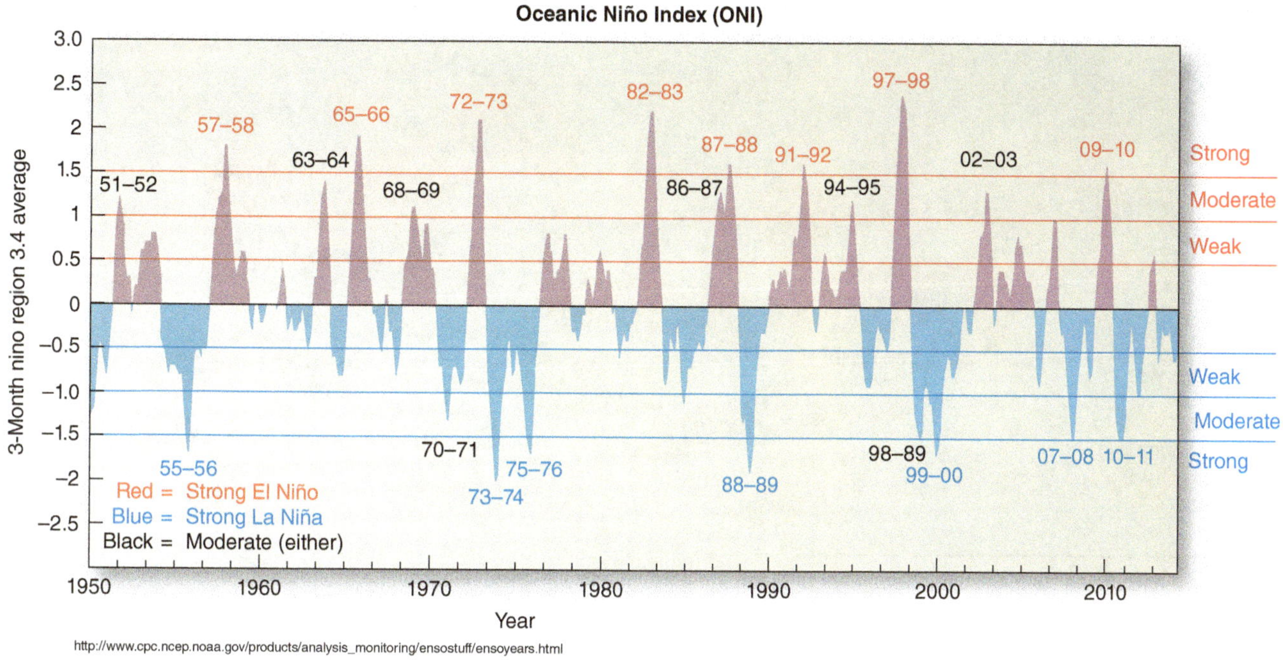

▲ **FIGURE 8–38 Time Series Plot of the Oceanic Niño Index (ONI).** Large positive values indicate El Niño conditions; extreme negative values are La Niñas.

La Niñas from occurring in succession. Data collected on a longer time scale show that although there have been some periods in which El Niños and La Niñas were more frequent or less frequent than others, it is clear that both have occurred with substantial regularity during the past century and a half. Moreover, close analysis of the data suggests that ENSO events have been more frequent in recent years than in the mid-1800s.

WEATHER IMPACTS OF EL NIÑO AND LA NIÑA Some useful observations have been made regarding the impact of El Niño events on seasonal weather. The central coast of California, for example, appears to have a greater likelihood of unusually high amounts of precipitation when El Niño is present, while the northwestern United States and Canadian Pacific coast regions tend to be unusually dry. The effects of El Niño are not restricted to the western coast of North America, however. As you have seen, the large-scale patterns of the atmosphere are strongly influenced by the position of Rossby waves. When high-pressure or low-pressure systems exist in some locations, they affect not only local weather conditions but also the overall size, shape, and position of the entire Rossby wave pattern. The establishment of an upper-level trough over the Pacific Coast, for example, will promote the development of a ridge farther to the east. For this reason, certain weather conditions in the eastern United States are set up through **teleconnections**, the relationships between weather or climate patterns at two widely separated locations. El Niños favor the formation of ridge-trough patterns that often bring rainy conditions to the southeastern United States and mild, dry conditions to the northeastern United States and eastern Canada.

These teleconnections also occur outside North America; El Niños tend to promote enhanced precipitation over coastal Ecuador and Peru, the central Pacific, parts of the Indian Ocean, and eastern equatorial Africa. The western tropical Pacific, Australia, India, southeastern Africa, and northeastern South America usually undergo decreased precipitation with the occurrence of El Niño events. Figure 8–39 depicts some of the worldwide climatic patterns generally associated with El Niños and La Niñas, and Figure 8–40 shows the average temperature and precipitation conditions that have occurred over the United States during recent ENSO episodes.

In the past few years, studies have found that El Niños have two variants and that these variants have different relationships with conditions elsewhere. We have been describing the conventional type, called an **Eastern Pacific (EP) type El Niño**. It gets its name from widespread warming of Eastern Pacific waters. But scientists have noticed that the warming in some El Niños is confined to the Central Pacific, and have named such events accordingly as **Central Pacific (CP) type El Niños**. In the past few decades, CP events have become more common than EP, leading some to conclude that the tropical system has changed in such a way that the CP type is the preferred El Niño state. If that is the case, this is significant because North American air temperatures are affected in different ways by the two types (Figure 8–41). In particular, EP events are correlated with warmer than normal winter temperatures in the U.S. northeast and colder than normal temperatures in the southwest. By contrast, CP events lead to wintertime warmth in the northwest and cold in the southeast. If CP events remain more common, El Niño impacts would more often show

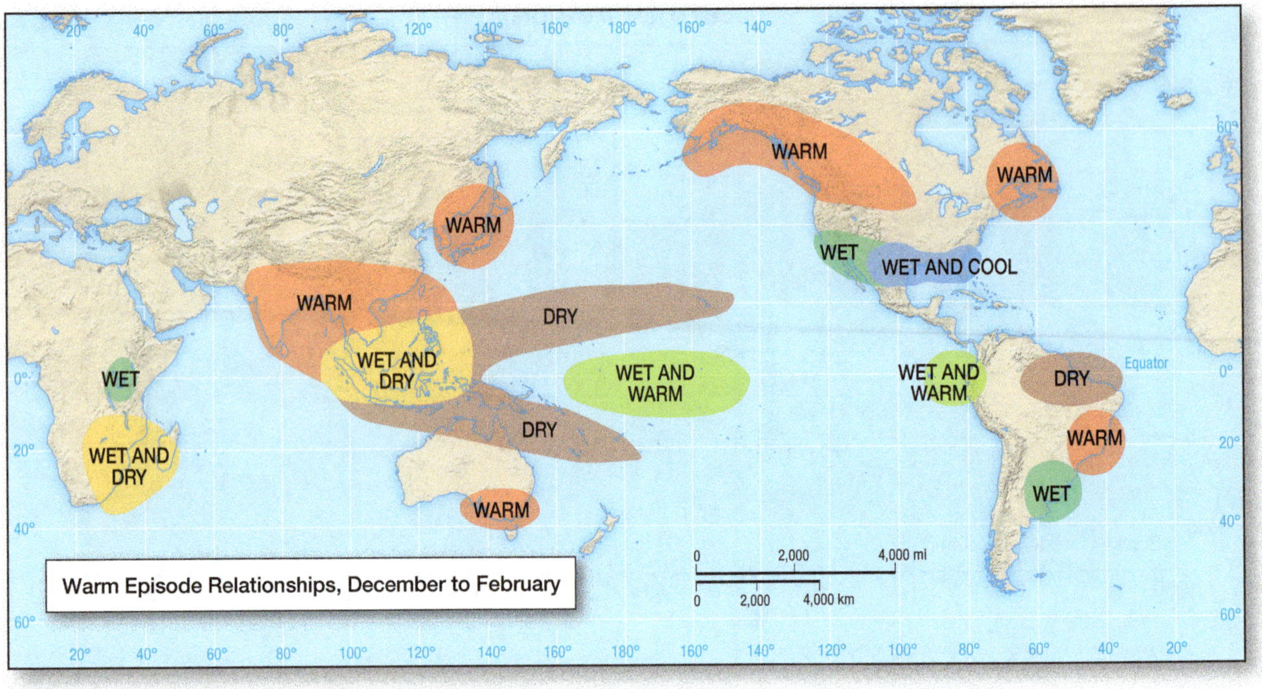

(a)

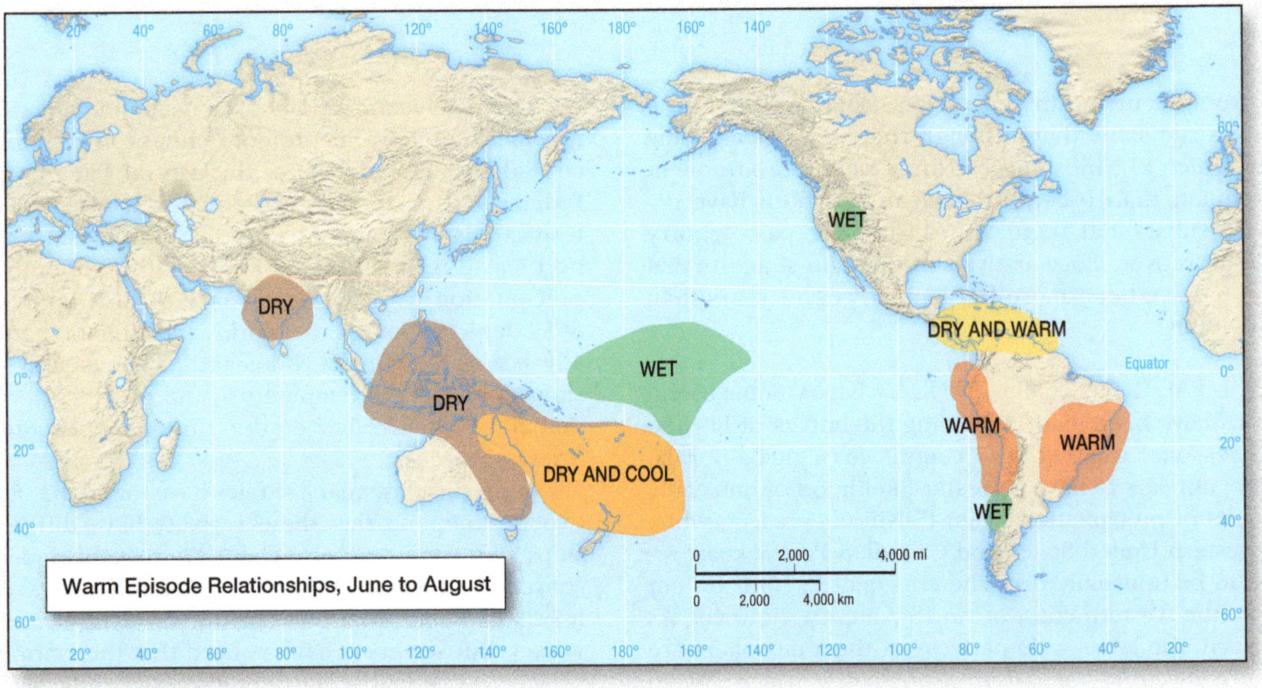

(b)

▲ **FIGURE 8–39 Effects of El Niño and La Niña.** The maps show worldwide climatic conditions associated with the occurrence of El Niño—(a) and (b)—and La Niña—(c) and (d)—events, for the periods of December–February and June–August (Maps (c) and (d) are on the following page.) While the indicated conditions often appear during ENSO events, they do not occur with all El Niños or La Niñas.

that pattern. This idea of distinct El Niño types is somewhat controversial. Some climatologists prefer to think of El Niños as existing along a continuum rather than falling into discrete categories.

Like its warm-water counterpart, a La Niña tends to favor distinct (but different) regional and teleconnection patterns, including dry conditions along the California coast, the southern United States, and Peru. Enhanced precipitation tends to occur in the Pacific Northwest and western Canada, the Caribbean, south Asia, and Indonesia. The 2010–2011 La Niña was the strongest on record. No fewer than eight powerful large tropical storms battered Northern Australia in the first four months of 2011, setting rainfall records and causing widespread flooding (Figure 8–42).

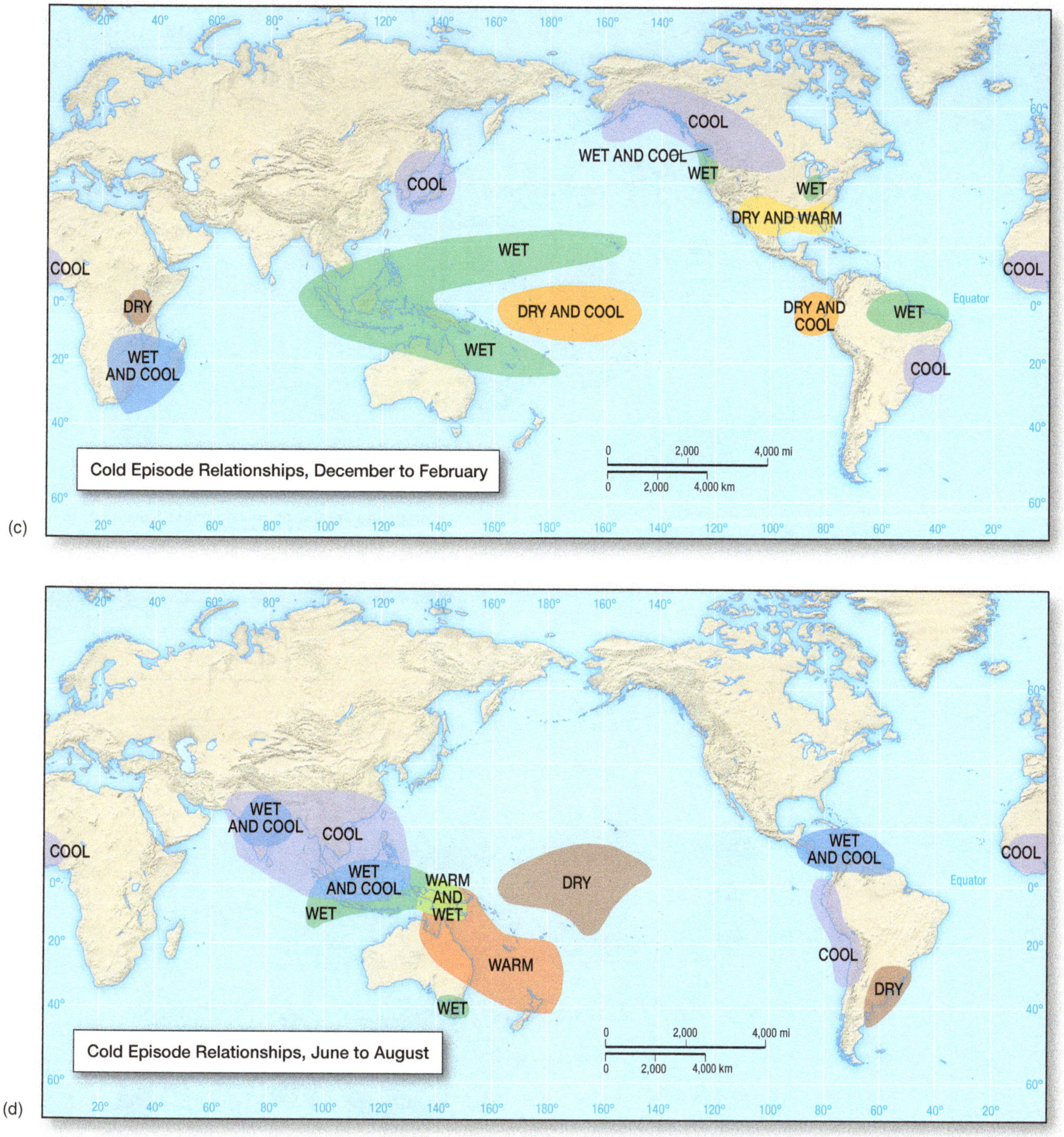

▲ **FIGURE 8–39** *Continued*

DID YOU KNOW?

Coral reefs provide information on ocean water temperatures that can be used to analyze the frequency and intensity of El Niño events of the distant past. One such study has suggested that the past century or so has witnessed the most intense El Niños of the past 130,000 years, with those of 1982–1983 and 1997–1998 possibly being the greatest to have occurred during that time span. It is possible, though not conclusively proven, that this change could be a response to global warming during the past century.

The fact that certain types of weather tend to occur in the presence of El Niños or La Niñas should not be taken to mean that some particular type of weather will always accompany the event. If the relationship between sea-surface temperatures and atmospheric patterns were that well defined, long-range weather forecasting would be very much easier than it is. Similarly, we should resist the tendency to "blame" El Niños and La Niñas for the occurrence of an

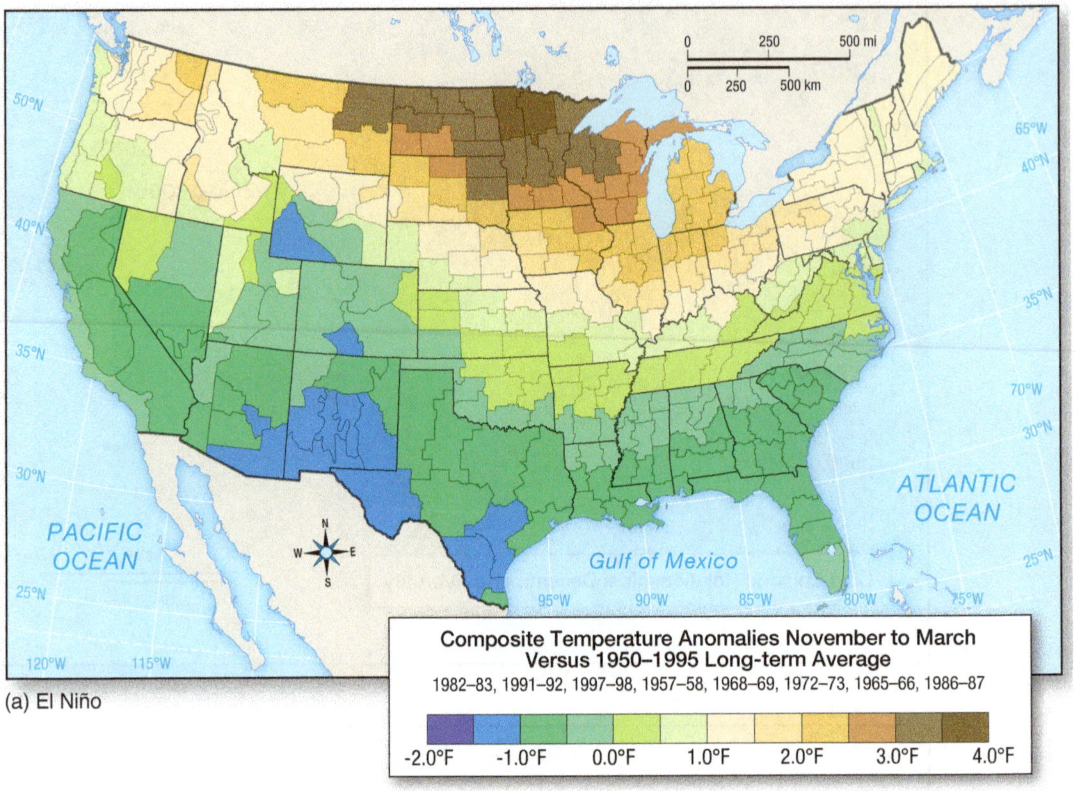

(a) El Niño

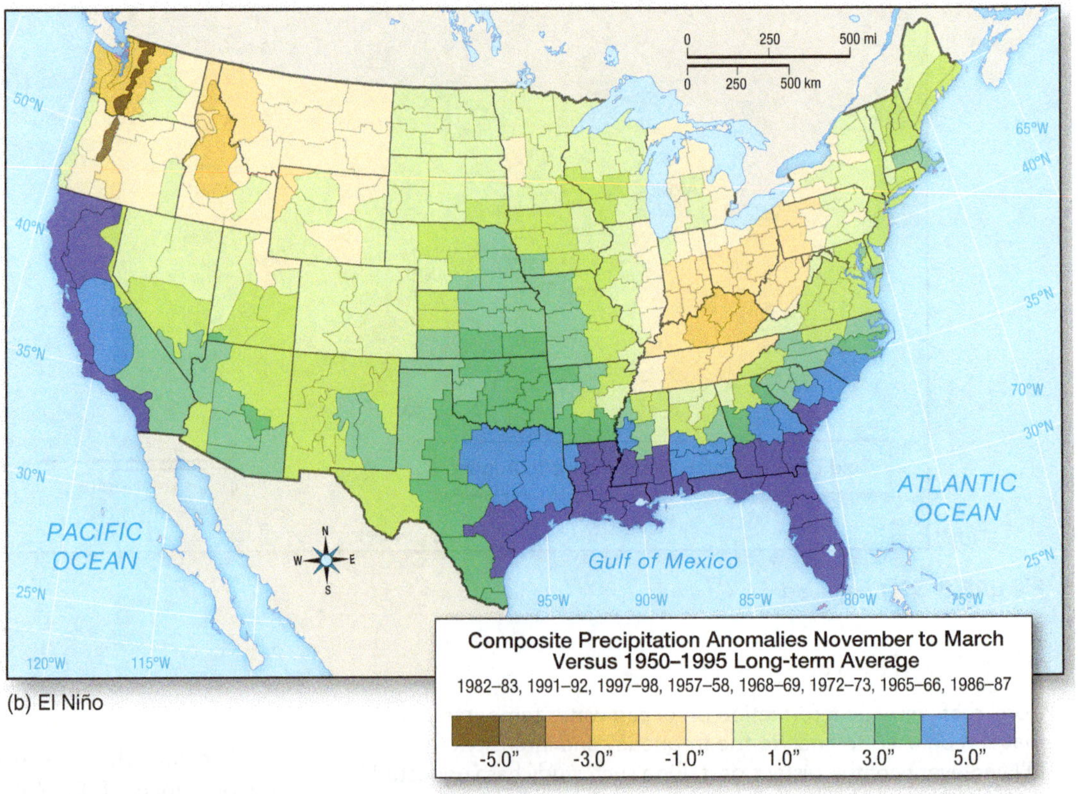

(b) El Niño

MapMaster
North America
Physical Environment
Temperature Anomalies
El Niño & La Niña Years (U.S)

▲ **FIGURE 8–40 Effects of El Niño and La Niña in the United States.** El Niño and La Niña events tend to promote temperature and precipitation responses across the conterminous United States. During eight recent El Niño events, November–March temperatures have tended to be higher than normal in the north-central United States and lower than normal across the southern tier of states (a). At the same time, coastal California and the southeast generally have wet conditions, while relative drought tends to occur in the Pacific Northwest and over some of the Ohio River Valley (b). As seen in the maps on the next page, La Niñas tend to promote warm winter conditions over much of the southeast (especially coastal Texas) and parts of the northwest and the north-central United States (c), along with wetness in the northwest and relative drought in the southeast (d).

Video
La Niña
http://goo.gl/E4S5U

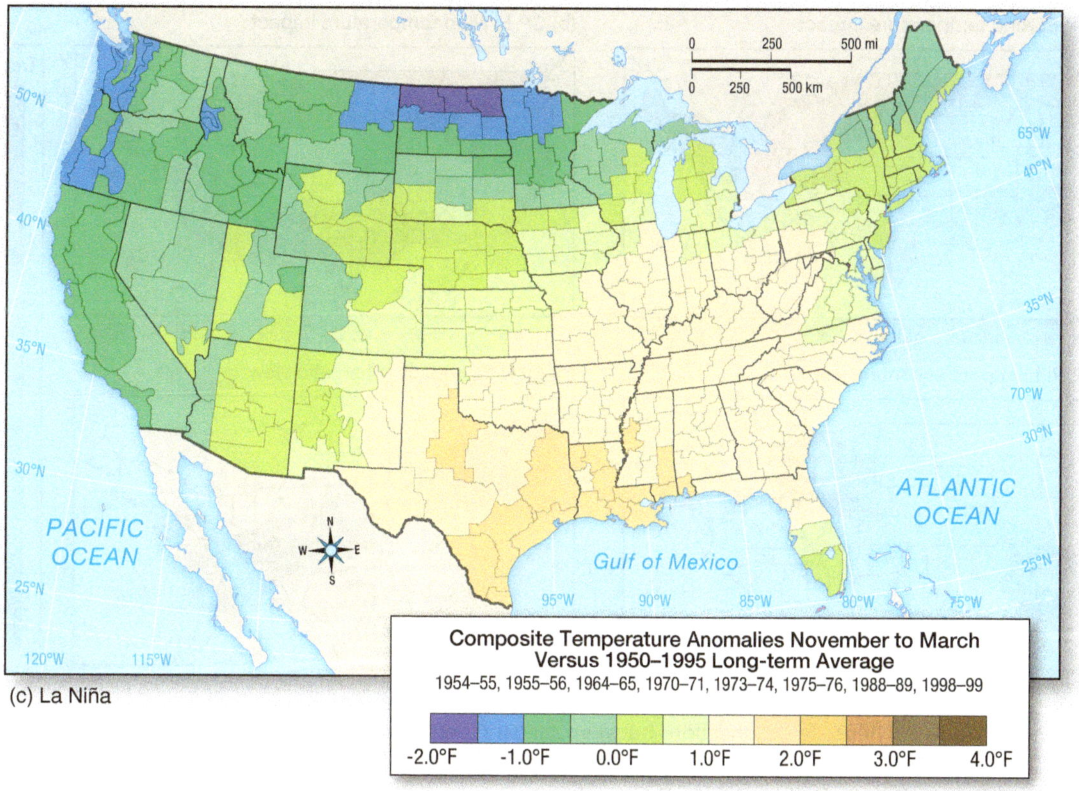

(c) La Niña

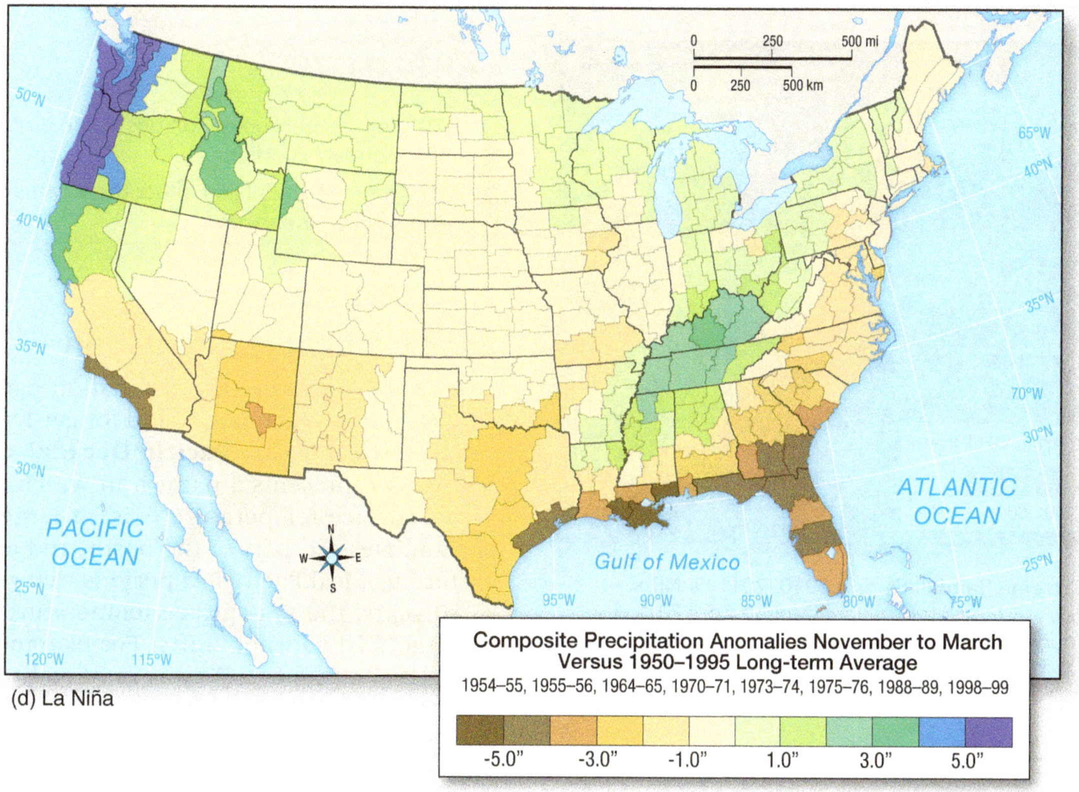

(d) La Niña

▲ **FIGURE 8–40** *Continued*

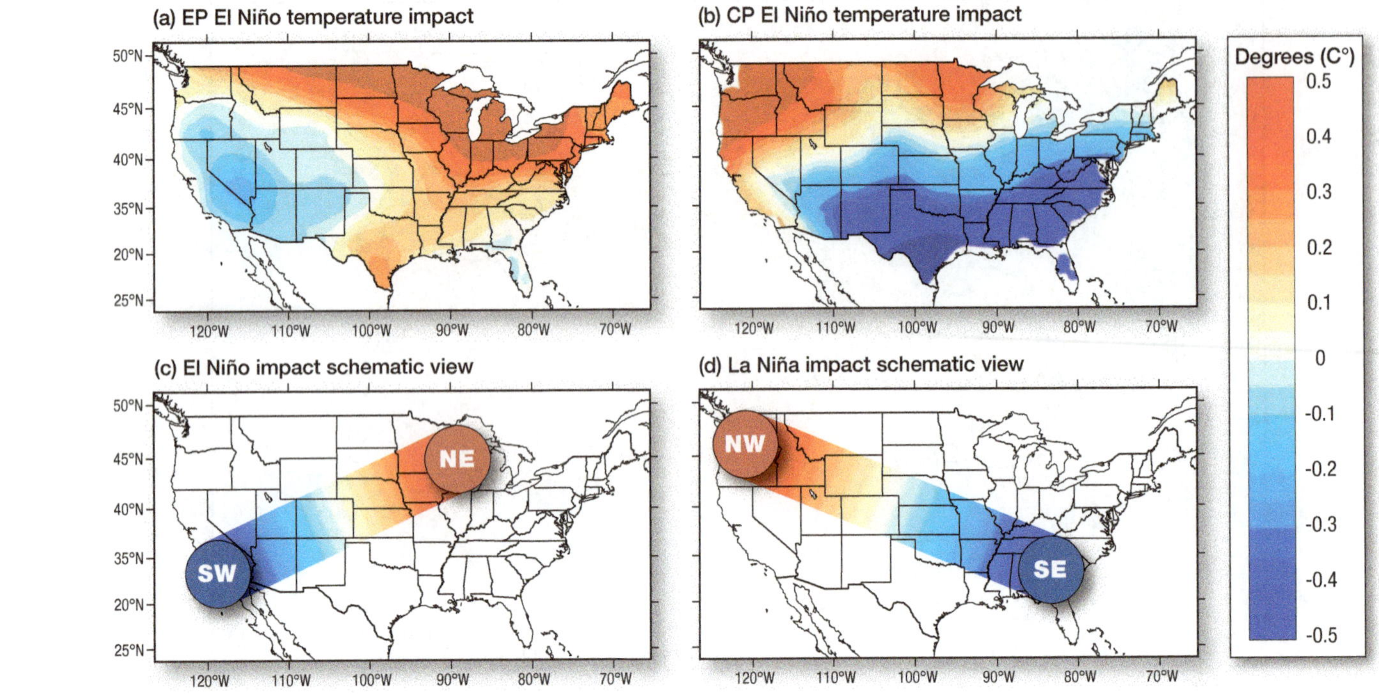

▲ **FIGURE 8–41 EP and CP Type El Niños.** Teleconnections between January temperatures in the United States and Eastern Pacific and Central Pacific type El Niños. The El Niño pattern has a Northeast-Southwest axis (a and c), whereas the La Niña pattern is aligned Northwest-Southeast (b and d).

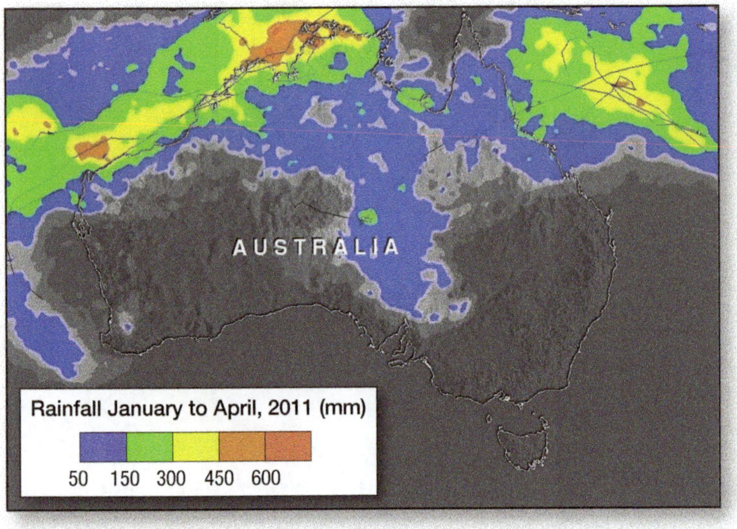

▲ **FIGURE 8–42 Extreme Rainfall During 2010–2011 La Niña.** Extreme rainfall totals are shown for January through April 2011 in Northern Australia during the strongest La Niña on record. Near the end of the La Niña, parts of Northern Australia received more than 60 cm (24 in.) of rain, and flooding was widespread.

individual event. Although the occurrence might be consistent with the teleconnections discussed above, there is no way to say it might not have occurred independently, with no assistance from the tropical Pacific.

CHECKPOINT

8.20 How do El Niño conditions in the eastern Pacific differ from "normal" conditions?

8.21 How can an El Niño affect the weather? Include teleconnections in your answer.

Pacific Decadal Oscillation

The ENSO is not the only oscillation pattern across the Pacific Ocean. A much larger and longer-lived reversal pattern exists, known as the **Pacific Decadal Oscillation (PDO)**. The PDO represents a pattern in which two primary nodes of sea-surface temperature exist, a large one in the northern and western part of the basin, and a smaller one in the eastern tropical Pacific. At periods that range from about 20 to 30 years, the sea-surface temperatures in the two zones undergo fairly abrupt shifts. For example, the period from about 1947 to 1976–1977 was marked by generally low temperatures in the northern and western part of the ocean and high temperatures in the eastern tropical Pacific. Then an abrupt transition occurred and the reverse temperature pattern became established until the late 1990s.

In the late 1990s TOPEX/Poseidon and Jason 1 satellite data suggested that another reversal was under way, but by the early 2000s this shift seemed to have halted. This phenomenon, along with other evidence, suggests that the

8–5 PHYSICAL PRINCIPLES

What Causes El Niños and La Niñas?

There are actually two parts to this question: (1) Why do these anomalies arise? That is, what causes an El Niño or La Niña to become established? and (2) What causes the reversals, or oscillations, between the two extremes?

The Role of Positive Feedback

An answer to the first question was proposed more than 40 years ago in a pioneering study by Jacob Bjerknes, and has been essentially confirmed by more recent analysis and data. Unlike some other types of climate change, both anomalies arise without any external forcing. That is, they do not result from a nonclimatic event such as an asteroid impact, a volcanic eruption, or a change in the Sun's radiation. Rather, they result from processes internal to the ocean–atmosphere system. They are, in fact, classic examples of a *positive feedback* mechanism, in which relatively small changes are continually reinforced and amplified.

Consider, for example, the La Niña, where the strengthening trades push more warm water westward. In the western Pacific, the growing pool of warm water promotes more uplift, precipitation, and lower pressure. Lower pressure in the west intensifies the pressure gradient and strengthens the trades, which in turn lead to further transport of warm water. At the same time, export of warm water cools the eastern Pacific, enhances upwelling, and raises the thermocline. All of these processes promote further growth of the anomaly. Similar reasoning, but in reverse, explains why El Niños develop. In this case warmer water in the eastern Pacific weakens the trades, which leads to less heat export by westward-moving water. With less heat exported, warmth in the eastern Pacific intensifies, further reducing the pressure gradient and reducing export of heat. Thus, we see that once the system begins to drift toward either condition, there will be a tendency for the anomaly to intensify and become firmly established. This line of reasoning favors two very different states. In the El Niño state, there is a thick pool of anomalously warm water in the eastern Pacific lying above a relatively deep thermocline. At the other extreme, the eastern Pacific is anomalously cold, the thermocline is shallow, and the trades are strong.

Breakdown of El Niños and La Niñas

If positive feedback explains the appearance of El Niños and La Niñas, what leads to their breakdown? That is, once established, why don't they persist indefinitely? Why is there oscillation from one state to the other? Much less is known with certainty about this phenomenon, and the topic remains a matter of intense interest within the scientific community. However, there is little doubt that for oscillations to arise naturally in the system, delayed negative feedback processes must be occurring that are out of phase with the positive feedbacks described above. That is, one or more restorative processes must be lagging behind processes driving the system toward the extreme states. As an El Niño or La Niña builds, the delayed restoring processes also grow and eventually become large enough to overwhelm the amplifying processes. When that happens, the system moves away from the extreme state (El Niño or La Niña) toward more normal conditions. But the positive feedbacks are still in play, and if they are large enough, the system moves to the other extreme. So, in general terms, an oscillation between the extremes arises from the joint effects of positive and delayed negative feedbacks.

A number of candidate processes have been proposed as delayed negative feedbacks. One possibility is that during an El Niño Rossby waves generated in the east-central Pacific travel westward and are reflected backward from the western ocean boundary as small waves on the ocean surface. These so-called Kelvin waves bring a deeper thermocline to the eastern Pacific, counteracting the El Niño. Similar waves form with La Niña, but they bring a shallower thermocline to the eastern Pacific, and thus similarly move the system toward more normal conditions.

Another possibility, termed the *recharge oscillator*, contends that the buildup and discharge of warm water from the tropical Pacific gives rise to the oscillation. The idea is that while an El Niño builds, the entire tropical Pacific experiences a gradual "recharge" of heat as the thermocline deepens. Some time during the El Niño proper, the excess warmth is discharged to higher latitudes, the thermocline becomes shallower, and the system moves toward the other extreme. This institutes another cycle of recharge, followed by another flushing of heat to the extratropics (the middle and high latitudes) with the ensuing El Niño.

Another possibility is that during an El Niño the central Pacific atmosphere warms because of enhanced convection and condensation, and this in turn generates low-pressure cells on either side of the equator. Initially these cyclones amplify the El Niño, but eventually they pump cold water eastward and thereby destroy the event they helped construct. Still other mechanisms have been suggested as important in the oscillations, including forcing by disturbances unrelated to the ENSO. No single process has been shown to be responsible for ENSO events, and many experts believe that multiple processes are involved to varying degrees from one episode to the next.

Support for all these ideas comes from computer models that are able to reproduce ENSO-like cycles. But there is no agreement about what controls the period of oscillation, why it averages about four years instead of some other period, or even why the period of oscillation is variable, with ENSOs appearing every two to seven years. Answers to these and other questions about ENSO events are important both for understanding our present climate and for anticipating the nature and impact of future climate changes.

8–5–1 What role does positive feedback play in La Niñas?

8–5–2 If you wanted to prove that a process is not involved in the development of El Niños or La Niñas, how might you go about that? What tools would you use?

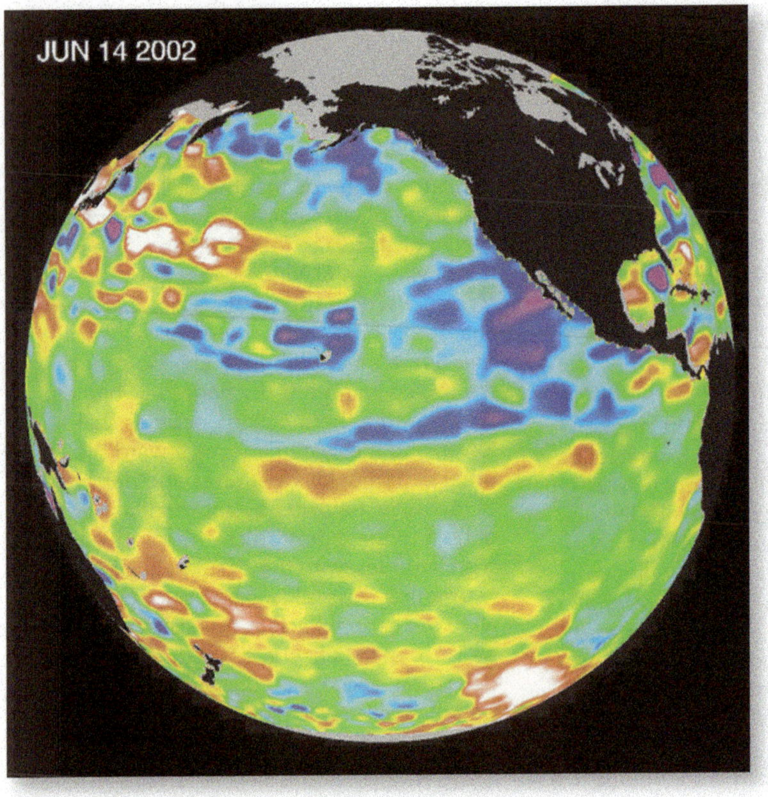

▲ **FIGURE 8–43 Pacific Decadal Oscillation.** Pacific sea-surface temperatures obtained in June 2002 suggest that a reversal in the Pacific Decadal Oscillation has begun. If so, this represents a major shift from the pattern in existence between 1976 and the late 1990s.

PDO pattern may in fact be more complex than originally believed. The image in Figure 8–43 shows a pattern of warm water extending off the southern coast of Japan in June 2002. A wedge of warmer waters also lies in the eastern tropical Pacific. Mostly cooler waters exist off the west coast of North America. Because of the enormous heat content of the oceans, the PDO may exert a direct, major impact on the pressure distribution of the atmosphere. A history of the PDO is shown in Figure 8–44, plotting a useful indicator of the phenomenon, the *PDO Index*, during the past century. Positive values indicate an episode with warm waters in the east tropical Pacific, and negative values representing cold waters in that area.

Recent research has also suggested that the phase of the PDO influences the impact that El Niño events have on climate. When the PDO is in a "warm phase" (high temperatures in the eastern tropical Pacific), an El Niño's impacts on weather are more pronounced than when the PDO is in a "cold phase" (low temperatures in the eastern tropical Pacific).

Arctic Oscillation and North Atlantic Oscillation

Multiyear oscillations are not restricted to the Pacific Ocean. Such patterns also exist in the Atlantic. One is called the **Arctic Oscillation (AO)**, which is closely related to another observed pattern called the **North Atlantic Oscillation (NAO)**. Refer back to Figure 8–6 and you will see that the atmosphere over the North Atlantic Ocean is dominated by the Bermuda–Azores high and the Icelandic low. Figure 8–6 depicts the average pressure pattern for summer and winter, but of course this pattern is variable and some years are marked by enhanced or reduced differences in pressure between the two semipermanent cells. The NAO is said to be in a *positive phase* when the pressure gradient is greater than normal (Figure 8–45a), and negative when it is less than normal (Figure 8–45b).

When the NAO is in a positive phase, there is an enhanced pressure gradient from south to north that causes both an intensification and a northward shift of the polar jet stream. The more vigorous jet stream brings increased storminess to Northern Europe in the winter months along with generally mild temperatures, as the storms pass over the warm ocean current in the North Atlantic. The eastern United States likewise experiences mild, wet conditions with a positive NAO, but eastern Canada and Greenland, north of the jet stream, usually have cooler and drier winter conditions.

Negative NAO phases occur when the high-pressure and low-pressure systems of the North Atlantic are less well

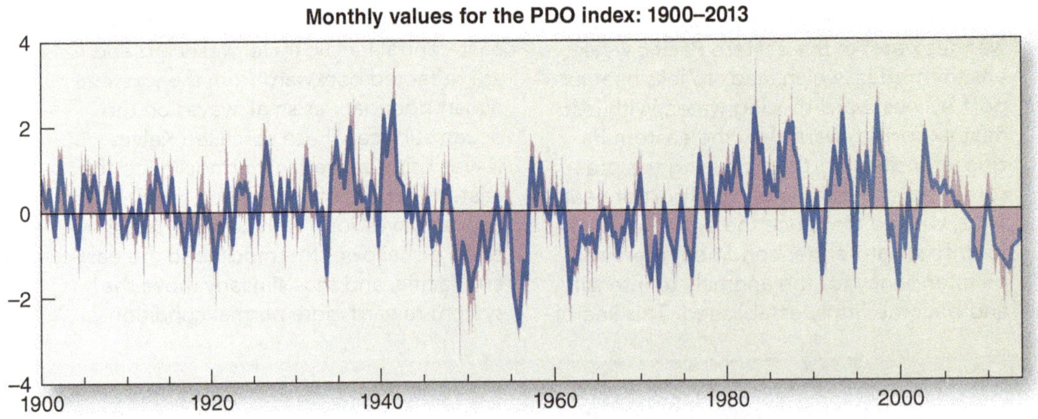

▶ **FIGURE 8–44 PDO Index.** The PDO Index is a commonly used indicator of the strength and mode of the Pacific Decadal Oscillation. Positive values of the index, plotted for the years 1900 through April 2014, indicate warm waters in the eastern tropical Pacific. In the late 1990s it appeared that a shift from positive to negative was occurring, but this shift seems to have abruptly ended in the early 2000s.

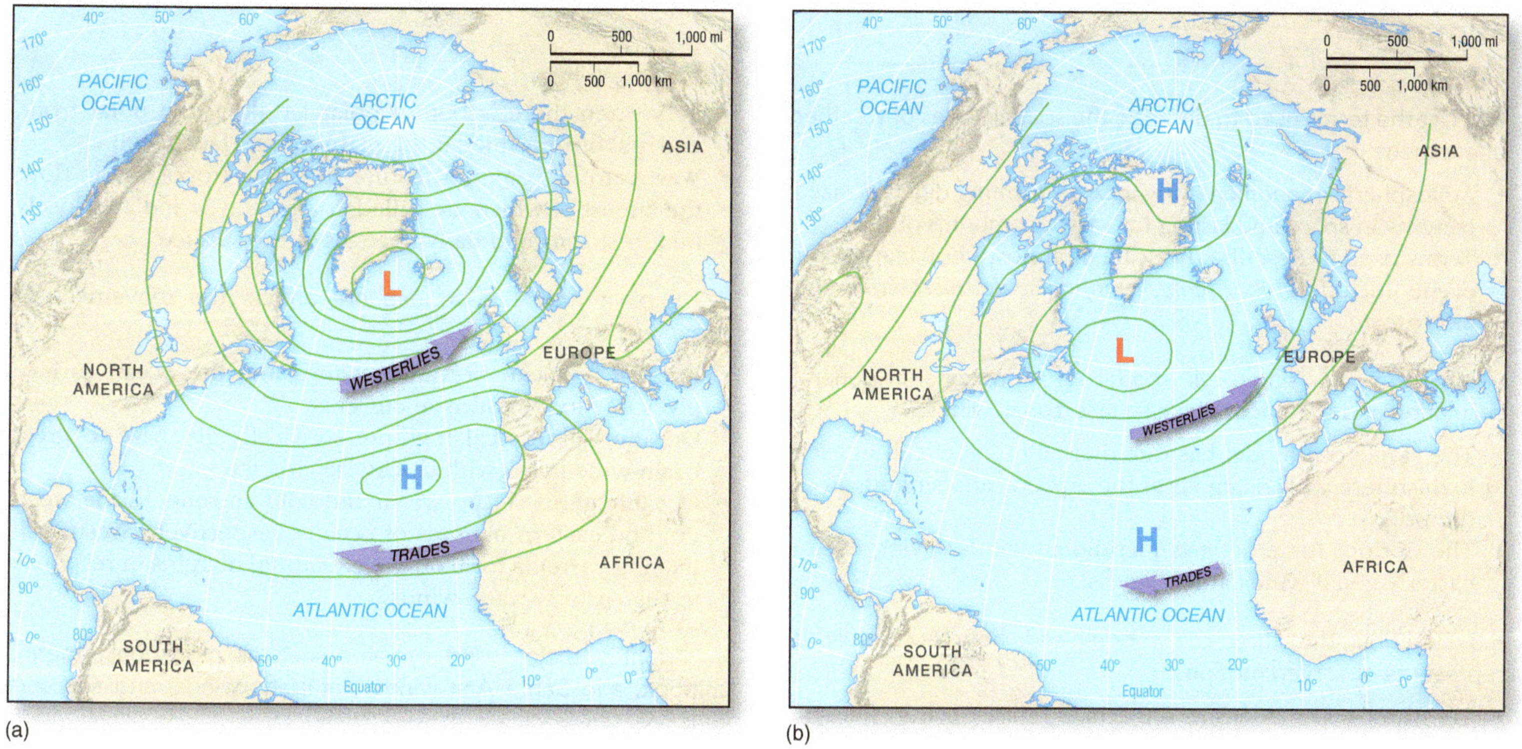

▲ **FIGURE 8–45 North Atlantic Oscillation.** The North Atlantic Oscillation relates to changes in the intensity of the high-pressure and low-pressure systems over the North Atlantic. (a) When the two pressure cells are more intense than normal, the jet stream becomes more intense and storms vigorously track toward northern Europe. (b) Negative NAO phases mean that the pressure cells are weaker than normal, leading to a less vigorous jet stream than normal. Northern Europe becomes cold and dry, while southern Europe and the Mediterranean tend to have increased winter storm activity.

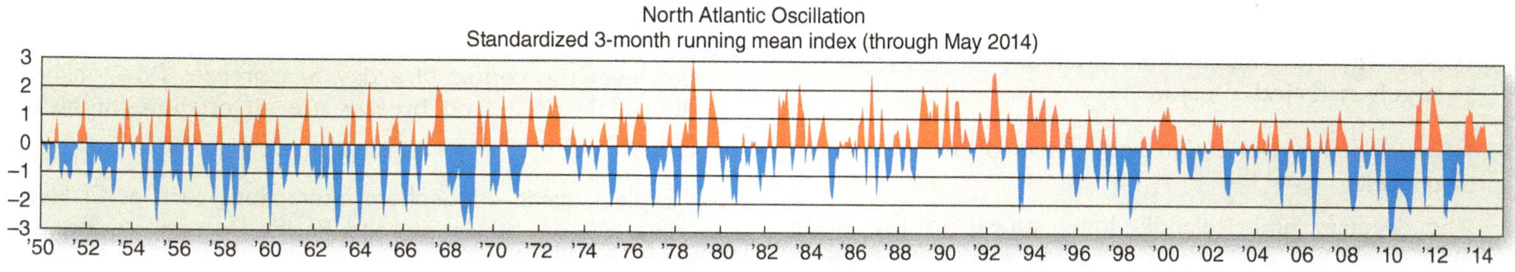

▲ **FIGURE 8–46 Positive and Negative Phases of NAO.** The NAO tends to remain dominated by a positive or negative phase for years at a time.

developed, leading to a reduced pressure gradient. This weakens the strength of the jet stream, which is also directed more toward southern Europe and the Mediterranean. This usually leads to wet winters over southern Europe while the northern part of the continent is cold and relatively dry. The eastern United States experiences more frequent outbreaks of cold air surging southward from interior Canada. Figure 8–46 plots a numerical index of the NAO. The 1990s were dominated by positive NAO conditions; a shift to negative NAO dominance took place just before the beginning of the new millennium and lasting into the beginning of its second decade. The extreme winter of 2010–2011 over the eastern United States occurred during a negative NAO regime.

CHECKPOINT

8.22 What is the PDO Index?

8.23 How do changes in the NAO influence weather conditions across Europe and eastern North America?

Summary

8.1 List the terms used to describe the spatial scales of weather phenomena.

- Atmospheric phenomena vary in scale, forming different-sized patterns in space and lasting for varying amounts of time.
- Terms used to describe the scale of atmospheric features or events include *global scale, synoptic scale, mesoscale*, and *microscale*.

8.2 Explain the single-cell model of atmospheric general circulation and the origin of east–west (zonal) winds.

- The single-cell model has one thermally direct cell in each hemisphere with rising air at the equator and sinking air at the poles.
- The Coriolis force deflects meridional winds, giving rise to easterly zonal winds at low levels.

8.3 Describe the three-cell model and how accurately it represents observed motions.

- The three-cell model has a thermally direct cell in the low latitudes, a friction-driven cell in the middle latitudes, and a thermally direct cell in the high latitudes. These are known as the Hadley cell, Ferrel cell, and polar cell, respectively.
- Winds in the lower troposphere in the three-cell model are easterly (Hadley cell), westerly (Ferrel cell), and easterly (polar cell).
- Of the three cells, the Hadley cell comes closest to matching observed circulation. The middle latitudes are better described as having westerly winds throughout the troposphere with wave-like motions predominant. The polar cell is largely statistical in nature.

8.4 Describe the distribution and effects of semipermanent pressure cells.

- The Aleutian and Icelandic lows appear in the North Pacific and North Atlantic winter and are breeding grounds for storms. The Tibetan high is another winter feature in the Northern Hemisphere, but with its descending air, it is not a storm-producer. All of these weaken or vanish in the summer.
- Subtropical high pressure appears in the summer in both the Northern Pacific (Hawaiian high) and Atlantic (Bermuda-Azores high). Clockwise circulation around these cells of high pressure affects surrounding areas on all sides.
- In summer the Tibetan low replaces the wintertime high.

8.5 Explain the distribution of wind and pressure in the upper troposphere.

- Because temperatures generally decrease poleward, upper-troposphere pressure decreases poleward.
- In the absence of friction, the Coriolis force and pressure gradient force give rise to generally westerly motions.
- Wind speeds are strongest in the upper troposphere, often concentrated in a narrow band called jet streams.
- Very often there is both a polar jet along the polar front, and a subtropical jet at the margin of a Hadley cell.
- Waves are a common feature in upper-tropospheric flow, the largest of which are called Rossby waves and are closely tied to storms and daily changes at the surface.

8.6 Explain how the atmosphere affects the movement of ocean waters.

- Friction between prevailing winds and the ocean surface gives rise to wind-driven currents.
- Oceanic subtropical high-pressure cells give rise to great basin-wide gyres with clockwise circulation.
- Oceanic gyres set up warm and cold currents in the western and eastern margins of oceans, respectively. Westward-moving currents form on the equatorial margins in response to the easterly trade winds.

8.7 Identify major wind systems such as monsoons; foehn, chinook, and Santa Ana winds; katabatic winds; and sea and land breezes.

- Monsoons are seasonal reversals of pressure and wind. In summer winds blow onto the continent, bringing heavy precipitation to many monsoon climates. In winter the flow is offshore and dry conditions prevail.
- Foehn, chinook, Santa Ana, and katabatic winds are all influenced by topography to some degree, with air flowing downslope and warming.
- Land and sea breezes are thermally direct circulations that form immediately along coastlines and switch direction over the course of a day. Sea breezes flow onshore during the day, land breezes are an offshore nighttime phenomenon.

8.8 Explain ocean–atmosphere interactions such as El Niño/La Niña, Walker circulation, and Pacific Decadal, Arctic, and North Atlantic oscillations.

- Walker cell (or Walker circulation) refers to a zonal cell in the tropical Pacific with easterly flow at low levels and westerly flow aloft. Variation in the Walker cell is known as the Southern Oscillation.
- Intensification of the Walker cell leads to a La Niña and anomalously cold Eastern tropical Pacific sea-surface temperatures.
- An El Niño forms if the Walker cell weakens sufficiently, allowing a pool of unusually warm water to form in the Eastern tropical Pacific.
- The Pacific Decadal Oscillation (PDO) is another form of ocean variation, involving a seesaw between temperatures in the north/western Pacific and the tropical Pacific.
- La Niñas, El Niños, and the PDO are all connected in different ways to weather and climate in other parts of the world.

Key Terms

Aleutian low *p. 250*
angular momentum *p. 249*
Arctic Oscillation (AO) *p. 282*
atmospheric rivers *p. 259*
Bermuda–Azores high *p. 250*
Canary Current *p. 262*
Central Pacific (CP) type El Niño *p. 275*
chinook winds *p. 266*
downwelling *p. 264*
East Greenland Drift *p. 263*
Eastern Pacific (EP) type El Niño *p. 275*
Ekman spiral *p. 261*
El Niño *p. 272*
ENSO events *p. 272*
Equatorial Countercurrent *p. 261*
equatorial low *p. 247*
Ferrel cell *p. 246*
foehn winds *p. 265*
frictional dissipation *p. 249*
general circulation *p. 244*
global scale *p. 245*
Gulf Stream *p. 261*
gyres *p. 263*
Hadley cell *p. 246*
Hawaiian high *p. 250*
horse latitudes *p. 247*
Icelandic low *p. 250*
intertropical convergence zone (ITCZ) *p. 247*
katabatic winds *p. 268*
La Niña *p. 272*
Labrador Current *p. 263*
lake breeze *p. 270*
land breeze *p. 270*
meridional winds *p. 246*
mesoscale *p. 245*
microscale *p. 245*
monsoon *p. 264*
monsoon depressions *p. 264*
mountain breeze *p. 271*
North Atlantic Drift *p. 261*
North Atlantic Oscillation (NAO) *p. 282*
North Equatorial Current *p. 261*
northeast trade winds *p. 247*
ocean currents *p. 261*
Oceanic Niño Index (ONI) *p. 273*
oceanography *p. 261*
Pacific Decadal Oscillation (PDO) *p. 280*
polar cell *p. 246*
polar easterlies *p. 250*
polar front *p. 254*
polar highs *p. 250*
polar jet stream *p. 254*
Rossby wave *p. 256*
Santa Ana winds *p. 267*
sea breeze *p. 269*
sea breeze front *p. 269*
semipermanent cells *p. 250*
Siberian high *p. 250*
single-cell model *p. 246*
South Equatorial Current *p. 261*
southeast trade winds *p. 247*
Southern Oscillation *p. 272*
subpolar lows *p. 250*
subtropical highs *p. 247*
subtropical jet stream *p. 255*
synoptic scale *p. 245*
teleconnections *p. 275*
three-cell model *p. 246*
Tibetan low *p. 250*
upwelling *p. 263*
valley breeze *p. 271*
Walker cell *p. 271*
West Greenland Drift *p. 263*
westerlies *p. 250*
zonal wind *p. 246*

Review Questions

1. Describe the single-cell and three-cell models of the general circulation.
2. What is the Hadley cell and where is it found?
3. Of the pressure and wind belts described in the three-cell model, which have the strongest basis in reality?
4. What are the Ferrel and polar cells?
5. Why do the trade winds flow from the northeast and southeast instead of directly from the east?
6. Explain how atmospheric motion is maintained in terms of potential energy, kinetic energy, and frictional dissipation.
7. Describe the distribution of semipermanent cells and their seasonal changes in location and size.
8. What is the Sahel? Describe its seasonal cycle of rainfall and explain its origin.
9. Describe the average patterns of the 500 mb level for January and July. What causes the patterns?
10. What are atmospheric rivers and why are they important?
11. Explain why upper-atmospheric winds outside the tropics have a strong westerly component, on average.
12. Why does the equatorward bending of height contours for the 500 mb level imply the presence of a trough?
13. Explain how temperature patterns lead to the development of the polar jet stream.
14. Describe the distribution of Rossby waves and their impact on daily weather.
15. What is upwelling? How is it caused?
16. What is the Ekman spiral?
17. Describe the scope of global, synoptic, mesoscale, and microscale wind systems.
18. Describe the wind patterns associated with the monsoon of southern Asia.
19. Describe foehn winds.
20. How do katabatic winds differ in origin from foehn winds?
21. What causes sea/land and mountain/valley breezes to develop?
22. Describe the Pacific Decadal Oscillation and the Arctic Oscillation.
23. What is an El Niño and how is it related to the Walker circulation?

Critical Thinking

1. Figure 8–4a depicts the classic description of the single-cell model. Can you think of any reason the arrows should not be directed in a straight line toward the equator?
2. The three-cell model of circulation places the center of the equatorial low right along the equator. Do you think that the varying solar declination through the course of a year would be able to shift the center of the low all the way to the tropics of Cancer and Capricorn?
3. Which of the belts depicted in the three-cell model is likely to exhibit the greatest temperature gradients?
4. Examine Figure 8–6 and determine which of the semipermanent cells have the highest and lowest surface air pressures. How do the strengths of the winter and summer cells in the Northern and Southern Hemispheres compare to each other?
5. Figure 8–6 shows that pressure gradients, and therefore wind speeds, are strong in the middle-high latitudes of the Southern Hemisphere (such as at the southern tip of South America). Why doesn't a similar feature exist in the Northern Hemisphere?
6. If the western Great Plains were to have very low temperatures and the East Coast relatively warm conditions, what inferences would you make about the position and amplitude of the Rossby wave pattern?
7. Weather forecasters often use a type of weather map depicting the thickness of the 1000–500 mb layer of the atmosphere. What information would this map convey to a meteorologist?
8. El Niño and La Niña conditions help meteorologists make seasonal forecasts for the southeastern United States—thousands of kilometers away. What atmospheric mechanism permits such extrapolations?
9. Why isn't upwelling an important process of the Gulf Stream?
10. Why do ocean surface temperature patterns change so slowly, especially when compared to atmospheric patterns?

Problems & Exercises

1. Go to http://weather.noaa.gov/fax/nwsfax.html. In the first group of maps (labeled "standard barotropic levels"), select 500 mb. Then click on the link under *4a*, called *Height/temp*. The map that will appear is a map of the 500 mb pattern, with solid lines depicting the height of the 500 mb level (in tens of meters).
   a. Locate the position of the major troughs and ridges.
   b. Is the current pattern a zonal or meridional one?
   c. Where is the jet stream most prominent?
   d. Does today's jet stream appear across the entire region mapped?
   e. How does today's map compare to the mean distributions shown in Figure 8–8? Why is the map for today (in all likelihood) less zonal than those in Figure 8–8?
2. If you have a high-speed connection to the Web, go to http://weather.unisys.com/nam/nam.php?plot=500&inv=0&t=1 to see an animation depicting the change in the 500 mb pattern as predicted by a numerical model.
   a. Describe the predicted movement of the Rossby waves.
   b. How does the movement of the Rossby waves respond to their wavelengths and internal wind speeds (as implied by the spacing of the contours)?

Visual Analysis

This map shows sea-surface temperature anomalies for early June 2014.

8.1. Is it suggestive of El Niño or La Niña conditions?

8.2. What part of the world ocean has the highest mapped values (above 4°C)? Is the sea surface warmer there than any place else? (Obviously not.) Please explain.

8.3. Considering your answer to Question 2, why are anomaly maps used to depict El Niños and La Niñas?

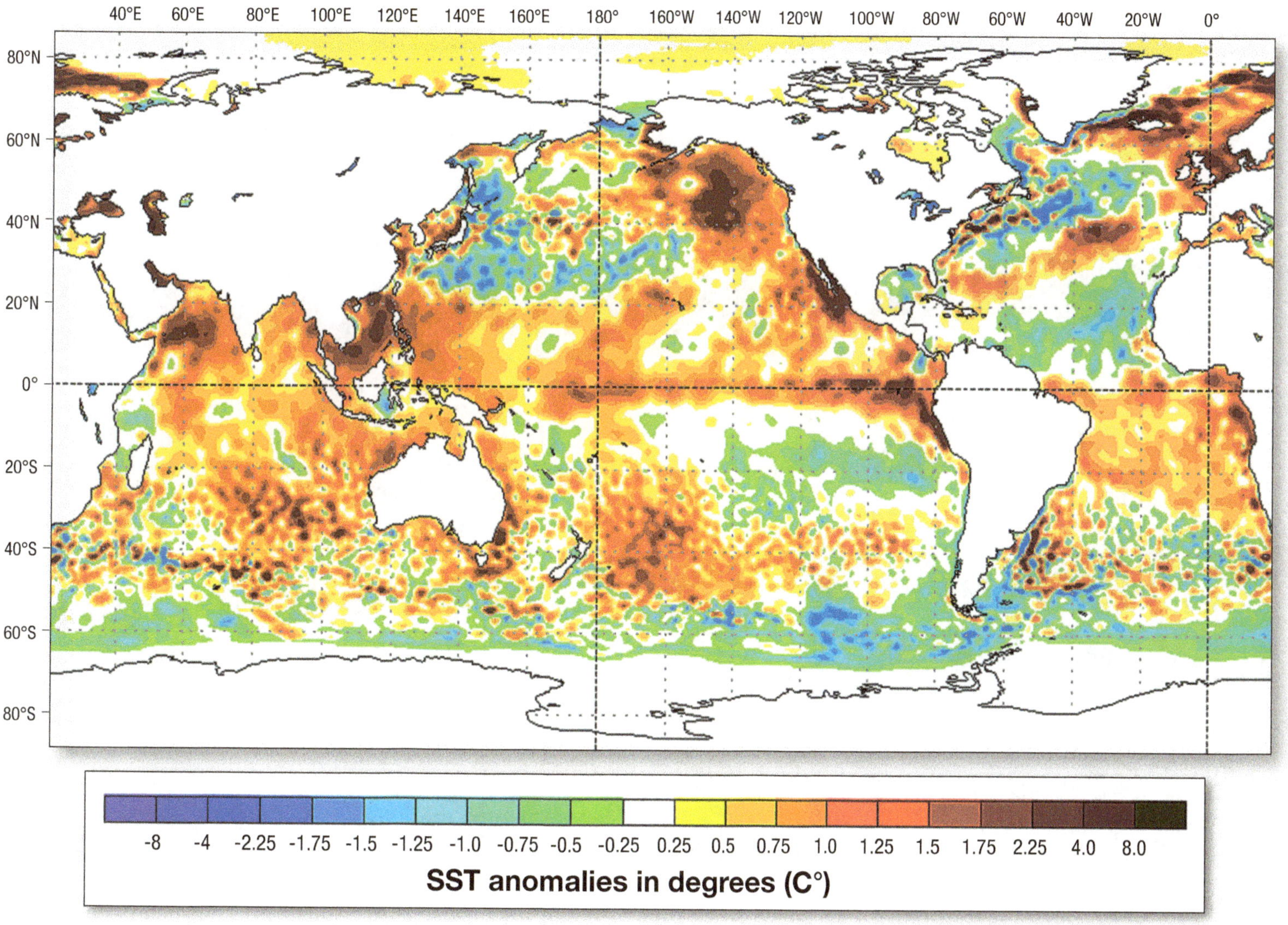

CHAPTER

9 Air Masses and Fronts

Residents of the Northeastern United States know that winter temperatures can be extreme and highly variable. Those that forgot this bit of wisdom were well reminded of it in January 2013. Cleveland, Ohio, for example, reached a low temperature of −11.7 °C (11 °F) on January 22. Pittsburgh and Syracuse had similar lows of −10.6 °C (13 °F) and −13.3 °C (8 °F), respectively. But eight days later things changed substantially and temperatures became spring-like, with a high of 17.2 °C (63 °F) in Cleveland.

Contrast those cold January conditions with the high temperatures experienced across the southwestern United States as record-breaking heat overtook much of the West in July of that same year. During the heat wave, Death Valley, California, got up to 53.3 °C (128 °F), Phoenix hit 48.3 °C (119 °F), and Las Vegas reached 47.2 °C (117 °F).

Have you ever wondered about episodes such as the two described above, in which large areas experience more or less similar weather? At times like these, broadcasters use phrases such as "throughout the Midwest" or

◀ *Extreme heat at Death Valley National Park on July 14, 2013. Four days earlier marked the 100th anniversary of the all-time maximum temperature ever recorded of 134 F.*

"across the eastern seaboard" or "Today the Pacific Northwest experienced. . . ." By way of contrast, it often happens that places within an hour's drive of each other have very different weather, with essentially nothing in common. What is the explanation for these conditions? Why does the atmosphere sometimes organize itself into broad uniform patches, and at other times show extreme variation over short distances? This chapter addresses these and related questions using some very simple, powerful concepts.

Learning Outcomes

After reading this chapter, you should be able to:

- **9.1** Describe the importance of source regions in air mass formation.
- **9.2** Describe the types of air masses and explain how they form.
- **9.3** Identify the different types of fronts and their characteristics, including cloud conditions.

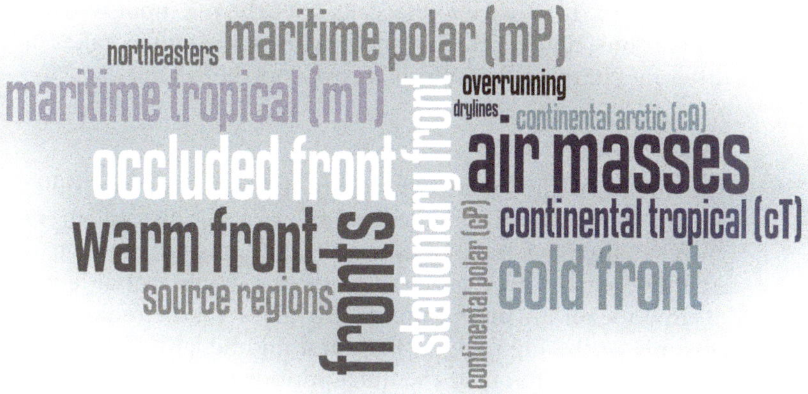

The January and July 2013 situations described in the chapter opener represent two extreme instances in which large regions are covered by a body of air having more or less uniform temperature and moisture. These large volumes of air are called **air masses**. Often an area the size of North America will be covered by several air masses at the same time so that, for example, the northeastern United States and southeastern Canada may experience cold, dry conditions, while the southern United States is dominated by warm, moist air. Consequently, a person might board an airplane in Nashville feeling perfectly comfortable in short sleeves, only to end up shivering in Boston. Moreover, these air masses are commonly separated from each other by fairly narrow boundary regions, called **fronts**, across which conditions change rapidly. The passage of a front is a significant weather event, because fronts often bring abrupt changes in temperature, humidity, and wind. They also provide a lifting mechanism that can lead to the formation of clouds and precipitation.

In this chapter we describe the formation and nature of air masses, the fronts that separate them, and their influences on local weather.

Air Masses and Their Source Regions

9.1 Describe the importance of source regions in air mass formation.

The temperature, pressure, and moisture characteristics of the atmosphere arise in large part from the continuous exchange of energy and water vapor between the air and surface. When energy inputs exceed energy losses, the temperature of the air increases. In the same way, when more evaporation than precipitation takes place, the moisture content of the atmosphere increases. Because heat and water are not uniformly distributed across the globe, the cooling and warming of the atmosphere vary from place to place, as does the net input of water vapor. Thus, air over the tropical Pacific, for example, takes on different characteristics from air over northern Canada.

Source Regions

The areas of the globe where air masses form are called **source regions**. Heating or cooling large bodies of air requires many days, as do changes in the moisture content, so air must remain over a source region for a substantial length of time for an air mass to form. Air mass source regions occur only in the high or low latitudes; middle latitudes are too variable and do not have the quiet periods necessary for an air mass to take on the characteristics of the underlying surface. Also, an area must be quite large—many tens of thousands of square kilometers—to act as a source region. Iceland is too small, for example, to allow the formation of air masses.

Although air masses have fairly uniform temperature and moisture content in a horizontal direction, they are not uniform from the surface to the upper atmosphere. In fact, large vertical gradients in temperature can easily occur in an air mass. These vertical differences in temperature affect the stability of the atmosphere (Chapter 6), which greatly affects the likelihood of precipitation. Some air masses, by their very nature, are more likely than others to yield precipitation.

CLASSIFYING AIR MASSES Air masses are classified according to the temperature and moisture characteristics of their source regions. Based on moisture content, air masses can be considered either *continental* (dry) or *maritime*[1] (moist). According to their temperature, they are either *tropical* (warm), *polar* (cold), or *arctic* (extremely cold). Meteorologists use a two-letter shorthand scheme for categorizing air masses. A small letter *c* or *m* indicates the moisture conditions, followed by a capital letter *T*, *P*, or *A* to represent temperature. Continental polar air, for example, is designated cP, and maritime tropical air is mT. While this combination of letters could theoretically yield six possible types of air masses, maritime arctic (mA) air masses do not occur in nature because water bodies do not get cold enough to foster arctic air (they freeze at arctic temperatures, which largely removes their maritime character). Thus, there are five types of air masses, as described in Table 9–1. The major source regions of North America are shown in Figure 9–1.

MOVEMENT OF AIR MASSES Air masses are not permanently confined to their source regions; they are able to migrate to regions marked by less extreme weather conditions. The movement of an air mass away from its source causes two things to happen: (1) The region to which the air mass moves undergoes a major change in temperature and humidity, and (2) the air mass becomes more moderate. For example, very cold air migrating southward across the U.S.–Canadian border may be warmed from below as it passes over surfaces exposed to greater solar heating.

We now examine the various types of air masses, the weather they bring, and the transformations they experience as they travel.

Video
Effects of the 2011 Groundhog Day Blizzard
http://goo.gl/500DwD

TABLE 9–1
Air Masses

Type	Source Regions	Properties at Source
Continental polar (cP)	High-latitude continental interiors	Cold and dry. Very stable. Minimal cloud cover.
Continental arctic (cA)	Highest latitudes of Asia, North America, Greenland, and Antarctica	Extremely cold and very dry. Extremely stable. Minimal cloud cover.
Maritime polar (mP)	High-latitude oceans	Cold, damp, and cloudy. Somewhat unstable.
Continental tropical (cT)	Low-latitude deserts	Hot and dry. Very unstable.
Maritime tropical (mT)	Subtropical oceans	Warm and humid.

[1]Sometimes called *marine*.

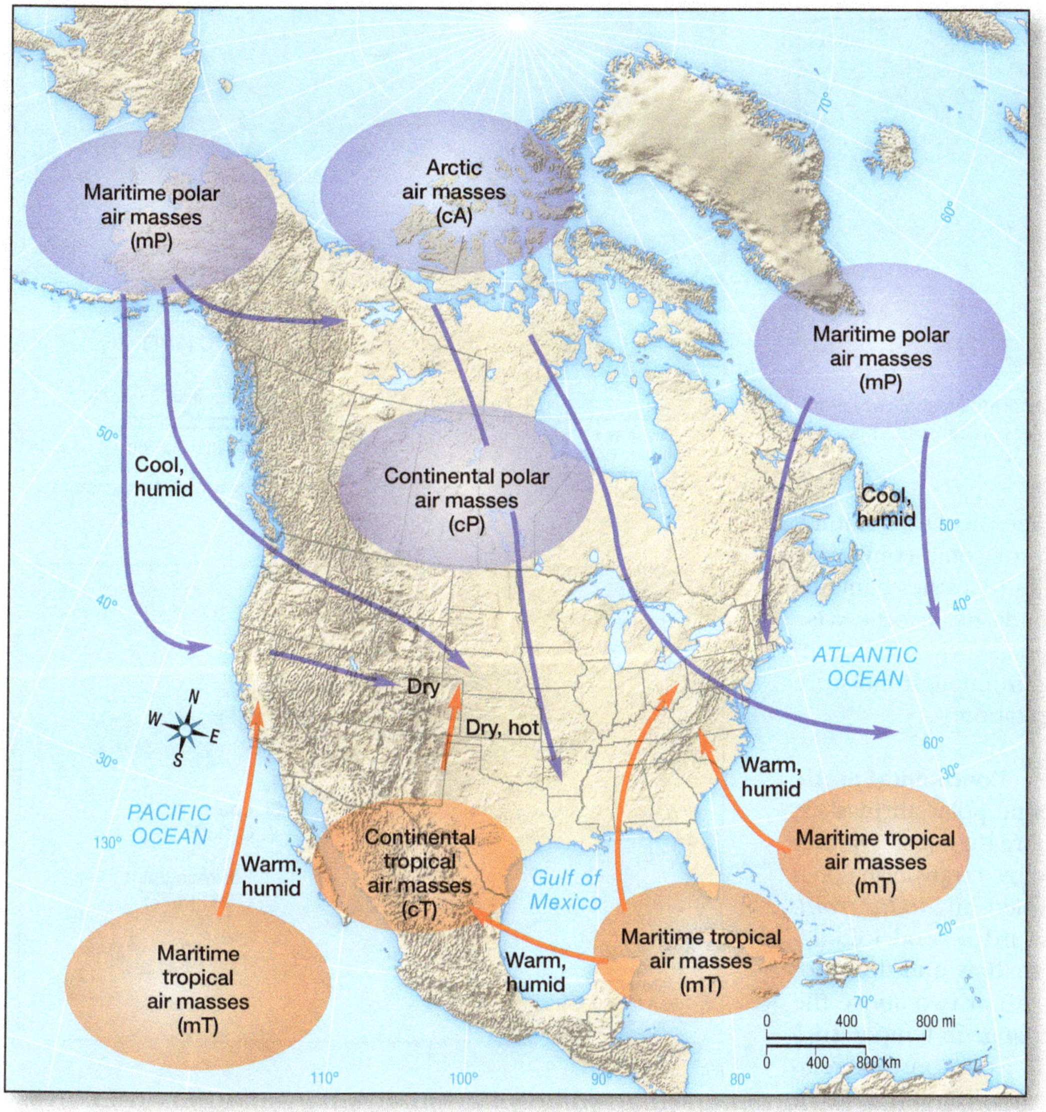

▲ **FIGURE 9–1** Source Regions for North American Air Masses.

CHECKPOINT

9.1 Ireland is a medium-sized island in the middle latitudes. Could Ireland serve as the source region for an air mass? Why or why not?

9.2 What are the two main criteria by which air masses are classified?

Air Mass Formation

9.2 Describe the types of air masses and explain how they form.

As described in Table 9–1, the five types of air masses are continental polar (cP), continental arctic (cA), maritime polar (mP), continental tropical (cT), and maritime tropical (mT). Factors affecting air mass formation include the characteristics of the source region and the season during which formation occurs.

Continental Polar (cP) and Continental Arctic (cA) Air Masses

Continental polar (cP) air masses form over large, high-latitude land masses, such as northern Canada and Siberia.

WINTER cP AIR MASSES In winter, these regions have short days and low solar angles. They also are usually snow covered during the winter and therefore reflect much of whatever solar radiation does reach the surface. This combination of circumstances virtually guarantees that the air will lose more radiant energy in the winter than it receives. The cooling of the air from below leads not only to low temperatures but also to radiation inversions and highly stable conditions.

In addition to having very low temperatures, winter cP air masses are extremely dry. Recall from Chapter 5 that little water vapor can exist in cold air and that the dew point (or frost point) temperature is always equal to or less than the air temperature. If the air temperature is very low—say −30 °C (−22 °F)—a kilogram of air at sea level can contain a maximum of only 0.24 g of water vapor. If the air is not saturated, the actual amount of moisture will be even lower.

The combination of dry air and stable conditions ensures that few if any clouds will form over a cP source region. Furthermore, the lack of water vapor reduces the absorption of incoming solar radiation by the atmosphere. Thus, despite their low temperatures, cP air masses over their source regions are usually bright and sunny. On the other hand, the stability of the atmosphere inhibits vertical mixing, so pollutants introduced at the surface remain concentrated near the ground. Because cold conditions lead to increased consumption of heating fuel (often coal and fuel oil), it isn't surprising that cP air is often associated with poor air quality over urban areas.

SUMMER cP AIR MASSES Summer cP air masses are similar, but much less extreme; they are both warmer and more humid than the winter version. They tend to remain at higher latitudes than winter cP air masses, so they do not influence as much of the globe as their winter counterparts. Daytime radiation inversions do not form because the air mass develops

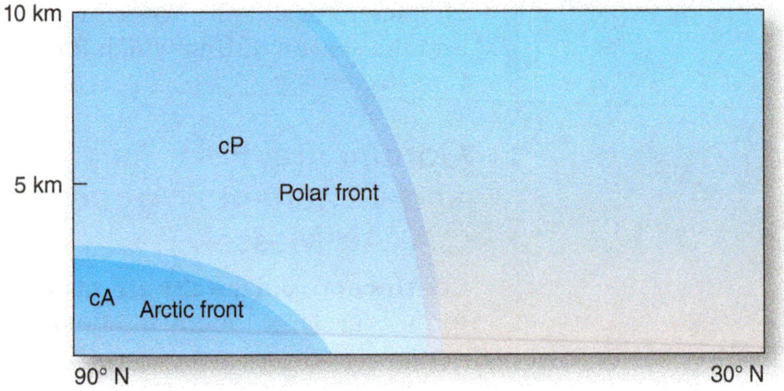

▲ **FIGURE 9–2 Arctic Front.** An arctic front separates a shallow layer of extremely cold arctic air from cold polar air. As you can see, it is much shallower than a polar front.

Video
Radar Reflectivity and Air Masses
http://goo.gl/MUX5nR

over a snow-free surface when the days are long. In fact, it's not uncommon for some convective uplift to occur, resulting in fair-weather cumulus clouds (scattered puffy clouds in an otherwise blue sky). Still, these air masses form over a continent, so they do not contain enough moisture to sustain significant precipitation.

CONTINENTAL ARCTIC (cA) AIR MASSES **Continental arctic (cA) air masses** are colder than continental polar air masses, but the distinction between the two is more than just a matter of degree. cA air and cP air are separated by a transition zone, similar to a polar front (Chapter 8), called an *arctic front.* Although a polar front can extend upward several kilometers from the surface, an arctic front is shallow and does not usually extend beyond a kilometer (0.6 mi) or two above the surface (Figure 9–2). Thus, we feel the change in temperature associated with the passage of an arctic front, but the relatively shallow front does not produce a great deal of uplift that might promote snowfall. On rare occasions, cA air can extend as far southward as the U.S.–Canadian border region. Some meteorologists believe that the distinction between cA and cP is rather minor, and therefore restrict their temperature categorization to just polar and tropical air masses.

DID YOU KNOW?

During much of the second week of January 2010 one could walk the streets of the French Quarter in New Orleans at night and see hardly anybody walking about. Why such little foot traffic in a normally bustling area? It was because of an unusual outbreak of continental polar air that surged down from northern North America bringing subfreezing temperatures as low as −5 °C (23 °F). This particular cP air mass caused very low temperatures to occur over the vast majority of North America east of the Rockies. This occurrence demonstrates the impact that the movement of air masses can have far away from their source regions.

MODIFICATION OF cP AND cA AIR MASSES Leaving their source region, cP air masses bring cold weather to more temperate latitudes. Figure 9–3 illustrates the movement of a hypothetical mass of continental polar air. In (a) the frontal

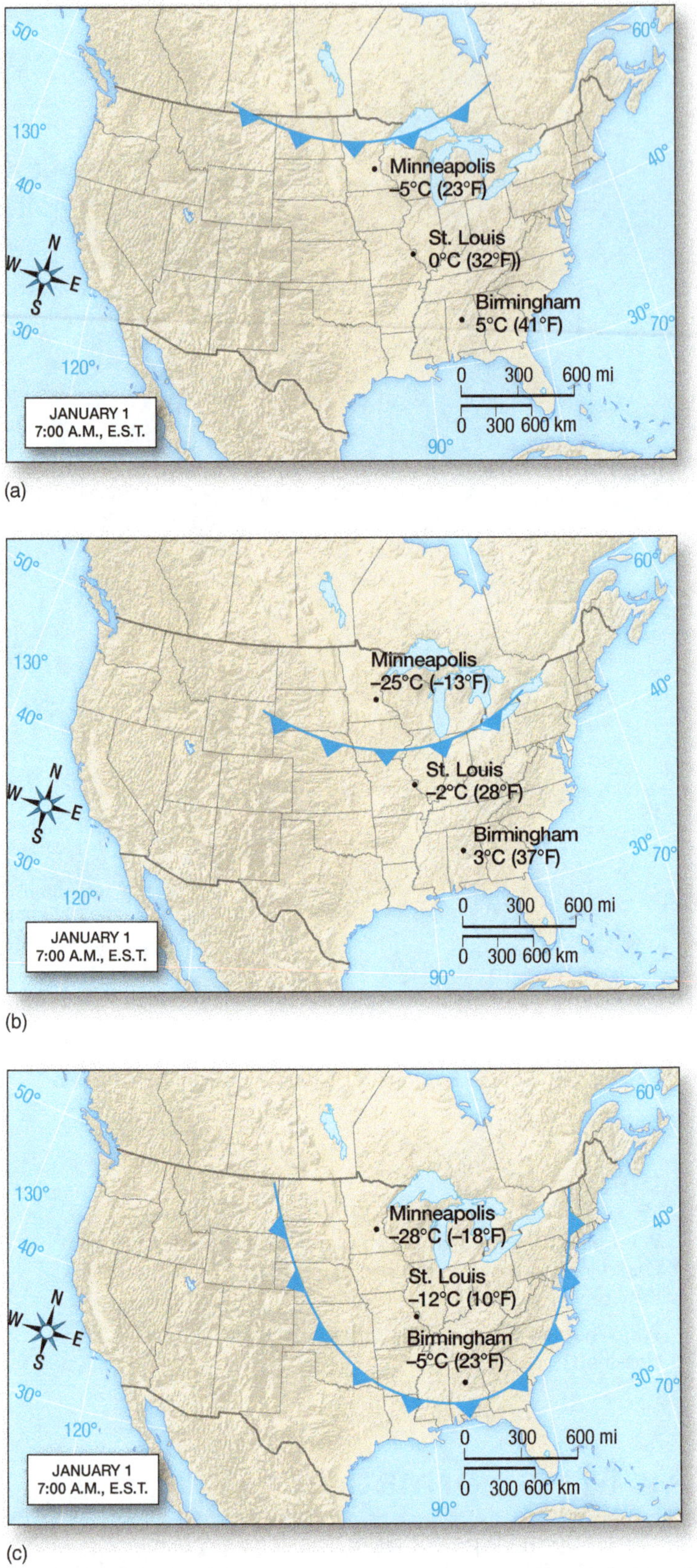

▲ **FIGURE 9–3 Air Mass Movement.** This sequence of surface weather maps shows the southward movement of continental polar air. In (a) the cold front is near the Canadian—United States border. In (b) it has moved to the southern Great Lakes region, dropping temperatures in its wake. In (c) the front has brought cold air to most of the rest of the eastern United States.

boundary is north of Minneapolis, and moderate temperatures exist throughout the central United States. Twenty-four hours later (b), the boundary of the continental polar air has passed over Minneapolis, causing a 20 °C (36 °F) drop in temperature. Farther to the south, St. Louis and Birmingham experience little temperature change from the previous day because the cold air has not yet extended that far southward. By the third day (c), however, the cold air has penetrated to the Gulf Coast and caused noticeable temperature declines at St. Louis and Birmingham, while Minneapolis remains under the influence of extremely cold air.

Notice that at each successive location the decline in temperature associated with the passage of the front is less pronounced than at the next most northerly site. Minneapolis experienced the most severe drop in temperature, while St. Louis and Birmingham underwent less extreme cooling. This occurs because of the gradual moderation of the cP air as it leaves its source region.

Continental polar and arctic air masses can also be modified by their passage over different terrain. The lake-effect and lake-enhanced snowfalls discussed in Chapter 7 are the result of such modification of the air mass, as dry continental air gains substantial moisture and warmth sufficient to yield copious snowfalls.

CHECKPOINT

9.3 What is the source region of a continental polar (cP) air mass? What are its properties?

9.4 Would you expect to find cP air masses in the Southern Hemisphere? Explain.

Maritime Polar (mP) Air Masses

Maritime polar (mP) air masses are similar to continental polar air masses but are more moderate in both temperature and dryness. Maritime polar air forms over the North Pacific as cP air moves out from the interior of Asia. The warm Japan current adds heat and moisture to the cold, dry air and converts it from cP to mP. The developing air masses migrate eastward across the Pacific; most of them pass across the Gulf of Alaska before reaching the west coast of North America. They approach the northern west coast of North America throughout the year but influence the California coast mainly in the winter.

Maritime polar air also affects much of the East Coast, but the manner in which it approaches is different. When midlatitude cyclones (rotating counterclockwise) pass over a region, the circulating air sweeps around the low-pressure system and approaches the coast from the northeast. The resultant winds are the famous **northeasters** (or *nor'easters*) that can bring cold winds and heavy snowfall.

MODIFICATION OF mP AIR MASSES Maritime polar air masses can be modified as they pass over major mountain barriers such as the Sierra Nevada, Rockies, or Appalachians. Recall from Chapter 6 that the passage of air over mountain barriers creates an orographic effect that favors the formation of clouds and precipitation. So consider what might happen to an mP air mass as it moves across North America from the eastern Pacific (Figure 9–4). The air mass would first pass over the Coast Ranges, which extract considerable moisture in the form of rain or snow (observe the enhancement of precipitation along the immediate west coast of North America, as shown in Figure 7–10). Descent along the eastern side of the Coast Ranges leads to dryer air on the leeward slopes, but uplift across the Sierra Nevada leads to further precipitation and drying of the air mass, which occurs yet again as it passes over the Rockies. By the time the air mass reaches the Great Plains, the maritime nature of the original air mass has been fully altered.

Continental Tropical (cT) Air Masses

Continental tropical (cT) air masses form during the summer over hot, low-latitude areas, such as the southwestern United States and northern Mexico. The desert areas where cT air masses form have little if any available surface water

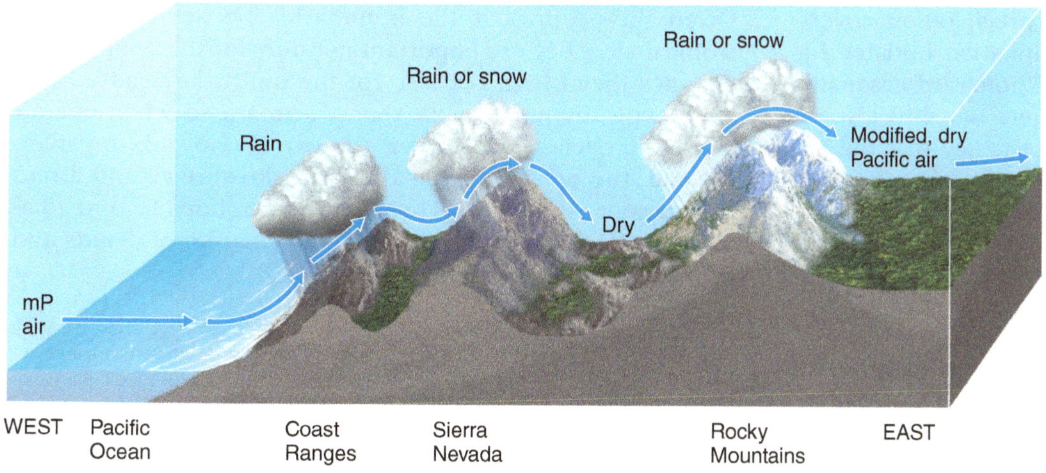

◄ **FIGURE 9–4** Modification of mP air masses as they move across the western United States.

and a minimal amount of vegetation to extract the water below the surface. Solar radiation inputs are extremely high during the summer. That, coupled with the lack of moisture, makes for very high ground temperatures that warm the overlying air by the transfer of sensible heat. The result is that these air masses are extremely hot and dry and, often, cloud free.

The extremely high surface temperatures cause intense heating of the air nearest the ground, which brings a steep temperature lapse rate and unstable conditions. Despite the instability, cT air masses often remain cloud free because of their inherent dryness. Consider, for example, a hot, dry air mass near the surface with a temperature of 45 °C (113 °F) and a dew point of 0 °C (32 °F). Because the temperature of a rising parcel of unsaturated air approaches the dew point at a rate of 0.8 °C/100 m (5 °F/1000 ft) of ascent, the air must be lifted 5.6 km (3.4 mi) before it can become saturated.[2] The unstable layer tends to be much lower than this height, which means that parcels at the surface may not be lifted sufficiently for clouds to develop. On the other hand, if the unstable layer is deep or if some moisture exists in the air, intense thunderstorms can develop. Thunderstorms can also develop near mountain peaks where converging valley breezes promote uplift.

Maritime Tropical (mT) Air Masses

Maritime tropical (mT) air masses develop over warm tropical waters. They are warm (though not as hot as cT), moist, and unstable near the surface—ideal for the development of clouds and precipitation.

Maritime tropical air masses have an enormous influence on the southeastern United States, especially during the summer. These air masses form over the Atlantic and the Gulf of Mexico and migrate into North America. As the air flows inland, heating from the warm surface increases the environmental lapse rate and makes the air even more unstable. The combination of high moisture content and increased instability favors the development of heavy but short-lived precipitation, often thunderstorms over relatively small areas.

The passage of a midlatitude cyclone can also trigger precipitation in mT air. This kind of precipitation covers larger areas and lasts longer than precipitation initiated by localized uplift. An mT air mass moving poleward gradually loses moisture by precipitation. Thus, there is a shift toward lower dew points northward. This is not to say that the air becomes dry as it reaches Chicago or Toronto, but it is unmistakably less oppressive than over New Orleans or Miami. An example of how transported mT air can make life more difficult is presented in *Box 9–1, Focus on the Environment and Societal Impacts: Air Mass Weather: Oppressive Heat and Humidity*.

The southwestern United States occasionally experiences the effects of mT air advected inland from the eastern tropical Pacific. Off the coast of southern California, the cold ocean current moderates the overlying air so that it does not have either the extreme moisture or the high temperature associated with air over the southeastern United States. In the late summer, however, moisture sometimes moves northeastward in the form of high-level outflow from tropical storms or hurricanes off the west coast of Mexico. This can lead to a layer of high clouds and increased humidity over southern California. It can also produce local thundershowers. These usually occur over the inland mountains and deserts, but occasionally interrupt the normal summer drought over the coastal region as well. Farther to the east in Arizona and eastern California, the introduction of mT air over the deserts causes what is locally (but not very accurately) called the *Arizona monsoon*. Another type of maritime airflow that can affect the western United States is the Pineapple Express, which is described in *Box 9–2, Focus on the Environment and Societal Impacts: The Pineapple Express*.

Boundary zones ranging from tens to a couple of hundred kilometers wide separate air masses. These boundary zones are categorized into four types of fronts, as described in the next section.

CHECKPOINT

9.5 What is the source region of a maritime tropical (mT) air mass? What are its properties?

9.6 Suppose an mT air mass moves inland from the Gulf of Mexico north to Chicago. How would it change?

Fronts

9.3 Identify the different types of fronts and their characteristics, including cloud conditions.

Fronts are defined as boundaries that separate air masses with differing temperature and moisture characteristics.[3] Often they represent the boundaries between polar and tropical air. They are important not only for the temperature changes they bring, but also for the uplift they cause. Cold air is typically more dense than warm air, so when one air mass encroaches on another, the two do not mix together. Instead, the denser air remains near the surface and forces the warmer air upward. These upward motions lead to adiabatic cooling and sometimes to the formation of clouds and precipitation.

[2]Recall that the DALR = 1.0 °C/100 m, but at the same time the dew point lapse rate = 0.2 °C/100 m. Thus, a rising parcel of unsaturated air approaches the dew point at a rate of 0.8 °C/100 m.

[3]Though this is essentially correct, one should understand that fronts are actually transition zones between two air masses—often with rapid changes in air temperature and humidity over very short distances.

Air Mass Weather: Oppressive Heat and Humidity

Maritime tropical air can extend far beyond its source region to bring hot humid conditions far inland and away from regions that we would normally consider tropical. The second half of August 2013 saw such an event occur. As shown in Figure 9–1–1, on August 27 temperatures exceeded 37.8 °C (100 °F) in Nebraska and South Dakota, with temperatures nearly as high across much of the north-central United States. More impressively perhaps is the way temperatures remained very warm into the morning hours, with Minneapolis, Minnesota, having a minimum temperature of 26.1 °C (79 °F)—10.6 °C (19 °F) above normal for that date. At the same time, coastal Southern California experienced humidities that reminded many residents of the weather they experienced in the more humid East.

Why such muggy conditions? The answer is easily explained by referring to the 500 mb map in Figure 9–1–2. Recall from Chapter 4 that airflow in the midtropospshere moves nearly parallel to the height contours. Thus, in late August midtropospheric air circulated clockwise around the large high-pressure system over the central United States. The air gained heat and moisture over the Gulf of Mexico, giving it the classic properties of maritime tropical (mT) air. The clockwise circulation transported the warm humid air as far west as southern California before turning it northward and eastward toward the north-central United States. In addition to the high daytime temperatures, little cooling occurred at night because water vapor is an effective absorber of outgoing longwave radiation (Chapter 3).

Events such as these do more than cause physical discomfort; they can have significant impacts on normal day-to-day activities and long-term conditions. This particular heat wave occurred as many city schools across the central United States were set to open. Some without air conditioning cancelled classes for several days. The heat also put additional stress on crops growing in a region that had already been under drought conditions for some time. The resultant increase in the intensity and areal extent of an ongoing drought led to an increase in corn and soybean prices on the commodity exchanges. But of course these consequences pale in comparison to the significant increase in mortality that occurs both in the United States and overseas during extended outbreaks of hot, humid conditions accompanying summer air masses.

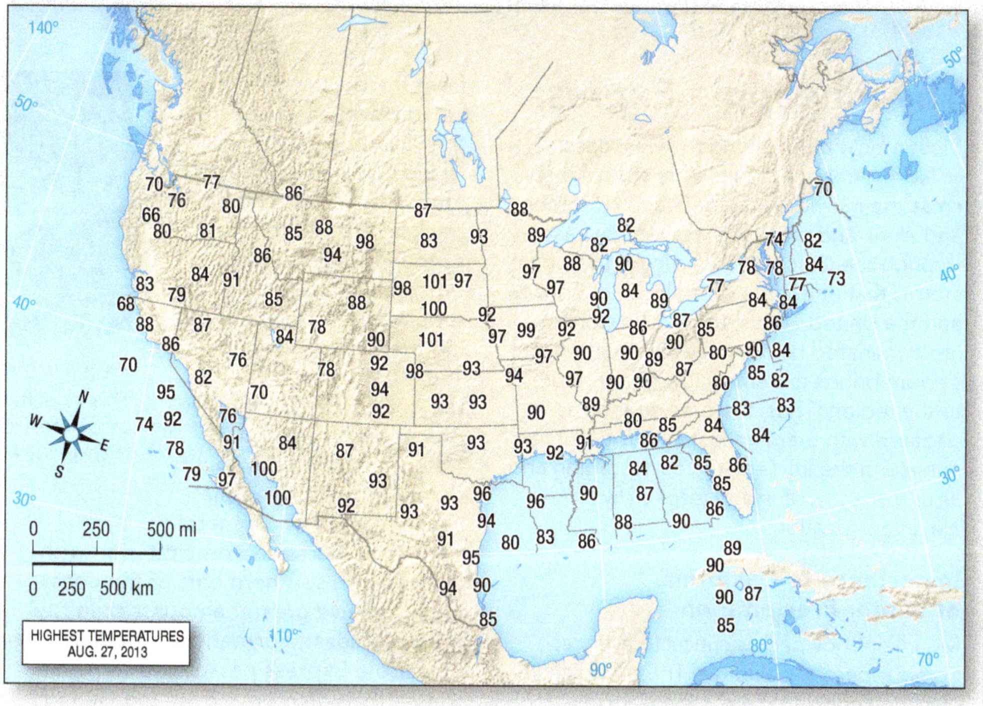

▲ **FIGURE 9–1–1 Oppressive Heat.** Maximum air temperatures (°F) on August 27, 2013.

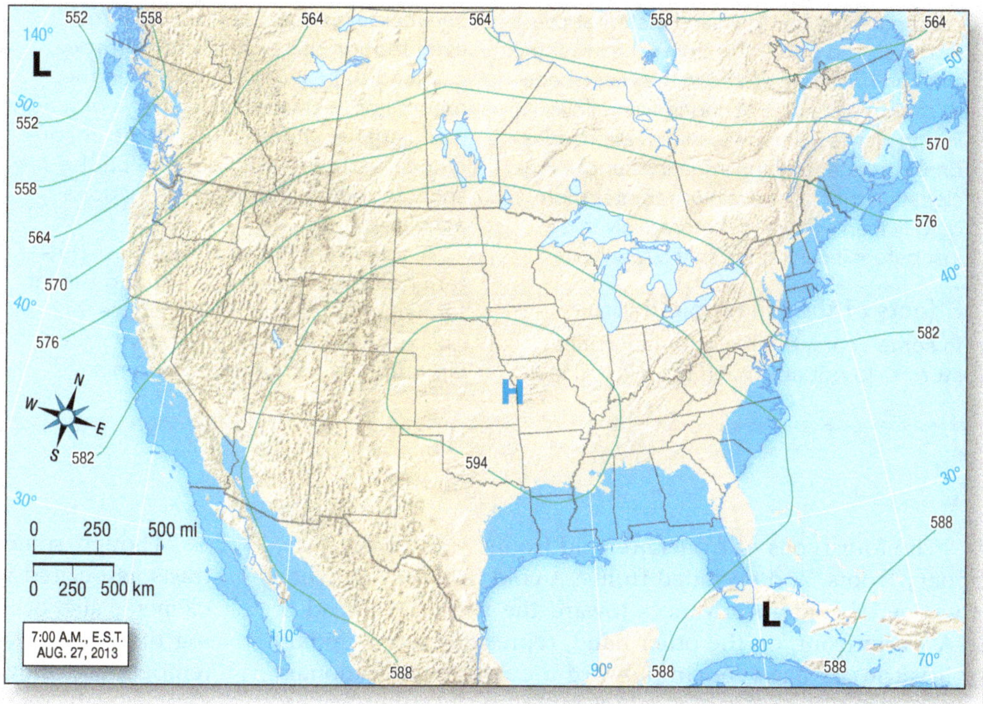

▲ **FIGURE 9–1–2 Warm Humid Air.** A 500 mb map for August 27, 2013.

9–1–1 Explain how the air mass over the central United States in August 2013 became so humid and how the hot, humid conditions came to extend as far west as southern California.

9–1–2 Describe an environmental impact of this weather event that also had a significant economic impact.

9–2 FOCUS ON THE ENVIRONMENT AND SOCIETAL IMPACTS

The Pineapple Express

Most precipitation along the west coast of North America results from storms that cross the Pacific Ocean between November and April, and most of these storms pass through the Gulf of Alaska. Unlike the winter storms that affect the central part of Canada and the United States, the Pacific storms are dominated by maritime polar air, which seldom brings extremely low temperatures to the region. Thus, the precipitation associated with them almost always occurs as rain along the immediate coastal region and as rain or snow at inland (especially high-elevation) locations.

Importance to California of Winter Precipitation

Mountain snowpack is critical to every economic aspect of California. The snow line (the elevation marking the change from rain to snow) for winter storms depends on the temperature. It usually varies between several hundred and a few thousand meters above sea level. The colder the storm, the lower the snow line. Water is stored in the watersheds as snow melts in the springtime and flows into the large reservoir system that provides drinking water and the irrigation supply. Precipitation that falls as rain evaporates more quickly than snow and is more likely to be lost before it can replenish the reservoirs.

Effects of the Pineapple Express

In some years, the majority of Pacific storms do not approach the West Coast from the northwest but instead travel eastward from the vicinity of the Hawaiian Islands along a plume of moist, humid air. This storm track is referred to locally as the *Pineapple Express*. When storms take this path repeatedly, as they did during the El Niño winters of 1982–1983 and 1992–1993, the greatest amounts of precipitation are concentrated farther to the south than during more normal years. Instead of a general pattern of increasing precipitation with latitude, the southern part of the coast might receive greater amounts than the northern coast. Furthermore, because the Pineapple Express passes over warmer waters than do storms from the Gulf of Alaska, the air tends to be warmer and more humid, and the height of the snow line increases considerably. As a result, even though this type of storm track can lead to heavy precipitation in the mountains, a smaller percentage of it accumulates as snow—much to the chagrin of local residents who depend on the snow for skiing and other recreational activities.

A significant Pineapple Express pattern had established itself during the latter half of December 2010 (Figure 9–2–1). This resulted in unusually heavy precipitation over southern California.

▲ **FIGURE 9–2–1 Pineapple Express, December 21, 2010.** This satellite image shows how plumes of warm moist air and associated storms follow a path generally from the Hawaiian Islands toward the west coast of North America.

For example, Pasadena, which averages 7.95 cm (3.13 in.) of rain for the entire month of December, received 31.2 cm (12.28 in) over a five-day period. Thus, there are both immediate and long-term effects of a Pineapple Express: the potential for heavy rains and a seemingly paradoxical potential decrease in water supply.

9–2–1 Compare a Pineapple Express storm with the typical Pacific winter storms that strike California. How are they similar? How are they different?

9–2–2 Describe how a Pineapple Express affects the snowpack in the Sierra Nevadas. Why is this a matter of tremendous importance to the residents of California?

The four types of fronts are cold fronts, warm fronts, stationary fronts, and occluded fronts. A cold front occurs when a wedge of cold air advances toward the warm air ahead of it. A warm front, on the other hand, represents the boundary of a warm air mass moving toward a cold one. A stationary front is usually similar to a cold front in structure but has not recently undergone substantial movement (in other words, it remains stationary). Unlike the other three types of fronts, occluded fronts do not separate tropical from polar air masses. Instead, they appear at the surface as the boundary between two polar air masses, with a colder polar air mass usually advancing on a slightly warmer air mass ahead of it. Figure 9–5 includes the symbols used to show the location of these fronts on surface weather maps.

Though some fronts are named for the temperature contrasts associated with them, distinct changes in other weather elements also occur along these boundaries. People experiencing the passage of a front not only notice a rapid change in temperature, they may also note a shift in wind speed and direction, a change in moisture content, and an increase in cloud cover.

The weather can change markedly for the worse when a cold front approaches a warm front and a type of storm called a *midlatitude cyclone* develops. Figure 9–6 illustrates the overall structure of typical Northern Hemisphere midlatitude cyclones, of which cold, warm, and (sometimes) occluded fronts are an important part. (We discuss midlatitude cyclones in detail in Chapter 10; in this chapter

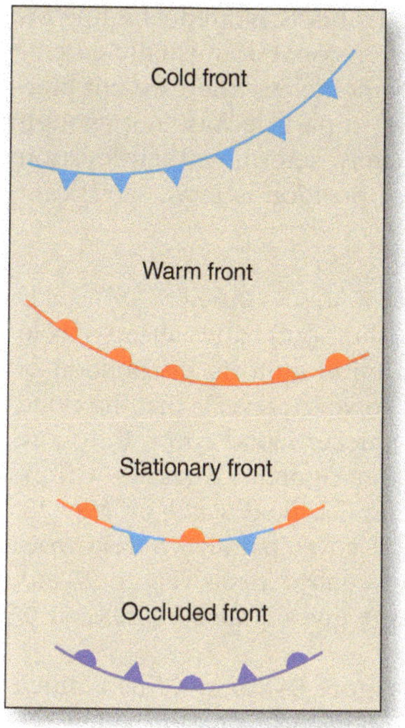

◂ **FIGURE 9–5 Front Symbols.** These symbols represent the four types of fronts on weather maps. Note that the direction in which the triangles and semicircles point indicates the direction of movement of the fronts.

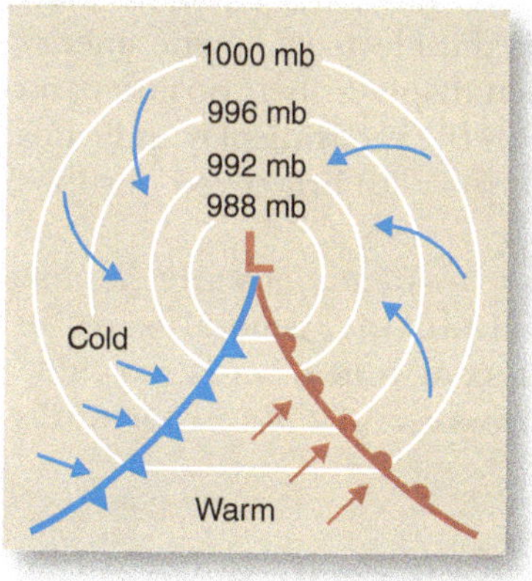

Animation
Cold Fronts
http://goo.gl/XIlinH

▴ **FIGURE 9–6 Structure of a Midlatitude Cyclone.** In a typical midlatitude cyclone, cold and warm fronts separated by a wedge of warm air meet at the center of low pressure. Cold air dominates the region outside the wedge separating the cold and warm fronts.

we discuss some of the more salient features of the particular fronts.)

Cold Fronts

We are all familiar with the dramatic shifts in weather that can occur quickly with the approach and passage of a **cold front**. During the winter, they can disrupt life for millions of people as they cause intense flooding from heavy rain or whiteout conditions from blowing snow.

Figure 9–7 shows the structure of a typical cold front, with its upper boundary sloping back in the direction of the cold air. As polar air surges away from its source region, it does not do so as a vertical wall of cold air; in fact, the boundary of the front slopes strongly backward with distance from the ground. You should note that the vertical scale in this figure is highly exaggerated and makes the slope of the front appear far steeper than it really is. In reality, the typical cold front has a surface slope of about 1:100, meaning that its surface rises only 1 m for every 100 m of horizontal extent.

EFFECT OF FRICTION ON A COLD FRONT BOUNDARY Observe also in Figure 9–7 that the frontal boundary is not flat but curved; close to the ground the slope of the front is steeper than it is aloft. This is due to the varying effect of friction. Friction is strongest at the ground but decreases upward. Close to the ground there is a rapid decrease in friction with height. This allows the air farther from the ground to advance more rapidly than the air below, advancing the air farther from the ground to the position of the surface front, and giving the lower portion of the frontal boundary a very steep profile. Farther away from the ground, the

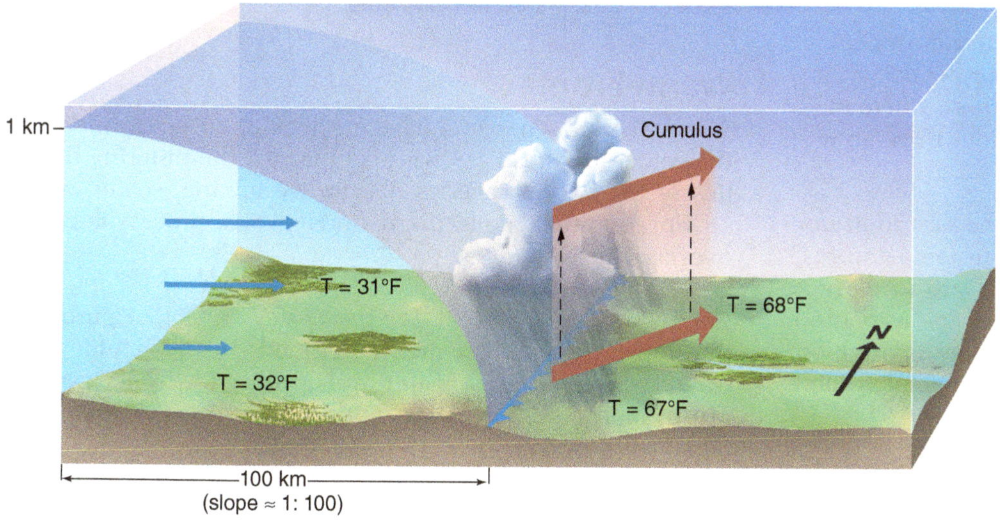

◂ **FIGURE 9–7 Cold Front.** Cold fronts typically move more rapidly and in a slightly different direction from the warm air ahead of them. This causes convergence ahead of the front and uplift of the warm air that can lead to cumuliform cloud development and precipitation. In this example, the cold air (in blue) advances from west to east (notice that the wind speed depicted by the thin arrows increases with height). The warm air (in red) is blowing toward the northeast. The cold air wedges beneath the warm air and lifts it up.

Video
Tornadoes
Ahead of a Cold
Front
http://goo.gl/UITBgL

effect of friction is fairly negligible, so there cannot be a large difference in friction with height. The result is that the wedge of cold air moves forward more uniformly and the highest portion of the cold air mass does not surge forward. The steep slope near the ground and the gentler slope aloft results in the convex profile. Cold fronts move at widely varying speeds, ranging from a virtual standstill to about 50 km/hr (30 mph).

CLOUD COVER ALONG COLD FRONTS The cloud cover we usually observe with the passage of a cold front results from the convergence of the two opposing air masses. Differences in wind speed and direction allow cold northwesterly wind to converge on the warm air ahead of it and displace it upward. Furthermore, the air ahead of a cold front tends to be unstable and therefore easily lifted. This promotes development of cumuliform clouds along these boundaries. With their large vertical extent, cumuliform clouds can often produce intense precipitation. However, because of the limited horizontal extent and rapid movement of the frontal wedge, such precipitation is often of short duration.

DETERMINING THE POSITION OF A COLD FRONT The production of surface weather maps is now almost entirely automated and performed by computer. But meteorologists still plot the location of fronts manually because identifying them is often a subjective process. Meteorologists will look for the following five features to determine cold front positions:

1. Significant temperature differences between adjacent regions, with lower temperatures behind the cold front
2. Dew point differences, with drier air in the cold sector
3. Bands of cloud cover and precipitation, which nearly coincide with the position of the front
4. Narrow zones where wind direction changes, typically northwesterly in the cold sector to southwesterly in the adjacent warm region
5. Boundaries separating regions where the atmospheric pressure has decreased over a three-hour period (typically ahead of the cold front) and those where the pressure has increased during the previous three hours (behind the cold front).

While one might logically expect that the best indicator of the position of a cold front is the shift in temperatures, that is not always true. When a cold front advances rapidly, the temperature does not decrease as rapidly behind the front as does the humidity. Therefore, the distribution of dew points is sometimes the better indicator of frontal position.

Though the position of cold fronts is plotted as a line on surface maps, they are three dimensional and usually extend upward past the 500 mb level. Nonetheless, they are not plotted on upper-level weather maps, in part because not enough data are available aloft to adequately determine their location and in part because their surface position is more significant than their position aloft.

INTERPRETING SATELLITE AND RADAR IMAGES OF COLD FRONTS When seen from space, the cloud bands ahead of cold fronts often appear to contain a fairly uniform distribution of cloud cover. Closer examination, however, reveals that the cloud bands often contain pockets of thicker cloud cover and more intense precipitation. For example, Figure 9–8 shows a frontal zone extending across the central United States on May 10, 2013. An extensive band of cloud cover parallels a cold front that extends northeastward from central Texas (Figure 9–8a). But as is often the case, this image gives a false impression of uniform precipitation.

The radar composite map (Figure 9–8b) gives us a much better description of the intensity of precipitation along the front. Weather radar imagery is obtained by emitting energy with wavelengths of about 10 cm (4 in.). Precipitating droplets and ice crystals scatter some of the electromagnetic energy back toward the radar unit, which displays their position and other characteristics. The color on the radar display indicates the intensity of the returned energy, with bright red portions indicating heavy precipitation.

In this instance, several narrow bands of more intense precipitation are embedded within the general zone of cloud cover. An example of a recent cold front is provided in *Box 9–3, Focus on Severe Weather: A Cold Front Chills Louisiana*.

CHECKPOINT

9.7 Describe the slope of the boundary along a cold front and the typical position of clouds relative to the boundary.

9.8 What types of information do meteorologists consider when plotting the location of a cold front?

Warm Fronts

Warm fronts separate advancing masses of warm air from the colder air ahead. As with cold fronts, the differing densities of the two air masses discourage mixing, so the warm air flows upward along the boundary. This process is called **overrunning**.

CLOUD FORMATION ALONG A WARM FRONT Figure 9–9 illustrates the typical sequence of clouds that form along the frontal surface. The warm air flows up along the frontal boundary in much the same way that air rises above a mountain

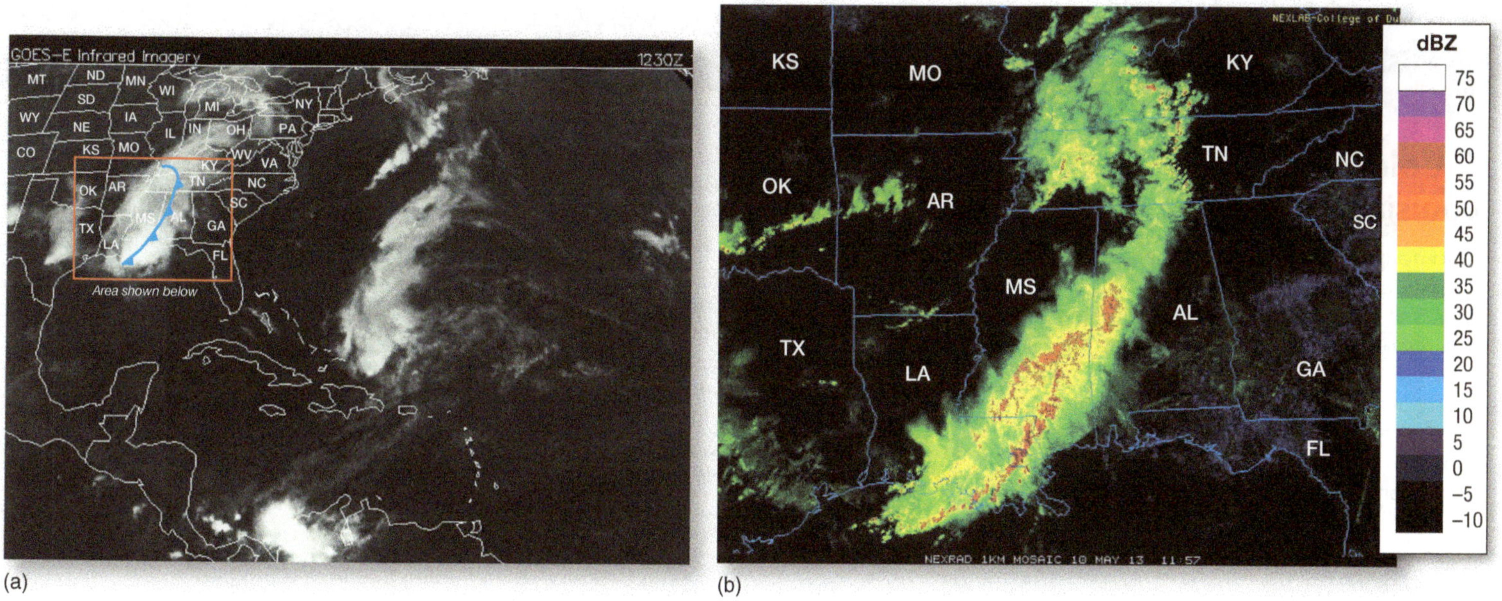

▲ **FIGURE 9–8 Midlatitude Cyclone Cloud Cover and Precipitation.** (a) The cloud cover in the satellite image, appears to be a continuous, uniform band. (b) The radar composite map reveals that the cloud cover is, in fact, marked by areas of varying precipitation intensities.

slope. But the gradual slope of the frontal surface leads to a very gradual progression of cloud types. As the air rises along the boundary, adiabatic cooling first leads to formation of low-level stratus clouds. As the air continues to rise, a sequence of higher clouds develops, with nimbostratus, altostratus, cirrostratus, and, finally, cirrus occurring in that order. As the front moves eastward or northeastward, the leading segment of the cloud sequence (the cirrus) is seen first, followed by the continually thickening and lowering cloud cover. Thus, even an amateur forecaster can predict an episode of continuous light rain from a warm front a day or two in advance, simply by observing the sequence of clouds as they pass overhead.

WARM FRONTS AND PRECIPITATION Warm fronts are about half as steep as cold fronts (their slopes are about 1:200), which causes the lifting of the warm air to extend for greater horizontal distances than for cold fronts. Although the clouds above the warm front cover a greater horizontal extent than the clouds above a cold front, they usually consist of smaller droplets and have a lower liquid water content. This results in less intense precipitation than is usually associated with cold fronts. For this reason, warm fronts in the summer are considered a friend to farmers; they provide a light, steady rain that nourishes crops but does not produce flash floods or severe storms.

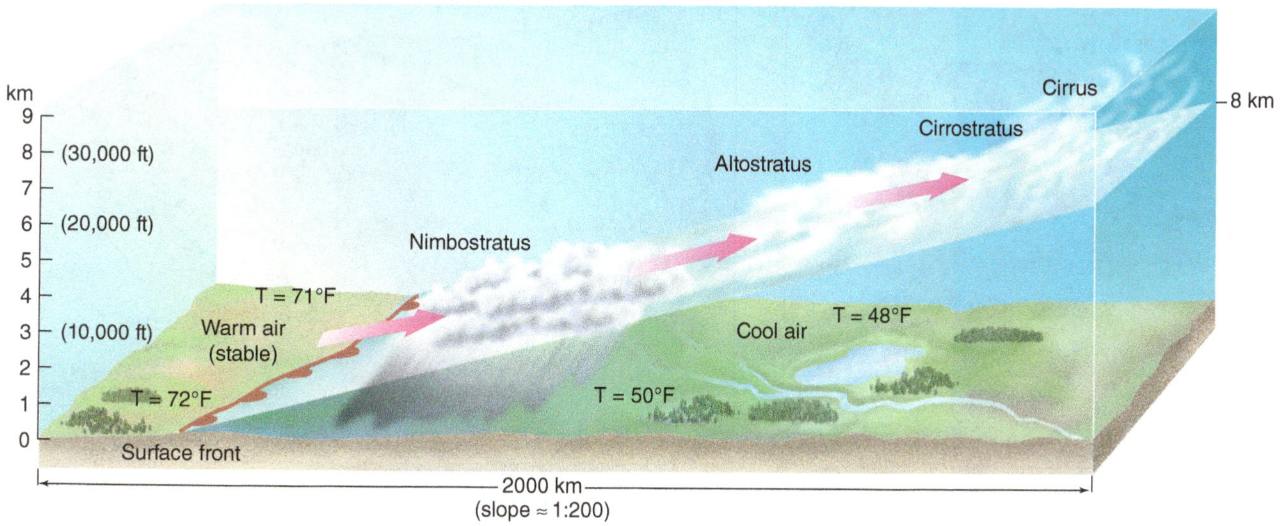

Animation
Warm Fronts
http://goo.gl/hdT4BM

▲ **FIGURE 9–9 Warm Front.** Overrunning leads to extensive cloud cover along the gently sloping surface of cool air.

9-3 FOCUS ON SEVERE WEATHER

A Cold Front Chills Louisiana

Cold fronts can lead to major changes in weather over short time periods. A cold front that moved across the United States in January 2013 provides an excellent example. Figure 9–3–1 shows the position of an advancing cold front on the morning of January 11 and corresponding observed minimum temperatures. A cold front extends from northwest Mexico to the Dakotas (Figure 9–3–1a) and continues northeastward as a stationary front. Behind the front, daily minimum temperatures were low, especially across Montana and northern Wyoming (Figure 9–3–1b).

By the next morning the cold front had advanced well to the southeast and extended from North Texas to the western Great Lakes (Figure 9–3–2a). As expected, the central part of the country experienced a substantial drop in temperatures and extreme cold covered much of the western Great Plains (Figure 9–3–2b). Twenty-four hours later, the cold front continued its eastward migration but at a slower rate than it had been moving previously (Figure 9–3–3a), and the map of minimum temperatures (Figure 9–3–3b) reflects this movement. Figure 9–3–4 shows the position of the front and the corresponding minimum temperatures on January 14.

Figure 9–3–5 shows three 24-hour plots of temperature illustrating how temperatures and dew points can change with the

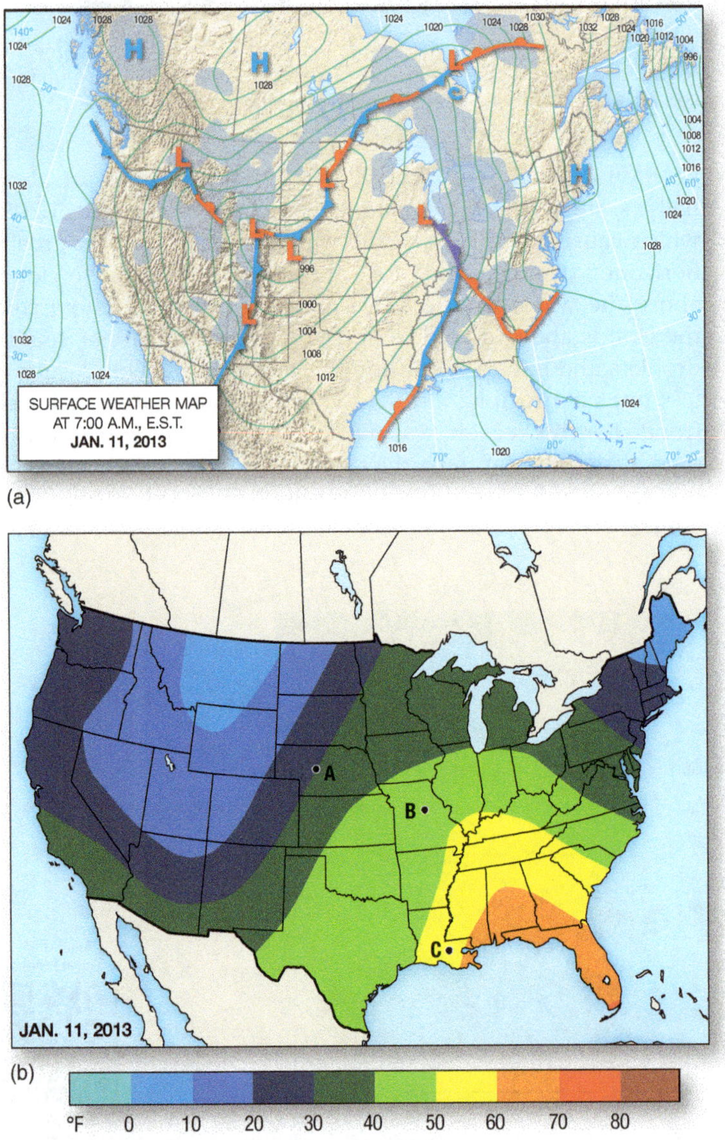

▲ **FIGURE 9–3–1 Cold Front: January 11.** (a) Surface weather and (b) morning minimum temperatures, January 11, 2013, 7 A.M. EST. Positions A, B, and C, mark the locations of North Platte, Nebraska; Columbia, Missouri; and Baton Rouge, Louisiana; respectively.

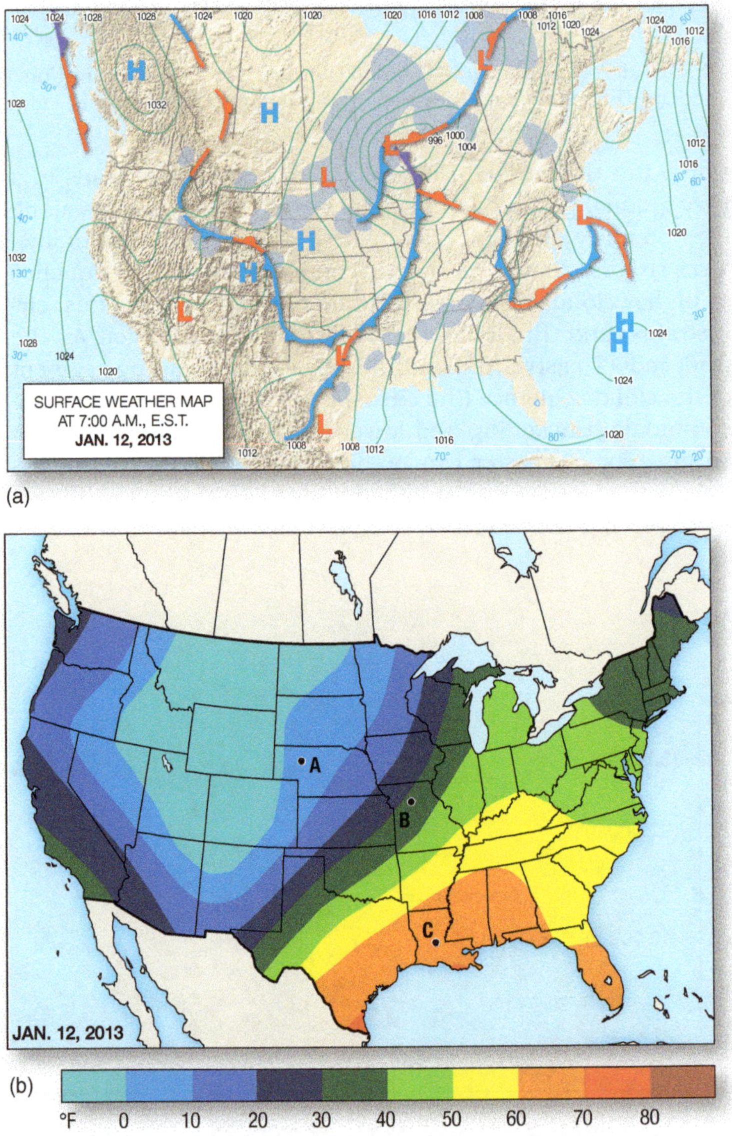

▲ **FIGURE 9–3–2 Cold Front: January 12.** (a) Surface weather and (b) morning minimum temperatures. Positions same as Figure 9–3–1.

passage of a cold front.[4] Figure 9–3–5a plots the temperature and dew points on January 11 at North Platte, Nebraska (shown as position A on Figure 9–3–1a). Both the temperature and dew point hovered in the 30s Fahrenheit for most of the day prior to the passage of the front. But a significant drop in both values occurs after 6 P.M. By 11 P.M. both values had dropped into the 20s. On January 12 a similar drop in temperature and moisture occurred as the front passed over Columbia, Missouri, at about 11 A.M. Notice that the temperatures continued to drop in the afternoon even as a rising afternoon sun would otherwise be expected to cause daytime warming. The following evening the front caused an abrupt cooling and drying of the air as it passed over Baton Rouge, Louisiana, transforming a warm winter evening in the 70s to much cooler and drier conditions.

[4]This plot of temperature and dew point is part of what is known as a *meteogram*, a useful graphical tool for illustrating changing weather conditions at a location. Meteograms are discussed further in Chapter 13.

9–3–1 What was the overall effect of the cold front passage described here?

9–3–2 The situation described here illustrates how abrupt decreases in temperature can occur. Do you think Louisiana experienced just as sudden a warm-up afterward? Why or why not?

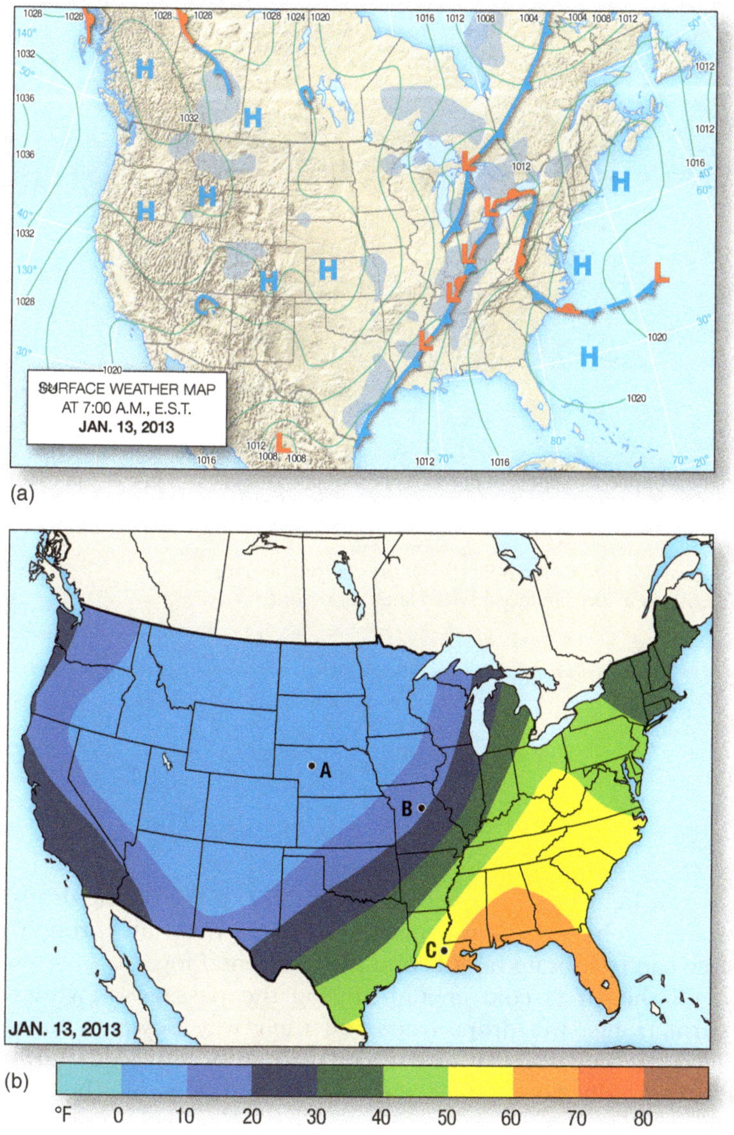

▲ **FIGURE 9–3–3 Cold Front: January 13.** (a) Surface weather and (b) morning minimum temperatures. Positions same as Figure 9–3–1.

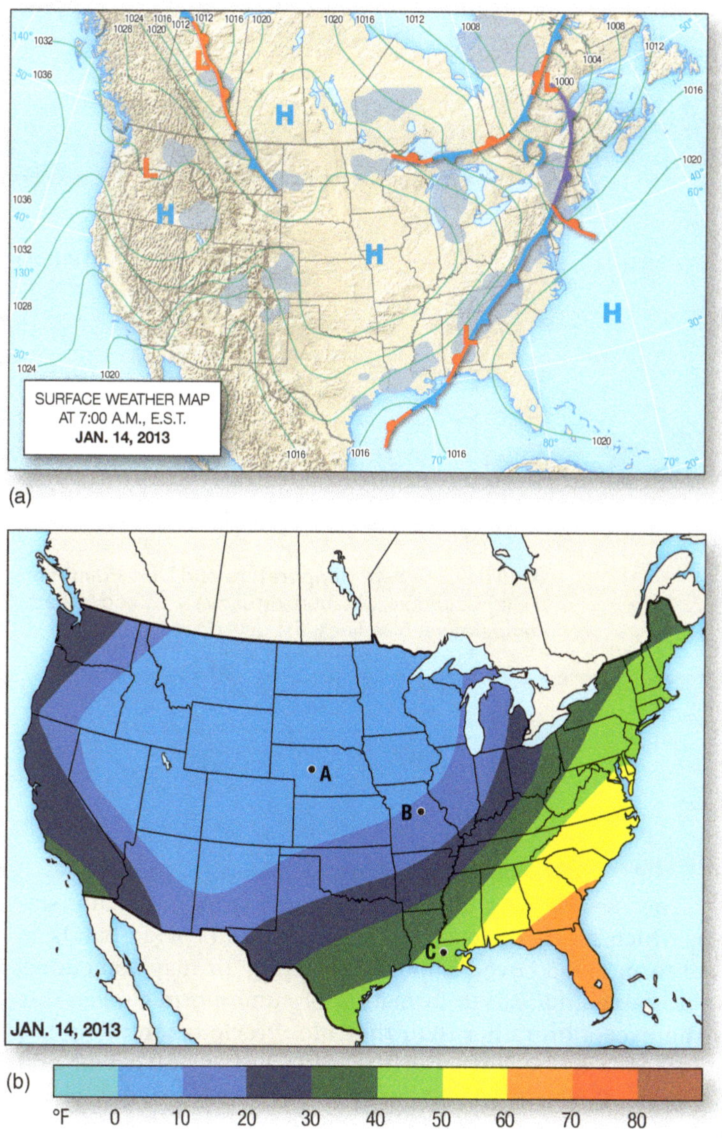

▲ **FIGURE 9–3–4 Cold Front: January 14.** (a) Surface weather and (b) morning minimum temperatures. Positions same as Figure 9–3–1.

9–3 FOCUS ON SEVERE WEATHER *(continued)*

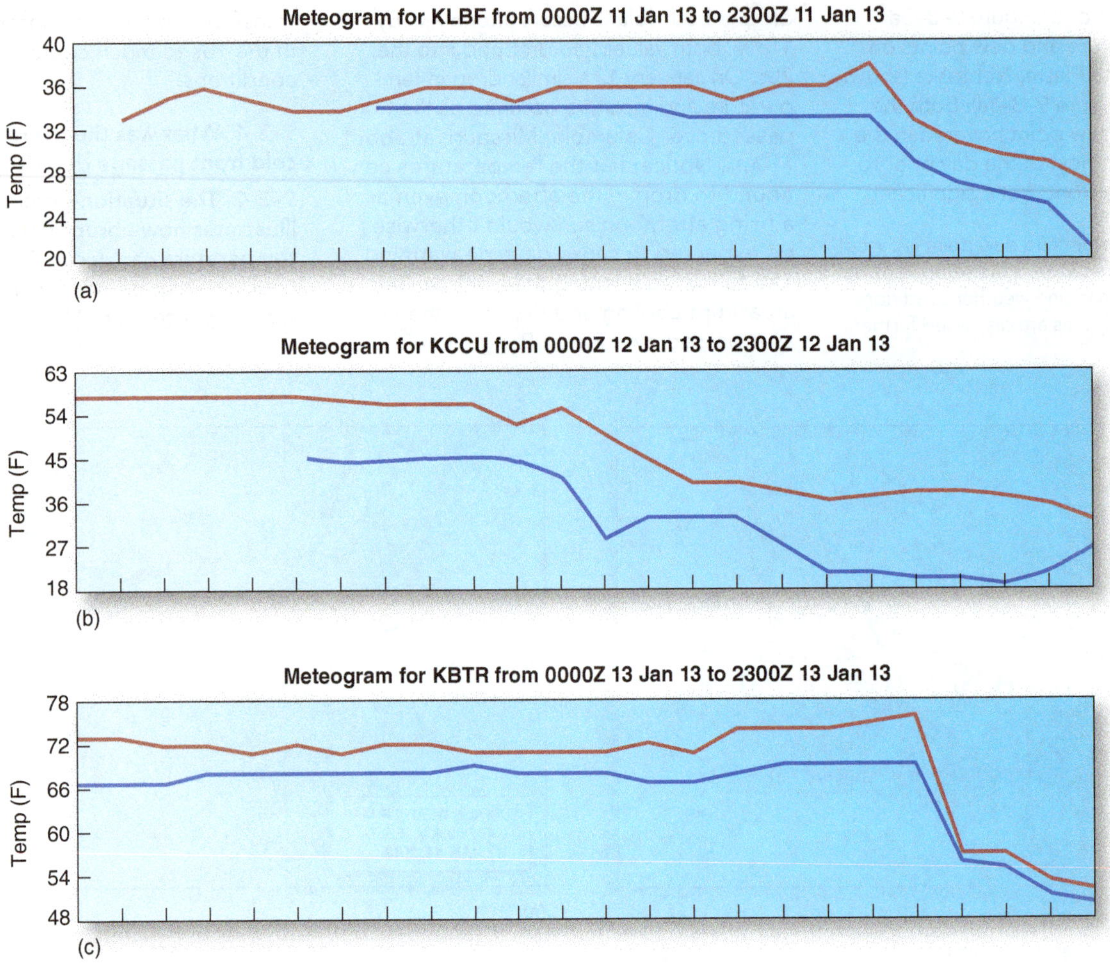

▲ **FIGURE 9–3–5 Temperature and Dew Point Plots.** Plots shown are for (a) North Platte, Nebraska, on January 11; (b) Columbia, Missouri, on January 12; and (c) Baton Rouge, Louisiana, on January 13. Red lines represent temperature, blue lines, the dew point.

The overrunning air above a warm front is usually stable, which normally leads to the formation of wide bands of stratiform cloud cover. Though the rising air may sometimes be unstable and lead to formation of cumuliform clouds, this is the exception rather than the rule. Precipitation along the front tends to be light, but the wide horizontal extent and typically slow movement of warm fronts (usually about 20 km/hr—12 mph) allow rain to persist over an area for up to several days.

The clouds along a warm front exist in the warm air above the wedge of dense, cold air. Thus, when rain falls from the base of these clouds, the falling droplets pass through the cold air below on their way to the surface. If the falling droplets are substantially warmer than the air they fall through, they can evaporate rapidly and form a frontal fog.

If the air is cold enough during the passage of winter warm fronts, the drops freeze on their way down to form sleet, or they solidify on contact with the surface to form freezing rain. So summer warm fronts are generally beneficial, while in winter they can cause widespread problems. (See *Box 9–4, Focus on Severe Weather: Interacting Air Masses Spawn a Powerful Nor'easter,* for an example of how a warm front might combine with other patterns to create a very notable weather event.)

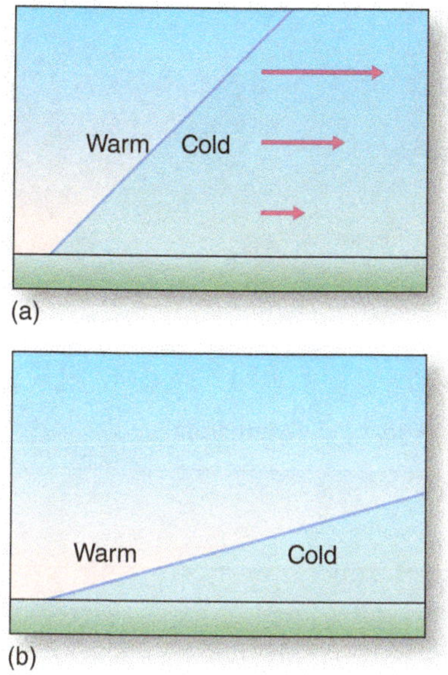

◀ **FIGURE 9–10 Slope of Warm Fronts.** Warm fronts have gentler sloping surfaces and do not have the convex-upward profile of cold fronts. (a) Surface friction decreases with distance from the ground, as indicated by the longer wind vectors away from the surface. (b) This causes the surface of the front to become less steep through time.

CHANGE IN SLOPE ALONG A WARM FRONT Figure 9–10 shows the typical change in the slope of a warm front over time. Because wind speeds are greater aloft than near the surface ahead of the front, the upper portion of the wedge advances more rapidly than does the part near the surface, and the slope of the frontal boundary becomes less steep.

DETERMINING THE POSITION OF A WARM FRONT Meteorologists use much the same rules for identifying warm front positions that they do for cold fronts, but with some differences. One must first look for a zone where warmer air advances toward cooler air (the opposite of cold fronts). Dew points typically increase behind the position of the warm front; winds commonly shift from southeasterly ahead of the front to southwesterly behind it; and cloud cover and precipitation bands are common. Also, the zone ahead of the warm front generally undergoes decreasing air pressures while the area immediately behind the front typically has stable air pressure.

CHECKPOINT

9.9 What is overrunning?

9.10 What cloud types typically form at different altitudes along the leading surface of a warm front? Explain.

Stationary Fronts

More often than not, the boundaries separating air masses move gradually across the landscape; at other times, they stall and remain in place for extended time periods. Nonmoving boundaries are called **stationary fronts**. They are identical to cold or warm fronts in terms of the relationship between their air masses. As always, the frontal surface is inclined, sloping over the cold air.

Determining whether a front is stationary is somewhat subjective. For example, if the front advances at a rate of 1 km/hr (0.6 mph), does that constitute enough movement for it to be nonstationary? If not, what about 2 km/hr? In practice, a meteorologist will make the designation by looking at the previous one or two surface weather maps (compiled at three-hour intervals). If there has been no movement during that period, the front is considered stationary.

If this sounds simple, keep in mind that the exact position of a front at any point in time is difficult to pinpoint. This is in part because fronts are zones of transition rather than abrupt boundaries; they are located at no precise line. Furthermore, maps are compiled from a finite number of weather stations. A front may undergo some movement that is not detected because it does not move across an observing station.

Stationary fronts can sometimes present significant flooding problems because their persistence over a particular area may yield copious precipitation, even if the intensity of precipitation is not particularly great.

DID YOU KNOW?

Although the concepts behind the location of fronts on weather maps may seem simple, there are occasional ambiguities, and sometimes what appears to be a frontal boundary may in fact be something else. For example, local temperature contrasts may arise due to elevation differences, with high plateaus having lower temperatures (especially at night) than adjacent areas at lower elevations. Ranges such as the Appalachian Mountains sometimes form a barrier between warm Atlantic air and colder cT air to the west. Snow-covered areas might also have lower temperatures than nearby areas, producing temperature contrasts not associated with fronts.

Occluded Fronts

The most complex type of front is an **occluded front** (sometimes called an *occlusion*). The term *occlusion* refers to closure—in this case, a warm air mass closed off from and thus not in contact with the surface. According to the traditional description proposed by Norwegian meteorologists in the early 20th century, occluded fronts may form when a faster-moving cold front "catches up" to the warm front ahead. The warm air rises. At the surface, one mass of cold air becomes adjacent to another, so there is a smaller difference in temperature (and likewise dew point) from one side of an occluded front to the other. A warm air mass is present, but it is aloft, pinched off from the surface, and therefore not reflected in surface temperature.

Figure 9–11a shows a typical midlatitude cyclone prior to occlusion according to the traditional model, with the cold and warm fronts intersecting at the center of low pressure. When the cold front meets the warm front ahead of it, the segments of the two fronts closest to each other become occluded, as shown in part (b). The warm air does not disappear but instead gets lifted upward, away from the surface. The occluded front becomes longer as more and more of the cold front converges with the warm front. Eventually, the cold front completely overtakes the warm front, as shown in part (c), and the entire system is occluded. According to the classical model, temperature differences between the cold air masses near the surface give

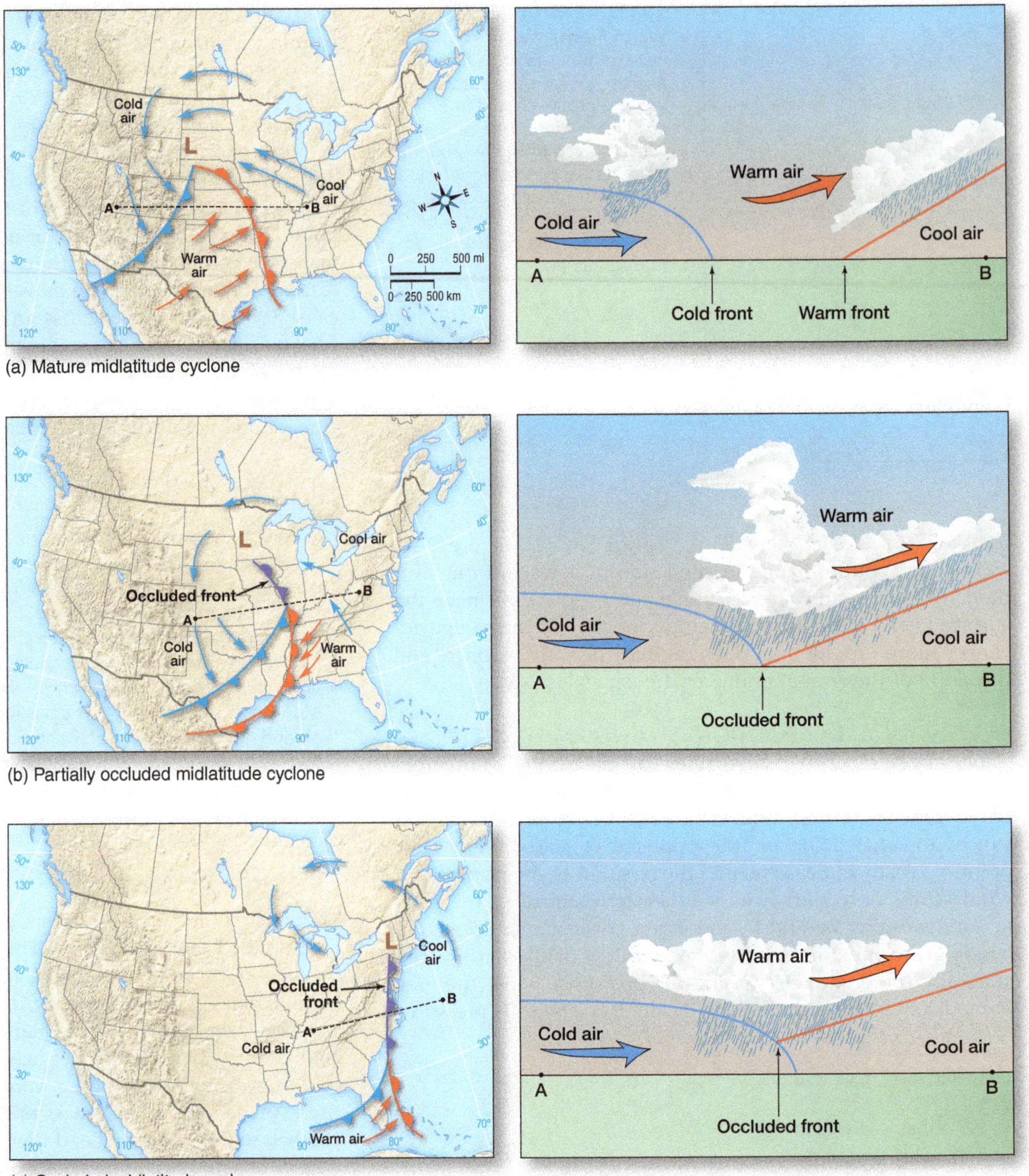

(a) Mature midlatitude cyclone

(b) Partially occluded midlatitude cyclone

(c) Occluded midlatitude cyclone

▸ **FIGURE 9–11 Formation of an Occluded Front.** The traditional explanation of the occlusion process, with a cold front overtaking a warm front. The panels on the left show the placement of the fronts relative to the surface, along with the location of a sample cross section (the dashed line with the letters *A* and *B* at the end of the line). The panels on the right show the profiles of the fronts across the transect shown on the corresponding panel on the left. Figures (a), (b), and (c) represent the mature, partially occluded, and almost totally occluded stages, respectively.

rise to a low-level front. If the overtaking air mass is colder, it produces a *cold-type occlusion*. If the trailing air mass is warmer, the model suggests a *warm-type occlusion* will form.

ALTERNATIVE OCCLUSION PROCESSES The occlusion process described above is essentially what was proposed by the leading meteorologists of the early 20th century, and it is consistent with many observed features of occluded fronts. However, it falls short in a number of ways. Most importantly, although the "catch-up" phenomenon is very common, it does not cause the occlusion. Second, the model incorrectly predicts that cold-type occlusions should be more common, when in fact they are rarely observed. Third, stability differences between the air masses determine the type of occlusion, not temperature differences.[5] Finally, the warm air aloft does not arise by "pinching" as asserted in the traditional view. Instead, as will be described in the next chapter, it is fed by an ascending stream of warm air entering the cyclone from the south.

[5]Whether an occlusion is of the cold or warm type is of little significance for our purposes, and indeed the types are not distinguished on maps produced by professional forecasters.

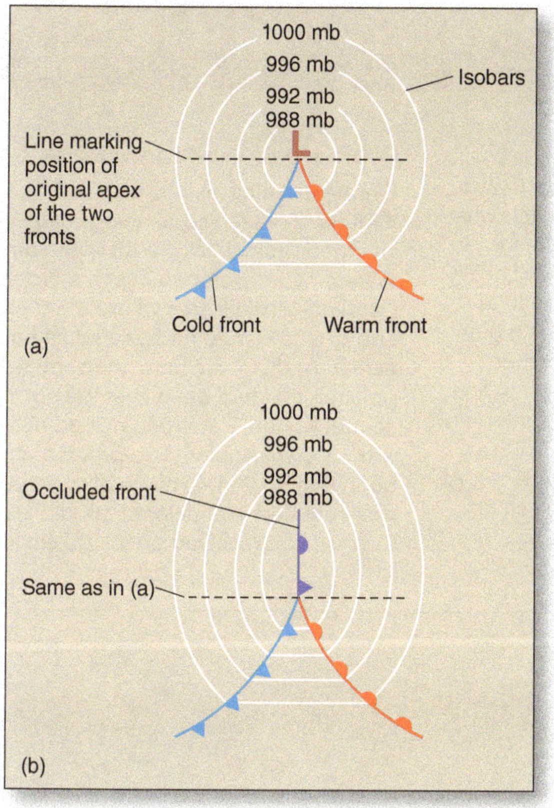

▲ **FIGURE 9–12 Occluded Front Formation Through Elongation of Surface Low.** Some occluded fronts form when the surface low elongates and moves away from the junction of the cold and warm fronts. In (a) we see the initial pattern. In (b) the elongation of the low creates the occlusion.

Occlusions arise from deformations in the flow around a cyclone. For example, occlusions sometimes occur when the circular core of low pressure near the junction of the cold and warm fronts changes shape and stretches backward, away from its original position at the junction of the cold and warm fronts (Figure 9–12). In (a), the cold and warm fronts are joined at the dashed line, and a rather uniform mass of air exists north of that line. At some later time (b), the cold and warm fronts have the same orientation with respect to each other as they did in (a), but the altered pressure pattern associated with the elongated zone causes a disruption in the airflow such that slightly different air masses are found on either side of the occluded front. The circular isobar pattern of (a) becomes elongated to form a trough over the occluded region. Thus, the structure of the occluded front is the same as that proposed in the traditional model, but the process responsible for its formation is somewhat different. In other cases, the cold front moves eastward relative to the warm front, so that the point where the fronts intersect moves progressively farther east (Figure 9–13). The occluded front appears as a relic portion of the warm front, situated to the west of the new intersection of the cold and warm fronts.

Video
An Infrared View of the 2011 Groundhog Day Blizzard
http://goo.gl/hSLiub

Yet another way occluded fronts can form is by a combination of distinct surface and upper-level processes. At low levels, a cold front overtakes and merges with a warm front, while aloft a rapidly moving upper-level frontal zone overtakes and joins with the surface front. Again the result is an occlusion, but not for reasons outlined in the classical description.

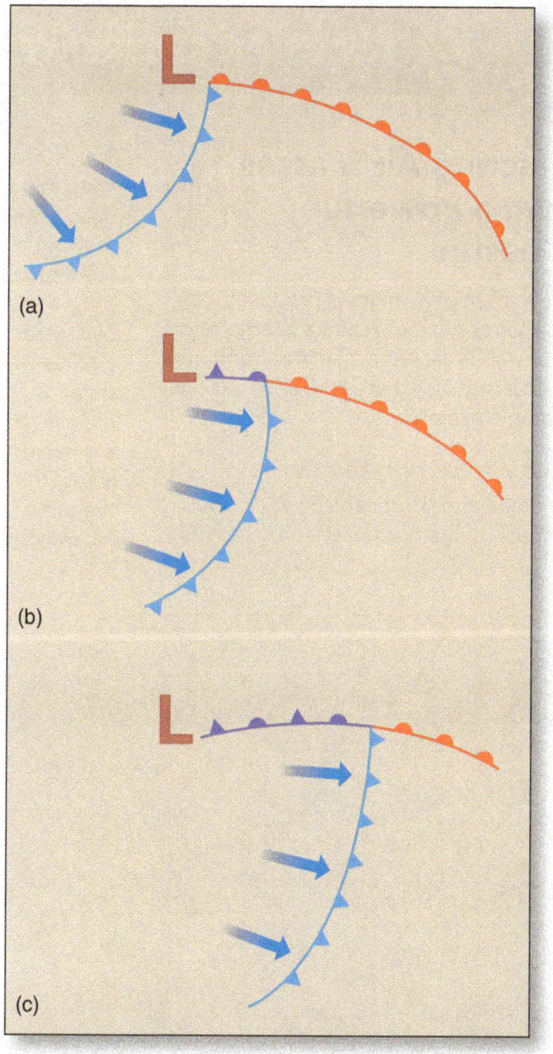

▲ **FIGURE 9–13 Occluded Front Formation Along Sliding Frontal Intersection.** Some occlusions occur when the intersection of the cold and warm fronts slides along the warm front. Parts (a), (b), and (c) show three stages in the process.

The types of fronts described above, like just about every other kind of natural feature on Earth, are subject to much variability. For example, temperatures might drop more strongly across some cold fronts than other cold fronts, and some warm fronts might move more rapidly than other warm fronts. Experienced weather watchers tend to observe these variations whenever they look at weather maps.

Drylines

Though fronts are the most common type of boundary separating air masses, we find another type of boundary in the spring and summer in Texas and the southern Great Plains. **Drylines** are areas where mT and cT air masses reside next

9–4 FOCUS ON SEVERE WEATHER

Interacting Air Masses Spawn a Powerful Nor'easter

The movement of air masses produces many abrupt changes in the weather. Examples of such transitions are almost limitless, and new ones occur regularly. Let's take a look at one very notable example.

February 7: Setting the Stage

Four months after Hurricane Sandy inundated much of the northeastern United States (discussed in Chapter 12), the last thing that part of the country needed was a major snowstorm. Unfortunately, that is what they got. On February 8 and 9, 2013, the Northeast was hit by a combination of heavy snow and strong winds that developed as two systems came together over the region. Figures 9–4–1a and b show the weather situation the day before the storm hit. A line consisting of two cold fronts and an intervening warm front extended from south Texas, across the Southeast and into the Atlantic. At the same time, cP air was advancing southeastward just west of the Great Lakes.

February 8: The Storm Forms

On the morning of February 8 (Figures 9–4–2a and b), things had changed dramatically. A low-pressure system had developed over the North Carolina–Virginia area where the trailing end of a cold front had existed 24 hours earlier. At the same time, the continental polar air that had been over the north-central United States had moved eastward with its own incipient low-pressure system. Soon (Figure 9–4–3) much of the country was engulfed in a major blizzard. Parts of Connecticut received up to 100 cm (40 in.)

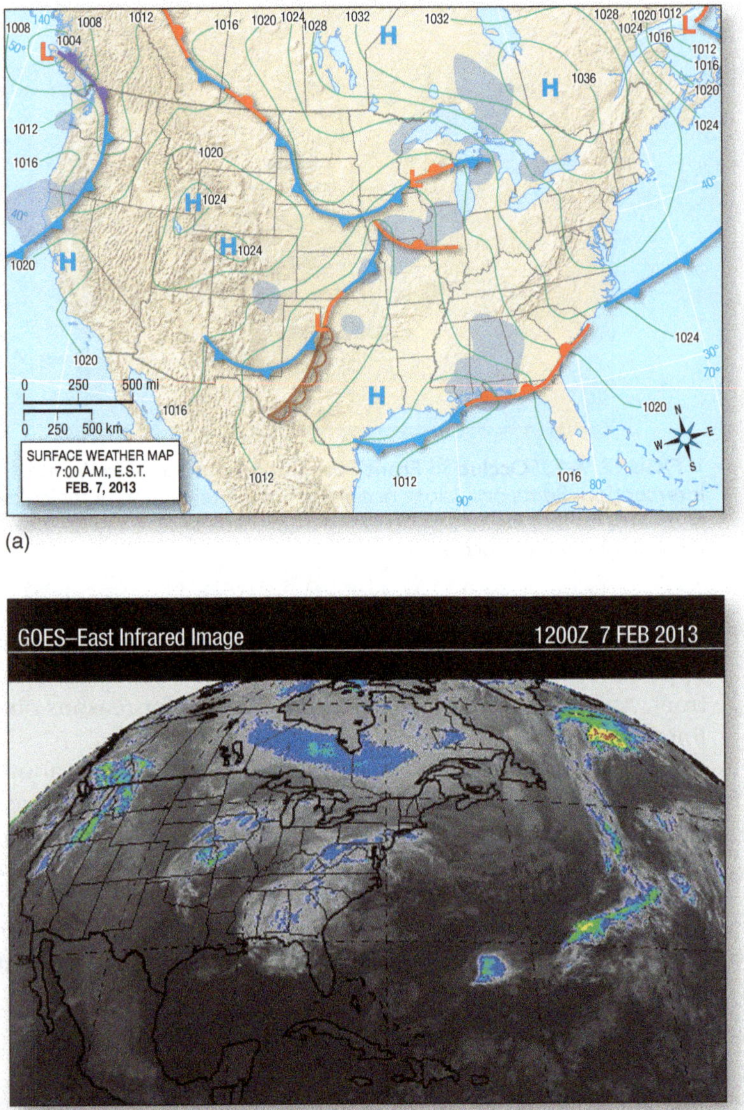

(a)

(b)

▲ **FIGURE 9–4–1 Powerful Nor'easter: February 7.** (a) Surface weather map and (b) infrared satellite image.

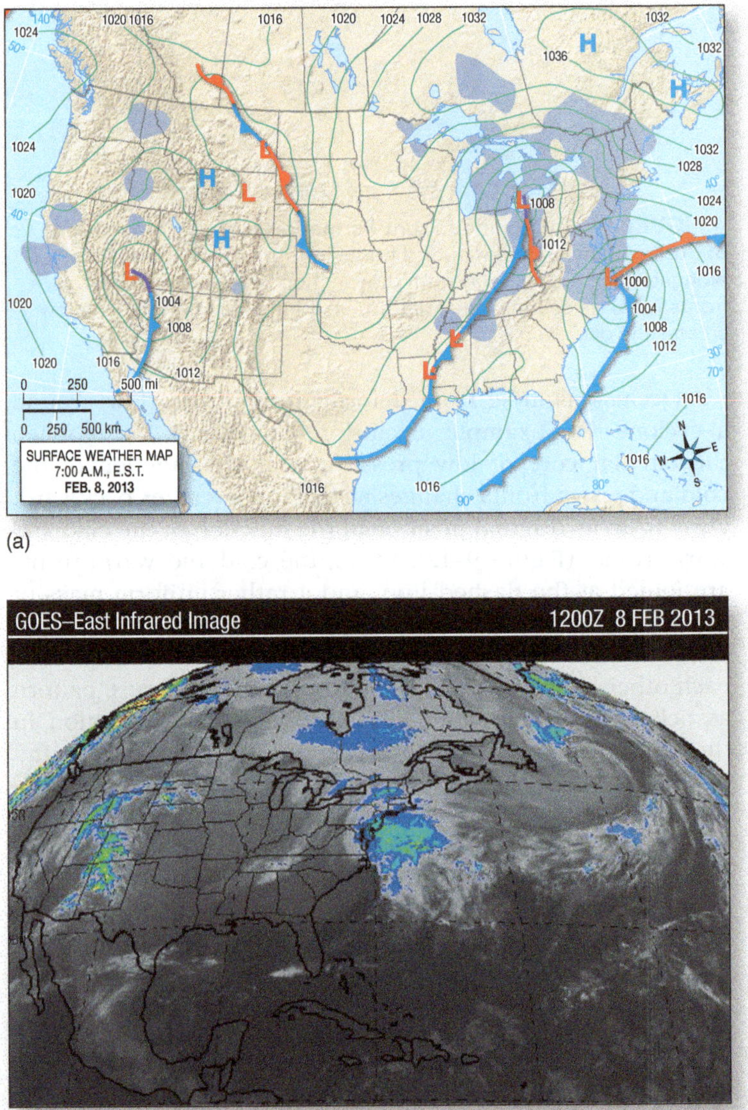

(a)

(b)

▲ **FIGURE 9–4–2 Powerful Nor'easter: February 8.** (a) Surface weather map and (b) infrared satellite image.

of snow (Figure 9–4–3). Hundreds of motorists were stranded on the Long Island Expressway east of New York City, and major airports closed across the region, leading to thousands of cancelled flights.

February 9: The Storm Intensifies

By February 9 (Figures 9–4–4a and b) the two storms had merged to form a single, very intense low-pressure system off the coast of eastern Canada with a pressure gradient usually associated with hurricanes (though, unlike hurricanes, this system was not tropical in origin). These winds kicked up waves to 9 m (30 ft) in height and created a storm surge that caused coastal flooding in many towns along Cape Cod. In Canada heavy snow extended as far inland as Toronto, Ottawa, and Montreal, while New Brunswick and Nova Scotia experienced major coastal flooding in addition to the heavy wind and snow.

9–4–1 What event occurred between February 8 and 9 that was crucial for the development of the major blizzard discussed? Explain.

9–4–2 Summarize the effects of the blizzard on inland and coastal areas as the storm moved northeast along the eastern seaboard of the United States and Canada.

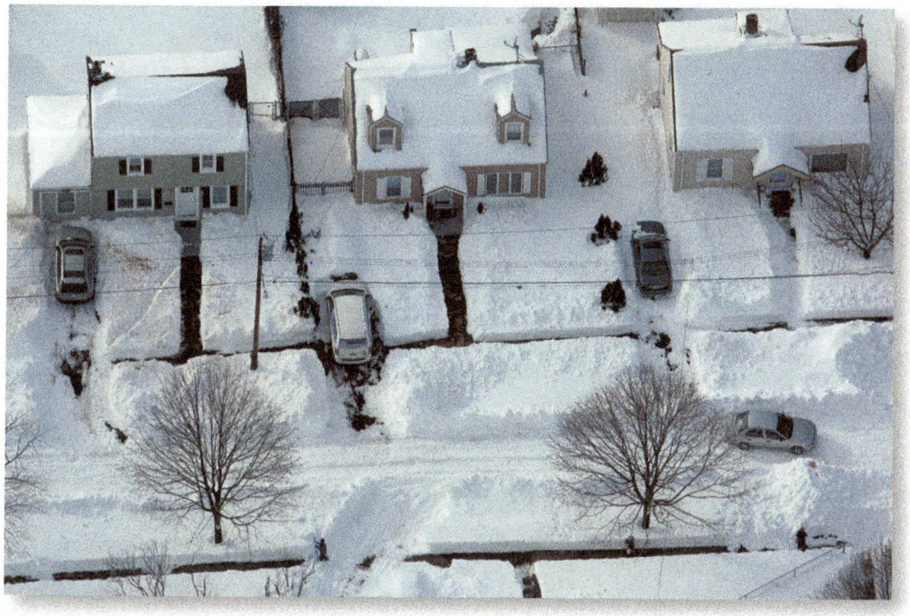

◂ **FIGURE 9–4–3 New Haven, Connecticut Snow.**

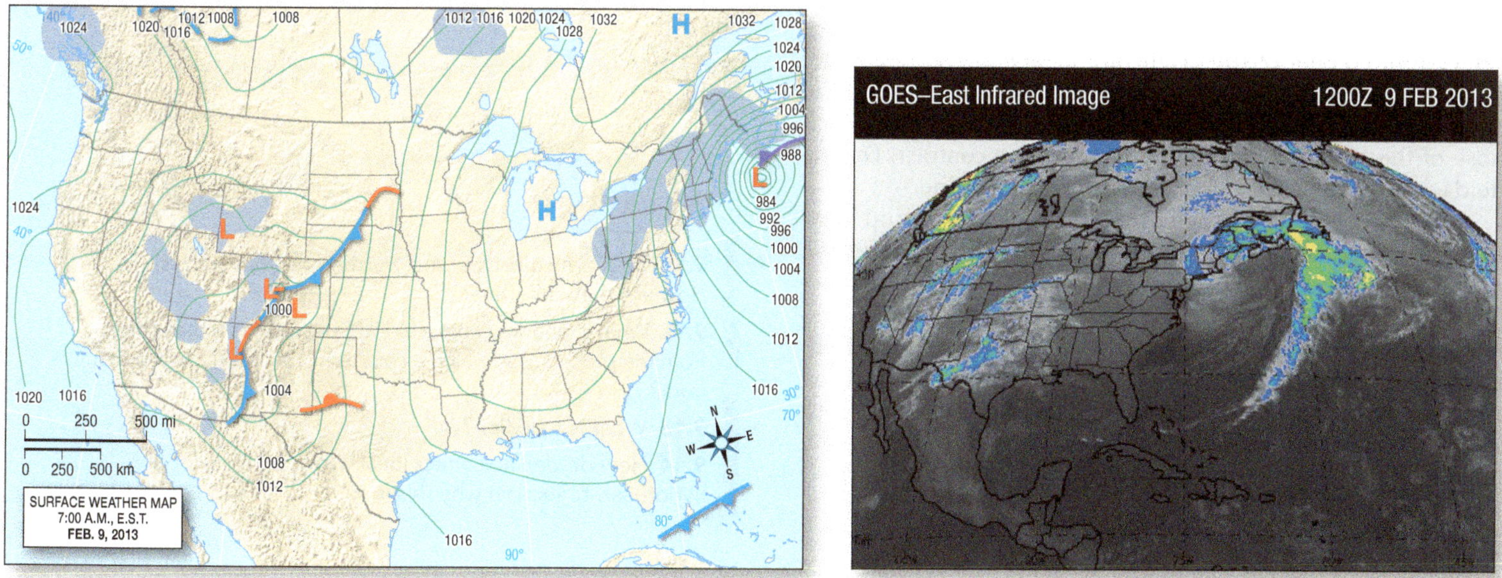

(a) (b)

▴ **FIGURE 9–4–4 Powerful Nor'easter: February 9.** (a) Surface weather map and (b) infrared satellite image.

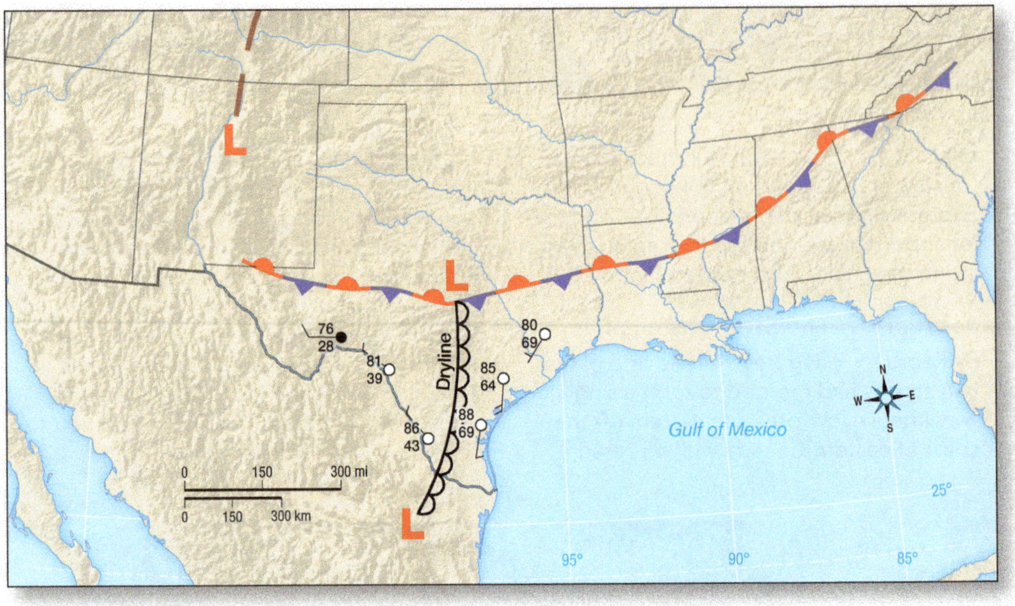

▸ **FIGURE 9–14 Dryline over Texas.** On both sides of the dryline the air is hot, but west of the line the humidity is low, with dew points ranging from the upper 20s to the 40s (°F). East of the line the humidity is greater, owing to the flow of air from the Gulf of Mexico.

Video
Hurricanes and Air Masses
http://goo.gl/evw00

to each other. Though the temperatures on either side of the boundary may not differ enough to create a front, the changes in water vapor content across the dryline can be substantial.

The dryline forms as warm moist air migrates westward and meets up with cT air moving in from the west or southwest. Figure 9–14 shows a dryline moving across Texas at noon CST on March 30, 2002. The temperatures and dew points (in °F)[6] at the selected weather stations are shown on the top left and bottom left of the station models, respectively. The dryline clearly demarcates the moist air from the Gulf and the drier air to the west.

Drylines tend to have greater temperature differences across their boundaries during the day than at night. The difference in temperature usually gets smaller overnight as the lower water vapor content of the cT air promotes more rapid cooling than occurs for the mT air to the east. By the early morning the greater cooling rate allows the cT air temperatures to approach those of the mT air, reducing the temperature contrast. During the day, more rapid warming occurs west of the line.

Table 9–2 shows the change in temperature and dew point observed at an automated weather-observing system near Austin, Texas. A substantial drop in the humidity occurred as the dryline passed between 2 and 3 P.M., though no increase in temperature occurred. It is noteworthy that thunderstorms had been going on in the area during the morning hours because drylines often promote thunderstorms with the potential for tornadoes or other severe weather. The mechanism by which this happens will be discussed further in Chapter 11.

TABLE 9–2

Temperatures and Dew Points Near Austin, TX, Prior to and Following the Passage of a Dryline

Time (CST)	Temperature °F (°C)	Dew Point °F (°C)
10 A.M.	66 (19)	64 (18)
11 A.M.	66 (19)	62 (17)
12 P.M.	71 (22)	64 (18)
1 P.M.	77 (25)	66 (19)
2 P.M.	78 (26)	62 (17)
3 P.M.	78 (26)	46 (8)
4 P.M.	80 (27)	46 (8)
5 P.M.	75 (24)	39 (4)

CHECKPOINT

9.11 How are cold-type and warm-type occlusions similar? How are they different?

9.12 What is one way an occluded front can form that differs from the processes in which a cold front catches up to a warm front?

9.13 What is a dryline?

9.14 How do drylines affect local weather? Develop a hypothesis to explain why this is so.

[6] We are using °F rather than °C in this instance because U.S. surface weather maps still rely on the Fahrenheit scale.

9-5 FOCUS ON AVIATION

Density Altitude and Aircraft Performance

In June, 2013, temperatures at Sky Harbor Airport in Phoenix, Arizona, edged toward the record high of 122 °F set in 1990. The 2013 heat wave peaked at 119 °F and did not break the record, but when the temperature reached 118 °F, some flights were canceled. Why? High temperatures can affect aircraft performance. As temperature rises, air becomes less dense, with fewer air molecules per unit of volume, and therefore provides less lift. Under hot conditions, planes require longer runways for takeoffs and landings, and pilots must fly the plane in a way that compensates for reduced lift. Failure to make these adjustments could result in an accident.

The relationship between air pressure, temperature, and altitude is key to understanding the effects of high temperature on aircraft performance.[7] Recall that air pressure always decreases with altitude. However, at higher temperatures, the air is less dense and air pressure is lower than it would otherwise be at a given altitude. In aviation, this relationship is expressed as *density altitude*, shown in Figure 9–5–1. As you can see from the chart, the higher the temperature, the higher the density altitude. For example, an airport that is 2,000 feet above mean sea level and 100 °F would have a density altitude of 5000 feet.

How do pilots correct aircraft performance for the effects of an airports increased density altitude? Aircraft manufacturers often provide a manual containing aircraft performance charts for density altitude. Based on these charts, the pilot can determine how much additional runway will be needed for takeoff, how much the rate of climb will be reduced, and how much air speed should be increased. The pilot may decide to reduce the aircraft's weight before takeoff by carrying less cargo or fuel. The Koch Chart in Figure 9–5–2 provides a quick estimate of how temperature affects takeoff distance and rate of climb. To use it, draw a straight line between the expected temperature on the left axis and the airport altitude on the right axis, and then observe the information at the point of intersection on the inset.

Density Altitude Chart

Std temp	Elev/temp	80 °F	90°F	100 °F	110 °F	120 °F	130 °F
59 °F	Sea level	1,200	1,900	2,500	3,200	3,800	4,400
52 °F	2,000	3,800	4,400	5,000	5,600	6,200	6,800
45 °F	4,000	6,300	6,900	7,500	8,100	8,700	9,400
38 °F	6,000	8,600	9,200	9,800	10,400	11,000	11,600
31 °F	8,000	11,100	11,700	12,300	12,800	13,300	13,800

▲ **FIGURE 9–5–1 Density Altitude and Temperature.**

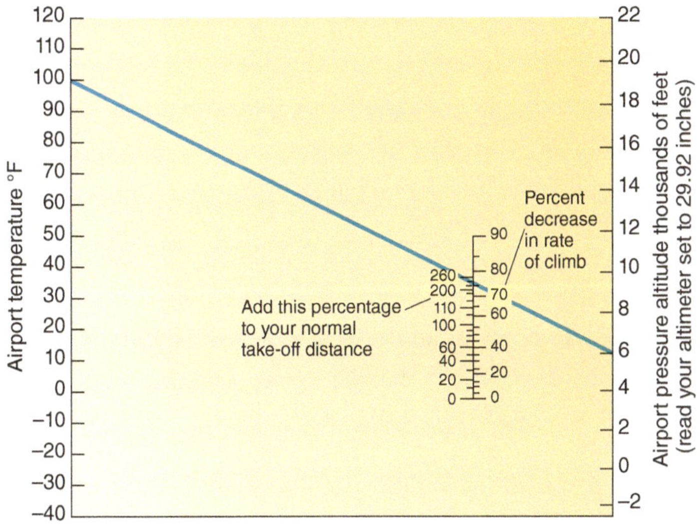

▲ **FIGURE 9–5–2 Koch Chart.** The blue line shows an example in which the expected airport temperature is 110 °F and the airport elevation is 6,000 feet.

Airlines may schedule flights so as to reduce aircraft exposure to high density altitude. For example, a flight heading to Phoenix in the summer could get priority for takeoff out of New York, so they can arrive before the hottest part of the day. Even then, the plane will need to be prepared for landing at a higher speed.

9–5–1 What are the factors that determine the density altitude of an airport?

9–5–2 Summarize how a pilot can compensate for the effects of density altitude on aircraft performance. Would it be a good idea to rely solely on the Koch chart? Explain.

[7] Humidity is only a very minor factor in influencing air density (Chapter 5), but more significantly, high humidity can reduce engine power.

Summary

9.1 Describe the importance of source regions in air mass formation.

- The atmosphere tends to arrange itself into large masses with relatively little horizontal change in temperature and humidity.
- Air masses form over particular source regions—large areas of high or low latitudes of either continental or maritime locations.

9.2 Describe the types of air masses and explain how they form.

- The five major types of air masses are continental polar, continental arctic, maritime polar, continental tropical, and maritime tropical.
- Air masses form when air remains over a large area with characteristic temperature and humidity for extended periods.

9.3 Identify the different types of fronts and their characteristics, including cloud conditions.

- The four types of fronts are cold, warm, stationary, and occluded. They are components of weather systems called *midlatitude cyclones*, which we discuss in the next chapter.
- Cold fronts tend to promote cumuliform clouds while warm fronts are more inclined to be accompanied by stratiform clouds.
- Stationary fronts are similar in structure to cold and warm fronts but remain in place with little movement.
- Occluded fronts are formed with the merger of cold and warm fronts.
- Drylines are similar to fronts in that they separate air masses, though the dryline marks a humidity boundary rather than a temperature boundary.

Key Terms

air masses *p. 290*
cold front *p. 297*
continental arctic (cA) air masses *p. 292*
continental polar (cP) air masses *p. 291*
continental tropical (cT) air masses *p. 293*
drylines *p. 305*
fronts *p. 290*
maritime polar (mP) air masses *p. 293*
maritime tropical (mT) air masses *p. 294*
northeasters *p. 293*
occluded front *p. 303*
overrunning *p. 298*
source regions *p. 290*
stationary front *p. 303*
warm front *p. 298*

Review Questions

1. What are the requirements for an area to serve as a source region?
2. Where are the primary air mass source regions in North America?
3. Describe the characteristics of cA, cP, cT, mT, and mP air masses.
4. Of the five types of air masses, which are the hottest, driest, coldest, and most damp?
5. Which of the air mass types are likely to be stable or unstable?
6. What is the primary difference between arctic and polar air masses?
7. Describe the structure of cold, warm, stationary, and occluded fronts.
8. Describe the changes that occur when a continental air mass migrates out of its source region.
9. What is overrunning?
10. Why do cold fronts have steeper slopes than warm fronts?
11. How do the alternative models of the occlusion process differ from the traditional model?
12. What are drylines and why are they important?
13. What is the difference between warm-type and cold-type occlusions?

Critical Thinking

1. Contrast the formation of air masses in the Northern and Southern Hemispheres.
2. Explain the limitations and benefits of classifying air into distinct masses.
3. What parts of North America are likely to experience the most frequent changes in air mass during the summer and winter?
4. Continental polar air masses can migrate into Florida during the winter but not into northern India. Why not?
5. Distinct temperature changes can be detected across narrow regions that are not associated with fronts. What can cause such conditions to exist?
6. Warm fronts are extremely rare over southern California. Why?
7. The southwestern United States experiences what is locally referred to as a monsoon. Can we say the same thing about Florida or Texas?
8. What does the presence of a continental polar air mass tell you about the height of the 500 or 300 mb levels?
9. Where in North America might you expect to see the collision of mT and cT air masses? What time of year would this be most likely to happen?
10. Which types of fronts are most likely to have inversions and which least likely?

Problems & Exercises

1. The following temperatures and dew points are observed. What are the likely types of air masses present for each?
   - **a.** T = 29 °C (85 °F) Dew Pt = 19 °C (66 °F)
   - **b.** T = −18 °C (0 °F) Dew Pt = −21 °C (−5 °F)
   - **c.** T = 3 °C (38 °F) Dew Pt = 0 °C (32 °F)
   - **d.** T = 38 °C (100 °F) Dew Pt = −4 °C (25 °F)
2. Assume that a cold front has a slope of 1:100 and that the height of the 700 mb level is 1500 m. How far behind the surface position of the front will the 700 mb position be?

Visual Analysis

This surface weather map from January 29, 2013, plots the temperatures in degrees Fahrenheit (top number in red) with the dew points (bottom number in blue) for cities across the United States. Examine the distribution of these values on the map to answer the following questions:

9.1. Where would a cold front likely be found?

9.2. Where would a warm front exist?

9.3. What kind of changes in temperature and dew point would you expect to find in the southeastern part of the United States during the next 24 hours?

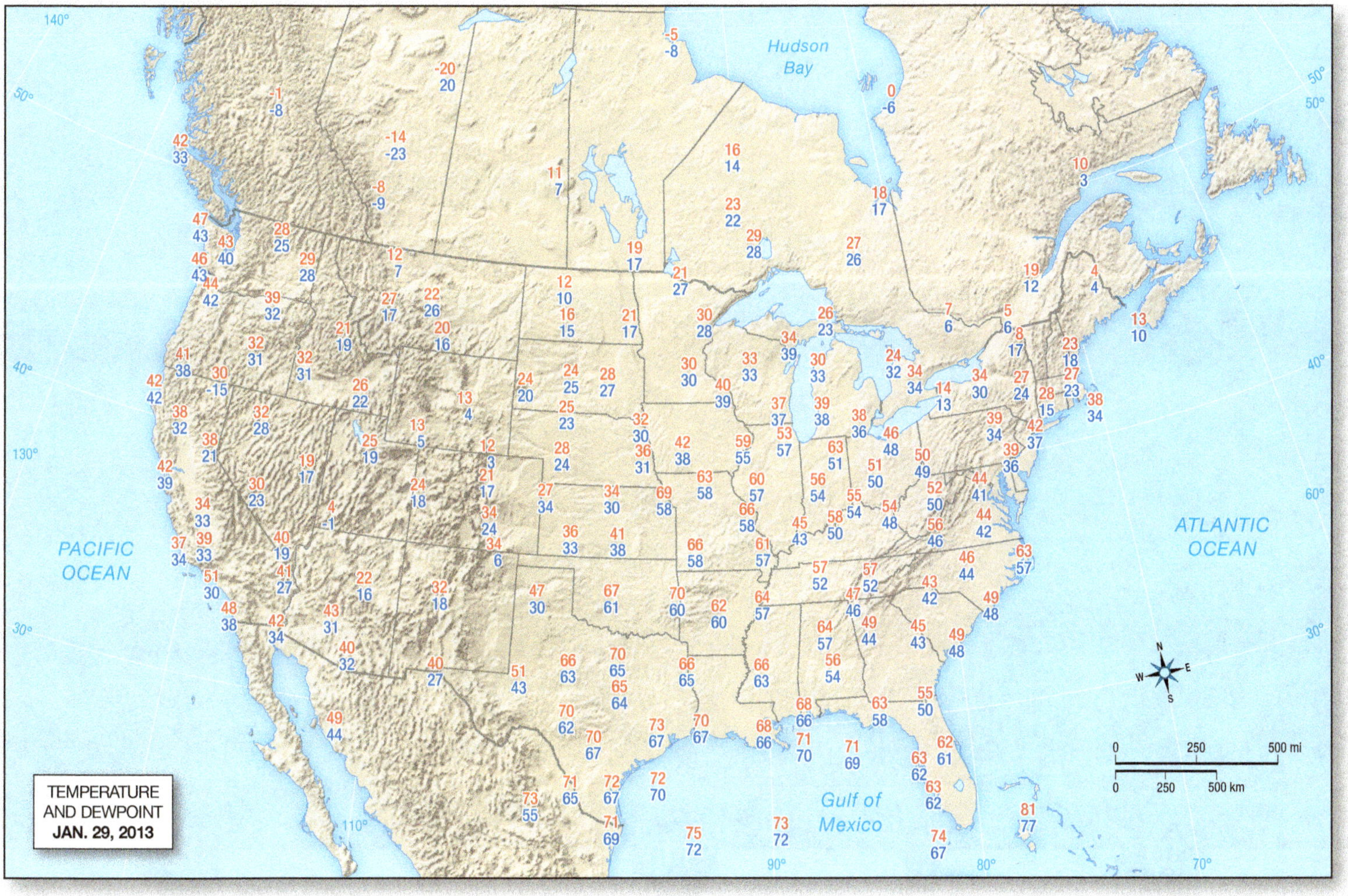

PART

4 Disturbances

CHAPTER

10

Midlatitude Cyclones

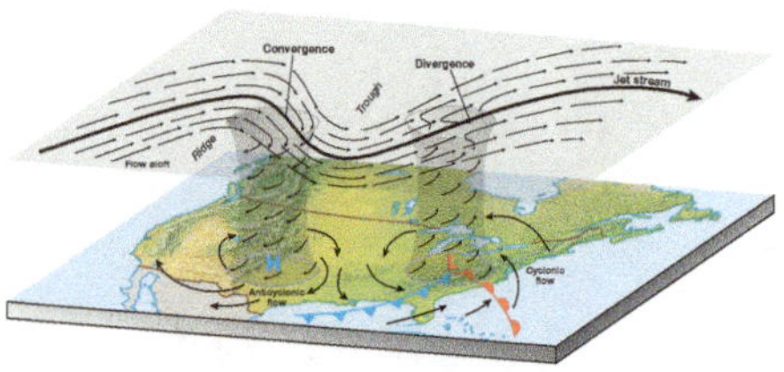

CHAPTER

11

Lightning, Thunder, and Tornadoes

CHAPTER

12

Tropical Storms and Hurricanes

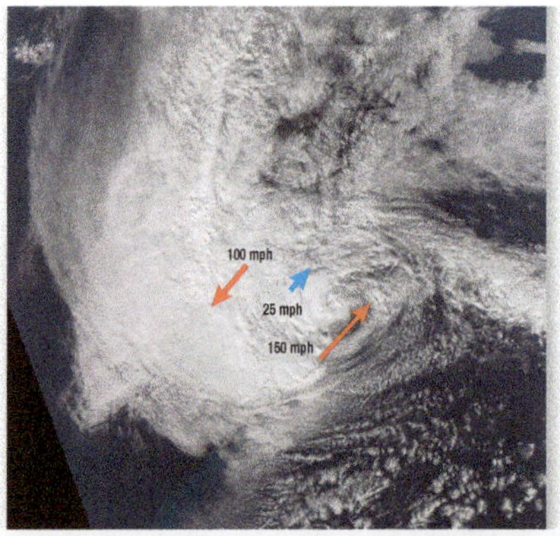

◄ *A rainbow over Banff National Park, Alberta, Canada.*

We are subject to a wide variety of weather conditions from calm to severe—the most severe of which can bring widespread death and destruction. The settings for these storms may be the passage of midlatitude cyclones, localized thunderstorm activities, organized clusters of thunderstorms, or tropical storms and hurricanes. This section of the book looks at the conditions associated with such activity and the processes involved in their formation and development.

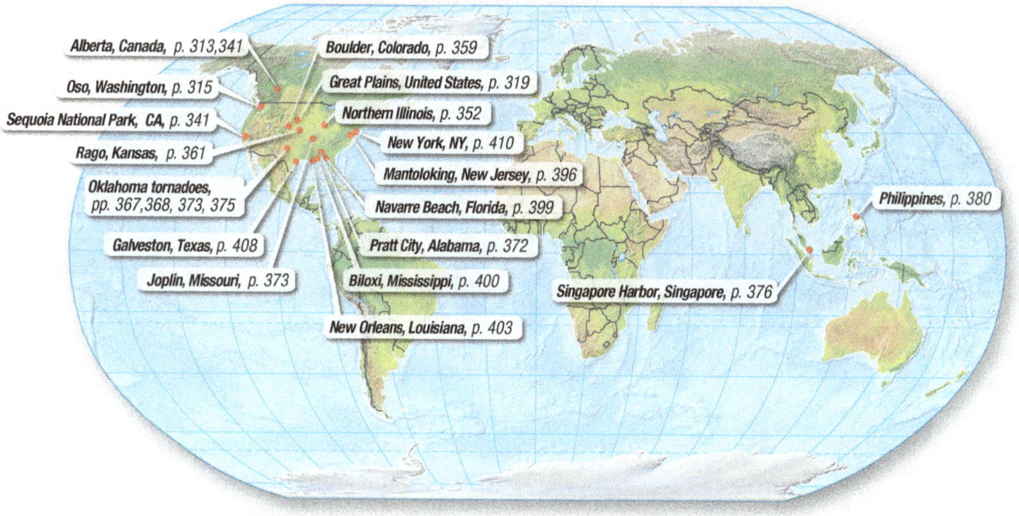

CHAPTER

10 Midlatitude Cyclones

Though the Pacific Northwest is known for its rain, September precipitation tends to be light when it occurs. The precipitation season normally begins in earnest in November and continues through the winter. But September 2013 witnessed a couple of very notable storms that produced precipitation that would be considered very heavy even in the peak of the season. On September 28 a passing cold front brought record rain across Oregon and Washington States. Seattle-Tacoma International Airport more than doubled the record for that date, with 1.71 inches. Eugene and Portland, Oregon, likewise received record-breaking rainfalls over September 28–29. These values paled in comparison to precipitation amounts received in the Cascade Mountains, where two-day totals exceeded 5 inches.

But the heavy rains were only part of the story, as strong winds followed in the wake of the heaviest precipitation. Gusts as high as 137 km/hr (85 mph) were observed at Hurricane Ridge in Washington, and widespread windiness caused power outages for tens of thousands of

◀ *Aftermath of a mudslide at Oso, Washington, March 2014. Such slides had occurred previously at this location, with this one triggered by heavy rainfall. The slide killed 43 people.*

people. Even more noteworthy was the occurrence of a tornado in Frederickson, Washington, with estimated winds of 177 km/hr (110 mph) that tore a large hole in the roof of a garage-door factory.

The storm system that brought this anomalous early-autumn event was a midlatitude cyclone (also referred to as an extratropical cyclone). This chapter discusses the processes that lead to the formation, growth, and eventual demise of these very important weather occurrences.

Learning Outcomes

After reading this chapter, you should be able to:

10.1 Summarize the polar front theory.

10.2 Outline the life cycle of a midlatitude cyclone as described in Bjerknes's model.

10.3 Explain processes in the middle and upper troposphere that relate to midlatitude cyclones.

10.4 Analyze how surface fronts and upper-level patterns are related.

10.5 Describe the behavior of a typical midlatitude cyclone as it crosses North America.

10.6 Explain how flow patterns and large-scale weather patterns affect the development, steering, and dissipation of midlatitude cyclones.

10.7 Evaluate the modern, conveyor belt model of midlatitude cyclones.

10.8 Describe anticyclones and the weather associated with them.

10.9 State how scientists think climate change may affect midlatitude cyclones.

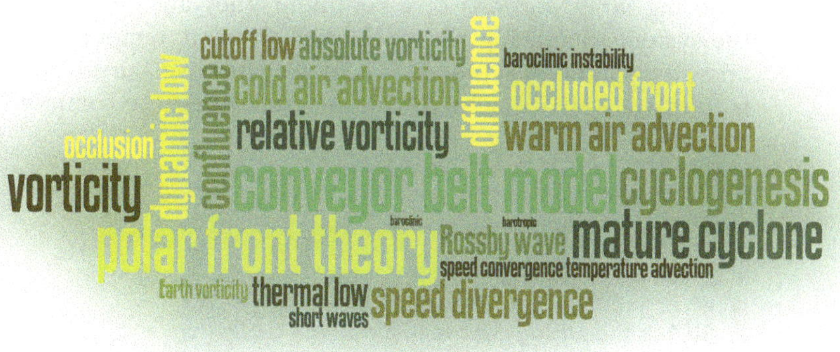

Polar Front Theory

10.1 Summarize the polar front theory.

From 1914 to 1918, the world experienced one of history's great catastrophes, World War I. The advent of new weapons, including the machine gun and mustard gas, made it nearly impossible for opposing armies to gain large tracts of ground from their opponents. Until then, an army attacking with rifles and bayonets could charge its opponents with some reasonable chance of success. However, against a foe dug into trenches and armed with the latest weapons, such maneuvers were almost doomed to failure. Thus, the war zone remained stagnant for long periods, since neither army could advance across the *front*, or battle line.

While the war was going on in western Europe, Vilhelm Bjerknes (pronounced *bee-YURK-ness*) established the Norwegian Geophysical Institute in the city of Bergen. With several colleagues,[1] including his son Jacob, Bjerknes developed a modern theory of the formation, growth, and dissipation of midlatitude cyclones, storms that form along a front in middle and high latitudes. Bjerknes observed the systems forming along a boundary separating polar air from warmer air to the south. Comparing that boundary to the one separating the opposing armies in western Europe, he called his model the **polar front theory** (also called the *Norwegian cyclone model*). This theory has stood the test of time remarkably well. Though we now have far more observational information available than did Bjerknes (especially for the middle and upper troposphere), we still describe the life cycle of a midlatitude cyclone in much the same way that the scientists of the Bergen school did decades ago.

Midlatitude cyclones are large systems that travel great distances and often bring precipitation—and sometimes severe weather—to wide areas. Lasting a week or more and covering large portions of a continent, they are familiar as the systems that bring abrupt changes in wind, temperature, and sky conditions. Indeed, all of us who live outside the tropics are well acquainted with the effects of these common events.

CHECKPOINT

10.1 Is the analogy between a wartime battle front and a weather front a good one? Why or why not?

10.2 What observations led to the development of the polar front theory?

[1]Tor Bergeron, who discovered the ice crystal process for the formation of precipitation (Chapter 7), was among the scientists in this group.

The Life Cycle of a Midlatitude Cyclone

10.2 Outline the life cycle of a midlatitude cyclone as described in Bjerknes's model.

The Bergen meteorologists were perfectly placed to witness the atmosphere along the polar front, and they used their observations to describe the formation of midlatitude cyclones, a process called **cyclogenesis**, along this boundary. Although many cyclones do originate along the polar front, they also form in other areas, especially downwind of major mountain barriers. We first discuss the formation of midlatitude cyclones at the polar front, as described in Bjerknes's classical model. Later in this chapter we incorporate more recent insights into cyclogenesis.

Cyclogenesis

Figure 10–1 illustrates the classical description of the life cycle of a typical midlatitude cyclone. Initially, the polar front separates the cold easterlies and the warmer westerlies (a). As cyclogenesis begins, a minor "kink" (b) develops along the boundary, and a low-pressure center begins to form. The cold air north of the front begins to push southward behind the cold front, and air behind the warm front advances northward. This creates a counterclockwise rotation (in the Northern Hemisphere) around the developing low-pressure system. With further intensification (c), the low pressure deepens and distinct warm and cold fronts emerge from the original polar front. Convergence associated with the low pressure can lead to uplift and cloud formation, while linear bands of deeper cloud cover develop along the frontal boundaries, as described in Chapter 9. Cyclogenesis along the polar front represents the initial stage in the development of a system that may affect millions of people thousands of kilometers away several days later as it progresses toward the occluded stage, parts (d) and (e).

The Bergen scientists could not explain *why* cyclogenesis occurs, but they did observe that it commonly happened near zones of thermal contrasts (such as along coastal regions or at the boundaries between warm and cold ocean currents) or where topographic features (such as major mountain chains) disrupt the normal airflow.

Mature Cyclones

Figure 10–2a illustrates the cloud patterns, wind, uplift processes, and precipitation patterns associated with a **mature cyclone**. (The precipitation probabilities, listed as percentages in the figure, should not be interpreted too literally;

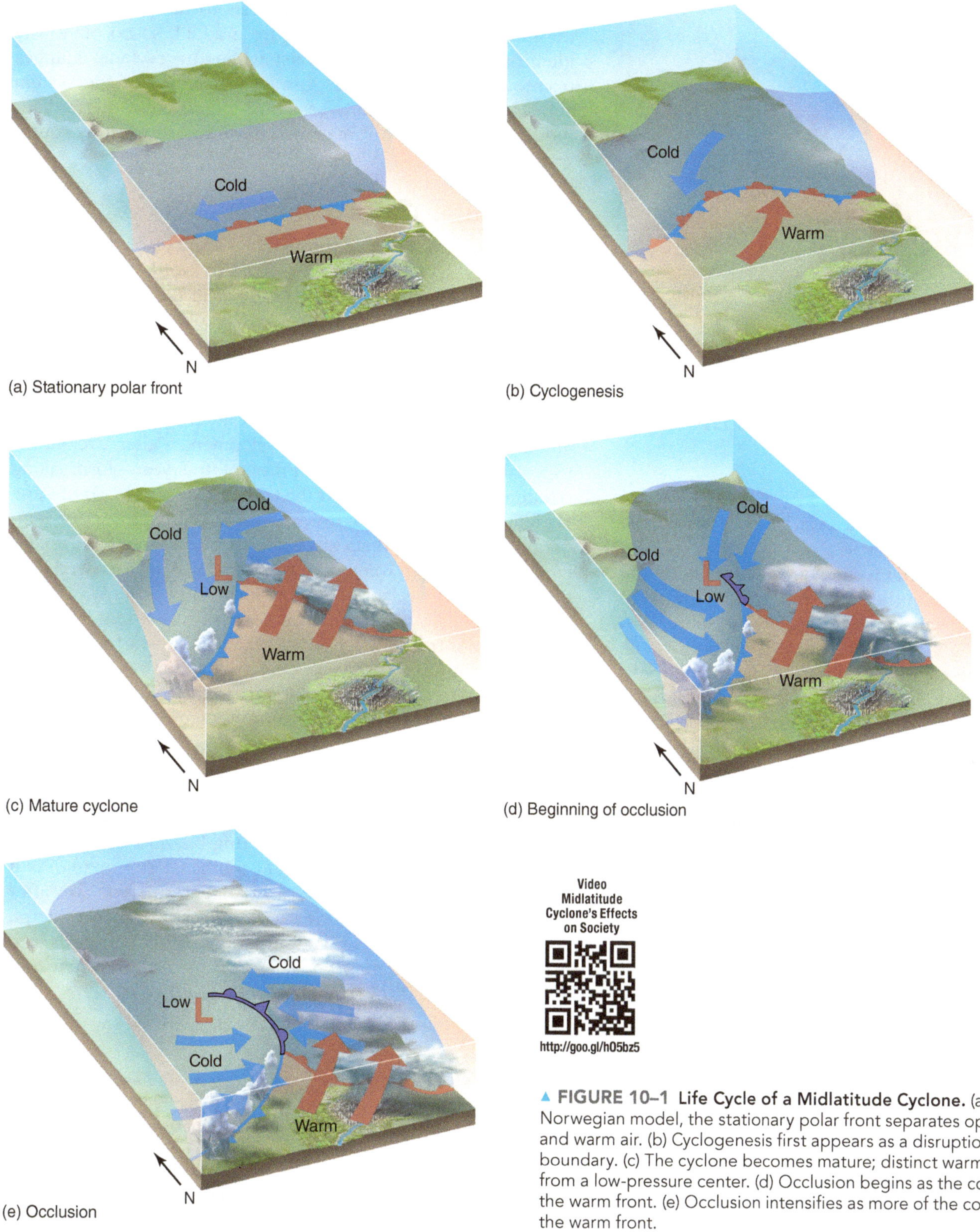

Video
Midlatitude
Cyclone's Effects
on Society

http://goo.gl/h05bz5

Animation
Midlatitude
Cyclone

http://goo.gl/mah9eC

▲ **FIGURE 10–1 Life Cycle of a Midlatitude Cyclone.** (a) According to the Norwegian model, the stationary polar front separates opposing masses of cold and warm air. (b) Cyclogenesis first appears as a disruption of the linear frontal boundary. (c) The cyclone becomes mature; distinct warm and cold fronts extend from a low-pressure center. (d) Occlusion begins as the cold front catches up to the warm front. (e) Occlusion intensifies as more of the cold front catches up to the warm front.

they merely give a general picture.) A band of mostly cumuliform cloud cover runs along and ahead of the cold front, caused by denser, cold air displacing warm air. The likelihood of precipitation along the front increases toward the center of the low pressure, where large-scale convergence adds to the uplift caused by the meeting of the two air masses. Because of the high moisture content and generally unstable conditions typically found ahead of a cold front, precipitation—in the form of rain, snow, or even sleet or hail—can be intense. But the band of cloud cover and precipitation is relatively narrow,

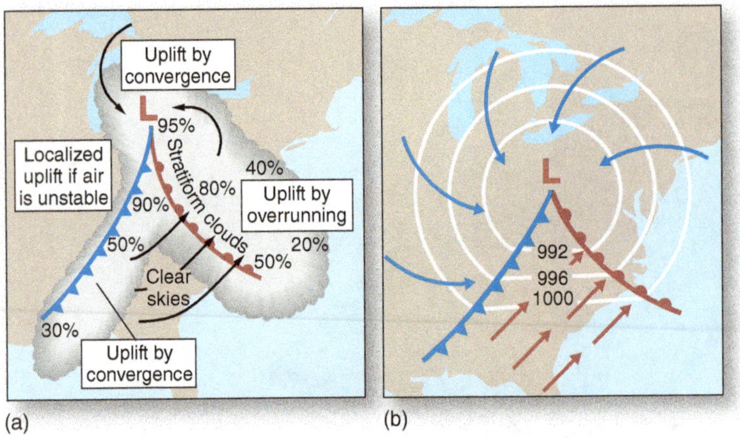

▲ **FIGURE 10–2 Structure of a Midlatitude Cyclone.** The diagrams show the typical structure of a mature midlatitude cyclone and the processes causing uplift. Shaded areas represent the presence of cloud cover. (a) The numbers represent an approximation of the precipitation probability. (b) The isobar pattern is shown.

so precipitation may last for only a brief period before the frontal zone moves on.

A wider band of mostly stratiform clouds lies ahead of the warm front. As we found with the cold front, the likelihood of precipitation increases toward the center of low pressure. Precipitation tends to be light along the warm front because its more gradual slope leads to slower uplift. But the warm front's greater horizontal extent and generally slower forward motion allow clouds and precipitation to last longer. Clear skies characteristically occur over the warm sector between the cold and warm fronts, although squall lines and other disturbances (discussed in Chapter 11) develop under certain conditions.

The isobar pattern, depicting the distribution of pressure within the cyclone (Figure 10–2b), is interrupted along the two fronts. This causes abrupt transitions in wind direction along the boundaries. The isobars are nearly straight in the warm sector but become curved in the larger, cold region. Looking at the warm front, the winds shift from southeasterly on the cold side to southwesterly in the warm sector. Across the cold front, the winds shift from southwesterly in the warm sector to northwesterly on the cold side.

Occlusion

Occlusion represents the end of the cyclone's life cycle, with an **occluded front** having replaced the warm and cold fronts of the mature phase. Refer back to Figure 10–1 and observe the latter stages of a midlatitude cyclone shown in parts (d) and (e). Although a temperature contrast exists across the occluded front, temperature differences here are not as great as those along the original cold or warm fronts. West of the frontal boundary, the air flows out of the northwest and is extremely cold. Slightly warmer air approaches the occluded front from the east, but this air originates in the cold sector of the cyclone. Thus, the temperature difference is less than where the fronts separate warm, tropical air from cold, polar air.

The transitions from cyclogenesis to maturity, and from the mature phase to occlusion, are gradual, so no obviously identifiable points in time exist when the cyclone changes from one stage to another. Also, the evolution of the system coincides with a generally eastward migration of the midlatitude cyclone, though it may also have a northward or southward component.

Location and Movement of Cyclones

Midlatitude cyclones may form over a number of different regions. Many approach the west coast of North America after forming off the coast of Japan. Many form downwind of large mountain ranges such as the Rockies. As columns of air descend the leeward slopes of the Rockies, they are vertically stretched and narrowed, which imparts a cyclonic (counterclockwise) spinning to the air.[2] Thus, the eastern foothills of the Rockies in south-central Colorado and in southern Alberta, Canada, are favored sites for cyclone development. Other cyclones—especially during the colder times of the year—form along coastlines, most notably near the Gulf of Mexico and off of Cape Hatteras, North Carolina, in response to the large surface temperature differences that exist between warm waters and colder land bodies. The movement of these cyclones is discussed later in this chapter in relation to upper-level wind patterns.

Let's look at a hypothetical scenario consistent with the original polar front theory, in which a weak disturbance in the airflow begins off the east coast of Japan. As the system matures and moves eastward, it can bring rain to the coastal portions of western North America and snow to the coastal mountains. If this storm occurs in the winter, upper-level winds may guide the storm southward into central and southern California and then eastward into the Rocky Mountain states. Passing over the lee of the mountains, the midlatitude cyclone may intensify and then barrel northeastward to bring blizzard conditions to the northeastern United States and southeastern Canada. As it moves offshore into the western Atlantic, a week or two after its formation in the western Pacific, the storm may undergo complete occlusion. (For an example of how blizzards can affect people, see *Box 10–1, Focus on the Environment and Societal Impacts: 2013 Great Plains Blizzard.*)

DID YOU KNOW?

Scientists at the Goddard Institute for Space Studies (GISS) have compiled an online atlas of midlatitude cyclones across the globe. On average, 1071 midlatitude cyclones form each year, 578.6 in the Northern Hemisphere and 492.1 in the Southern Hemisphere. Thus, in each hemisphere an average of about one and a half new midlatitude cyclones forms each day. Midlatitude cyclones of the Southern Hemisphere average lower minimum pressures at their most intense stage (972 mb) than do those of the Northern Hemisphere (988 mb). Monthly and seasonal maps of cyclone tracks for almost 40 years are available for download at **http://data.giss.nasa.gov/stormtracks.**

[2]The concept behind this occurrence is known as the conservation of potential vorticity, and is beyond the scope of this text.

10–1 FOCUS ON THE ENVIRONMENT AND SOCIETAL IMPACTS

2013 Great Plains Blizzard

In early October 2013 a midlatitude cyclone created a large and unusually early blizzard across the Great Plains. In Figure 10–1–1 observe the strong pressure gradient centered over South Dakota on the morning of October 5, which produced the strong winds. At the same time, a flow of warm moist air from well to the southeast was drawn into the low-pressure system, providing the uplift required for heavy precipitation. In this case the results were tragic, as tens of thousands—perhaps as many as 100,000—cattle died in the blizzard. Because the storm arrived so early in the season, the cattle had yet to grow their winter coats, which offer some protection. First came a cold rain that lasted for half a day, followed by heavy snowfall that measured up to 1.2 m (4 ft) in depth through winds up to 97 km/hr (60 mph). Many of the cattle froze to death while others were suffocated beneath the snow pack (Figure 10–1–2).

The economic impact was enormous for many ranchers who lost significant portions of their herds. At the same time, the emotional trauma was universal among the ranchers—a group not generally known for outward displays of sentiment. The immediate task for them was the burial of the dead cattle (in some cases, the removal of carcasses from streams where the cattle had died). Not unexpectedly, the ranchers worked together and helped their neighbors—sometimes people previously unknown to them. But the situation had one complicating factor: the U.S. government had "shut down" due to an impasse over certain aspects of the national budget. As a result, ranchers who called the U.S. Department of Agriculture for disaster assistance got nothing more than a recorded message.

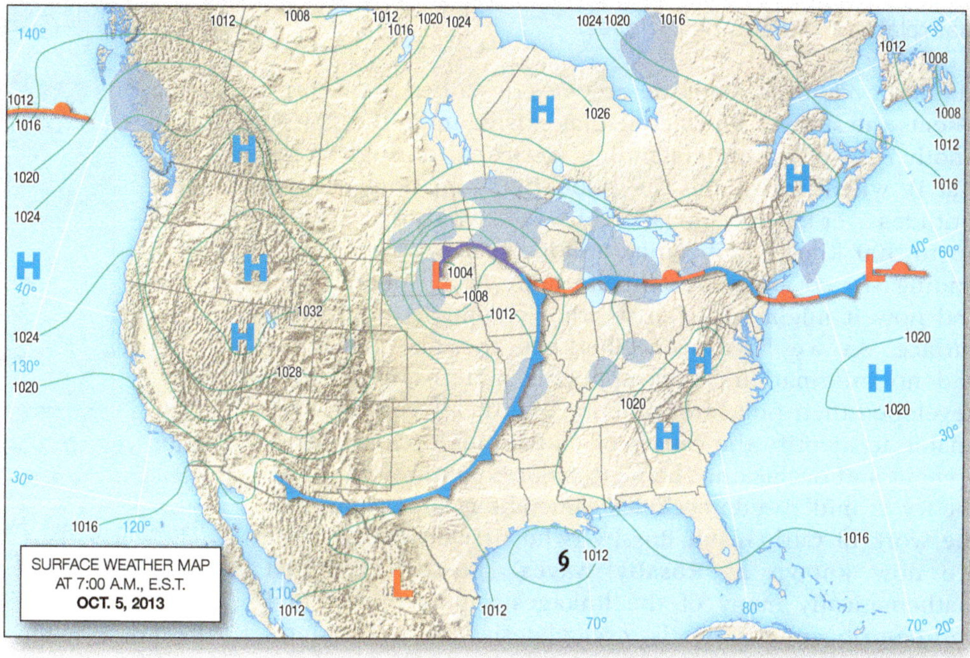

▲ **FIGURE 10–1–1 Surface Weather Map.** Surface weather map depicting the weather situation on October 5, 2013, at about the time of blizzard conditions over South Dakota.

10–1–1 What features of the weather map would alert you to the possibility of a blizzard over South Dakota?

10–1–2 Why would the same exact weather pattern have had different societal and environmental impacts if it had occurred later in the year?

▲ **FIGURE 10–1–2 Impact of Storm.** Numerous cattle such as this one in Sturgis, South Dakota, died from the Great Plains blizzard of 2013.

Video
Water Vapor Transport by Midlatitude Cyclones
http://goo.gl/aRLhHB

At a particular place (say, Kansas City, Missouri), the passage of the system brings predictable effects. As a warm front approaches, cloud cover usually deepens and increases and light to moderate precipitation is possible. The rain or snow eventually gives way to warmer, sunny conditions with the passage of the warm front, and the wind shifts from a southerly to southwesterly direction. Clear, warm conditions may then persist for a day or so. But as the cold front approaches, a strong, fast-moving band of heavy clouds and precipitation can cause major snowfall or rainfall. Afterward, the cold air behind the front brings cold, clear conditions.

CHECKPOINT

10.3 What initial conditions are required for the formation of a midlatitude cyclone?

10.4 In a mature cyclone, which area of precipitation might deliver the greatest rainfall totals: the area ahead of the cold front or the area ahead of the warm front? Explain.

Processes of the Middle and Upper Troposphere

10.3 Explain processes in the middle and upper troposphere that relate to midlatitude cyclones.

The life cycle of midlatitude cyclones described above represents the state of knowledge that existed up to the 1940s. Another leap in understanding occurred during World War II, when British and U.S. pilots flying missions at high altitudes over Europe and Japan observed winds with speeds up to 400 km/hr (250 mph). Among meteorologists, this finding stirred an interest in the upper-tropospheric flow and how it might relate to weather conditions on Earth's surface. As we have seen, Bjerknes and his colleagues had no information about upper-air patterns when they developed their polar front theory, and therefore they were unable to identify the causes of midlatitude cyclone development and occlusion. The next major breakthrough in the theory of midlatitude cyclones came about largely through the work of Carl-Gustaf Rossby (who first described what are now known as **Rossby waves**). Rossby explained mathematically many of the linkages among upper- and middle-tropospheric winds, cyclogenesis, and the maintenance of midlatitude cyclones.

Rossby Waves and Vorticity

In Chapter 8 we described the large Rossby waves of the upper troposphere. Figure 10–3 illustrates how the air turns left and right as it flows through these waves. Moving from points 1 to 3, the air rotates counterclockwise (as indicated at the bottom left of the figure). Between points 3 and 5, it rotates clockwise. The rotation of a fluid (such as air) is referred to as its **vorticity**.[3] The figure shows vorticity changes in the moving air, relative to the surface. Viewed from space, there is an additional component of vorticity arising from Earth's rotation on its axis. The overall rotation of air, or its **absolute vorticity**, thus has two components: **relative vorticity**, or the vorticity relative to Earth's surface, and **Earth vorticity**, which is due to Earth's daily rotation about its axis.

Relative vorticity depends on air motions with respect to Earth's surface, whereas Earth vorticity is a function solely of latitude—the higher the latitude, the greater the vorticity—with zero vorticity at the equator.[4] If the flow of air relative to the surface is in the same direction as the rotation of Earth itself (counterclockwise in the Northern Hemisphere), relative and Earth vorticity complement each other and increase the total or absolute vorticity (Figure 10–4). For this reason, counterclockwise rotation in the Northern Hemisphere is said to have *positive vorticity*, to be consistent with the convention used for the Coriolis force (see Chapter 4). Air rotating clockwise possesses *negative vorticity*.

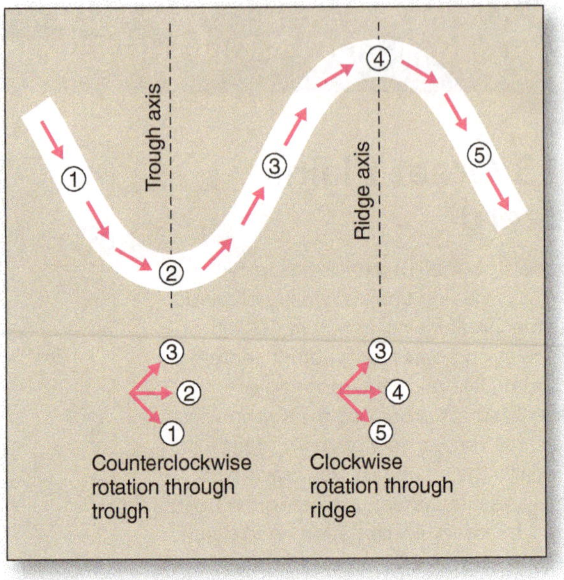

▲ **FIGURE 10–3 Vorticity Around a Rossby Wave.** As air flows from positions 1 to 3, it undergoes a counterclockwise rotation. Along the ridge, the air turns clockwise from positions 3 to 5. The bottom of the figure shows the wind vectors representing the flow at the five positions.

Figure 10–5 shows the trough from Figure 10–3 in greater detail so we can examine the air's vorticity. In segment 1, the air flows southeastward with no change in direction or speed. Because it undergoes no rotation, the air has zero relative vorticity. In segment 2, the air continuously turns to its left to yield positive relative vorticity. In segment 3, the air flows continuously toward the northeast and has no relative vorticity. Thus, the trough has three distinct regions: two with zero relative vorticity and one with positive relative vorticity. Two transition zones separate the regions of maximum and minimum (zero) relative vorticity. Across transition zone *A*, vorticity increases as the air flows, whereas across transition zone *B* the relative vorticity decreases. (The Rossby wave shown here covers a limited range of latitude. As a result, Earth vorticity changes only slightly in Figure 10–5, and changes in absolute vorticity correspond closely to changes in relative vorticity.)

[3]For this discussion, we are only concerned with rotation with respect to the local vertical (such as on a merry-go-round). *Vorticity* can also refer to rotation around a horizontal axis (such as with a horizontal roll of paper towels).

[4]Earth vorticity is proportional to the sine of the latitude. Thus, Earth vorticity is only 16 percent greater at latitude 55° (sin 55° = 0.819) than at 45° (sin 45° = 0.707). This means that over the midlatitudes, where Rossby waves are most likely to occur, the changes in Earth vorticity with latitude are fairly small.

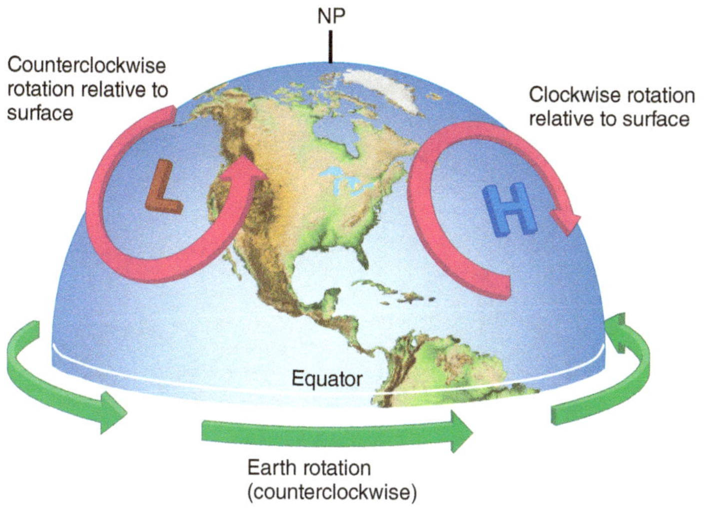

▲ **FIGURE 10–4 Earth Vorticity and Relative Vorticity of Air.** As the Northern Hemisphere rotates counterclockwise, it generates Earth vorticity. Relative vorticity is the rotation of air relative to the surface, without regard to the planet's rotation. Absolute vorticity is the sum of the two.

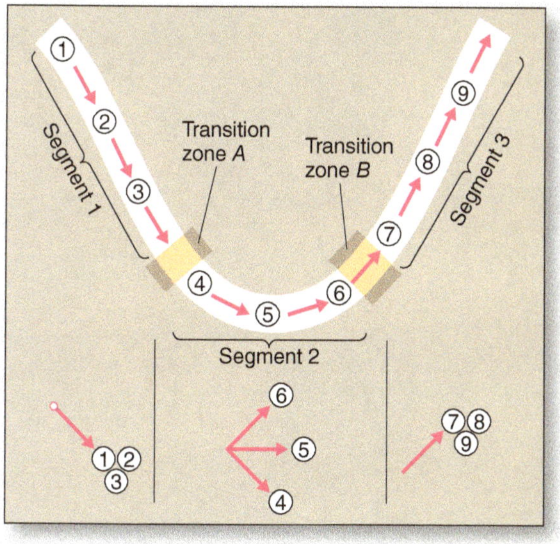

▲ **FIGURE 10–5 Change in Vorticity Along a Rossby Wave Trough.** As the air flows from positions 1 to 3, it undergoes little change in direction and thus has no relative vorticity. From positions 4 to 6, it turns counterclockwise and thus has positive relative vorticity. The air flows in a constant direction from positions 7 to 9. Thus, the trough has three segments based on vorticity, separated by two transition zones.

At this point you might logically ask, "So what?" The answer is that vorticity changes in the upper troposphere lead to pressure changes near the surface. Let's look at how. As you learned in Chapter 8, angular momentum is conserved in the absence of any outside forces. As a cowboy's twirling rope is pulled in, the reduction in the area swept by the rope causes it to twirl faster. The same thing happens when the vorticity or rotation of a parcel of air changes. That is, as the horizontal area occupied by an air parcel decreases by convergence, its vorticity or spin must increase, as in transition

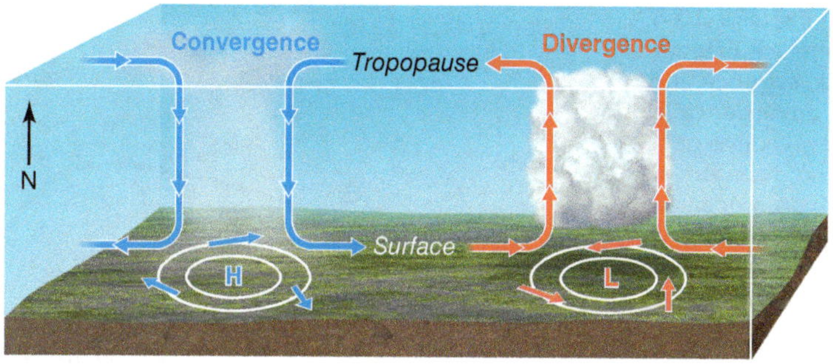

▲ **FIGURE 10–6 Divergence and Convergence.** The diagram shows the relationship between upper-level convergence and divergence and surface highs and lows.

zone *A*. Decreasing vorticity, as in transition zone *B*, leads to divergence. Conversely, an increase in absolute vorticity with respect to time leads to convergence. (For simplicity, we limit our discussion of divergence and convergence to horizontal changes in area.)

Divergence in the upper atmosphere, caused by decreasing vorticity, draws air upward from the surface and provides a lifting mechanism for the intervening column of air. This, in turn, can initiate and maintain low-pressure systems at the surface (Figure 10–6). The situation associated with the upper-level divergence and the surface low is shown in Figure 10–7. Conversely, increasing upper-level vorticity leads to convergence and the sinking of air, which creates high pressure at the surface. Surface low-pressure systems resulting from upper-tropospheric motions are referred to as **dynamic lows** (also called *cold core lows*), which are distinct from the **thermal lows** (*warm core lows*) caused by localized heating of the air from below. Cold core lows at the surface typically exist beneath regions of decreasing vorticity in the upper atmosphere, just downwind of trough axes. They are therefore associated with low pressure aloft, although the center of the low is not located directly above the surface low—instead the surface low is generally located to the east or southeast of the center of the upper-level trough. In contrast, warm core lows have high pressure aloft. This arises because they have a relatively weak decline in pressure with altitude due to their higher temperatures. Thus, at some height in the middle troposphere their pressure becomes greater than that of the surrounding air, and a zone of high pressure exists. (Hurricanes, discussed in Chapter 12, are classic examples of warm core lows.)

Figure 10–8 shows a typical relationship between the distribution of 500 mb heights (representative of the pressure pattern in the middle troposphere) and absolute vorticity. The areas of greatest vorticity (shaded blue) occur along the two trough axes (in this case, over northern California and the lower Mississippi Valley). Downwind of these zones, vorticity decreases very rapidly. Thus, as air flows away from the vorticity maxima, upper-level divergence occurs, which in turn promotes low pressure at the surface—the central component of the midlatitude cyclone.

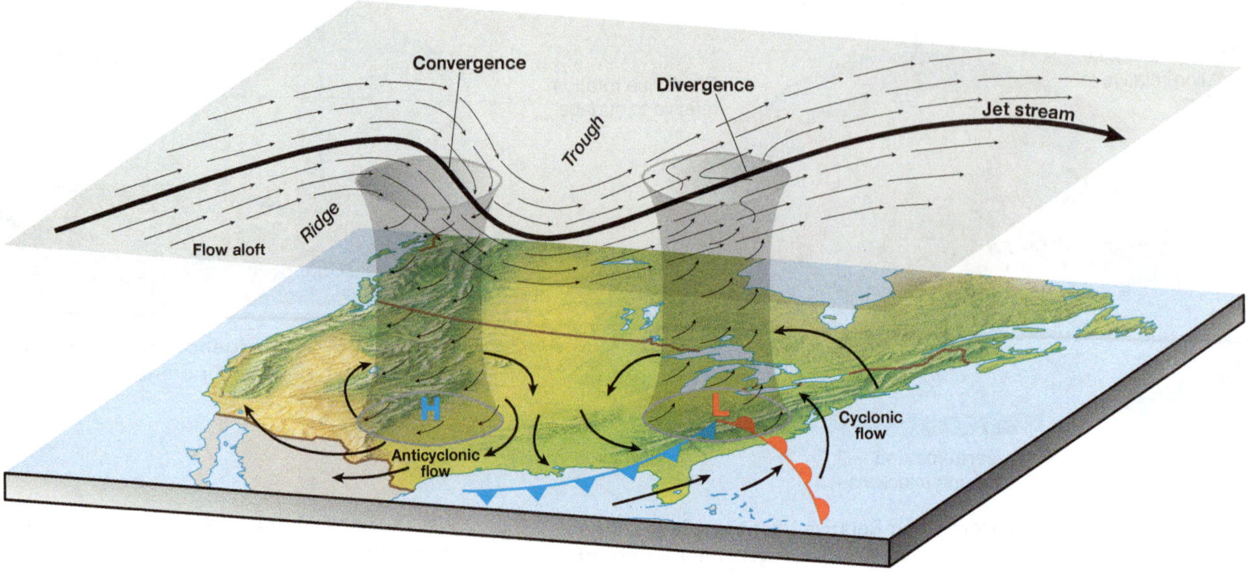

▸ **FIGURE 10–7 Surface Effects of Divergence and Convergence.** Upper-level convergence and divergence along favored positions on a Rossby wave create high and low pressures at the surface.

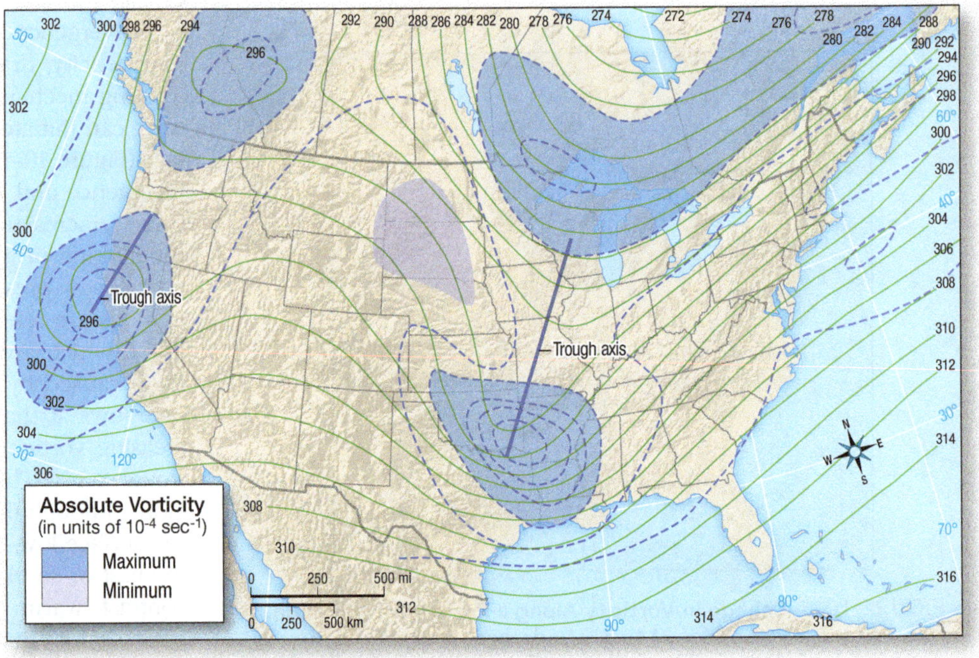

▸ **FIGURE 10–8 Absolute Vorticity.** Values of absolute vorticity are shown in dashed lines on a hypothetical 500 mb map. Notice that the greatest values appear near trough axes (vorticity in units of 10^{-4} sec^{-1}). As air flows away from the areas of maximum vorticity, the decrease in that property results in upper air divergence. In this instance, that would occur near the Northern California–Nevada border and Louisiana–Arkansas regions.

The region of lowest absolute vorticity (shaded in purple) occurs near the ridge axis, centered over the Dakotas. An area of increasing vorticity that is a likely center of high pressure at the surface exists just downwind of this region. (See *Box 10–2, Physical Principles: Vorticity and the Maintenance of Rossby Waves,* for further information on vorticity patterns.)

CHECKPOINT

10.5 What are positive vorticity and negative vorticity?

10.6 Discuss the role of vorticity in the formation of the cyclone shown in Figure 10–7.

Surface Fronts and Upper-Level Patterns

10.4 Analyze how surface fronts and upper-level patterns are related.

At this point, we know that upper-level divergence causes the formation and intensification of surface midlatitude cyclones, whereas upper-level convergence causes high pressure at the surface. We have also seen that the airflow along Rossby waves can generate upper-level divergence and convergence. (Further information on divergence and convergence is presented in *Box 10–3, Physical Principles: A Closer Look at Divergence and Convergence.*)

Vorticity and the Maintenance of Rossby Waves

We have seen that the vorticity associated with airflow has two components: Earth vorticity and relative vorticity.

Earth Vorticity

Earth vorticity arises from the planet's 24-hour rotation. A person sitting on a chair at the North Pole undergoes one complete rotation each day relative to Earth's axis and thus has maximum Earth vorticity. Like the Coriolis force, Earth vorticity increases with latitude so that it is at its maximum at either pole and is nonexistent at the equator.

Relative Vorticity

We define the second source of rotation, *relative vorticity*, in terms of motions of air relative to the surface. Relative vorticity itself has two sources. The first is the *shear* that occurs when the speed of a fluid varies across the direction of flow. Figure 10–2–1a illustrates this process as water flows through a channel (the same applies to air, but we use water as an example simply because it is easier to visualize). The flow is faster along the right bank. If a paddlewheel was fixed in the middle of this stream, the faster-moving water to the right would exert a greater force on the wheel than would the slower-moving flow on the left. This would cause the wheel to rotate counterclockwise and thereby undergo vorticity due to shear.

The second source of relative vorticity depends on the *curvature* of the flow, as shown in Figure 10–2–1b, where the curved channel forces a fluid to turn counterclockwise. As air flows through a Rossby wave, it undergoes this type of curvature. As we said, relative vorticity is the sum of curvature and shear; both are signed quantities. When they have the same sign, they act in the same direction. When the signs are different, they offset one another, perhaps completely. In general, shear is much less important to the occurrence of relative vorticity than curvature.

Absolute Vorticity and Conservation of Angular Momentum

Absolute vorticity, the overall rotation of a fluid, like angular momentum, is conserved—that is, in the absence of intervening forces, it remains constant. Figure 10–2–2 shows a Rossby wave in the Northern Hemisphere. As the air flows southward, west of the trough axis, its Earth vorticity decreases (recall that Earth vorticity decreases toward the equator). But because absolute vorticity is conserved, an increase in relative vorticity compensates for the decrease in Earth vorticity, causing the air to turn counterclockwise. Then, as it starts to flow poleward, the Earth vorticity increases. Thus, the air turns back to its right and once again exhibits negative relative vorticity. (For the sake of clarity, we are assuming no shear exists.)

In short, there is a constant trade-off between Earth and relative vorticity. As the air moves poleward, it assumes a greater clockwise rotation relative to the surface; as it moves toward the equator, it turns in a counterclockwise manner. Such reversals in relative vorticity, along with the conservation of angular momentum, help maintain Rossby waves.

10–2–1 What are the two sources of relative vorticity?

10–2–2 Explain why there is a trade-off between Earth vorticity and relative vorticity as air flows along a Rossby wave.

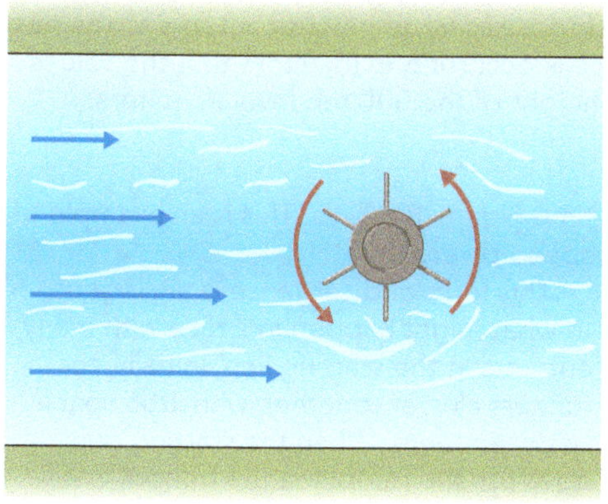

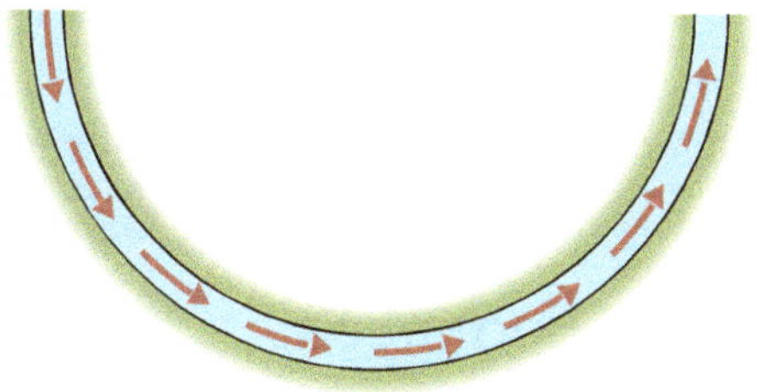

FIGURE 10–2–1 Relative Vorticity. (a) Relative vorticity by shear, occurs when a fluid moves at a differential speed across the direction of flow. A paddle wheel fixed across the fluid rotates counterclockwise as a greater forward stress is exerted to the right of the direction of flow. (b) Curvature in the direction of flow can also produce relative vorticity.

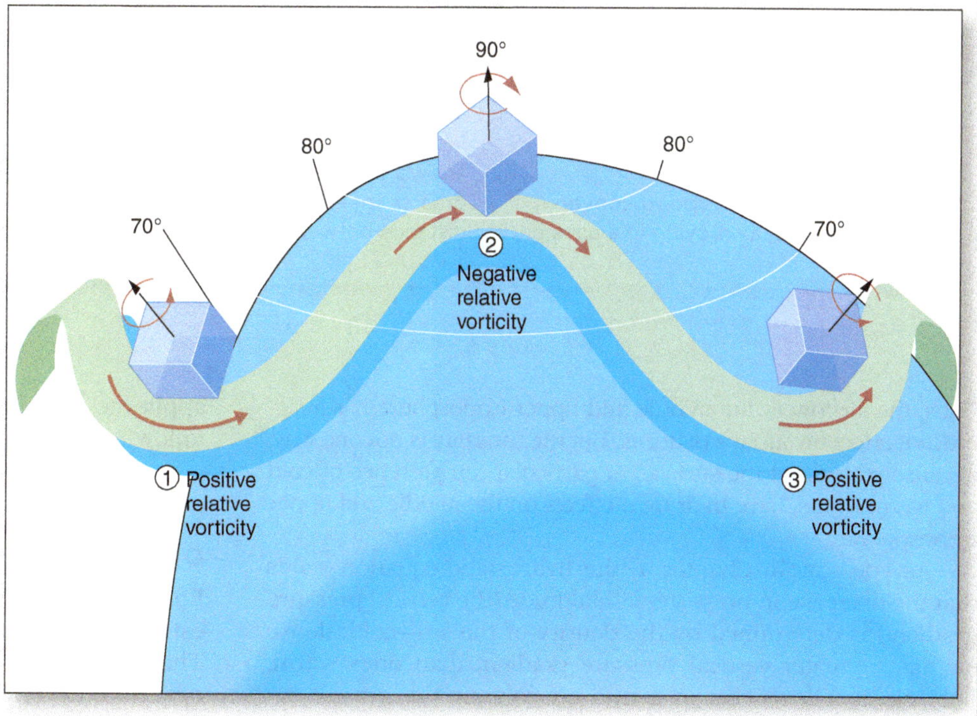

FIGURE 10–2–2 Maintenance of a Rossby Wave. As the air moves poleward from position 1, the gain in Earth vorticity is compensated for by a decrease in relative vorticity. As the parcel approaches position 2, the relative vorticity becomes negative and the parcel turns back equatorward. At position 3, the reduction of Earth vorticity causes an increase in relative vorticity, and the air again turns poleward.

10–3 PHYSICAL PRINCIPLES

A Closer Look at Divergence and Convergence

Upper-level divergence and convergence are changes in the horizontal area occupied by an air parcel and result from changes in vorticity as air flows. This relationship between divergence and vorticity can be summarized in the simple equation

$$-\frac{1}{\zeta}\frac{\Delta\zeta}{\Delta t} = div \quad \text{vorticity and divergence}$$

where $-\frac{1}{\zeta}\frac{\Delta\zeta}{\Delta t}$ is the standardized change (here, the decrease) in absolute vorticity (ξ) with respect to time (*t*), and *div* = divergence. If absolute vorticity increases, convergence must result. If absolute vorticity decreases, divergence must result.

Divergence and convergence can occur in two ways. The first is by an increase or a decrease in the speed of air as it flows. The second is by a stretching out or pinching inward of the air, in a direction perpendicular to the direction in which it is moving. The divergence and convergence described earlier in this chapter can take either form.

Speed Divergence and Speed Convergence

Speed divergence and **speed convergence** occur when air moving in a constant direction either speeds up or slows down. Consider the two parcels of air, *A* and *B*, in Figure 10–3–1a. Both parcels are moving in the same direction, but parcel *B* moves faster, as indicated by the length of the arrows. Because the leading parcel has greater speed than the one behind it, the distance between the two increases with time Figure 10–3–1b. This is an example of speed divergence.

This form of divergence is analogous to what might happen in a race with many entrants at the starting line. Initially, the runners cluster together, with little space between them. When the starting gun goes off, the people at the front of the

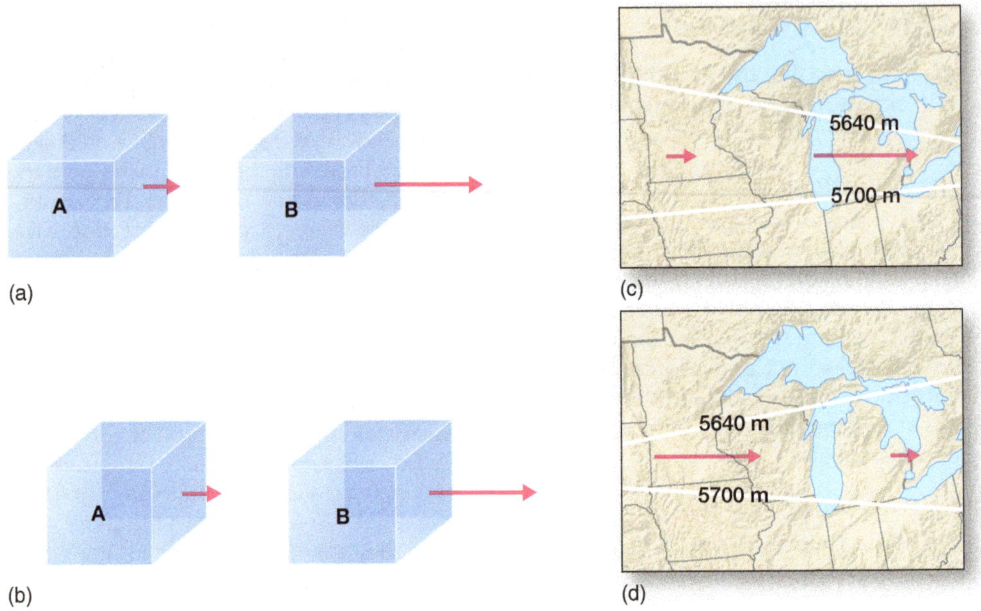

▲ **FIGURE 10–3–1 Speed Divergence and Convergence.** (a) Two hypothetical parcels of air are moving in the same direction, with the one in front moving faster. (b) At some interval of time later, the leading parcel has moved even farther ahead, creating speed divergence. This is also illustrated in (c), with the tighter spacing of height contours to the east creating speed divergence. Speed convergence is occurring in (d). Note that the values shown on the lines in (c) and (d) represent the height of the 500 mb level in meters.

Thus, the airflow in the middle and upper troposphere has a significant effect on surface patterns. But the causality is not one way, because patterns at the surface, in particular the presence of cold and warm fronts, have their own effects on the middle and upper troposphere.

As you saw in Chapter 4, the hydrostatic equation states that the decrease in pressure with altitude (the vertical pressure gradient) is determined by the density of the air—cold, dense air has a greater vertical pressure gradient than does warm, light air. It follows that in a cold air column, the greater decrease in pressure with altitude should lead to lower pressure aloft compared to warm air. The differences in temperature on either side of a cold front must therefore lead to significant differences in upper-level pressure there (the same reasoning applies to warm fronts). We develop this idea more fully in the following section.

Cold Fronts and the Formation of Upper-Level Troughs

Rossby waves consist of large, alternating troughs and ridges that establish patterns of upper-level divergence and convergence. The troughs in the waves normally develop behind the position of surface cold fronts, not by some grand coincidence but in response to the presence of the fronts. Figure 10–9 illustrates how this happens by showing the temperature and pressure changes in a 1 km thick layer of air on either side of a cold front. The air above point *A* lies entirely within the warm sector ahead of the pack dash away from those farther back, who shuffle along as they wait for the crowd to move forward. The cluster of people gradually thins out as the faster runners pull away from the slower ones, and the same number of people now occupy a greater area.

Because wind speed on an upper-level weather map is directly proportional to the spacing of height contours, we can use these maps to identify regions of speed divergence. Specifically, speed divergence occurs where contour lines come closer together in the downwind direction. In Figure 10–3–1c, the wind speed, indicated by the length of the arrows, increases in the direction of flow and causes speed divergence.

Speed convergence occurs when faster-moving air approaches the slower-moving air ahead. In the example of runners in the race, convergence might occur behind a muddy part of the track that slows the runners. The fastest runners, who have pulled ahead of the others, are the first to encounter the muddy spot. As they slog through the muck, the trailing runners have the opportunity to catch up. The entire pack bunches up in a smaller area and convergence occurs. A similar phenomenon is illustrated in Figure 10–3–1d.

Diffluence and Confluence

A second type of divergence and convergence, **diffluence** and **confluence**, occurs when air stretches out or converges horizontally due to variations in wind direction.

In Figure 10–3–2a, a certain amount of air is contained in the shaded area between points 1 and 3. As it passes to the region between points 2 and 4, the same amount of air occupies a greater horizontal area. This is diffluence, a pattern that commonly appears wherever vorticity changes cause divergence. Confluence is shown in Figure 10–3–2b.

Close inspection of Figure 10–3–2 reveals an interesting relationship between the different types of convergence and divergence. Diffluence occurs where height contours on an upper-level map spread apart in the upwind direction (confluence occurs where they converge). But where height contours are spread farther apart, there is a lesser horizontal pressure gradient, and therefore weaker winds. So diffluence (a type of divergence) in the upper figure occurs at the same time that speed convergence is occurring; that is, two opposite processes are occurring at once. We know that divergence occurs downwind of a trough axis. How does divergence actually occur downwind of a trough axis when diffluence and speed convergence occur simultaneously? The answer is that in most instances the diffluence is greater in magnitude than the accompanying speed convergence, so the sum of the two yields a net divergence.

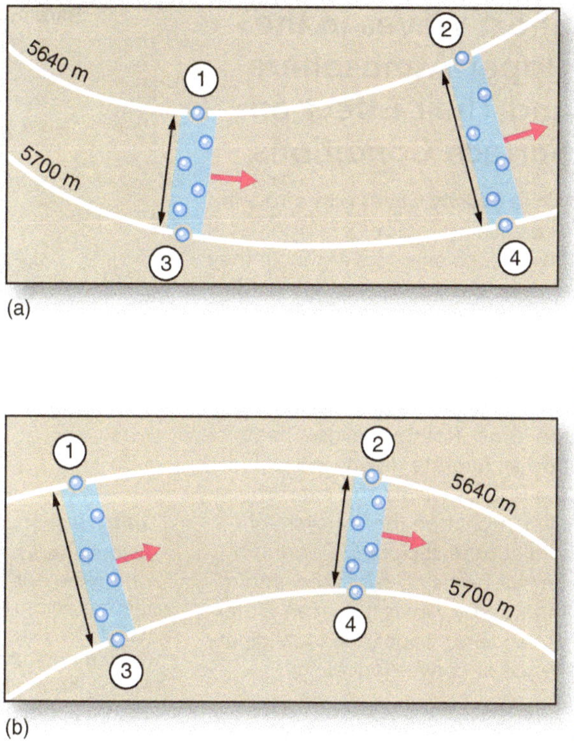

▲ **FIGURE 10–3–2 (a) Diffluence and (b) Confluence.**

10–3–1 What are diffluence and speed divergence?

10–3–2 Explain why you can have one kind of divergence and one kind of convergence occurring at the same location.

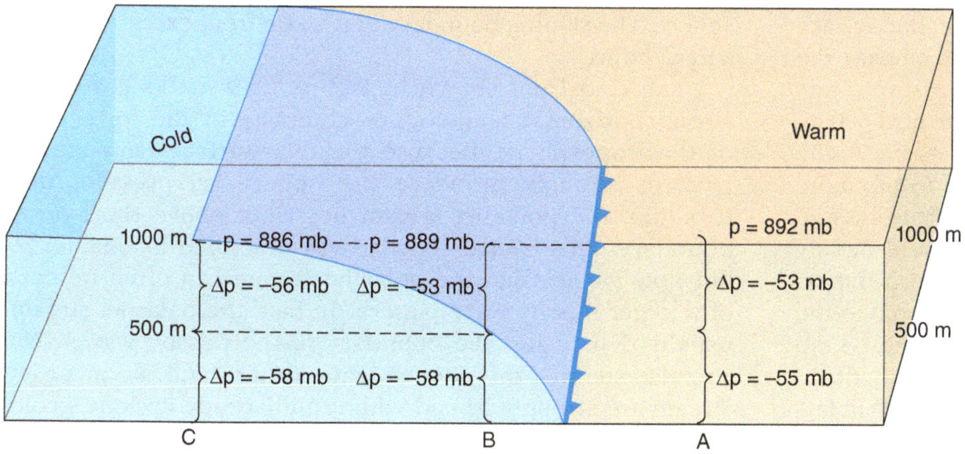

◀ **FIGURE 10–9 Effects of Temperature Distributions on Air Pressure.** Temperature distributions in the lower atmosphere lead to variations in upper-level pressure. Above *A*, the entire column of air in the lower atmosphere is warm, so the pressure drops relatively slowly with height. At *B*, cold air occupies the lowest 500 m, with warmer air aloft. This leads to a slightly lower pressure at the 1 km level. At *C*, cold air occupying the lowest 1000 km causes a greater rate of pressure decrease with altitude and, as a result, a lower pressure at the 1 km level. In this way, the existence of sloping frontal boundaries establishes horizontal pressure gradients in the upper and middle atmosphere.

10–4 FORECASTING

Short Waves in the Upper Atmosphere and Their Effect on Surface Conditions

While Rossby waves play a role in establishing regions of upper-level divergence and convergence, they are not the only waves in the atmosphere that do so. The atmosphere also contains smaller eddies. Some of these, called **short waves**, are smaller ripples superimposed on the larger Rossby waves. These eddies migrate downwind within the Rossby waves and exert their own impact on the life cycle of midlatitude cyclones. Depending on where they are located within the Rossby waves, they can either enhance or reduce the local divergence or convergence.

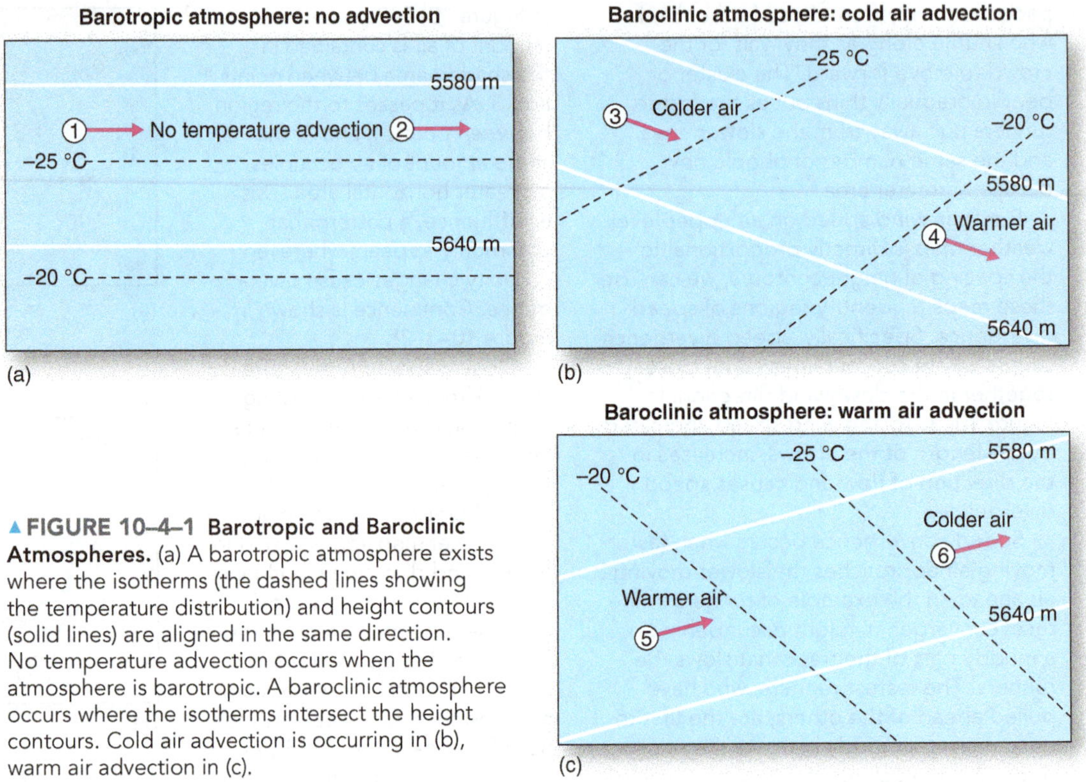

▲ **FIGURE 10–4–1 Barotropic and Baroclinic Atmospheres.** (a) A barotropic atmosphere exists where the isotherms (the dashed lines showing the temperature distribution) and height contours (solid lines) are aligned in the same direction. No temperature advection occurs when the atmosphere is barotropic. A baroclinic atmosphere occurs where the isotherms intersect the height contours. Cold air advection is occurring in (b), warm air advection in (c).

The Role of Temperature Advection

The formation of short waves depends on **temperature advection**, the horizontal transport of warm or cold air by the wind. Because air in the upper troposphere is well removed from the direct source of atmospheric heating (the surface), the temperature changes we experience from day to night are barely perceptible in the upper atmosphere. Upper-level air changes temperature very slowly as it moves from one region to another, and air flowing horizontally from a warm region can retain its high temperature as it moves into a region otherwise occupied by cold air. We refer to the horizontal movement of relatively warm air as **warm air advection**. The opposite, of course, is called **cold air advection**. Both types of temperature advection appear on Rossby waves and, as we will see, affect the development of surface cyclones.

Video
Short Waves and Long Waves
http://goo.gl/WLIc81

front. The frontal boundary slopes backward so that at *B* the air is cold in the lowest 500 m and warm in the upper 500 m. Cold air occupies the entire kilometer-thick layer above *C*.

Now let's assume (for simplicity's sake) that the surface pressure is 1000 mb at all three locations and compare the vertical pressure distributions. Above *A*, the pressure drops 55 mb in the lowest 500 m and another 53 mb in the next 500 m, to yield a pressure of 892 mb at the 1 km level. Above *B*, the pressure drops 58 mb in the cold, lowest 500 m (3 mb more than that above *A*). (These different rates of vertical pressure decrease occur because air pressure decreases more rapidly with height through cold air than through warm air.) But, as in the upper 500 m above *A*, the pressure drops 53 mb to the 1 km level. Thus, the pressure at that height is 889 mb—3 mb less than the pressure at *A*. Over *C*, the pressure drops 58 and 56 mb, respectively, in the lower and upper 500 m layers, so that the pressure at 1 km is 886 mb. Thus, the pressure at 1 km decreases from 892 to 889 to 886 mb across the frontal boundary. This is how the differing temperature characteristics cause a horizontal pressure gradient in the middle atmosphere. The sloping boundary of a warm front exerts the same sort of effect.

The cold front shown in Figure 10–9 marks a zone of strong horizontal temperature contrasts in the lower half of the troposphere. Because the near-surface temperature patterns strongly influence the upper-level pressure distribution, the polar jet stream lies right above the frontal boundary. This is exactly what was described in Chapter 8, but now we see that the polar jet stream can often be part of a larger Rossby wave pattern. In fact, the polar jet stream usually demarcates the boundary of the Rossby waves. We can also see that this Rossby wave pattern and the polar jet stream are strongly linked with a midlatitude cyclone's cold and warm fronts.

Detecting Warm and Cold Air Advection

A useful method for detecting warm and cold air advection on a map of the upper atmosphere is to compare the orientation of height contours and isotherms. Figure 10–4–1 illustrates three possible patterns of 500 mb height levels and temperature advection. In Figure 10–4–1a, parallel height contours (solid lines) are aligned in a west-to-east direction so that a geostrophic wind flows from west to east. The isotherms (the dashed lines showing the temperature distribution at the 500 mb level) run parallel to the height contours and indicate a northward decrease in temperature. Because the airflow is parallel to the height contours (and in this case the isotherms), the temperature is the same (just less than –25 °C) at positions 1 and 2, and there is neither cold nor warm air advection. When the height contours and isotherms are in alignment, the atmosphere is said to be **barotropic**.

In Figure 10–4–1b, the height contours are parallel to each other, as are the isotherms. But in this instance the height contours and isotherms intersect each other, and the temperature increases from position 3 (below –25 °C) to position 4 (above –20 °C). This is an example of cold air advection, wherein a parcel of colder air is transported from 3 to 4. The opposite situation, warm air advection, occurs in Figure 10–4–1c, where the temperature decreases in the direction of airflow. When the height contours and the isotherms intersect, as in both parts (b) and (c), the atmosphere is said to be **baroclinic**.

Effects of Advection in a Rossby Wave

Refer to Figure 10–4–2 and observe the two baroclinic zones at positions 1 (cold advection) and 2 (warm advection). Where cold advection exists, the entering air is denser than the air ahead of it because of its lower temperature. This gives it a negative buoyancy that causes it to sink downward, bringing cold air toward the surface. Cold air advection typically occurs behind a cold front, thereby enhancing the temperature contrast found on either side of the front. Where warm advection occurs, entering air is warmer and more buoyant than the air ahead of it and therefore rises. The warm and cold air advection thus cause vertical motions similar to those associated with statically unstable air (Chapter 6). This situation is called **baroclinic instability**, an important mechanism in creating low pressure at the surface.

In addition to undergoing rising or sinking motions, the air in areas of warm or cold air advection undergoes a slight turning—to the right in areas of cold air advection and to the left in regions of warm air advection. These motions are what cause the ripples (short waves) to form on the Rossby waves. When a short wave is located downwind of a Rossby wave trough axis, the divergence is enhanced and surface cyclones intensify.

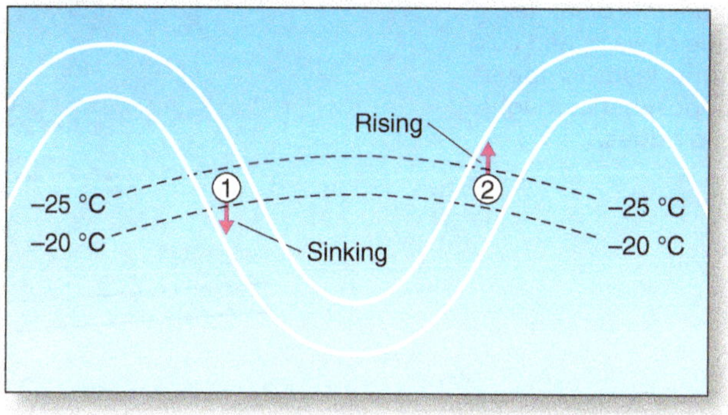

▲ **FIGURE 10–4–2 Warm and Cold Air Advection Around a Rossby Wave.** The air at position 1 flows from colder to warmer air, resulting in cold air advection. Along a zone of cold air advection, sinking motions and a turning of the air to the right tend to take place. Warm air advection occurs at position 2 along with a rising of the air.

10–4–1 What are warm and cold air advection?

10–4–2 Explain what baroclinic instability is and where it would be identified on a 500 mb map.

Interaction of Surface and Upper-Level Conditions

Despite the fact that we commonly refer to the "surface level" and the "upper level" of the atmosphere, they are *not* separate entities; they are simply different parts of a single atmosphere that are fully connected and intertwined with each other. The upper-level winds influence surface conditions by generating divergence and convergence that lead to the formation of surface cyclones and anticyclones (Chapter 4). At the same time, the temperature patterns in the lower atmosphere affect the rate at which pressure decreases with altitude and thereby influence the upper-level wind flow. More specifically, upper-level patterns with strong north–south components (meridional flow) cause the formation of midlatitude cyclones downwind of trough axes (where vorticity decreases). At the same time, the presence of fronts can cause Rossby waves in the upper atmosphere. The usual juxtaposition of surface cold fronts and upper-level troughs and ridges is shown in Figure 10–10.

The bottom line is that the interconnectedness between surface patterns and those aloft provides the true foundation for understanding the life cycle of midlatitude cyclones. While the model of cyclogenesis, maturity, and occlusion described by Bjerknes and his colleagues provided an excellent *description* of the life cycle of midlatitude cyclones, the *explanation* behind the processes eluded the early scientists. Today we know that upper-level patterns and their associated divergence and convergence affect pressure distributions (and hence the changes in midlatitude cyclones) at the surface. Furthermore, we know that the upper-level patterns are influenced in turn by temperature conditions near the surface. (See *Box 10–4, Forecasting: Short Waves in the Upper Atmosphere and Their Effect on Surface Conditions,* for a more detailed analysis of upper-level and surface phenomena.)

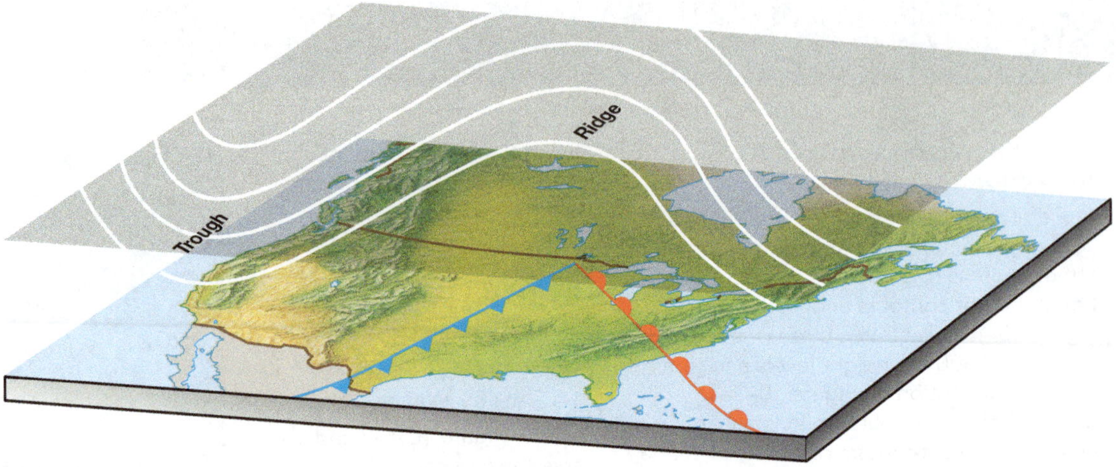

▶ **FIGURE 10–10 Troughs and Ridges in Relation to Fronts.** The effect of differing vertical pressure gradients on either side of warm and cold fronts leads to upper-tropospheric troughs and ridges.

CHECKPOINT

10.7 Assuming the same pressure at the surface, which would have higher pressure at an altitude of 1 km: a cold air column or a warm air column? Explain.

10.8 Why does the polar jet stream lie above the frontal boundary?

DID YOU KNOW?

People are often surprised to learn that the National Oceanic and Atmospheric Administration (NOAA) routinely flies planes into active hurricanes as part of their tracking and forecasting procedures. This sounds risky indeed. Yet the maximum winds in those hurricanes are often only half as strong as those of the polar jet stream that lies above the surface fronts—and commercial aircraft fly through these jet streams every day!

An Example of a Midlatitude Cyclone

10.5 Describe the behavior of a typical midlatitude cyclone as it crosses North America.

The descriptions and examples of troughs, ridges, cyclones, anticyclones (each described in Chapter 4) and fronts (Chapter 9) represent idealized patterns. But the real atmosphere seems not to have learned the lessons presented in meteorology textbooks, and real-life conditions often depart from these idealized examples. The following example from 1994 is a fairly typical midlatitude cyclone making its trek across North America.

April 15

Figure 10–11 shows the surface and 500 mb weather maps and a satellite image of North America on April 15. Clear skies, low humidities, and light winds dominate the western United States and southwest Canada. In contrast, the midlatitude cyclone over the north-central part of the United States has brought strong winds, overcast skies, and heavy rain showers. The surface wind in the warm sector flows northward out of the southern states and turns somewhat to the northwest as it approaches the center of low pressure. North and west of the system, the air rotates counterclockwise around the low. Temperatures in the warm sector are typically in the 60s to low 70s Fahrenheit, considerably greater than those to the west of the cold front.[5]

At the 500 mb level (Figure 10–11b), a well-defined trough exists to the west of where the surface cold front is located. Strong westerly winds in excess of 100 km/hr (60 mph) occur over the western portion of the Canadian–United States border and then become northwesterly as they enter the trough. Downwind of the trough axis, the air again flows eastward toward the North Atlantic.

Just downwind of the trough axis, the vorticity decreases over northern Missouri, Iowa, and Minnesota. As we would expect, diffluence in this region leads to net divergence aloft and low pressure at the surface. Out west, downwind from the ridge axis, upper-level convergence leads to high pressure at the surface centered over central Idaho.

April 16

During the next 24 hours (Figure 10–12), the midlatitude cyclone migrates some 800 km (500 mi) to the northeast. It also intensifies, with about a 5 mb decrease in sea-level pressure from the previous morning. For the first time, a portion of the cyclone becomes occluded north of the Great Lakes and rain showers cover southeast Canada and the northeastern

[5]Observed temperatures, dew points, wind velocities, and sea-level pressures are depicted on the surface map for a few stations. Note that the temperatures and dew points (°F) are shown on the upper and bottom left, respectively, of the so-called *station models*. A line coming out of the circle shows the direction from which the wind is blowing, and the number of short and long tick-marks attached near the end of the line represents the wind speed. *Appendix C: Weather Map Symbols* provides more detailed information.

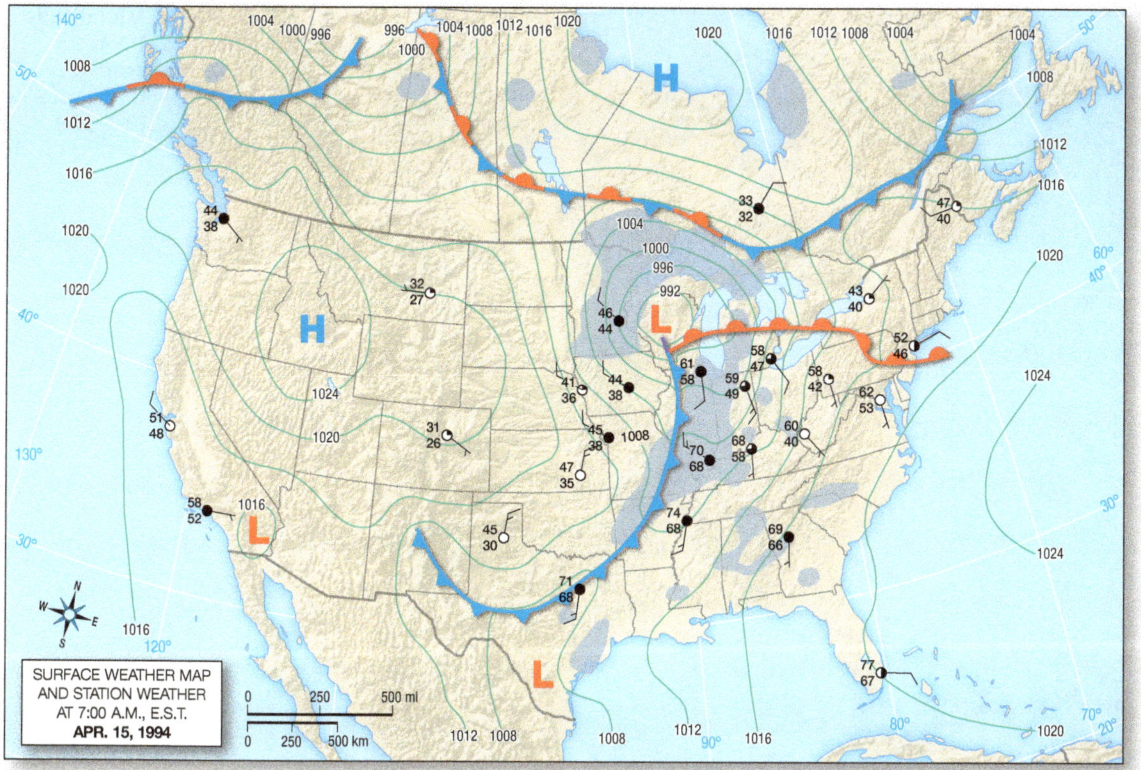

(a)

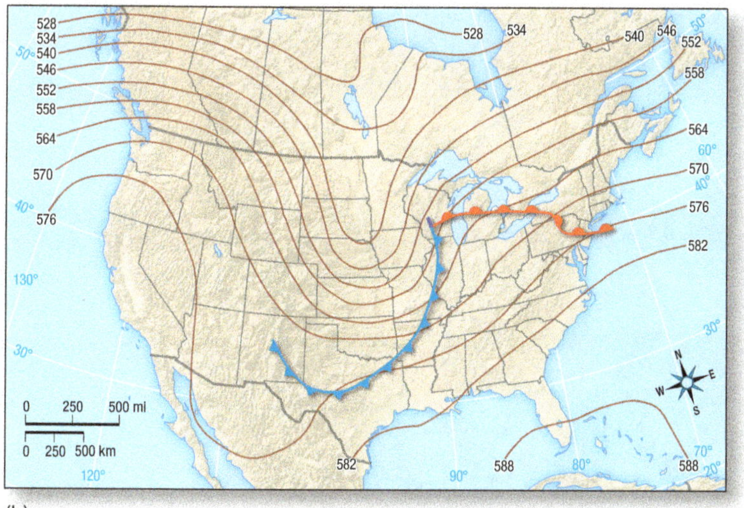

(b)

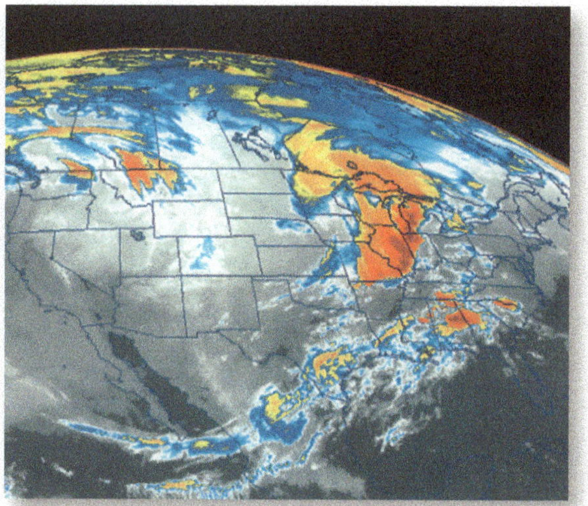

(c)

◀ **FIGURE 10–11 Midlatitude Cyclone: April 15, 1994.** (a) Weather maps of the surface and (b) 500 mb level, and (c) a satellite image for April 15, 1994, 7 A.M. EST. Note that the positions of the surface fronts associated with the midlatitude cyclone have been superimposed on the 500 mb map. Precipitation occurs in shaded areas.

United States. The anticyclone over the western United States expands southeastward and intensifies slightly to a maximum sea-level pressure of about 1027 mb.

The 500 mb pattern (Figure 10–12b) shows considerable change from the day before. The trough has moved east, and its axis changed from the north–south orientation of the day before to a northwest–southeast orientation. The trough has also intensified, with peak winds now reaching about 150 km/hr (90 mph) over the east-central United States. Another interesting change during the past 24 hours is the closing of the 5280 m contour over southern Ontario to form what is called a **cutoff low**. Although the main flow of air loops around the low and eventually flows off the map, a circular rotation of the air is also embedded in the trough.

April 17

During the 24 hours preceding the morning of April 17, the surface low-pressure center migrated a short distance to the northeast, with no change in central pressure (Figure 10–13). At the same time, the occluded front swept northward so that it is situated in a nearly west-to-east direction over southeast Canada. Precipitation,

◂ **FIGURE 10–12 Midlatitude Cyclone: April 16, 1994.** (a) Surface weather map, (b) 500 mb map, and a (c) satellite image as in Figure 10–11, for April 16.

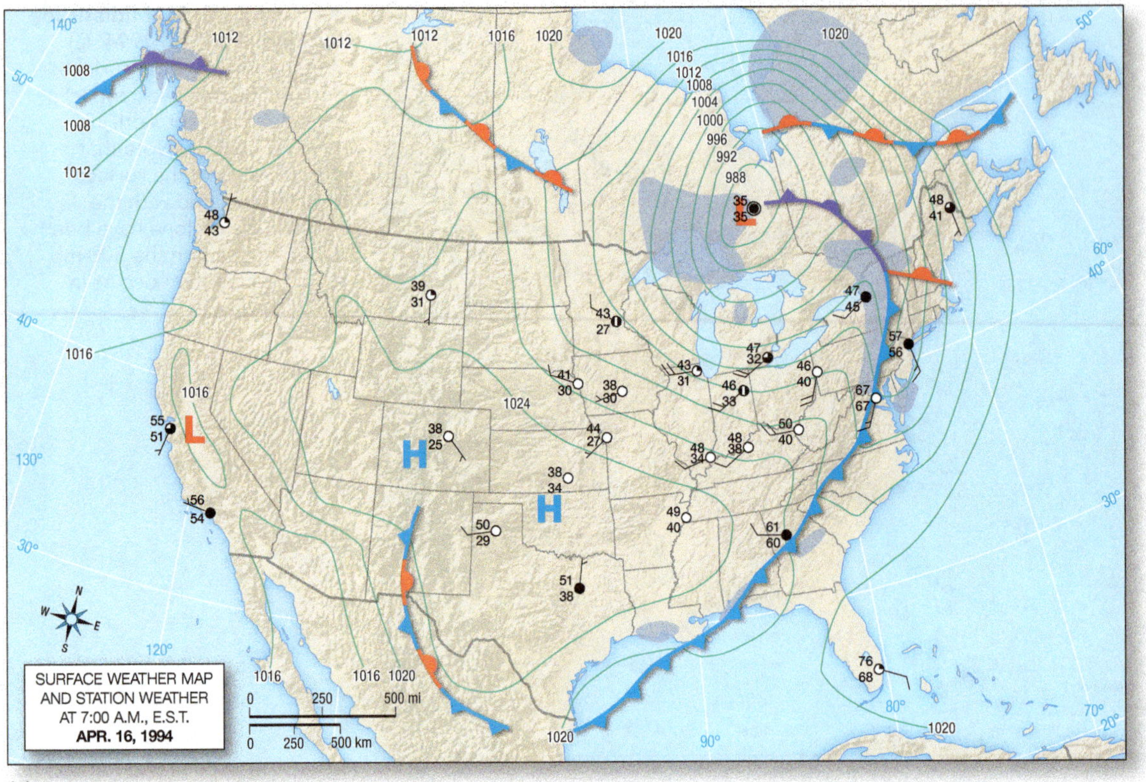

(a)

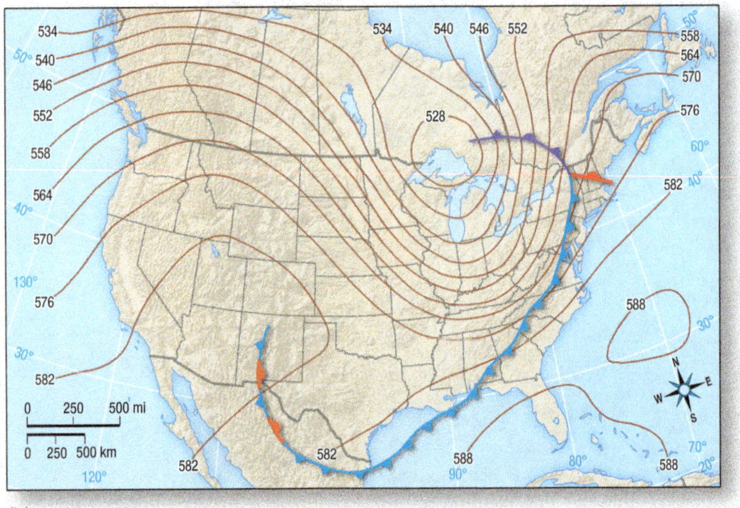

(b)

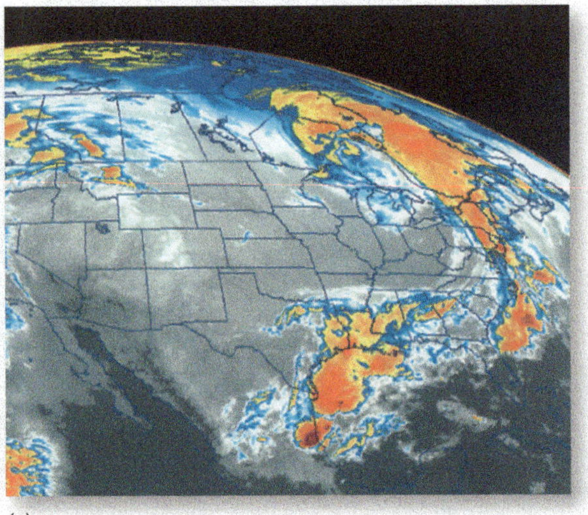

(c)

mainly as snow, is scattered along the frontal boundary and concentrated near the low-pressure center east of Hudson Bay.

The upper-level low (Figure 10–13b) has continued to deepen from the day before, as indicated by the lower heights of the 500 mb level. Not only has the height of the 500 mb level at the center of the cutoff low decreased by about 90 m, but the number of closed height contours has increased to four (from the previous one).

April 18

By April 18 (Figure 10–14), the center of the low-pressure system and most of the frontal boundaries have migrated off the surface map, but the system still is evident on the 500 mb map as the large trough extending to the southeast. As far as most of the population of eastern Canada and the United States is concerned, the midlatitude cyclone no longer exerts any direct influence on the weather.

CHECKPOINT

10.9 Referring to Figure 10–11, describe the situation that existed on the morning of April 15 at the surface and at the 500 mb level.

10.10 How closely does the polar jet stream correspond to the location of the major fronts of the storm over the entire period depicted in Figures 10–11 through 10–14?

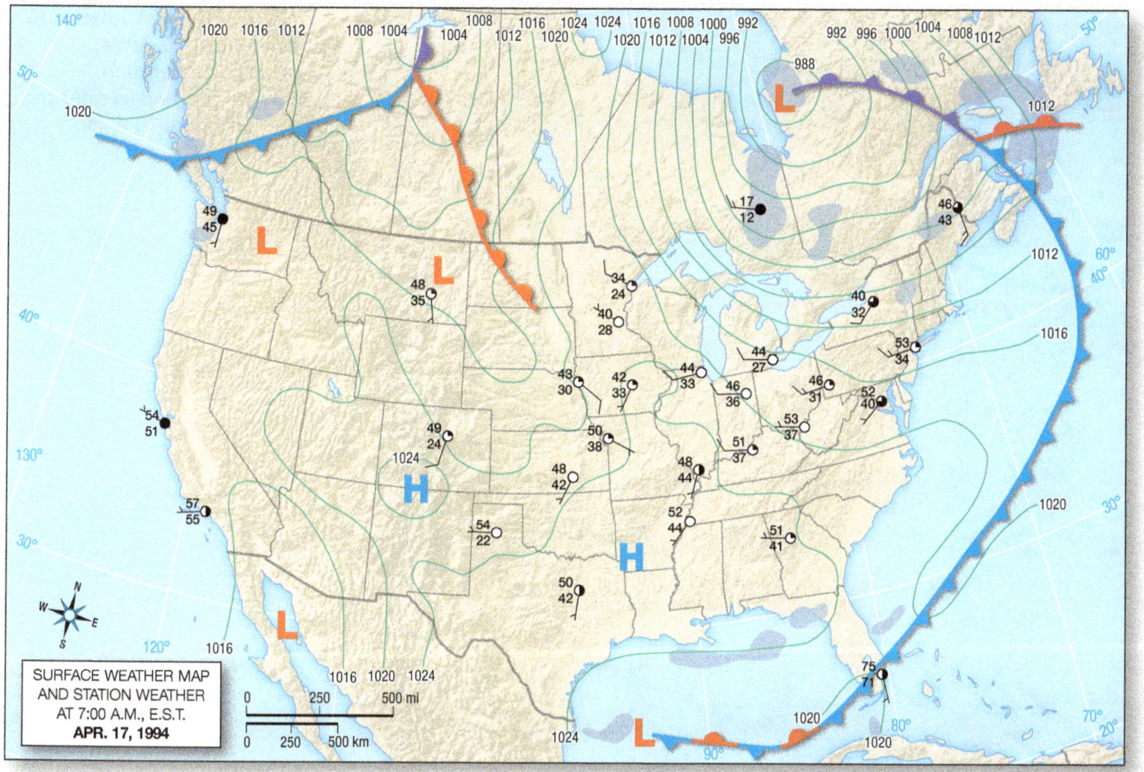

(a)

◂ **FIGURE 10–13 Midlatitude Cyclone: April 17, 1994.** (a) Surface weather map, (b) 500 mb map, and (c) satellite image as in Figure 10–11, for April 17.

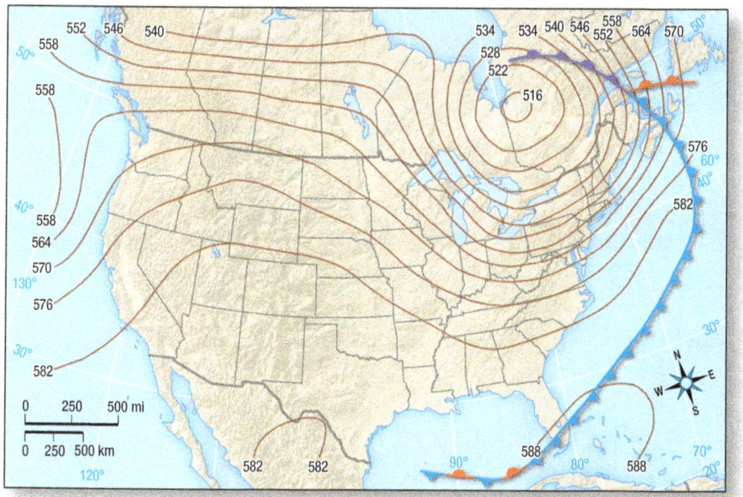

(b)

(c)

Flow Patterns and Large-Scale Weather

10.6 Explain how flow patterns and large-scale weather patterns affect the development, steering, and dissipation of midlatitude cyclones.

We have seen that changes in upper-level vorticity create divergence and convergence patterns that influence the formation, intensification, and dissipation of surface cyclones and anticyclones. Thus, strong looping motions of upper-level air are likely to create distinct regions of high and low pressure at the surface. Conversely, air over a large area such as North America flowing in a straight westerly direction will have uniform vorticity, with little chance for significant upper-level divergence or convergence. Without upper-level divergence or convergence, no major areas of high or low pressure develop at the surface.

Compare the 500 mb maps in Figure 10–15. In (a), the height contours exhibit a zonal pattern with a minimum of north–south displacement. In contrast, the pattern in (b) shows a strong meridional component.

Because they have no pronounced vorticity changes, zonal patterns (Chapter 8) hamper the development of intense cyclones and anticyclones. They are therefore more often associated with a large-scale pattern of light winds, calm conditions,

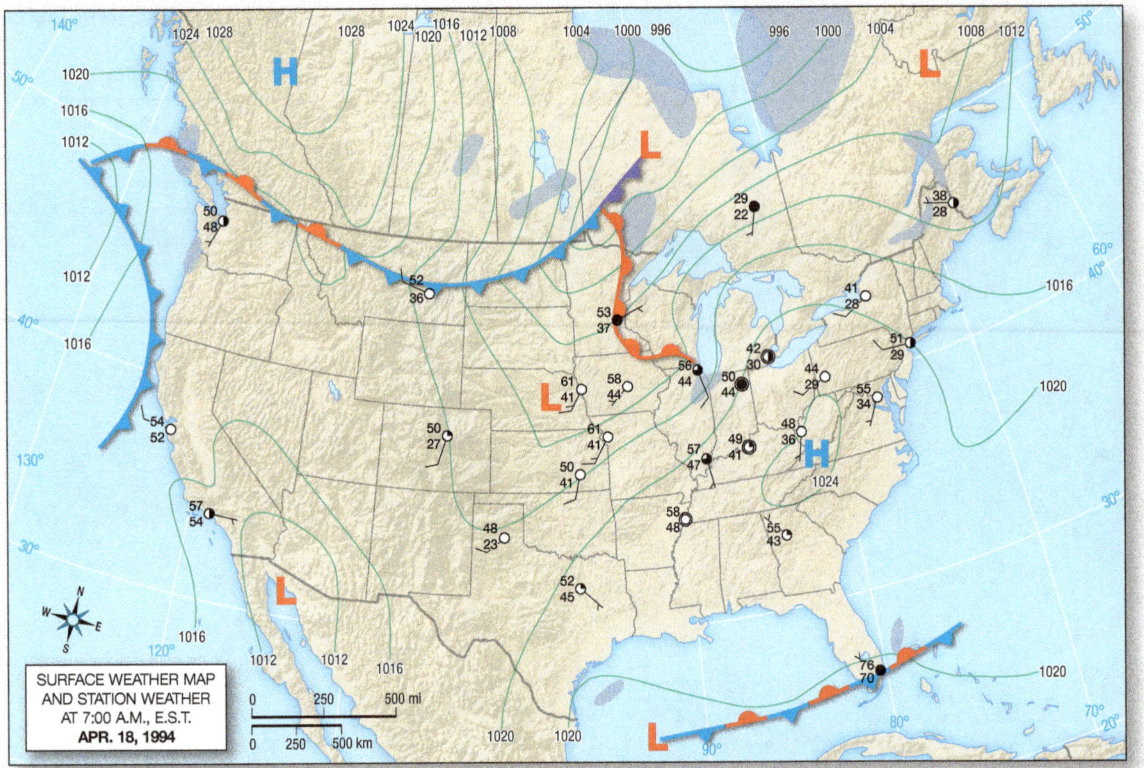

(a)

◂ **FIGURE 10–14 Midlatitude Cyclone: April 18, 1994.** (a) Surface weather map, (b) 500 mb map, and (c) satellite image as in Figure 10–11, for April 18.

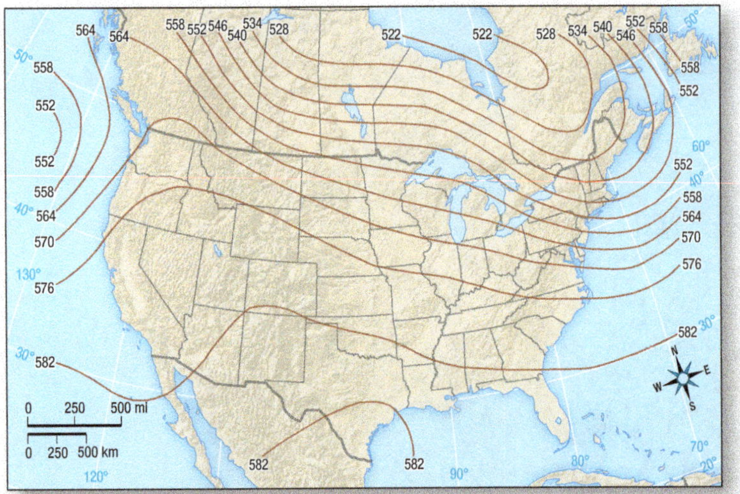

(b)

(c)

and no areas of widespread precipitation. Certainly there may be areas of localized precipitation—and the precipitation may even be quite heavy, as when orographically produced—but this activity will be spotty and widely scattered. Meridional flow, in contrast, can lead to the formation of major cyclones and anticyclones. If you look at an upper-level weather map and see strongly meridional flow, you can expect that some areas are experiencing cloudy and wet conditions while others are calm and dry. If you see a zonal pattern, it is less likely that large temperature contrasts exist from place to place or that there are large areas of heavy precipitation.

Experience shows that large-scale wind patterns in the upper atmosphere often persist, with one general type of pattern dominating for weeks or longer at a time. Such persistence of a zonal or meridional pattern can lead to droughts or episodes of heavy precipitation. A persistent zonal pattern can cause very widespread droughts due to the lack of vorticity. Regional droughts can also occur if a meridional pattern remains in place, with the zone of upper-level convergence downwind of a ridge axis persisting over a particular region.

The Steering of Midlatitude Cyclones

Upper-level winds have another important effect on surface conditions by governing the direction and speed at which the surface systems move. Outside of the tropics, the upper atmosphere includes a strong component of west-to-east flow. Likewise, experience tells us that both cyclones and anticyclones outside of the tropics typically migrate eastward. These

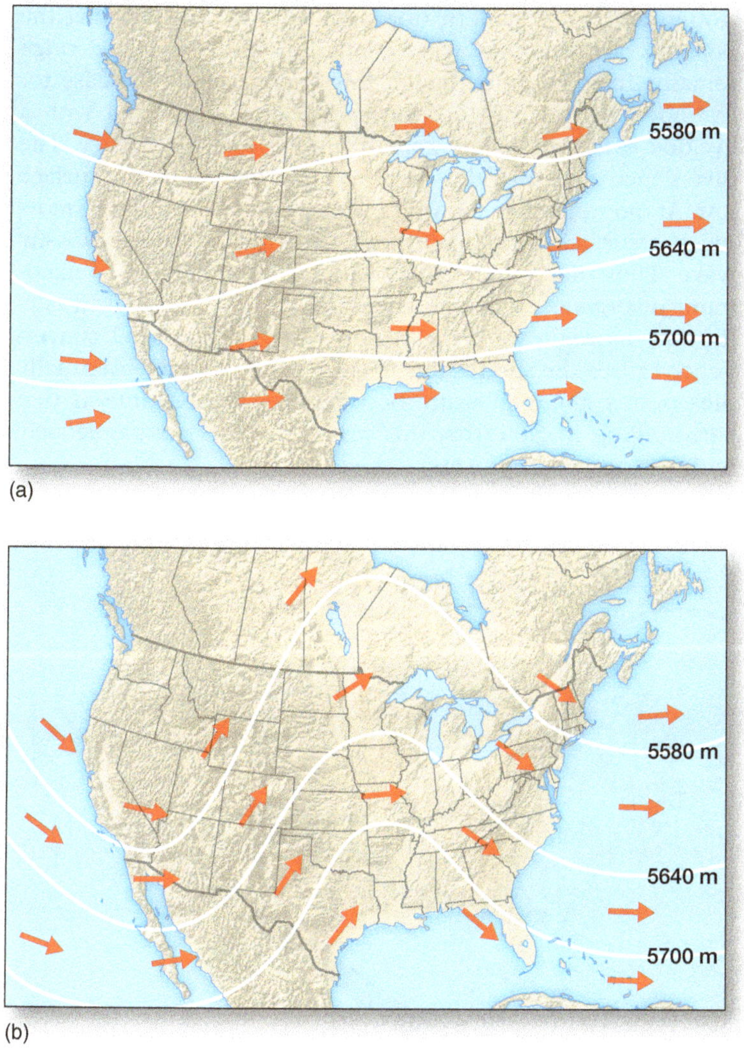

▲ **FIGURE 10–15 Zonal and Meridional Flow Patterns.** (a) Zonal flow is associated with light winds and relatively little precipitation. (b) Meridional airflow has a greater likelihood of generating large-scale precipitation.

two facts are not mere coincidence. In fact, the movement of surface systems can be predicted by the 500 mb pattern, with the surface systems moving in about the same direction as the 500 mb flow, at about one-half the speed. Keep in mind, however, that the 500 mb level wind pattern changes through time, so predicting the track of a cyclone involves more than just examining the current upper-level flow and assuming a constant movement parallel to the current height contour pattern; one must also predict the change in the 500 mb pattern.

Many midlatitude cyclones have their origin over the north Pacific off the coast of Japan. Upon reaching the Aleutian Islands of Alaska, the systems can die out, migrate toward the southeast, or continue on an eastward path across British Columbia. The most likely path the cyclones take upon reaching North America varies with the season, with northern treks favored in the summer, and movement toward the southeast more likely in the winter.

Upper-level winds are about twice as vigorous on average in the winter than in the summer. During the winter, net radiation decreases rapidly with increasing latitude, giving rise to a stronger latitudinal temperature gradient than in summer

DID YOU KNOW?

At one point the famous "storm of the century" that hit the eastern United States and Canada in March 1993 recorded a central barometric pressure of 960 mb—lower than that associated with 80 percent of the hurricanes and tropical storms that make landfall in the United States. It was also an unusually large midlatitude cyclone that affected 26 of the 50 United States, spawning blizzards that shut down virtually every airport on the eastern seaboard.

Video
Winds During the Floods of 1993
http://goo.gl/AJBtvB

(Chapter 3). This results in greater pressure gradients (and winds) in the upper atmosphere. It is no surprise that midlatitude cyclones generally move faster in the winter.

Though a winter midlatitude cyclone can take many different paths across North America, two are particularly common: the Alberta Clipper and the Colorado Low (Figure 10–16). The Alberta Clipper is associated with zonal flow and a polar jet stream that sweeps across southern Canada and the northern United States. Though it can bring frigid conditions, snowfall is usually light. In contrast, some midlatitude cyclones passing farther to the south over western North America spawn new centers of low pressure as they pass over the central Rocky Mountains. They then follow a path from the southern Plains toward the northeastern United States and eastern Canada. These storms, usually warmer and containing greater amounts of water vapor in the air, often produce extremely heavy snowfall.

Storms occasionally have their genesis well to the south of the polar front (in contrast to the original model of the early Norwegian meteorologists) and can track northward along the eastern United States. Such storms often have strong uplift and high water vapor contents—conditions favorable for the development of extremely heavy snowfalls. Such conditions led to the "storm of the century" in March 1993, an enormously powerful nor'easter (Chapter 9) that produced strong winds and record-breaking snow accumulations over the eastern third of the United States.

▲ **FIGURE 10–16 Typical Winter Storm Paths.** The map shows the paths of the Colorado Low and Alberta Clipper across North America.

Migration of Surface Cyclones Relative to Rossby Waves

For a midlatitude cyclone to form, upper-level divergence must occur. If there is more divergence aloft than convergence near the surface, the surface low deepens and a cyclone forms. If the convergence at the surface exceeds the divergence aloft, more air flows into the low than exits it, and the low weakens and dies out.

Although the optimal place for midlatitude cyclones to develop is just beneath the zone of decreasing vorticity aloft, they don't usually remain in a fixed position relative to the upper-level trough. Instead, they are usually pushed along so that they migrate in the same direction (and at about half the speed) as the winds at the 500 mb level. Figure 10–17 shows the movement of a surface cyclone relative to an upper-level trough and ridge pattern (for simplicity, we assume that the position of the Rossby wave containing the trough and ridge remains fixed in time, though this is not usually the case for extended periods). In (a), the surface low associated with a cyclone exists under the zone of upper-level divergence. The divergence aloft provides the uplift that maintains the surface low. At the same time, however, the upper-level winds are moving the surface system northeastward relative to the Rossby wave. Thus the cyclone moves away from the region of maximum divergence aloft and therefore weakens (b). Eventually the low moves toward the region where upper level convergence causes the surface low to fill, and the cyclone eventually dies out as its air pressure becomes nearly the same as that surrounding it. Note that this process occurs simultaneously with the occlusion process.

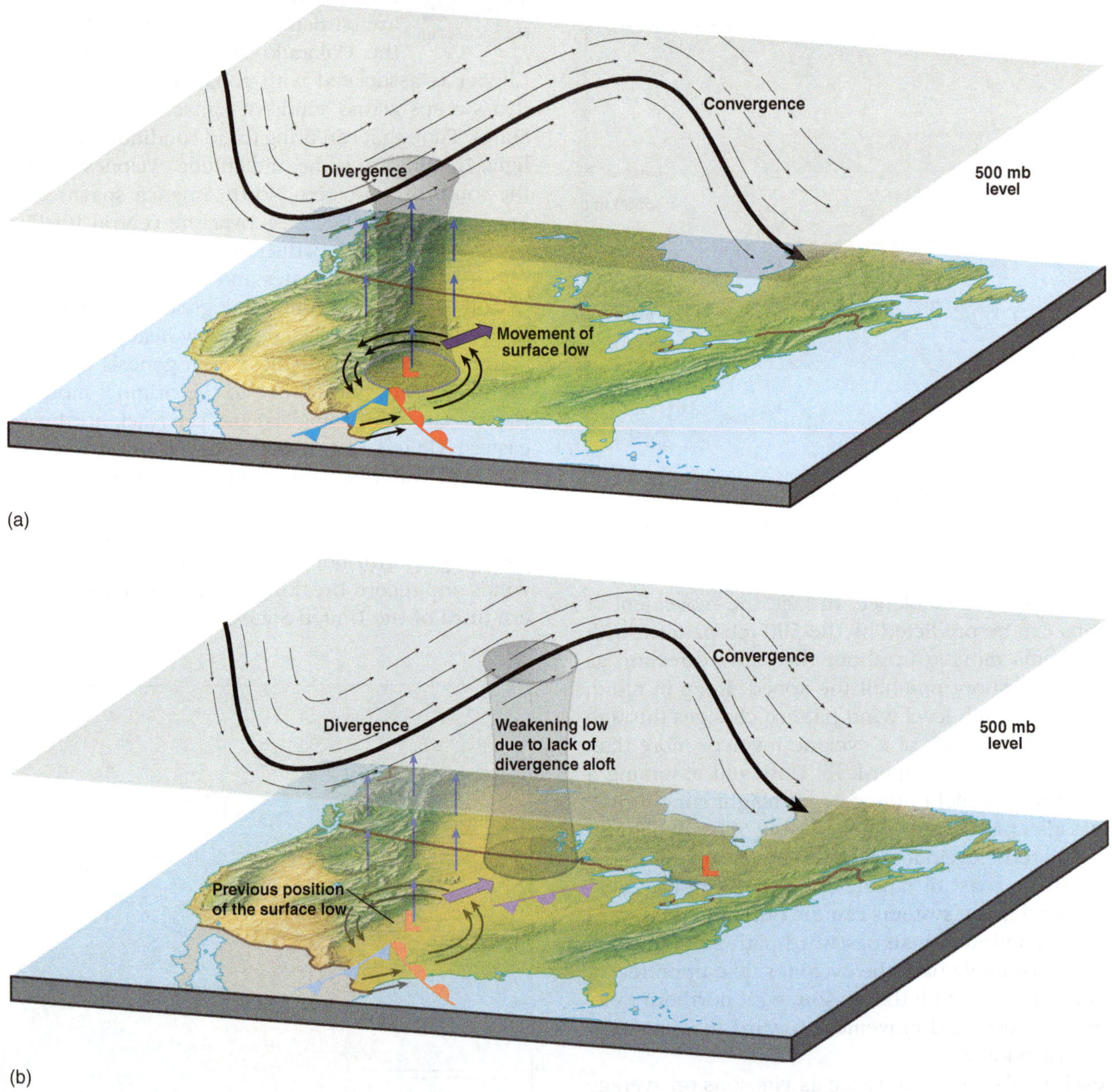

▲ **FIGURE 10–17 Surface Highs and Lows in Relation to Rossby Waves.** Surface high- and low-pressure centers can migrate relative to the Rossby wave aloft. (a) The surface low exists below the area of upper-level divergence, and the resultant uplift maintains or strengthens the cyclone. (b) Surface systems are generally guided by upper-level winds, so eventually the center of the low might be displaced to the zone where upper-level convergence occurs. The sinking air fills into the low, causing its demise. This process occurs as the cyclone undergoes the transition from its mature to occluding stages.

CHECKPOINT

10.11 How does an Alberta Clipper differ from a Colorado Low? Explain.

10.12 What role do convergence and divergence in Rossby waves play in the eventual dissipation of a midlatitude cyclone?

The Modern View: Midlatitude Cyclones and Conveyor Belts

10.7 Evaluate the modern, conveyor belt model of midlatitude cyclones.

We have examined the structure of midlatitude cyclones at the surface and in the middle troposphere, but we should not overlook the fact that cyclones are three-dimensional entities. In other words, the 500 mb level (or any other level) is not separate from the surface portion—it is simply a different part of the same system. The **conveyor belt model** (Figure 10–18), which we look at next, provides a better depiction of the three-dimensional nature of midlatitude cyclones.

This model describes the midlatitude cyclone in terms of three major flows. The first flow, the *warm conveyor belt*, originates near the surface in the warm sector and flows toward and over the wedge of the warm front. As the warm belt flows up the frontal surface, adiabatic cooling leads to condensation and precipitation. Moreover, as the air rises into the middle troposphere, it begins to turn to its right and become incorporated into the general westerly flow downwind of the upper-level trough. The cloud cover associated with the rising warm conveyor belt appears prominently as the bright, wide band extending from south to north in Figure 10–18b.

The *cold conveyor belt* lies ahead (north) of the warm front. It enters the storm at low levels as an easterly belt flowing westward toward the surface cyclone. But like the warm conveyor belt, it too ascends as it flows, turns anticyclonically (clockwise in the Northern Hemisphere), and becomes incorporated into the general westerly flow aloft. Although it originates as cold (and therefore relatively dry) air, the cold conveyor belt gains moisture from the evaporation of raindrops falling from the warm conveyor belt above. The cold conveyor belt extends in Figure 10–18b from northern Michigan to eastern South Dakota.

The final component of the three-dimensional circulation is the *dry conveyor belt* that originates in the upper troposphere as part of the generally westerly flow. This broad current brings the coldest air into the cyclone, and it is important in maintaining the strong temperature contrast across the cold front. The upper-level air sinks slightly as it approaches the cold front from the west, but then it rises and merges with the general upper-level flow. The dry conveyor belt separates the cloud bands from the warm and cold conveyor belts, creating the relatively cloud-free area seen between the brightly

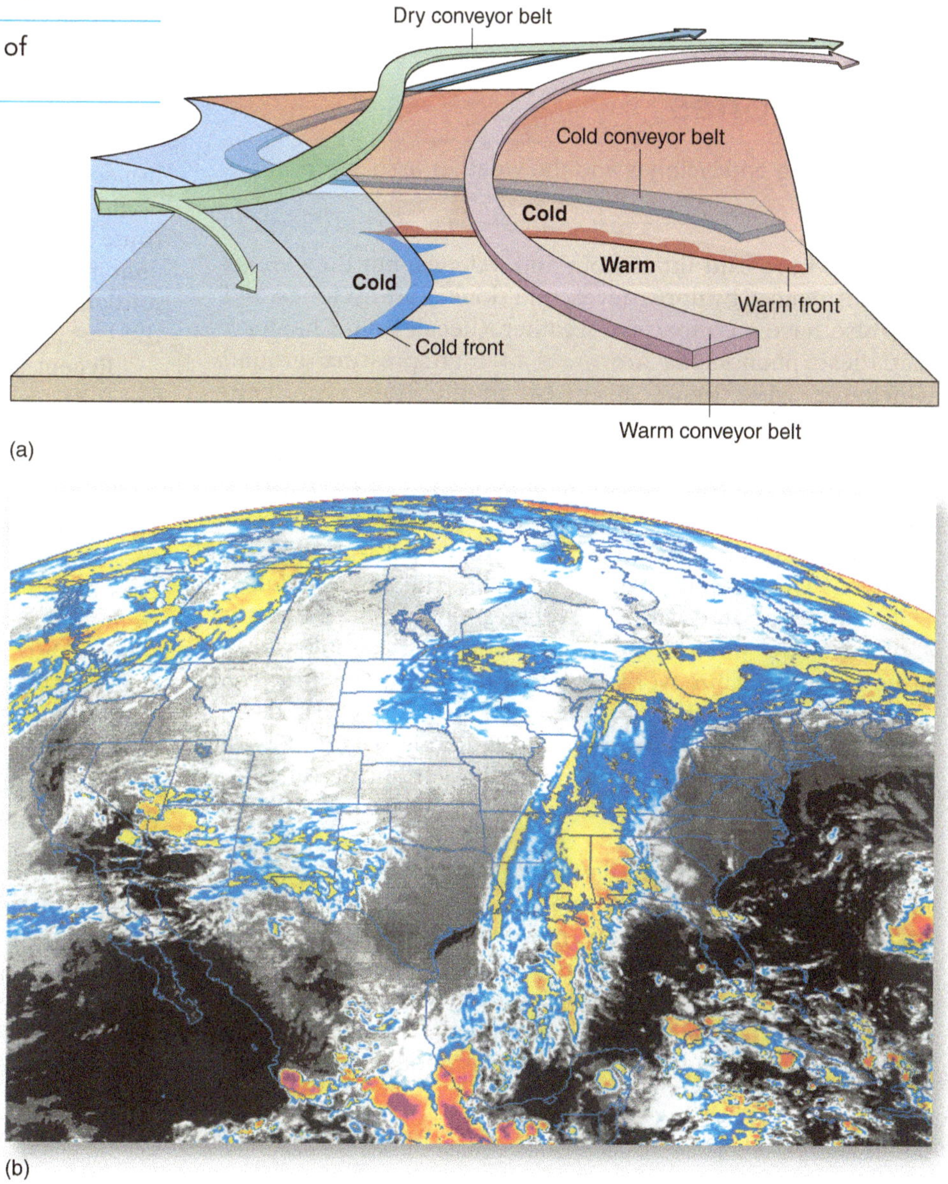

▲ **FIGURE 10–18 Conveyor Belt Model of Midlatitude Cyclones.** (a) In the conveyor belt model, the warm conveyor belt between the cold and warm fronts is responsible for much of the large band of cloud cover evident in the satellite image (b). The cold conveyor belt flows toward the center of the low-pressure system at the surface and rises as it does so, producing the cluster of clouds over the Dakotas and Minnesota. The dry conveyor belt flows in the upper troposphere, with some of the flow sinking as it approaches the cold front. The rest of the conveyor belt remains in the upper atmosphere and remains dry and relatively cloud free. That is evident in (b) as the gap over Wisconsin separating the major cloud regions.

Video
Midlatitude Cyclone's Dry Slot
http://goo.gl/LbiVlX

colored area of clouds associated with the warm and cold conveyor belts in Figure 10–18b.

CHECKPOINT

10.13 Describe the positioning and flow of the warm and cold conveyor melts.

10.14 What effect does the dry conveyor belt have on the temperature contrasts across a cold front?

Anticyclones

10.8 Describe anticyclones and the weather associated with them.

So far we have said little about anticyclones, but they are as much influenced by upper-level conditions as are cyclones, and they also have an impact on weather. (Recall from Chapter 4 that these phenomena are areas of high pressure around which the wind blows clockwise in the Northern Hemisphere.) While cyclones can bring heavy precipitation and strong winds, anticyclones foster clear skies and calm conditions because the cool air within them slowly sinks toward the surface. That should not be taken to mean that anticyclones are always associated with wonderful weather. Indeed, outbreaks of continental polar (cP) air over the eastern United States are associated with anticyclones behind southward- or southeastward-moving cold fronts. Furthermore, anticyclones often tend to remain over a region for a long time, which can lead to droughts. Finally, anticyclones over the Great Basin can lead to hot Santa Ana wind conditions over the West Coast or compressional heating elsewhere. Ironically, at other times high pressure is the result of the incursion of cold, dense air.

This chapter has presented the characteristics of midlatitude cyclones and anticyclones that influence weather outside the tropics. We know, however, that the atmosphere often undergoes violent types of weather, usually on smaller time and space scales than those associated with midlatitude cyclones. In Chapter 11 we will meet thunderstorms and tornadoes, phenomena that can cause considerable damage and loss of life.

CHECKPOINT

10.15 What kind of cloud and wind conditions are usually associated with anticyclones?

10.16 List three examples of important weather events associated with anticyclones.

Climate Change and Midlatitude Cyclones

10.9 State how scientists think climate change may affect midlatitude cyclones.

Midlatitude cyclones are intricately associated with the polar front and polar jet stream, which situate farther poleward in the summer months and migrate to lower latitudes during the cold season. This is particularly evident over the North Pacific Ocean, where summer storms routinely approach North America along western Canada and Alaska, but commonly pass southward to southern California and Baja California during the winter.

We have seen in earlier chapters that increasing greenhouse gas concentrations have affected global temperatures, cloud conditions, and precipitation, and will very likely continue to do so. We also need to consider whether climate change might also cause a shift in the average tracks of midlatitude cyclones; the evidence suggests this is indeed the case.

Recent studies have shown that since the middle of the 20th century, the average path of wintertime midlatitude cyclone tracks has shifted poleward. This change has been strongest in the Southern Hemisphere but also shows up in the Northern Hemisphere. This could have dire consequences for some regions of the globe. Take the case of the arid and semiarid western United States, where some very extensive and severe droughts have contributed to extremely destructive and widespread fires in recent years. If this northward shift in storm tracks becomes permanent, the region will witness fewer storms each year and be forced to deal with reduced water supplies and an increasing vulnerability to major burns.

The highly influential Intergovernmental Panel on Climate Change (IPCC), whose Fifth Assessment Report was released in 2013, has concluded that continued warming will likely lead to a further poleward displacement of storm tracks. Some of the regional patterns identified included a reduction of Atlantic midlatitude cyclones approaching the Mediterranean as storms become more frequent in extreme Northern Europe, increasing aridity over southern Australia, and further drying over much of the western United States.

CHECKPOINT

10.17 What kind of shift has occurred in the wintertime midlatitude cyclone tracks during the 20th century?

10.18 What impacts of climate warming have been projected for the path of winter storm tracks?

Summary

10.1 Summarize the polar front theory.

- Early in the 20th century, Norwegian scientists Vilhelm and Jacob Bjerknes developed the polar front theory explaining the formation, growth, and dissipation of midlatitude cyclones.
- Their "Norwegian cyclone model" remains the standard explanation for the life cycle of midlatitude cyclones.

10.2 Outline the life cycle of a midlatitude cyclone as described in Bjerknes's model.

- According to the Norwegian model, midlatitude cyclones undergo an evolution that begins with their incipient stage (cyclogenesis) and continues through the mature stage and eventual dissipation. The dissipation phase is associated with occlusion of the fronts.

10.3 Explain processes in the middle and upper troposphere that relate to midlatitude cyclones.

- Midlatitude cyclones and anticyclones both depend on a close interaction between processes occurring in the upper and lower troposphere.
- The decreasing vorticity immediately downwind of the Rossby wave trough axis causes divergence, which leads to the formation of cyclones at the surface. At the same time, the fronts associated with midlatitude cyclones help form the Rossby waves that create upper-level convergence and divergence. Thus, a constant interaction takes place between the upper and lower atmosphere.

10.4 Analyze how surface fronts and upper-level patterns are related.

- The dense air behind cold fronts favors the existence of troughs in the middle and upper troposphere, while patterns of divergence and convergence create surface areas of low and high pressure.

10.5 Describe the behavior of a typical midlatitude cyclone as it crosses North America.

- As a midlatitude cyclone migrates eastward there is a reduction in the distance between the cold and warm fronts, with occlusion eventually developing.

10.6 Explain how flow patterns and large-scale weather patterns affect the development, steering, and dissipation of midlatitude cyclones.

- Surface cyclones and anticyclones migrate in the direction of the midtropospheric (700 mb) winds of Rossby waves.
- Cyclones move from regions of upper-level divergence (which help maintain or intensify the surface low pressure) to regions of upper-level convergence (which weaken the cyclones).

10.7 Evaluate the modern, conveyor belt model of midlatitude cyclones.

- The modern, conveyor belt approach explains the midlatitude cyclone airflow as the result of three separate streams, or conveyor belts.
- A cold conveyor belt flows toward the center of low pressure ahead of the warm front. As it approaches the low pressure, it rises, and adiabatic cooling produces the wide band of cloud cover. Similarly, a rising conveyor belt of warm air flows ahead of the cold front to provide another band of cloud cover.
- Both the warm and cold conveyor belts turn anticyclonically near the center of low pressure and join the upper-level westerly flow.
- A dry conveyor belt in the middle atmosphere flows above the cold front.

10.8 Describe anticyclones and the weather associated with them.

- Anticyclones are marked by slowly descending air, which favors clear skies.

10.9 State how scientists think climate change may affect midlatitude cyclones.

- Climate change has apparently caused a poleward shift in the mean position of storm tracks in recent decades, which is likely to continue.
- This shift will alter the precipitation distribution in some parts of the world, which may have very significant social impacts, such as increased water shortages and greater vulnerability to wildfires.

Key Terms

absolute vorticity *p. 320*
baroclinic *p. 327*
baroclinic instability *p. 327*
barotropic *p. 327*
cold air advection *p. 326*
confluence *p. 325*
conveyor belt model *p. 335*
cutoff low *p. 329*
cyclogenesis *p. 316*
diffluence *p. 325*
dynamic low *p. 321*
Earth vorticity *p. 320*
mature cyclone *p. 316*
occluded front *p. 318*
occlusion *p. 318*
polar front theory *p. 316*
relative vorticity *p. 320*
Rossby wave *p. 320*
short waves *p. 326*
speed convergence *p. 324*
speed divergence *p. 324*
temperature advection *p. 326*
thermal low *p. 321*
vorticity *p. 320*
warm air advection *p. 326*

Review Questions

1. Describe the isobar and wind patterns associated with mature midlatitude cyclones.
2. Define *cyclogenesis*. Where does it most commonly occur, according to the original polar front theory?
3. Where is precipitation most likely to be found within midlatitude cyclones?
4. Where within a midlatitude cyclone are overrunning, convergence, and instability likely to serve as precipitation-inducing processes?
5. What are Earth, relative, and absolute vorticity?
6. Why is counterclockwise rotation in the Northern Hemisphere said to have positive vorticity?
7. Where are the zones of positive, negative, and zero vorticity in a typical ridge and trough pattern?
8. In what part of a ridge and trough system do you find the areas of decreasing and increasing vorticity? Why is their existence important?
9. Where are upper-level divergence and convergence most likely to occur?
10. How do dynamic and thermal lows differ from each other?
11. What type of upper-level condition typically lies above and behind a cold front?
12. Where are upper-level ridges generally located relative to midlatitude cyclones?
13. What are diffluence and speed divergence? How do they differ from confluence and speed convergence?
14. Which type of general upper-level airflow is more likely to occur if minimal temperature variations exist across North America?
15. If the upper-level airflow over North America is zonal, what can you infer with regard to widespread precipitation conditions?
16. Describe the three conveyor belts of the conveyor belt model of midlatitude cyclones. How does the conveyor belt model differ from the description of airflow presented in the original polar front model?
17. Explain how the movement of midlatitude cyclones relative to the upper-level airflow contributes to the demise of cyclones.

Critical Thinking

1. Why is the term *polar front theory* probably a misnomer in view of current knowledge about cyclogenesis?
2. A commercial aircraft is flying at the 300 mb level and goes across a midlatitude cyclone over both the cold and warm fronts. What kind of weather changes might the aircraft encounter?
3. After a front is fully occluded, its demise is imminent. Why can't the occluded front persist for several more weeks?
4. Why can't systems similar to midlatitude cyclones develop over the tropics?
5. Vorticity is usually discussed with reference to a vertical axis. What types of Earth, relative, and absolute vorticity conditions would you expect to exist with regard to a horizontal axis?
6. Why do forecasters take particular interest in the distribution of vorticity on 500 mb weather maps?
7. Why don't thermal lows migrate like dynamic lows do?
8. Clear skies often portend warm conditions during the summer but are often associated with very cold conditions in the winter. Explain why this is true.
9. Are Rossby waves likely to have greater representation in the Southern Hemisphere or the Northern Hemisphere? Why?

Problems & Exercises

1. Visit **http://weather.uwyo.edu/surface/front.html** on a daily basis and keep track of existing midlatitude cyclones and their associated fronts (click the most recent map available). Do the systems correspond well to the life cycle described in this chapter?
2. Go to **http://weather.uwyo.edu/upperair/uamap.html** and click the Get Map button to view the current 500 mb map for North America. Without referring to a surface map, make an educated guess about the position of surface midlatitude cyclones, cold fronts, and warm fronts. Then go to **http://weather.uwyo.edu/surface/front.html** to see how well your educated guess worked out (be sure to highlight Analysis for image type).
3. Using the same website as for Problem 2, print the maps of the surface, 300, 500, 700, and 850 mb levels. What patterns emerge as you move upward?

Visual Analysis

This weather map from October 14, 2013, plots the height of the 500 mb level. Examine distribution of these values on the map to answer the questions below. (*Hint:* Figures 10–7 and 10–8 provide useful background for answering the questions.)

10.1. Identify the location and orientation of the major troughs and ridges.

10.2. Which location over the United States has the highest vorticity?

10.3. Where do you anticipate the zone of maximum divergence?

CHAPTER

11

Lightning, Thunder, and Tornadoes

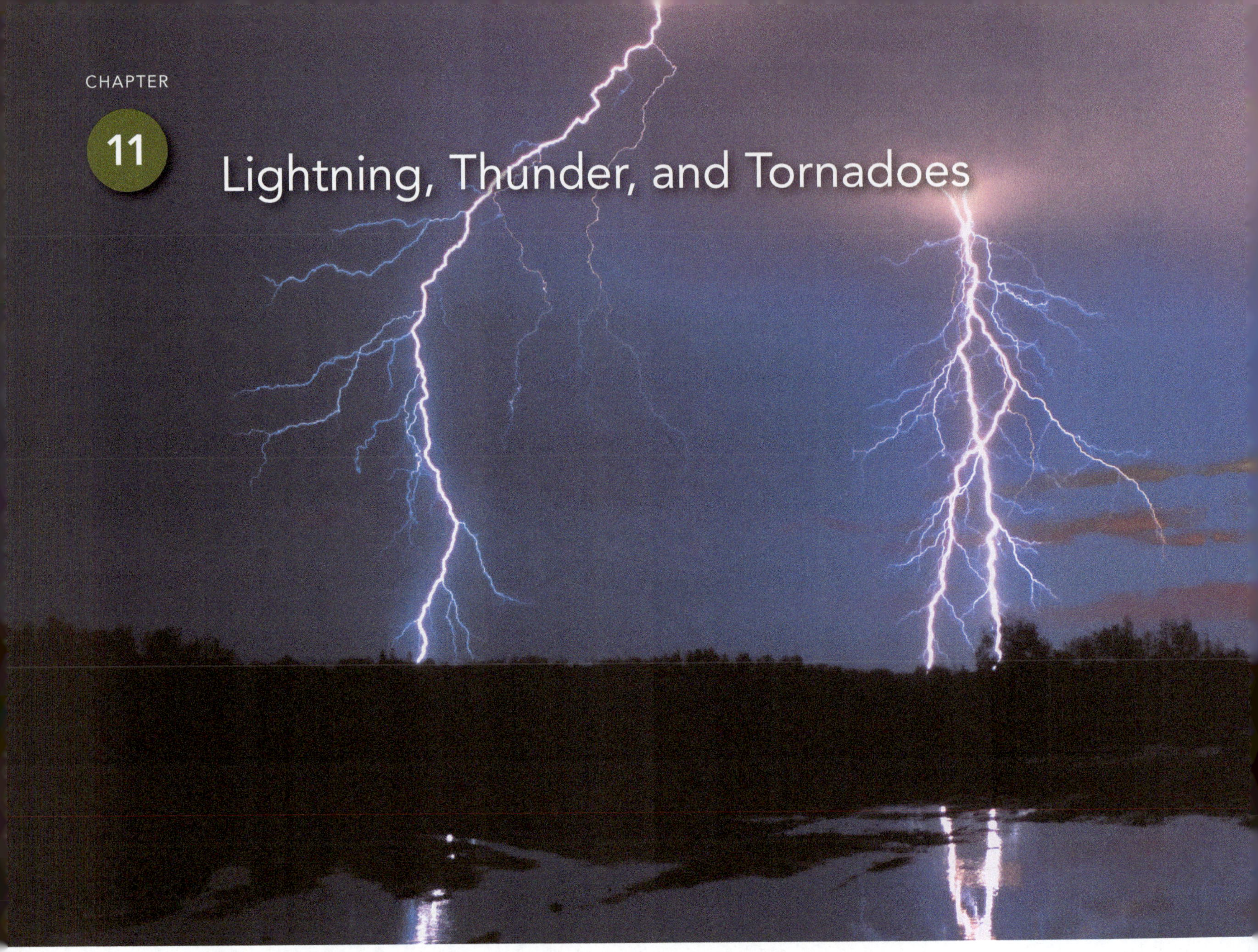

It is hard to imagine anybody failing to be impressed by the beauty—as well as the danger—brought about by a thunderstorm. Spectacular though they may be, such storms occur about 40,000 times each day. Their frequency varies greatly from place to place, yet virtually every location on Earth is vulnerable to thunder and lightning from time to time.

Lightning can create inconveniences—such as blowing out all the electrical appliances in a house. It can also do considerable damage, such as starting forest fires. And, of course, it can kill; during an average year, about 70 to 80 people are killed by lightning in the United States and Canada. But considering that the population of these two countries exceeds 300 million people, it is easy to see that your chances of being struck are extremely remote.

Consider, however, the experience of the McQuilken family on their trip to Sequoia National Park, California, in August 1975. As the sky began to darken, Sean, Michael, and their sister Mary noticed their hair standing on end. Recognizing the apparent comedy of the situation, the boys posed for the photograph shown in Figure 11–1. Hail followed almost immediately. Then lightning struck—literally—and Sean was knocked unconscious. Michael quickly administered artificial respiration, which probably saved Sean's life. However, another victim was less fortunate. The lightning had apparently forked off, with another branch hitting two nearby people, one of whom was killed.

The effects of lightning and thunder are eclipsed by an even greater menace—tornadoes. We now examine how, where, and why violent weather occurs, and we look at the situations that cause some storms to be weak and others to become destructive and deadly.

◄ *Lightning in Alberta, Canada.*

Learning Outcomes

After reading this chapter, you should be able to:

11.1 Sequence the processes involved in lightning formation and identify the different types of lightning.

11.2 State measures that can be taken to ensure safety from lightning.

11.3 Explain how the different types of thunderstorms form and describe the strong winds that can accompany them.

11.4 Summarize the characteristics and processes involved in the formation of floods and flash floods.

11.5 Analyze the geographic and temporal distribution of thunderstorms.

11.6 Explain how, where, and when tornadoes form.

11.7 Identify aspects of tornadoes as a natural hazard, including areas most at risk for tornadoes, damage, and fatalities, and explain how tornadoes are rated and forecasted.

11.8 Explain how a waterspout forms.

▲ **FIGURE 11–1 Electrical Fields.** Sean and Michael McQuilken in a strong electric field just prior to a lightning strike.

Processes of Lightning Formation

11.1 Sequence the processes involved in lightning formation and identify the different types of lightning.

About 80 percent of all **lightning** results from the discharge of electricity *within* clouds, as opposed to discharge from cloud to surface. This **in-cloud lightning** occurs when the voltage gradient within a cloud overcomes the electrical resistance of the air. The result is a very large and powerful spark that partially equalizes the charge separation. Lightning between clouds, called **cloud-to-cloud lightning**, causes the sky to light up more or less uniformly. Because the flash is obscured by the cloud itself, it is commonly called **sheet lightning**.

The remaining 20 percent of lightning strokes are the more dramatic events in which the electrical discharge travels between the base of the cloud and Earth's surface. Most of this **cloud-to-ground lightning** occurs when the negative charges accumulate in the lower portions of the cloud. Positive charges are attracted to a relatively small area in the ground directly beneath the cloud. This establishes a large voltage difference between the ground and the cloud base. The positive charge at the surface is a local phenomenon; it arises because the negative charge at the base of the cloud repels electrons on the ground below. Although the term *cloud-to-ground* is used, the same effect occurs in water—and lightning often strikes lakes, rivers, and oceans. Another form of lightning, *cloud-to-air lightning*, discharges from a cloud to the surrounding air. Cloud-to air lightning most commonly occurs near the positively charged, upper part of the cloud and the surrounding air. This type of lightning is less common at the bottom portion of the cloud, where discharge with the ground is easier to achieve due to greater charge differences.

Although a stroke of lightning may come and go in just a few moments, a regular sequence of events must occur for the event to take place. Electrification of a cloud is the initial stage in all lightning. After that, a path must develop through which electrons can flow. Only then is electricity actually discharged to produce a lightning stroke.

Charge Separation

All lightning requires the initial separation of positive and negative charges into different regions of a cloud. Most often the positive charges accumulate in the upper reaches of the cloud, negative charges in lower portions. Small pockets of positive charges may also gather near the cloud base (Figure 11–2a). Now the question is: How does this **charge separation** occur in the first place? Nobody knows for sure, because clouds that produce lightning and thunder happen to be particularly inhospitable laboratories. But we do know several facts from which we can get some idea of how charges separate. Lightning occurs only in clouds that extend above the freezing level, and it is also restricted to precipitating clouds. Thus, the ice crystal processes responsible for precipitation must also influence charge separation. Laboratory experiments with artificial clouds and numerical calculations suggest that electrification results from collisions between ice crystals and graupel surrounded by cloud droplets. It seems likely that charges are transferred across thin films of water present on ice crystals and soft hail.

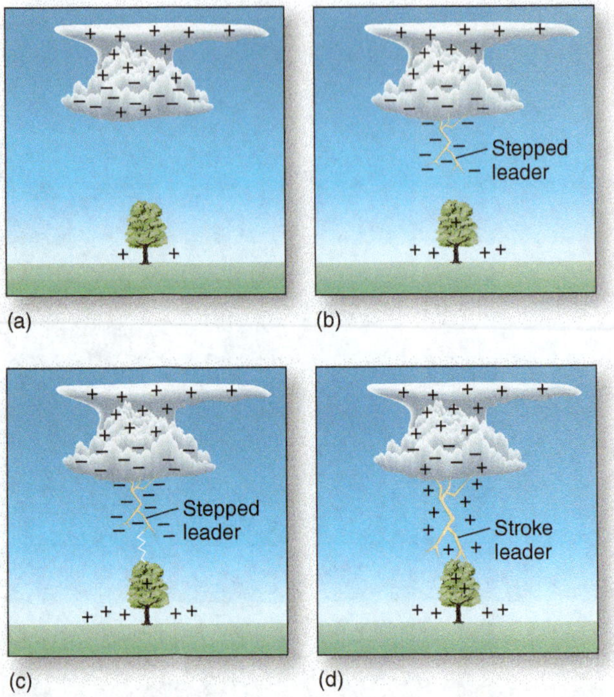

▲ **FIGURE 11–2 Formation of Lightning.** After charge separation is established in a cloud (a), the first step in the formation of lightning is the development of a stepped leader. The leader approaches the surface as a very rapid sequence of steps (b and c), until contact is made with an object at the ground. The flow of electricity produces the lightning stroke (d).

Though we normally don't notice it, solids are often coated with a liquid surface layer just a few molecules thick. This layer is present even at temperatures well below the freezing point. (This layer helps to explain why ice is so slippery at temperatures well below 0 °C.[1]) In a cloud, when an ice crystal and soft hail collide, some of the liquid-water molecules on the hailstone's surface migrate to the ice. A transfer of positive charge usually accompanies the movement of water from the hailstone to the crystal or, equivalently, a negative charge is transferred from the crystal to the hailstone. In this way ice crystals surrender negative ions to the much larger hailstones, which then fall downward toward the cloud base.

Depending on the liquid-water content and temperature of the cloud, a *positive* charge is occasionally transferred to the hailstone, leading to an accumulation of negative charges aloft. Very recent research suggests that surface defects also regulate the charge transfer between particles. (See *Box 11–1, Physical Principles: Electricity in the Atmosphere*, for more information on electrical charges in the air.)

DID YOU KNOW?

Worldwide, there are about 4 million lightning discharges per day, resulting in about 2000 injuries and 600 deaths per year.

[1]You may have heard that pressure from an ice skate's blade melts enough ice to create a slippery film of water, but that's not correct. The ice is slippery whether you press hard or not.

11-1 PHYSICAL PRINCIPLES

Electricity in the Atmosphere

Lightning is, of course, an electrical disturbance, much of which can be explained by the basic principles of atmospheric electricity. You know from Chapter 1 that ions (charged particles) are most abundant high in the atmosphere (in the ionosphere, from about 80 to 500 km, or 50 to 300 mi). The upper atmosphere has a positive charge, just as we find near the positive pole of a battery. In the same way that a battery stores energy, electrical charges in the atmosphere represent stored energy and have the potential to do work. For both batteries and the atmosphere, this electrical potential is expressed by voltage, which is simply the energy per unit charge. For example, if a battery is rated at 1.5 volts (V), it means that 1.5 joules are available per coulomb of charge (1.5 J/C). A coulomb (C) is equivalent to the charge carried by about 6×10^{19} electrons. The higher the voltage, the greater the energy release for each coulomb transferred.

In the case of Earth, a huge voltage difference exists between the surface and the ionosphere—about 400,000 volts! This voltage gradient sets up what we call the **fair-weather electric field**. The fair-weather field is always present, even in bad weather, so a better name might be the **mean electric field**. The fair-weather field can be thought of as the background situation on which extreme events such as lightning are superimposed.

Does electricity flow in response to the voltage gradient of the fair-weather field? Yes, but because air is a good insulator, the current is weak, about 2000 coulombs per second (2000 Ampere) for the entire planet. In North America, individual houses are typically wired for 200-Ampere service, so we see that the atmospheric current is truly very small. Nevertheless, it does represent a continuous leakage, whereby electrons are transferred from the surface or, equivalently, positive charges are transferred from the atmosphere. This implies that for the mean electric field to be maintained, it must be continuously replenished. As a matter of fact, lightning discharges in thunderstorms are thought to be the primary recharge mechanism. In other words, cloud-to-ground lightning discharges transfer electrons to the surface, maintaining the voltage difference and the resulting electric field.

In the lower atmosphere, the fair-weather electric field gradient is on the order of 100 V per meter. (Although this might sound impressive, remember that few ions are present, so the total available energy is very low.) Of course, for lightning to occur, the field strength must be greatly intensified above the background value. How this happens can be only partly explained today, nearly 250 years after Ben Franklin performed his famous kite experiment.

11-1-1 What is the fair-weather electric field?

11-1-2 What role does cloud-to-ground lightning play in regard to the mean electrical field?

Leaders, Strokes, and Flashes

In cloud-to-ground lightning, the actual lightning event is preceded by the rapid and staggered advance of a shaft of negatively charged air called a **stepped leader** (Figure 11-2b) from the base of the cloud. The leader is not a single column of ionized air; it branches off from a main trunk in several places. Only about 10 cm (4 in.) in diameter, each section of the column first surges downward about 50 m (165 ft) from the base of the cloud in about a millionth of a second (or microsecond). This invisible leader pauses for about 50 microseconds, as electrons pile up at the tip and generate a strong electric field in the surrounding area. The field generates more runaway electrons that surge downward another 50 or so meters in the next step. The downward movement in a rapid sequence of individual steps gives the stepped leader its name. In each step, the newly created runaway electrons collide with air molecules and trigger a flare of X-rays.

When the leader approaches the ground (Figure 11-2c), a spark surges upward from the ground toward the leader. When the leader and the spark connect, they create a pathway for the flow of electrons that initiates the first in a sequence of brightly illuminated **strokes**, or **return strokes**. The electrical current, flowing at about 20,000 amperes (A), appears to work its way downward from the base of the cloud, but the stroke actually propagates upward (Figure 11-2d). The conducting path is completed at the surface, from which a positive charge surges upward toward the cloud. The current heats the air in the conducting channel to temperatures up to 30,000 K (54,000 °F), or five times that of the surface of the Sun! (Figure 11-3).

The electrical discharge of the first stroke neutralizes some, but not all, of the negatively charged ions near the base of the cloud. As a result, another leader (called a **dart leader**) forms within about a tenth of a second, and a subsequent stroke emerges from it. We call the combination of strokes a lightning **flash**, the net effect of which is to transfer electrons

▲ FIGURE 11-3 Lightning over Indianapolis, Indiana.

from the cloud to the ground. Although most flashes consist of only 2 or 3 strokes, some consist of as many as 20 individual strokes. Because they occur in such rapid succession, they appear to be a single stroke that flickers and dances about.

The total transfer of electrons is not large, only about as many electrons as we use in burning a 100-watt lightbulb for about half a minute. So how is lightning able to split trees and perform other dramatic work? For one thing, in lightning the charge transfer is rapid, so the electric current is discharged many thousands of times faster than in a household current. (Think of a lightbulb as having low current flowing for a relatively long time. The total charge transferred is the same as in lightning, where a huge current flows for just an instant.) Another factor is that the voltage is much larger than in a household circuit, so the energy release is much larger for each electron transferred. Taken together, these facts mean that a huge amount of energy gets released over a very brief period of time, making each stroke extremely powerful. A lightbulb would need a month or two to release the same amount of energy.

Positive charges found at the top of thunderstorm clouds can also lead to lightning. When high-level winds are strong, thunderstorm clouds become tilted, with positive charges carried ahead of the storm (Figure 11–4). These positive charges induce negative charges at the surface, resulting in a lightning strike that shoots positive charges to the surface. As a result, it often happens that the first of a storm's lightning strikes are positive, and measurements reveal that they can be many times stronger than the negative strokes that follow. This positive form of lightning is therefore particularly dangerous. It can occur several miles away from the storm, where people do not feel threatened; it tends to have larger peak electrical currents; and it typically lasts longer, making fires more likely. Although an estimated 9 percent of all cloud-to-ground flashes in the United States and Canada originate this way, considerable variability is seen across North America. Strokes originating from positive polarity are more common in the upper Midwest of the United States and throughout much of Canada, where they represent perhaps 20 percent of all occurrences.

CHECKPOINT

11.1 What role do leaders play in the formation of lightning?

11.2 Why do lightning flashes often appear to be single bolts of electricity that move around, when in fact they are not?

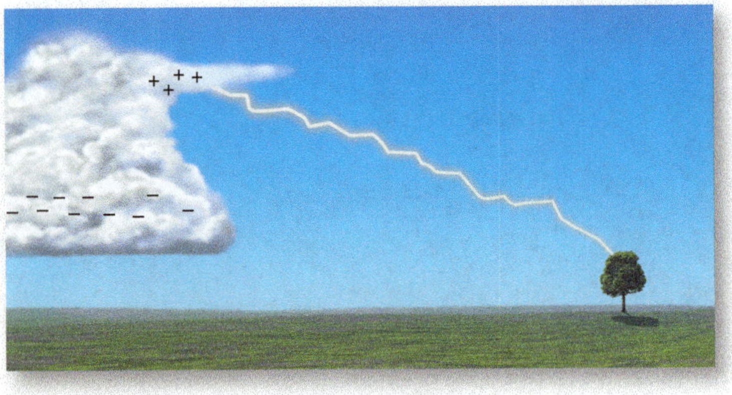

▲ **FIGURE 11–4 Positive Lightning Stroke.**

▲ **FIGURE 11–5 Ball Lightning.** A 19th-century engraving.

Types of Lightning

Far less common than strokes and leaders is the bizarre type of electrification called **ball lightning** (Figure 11–5). Ball lightning appears as a round, glowing mass of electrified air, up to the size of a basketball, that seems to roll through the air or along a surface for 15 seconds or so before either dissipating or exploding. One form is a free-floating, reddish mass that tends to avoid good electrical conductors and flows into closed spaces or through doorways and windows. Another form is considerably brighter and is attracted to electrical conductors (including people). Various explanations for ball lightning have been offered for at least 150 years, but until recently none could account for all aspects of the phenomenon, and most had glaring weaknesses. The situation improved significantly early in 2000, with the report of experiments involving artificial lightning strikes on soil. Researchers observed that lightning reduces silicon compounds in the soil to tiny nanoparticles of silicon carbide (SiC), silicon monoxide (SiO), and metallic silicon (Si). Unlike the original silicon compounds, these contain significant chemical energy and are unstable in an oxygen environment. They are ejected into the air, where they cool rapidly and condense into filmy chains and networks. The networks are light, so they float easily in the atmosphere. Most important, they burn brightly as they oxidize, releasing the stored energy in the form of visible light. (See *Box 11–2, Focus on Severe Weather: A Personal Account of Ball Lightning,* for a firsthand description of this phenomenon.)

DID YOU KNOW?

Lightning strokes can extend more than 16 km (10 mi) from the side of clouds. Thus, lightning can hit in an area where clear skies prevail.

St. Elmo's fire (Figure 11–6) is another rare and peculiar type of electrical event. Ionization in the air—often just before the formation of cloud-to-ground lightning—can cause tall objects such as church steeples or ships' masts to glow as

11–2 FOCUS ON SEVERE WEATHER

A Personal Account of Ball Lightning

I saw ball lightning during a thunderstorm in the summer of 1960. I was 16 years old. It was about 9:00 P.M., very dark, and I was sitting with my girlfriend at a picnic table in a pavilion at a public park in upstate New York. The structure was open on three sides and we were sitting with our backs to the closed side. It was raining quite hard. A whitish-yellowish ball, about the size of a tennis ball, appeared on our left, 30 yards away, and its appearance was not directly associated with a lightning strike. The wind was light. The ball was 8 feet off the ground and drifting slowly toward the pavilion. As it entered, it dropped abruptly to the wet wood plank floor, passing within 3 feet of our heads on the way down. It skittered along the floor with a jerky motion (stick-slip), passed out of the structure on the right, rose to a height of 6 feet, drifted 10 yards further, dropped to the ground and extinguished nonexplosively. As it passed my head, I felt no heat. Its acoustic emission I liken to that of a freshly struck match. As it skittered on the floor it displayed elastic properties (a physicist would call them resonant vibrating modes). Its luminosity was such that it was not blinding. I estimate it was like staring at a less than 10-watt light bulb. The whole encounter lasted for about 15 seconds. I remember it vividly even today, as all eyewitnesses do, because it was so extraordinary. Not until 10 years later, at a seminar on ball lightning, did I realize what I had witnessed.

11–2–1 Based on the writer's description, how would you define ball lightning?

11–2–2 If the recent hypothesis about ball lightning's formation described in the text is correct, how can you explain the way the ball lightning Hubler saw "extinguished nonexplosively"?

Source: Graham K. Hubler, reprinted by permission from *Nature*, Copyright 2000, Macmillan Magazines Ltd.

▲ **FIGURE 11–6 St. Elmo's Fire.** From the cockpit of a military aircraft.

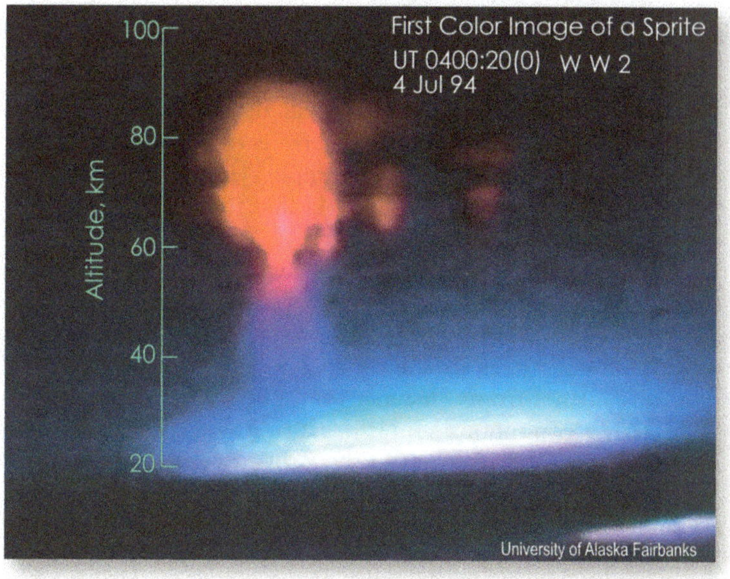

▲ **FIGURE 11–7 A Sprite.** First color image of a Sprite ever taken above Oklahoma City.

they emit a continuous barrage of sparks. This often produces a blue-green tint to the air, accompanied by a hissing sound.

Observations and photographs from space shuttle missions have revealed the existence of previously unknown electrical phenomena at the tops of thunderstorms. **Sprites** (Figure 11–7) are very large but short-lived electrical bursts that rise from cloud tops as lightning occurs below. A sprite looks somewhat like a giant red jellyfish, extending up to 95 km (57 mi) above the clouds, with blue or green tentacles dangling from the reddish blob. Sprites accompany only about 1 percent of all lightning events. (Interestingly, military and commercial pilots now admit to having seen sprites before they were observed from shuttle missions, but they did not often report them lest they be accused of having hallucinations.)

Blue jets (Figure 11–8) are upward-moving electrical ejections from the tops of the most active regions of thunderstorms. They shoot upward at about 100 km/sec (60 mi/sec) and attain heights of up to 50 km (30 mi) above the surface. In all likelihood, other types of electrical activity above thunderstorms remain to be discovered.

Thunder

The tremendous increase in temperature during a lightning stroke causes the air to expand explosively and produce the familiar sound of **thunder**. Although sound travels rapidly—about 0.3 km (0.2 mi) per second—it is much slower than the speed of light (300,000 km, or 186,000 mi, per second). This difference creates a lag between the flash of light and the sound of thunder; the farther away the lightning, the longer the time lag. You probably know the familiar

▲ **FIGURE 11–8 Blue Jets are Optical Ejections from the Tops of Thunderstorms.**

rule of thumb for estimating the distance of a lightning stroke: Simply count the number of seconds between the stroke and the thunder and divide by three to determine the distance in kilometers (divide by five for the distance in miles).

This method does not work for very distant strokes, those more than about 20 km (12 mi) away. The decrease in the density of air with height causes sound waves to bend upward. At relatively short distances, the amount of bending is negligible. But beyond about 20 km, it is sufficient to displace the sound waves so that they cannot be heard at ground level. Lightning that seems to occur without thunder is sometimes called **heat lightning**, though this term is misleading in that it implies there is something unusual about it. The only oddity is that the sound of distant thunder does not reach the listener.

You have probably noticed that nearby thunder sounds like a loud, brief clap, while more distant thunder often occurs as a continuous rumble. A lightning stroke producing thunder may be several kilometers in length, so one part of it may be significantly farther from a listener than other parts. Thus, thunder makes a continuous sound because it takes longer for the sounds of more distant parts of the stroke to reach the listener. At greater distances, the echoing of sound waves off buildings and hills can cause the thunder to make a rumbling sound.

CHECKPOINT

11.3 Describe four types of electrification that occur in the atmosphere other than lightning.

11.4 Explain how measuring the elapsed time between the appearance of lightning and hearing thunder can provide information on the distance to the lightning.

DID YOU KNOW?

On June 29, 2005, Jeff Johnson was struck by lightning as he worked in his office near Des Moines, Iowa. The lightning appears to have been carried into the office via electrical cabling, disabling the computer equipment and hitting Johnson. What made this event noteworthy was that Johnson was working at his job as a meteorologist at a National Weather Service forecast office and was following the progress of a lightning storm—the same one that hit him. After a visit to the hospital, Johnson took the rest of the day off at home—in good physical condition.

Lightning Safety

11.2 State measures that can be taken to ensure safety from lightning.

MM **MapMaster** World and North America
Physical Environment: Lightning Strikes

Despite its splendor, we must not forget that lightning can be lethal, killing an average of 69 people each year in the United States and 7 in Canada. Fortunately, our current understanding of lightning suggests some safety rules.

First and foremost, in the presence of lightning, always take cover in a building, being careful not to make contact with any electrical appliances or landline telephones. Do not stand under a tree or other tall object that is likely to serve as a natural lightning rod. Lightning hitting a tree can easily flow through it and electrocute or burn those near this particularly vulnerable location. Avoid standing on rooftops, hill crests, or other high areas where lightning requires a shorter path from the cloud base. And of course, stay out of the water. Do not watch the lightning storm from a pool, lake, or hot tub!

Automobiles (other than convertibles) are relatively safe, but not because the rubber tires provide insulation against grounding (as believed by many people). The real reason is that if a car is hit, the electricity will flow around the car body rather than through the interior (or its occupants). The same fact explains why lightning seldom brings down airplanes, even though any particular commercial aircraft is hit on an average of once a year (see *Box 11–3, Focus on Aviation: Lightning and Aircraft*).

We often associate deadly lightning strikes with golfing during a thunderstorm and other foolish behaviors, but the danger is not always easy to avoid. On January 1, 2000, for instance, a single lightning bolt killed a family of six near Mount Darwin, Zimbabwe, in a tragedy eerily similar to one that had occurred a few months earlier in Zimbabwe, when a single strike killed six persons near the city of Gokwe. Lightning deaths are becoming ever more frequent throughout that region as forests are cleared, leaving villages more exposed in open areas. The problem is greatly exacerbated by the use of dry thatch as a roofing material. Soot from cooking fires impregnates the thatch with carbon, making the roof highly conductive and attractive for lightning.

DID YOU KNOW?

Only about 10 percent of lightning victims die after being hit. By far the greatest number of those who are killed die from cardiac arrest; victims only rarely suffer external burns. Though most people are not killed when struck by lightning, survivors often suffer long-term psychological problems, including irritability, personality changes, chronic fatigue, and depression.

11–3 FOCUS ON AVIATION

Lightning and Aircraft

Lightning strikes on aircraft are not uncommon (Figure 11–3–1). Fortunately, they seldom result in serious damage, injury, or fatalities. During the 57-year period prior to 2009, there were 18 reported air disasters due to lightning worldwide—less than one every 3 years. The last confirmed U.S. civilian plane crash attributed to lightning was reported in 1967, the result of a fuel tank explosion.

Effects of Lightning on Aircraft

Aircraft are protected in part by their highly conductive skins, which conduct the electrical current around the fuselage and extremities of the plane. In the United States, the Federal Aviation Administration (FAA) mandates high levels of additional protection against the potential direct and indirect consequences of lightning hits. For example, all electronic equipment must be shielded against power surges. The nose cones of commercial aircraft that house the radar (the radome) must be surrounded by lightning diverter strips that act much like lightning rods do on homes. The fuel tank and lines transporting fuel to the engines must be protected against the possibility of any sparks that could ignite the fuel, and the fuel itself is now designed to produce less explosive vapors.

Sometimes an existing lightning stroke will make contact with a plane, and passengers and crew will be greeted by a bright flash and a loud noise. Pilots may observe a flickering of lights in the cockpit. At other times the plane itself will cause a lightning event as it passes through a heavily ionized part of a cloud.

Lightning-Strike Incidents

Though it is rare for more than one plane to get hit by lightning at any given time or place, such events do occasionally happen. On February 24, 1987, at least 6 planes were struck near Los Angeles (a location that sees relatively few lightning events). On October 19, 2011, at least 10 planes were hit near Helsinki, Finland.

And it isn't just planes that get struck. During the afternoon of September 12, 2013, the control tower at Baltimore-Washington International Thurgood Marshall Airport (BWI) was struck by lightning—one of five strikes to hit the airport grounds that day. One FAA employee was injured by the strike as he attempted to turn on a generator to power backup runway lights. The airport was forced to suspend landings and takeoffs over a 3-hour period, ultimately cancelling a total of 118 flights, impacting air traffic across the country.

Ball Lightning and Aircraft

Scientists have begun to examine the incidence of ball lightning in and near planes. A recent analysis of 87 such events from 1938 to 2007 indicated that about half of these occurred inside the aircraft and the other half outside the airframe. Sometimes the ball lightning appeared to be a side effect of a regular lightning strike, but it can also occur independently. Although no damage or minimal damage is usually incurred, there have been reports of major damage and even three aircraft downings due to ball lightning.

▲ **FIGURE 11–3–1 Cloud-to-air l.** Lightning strikes a commercial aircraft in Germany.

11–3–1 Why do lightning strikes seldom bring down planes, despite the frequency with which planes are hit?

11–3–2 Describe what passengers and crew experience when a plane is struck by lightning.

CHECKPOINT

11.5 Describe the best safety measures to take during lightning storms.

11.6 Why do automobiles provide relative safety in thunderstorms?

Thunderstorms: Air Mass, Multicell, and Supercell

11.3 Explain how the different types of thunderstorms form and describe the strong winds that can accompany them.

Most lightning events are associated with localized, short-lived storms that dissipate within tens of minutes after forming. These storms, called **air mass thunderstorms**, actually extinguish themselves by creating downdrafts that cut off the supply of moisture into the precipitating clouds. For that reason, they do not normally produce severe weather. On other occasions, however, downdrafts from heavy precipitation do not impede the replenishment of moisture into the storm from updrafts. Those storms can become severe and produce serious damage and loss of life.

Air Mass Thunderstorms

Air mass thunderstorms are the most common and least destructive of thunderstorms. They also have very limited life spans, usually lasting for less than an hour. Despite the name, which implies that these thunderstorms might occupy entire air masses (which are very large), air mass thunderstorms are very localized. But the term does make sense when you consider that air mass thunderstorms occur within individual air masses and are well removed from frontal boundaries. Think of it this way: Air mass thunderstorms are contained *within* uniform air masses, but they do not occupy the *entire* air mass.

Air mass thunderstorms will not occur unless certain conditions exist: The air must contain sufficient moisture with the temperature being not too much greater than the dew point, as that permits condensation to occur fairly close to the ground. The air also must be somewhat unstable so that uplift can be sufficient to produce a deep cumulus cloud. Another characteristic of air mass thunderstorms is that they form in an environment where there is little *wind shear*, a change in wind velocity with altitude. And, of course, there must be some initial uplift mechanism. Such mechanisms include:

1. Localized heating of the ground to trigger the lifting of air parcels
2. An orographic effect induced when a hill or mountain forces air upward
3. Convergence at the surface, which may be caused by air being slowed down as it encounters regions of rougher terrain
4. Divergence in the upper atmosphere
5. Lifting along a warm, cold, stationary or occluded front.

Our current understanding of air mass thunderstorms is based on the Thunderstorm Project, which examined such events in Ohio and Florida during the late 1940s. An air mass thunderstorm normally consists of a number of individual updrafts (*cells*), each undergoing a sequence of three distinct stages—cumulus, mature, and dissipative (Figure 11–9).

CUMULUS STAGE The first stage of an air mass thunderstorm begins when unstable air begins to rise, often by the localized convection that occurs as some surfaces undergo more rapid heating than others. Because air mass thunderstorms frequently occur during the late evening when the air is cooling, we know that other processes can also trigger uplift. Regardless of which process causes uplift, the rising air cools adiabatically to form fair-weather cumulus clouds. These initial clouds may exist for just a matter of minutes before evaporating. Although they do not directly lead to any precipitation, the initial clouds play an important role in thunderstorm development by moving water vapor from the surface to the middle troposphere. Ultimately, the atmosphere becomes humid enough that newly formed clouds do not evaporate but instead undergo considerable vertical growth. This growth represents the **cumulus stage** in the air mass thunderstorm.

Clouds in the cumulus stage grow upward at 5 to 20 m/sec (10 to 45 mph). Within the growing clouds, the temperature decreases with height at roughly the saturated adiabatic lapse rate, and a portion of the cloud extends above the freezing level. Ice crystals begin to form and grow by the Bergeron process. The sky rapidly darkens under the thickening cloud; when precipitation begins to fall, the storm enters its next stage of development.

MATURE STAGE The **mature stage** of the air mass thunderstorm begins when precipitation—as heavy rain or possibly hail—starts to fall. As the falling rain or hail drags air toward the surface, downdrafts form in the areas of most intense precipitation. You can observe this process in your own yard. Simply turn on a garden hose full blast and put your hand just outside the stream of water; you will notice a breeze in the direction in which the hose is pointed. The downdrafts are strengthened by the cooling of the air—by as much as 10 °C (18 °F)—that occurs as the precipitation evaporates.

The mature stage marks the most vigorous stage of the thunderstorm, when precipitation, lightning, and thunder are most intense. The top of the cloud extends to an altitude where stable conditions suppress further uplift. Strong winds at the top of the cloud push ice crystals forward and create the familiar anvil shape extending outward from the main part of the cloud.

During the cumulus and mature phases of the storm, an abrupt transition exists between the edge of the cloud and the surrounding unsaturated air. Updrafts dominate the interior of the cloud, while downdrafts occur just outside it. This sets up a highly turbulent situation that encourages entrainment (Chapter 6). The entrainment of unsaturated air causes the droplets along the cloud margin to shrink and cool the cloud by evaporation. The outer part of the cloud becomes more dense and less buoyant, thus suppressing further uplift.

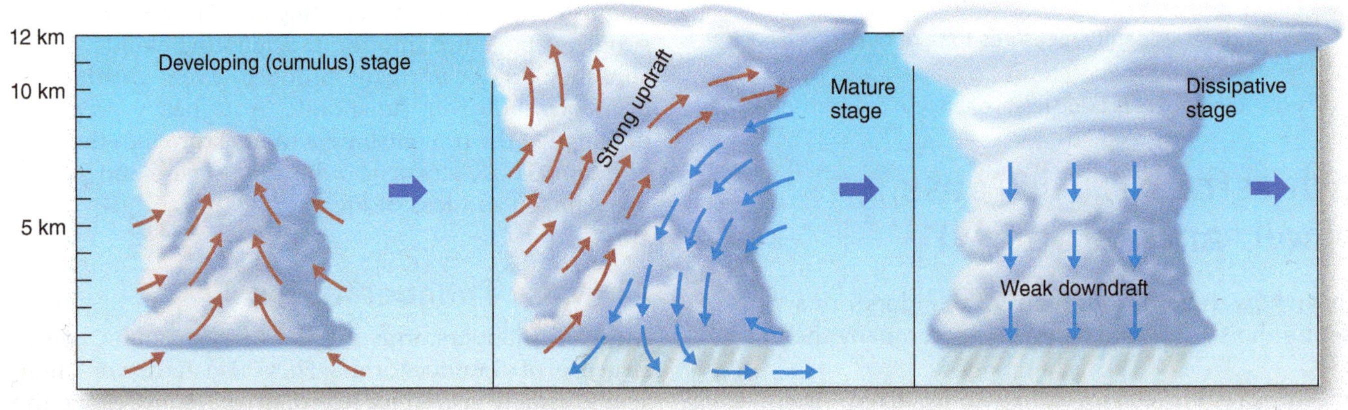

▲ **FIGURE 11–9 Air Mass Thunderstorm.** These storms have three stages: (a) cumulus, (b) mature, and (c) dissipative.

▲ **FIGURE 11–10 Mature Air Mass Thunderstorm.** The part of the cloud on the right of the photo that has a washed-out appearance has become glaciated.

Figure 11–10 shows an air mass thunderstorm. As is typical for thunderstorms in the mature phase, each tower consists of an individual cell and is in a different part of its life cycle. Notice in particular that some of the storm cloud appears washed-out and less well-defined than the rest. Such areas consist entirely of ice crystals, with no liquid droplets, and are said to be *glaciated*. They are not necessarily colder than other parts of the cloud; they are merely old enough so that all the supercooled droplets have had a chance to freeze.

DISSIPATIVE STAGE As more and more of the cloud yields heavy precipitation, downdrafts occupy an increasing portion of the cloud base. When they occupy the entire base, the supply of additional water vapor is cut off and the storm enters its **dissipative stage**. Precipitation diminishes and the sky begins to clear as the remaining droplets evaporate. Only a small portion—perhaps 20 percent—of the moisture that condenses within an air mass thunderstorm actually falls as precipitation. The greatest amount simply evaporates from the cloud.

CHECKPOINT

11.7 What are the necessary conditions for the formation of an air mass thunderstorm?

11.8 What is the series of steps that leads to the formation of a mature air mass thunderstorm?

Multicell and Supercell Storms

Sometimes thunderstorms develop into clusters referred to as **multicell thunderstorms**. These organized groups of thunderstorms are generally referred to as **mesoscale convective systems (MCSs)**. In some cases, MCSs occur as linear bands called **squall lines**. At other times, they appear as oval or roughly circular clusters called **mesoscale convective complexes (MCCs)**. Regardless of how they are arranged, the individual storm cells of an MCS form as part of a single system: They are not just a grouping of individual storms that happen to be near each other. The storm cells develop from a common origin or exist in a situation in which some cells directly lead to the formation of others.

MCSs can bring intense weather conditions to areas covering several counties. They often have life spans of up to 12 hours, but in some cases they can exist for as long as several days. They are fairly common in North America, and in some parts of the central United States and Canada they account for as much as 60 percent of the annual rainfall. Because the surrounding circulation supports an MCS, they lead to much stronger winds and heavier precipitation than is normally found in an air mass thunderstorm.

Even more severe at times are **supercells**, intensely powerful storms that contain a single updraft zone. Although supercells often appear in isolation, they can also occur as a part of an MCS.

Mesoscale convective systems and supercells are capable of producing **severe thunderstorms**. By definition, severe thunderstorms have wind speeds that exceed 93 km/hr (58 mph),[2] have hailstones larger than 2.4 cm (1 in.) in diameter, or spawn tornadoes. The downdrafts and updrafts in severe storms reinforce one another and thereby intensify the storm. This reinforcement usually requires suitable conditions over an area from 10 to 1000 km (6 to 600 mi) across. In other words, most severe thunderstorms get a boost from a mesoscale atmospheric pattern that allows the wind, temperature, and moisture fields to "cooperate" and thereby create very strong storms.

Certain conditions are necessary for the development of all severe thunderstorms. Among these are wind shear, high water vapor content in the lower troposphere, some mechanism to trigger uplift, and a situation called *potential instability*, described earlier in *Box 6–2, Forecasting: Potential Instability.* Note that the requisite wind shear for these storms is a major factor that allows the development of multicell and supercell storms, as opposed to the less intense air mass thunderstorms.

We now briefly describe MCCs, squall lines, supercells, and drylines. Keep in mind that the descriptions of these storm types are quite general, and individual systems might not be easy to categorize. It is also common for storm systems to evolve from one type of system to another.

[2]This seemingly odd value was originally designated as 50 nautical miles per hour, or knots.

MESOSCALE CONVECTIVE COMPLEXES In the United States and Canada, severe weather often arises from MCCs (Figure 11–11). In the most general sense, MCCs are defined as oval or roughly circular organized systems containing several thunderstorms.[3]

Although not all MCCs create severe weather, they are self-propagating in that their individual cells often create downdrafts, leading to the formation of new, powerful cells nearby. To understand how this occurs, imagine a large cluster of thunderstorms. At the surface a flow of warm, humid air comes from the south, and in the middle troposphere the wind flows from the southwest. This setting provides the wind shear necessary for a severe thunderstorm. As we have already seen, the precipitation from each thunderstorm cell creates its own downdraft. The downdraft is enhanced by the cooling of the air as the rain evaporates and consumes latent heat. Upon hitting the ground, the downdraft spreads outward and converges with the warmer surrounding air to form an **outflow boundary** (Figure 11–12).

Figure 11–13 illustrates the progressive movement of cells in an MCC. Initially, at time $t = 0$, five cells labeled *A, B, C, D,* and *E* are moving toward the northeast. Somewhat later, at $t = 1$, cells *B* through *E* have migrated to the northeast. Cell *A,* located farthest to the north, has died out, while along the southern margin of the complex a new cell, *F,* has formed. At $t = 2$, cell *B* has died out, to be replaced by cell *G* along the south. In this manner, the complex of thunderstorms moves eastward, though each individual cell moves to the northeast.

Near the southern side of the MCC, the cold outflow collides with the large-scale southerly surface flow and lifts it upward. The warm, humid air is drawn into the southern edge of the MCC, where it forms new cells. At the same time, the older cells on the northern side of the MCC dissipate because they lack the updrafts needed for replenishment.

SQUALL LINE THUNDERSTORMS Squall line thunderstorms consist of a large number of individual violent storm cells arranged in a linear band, typically about 500 km (300 mi) in length (Figure 11–14). Although squall lines may be as much as 100 km (60 mi) in width, the embedded band of intense showers is much narrower, usually about 5 km (3 mi), and located near the leading edge of the advancing band of cloud cover and precipitation. A wide area of continuous precipitation normally exists in the central and western portions of the squall line, immediately behind the band of heavy showers.

Squall lines usually form in the warm sector of a midlatitude cyclone immediately ahead of the cold front but then advance more rapidly than the front itself. They may eventually reside about 300 to 500 km (180 to 300 mi) ahead of the front. Squall lines are most common over the southern and central United States during spring and summer. The average squall line has a life span on the order of 10 hours, though some have lasted up to four days.

As is the case for all intense thunderstorms, strong vertical wind shear is an important component of squall line

[3]Meteorologists have some precise criteria for classifying a system as an MCC, based on its signature on satellite imagery. We will simply apply the term to organized systems of thunderstorms clustered in a pattern that is closer to circular than linear.

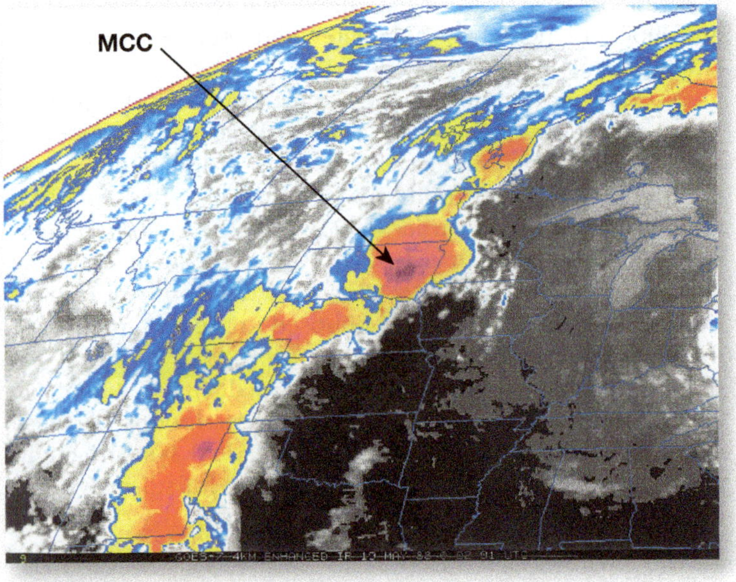

▲ **FIGURE 11–11 Mesoscale Convective Complex.** Satellite image of MCC over eastern South Dakota.

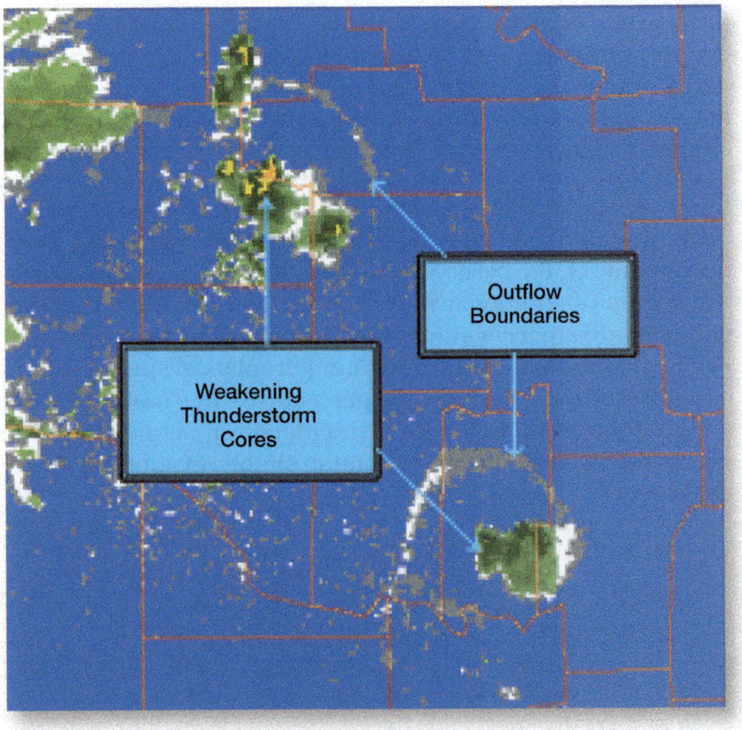

▲ **FIGURE 11–12 Outflow Boundaries.** A radar image highlights two outflow boundaries in relation to weakening thunderstorm cores.

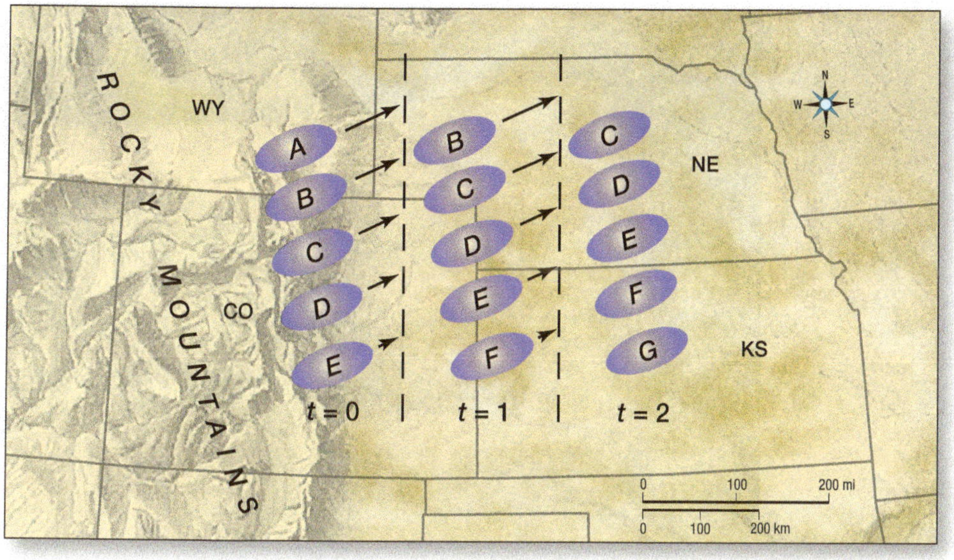

▲ **FIGURE 11–13 Movement of Thunderstorm Cells.** Initially (at $t = 0$) in a mesoscale convective complex, all the cells are moving toward the northeast. Cell A is the oldest, and cell E is the most recently formed. Later ($t = 1$), cell A has dissipated, but a new cell (F), has formed along the southern margin of the complex. At $t = 2$, cell B has dissipated while a new cell (G) has formed.

thunderstorms. As shown in Figure 11–15a, wind velocities in the direction of storm movement typically increase with height. The strong winds aloft push the updrafts ahead of the downdrafts and allow the rising air to feed additional moisture into the storm. As the downdrafts reach the ground, they surge forward as a wedge of cold, dense air, called a **gust front** (Figure 11–15b). Gust fronts act in much the same way as advancing cold fronts by displacing warm air upward. They are often discernible by the cloud of dust picked up from the ground transported by the heavy winds. They can also lift the warm air ahead to form a **shelf cloud** (sometimes called an *arcus*) just above the gust front and ahead of the main portion of the thunderstorm (Figures 11–16 and 11–17). Beneath the leading edge of the gust front, horizontally rotating air can produce a **roll cloud** (see Figure 11–17).

SUPERCELL STORMS Few weather systems are as awesome as a supercell storm (Figure 11–18). With diameters that range from about 20 to 50 km (12 to 30 mi), they are smaller than either squall lines or MCCs. On the other hand, they are usually more violent and provide the setting for most very large tornadoes. Unlike MCCs and squall lines, a supercell storm consists of a single, extremely powerful cell rather than a number of individual cells.[4] Supercell storms also undergo a large-scale rotation absent from squall lines and MCCs. The typical life span of a supercell is 2 to 4 hours.

Despite their single-cell structure, supercell storms are remarkably complex, with the updraft and downdraft bending

[4]Supercells usually occur as isolated storms, though "squall lines" of supercells have been observed.

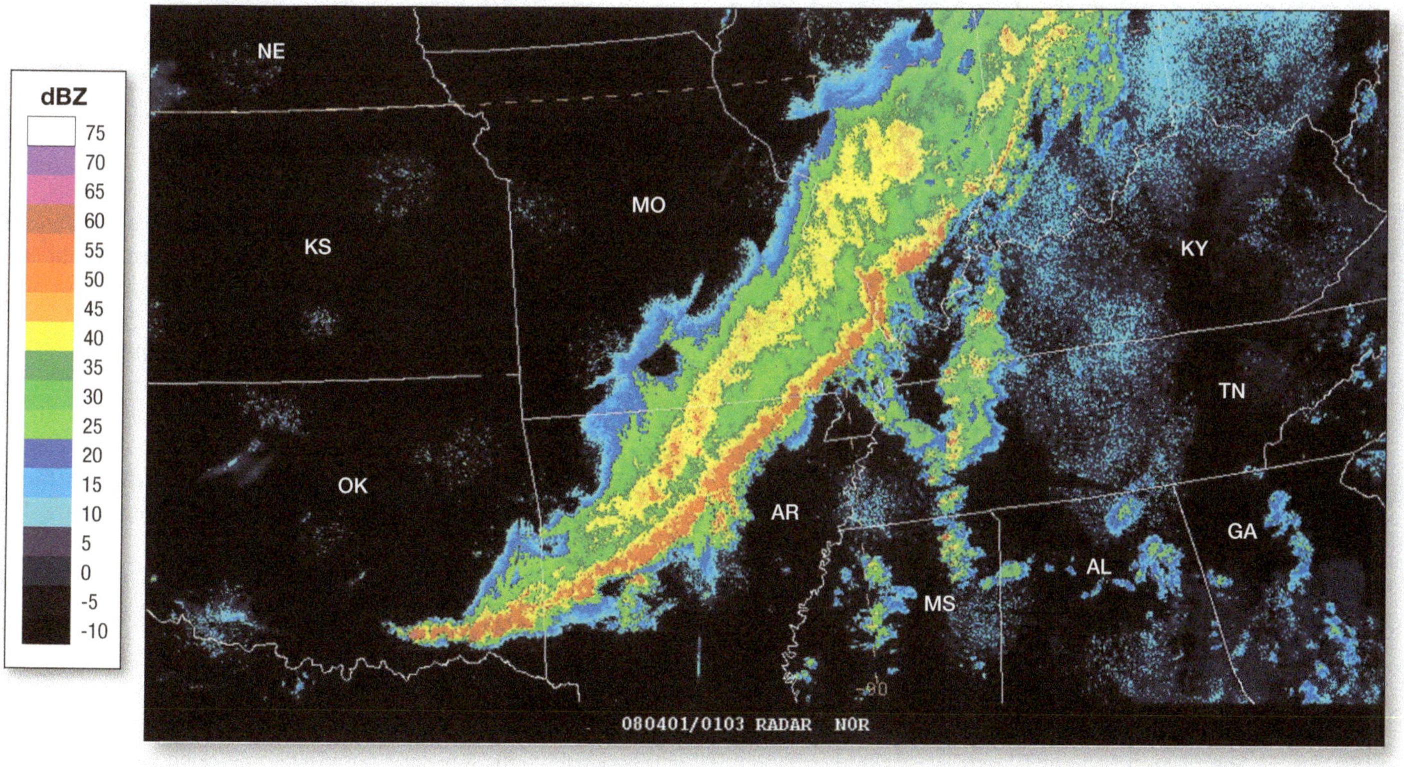

▲ **FIGURE 11–14 Squall Line.** Radar image of a squall line extending from southeastern Oklahoma to Illinois.

▶ **FIGURE 11–15 Squall Line Thunderstorms and Wind Shear.** Squall lines require the presence of wind shear. The arrows in (a) represent the wind speeds with respect to the movement of the storm. The movement of the air within the cumulonimbus is shown in (b). The upper part of the cloud is pushed forward more rapidly than the lower part, which helps to draw in warm, moist air. Note the gust front near the ground ahead of the rain shaft.

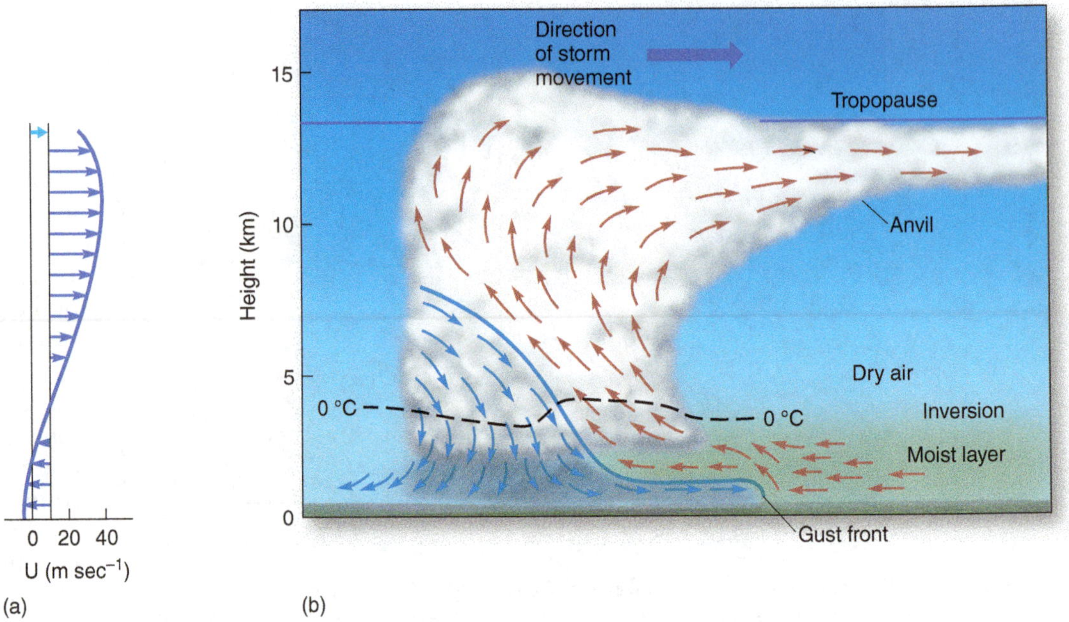

▲ **FIGURE 11–16 Shelf Cloud in Northern Illinois.**

▲ **FIGURE 11–18 Supercell in Oklahoma.** Note the tornado at the far left.

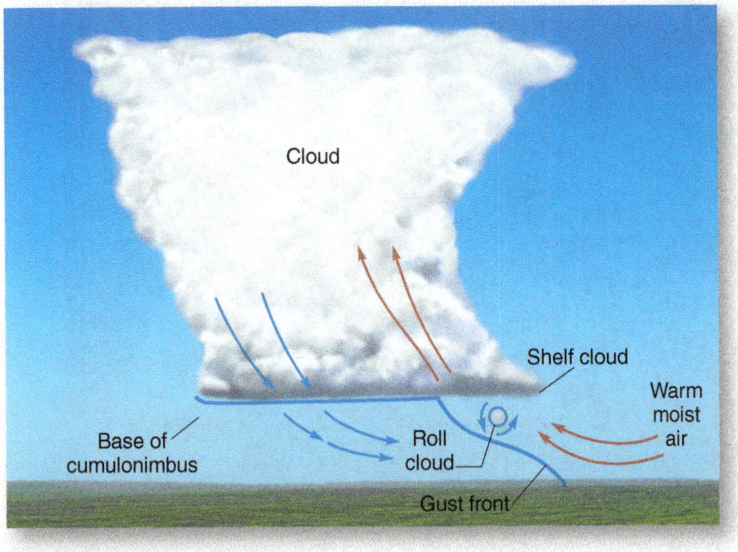

▲ **FIGURE 11–17 Shelf Cloud and Roll Cloud.** These clouds occur along gust fronts.

and wrapping around each other due to strong wind shear (Figure 11–19). As in any other weather system that spawns severe weather, the downdrafts serve to amplify the adjacent updrafts.

Meteorologists follow supercells with tremendous interest. Fortunately, they have at their disposal an extremely useful tool in the form of weather radar. Radar can reveal one of the most noteworthy features of a supercell, called a **hook** (or **hook echo**), which looks like a small appendage attached to the main body of the storm on the radar image (Figure 11–20).[5]

[5]Echoes refer to the way radar waves reflect off cloud constituents, as sound waves echo off canyon walls.

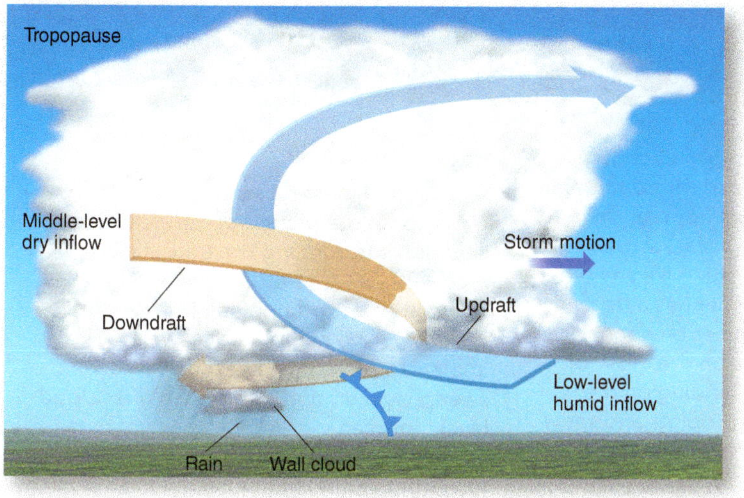

▲ **FIGURE 11–19 Internal Structure of a Supercell.**

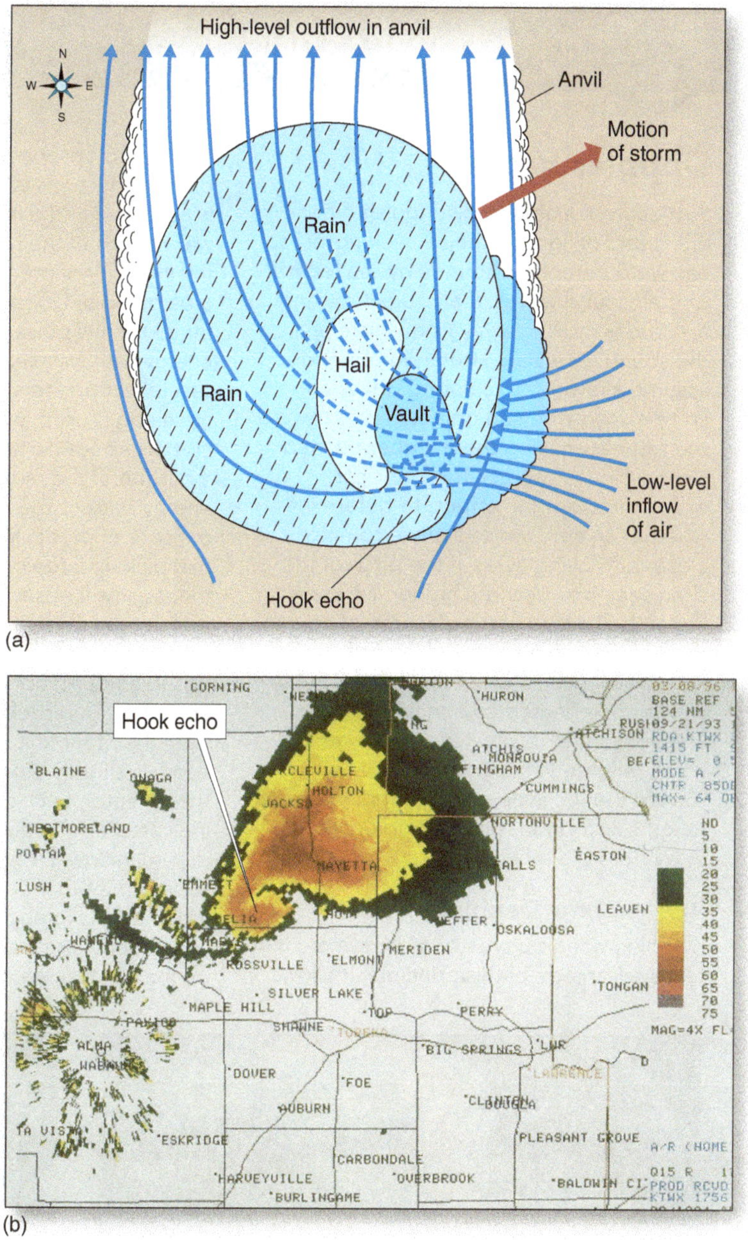

▲ **FIGURE 11–20 Supercells on Radar.** These display certain features, such as a hook echo and vault (a). The vault is a zone of substantial cloud cover that appears on a radar image as a cloud-free zone because of the small size of the cloud droplets. The hook echo extends around the vault. An actual radar image showing a hook is seen in (b).

The zone with no radar return between the main part of the supercell and the hook echo, known as a **vault**, also called the **Bounded Weak Echo Region** (BWER). is where the inflow of warm surface air enters the supercell. The air entering the vault rises, and water vapor condenses to form a dense concentration of water droplets. But the newly formed droplets in the vault are too small to effectively reflect radar waves. Thus, this zone does not show up on the radar image despite the dense concentration of water droplets.

The appearance of hook echoes has long been significant to forecasters, as they often are followed by tornadoes. Today **Doppler radar** can allow us to observe rotation within clouds, providing an even better prediction of tornadoes (see *Box 11–4, Forecasting: Doppler Radar*).

DRYLINES Drylines were briefly described in Chapter 9 as boundaries separating mT and cT air masses. They are most likely to form in spring and early summer over Texas and Oklahoma, where they can be the site of severe weather. Drylines develop as moist air near the surface from the Gulf of Mexico flows northward, separated from the dry air to the west. Terrain effects are important, as the elevation of the surface gradually increases westward from east Texas and Oklahoma toward the Rocky Mountains. Dry air flowing eastward out of the high plains overrides the moist air below, creating a situation called *potential instability* (described in Chapter 6). Potential instability can lead to severe storm activity if enough lifting occurs below the boundary of the warm and moist air. This can occur due to surface convergence as the mT air from the Gulf of Mexico and the westerly flow of cT air collide along the dryline.

CHECKPOINT

11.9 Identify the two types of mesoscale convective systems and explain how they differ from each other and from supercells.

11.10 What are drylines and why are they important?

Downbursts, Derechos, Microbursts, and Haboobs

Video
Identifying Tornadic Thunderstorms Using Radar Reflectivity Data
http://goo.gl/Sbbm0j

We have seen how downdrafts are an important feature of thunderstorms, especially in the maintenance of severe thunderstorms. Strong downdrafts may also create **downbursts**, potentially deadly gusts of wind that can reach speeds in excess of 270 km/hr (165 mph). When strong downdrafts reach the surface, they can spread outward in all directions to form intense horizontal winds capable of causing severe damage at the surface. In fact, damage attributed to tornadoes may in some cases be the result of downbursts.

11–4 FORECASTING

Doppler Radar

Just as we are able to distinguish different colors of light by their wavelengths, so can we differentiate sounds by the length of their sound waves. If an object making a sound is moving away from a listener, the sound waves are stretched out and assume a lower pitch. Sound waves are compressed when an object moves toward the listener, making them higher pitched. Unconsciously, we use this principle, called the **Doppler effect**, to determine whether an ambulance siren is coming closer or moving away. If the pitch of the siren seems to become higher, we know the ambulance is getting nearer (of course, the siren would also sound louder). A similar process occurs when electromagnetic waves are reflected by a moving object: The light shifts to shorter wavelengths when reflected by an object moving toward the receiver and to longer wavelengths as it bounces off an object moving away from the receiver.

Applying the Doppler Effect

Doppler radar is a type of radar system that takes advantage of this principle. It allows the user to observe the movement of raindrops and ice particles (and thus determine wind speed and direction) from the shift in wavelength of the radar waves, as well as the intensity of precipitation. Like any other type of radar, Doppler radar has a transmitter that emits pulses of electromagnetic energy with wavelengths on the order of several centimeters. Depending on the wavelength used, water droplets and snow crystals above certain critical sizes reflect a portion of the radar's electromagnetic energy back to the transmitter/receiver. In the case of particularly violent tornadoes that pick up large objects from the ground, the radar will observe this airborne material and display it as a *debris ball*.

Doppler radar is special in its ability to observe the motion of the cloud constituents. If a cloud droplet is moving away from the radar unit, the wavelength of the beam is slightly elongated as it bounces off the reflector. Such reflections are normally indicated on the display monitor as reddish to yellow. Likewise, a droplet moving toward the radar unit undergoes a shortening of the wavelength. Echoes from these constituents are displayed as blue or green on the radar screen.

Radar Scans

A radar unit must rotate 360 degrees to get a complete picture of the weather situation surrounding the transmitter/receiver unit. When the transmitter makes one complete rotation at a fixed angle, it is said to have completed a **sweep**. The angle can then be increased as a second sweep is taken that depicts a higher cloud level. This can be repeated several times so that the radar can peer into multiple levels of the cloud. The compilation of all the individual sweeps takes approximately 5 to 10 minutes and produces a **volume sweep**.

Figure 11–4–1 shows a pair of Doppler radar images of a major storm near Dallas–Fort Worth, Texas, on March 29, 2000. Figure 11–4–1a shows the reflectivity of the storm, with redder regions indicating intense precipitation and green areas representing less intense precipitation. The white arrows point toward a hook echo (described in the main text of this chapter). Figure 11–4–1b displays the *storm radial velocity (SRV) pattern*, which describes the motions taking place within the cloud. SRV displays use redder colors to represent winds blowing away

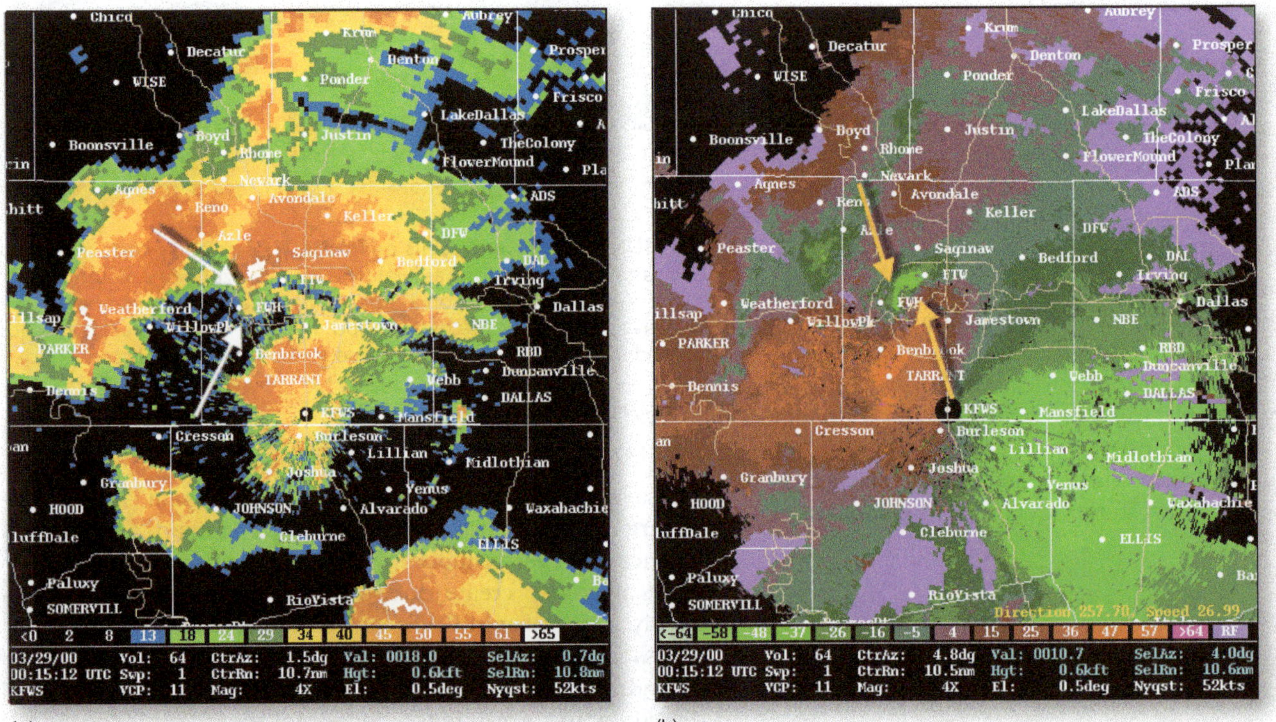

(a) (b)

Video Identifying Tornadic Thunderstorms Using Radar Velocity Data
http://goo.gl/szdHJQ

▲ **FIGURE 11–4–1 Doppler Radar Images.** A storm near Dallas–Fort Worth, Texas, on March 29, 2000. Part (a) depicts the intensity of precipitation; part (b) shows the storm radial velocity (SRV) pattern, which is the movement of different parts of the storm toward or away from the radar unit.

from the radar and green to indicate movement toward the radar. The yellow arrows on this image highlight a region of counterclockwise rotation. As we discuss later in the chapter, this pattern, called a *mesocyclone,* sometimes precedes the formation of a tornado. After the onset of rotation, it takes only 30 minutes or so for the tornado to form, which allows meteorologists to give warnings in advance. In this particular case, a tornado did hit the city of Fort Worth. A study published in 2005 concluded that the implementation of Doppler radar units across the United States has prevented 79 fatalities and 1050 injuries per year. Doppler radar has increased the average warning time for all tornadoes to 13 minutes. Prior to Doppler radar, warnings typically did not go out until 4 minutes after touchdown.

MapMaster

North America Physical Environment

Doppler Radar Sites (U.S.)

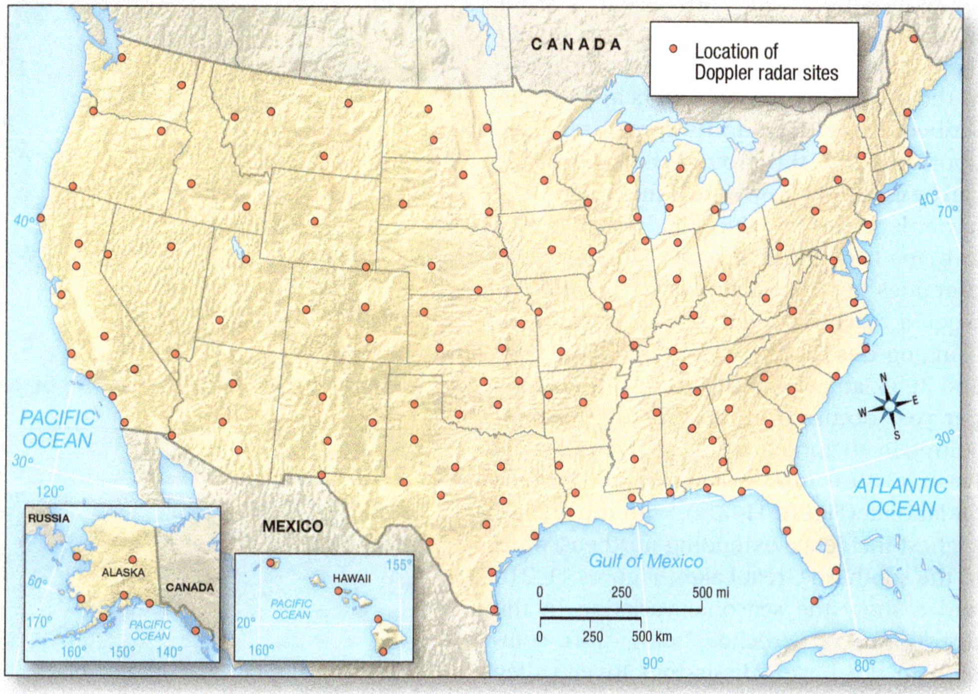

▲ **FIGURE 11–4–2 Doppler Radar Sites in the United States.**

The NEXRAD Network

Today about 160 Doppler sites are scattered across the United States as part of the NEXRAD network (Figure 11–4–2). The National Weather Service operates 113 of these sites; the rest are owned by the Federal Aviation Administration and the Department of Defense. In part because of budgetary cutbacks, the Atmospheric Environment Service of Canada has just a handful of Doppler radar installations. Because both sides of the border area tend to be heavily populated, Doppler radar from the United States provides extensive coverage of severe storms that could affect many large Canadian urban centers.

NEXRAD is also useful for flood forecasting, providing continual precipitation estimates over large areas. Doppler radar can sometimes observe wind movements even when no clouds exist, as large clusters of flying bugs or heavy dust concentration scatter radar waves back toward the transmitter. The resultant echoes are called *clear air echoes.*

Just as the introduction of Doppler radar represented a major breakthrough for monitoring severe weather, the introduction of dual-polarization technology (Chapter 7) marks a revolutionary breakthrough for forecasting. Unlike earlier versions of Doppler radar, dual-polarization radar gives information about both the vertical and horizontal dimensions of ice particles and water droplets in clouds, thus allowing a better identification of hail. In addition, it is particularly useful for identifying flying tornado debris, allowing more precise identification of the location and movement of tornadoes.

Portable Doppler Radar

The NEXRAD network of Doppler radar, installed primarily for forecasting purposes, provides information on tornadoes that has proven useful to researchers. But NEXRAD radar units are spaced too far apart to provide close scrutiny of most passing tornadoes, so researchers have looked to transportable radar to observe tornadoes at close range. For this reason, tornado researchers have come to rely on portable Doppler radar units called *Doppler on wheels* that have units mounted to a flatbed truck (Figure 11–4–3). These units have been instrumental in acquiring new information on tornado dynamics.

▲ **FIGURE 11–4–3 Doppler on Wheels in Nebraska.**

11–4–1 What is the guiding principle behind Doppler radar?

11–4–2 Explain how the use of Doppler radar allows forecasters to provide early warnings of tornado development.

Strong downdrafts associated with mesoscale convective systems can produce very powerful, larger-scale horizontal winds called **derechos** (a Spanish word meaning "straight ahead"). Such winds may last for hours at a time and achieve speeds higher than 200 km/hr (120 mph)—comparable to those of many tornadoes. As already described, these downdrafts spread outward upon reaching the ground. But the winds can be especially powerful if the descending upper-level air has high wind speeds prior to being brought downward. This momentum does not disappear as the air sinks, so when it reaches the surface, it is capable of bringing destructive winds. Between 1986 and 2003, an average of nearly 21 derechos per year occurred in the United States, resulting in an annual average of 8.5 fatalities. Derechos are most common over eastern Oklahoma (Figure 11–21a), with the area of highest incidence extending northeastward to the southern Great Lakes. Figures 11–21b and c show the seasonal variation in the distribution of derechos, being more common in the lower Mississippi River valley between September and April. Over the entire United States, derechos are most prevalent between May and July (Figure 11–22).

On June 29, 2012, a particularly devastating storm carried derecho winds across a 965 km (600-mi) zone extending from Northern Indiana to the middle Atlantic states. Winds consistently exceeded some 97 km/hr (60 mph), with isolated areas experiencing gusts in excess of 161 km/hr (100 mph). In all, 13 people were killed, mostly by falling trees, and power was cut off to as many as 4 million people for up to one week. Figure 11–23 shows the path of this storm.

Downbursts with diameters of less than 4 km (2.5 mi), called **microbursts**, can produce a particularly dangerous problem when they occur near airports (see *Box 11–5, Focus on Aviation: Microbursts*).

Very strong horizontal winds created by downdrafts over desert regions can dislodge sand and dust from the surface and incorporate them into the wind gusts. These winds can be remarkably turbulent and disburse the sand and dust to heights as great as 3 km (10,000 ft) above the ground. These winds can persist for hours at a time and typically advance at speeds of about 50 km/hr (30 mph), creating an ominous wall of loose material that appears like a very dense, dark cloud.

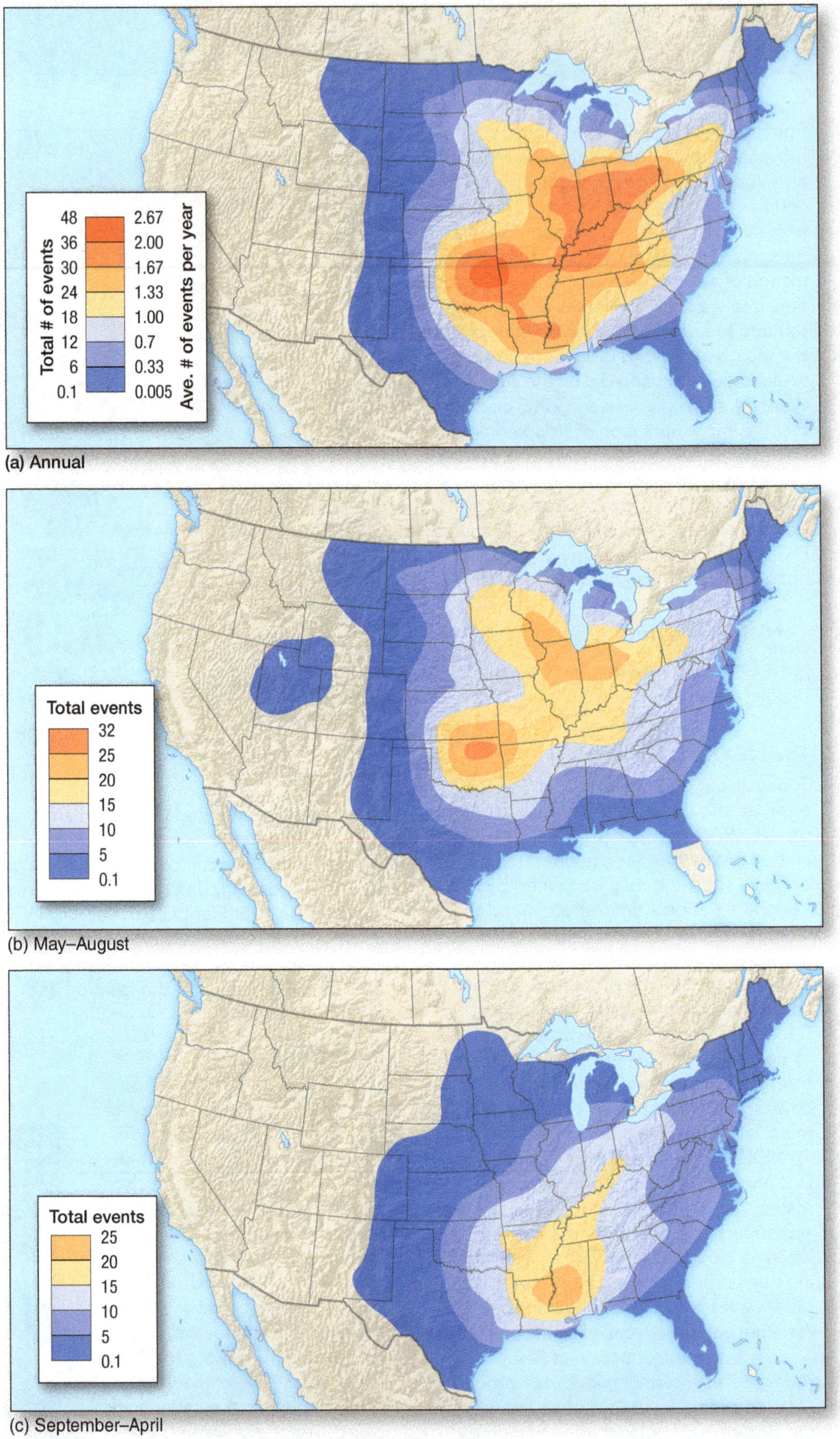

▲ **FIGURE 11–21 Derechos Across the United States.** Frequency of derechos (a) annually, (b) between May–August, and (c) between September–April.

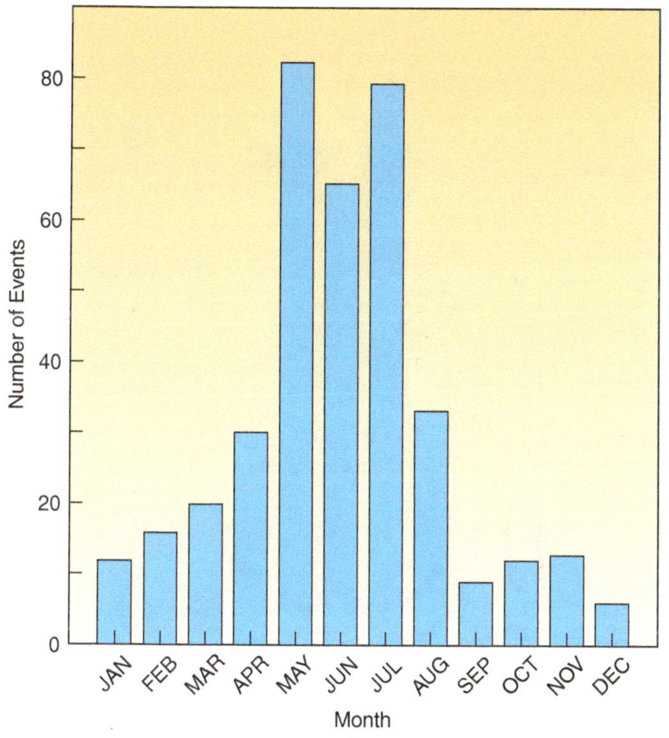

▲ **FIGURE 11–22 Derechos.** Frequency of derechos across the United States by month.

These sandstorms are called **haboobs**, from the Arabic word meaning "wind." They are most common in northern and central Sudan, occurring perhaps two dozen times each year. A very dramatic haboob occurred over central Arizona in July 2011 and attracted considerable media attention.

CHECKPOINT

11.11 Briefly define each of the following terms in your own words: MCCs, squall line, supercell.

11.12 How are downbursts, derechos, microbursts, and haboobs similar? How are they different?

Floods and Flash Floods

11.4 Summarize the characteristics and processes involved the formation of floods and flash floods.

Sometimes thunderstorms can create enormous problems simply by delivering excessive amounts of water that create **floods** and **flash floods**. Floods can occur under a lot of different conditions. Sometimes a deep snowpack will melt

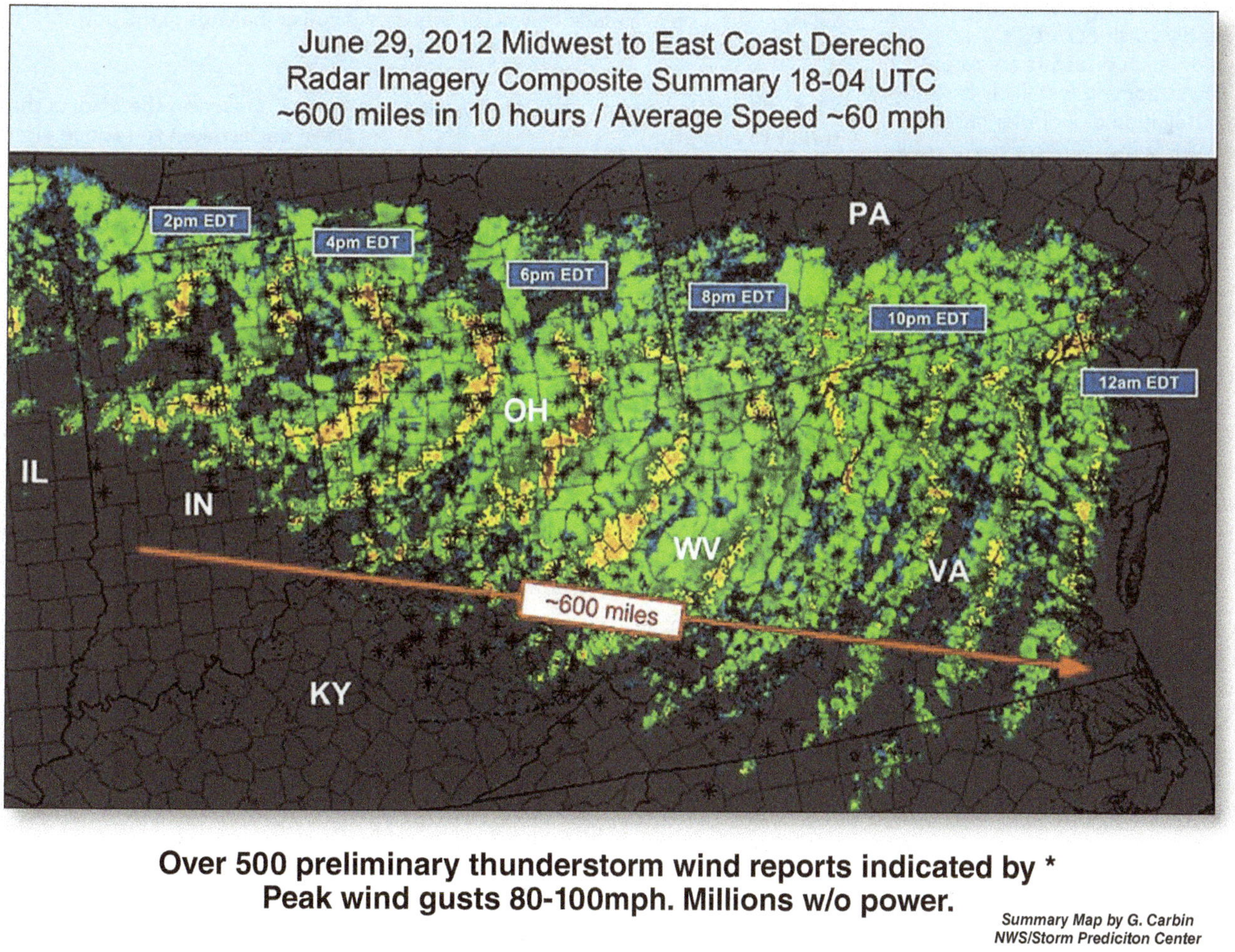

▲ **FIGURE 11–23** 2012 Derecho Path.

11–5 FOCUS ON AVIATION

Microbursts

Microbursts can pose a serious threat to aircraft, especially during takeoffs and landings (Figure 11–5–1). The horizontal spreading of a microburst creates strong wind shear when it reaches the surface. For example, air may flow westward on one side of the microburst while spreading eastward on the opposite side. Imagine what this might do to an aircraft attempting to land in a microburst. As the plane enters the microburst, a headwind provides lift, to which the pilot might respond by turning the aircraft downward. As soon as the plane passes the core of the downdraft, however, the headwind not only disappears, it is replaced by a tailwind, decreasing lift. Coming after the pilot's earlier downward adjustment, this causes the plane to abruptly drop in altitude. Because the plane is not far above the ground when these events occur, the pilot may not have time to compensate before a deadly crash occurs.

Fortunately, such disasters are rare. They are also becoming less likely because the installation of Doppler radar at about 40 U.S. airports has proven highly effective at detecting microbursts, with a detection rate of about 95 percent.

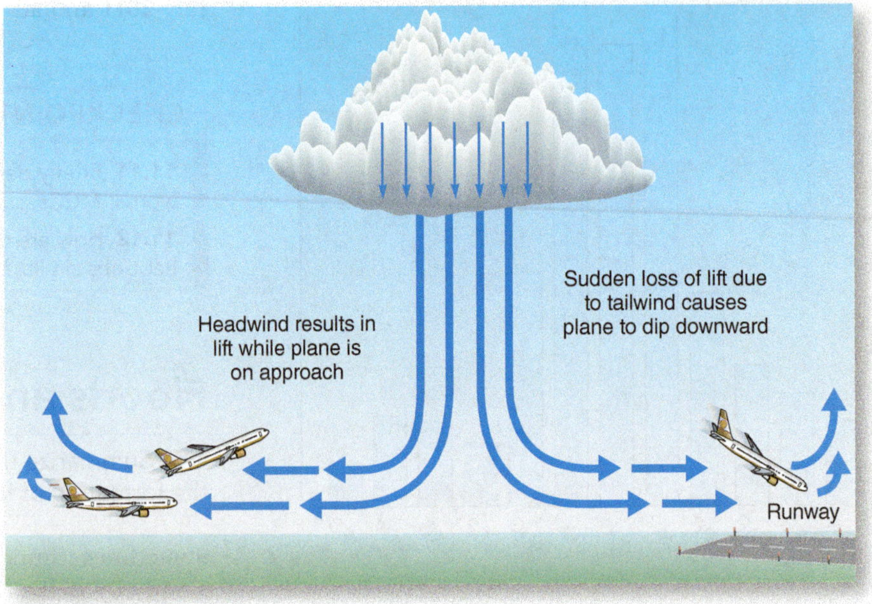
Headwind results in lift while plane is on approach
Sudden loss of lift due to tailwind causes plane to dip downward
Runway

▲ **FIGURE 11–5–1** Microbursts can make aircraft landing and takeoff perilous. A plane flying into the headwinds of a microburst gets a sudden increase in lift. This lift suddenly disappears and is replaced by a tailwind as it exits the downdraft, thereby reducing the lift. If the pilot overcompensates and guides the plane downward while entering the downdraft, a dangerous drop in altitude may occur. Notice the curl at the ends of the downdrafts, which mark the outer limit of the microburst at the ground.

11–5–1 Explain how microbursts create a threat to aircraft.

11–5–2 Describe the efforts that have been undertaken to reduce aircraft vulnerability to microbursts.

rapidly as warm rain falls, and the combination of meltwater and rain is rapidly routed into a stream system. Flooding might also occur as repeated or prolonged rainfall occurs over an area. Initially the rainfall might be absorbed into the soil, but if the soil becomes nearly saturated it may no longer be able to take in further moisture, and the excess runs into streams and rivers, which then overflow their banks.

Floods like these may or may not be associated with thunderstorms. Flash floods, on the other hand, are far more likely to occur with the high-intensity precipitation associated with thunderstorms. Flash floods are most likely to occur when a strong thunderstorm remains over an area sufficiently long that heavy runoff immediately occurs and flows into stream channels and other low-lying areas. They are sometimes observed in areas that receive very little precipitation on average, such as the desert Southwest of the United States. Over a very short period of time, dry channels can become choked with rapidly moving water capable of carrying away vehicles whose drivers decided to take a risk and attempt a crossing.

Sometimes flash floods and longer-lived floods can occur as part of the same weather system. This type of situation was exemplified by flooding in September 2013 along the Front Range of Colorado, extending east along river banks into the Great Plains (Figure 11–24). The flooding began on September 9 as a passing cold front triggered heavy thunderstorms over the Denver area. During the next two days the rainfall became more intense and widespread, as very moist upper-level flow combined with a stationary front and orographic uplift to produce heavy rainfall along much of the Front Range of the Rockies. Between the late afternoon on September 11 and the early morning of September 12, as much as 23 cm (9 in.) of rain were locally recorded, with Boulder and nearby counties hardest hit. River flow through Boulder Canyon was so intense that classes at the University of Colorado in Boulder were cancelled for the last two days of the week, and some parts of the campus had substantial flood damage.

Within about a week's time, a large area ended up with an average year's worth of precipitation. Some areas fared better than others. For example, the city of Boulder had already taken to heart many of the lessons taught in University of Colorado classrooms and had incorporated safety measures, such as buying and removing buildings from particularly

▲ **FIGURE 11–24 Damage from 2013 Floods along Thompson River near Boulder, Colorado.**

vulnerable locations and elevating bridges to remain above unusually high river flows. Over the entire 17-county area, the death toll included at least seven victims. At least 1800 homes were destroyed and another 16,000 damaged.

CHECKPOINT

11.13 What are flash floods and how are they different from other floods?

11.14 Describe why flash floods and other floods are potentially dangerous, and how humans can minimize their risk from these events.

Distribution of Thunderstorms

11.5 Analyze the geographic and temporal distribution of thunderstorms.

Thunderstorms are extremely common across much of the globe, numbering some 14.5 million per year. They are most likely to develop where moist air is subject to sustained uplift; not surprisingly, such conditions occur most commonly in the tropics. Until recently the occurrence of lightning in low-populated or economically underadvantaged areas precluded the gathering of reliable statistics across the globe. Fortunately, satellites have provided high-quality observations of lightning incidence since the mid-1990s (Figure 11–25). Lightning strikes most frequently over the Congo basin of central Africa and occurs frequently over many other regions of the world. Colder and less humid regions typically have a lower incidence of lightning. Outside of the tropical regions, lightning is most common during the summer months when midday sun angles are greatest, while equatorial regions usually have their peak activity near the equinoxes (Chapter 2), due to abundant solar radiation.

Detailed data on cloud-to-ground lightning flashes have been assembled using ground-based data from the Canadian Lightning Detection Network (CLDN) and the National Lightning Detection Network (NLDN) of the United States.

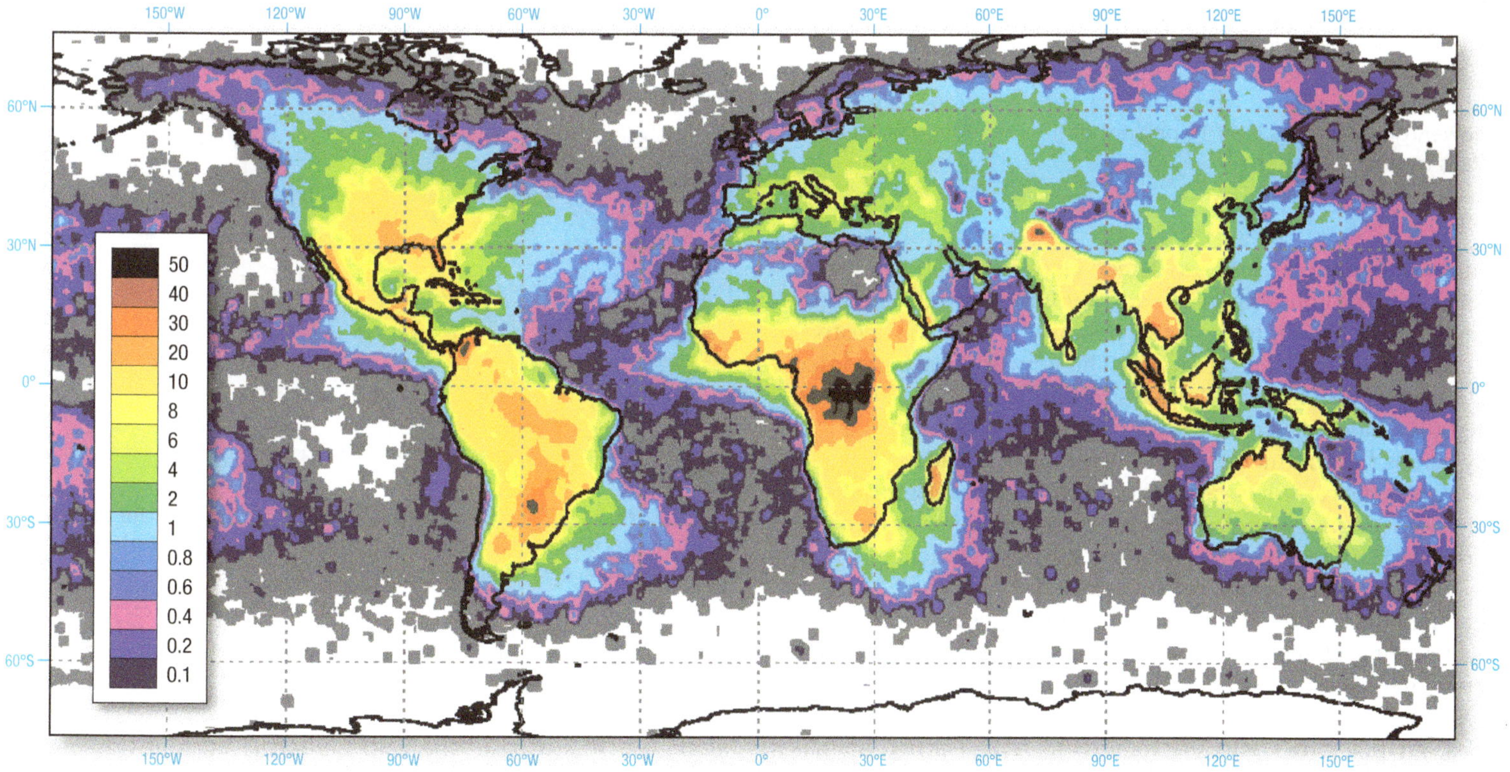

▲ **FIGURE 11–25 Worldwide Lightning Flashes.** Data from space-based optical sensors reveal the uneven distribution of lightning flashes. Units: flashes/km^2/yr.

MM MapMaster World Physical Environment: Lightning Strikes

Video
Thundersleet and Thundersnow
http://goo.gl/GOm7IV

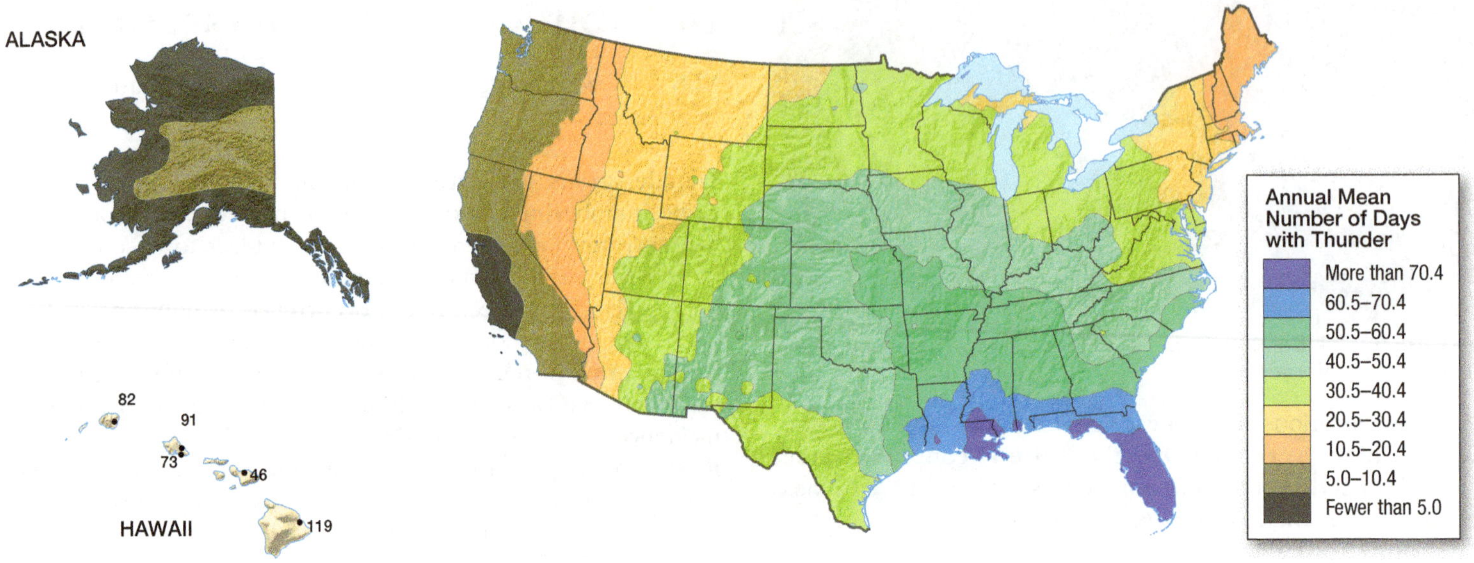

▲ **FIGURE 11–26 Thunder.** The average number of days annually in which thunder is heard in the United States.

MapMaster World Physical Environment: Thunderstorm Occurrences per Year

DID YOU KNOW?

Thunderstorms account for a large part of the total rainfall that occurs over much of North America. Thunderstorms provide almost half the total rainfall that occurs over the Mississippi River Valley, which covers 41 percent of the conterminous 48-state land area. In the south-central United States they account for as much as 70 percent of total precipitation. Across the western portion of North America, the percentages are far lower; generally less than 10 percent of the rainfall in that area falls from thunderstorms.

The two networks' sensors detect electromagnetic radiation emitted during lightning strikes and automatically relay information on the location, timing, and polarity (positive or negative) of each stroke to a central facility. The networks have recorded an average of 28 million cloud-to-ground lightning flashes and 100,000 thunderstorms (280 flashes per thunderstorm) annually across the continental United States and Canada.

Across the eastern two-thirds of North America (Figure 11-26) there is a general pattern of decreasing thunderstorm activity northward. By far the state with the greatest incidence is Florida, where thunderstorms occur on average more than 100 days per year. Much of the necessary lifting of air results from strong solar heating of the surface. But the situation over the Florida peninsula is unique within the continental United States, because the area is almost completely surrounded by warm water. Thus, air that flows into the interior to replace lifted air has a very high moisture content, which in turn supports heavy precipitation and the development of thunderstorms. The area of the United States and Canada west of the Rocky Mountains (except for Hawaii) experiences considerably fewer thunderstorms than the eastern two-thirds of the continent. This is largely explained by the presence of the cold ocean current along the West Coast. The cold California current (Chapter 8) keeps humidities lower than they typically are in the East and also tends to keep the air more stable (Chapter 6). Both the availability of moisture and easily uplifted, unstable air are important for the formation of thunderstorms.

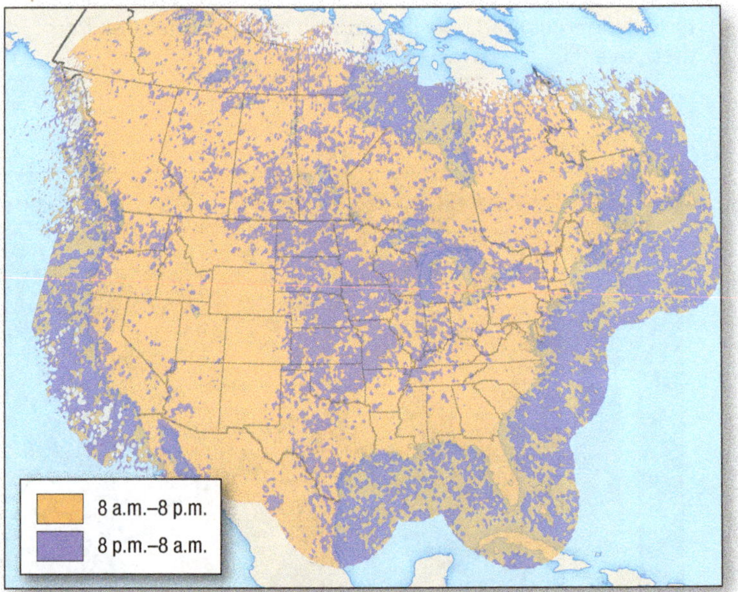

▲ **FIGURE 11–27 Daytime and Nighttime Lightning.** The spatial distribution of areas where the majority of lightning takes place from 8 A.M. to 8 P.M. and 8 P.M. to 8 A.M.

MapMaster North America Physical Environment: Lightning Strikes

Figure 11–27 illustrates the areas that have the maximum number of lightning strikes during the daytime

(8 A.M. to 8 P.M. local standard time) and those occurring during the 12 hours following 8 P.M. Much of North America exhibits a daytime maximum in lightning flashes, though much of the midcontinent has a greater likelihood of nighttime events.

CHECKPOINT

11.15 From a global perspective, where are the conditions necessary for the formation of a thunderstorm most commonly found? Explain.

11.16 Look at Figure 11–26. What can you infer about climate in the United States from the distribution of annual mean number of days with thunder?

Tornadoes

11.6 Explain how, where, and when tornadoes form.

The large hail and strong winds of a severe thunderstorm can bring widespread destruction, but even hailstorms are relatively tame compared to **tornadoes** (Figure 11–28). Tornadoes are zones of extremely rapid, rotating winds beneath the base of cumulonimbus clouds. Though the overwhelming majority of tornadoes rotate cyclonically (counterclockwise in the Northern Hemisphere), a few spin in the opposite direction. Some appear as very thin, rope-shaped columns, while others have the characteristic funnel shape that narrows from the cloud base to the ground. Regardless of their shape or spin, tornadoes are extremely dangerous.

Strong tornadic winds result from extraordinarily large differences in atmospheric pressure over short distances. Over just a few tenths of a kilometer, the pressure difference between the core of a tornado and the area immediately outside the funnel can be as great as 100 mb. To put this in perspective, on a typical day the highest and lowest sea-level pressure across all of North America may differ by only about 35 mb—and this difference exists over horizontal distances of up to thousands of kilometers.

Tornado Characteristics and Dimensions

It is difficult to generalize about tornadoes because they occur in a wide variety of shapes and sizes. While most have diameters about the length of a football field (100 yards or so), some are 15 times as large. Usually they last no longer than a few minutes, but some have lasted for several hours. Tornadoes normally travel northeastward at speeds comparable to a car driving down a city street—about 50 km/hr (30 mph). A typical tornado covers about 3 or 4 km (2 to 2.5 mi) from the time it touches the ground to when it dies out.

Estimates of wind speeds within tornadoes are based primarily on the damage they have produced. The weakest have wind speeds as low as 105 km/hr (65 mph); the most severe are in excess of about 450 km/hr (280 mph).

Tornado Formation

Tornadoes can develop in any situation that produces severe weather—frontal boundaries, squall lines, mesoscale convective complexes, supercells, and tropical cyclones (see Chapter 12). The processes that lead to their formation are not very well understood. Typically, the most intense and destructive tornadoes arise from supercells.

SUPERCELL TORNADO DEVELOPMENT In a supercell storm, the first observable step in tornado formation is the slow, horizontal rotation of a large segment of the cloud (up to 10 km—6 mi—in diameter). Such rotation begins deep within the cloud interior, several kilometers above the surface. The resulting large vortices, called **mesocyclones**, often precede the formation of the actual tornado by some 30 minutes or so.

(a)

(b)

▶ **FIGURE 11–28 Tornadoes.** Tornadoes come in a wide range of shapes and sizes. (a) A tornado in Mulvane, Kansas on June 12, 2004 with considerable tilt makes a sharp bend near the cloud base. (b) A more nearly vertical and symmetrical tornado near Rago, Kansas on May 19, 2012.

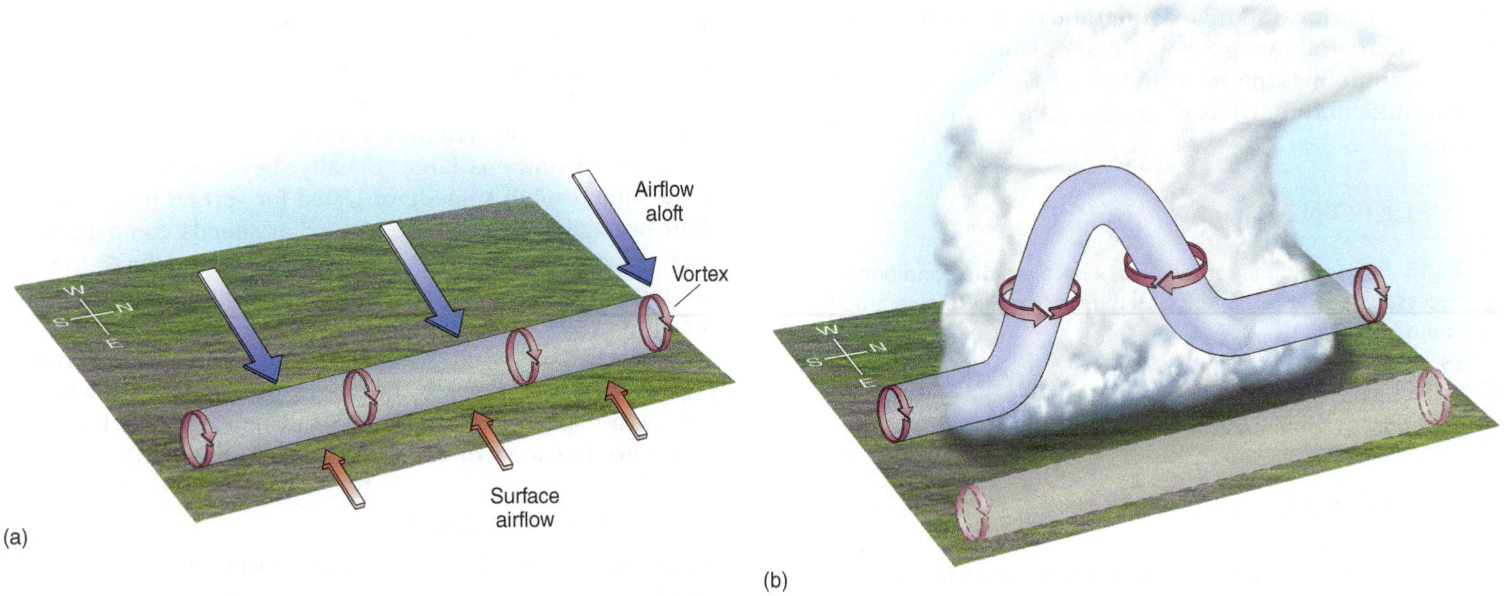

▲ **FIGURE 11–29 Mesocyclones.** These zones of rotation can form when a horizontal vortex of air (a) becomes tilted upward, (b) yielding vertical counterclockwise and clockwise rotating vortices.

▶ **FIGURE 11–30 Wall Clouds in Oklahoma.** The wall clouds in (a) and (b) protrude below the base of the cloud.

The formation of a mesocyclone depends on the presence of vertical wind shear. Moving upward from the surface, the wind shifts direction and its speed increases. This wind shear causes a rolling motion about a horizontal axis, as shown in Figure 11–29a. Under the right conditions, strong updrafts in the storm tilt the horizontally rotating air so that the axis of rotation becomes approximately vertical (Figure 11–29b). This provides the initial rotation within the cloud interior.

Intensification of the mesocyclone requires that the area of rotation decrease, which leads to an increase in wind speed.[6] The narrowing column of rotating air stretches downward, and a portion of the cloud base protrudes downward to form a **wall cloud** (Figure 11–30). Wall clouds form where cool, humid air from zones of precipitation is drawn into the updraft feeding the main cloud. The cool, humid air condenses at a lower height than does the air feeding into the rest of the cloud. Wall clouds most often occur on the southern or southwestern portions of supercells, near areas of large hail and heavy rainfall (Figure 11–31). They are particularly noteworthy because the most significant tornadoes associated with supercells usually form within or near wall clouds. For this reason professional storm chasers[7] often try to position

[6]This is another application of the conservation of angular momentum, initially described in Chapter 8.

[7]Storm chasing can be extremely dangerous if undertaken without the supervision of a professional. Do not ever attempt it on your own—*always* treat tornadoes with the utmost caution.

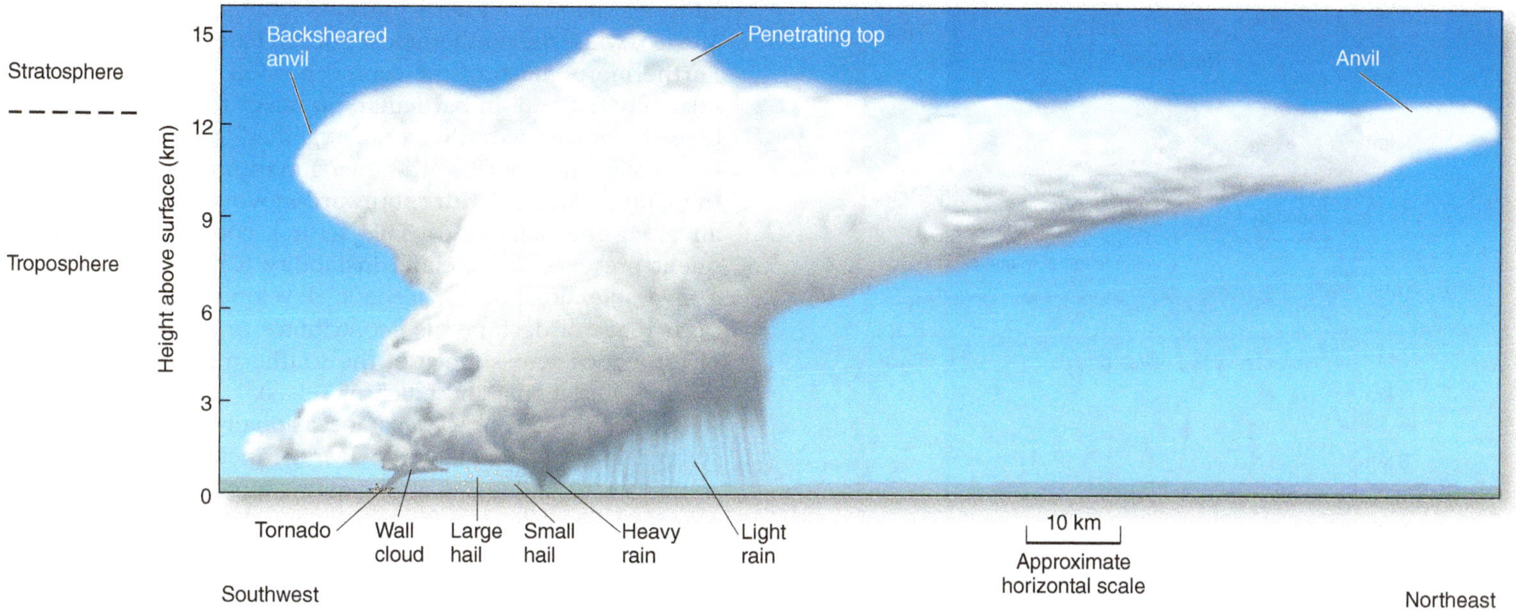

▲ **FIGURE 11–31 Idealized Supercell.** Typical pattern of features with a wall cloud on the southwestern portion.

themselves ahead and to the right of the position of the wall cloud, relative to the storm's motion. From this vantage point they maximize their chances of not having the tornado obscured by rainfall while minimizing their chances of being hit.

Funnel clouds form when a narrow, rapidly rotating vortex emerges from the base of the wall cloud. As air is drawn upward into the zone of rotation, condensation and the importation of dust and debris from the surface may give the funnel a dark, ominous appearance. A funnel cloud has all the characteristics and intensity of a true tornado; the only difference between the two is that the strong rotation has not reached the ground in the case of a funnel cloud.

Animation
Tornado Wind Patterns
http://goo.gl/luOPiY

Because Doppler radar enables forecasters to observe the rotating winds of mesocyclones, the network has greatly increased lead times in issuing tornado warnings (refer again to *Box 11–4, Forecasting: Doppler Radar*). Only about 20 percent of all mesocyclones actually spawn tornadoes, however. Exactly why some mesocyclones produce tornadoes and others do not is unknown; for that reason, forecasters cannot tell in advance which mesocyclones will produce tornadoes. Despite these uncertainties, Doppler radar has proven itself to be the most important tool yet developed in "nowcasting" (making short-term predictions about) the formation of tornadoes.

NONSUPERCELL TORNADO DEVELOPMENT The exact mechanisms that lead to nonsupercell tornadoes are also poorly understood, though recent research indicates that these tornadoes have their origins nearer to the surface than do those that begin as mesocyclones. Figure 11–32 illustrates one situation that may lead to nonsupercell tornadoes. The arrows show the outflow of air from two thunderstorm regions: at the top left and bottom right of the figure. From the bottom left to the top right of the figure is a zone of convergence between the two masses of air. At certain areas along the convergence zone (the circled areas), strong rotation develops. Another possible mechanism is shown in Figure 11–33, where strong convection along the convergence zone causes uplift and the formation of a cumulus cloud. In Figure 11–33c, the cloud develops into a cumulonimbus, and the strong rotation stretches down from the cloud base to form a funnel cloud.

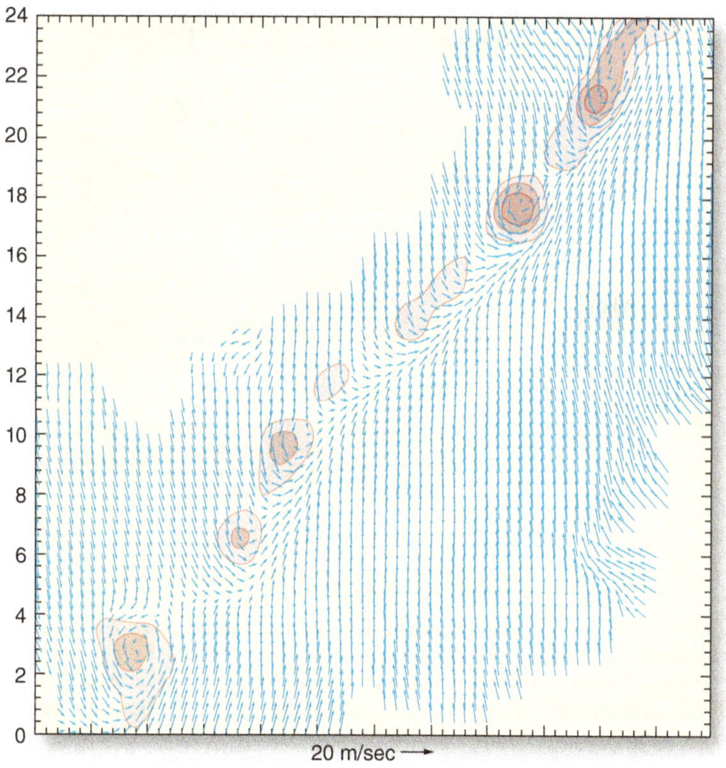

▲ **FIGURE 11–32 Nonsupercell Tornadoes.** These are weaker tornadoes that may develop where the outflow from separate storm downdrafts causes convergence. The arrows depict the wind flow over a region of about 400 sq km. The circled areas, organized from the bottom left to upper right, represent zones of rapid rotation where tornadoes appeared.

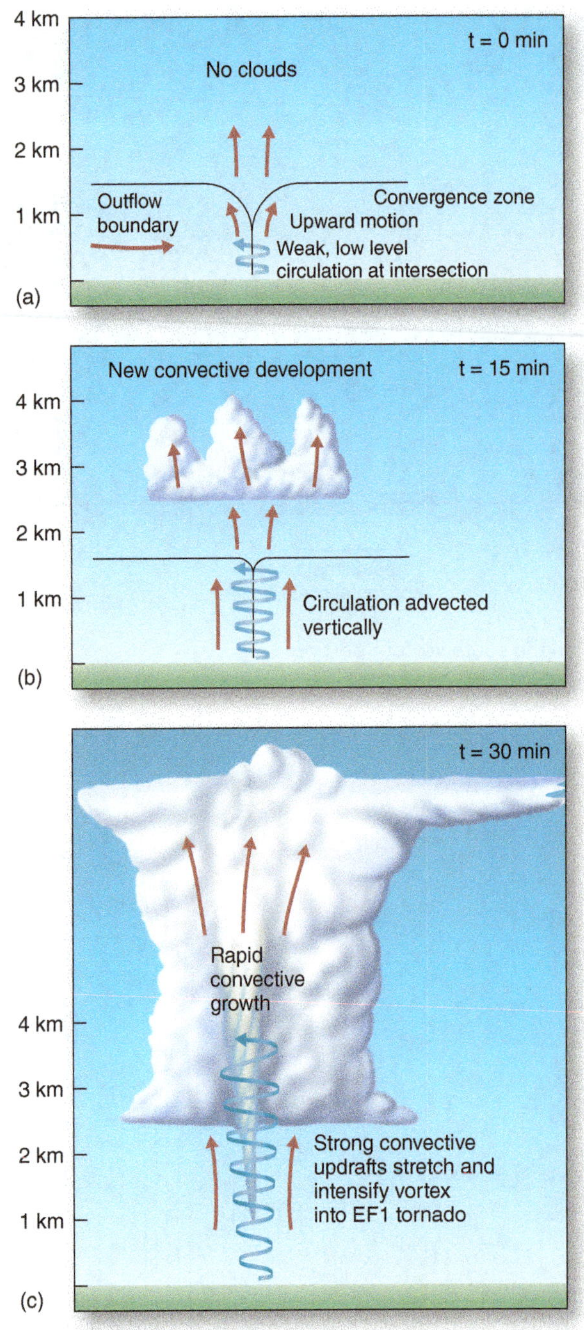

▲ **FIGURE 11–33 Evolution of a Tornado Along a Convergent Boundary.** Spinning motions along the boundary (a) can be carried upward if there is sufficient convection (b). Once the cumulonimbus develops (c), the downward movement of the strong rotation can lead to tornadoes.

The Location and Timing of Tornadoes

Video
Weather in Motion NSSL in the Field
http://goo.gl/fttNOK

Tornadoes are a very American phenomenon; no other country in the world has nearly as many as does the United States (Figure 11–34). Several factors combine to make North America a haven for tornadoes. The continent covers a wide range of latitudes: Its southeastern portion borders the warm Gulf of Mexico, while the northernmost portion extends into the Arctic. Furthermore, much of the central portion of the continent is relatively flat and, in particular, no major mountain range extends in an east–west direction. Together, these features allow for a collision of northward-moving maritime tropical air from the Gulf of Mexico with southward-moving continental polar air along the polar front. This setting, coupled with the frequent presence of potential instability (Chapter 6), provides a favorable situation for tornado development. The frequent occurrence of drylines also contributes to the high incidence of tornadoes across much of the southern Great Plains states (such as Oklahoma and much of Texas).

Tornadoes occur at least occasionally in almost all 50 states. Figure 11–35 maps the average incidence and concentration of tornadoes. A great many tornadoes touch down along a wide strip extending from east Texas to eastern Kansas and Nebraska, commonly called *Tornado Alley* (though the incidence of tornadoes is also high up through the Mississippi River Valley and the lower Great Lakes region). A secondary zone of high tornado incidence runs from east Texas to Georgia that is sometimes called *Dixie Alley*.

Ever since Dorothy was swept up by a tornado and dropped into the Land of Oz, many people have believed that Kansas leads the nation in tornado incidence. In fact, that state ranks only third. Texas easily leads the rest of the states in total number of tornadoes, with considerably more than 100 annually. But when the large size of that state is taken into account, it ranks only ninth in number of tornadoes per unit area. Florida (which happens to be well removed from Tornado Alley) has the highest tornado density (number of tornadoes per unit area) of all the states, followed by Oklahoma. Unlike the twisters in Tornado Alley, many of Florida's tornadoes are embedded in passing hurricanes and tropical storms, and many occur offshore as relatively weak waterspouts (described later in this chapter). Oklahoma is especially dominant with regard to what are considered strong tornadoes—those categorized as EF-2 or higher on the Fujita scale, which is discussed later in this chapter—so that state is the overall leader with regard to the density and severity of tornadoes.

Tornadoes can occur at any time of the year. But, as shown in Figure 11–36, a strong concentration of tornadoes occurs during the spring, when air mass contrasts are especially strong. May has the greatest number of tornadoes, with June a close second. Tornadoes are most likely to occur in the afternoon and early evening between 3 and 8 P.M., but many also occur well into the night.

Tornadoes are far less common in Canada, with an annual average of only about 100. The greatest concentration is in the extreme southern part of Ontario, between Lake Huron and Lake Erie. Most Canadian tornadoes outside Ontario occur in the southern region of the Prairie provinces and southwestern Quebec. The Canadian tornado season extends from April through October, with the greatest frequency in June and July. Despite the fact that the greatest concentration of tornadoes

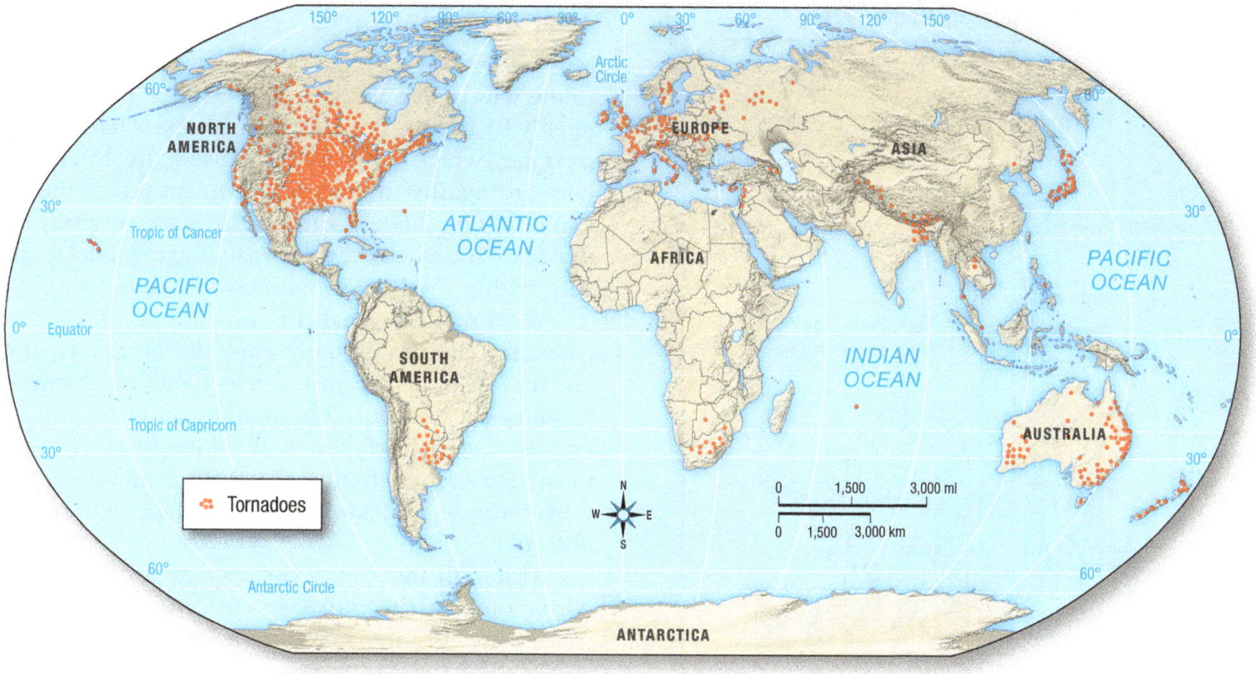

▲ **FIGURE 11–34 Tornadoes Around the Globe.** The areas of greatest dot concentration correspond to those of greatest tornado frequency.

MapMaster World Physical Environment: Tornadoes

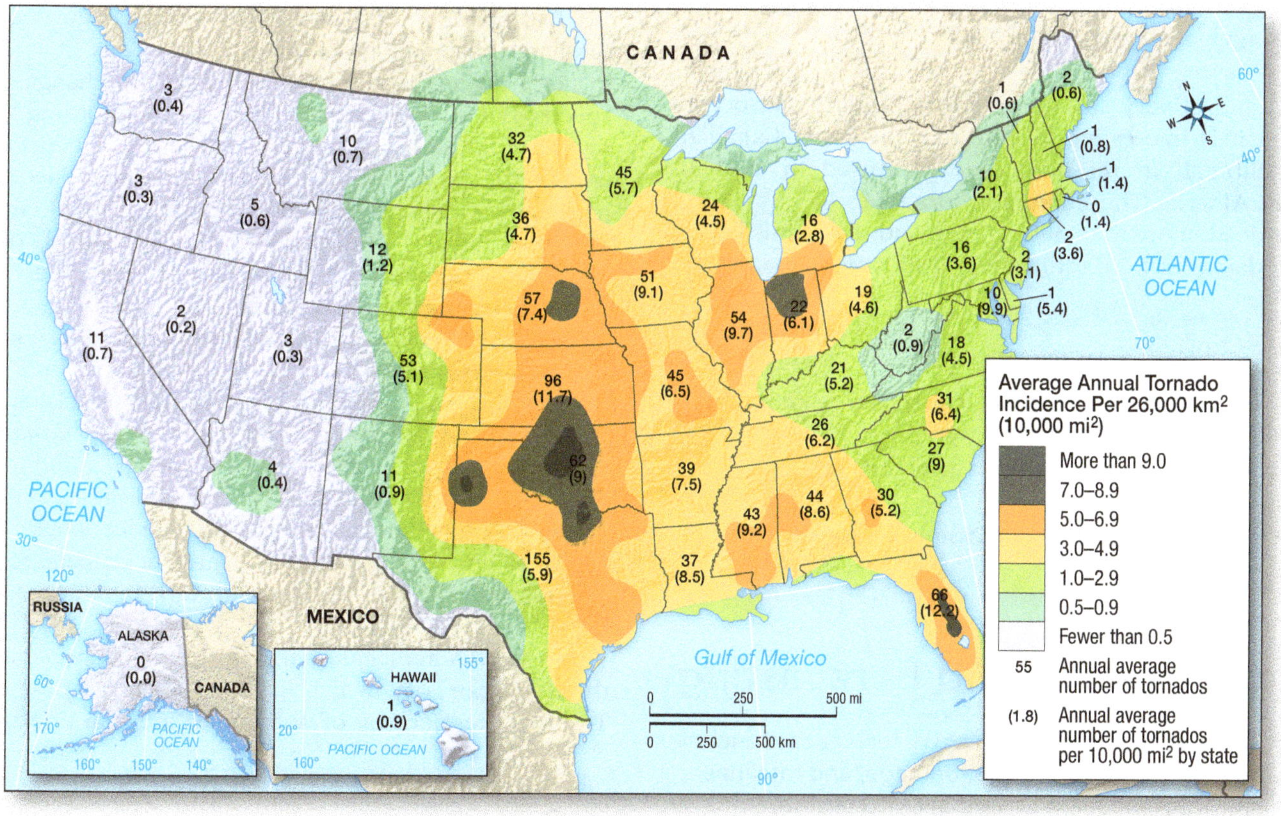

▲ **FIGURE 11–35 Annual Average Number of Tornadoes.** Total between 1953 and 2004 (top number) and the annual average number of tornadoes per 10,000 square miles (lower numbers in parentheses).

MapMaster North America Physical Environment: Tornado Incidence/Tornado Alley (U.S.)

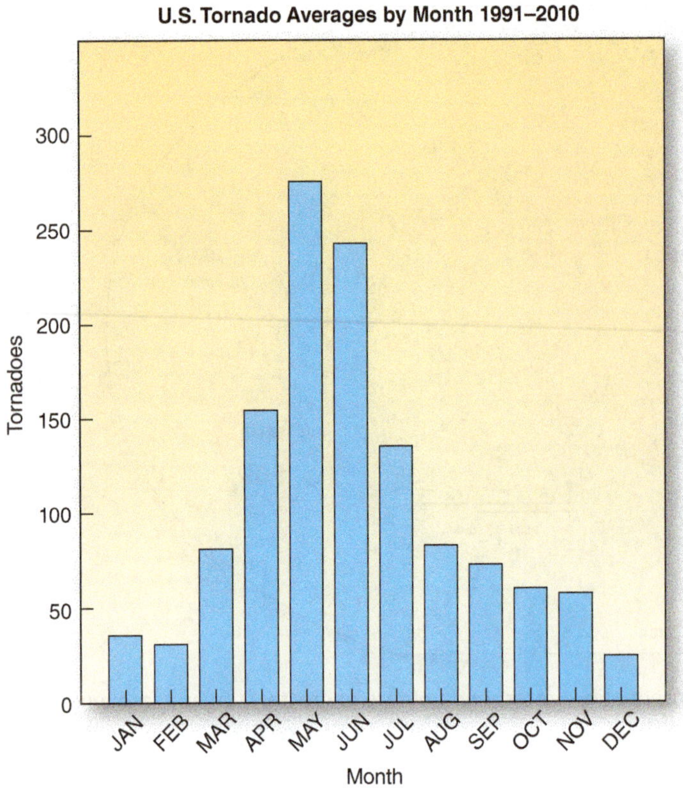

▲ **FIGURE 11–36 Seasonal Tornado Distribution.** U.S. tornadoes occur with greatest frequency in May and June. Values represent monthly averages for the years 1991–2010.

in general and of very strong tornadoes in particular occurs in Ontario, it is interesting to note that the worst Canadian tornado outbreak in recent decades occurred in 1987 in Edmonton, Alberta. The Edmonton tornadoes of July 31 left 27 people dead, more than 300 injured, and thousands homeless. Total damages were estimated at $300 million.

CHECKPOINT

11.17 What type of weather situation most often leads to the formation of a large, destructive tornado?

11.18 In what way are mesocyclones associated with tornadoes?

11.19 What factors account for the distribution of tornadoes in Figure 11–34?

Tornado Hazards and Forecasting

11.7 Identify aspects of tornadoes as a natural hazard, including areas most at risk for tornadoes, damage, and fatalities, and explain how tornadoes are rated and forecasted.

Having seen the characteristics of tornadoes, we now turn to what they can do, how their threat has changed through time, and what people can do to protect against these most powerful and dangerous of storms.

Tornado Damage

Most structural damage from tornadoes results from their extreme winds. People once believed that homes were destroyed mostly by the pressure differences associated with a tornado's passage, which supposedly caused the interior air to push outward against the walls so violently that the house would explode. For this reason, people were advised to open their windows if they saw an approaching tornado, so that the pressure within the house could be reduced.

We now know that this was not good advice, in part because few homes actually explode. Moreover, though winds are the major factor in tornado damage, flying debris is the primary cause of tornado injuries, and opening a window increases the risk of personal injury from flying debris. (We must also suspect that opening the windows is useless in any case, because they are likely to be "opened" anyway by flying objects.)

Although most tornadoes rotate around a single, central core, some of the most violent ones have several relatively small zones of intense rotations (about 10 m—30 ft—in diameter) called **suction vortices** (Figure 11–37). It is these small vortices that probably cause the familiar phenomenon of one home being totally destroyed while the next one remains relatively unscathed. Sometimes the path of a tornado is remarkably well defined after a sweep across the landscape (Figure 11–38).

Except for those rare times when tornado chasers make firsthand observations of passing tornadoes, it is impossible to get a precise reading on their pressure changes and wind speeds. But it is possible to classify them according to the magnitude of the damage they cause. The **enhanced Fujita scale** (named for the late eminent tornado specialist Theodore Fujita) provides a widely used system for ranking tornado intensity. As shown in Table 11–1, tornadoes fall into six levels of intensity, with each assigned a particular EF-value (called an F-value on the original Fujita scale) ranging from 0 to 5. Fujita values are determined by trained observers who assess a tornado's impact on any of 28 categories of natural objects or built structures (softwood trees, service station canopies, motels, small barns, single-wide mobile homes, and so on) called *damage indicators.* The observers then assign a numerical value to represent the *degree of damage* to each damage indicator. A certain minimum wind gust speed lasting 3 seconds or longer is then assumed to create any particular degree of damage to a damage indicator, and this is translated to an EF value based on the values shown in Table 11–1.[8]

In the United States, the majority (69 percent) fall into the *weak* category, which includes EF-0 and EF-1 tornadoes. Twenty-nine percent of tornadoes are classified as *strong*

[8]Those familiar with the original Fujita scale will note that the wind speeds for the larger categories have been reduced in the enhanced version, based on better knowledge of the wind speeds necessary to incur certain types of damage. The damage that would have resulted in the classification of, say, an F4 on the original scale is the same as that which would be considered an EF-4 on the new scale.

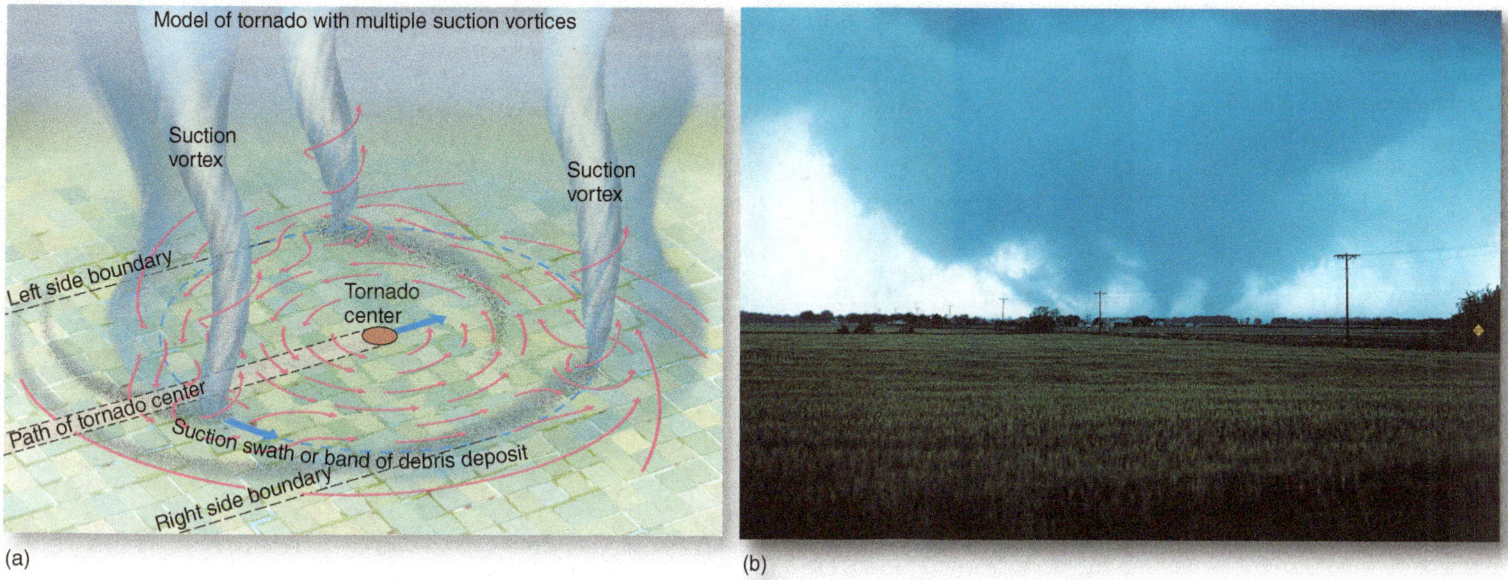

▲ **FIGURE 11–37 Suction Vortices.** (a) These zones of intense rotation sweep around a tornado center and often cause major destruction. (b) A multiple vortex tornado over Friendship, Oklahoma, on May 11, 1982.

(EF-2 and EF-3), which makes them capable of causing major structural damage even to well-constructed homes. Fortunately, only 2 percent of tornadoes are *violent* (EF-4 and EF-5). Those tornadoes are capable of wreaking incredible destruction. Cars can be picked up and carried tens of meters, pieces of straw can be driven into wooden beams, and freight cars can be carried off their tracks. Indeed, these storms are the true stars in all the movies and videos about tornadoes.

Between 1950 and 2013, 59 EF-5 tornadoes occurred in the United States and none in Canada. Thus, they happen about once a year in all of North America. Texas holds the lead for the greatest number of EF-5 tornadoes (6) during that period, while Alabama, Iowa, Kansas, and Oklahoma each had 5.

Tornado Outbreaks

Tornados are often particularly devastating because they can occur as part of a larger **tornado outbreak**—an event in which a single weather system produces at least six tornadoes. The most devastating of all outbreaks in the United States was the Tri-State tornado that rolled through eastern Missouri, southern Illinois, and southwest Indiana on March 18, 1925. The tornado killed about 700 people (roughly 600 of them in Illinois) and leveled several towns. To this day we do not know for sure if the Tri-State tornado was a single, incredibly large tornado, or part of an outbreak.

Though the Tri-State was the deadliest of all tornado events, it was not the largest. An outbreak on April 3-4, 1974 ranked as the largest until it was overtaken by that of April 2011, discussed later in this chapter.

DID YOU KNOW?

Between 1950 and 2005 only 27 percent of U.S. tornadoes occurred at night, but these nocturnal tornadoes were responsible for 39 percent of the fatalities. This is likely due to their being harder to see at night and the fact that many potential victims are asleep and therefore unable to take protective measures.

TABLE 11–1

Enhanced Fujita Intensity Scale

Intensity	Maximum 3-Second Wind Gust (km/hr)	Maximum 3-Second Wind Gust (mph)	Typical Amount of Damage
EF-0	105–137	65–85	Light: Branches broken, shallow trees uprooted, signs and chimneys damaged.
EF-1	138–177	86–110	Moderate: Roofs damaged, moving autos swept off road, mobile homes overturned.
EF-2	178–217	111–135	Considerable: Roofs torn off homes, mobile homes completely destroyed, large trees uprooted.
EF-3	218–266	136–165	Severe: Trains overturned, roofs and walls torn off well-constructed houses.
EF-4	267–322	166–200	Devastating: Frame houses completely destroyed, cars picked up and blown downwind.
EF-5	Over 322	Over 200	Incredible: Steel-reinforced concrete structures badly damaged.

Note: EF-0 and EF-1 tornadoes are collectively called weak, EF-2 and EF-3 strong, and EF-4 and EF-5 violent.

▸ **FIGURE 11–38 Tornado Damage.** Swath of damage from the tornado that hit Moore, Oklahoma, on May 23, 2013.

Trends in U.S. Tornado Occurrence

A plot of tornado occurrences in the United States, based on a log of events collected by the National Severe Storms Laboratory of NOAA, shows a doubling of the tornado frequency from 1950 to the early 2000s (Figure 11–39). Taken at face value, this would be a very disturbing trend. But there are reasons to believe that this increase might only partially reflect an actual increase in tornado activity, and that the trend may be primarily due to a greater likelihood that a given tornado will actually be observed. One possible explanation relates to population increases. As population centers have expanded out into formerly rural areas, there is a greater probability that a tornado will hit a structure or be observed directly. Furthermore, installation of the Doppler radar network has undoubtedly contributed to improved detection and an increase in the number of participants in the national storm spotter network. Changes in the way we classify tornadoes may also have played a role in the apparent increase.

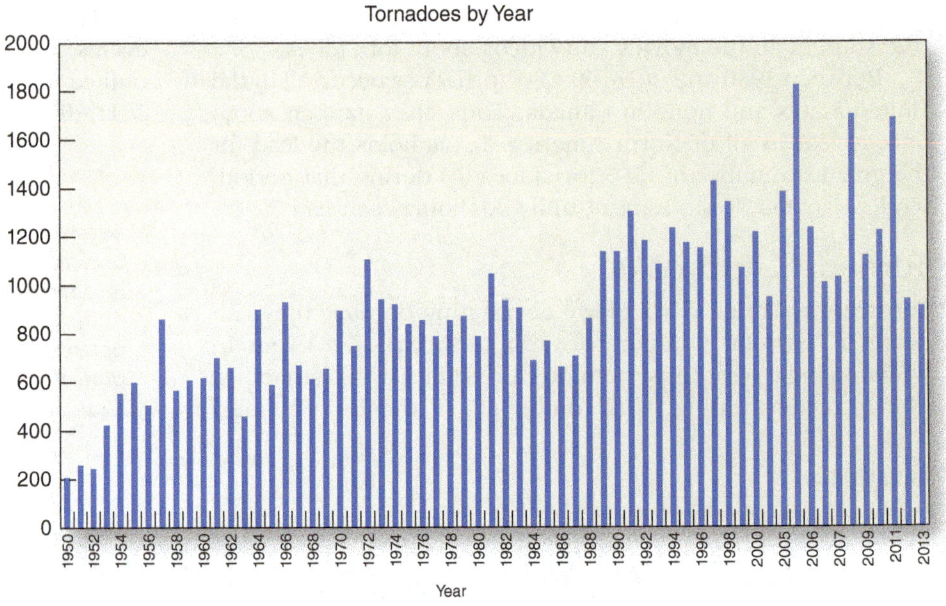

▴ **FIGURE 11–39 Increase in U.S. Reported Tornadoes.** The number of observed tornadoes in the United States has shown an apparent doubling since the 1950s, but this increase may be mostly due to better observation rather than to a genuine increase in tornado activity.

In contrast to the question of increasing frequency of all tornadoes is that of whether the incidence of *violent* tornadoes (EF-4 and EF-5) has changed through time. Interestingly, the incidence of these most deadliest of storms has decreased in recent decades, as indicated in Figure 11–40.

The discussion on recent global warming and the importance of human activities to that warming has led to concern about a future rise in tornado frequency. At this point in time, however, these ideas are still speculative.

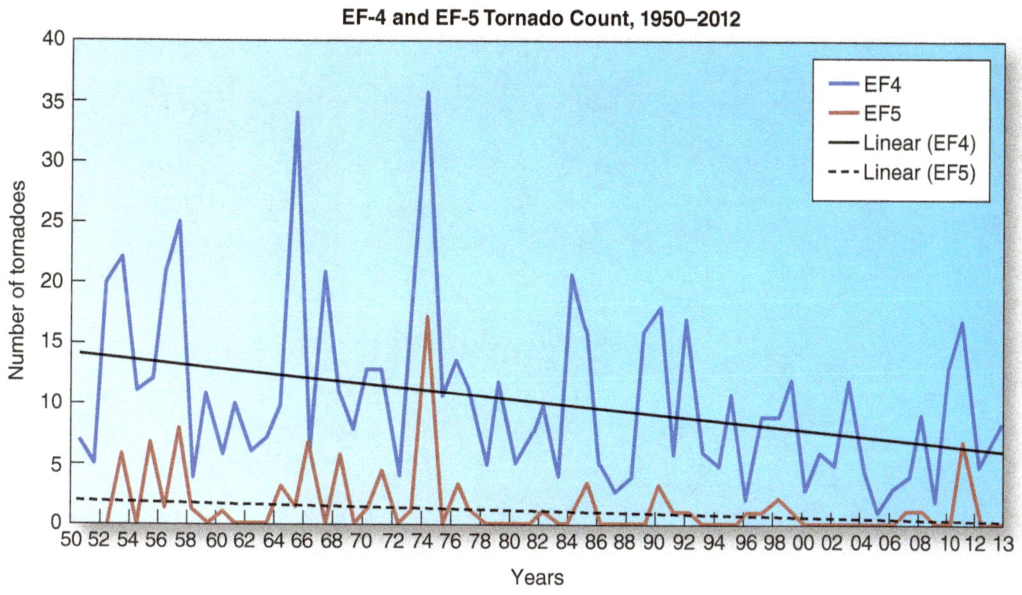

FIGURE 11–40 Major Tornadoes. The average number of violent (EF-4 and EF-5) tornadoes in the United States decreased between 1950 and 2012. Actual annual values of EF-4 tornadoes are shown in blue, EF-5 tornadoes in red. The thin black lines depict the statistical trends.

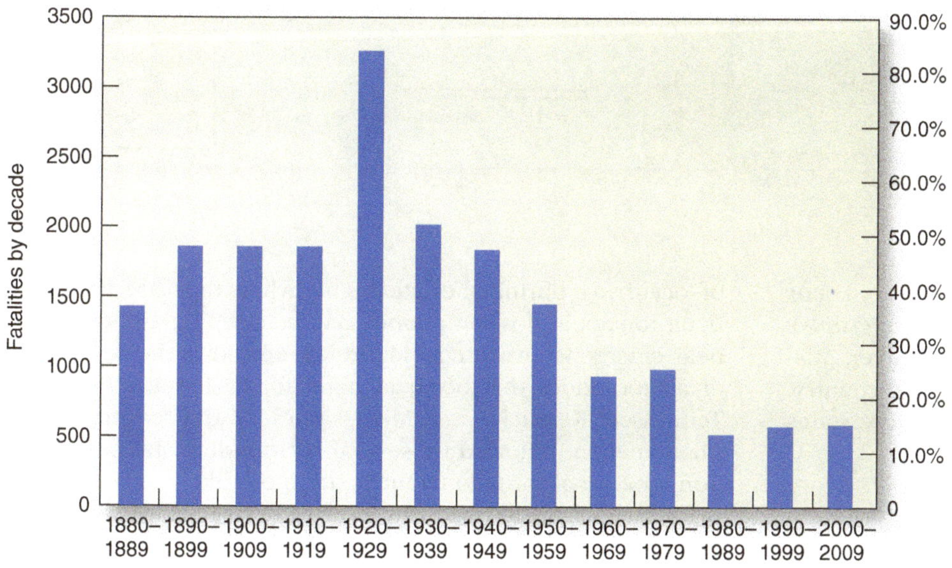

FIGURE 11–41 Tornado Fatalities. The number of reported U.S. tornado fatalities by decade.

Fatalities

Because they are small and last for such a short time, the overwhelming majority of tornadoes kill no one. A summary of data compiled by the University of Nebraska shows that the more than 48,000 tornadoes reported in the United States from 1950 to 2005 resulted in some 4700 fatalities. Thus, well over 43,000 (90 percent) of those reported tornadoes killed no one. And the actual proportion of fatality-free tornadoes must have been even higher, for at least two reasons. First, tornadoes in which no one dies are preferentially undercounted, because a storm that kills is almost certain to be reported. Second, most fatalities result from a few very large storms that kill up to dozens of people. According to the National Severe Storms Laboratory (NSSL), fewer than 5 percent of all U.S. tornado deaths are associated with weak (EF-0 or EF-1) tornadoes, nearly 30 percent with strong tornadoes (EF-2 and EF-3), and about 70 percent with violent tornadoes (EF-4 and EF-5). Thus, only 2 percent of all tornadoes are responsible for more than two-thirds of all fatalities. Although the number of reported tornadoes has increased over the years, the number of resulting fatalities has dropped markedly since the 1930s, with much of that reduction occurring since the 1970s (Figure 11–41). There are several reasons for this drop, including better building construction, vastly improved technology for predicting and tracking tornadoes, and a better network for broadcasting emergency information to the public.

While Figure 11–35 shows that the region of maximum tornado activity in the United States is the southern Great Plains, Figure 11–42 indicates that the distribution of tornado fatalities is surprisingly different. For the period of January 2000 through June 2011, Alabama easily wins the dubious distinction of leading the country in tornado fatalities, followed by Missouri. It is noteworthy that both these states suffered extremely devastating killer tornadoes in the first half of 2011

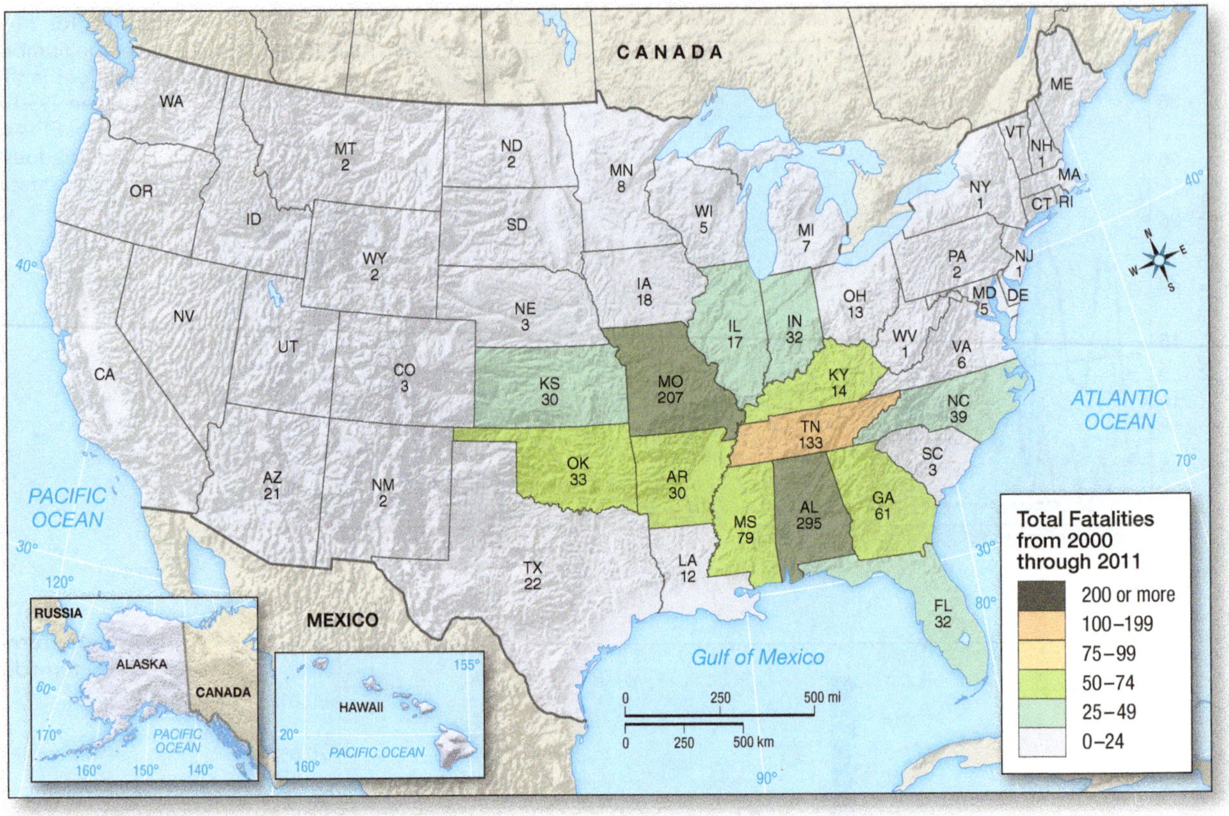

▸ **FIGURE 11–42 Tornado Fatalities.** Total number of tornado fatalities by state, 2000–2011.

that greatly inflated their fatality totals (see *Box 11–6, Focus on Severe Weather: Deadly 2011 and 2013 Tornado Seasons*). Even excluding the anomalous events of 2011, however, Missouri and Alabama still ranked two and three in the country, following Tennessee, for fatalities. None of those three states ranks in the top 10 for total tornado incidence.

DID YOU KNOW?

Social media is now being used to learn how far personal items can be transported by tornadoes. After the tornado outbreak of April 2011, people used the Internet to report items they found far from areas hit by the tornadoes. Such reports aid in the return of items, and also contribute to a knowledge of just how far things can be carried before being deposited. The record is 353 km (219 mi) from Alabama to Tennessee.

Without question, large populations living in mobile homes increase the susceptibility to tornado fatalities. A disproportionately large portion—40 percent—of tornado-related deaths do in fact occur in mobile homes, in contrast to the 31 percent in permanent houses. This increases the vulnerability in the southeastern United States, which has the highest percentage of mobile homes east of the Rockies. Trailers offer little protection to their occupants because they are easily blown off their foundations and tossed about by strong winds.

Another likely factor in the disproportionate number of tornado fatalities in much of the South is their greater likelihood of occurring during the late night, when they are more difficult to spot and when people may be sleeping and unable to hear emergency warnings. Figure 11–43 shows the percentage of all tornadoes that occur at night for each state. Arkansas, Tennessee, Kentucky, and Mississippi lead the country in this phenomenon, followed by several surrounding states. In addition, nocturnal tornado fatalities are more likely to occur from November to April, prior to the peak of tornado season when people are more complacent and also during the time of year when nights are longest.

It is also possible there is less of a public awareness of tornadoes in the Deep South than there is in Tornado Alley, where residents are particularly mindful of the threat.

According to the National Weather Service's Storm Prediction Center, between 1997 and 2007, 49 of the 705 U.S. tornado fatalities (7 percent) were in motor vehicles. These victims were often those who panicked and made the fatal mistake of leaving the relative safety of their homes to outrun the storm in their vehicles. For the majority of tornadoes, the safest place you can be is inside a well-constructed building, preferably in the basement and away from the windows.

A famous video shot in 1991 by a television news crew has led many people to falsely believe that highway overpasses provide good shelter from a tornado. The video shows a film crew along a Kansas highway fleeing an approaching tornado and taking shelter under the girders of a highway overpass. The news crew continued to film the tornado as it passed directly over them—with nobody being

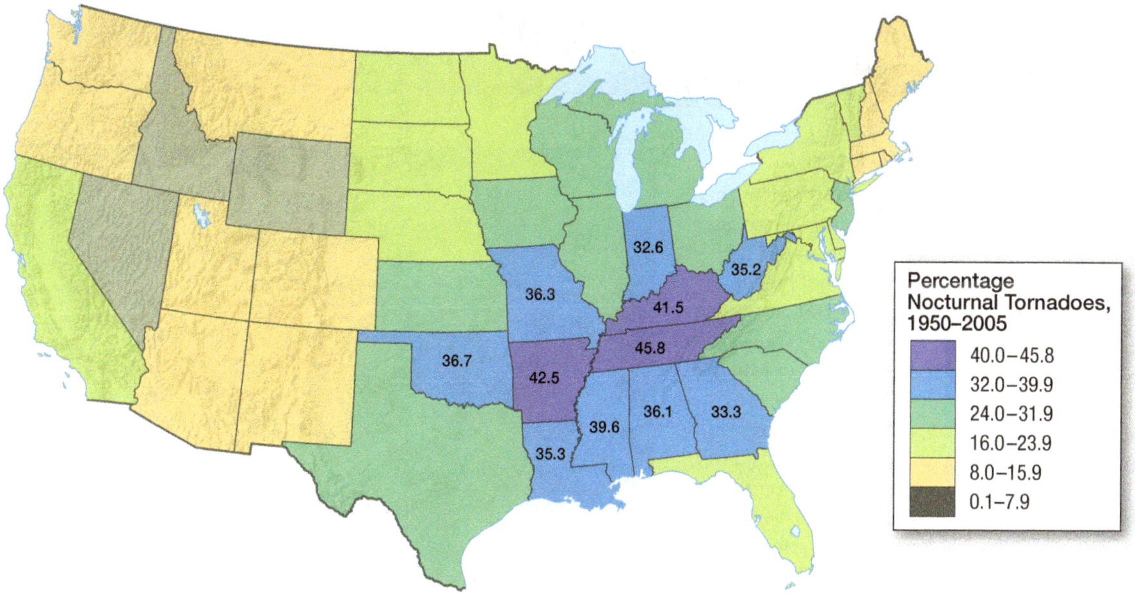

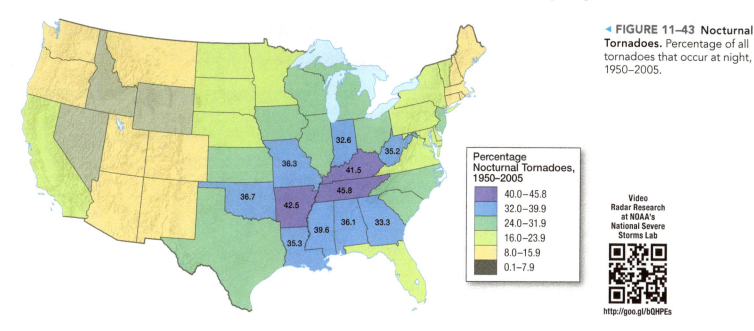

FIGURE 11–43 Nocturnal Tornadoes. Percentage of all tornadoes that occur at night, 1950–2005.

Video
Radar Research at NOAA's National Severe Storms Lab

http://goo.gl/bQHPEs

hurt. Unfortunately, the video did not issue a disclaimer about the relative lack of safety provided by overpasses in most instances.[9]

The primary mission of the National Weather Service is the protection of life and property, and the issuance of tornado warnings is one of the foremost methods of carrying out this task. If a warning is issued, one should take action immediately and not wait until the tornado is actually seen, as far too often happens. The National Weather Service and American Red Cross offer the following safety rules for when tornadoes threaten:

1. Stay indoors and seek shelter in a basement.
2. If you are in a building with no basement, move to an interior portion of the lowest floor and crouch to the floor. **Get as low as possible and put as many walls between you and the tornado as possible.** If possible, cover yourself with a mattress or some other form of padding to protect yourself from falling or flying debris. Despite what is sometimes said, the southwest corner of a building does not offer additional protection.
3. If you are in a mobile home, you should immediately vacate it for a nearby shelter or sturdy structure.
4. If you are caught outdoors and are unable to quickly get to a shelter or sturdy building, you should get into your vehicle, fasten your seat belt, and drive to the nearest available place of shelter. If you encounter flying debris while in your vehicle, you should do either of the following: (a) pull over with your seat belt still fastened and lower your head to below window level (covering your head with your hands or other covering), or (b) leave the vehicle for a noticeably lower-lying area, lie down, and cover your head.

The above advice is based on considerable research and prior experience. Though some debate arose about this after the 2013 Oklahoma tornadoes, the leading agencies involved in tornado safety agree that this is still the best advice that people can follow. See *Box 11–7, Focus on the Environment and Societal Impacts: Life and Death Decisions with the Threat of Tornadoes.*

DID YOU KNOW?

Between 1985 and 2003, 40 percent of all tornado fatalities in the United States occurred in mobile homes. This is greater than the 31 percent of fatalities that occurred in permanent homes, and it is very noteworthy in view of the relatively small proportion of people who live in mobile homes.

CHECKPOINT

11.20 In your own words, describe the level of damage associated with each level of the enhanced Fujita scale.

11.21 Look at the map in Figure 11–42. What factors might account for the state-to-state differences in tornado fatalities per year? Explain.

Saving Lives: Watches and Warnings

Without question, one of the key responsibilities of the National Weather Service is to issue severe weather advisories. These take two forms, watches and warnings, either of which can be issued for severe storms or tornadoes.[10] The declaration

[9] The Norman, Oklahoma, office of the National Weather Service has an interesting slide presentation on exactly why the overpass provided shelter despite the usual lack of safety such structures offer (http://www.srh.noaa.gov/oun/?n=safety-overpass-slide01).

[10] Warnings and watches also can be issued for other types of threatening weather, such as hurricanes and flash floods.

11–6 FOCUS ON SEVERE WEATHER

Deadly 2011 and 2013 Tornado Seasons

Tornado Alley has been hit by some particularly deadly events in recent years, with particularly notable tornadoes in 2011 and 2013. These events not only illustrated the destructive potential of tornadoes but also highlighted the effectiveness with which forecasters can warn the public about their imminent formation and provide advice on what lifesaving measures should be taken.

2011 Tornado Season

The 2011 tornado season was remarkably deadly, with 553 fatalities—more than seven times the average over the 30-year period. Many of the deaths occurred from two particularly deadly outbreaks.

Tuscaloosa and Birmingham, Alabama

April 2011 had more tornadoes than any other month in recorded U.S. history. The deadliest of these was an EF-4 that hit the cities of Tuscaloosa and Birmingham, Alabama (Figure 11–6–1) on the 27th. With winds that peaked at 306 km/hr (190 mph) and a width of 2.4 km (1.5 mi), the tornado cut a path of 129 km (80 mi), killed at least 71 people, and injured at least another 1000. Over the entire region on the 27th alone there were 317 fatalities, making it the worst single day for tornado fatalities since 1925.

Video
The Deadliest U.S. Tornado Event Since Modern Recordkeeping Began
http://goo.gl/EA3ry7

Joplin, Missouri

Less than a month after the deadly Alabama tornadoes, a supercell generated an EF-5 tornado that hit the heavily populated, southern part of Joplin, Missouri, during the afternoon of May 22 (Figure 11–6–2). The tornado, which had a maximum diameter of 1.2 km (0.75 mi), cut a nearly 10 km (6 mi) path through the town with maximum winds exceeding 320 km/hr (200 mph). About 171 people died in Joplin, making the tornado the deadliest during the modern period of record keeping that began in 1950. The Joplin tornado ranks as the seventh deadliest in U.S. history.

2013 Tornado Season

Though 2013 did not have nearly the number of tornadoes as did 2011, the month of May 2013 did have two very deadly tornadoes—both in Oklahoma and only days apart.

Newcastle–Moore, Oklahoma

At 2:56 P.M. on May 20 an EF-5 tornado touched down south of Oklahoma City in the town of Newcastle. The tornado cut a path to the east-northeast, staying on the ground about 40 minutes and cutting a swath as much as 2.1 km (1.3 mi) wide and 27 km (17 mi) long. The storm killed 25 people and injured 377 others. The winds, peaking at 340 km/hr (210 mph), destroyed 1150 homes and brought about some $2 billion in damages. The Newcastle–Moore tornado was the deadliest since the 2011 Joplin tornado, but it was not the last to hit Oklahoma that month—another powerful and deadly twister hit the town of El Reno 11 days later.

▲ **FIGURE 11–6–1 Alabama Tornado Destruction.** Dozens of homes lay in ruin in Pratt City, near Birmingham, Alabama, following a devastating April 28, 2011, tornado.

Video
A Satellite View of the Joplin, Missouri Torando
http://goo.gl/R9Clf6

El Reno, Oklahoma

It took only 11 days for a second EF-5 tornado to hit Oklahoma, the shortest time span in recorded history. The storm hit the town of El Reno on May 31. Although the death toll at El Reno was slightly

of a watch does not mean that severe weather has developed or is imminent; it simply tells the public that the weather situation is conducive to the formation of such activity. Most watches are issued for a period of 4 to 6 hours, for an area that normally encompasses several counties—about 50,000 to 100,000 sq km (20,000 to 40,000 sq mi).

The Storm Prediction Center (SPC) of the U.S. Weather Service in Norman, Oklahoma, has responsibility for putting out **severe storm and tornado watches** for the entire country. Operating 24 hours a day, every day, the center constantly monitors surface weather station data, information from weather balloons and commercial aircraft, and satellite data for all of the United States. Forecasters also have a large number of analytical tools for assessing the threat of severe weather. Some of these are in the form of indices that describe the stability of a large portion of the overlying atmosphere, with numerical values indicating the level of danger. Such indices, including the lifted index and K-index, are discussed in Chapter 13. If any particular part of the country appears vulnerable to impending severe storm activity, the SPC issues a severe storm or tornado watch. The advisory then goes to the local office of the National Weather Service, which notifies local television and radio stations. The broadcast media then relay the information to the public. Watches for the entire country are also available from a number of sources, many of which are posted on the Internet. Here is a portion of the text of the tornado watch

▲ **FIGURE 11–6–2 Joplin, Missouri Tornado Destruction on May 24, 2011.**

▲ **FIGURE 11–6–3 Oklahoma Tornado Destruction.** Satellite image showing the swath of the El Reno tornado.

less than that of the Moore tornado—19 fatalities—the storm was noteworthy for setting a record for the widest path for an American tornado, measuring 4.2 km (2.6 mi) at its widest (Figure 11–6–3).

Touching west of Oklahoma City at 6:03 P.M., the tornado swept a 25.7 km (16 mi) path over the next 40 minutes. The storm produced 475 km/hr (295 mph) winds at its maximum. It was also noteworthy in that prominent storm chasers Tim Samaras, Carl Young, and Samaras's son, Paul, were killed while making a scientific reconnaissance of the tornado. Given the severity and similarities of the two events, the Newcastle–Moore and El Reno tornadoes provoked intense discussions about the reevaluation of recommended safety procedures. This issue is discussed in *Box 11–7, Focus on the Environment and Societal Impacts: Life and Death Decisions with the Threat of Tornadoes*.

The Role of National Weather Service Alerts

As horrific as the death and injury statistics were for these tornadoes, they almost certainly would have been far worse without the alerts given by the National Weather Service. When a roof was partially blown off one of the terminals at Lambert International in St. Louis on April 22, 2011, there were no fatalities, because much of the airport had already been evacuated following a tornado warning. Similarly, the death toll relative to the degree of destruction during the 2013 Oklahoma tornadoes was nothing short of remarkable.

The National Weather Service played a lifesaving role for all of these outbreaks, providing residents time to take cover prior to being hit by the tornadoes. For example, a tornado warning was issued for Joplin, Missouri, at 5:17 P.M. local time by the forecast office in Springfield. The first report of tornadoes touching ground occurred at 5:41 P.M. A tornado watch had also been issued some 4 hours prior to the tornado, giving the public additional advanced safety information.

11–6–1 How do the four tornadoes described here compare in terms of damage and lives lost?

11–6–2 Given the size and strength of the tornadoes, how would the outcome have been different had no warnings or other advisories been issued?

issued hours before the devastating Joplin, Missouri, tornado of May 22, 2011:

URGENT - IMMEDIATE BROADCAST REQUESTED

TORNADO WATCH NUMBER 325

NWS STORM PREDICTION CENTER NORMAN OK

130 PM CDT SUN MAY 22 2011

THE NWS STORM PREDICTION CENTER HAS ISSUED A

TORNADO WATCH FOR PORTIONS OF

NORTHWEST ARKANSAS

SOUTHEAST KANSAS

SOUTHWEST AND CENTRAL MISSOURI

EASTERN OKLAHOMA

EFFECTIVE THIS SUNDAY AFTERNOON AND EVENING FROM 130 PM UNTIL 900 PM CDT.

TORNADOES . . . HAIL TO 4 INCHES IN DIAMETER . . . THUNDERSTORM WIND GUSTS TO 70 MPH . . . AND DANGEROUS LIGHTNING ARE POSSIBLE IN THESE AREAS.

THE TORNADO WATCH AREA IS APPROXIMATELY ALONG AND 75 STATUTE MILES EAST AND WEST OF A LINE FROM 30 MILES WEST NORTHWEST OF JEFFERSON CITY MISSOURI TO 30 MILES SOUTH OF MUSKOGEE OKLAHOMA. FOR A COMPLETE DEPICTION OF THE WATCH SEE THE ASSOCIATED WATCH OUTLINE UPDATE (WOUS64 KWNS WOU5).

REMEMBER . . . A TORNADO WATCH MEANS CONDITIONS ARE FAVORABLE FOR TORNADOES AND SEVERE

THUNDERSTORMS IN AND CLOSE TO THE WATCH AREA. PERSONS IN THESE AREAS SHOULD BE ON THE LOOKOUT FOR THREATENING WEATHER CONDITIONS AND LISTEN FOR LATER STATEMENTS AND POSSIBLE WARNINGS.

AVIATION . . . TORNADOES AND A FEW SEVERE THUNDERSTORMS WITH HAIL SURFACE AND ALOFT TO 4 INCHES. EXTREME TURBULENCE AND SURFACE WIND GUSTS TO 60 KNOTS. A FEW CUMULONIMBI WITH MAXIMUM TOPS TO 600. MEAN STORM MOTION VECTOR 27025.

If a severe thunderstorm has already developed, the public is notified by a **severe thunderstorm warning**. Likewise, **tornado warnings** alert the public to the observation of an actual tornado (usually by a trained weather spotter) or the detection of tornado precursors on Doppler radar. Unlike watches, which are issued by the SPC, warnings are given by local weather forecast offices. They warn the public to take immediate safety precautions, such as finding shelter in a basement. The information is broadcast immediately by television and radio stations, and civil defense sirens are sounded.

Sometimes Doppler radar enables meteorologists to give warning of an impending tornado about half an hour before it actually forms. Prior to the advent of Doppler radar, warnings were usually issued only after a funnel cloud had been seen. Many tornado warnings are still based on *in situ* observations, but the use of Doppler radar provides better lead time in many instances. When tornado warnings are based on visual sightings, observers must report their findings to the local weather service office, where the meteorologist in charge then passes on the information to the broadcast media. The procedure normally takes several minutes. But most tornadoes have life spans of about 3 to 4 minutes, so in many instances the warnings are issued to the public only after the tornado has died out. Though this might make the procedure seem pointless, that really is not the case. First of all, one tornado is often followed by others within the same storm system. Consequently, the precautions taken for the defunct tornado might still save lives when subsequent ones appear. Second, rare killer tornadoes tend to have considerably longer lifetimes and can remain in existence long after the warning has been issued. Thus the warning goes out while the tornado still presents a serious risk to people.

Tornado Forecasting: Convective Outlooks

The Storm Prediction Center issues **convective outlook** maps on its website (listed at the end of this chapter). These maps are issued several times each day, showing the probability of severe weather hitting parts of the United States for the current day, the next day, or the day after. Figure 11–44 shows such a series of maps for May 15, 2003, with part (a) plotting the threat of severe weather of any kind across the country. Here the greatest possibility of severe weather occurs around the Texas and Oklahoma panhandles, with much of the Ohio River Valley and Southeast having a slight risk. Parts (b), (c), and (d) give more specific probabilities for the occurrence of EF-2 or higher tornadoes, damaging winds, and large hail, respectively, occurring within 25 miles of any point within the outlined areas. So, for example, a resident of north Texas or western Oklahoma has about a 25 percent probability that a tornado or dangerous winds will occur within 25 miles of home on that day, with an even greater probability (35 percent) of large hail. Residents of Tennessee and Kentucky, on the other hand, have less than a 2 percent probability of tornadoes hitting, but a much greater likelihood (25 percent) of strong winds or large hail. In addition, the SPC probability maps are accompanied by narratives further describing the threats.

Despite the excellent warning system, disasters do indeed happen. Sometimes, sheer luck is all that stands in the way of major loss of life. On September 2, 2002, at 4:20 P.M. CDT, an F3 tornado hit the town of Ladysmith, Wisconsin, destroying 26 businesses and 17 homes. This time no tornado warning was issued until 17 minutes *after* the tornado hit the town, although a severe thunderstorm warning had been in effect for more than half an hour. (The fact that the nearby Doppler radar unit at La Crosse, Wisconsin, had been temporarily out of order may have contributed to the absence of a more timely warning.) Although most residents received no warning about the impending tornado, somehow the town escaped with no fatalities.

CHECKPOINT

11.22 In terms of assessing the risk of a tornado, what is the difference between a severe thunderstorm watch and a tornado warning?

11.23 What is the value of convective outlook maps in predicting the risk of tornadoes for an area on a given day? Explain.

Waterspouts

11.8 Explain how a waterspout forms.

So far we have discussed tornadoes over land. Similar features, called **waterspouts** (Figure 11–45), occur over warm-water bodies. Waterspouts are typically smaller than tornadoes, having diameters between about 5 and 100 m (17 to 330 ft). Though they are generally weaker than tornadoes, they can have wind speeds of up to 150 km/hr (90 mph), which makes them strong enough to damage boats.

Some waterspouts originate when land-based tornadoes move offshore. Most, however, form over the water itself. These "fair-weather" waterspouts develop as the warm water heats the air from below and causes it to become unstable. As air rises within the unstable atmosphere, adiabatic cooling lowers the air temperature to the dew point, and the resultant condensation gives the waterspout its ropelike appearance. Waterspouts form in conjunction with cumulus congestus clouds, those having strong vertical development but not enough to form the anvil that characterizes cumulonimbus.

11-7 FOCUS ON THE ENVIRONMENT AND SOCIETAL IMPACTS

Life and Death Decisions with the Threat of Tornadoes

No decision can be more critical than the one about whether or not to evacuate an area if a powerful tornado is on the way. Most safety organizations, including the National Weather Service and the American Red Cross, recommend taking shelter in a basement, storm shelter, or interior portions of a home if you are already there. But is this always the best course of action in the face of a particularly large and powerful tornado that could completely devastate the structure? Regrettably, there may be no one-size-fits-all answer. An examination of what happened during the Moore and El Reno tornadoes of 2013 provides some background on this issue.

The May 20 tornado in Moore, Oklahoma, occurred at 2:56 P.M.—before the beginning of rush hour (Figure 11–7–1). The massive tornado had already touched down and was being tracked by the National Weather Service as well as local media. One highly respected television meteorologist went on air and, using information from the station's helicopter, gave current information not only on the position of the tornado but on traffic conditions in various areas. Viewers were able to put this information together and make a somewhat informed decision on whether to take shelter within their homes or to try and escape from the path of the storm. Were lives saved because of this? Possibly.

But the El Reno tornado a week and a half earlier highlights the danger involved in leaving a fixed structure and attempting to flee by car. That tornado occurred after 6 P.M., during rush hour. Moreover, the largest interstate highway not only carried local commuter traffic but substantial long-distance travel as well. The result: Some people got on the highway only to find themselves in heavy traffic and totally defenseless, while many others stayed in their homes and survived despite serious building damage. Figure 11–7–2 shows the locations at which the fatalities occurred. Clearly the roads were not a safe place to be. Bad driving conditions due to normal rush hour traffic compounded by heavy rainfall made people in vehicles extremely vulnerable. In addition to the death of the experienced storm chasers, a mother and daughter were killed as their car was picked up off the highway. Moreover, the fatality total on the highway could have been even worse had the oncoming tornadoes not dissipated. Clearly, the decision to flee depends not only on the amount of time available, but also on the strength of the structure one is in, the type of road network available, and even the time of day.

Tornado experts continue to evaluate all evidence as to what is the best course of action to take as a tornado threatens, and they remain firm in their expert opinion. They advise people that the best thing one can do is remain in a fixed building and take cover in a basement, shelter, or interior room with as many walls as possible separating that room from the outside.

▲ **FIGURE 11–7–1 Approaching Tornado on May 20, 2013 in Moore, Oklahoma.**

11–7–1 What do tornado safety experts advise as the best way to protect yourself from an approaching tornado?

11–7–2 Based on the experiences of Moore and El Reno, Oklahoma, in 2013, how would you assess the wisdom of attempting to flee an approaching tornado in your car?

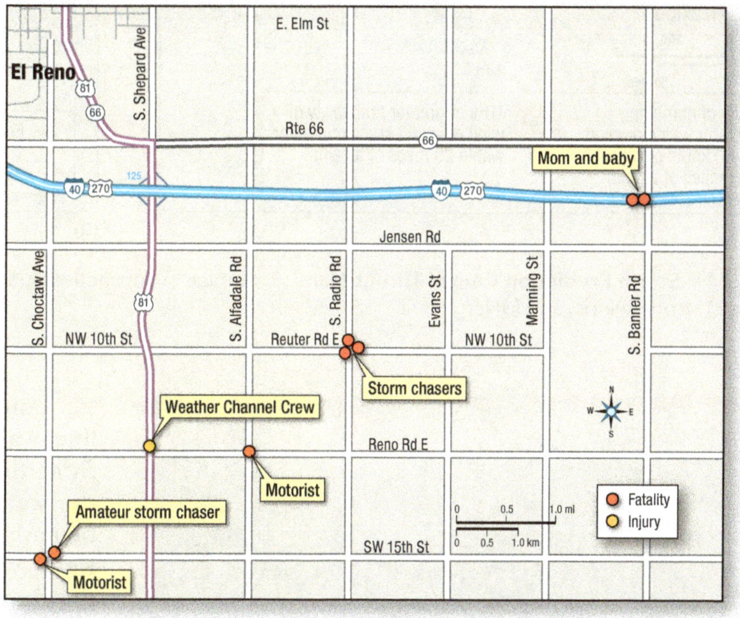

▲ **FIGURE 11–7–2 Locations of Fatalities During El Reno Tornado.** Notice the southwest to northeast pattern reflecting the path of the tornado.

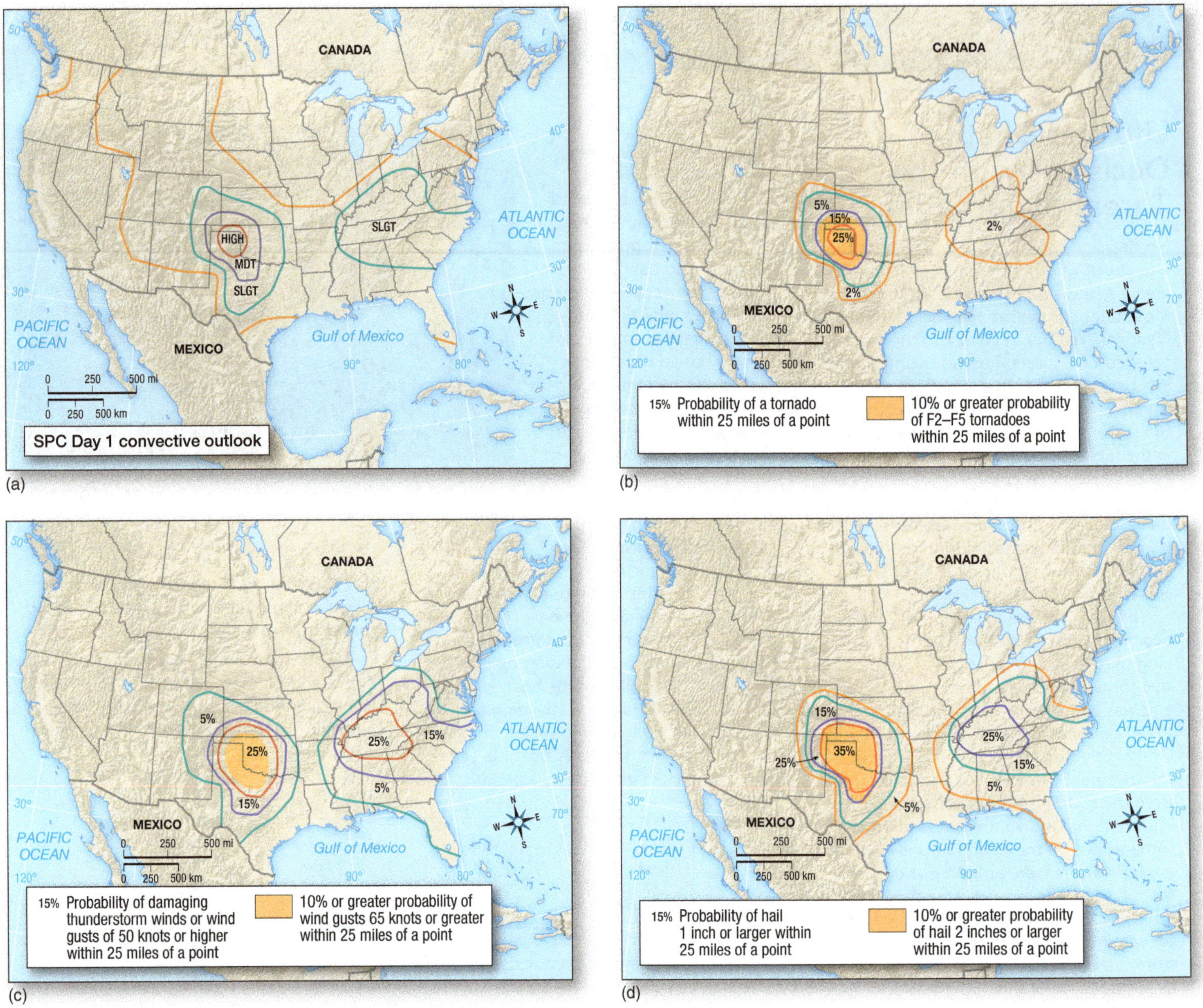

▲ **FIGURE 11–44 Storm Prediction Center Threat Maps.** Examples of convective outlook maps indicating the threat of any type of (a) severe weather, (b) tornadoes, (c) strong winds, and (d) large hail.

Contrary to what we might assume, the visible water in the waterspout is not sucked up from the ocean below; it actually comes from the water vapor in the air. Waterspouts are particularly common in the area around the Florida Keys, where they can occur several times each day during the summer.

Video
Waterspouts
http://goo.gl/456i47

CHECKPOINT

11.24 What differentiates waterspouts from tornadoes?

11.25 How do waterspouts form?

◀ **FIGURE 11–45 Waterspout in Singapore Harbor.**

Summary

11.1 Sequence the processes involved in lightning formation and identify the different types of lightning.

- Lightning begins when negative electrical charges build up near the base of a cloud and positive charges gather at the top.
- In cloud-to-ground lightning, rapidly growing leaders extend downward from the base of the cloud. When they connect with some object at the surface, a visible stroke develops.
- Most often a rapid sequence of multiple strokes follows the initial one to produce a lightning flash. Extreme heating of the air within the stroke causes the air to expand explosively and create the sound of thunder.

11.2 State measures that can be taken to ensure safety from lightning.

- If lightning threatens, even at a distance, it is important to take shelter indoors and to stay away from electrical appliances.
- Cars with metal exteriors and hard tops are relatively safe.

11.3 Explain how the different types of thunderstorms form and describe the strong winds that can accompany them.

- In air mass thunderstorms, downdrafts eventually destroy the storm. They undergo a sequence from the cumulus to the mature to the dissipative stages in tens of minutes.
- Severe storms—those that produce large hail, damaging winds, or tornadoes—develop when downdrafts reinforce the storm. Such storms occur within squall lines, in multicellular mesoscale convective complexes, or as supercells.

11.4 Summarize the characteristics and processes involved in the formation of floods and flash floods.

- Floods can result from prolonged or repeated rain events, especially if they occur over existing snowpacks. This will often take the form of streams or rivers overtopping their banks.
- Flash floods are those that occur when a strong thunderstorm remains over an area sufficiently long to create heavy runoff.

11.5 Analyze the geographic and temporal distribution of thunderstorms.

- Worldwide, thunderstorms are most common in the tropics, especially over interior equatorial Africa. Within the United States they are most common in the extreme southern states.

11.6 Explain how, where, and when tornadoes form.

- Tornadoes are most likely to form in the spring or early summer, but they can occur at any time of the year.
- Many tornadoes (especially those that emerge from supercells) follow the formation of large rotating areas within storm clouds called *mesocyclones*. However, other unexplained processes can also lead to their formation.

11.7 Identify aspects of tornadoes as a natural hazard, including areas most at risk for tornadoes, damage, and fatalities, and explain how tornadoes are rated and forecasted.

- The majority of tornadoes are classified as weak and cause no fatalities, but relatively rare, extremely large tornadoes can cause tremendous devastation and loss of life.
- Tornadoes hit the United States more than any other part of the world, and are most common in a swath from Texas northeast toward the Great Lakes.
- The enhanced Fujita scale is a ranking system that provides information on a tornado's strength by the type of damage done to particular types of structures.
- The U.S. Weather Service has a system of tornado watches and warnings that alerts the public to threatening weather. Watches cover fairly large areas where conditions are favorable for the formation of tornadoes. Warnings mean that a tornado has touched down or is probably imminent.
- Tornado outbreaks are weather systems that spawn multiple tornadoes.

11.8 Explain how a waterspout forms.

- Waterspouts are similar in structure to tornadoes over land. They typically form over warm bodies of water, with the lower atmosphere being heated from below to create unstable conditions.

Key Terms

air mass thunderstorms *p. 347*
ball lightning *p. 344*
blue jets *p. 345*
charge separation *p. 342*
cloud-to-cloud lightning *p. 342*
cloud-to-ground lightning *p. 342*
convective outlooks *p. 374*
cumulus stage *p. 348*
dart leader *p. 343*
derechos *p. 356*
dissipative stage *p. 349*
Doppler effect *p. 354*
Doppler radar *p. 353*
downbursts *p. 353*
enhanced Fujita scale *p. 366*
fair-weather electric field *p. 343*
flash *p. 343*
flash floods *p. 357*
floods *p. 357*
funnel clouds *p. 363*
gust front *p. 351*
haboobs *p. 357*
heat lightning *p. 346*
hook or **hook echo** *p. 352*
in-cloud lightning *p. 342*
lightning *p. 342*
mature stage *p. 348*
mean electric field *p. 343*
mesocyclones *p. 361*
mesoscale convective complexes (MCCs) *p. 349*
mesoscale convective systems (MCSs) *p. 349*
microbursts *p. 356*
multicell thunderstorms *p. 349*
outflow boundary *p. 350*
roll cloud *p. 351*
severe storm and tornado watches *p. 372*
severe thunderstorm warning *p. 374*
severe thunderstorms *p. 349*
sheet lightning *p. 342*
shelf cloud *p. 351*
sprites *p. 345*
squall lines *p. 349*
St. Elmo's fire *p. 344*
stepped leader *p. 343*
stroke or **return stroke** *p. 343*
suction vortices *p. 366*
supercells *p. 349*
sweep *p. 354*
thunder *p. 345*
tornado outbreak *p. 367*
tornado warning *p. 374*
tornadoes *p. 361*
vault or Bounded weak Echo Region *p. 353*
volume sweep *p. 354*
wall clouds *p. 362*
waterspouts *p. 374*

Review Questions

1. How common is cloud-to-ground lightning relative to cloud-to-cloud lightning?
2. What is the difference between a lightning stroke and a lightning flash?
3. Describe the current theories regarding the formation of charge separation.
4. Describe the sequence by which electrical imbalances lead to lightning strokes.
5. Briefly describe the following phenomena:
   a. ball lightning
   b. St. Elmo's fire
   c. sprites
   d. blue jets
6. What causes thunder?
7. Why is the term *heat lightning* misleading?
8. How big are air mass thunderstorms and how long do they usually persist?
9. What are the three stages of an air mass thunderstorm?
10. Explain why some thunderstorms have short life spans and yield little damage and others are able to develop into severe thunderstorms.
11. Describe the following types of storm systems:
    a. mesoscale convective systems
    b. squall lines
    c. mesoscale convective complexes
    d. supercells
12. How are outflow boundaries formed, and what effect do they have?
13. Describe the processes that lead to tornado development in supercell and nonsupercell storms.
14. What features of Doppler radar make it an effective tool for severe storm forecasting?
15. Explain how microbursts form and why they present a serious threat to aviation.
16. What are hook echoes and vaults, and why are they important?
17. Describe the location and timing of tornadoes in North America.
18. Describe the process of tornado formation from supercell storms.
19. What are wall clouds, and why is their appearance a cause for concern?
20. What is the leading threat to human safety when tornadoes hit?
21. Describe the enhanced Fujita scale for classifying tornadoes. Which category is most common? What is the highest EF-value that can actually occur in nature?
22. How do waterspouts compare to tornadoes, on average, in terms of intensity?
23. What is the difference between a tornado watch and a tornado warning?

Critical Thinking

1. Is charge separation necessary for sheet lightning and/or ball lightning?
2. Why does the environmental lapse rate affect the distance at which "heat lightning" can be observed?
3. You are outside on a sunny afternoon and observe a thunderstorm far to the west. An hour later, the storm passes over you. Is this more likely to have been an air mass thunderstorm or some sort of mesoscale convective system?
4. Why is the incidence of thunderstorms much lower near the Pacific coast than at the Atlantic coast?
5. In what fundamental ways are gust fronts different from passing cold fronts?
6. What conditions east of the Rocky Mountains promote a much greater incidence of tornadoes than exists in western North America?
7. Why is it not possible for a mesocyclone to occur within an air mass thunderstorm?
8. Why is it extremely unlikely that a tornado will move from east to west?
9. Other than their location with respect to land versus water, how do waterspouts differ from tornadoes?

Problems & Exercises

1. A tornado has a ring of uniform winds of 200 km/hr (120 mph) around the vortex 20 m (66 ft) away from its center. If the tornado moves to the northeast at 50 km/hr (30 mph), what is the effective wind speed on the northwestern and southeastern portions of the tornado?
2. Compare the map of annual hailstorms (Figure 7–20) to the map depicting the distribution of lightning flashes (Figure 11–25). Identify the regions where both are frequent, and those (if any) where only hail or thunderstorms frequently occur.

Visual Analysis

These maps show tornado concentration (the number of tornadoes per 10,000 square miles of area) over the 20-year period from 1991 to 2001. Map (a) shows the concentration of all tornadoes over the period. Map (b) plots only the larger tornadoes—EF-3 and above.

11.1. Compare the patterns for Tornado Alley and Dixie Alley in maps (a) and (b). Do the states that have the greatest concentration of all tornadoes also have the highest concentration of EF-3 to EF-5 tornadoes? Which states, if any, stand out? Explain your answers.

11.2. Why do you think Florida's incidence of strong tornadoes is so different from its incidence of all tornadoes? (*Hint:* Think of the conditions in which waterspouts form.)

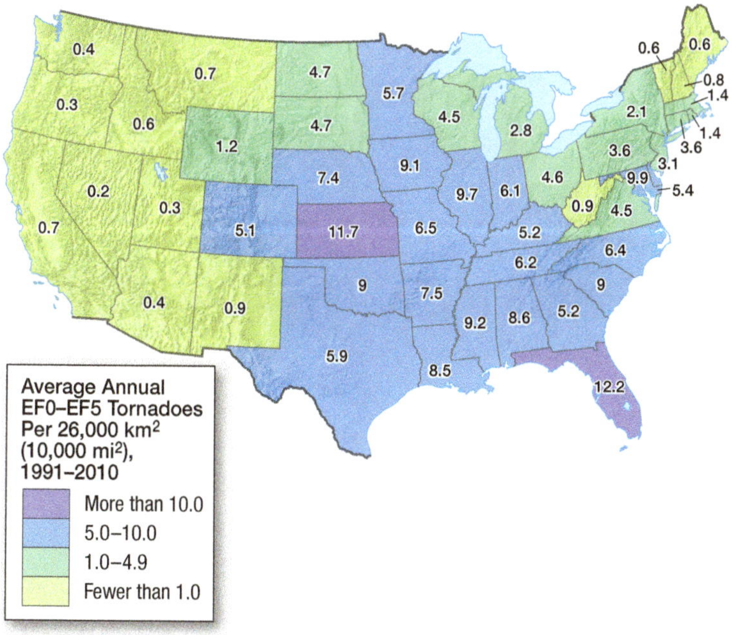

(a)

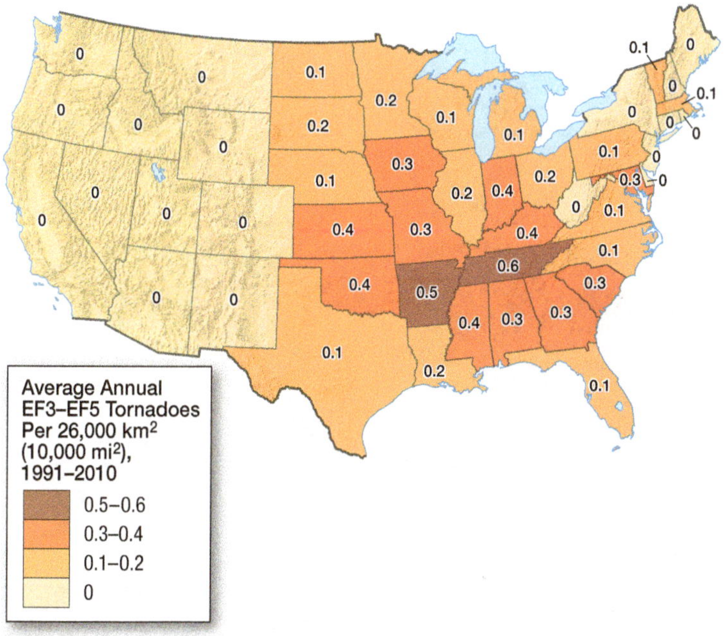

(b)

CHAPTER

12

Tropical Storms and Hurricanes

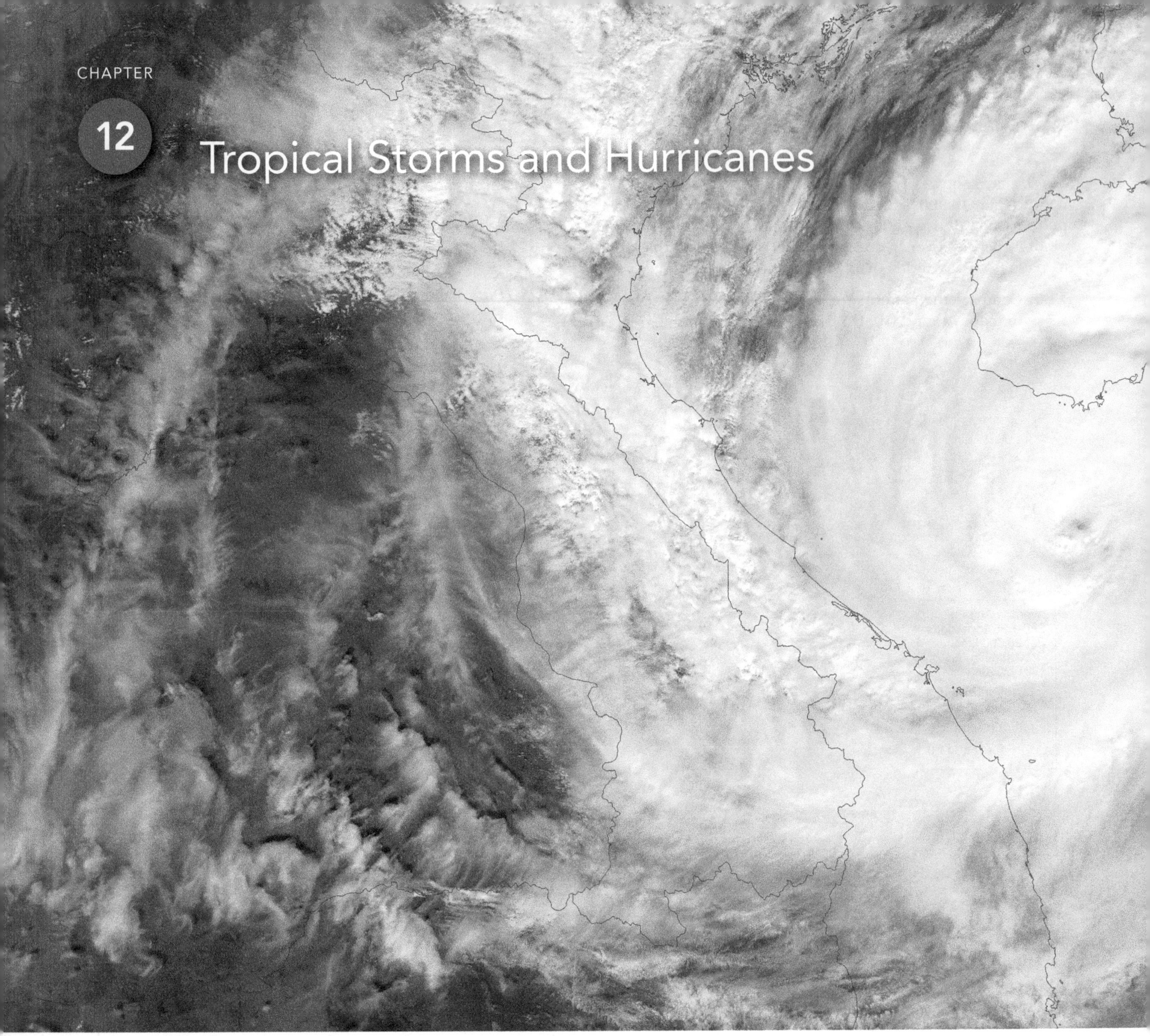

The year 2013 was a relatively quiet one with regard to hurricanes making landfall across the United States, but that doesn't mean that other parts of the world were not severely hit, especially islands in the Western Pacific Ocean. This shouldn't come as a surprise because tropical storms are generally more common, larger, and more powerful in that part of the world than they are in the Atlantic and Caribbean.

The most devastating tropical event of the season was Typhoon Haiyan (called Typhoon Yolanda in the Philippines), which hit numerous countries across the Pacific and Southeast Asia in November 2013. It is not easy to determine wind speeds accurately in such storms, but this one ranks as the most powerful storm ever to have made landfall, with estimated maximum one-minute sustained winds of 315 km/hr (195 mph).

Haiyan made landfall in five locations, but the most devastated country was the Philippines, where more than 6000 people were killed. Tacloban City may lay

◀ *With some of the strongest winds ever recorded for a tropical cyclone, Typhoon Haiyan swept across the Philippines in November, 2013.*

claim to having witnessed the greatest destruction, where 90 percent of all structures were destroyed

While Haiyan's winds were of historic magnitude, they were not the main cause of death and destruction. Extremely high waves combined with an elevated sea level—a so called *storm surge*—produced flooding that drowned thousands and completely leveled coastal communities. In this chapter we will see how such storms form and how they behave.

Learning Outcomes

After reading this chapter, you should be able to:

12.1 Identify the geographical settings where most hurricanes occur.

12.2 State the major characteristics of hurricanes.

12.3 Describe the structural features of a hurricane.

12.4 Explain the process of hurricane formation.

12.5 Describe hurricane movement and dissipation.

12.6 Describe how hurricanes cause destruction and fatalities.

12.7 Explain how meteorologists develop hurricane forecasts and advisories.

12.8 Describe efforts to identify trends in recent hurricane activity and project future trends.

▲ **FIGURE 12–1 Typhoon Haiyan.** This satellite image shows the typhoon on November 5, 2013 as it heads towards the Philippines.

Video
Improving Hurricane Predictions
http://goo.gl/DLMR8T

Hurricanes Around the Globe: The Tropical Setting

12.1 Identify the geographical settings where most hurricanes occur.

Extremely strong tropical storms go by a number of different names, depending on where they occur. Over the Atlantic and the eastern Pacific they are known as **hurricanes**. Those over the extreme western Pacific are called **typhoons**; those over the Indian Ocean and Australia, simply **cyclones**. In structure, the three kinds of storms are essentially the same, although typhoons tend to be larger and stronger than the others. We will use the term *hurricane* for the general class of storm, regardless of location.

Most U.S. residents associate hurricanes with storms that form in the Atlantic Ocean or the Gulf of Mexico. Yet other parts of the world have a much greater incidence of hurricanes (Table 12–1 and Figure 12–2). The Atlantic and Gulf of Mexico receive an average of 6.4 hurricanes each year, while the eastern North Pacific off the coast of Mexico has an average of 8.9. Most tropical storms in the east Pacific move westward, away from population centers, and so they receive little public attention. Sometimes, however, they migrate to the northeast and bring severe flooding and loss of life to western Mexico.

The region having the greatest number of these events—by far—is the western part of the North Pacific. In a typical year, 16.5 typhoons hit the region. At the other extreme, no hurricanes form in the Southern Hemisphere Atlantic, even at tropical latitudes (except for a very unusual event off the coast of Brazil in 2004). As you will see later, hurricanes depend on a large pool of warm water, a condition that does not arise in the relatively small South Atlantic basin.

DID YOU KNOW?

For the first time in more than half a century both the east and west coasts of Mexico were hit by a tropical storm and/or hurricane at the same time in September 2013. Tropical storm Manuel first hit the west coast of Mexico before moving into the Gulf of California. It then reintensified to a Category 1 hurricane and made a second landfall in the state of Sinaloa. Meanwhile, on September 14 Tropical Storm Ingrid hit the east coast of Mexico near La Pesca, Tamaulipas. Together the two storms killed over 150 people—many from flooding—and left billions of dollars worth of damage.

In Chapter 8 we saw that during much of the year air spirals out of massive high-pressure cells that occupy large parts of the Atlantic and Pacific Oceans. Middle- and upper-level air along the eastern side of these anticyclones sinks as it approaches the west coasts of the adjacent continents. Because the air does not descend all the way to the surface, a subsidence inversion (see Figure 6–14, page 200) forms above the surface. This particular subsidence inversion is called the **trade wind inversion**. The air below the inversion, called the **marine layer**, is cool and relatively moist.

The thickness of the marine layer and the height of the inversion base vary across the tropical oceans (Figure 12–3). The inversion is lowest along the eastern margins of the oceans, where upwelling and cold ocean currents maintain a relatively cool marine layer. Here the inversion may be only a few hundred meters above the surface. Farther to the west, the warmer surface waters heat the marine layer and cause it to expand to a greater height. Over the eastern part of the oceans, the low inversion inhibits vertical cloud growth, and low stratus clouds often occupy the region. Farther to the west, the greater height of the inversion (or even its total disappearance) allows for more convection, and deep cumulus clouds are more likely to form. For this reason, more hurricanes occur along the western portion of the ocean basins.

CHECKPOINT

12.1 What are hurricanes called in the Pacific Ocean? The Indian Ocean?

12.2 What is the role of the trade wind inversion in the development of hurricanes along the western parts of ocean basins? Explain.

TABLE 12–1

Number of Storms by Basin. Calculated for 1981–2010 in Northern Hemisphere, 1981–1982 to 2010–2011 in Southern Hemisphere

	Tropical Storm (63–117 km/hr (39–73 mi/hr))			Hurricane/Typhoon/Severe Tropical Cyclone (above 117 km/hr (73 mi/hr))		
Basin	Most	Least	Average	Most	Least	Average
Atlantic	28	4	12.1	15	2	6.4
Northeast Pacific	28	8	16.6	16	3	8.9
Northwest Pacific	35	14	26.0	23	7	16.5
Northern Indian Ocean	10	2	4.8	5	0	1.5
Southwest Indian Ocean	14	4	9.3	8	1	5.0
Australia/Southeastern Indian Ocean	16	3	7.5	8	1	3.6
Australia/Southwestern Pacific	20	4	9.9	12	1	5.2
Globally	102	69	86.0	59	34	46.9

Source: Atlantic Oceanographic and Meteorological Laboratory, NOAA.

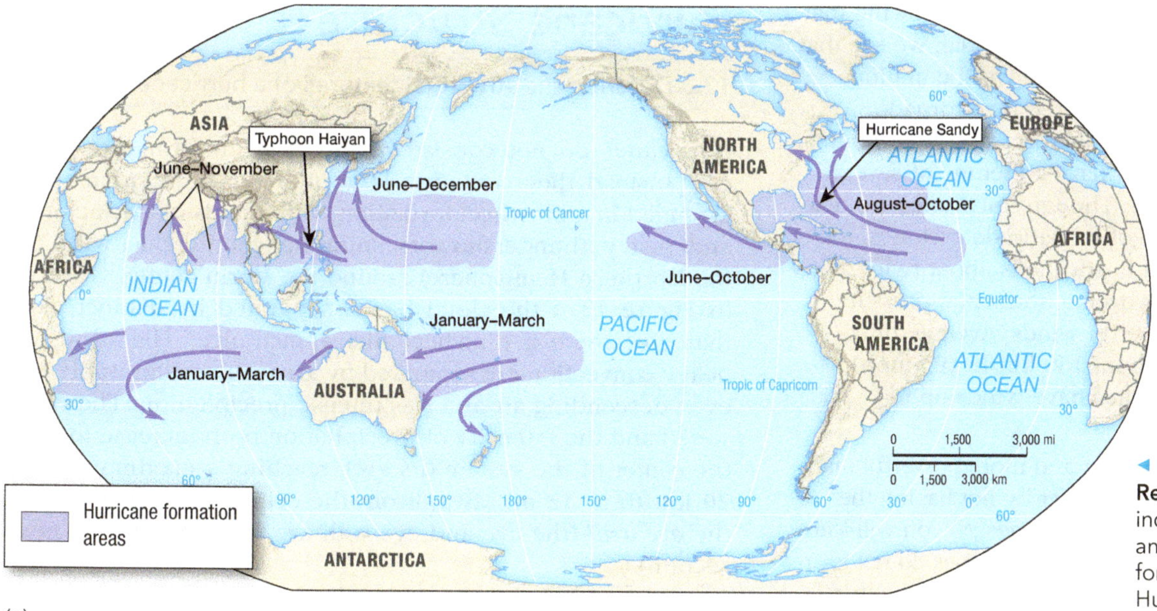

(a)

(b)

◂ **FIGURE 12–2 Hurricane Formation Regions Around the Globe.** (a) Shading indicates pools of warm surface water and favored regions for hurricane formation. (b) Damage caused by Hurricane Sandy along the New Jersey coast in 2012.

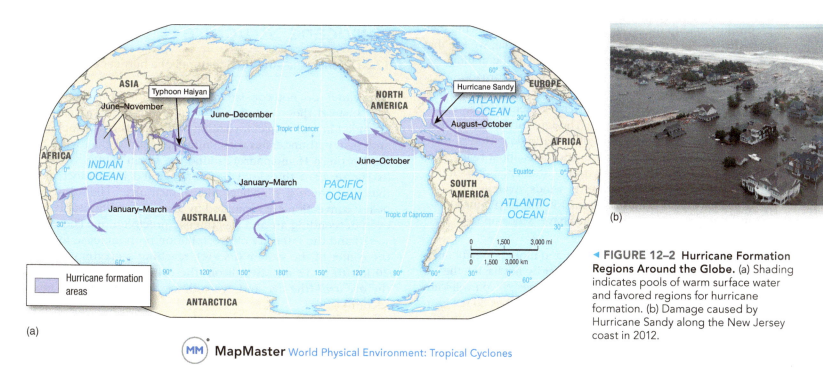

MapMaster World Physical Environment: Tropical Cyclones

Hurricane Characteristics

12.2 State the major characteristics of hurricanes.

Hurricanes are the most powerful of all storms. As such, they have fascinated meteorologists, who have exhaustively studied hurricanes' powerful winds, source of energy, structure, process of formation, and movements. The energy unleashed by just a single hurricane can exceed the annual electrical consumption of the United States and Canada. Here are their basic characteristics:

- **Wind speed** By definition, hurricanes have sustained wind speeds of 119 km/hr (74 mph) or greater. Though their wind speeds are less than those of tornadoes, hurricanes are very much larger and have far longer life spans.
- **Pressure** Sea-level pressure near the center of a typical hurricane is around 950 mb, but pressures as low as 870 mb have been observed for extremely powerful hurricanes. The weakest hurricanes have central pressures of about 990 mb.
- **Size** In contrast to tornadoes, whose diameters are typically measured in tens of meters, hurricanes are typically about 600 *kilo*meters (373 mi) wide. Thus, the typical hurricane has a diameter thousands of times greater than that of a tornado. Remembering that the area of a circle is

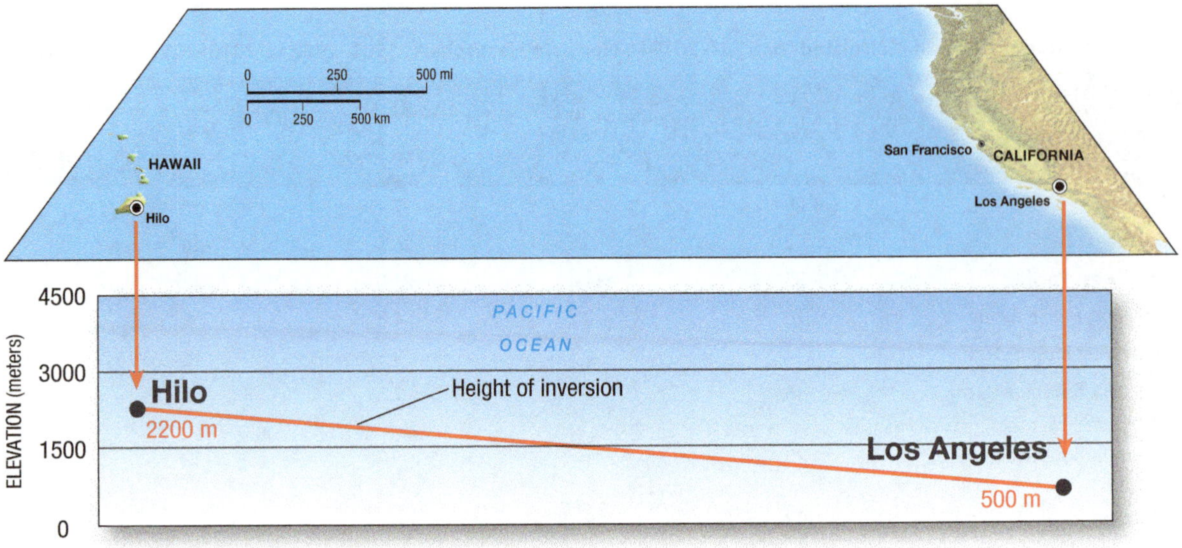

▶ **FIGURE 12–3 Trade Wind Inversion.** The height of the base of the trade wind inversion tends to increase moving westward across the ocean. In this example, the height of the inversion layer increases from Los Angeles, California, to Hilo, Hawaii.

proportional to the square of its radius, and knowing that tornadoes and hurricanes are roughly circular, we see that the area covered by a hurricane is likely to be millions of times greater than the area covered by a tornado. Furthermore, a tornado exists only for a couple of hours at most, while a hurricane can have a lifetime of several days or even a week or more. Though hurricanes are usually about one-third the size of midlatitude cyclones, the pressure difference across a hurricane is about twice as great. They therefore have extreme horizontal pressure gradients that generate powerful winds: Average hurricanes have peak winds of about 150 km/hr (93 mph), and the most intense hurricanes have winds up to 350 km/hr (218 mph).

- **Form** In addition to being smaller and more powerful than midlatitude cyclones, hurricanes differ by not having the fronts that characterize midlatitude cyclones. As you will learn in the section on hurricane formation, hurricanes do not form in regions having dissimilar air masses abutting each other.
- **Energy source** Because hurricanes obtain most of their energy from the latent heat released by condensation, they are most common where a deep layer of warm water fuels them. Given that tropical oceans have their highest surface temperatures and evaporation rates in late summer and early fall, it is not surprising that August and September are the prime hurricane months in the Northern Hemisphere, with January to March the main season in the Southern Hemisphere.[1]

CHECKPOINT

12.3 What is a hurricane's source of energy?

12.4 What two factors explain the spatial and temporal distribution of hurricanes?

[1]The U.S. National Hurricane Center defines the hurricane season as the period from June 1 to November 30. Tropical storms in other months are rare events—from 1871 to 1996, only six storms formed in December.

Hurricane Structure

12.3 Describe the structural features of a hurricane.

Hurricanes do not consist of only one uniform convective cell. Instead they contain a large number of thunderstorms arranged in a pinwheel formation, with bands of thick clouds and heavy thundershowers spiraling counterclockwise (in the Northern Hemisphere) around the storm center. Observe in Figure 12–4 the cloud bands separated by distinct gaps that give them a pinwheel like appearance. The bands of heavy convection are separated by areas of weaker uplift and even descending air and less intense precipitation. The wind speed and the intensity of precipitation both increase toward the center of the system (its *eye*), reaching a maximum 10 to 20 km (6 to 12 mi) away from the center, at what is called the *eye wall* (the eye and eye wall are described in the next section).

Figure 12–5 depicts generalized cross sections through a hurricane. The thickness of the cloud bands (a) corresponds well to the intensity of rainfall (b). On the other hand, the distributions of pressure (c) and wind speed (d) do not exhibit a similar banding. Both the pressure gradient and wind speed increase gradually toward the center of the storm and then increase rapidly in the vicinity of the eye wall.

Temperature Gradients Within a Hurricane

Pressure gradients and temperature differences within a hurricane are key to understanding the tremendous power of these storms. As air flows inward toward lower pressure, the warm ocean surface supplies large amounts of latent and sensible heat to the overlying air. Because pressure within the moving air decreases as it flows toward the low, adiabatic expansion keeps the temperature from increasing dramatically, with the result that there is little temperature difference across the base of the storm.

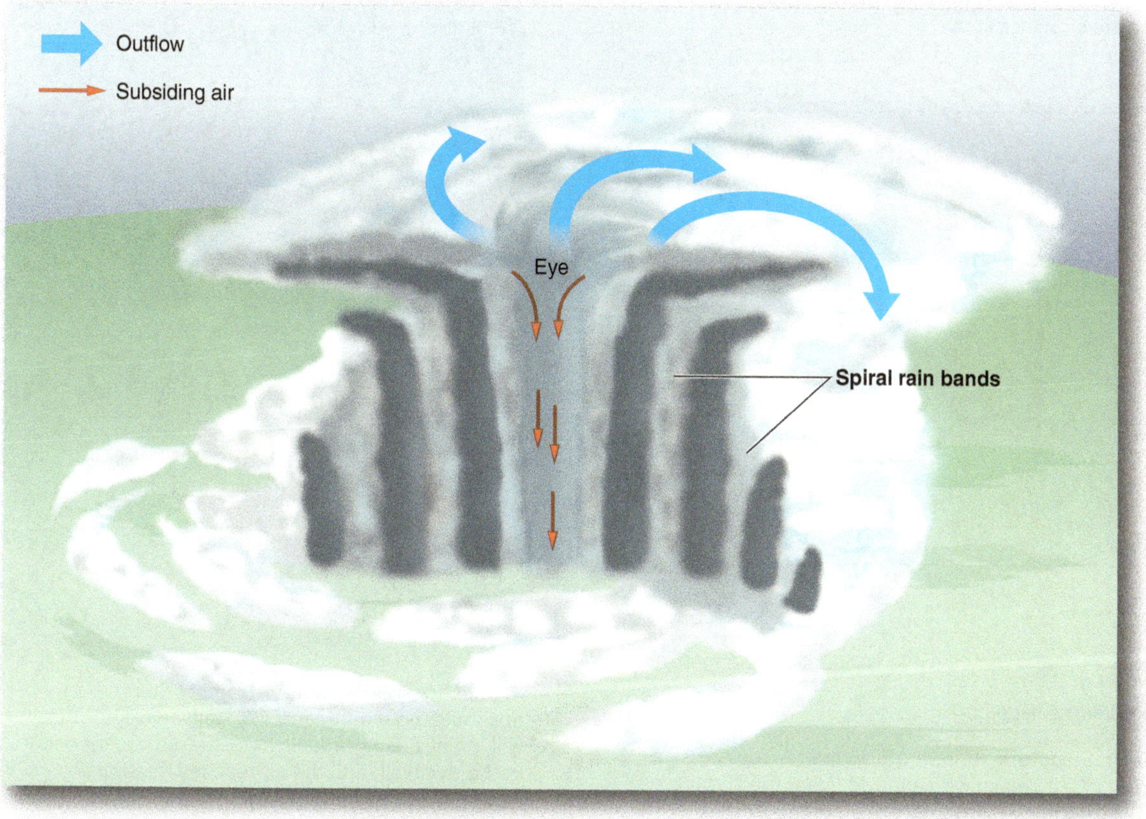

◄ **FIGURE 12–4 Inside a Hurricane.** Looking into the hurricane there are parallel bands of cloud cover that spiral in toward the hurricane center. Cloud bands tend to be deeper toward the center of the hurricane until they meet the *eye wall*, which separates the intense part of the storm with strong winds and heavy showers from the relatively calm *eye*.

Animation
Hurricane Wind Patterns

http://goo.gl/gqBLv2

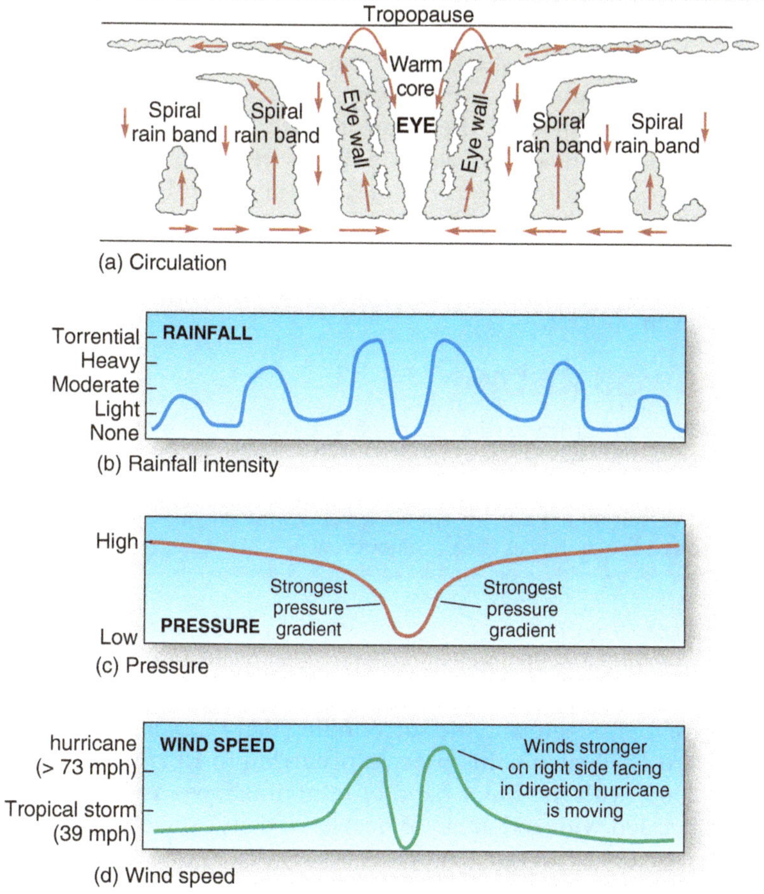

▲ **FIGURE 12–5 Hurricane Cross Section.** (a) Cloud bands become deeper toward the center of the hurricane. (b) Rainfall becomes more intense under each cloud band. Air pressure decreases gradually in the outer portions of the hurricane but intensifies rapidly toward the center of the hurricane (c), as does wind speed (d).

Nevertheless, much thermal energy is added, resulting in a "warm" central core. Aloft, after condensation and the release of latent heat, the warmth is reflected in temperature, so that temperatures near the center are much higher than those of the surrounding air (Figure 12–6).

As a warm-core low, pressure within a hurricane decreases relatively slowly with increasing altitude (see Chapter 4). Thus, the horizontal pressure gradient within the storm gradually decreases with altitude. At about 7.5 km (25,000 ft)—about the 400 mb level—the air pressure is the same as that immediately outside the storm. From this height to the lower stratosphere, the hurricane has relatively high pressure. So unlike the lower part of the hurricane in which the air rotates cyclonically, much of the air in its upper portion spirals anticyclonically outward from its center (clockwise in the Northern Hemisphere). At the same time, some of the rising air along the eye wall flows inward toward the eye wall, which accounts for the slowly descending air encountered in that region. Despite the introduction of this air into the upper reaches of the eye, the increase in mass is not sufficient to diminish its characteristic low pressure. Figure 12–7 shows a schematic of how surface winds typically flow inward, then rapidly rotate around the eye wall as they ascend to higher levels and then spiral outward. The outflow from a hurricane can extend outward for many hundreds of kilometers.

In the upper reaches of the storm, the low temperatures cause water droplets to freeze into ice crystals. As the ice crystals spiral out of the storm center, they create a blanket of cirrostratus clouds that overlies and obscures the pinwheel-like structure of the storm. That explains why

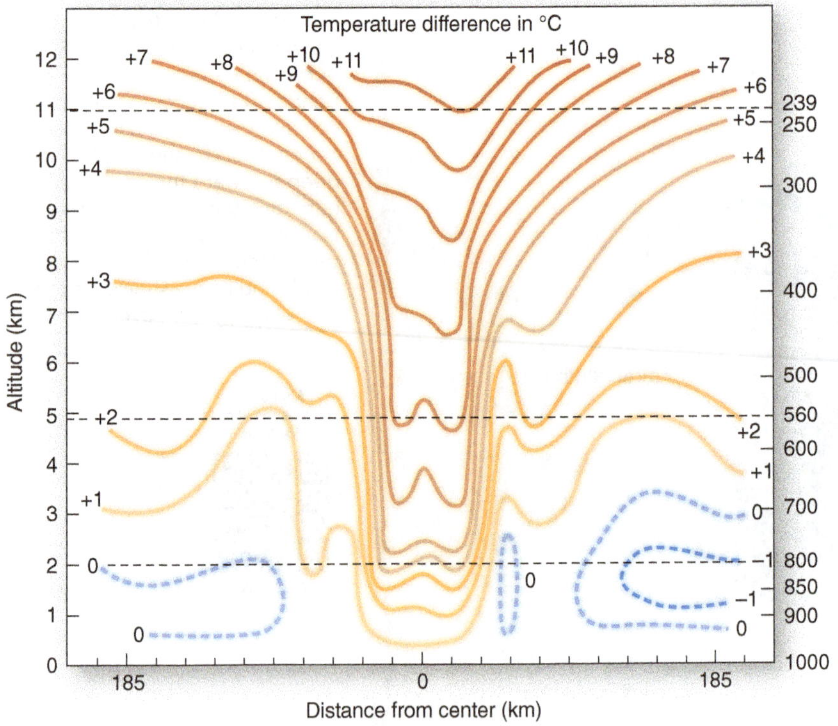

▲ **FIGURE 12–6 Vertical Temperature Structure.** Temperature differences across a hurricane relative to the surrounding air (°C) increase with altitude. Near the surface, temperatures increase only slightly toward the eye. Aloft, however, temperatures exceed those of the surroundings by about 10 °C (18 °F).

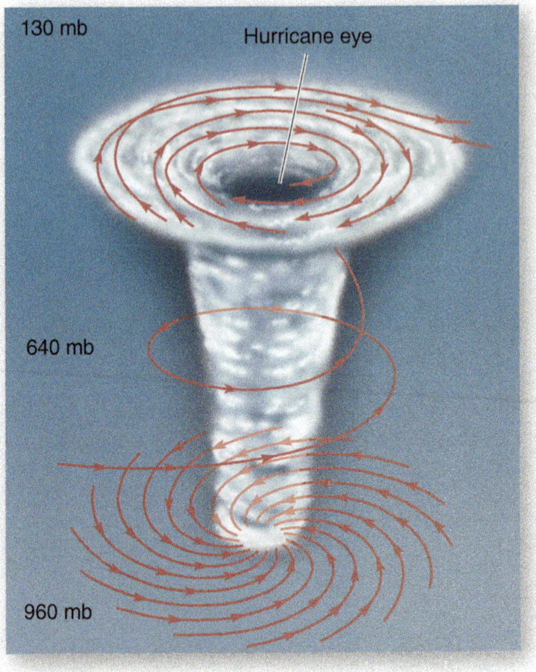

▲ **FIGURE 12–7 Airflow Trajectories.** When approaching a hurricane eye wall, air trajectories will initially rise gradually. Most rapid ascent occurs in the eye wall until the air reaches the upper troposphere and flows outward anticyclonically.

hurricanes on satellite images often appear to have a uniform thickness and intensity, when in fact they are strongly banded.

The Eye and the Eye Wall—A Closer Look

One of the most distinctive characteristics of a hurricane is its **eye**, a region of relatively clear skies, slowly descending air, and light winds. Hurricane eyes average about 30 km (19 mi) in diameter, with most ranging from 20 to 50 km (15 to 31 mi). Eye diameters vary considerably among individual storms, with some as small as 6 km (3.5 mi) and others almost as large as 100 km (60 mi). The change in the size of an eye through time gives some indication of whether the hurricane is intensifying or weakening. Generally, a shrinking eye indicates an intensifying hurricane.

EYE WALL CHARACTERISTICS Along the margin of the eye lies the **eye wall**, the zone of most intense storm activity. The eye wall contains the strongest winds, thickest cloud cover, and most intense precipitation of the entire hurricane. Directly beneath the eye wall, rainfall rates of 2500 mm/day (100 in./day) are not uncommon. The abrupt transition from an eye wall to the eye causes a strong and rapid change in weather. Imagine a hurricane about to make a direct hit on a small island. As the hurricane center approaches the island, the intensity of the wind and rain steadily increases, becoming most intense as the eye wall arrives. But when the eye reaches the island, the storm seems to suddenly dissipate, as blue skies and calm conditions take hold. Of course, the storm has not dissipated at all. Rather, there is just a brief lull until the opposite side of the eye wall covers the island and storm conditions resume. Because the average hurricane eye is about 30 km (18 mi) in diameter and travels at about 20 km/hr (12 mph), the calm associated with passage of the eye lasts about an hour or two. Clearly, if the eye just grazes the island, the break in the storm will be even shorter.

DID YOU KNOW?

Although hurricane wind speeds are, by definition, extremely fast, it still takes a parcel of air an average of 8 days to flow into and out of a hurricane. This is due to the great distance traversed by the air as it spirals upward, and the relatively slow speed the air has as it exits as outflow.

FORMATION OF A DOUBLE EYE WALL Some intense hurricanes develop **double eye walls** as surrounding rain bands contract and intensify. As the interior eye wall contracts, it can begin to dissipate and the surrounding band can take its place. Figure 12–8 shows this process occurring in Hurricane Emily in 2005, with the outer band of rainfall (shown as green arcs embedded with red areas of heavier activity) surrounding the small eye wall in the center.

EYE WALL REPLACEMENT Sometimes, especially with very strong hurricanes, an inner eye wall will contract in size so that the eye's diameter becomes relatively small. If

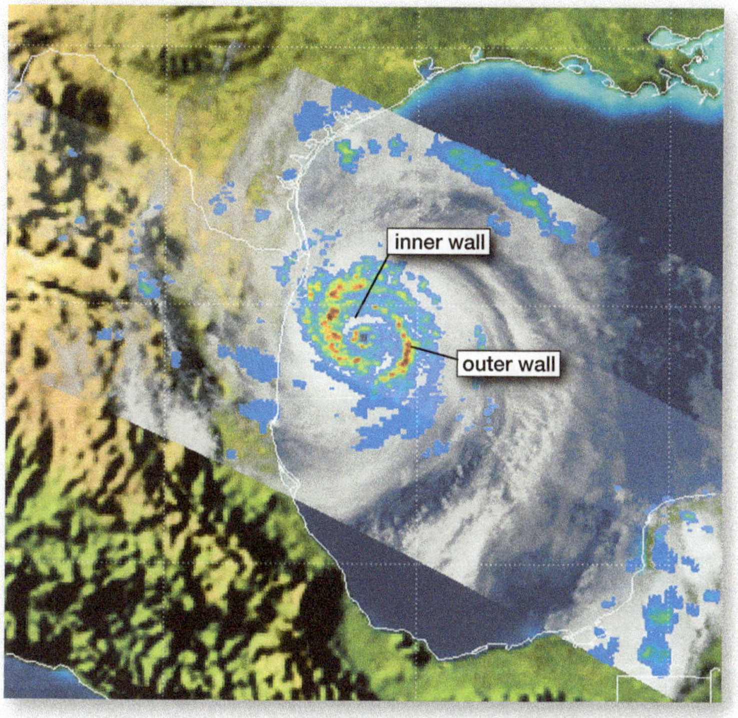

▲ **FIGURE 12–8 Double Eye Walls.** Some hurricanes develop double eye walls, such as Hurricane Emily in 2005. Usually occurring as the hurricane achieves maximum strength, both eye walls contract, with the innermost eye wall eventually dissipating and giving way to the outer eye wall.

that happens, there can be a major reduction in the inflow of moisture to the inner eye wall and it can begin to lose its intensity and eventually die out. At the same time, what had been an outer wall can intensify and contract toward the center of the hurricane and take over as the single hurricane eye wall. This process, called **eye wall replacement**, often leads to an initial weakening of the hurricane winds (as the inner eye wall begins to die out), followed by renewed strengthening as the outer wall contracts and intensifies. Some hurricanes undergo this process more than once, with each cycle occurring on the order of about a half a day to two or three days. This is illustrated by Figure 12–9, showing satellite images of Hurricane Wilma over a 48-hour period, October 19–21, 2005.

HOT TOWERS NASA scientists have recently uncovered the existence of **hot towers** (Figure 12–10) embedded in some eye walls that last between 30 minutes and 2 hours. Hot towers are localized portions of eye walls that rise to greater heights (up to 12 km, or 7 mi) than the rest of the eye wall. The researchers found that development of hot towers indicates a greater likelihood that the hurricane will intensify within the next 6 hours.

Video
Hot Towers and Hurricane Intensification
http://goo.gl/jJmpo

The air temperature at the storm's surface within an eye is several degrees warmer than outside the eye because compression of the sinking air causes it to warm adiabatically. The air is also drier, because warming the unsaturated air lowers its relative humidity. Contrary to common belief, however, the air is not entirely cloud free within the eye; instead, fair-weather cumulus clouds are scattered throughout the otherwise blue sky.

CHECKPOINT

12.5 Where in a hurricane would you find ascending air? Descending air?

12.6 What are some of the major changes that can occur in a hurricane that are associated with weakening or intensification?

Steps in the Formation of Hurricanes

12.4 Explain the process of hurricane formation.

Hurricanes do not suddenly appear out of nowhere; they begin as much weaker systems that often migrate large distances before turning into hurricanes. In this section we examine hurricane development, with particular emphasis on how it occurs in the Atlantic.

Tropical Disturbances

Although most tropical storms attain hurricane status in the western portions of the oceans, their earliest origins often lie far to the east as small clusters of small thunderstorms called **tropical disturbances**. Tropical disturbances are disorganized groups of thunderstorms having weak pressure gradients and little or no rotation.

Tropical disturbances can form in several different environments. Some form when midlatitude troughs migrate into the tropics; others develop as part of the normal convection associated with the intertropical convergence zone (ITCZ). But most tropical disturbances that enter the western Atlantic and become hurricanes originate in **easterly waves**, large undulations or ripples in the normal trade wind pattern. Figure 12–11 illustrates a sequence of easterly waves. Because pressure gradients in the tropics are normally weaker than those of the extratropical regions, the easterly waves are better shown by plotting lines of wind direction (called *streamlines*) rather than isobars. The air in the wave initially flows westward, turns poleward, and then flows back toward the equator, with the entire wave pattern extending 2000 to 3000 km (1242 to 1864 mi) in length. On the upwind (eastern) side of the axis, the streamlines become progressively closer together (convergence). With convergence there is low pressure and rising motion (see Chapter 6); thus the tropical disturbance is located upwind of the wave axis (the dashed line) of the easterly wave. Surface divergence downwind of the wave axis leads to clear skies. (An explanation for why the streamlines are convergent and divergent is somewhat complicated; the main factor involves changes in relative vorticity that occur as the air moves poleward and equatorward.)

The tropical disturbances that affect the Atlantic Ocean, Caribbean, and the Americas mostly form over western Africa,

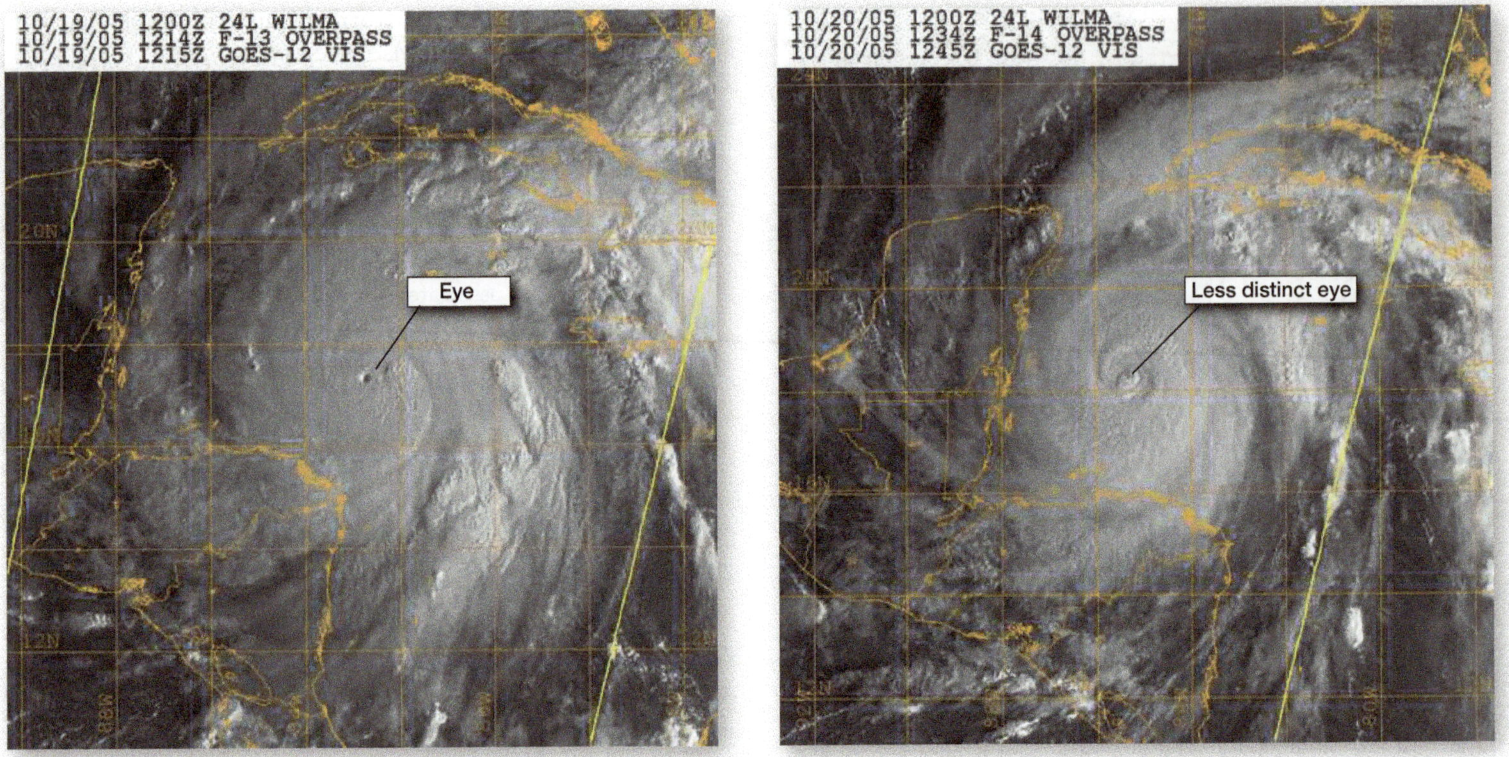
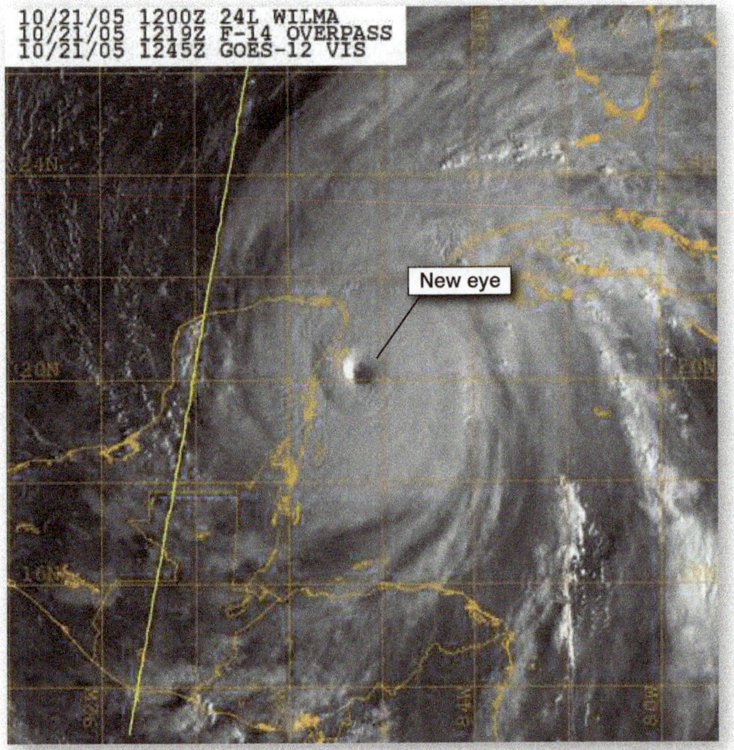

▲ **FIGURE 12–9 Eye Wall Replacement.** Hurricane Wilma underwent eye wall replacement between October 19 and 21, 2005. (a) Initially a very tight eye wall is evident. (b) As the inner wall weakened, the eye became less distinct and the maximum wind speeds weakened. (c) A day later the outer wall has completely replaced the original eye wall.

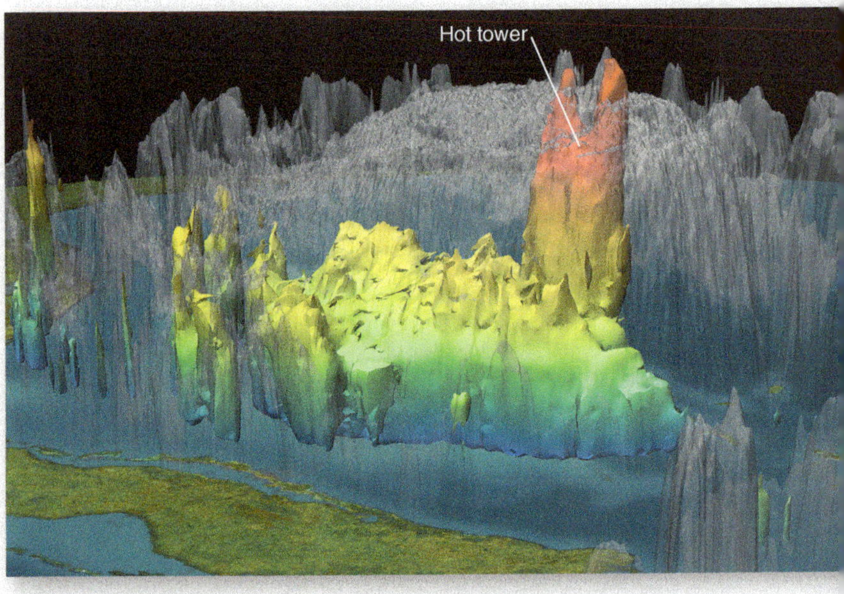

▲ **FIGURE 12–10 Hot Towers.** This hot tower in Hurricane Rita in 2005 was observed by radar aboard NASA's Tropical Rainfall Measuring Mission (TRMM) spacecraft. These localized areas of deep cloud cover and intense precipitation often precede hurricane intensification.

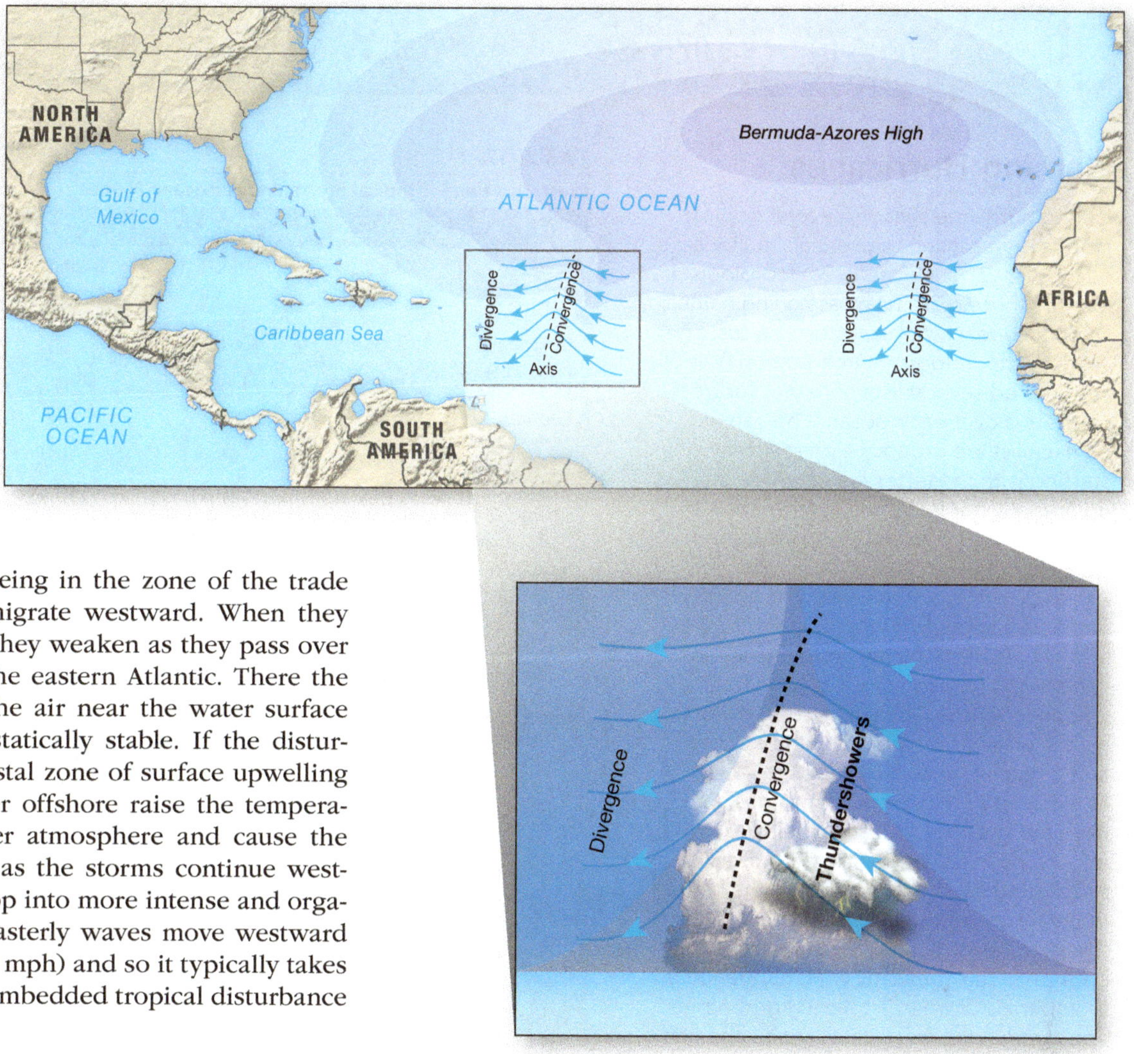

▶ **FIGURE 12–11 Easterly Waves.** East of the axis, easterly waves have cloud cover and surface convergence. While west of the axis, they diverge. They represent troughs of low pressure, embedded in the trade winds, that migrate westward. They sometimes develop into tropical depressions, storms, or hurricanes.

south of the Saharan desert. Being in the zone of the trade winds, these storms tend to migrate westward. When they reach the west coast of Africa, they weaken as they pass over the cold Canary current over the eastern Atlantic. There the low water temperatures chill the air near the water surface and cause the air to become statically stable. If the disturbances migrate beyond the coastal zone of surface upwelling however, warmer waters farther offshore raise the temperature and humidity of the lower atmosphere and cause the air to become unstable. Then, as the storms continue westward, a small percentage develop into more intense and organized thunderstorm systems. Easterly waves move westward at about 15 to 35 km/hr (9 to 22 mph) and so it typically takes about a week to 10 days for an embedded tropical disturbance to migrate across the Atlantic.

Tropical Depressions and Tropical Storms

The vast majority (probably more than 90 percent) of tropical disturbances die out without ever organizing into more powerful systems. But some undergo a lowering of pressure and begin to rotate cyclonically. When a tropical disturbance develops to the point where there is at least one closed isobar on a weather map, the disturbance is classified as a **tropical depression**. If the depression intensifies further and maintains wind speeds above 63 km/hr (39 mph), it becomes a **tropical storm**. (At this point the system is named. See *Box 12–1, Forecasting: Naming Hurricanes,* for more information on this practice.) A further increase in sustained wind speeds to 119 km/hr (74 mph) creates a true hurricane. While only a small fraction of tropical disturbances ever become tropical depressions, a larger proportion of depressions become tropical storms, and an even greater percentage of tropical storms ultimately become hurricanes.

Hurricanes

The location at which hurricanes are most likely to form varies seasonally. Early in the Atlantic season, dissipating midlatitude cyclones (see Chapter 10) in the western ocean extend southward over warm tropical water. Such systems can provide the rotation favorable for hurricane development. Later in the season, fronts are confined to higher latitudes and no longer play a role in the formation of hurricanes. Instead, warm waters are found progressively farther to the east, so that disturbances leaving the African continent can grow into full-scale hurricanes. (Systems that become tropical storms in the tropical waters just off western Africa and become hurricanes before reaching the Caribbean are often referred to as *Cape Verde hurricanes*, so named for the islands near which they originate.) The net effect is that the birthplace of tropical cyclones moves from west to east across the tropical ocean during the first half of the season. In the late fall, the breeding ground moves westward as frontal activity again emerges as a primary agent of cyclone genesis.

Video
A Hurricane in the Middle Latitudes
http://goo.gl/r359VH

As with their Atlantic counterparts, Pacific hurricanes move westward during their formative stages. Many come near Hawaii, but most bypass the islands or die out before reaching them. Unfortunately, this is not always the case. In September 1992, Hurricane Iniki battered the

12–1 FORECASTING

Naming Hurricanes

During hurricane season, several tropical storms or hurricanes can arise simultaneously over various oceans. Meteorologists identify these systems by assigning names when they reach tropical storm status. The World Meteorological Organization (WMO) has created several lists of names for tropical storms over each ocean. The names on each list are ordered alphabetically, starting with the letter *A* and continuing up to the letter *W*. When a depression attains tropical storm status, it is assigned the next unused name on that year's list. At the beginning of the following season, names are taken from the next list, regardless of how many names were unused in the previous season. Six lists have been compiled for the Atlantic Ocean, and the names on each list are used again at the end of each 6-year cycle (Table 12–1–1). English, Spanish, and French names are used for Atlantic hurricanes. If all the available names on a season's list are used, subsequent storms will be given the names of the letters of the Greek alphabet. Thus, the 22nd named storm for a season would be Alpha. This is exactly what happened in October 2005 when Tropical Storm Alpha appeared in the western Atlantic—the first time ever that the list of names was unable to accommodate all of the tropical storms in a single season.

TABLE 12–1–1

Western Atlantic Tropical Storm and Hurricane Names

2014	2015	2016	2017	2018	2019
Arthur	Ana	Alex	Arlene	Alberto	Andrea
Bertha	Bill	Bonnie	Bret	Beryl	Barry
Cristobal	Claudette	Colin	Cindy	Chris	Chantal
Dolly	Danny	Danielle	Don	Debby	Dorian
Edouard	Erika	Earl	Emily	Ernesto	Erin
Fay	Fred	Fiona	Franklin	Florence	Fernand
Gonzalo	Grace	Gaston	Gert	Gordon	Gabrielle
Hanna	Henri	Hermine	Harvey	Helene	Humberto
Isaias	Ida	Ian	Irma	Isaac	Ingrid
Josephine	Joaquin	Julia	Jose	Joyce	Jerry
Kyle	Kate	Karl	Katia	Kirk	Karen
Laura	Larry	Lisa	Lee	Leslie	Lorenzo
Marco	Mindy	Matthew	Maria	Michael	Melissa
Nana	Nicholas	Nicole	Nate	Nadine	Nestor
Omar	Odette	Otto	Ophelia	Oscar	Olga
Paulette	Peter	Paula	Philippe	Patty	Pablo
Rene	Rose	Richard	Rina	Rafael	Rebekah
Sally	Sam	Shary	Sean	Sara	Sebastien
Teddy	Teresa	Tobias	Tammy	Tony	Tanya
Vicky	Victor	Virginie	Vince	Valerie	Van
Wilfred	Wanda	Walter	Whitney	William	Wendy

Particularly notable hurricanes can have their names "retired" by the WMO if an affected nation requests the removal of that name from the list. All replacements are made with names of the same gender, first letter and language. As of the end of the 2013 season, 79 names had been retired from the Atlantic hurricane list. If a hurricane with a Greek letter merits special designation, it would do so with its year appended to the letter. Thus, if in 2015 Alpha is very destructive, it can be noted as Alpha 2015.

The practice of naming tropical storms and hurricanes appears to have begun during World War II when meteorologists in the Pacific assigned female names (possibly after wives and girlfriends) to tropical storms and typhoons. This practice was adopted by the U.S. National Weather Service (then called the Weather Bureau) in 1953 and maintained until 1979, when male names were added to the lists.

12–1–1 At what point in their evolution do hurricanes get assigned names?

12–1–2 How have the conventions for assigning names changed since World War II?

island of Kauai with wind gusts up to 258 km/hr (160 mph) and brought heavy flooding to the beach resort areas. The hurricane destroyed or severely damaged half of the homes on the island and devastated most of the tourist industry.

Conditions Necessary for Hurricane Formation

Although the dynamics of hurricanes are extremely complex, meteorologists have long recognized the conditions that favor their development. Great amounts of heat are needed to fuel hurricanes, and the primary source of this energy is the release of latent heat supplied by evaporation from the ocean surface.

WARM SURFACE LAYER TEMPERATURES Because high evaporation rates depend on the presence of warm water, hurricanes form only where the ocean has a deep surface layer (several tens of meters in depth) with temperatures above 27 °C (81 °F). The need for warm water precludes hurricane formation poleward of about 20 degrees latitude; sea surface temperatures are usually too low there. Hurricanes develop most often in late summer and early fall, when tropical waters are warmest.

HUMIDITY Also related to the presence of deep layers of warm water is the need for a high moisture content in the air. As water evaporates from the sea surface, putting latent heat into the atmosphere, it also feeds enough water vapor into the air to help it maintain a deep layer of air with high humidity. This provides the necessary water vapor for the growth of deep cumulus clouds.

CORIOLIS FORCE Hurricane formation also depends on the Coriolis force, which must be strong enough to prevent filling of the central low pressure. The absence of a Coriolis effect at the equator prohibits hurricane formation between 0° and 5° latitude. This factor and the need for high water temperatures explain the pattern shown earlier in Figure 12–2, in which tropical storms attain hurricane status mostly between the latitudes of 5° and 20° North and South latitude.

UNSTABLE CONDITIONS Stability is also important in hurricane development, with unstable conditions throughout the troposphere an absolute necessity. Along the eastern margins of the oceans, cold currents and upwelling cause the lower troposphere to be statically stable, inhibiting uplift. Moreover, the trade wind inversion puts a cap on any mixing that might otherwise occur. Moving westward, water temperatures typically increase and the trade wind inversion is found farther up or disappears altogether, and so hurricanes become more prevalent. Finally, hurricane formation requires an absence of strong vertical wind shear (a substantial change in wind speed or direction with height), which disrupts the vertical transport of latent heat.

Once formed, hurricanes are self-propagating (just as severe storms outside the tropics are self-maintaining). That is, the release of latent heat within the cumulus clouds causes the air to warm and expand upward. The expansion of the air supports upper-level divergence, which draws air upward and promotes low pressure and convergence at the surface. This leads to continued uplift, condensation, and the release of latent heat.

So if hurricanes are self-propagating, can they intensify indefinitely until they attain supersonic speeds? No, because they are ultimately limited by the supply of latent heat, which in turn is constrained by the temperature of the ocean below and by the processes underlying evaporation and convection. The importance of ocean temperature suggests that if the oceans were to become warmer, hurricanes would theoretically become more intense. This topic has received considerable attention lately because of climatic warming, which could be accompanied by higher ocean temperatures.

CHECKPOINT

12.7 What initial conditions are necessary to begin the process of hurricane formation?

12.8 What makes a hurricane self-propagating?

12.9 What factors could cut off or limit the intensification of a developing hurricane? Explain.

Hurricane Movement and Dissipation

12.5 Describe hurricane movement and dissipation.

The movement of tropical systems is related to the stage in their development. Tropical disturbances and depressions are guided mainly by the trade winds and, therefore, tend to migrate westward. The influence of the trade winds often diminishes after the depressions intensify into tropical storms. Then the upper-level winds and pressure and their interactions with the storm's internal motions more strongly determine their speed and direction.

Hurricane Paths

Once fully developed, tropical storms become more likely to move poleward, as shown in Figures 12–2 and 12–12. Figure 12–13 shows the locations of all continental United States hurricane landfalls from 1950 through 2011. Hurricanes and tropical storms can move in wildly erratic ways—for example, moving in a constant direction for a time, then suddenly changing speed and direction, and even backtracking along its previous path. Figure 12–14 plots some particularly erratic paths along the east coast of North America. *Box 12–2, Focus on Severe Weather: 2005—A Historic Hurricane Season,* describes the paths and damage done by two notable hurricanes in 2005.

While hurricanes can hit any part of the Gulf of Mexico or Atlantic coasts at any time during hurricane season, they have a greater likelihood of taking particular paths during different months. Figure 12–15 shows the likelihood of hurricane passage for various months:

- In August, the most likely path of hurricanes tracks over the West Indies (Figure 12–15a). From there, hurricanes are about equally likely to track toward the Texas coast or along the Atlantic coast from Florida to North Carolina.
- Two prominent paths dominate in September (Figure 12–15b). One goes from between the Yucatan Peninsula and western Cuba, northward toward the central Gulf of Mexico coast; the other moves northward from around Haiti, the Dominican Republic, and Puerto Rico into the western Atlantic. Hurricanes taking the more easterly track are most likely to hit the middle Atlantic states if they make landfall.
- October paths exhibit a greater tendency to track from the Gulf of Mexico and Caribbean Sea northward to Florida and the rest of the southeastern United States (Figure 12–15c).

Hurricane Dissipation

Atlantic tropical storms and hurricanes can travel great distances along the North American east coast, but they usually

▲ **FIGURE 12–12 Hurricane Tracks.** The tracks of all western Atlantic and eastern Pacific major hurricanes (category 3 or higher), from 1851–2013 for the Atlantic and 1949–2013 in the Pacific. If plotted, the paths of weaker hurricanes would be very similar to those shown (and are so numerous would render the map illegible).

▶ **FIGURE 12–13 Hurricane Landfalls.** The locations of all hurricane landfalls over the continental United States from 1950 through 2011. Saffir-Simpson categories are those at the time of landfall. *Source:* http://www1.ncdc.noaa.gov/pub/data/images/2011-Landfalling-Hurricanes-11x17.pdf

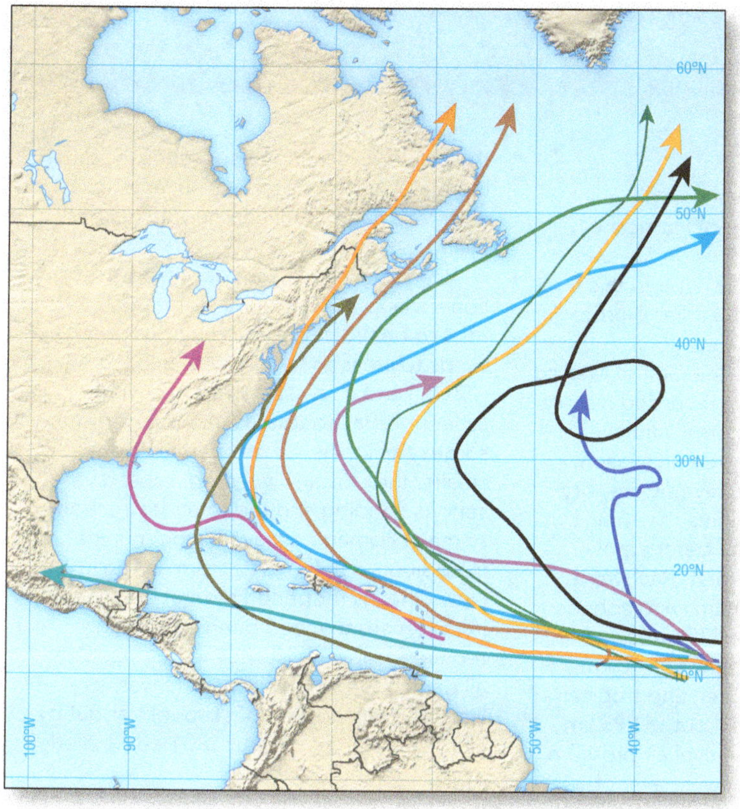

▲ **FIGURE 12–14 Erratic Hurricane Paths.** Tropical storms and hurricanes have a tendency to move north or northeast out of the tropics along the southeast coast of North America. Their paths are often erratic, as seen in these examples.

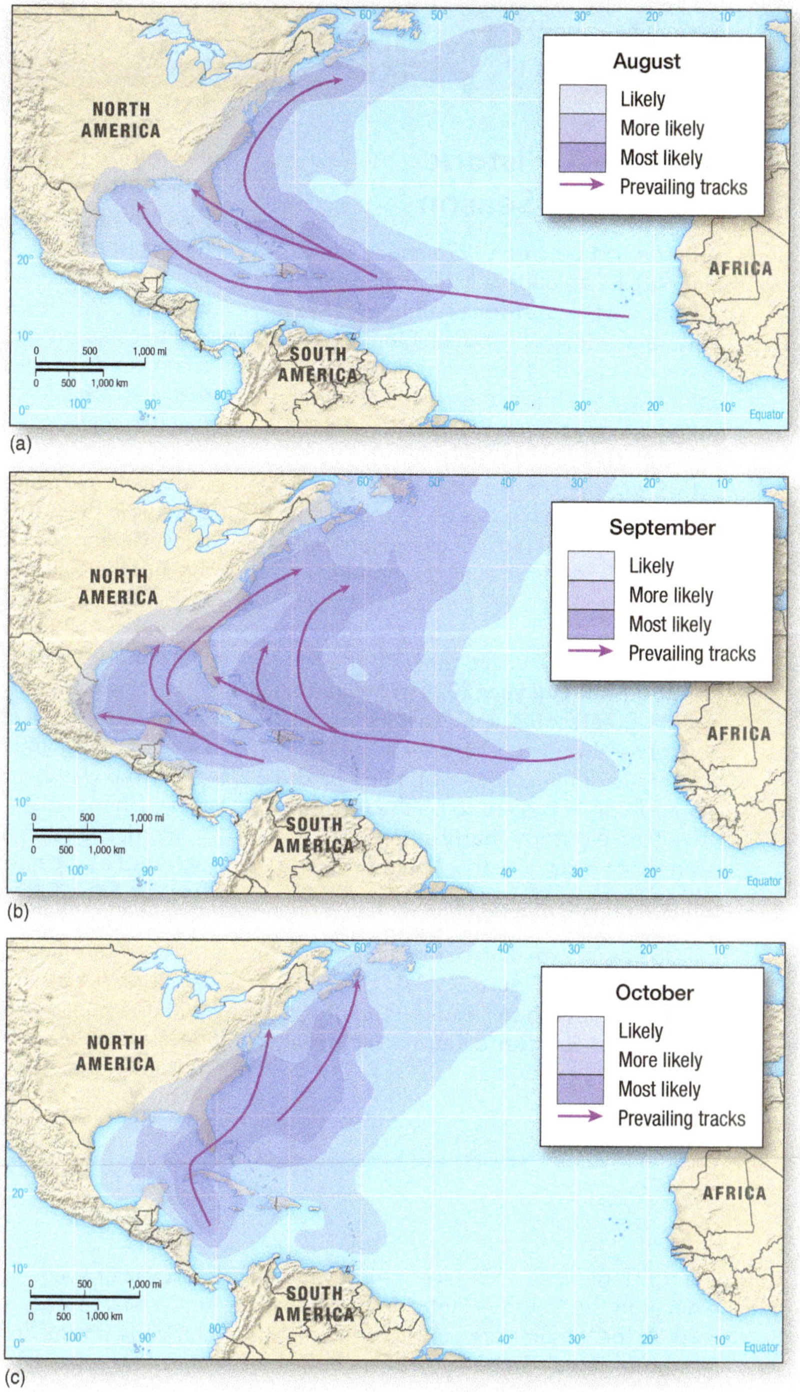

▲ **FIGURE 12–15 Predominant Hurricane Regions.** The predominant paths of Atlantic hurricanes vary in different months from early to late in the hurricane season. (a) In August hurricanes may enter the Gulf of Mexico or sweep up the East Coast, but are more likely to remain in the Caribbean. (b) By September the likelihood of hurricanes making landfall along the Gulf of Mexico or East Coast increases considerably. (c) Late season hurricanes occurring in October are more confined to the far western Atlantic and Caribbean, but it is important to note that the overall frequency of hurricanes is less than it is in earlier months.

weaken considerably as they approach the northeastern United States and the Maritime Provinces of Canada. These storms usually lack the strong winds that characterize hurricanes in the low latitudes but can still bring intense rainfall and flooding. On rare occasions, the storms can maintain their strong winds even as they move considerable distances from the subtropics. For example, an intense September 1938 hurricane brought 200 km/hr (120 mph) winds to Long Island, New York, as it moved toward New England. Its estimated 600 fatalities made it the fourth deadliest of all U.S. hurricanes. More recently, in October 2012, Hurricane Sandy brought enormous damage to the northeastern United States (see *Box 12–3, Focus on the Environment and Societal Impacts: Superstorm Sandy, 2012*).

Although hurricanes and tropical storms can move into the northeastern United States, along the West Coast they do not migrate nearly as far north without weakening to tropical depressions. The reason for this is the difference in water temperatures along the two coasts. The Pacific Coast is dominated by upwelling and the cold California current, while the warm Gulf Stream flowing along the East Coast provides a greater supply of latent heat. Sometimes, storms off the coast of Mexico move to the northeast across Baja, California, and into southern California. These storms lose their supply of latent heat and lose their intensity as they move inland, but they can still bring heavy rains and flooding.

12–2 FOCUS ON SEVERE WEATHER

2005—A Historic Hurricane Season

The period between 1995 and 2005 was marked by an unusually high number of Atlantic tropical storms and hurricanes, with many making landfall. Of the 10 seasons during this time span, all but two had at least eight hurricanes—well above the annual average of 5.9. But 2004 and 2005 were particularly noteworthy. The year 2004 was the costliest hurricane season to hit the United States up to that time, bringing $42 billion in losses. Hurricanes damaged one-fifth of all the homes in Florida and killed 117 people that year. Incredibly, that devastation was dwarfed by the hurricane season that hit the following year. The year 2005 produced one of the greatest natural disasters in U.S. history—Hurricane Katrina—and several other major hurricanes that made landfall; in addition, it proved to be the most active tropical storm–hurricane season in American history, with 27 named storms (breaking a record set in 1933).

Two noteworthy hurricanes of the 2005 season are summarized here and plotted in Figure 12–2–1. Given its enormous death toll and destruction, we discuss Hurricane Katrina separately later in this chapter.

Hurricane Rita, September 18–25, 2005

Following the devastation of Hurricane Katrina (see *Box 12–4, Focus on Severe Weather: Hurricane Katrina,* later in this chapter) only a few weeks earlier, Hurricane Rita became the second Category 5 hurricane to develop in the Gulf of Mexico that year—the first time in recorded history that two such powerful storms had ever occurred in the Gulf in the same season. (Hurricane Wilma also achieved Category 5 status in the Caribbean Sea that year, but it lost Category 5 intensity before moving into the Gulf.) At its worst, Rita was a monster storm. Its maximum wind speeds topped out at 280 km/hr (175 mph), hurricane-force winds extended 110 km (70 mi) away from the center, and tropical storm-force winds reached out to 295 km (185 mi). Its minimum sea-level air pressure of 897 mb was the third lowest ever observed in the Atlantic Ocean.

Hurricane Rita's first brush with land occurred as it passed south of the Florida Keys on September 20 as a Category 2 hurricane. As it moved westward, the hurricane rapidly intensified and reached Category 5 status on the afternoon of September 21. With the public very much aware of the devastation wrought by Hurricane Katrina, citizens along much of the Gulf Coast took the call for evacuation very seriously, causing massive traffic problems in east Texas as an estimated 3 million people headed inland.

Rita struck the Texas coast just west of Louisiana as a Category 3 storm on September 23. It delivered hurricane-force winds to areas as far inland as 240 km (150 mi) and tropical storm-force winds as far north as southern Arkansas. Storm surges as high as 4.6 m (15 ft), though smaller than originally feared, caused very serious flooding and the total destruction or major damage of several communities. In addition, a 2.4 m (8 ft) storm surge in New Orleans reopened several breaches in the levees that had been temporarily repaired following Hurricane Katrina. Overall damage from Hurricane Rita was very large but far short of that brought about by Hurricane Katrina. About 120 people died as a result of the storm.

Hurricane Wilma, October 17–25, 2005

Wilma was the 12th hurricane to hit the western Atlantic or Caribbean in 2005, having grown explosively from tropical storm to Category 5 hurricane in less than 24 hours! Its sea-level air pressure of

Video
The 2005
Hurricane Season
http://goo.gl/ZvfcZ

DID YOU KNOW?

A quick glimpse of Figure 12–2 reveals that hurricanes never occur in the South Atlantic. Well, almost never. On March 27, 2004, for the first time in recorded history a hurricane hit the coast of Brazil. Tropical cyclones have hit the Brazilian coast twice before, but "Catarina" (so named for the Brazilian state of Santa Catarina, which was hit hardest) packed 147 km/hr (90 mph) winds—making it a true hurricane. This storm did not develop in exactly the same manner as do most of those in the North Atlantic. It appears to have been a hybrid between a tropical system and a midlatitude cyclone, having initially formed along a cold front in the South Atlantic. Considerable damage occurred along the coastal region, but effective warnings to the public minimized the loss of life.

Effect of Landfall

After making landfall, a tropical storm may die out completely within a few days. Even as the storm weakens though, it can still import huge amounts of water vapor and bring very heavy rainfall hundreds of kilometers inland. This is especially true when the remnant of a hurricane moving poleward joins with a midlatitude cyclone moving eastward. Exactly this happened in 1969 when one of North America's most notorious hurricanes, Camille, moved northward from the coast of Mississippi (Figure 12–16). After its high winds and tidal flooding brought extreme damage to the Gulf Coast, the storm moved northeastward toward the western slopes of the Appalachians. There, orographic uplift coupled with low pressure and the high water vapor content of the remnant hurricane could easily have produced serious flooding. But to make matters worse, an eastward-moving cold front reached the mountains at the same time as the former hurricane. The combination of moist air, low pressure, an approaching front, and the orographic effect set the stage for intense rains and flash flooding that killed more than 150 people.

884 mb was the lowest ever observed in the Atlantic.

Wilma scored a direct hit on the Yucatan Peninsula, where an estimated 22,000 tourists and many more residents sought shelter from the 200 km/hr (125 mph) winds and heavy downpours. Meanwhile, 700,000 people had been evacuated from the west side of Cuba in anticipation of the heavy flooding that hit that portion of the island. Wilma moved rapidly northeastward across the Gulf of Mexico toward the southwestern coast of Florida. It hit the western Gulf Coast as a Category 3 hurricane, causing storm surges up to 2.75 m (9 ft) and extensive flooding, especially in Key West. The Miami and Ft. Lauderdale areas experienced considerable wind damage with millions of people losing electric power for extended time periods. In all, over 60 people were killed and damages were estimated at $29 billion.

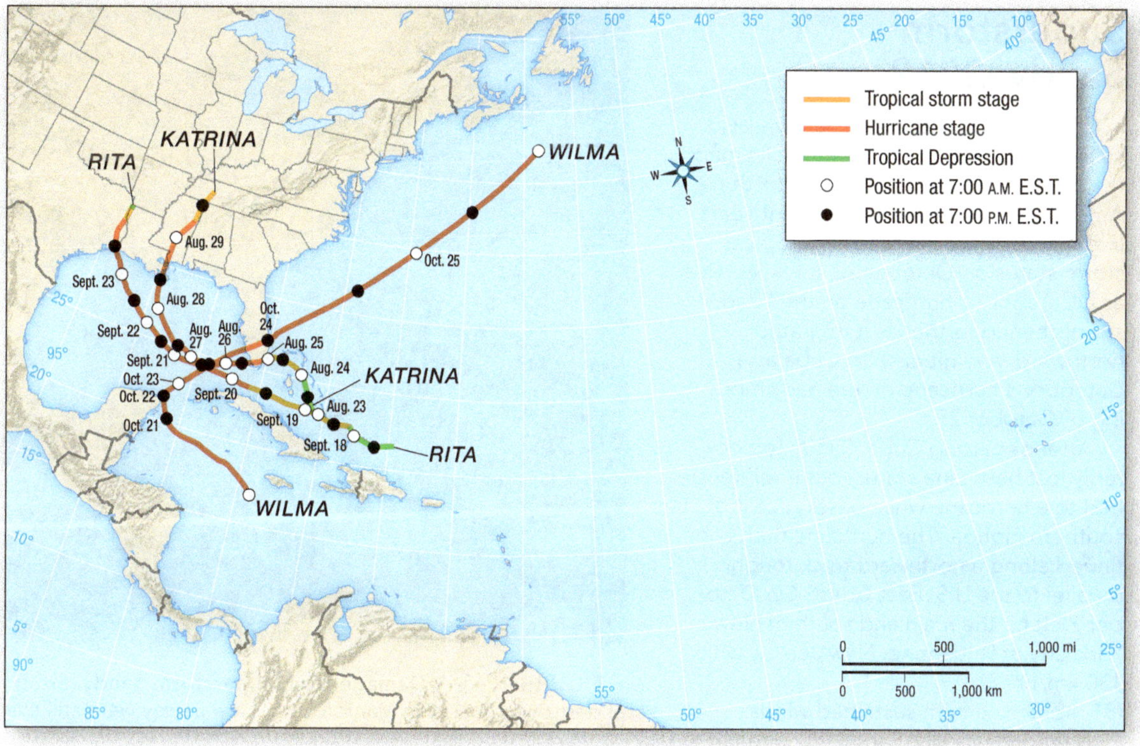

▲ **FIGURE 12–2–1 2005 Hurricanes.** The paths of Hurricanes Katrina, Rita, and Wilma in 2005.

12–2–1 What are two aspects of the 2005 hurricane season that made it particularly noteworthy?

12–2–2 Which areas received the worst effects of Hurricanes Rita and Wilma?

Hurricane Destruction and Fatalities

12.6 Describe how hurricanes cause destruction and fatalities.

Hurricanes can bring death and devastation in several ways, through any combination of strong winds, heavy rain, hurricane-spawned tornadoes, and the elevation of coastal waters combined with heavy surf.

Wind

By definition, hurricane winds exceed 119 km/hr (74 mph), and many are much faster. It is not surprising then, that hurricane-force winds can cause extensive damage and even destroy well-built homes. An example of this happened on August 24, 1992, when Hurricane Andrew devastated much of southern Florida. It approached southern Florida almost directly from the east and cut westward across the state, with its eye passing about 40 km (25 mi) south of Miami Beach. When the hurricane made landfall it had sustained winds of up to 233 km/hr (145 mph) with gusts in excess of 281 km/hr (175 mph). There was relatively little damage due to coastal flooding or heavy rainfall, but winds completely leveled the town of Homestead, killed 24 people, and left 180,000 homeless in the state before crossing into the Gulf of Mexico.

Heavy Rain

Hurricanes produce exceedingly intense rainfall, with rates on the order of meters per day found beneath the eye wall. The rate for a fixed location beneath a passing storm is smaller, but still huge—on the order of 25 cm/day (10 in./day).

Heavy rain made Hurricane Mitch the deadliest hurricane to hit the Western Hemisphere in the past 200 years, killing thousands of people in Central America in October 1998. Mitch hit Central America from the east. Although Mitch's winds weakened substantially, the remnant system

12–3 FOCUS ON THE ENVIRONMENT AND SOCIETAL IMPACTS

Superstorm Sandy, 2012

In October 2012 Hurricane Sandy—also called Superstorm Sandy—made its mark as one of the most devastating North American weather events in recent years (Figure 12–3–1). First reaching tropical storm status on October 22, it hit Jamaica as a Category 1 hurricane on the 22nd, strengthened further as it migrated northward, and hit eastern Cuba as a Category 3 hurricane on the early morning of October 25.

After departing Cuba, Sandy temporarily lost hurricane status, but it reintensified to a hurricane while passing east of southern Florida. The hurricane then continued along a northward track roughly parallel to the U.S. East Coast. On October 29 it hit the mainland south of New York City at Brigantine, New Jersey, with 130 km/hr (81 mph) maximum sustained winds.

Meteorologically, Sandy was a remarkable event for a number of reasons:

- It was an unusually large hurricane in terms of area covered.
- The hurricane brought record-breaking low pressure across much of the East Coast.
- Sandy traveled farther northward than most Atlantic hurricanes, especially for that time of year.
- The hurricane merged with a midlatitude cyclone that swept along the United States to become a hybrid system with tropical origins but well-defined cold and warm fronts.
- It made landfall near the most densely populated area of the United States.
- Landfall occurred during a large high tide.

▲ **FIGURE 12–3–1 Damage from Superstorm Sandy.** Six months after the storm, the remains of destroyed houses in Mantoloking, New Jersey were still evident.

Hurricane Path

As the hurricane paralleled the East Coast (Figure 12–3–2) it gained considerable latent heat as it passed over the warm Gulf Stream, where water temperatures were anomalously high for that time of year (Figure 12–3–3). This would not be a huge problem on land under normal conditions, because most hurricanes move to the northeast out to sea. But a combination of an upper-level trough to the west and high surface pressure to the east forced an unusual westward turn toward the mainland on October 28, the day before eventual landfall on the U.S. mainland.

Transition from Hurricane to Midlatitude Cyclone

One very interesting aspect of Sandy was how it evolved from a purely tropical system to one that joined with a midlatitude cyclone as it approached the eastern shore. Refer to Figure 12–3–2 to observe this development. On the 27th and 28th, the two systems are completely distinct from each other (Figures 12–3–2a and b), but by the 29th and 30th the storms had become one (Figures 12–3–2c and d.).

Winds and Storm Surge

Though the winds decreased rapidly after landfall, the storm—which by this time had lost its tropical characteristics—continued across much of the mid-Atlantic region, as far inland as Ontario and eastern Ohio. Many areas of New York City and the New Jersey shoreline were completely inundated by a storm surge of up to 3.86 m (12.65 ft) above normal level at Kings Point on western Long Island. Tourist areas along the shores of New York City and New Jersey were severely damaged by the flood surge. The town of Breezy Point on Staten Island lost over 100 homes as a fire spread and the volunteer fire department was unable to immediately get to the burning structures. Ironically, when they did arrive, a lack of water pressure prevented them from extinguishing the flames.

Fatalities and Damage

About 150 direct fatalities such as drowning and trauma from flying debris were attributed to the storm across the Caribbean and the United States (about 70 of these in the United States), with about an equal number of deaths resulting from indirect causes such as heart attacks. The United States incurred about $70 billion in damages—second only to Hurricane Katrina.

12–3–1 What were some of the unusual features of Superstorm Sandy?

12–3–2 What factors led to the huge amount of damage from Sandy?

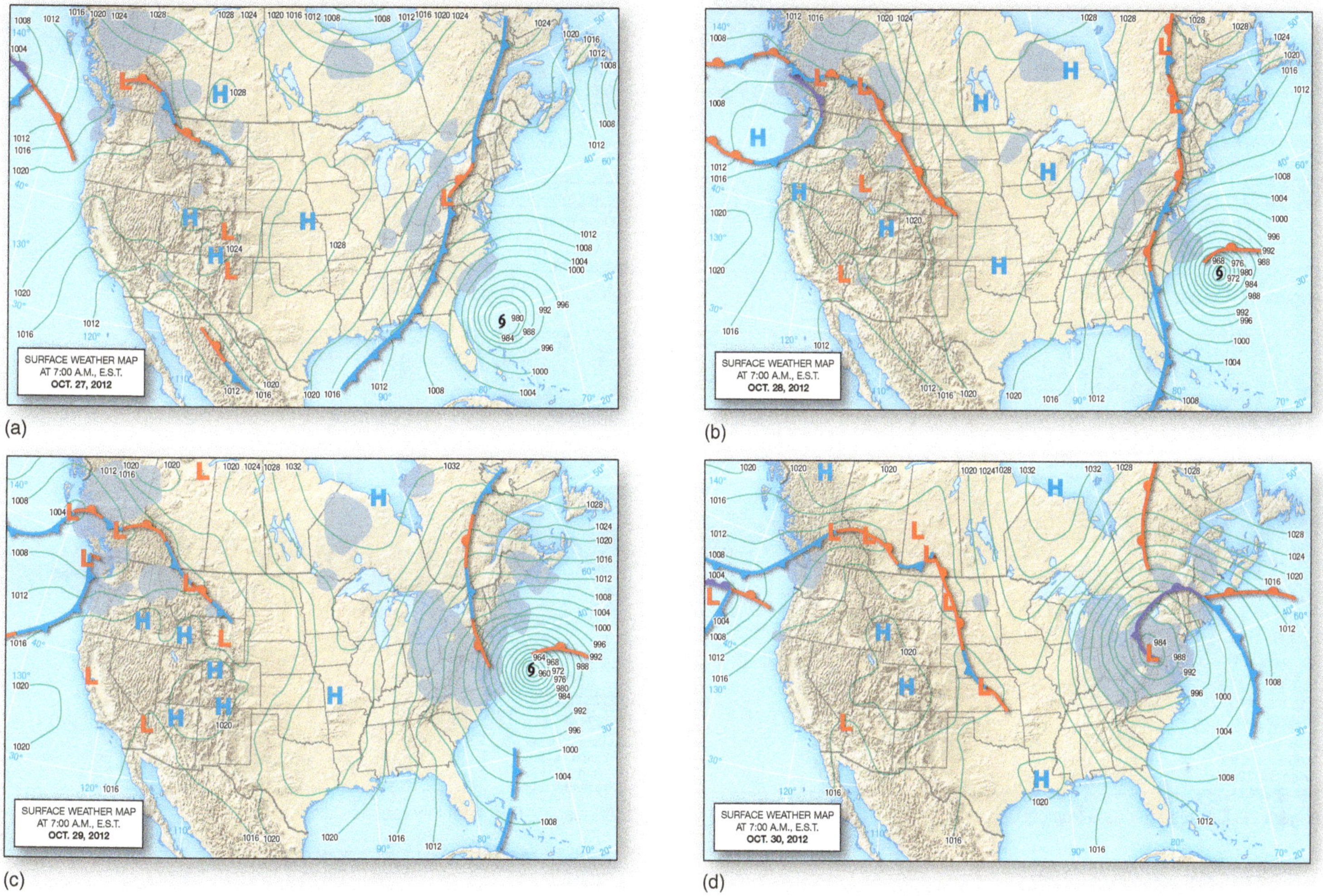

▲ **FIGURE 12–3–2 Sandy's Path.** The maps show the progress of Hurricane Sandy, October 27–30, 2012.
Source: http://www.hpc.ncep.noaa.gov/dailywxmap/.

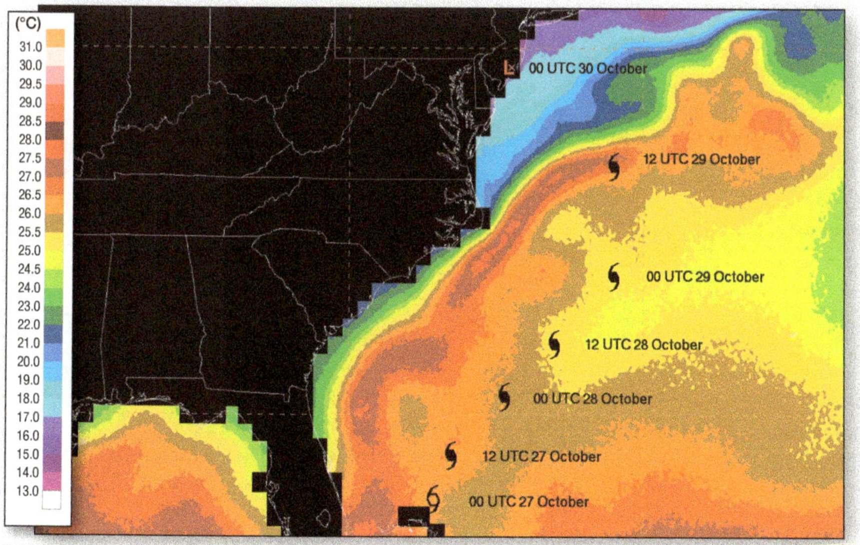

SST's (°C) on 28 Oct 2012 with the best track of Sandy plotted at 12 hour intervals

◄ **FIGURE 12–3–3 Sandy's Source of Heat.** The map shows Hurricane Sandy's path in relation to Gulf Stream sea surface temperatures. For several days, Sandy gained latent heat over the Gulf Stream's warm waters.
Source: http://www.hpc.ncep.noaa.gov/dailywxmap/.

Video
The Making of a Superstorm

http://goo.gl/4dm6PN

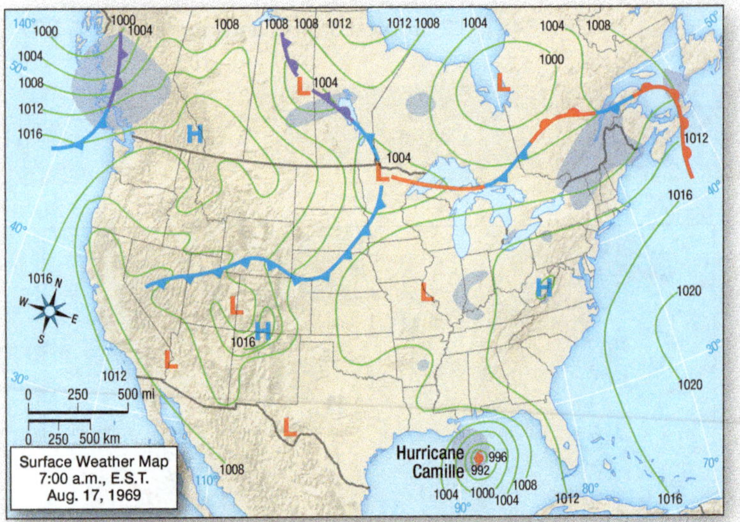

(a) August 17: Comes ashore as a Category 5 hurricane.

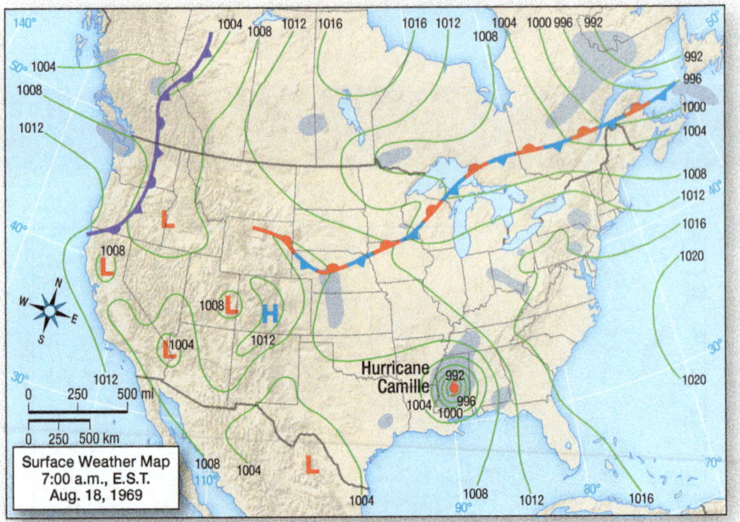

(b) August 18: Moves north over Mississippi.

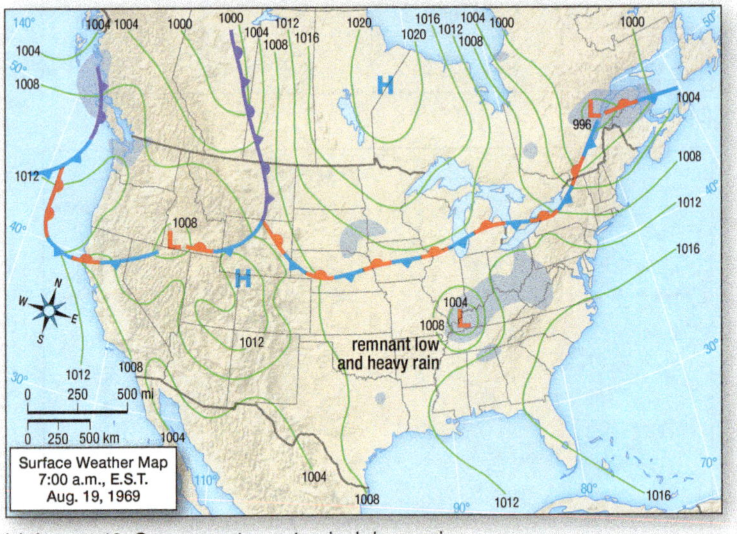

(c) August 19: Curves east as a tropical depression.

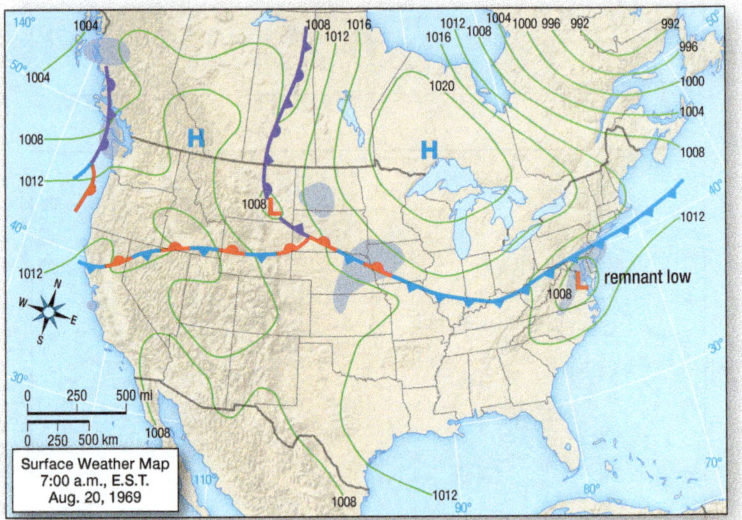

(d) August 20: Remnant low reaches Virginia coast.

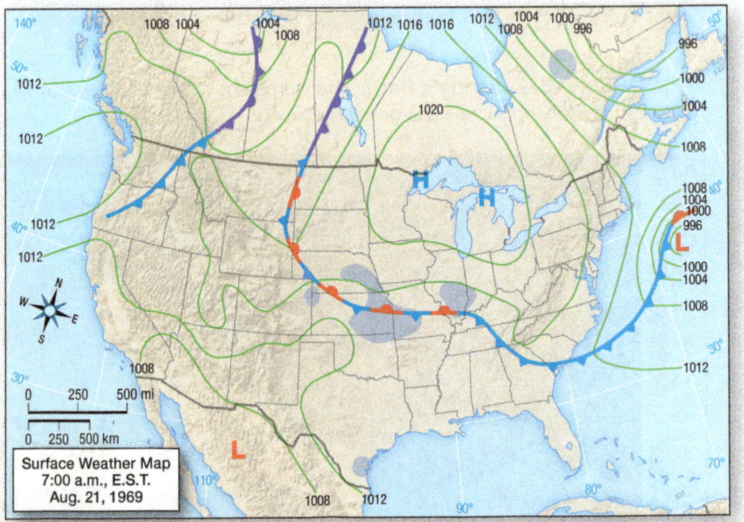

(e) August 21: Low moves northeast in the Atlantic.

▲ **FIGURE 12–16 Hurricane Camille.** Camille hit the Gulf Coast as a major hurricane and moved north into Mississippi (a and b). Winds diminished considerably after advancing inland but heavy rains continued, especially over the Appalachians (c). Very serious flooding occurred as the passage of a cold front intensified rainfall (c and d). Eventually the remnants of the storm passed eastward into the Atlantic (e).

brought heavy rainfall across the region as it tracked northward toward the Gulf of Mexico. Intense rains lasted for several days, and parts of Honduras and Nicaragua received estimated precipitation totals of 85 cm (35 in.), causing extensive flooding and mudslides in this mountainous region.

In 2002 Tropical Storm Allison demonstrated that a tropical storm need not attain hurricane status to become a major disaster. Allison hit the south Texas coast on June 5 and hovered over the area for nearly a week. Its heavy rainfall (as much as 96 cm, or 40 in.) caused major flooding across Texas and Louisiana, killing 24 people and flooding more than 46,000 homes and businesses.

Tornadoes

Many hurricanes also contain clusters of tornadoes, most often in the right-forward quadrant (Figure 12–17). They usually occur far enough away from the center that they are surrounded by relatively weak winds. It appears that slowing of the wind by friction at landfall contributes to tornado formation. Hurricane-spawned tornadoes tend to have shorter life spans than tornadoes in the central United States.

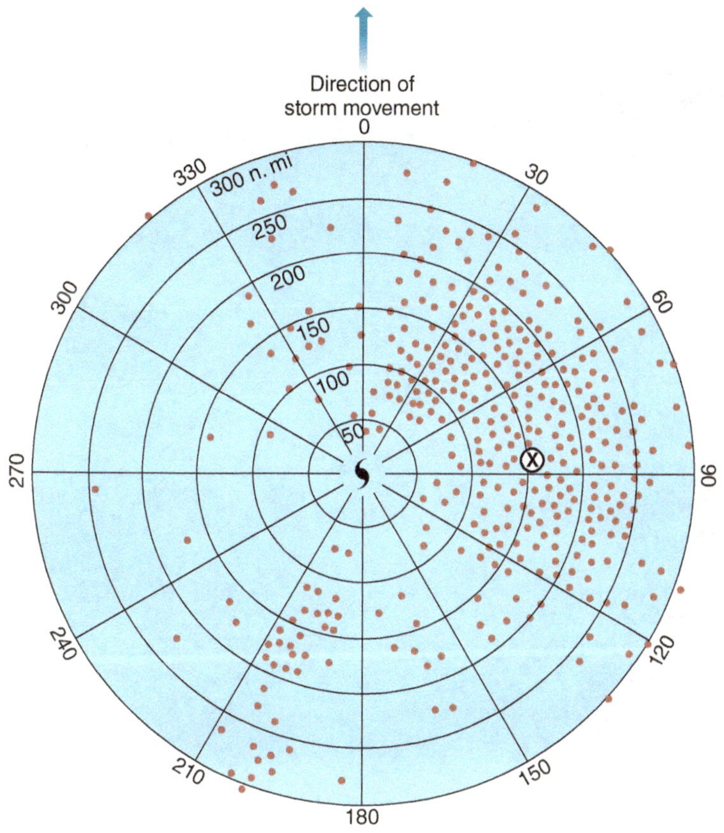

▲ **FIGURE 12–17 Embedded Tornadoes.** Tornadoes most often form in the right-forward quadrant of hurricanes (based on the direction in which the storm is moving). This figure is based on data from 373 hurricane-embedded tornadoes between 1948 and 1972 in the Northern Hemisphere. Each dot represents a tornado. The circled X indicates the mean position of tornadoes relative to the storm center.

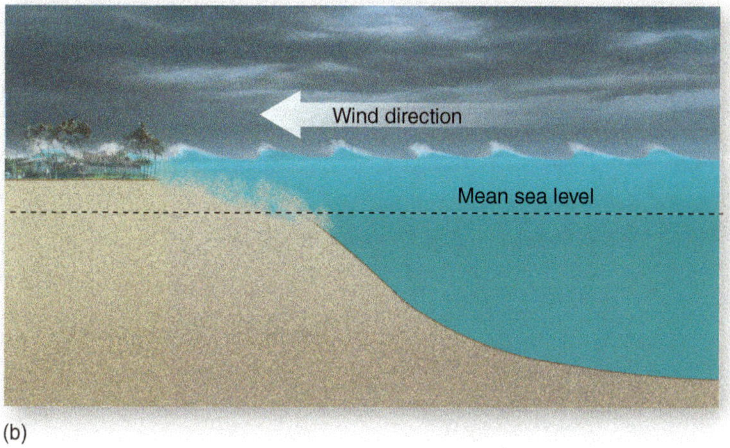

▲ **FIGURE 12–18 Storm Surge.** (a) A normal high tide presents no problems. (b) A storm surge is created primarily by strong winds blowing ocean water ashore, the effect of which is compounded by heavy wave activity. The reduction in air pressure associated with a hurricane also contributes to the rise in sea level but is not the primary factor.

Storm Surges

In addition to the threat of heavy rain, strong winds, and tornadoes, coastal regions are vulnerable to a special problem called a **storm surge**, a rise in water level induced by a hurricane (Figure 12–18). Two processes create a storm surge, the major one being the piling up of water as heavy winds drag surface waters forward. Strong winds blowing toward a coast force surface waters landward and thereby elevate sea level, while also bringing heavy surf. The low atmospheric pressure in a hurricane also contributes to the storm surge, in the same way that the height of a column of mercury in a barometer responds to variations in atmospheric pressure. For every millibar the pressure decreases, the water level rises 1 cm (0.4 in.). For most hurricanes along a coastal zone, the combined effects elevate the water level only a meter or two. But in extreme circumstances, storm surges can increase the water level by as much as 7 m (23 ft), as was the case for Hurricane Camille along the coast of Mississippi in 1969.

DID YOU KNOW?

Hurricane damage is not all water and wind. For example, in Navarre Beach, Florida, Hurricane Katrina left behind meters of sand along a 20-mile (12 km) stretch of road (Figure 12–19). Months of work were needed to clear these storm-produced drifts.

▼ **FIGURE 12–19 Transported Sand.** A stretch of road at Navarre Beach, Florida, cut off by blowing sand from Hurricane Katrina.

Storm surges along low-lying coastal plains can be extremely devastating where the rise in sea level brings waters far inland. Furthermore, the heavy waves generated by the strong winds pound away at structures, with debris carried by the waves adding to the problem. Figure 12–20 illustrates the type of devastation that can result from a storm surge, in this case the result of Hurricane Katrina. Storm surges are most destructive when they coincide with high tides, especially over bays and inlets that have an extreme range of height between high and low tide. The danger storm of surges is so great that in 2014 the National Hurricane center began issuing storm surge flooding maps on an experimental basis for the tropical storms predicted to impact the East and Gulf Coasts of the United States.

LEFT SIDE–RIGHT SIDE WINDS AND STORM SURGES Hurricane winds and storm surges are most intense on the right-hand side of the storm relative to the direction it is moving.

(a)

(b)

▲ **FIGURE 12–20 Damage from Hurricane Katrina.** The red arrows in (a) point to a pierhouse and antebellum house in Biloxi, Mississippi. They, along with the pier in the foreground, were completely destroyed by the storm (b).

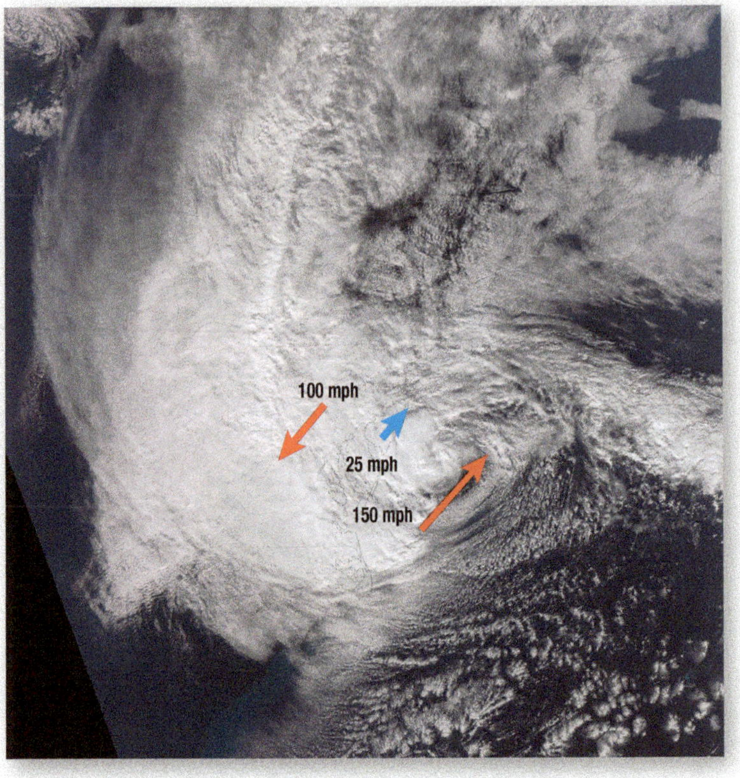

▲ **FIGURE 12–21 Varying Winds of Hurricane Charley.** This figure shows the movement of the storm itself (indicated by blue arrow), 40 km/hr (25 mph) to the northeast. The wind speed on the right-hand side of the storm was 241 km/hr (201 km/hr + 40 km/hr) [150 mph (125 mph + 25 mph)]; the wind speed on the left-hand side was 161 km/hr (201 km/hr − 40 km/hr) [100 mph (125 mph − 25 mph)]. The red arrows indicate the net wind speed on the left and right sides of the hurricane.

Figure 12–21 shows Hurricane Charley as it approaches the west coast of Florida. The storm moved to the northeast at 40 km/hr (25 mph), while winds relative to the eye were moving at about 201 km/hr (125 mph). To the right of the eye relative to its direction of motion, the 201 km/hr (125) mph wind speeds were supplemented by the 40 km/hr (25 mph) movement of the storm in the same direction, for a net speed of 241 km/hr (150 mph). To the left of the eye, the counterclockwise rotating air was toward the southwest—the opposite direction of the storm itself. Thus, the net speed there was 161 km/hr (100 mph).

Hurricane Fatalities

Though storm surges present the greatest potential for catastrophic coastal destruction and have claimed thousands of American lives during the past few centuries, they do not account for the majority of American hurricane fatalities. A study published by a researcher at the National Hurricane Center of the National Weather Service revealed that between 1970 and 2002, more than half the fatalities from tropical storms and hurricanes in North America resulted not from storm surges, but rather from freshwater flooding from heavy rain. Only about one-quarter of the fatalities associated with tropical storms and hurricanes (or their remnants) occurred in coastal counties. For several decades prior to 2005 and Hurricane Katrina, there had been a decrease in the incidence of storm surge fatalities. (See *Box 12–4, Focus on Severe Weather: Hurricane Katrina,*

12–4 FOCUS ON SEVERE WEATHER

Hurricane Katrina

At the end of August 2005 we witnessed one of the major natural catastrophes in American history: Hurricane Katrina. Katrina was the first of three Category 5 hurricanes to form in the Gulf of Mexico or the Caribbean in 2005, bringing with it substantial flooding in southern Florida, the inundation of New Orleans, and a devastating storm surge in coastal Louisiana, Mississippi, and Alabama. Two months later, the storm's death toll was complete: More than 1300 people had perished in the United States. Damage estimates exceeded $100 billion. In this section we present a chronology of what happened meteorologically.

Wednesday, August 24, 2005: Tropical Storm East of South Florida

At 11 P.M. EDT, Tropical Storm Katrina was located to the east of south Florida. Figure 12–4–1a shows the location of the storm, as well as the official National Hurricane Center (NHC) predictions for future movements and status. Hurricane warnings had recently been issued for the southeast Florida coastline, with anticipated landfall near Miami. The storm was forecast to be a Category 1 hurricane by 8 P.M. the following day very near the shoreline. Though the cone depicted a fairly extensive range of possible positions, the forecast models the center used were in close agreement, and the storm moved much as predicted over the short term.

Thursday, August 25, 2005: Hurricane Katrina Landfall

At 5 P.M. EDT on August 25 (Figure 12–4–1b), Katrina had developed into a hurricane and was near landfall at the position forecast the night before (though it had moved somewhat faster than expected). Landfall occurred at about 6:30 P.M. near the Broward/Miami–Dade County state line (Figure 12–4–2). At this point, forecasters officially called for the cyclone to move directly westward and, upon entering the Gulf Coast the following afternoon, to turn toward the northwest and follow a track somewhere in the eastern Gulf along the west coast of Florida.

The hurricane moved rapidly across the state, which reduced the amount of weakening normally undertaken as a hurricane passes over land. So even though Katrina traveled across southern Florida mainly as a tropical storm, it was able to reintensify into a hurricane when it reached the Gulf.

Katrina did considerable damage to Florida mainly due to heavy rain that exceeded 26 cm (10 in.) in places. The heavy rainfall led to major flooding and trees toppled by the combination of saturated soils and strong winds. Six people died in Florida from Katrina, which also caused $100 million in damages and $423 million in agricultural losses.

Friday, August 26, 2005: In the Eastern Gulf of Mexico

Shortly after midnight, Katrina entered the Gulf of Mexico. Early morning forecasts called for the storm to initially move to the west and then begin to arc northwestward toward the coast anywhere from the Florida panhandle to extreme eastern Louisiana (Figure 12–4–1c). But by late evening, the system had moved farther to the southwest than anticipated, and by 11 P.M. a very different track was predicted—one that would put Katrina on a collision course with New Orleans and the Mississippi coast (Figure 12–4–1d). The hurricane was now set to pass over a region of very warm Gulf waters, which the NHC described as ". . . like adding high octane fuel to the fire." All the forecast models predicted further intensification of Katrina, one of them calculating that wind speeds would top out at more than 243 km/hr (151 mph)—a strong Category 4.

Saturday, August 27, 2005: On Course

By Saturday night (Figure 12–4–1e) the previous day's forecasts had proven very accurate, and there was little change in the expected path of the storm. The NHC issued a hurricane warning that included some of the following text:

> . . . POTENTIALLY CATASTROPHIC HURRICANE KATRINA MENACING THE NORTHERN GULF COAST . . .
>
> A HURRICANE WARNING IS IN EFFECT FOR THE NORTH CENTRAL GULF COAST FROM MORGAN CITY LOUISIANA EASTWARD TO THE ALABAMA/FLORIDA BORDER . . . INCLUDING THE CITY OF NEW ORLEANS AND LAKE PONTCHARTRAIN.
>
> MAXIMUM SUSTAINED WINDS ARE NEAR 175 MPH . . . WITH HIGHER GUSTS. KATRINA IS A POTENTIALLY CATASTROPHIC CATEGORY FIVE HURRICANE ON THE SAFFIR-SIMPSON SCALE. SOME FLUCTUATIONS IN STRENGTH ARE LIKELY DURING THE NEXT 24 HOURS.
>
> HURRICANE FORCE WINDS EXTEND OUTWARD UP TO 105 MILES FROM THE CENTER . . . AND TROPICAL STORM FORCE WINDS EXTEND OUTWARD UP TO 205 MILES.
>
> COASTAL STORM SURGE FLOODING OF 18 TO 22 FEET ABOVE NORMAL TIDE LEVELS . . . LOCALLY AS HIGH AS 28 FEET ALONG WITH LARGE AND DANGEROUS BATTERING WAVES . . . CAN BE EXPECTED NEAR AND TO THE EAST OF WHERE THE CENTER MAKES LANDFALL. SIGNIFICANT STORM SURGE FLOODING WILL OCCUR ELSEWHERE ALONG THE CENTRAL AND NORTHEASTERN GULF OF MEXICO COAST.

The most vulnerable city in the United States, New Orleans, Louisiana, was about to be hit by a catastrophic storm. Bounded by Lake Pontchartrain to the north and surrounded by the winding Mississippi River, the city—much of which is below sea level—had long been known to be extremely vulnerable. The levees protecting much of the city were believed to be able to withstand a direct hit from a Category 3 hurricane, but they had never been seriously tested, as they would be soon. Meanwhile, coastal Mississippi and Alabama were lined up on the right-hand side of an enormously powerful hurricane track.

Sunday, August 28, 2005: Landfall Imminent

By late Sunday night the hurricane was just offshore (Figure 12–4–1f). Those who had set out to evacuate were long gone. Those unable to leave were directed to shelters or hunkered down to take their chances at home. That morning, winds had easily exceeded the threshold for a Category 5 hurricane, with sustained winds of 282 km/hr (175 mph). The hurricane had become as intense as Hurricane Camille, which devastated coastal Mississippi in 1969, and was even larger (Figure 12–4–3). Katrina's

(continued)

12–4 FOCUS ON SEVERE WEATHER *(continued)*

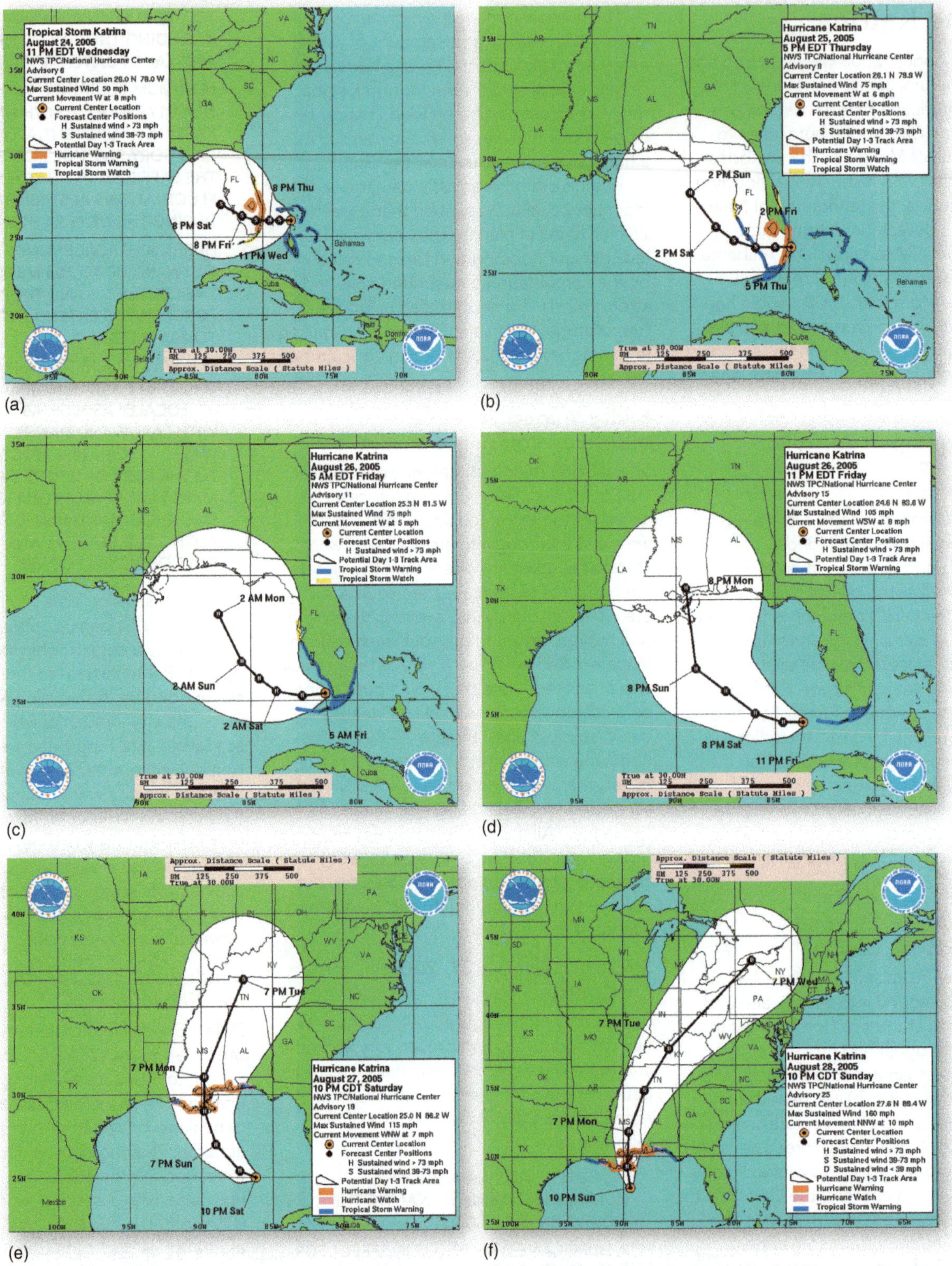

▲ **FIGURE 12–4–1 Katrina Forecasts and Actual Positions.** Maps (a) through (f) represent a four-day sequence from August 24 to August 28. The position and forecasted movements of Hurricane Katrina at various times. The area in white shows the range of possible locations forecast by the National Hurricane Center at the times shown.

▲ **FIGURE 12–4–2 Katrina Approaches Miami.** Radar image of Hurricane Katrina as it approached Miami. This is shown as a movie in the book's online resources.

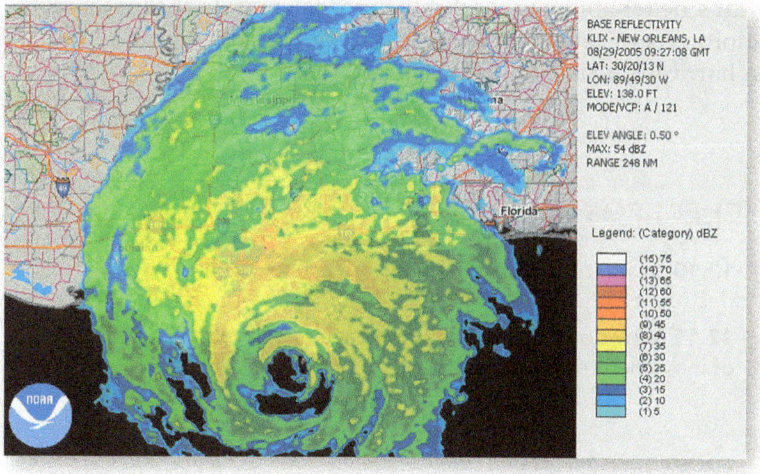

▲ **FIGURE 12–4–3 Katrina Approaches the Louisiana Coast.** Radar image of Hurricane Katrina as it approached the Louisiana coast. This is shown as a movie in the book's online resources.

minimum air pressure of 902 mb was the fourth lowest ever observed for an Atlantic storm up to that time.

Monday, August 29, 2005: Landfall

Hurricane Katrina made landfall early Monday morning. Coastal Louisiana was battered first as a huge storm surge overtook the area. Though the storm had weakened enough to be a strong Category 4, everybody knew that a historic disaster was occurring. The eye tracked just to the northeast of New Orleans, putting the city on the less-threatening "left side" of the storm. This initially gave the false impression that the city had narrowly escaped a disaster. This, of course, proved entirely incorrect. As New Orleans was subjected to wind speeds on the order of 160 km/hr (100 mph), Lake Pontchartrain waters rose along the levees, which were unable to hold them back. Eighty percent of the city came underwater (Figure 12–4–4).

In addition to the flooding in and near New Orleans, a huge storm surge and heavy winds devastated much of the Louisiana and Mississippi coast, helping to make Katrina one of the major natural disasters in U.S. history.

12–3–1 How would you assess the quality of the forecasts leading up to Katrina's landfall over Louisiana and Mississippi?

12–3–2 Why was flooding so severe in New Orleans despite the fact that the right side of the storm made landfall east of the city?

▲ **FIGURE 12–4–4 Katrina.** Downtown New Orleans under water.

for more information on this historic event.) The reduction in deaths associated with storm surges is partially the result of an episodic decrease in the number of strong hurricanes hitting populated coastal regions during the 30-year period, along with a better ability to predict the movement of hurricanes and improved evacuation procedures. (We discuss trends and cycles in hurricane activity later in this chapter.)

CHECKPOINT

12.10 In what ways do hurricanes cause fatalities and destruction?

12.11 Why does the side of a hurricane relative to its direction of movement matter to coastal communities?

Hurricane Forecasts and Advisories

12.7 Explain how meteorologists develop hurricane forecasts and advisories.

Responsibility for tracking and predicting Atlantic and east Pacific hurricanes lies with the National Hurricane Center (NHC) in Miami, Florida. During hurricane season, this office of the National Weather Service obtains constantly updated surface reports and satellite data to determine current storm conditions. Sophisticated numerical models run on a supercomputer predict the formation, growth, and movement of tropical storms and hurricanes. When active hurricanes approach land, specially equipped aircraft fly into the storms and provide reconnaissance data from airborne radar and *dropsondes*, packages containing temperature, pressure, and moisture sensors and transmitters released from the plane into the storm.

The NHC uses the standard computer models for conventional weather forecasting (discussed in Chapter 13), as well as others developed specifically for hurricanes. The latter fall into three categories: *statistical, dynamical*, and *hybrid*. Statistical models apply information on past hurricane tracks and use those tracks as predictors for current storms. Dynamical models take information on current atmospheric and sea surface conditions and apply the governing laws of physics to current data. Hybrid models combine elements of statistical and dynamical models. The models repeatedly forecast the movement and internal changes of hurricanes for short time increments and then print information on projected storm positions, air pressure, and wind at 6-hour intervals. Not surprisingly, model forecasts become less accurate as lead time increases and are unreliable for more than about 72 hours.

DID YOU KNOW?

Meteorologists once believed that seeding clouds with silver iodide (see Chapter 7) could be an effective way to reduce hurricane strength. The idea was to seed parts of the cloud outside the eye wall. If the seeding successfully led to enhanced convection outside the eye wall, the new zone of heavy activity could compete with and thereby weaken the strength of the eye wall. Seeding was undertaken sporadically between 1962 and 1983 on 8 days in four different hurricanes. Although the results once appeared promising, the method ultimately proved unviable.

Hurricane forecasting requires a tremendous amount of data and places a great demand on computer hardware. The *National Oceanic and Atmospheric Administration (NOAA)* has deployed geostationary orbiting GOES satellites (meaning they remain over fixed locations on Earth) that provide improved data acquisition from space. NOAA also flies aircraft into hurricanes and observes conditions from on-board radar systems and the use of GPS dropsondes (Figure 12–22). Dropsondes are instrument packages that observe pressure, temperature, and moisture conditions as they descend through a hurricane. They are dropped from the aircraft, and they have parachutes that allow them to slowly descend through the hurricane and transmit their observations. NOAA relies heavily on supercomputers to run complex forecast models designed specifically for hurricanes.

The improvement in hurricane forecasting in recent years has been substantial. For example, the average error in 48-hour forecasted positions for Atlantic tropical storms and hurricanes was typically above 370 km (230 mi) in the mid-1980s. Over the 5-year period 2008 through 2012 it was 142 km (88 mi).

Hurricane Watches and Warnings

Improvements in forecasting capabilities and the need to give people as much notice as possible in anticipation of a hurricane led the National Hurricane Center to change its criteria for issuing hurricane watches and warnings. A **hurricane watch** is issued when hurricane conditions are *possible* for a particular coastal region. These watches are issued 48 hours ahead of the predicted onset of tropical storm–force winds for that location. A **hurricane warning** means that hurricane conditions are *expected* somewhere within the identified area. Even though hurricanes are expected when such warnings go out, they are issued 36 hours ahead of the expected onset of *tropical storm–force* winds. This is done because evacuation and other measures are very difficult to undertake once tropical storm–force winds have already arrived. If the winds are not expected to achieve hurricane status, then the NHC issues **tropical storm watches** and **tropical storm warnings** with the same anticipated lead time.

(a)

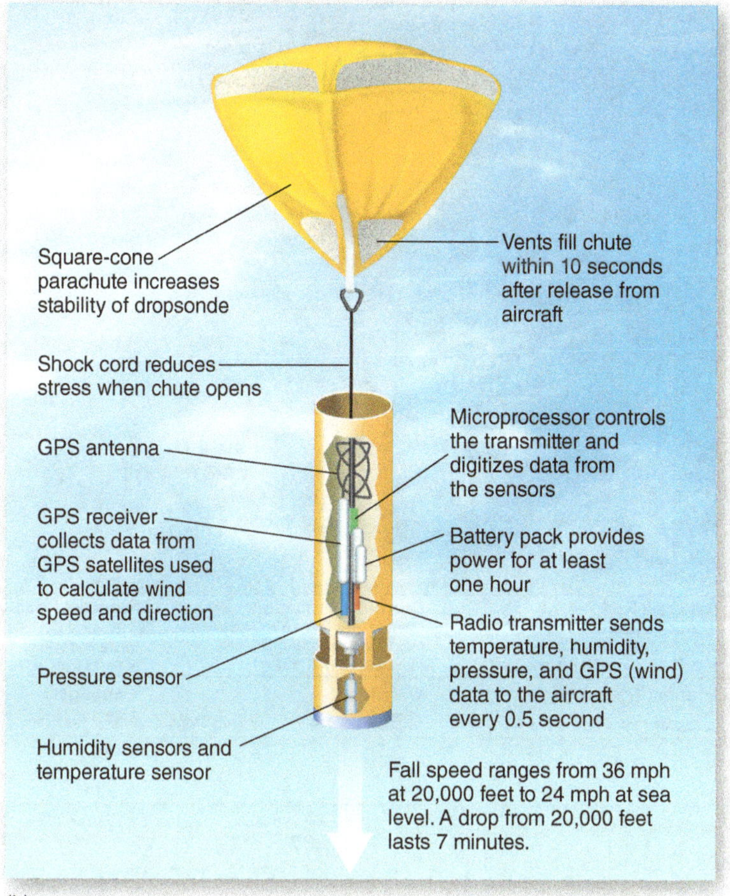

(b)

▲ **FIGURE 12–22 Dropsonde.** (a) A photo of a typical dropsonde used for hurricane surveillance. These packages are dropped into hurricanes and record storm characteristics as they descend via parachute. (b) A schematic of a dropsonde instrument package.

Figure 12–23 provides an example of how watches and warnings evolve as a hurricane approaches land. In Figure 12–23a, Hurricane Earl's position is marked by the bull's-eye near the Bahamas at 8 A.M. EDT on September 1, 2010. The map shows the forecasted movement of the hurricane along with a cone indicating the range of positions at which the storm may be located at indicated times. A hurricane watch has been issued for much of the North Carolina coast up to eastern Maryland and Virginia. Twenty-four hours later (Figure 12–23b) the hurricane has come closer to land and most of the North Carolina coast has come under a hurricane warning. In addition, a short stretch of coast to the southwest of the warning area and a large area extending northeast to southern New England is under a tropical storm warning, with a small hurricane watch area farther up the coast. In the end, the Mid-Atlantic coast of the United States experienced only tropical storm winds from Earl. However, the storm, whose intensity had been weakening as it passed east of the Mid-Atlantic states, reintensified back to a hurricane and hit the coast of southern Nova Scotia on the morning of September 4.

DID YOU KNOW?

Between 1992 and 2001 the average forecast error for the 24-hour hurricane landfall position was 149 km (92 mi). If you live near a threatened coastal zone, you must keep in mind that the endangered area extends far beyond the point of landfall. Hurricane-force winds can occur for distances well exceeding 160 km (100 mi) from the eye in any direction. So even if the point of expected landfall is far away or if the eye of the storm is still offshore, you may be subject to extremely hazardous hurricane conditions.

The erratic nature of hurricanes makes them notoriously difficult to predict. When predicting hurricane movements, forecasters must weigh the effects of issuing watches or warnings for hurricanes that never make landfall versus the consequences of failing to issue a watch or warning for a storm that ultimately does hit. Obviously, the failure to warn people to evacuate may lead to unnecessary loss of life and property. On the other hand, false warnings have serious ramifications, especially if they occur repeatedly. Repeated false warnings

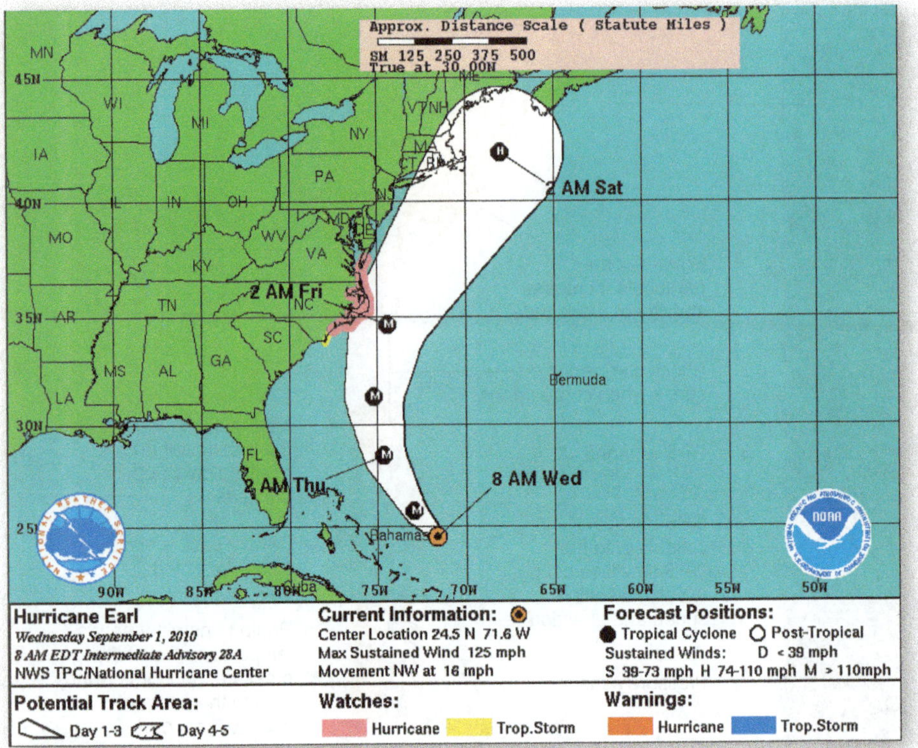

(a)

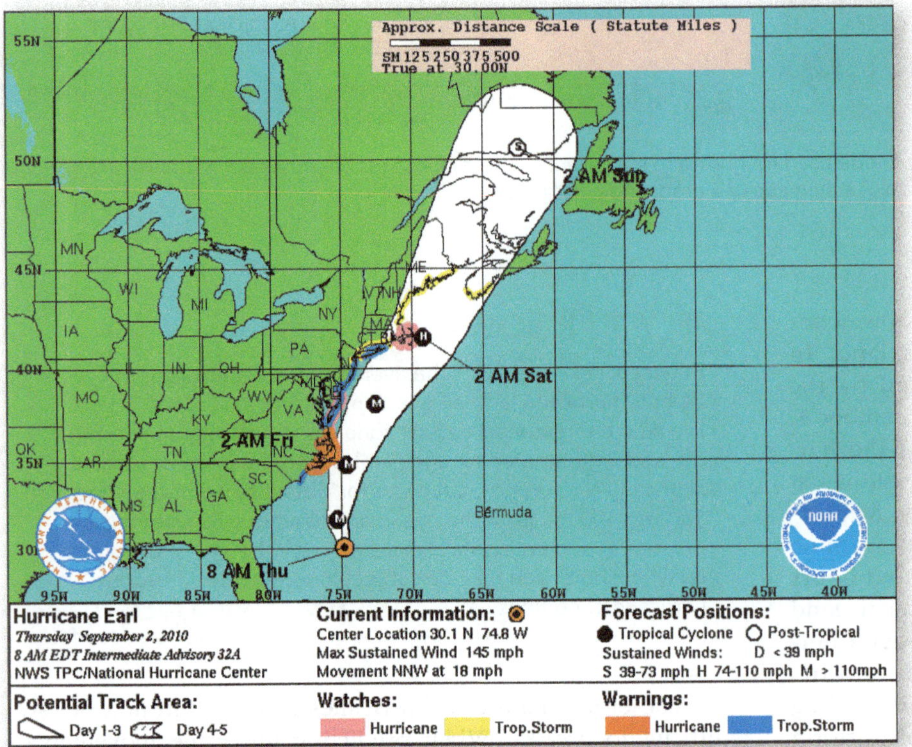

(b)

▲ **FIGURE 12–23 Warnings and Watches.** Hurricane warnings and watches were issued as a result of Hurricane Earl on September 1 and 2, 2010. (a) On September 1, Earl was expected to move from its position near the Bahamas to the North Carolina coast, for which a hurricane watch was issued. (b) By the next morning, the storm had advanced toward the coast and a hurricane warning had been issued for much of the North Carolina coast, with an extensive band of shoreline under a tropical storm warning.

can make the public so complacent that people will eventually disregard warnings that prove accurate.

Evacuations based on these advisories have immense economic costs for the general public, government agencies, and industry. Local residents and small businesses have their lives thoroughly disrupted as they board up windows and prepare to evacuate or take shelter, while large industries (such as petroleum mining and processing) incur costs measured in tens of millions of dollars from having to shut down and reopen their plants.

Hurricane Intensity Scale

In addition to alerting the public to the location and projected movement of hurricanes, meteorologists use a simple scale to categorize their intensity. The **Saffir-Simpson scale** (Table 12–2) classifies hurricanes into five categories based on the highest 1-minute average winds in the hurricane. Generally, higher-category hurricanes have lower central pressures and larger storm surges. Figure 12–24 gives a general idea of how high water levels would be for various categories of landfall hurricanes. Of course, the water would not rise up gently like a bath tub, but would hit the area with violent wave activity, intense rainfall, and extreme wind conditions. Though storm surges are usually the more destructive element of hurricanes upon landfall, the scale is based on wind speeds because storm surges are affected by nonmeteorological factors such as coastal configuration and the steepness of the offshore continental shelf. Extremely violent hurricanes are rare, with only three Category 5 and 16 Category 4 hurricanes having hit the mainland United States between 1900 and 2010.

Of course, Category 4 and 5 hurricanes are far more deadly and devastating than lower-category hurricanes. The effects of Hurricane Camille (Category 5) in 1969 have already been described. Since then, 16 Category 5 hurricanes have occurred in the Gulf of Mexico, Caribbean Sea, or western Atlantic, and only three of those struck land at full Category 5 intensity: Andrew in 1992, and Dean and Felix, both in 2007. But even though only a handful of Category 5 hurricanes made landfall at that maximum level, the majority of these extreme hurricanes have made landfall at some point in their lifetimes, often with catastrophic consequences. The great Galveston hurricane of 1900 that killed 6000

Video
Interview with a Chase Pilot
http://goo.gl/dNcCfB

TABLE 12–2
The Saffir-Simpson Scale

Category	Pressure (mb)	Wind Speed (km/hr)	Wind Speed (mph)	Storm Surge (m)	Storm Surge (ft)	Damage
1	≥980	119–153	74–95	1–2	3–7	*Minimal.* No major damage to most building structures.
2	965–979	154–177	96–110	2–3	7–10	*Moderate.* Some roof, door, and window damage. Some trees blown down. Considerable damage to mobile homes.
3	945–964	178–208	111–129	3–4	10–13	*Extensive.* Some structural damage to small residences. Some large trees blown down. Some mobile homes destroyed.
4	920–944	209–251	130–156	4–6	13–20	*Extreme.* Some complete roof structure failures on small residences. Many shrubs, trees, and all signs are blown down. Complete destruction of mobile homes.
5	<920	>251	>156	>6	>20	*Catastrophic.* Some complete building failures. All shrubs, trees, and signs blown down.

▲ **FIGURE 12–24 Storm Surges.** Approximate level of storm surge height relative to coastal houses for different categories of hurricanes. Note the placement of houses on stilts, as is done on much beachfront property along the Gulf of Mexico coast.

people (see *Box 12–5, Focus on Severe Weather: The Galveston Hurricane of 1900*) is believed to have been a Category 4 hurricane. *Box 12–6, Focus on Severe Weather: Recent Deadly Cyclones* describes some noteworthy cyclones elsewhere.

CHECKPOINT

12.12 When are hurricane watches and warnings issued?

12.13 Describe the scale used to categorize the intensity of hurricanes.

Recent (and Future?) Trends in Hurricane Activity

12.8 Describe efforts to identify trends in recent hurricane activity and project future trends.

The extremely destructive seasons of 2004 and 2005 greatly heightened the public's awareness of the danger hurricanes pose. Florida made repeated headlines as four hurricanes hit it in 2004, and the Gulf Coast witnessed destruction of historic proportions in 2005, compliments of Hurricanes Katrina and Rita. These events stimulated public and media interest about whether the recent upsurge in Atlantic hurricane activity was related to natural periodic cycles, global warming, or both.

Effects of Atlantic Multidecadal Oscillation

One thing we know for certain is that the **Atlantic Multidecadal Oscillation (AMO)**, a 25- to 40-year oscillation in water temperatures (Figure 12–25), has been a major factor in the increase in Atlantic hurricane activity—and especially in strong hurricanes. Figure 12–26 plots the annual number of Atlantic-named systems (tropical storms and hurricanes), hurricanes, and Category 3 through 5 hurricanes from 1851 through 2010. The period between the early 1970s and mid-1990s was one of relatively low Atlantic sea surface temperatures and hurricane activity. In fact, the years 1991 through 1994 had less Atlantic hurricane activity than any other 4-year period on record (despite the Category 5 Hurricane Andrew in 1992). Then came an abrupt transition to a very active period beginning in 1995 that coincided with a shift in the AMO; the 1995–1999 period proved to be the most active 5-year period on record (41 hurricanes)—at least until that record was surpassed in 2001–2005 (44 hurricanes).

Effects of Climate Change

Has global warming also influenced the increase in hurricane activity? Currently there is little, if any, evidence to indicate that the increase in Atlantic tropical storms and hurricanes is due to a longer-term trend in sea surface temperatures. In fact, the increase in Atlantic tropical storm and hurricane activity has not been observed in the other ocean basins around the globe—despite the fact that worldwide tropical ocean temperatures have increased by about 0.5 °C (1 °F) between 1970 and 2004.

The situation might be different with regard to the effect on *intense* hurricanes of long-term sea surface warming. Research published in 2005 showed a near doubling in

12–5 FOCUS ON SEVERE WEATHER

The Galveston Hurricane of 1900

Some natural disasters are so embedded in our folklore that virtually everybody knows about them. We have all heard about the San Francisco earthquake of 1906 and the Great Chicago Fire of 1871. Yet the single deadliest natural disaster in U.S. history, the Galveston (Texas) hurricane of 1900, seems to have been lost from the national memory. In just a few hours, rising sea waters and heavy surf drowned 6000 people on Galveston Island—a narrow strip of land that peaks at less than 3 m (9 ft) above sea level (Figure 12–5–1).

The loss of life resulted not from lack of warning but rather from a failure to take the threat seriously. Two days earlier, a strong storm was reported moving westward into the Gulf of Mexico off Cuba, and ships returning from the Gulf of Mexico reported encountering the storm offshore the day before it made landfall. Furthermore, the local weather forecaster, Isaac Cline, observed the combination of winds and heavy surf along the local beach and deduced that the storm would move onshore. But evidence of the impending landfall seems to have been largely unheeded, in part because some meteorologists erroneously believed it was virtually impossible for a storm in the Caribbean to track across the Gulf. Scientists (including Cline) were also erroneously convinced that the gently sloping seafloor offshore would protect Galveston from major flooding in the event of a hurricane.

There is some uncertainty as to when people started to take the hurricane seriously. According to Cline's account of the disaster, he rode through Galveston Island urging residents to evacuate, but recent research casts doubt about the degree to which he actually warned the populace. Regardless of how urgent Cline's warnings were, however, few people evacuated, and some residents even rode to the beach to watch the heavy waves crash against the shore.

When the hurricane arrived, the people of Galveston had no way to escape. Within hours the rising seas completely covered the island so that the only potential shelter was in taller, well-built structures. Even these failed to withstand the pounding of waves and debris. Cline later gave the following account of his ordeal:

> By 8 P.M. a number of houses had drifted up and lodged to the east and southeast of my residence, and these with the force of the waves acted as a battering ram against which it was impossible for any building to stand for any length of time, and at 8:30 P.M., my residence went down with about fifty persons who had sought it for safety, and all but eighteen were hurled into eternity. Among those lost was my wife, who never rose above the water after the wreck of the building.

Cline and his brother were luckier and grabbed onto floating debris that helped them stay afloat. After three hours, the floodwaters subsided and the Clines were on solid land, among the survivors.

The horror did not end with the passage of the hurricane. There were still 6000 bodies to deal with. Some were taken out to sea on barges, but many washed back to shore. Ultimately, most of the bodies were cremated where they were found.

With our current ability to track and forecast the movement of approaching hurricanes, there is no reason for a repeat of this type of loss of life in North America. When Hurricane Ike (Figure 12–5–2) struck the Galveston area in 2008 the loss of life was tragic, but at levels far below those of the 1900 hurricane. But hurricanes will always present a threat to Gulf and Atlantic coasts that must be respected.

12–5–1 What human errors contributed to the loss of life from the 1900 hurricane?

12–5–2 Why is such a loss of life from a hurricane no longer likely in the United States?

▲ **FIGURE 12–5–1 Galveston Hurricane.** The Galveston hurricane of 1900 was the deadliest natural disaster in U.S. history.

▲ **FIGURE 12–5–2 Hurricane Ike.** Damage to Galveston after Hurricane Ike in 2008.

12-6 FOCUS ON SEVERE WEATHER

Recent Deadly Cyclones

Hurricanes that make landfall over the United States can be deadly, but nothing on the scale of what has been witnessed in other parts of the world due to cyclones and typhoons. In 1970 Bangladesh (then part of Pakistan) was hit by a tropical cyclone that killed between 300,000 and 500,000 people. This disaster led to the construction of more than 2500 concrete shelters on pillars (Figure 12–6–1) to protect residents against future cyclone hits. These shelters have undoubtedly saved hundreds of thousands of lives, first in 1991 when a cyclone hit the country with heavy rain, Category 5 winds, and a 9 m (30 ft) storm surge. Despite the fact that this was the strongest cyclone to hit the country in more than a century, the death toll was about 70,000—a horrific number but far smaller than that of the 1970 cyclone. The shelters once again saved tens of thousands of lives in November 2007 when another Category 5 cyclone, Sidr, hit the country. Estimates of the number of fatalities have varied between 3000 and 10,000—another terrible number but far less than that which would have occurred without the shelters.

In May 2008 Tropical Cyclone Nargis (Figure 12–6–2) made international news when it hit Myanmar (formerly called Burma) at its peak strength, packing peak winds estimated at 213 km/hr (132 mph) and producing heavy rains and a 3.7 m (12 ft) storm surge. The disaster was intensified by the ruling military dictatorship that denied entry to international relief workers trying to bring food and medical supplies to the country. At least 77,000 people died from Nargis—perhaps as many as 100,000—which would make it the deadliest cyclone to hit Asia since the 1991 storm. Two to three million people were left homeless.

Much of the coastline of South Asia is perfectly situated for disasters such as these. Low-lying coastal areas, often near rivers that can overflow their banks, and a poorly developed infrastructure for shelter and evacuation put the lives of millions of people in great danger. It is a tragic fact that events such as the ones described here are likely to be repeated at great human cost.

▲ **FIGURE 12–6–1 Cyclone Center.** One of more than 2500 cyclone shelters set up in Bangladesh after the catastrophic cyclone of 1970.

12–6–1 What measures have been undertaken to mitigate the threat from tropical cyclones in South Asia?

12–6–2 Why are parts of South Asia so vulnerable to tropical cyclones?

▲ **FIGURE 12–6–2 Nargis.** Satellite image of Tropical Cyclone Nargis approaching Myanmar.

12–7 FOCUS ON AVIATION

Airports' and Airlines' Response to Hurricanes

Hurricanes pose complex challenges to our air transportation system. Airlines and airports must be ready to respond quickly, even as the hurricane's precise path and the magnitude of its effects remain uncertain. How can airlines and airports plan to cope with the massive disruptions a hurricane can bring?

Airport (from south to north)	Sun., 10/28	Mon., 10/29	Tues., 10/30	Wed., 10/31
Miami	9%/2%	32%/28%	26%/31%	13%/19%
Washington (Dulles)	25%/38%	99%/100%	98%/85%	17%/12%
Philadelphia	2%/27%	100%/100%	97%/93%	26%/6%
Newark	21%/46%	100%/100%	100%/100%	99%/85%
JFK International	6%/34%	100%/100%	100%/97%	82%/56%
La Guardia	14%/38%	100%/100%	100%/100%	100%/99.7%

▲ **FIGURE 12–7–1** Percent departures/arrivals cancelled at East Coast airports, October 28–31, 2012.

▲ **FIGURE 12–7–2** La Guardia airport in New York City, shortly after it was closed down by Hurricane Sandy.

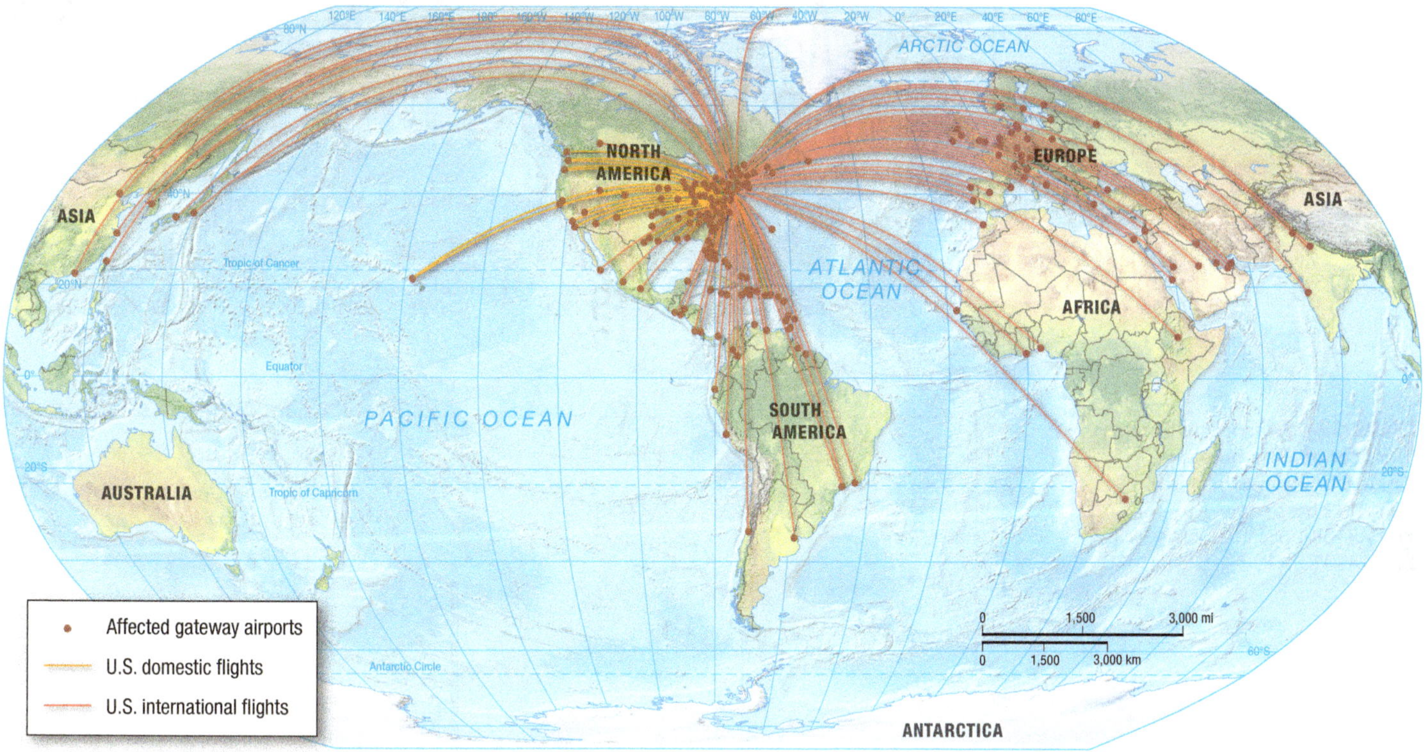

▲ **FIGURE 12–7–3** Global view of flight cancellations due to Hurricane Sandy.

Most airports have emergency management plans for severe weather. Before the storm arrives, employees secure buildings and equipment against expected winds and rain. Airlines must manage their own fleets to minimize the hurricane's impact on flight schedules and risks to aircraft. As the storm approaches, airlines may route planes in their fleet to airports out of the path of the storm. For example, the Pittsburgh airport is 250 miles inland from the Atlantic coast and has excess capacity for parked planes, so during coastal storms, airlines have diverted planes to Pittsburgh to keep them out of harm's way. Planes remaining at an airport can be tied down to protect them from wind or moved inside hangars.

Even with advance preparations, a major storm like Sandy triggers a "ripple effect" of delays and cancellations at other airports. For example, when Hurricane Sandy slammed into New Jersey on Monday night, October 29, 2012, its wind, rain, and flooding closed some of the busiest airports in the United States for two days or more, causing near-record levels of flight cancellations and delays (Figure 12–7–1). Sandy's storm surge flooded runways at New York City's JFK International and LaGuardia (Figure 12–7–2). The closure of so many major airports disrupted air travel worldwide (Figure 12–7–3). According to the International Air Transport Association, almost 17,000 flights were cancelled—nearly 10,000 from the New York area airports alone—and airline losses totaled approximately 0.5 billion dollars.

Once a hurricane has passed, airline dispatchers must work to "reboot" the air transportation system. The airlines fly planes and flight crews back to airports from which they had been evacuated, and incoming flights resume. In the case of Sandy, incoming but mostly empty aircraft began to return to airports by Wednesday, October 31st. The process of getting passengers from canceled flights to their destinations took several days. But the backlog may not have been quite as bad as one might expect, because forecasts of the impending storm caused many passengers to cancel their travel plans into and out of the area days in advance.

12–7–1 How do airlines respond to the threat of a hurricane and restore service after the storm has passed?

12–7–2 What aspect of Hurricane Sandy had the greatest impact on airports? What could be done to mitigate the risks and costs of this hazard?

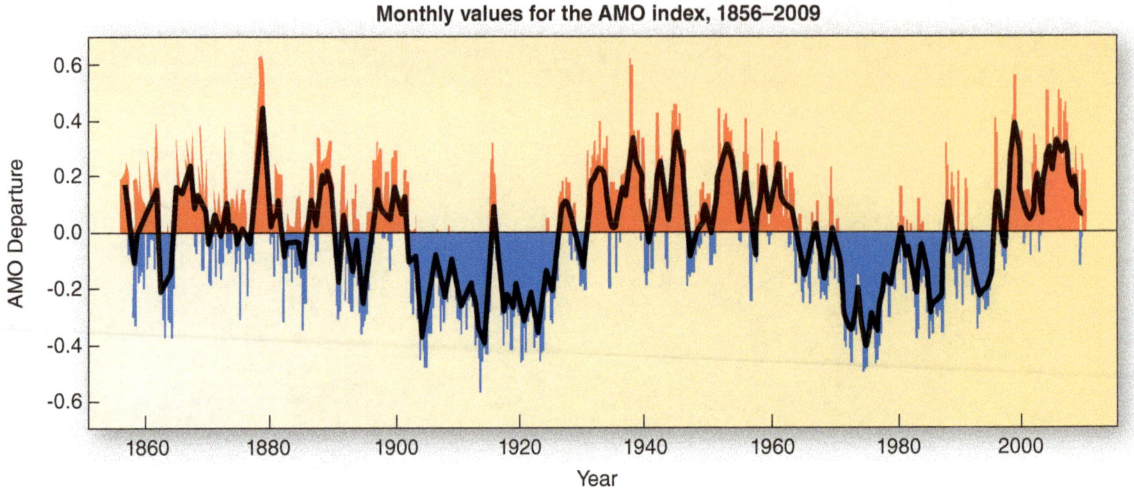

▸ **FIGURE 12–25 Atlantic Multidecadal Oscillation.** A time series of an index representing the Atlantic Multidecadal Oscillation from 1856 to 2009.

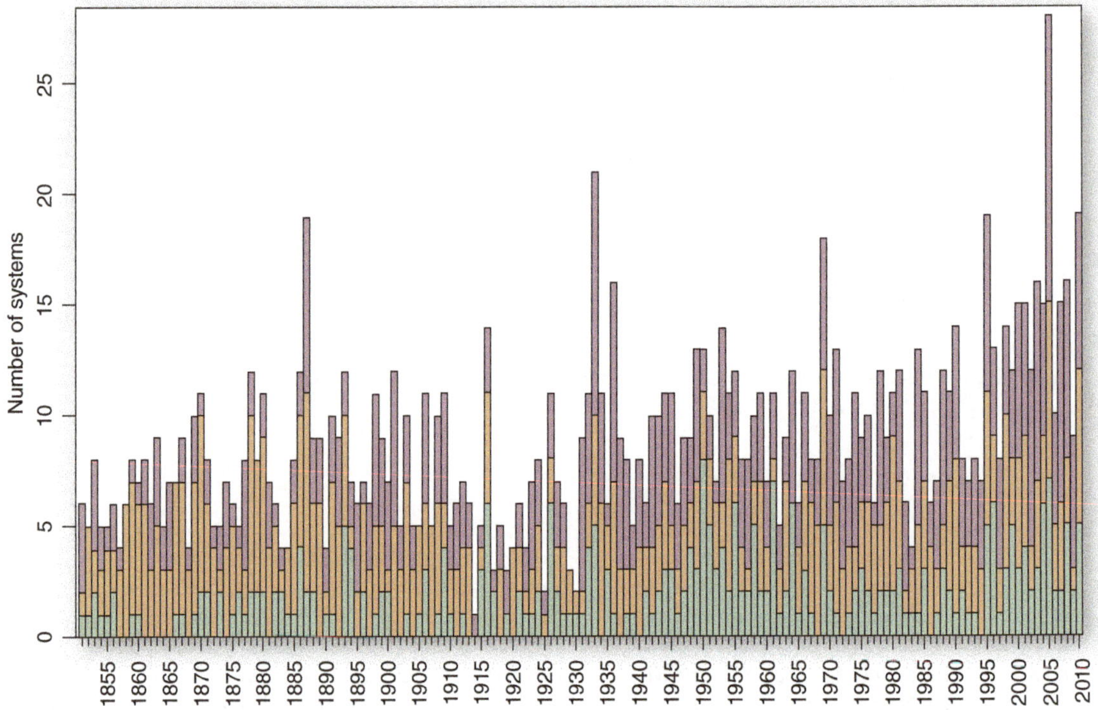

▸ **FIGURE 12–26 Tropical Storms and Hurricanes.** The number of Atlantic tropical storms (red), Category 1–2 hurricanes (green), and Category 3–5 hurricanes (orange) by year, 1851–2010.

the number of Category 4 and 5 hurricanes in the western North Pacific, western South Pacific, Indian, and Atlantic oceans since 1970 that coincides with increasing water temperatures. Interestingly, the North Atlantic experienced the smallest increase in major hurricanes among those basins. Another highly cited 2005 study pointed to a substantial increase in the overall energy released by Atlantic hurricanes in recent decades, reflected in both hurricane intensity and duration. That article noted, however, that only part of the increase in Atlantic hurricane activity could be ascribed to increasing sea surface temperatures (an assertion that was not widely reported by the popular press). And yet another 2005 article used the results of computer simulations to obtain the same conclusions based on empirical observations. Thus, there is better reason to suspect that global warming might influence the *intensity* of hurricanes than the *number* of hurricanes.

Factors Affecting Future Trends

Based on the preceding considerations, it appears that an increase in hurricanes and intense hurricanes may be a fact of life for residents along the Gulf of Mexico and the Atlantic coast in the years to come—at least for a few decades. Whether we are witnessing a longer-term trend is currently unknown. Factors beyond sea surface temperatures could affect the intensity of hurricanes in a world with higher air and sea surface temperatures, and some of these factors could make it harder for hurricanes to develop in a warmer world.

We are more vulnerable than ever to the destructive potential of hurricanes for the simple reason that there has been enormous

Video
2013 Atlantic
Hurricane
Season Outlook

http://goo.gl/FwNuYu

population growth along the Atlantic and Gulf coasts in recent decades. Moreover, increasing temperatures have led to a 19 cm (7.5 in) increase in sea level during the past century. These factors will continue to make evacuations of large populations more difficult in response to approaching hurricanes (as witnessed during the evacuation of southeast Texas as Hurricane Rita approached in 2005).

CHECKPOINT

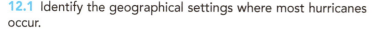

12.14 What multiple-year oscillation pattern of Atlantic sea surface temperatures may have impacted hurricane activity in recent decades?

12.15 What is the current consensus on how global warming might be changing hurricane characteristics?

Summary

12.1 Identify the geographical settings where most hurricanes occur.

- Hurricanes (and their counterparts such as typhoons and tropical cyclones) are extremely powerful storms that originate in tropical regions and migrate into the middle latitudes.
- They are called typhoons in the western Pacific and cyclones when found in the Indian Ocean.

12.2 State the major characteristics of hurricanes.

- Hurricanes are smaller than midlatitude cyclones but much larger than tornadoes.
- They can last for a week or more and travel thousands of kilometers before dissipating.

12.3 Describe the structural features of a hurricane.

- The heaviest thunderstorm activity in hurricanes occurs within bands of thick cloud cover that spiral toward the center of the system in a pinwheel pattern.
- The intensity of a storm increases toward its center until reaching the eye wall, the concentric zone of maximum activity that surrounds the eye.
- The eye of a hurricane is strikingly different from the rest of the hurricane because it is marked by generally clear skies, light winds, and higher air temperatures.

12.4 Explain the process of hurricane formation.

- Hurricanes begin their life cycles as uneventful tropical disturbances, or small clusters of thunderstorms. When they intensify and organize into a rotating band of cloud cover and thunderstorm activity, they are called *tropical depressions*.
- Further intensification results in their being classified as tropical storms. Some tropical storms achieve greater wind speeds to become hurricanes.

12.5 Describe hurricane movement and dissipation.

- Hurricanes initially move westward as they are steered by the trade winds, but may then move poleward and take on highly meandering paths.

12.6 Describe how hurricanes cause destruction and fatalities.

- Copious amounts of rain can bring intense floods, and strong winds can bring down structures.
- The most serious threat posed by a hurricane is the storm surge, the elevated rise in sea level due to low atmospheric pressure and the piling up of water by strong winds.

12.7 Explain how meteorologists develop hurricane forecasts and advisories.

- The National Hurricane Center of the National Weather Service uses a sophisticated network of satellites, research aircraft, and computer hardware and software to issue advisories on the likelihood of hurricane landfall.
- The erratic nature of hurricanes makes predicting them particularly difficult, but recent modernization at the National Weather Service has substantially increased forecast accuracy.

12.8 Describe efforts to identify trends in recent hurricane activity and project future trends.

- Hurricanes have become more frequent in the Atlantic Ocean since the mid-1990s.
- Although the potential impact of global warming on hurricanes is not entirely understood, it is likely that global warming will lead to more intense hurricanes rather than more frequent hurricanes.

Key Terms

Atlantic Multidecadal Oscillation (AMO) *p. 407*
cyclone *p. 382*
double eye walls *p. 386*
easterly waves *p. 387*
eye *p. 386*
eye wall *p. 386*
eye wall replacement *p. 387*
hot towers *p. 387*
hurricane *p. 382*
hurricane warning *p. 404*
hurricane watch *p. 404*
marine layer *p. 382*
Saffir-Simpson scale *p. 406*
storm surge *p. 399*
trade wind inversion *p. 382*
tropical depression *p. 389*
tropical disturbance *p. 387*
tropical storm *p. 389*
tropical storm warning *p. 404*
tropical storm watch *p. 404*
typhoon *p. 382*

Review Questions

1. Describe the geographic distribution of hurricanes, typhoons, and cyclones. What environmental conditions at these locations favor the development of such storms?
2. What is the trade wind inversion, and what impact does it have on the formation of hurricanes?
3. Which region has the greatest incidence of major tropical storms?
4. Describe the size, sea-level air pressure, and wind speed of a typical hurricane.
5. When are hurricanes most likely to form?
6. Describe the cloud and precipitation patterns associated with hurricanes, including those associated with the eye and eye wall.
7. What are tropical disturbances, and how do easterly waves influence them?
8. Describe the characteristics that distinguish tropical disturbances, tropical depressions, tropical storms, and hurricanes from one another.
9. Describe the various ways in which hurricanes differ from midlatitude cyclones.
10. What ocean surface characteristics are required for the intensification of storms into hurricanes and the maintenance of hurricanes?
11. Is there a "typical" path that hurricanes take after forming? Explain.
12. What feature associated with hurricanes causes the greatest destruction to coastal regions? Is this also true of inland regions?
13. Why is the right-hand side of a hurricane (relative to its direction of movement) the most dangerous?
14. Where are tornadoes most likely to be embedded in a hurricane?
15. What are hurricane watches and warnings? Are they exact corollaries to tornado watches and warnings?
16. What is the highest hurricane category on the Saffir-Simpson scale? How frequently do hurricanes of that magnitude occur?
17. Why are forecasters concerned with issuing hurricane advisories for areas that do not eventually get hit?

Critical Thinking

1. Why don't hurricanes cross the equator?
2. If two hurricanes pass just to the west of Cuba over a two-week period, what reasons might one have for expecting the second one to be weaker than the first?
3. How might previous drought conditions affect the intensity of a former hurricane as it passes over the southern United States?
4. El Niño conditions are believed to suppress hurricane development in the Atlantic. How might the phenomenon affect hurricane formation and movement in the Pacific?
5. If global warming continues, thermal expansion of the oceans and the melting of glaciers and ice sheets will lead to a higher sea level. How would this affect the threat of storm surges relative to wind damage and flooding?
6. It has been postulated that an increase in global temperatures could lead to an increase in the number and intensity of tropical storms and hurricanes. Global temperatures were particularly high during the 1990s and early 2000s, and there has been an increase in Atlantic hurricane activity since 1995. Does this association prove the connection between temperature and hurricane activity? Explain why or why not.
7. Experts believe that New York City is the third most dangerous city in the United States with regard to hurricanes, despite the fact that there has been no major hurricane-inflicted damage on the area (other than some wind damage and coastal erosion from Hurricane Gloria in 1985) and the more recent damage from Sandy. What factors could be responsible for this vulnerability? After answering this question, refer to **www2.sunysuffolk.edu/mandias/38hurricane** for an informative discussion of this issue.

Problems & Exercises

1. Compare the area of a hurricane that measures 600 km in diameter to a midlatitude cyclone having a diameter of three times greater (1800 km).
2. Refer to the forecast for the upcoming tropical storm season at **http://tropical.atmos.colostate.edu**. What existing conditions have led the forecast team to make its prediction? Also, use this website to determine how successful last year's forecast was.
3. Refer to **www.ncdc.noaa.gov/oa/climate/severeweather/hurricanes.html** and read the Special Reports on the past year's hurricane activity. Were any tropical storms particularly noteworthy?

Visual Analysis

Observe the maps showing the areas of maximum sustained wind speeds (Figure VA–12–1) and height of flooding (Figure VA–12–2) as Superstorm Sandy made landfall along the East Coast.

12.1. How did the wind speeds vary on either side of Sandy's path?

12.2. Which areas of the Mid-Atlantic coast had the greatest inundation?

12.3. What factors contributed to the patterns of wind speed and flooding height you observe on the maps?

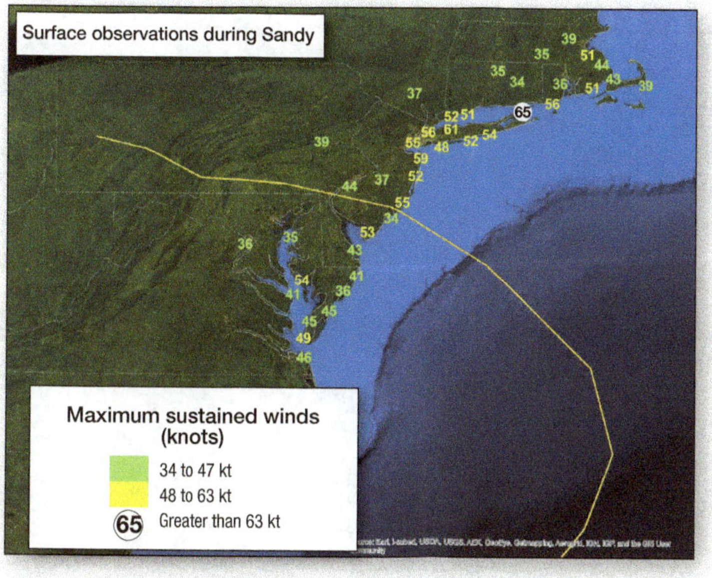

▲ **FIGURE VA–12–1 Sandy's Winds.** The map shows the maximum sustained wind speeds measured during Hurricane Sandy's northward progress along the coast.

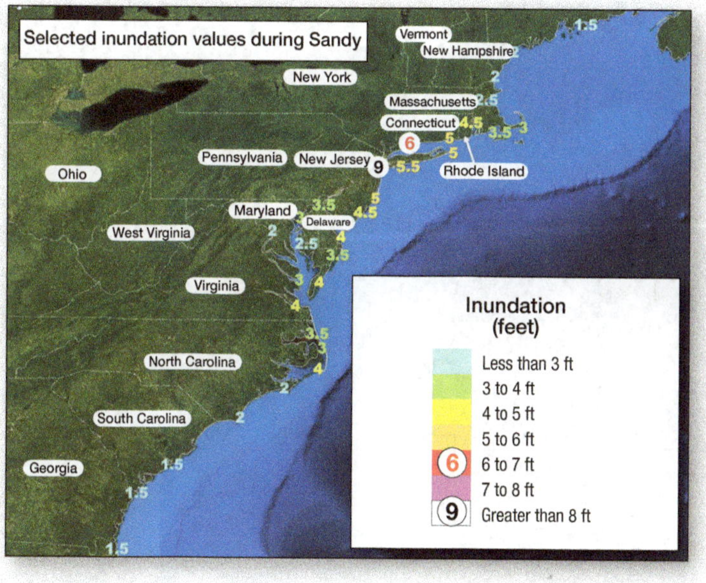

▲ **FIGURE VA–12–2 Sandy's Storm Surge.** The map shows the height of the storm surge as Hurricane Sandy made landfall.

PART

5 Weather Forecasting and Human Impacts on the Atmosphere

CHAPTER

13 Weather Forecasting and Analysis

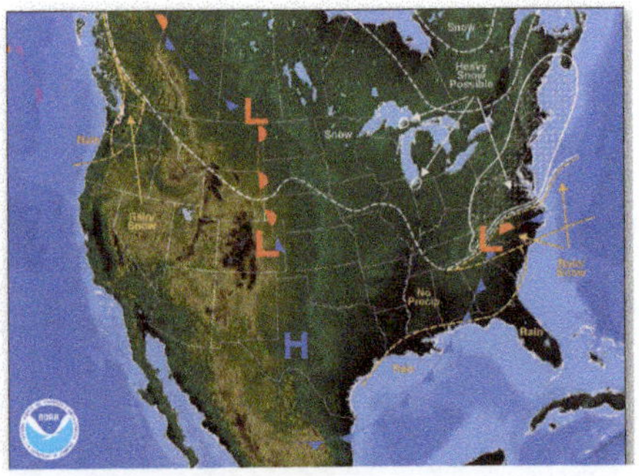

CHAPTER

14 Human Effects on the Atmosphere

◀ *Floods in southwest United Kingdom, February 2014.*

Weather can have a profound impact on our lives, but people are not defenseless against it. One of the best tools people have in dealing with weather is a sophisticated system of forecasting weather days in advance. But people interact with the atmosphere in other ways as well. For centuries we have polluted the air in the course of industrial and even agricultural activities. We also alter the climate by building large cities that modify their own climate. This section looks at several ways people are connected to their atmospheric environment.

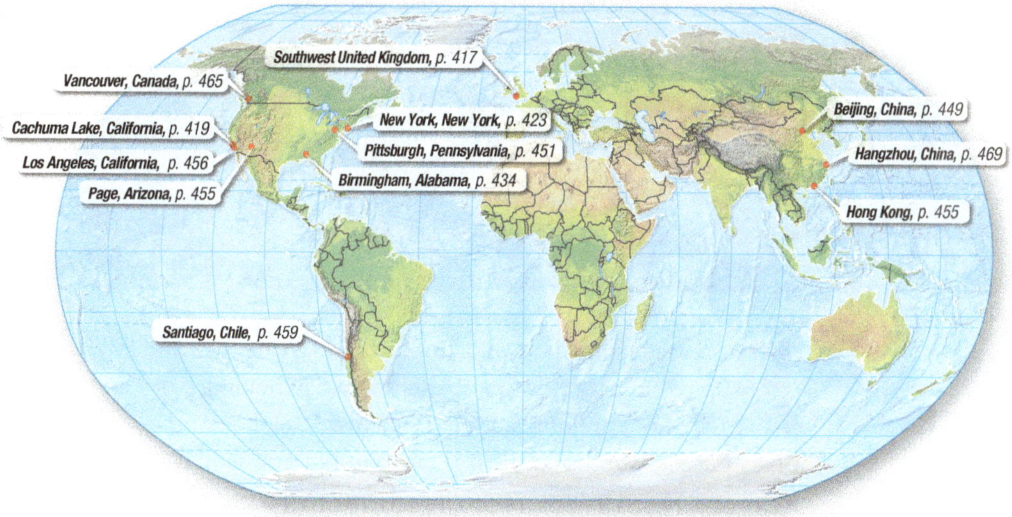

CHAPTER

13

Weather Forecasting and Analysis

Forecasters for the National Weather Service (NWS) spend a large portion of each workday fielding telephone inquiries about upcoming weather. Sometimes people want to know if their round of golf is likely to be rained out or if their outdoor plants might be vulnerable to overnight frost damage. But the most important aspects of a forecaster's job do not involve matters of simple convenience but questions that deal with life and death.

Such was the case for Mark Moede, a meteorologist with the NWS Forecast Office in San Diego. On one late August day much of southern California was threatened by brush fires triggered by unusually heavy thunderstorm activity. Firefighters battling the blazes received constant updates on weather conditions that could either suppress or enhance the spread of the fire. A few days later the situation reached its most critical stage. Pete Curran of the Orange County Fire Department maintained close contact with Moede for constant updates on a line of approaching thunderstorms. At issue was whether the thunderstorms would continue to strengthen and whether they would pass through the areas where the firefighters were working. If the storms passed that way, then strong winds, deadly lightning, and blinding rains would place Curran's crews in jeopardy. With the aid of Doppler radar and information from automated weather stations, Moede made the right call. He advised Curran to evacuate his crews

◂ *Cachuma Lake, California in February of 2014. The lake was at less than 20 percent of its normal level, imperiling the main source of drinking water for 200,000 people in coastal Santa Barbara County. Despite great advances over the last few decades in forecasting individual weather events, the ability to predict droughts such as this remains elusive.*

from the eventual path of the storm, where winds in excess of 95 km/hr (60 mph) created an uncontrollable firestorm. Afterward, firefighters reported that the fire line they had evacuated had been completely burned over. If there had been no call to evacuate, in all likelihood a number of firefighters would have been killed.

This chapter discusses the methods by which forecasters perform their job. We first look at some important issues regarding the general concept of weather forecasting and then discuss ways in which necessary data are obtained and processed. We then study the various types of weather maps and how they are used in weather analysis.

Learning Outcomes

After reading this chapter, you should be able to:

13.1 Summarize the basic procedures of weather forecasting.

13.2 Explain why weather forecasting is imperfect.

13.3 Describe the main methods of weather forecasting

13.4 Describe the general types of forecasts.

13.5 Explain how meteorologists assess the quality of weather forecasts.

13.6 List some major sources of data for weather forecasting.

13.7 Describe forecast procedures, including numerical modeling, and explain how meteorologists develop short-, medium-, and long-range forecasts and seasonal outlooks.

13.8 Explain how meteorologists use weather maps and satellite images in forecasting.

13.9 Explain the use of thermodynamic diagrams in forecasting.

13.10 Describe the characteristics of numerical models, including scale considerations and horizontal representation.

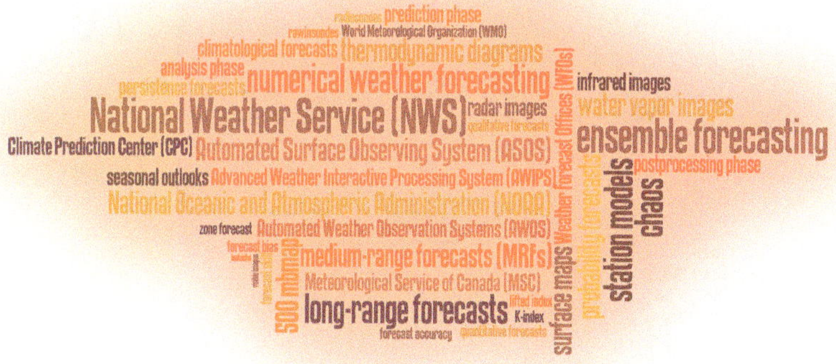

Weather Forecasting—Both Art and Science

13.1 Summarize the basic procedures of weather forecasting.

Weather forecasting is a highly sophisticated and data-intensive endeavor. Forecasters routinely employ state-of-the-art computer hardware and software systems that perform billions of calculations, based on a huge amount of input data and displayed at sophisticated workstations. They analyze information on variables such as temperature, dew point, and wind at multiple pressure levels of the atmosphere. The meteorologists who work with the information come armed with rigorous university training that includes intensive coursework in mathematics, physics, chemistry, and, of course, meteorology. But the scientists who work at the public and private forecasting agencies also rely heavily on their ability to subjectively analyze the current weather situation, thus making their task a blending of scientific principle and art.

The official agency for weather forecasting in the United States is the **National Weather Service (NWS)**. Weather forecasting by the U.S. government began in the 1870s, when Congress established a National Weather Service under the authority of the Army Signal Corps. In 1890 the Weather Service was renamed the National Weather Bureau and transferred to the Department of Agriculture. There it remained until it was shifted to the Department of Commerce in 1942. In 1970 the agency reverted to its earlier name and became part of the **National Oceanic and Atmospheric Administration (NOAA)**.[1] Other agencies under NOAA include the National Fisheries Service, National Environmental Satellite, Data, and Information Service, and the National Ocean Service. The **Meteorological Service of Canada (MSC)**, located in Downsview, Ontario, assumes all forecasting duties for that country and provides local and regional information to its 14 **regional weather centres**.

Forecasters for the NWS work eight-hour shifts in one of 122 **Weather Forecast Offices (WFOs)** offices that are open 24 hours a day, seven days a week. The forecasters at each office must issue a variety of products that include basic forecasts and weather statements for specific applications (such as aviation forecasts) for a designated region called a *county warning area* (CWA). Figure 13–1 shows the WFOs and associated CWAs for the conterminous United States. Not shown are offices in Alaska (three), Hawaii and Guam (one each).

[1]Employees of NOAA sometimes claim that the letters NOAA stand for National Organization for the Advancement of Acronyms.

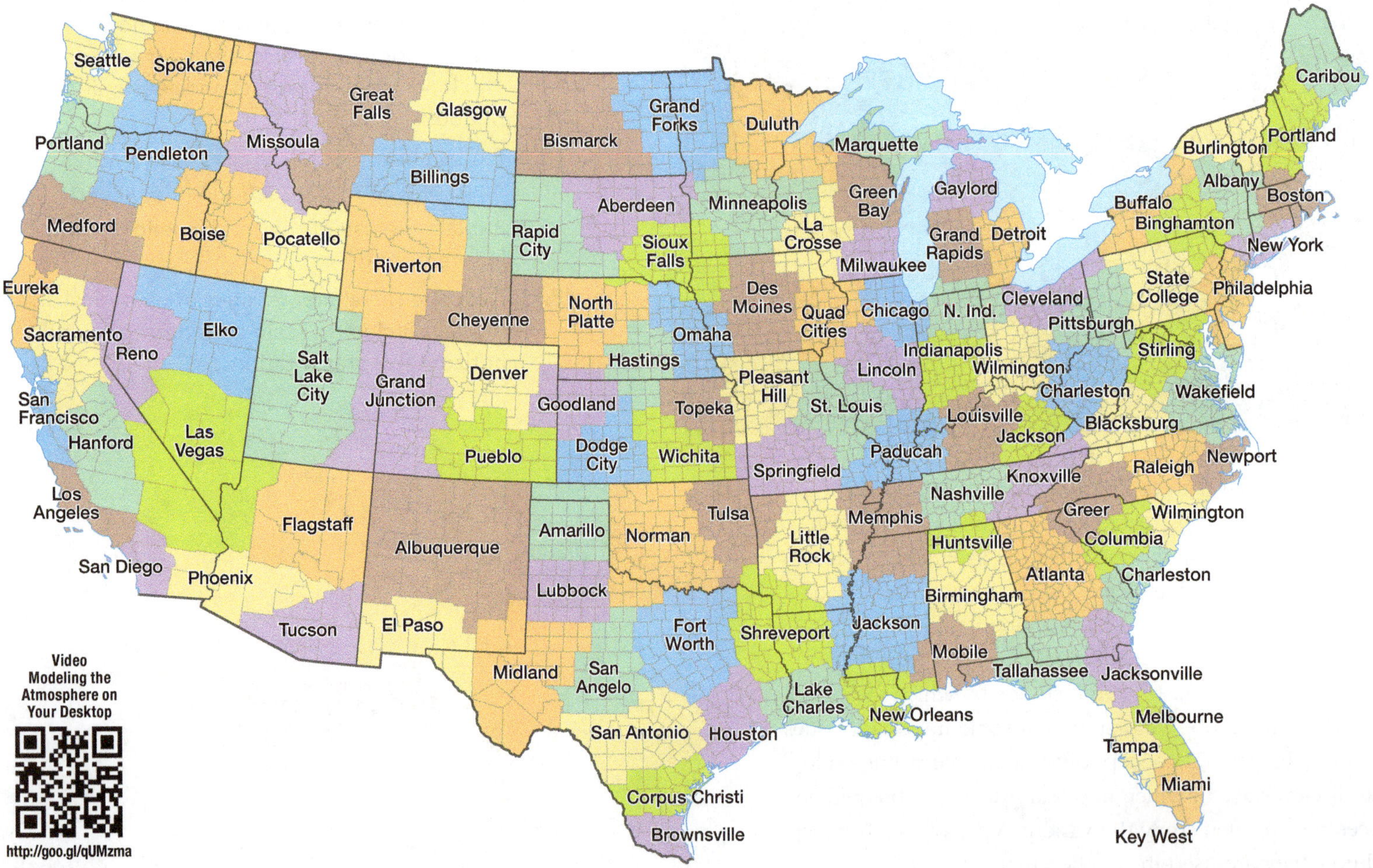

Video
Modeling the
Atmosphere on
Your Desktop
http://goo.gl/qUMzma

▲ **FIGURE 13–1 County Warning Areas.** Each CWA contains a Weather Forecast Office with staff responsible for surrounding counties.

▲ **FIGURE 13–2 Advanced Weather Interactive Processing System.** A meteorologist working on his forecast at the San Diego office of the National Weather Service. The monitors are part of the AWIPS workstation.

The Work of a Forecaster

All forecasters apply the same scientific underpinnings to their work and make use of the same types of resources, but there is no precisely defined routine by which a forecast is made. Upon starting his or her shift, a forecaster will receive a verbal briefing about the recent and current weather from the meteorologists whose shifts are about to end. What follows largely depends on the type of situations that the local area is subject to and which particular weather elements are of concern at that time. A forecaster in Indianapolis may be most interested in the possible development of severe thunderstorms on a June day, but have to decide whether to issue blizzard warnings six months later. On the other hand a forecaster for the Seattle, Washington, area might be tracking the movement of a major storm approaching from the Gulf of Alaska.

AWIPS COMPUTER WORKSTATIONS Weather information comes to the forecaster from a huge bank of observation platforms, satellites, radar facilities, and other sources of data. The forecaster accesses that information from a workstation. Figure 13–2 shows a typical National Weather Service office workstation called the **Advanced Weather Interactive Processing System (AWIPS)**. Each AWIPS allows forecasters to simultaneously display maps of current weather conditions, output of computer forecast models, satellite and radar images, forecast advisories and discussions from other weather facilities, and a great deal more. All maps and satellite images can be presented as movie loops. Typically several employees will work together at any particular shift, each handling different aspects of daily operations. For example, one forecaster may be working on the general forecast while another analyzes aviation-related matters.

USING COMPUTER MODELS Forecasters rely heavily on the output of computer models. These highly complex programs apply the fundamental laws of atmospheric physics to a huge amount of data and produce maps showing the expected distribution of air pressure, temperature, and other variables at the surface and upper atmosphere. The meteorologist interprets the overall patterns depicted on the maps, infers the overall type of weather that can be expected at a particular location, and issues a forecast that can be used by the general public. But the computer output is not infallible, nor does it tell us everything that is relevant to the issuance of a forecast. Thus, the forecaster has to interpret the so-called model guidance using her or his professional training and experience. Sometimes forecasters in different offices observing the same weather systems will consult with each other to exchange ideas on how to interpret the current situation.

Although much of model output is specialized and of primary use to forecasters, twice a day a forecast map will be produced by the Weather Prediction Center summarizing expected conditions across the conterminous United States. These maps give the general public an idea of what kinds of weather conditions can be expected across the country. For example, Figure 13–3 plots the 24-hour predicted conditions as of the morning of January 21, 2014.

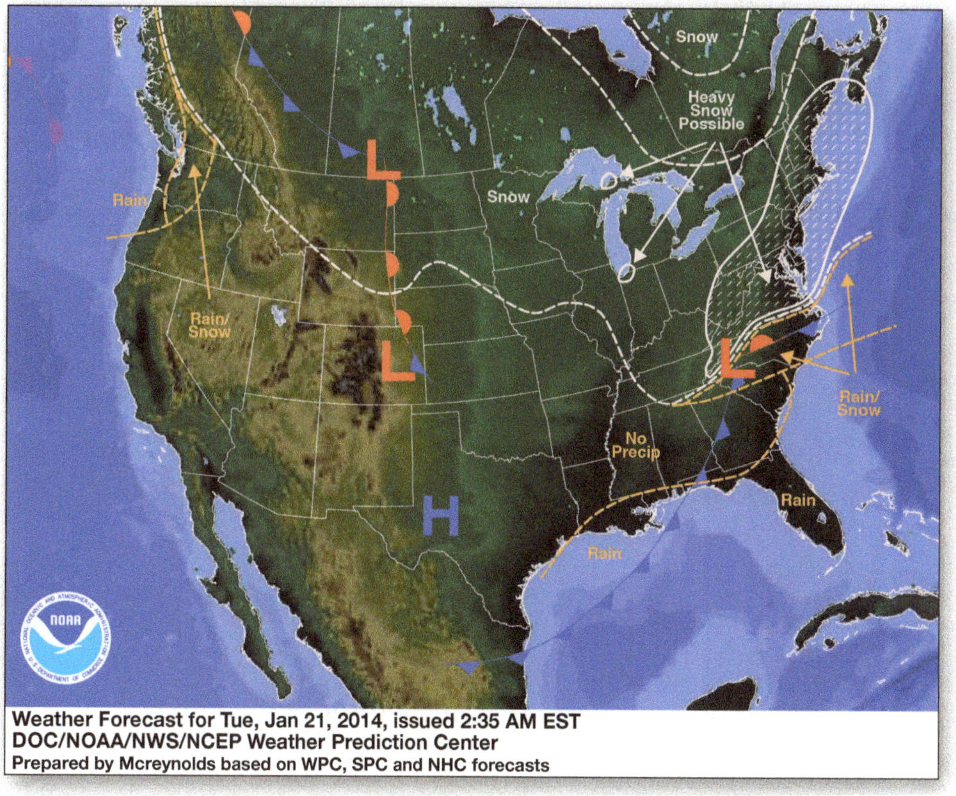

▲ **FIGURE 13–3 Simplified Forecast Map.** The NWS issues simplified forecast maps twice a day based on interpretation of numerical models. This one was issued during the early morning hours of January 21, 2014, and highlighted the possibility of heavy snow in the Northeast.

In particular, an area of potentially heavy snow is depicted across much of the northeastern United States, along with a narrow belt of lake-effect snow in northwest Indiana.

Forecast maps produced by the computer models often depict the predicted distribution of weather variables directly as an application of physical laws. These maps may depict predictions of certain characteristics of the atmosphere at particular levels, such as the 500 mb level (examples of such maps are given later in this chapter).

Other models rely on statistical methods to assist the forecaster. These products rely on output from the physical models. Thus, the physical models predict the state of the entire atmosphere over a wide area (such as North America), and the statistical models use that information to predict values such as maximum and minimum temperature at particular locations. Though the temperatures predicted by the computer can provide the forecaster with useful guidance, the professional meteorologist knows when to overrule such output by using his or her professional judgment.

ISSUING NWS FORECASTS Forecasters make more than a single forecast and various weather statements over the course of a shift, each intended to serve a particular person or constituency. Numerous types of forecasts are routinely issued at particular times during each shift, and a forecaster's activities ensure that these reports are disseminated on time. Scheduled reports include short-term forecasts, more detailed forecast discussions intended for people with a more advanced understanding of meteorology, local aviation advisories, hydrologic information (such as information on potential flooding), and numerous other statements.

Perhaps the most familiar type of weather statement to many people is the **zone forecast**, issued at designated times each day and extending out to a week into the future. The following forecast from New York City serves as a good example[2]:

ZONE FORECAST PRODUCT

NATIONAL WEATHER SERVICE NEW YORK NY

421 AM EST TUE JAN 21 2014

NYZ072-212130-

NEW YORK (MANHATTAN)-

421 AM EST TUE JAN 21 2014

. . . WINTER STORM WARNING IN EFFECT FROM NOON TODAY TO 6 AM EST

WEDNESDAY . . .

.TODAY . . . CLOUDY . . . SNOW . . . MAINLY THIS AFTERNOON. SNOW MAY BE HEAVY AT TIMES THIS AFTERNOON. SNOW ACCUMULATION OF 3 TO 5 INCHES. BRISK AND MUCH COOLER WITH HIGHS IN THE MID 20S. TEMPERATURE FALLING TO AROUND 17 THIS AFTERNOON. NORTH WINDS 15 TO 20 MPH WITH GUSTS UP TO 30 MPH. CHANCE OF SNOW 80 PERCENT. WIND CHILL VALUES AS LOW AS ZERO.

.TONIGHT . . . SNOW. BLOWING SNOW. SNOW MAY BE HEAVY AT TIMES IN THE EVENING WITH VISIBILITY ONE QUARTER MILE OR LESS AT TIMES. TOTAL SNOW ACCUMULATION OF 6 TO 10 INCHES. WINDY WITH LOWS AROUND 10 ABOVE. NORTH WINDS 20 TO 25 MPH WITH GUSTS UP TO 40 MPH. CHANCE OF SNOW NEAR 100 PERCENT. WIND CHILL VALUES AS LOW AS 10 BELOW.

.WEDNESDAY . . . PARTLY SUNNY IN THE MORNING . . . THEN CLEARING. BLUSTERY AND COLD WITH HIGHS AROUND 17. NORTHWEST WINDS 15 TO 20 MPH WITH GUSTS UP TO 35 MPH. WIND CHILL VALUES AS LOW AS 10 BELOW IN THE MORNING.

.WEDNESDAY NIGHT . . . PARTLY CLOUDY IN THE EVENING . . . THEN BECOMING MOSTLY CLOUDY. LOWS AROUND 9 ABOVE. NORTHWEST WINDS 10 TO 15 MPH. WIND CHILL VALUES AS LOW AS 1 BELOW.

.THURSDAY . . . MOSTLY CLOUDY IN THE MORNING . . . THEN BECOMING PARTLY SUNNY. COLD WITH HIGHS IN THE MID 20S. WEST WINDS 5 TO 10 MPH. WIND CHILL VALUES AS LOW AS 1 BELOW.

.THURSDAY NIGHT . . . PARTLY CLOUDY. LOWS AROUND 10 ABOVE. WIND CHILL VALUES AS LOW AS 5 BELOW.

.FRIDAY . . . MOSTLY SUNNY. COLD WITH HIGHS IN THE LOWER 20S. WIND CHILL VALUES AS LOW AS 5 BELOW IN THE MORNING.

.FRIDAY NIGHT . . . MOSTLY CLOUDY. LOWS AROUND 20.

.SATURDAY . . . MOSTLY CLOUDY. HIGHS IN THE MID 30S.

.SATURDAY NIGHT . . . MOSTLY CLOUDY IN THE EVENING . . . THEN BECOMING PARTLY CLOUDY. LOWS IN THE LOWER 20S.

.SUNDAY . . . PARTLY SUNNY. HIGHS IN THE UPPER 20S.

.SUNDAY NIGHT . . . PARTLY CLOUDY IN THE EVENING . . . THEN BECOMING MOSTLY CLOUDY. LOWS 15 TO 20.

.MONDAY . . . PARTLY SUNNY IN THE MORNING . . . THEN BECOMING MOSTLY CLOUDY. HIGHS IN THE UPPER 20S.

The above forecast informed users about the likelihood of heavy snow. But given the disruption that was imminent on that particular day in January 2014, the local NWS soon released an updated winter weather message:

URGENT—WINTER WEATHER MESSAGE

NATIONAL WEATHER SERVICE NEW YORK NY

452 AM EST TUE JAN 21 2014

. . . WINTER STORM TO IMPACT THE REGION TODAY INTO TONIGHT . . .

[2]NWS policy mandates that forecasts be disseminated in all-caps lettering, much to the chagrin of many employees—not to mention the general public.

NYZ072>075-176>179-212300-

/O.CON.KOKX.WS.W.0002.140121T1700Z-140122T1100Z/

NEW YORK (MANHATTAN)-BRONX-RICHMOND (STATEN ISLAND)-

KINGS (BROOKLYN)-NORTHERN QUEENS-NORTHERN NASSAU-SOUTHERN QUEENS-

SOUTHERN NASSAU-

452 AM EST TUE JAN 21 2014

. . . WINTER STORM WARNING REMAINS IN EFFECT FROM NOON TODAY TO 6 AM

EST WEDNESDAY . . .

* LOCATIONS . . . NEW YORK CITY AND WESTERN LONG ISLAND.

* HAZARD TYPES . . . HEAVY SNOW AND BLOWING SNOW.

* ACCUMULATIONS . . . SNOW ACCUMULATION OF 8 TO 12 INCHES.

* WINDS . . . NORTH 15 TO 25 MPH WITH GUSTS UP TO 35 MPH.

* WIND CHILLS . . . AS LOW AS 10 BELOW ZERO LATE TONIGHT.

* TEMPERATURES . . . HIGHS IN THE MID 20S THIS MORNING . . . DROPPING INTO THE LOWER TEENS TONIGHT.

* VISIBILITIES . . . ONE QUARTER MILE OR LESS AT TIMES THIS AFTERNOON INTO TONIGHT.

* TIMING . . . SNOWFALL WILL BEGIN LATE THIS MORNING AND WILL CONTINUE THROUGH TONIGHT . . . TAPERING OFF EARLY WEDNESDAY MORNING. THE HEAVIEST SNOWFALL WILL OCCUR LATE THIS AFTERNOON INTO TONIGHT.

* IMPACTS . . . FALLING . . . BLOWING . . . AND DRIFTING SNOW WILL CAUSE HAZARDOUS TRAVEL AND WALKING CONDITIONS ACROSS THE AREA . . . IMPACTING THE LATE AFTERNOON AND EVENING COMMUTE. SNOW REMOVAL WILL BE DIFFICULT THIS AFTERNOON THROUGH TONIGHT. PROLONGED EXPOSURE TO FRIGID COLD AND LOW WIND CHILLS COULD CAUSE FROST BITE. DRESS APPROPRIATELY.

PRECAUTIONARY/PREPAREDNESS ACTIONS . . .

A WINTER STORM WARNING FOR HEAVY SNOW MEANS SEVERE WINTER WEATHER CONDITIONS ARE EXPECTED OR OCCURRING. SIGNIFICANT AMOUNTS OF SNOW ARE FORECAST THAT WILL MAKE TRAVEL DANGEROUS. ONLY TRAVEL IN AN EMERGENCY. IF YOU MUST TRAVEL . . . KEEP AN EXTRA FLASHLIGHT . . . FOOD . . . AND WATER IN YOUR VEHICLE IN CASE OF AN EMERGENCY.

You might have noticed that the two text forecasts shown above were for the same event predicted in the map of Figure 13–3.

▲ **FIGURE 13–4 Snowy Winter.** Central Park, New York, covered in snow from one of the many storms during the winter of 2013–2014.

Although this was a major snow event in the Northeast, it was hardly the only one that season, as a seemingly endless string of storms swept through the eastern two-thirds of North America in the winter of 2013–2014 (Figure 13–4). For the most part these storms were forecast with very good accuracy and allowed people, businesses, and governments to prepare.

DID YOU KNOW?

Meteorology offers a wide range of career possibilities to students who enjoy the physical sciences. The American Meteorological Society's website has several useful links for budding professionals. The web page http://ametsoc.org/atmoscareers/ provides some background on the training required to be a professional meteorologist. Also, the site at www.ametsoc.org/amsucar_curricula/curriculaAlpha.cfm offers a listing of all the programs in the United States that offer meteorology or atmospheric science degrees.

From the above discussion you can see that not all meteorologists have the same set of procedures to follow on a given day. Variations from one forecast office to another affect a meteorologist's schedule, and differing weather conditions may alter the usual routine.

CHECKPOINT

13.1 What is a zone forecast?

13.2 What are the roles of physical models and statistical models in weather forecasting? Explain.

Why is Weather Forecasting Imperfect?

13.2 Explain why weather forecasting is imperfect.

We've all had careful plans upset by a bad weather forecast and are understandably quick to find fault when actual conditions depart from the forecast. Implicit in such criticism is the premise that forecasts ought to be accurate and that there is no excuse for a miss. So why are forecasts sometimes so far from correct? After all, with powerful computers, satellites, weather radar, and global communication networks, it seems as if making a good forecast ought to be easy. But as much as the public might think so, this is definitely not the case—in fact, accurate weather forecasting is extraordinarily difficult.

To see why, imagine you want to forecast tomorrow's temperatures and think about just a few of the factors you must consider. First, remember that the temperature structure of the atmosphere depends in part on absorption and emission of radiation (shortwave and longwave), which itself depends on the vertical and horizontal distribution of atmospheric gases, clouds, and so on. So, to compute the temperature at a point, you need to begin with detailed information about the composition of the atmosphere in three dimensions.

Of course, with water constantly shifting between the liquid, solid, and vapor phases, atmospheric composition is hardly constant, so you will need to forecast those changes over time. Remember also that as water changes phase, latent heat is added or removed from the atmosphere; thus, you will have to keep track of that as well as radiation transfer. But the phase changes are influenced by small-scale updrafts and downdrafts, so you will need to somehow forecast vertical motions as part of your overall effort. Furthermore, horizontal motion can't be neglected—you will need to allow for warm or cold air advection (see Chapter 10). With regard to wind, near the surface you face the problems of flow around complex terrain and somehow quantifying frictional effects between the atmosphere and Earth's surface.

Obviously, weather forecasting involves a set of interlocking problems, each difficult to solve in isolation, let alone in combination. In light of all these difficulties, it's remarkable that forecasts show any accuracy at all. Rather than wondering why they fail, we're more likely to wonder how they are able to succeed!

CHECKPOINT

13.3 What factors complicate the prediction of temperature at a given point?

13.4 Why is the prediction of water vapor in the atmosphere so complicated?

Forecasting Methods

13.3 Describe the main methods of weather forecasting.

There is no single "correct" way to forecast the weather. Depending on the length of forecast, the type of information desired, and how much is known about the current state of the atmosphere, any one of a number of methods might be used. One can even attempt a forecast in the absence of any data about current weather, provided that long-term information is available. For instance, a forecast of hot, muggy conditions with a chance of afternoon thundershowers in Orlando, Florida, in mid-August has a reasonably good chance of proving correct. Such prognoses based on long-term averages are known as **climatological forecasts**. Obviously, the reliability of a climatological forecast depends on year-to-year variability in weather conditions for the forecast day. Thus, a forecast based on climatology might have a reasonable chance of being accurate in Orlando during the summer, but only the truly daring would try this for Chicago in April, when almost any kind of weather can occur.

Another type of forecast, called a **persistence forecast**, relies completely on current conditions with no reference to climatology. Actually, a special case of persistence forecasting is used by all of us in everyday life. When we see clear skies and leave the umbrella behind, we are betting that the prevailing conditions will continue and are making a short-term forecast on that basis. This simple procedure might work for a little while, but it will eventually fail to catch changes in weather. A more sophisticated version might assume that a decrease in pressure during the past few hours will continue through time and serve to predict the approach of a low-pressure system and its associated increase in cloud cover. Here, too, one extrapolates present behavior forward in time, again with certain knowledge that the forecast will fail when that behavior ceases. Of course, it is often precisely those breaks from past behavior we want to forecast; a method that can't provide such information is not terribly useful.

Until the 1950s, weather analysis and forecasting depended entirely on the experience of meteorologists and their interpretation of current conditions and the recent past. A meteorologist would base a forecast largely on the comparison of the current situation to similar conditions encountered previously (but not necessarily in the immediate past). This approach led to development and use of numerous "rules of thumb," which attempted to capture repeatable patterns and relations among various weather elements. For example, winter precipitation over the eastern United States and Canada tends to be snow north of the 5400 m contour for the 500 mb level and rain south of that curve. In this so-called **analog approach**, one tries to recognize similarities between current conditions and similar well-studied patterns from before. There are many variations of the analog approach, some subjective (depending on the forecaster's expertise), others objective (depending on statistical relations). All assume that what happened sometime in the past holds a clue about the future.

In the past few decades, **numerical weather forecasting** has come to occupy a dominant position. The term is somewhat misleading, because the methods just described all produce numerical forecasts. What is different is that this method is based on computer programs that attempt to mimic the actual behavior of the atmosphere. Numerical weather models explicitly compute the evolution of wind, pressure, temperature, and other elements over time. By examining the output for a given point in time, one obtains a depiction of the three-dimensional state of the atmosphere for that moment.

Video Forecasting Precipitation

http://goo.gl/DZxFLZ

(This is in contrast to forecasting the surface values for just a few weather elements, as might be done with other methods.) The numerical models typically used in weather forecasting are very large and can only be run on supercomputers. As reflects their importance in modern forecasting, these models are described in more detail in the later section called *Forecast Procedures and Products*. For now, we simply want to draw a contrast between other methods and this physically based ("numerical") approach.

CHECKPOINT

13.5 Define climatological forecasts and persistence forecasts.

13.6 How do analog approaches and numerical weather forecasting differ?

Types of Forecasts

13.4 Describe the general types of forecasts.

The product, or result, of a forecast method can take a variety of forms, which we are calling the *type* of forecast. We are all familiar with quantitative forecasts, in which the "amount" of the forecast variable is specified. For example, a forecast that says "An inch of rain is expected" is a **quantitative forecast**. Figure 13–5 maps the expected value of 24-hour precipitation beginning on the morning of January 21, 2014. In this case the greatest precipitation is forecast to extend across a swath from West Virginia northeastward into the Atlantic. Similarly, forecasts of the expected high or low temperature are quantitative, because a value for the forecast variable is provided.

In contrast, **qualitative forecasts** provide only a categorical value for the predicted variable. Examples of this are "rain/no rain," "hurricane/no hurricane," "above/below normal," or "cloudy/partly cloudy/mostly clear." In these examples, the predicted variable is assigned to a particular class, or category; hence, it is a qualitative forecast.

In the preceding examples, the forecasts were provided without qualification. **Probability forecasts** are an alternative, in which the chance of some event is stated. For example, a categorical hurricane forecast might be stated as a probability, rather than as a certainty. Probability forecasts can take a variety of forms, the most common of which is undoubtedly the probability-of-precipitation (PoP) forecast. When the broadcaster says "The rain chance today is 70 percent" or "There is a 60 percent chance of afternoon showers," he or she is reporting a PoP forecast. Note that these forecasts don't specify an amount of rain. Rather, a PoP forecast means that a randomly chosen point in the forecast area has the stated probability of receiving measurable precipitation. For example, a 75 percent PoP forecast means that the odds of precipitation are 3:1, or equivalently, you have only one chance in four (25 percent) of staying dry throughout the forecast period. Figure 13–6 maps the probability of snow accumulations greater than 4 inches during the next 24 hours. Consistent with Figure 13–3, heavy snowfall was most likely to occur over the Northeast on the morning of January 21, 2014.

CHECKPOINT

13.7 What is a quantitative forecast?

13.8 What are probability forecasts?

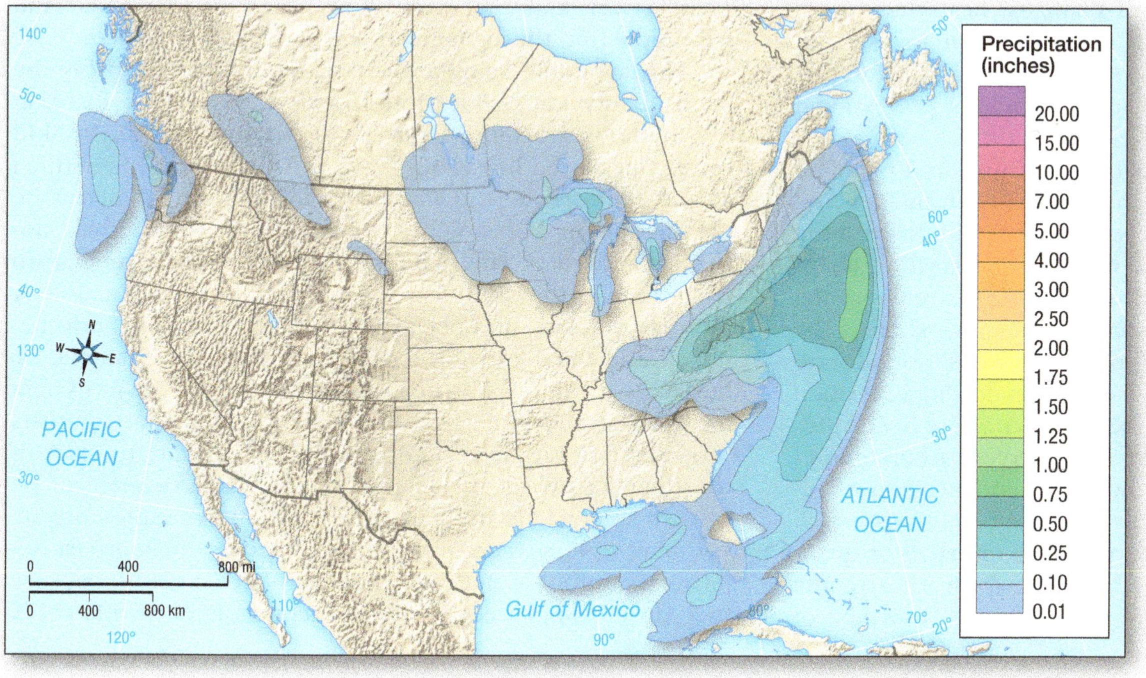

◄ **FIGURE 13–5 Quantitative Precipitation Forecast.** This is the predicted distribution of precipitation for the next 24-hour period issued on January 21, 2014.

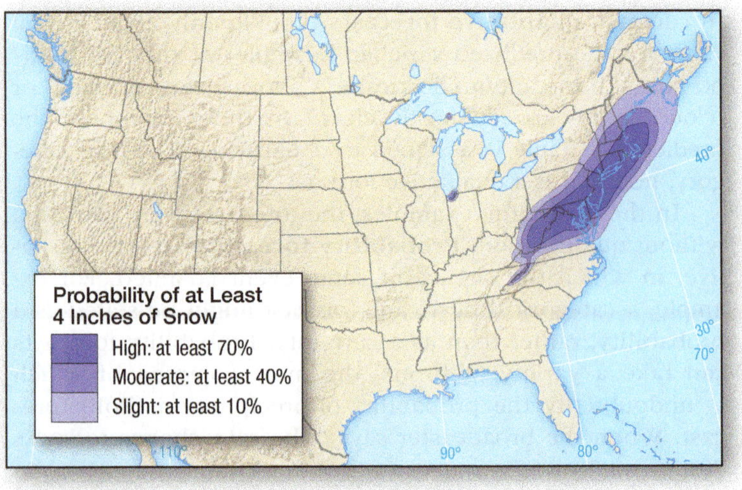

▲ **FIGURE 13–6 Probability of Precipitation.** PoP maps can take various forms. The simplest would be the likelihood of any kind of precipitation occurring over a particular time period. In this case we see the probability of snowfall greater than 4 inches in the 24-hour period beginning on the morning of January 21, 2014.

Assessing Forecasts

13.5 Explain how meteorologists assess the quality of weather forecasts.

Regardless of which method is used, or what form it takes, we obviously need some way of deciding how good a forecast is likely to be. Measures are needed, for example, to compare one forecast method to another or to decide how much to consider a forecast when making plans. Most importantly, assessment measures are needed by those responsible for developing and administering forecasting programs. As attempts are continually made to improve data-gathering and forecast procedures (at an ever-increasing cost), methods are needed for judging the value of changes, justifying future expenditures, and determining the return on investment. Although as consumers we don't hear much about forecast assessment, it is a routine and integral part of professional forecasting.

Over the years, many evaluation measures and practices have been developed, each with particular advantages and disadvantages. To sort them out, we must first think about the purpose of the assessment. Do we want information about **forecast quality** or **forecast value**? The former refers to the agreement between forecasts and observations, whereas the latter refers to the usefulness of a forecast. These sound similar but are quite different. Because there is no simple relation between quality and value, different measures are needed for each. For example, a high-quality, 100 percent accurate forecast of rain might have zero value for scheduling crop irrigation, if knowing the total amount of rainfall is also essential. A second issue is the type of forecast: quantitative or qualitative, probability or unqualified, and so forth. Clearly, the appropriate assessment measure will vary with the type of forecast variable. Finally, do we want an absolute measure of performance or a relative, comparative measure?

Forecast *value* necessarily depends on applying a forecast to a particular problem or decision. Most measures of value are based on loss/payoff tables, which attempt to capture the risks and rewards associated with various forecasts and responses to those forecasts. For example, knowing the cost of a ruined concrete job, the money earned when things go right, and the probability of a correct forecast, you could assign monetary value to the forecast. The details involve probability concepts that are beyond an introductory text; therefore, we won't discuss forecast value any further, except to note that a single forecast can have widely varying value, depending on how it is used.

With regard to the *quality* of a forecast, an obvious question to ask concerns **forecast accuracy**. That is, on average, how close is the forecast value to the true value? There are many ways to answer this simple question, each leading to a different accuracy measure. At the broadest level, we might want information about **forecast bias**, which concerns systematic over- or underprediction. A biased forecast method is one whose average forecast is above or below the true average. An unbiased method, by contrast, shows no tendency for over- or underprediction. Of course, that is not to say the method is perfect—it simply means the average overprediction is as large as the average underprediction, yielding an average error of zero. For example, if you wanted to predict the number of spots showing after the roll of a die, you could forecast 3.5 dots on every roll. Over many rolls (of a fair die), the average would be 3.5, agreeing with your predictions—the predictions are therefore unbiased. Of course, it is impossible for this method to yield even a single correct prediction. It is clear that although bias is a useful measure, we also need accuracy measures that don't allow positive errors to cancel negative errors. The simplest is the *mean absolute error* (MAE), which ignores the sign (positive or negative) of the errors. That is, over- and underpredictions are treated the same, and we simply find the average error without regard to sign.

For laypeople using a forecast, accuracy is probably the main quality issue. But professionals who develop forecasting methods are more likely to report **forecast skill**. Skill can be measured in various ways, but the concept is defined as the *improvement* a method provides over what can be obtained using climatology, persistence, or some other "no-skill" standard. If the method is no better than relying on past climate, the skill score would be zero—a climatology forecast requires no special knowledge of atmospheric behavior and thus does not contain any skill. For example, no measurable rainfall has been recorded for July in Jerusalem for the last 100 years. Any "no-skill" method that forecasts "no rain" for July is certainly going to be accurate most of the time. Likewise, if you predict that a hurricane will be present somewhere in the Atlantic Ocean next September 10, you have a 90 percent probability of success (using past behavior as a guide). But these forecasts have no skill, because they do not improve on "chance" (climatology in this case).

Only if predictions were correct more than 90 percent of the time would we say that the forecast method possesses any skill. In the case of air temperature, we might compare the MAE of the forecast method to the MAE obtained using the climatological mean temperature.

CHECKPOINT

13.9 What are two important elements of forecast quality?
13.10 What is forecast skill?

Data Acquisition and Dissemination

13.6 List some major sources of data for weather forecasting.

The starting point for almost all weather forecasting is information about the current state of the atmosphere. To know the future, we begin with information about the present. Thus, the first process in operational weather forecasting is acquiring the necessary data. For reasons that will become clear, this requires an international effort, even when making forecasts for "small" areas, such as individual countries.

The **World Meteorological Organization (WMO)**, under the auspices of the United Nations, oversees the collection of weather data across the globe from its 179 member nations. The WMO collects data from about 10,000 land observation stations, 7000 ship stations, 300 moored and drifting buoys with automatic weather sensors, and several weather satellites. It also obtains upper-air data from about 800 weather balloon sites twice a day and on a continuous basis from instruments aboard wide-bodied commercial aircraft. The data from all countries are sent to the three **World Meteorological Centers** at Washington, D.C.; Moscow, Russia; and Melbourne, Australia. These centers, in turn, disseminate the data to all member countries.

The member nations of the WMO maintain their own meteorological agencies that obtain and process the data and issue regional and national forecasts. In the United States, the **National Centers for Environmental Prediction (NCEP)** of the Weather Service performs these tasks, while in Canada they are handled by the **Canadian Meteorological Centre** of the **Atmospheric Environment Service (AES)**.

Not surprisingly, the United States has a relatively dense network of surface observation stations. Of the approximately 1000 locations where surface conditions are recorded, about 120 are National Weather Service offices; the rest are Federal Aviation Administration (FAA) airport sites. The Canadian AES operates about 270 surface stations. Together, these sites record temperature, humidity, pressure, cloud conditions (including type and height, and the percentage of sky obscured by cloud), wind direction and speed, visibility, the presence of significant weather, such as fog or rain, and accumulated precipitation readings at ground level. The FAA and NWS maintain a network of more than 800 automated sensors, called **Automated Surface Observing Systems,** or **ASOSs**, for measuring and recording these variables (Figure 13–7). The FAA also maintains some similar but older units called **Automated Weather Observation Systems** (AWOSs). In Canada, the AES operates more than 100 AWOSs.

In addition to the surface observations made at these sites, the NWS launches hydrogen-filled balloons[3] carrying weather instrument packages called **radiosondes** (Figure 13–8). Twice a day—at 0000 and 1200 UTC (Universal Coordinated Time[4])—about 800 radiosondes are launched worldwide, about 80 within the United States and Canada. Radiosondes continually observe and transmit to ground recording stations the pressure, air temperature, and wet bulb temperature (from which dew point temperatures are determined). Some radiosondes are tracked by radar as they ascend through the atmosphere, which enables a determination of the wind speed and direction of the middle and upper atmosphere. Radiosondes tracked by radar are called **rawinsondes**.

Most radiosondes rise high into the stratosphere to about the 4 mb level—or about 30 km (19 mi)—whereupon the balloon bursts and the package parachutes back to the surface. Usually a complete radiosonde ascent will take about 1 hour and 50 minutes, by which time the instrument package may have been carried many kilometers downwind from the launch site. Interestingly, radiosondes are often found by nonmeteorologists who happen to stumble across them. Many people who find them follow the instructions on the packages requesting that the package be dropped in a mailbox for return to the Weather Service, and the recovered radiosondes are then refurbished for subsequent reuse. Most of them are never recovered, however, landing in oceans or remote areas on land.

Valuable information is also obtained from geostationary orbiting Earth satellites called GOES. There are usually two active GOES satellites that provide images and digital information for multiple layers of the atmosphere for the eastern and western parts of North America. These satellites are replaced as they become inoperable or obsolete. Currently GOES 13 covers eastern North America and GOES 15 the west.

Upper-level information is obtained from other sources as well. Many commercial aircraft are equipped with weather sensors that automatically monitor the atmosphere throughout their flights, and weather satellites supplement the upper-level database by determining temperatures and humidities at several depths throughout the atmosphere. Together, the aircraft and satellites provide data from locations far away from any radiosonde sites and play a particularly important role in gathering information over the oceans. In addition, surface data are collected by sensors on buoys and relayed to land via satellites.

---
[3]Rather than using helium in the balloons, the Weather Service uses hydrogen, which it extracts locally, directly from the surrounding air.
[4]These times correspond to 1900 (7:00 P.M.) and 0700 (7 A.M.) Eastern Standard Time.

▸ FIGURE 13–7 A Typical Automated Surface Observation Unit (ASOS).

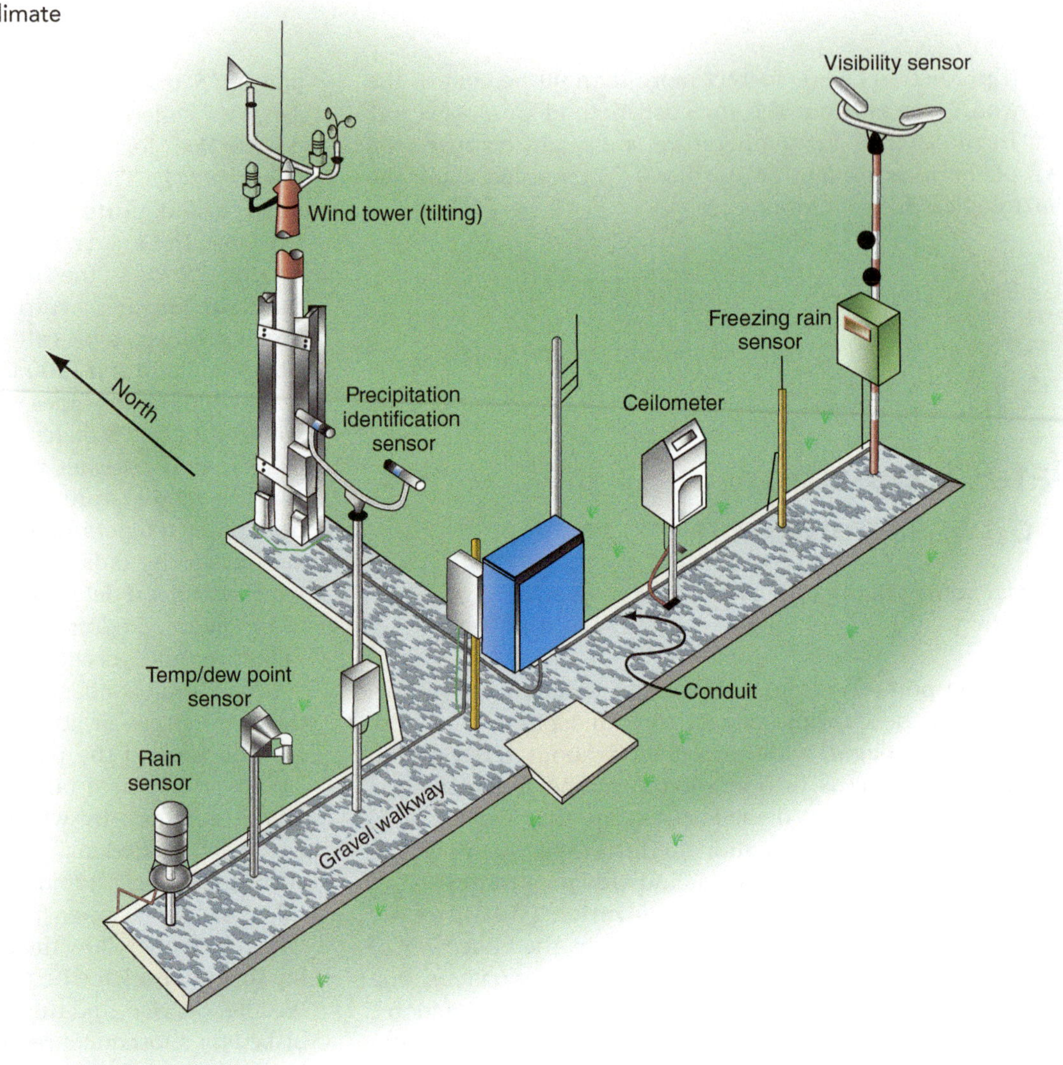

▸ FIGURE 13–8 Radiosonde. One of twice-daily launches of a weather balloon and radiosonde outside the Topeka, Kansas, Weather Service Office.

CHECKPOINT

13.11 Describe the ground-based instruments for obtaining automated weather information.

13.12 What types of systems gather information about weather variables away from the surface?

Forecast Procedures and Products

13.7 Describe forecast procedures, including numerical modeling, and explain how meteorologists develop short-, medium-, and long-range forecasts and seasonal outlooks.

As mentioned earlier, numerical models are the preeminent tool of modern weather forecasting. Weather agencies around the world develop their own models and typically maintain a suite of models rather than a single program. The models are updated continually, and from time to time new models are introduced and older ones retired.

Phases in Numerical Modeling

Although there are large differences among models, the general procedure is the same for all numerical models. The three phases include analysis, prediction, and postprocessing.

ANALYSIS PHASE First is the **analysis phase**, in which observations are used to supply values corresponding to the starting ("current") state of the atmosphere for all the variables carried in the model.

Unfortunately, the network of weather stations and radiosonde launches is highly irregular and does not come close to providing even coverage. Part of what analysis accomplishes is converting those irregular observations into "uniform" initial values. Though only a preparatory step, this is a difficult task. There are millions of data values from a variety of sources (satellite, ships, and so on), representing various moments in time. None of the measurements is completely free of error, and many are subject to large error.

PREDICTION PHASE Fundamentally, the job of a numerical model is to solve the basic equations describing atmospheric behavior. Beginning with values delivered by the analysis phase, the model uses physical equations to obtain new values a few minutes into the future. The process is then repeated, using the output from the first step as input for the next set of calculations. This procedure is performed over and over as many times as necessary to reach the end of the forecast period (24 hours, 48 hours, or whatever). This is called the **prediction phase** of the model run. This involves many billions of calculations for each time step, despite the fact that there are just a handful of fundamental atmospheric variables (temperature, pressure, wind velocity vector, density, and moisture).

POSTPROCESSING PHASE The conditions forecast by the model at regular intervals (for example, every 12 hours) are represented on a grid for mapping and other display purposes. For example, in this **postprocessing phase**, a series of maps might be produced for each of the 12-, 24-, 36-, and 48-hour periods, depicting the forecast distributions for the following:

1. Sea-level pressure and 1000 to 500 mb thicknesses
2. 850 mb heights and temperature
3. 700 mb heights and vertical velocities (such as the speed at which air rises or sinks)
4. 500 mb heights and absolute vorticity values
5. Precipitation amounts.

These products are used in various ways, some of which are described later in this chapter. Forecasters study maps for each period and interpret the conditions that would probably be associated with such patterns. They compare the maps with one another and with corresponding maps from other models for the same time periods. Of course, the model forecasts differ from one another—the forecaster uses model output as guidance, weighting the results differentially according to what is known about each model's strengths and weaknesses and supplementing model guidance with other analyses and observations.

Often the actual (final) forecast will not match any of the models exactly. Despite the ever-improving and now very impressive quality of the numerical models, they will never be perfect. This is where the training and experience of professional forecasters comes in. For this reason NWS offices routinely critique the accuracy of recent forecasts yielded by each model and by individual forecasters. The goal of this exercise is not to create undue stress and premature aging of forecasters, but rather to improve future forecasts by noting which models have been providing the optimal output for a given season and in which meteorologists have a higher level of confidence.

In addition to maps of model variables, forecasts for a number of secondary variables are produced. Examples include maximum and minimum temperature, dew point, wind conditions, and the probability of precipitation. These forecasts are constructed using statistical relationships between model output and observed surface conditions from the past. The output products are called *model output statistics* (MOS) and are designed to capture the effect of topography and other factors that influence local weather conditions. Numerical models have only a limited ability to represent processes occurring near the surface, and they provide a rather generalized picture of the atmosphere. Thus, a statistical approach has considerable appeal. MOS is most effective for the places from which statistical relations were derived and is somewhat less effective at intermediate locations having a different topographic setting.

DID YOU KNOW?

The National Weather Service has implemented a *National Digital Forecast Database* (NDFD). Unlike conventional forecasts that focus only on large, well-known sites (such as Indianapolis, Indiana) and give only general information about surrounding areas, the system sets up a grid and extrapolates weather forecasts for all individual points on the grid—even if they are not part of a large metropolitan area (for example, Beanblossom, Indiana, with a population of 2740). You can use this system by going to http://graphical.weather.gov/ and scrolling on a sequence of maps. (You may have to experiment a bit at first, but the results are well worth it.) More details on this emerging technology can be found at www.nws.noaa.gov/ndfd.

Advances in theory, computer technology, observing networks (especially satellites), and data integration have all contributed to steadily increasing forecast skill. Consider Figure 13–9a, which shows the trend in skill scores for NCEP forecasts of 2.5 cm (1 in.) rainfall events. Skill for one-day forecasts has nearly doubled since 1965 and has almost quadrupled over the same period for two-day forecasts. Although the record is much shorter, skill for three-day forecasts has been increasing even faster. Despite technological advances, there remains a large role for human judgment in forecasting. In fact, the forecasts underlying Figure 13–9a have a large human component as forecasters blend guidance from various models with their own experience when producing a forecast. Figure 13–9b exposes the role of humans by comparing actual forecasts with those that would be obtained from model guidance alone. Exactly how much skill a human adds to a forecast depends on the model used for comparison, but it is interesting to see that the gap has remained fairly constant

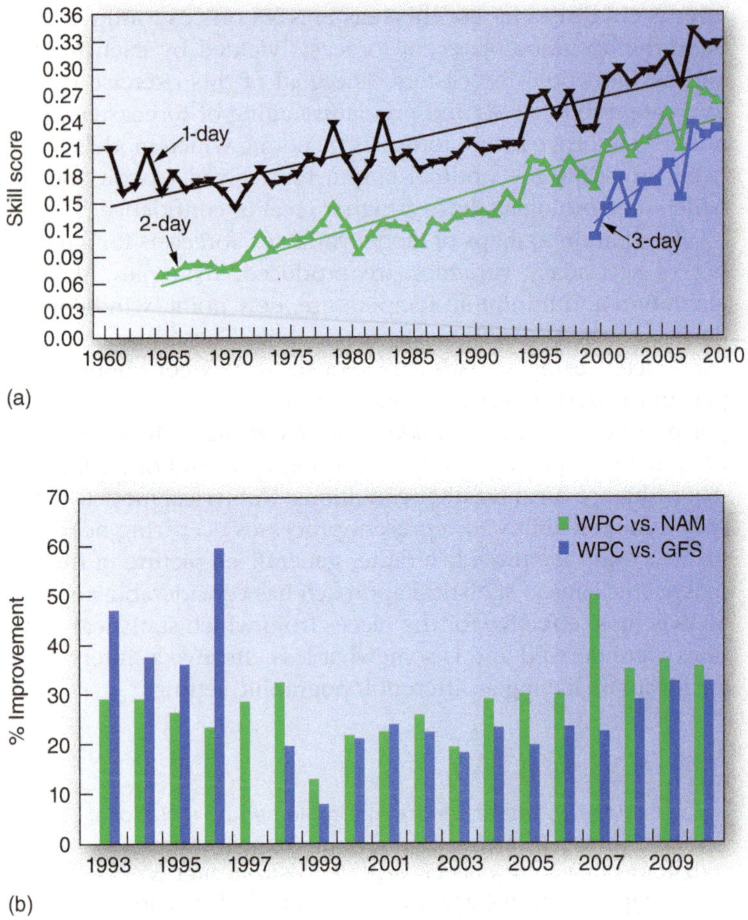

▲ **FIGURE 13–9 Skill Scores.** (a) Annual scores for 2.5 cm (1 in.) precipitation forecasts issued by the Weather Prediction Center of NCEP. (b) Percentage difference in skill between WPC forecasts in (a) and two numerical weather prediction models. WPC forecasts result from subjective forecaster judgments applied to a combination of model output and other data.

during the past 15 years or so, with judgment contributing about 30% to forecast skill. Clearly, local knowledge (and experience regarding model failings) continues to be useful and adds significant value to purely objective forecasts.

CHECKPOINT

13.13 What are the three phases of numerical modeling?

13.14 What is the role of a forecaster with regard to the use of numerical modeling?

Models Employed

The National Weather Service uses output from multiple numerical models every time a forecast is made. One important one is the **North American Mesoscale Forecast System (NAM)**. This model, run four times each day, uses multiple grid systems in which large areas are overlaid by widely spaced grid points; superimposed on these are smaller areas with a denser network of grid points.

Changing weather conditions can necessitate a rapid update of model output. To do this, NWS employs the **Rapid Refresh (RAP) model**, which was implemented on May 1, 2012. One version of the RAP interpolates detail from other models onto a fine-scale grid that yields much greater spatial detail over local regions. For example, forecasters in the Los Angeles region need output that shows the difference in wind and other variables over a region that extends from a coastal plain to a mountainous region and an open desert. RAP forecasts are rerun every hour and provide forecasts going out 18 hours.

Going beyond "short-term" forecasts (72 hours or less), so-called **medium-range forecasts (MRFs)** are receiving considerable attention. One such system is the Global Forecast System (GFS), which is used to formulate forecasts up to 16 days into the future. Since weather systems can form, move great distances, and die out over that period, the GFS requires a grid that covers the entire world. The GFS is, in fact, an amalgam of four models that bring together information on the atmosphere, oceans, soils, and sea ice.

Another medium-range model is from the **European Center for Medium-range Weather Forecasting (ECMWF)**, which generates forecasts for up to 15 days. It can also be used for the generation of monthly and seasonal forecasts. The ECMWF is perhaps the single best forecast model in operation and received much recognition for its ability to forecast the development and movement of superstorm Sandy in 2012 (Chapter 12).

Other models with more specific objectives are also widely used. Such models would include those for forecasting hurricanes and local severe storms.

ENSEMBLE FORECASTS Rather than making just a single forecast, **ensemble forecasting** is widely employed, in which a number of different computer-generated forecasts (runs) are performed for the same period. The reasons have to do with something we mentioned earlier: that small disturbances can grow into large disturbances. If two model runs are made with slightly different initial values, the results might be very different after a week or so. Recognized in the early 1960s by E. N. Lorenz, this behavior is now known to be typical of many natural and human systems and is usually called *chaotic* behavior.

Chaos is a condition that occurs in physical systems that makes it impossible to precisely predict how a system (such as the state of the atmosphere) will appear some time in the future. Chaos arises because small errors in the input value of a variable (or even round-off error) can become magnified in time. This situation is referred to as sensitive dependence on initial conditions. It is a serious problem for weather forecasting because the initial conditions are never known exactly. For example, if an upper-level trough appears in the 15-day forecast, there is no way to know if it is "real" or if it arose because of errors in the initial data. Ensemble forecasting uses multiple runs starting with slightly different initial values. The different initial fields are created by introducing small changes (perturbations) in the best-guess field. Several methods are used for assigning perturbations, but all attempt to mimic errors that might reasonably appear in the data. From this ensemble of initial conditions, an ensemble of forecasts is produced, each different.

Figure 13–10 shows a 10-day ensemble (sometimes referred to as a *spaghetti diagram*) for the Northern Hemisphere

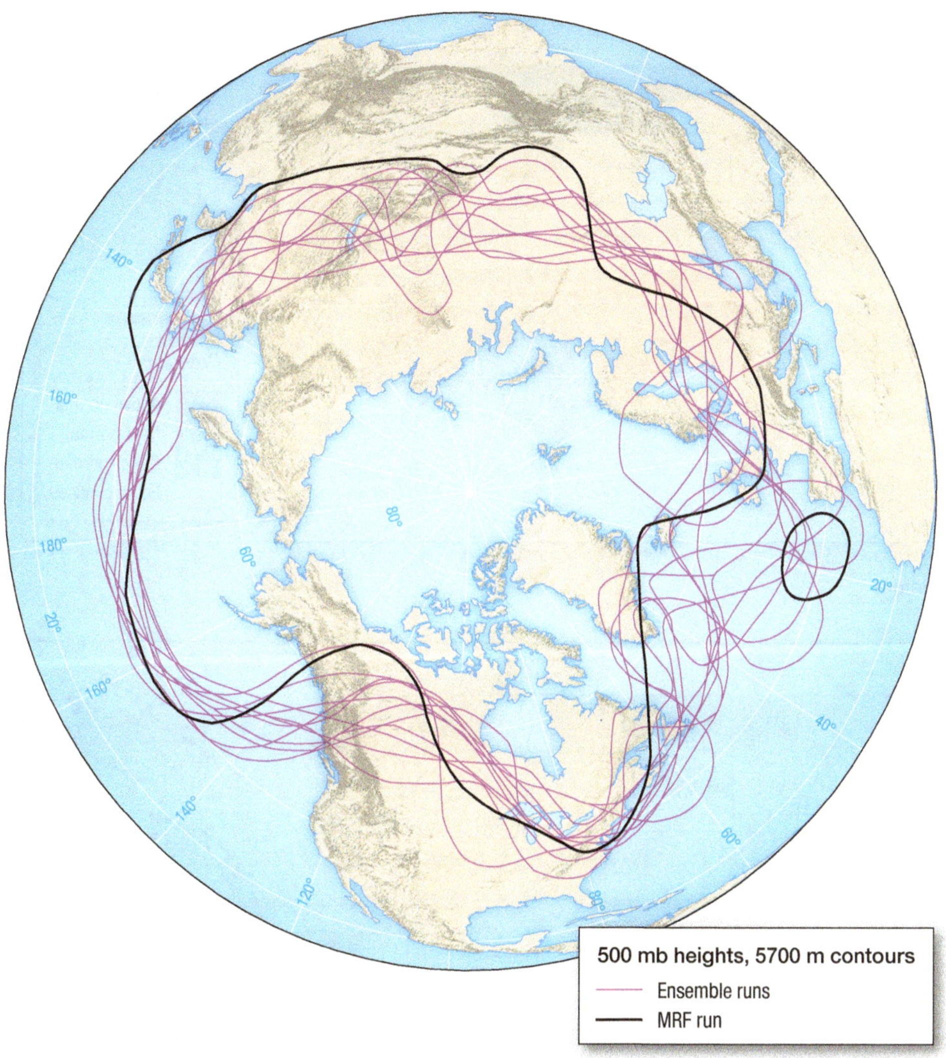

FIGURE 13–10 Spaghetti Map. Ten-day ensemble forecast from the NCEP MRF model. The 5700 m contour is plotted for all 17 ensemble members. Each of the contours falls into somewhat different positions because of the effect that minor changes in the initial conditions of each model run exert on the model output.

from NCEP. The map is a kind of 500 mb map, except that only one contour is shown: the 5700 m contour. Plotted are the 17 NCEP ensemble members, plus the unperturbed (control) run. Notice that all ensemble members call for a trough in the eastern Pacific Ocean. The implication is that whatever errors might exist in the data, they don't affect the forecast there. Thus, we might have some confidence that a real trough will develop. But over central Asia, western Europe, and the north Atlantic, ensemble spread is much larger: Ensemble members show much less consistency with one another, suggesting we should be less confident about the forecast for these regions. (Of course, we would not base the final assessment on a single contour, and we would look at other variables in addition to 500 mb height.)

Figure 13–10 illustrates the primary use of ensembles, which is to provide information about forecast uncertainty. With that, one has an estimate of reliability to go along with the forecast and won't pay much attention to forecasts deemed unreliable. Ensembles also can be used in other ways, including to generate the forecast itself. In particular, the mean of all ensembles can be treated as a forecast, even though it doesn't arise from a model run. Moreover, we might expect this forecast to be reasonably good, on the grounds that averaging will smooth away features found in just one or two members.

LONG-RANGE FORECASTS AND SEASONAL OUTLOOKS

Forecasts at still longer lead times are called **long-range forecasts**. In the United States, the **Climate Prediction Center (CPC)** of NCEP is charged with preparing forecasts for periods ranging from a week to the limits of technical feasibility. The methods used include climatology, statistics, numerical models, and subjective judgment. For example, because of its role in the global climate system, sea surface temperature (SST) in the tropical Pacific is routinely forecast for up to a year in advance. The SST forecasts are based on a combination of three models:

1. An analog statistical model
2. A "canonical" statistical model based on correlations of temperature and precipitation patterns with prior sea surface and atmospheric conditions measured over space and time
3. A numerical model.

In the numerical model, ocean and atmosphere are coupled so that the ocean responds to changes in the atmosphere, and vice versa. Output from the three models is combined statistically to yield the final forecast.

▸ **FIGURE 13–11 Seasonal Forecast.** Predictions for (a) temperature and (b) precipitation for the upcoming December to February period issued by the Climate Prediction Center in November 2013.

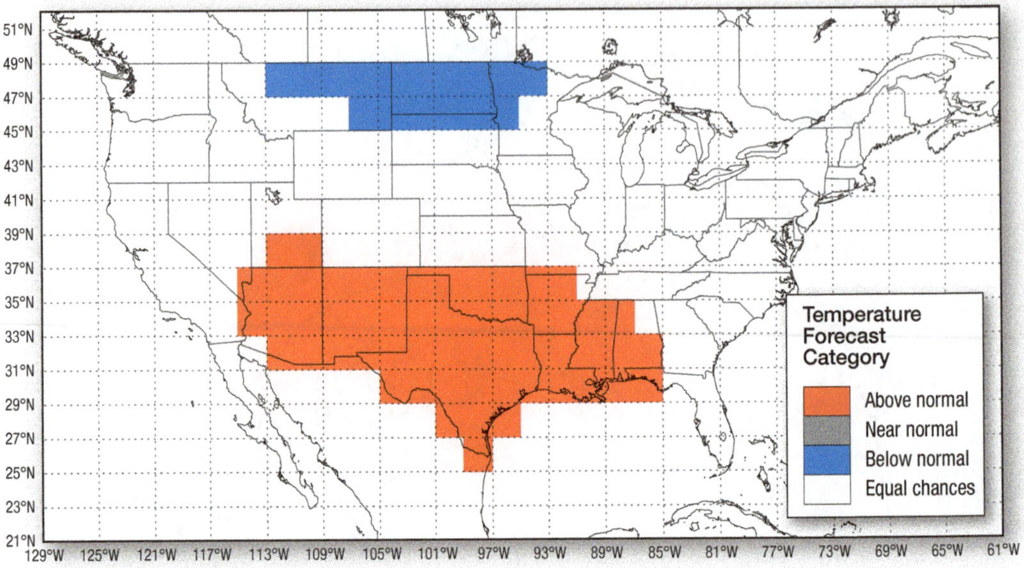

(a) Temperature

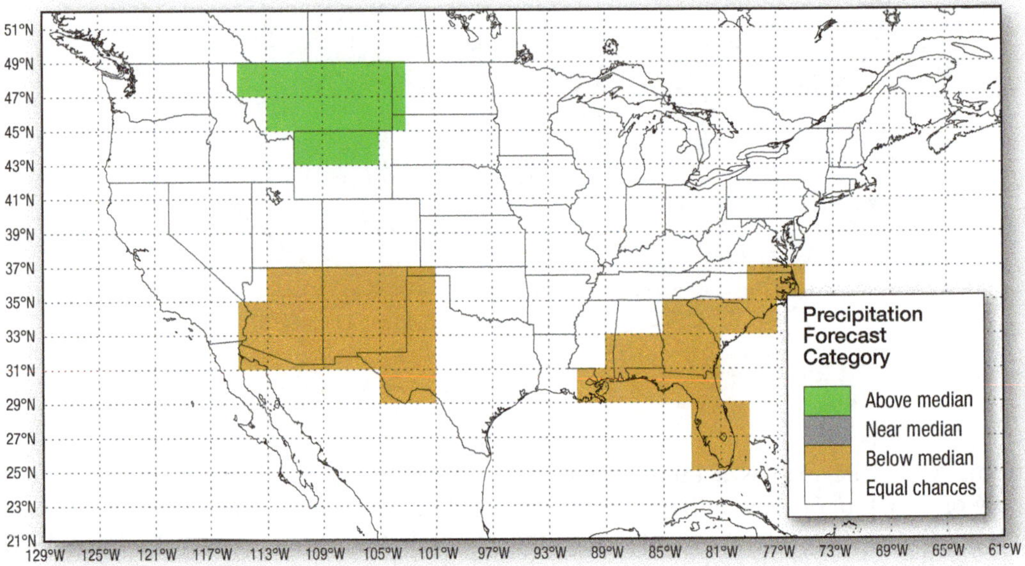

(b) Precipitation

Another CPC product is the **seasonal outlook**, a kind of forecast for an entire season, often misunderstood. In contrast to long-range forecasts that predict conditions *for particular days*, seasonal outlooks predict *average conditions* for an entire season. Figure 13–11 shows the format, in which seasons are classed as "above normal," "near normal," or "below normal for temperatures, and "above median," "near median" and "below median" for precipitation.[5] "Above normal" includes the upper third of the distribution; it occurs about 33 percent of the time, when a variable is at the 66th percentile or higher. "Near normal" and "below normal" are defined similarly, as the middle and bottom thirds, respectively. Thus, without any special knowledge we could forecast "above normal" with 33 percent accuracy and likewise for "near normal" and "below normal." Suppose we believe "near normal" conditions are quite likely, say, 45 percent likely. In that case we would issue a "near normal" forecast. But if we believe the chances are only 34 percent, we would probably refuse to make a forecast on the grounds that above or below normal conditions are just about as likely. Looking at the maps of Figure 13–11, you see that the CPC predicted cold conditions (a) along much of the northern Plains and a warm season over much of the South and Northeast. As for precipitation (b), the CPC anticipated dryness in some of the Southwest and Southeast, and wetter than normal conditions around Montana and environs.

How did the forecasts in Figure 13–11 work out? Not so good, as shown in Figure 13–12. The winter of 2013–2014 was marked by wave after wave of snowstorms across much of the United States east of the Rockies, and conditions were anything but warm. And the forecast failed to anticipate the extreme drought that hit California. Fortunately, this weak performance was an anomaly, with most seasonal forecasts

[5] For simplicity we will drop the term *median* in the rest of the discussion and instead describe conditions as above, near, or below normal.

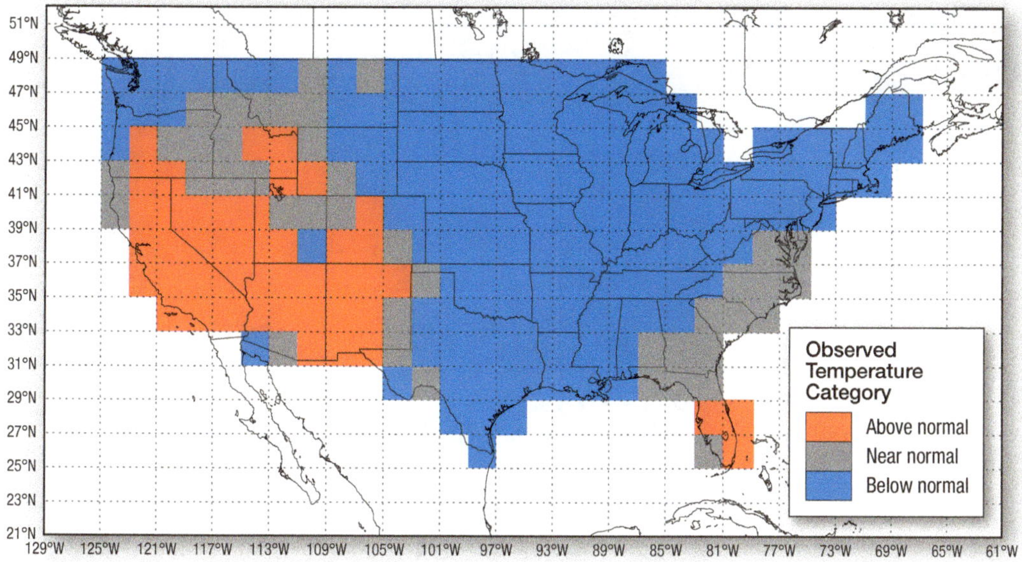

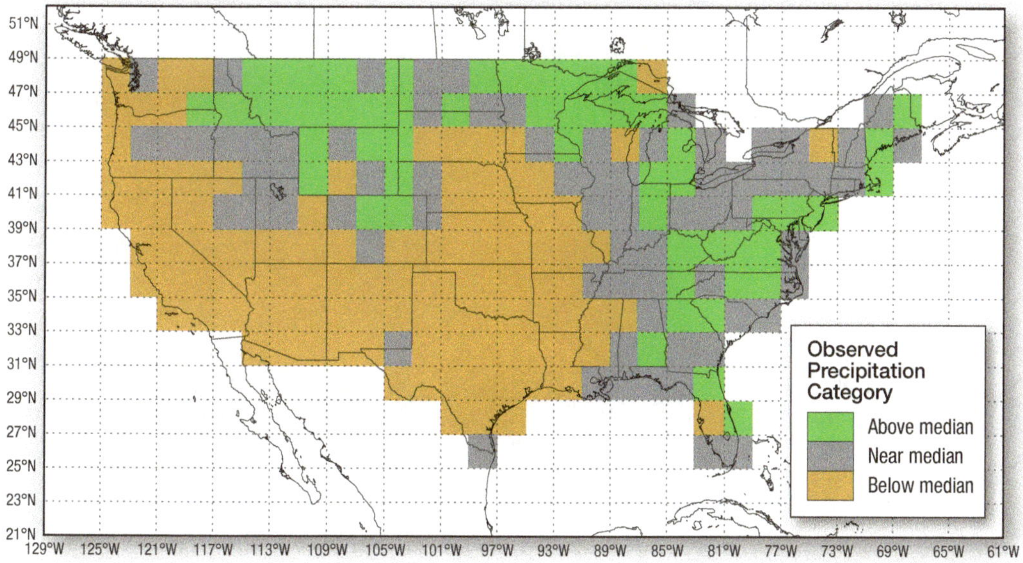

FIGURE 13–12 Observed Weather. (a) actual temperature and (b) precipitation disturbances experienced from December 2013 through February 2014; compare to Figure 13-11.

having better success. (Further analysis of the successes and failures of the forecasting efforts that season are discussed in *Box 13–1, Focus on the Environment and Societal Impacts: Forecasting the Winter of 2013–2014*).

How are these forecasts obtained? Again, forecasters use a combination of methods, both numerical and statistical. For example, SST forecasts from the coupled ocean–atmosphere model are used as boundary conditions for repeated runs of an atmospheric general circulation model. Another set of runs is made using SST based on persistence rather than physical principles. The result, once again, is an ensemble of atmospheric forecasts. The statistical techniques are mainly based on persistence in anomaly patterns. That is, departures from average are analyzed for stability and/or consistent patterns of evolution. Although the details are

DID YOU KNOW?

In 1961, Edward Lorenz of MIT discovered that his primitive prediction model was infuriatingly sensitive to initial conditions. In other words, just like the real atmosphere, future states (predictions) were easily contaminated by seemingly insignificant differences in starting values. This became evident when he reran a primitive forecast model with input data slightly different from those of a previous run. To his surprise, the model output was very different from that which derived from the initial input. What he discovered was the concept of *sensitive dependence on initial conditions*, which tells us that over time, minor perturbations in model input can compound to produce very different outputs. This factor inevitably defeats any attempt to numerically forecast weather beyond a certain time period—currently believed to be about 14 days.

13–1 FOCUS ON THE ENVIRONMENT AND SOCIETAL IMPACTS

Forecasting the Winter of 2013–2014

Usually we think of weather forecasts in terms of whether tonight's baseball game might be cancelled or if a stroll through the park is a good idea or not. But forecasting the arrival of major weather events is the primary mission of a NWS forecaster—the agency's number one responsibility is the protection of life and property. We saw earlier in this chapter that the seasonal forecast did not anticipate the kind of winter that actually occurred. The societal ramifications of a failed seasonal forecast are probably fairly minor, however, compared to a failed short-term forecast.

Consider, for example, the major snow and ice storm that hit the Southeast on January 28, 2014. Many residents of Atlanta, Georgia, and—most importantly—numerous city officials believed that heavy snow would hit the general area but fall mostly to the south of the city. As a result, schools and businesses remained open and people took the roads for their homeward journey in the afternoon as usual. But the snow proved to be quite heavy and motorists found themselves stuck on highways made impassable by the snow. Some people remained in their cars for up to 11 hours. The hardship wasn't limited to the Atlanta area, as snow across the Deep South totally shut down traffic (Figure 13–1–1). In Birmingham, Alabama, hundreds of schoolchildren spent the night at school and did not return home until the next day.

Was the Atlanta experience the fault of an errant forecast? Not really. The previous evening forecasters had called off a winter storm warning for Atlanta, but advised that a small change in overall conditions could cause a major change in where and when snow would arrive. Then during the predawn hours the local NWS office reissued a winter storm warning and advised that snow would likely begin in the midmorning and continue through the following morning. They also specifically indicated that snow-covered roads and reduced visibility would make travel difficult.

This weather event highlighted one of the biggest difficulties in weather forecasts: dealing with precipitation amounts that may vary a lot over short distances. In some cases the transition from heavy snow to a light dusting may occur over a few tens of miles. Thus, a forecast that is generally correct may be off the mark for people on either side of the transition zone. Similarly, a very minor error in forecasted temperature can mean that people expecting a moderate rainfall may instead find themselves slogging through deep snow. Forecasters generally try to convey these uncertainties to the public and emergency managers.

▲ **FIGURE 13–1–1 Ice Storm, January 28, 2014.** Parents in Birmingham, Alabama leave their cars and walk to pick up high school students stranded overnight.

13–1–1 What could residents and officials have done differently to reduce the impact of the January 2014 snowstorm?

13–1–2 What factors can make the prediction of heavy snow particularly difficult?

beyond an introductory text, the basic idea is not: The past is the key to the future.

Keep several things in mind when interpreting seasonal forecasts. First, they are forecasts for the entire period (season), not a particular day. One certainly does not expect above-average conditions throughout the forecast period, no matter what the "above average" probability might be. Also, at the present time, these forecasts do not exhibit much skill, and what skill exists varies by season, location, and the forecast variable (temperature vs. precipitation). Finally, even if the skill is relatively high, the associated probability is not likely to be large. In fact, it is unusual to have probabilities much above 50 percent, even in areas of highest confidence.

CHECKPOINT

13.15 What is ensemble forecasting and how is it used?

13.16 Which type of forecasting does the production of a seasonal outlook most resemble: climatological forecasting, persistence forecasting, or analog forecasting? Explain.

Weather Maps and Images

13.8 Explain how meteorologists use weather maps and satellite images in forecasting.

Although computers play a critical role in weather analysis, ultimately the meteorologist applies his or her knowledge to produce the forecast that is issued to the general public. Probably no tool is as valuable to a forecaster as a weather map. Newspapers and television news segments often just show surface maps, but weather forecasting requires analyzing conditions in the middle and upper atmosphere as well. Not only do clouds exist well above the surface, but the middle and upper atmosphere are closely intertwined with the air near the surface. As a result, accurate weather analysis requires using a series of maps representing different layers of the atmosphere. In this section we review resources commonly used in general weather forecasting. See *Box 13–2, Focus on Aviation: Special Forecasts and Observations for Pilots,* to see some resources of particular interest to pilots.

13–2 FOCUS ON AVIATION

Special Forecasts and Observations for Pilots

The National Weather Service offers comprehensive weather information and forecasts for aviators that is readily accessible from the Aviation Weather Center's Aviation Digital Data Service (ADDS). These data are readily accessible at www.aviationweather.gov/adds (Figure 13–2–1). Some of the more commonly used resources are described below. In some cases information is presented using abbreviations and codes. For a complete description of the codes and conventions used in these reports, you can download at no cost a copy of *Aviation Weather Services, Advisory Circular 00-45G Change* 1 (the URL is a long one, so it's easiest to access if you do a web search for the document and click on the link).

Aviation Routine Weather Reports (METARs)

METARs are issued by airports on an hourly basis and provide basic weather observations. These reports present observations in a string of "groups." Each airport is represented by a four-digit identifier, followed by the time of observation and a modifier report describing the data source. The coding system used is somewhat archaic but usually easy to decipher; at other times it is less obvious. But this problem is now remedied by the option to click on a "translation" of the data that presents the information in real words. The ADDS website also has an interactive Java tool that allows the user to scroll over a map and have the data immediately displayed.

Aviation Selected Special Weather Reports (SPECIs)

Though METARs are issued every hour, weather conditions can sometimes change dramatically between reports to a level that might be important for aviators. When this happens, the AWC issues Aviation Selected Special Weather Reports, or SPECIs. These are formatted the same way as METARS and are differentiated from them by the use of SPECI as the type of report preceding the station identifier.

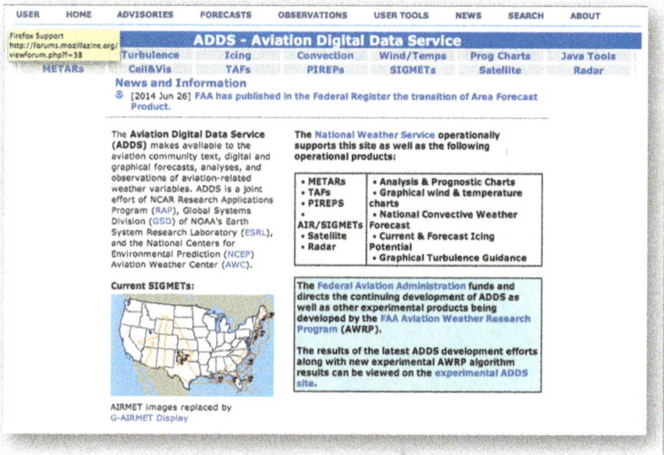

▲ **FIGURE 13–2–1 Aviation Weather.** The home page for the Aviation Weather Center's Aviation Digital Data Service.

Pilot Weather Reports (PIREPs)

Weather data from airports help pilots prepare for takeoffs and landings, but aviators also need information on flight conditions along their flight route. This type of information is provided to the Aviation Weather Center by pilots themselves and then disseminated to the rest of the aviation community. PIREPs report on icing, turbulence, wind, or other conditions at designated flight levels. As with METARs and SPECIs, the ADDS website offers decoded versions of the reports and a Java tool for easy access.

Significant Meteorological Information (SIGMETs)

SIGMETs alert pilots to existing or expected weather conditions that could impact aircraft safety. The threats could involve severe icing, heavy turbulence, sandstorms or dust storms, or airborne volcanic ash. These are issued by regional Meteorological Watch Offices (MWOs) and remain valid for a periods of up to four hours (they can be reissued an unlimited number of times). Most SIGMETs are classified as nonconvective; convective SIGMETs are issued when thunderstorms may lead to extreme turbulence, icing, or low-level wind shear.

Short-Range Surface Prognostic Charts (PROG) and Low-Level, Midlevel, and High-Level Significant Weather (SIGWX) Charts

You have seen numerous examples of surface and upper-air maps in this and previous chapters of this text, all of which are useful to pilots. In addition to these, the Aviation Weather Center compiles specialized maps for the aviation community—both for current conditions and forecasts up to 48 hours. These maps are tailored to the needs of pilots and highlight weather conditions of special importance to aviators.

13–2–1 What NWS agency disseminates products of use to pilots?

13–2–2 Describe five forecasting products issued primarily for aviation purposes.

Surface Maps

Surface maps of prevailing conditions (such as that shown in Figure 13–13) present a general depiction of sea-level pressure distribution and the location of frontal boundaries. The pressure is shown by isobars drawn every 4 mb, with zones of locally highest and lowest pressure labeled *H* and *L*, respectively.

LARGE-SCALE FEATURES Even a nonprofessional can make a number of inferences from briefly inspecting a surface weather map. General wind speeds and directions obey the rules discussed in Chapter 4. That is, wind speed varies according to the spacing of the isobars, and Northern Hemisphere winds rotate clockwise out of high-pressure systems and counterclockwise into lows. High-pressure systems favor downward vertical motions that promote clear skies, whereas low pressure promotes updrafts that lead to adiabatic cooling and formation of cloud cover. Surface maps become even more valuable when viewed in sequence. Because new maps are compiled every three hours, one can easily track the movement of individual weather systems as they move cross-country. By assuming that the systems will continue to behave as they have in the past few hours, we can infer their movement and intensification or dissipation over the next few hours.

STATION MODELS More detailed knowledge of the conditions at particular locations can be obtained from **station models**. Well over a dozen weather elements, including temperature, dew point, and pressure, are represented on each station model. A complete station model contains information beyond the needs of most readers, so Figure 13–14 describes only the most important symbols. A more complete description of the surface station model, along with the basics of the 500 mb station model, is presented in Appendix C. (For simplicity the surface weather map in Figure 13–13 does not use complete station models.)

The surface station model indicates cloud coverage by the amount of shading inside the central circle. A completely open circle indicates cloud-free conditions, a fully darkened one represents complete overcast, and intermediate amounts of shading correspond to varying fractions of cloud cover. A line extending from the circle with tick marks or flags at the end is a wind "arrow," or "barb," showing the wind speed and direction.

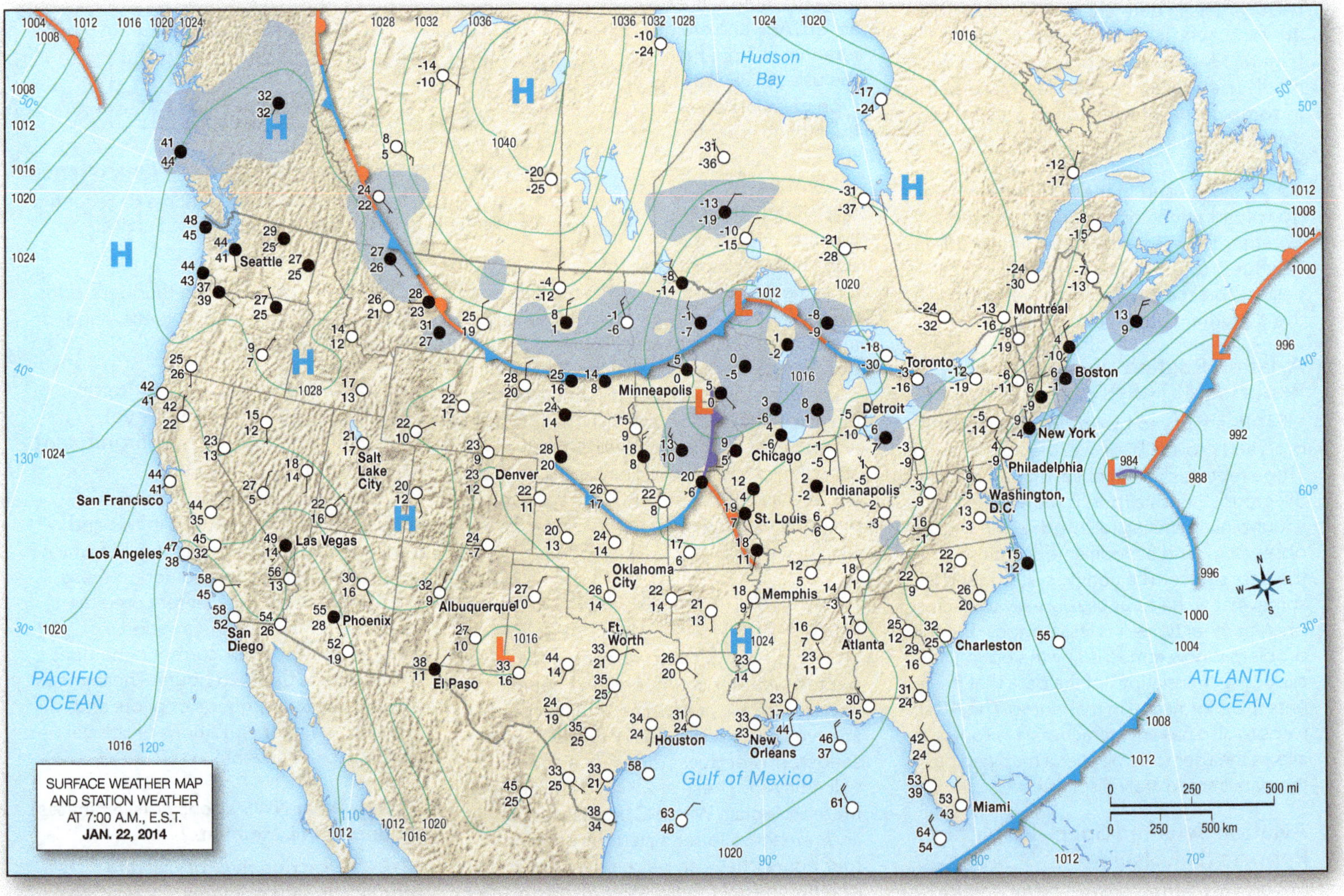

▲ **FIGURE 13–13 Surface Map.** A typical surface weather map, for January 22, 2014.

Simplified surface station model

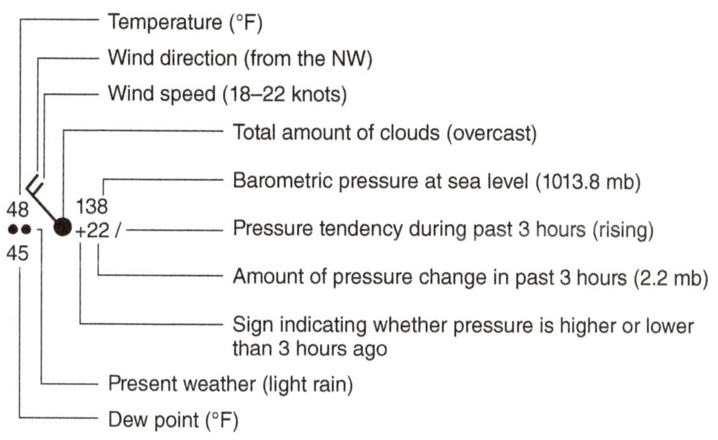

Total sky cover

- No clouds
- 1/8
- Scattered
- 3/8
- 4/8
- 5/8
- Broken
- 7/8
- Overcast
- Sky obscured

Wind entries

	Miles (statute) per hour	Knots	Kilometers per hour
	Calm	Calm	Calm
	1–2	1–2	1–3
	3–8	3–7	4–13
	9–14	8–12	14–19
	15–20	13–17	20–32
	21–25	18–22	33–40
	26–31	23–27	41–50
	32–37	28–32	51–60
	38–43	33–37	61–69
	44–49	38–42	70–79
	50–54	43–47	80–87
	55–60	48–52	88–96
	61–66	53–57	97–106
	67–71	58–62	107–114
	72–77	63–67	115–124
	78–83	68–72	125–134
	84–89	73–77	135–143
	119–123	103–107	144–198

Pressure tendency

- Rising, then falling
- Rising then steady; or rising, then rising more slowly
- Rising steadily or unsteadily
- Falling or steady, then rising; or rising, then rising more quickly

(Barometer now higher than 3 hours ago)

- Steady, same as 3 hours ago
- Falling, then rising, same or lower than 3 hours ago
- Falling, then steady; or falling, then falling more slowly
- Falling steadily, or unsteadily
- Steady or rising, then falling; or falling, then falling more quickly

(Barometer now lower than 3 hours ago)

Common weather symbols

- Light rain
- Moderate rain
- Heavy rain
- Light snow
- Moderate snow
- Heavy snow
- Light drizzle
- Ice pellets (sleet)
- Freezing rain
- Freezing drizzle
- Rain shower
- Snow shower
- Fog
- Haze
- Smoke
- Thunderstorm
- Hurricane

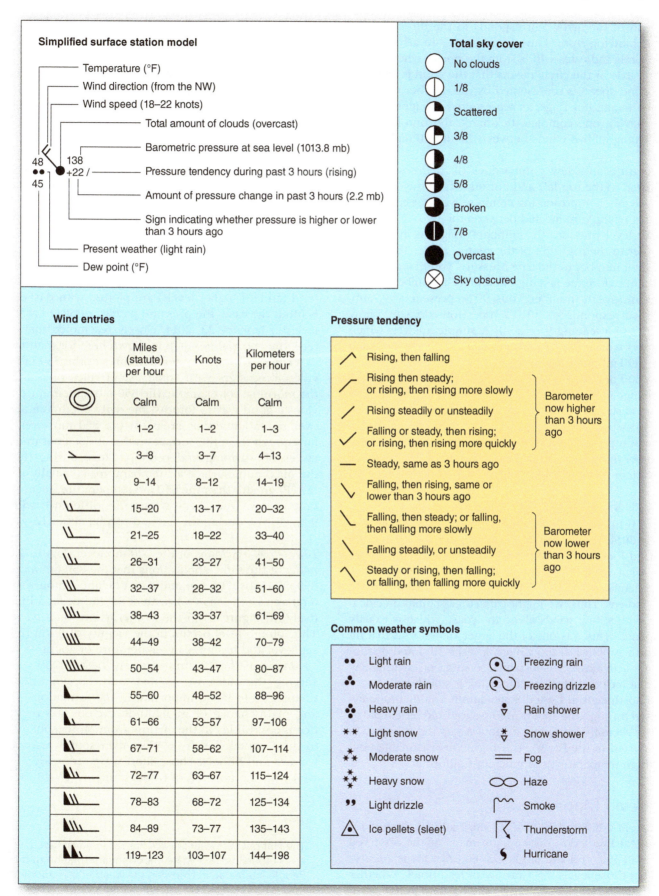

▲ **FIGURE 13–14 Station Models.** The arrangement of a surface station model and some important symbols.

The free end of the arrow corresponds to the direction that the wind is blowing *from*. Thus, for instance, an arrow at the top of the circle indicates winds from the north, whereas one on the right side of the circle means that the wind is from the east. The wind speed is represented by the number of full or half tick-marks, and/or flags, as represented in Figure 13–14. An arrow having one complete tick-mark and one half tick-mark, for example, has winds between 15 and 20 mph (20 to 32 km/hr).

Temperature and dew point values (in degrees Fahrenheit) are listed to the top left and bottom left of the circle, respectively. Symbols representing common weather conditions (such as rain or fog) are located between the two.

The sea-level pressure (in millibars) is given in a shorthand manner to the top right of the central circle. To convert the three-digit number to the true pressure, you first assume a decimal point before the last digit and place a number 9 or 10 at the beginning of the number. Thus, 997 represents 999.7 mb, while 104 corresponds to 1010.4 mb. How do you know whether to add a leading 9 or a 10? A simple answer, which nearly always works, is to add the number that yields a value closest to 1000 mb (in other words, add 9 if the coded value is more than 500 or add 10 if it is less than 500).

The change in pressure (pressure tendency) during the past three hours is shown just to the right of the circle. Again, it is necessary to assume a decimal point before the last digit so that −10 indicates a pressure drop of 1.0 mb. A symbol to the right of the number indicates in a qualitative sense how the pressure has been changing (for example, initially rising but then falling).

INTERPRETING A SURFACE WEATHER MAP Let's refer back to Figure 13–13, which shows the surface weather map for January 22, 2014. Earlier in this chapter Figure 13–3 presented a generalized forecast for the country for that date; this figure shows us what really happened. A very well developed low-pressure system is shown off the Atlantic coast with a fairly strong pressure gradient. Thus, we might expect substantial precipitation along with strong winds along the coast, which is exactly what happened. This was one of the larger snowstorms that hit eastern North America in the winter of 2013–2014 and disruption of daily activities was substantial, coming off a three-day weekend that included Martin Luther King Jr.'s birthday.

In the north-central United States another midlatitude cyclone passed through the area bringing snow and bitter cold to much of the region. Meanwhile a Santa Ana wind (Chapter 8) situation existed in the Far West, bringing unseasonable (and enviable) high temperatures to much of California.

Upper-Level Maps

Twice a day, at 0000 and 1200 UTC, the National Centers for Environmental Prediction disseminate maps of the observed 850, 700, 500, 300, and 200 mb levels. Forecast maps are also produced for those levels for a variety of lead times. Whether the maps represent current or predicted conditions, each provides its own unique combination of advantages for the weather analyst.

850 mb MAPS The 850 mb level resides at an average height of 1.5 km (1 mi) above sea level. Because friction is often considered to be negligible at heights of about 1.5 km above the surface, gradient or geostrophic flow can exist at this level over terrain with elevations near sea level. On the other hand, much of the Rocky Mountain region of the western United States and Canada has elevations above this height, so in those areas the **850 mb map** actually represents near-surface conditions. At high elevations, friction retards the wind and the air flows somewhat across the height contours.

Heights of the 850 mb level are plotted with solid lines, analogous to the isobars found on surface maps. Contours are spaced at 30 m intervals and labeled in units of decameters (10 m). Thus, a value of 150 represents a height of 1500 m. Though not specifically drawn, frontal boundaries are distinguishable at the 850 mb level, where the height contours are packed closely together. Air temperatures (in °C) at the 850 mb level (and all higher levels) are plotted with dashed lines. As is often the case, the pressure pattern depicted at the 850 mb level on January 22, 2014, closely resembled that at the surface; thus no 850 mb map is shown here for that date.

The distribution of temperatures at the 850 mb level provides forecasters with some useful rules of thumb. During the morning, for example, the 850 mb temperature often provides a good way to forecast the daily maximum temperature over nonmountainous areas. At the 850 mb level, the air is far enough from the surface so that it does not undergo daily cycles of warming and cooling. Thus, during the summer, the maximum surface air temperature is usually about 15 °C (27 °F) greater than the 850 mb temperature, regardless of the time of day. During the winter, the difference will be about 9 °C (16 °F); during the fall and spring, about 12 °C (22 °F).

700 mb MAPS Maps of the 700 mb level (Figure 13–15) have many of the same applications that the 850 mb maps do. Like 850 mb maps, **700 mb maps** plot height contours (in decameters) and isotherms (°C), with solid and dashed lines, respectively (in Figure 13-15 the isotherms are omitted for clarity). The 700 mb level occurs in the vicinity of 3 km (2 mi) above sea level and has a mean temperature of about −5 °C (23 °F).

Maps of the 700 mb level are best for observing the short waves that were shown in Chapter 10 to be so important in the formation and maintenance of midlatitude cyclones. Notice the wrinkles in the airflow from Minnesota to the Dakotas and from Wisconsin southward to Iowa and Illinois. Those short waves coincide well with the position of fronts in Figure 13–13.

Video
Forecasting Upper-Level Winds
http://goo.gl/7goQvk

DID YOU KNOW?

The National Weather Service has taken advantage of social media to help disseminate important information to the public. People interested in following the weather can track local NWS offices and other agencies such as the Storm Prediction Center via Facebook and Twitter.

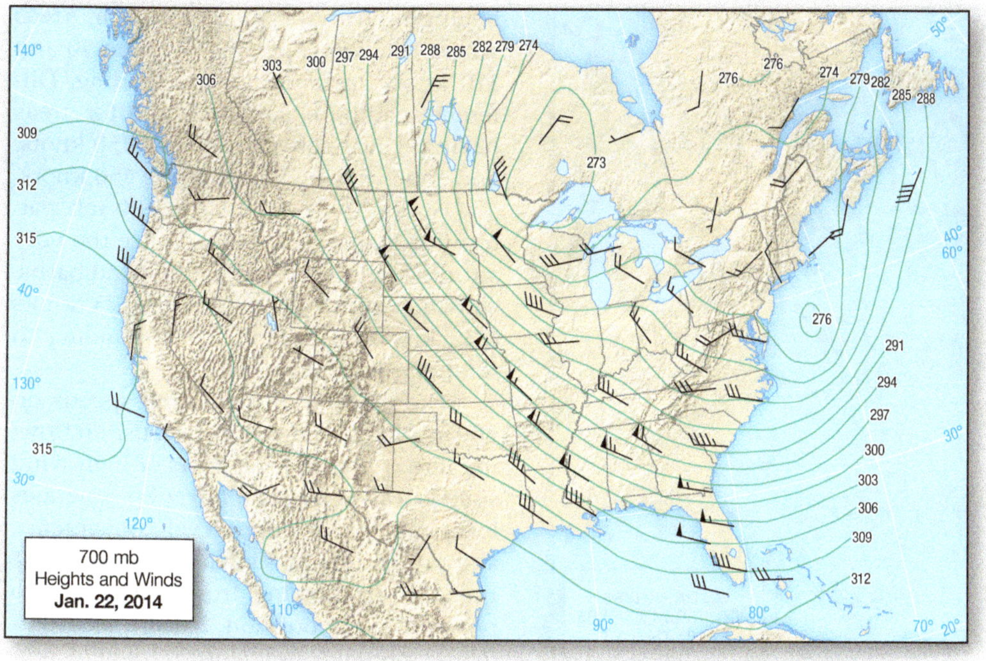

▲ **FIGURE 13–15** **700 mb Map.** The solid lines represent the altitude of the 700 mb level in tens of meters above sea level on January 22, 2014. Wind speeds are indicated by barbs as shown in Figure 13-14.

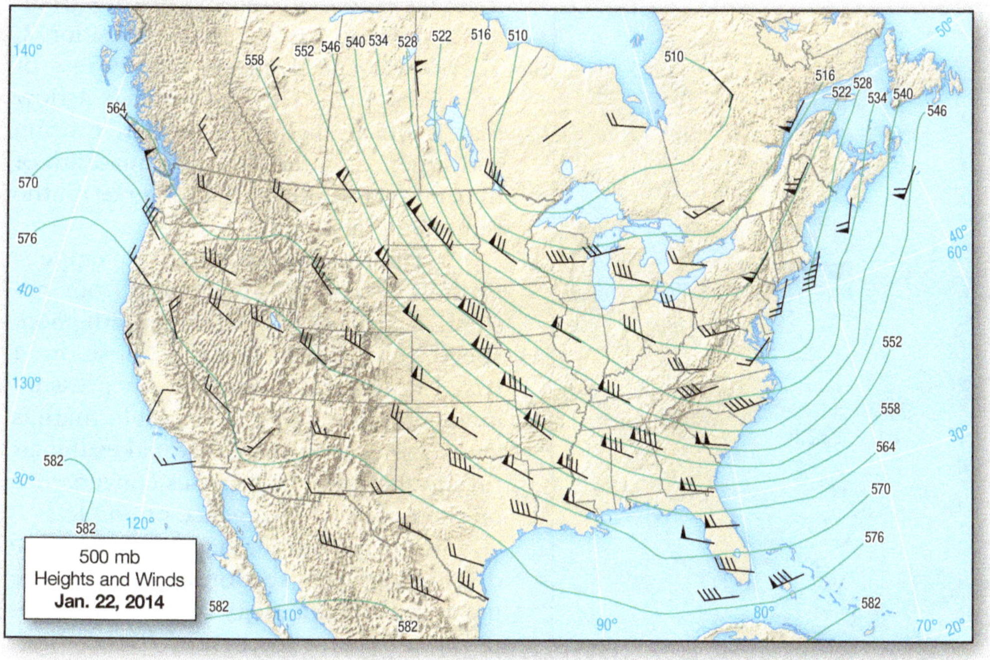

▲ **FIGURE 13–16** **500 mb Map.** Maps of the 500 mb level are often used to observe short waves and to predict the movement of air mass thunderstorms. On this day two short waves can be seen in the north-central United States.

Maps of the 700 mb level are also particularly valuable in predicting the movement of air mass thunderstorms, which usually move with about the same velocity as the 700 mb winds.

500 mb MAPS The **500 mb map** (Figure 13–16) is commonly used to represent conditions in the middle atmosphere. The globally averaged height of the 500 mb level is about 5.6 km (18,000 ft) above sea level, and the mean temperature is about −20 °C (−4 °F). Because the mean pressure at sea level is nearly 1000 mb, about half the mass of the atmosphere exists below the 500 mb level and half above.

Compare Figures 13–13, 13–15, and 13–16 and observe how the low-pressure systems off the middle Atlantic states and the northern Plains seen at the surface gradually have faded out at the 500 mb level. This is a typical pattern in which surface maps often feature the presence of cyclones and anticyclones but at higher levels we usually see more general patterns of large troughs and ridges. Notice also in the 500 mb map the trough axis (Chapter 10) extending southeastward from about Chesapeake Bay. As air flows around the axis it undergoes a decrease in vorticity, thus helping create the low-pressure system observed near the surface in Figure 13–13. In other words, nature was doing exactly what we expected it to do in this instance. Out west a ridge exists over the eastern Pacific along the West Coast. Downwind of the ridge axis there is an increase in vorticity causing upper-level convergence and the creation of the surface high-pressure system over the Great Basin. Hence the 500 mb pattern can be associated with the Santa Ana winds over southern California.

For decades meteorologists have made special use of 500 mb maps. For example, there is a certain type of pattern at the 500 mb level called an *omega high*. This feature, so named for its resemblance to the Greek letter Ω, often signals that the upper-level pattern is likely to change only slowly for several days. This pattern is clearly evident in Figure 13–17, with a single omega high covering the central part of North America and troughs over the western and eastern regions. The height contours of the 500 mb maps are spaced at 60 m intervals instead of the 30 m intervals used for the 850 and 700 mb maps.

300 AND 200 mb MAPS Lying near the tropopause, the 300 mb (approximately 9 km above sea level and having mean temperatures of about −45 °C, or −50 °F) and 200 mb levels (12 km; −55 °C, or −65 °F) have the strongest jet streams. During the colder months, the **300 mb map** works best for identifying the jet stream; during the summer, the **200 mb map**

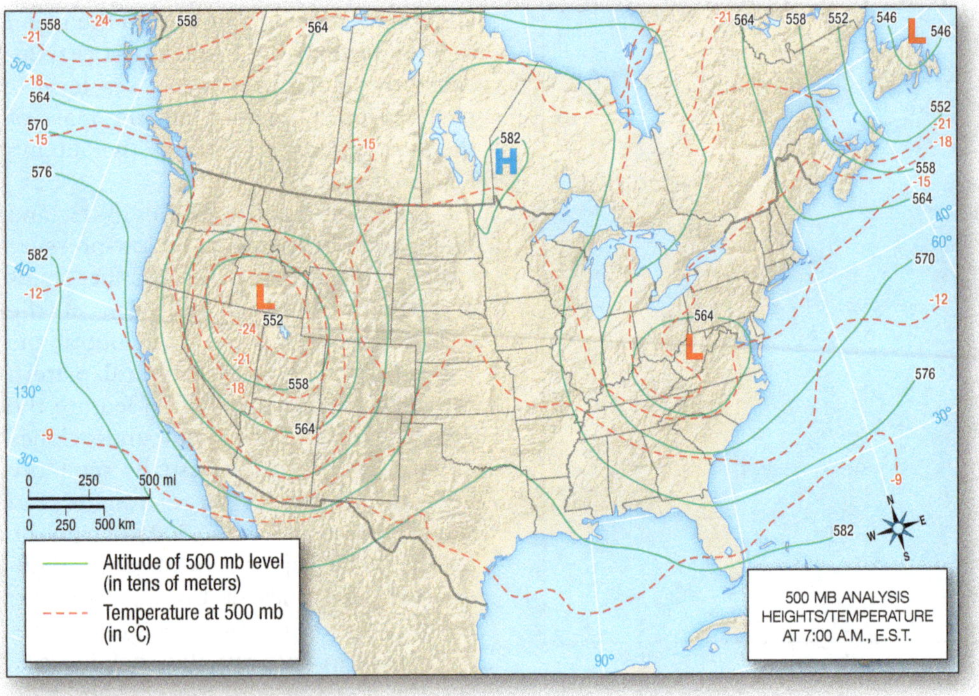

▲ **FIGURE 13–17 Omega Highs.** These 500 mb patterns resembling the Greek letter omega often indicate unchanging conditions in the near future.

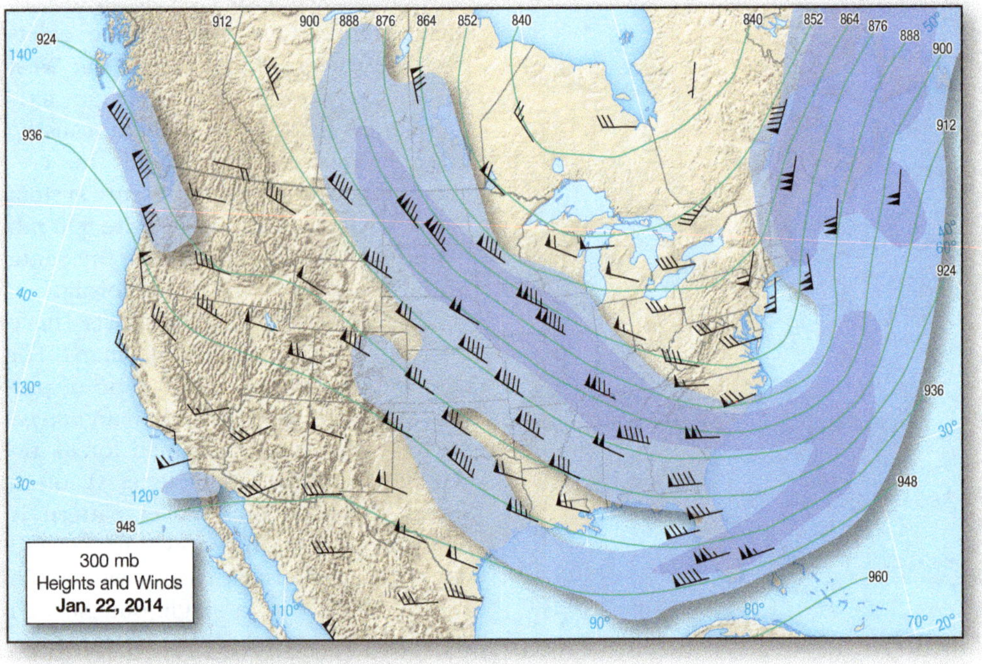

▲ **FIGURE 13–18 300 mb Map.** Maps of this level are useful for observing conditions near the top of the troposphere. Up to three levels of color shading are used to indicate wind speeds in excess of 70, 100, and 150 knots.

is best. (In this section, all references to the 300 mb level pertain equally to the 200 mb level.) Rossby waves show up best on the 300 mb maps, which makes the charts useful for determining the rate at which the waves are likely to migrate downwind (or in rare instances, upwind).

In addition to height contours (at 120 m intervals) and isotherms, 300 mb maps also plot **isotachs**, which are lines of equal wind speed. These are drawn at intervals of 20 knots, beginning with the 10-knot isotach. Areas where the wind speed is between 70 and 110 knots are indicated by shading. Different shading is superimposed for areas having winds between 110 and 150 knots, and zones having winds above 150 knots.

Figure 13–18 shows the 300 mb pattern for January 22, 2014. Notice the very strong winds indicated by the wind barbs, showing winds in excess of 125 knots (232 km/hr, 144 mph) corresponding to the polar jet stream.

The plotting of high wind speeds on 300 mb maps is important because air flowing into or out of local areas of high wind speeds (called *jets,* or *jet streaks*) generates local regions of upper-level convergence and divergence, as shown in Figure 13–19. Imagine air flowing from position 1 to position 2. At position 2 there is a particular pressure gradient that would normally cause a particular wind speed. But a parcel entering that point comes in at a slower speed than expected at position 2 because the south–north pressure gradient at position 1 is less than at position 2. The slower speed means slightly less of a Coriolis force. The end result is a flow that is offset slightly to its left, causing convergence on the left-hand quadrant of the entrance region and divergence on the right-hand quadrant.

In Figure 13–19 the opposite happens in the exit region. Air moving from position 3 arrives at position 4 with more speed than would exist at position 4 based only on the south–north pressure gradient at position 4. As a result, there is an excess Coriolis force that makes the air turn to the right. This causes convergence on the right and divergence on the left.

Studying weather maps such as those discussed in this section is probably the most important activity involved in compiling a general forecast. What such an exercise in this case would tell us is that on January 22, 2014, a very large trough extended down from Canada to Florida. This trough imported cold air to virtually all of North America east of the Rocky Mountains and created patterns that fostered the development of midlatitude cyclones over the north-central and northeastern United States. Along the Pacific Coast a ridge led to the development of warm Santa Ana wind conditions. Forecasters across the country all saw the same patterns, but adapted their analysis to relate those patterns to explaining and predicting the weather for their local areas. While these forecasting activities may have

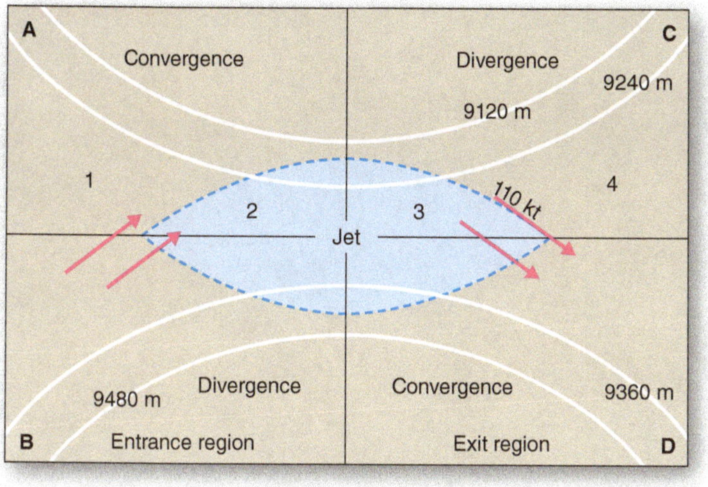

▲ **FIGURE 13–19 Jet Streaks.** Map view of a jet streak. Air flowing into and out of jet streaks (areas of locally fastest winds) creates patterns of upper-level convergence and divergence. Air entering the jet streak sets up convergence along its left flank (quadrant A), and divergence on its right (B). Exiting the jet streak, there is divergence to the left (quadrant C) and convergence to the right (D).

centered on fairly extreme conditions during this period, they were all based on the principles and resources normally used by meteorologists.

CHECKPOINT

13.17 Draw a station model, showing where the temperature, dew point, wind speed and direction, cloud conditions, and sea-level pressure are indicated.

13.18 How are maps of the 850 mb level used? The 700 mb level? The 500 mb level?

Satellite Images

Anybody who watches the weather segment of any news broadcast is certainly familiar with radar and satellite imagery. **Visible images** (Figure 13–20a) view the atmosphere the way an astronaut in space would, simply by registering the intensity of reflected shortwave radiation. Obviously, these images are available only during the daytime.

Infrared images (Figure 13–20b) are based on measurements of longwave radiation *emitted* (not reflected) from below. If dense clouds are present, the source of the radiation will be the cloud top; otherwise, the surface and lower atmosphere supply most of the upwelling radiation. Cumuliform clouds result from condensation associated with the adiabatic cooling of rising air. A deep cloud formed by air parcels rising great distances will therefore have a lower cloud-top temperature than will a midlevel cloud formed by just a kilometer or two of ascent. It follows that a satellite sensor will receive

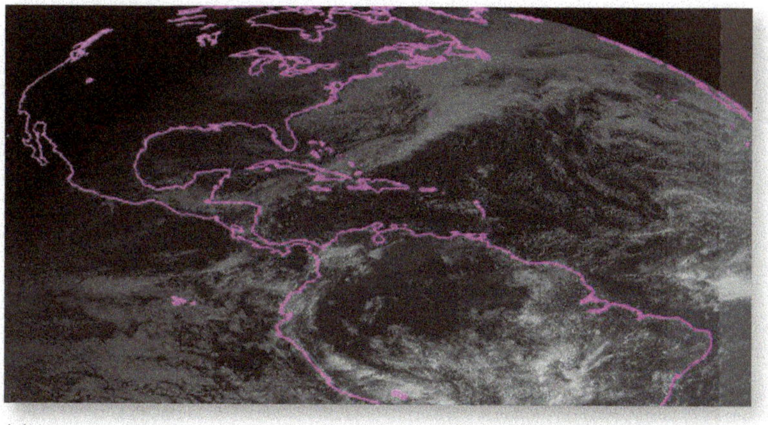

(a)

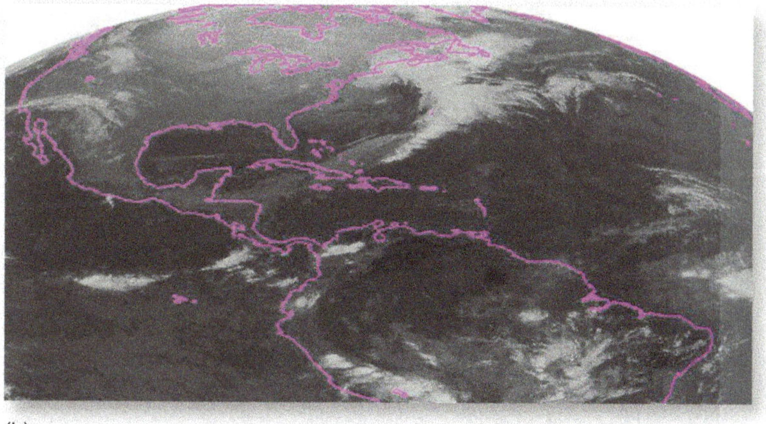

(b)

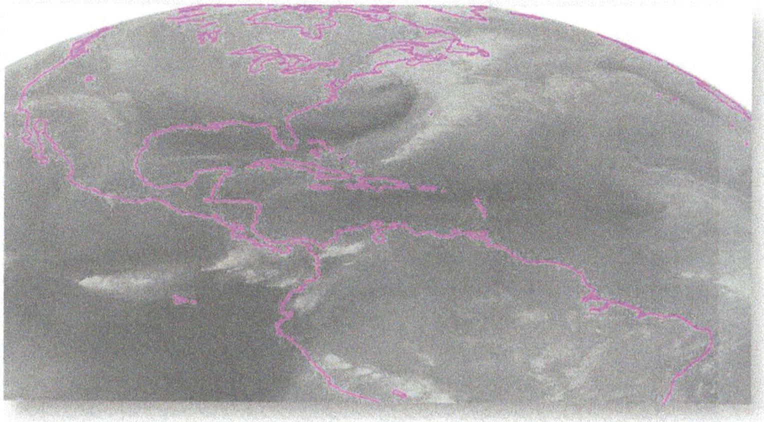

(c)

▲ **FIGURE 13–20 Satellite Images.** A (a) visible, (b) infrared and (c) water vapor image obtained from the GOES-13 satellite at 8:45 a.m. Eastern Standard Time on January 22, 2014. Visible images are based on sunlight reflected off cloud tops, and can therefore only be obtained for regions having daytime conditions. Infrared images are based on the intensity and wavelengths of radiation actually emitted by clouds. Water vapor images show whiter areas where either substantial cloud cover or high water vapor contents exist.
Source: Space Science and Engineering Center, University of Wisconsin-Madison.

less radiation from a tall cloud than a low cloud and less from a low cloud than from a cloud-free region. Satellite images show the colder (higher) cloud top as whiter than warmer (low) clouds. Clear areas are very dark on infrared images. In Figure 13-20b we see the extensive area of precipitation for the

January 22, 2014, storm. The storm left large snow accumulations in the eastern states, but at this point in time the most intense precipitation activity is off the Atlantic coast.

Water vapor images (Figure 13–20c) provide a unique and often beautiful perspective on the atmosphere. Water vapor is a very effective radiator at wavelengths near 6.7 μm. Relative humidities above about 50 percent result in a high output of radiation in this part of the spectrum, and the sensors translate high values of this radiation into bright regions on the imagery. Water vapor images are particularly useful for tracking the flow of moisture across wide regions and helping identify the location of frontal boundaries.

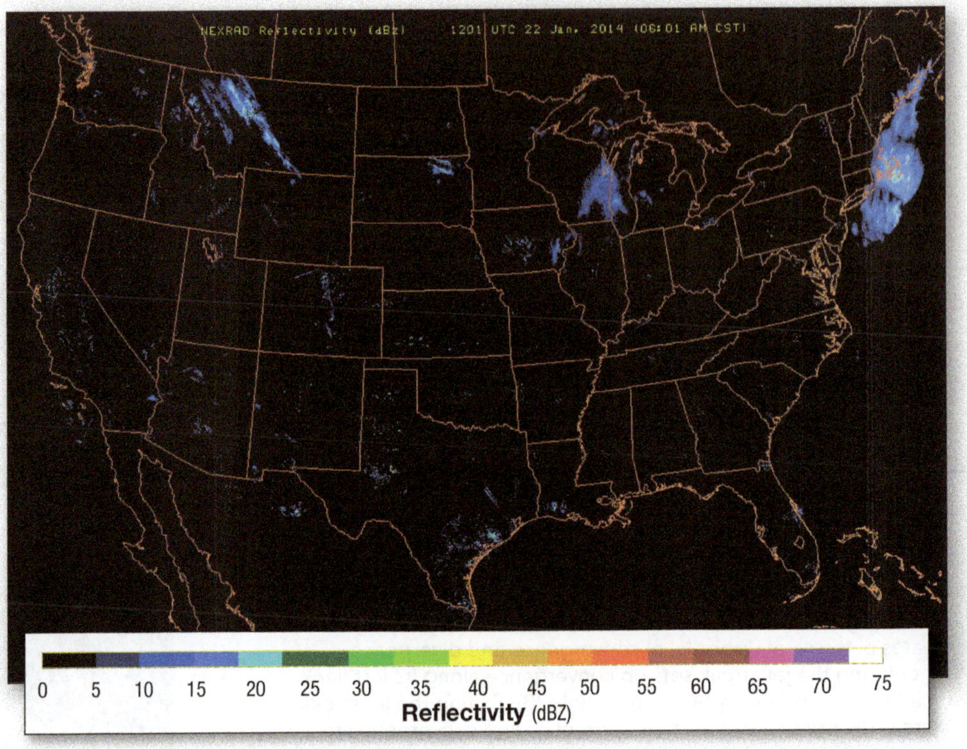

▲ **FIGURE 13–21 Radar Map.** Composite maps like this assemble data obtained from regional radar units across the country and combine the information onto a single map.

Radar Images

Radar images generated by conventional systems have been extremely valuable for routine weather analysis for decades. Radar observes the internal cloud conditions by measuring the amount of radiation backscattered by precipitation (both liquid and solid). A transmitter sends out brief pulses of electromagnetic energy with wavelengths on the order of several centimeters. A receiver records the intensity of the echoed pulses (indicative of the number and size of droplets and crystals) and the time elapsed between pulse transmission and return (which indicates the distance to the scattering agent).

The radar continually emits these pulses as the transmitter/receiver rotates 360 degrees, giving a two-dimensional representation of the precipitation pattern surrounding the unit. After each rotation, the transmitter angle is increased slightly, and the radar scans a higher slice of the atmosphere. This procedure is repeated until a large volume of the surrounding atmosphere has been scanned, and the meteorologist can observe the distribution and the relative intensity of precipitation. The information is then displayed in color-coded map form. A sequence of these images can be put together into a loop that shows the movement and changes in weather activity.

Each radar unit covers distances up to about 400 km (250 mi). The National Weather Service assembles the returns from all the radar sites in the national network and compiles the information onto composite maps of the 48 contiguous United States and southern Canada (Figure 13–21).

Doppler radar has been a staple for forecasters since the 1980s. As discussed in Chapters 7 and 11, all U.S. radar stations have now updated to the newest version, *dual-polarization radar*. These units transmit pulses of energy with both horizontal (i.e., nearly parallel to the surface) and vertical pulses that provide additional information on precipitation characteristics, such as droplet size, the relative mixture of rain and snow, and precipitation rate.

While it is relatively easy to determine the big picture of local activity from a radar image, a meteorologist has to beware of potential errors in its use. For instance, the curvature of Earth's surface causes a horizontally emitted beam of radiation to assume greater heights above the surface as it moves farther from the transmitter. That is why the large area of precipitation evident in Figure 13-20b is mostly missing from Figure 13–21. In essence, the radar beams are overshooting the tops of the storm clouds due to Earth's curvature.

CHECKPOINT

13.19 Which type of satellite imagery would be most useful in detecting differences in the altitude of cloud tops?

13.20 How is a radar composite map produced and how could it be useful in forecasting?

Thermodynamic Diagrams

13.9 Explain the use of thermodynamic diagrams in forecasting.

The maps and images previously described provide two-dimensional views of atmospheric conditions, but they fail to provide detailed vertical information. Vertical profiles of temperature and dew point data observed by radiosondes are plotted on **thermodynamic diagrams** (also called *pseudo-adiabatic charts*). The simplest thermodynamic diagram is the Stuve chart (Figure 13–22), on which the air temperature is scaled along the horizontal axis and the pressure on a nearly logarithmic vertical axis. The straight, solid lines that slant upward to the left are called *dry adiabats*. These show the rate of temperature change for an unsaturated parcel of air that is lifted or lowered (in other words, they plot the dry adiabatic

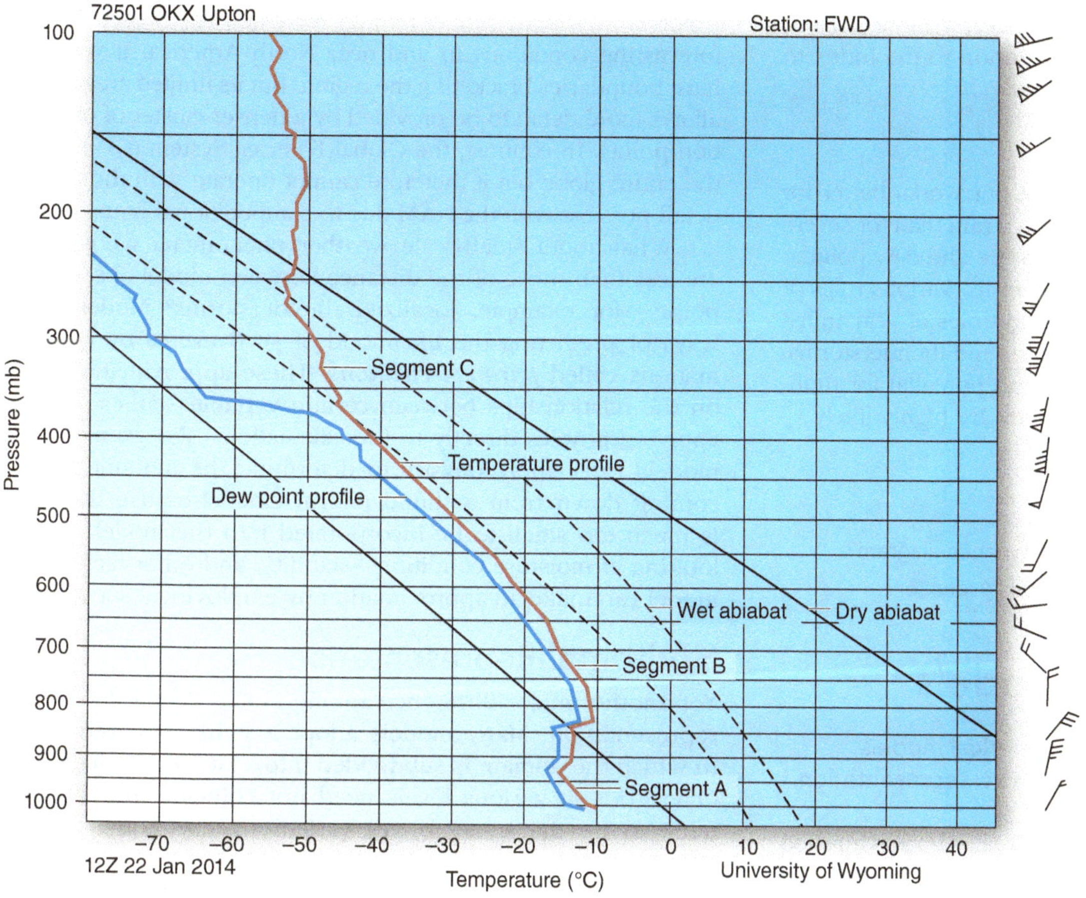

◂ **FIGURE 13–22 Thermodynamic Diagram.** An example of a sounding on a Stuve diagram. The heavy red line on the right plots temperature; the heavy blue line on the left shows the dew point profile. Wind barbs on the right of the diagram depict wind velocities at various levels. Dashed lines representing the mixing ratio are included, as are dry and wet adiabats.

Video
Forecasting
Thunderstorms

http://goo.gl/T26p2r

lapse rate). The dashed, slightly curved lines are called *wet adiabats,* showing temperature changes experienced by a rising saturated parcel. Lines depicting the mixing ratio are also provided, as previously discussed in *Box 5–3, Forecasting: Vertical Profiles of Moisture.* Wind barbs at the right of the diagram tell us about wind speed and direction at different altitudes.

Figure 13–22 shows an actual sounding for Upton, New York, on the morning of January 22, 2014. The heavy lines represent the temperature and dew point profiles (the temperature profile is the one on the right—which must be the case because the temperature is always equal to or greater than the dew point).

In this case the temperatures and dew points are nearly equal to each other from near the surface up to about the 400 mb level. When the two are equal, we know the air is saturated. In this case the two lines don't touch each other but come close to doing so. Given the fact that heavy snow was falling at this time, we can assume the air was in fact saturated for much of the atmosphere and that the difference in air and dew point temperature plotted was the result of minor instrument error.

The steepness of the temperature profiles relative to the dry and wet adiabats indicates the static stability for any portion of the atmosphere. Where the temperature decreases more rapidly with height than the DALR, the air is absolutely unstable. Where the temperature drops less rapidly with height than adjacent wet adiabats (segment C), the air is absolutely stable. Temperature lapse rates intermediate between the two adiabatic lapse rates (segment B) indicate conditionally unstable conditions, as we discussed in Chapter 5. But as this sounding shows, different stability conditions can exist at different levels. For that reason several numerical indices have been developed that incorporate temperatures and dew points at various levels to indicate the overall likelihood that air will behave as either a stable or unstable layer. The lifted index and the K-index are two such summary measures.

Lifted Index

The **lifted index** was developed in the mid-1950s. The index is computed by imagining a parcel at the predicted high temperature for the day is lifted adiabatically to the 500mb level. The lifted index is the difference between the sounding temperature at 500 mb and the calculated parcel temperature. If the parcel is colder, the index is positive, reflecting stable conditions. Unstable conditions will have negative index values because the parcel is warmer. The magnitude and sign of the values together indicate the potential for thunderstorms; negative values indicate sufficient water vapor and instability to trigger thunderstorms. More specifically, lifted index values between −2 and −6 indicate a high potential for thunderstorms, whereas values lower than −6 suggest a threat of severe thunderstorms. The lifted index calculated for Upton, New York, on January 22, 2014, was a positive value. Of course, there was no likelihood of thunderstorms on this day, as was evident from the consistent snowfall occurring, so the forecasters probably never bothered to actually make note of the lifted index. On a warm sunny day a few months later, however, forecasters might pay close attention to the index to help determine the threat of such storms.

K-Index

The **K-index** is similar to the lifted index but works better for predicting air mass thunderstorms and heavy rain than for severe weather. The index uses values of temperature and dew point at the surface and the 850, 700, and 500 mb levels. Various rules of thumb for different geographic regions and times of year translate K-values to the probability of heavy rains and thunderstorms. In general, K-values less than 15 indicate no potential for thunderstorms; values above 40 suggest that they are highly likely.

CHECKPOINT

13.21 What data are shown on a thermodynamic diagram?

13.22 What are the lifted index and K-index useful for?

Numerical Forecast Models

13.10 Describe the characteristics of numerical models, including scale considerations and horizontal representation.

During the past half century, weather forecasters have relied on several generations of computers and a variety of different models for guidance. As computers have increased in speed and capacity, the models have become increasingly complex, always straining the limits of computer power in an effort to achieve greater realism and accuracy. Even with today's computers, numerous compromises and approximations are necessary. In fact, it might be said that the spectrum of models arises not so much from differences in purpose or theory, but rather from differing approaches to the basic problems of abstraction and simplification. The result is tremendous variety in both the details and major features of numerical models. In this section we first discuss the major features of numerical models, using the primary National Centers for Environmental Prediction (NCEP) operational models as examples. We then describe some methods for assessing forecast quality.

Scale Considerations

Thinking first about large-scale features, perhaps the most fundamental difference among models concerns the model **domain**, the region of the globe to be represented. The North America Mesoscale Model (NAM), for example, is used just for forecasting conditions in and near North America. It will thus have boundaries bracketing the region. But its limited areal extent allows more detail to be provided by a denser cluster of observation points. In contrast, the Global Forecast System (GFS) covers the entire globe but it therefore cannot operate with the level of detail possible with the NAM due to computational limitations.

What about small-scale weather phenomena whose sizes are less than those of the distance between model observation points—for example, localized thunderstorms? Models have ways of predicting the likelihood of such events by approximations called ***parameterizations***. These approximations rely on the relationships between certain variables rather than on applying physics directly to the calculations. For example, the models can't apply physics to determine the amount of rain coming down from a thunderstorm cloud because the scale is much too small to be incorporated into the model. But by looking at moisture conditions, stability, and other factors, the model can make an approximation by employing a set of rules.

Horizontal Representation

Yet another major difference among models is the horizontal representation. Many models adopt a **grid representation**, in which the domain is subdivided into a lattice of grid points. (The grids at various levels need not coincide, nor do they need to be the same for all variables.) The equations are solved only at the grid points.

The alternative type of horizontal representation is called a *spectral representation* (so named because the waves are analogous to those of the radiation spectrum.) Variables are represented as a series of "waves" in space, each having a characteristic wavelength. For example, the model has waves that correspond to Rossby waves, repeating just a few times around a parallel of latitude, extending across the globe. Superimposed on these are other waves representing smaller-scale and larger-scale variations. To obtain the value for a particular variable at a point, its various wave functions are summed over all wavelengths.

Video
Uncertainty in Numerical Models
http://goo.gl/NvqnYP

CHECKPOINT

13.23 Describe some of the important considerations with respect to model scale.

13.24 What are the two major types of representations of forecast models?

Summary

13.1 Summarize the basic procedures of weather forecasting.

- Forecasters rely on data from land- and sea-based stations, radiosondes, and satellites.
- Data from these sources are run through numerical models that produce output. Forecasters analyze the output and interpret it based on their experience and training.

13.2 Explain why weather forecasting is imperfect.

- Numerical models must analyze huge amounts of data that may be imperfect at the outset.
- Many additional factors make modeling efforts even more difficult. These factors include the absorption and emission of shortwave and longwave energy, phase changes of water, and the effects of friction near the surface.

13.3 Describe the main methods of weather forecasting.

- Climatological forecasts rely on past experience.
- Persistence forecasts are based on the notion that what is happening now will continue to happen in the short-term future.
- The analogue approach interprets current conditions in terms of what has happened previously under similar situations.

13.4 Describe the general types of forecasts.

- Quantitative forecasts try to predict the future value of some weather variable, such as temperature or accumulated precipitation.
- Probability forecasts tell us that some type of occurrence will happen with some level of probability, ranging from 0 to 100 percent.

13.5 Explain how meteorologists assess the quality of weather forecasts.

- Forecast quality differs from forecast value, in that the former expresses the degree to which the forecast resembled reality, whereas the latter indicates the usefulness of the forecast.
- Components of forecast quality include skill and bias.

13.6 List some major sources of data for weather forecasting.

- Agencies around the world obtain data from ground-based platforms, radiosondes, commercial aircraft, satellites, and radar.

13.7 Describe forecast procedures, including numerical modeling, and explain how meteorologists develop short-, medium-, and long-range forecasts and seasonal outlooks.

- Numerical modeling involves the assembling and correcting of data during the analysis phase.
- The major computation work is done during the prediction phase, which provides the output for the analysis phase.
- Different models are used, so forecasters must decide which is most likely to yield the most accurate results.
- Models are developed to predict conditions in the short term (several days), the medium-range (about two weeks), and for entire seasons.

13.8 Explain how meteorologists use weather maps and satellite images in forecasting.

- Maps are created that show conditions at the surface and particular pressure levels of the atmosphere. Each upper-level map has its own set of uses.
- The maps are supplemented by displays based on satellite and radar observations.

13.9 Explain the use of thermodynamic diagrams in forecasting.

- Thermodynamic diagrams are used to plot temperature and dew point data from the surface to well beyond the tropopause.
- They also provide the information needed to yield numerical indices that are helpful as a first step in predicting the likelihood of certain events, such as severe weather.

13.10 Describe the characteristics of numerical models, including scale considerations and horizontal representation.

- Numerical models differ from each other in terms of their general structure and the type of scales. For example, models used to predict further into the future might incorporate data from a much larger area than those used in shorter-term forecasts.

Key Terms

200 mb map *p. 439*
300 mb map *p. 439*
500 mb map *p. 439*
700 mb map *p. 438*
850 mb map *p. 438*
Advanced Weather Interactive Processing System (AWIPS) *p. 421*
analog approach *p. 424*
analysis phase *p. 429*
Atmospheric Environment Service (AES) *p. 427*
Automated Surface Observing System (ASOS) *p. 427*
Automated Weather Observation Systems (AWOS) *p. 427*
Canadian Meteorological Centre *p. 427*
chaos *p. 430*
Climate Prediction Center (CPC) *p. 431*
climatological forecasts *p. 424*
domain *p. 444*
ensemble forecasting *p. 430*
European Center for Medium-range Weather Forecasting (ECMWF) *p. 430*
forecast accuracy *p. 426*
forecast bias *p. 426*
forecast quality *p. 426*
forecast skill *p. 426*
forecast value *p. 426*
grid representation *p. 444*
infrared images *p. 441*
isotachs *p. 440*
K-index *p. 444*
lifted index *p. 443*
long-range forecasts *p. 431*
medium-range forecasts (MRFs) *p. 430*
Meteorological Service of Canada (MSC) *p. 420*
National Centers for Environmental Prediction (NCEP) *p. 427*
National Oceanic and Atmospheric Administration (NOAA) *p. 420*
National Weather Service (NWS) *p. 420*
North American Mesoscale Forecast System (NAM) *p. 430*
numerical weather forecasting *p. 424*
persistence forecasts *p. 424*
postprocessing phase *p. 429*
prediction phase *p. 429*
probability forecasts *p. 425*
qualitative forecasts *p. 425*
quantitative forecasts *p. 425*
radar images *p. 442*
radiosondes *p. 427*
Rapid Refresh (RAP) model *p. 430*
rawinsondes *p. 427*
regional weather centre *p. 420*
seasonal outlook *p. 432*
station models *p. 436*
surface maps *p. 436*
thermodynamic diagrams *p. 442*
visible images *p. 441*
water vapor images *p. 442*
Weather Forecast Offices (WFOs) *p. 420*
World Meteorological Centers *p. 427*
World Meteorological Organization (WMO) *p. 427*
zone forecast *p. 422*

Review Questions

1. Briefly describe some of the variables that complicate weather forecasting.
2. Describe the basic characteristics of climatological forecasts, persistence forecasts, the analog approach, and numerical forecasting.
3. What are the distinguishing characteristics of quantitative, qualitative, and probability forecasts?
4. Explain how weather data are obtained and disseminated to agencies across the globe.
5. What are radiosondes and rawinsondes? What other sources of upper-atmosphere information are available to forecasters?
6. What are model output statistics?
7. Describe the analysis, prediction, and postprocessing phases in numerical forecasting.
8. What are the primary characteristics of short-range, medium-range, and long-range forecasts? What types of information are needed to prepare the individual forecasts?
9. What is ensemble forecasting?
10. Describe the station model used for surface weather maps. How is the information presented on the station model? What measures must be used to convert the numerical data on the station model to real values?
11. Describe the characteristics of the 850 mb, 700 mb, 500 mb, 300 mb, and 200 mb maps that make each of them useful to forecasting.
12. Why are omega highs significant? Which map is best for identifying them?
13. Which upper-level weather map would you use to locate short waves?
14. What is an isotach?
15. Which maps are most useful for locating the polar jet stream? [Hint: recall that the polar jet is found at altitudes near 10km.]
16. Describe the three types of satellite images discussed in this chapter. What characteristics make them useful?
17. Explain what a thermodynamic diagram does and how it is constructed.
18. Describe how radar works and how its information is presented.
19. Describe the lifted index and the K-index. How are they valuable to forecasters?

Critical Thinking

1. What would have to happen to the data acquisition network for the analysis phase of forecasting to be bypassed?
2. Are further improvements in weather forecasting more likely to occur for large-scale phenomena or for smaller-scale events? Explain your answer.
3. Why will the climatological, persistence, and analog approaches never be entirely eliminated from the process of weather prediction?
4. If a forecast calls for a 70 percent chance of rain and no precipitation occurs, was the forecast actually wrong? What if this happens two days in a row? Ten days?

Problems & Exercises

1. During times of unusual or inclement weather, visit the websites listed below to obtain weather map, satellite, radar, and thermodynamic diagram information. Is the weather you are experiencing consistent with what you would have expected based on the information you obtained?
2. On a daily basis, make use of the available information from the websites listed below and make your own forecast (before reading the official forecast for your area). Then compare your forecast to that of the local weather office. Are your forecasts generally consistent with those of professional meteorologists?
3. Read the forecast discussion on the web page of your local weather service office each day. These discussions explain the meteorologists' reasons for making their particular forecasts. Determine how this elaborates on the zone forecast for the same area.
4. Go to the website www.nws.noaa.gov/organization.php and click on the National Weather Service office nearest to you. Make note of the 24-hour forecast and follow up the next day to see if the forecast was correct. Do this for extended forecasts as well. In general, do you find the forecasts to be accurate?

Visual Analysis

Examine the surface and 500 mb maps for January 6, 2014, and answer the following questions:

13.1. What kind of temperatures might be expected for the eastern and western United States? Explain.

13.2. Where are the areas most likely to experience precipitation currently and within the next 24 hours? Explain.

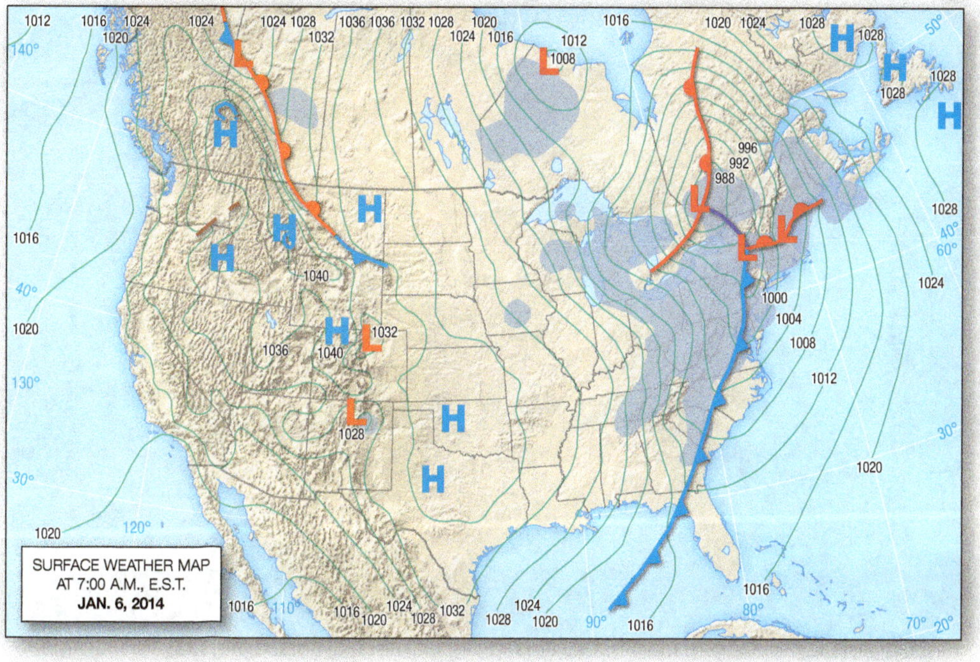

▲ Surface map for January 6, 2014.

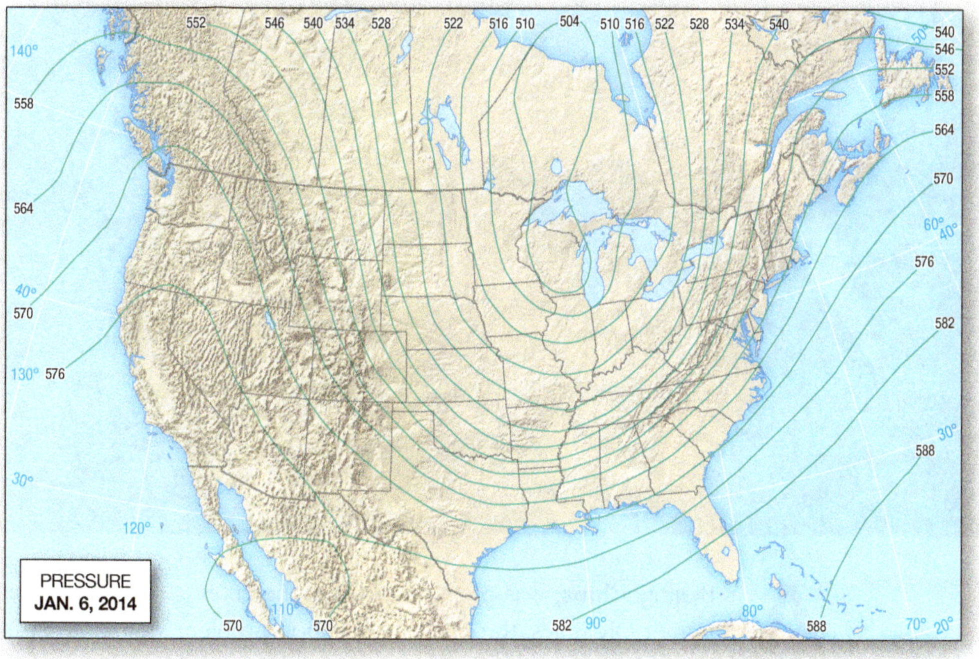

▲ 500 mb map for January 6, 2014.

CHAPTER

14 Human Effects on the Atmosphere

As Beijing, China, was preparing to host the 2008 Summer Olympics, the city's air pollution problem began to receive worldwide attention. Beijing's air situation has only gotten worse since then, so it is little surprise that its reputation for bad air has endured. But a 2014 report by the World Health Organization (WHO) has revealed that Beijing may not have the world's worst air—that dubious honor goes to New Delhi, India.

The WHO conclusion is based on the concentration of very small airborne particles, which are believed to be the most damaging type of pollutant to human health. While some officials in India question whether the air in New Delhi is indeed worse than that of Beijing, it is indisputable that the air quality is so bad that people are routinely seen on the streets wearing masks to filter out some of the material. New Delhi's air pollution is the result of both vehicular and industrial emissions. The situation is worse in winter, during the monsoon dry period (Chapter 8), and improves during the wet summer months when heavy rains wash out much of the pollution. New Delhi is not

◂ *Smog in Beijing, February 23, 2014. Air pollution levels in Beijing China often exceed air quality standards set by the World Health Organization. This episode had values nine times higher than recommended levels.*

alone among Indian cities for polluted air: 13 Indian cities are among the 20 cities having the world's worst air.

Worldwide, the WHO estimates that 3.7 million people under the age of 60 died from air pollution in 2012. And in many cities the situation is getting worse as industrialization increases, but few if any air pollution abatement measures are being put in place. Fortunately, the situation is not hopeless—at least for countries with the economic ability and determination to attack the problem—as exemplified by countries in North America and Europe.

Learning Outcomes

After reading this chapter, you should be able to:

14.1 Describe the different types of atmospheric pollutants.

14.2 Explain the role of atmospheric conditions in determining pollution concentrations.

14.3 Describe the efforts that have been made to reduce air pollution.

14.4 Explain the factors that produce urban heat islands and describe their effects on local weather.

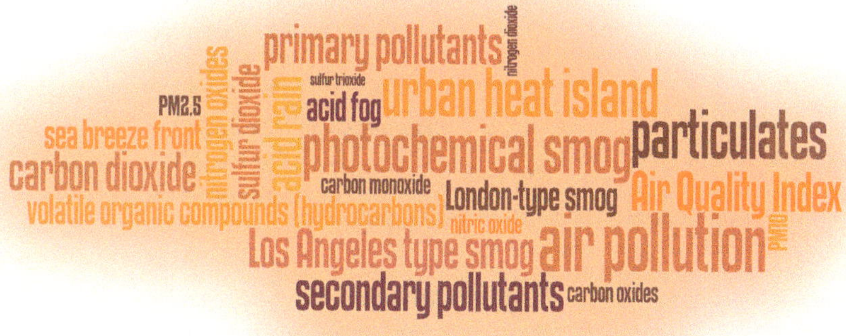

Atmospheric Pollutants

14.1 Describe the different types of atmospheric pollutants.

Video
Supercomputing the Climate
http://goo.gl/Wwl0o0

Nowhere is the air entirely pristine. Small suspended solids and liquids (called *particulates*) enter the atmosphere from natural and human sources. Likewise, many gases that are considered pollutants also arise naturally from processes such as lightning-induced forest fires and volcanic eruptions. Nonetheless, natural dilution and removal of these gases and particulates makes them relatively unimportant to the air quality experienced by most people. More important are the effects of human activity, especially in and around urban and industrial centers where anthropogenic emissions are concentrated into much smaller areas. In this chapter all references to **air pollution** will concern the introduction of undesirable gases and particulates by humans. The varying sources of particulates and other pollutants in the United States and their relative concentrations are shown in Figure 14–1. (We present some background on major pollution episodes past and present in *Box 14–1, Focus on the Environment and Societal Impacts: Severe Pollution Episodes.*)

In the most general sense, pollutants can be divided into two categories. Some, called **primary pollutants**, are emitted directly into the atmosphere. Others, called **secondary pollutants**, do not go directly into the atmosphere but result from one or more chemical transformations. Thus, a chemical emitted into the atmosphere may be innocuous in its original state but become a noxious gas or particulate after combining with other emissions or naturally occurring compounds. Several primary and secondary pollutants figure most prominently in the degradation of air quality.

Particulates

Particulates (also called *aerosols*) are solid and liquid materials in the air that are of natural or anthropogenic (human-made) origin. Though always small, particulates come in a wide range of sizes from about 0.1 to 100 μm (Figure 14–2). Some of the particulates are primary pollutants put directly into the atmosphere, whereas others are secondary pollutants formed by the transformation of preexisting gases or from the growth of smaller particulates into larger ones by coagulation.

SOURCES OF PARTICULATES Particulates introduced directly into the air can originate from natural fires, volcanic eruptions, the ejection of salt crystals by breaking ocean waves, and—as any sufferer of hay fever can tell you—by plants whose pollen is picked up and carried by winds. Human activities, especially those involving combustion, produce primary and secondary particulates.

Some secondary particulates form by the coagulation of gases. This process is most rapid when the humidity is high, which creates an interesting situation. Recall from Chapter 5 that water droplets in nature always form on condensation nuclei—with large, hygroscopic aerosols being particularly effective at attracting water and lowering the relative humidity needed for droplet formation. Thus, the introduction of particulates, especially large ones, promotes the formation of fog or cloud droplets. At the same time, high humidities favor the conversion of certain gases into secondary particulates, which in turn promote the condensation of water vapor into liquid droplets. As a result, humid areas with a high concentration of industrial activities can become foggy at relative humidities considerably below 100 percent. These processes worked together to make the London type of smog ubiquitous in eastern, industrial North American cities in previous years.

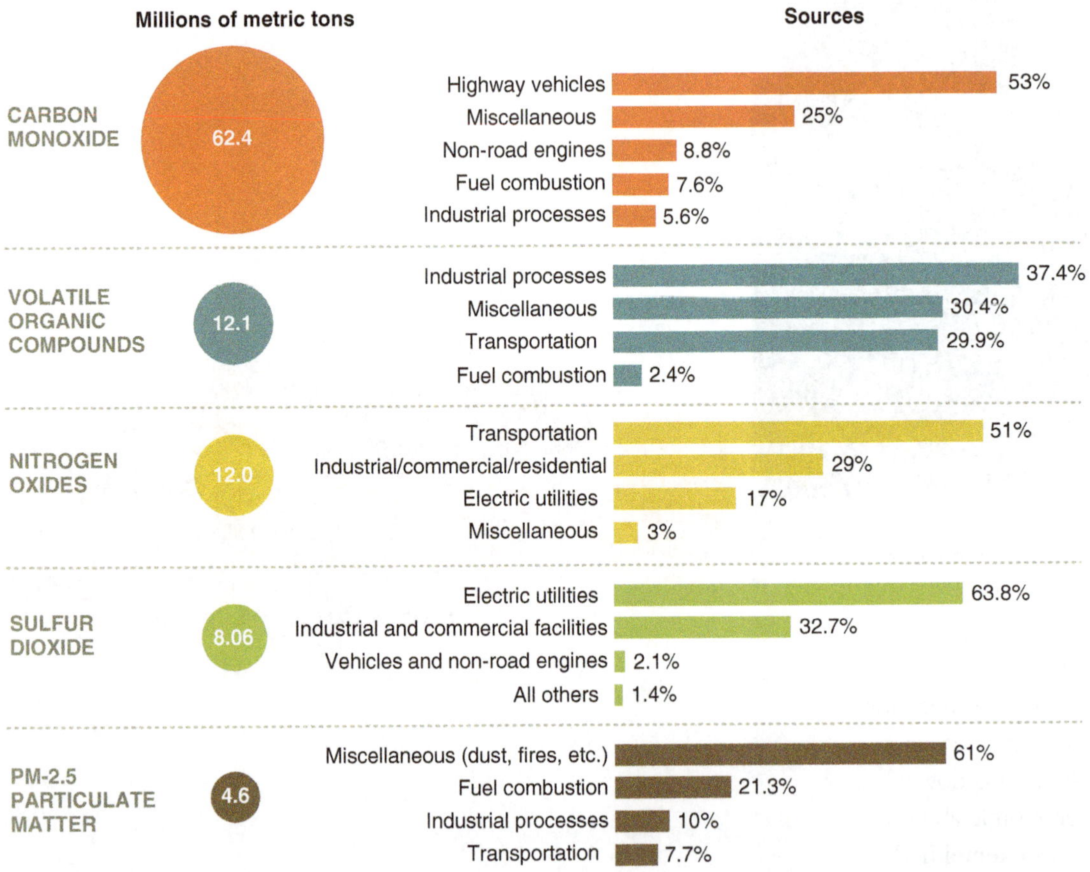

▲ **FIGURE 14–1 Pollution Sources.** The relative amounts of pollutants and their sources in the United States in 2011.

14–1 FOCUS ON THE ENVIRONMENT AND SOCIETAL IMPACTS

Severe Pollution Episodes

(a) (b)

▲ **FIGURE 14–1–1 Pittsburgh's Air Quality.** Like some other former industrial centers, Pittsburgh's air quality has undergone a huge improvement due to the closure of foundries and factories. These photos show (a) Pittsburgh in 1906 and (b) the same scene in 1986.

Though many of us live in places where poor air quality is a disturbing fact of life, much progress in solving the problem has been made in the developed world in recent decades, with the result that the most disastrous types of smog events are a thing of the past. Consider, for example, what is probably the most famous air pollution episode in recent history—the one that hit London, England, between December 5 and 9, 1952. In this five-day period, a combination of stagnant, damp air and the burning of low-quality coal produced a lethal mixture of smoke and fog. An estimated 3500 to 4000 people—mostly children, elderly, and the already infirm—died as a direct result of the episode.

The most famous air pollution disaster in North America occurred in Donora, Pennsylvania, 50 km (30 mi) from Pittsburgh. Between October 26 and 31, 1948, sulfur, carbon monoxide, and heavy metal dusts emitted from the American Steel & Wire's Zinc Works mixed with a dense radiation fog to create what has been called the "Hiroshima of air pollution." Four days of intense smog took on even greater proportions by Saturday, October 30. Fans at a high school football game were unable to see the events happening on the field. Others left the game early as word came that family members at home had died or were hospitalized from respiratory problems brought on by the smog. Those who tried to evacuate the town were unable to leave because near-zero visibility completely stalled traffic. By Sunday morning firefighters were bringing oxygen to people who were unable to breathe, but relief was only temporary as the departing firefighters felt their way over to the next victim requiring assistance. On Sunday morning, officials finally closed down the Zinc Works, and later that day the smog was finally washed away by rain—but only after 20 people had died and 7000 people had been hospitalized. It is widely believed that the Donora event was the principal catalyst in the enactment of antipollution legislation in the United States (see Figure 14–1–1).

14–1–1 What caused the 1952 London smog episode?

14–1–2 What event was most responsible for the enactment of air pollution abatement laws in the United States?

REMOVAL OF PARTICULATES Though particulates are always present in the air, no individual particulate stays in the atmosphere forever. As we saw in Chapter 7, terminal velocities increase with the size of falling objects. Thus, particulates, which are always small, can remain suspended in the atmosphere for considerable lengths of time. Larger ones remain in the air for perhaps just a few hours, while smaller particulates can exist for weeks.

Several different processes remove particulates from the air. *Gravitational settling*, the process wherein particulates fall from the air (even if very slowly), effectively removes larger particulates. The smaller ones are less susceptible to this process because even very small eddies can keep them in suspension. Precipitation, on the other hand, removes both large and small particulates in two ways. First, the particulates that served as condensation nuclei in clouds are removed when the droplets that they are part of fall as rain or snow. Other particulates are removed by *scavenging*, the process in which falling droplets and crystals collide with particulates in their path. Upon collision, the precipitation incorporates the particulate and carries it to the surface. The scavenging of particulates largely explains why the air is so much cleaner and visibility is enhanced after a rain shower.

EFFECTS OF PARTICULATES Particulates reduce visibility by increasing scattering of visible radiation, but their effect on visibility is of less importance than their impacts on health. Perhaps this is not surprising, given that we are bathed in these tiny objects every minute of the day. By 1987 it had become clear that a certain class of particulates—those smaller than 10 μm in diameter (called $\mathbf{PM_{10}}$)—most readily enters the lungs and brings about the most serious tissue damage.

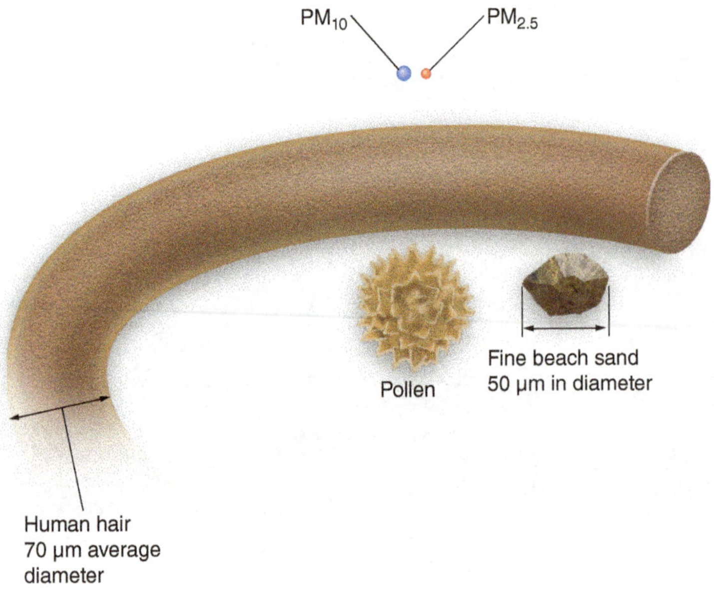

▲ **FIGURE 14–2 Particulates.** Microscopically small particles called $PM_{2.5}$ and PM_{10} are extremely small, even compared to the thickness of a human hair.

Although the lungs have cilia that can remove these small (less than one-tenth the diameter of a human hair) particulates, the particulate removal occurs very slowly—on the order of several months.

A large body of research analyzing the effects of particulates has shown that a more specific class of particulates—those smaller than 2.5 µm (called **$PM_{2.5}$**)—also presents serious health problems. For this reason, the U.S. Environmental Protection Agency (EPA) bases its regulations regarding particulates on these so-called *fine particles.* But the recent focus on $PM_{2.5}$ should not be taken to imply that larger particulates are not dangerous. For example, one study has shown a strong correlation between hospital admissions in the Los Angeles basin and the levels of PM_{10} particulates in the air. The increase in hospital admissions is nearly equally divided between patients with acute respiratory illness and those with cardiovascular disease.

CHECKPOINT

14.1 What are the major sources of particulates?

14.2 How are the processes of gravitational settling and scavenging similar? How are they different?

Carbon Oxides

Carbon oxides (also called *oxides of carbon*) include **carbon monoxide (CO)** and **carbon dioxide (CO_2)**. The latter has been discussed in Chapters 1 and 3 as one of the important variable gases that make up the atmosphere and for its role as a greenhouse gas in absorbing a portion of longwave radiation that would otherwise escape to space (refer back to Figure 3–17). The topic is discussed further in Chapter 16 where we expand on its possible role in climate change. Though current levels of CO_2 have no immediate health effects on people as they breathe carbon dioxide, its role as a greenhouse gas has led to a United States Supreme Court ruling in 2007 that it was indeed a pollutant subject to regulations of the Clean Air Act and enforceable by the EPA.

Carbon monoxide is a colorless, odorless gas. In the natural environment it is released as a primary pollutant by volcanic eruptions, forest fires, bacterial action, and other processes. Though natural processes emit far more CO into the environment than do human activities, soil microorganisms consume it effectively, and background values are very low. In cities, however, inputs can greatly exceed the rate of removal, and unsafe concentrations can accrue. In the United States, the most important source of CO is the automobile (see Figure 14–1), which releases the gas as a by-product of incomplete combustion. In well-maintained vehicles, carbon monoxide emissions are low, but poorly operating engines can cause CO concentrations to accumulate to unsafe levels. This is particularly true in confined areas, such as garages and tunnels. In the home, an improperly vented or malfunctioning furnace can release lethal doses of CO very quickly. Carbon monoxide is also released in home fires, where it probably is responsible for a high percentage of fire-related fatalities. Cigarette smoke also releases carbon monoxide as a by-product sufficient to greatly increase its concentration in the bloodstream.

Carbon monoxide is extremely toxic. Even low levels can cause a person to immediately experience slowed reflexes, drowsiness, and a reduction or loss of consciousness. Exposure for three hours at 400 parts per million (ppm) is life threatening, and at 1600 ppm death comes within an hour. Over the long haul, it can contribute to heart disease. Table 14–1 highlights some effects of varying CO levels.

Unlike other pollutants that act mainly on the pulmonary system, carbon monoxide's toxicity arises from its effects on the bloodstream. Hemoglobin (the agent that gives red blood

TABLE 14–1

Threshold Levels of Carbon Monoxide

Carbon Monoxide Concentration (ppm)	Comment
50	Maximum allowable Occupational Safety and Health Administration dose for 8-hour exposure
200	Headache, fatigue, dizziness, nausea in 2–3 hours
400	Headache in 1–2 hours, life threatening after 3 hours
800	Dizziness, nausea, and convulsions within 45 minutes; death in 2–3 hours
1600	Headache, dizziness, nausea in 20 minutes; death in 1 hour
3200	Headache, dizziness, nausea in 5–10 minutes; death in 25–30 minutes
6400	Headache, dizziness, nausea in 1–2 minutes; death in 10–15 minutes
12,800	Death within 1–3 minutes

cells their characteristic color) absorbs oxygen in the lungs and circulates it throughout the body. Under ideal conditions the hemoglobin releases oxygen to cells and then returns to the lungs, whereupon the process is repeated again and again. Carbon monoxide in the bloodstream completely disrupts this process. As it turns out, hemoglobin has a 200-fold greater affinity for carbon monoxide than O_2. In other words, with carbon monoxide and oxygen both available in the lungs, the blood will far more readily absorb the carbon monoxide. Thus, the inhalation of carbon monoxide reduces the cardiovascular system's ability to circulate oxygen to the rest of the body.

Video
Global Carbon Monoxide Concentrations
http://goo.gl/AAvWkd

Sulfur Compounds

Sulfur compounds in the atmosphere can occur in gaseous or aerosol form. The majority—roughly two-thirds—of all sulfur compounds emitted into the atmosphere originate from natural processes. Steam vents, such as those at Yellowstone National Park in Wyoming or Lassen Volcanic National Park in California, provide interesting examples of the emission of sulfur compounds. The most important of these is the bacterial release of hydrogen sulfide (H_2S), a particularly noxious gas that smells like rotten eggs. Volcanic eruptions and sea spray also play an important role in releasing sulfur compounds. Fortunately, sulfur gases are readily dispersed in the atmosphere, so background concentrations are extremely low (on the order of one-half a part per billion) and their environmental and health impacts are minimal.

Of the anthropogenic sulfur compounds released to the atmosphere, the most important are **sulfur dioxide (SO_2)** and **sulfur trioxide (SO_3)**. These oxides of sulfur fall under the collective designation of SO_x. Sulfur dioxide is a primary pollutant released mainly by the burning of sulfur-containing fossil fuels, particularly coal and oil used for heating and electric power. Other industrial activities, such as petroleum refining and ore smelting, also contribute SO_2 (see Figure 14–1). Unlike natural processes, human activities tend to be concentrated over relatively small areas, allowing SO_x to attain high values over urban and industrial areas.

Sulfur dioxide is a colorless but highly corrosive gas that irritates human respiratory systems. High concentrations are associated with a number of lung problems, and even low concentrations can cause asthmatic subjects to experience severe bronchial constriction during exercise. Though it is widely blamed for causing respiratory problems, scientists are not sure what role high SO_2 concentrations directly play in their onset. It may be that the occurrence of respiratory illness during high SO_2 episodes is not due directly to the presence of the gas, but rather to the particulates that often accompany high sulfur dioxide concentrations.

Sulfur trioxide can be put directly into the air as a primary pollutant, but it more commonly builds up as a secondary pollutant following reactions involving SO_2. Sulfur trioxide is not by itself a major component of air pollution. However, it readily combines with water droplets to form *sulfuric acid,* H_2SO_4. If this process occurs near the surface, it forms **acid fog**; if

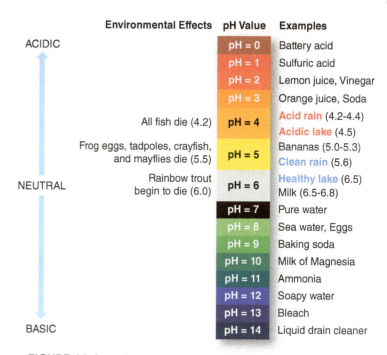

	Environmental Effects	pH Value	Examples
ACIDIC		pH = 0	Battery acid
		pH = 1	Sulfuric acid
		pH = 2	Lemon juice, Vinegar
		pH = 3	Orange juice, Soda
	All fish die (4.2)	pH = 4	Acid rain (4.2-4.4) Acidic lake (4.5)
	Frog eggs, tadpoles, crayfish, and mayflies die (5.5)	pH = 5	Bananas (5.0-5.3) Clean rain (5.6)
NEUTRAL	Rainbow trout begin to die (6.0)	pH = 6	Healthy lake (6.5) Milk (6.5-6.8)
		pH = 7	Pure water
		pH = 8	Sea water, Eggs
		pH = 9	Baking soda
		pH = 10	Milk of Magnesia
		pH = 11	Ammonia
		pH = 12	Soapy water
		pH = 13	Bleach
BASIC		pH = 14	Liquid drain cleaner

▲ **FIGURE 14–3 Acidity.** The lower the pH value of a substance, the greater its acidity. In the absence of pollution, normal rainfall is mildly acidic.

it occurs in clouds, subsequent precipitation of the acid compound produces **acid rain** or acid snow. In addition to these wet processes, **dry deposition** of acid occurs when acid chemical are incorporated in dust or other dry particulates and fall to surface. Dry deposition is an important component of what is often simply referred to as "acid rain".[1] Figure 14–3 shows typical values of acidity for various substances. Note that lower values denote greater acidity. Also note that rainwater unaffected by pollution is still somewhat acidic with values pH values ranging from 5.0 to 5.5 (due largely to the incorporation of atmospheric carbon dioxide into the rain water).

Not surprisingly, acid fog and rain are both capable of causing extreme environmental harm and through time can wear down human structures. Acid fog can be particularly dangerous to people because it is so easily inhaled. Buildings and monuments made of limestone are especially vulnerable to weathering from acid rain and fog (Figure 14–4a). Natural environments are likewise susceptible to the effects of acid fog and rain (Figure 14–4b).

Acid precipitation reaching the surface eventually feeds into the hydrologic system. Though some water falls directly onto lakes and rivers, the majority gets into them indirectly as soil water or groundwater. Despite the indirect input into the surface waters, however, the water retains its acidity as it flows beneath the surface and eventually enters lakes and rivers. Through time, the surface water system becomes so acidic that it is inhospitable to life. At its worst, acidification can render lakes and rivers completely devoid of fish and aquatic birds. Unfortunately, this problem is neither hypothetical nor abstract. In the eastern United States, nearly 1200 lakes and 4700 streams have become acidified—some to the point

[1]The term *acid deposition* refers to the acidification of the surface environment by wet or dry processes.

▲ **FIGURE 14–4 Acid Deposition.** (a) Sulfur oxides can increase the acidity of rain water, gradually wear down the surfaces of monuments and buildings, and (b) cause damage to vegetation as it has to this forest at Grayson Highlands State Park in Virginia.

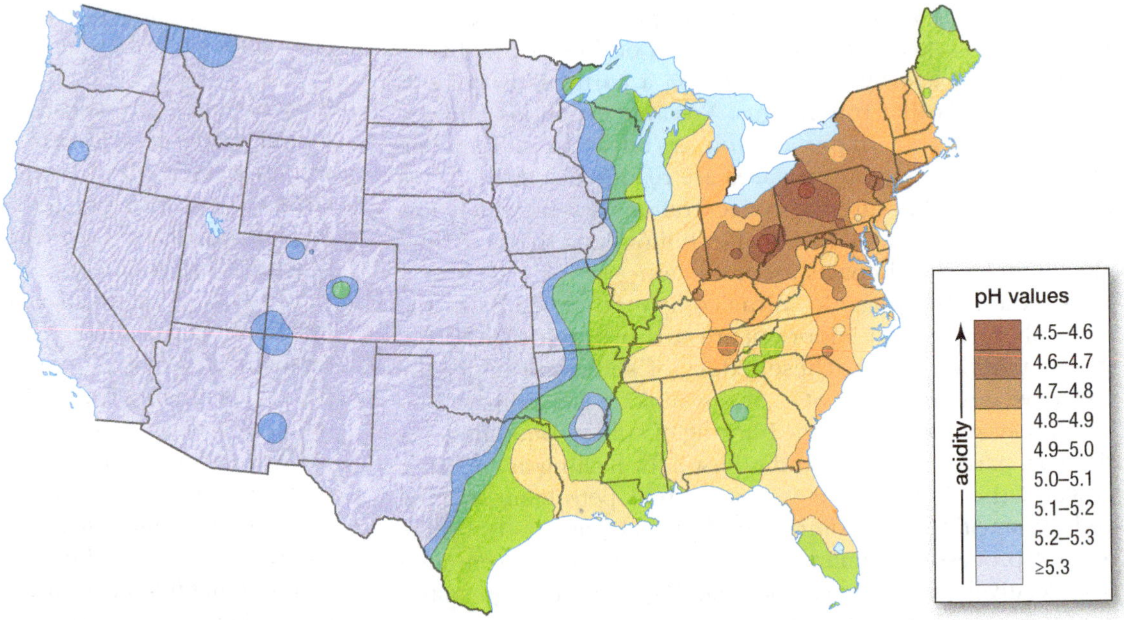

► **FIGURE 14–5 Precipitation Acidity.** The eastern United States is subject to acid precipitation because of human-source sulfur compounds in the atmosphere. Average values shown here obtained from Central Analytical Laboratory.

MapMaster
North America
Physical Environment
Acid Rain (U.S.)

where they no longer support any fish. In the Canadian province of Ontario, 1200 lakes are now essentially lifeless. Staggering as these figures may be, they pale in comparison to the 6500 lakes similarly impacted in Norway and Sweden.

As shown in Figure 14–5, acid precipitation is a much greater problem in the eastern United States and Canada than in the West, primarily because of the greater use of coal and heating oil. A huge proportion of the sulfur dioxide contributing to the acid rain originates from a relatively small number of sources. It is estimated that the 50 largest sulfur emitters in the region (all of which are power-generating plants) account for half of the acid deposition.

Interestingly, one of the measures undertaken to improve air quality near sulfur-emitting industries and power utilities may have exacerbated the acid deposition problem farther downwind. To help in the dispersion of sulfur oxides from industrial plants, many industries and utilities have built large smokestacks to release the pollutants well above ground level (Figure 14–6). The idea behind the stacks is that, by releasing the smoke far above the surface, sulfur compounds will be transported considerable distances downwind before settling near the ground. While the stacks have successfully reduced sulfur concentrations near their source, they have had the unintended consequence of allowing the sulfur compounds to be transported hundreds of kilometers downwind, where they react to form acid deposition. Thus, the acid problem over much of the eastern United States and Canada is due to transported, rather than locally generated, pollutants. This has led

▲ **FIGURE 14–6 Industrial Plant Emissions, at Navajo Nation Reservation in Page, Arizona.** Smokestacks on manufacturing and power plants are designed to keep emissions away from the ground near the source. Unfortunately, pollutants are carried downwind for hundreds of kilometers, where they can exacerbate acid deposition.

▲ **FIGURE 14–7 Nitrogen Dioxide.** This photo of Hong Kong, China illustrates the yellowish to reddish brown color caused by nitrogen dioxide.

to many years of litigation between states in the Midwest and Northeast and between the United States and Canada.

While most acid deposition in eastern North America is associated with sulfur compounds, this is not always the case in other areas. Some acid deposition, especially in the western United States and Canada, is related to compounds made of nitrogen and oxygen.

Nitrogen Oxides (NO_x)

Nitrogen oxides (also called *oxides of nitrogen*) are compounds consisting of nitrogen and oxygen atoms. The two most important of these from an air pollution viewpoint are **nitric oxide (NO)** and **nitrogen dioxide (NO_2)**. Together the two gases are commonly referred to as NO_x. Nitric oxide is a nontoxic, colorless, and odorless gas that forms naturally by biological processes in soil and water. While millions of tons of the material are introduced into the atmosphere each year, it is highly reactive and tends to break down very quickly. Nitric oxide also forms as a by-product of high-temperature combustion associated with automobile engines, industrial manufacturing, and electric power generation. Its primary importance from an air quality perspective is that it oxidizes to form nitrogen dioxide, a major component of smog in many places. In addition, about 25% of the acidity of rainfall is caused by nitric acid, which is formed from NO.

Nitrogen dioxide is a toxic gas that gives polluted air its familiar yellowish to reddish brown color (Figure 14–7), as well as a pungent odor. It is an important component in air pollution in that it is relatively toxic and corrosive and undergoes transformations that contribute to acid deposition and other secondary pollutants. As with nitric oxide, nitrogen dioxide breaks down very readily and, as a result, NO_2 concentrations in urban areas tend to rise and fall in accordance with vehicular traffic patterns. Furthermore, the rapid decay of nitrogen dioxide prevents large concentrations from occurring in rural areas surrounding source areas.

Like sulfur compounds, nitrogen oxides can cause serious pulmonary health problems. Clinical studies have shown that NO_2 easily passes through bronchial passages and irritates tissue deep within the lungs. Laboratory tests have shown animals to experience severe lung damage and reduced immunity to infection when exposed to high levels of NO_2.

Volatile Organic Compounds (Hydrocarbons)

Volatile organic compounds (VOCs) are materials that can rapidly evaporate at normal temperatures and include a vast number of different materials. Many VOC sources are found in our homes and for that reason tend to occur indoors at several times the concentrations found outdoors. Sources of VOCs include paint, solvents, disinfectants, air fresheners, and even dry cleaning residues in clothing (among many other sources). Just as there are many types of VOCs, there likewise are a number of health ramifications that can occur from breathing them, including dizziness, visual problems, and short-term memory loss. Some have been linked to cancer in laboratory animals.

An important class of VOCs found outdoors is **hydrocarbons**, materials made entirely of carbon and hydrogen atoms. These compounds, including methane, butane, propane, and octane, occur in both gaseous and particulate forms. Globally, the vast majority of hydrocarbons arrive in the atmosphere via natural processes, including plant and animal emissions and decomposition. In the United States, industrial activities account for the greatest proportion of anthropogenic hydrocarbons, with automobiles also contributing a major share. The emissions associated with automobiles arise primarily from incomplete fuel combustion and the evaporation of gasoline (often while filling gas tanks).

Except for industrial accidents, VOC concentrations in cities don't reach the level needed to produce the acute health effects mentioned above. Nonetheless, VOCs are extremely important because in the presence of sunlight they recombine with nitrogen oxides and oxygen to produce photochemical smog.

(a) (b)

▲ **FIGURE 14–8 Los Angeles-Type Smog.** (a) Dramatic views in the Los Angeles basin are commonly spoiled (b) by photochemical smog.

Photochemical Smog

If you have ever visited Los Angeles in the summer, you have probably heard the term **photochemical smog** and know what it feels like. Burning eyes, sore lungs, and a subtle but unpleasant odor accompany an atmosphere with poor visibility (Figure 14–8). Photochemical smog consists of secondary pollutants that include ozone (O_3), NO_2, peroxyacyl nitrate (PAN), formaldehyde, and other gases that occur in very minute quantities. As the name implies, this type of smog forms when sunlight triggers numerous reactions and transformations of gases and aerosols. Unlike the **London-type smog** found in many places where smoke combines with damp air (the word *smog,* in fact, originally derived from the terms *smoke* and *fog*), this **Los Angeles–type smog** usually involves dry air.

DID YOU KNOW?

Paris, France, is considered by many to be among the world's most beautiful cities, so an episode of unusually high atmospheric particulates in March 2014 was particularly noteworthy. For several days the air became unusually stagnant and reduced the normal dispersal of pollutants. So for only the second time since 1997 authorities restricted the use of vehicles with odd- and even-numbered license plates on alternating days and temporarily reduced speed limits in the region around that city. But the city did not totally shut down—people were allowed to ride public transportation and use rental bicycles for free during the event.

Ozone has been designated by the EPA as the most important agent of photochemical smog. It can cause serious physical and environmental harm at surprisingly low concentrations, so low that the EPA established a concentration of only 0.12 ppm averaged over a one-hour period as the maximum allowable without exceeding federal standards. Exposure to ozone causes inflammation of air passages that can reduce lung capacity by as much as 20 percent. The EPA estimates that perhaps 20 percent of all respiratory-related hospital visits in the northeastern United States during the summer result from exposure to ozone. Although acute symptoms usually subside fairly quickly after ozone concentrations decline, research has shown that long-term exposure to ozone can cause permanent damage to lung tissue and impairment of the body's ability to resist bronchitis, pneumonia, and other diseases. Of course, ozone causes even greater problems for people with asthma and other preexisting pulmonary problems. For those people, ozone constricts lung passages to the point where breathing becomes nearly impossible, and the gas may contribute considerably to the number of asthma-related fatalities that occur in the United States annually. Currently, an average of 5000 Americans die each year during acute asthma attacks.

DID YOU KNOW?

Severe air pollution is not a phenomenon restricted to large, urban centers. Indeed, even national parks can have major smog problems. At the Great Smoky Mountains National Park in North Carolina and Tennessee, pollutants—mostly from regional industry—have caused the average summer visibility to be reduced from a maximum summertime distance of 187 km (117 mi) to only 32 km (20 mi). Acid precipitation is also a problem, with rainfall acidity being as much as 10 times that of natural precipitation. The park has exceeded EPA standards for ozone an average of 30 days per year from 1991 to 2003.

The situation is even worse for Sequoia & Kings Canyon National Park in California. The smoggiest of all U.S. national parks, this otherwise magnificent natural resource routinely has more days in excess of EPA ozone standards (370 days in the period from 1999 to 2003) than do most large U.S. cities. Visitors and staff there are frequently advised to limit their physical activities because of unhealthful air.

TABLE 14–2

Air Quality Index

Air Quality Index (AQI) Values	Levels of Health Concern	Colors
When the AQI is in this range:	. . . air quality conditions are:	. . . as symbolized by this color:
0–50	"Good." Air quality is considered satisfactory, and air pollution poses little or no risk.	Green
51–100	"Moderate." Air quality is acceptable; however, for some pollutants there may be a moderate health concern for a very small number of people. For example, people who are unusually sensitive to ozone may experience respiratory symptoms.	Yellow
101–150	"Unhealthy for Sensitive Groups." Although general public is not likely to be affected at this AQI range, people with lung disease, older adults and children are at a greater risk from exposure to ozone, whereas persons with heart and lung disease, older adults and children are at greater risk from the presence of particles in the air.	Orange
151 to 200	"Unhealthy." Everyone may begin to experience some adverse health effects, and members of the sensitive groups may experience more serious effects.	Red
201 to 300	"Very Unhealthy." This would trigger a health alert signifying that everyone may experience more serious health effects.	Purple
301 to 500	"Hazardous." This would trigger a health warnings of emergency conditions. The entire population is more likely to be affected.	Maroon

Source: http://airnow.gov.

Not only are people directly harmed by ozone, but high levels also result in serious environmental degradation. Damage to agricultural crops by photochemical smog (mainly ozone) was identified in southern California grape fields in the late 1950s. Since that time, plant damage from oxidants such as ozone has been widespread across North America, with conifers being particularly vulnerable.

Air Quality Index

The Environmental Protection Agency (EPA) has created a uniform index that is useful for air pollution monitoring across the United States called the **Air Quality Index (AQI)**. The AQI is calculated each day for locations across the United States by applying a particular formula for ozone, particulates, carbon monoxide, sulfur dioxide, and nitrogen dioxide, and expressing each pollutant on a scale that ranges from 0 to 500. The official AQI for any place at a particular time and place is the highest of the five individual pollutant values. The EPA has also established a color scheme to represent different levels of AQI ranges for better communication to the public. Table 14–2, taken directly from the EPA, illustrates the color scheme used and summarizes the significance of AQI values within designated ranges. Current maps of AQI values and next-day forecast maps can be obtained at www.airnow.gov.

CHECKPOINT

14.3 Pick one of the following atmospheric pollutants and explain its causes and effects: carbon oxides, sulfur compounds, nitrogen compounds, volatile organic compounds.

14.4 Is acid precipitation a primary or secondary pollutant? Explain.

14.5 What are two important reasons why ozone levels in photochemical smog should be reduced? Explain.

Atmospheric Conditions and Air Pollution

14.2 Explain the role of atmospheric conditions in determining pollution concentrations.

As we have seen, a large portion of the chemicals that we consider pollutants occur naturally in the environment. These emissions do not create high concentrations, however, because their release into the atmosphere is over such a large area that they are immediately diluted. Urban emissions, on the other hand, are concentrated over much smaller areas and can thereby lead to significant pollution episodes.

What is more important than the amount of pollutants created in a region is their *concentration*. A given amount of pollution confined to a smaller depth and area of the atmosphere has a greater concentration than one in which it is allowed to disperse throughout a deeper and wider area, and its effects are therefore felt more strongly. Atmospheric conditions play a major role in determining pollution concentrations in several ways. Atmospheric stability and wind conditions control the vertical and horizontal dispersion of pollutants, and cloud conditions can influence the rate of photochemical reactions taking place. Furthermore, unusually cold or warm conditions encourage the increased use of heaters or air conditioners, which can increase emissions.

Effect of Winds on Horizontal Transport

Strong winds aid in the dispersal of pollutants in two ways. First, they rapidly transport emissions from their source and spread them over a wide horizontal extent. Figure 14–9 illustrates how the concentration of pollution is inversely proportional to the wind speed. (To make this easier to visualize, the figure depicts puffs of smoke being released from a stationary

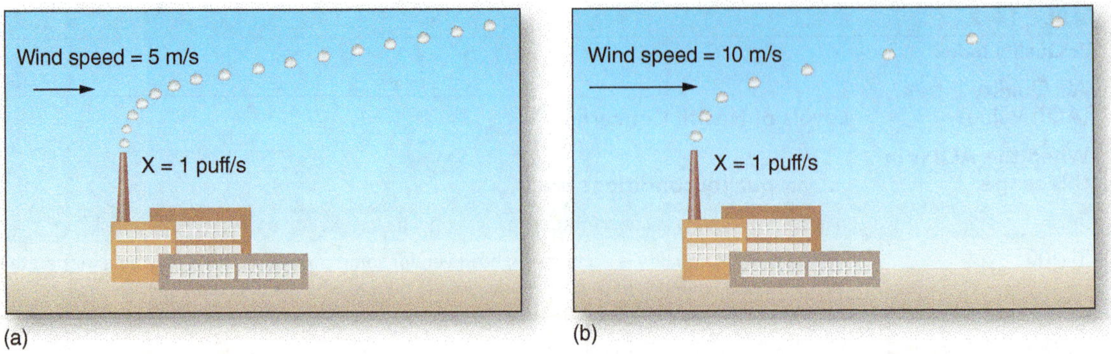

▸ **FIGURE 14–9 Wind Effect.** Wind speed is an important factor in pollutant dispersal. In (a), the 5 m/sec (11 mph) wind moves individual puffs of smoke downwind slowly, so each successive puff is only 5 m behind the previous one. In (b), the 10 m/sec (22 mph) wind causes twice as great a distance between puffs, and thus only one-half the concentration of smoke.

source every second, although in reality pollutants would be released continuously.) In (a), the wind blows at 5 m/sec (11 mph) so that each puff of smoke travels 5 m before the next one is released. In (b), the wind flows twice as fast as in (a), and the distance between successive puffs of smoke likewise doubles. Thus, the greater wind speed in (b) causes the same amount of pollution to be diluted within twice as large a volume of air.

Greater wind speeds also lower the pollution concentration indirectly. Recall from Chapter 3 that air does not flow uniformly in a given direction. Instead, it contains small, swirling motions, called *eddies,* that mix the air vertically. This forced convection increases with wind speed and, as a result, strong winds favor greater vertical dispersion.

Short-term variations in wind direction also affect dispersion. If the wind direction varies only slightly through time, pollution will be concentrated within a relatively narrow area downwind of the source. If wind directions are highly variable, the pollutants will spread out over a wider area. More people will be subjected to the pollutants, but the concentration will be lower than it would be under a more constant wind regime.

Effect of Atmospheric Stability

Just as the stability of the air (Chapter 6) influences lifting and cloud formation, it also affects the vertical movement of pollutants. Recall that when the air temperature decreases slowly with height (or if it increases with height), the air is said to be stable. Stable air resists vertical displacement and leads to higher pollutant concentrations near the ground. Unstable air, on the other hand, enhances vertical mixing, and any material introduced near the surface is easily displaced upward. This reduces pollution concentrations near the surface.

Inversions, the situation in which air temperature increases with height, make the air extremely stable and impose the greatest restraint on vertical mixing. Radiation inversions (described in Chapter 6) originate at the surface in response to cooling of the lower atmosphere (Figure 14–10a). These inversions usually dissipate in the late morning after the Sun has warmed the surface and lower atmosphere (Figure 14–10b). As a result, they tend to have the greatest impact on pollution concentrations in the early morning. These inversions are most important in areas subject to a London-type smog.

Subsidence inversions are often important where photochemical smog is the major problem (Figure 14–11). The base of a subsidence inversion marks the maximum height to which the air below can be easily mixed. An inversion with a base at 500 m (0.3 mi) above the surface will result in a *mixing depth* of 500 m, doubling the concentration of pollutants that would accompany a mixing depth of 1000 m (0.6 mi). Figure 14–12 shows how distinct that base of an inversion layer can be.

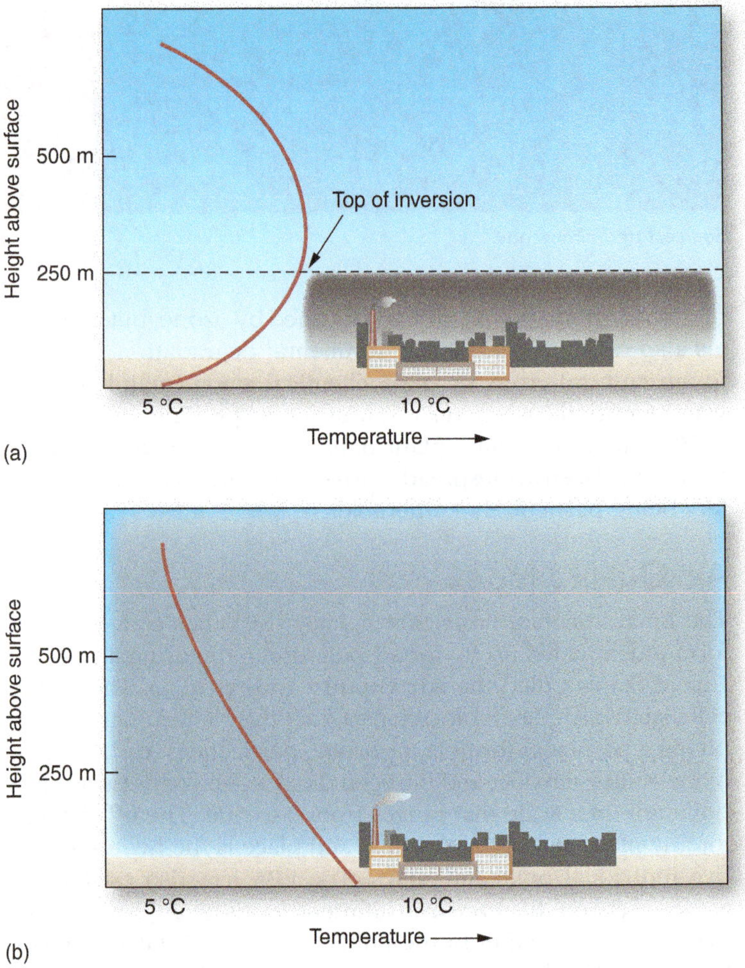

▲ **FIGURE 14–10 Radiation Inversions.** (a) Nighttime cooling lowers the air temperature more rapidly near the ground than aloft, creating an inversion. The stability associated with the inversion inhibits vertical mixing of air pollutants, keeping them close to the ground. (b) Daytime heating warms the air closest to the ground most rapidly and eventually breaks up the inversion.

Just as the semipermanent Hawaiian high-pressure system accounts for the mostly dry summers of southern California,

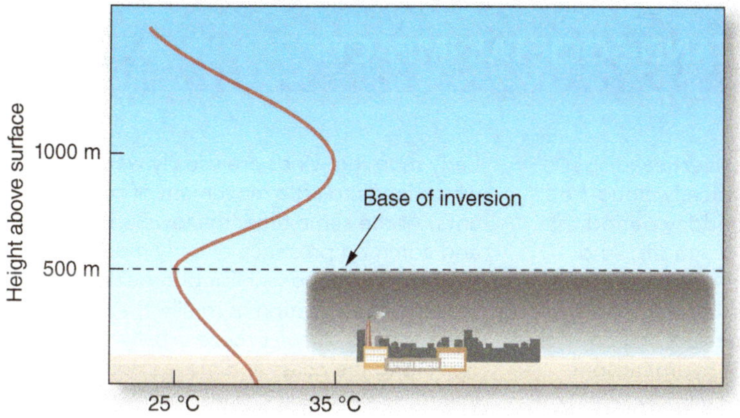

▲ **FIGURE 14–11 Subsidence Inversions.** Unlike radiation inversions that develop at the surface, subsidence inversions occur some distance above ground level. The strong stability of the inversion layer prevents air from mixing above its base and thereby confines pollutants to the layer of air below the inversion.

subsidence from the same system also plays an important role in the region's poor air quality. As air rotates clockwise out of the eastern margin of the high, air in the middle troposphere descends and creates an inversion. During the summer, the base of the inversion level over Los Angeles typically occurs at about 700 m (2300 ft) above sea level, but the base of the inversion can also occur at lower levels and lead to particularly bad smog events. (*Box 14–2, Focus on the Environment and Societal Impacts: Smog in Southern California*, presents further information on atmospheric and other controls on air pollution in the region.)

CHECKPOINT

14.6 What are two main factors that can disperse or concentrate air pollutants?

14.7 What are typical heights for the base of Los Angeles inversions? Why is the height of the inversion base important?

▼ **FIGURE 14–12 Effect of an Inversion.** The base of a subsidence inversion is clearly evident, where its stability effectively caps the level of pollutant dispersal in the smog layer in Santiago, Chile.

The Counteroffensive on Air Pollution

14.3 Describe the efforts that have been made to reduce air pollution.

Regulations designed to improve air quality have made a substantial impact on the lives of people in the United States and Canada. Although federal regulations regarding air pollution in the United States did not come into being until the 1950s, certain cities and states have had laws on the books regulating smoke emissions since the late nineteenth century. In some cases, such as in Pittsburgh, fairly stringent controls were in effect by the 1940s.

The first major U.S. initiative to clean up the nation's air was the 1970 Clean Air Act. This act charged the EPA with creating and enforcing air quality standards for certain classes of pollutants. Individual states have the task of ensuring compliance, at the risk of losing a portion of federal funding for noncompliance. Regions where air quality falls short of the standards are designated nonattainment areas. Significant expansion of the EPA's role in air pollution was authorized in 1977 and 1990.

Overall, the act and its amendments have resulted in substantial reductions in pollutant levels in urban areas, despite a large increase in motor vehicle miles driven each year. Figure 14–13 shows the reduction in aggregate pollution levels relative to 1990 values that occurred despite increases

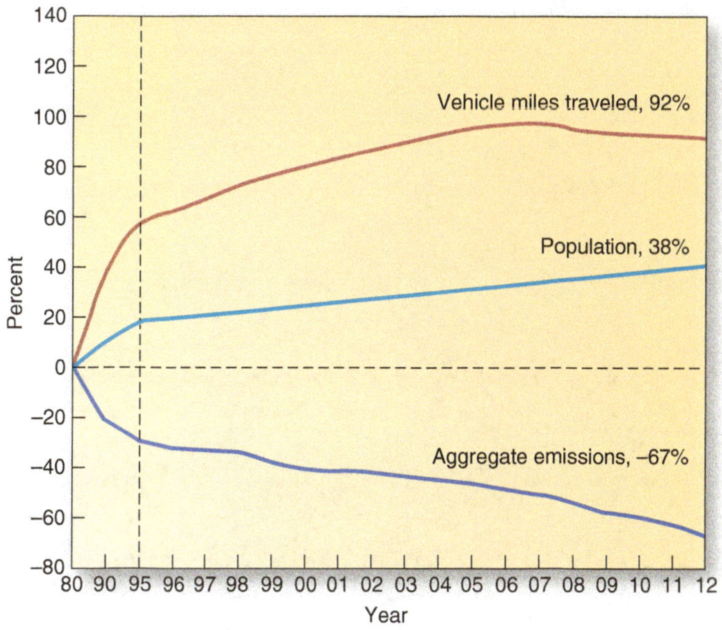

▲ **FIGURE 14–13 Changes since 1990.** The graph shows trends in aggregate pollutant levels, motor vehicle miles traveled annually, and population in the United States relative to 1990.

14-2 FOCUS ON THE ENVIRONMENT AND SOCIETAL IMPACTS

Smog in Southern California

Los Angeles has long had a reputation for extremely bad air quality—and for good reason. Of all the cities in the United States, Los Angeles is the only one classified by the Environmental Protection Agency as an "extreme area" of noncompliance with ozone standards. A number of factors work together to make the air quality bad enough to earn this dubious distinction. As shown in Figure 14–2–1, Los Angeles occupies part of a basin bounded by mountains to the north and east that block the free movement of pollutants by the sea breeze, while the presence of a subsidence inversion during the warmer months restricts vertical dispersion. Add to that the typically cloud-free conditions during the midday period that trigger photochemical reactions. And finally there is the city's well-known love affair with the automobile, which contributes much of the estimated 2 million kilograms (2200 tons) of hydrocarbons and 1 million kilograms (1200 tons) of NO_x released each day into the four-county South Coast Air Basin.

During the summer, daily concentrations of photochemical smog vary on a regular basis in the course of a day. Prior to the morning rush hour, residual primary and secondary pollutants from the previous day leave a background level of air pollution. As traffic increases during the morning, emissions increase substantially. Early morning winds are usually weak, which leads to little movement of pollutants. At the same time, the low Sun angle and common presence of early morning fog and low clouds inhibit photochemical activity. The situation normally changes by late morning. A sea breeze usually develops along the coast and moves pollutants inland, while clearing skies and increasing Sun angles increase photochemical conversions.

As the sea breeze develops, a boundary called a **sea breeze front** separates the relatively clean marine air from the more polluted, drier air ahead. As the sea breeze front moves inland, it pushes the emissions eastward or northeastward. This often creates a strong gradient in ozone concentrations near the sea breeze front, with relatively clean air

▲ **FIGURE 14–2–1 Topography of the Los Angeles Basin.**

behind it and increasing concentrations to the east or northeast (Figure 14–2–2). By late afternoon, the cities in the eastern portion of the basin get the full onslaught of the advected pollutants, while local commuters add their own contribution to the photochemical smog. As a result, pollution levels can become extraordinarily high in the areas downwind of Los Angeles.

Video
The Smog Bloggers
http://goo.gl/s6wVDV

The area to the south of Los Angeles—including Orange and San Diego counties—also has a significant air pollution problem. Usually the air pollution in San Diego, some 150 km (100 mi) to the south, is of local origin. During severe episodes, however, most of the pollution originates over Los Angeles and Orange counties and gets carried into San Diego by the wind. These episodes often occur as Santa Ana winds die out. During the Santa Ana, the easterly winds force the basin's pollutants offshore. As the Santa Ana begins to weaken, a thermally induced low-pressure system over the eastern desert stretches into the San Diego area. This creates a northwesterly flow that transports pollution originating over Los Angeles and Orange counties. A number of state-mandated rules have had a remarkable effect on air quality, such as biannual smog checks on vehicles, special pollution-reducing gasoline blends during the summer period, and rubber sleeves on gas pump nozzles that capture escaping fumes. Figure 14–2–3 illustrates the dramatic and consistent improvement in air quality in the Los Angeles area, showing both the number of days that ozone exceeded federal guidelines and the maximum eight-hour average of ozone. Though ozone pollution is still a serious problem, the number of days with excessive concentrations is about half of what it was during the late 1970s.

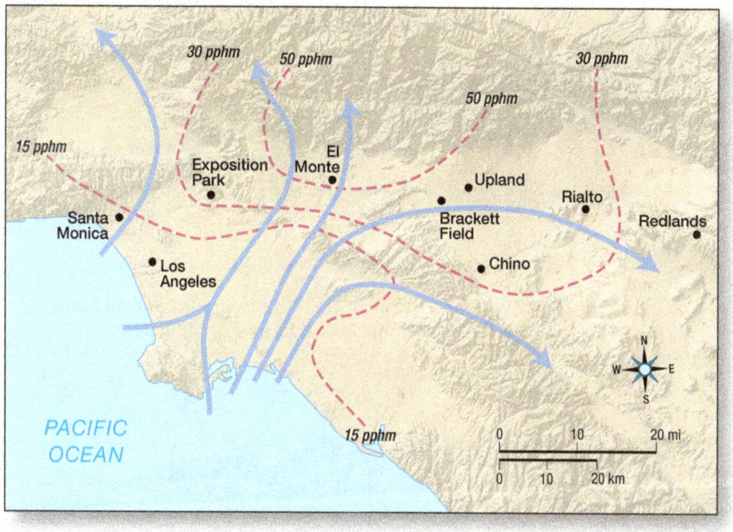

▲ **FIGURE 14–2–2 Smog Pattern.** The map shows a typical pattern of smog on a summer afternoon in the Los Angeles basin. Highest ozone concentrations (shown by dashed lines with units of parts per hundred million) occur in the northeast, ahead of the sea breeze front. Solid lines show the wind direction.

14–2–1 Describe some of the reasons why Los Angeles is so prone to heavy air pollution.

14–2–2 How does the sea breeze front influence Los Angeles air quality?

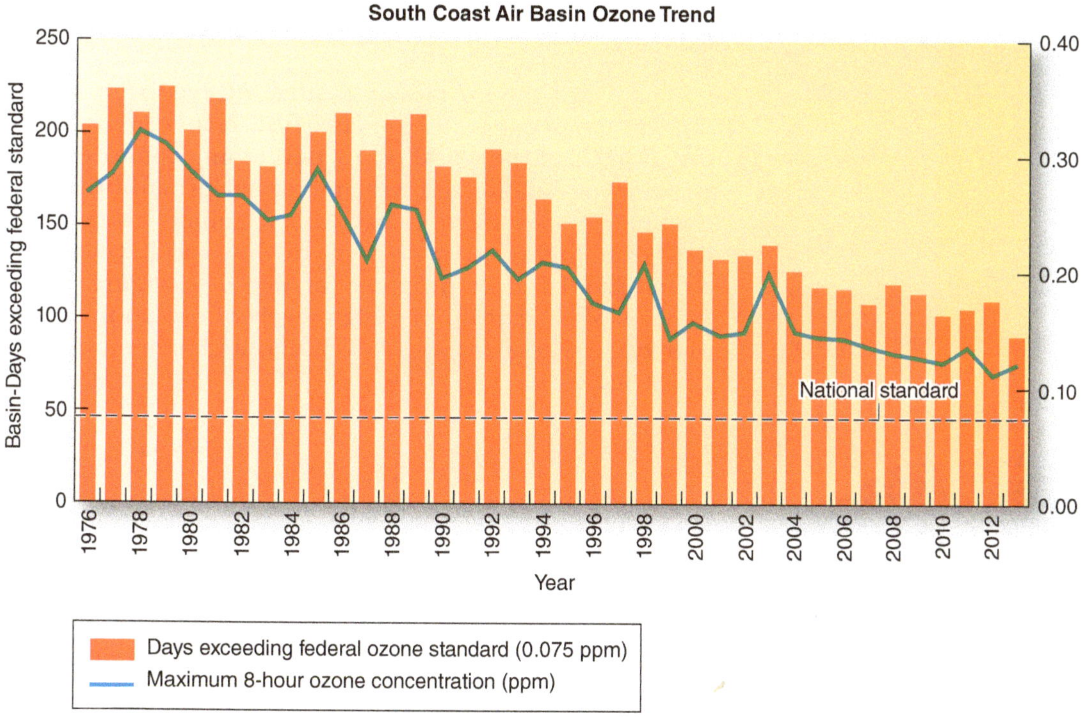

◂ **FIGURE 14–2–3 Ozone Trend.** Though still excessive, the number of days with an ozone level above federal standards (bars) and the highest eight-hour average ozone reading for each year (line) have decreased substantially.

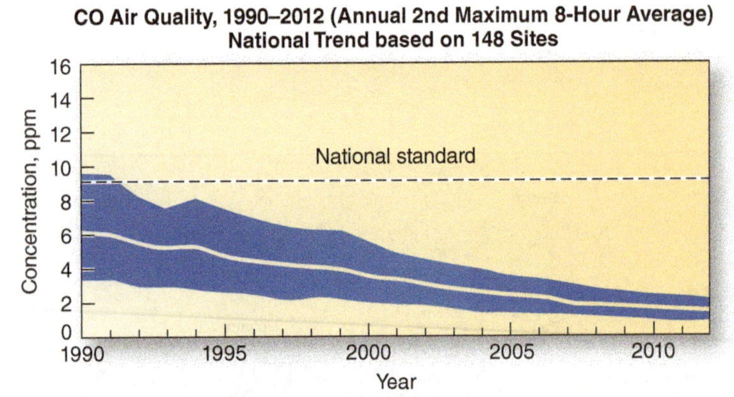

(a) 1990 to 2012: 75% decrease in CO national average

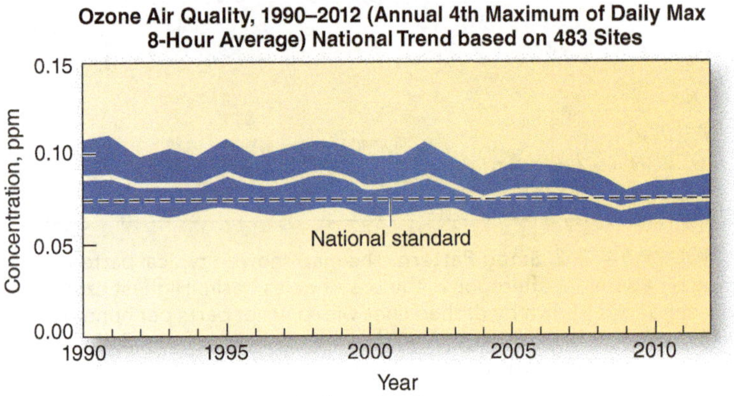

(b) 1990 to 2012: 14% decrease in O_3 national average

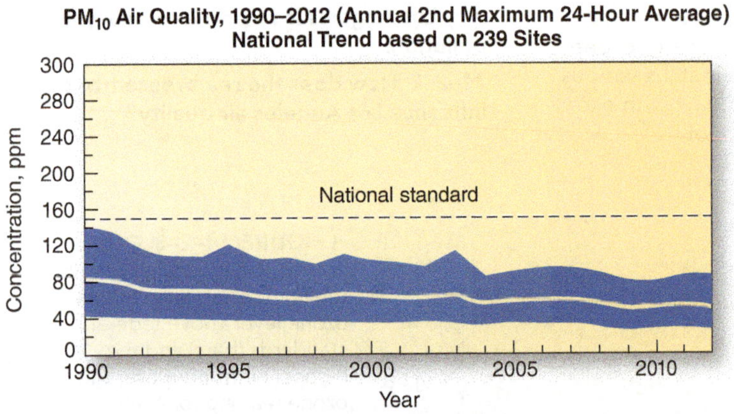

(c) 1990 to 2012: 39% decrease in PM_{10} national average

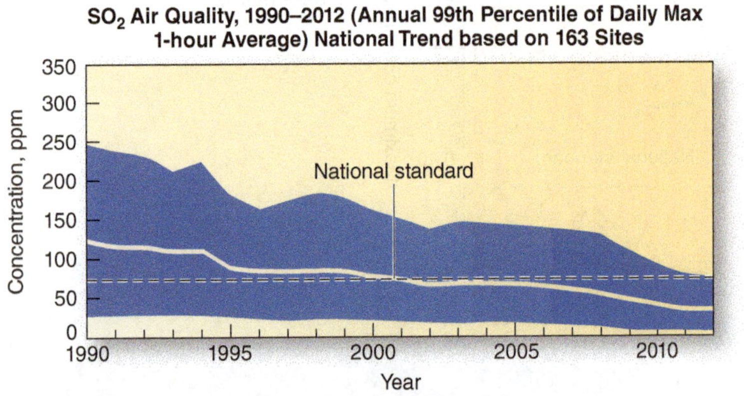

(d) 1990 to 2012: 72% decrease in SO_2 national average

▲ **FIGURE 14–14 Pollutant Trends.** Between 1990 and 2012, U.S. levels of (a) CO, (b) O_3, (c) PM_{10}, and (d) SO_x have dropped dramatically.

in population and vehicle miles traveled. Figure 14–14 plots national trends in air quality indices for carbon monoxide (a), ozone (b), PM_{10} (c), and sulfur dioxide (d) between 1990 and 2012. The decline in those pollutants ranges from substantial (ozone has had a 14% reduction) to dramatic (75% for carbon monoxide).

Because each pollutant has different primary sources, a variety of actions have been required to reduce the amount of each pollutant. Nitrogen oxide, hydrocarbon, and carbon monoxide emissions from trucks and automobiles have been reduced to a large extent by the use of catalytic converters that convert much of the by-product of gasoline consumption to less toxic gases. Improvements in engine efficiency from the computerization of engines and electronic fuel injection have also played an important role, as have overall improvements in gas mileage. Reductions in nitrogen oxide emissions from nonautomotive sources have also occurred through the implementation of regulations regarding power plant emissions.

The reduction in particulate concentrations has been the result of filters and other devices that reduce smokestack emissions. Sulfur emissions have been recently targeted by the Clean Air Interstate Rule (CAIR) of 2005 and the 2011 Cross-State Air Pollution Rule (CSAPR). The latter body of regulations was struck down by an appeals court in 2012, but in April 2014 this decision was reversed by the United States Supreme Court. Though progress has been dramatic, some counties still fail to meet minimum standards for at least one of the major pollutants, and millions of people still live in nonattainment areas (Figure 14–15). The impact of air pollution, though improved, is still immense. A 2013 MIT study has estimated that 200,000 premature deaths occur each year in the United States as a result of air pollution. The biggest culprit is pollution from cars and trucks, accounting for 53,000 of those deaths. Just behind is air pollution from power-generating plants, at 52,000 deaths.

Although the air pollution situation has been improving throughout the United States, the same cannot be said for all parts of the world. Many developing countries experiencing population growth, increased urbanization, and greater industrialization have been subject to pollution levels far higher than in parts of the world where mitigation has been in progress for decades (Figure 14–16). Reports of extraordinarily high levels of pollution are regular occurrences from China, India, and numerous other countries.

CHECKPOINT

14.8 How does the Clean Air Act mandate the reduction in air pollution?

14.9 Which air pollutants have undergone the most dramatic reductions since 1990?

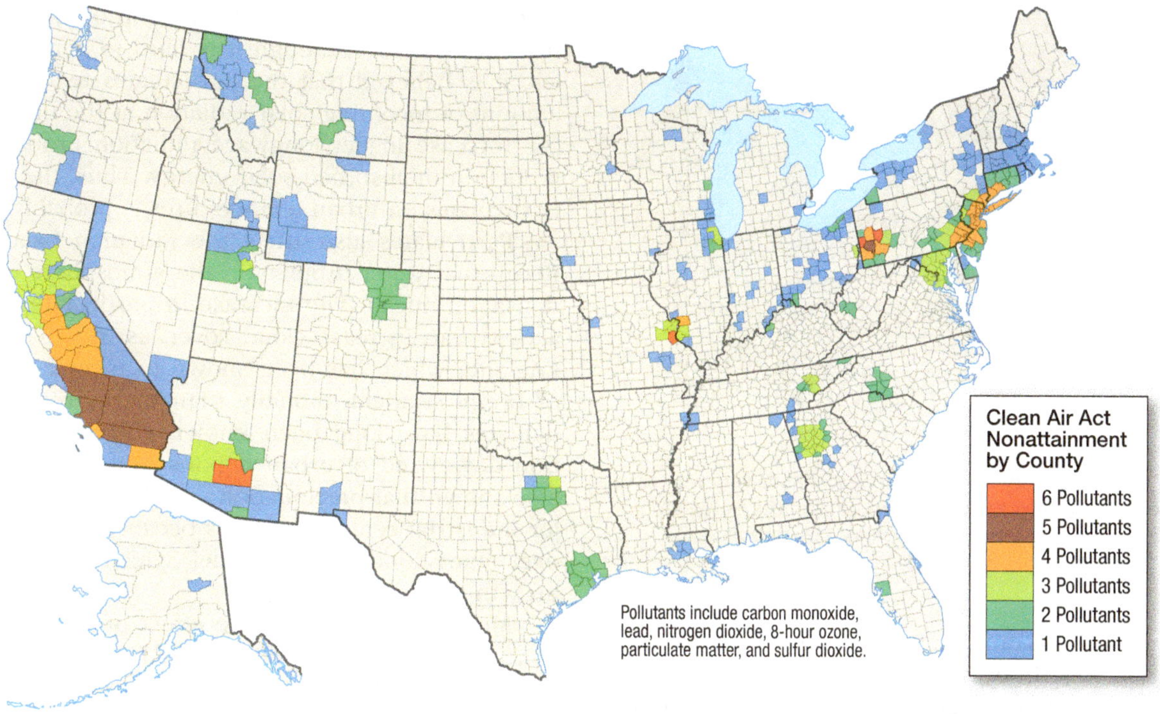

▲ **FIGURE 14–15 Nonattainment Areas.** The map shows counties that have failed to meet air quality standards as of December 2013.

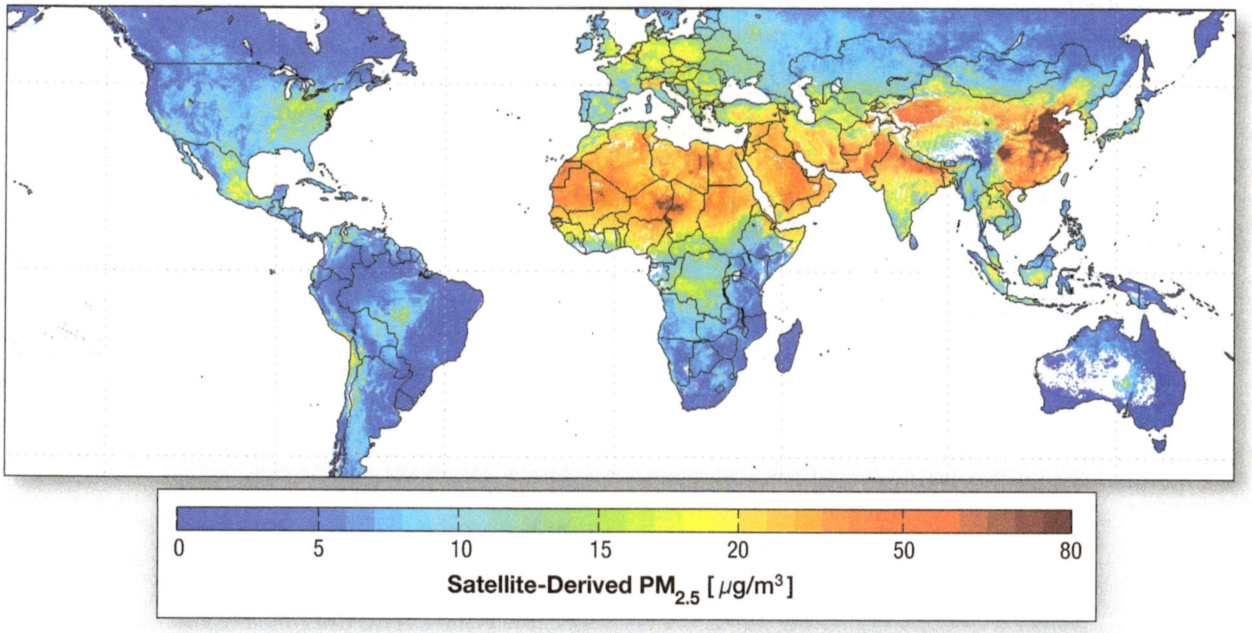

▲ **FIGURE 14–16 Worldwide Pollution.** Two Canadian scientists compiled this map of average $PM_{2.5}$ levels between 2001 and 2006 from NASA satellite data.

Urban Heat Islands

14.4 Explain the factors that produce urban heat islands and describe their effects on local weather.

Not all human impacts on the atmosphere are as dramatic as the pollution of the atmosphere. The well-known **urban heat island** is an excellent case in point. For centuries it has been known that urbanized areas often have higher temperatures than do adjacent countrysides. These differences can be quite dramatic, with temperatures in major metropolitan areas sometimes exceeding those of their hinterlands by as much as 12 °C (22 °F). Although the exact nature of the urban heat island varies from one city to another, in general the urban–rural temperature differences are greatest during the late evening and night and during the winter months.

Urban heat islands occur because of modifications to the energy balance (Chapter 3) that result when natural surfaces

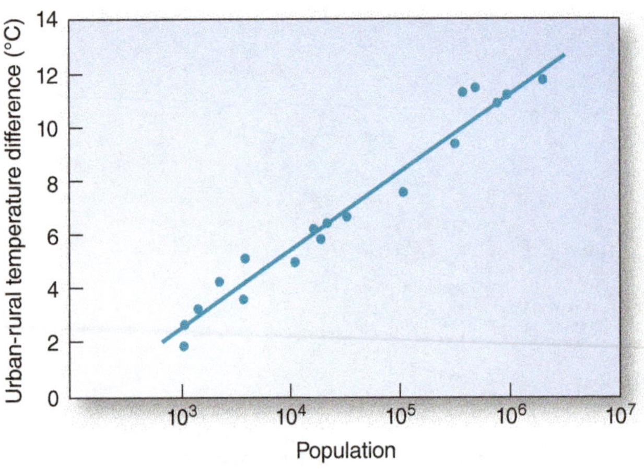

▲ **FIGURE 14–17 Effect of City Size.** Urban heat islands vary with city population. The vertical axis plots temperature differences between urban centers and surrounding areas in °C against city population. Note that the horizontal axis (population) is on a logarithmic scale.

Video Urban Heat Islands
http://goo.gl/M701Ba

are paved and built on and when human activities release heat into the local environment. Though it is not possible to generalize the relative importance of these processes for every city, all of them probably play some role in causing the phenomenon to occur.

Several variables influence the magnitude of the heat island effect. Some are related to the local setting, others to the activities undertaken within it. But the most important are the size of the city itself and the density of its population, with large, densely populated cities having the largest heat island effect. Figure 14–17 illustrates the close relationship between the population of North American cities and the maximum urban and rural temperature differences (note that the horizontal axis is plotted on a logarithmic scale).

The intensity of an urban heat island varies spatially across a city, with highest temperatures normally found within the city core. Figure 14–18 illustrates this effect by plotting temperatures over Vancouver, British Columbia, on a July evening. The downtown area is located on the southeastern part of the peninsula that juts into Burrard Inlet. Immediately to the northwest of downtown (on the northwestern part of the peninsula) lies Stanley Park, a wooded area with few buildings. As expected, temperatures are greatest over the downtown region and decrease substantially over less-populated areas. Temperatures on the peninsula decrease dramatically between downtown and the middle of the park, with a difference of about 9 °C (16 °F) occurring over a distance of only about 1.5 km (1 mi). Wind speeds also play an important role. During windy conditions, cooler air from the surrounding countryside displaces the warmer urban air and thus reduces the magnitude of the urban heat island.

Radiation Effects

Urban particulates can also affect the intensity of the urban heat island through their effect on the radiation balance. Increased particulates associated with urban activity can absorb and scatter incoming solar radiation and also increase the amount of absorption and reradiation of longwave energy in the atmosphere. Although it is hard to generalize for all cities, it is believed that the particulates tend to decrease the amount of incoming solar radiation at the urban surface, but the increase in net longwave radiation probably offsets the reduction in absorbed solar energy. Thus, the direct effect of particulates on urban temperatures is probably negligible.

Particulates also affect the radiation balance indirectly. Recall that water droplets in the atmosphere form onto condensation nuclei. The increase in particulates due to human activity can increase cloud cover, as in the case of London, England, which has been shown to receive 270 fewer hours of bright sunlight annually than the surrounding area. Particulates have long been known also to increase precipitation downwind of urban regions. Interestingly, studies have also shown that precipitation can decrease downwind of major industrial centers, perhaps because cloud water is spread over many condensation nuclei, which lessens the chance of growth to precipitation size.

More important than the effect of increasing particulate concentrations is the impact that buildings have on the radiation balance. Consider the impacts the construction of buildings with vertical walls might have on the receipt of solar radiation. When the sun is low in the sky—near sunrise and sunset, and during much of the day at high latitudes in the winter—direct sunlight that would otherwise reach the horizontal surface hits the vertical walls of buildings. This causes the angle of incidence to become closer to perpendicular and increases surface heating, which leads to a higher temperature.

The presence of buildings also affects the rate of heating by changing the surface albedo. Darker buildings, of course, absorb more sunlight than lighter ones, and, in general, urban surfaces (asphalt streets, roofing materials) have lower albedos than the natural surfaces they replace. The presence of buildings also affects the amount of absorption by causing multiple reflections to occur, as shown in Figure 14–19. As sunlight penetrates the urban landscape and hits the side of a building, some of the energy is absorbed and some is scattered back as diffuse radiation. Some of the scattered radiation strikes an adjacent building, where once again a portion is absorbed. This process goes on repeatedly, with each successive reflection at least partially absorbed upon contact with another wall. This increases the total absorption so the albedo of the urban area is actually lower than the albedo of the individual surfaces.

The presence of tall buildings also affects the transfer of longwave radiation in a way that favors higher nighttime temperatures. Essentially, the process is very similar to the multiple reflection of solar radiation just discussed. Longwave radiation emitted from an open, rural surface travels upward without being impeded by buildings. Urban areas, in contrast, reduce the amount of longwave radiation that freely escapes to space because walls absorb a portion of the outgoing radiation. The resultant reduction in longwave radiation loss slows the rate of nocturnal cooling and promotes higher daily minimum temperatures, which are more strongly affected by urban heat islands than are daily maximum temperatures.

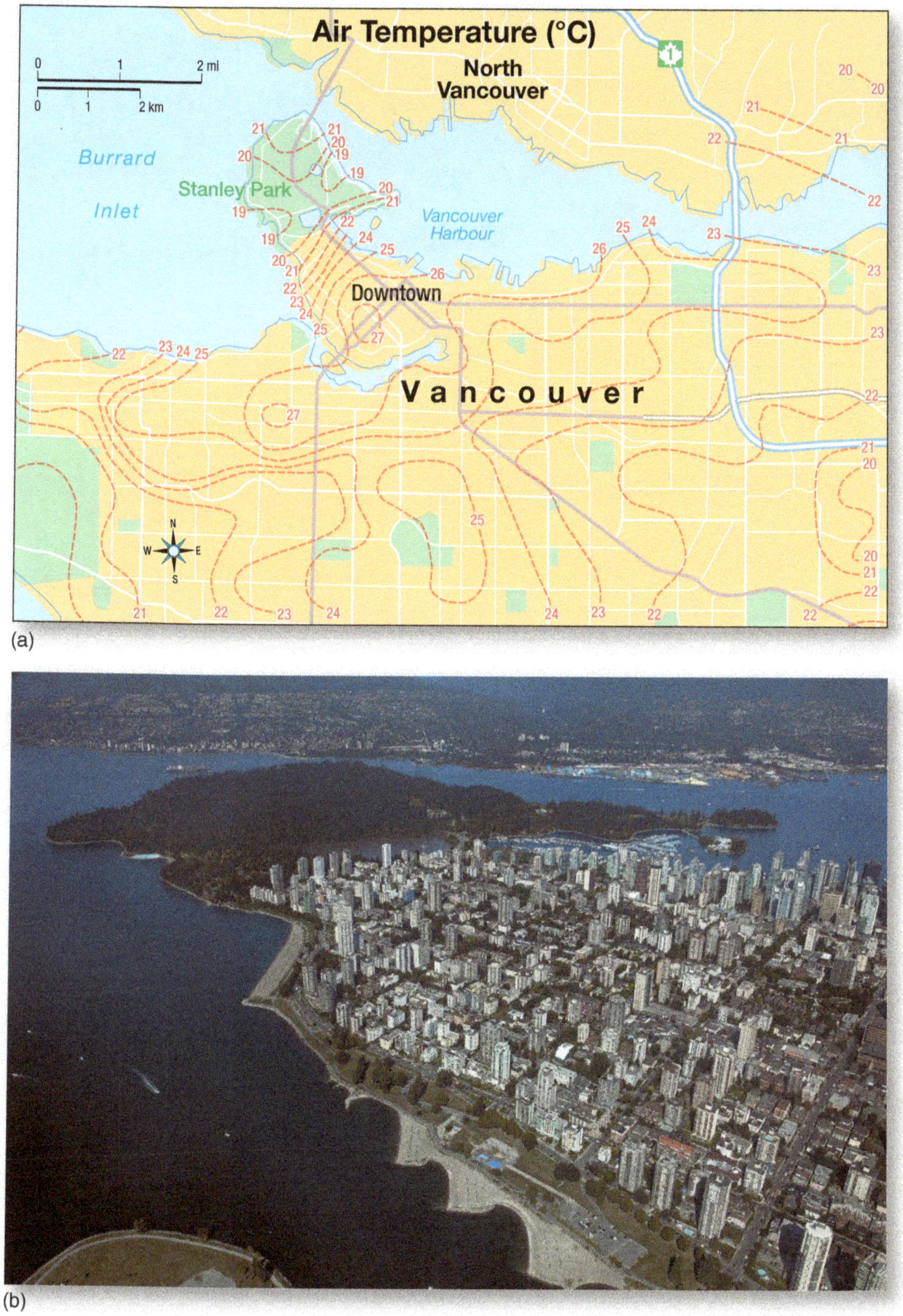

▲ **FIGURE 14–18 Example of Heat Island.** (a) Temperatures in Vancouver, British Columbia, at 9 P.M. on July 4, 1972 in °C. Note the sizable temperature gradient between downtown and Stanley Park. (b) The Vancouver downtown area with Stanley Park in the background.

Changes in Heat Storage

As explained in Chapter 3, radiation is not the only mechanism that transfers energy from one place to another; conduction and convection are also important heat transfer mechanisms. In the middle of a sunny day, absorption of solar radiation warms the surface of the ground. Conduction within a very thin layer of the atmosphere and convection transfer much of this energy to the air. At the same time, a gradient develops in which soil temperature decreases with depth, and heat is conducted downward.

During the late afternoon, the surface begins to cool when energy losses by radiation and convection exceed the absorption of shortwave and longwave radiation. The soil temperature profile eventually reverses, with temperature increasing with depth. Thus, during the evening hours, heat that has been stored within the soil is transferred to the surface.

The same processes just described occur in the walls and roofs of urban buildings. As the surface is warmed during the day, a temperature gradient develops that conducts heat toward the building interior. When cooling occurs in late afternoon, the stored heat is released to the surface. What is different from the rural setting is that materials used in building construction have a much greater capacity to store heat than most natural surfaces. As a result, more stored heat is available for transfer to the lower atmosphere during the evening and nighttime than for natural surfaces, and nocturnal temperatures are increased.

The release of heat from buildings just described is supplemented by the anthropogenic heat produced for comfort (e.g., space heating) or as a by-product of other activities (e.g., waste heat from a hot car engine). Anthropogenic heating is greatest during the winter, which partially explains why heat islands are most pronounced during the low-Sun season. Anthropogenic heat can be a surprisingly large component of the urban energy balance. In Vancouver, British Columbia (49° N), for example, the amount of anthropogenic heat released in the winter has been estimated to be nearly four times that received as net radiation. That estimate, based on 1970 data, probably understates the importance of anthropogenic heat currently released because the city has grown significantly since then.

Sensible and Latent Heat Transfer

In Chapter 3 we saw that most of the global surface has a surplus of net all-wave radiation on an annual basis. That surplus is transferred to the atmosphere as sensible and latent heat. When moisture is available near the surface, the transfer of energy as latent heat can exceed sensible heat transfer, indicating that most of the surplus is consumed by evaporation. On the other hand, if the surface is completely dry, surplus energy raises the surface temperature far above the air temperature, and sensible heat dominates. All other things being equal, the higher the ratio of sensible to latent heat, the greater the temperature.

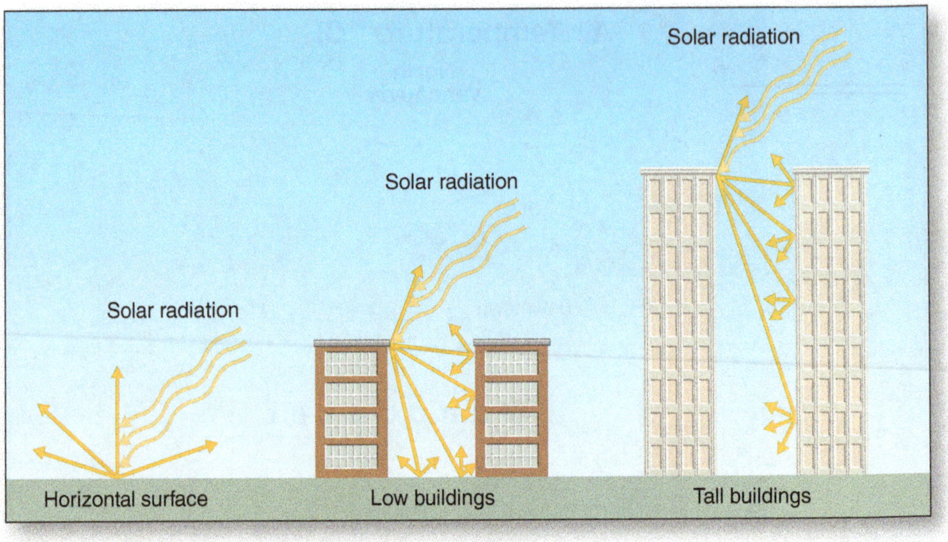

▶ **FIGURE 14–19 Effect of Buildings.** As incoming radiation contacts a building, some is scattered off in all directions and some is absorbed. The scattered radiation may in turn hit an adjacent building, where further absorption can take place. This lowers the urban albedo.

Video
Hello Crud
http://goo.gl/gDF9pS

DID YOU KNOW?

It has long been known that cities create their own urban heat islands. But cities can also affect precipitation patterns—in conflicting ways. The urban areas around the San Francisco Bay Area, Los Angeles, and San Diego increase the amount of small particulates in the air that are transported downwind. The increase in particulates causes clouds over the foothills of the mountains to the east to have their liquid or ice content distributed among a larger number of droplets or crystals. The smaller size of the cloud constituents makes it harder for precipitation to develop and may reduce mountain snowfalls by as much as 20 percent in some places.

But a separate NASA study shows that in cities more prone to thunderstorm activity, the urban heat island effect—combined with other urban factors such as surface convergence and increased aerosol concentrations—can *enhance* precipitation downwind by as much as 51 percent. Summertime precipitation increases have been observed downwind of numerous cities, ranging from Atlanta, Georgia, in the Southeast, to Dallas, Texas, in the South, and as far north as Chicago, Illinois.

Urbanization affects the routing of precipitation in a way that favors increased sensible heat transfer. Unlike natural surfaces that allow rainfall or snowmelt to permeate the soil and be retained below ground, city streets and sidewalks are almost impervious to water (Figure 14–20). So when precipitation occurs, most of the water runs off the surface and ultimately flows out through the flood-control system. The reduction in available water increases the input of sensible heat to the atmosphere at the expense of latent heat and helps increase the temperature of the urban environment.

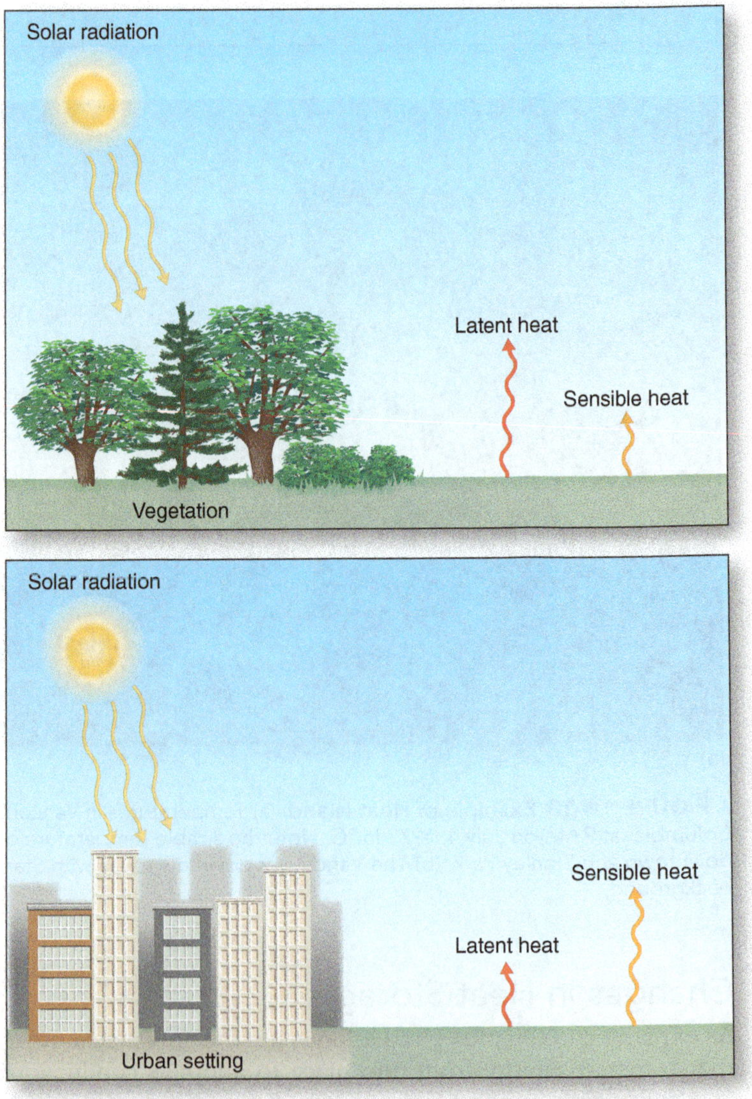

▲ **FIGURE 14–20 Latent and Sensible Heat.** Urban pavements greatly reduce the amount of precipitation that permeates the soil and becomes available for evapotranspiration. The result is an increase in sensible heat transfer and a decrease in latent heat transfer into the atmosphere.

Urban Heat Islands and the Detection of Climate Change?

By now everyone has heard much discussion about the possibility of impending climatic change resulting from the anthropogenic emission of greenhouse gases. Many atmospheric scientists believe this change has already begun to take place. To support this notion, they point to an overall warming of 0.3 to 0.6 °C (0.5 to 1 °F) occurring since the late nineteenth century for weather stations having records going back more than a century. However, we cannot use these recorded temperature changes at face value because many of the data come from urban weather stations, and most cities with long-running weather stations have undergone considerable growth during the past century. Thus, we must contend with the problem of enhanced urban heat islands influencing the data, which means that records from large urban areas are not representative of the surrounding region. Atmospheric scientists are well aware of this source of bias in temperature records—which happens to be relatively small—and routinely account for its effect, either by discarding contaminated data or by adjusting values downward for affected stations.

CHECKPOINT

14.10 Why does the intensity of an urban heat island vary spatially across a city?

14.11 What are some ways in which a city could reduce the heat island effect?

Summary

14.1 Describe the different types of atmospheric pollutants.

- Particulates are solid and liquid aerosols that can be produced as either primary or secondary pollutants.
- Pollutant gases include carbon oxides, sulfur compounds, nitrogen oxides, volatile organic compounds (including hydrocarbons), and photochemically formed gases (the most notable of which is ozone).

14.2 Explain the role of atmospheric conditions in determining pollution concentrations.

- If winds are strong and constantly shift direction, pollutants are distributed over a larger area and concentrations decrease.
- Statically unstable air also favors the dilution of gases and particulates by enhancing vertical mixing.
- On the other hand, stable conditions and, in particular, the presence of an inversion, can greatly restrict vertical motions and concentrate pollutants near the ground.

14.3 Describe the efforts that have been made to reduce air pollution.

- The original Clean Air Act has had a dramatic impact on air quality during the past several decades, despite the fact that the population and miles driven have increased nationwide.
- Although reductions in many pollutants have been dramatic, numerous cities in the United States still have not met clean air goals.

14.4 Explain the factors that produce urban heat islands and describe their effects on local weather.

- The urban heat island is a well-known phenomenon in which changes in the surface (such as the replacement of vegetated surfaces with concrete and asphalt), the existence of buildings with vertical walls, and the release of heat as a by-product of human activity combine to increase temperatures. These increases are most notable during the evening and nighttime hours and in the winter season.

Key Terms

Review Questions

1. Explain the distinction between primary and secondary pollutants.
2. What are particulates and how are they introduced into the atmosphere?
3. What are the two processes most responsible for removing particulates from the atmosphere?
4. What are PM_{10} and $PM_{2.5}$? Does one pose a greater health risk than the other?
5. List the most important gases that contribute to air pollution.
6. In what way does carbon monoxide harm the human body?
7. What are the primary sources of carbon monoxide in the atmosphere? If these sources are nonanthropogenic, why is CO considered a pollutant?
8. What are the primary sources of sulfur dioxide and sulfur trioxide in the atmosphere?
9. Would a person be more likely to notice the presence of high CO or high SO_2 contents in the ambient air?
10. Which primary pollutant is most likely to promote the formation of acid fog or acid rain?
11. Why is nitric oxide much less common in the atmosphere than nitrogen dioxide?
12. Describe the general composition of volatile organic compounds.
13. How do London-type and Los Angeles–type smog differ from each other?
14. Describe the various atmospheric controls that affect the concentration of air pollutants.
15. Which pollutant gases cause a noticeable coloration of the atmosphere? Which have a characteristic odor?
16. How does the construction of buildings in cities alter the radiation exchange near the surface and contribute to the urban heat island effect?
17. What effect does urbanization have on the exchange of sensible and latent heat?
18. Describe the way in which heat storage in cities differs from that of surrounding rural areas.
19. Does the urban heat island manifest itself equally during the day and night? Are there seasonal differences in the magnitude of the heat island effect?

Critical Thinking

1. There has been much improvement in air quality for North America as a whole. Further improvements will only arise from measures that may be costly, both directly through the application of technology and by reductions in certain economic activities. Do you personally believe that further improvements can be brought about at a cost that people are willing to bear?
2. It is believed that global warming during the past few decades might have been even greater were it not for the effect of certain types of air pollution. Explain how this could come about.
3. Visit the web page at http://earthobservatory.nasa.gov/GlobalMaps/ and form your own conclusions about whether air pollution is now primarily a local or a global process.

Problems & Exercises

1. Examine the map in Figure 14–15 or Figure 14–16. How is the air pollution situation in your area? Is the information on the map consistent with your perception of the local air quality? What factors do you think lead to the type of air quality that your area has?
2. If you live in or near an urban area, make note of the daily maximum and minimum temperatures in your region. Do you detect a significant urban heat island? How does the magnitude of this heat island compare for maximum and minimum temperatures? Are there seasonal differences?
3. Visit the web page at www.epa.gov/air/partners.html for a listing of EPA, state, and local agencies that provide data on air quality. Refer to one of the agencies for a daily report on the air quality in your area. How do the changes in air quality fluctuate with different weather patterns where you live?

Visual Analysis

Examine the photograph below of Hangzhou, a city in eastern China and answer the following questions?

14.1. What activities can you identify that are creating air pollution?

14.2. Which types of pollutants do you think are being emitted by each activity?

PART

6 Current, Past, and Future Climates

CHAPTER 15

Earth's Climates

CHAPTER 16

Climate Changes: Past and Future

◂ *Icebergs produced by the Greenland ice sheet.*

Climatic conditions are extremely varied across Earth's land surfaces, with large areas of tropical rainforest, extremely dry deserts, and cold polar and high-elevation regions. World oceans are almost equally variable. But variations through time have been (and will continue to be) equally dramatic. We now turn to an examination of climatic distributions across Earth's landmasses and oceans and the reasons behind their existence, followed by an examination of climatic changes and how atmospheric scientists infer past conditions and predict future changes—including climate changes resulting from the buildup of carbon dioxide and other greenhouse gases released by human activities.

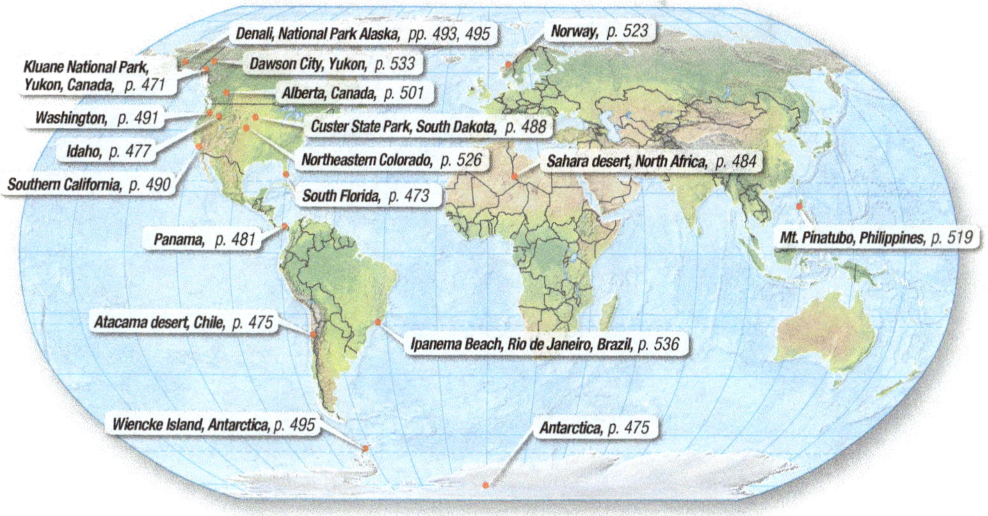

CHAPTER

15

Earth's Climates

Although we don't always know in advance what type of weather will occur at any particular place on a particular day, we do have some idea of what type of weather is considered normal for a given location. The scene in the opening photo probably fits with everyone's mental image of Miami, Florida's climate. The relative warmth of Miami in March makes it an attractive destination for countless students on spring break, whereas a vacationer planning to be there in August knows to expect heat and stifling humidity (along with correspondingly low hotel room rates). Along the same lines, the bitter cold usually experienced during a Minneapolis winter makes it a rare vacation travel destination during that time of year. However loosely formed, these are conceptualizations of climate, our focus for this chapter. The impression about a place's climate is based on many years. We would not revise our impression of Miami based on a single unusually mild summer day, nor that of Minneapolis based on a surprisingly mild day in January. Thus climate deals with what is typical for a location over a long time frame, as opposed to the passage of daily weather occurrences. In this chapter we examine the issues involved in classifying climate, and we describe the broad climatic regions across the globe.

◀ *South Florida has a humid subtropical climate. In winter, while much of North America shivers, the Miami area enjoys warmth and sunshine suitable for beachgoing or cruise-ship getaways.*

Learning Outcomes

After reading this chapter, you should be able to:

15.1 Define climate and explain the climatic normal and its limitations as the sole indicator of climate.

15.2 Identify the main climate groups in the Koeppen system.

15.3 Describe the characteristics of tropical climates.

15.4 Describe the characteristics of dry climates.

15.5 Describe the characteristics of mild midlatitude climates.

15.6 Describe the characteristics of severe midlatitude climates.

15.7 Describe the characteristics of polar and highland climates.

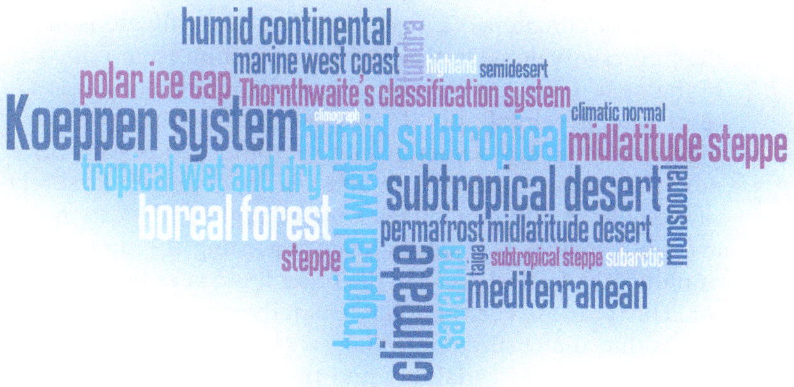

Climate and Controlling Factors

15.1 Define climate and explain the climatic normal and its limitations as the sole indicator of climate.

We define **climate** formally as the statistical properties of the atmosphere. By far the most common climatic statistic is an average calculated over a number of consecutive years. Note, however, that in using an average value we deliberately ignore differences from one year to the next. Using temperature as an example, one year in the averaging period might have been abnormally warm and another very cold, but the average does not reflect those differences nor does it represent the most extreme value observed. Rather, the average reflects the middle of the distribution of temperatures and is a valid definition of "typical" for the period. Two obvious questions arise with regard to climatic averages. First, how long a period should be used? Too short a record will fail to remove enough year-to-year variation, and too long a record runs the risk of missing any changes in climate that occurred over the period. Standard practice regards a 30-year record as the appropriate trade-off. A second question is: What time period? That is, what 30-year period do we select? It seems natural to use the most recent 30 years in order to capture the present climate, but that would call for yearly updates in climate statistics. Instead the convention is to use the most recent 30-year periods ending on a decade, such as the period 1981–2010. Taken together these two conditions define the **climatic normal**, which has been in use worldwide for about 75 years. When a TV reporter says "normal for the date is 73 °F," he or she is almost certainly referring to the climatic normal.

Statistical Properties of Climates

It is worth realizing that the climatic normal is merely a convention and may not be suitable for some uses. For example, electric energy companies trying to estimate power demands find the normal to be a relatively poor predictor of the future. Nevertheless, the climatic normal is preeminent and reinforces the idea of climate as "average weather." But climate is more than average values. It also includes variation from one year to the next. For example, two places might have similar long-term average rainfall. But if one place typically has both very high and very low values, while the other tends to receive nearly the same total year in and year out, we would surely say they have different climates. After all, in one location the average provides a good estimate regarding the amount of rain that will fall, whereas in the other it is a very poor indicator. This year-to-year variability is another statistical property and is thus a measure of climate.

In addition to variability and the occurrence of extreme events, we might be interested in the degree to which above-average or below-average values tend to come in runs or be clustered in successive years. For example, do periods of drought tend to be followed by wet periods? Or are extended episodes of extreme heat and cold a common occurrence? These are questions about another statistical property, namely, the correlation between values in successive years. Again, this is part of climate. The point we want to make is that climate consists of many statistical properties, not just average values. For our purposes, a consideration of average values will be sufficient, but a complete description of climate would include much more.

Climate Variables

As mentioned, the definition of climate is not limited to just one or two variables. Temperature and precipitation are surely important elements, but climate also includes wind, cloudiness, and net radiation, to name just a few others. Thus, the climate of a place involves a wide array of variables. It should be obvious that because no two places will have identical statistical properties across the full range of atmospheric variables, in a technical sense every location has a unique climate. We might speak of Chicago's climate, but its Miracle Mile along Lake Michigan has a different climate than the area surrounding O'Hare Airport just to the west. Likewise, no place on Earth has a climate exactly like that of the Miracle Mile. But for many purposes small differences in climate can be safely ignored. We know rain-fed corn can be grown across wide swaths of the Midwest United States with little risk of complete crop failure, and we know the same attempt would be well beyond foolhardy for the entire African Sahara. In this text we will therefore work with climates on a large spatial scale, encompassing areas that are typically hundreds to thousands of kilometers wide.

Controlling Factors

On large spatial scales the factors determining climates can be readily identified. The change of solar radiation input with latitude comes to mind immediately. Thus, to a first approximation, symmetric climatic bands centered on the equator is a reasonable guess and has some basis in reality. However, other controls exert their influence to create a considerably more complicated pattern, as discussed in Chapters 3 and 4:

- The uneven distribution of land and ocean
- The role of elevation, with high-elevation locations colder than would be expected solely on the basis of latitude
- Prevailing atmospheric and ocean circulations—for many places a dominant influence.

Notice that depending on location, controlling factors can interact with one another. Thus, prevailing winds blowing across a cold ocean surface toward land yield a very different climate than similar winds moving over a warm tropical sea (see Figure 15–1 for an example). Climatic gradients on the windward side of mountains are very different from those on the leeward side, but the effect varies by latitude. The variety of combinations and interactions is huge, so much so that a general discussion of climatic controls is not very useful. Therefore, throughout our discussion of individual climates we will make reference to underlying controlling factors. We will rely

▲ **FIGURE 15–1 Atacama Desert.** Elevation interacts with prevailing winds and cold sea-surface temperatures to create fogs that nourish desert vegetation in the bone-dry Atacama Desert of Chile.

▲ **FIGURE 15–2 Sea Ice in Antarctica.** Seasonally varying sea ice extent is one of the important factors in the way oceans influence the climate.

on the reader to be familiar with the relevant background material as presented in previous chapters.

Oceans as a Factor in Climate

Before proceeding, bear in mind that the importance of oceans goes beyond influencing coastal locations; oceans help shape climate throughout the Earth system. For example, they are the main supplier of water to the atmosphere. But as we saw in Chapter 8, some ocean basins have much higher evaporation rates than others, and thus rainfall on continents is partly constrained by how much water vapor can be transported from an oceanic source. As another example of the ocean's influence, consider the fact that sea ice comes and goes in the Arctic Ocean in winter and summer (Figure 15–2). However, seasonal changes in land ice are larger because the land changes temperature easily compared to water. Thus, albedo changes in the Arctic ocean are less than those over land, and the energy budget and climate throughout the high northern latitudes is different because of the Arctic Ocean. Looking ahead to Chapter 16 and the concern that the Arctic will become ice free in the summer, we see the disproportional influence of that particular ocean in the future.

Classifying Climates

Although the delineation of distinct climates may seem like a very straightforward endeavor, establishing the criteria by which climates are separated requires considerable subjectivity. Consider what you would do if called on to devise a scheme by which Earth's surface would be covered by distinct climatic zones, each having properties that set them apart from the others. The job would require you to set boundaries that separate one climate zone from another. Yet in nature such clear boundaries are rare. Thus, there is a considerable difference in temperature and precipitation along the east coast of North America from Florida to the Maritime Provinces of Canada, and you certainly would not want to put St. John's, Newfoundland, in the same climatic zone as Tallahassee, Florida. But exactly where would you draw the lines that separate the various climates? And how would you decide how many climates? Too many would make the system too complex; too few would fail to capture the patterns you would want to identify.

Climatologists over the years have made numerous attempts to establish useful climatic classification schemes. Some are based on the obvious properties of temperature and precipitation. Others use the frequency with which air mass types occupy various regions, differences in energy budget components, or seasonal characteristics of the water balance at the surface. Each has its own advantages, depending on the purpose of the classification. Agriculturalists, for example, would probably be most interested in using a classification that yields information on water availability relative to plant needs, reflecting gains and losses of water in the soil column (precipitation, evapotranspiration,[1] runoff, and losses to groundwater). For more on this topic, see *Box 15–1, Physical Principles: The Thornthwaite Classification System.*

CHECKPOINT

15.1 What is climate?

15.2 What considerations enter the decision regarding the length of period to use in a climatic average?

15.3 What are some of the factors that determine the climate of a location?

15.4 What are some of the problems encountered when trying to divide the Earth into climatic zones?

[1]Evapotranspiration is the combination of water directly evaporated at the ground surface and that absorbed by plants and evaporated through their leaves.

15–1 PHYSICAL PRINCIPLES

The Thornthwaite Classification System

As with any other climate system, Koeppen's has its shortcomings. One of the most important is that it is based on vegetation boundaries that have been associated with monthly values of mean temperature and precipitation. This is problematic because these two variables alone do not directly determine the geographic limits of natural vegetation. A superior system would also include potential evapotranspiration as a factor that plays a direct role in determining the geographic limits of vegetation.

In tandem, the opposing effects of evapotranspiration and precipitation determine the *water balance*, examples of which are shown in Figure 15–1–1. Wherever evapotranspiration exceeds precipitation, the amount of moisture in the soil is reduced, as its loss is not offset by soil moisture inputs. When precipitation exceeds evapotranspiration, the soil moisture is replenished until it reaches the maximum amount of water that can be retained by the soil against the force of gravity (field capacity).

Thornthwaite's classification system was developed to distinguish climates based on moisture availability. It is based on the principle of the water balance and evolved through decades of work that culminated in its final form in 1955. Its first criterion is a *moisture index* that compares the amount of average precipitation each month to the potential evapotranspiration. The latter value derives from a formula using mean temperatures and the monthly values of average period of daylight (a function of latitude) for each station to establish a monthly moisture index. These monthly values are then summed to produce an annual moisture index, the value of which distinguishes arid, semiarid, subhumid, humid, and perhumid (very humid) climates. These divisions are based on arbitrary percentage changes in the moisture index (20 percent changes for the humid climates, 33 percent changes for the dry climates), not from associations with plants or other nonclimatic phenomena.

The second criterion is the *thermal efficiency* of a location, or the total amount of *potential evapotranspiration* (the amount of evaporation that would occur under given climate conditions and assuming an unlimited supply of soil water). The remaining two criteria are based on the seasonality of precipitation and potential evapotranspiration. When combined, the four criteria create a more physically based climate classification scheme than that of Koeppen.

So why hasn't the Thornthwaite system supplanted Koeppen's as the most popular? This is partly explained by its greater complexity. Compared to Koeppen's system, the water-balance computations needed by Thornthwaite's method are laborious, and the resulting regions are therefore quite removed from the underlying climatic data. Thus, the pattern of climates that emerges is harder to interpret in terms of large-scale processes. Also, although the basic concept of potential evapotranspiration is widely accepted, the method developed by Thornthwaite does not follow from physical principles, but instead relies on data collected mainly in the eastern United States. The data were used to construct empirical (observation-based) equations for potential evapotranspiration. Unfortunately, Thornthwaite did not publish details regarding how the equations were developed, and there are questions as to how well they work in other areas. Thus, if Thornthwaite's method were applied to the entire globe, it is not clear how meaningful the resulting regions would be.

Of course, the Koeppen and Thornthwaite systems are not the only systems that have been devised. Most of the others are designed for more specific applications than either of these two. Some, for example, have identified regions of human comfort, whereas others consider the dominance of different types of air masses. The familiar map of plant-hardiness zones found on seed packages is another example. More sophisticated classifications have used energy budget considerations.

The Koeppen System

15.2 Identify the main climate groups in the Koeppen system.

For many people, a climatic classification scheme based on temperature and precipitation is useful because it yields information on the two meteorological variables of greatest general interest. The most widely used systems based on these variables have followed on the work of Vladimir Koeppen,[2] a German citizen of Russian ancestry. The **Koeppen system** was developed over the period from 1918 to 1936 in a process of almost continual revision and refinement. Koeppen looked at the world distribution of natural vegetation types, located the boundaries that separated them, and determined what combinations of monthly mean temperature and precipitation were associated with those boundaries. Thus, although its climatic types are determined by temperature and precipitation, Koeppen's system is inherently tied to natural vegetation. Contrary to what one might assume, it does not begin with the idea of "natural" temperature/precipitation regimes. The boundaries are associated with plant associations, which may or may not coincide with what seem to be obvious or striking gradients in temperature and precipitation.

This chapter uses a version of the Koeppen system as modified by Glenn Trewartha, American geographer. Various versions of the Koeppen system apply different names for each of the climates and may have slightly varying criteria for distinguishing them. Thus, our maps and descriptions are necessarily somewhat different from some other portrayals

[2]Also spelled Köppen.

Regardless of the premises on which they are based, each has its own set of advantages and disadvantages.

15–1–1 In what way is the Thornthwaite system superior to the Koeppen system?

15–1–2 Explain why the Thornthwaite system is not more widely used than the Koeppen system.

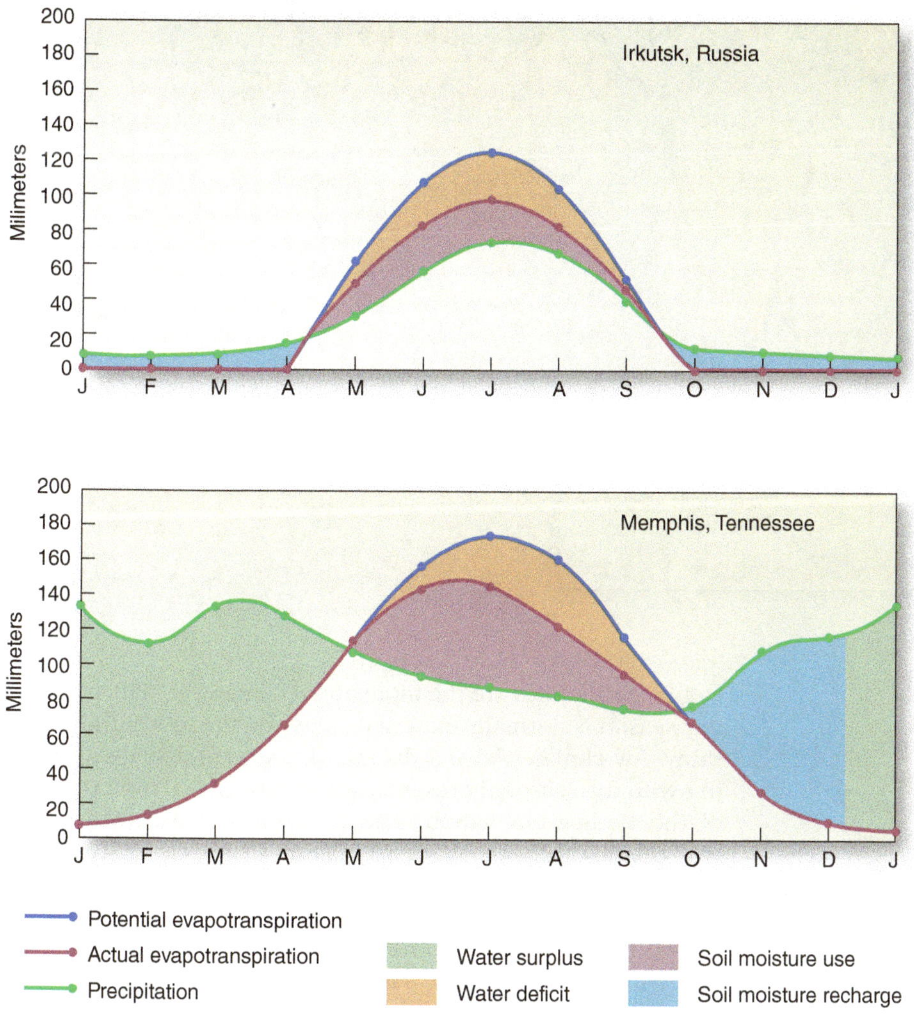

Video
Diurnal Variability in Global Precipitation

http://goo.gl/2rbyTZ

◀ **FIGURE 15–1–1 Water Balances.** The diagrams illustrate the monthly balances of precipitation, potential evapotranspiration, and actual evapotranspiration for Irkutsk, Russia, and Memphis, Tennessee.

of the Koeppen scheme. Moreover, this scheme is totally descriptive and does not attempt to explain the causes of the various climates. However, as seen in Figure 15–3, in most cases climates can be related to the various controlling factors mentioned above.

Koeppen used a multitiered classification system that delineated primary climates by capital letters ranging from A through E. These five broad categories tend to arrange themselves across Earth's surface in response to the latitude, degree of continentality, and location relative to major topographic features. In addition to these climates, the version we are using includes an additional one for mountain environments, designated by H. The main climate groups (designated by first letter only) can be briefly described as follows:

- A—**Tropical.** Climates in which the average temperature for all months is greater than 18 °C (64 °F). Almost entirely confined to the region between the equator and the tropics of Cancer and Capricorn.
- B—**Dry.** Potential evaporation exceeds precipitation.
- C—**Mild Midlatitude.** The coldest month of the year has an average temperature higher than −3 °C (27 °F) but below 18 °C (64 °F). Summers can be hot.
- D—**Severe Midlatitude.** Winters have at least occasional snow cover, with the coldest month having a mean temperature below −3 °C (27 °F). Summers are typically mild.
- E—**Polar.** All months have mean temperatures below 10 °C (50 °F).

Climatic zones A, C, D, and E are based on temperature characteristics. The A climates (tropical) tend to straddle the equatorial regions; C, D, and E climates usually occur

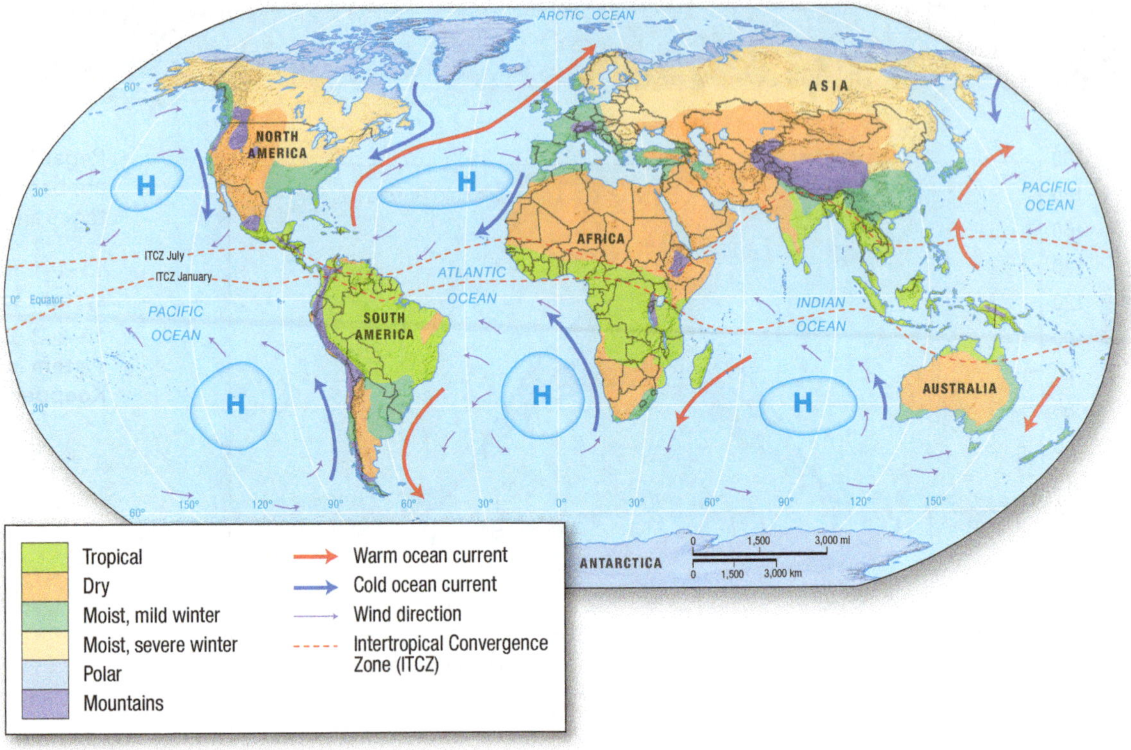

▸ **FIGURE 15–3 Factors Influencing Climates.** A generalized map of Koeppen climates for the continents and important controlling factors. Note the importance of prevailing pressure and winds, ocean currents, and topography.

Video
Supercomputing the Climate
http://goo.gl/Wwl0o0

DID YOU KNOW?

Climate classifications have been derived based on some unconventional but nonetheless significant climatological elements. For example, scientists have calculated and mapped the average apparent temperatures (an index of temperature discomfort due to a combination of high temperature and humidity, which was discussed in Chapter 5) for the hottest 12 days of the year for weather stations across the United States. The maps indicate that typically the most severe heat occurs in the region extending from south Texas northeastward to about St. Louis, Missouri. There the heat index on the hottest days of the year exceeds that found in much of southern Arizona or the Florida panhandle.

sequentially further from the tropics and toward the polar regions. The sole primary climate type that considers precipitation is the B climate, designating deserts and semideserts.

The A through E climates are subdivided into smaller zones represented by a second letter, which are further subdivided. We will use subdivisions only through the third letter; thus, each individual climate is represented by at most a three-letter combination describing its temperature and precipitation characteristics. The descriptions for each of the climates are presented along with their distributions in Figure 15–4. Although this system has been taught to countless numbers of students over the years, there is little insight to be gained by memorizing the exact temperature and precipitation values that define the climate boundaries. Thus, the critical values are omitted from the table.

For the A climates, the second letters *f*, *m*, and *w* indicate if and when a dry season occurs. Af climates have no dry season at all. Am climates are the monsoonal climates in which a short dry season is normally experienced while the rest of the year is rainy. Aw climates have a distinct dry season, usually coinciding with the seasonal presence of the subtropical high pressure of the Hadley circulation. The dry climates are divided into two classes: the true deserts (BW) and the semideserts (BS). The second letter of the C and D climates signifies the timing of the dry season. The letter *f* indicates no dry season at all (as with the A climates), while *s* and *w* represent dry summers and winters, respectively. The second letter of the E climates (capitalized) distinguishes polar tundra (ET) regions from areas covered by glaciers (EF). The third letter of each climate represents the temperature regime. As you can see, the meaning of the third symbol varies among the major groups.

Climate Regions over the Oceans

The Koeppen classification was created with vegetation and land areas in mind, and to this day most Koeppen maps confine themselves to continental conditions. But both our maps and discussion of Koeppen climates include ocean areas. This seems fitting considering the planet is more than 70 percent ocean—how could we claim to describe Earth's climate and leave out more than two-thirds of the planet? Clearly there are some differences between land and ocean that can be anticipated. Oceans will have a smaller annual temperature range than land at the same latitude. If there is upwelling, air temperatures will be colder. A warm ocean current will extend a more moderate climate poleward of where it would be found on land. All these things and more must be kept in mind. Perhaps more subtly, the partition of a particular climate type between land

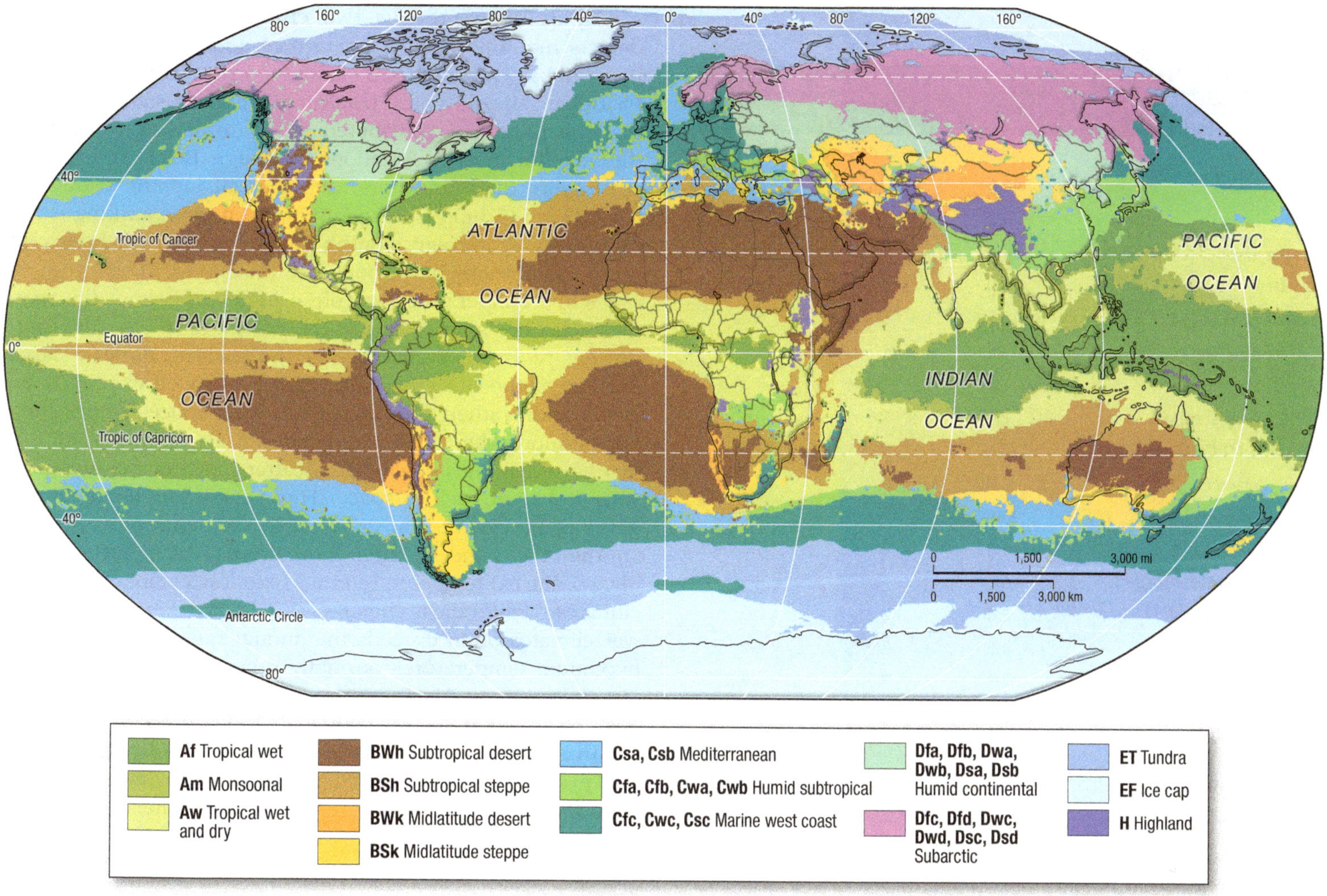

▲ **FIGURE 15–4 World Climates.** Map of Koeppen climates. Notice how some climates are much more extensive over ocean than land, whereas others are almost exclusively continental.

MapMaster World and North America
Physical Environment: Climate

and ocean is an important part of the Earth System. Table 15–1 provides the percentages of Earth's area with each climate and will be referenced repeatedly in the sections that follow.

Of course, determining climates over the ocean is challenging given the lack of weather stations. You may therefore wonder how Table 15–1 and the associated maps were created. Shipboard measurements are helpful, but far too uneven in time and space for our purposes. Therefore, we have turned to recently available satellite-derived estimates of temperature and rainfall for oceans. For the continents we used conventional data based on weather stations numbering in the many thousands.

CHECKPOINT

15.5 What are the five main climate groups in the Koeppen system and what distinguishes them?

15.6 What criteria are used to determine subclimates in each of the Koeppen groups?

Tropical Climates

15.3 Describe the characteristics of tropical climates.

The name of this climatic group could not be more straightforward or accurate. Tropical climates exist almost entirely between the tropics of Cancer and Capricorn (see Figure 15–4). Collectively they are the largest group, covering almost 30 percent of the planet (see Table 15–1). This isn't surprising considering how much of Earth's surface area is found at tropical latitudes (50 percent between 30°S and 30°N). Most of the tropics are ocean covered, and most tropical climates are oceanic rather than continental. According to the table, the ratio is more than 4 to 1. The tropical group consists of three climates, each of which is warm year-round, with only minor—and in some cases, minimal—variation in temperature throughout the year. The three climates are distinguished by their different degrees of precipitation seasonality. The **tropical wet** climate has significant rainfall every month of the year, the **tropical wet and dry** climate has a pronounced dry season, and the **monsoonal** climate undergoes relative

TABLE 15-1
Climate Type Coverage

Climate Type	Percentage of Earth Area Total	Land	Ocean
A—Tropical	29.4	5.6	23.9
Tropical wet	10.5	1.3	9.2
Monsoon	4.3	1.0	3.4
Tropical wet and dry	14.6	3.3	11.3
B—Dry	25.0	8.3	16.7
Subtropical desert	10.5	3.9	6.6
Subtropical steppe	10.8	1.9	9.0
Midlatitude desert	1.3	1.0	0.3
Midlatitude steppe	2.4	1.6	0.8
C—Mild Midlatitude	20.1	3.9	16.2
Mediterranean	4.5	0.6	3.9
Humid subtropical	13.8	3.2	10.6
Marine west coast	1.8	0.1	1.7
D—Severe Midlatitude	6.5	6.0	0.6
Humid continental	2.3	2.2	0.2
Subarctic	4.2	3.8	0.4
E—Polar	17.8	2.9	14.9
Tundra	11.4	1.2	10.2
Ice cap	6.4	1.7	4.7
H—Highland	1.2	1.2	0.0

dryness for one to three months but receives sufficient moisture that vegetation need not be adapted to seasonal drought. All three tropical climates are dominated by the seasonal movement of the Hadley cells, described in Chapter 8.

Tropical Wet (Af)

The three largest tropical wet climates are found in the Amazon Basin of South America, over western equatorial Africa, and on the islands of the East Indies (Figure 15–5). As the map shows, the majority of these locations exist within about 10° on either side of the equator, although some are found as far poleward as 20°. Tropical wet climates have no dry period because their position near the equator puts them under the constant influence of the intertropical convergence zone. For this reason, precipitation is almost always convective, with strong solar heating of the surface triggering brief but heavy thundershowers in the mid- to late afternoon. In the more poleward areas in which Af climates occur, rain often results from orographic uplift of the predominant trade winds. The tropical wet climate of the Atlantic coast of Central America serves as an excellent example of this phenomenon.

Figure 15–6 presents two **climographs** (a climograph depicts monthly mean temperatures and precipitation, with line and bar graphs plotted simultaneously) for Singapore and for Belém, Brazil, typical tropical wet climate stations. Notice that the rainfall is distributed nearly uniformly throughout the year for most Af locations (although Belém has greater seasonality than Singapore) and that all months average at least 22 cm (5 in.) of precipitation. Even more striking is the uniformity of average monthly temperatures, which vary in these examples by only about 2 °C (4 °F).

While temperatures are often high throughout the year, these climates are not among the hottest on Earth. The ever-present moisture availability at the surface allows a large portion of the incoming solar radiation to be expended on evaporation rather than increasing the surface temperature. Furthermore, the convection of humid air promotes the formation of cumulus clouds that scatter much of the incoming solar radiation back to space. Thus, maximum temperatures never come close to those found in the subtropical deserts. On the other hand, the high humidity retards nighttime cooling, so diurnal temperature ranges are low compared to drier climates. Unlike most climates, the diurnal range in tropical wet climates often exceeds the annual range. Minimum and maximum temperatures normally range from about the low 20s Celsius (low 70s Fahrenheit) in the morning to the low 30s Celsius (high 80s Fahrenheit) in the afternoon.

In addition to the unsurpassed consistency in temperature and precipitation, areas having tropical wet climates tend to have the same type of natural vegetation—the tropical rain forests—which house a very dense canopy of tree cover and a tremendous species diversity in both the plant and animal kingdoms (Figure 15–7).

OVER THE OCEANS: TROPICAL WET Notice in Figure 15–5 the large area that tropical rain forest climates occupy over the

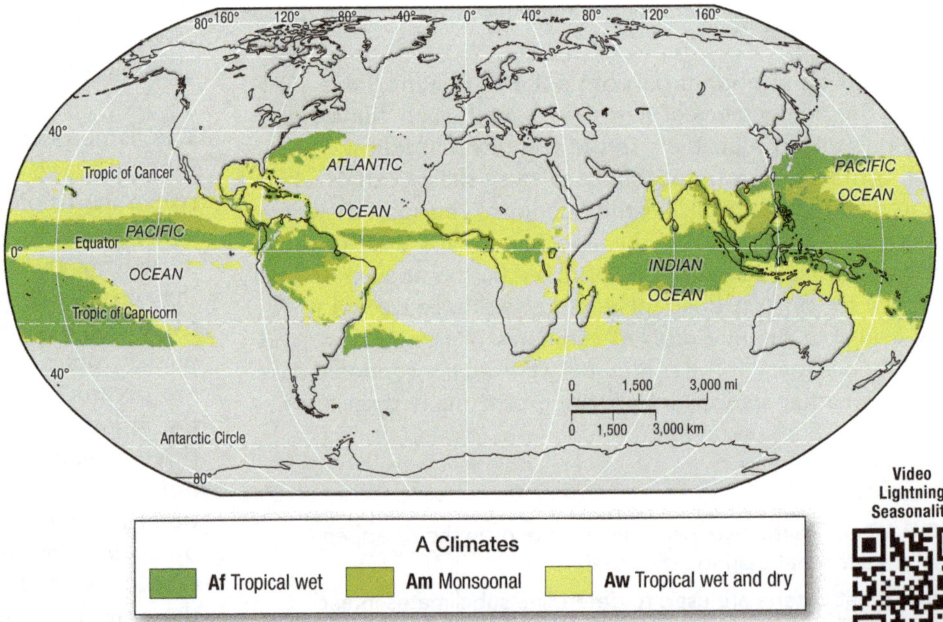

Video Lightning Seasonality
http://goo.gl/Qq8sfA

▲ **FIGURE 15–5 A Climates.** The distribution of tropical climates across the globe.

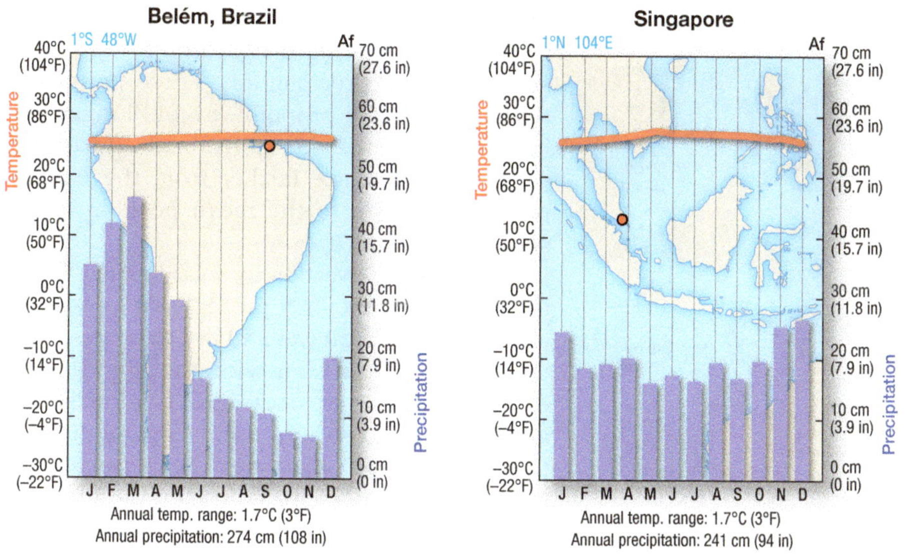

▲ **FIGURE 15–6 Tropical Wet Climates.** Climographs for Belém, Brazil, and Singapore, representative of tropical wet climates. The bars plot the monthly mean precipitation (scaled on the right vertical axis). The lines represent mean monthly temperature (scaled on the left).

▲ **FIGURE 15–7 Tropical Wet Climate Vegetation.** This rain forest in Panama is a response to the year-round warmth and rainfall of tropical moist climates.

equatorial oceans, particularly in the western regions. This is, in part, the result of the high water temperatures typically observed where the westward-moving equatorial current transports heat westward, fostering increased evaporation rates. With seven times as much area over ocean than land, this subtype is clearly dominated by ocean.

Monsoonal (Am)

The monsoonal climate can be thought of as a transitional climate between the tropical wet and the tropical wet and dry climates. It the smallest of the tropical climates, covering only 4 percent of the planet. Monsoonal climates usually occur along tropical, coastal areas subjected to predominant onshore winds that supply warm, moist air to the region throughout most of the year. Such areas are found along northeastern South America; southwest India, near the eastern Bay of Bengal; and in the Philippines.[3] These areas do not extend nearly as far inland as the tropical wet climates, because their existence depends largely on the effect of speed convergence that occurs as offshore air reaches the coast. Rainfall in these climates is also enhanced by orographic uplift. Thus, localized convergence from surface heating is much less a factor in causing precipitation than it is in the tropical wet climates. During the low sun season, some precipitation may occasionally result from the passage of midlatitude cyclones migrating unusually far equatorward. Near the end of summer and into early fall, tropical cyclones and hurricanes can also bring heavy deluges.

As shown in Figure 15–8, precipitation does not occur nearly as steadily throughout the year in a monsoonal climate as it does in a tropical wet climate. Some months can experience exceedingly heavy rainfall while others are nearly dry. In many cases, the wet months in monsoonal climates yield far more rain than does the wettest month for tropical wet climates. In fact, annual precipitation totals in some monsoonal regions are among the highest in the world, with monthly precipitation values during the peak rainfall periods easily exceeding 80 cm (33 in.). Seasonal totals can even surpass 10 m (400 in.)!

Despite the presence of a brief dry season, monsoonal climates usually support dense forests. In these environments, the soil retains sufficient moisture to maintain the lush vegetation even in the absence of heavy rain for part of the year. In other words, the annual total precipitation is large enough that plants do not experience pronounced moisture stress during the dry season; therefore, vegetation generally does not require adaptation to drought. Thus, the Am climate is said to have a "noncompensated" dry season, unlike the tropical wet and dry climate. While not as luxuriant or as abundant in species diversity as the tropical wet environments, these locales contain much more living matter than those of the drier, tropical wet and dry climates.

OVER THE OCEANS: MONSOONAL Over the tropical oceans, areas of monsoonal climates flank the tropical wet climates. Collectively, they amount to 77 percent of the total monsoon area, but these zones tend to be small compared to those of the other two tropical climate types because monsoon areas are not as extensive generally.

[3]Note that the designation *monsoonal climate* is not synonymous with locations subject to the reversal of the winds, as discussed in Chapter 8.

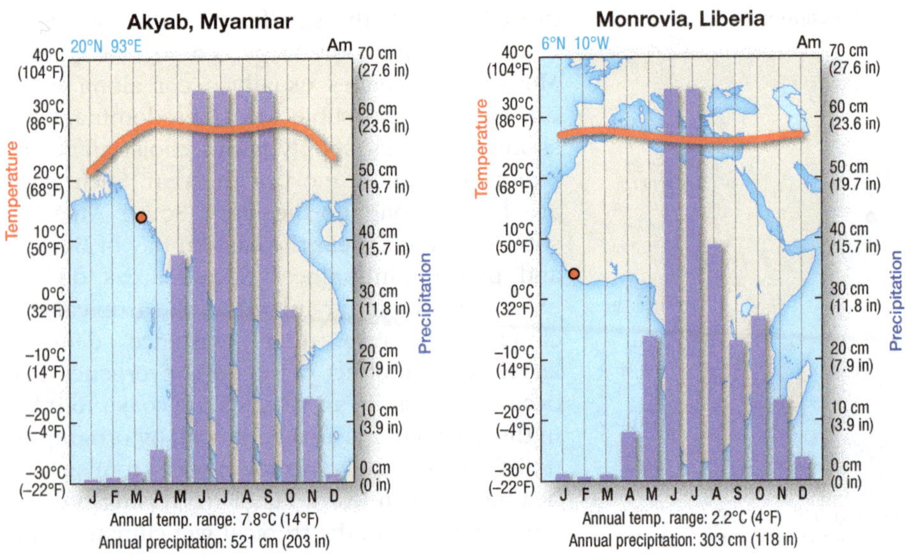

▲ **FIGURE 15–8 Monsoon Climates.** Climographs for Akyab, Myanmar, and Monrovia, Liberia, representative of monsoonal climates.

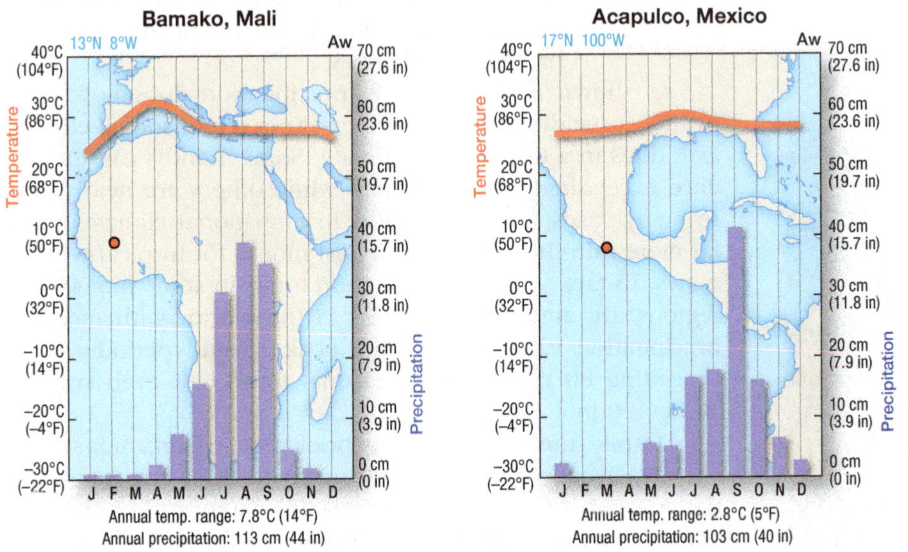

▲ **FIGURE 15–9 Tropical Wet and Dry Climates.** Climographs for Bamako, Mali, and Acapulco, Mexico, representative of tropical wet and dry climates.

Tropical Wet and Dry (Aw)

Tropical wet and dry climates often occur along the poleward margins of the tropics and border dry climates on one side and tropical wet climates on the other. With area equal to about 15 percent of the planet, this is the largest of the tropical climates. They are most extensive in South and Central America and southern Africa. Because they are farther from the equator, they undergo much greater seasonality in precipitation than do the tropical wet and the monsoonal climates, and they have somewhat greater seasonal variation in temperature as well.

As with the other two climates of the humid tropics, tropical wet and dry climates owe their existence largely to the Hadley cell. During the high sun season, the intertropical convergence zone (ITCZ) favors the formation of afternoon thundershowers. As the position of the overhead sun shifts to the opposite hemisphere, however, the subtropical high arrives to bring descending air and a resultant lack of precipitation. These periods of dryness are more pronounced and longer lasting than those of the monsoonal climate because their distance farther from the equator puts them closer to the mean position of the subtropical high.

Localized convection by solar heating within the ITCZ is not the only process that brings precipitation to tropical wet and dry climates. Tropical depressions can bring widescale precipitation. Along coastal areas, occasional tropical storms and hurricanes can increase average accumulations. Figure 15–9 illustrates the seasonality of temperature and precipitation for typical tropical wet and dry climates. In Acapulco, for example, each of the months between May and October receives an average of at least 12 cm (5 in.) of rain. September is by far the wettest month, with an average of about 36 cm (15 in.) of rain. This is largely due to the occasional passage of tropical storms and hurricanes that can dump huge amounts of precipitation. Although most years go by without any of these storms happening, their occasional occurrence increases the monthly average. During the fall, the monthly precipitation decreases, until the dry months of February through April. On an annual basis, these climates receive less precipitation than do either the tropical wet or the monsoonal climates.

Unlike the other two tropical humid climates, the tropical wet and dry areas undergo considerable year-to-year variability. Thus, drought episodes can reduce even further the amount of precipitation received in the dry season—often with fatal consequences, as we have witnessed in the Sahel of Africa (see Chapter 8). Likewise, unusually wet rainy seasons can lead to severe flooding and erosion.

Within the year, monthly mean temperatures exhibit more variability than is found in the other tropical climates, but the variability is nonetheless low compared to most others. The annual temperature range is usually between about 3 °C and 10 °C (5 °F and 18 °F). Diurnal variations are likewise greater in these parts of the tropical humid environment than in the wetter regions. This is especially true during the dry season, when the absence of clouds facilitates greater daytime heating and nighttime cooling, and daily lows and highs might range between 15 °C and 30 °C (59 °F and 86 °F). During the rainy season, the combination of high humidity and cloud cover reduces the diurnal ranges to values similar to those of the tropical wet regions (about 10 °C, or 18 °F).

Tropical wet and dry climates are associated with a natural vegetation type unique among the tropical regions—the **savanna**. This vegetation consists mainly of grasses interspersed with widely separated trees or clumps of trees (Figure 15–10). While it is tempting to attribute the lack of forest

▲ **FIGURE 15–10 Savanna Vegetation in Africa.** Forests of tropical moist climates give way to savannas in the tropical wet and dry climate.

to the presence of the dry season, many ecologists doubt the causality of this relationship. Instead they believe that the vegetation complex results from numerous factors, including recurrent fire, waterlogged soils, and the development of hard layers within the soil.

OVER THE OCEANS: TROPICAL WET AND DRY Tropical wet and dry climates are relatively abundant over the tropical ocean. The ocean fraction represents about 77 percent of the climate. This shows that despite being comparatively resistant to temperature change, the seasonal migration of pressure and winds applies to the ocean just as it does to continents.

CHECKPOINT

15.7 What data are shown in a climograph?

15.8 In what ways are tropical moist climates and tropical monsoons similar? How do they differ?

15.9 What is a savanna, and with what type of climate is it associated?

Dry Climates

15.4 Describe the characteristics of dry climates.

It may surprise many people to learn that the definition of a dry climate is not based on precipitation alone. In other words, there is no set value of annual precipitation (for example, 10 cm) that would mean a region is considered to be arid. Instead, aridity also depends on the potential evapotranspiration, as well as the timing of the precipitation relative to the period of peak potential evapotranspiration. For our purposes, it is sufficient to state that climates are dry when annual potential evaporation exceeds the annual precipitation. These climates occupy about 30 percent of Earth's land surface,[4] which is considerably more than any other climate group (Figure 15–11). About two-thirds of this group is found over oceans.

The dry regions of the world can be divided two ways: by the degree of aridity (that is, true deserts versus the less arid semideserts) and by their latitudinal position (hot versus cooler dry areas). **Semideserts** are transitional zones that separate the true deserts from adjacent climates. They are also called **steppe** climates, with reference to the associated vegetation type consisting of short grasses. True deserts are so dry that only a sparse vegetation consisting entirely of xerophytic species (that is, species adapted to drought conditions) can take hold.

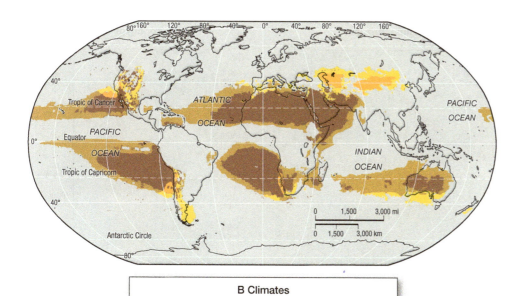

Deserts and semideserts can be classified as either *subtropical* or *midlatitude*. Subtropical dry regions extend across wide expanses of land between the latitudes of about 10° to 30° in either hemisphere and result from large-scale sinking air motions during most of the year. Notable examples include

[4]To see this, realize that only 28 percent of Earth is land. The land portion of B climates is 8.3 percent of the whole Earth, which means it covers about 30 percent of the continents (8.3/28 × 100).

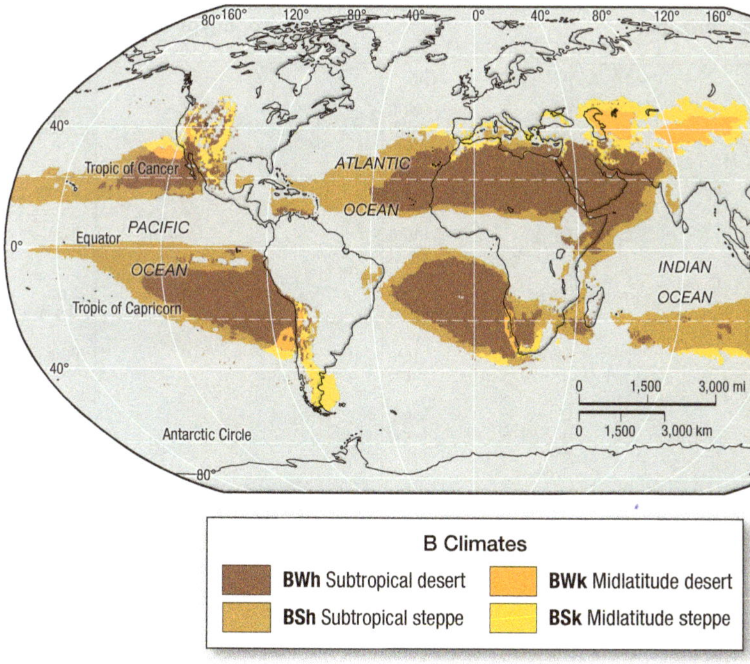

◄ **FIGURE 15–11 B Climates.** The distribution of desert and steppe climates.

Video
Studying Fires
Using Multiple
Satellite Sensors

http://goo.gl/qKVJa

the desert of southern California–Arizona–Baja California, the Sahara Desert of North Africa and the Arabian Peninsula, and most of the Australian interior.

Midlatitude deserts and semideserts usually occur to the east of major topographic barriers that create a strong rain shadow or over interior continental regions well removed from moisture sources. They are found over large portions of the western United States and interior Asia.

The two-tiered system of categorization yields four types of dry climates: **subtropical desert**, **subtropical steppe**, **midlatitude desert**, and **midlatitude steppe**.

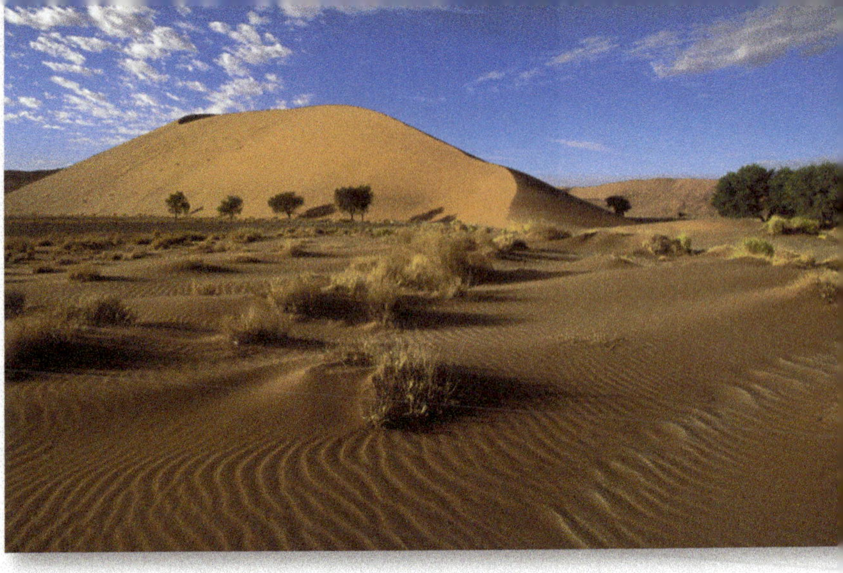

▲ **FIGURE 15–12 Desert Environment.** A hot subtropical desert scene from the Sahara desert of North Africa.

DID YOU KNOW?

One of the driest desert climates of the world is separated from a tropical wet climate by a mere 250 km (150 mi). The coastal area of northern Peru occupies part of the Atacama Desert and receives on average less than 2.5 cm (1 in.) of precipitation annually. The desert is abruptly truncated by the rugged Andes Mountains, which separates it from the large Af climate to the east, where precipitation exceeds 285 cm (114 in.) per year.

DID YOU KNOW?

In some parts of South America and Baja California, villagers in remote desert areas use a simple but ingenious system for capturing drinking water. Nylon mesh screens, each about one square meter in area, are extended between supporting posts so that passing advection fogs condense water onto their surfaces. As the fog accumulates on the mesh screens, the surplus drips down into collecting systems and provides enough drinking water to sustain small communities.

Subtropical Desert (BWh)

As shown earlier in Figure 15–11, the most extensive areas of desert exist in the subtropical regions, particularly within the western portions of the continents. The most important factor in the formation of the subtropical deserts is the subsidence associated with the Hadley cell. As we saw in Chapter 8, the subtropical highs of the Hadley cell do not appear as continuous bands at the surface, but rather as semipermanent cells. Although the Hawaiian and Bermuda–Azores high-pressure systems at the surface contract and weaken during the winter, subsidence nonetheless dominates within the middle troposphere. This causes stable conditions to exist in the middle troposphere, which restricts uplift and inhibits precipitation, creating the major subtropical deserts exemplified by the scene in Figure 15–12.

Some deserts occur in subtropical regions as narrow strips along the west coast of continents, adjacent to cold ocean currents. The Atacama Desert along the west coast of Chile has the lowest average precipitation on Earth and provides an excellent example of this phenomenon (Figure 15–1). As air flows out of the subtropical high-pressure system in the eastern part of the South Pacific, it flows over the cold Peru current, and the lower portion of the atmosphere is cooled. This cooling can bring the air temperature down to the dew point so the air becomes damp and foggy. It also lowers the environmental lapse rate and causes the air to be extremely stable. Although the air advected off the coast is damp, stable conditions suppress uplift so several years can go by without any precipitation at all in some areas. Similar west coast deserts are found in Baja California, Namibia, and Morocco.

Over many subtropical deserts, the precipitation that does occur often comes in the form of localized showers from summertime convectional activity. This is not the case for all subtropical deserts, however, as illustrated in Figure 15–13. Yuma, Arizona, and Cairo, Egypt, are both located at about the same latitude, in the midst of subtropical deserts. Although both receive scant precipitation over the course of the year, Cairo's precipitation occurs mainly in the winter, whereas Yuma's is

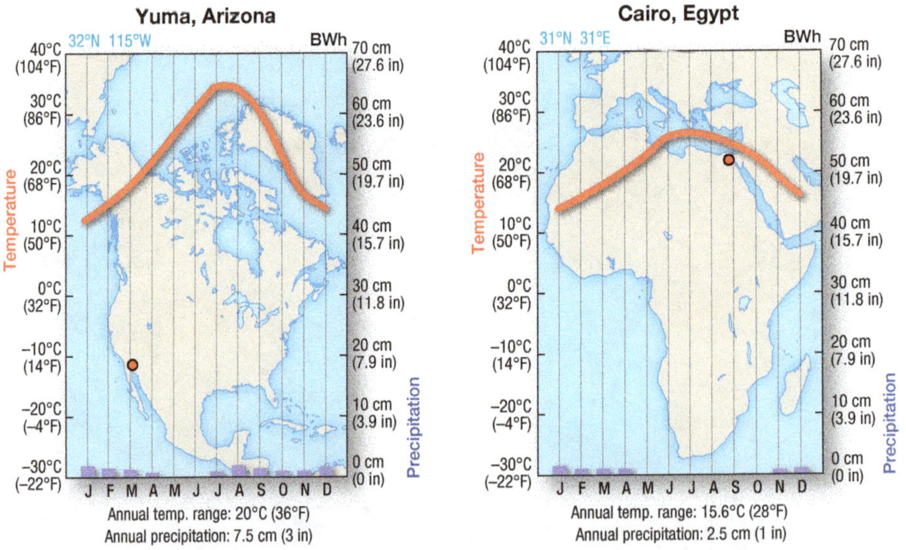

▲ **FIGURE 15–13 Tropical Deserts.** Climographs for Yuma, Arizona, and Cairo, Egypt, representative of subtropical deserts.

about equally divided between the late summer and winter months. As is the case over much of southern Arizona, August usually marks the peak of what is locally known as the *Arizona monsoon.* Although it has little similarity to the wet season of the Asian monsoon, the Arizona version does undergo a shift in the airflow that brings damp air into the area in late summer. This influx of moisture is subject to strong surface heating that can lift the air sufficiently to trigger isolated thunderstorms. While these thunderstorms can be intense and cause flash flooding, they are neither frequent nor strong enough to make the area anything other than the true desert that it is. The winter precipitation at Yuma results from the passage of midlatitude cyclones. While these systems can bring precipitation across a wide swath of the country, often in the form of heavy rain showers or snowstorms, over the desert the moisture supply is usually too low to allow much precipitation to fall. This is because Pacific moisture is blocked by the western mountains, and Gulf of Mexico air does not usually flow this far westward.

Daytime summer temperatures in subtropical deserts can be extremely high. In fact, the hottest locations in the world are all found in these regions. During the summer, the combination of low humidities, high sun, and clear skies allows high inputs of solar radiation to be absorbed at the surface. Furthermore, the lack of soil moisture (except after recent rain showers) causes the ground temperatures to become extremely high because little heat is expended in the evaporation of water. Thus, it is not uncommon for daytime temperatures to reach as high as 45 °C (113 °F) or higher. After the Sun sets, the clear skies and low humidities that led to rapid heating also allow the air to cool considerably. As a result, diurnal temperature ranges can be very large.

The same applies to the annual temperature ranges. At Baghdad, Iraq, for example, the mean monthly temperature in August is 35 °C (95 °F), which is 25 °C (45 °F) higher than the January mean of 10 °C (50 °F). This type of temperature range is greater than that found in any other climate of the tropics or subtropics. As we will see, however, midlatitude deserts commonly have even greater temperature ranges.

There is a widely heard axiom in regional climatology that areas of low annual precipitation also have the greatest amount of interannual variability. This certainly applies to subtropical deserts, where many years can go by with only minimal rainfall, only to be followed by a season having numerous rain showers, delivering several years' worth of precipitation. Thus, the annual average is a poor indicator of how much rain is likely to fall in a given year. Figure 15–14 plots annual rainfall at Borrego Springs, California, using a 45-year period as an example.

OVER THE OCEANS: SUBTROPICAL DESERT Interestingly, large subtropical desert climates over land areas extend considerable distances over adjacent oceans, as shown in Figure 15–11. In fact, the ocean fraction is about 62 percent, which is testament to the large oceanic subtropical highs that emerged so prominently when we considered the general circulation of the atmosphere. While dry, these oceanic regions do not attain the extreme high temperatures that occur over land.

Subtropical Steppe (BSh)

All the conditions that distinguish a subtropical desert also apply to subtropical steppe climates—though to a lesser degree. Like their more extreme desert counterparts, subtropical steppes are marked by aridity, high year-to-year variations in

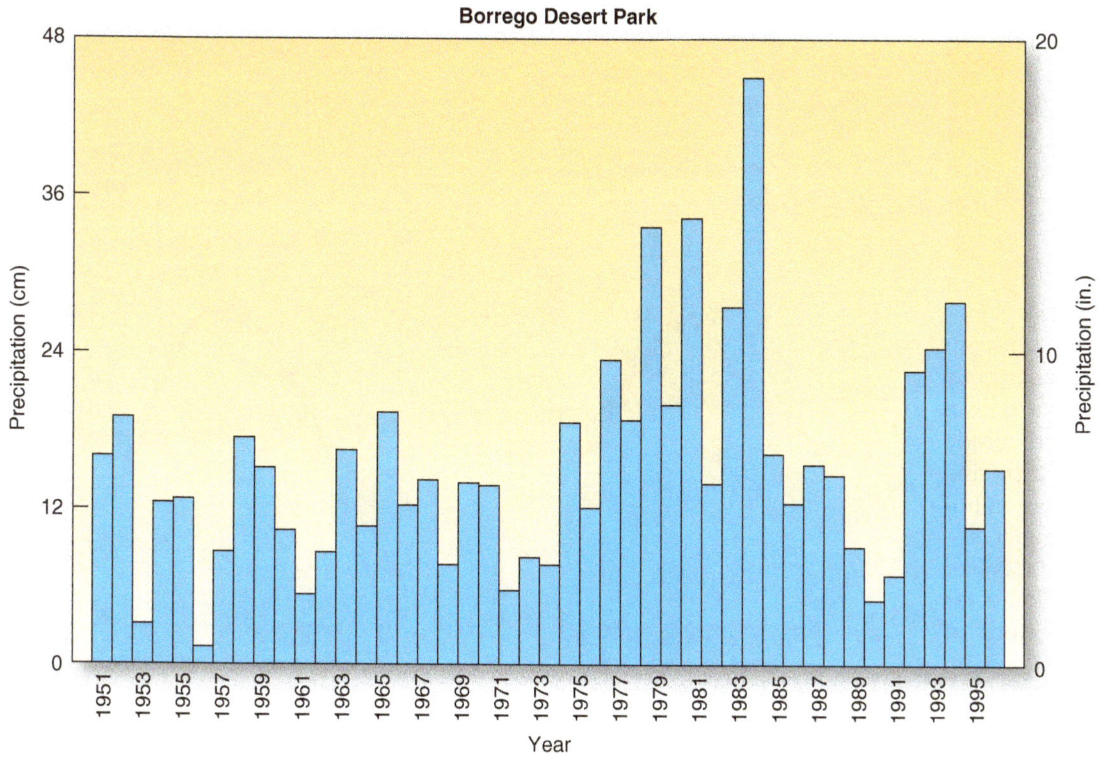

FIGURE 15–14 Desert Precipitation Variability. Annual precipitation at Borrego Springs, California, 1951–1995. Dry climates such as this one normally have a wide year-to-year variability in precipitation.

precipitation, extreme summer temperatures, and large annual and daily temperature ranges. It is therefore not surprising that subtropical steppe climates commonly border the subtropical deserts. And though they are transitional between deserts and nonarid regions in terms of their climatological characteristics, they are not necessarily narrow buffer zones. Much of the southwestern United States and northern Mexico, for example, is occupied by subtropical steppe. Because they appear as fringes of other climates many places, they can appear smaller on maps than they are. Actually, these climates are about the same size as subtropical deserts, which means they are large (covering 11 percent of Earth).

Cloncurry, Australia, and Monterrey, Mexico (Figure 15–15), illustrate the greater precipitation totals, somewhat cooler conditions, and lower annual temperature ranges associated with these climates. These two examples also reveal a distinct seasonality in the precipitation regime typical of subtropical steppes located on the equatorward side of deserts. In these regions, precipitation occurs more often during the summer months than during the winter, as a result of localized convection and tropical disturbances. In contrast, steppe regions on the poleward side of subtropical deserts experience most precipitation in the winter in response to the passage of midlatitude cyclones. Subtropical steppe climates over oceans generally occur on the fringes of the true subtropical deserts, and often serve as a boundary zone between desert climates and tropical wet and dry climates (see Figure 15–11).

OVER THE OCEANS: SUBTROPICAL STEPPE

Most tropical steppe climates are oceanic (83 percent of the total). The largest areas are in the South Indian, South Atlantic, and South and North Pacific. Like the oceanic deserts, they are mainly found in the eastern parts of oceans. They do not appear at all in the North Indian Ocean, as summer rains associated with monsoon circulation prevent even a semiarid climate from developing.

Midlatitude Desert (BWk)

Midlatitude deserts result from extreme continentality. Such regions occur deep within continental interiors or downwind of orographic barriers that cut off the supply of moisture from the ocean. The greatest expanse of midlatitude desert occurs in Asia—which is not surprising considering the immense size of that continent. The two major areas of midlatitude desert in Asia are found just east of the Caspian Sea and north of the Himalayas. Both are far to the east of the Atlantic Ocean, so much of the moisture associated with eastward-moving cyclones is depleted before reaching either area. Mountains to the south block the northward flow of moisture out of the Indian Ocean.

The second greatest expanse of midlatitude desert occurs in the western United States. The midlatitude desert extends southward and merges with the subtropical desert of the Southwest. Although the midlatitude desert lies poleward of the subtropical desert, it should not be assumed that it is neither as dry nor as hot in the summer as the desert to the south. In fact, Death Valley, California, one of the hottest and driest places in the world, lies within the midlatitude desert.

Midlatitude desert in the Southern Hemisphere is confined to a narrow strip in South America, east of the Andes. A quick look at Figure 15–3 reveals that the midlatitudes of the Southern Hemisphere are almost completely covered by ocean. Thus, it is rare to see any type of land climate in this region, let alone a midlatitude desert.

Figure 15–16 shows two typical climographs for midlatitude deserts. Midlatitude deserts have a greater range of

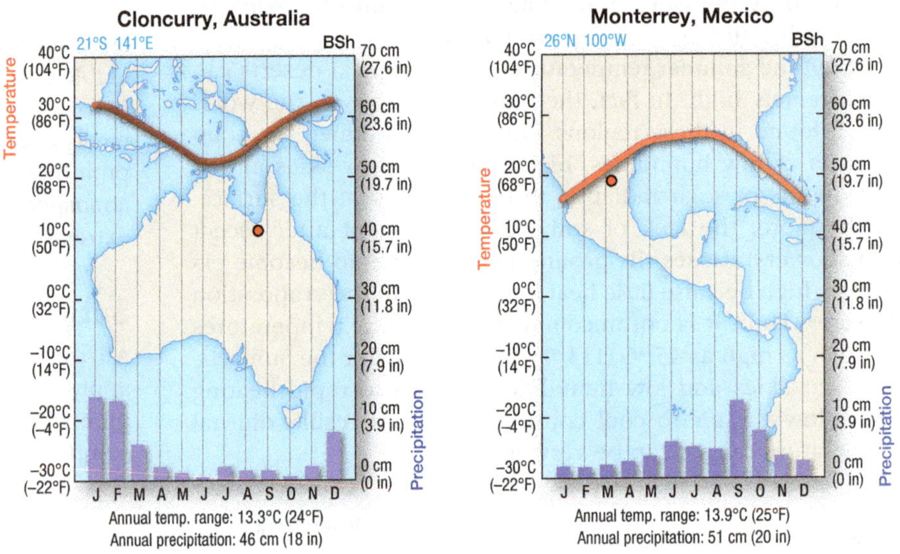

▲ **FIGURE 15–15 Subtropical Steppe Climates.** Climographs for Cloncurry, Australia, and Monterrey, Mexico, representative of subtropical steppe climates.

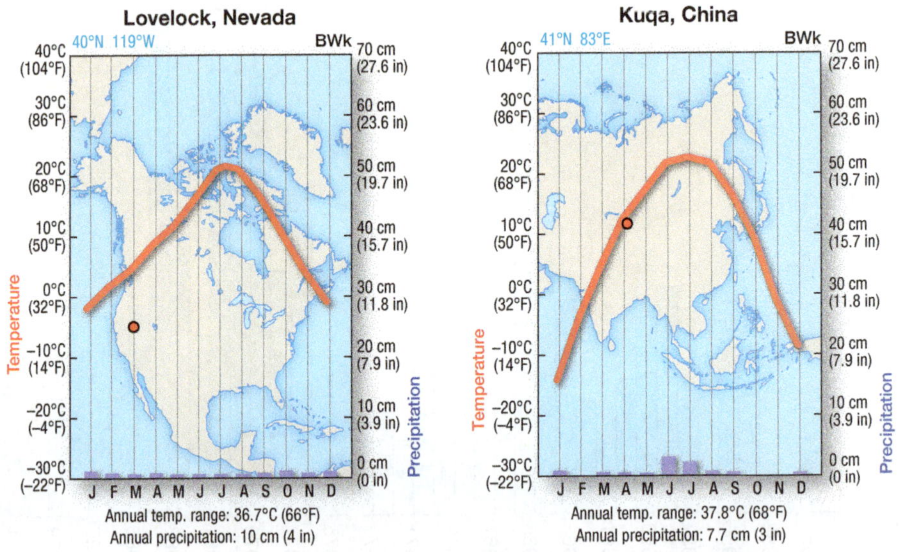

▲ **FIGURE 15–16 Midlatitude Deserts.** Climographs for Lovelock, Nevada, and Kuqa, China, representative of midlatitude desert climates.

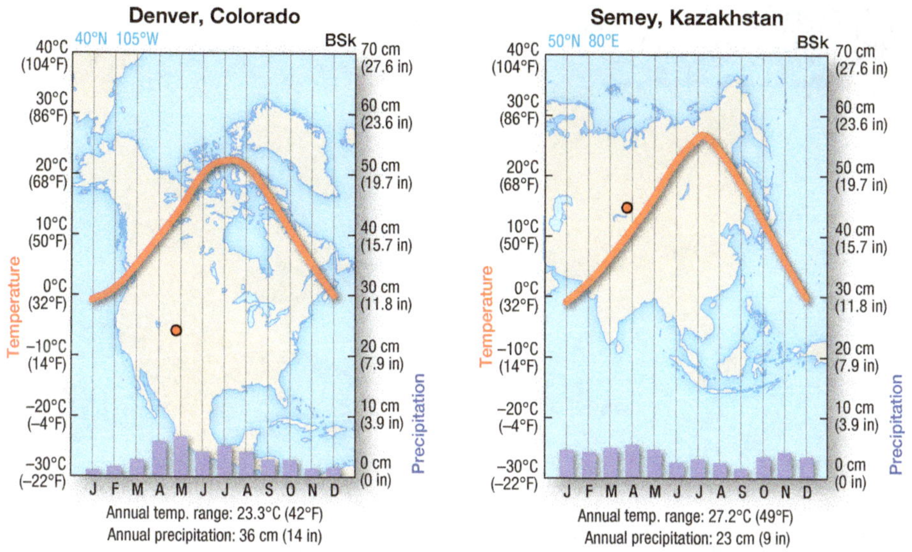

▲ **FIGURE 15–17 Midlatitude Steppe Climates.** Climographs for Denver, Colorado, and Semey, Kazakhstan, representative of midlatitude steppe climates.

temperatures, both on a daily and annual basis, than do their subtropical counterparts. While both types of deserts become extremely hot during summer days, midlatitude deserts have more rapid nighttime and winter cooling. With more precipitation and lower potential evapotranspiration, they are generally more humid than subtropical deserts. Thus, although vegetation is adapted to dry conditions, ground cover is likely to be continuous, without large patches of bare ground common in hotter (subtropical) desert regions.

Overall, this is a small climate type, only a little more than 1 percent of Earth. Three-fourths of that is on land, meaning that oceanic midlatitude deserts are uncommon indeed. But one would expect this, considering that they arise from continental effects!

Midlatitude Steppe (BSk)

The midlatitude steppe accounts for most of the arid regions of western North America. It flanks the northern portion of the midlatitude desert and merges with the subtropical steppe to the south. A large swath of steppe also extends from the Great Plains, east of the Rocky Mountains, all the way from northeast Mexico into western Canada (see Figure 15–11). The midlatitude steppe region of Asia is in many places a fairly narrow strip of land surrounding the true deserts.

▼ **FIGURE 15–18 Midlatitude Steppe Vegetation in Idaho.**

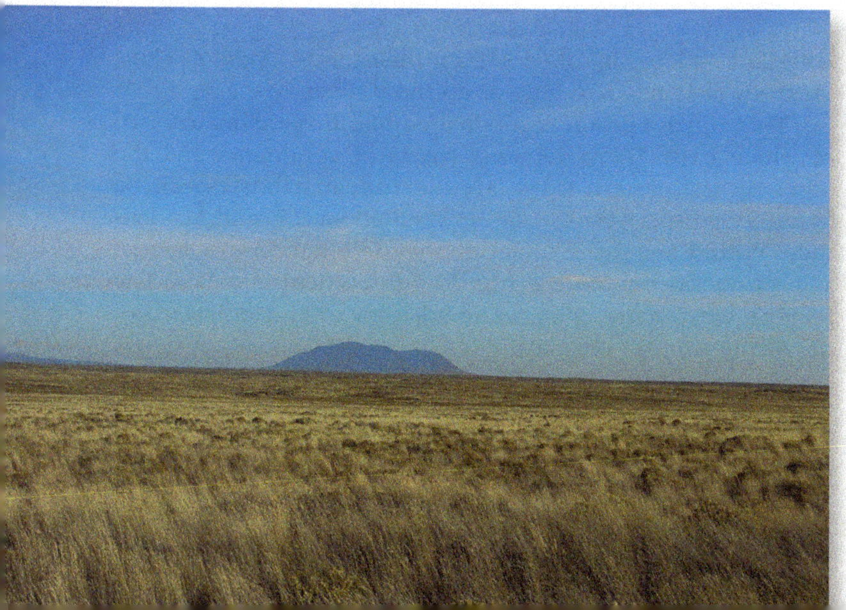

Midlatitude steppes have the same temperature characteristics as the midlatitude deserts. The primary difference between the two is the greater amount of precipitation in the steppes, which commonly totals about 50 cm (20 in.) annually (Figure 15–17). Typical steppe vegetation is shown in Figure 15–18, and a discussion of the natural ecosystem is given in *Box 15–2, Focus on the Environment and Societal Impacts: North American Prairies.*

Overall, the fraction of Earth with a midlatitude steppe climate is a little more than 2 percent. As with midlatitude deserts, it is mostly a continental climate and for the same reason.

CHECKPOINT

15.10 Explain why low rainfall is an inadequate indicator of dry climates.

15.11 What accounts for the band of subtropical deserts found between the north and south latitudes of 10° and 30°?

15.12 What accounts for midlatitude deserts and steppes?

Mild Midlatitude Climates

15.5 Describe the characteristics of mild midlatitude climates.

The mild midlatitude climates are located in parts of the latitude range between 30° and 60° in either hemisphere (Figure 15–19). They cover about 20 percent of Earth, just a little less than the 25 percent we saw in the B group (see Table 15–1). About 80 percent of the group is oceanic.

Over land the C climates occur as long, narrow strips of land along the west coasts of North and South America and southern Australia. In North and South America, the eastern border of the climate is delimited by mountains. Another west coast mild midlatitude climate—in fact, the largest—surrounds the Mediterranean Sea.

Mild midlatitude climates also cover large areas of the eastern portions of the continents, especially in North and South America and Asia. While these areas extend farther inland than do their counterparts along the west coasts, these mild midlatitude climates have narrower latitudinal extents. While west coast midlatitude climates can be found as far poleward as just a few degrees shy of the Arctic Circle, most on the eastern sides of continents do not extend poleward of 40° latitude. C climates occupy a very large part of the midlatitude Atlantic and Pacific Oceans.

15–2 FOCUS ON THE ENVIRONMENT AND SOCIETAL IMPACTS

North American Prairies

We have seen that climatic boundaries generally coincide with vegetation boundaries, but it is noteworthy that humans can greatly modify extant vegetation. The great prairies once reached from the southern parts of the Canadian Plains southward to Texas and westward to the Rocky Mountain foothills and extended to the east into the Great Lakes region. Prairies consist primarily of grasses, along with some herbs and shrubs and the exclusion of trees.

Many factors are involved in creating prairie environments. Large temperature differences between summer and winter are common and precipitation is lower than that of most woodland areas. Precipitation across North America generally increases eastward from the foothills of the Rockies, and natural prairie populations have reflected this by having taller grasses eastward.

Nonclimatic factors also play a role in the existence of prairies. The existence of periodic fire is important because it prevents the incursion of trees that, unlike prairie grasses, do not quickly regenerate in the wake of burns. Like other ecosystems, prairies maintain an interaction between the vegetation and animal species. In North America bison fed on unlimited grass and at the same time helped maintain the grass by naturally tilling the soil with their very narrow hooves (Figure 15–2–1).

▲ **FIGURE 15–2–1 North American Prairies.** Bison grazing in Custer State Park, South Dakota.

The extent of this major ecosystem has been reduced greatly in North America since the westward expansion. The reduction has been particularly acute where the tall grass prairies once ruled; in Illinois only an estimated 0.1 percent of the original prairie still exists. An estimated 20 to 25 percent of the short grass prairie still exists farther to the west.

The biggest culprit in the demise of prairielands is expansion by humans. Prairie soils are particularly fertile and valuable for agriculture. Urban expansion has also played a role. Even where people have not taken over prairielands, fire suppression has facilitated the expansion of trees into former grasslands. There is also a threat from the spread of nonnative species.

As we will see in Chapter 16, climates have been changing and will continue to do so. These changes may have significant effects on the plants and animals from all ecosystems, including North American prairies.

15–2–1 What climatic factors explain the existence of the North American prairie?

15–2–2 Give an example of a natural factor that helped to maintain the prairie ecosystem.

The individual climates within this general group do not share similar precipitation patterns. In fact, the precipitation regime can vary greatly from one region to another. To get an idea of the extent of this variability, consider the fact that along the west coast of the United States, the mild midlatitude region includes San Diego, California, which receives about 25 cm (10 in.) of annual precipitation, and the Olympic rain forest in Washington State, where the precipitation locally exceeds 375 cm (150 in.) per year.

The term *mild* refers to the winter temperatures and not necessarily those of the summer. In North America, for example, this climate group is found over inland areas of California where summer temperatures routinely exceed 38 °C (100 °F), as well as in the Gulf states of Florida, Alabama, Mississippi, and Louisiana—places hardly noted for their mild summers. Thus, in effect, *mild* refers to little or no snow cover.

This climate group is subdivided into three climates, two of which exist along the west coasts of continents and the other on the eastern sides. **Mediterranean** climates can be found along the west coasts between about 25° and 40° latitude. Within about the same range of latitude on the eastern side of continents are the **humid subtropical** climates. The **marine west coast** climates lie adjacent to and poleward of the Mediterranean climates.

OVER THE OCEANS: MILD MIDLATITUDE CLIMATES

Collectively this group is 80 percent oceanic. Some details will be given for each subtype below, but one general feature

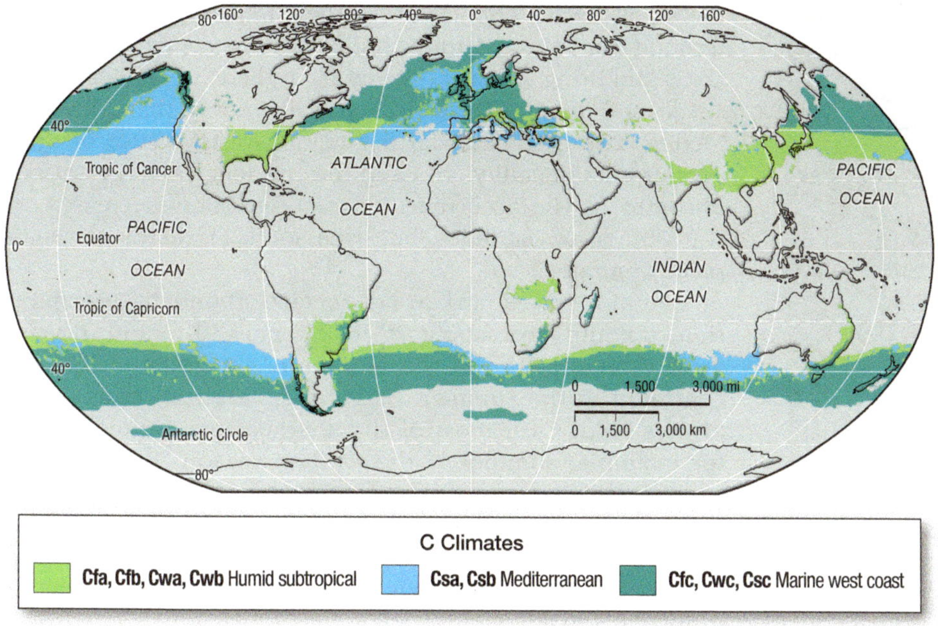

▲ **FIGURE 15–19 C Climates.** Map of the mild midlatitude climates.

deserves mention here. Note that mild climates are found far to the north in the Atlantic and Pacific Oceans. This is evidence of how much energy is carried poleward by the North Atlantic drift and the North Pacific Current (see Figure 8–19). The North Sea and the North Pacific are certainly not warm places, but they would be far colder without warmth brought in from more southerly locales. To a lesser degree, warm currents in Southern Hemisphere oceans accomplish a similar poleward extension of mild climates. In the Southern Ocean the strong Antarctic Circumpolar Current interrupts the southward transport.

Mediterranean (Csa, Csb)

Mediterranean climates are the only extensive climates with a distinct summer dry season and a concentration of precipitation in the winter. Figure 15–20 clearly depicts this pattern for two typical sites, Los Angeles, California, and Athens, Greece. The summer aridity in Mediterranean climates is attributable to the presence of semipermanent subtropical high-pressure systems offshore. For North America, the Hawaiian high-pressure system to the west "blocks" the eastward migration of summertime midlatitude cyclones and deflects them to the north. This deflection of the storms, coupled with subsidence along the eastern portion of the subtropical high, deprives coastal southern California of the uplift mechanisms necessary for precipitation. During the winter months, the Hawaiian high weakens, shrinks in size, and migrates toward the equator. This opens the way for the arrival of Pacific cyclones.

Annual precipitation increases with latitude and with elevation along windward slopes in Mediterranean climates. The latitudinal gradient results from the more frequent passage of midlatitude cyclones at the higher latitudes. At the same time, the mountain slopes induce orographic uplift of the predominantly westerly airflow.

While the climographs in Figure 15–20 reveal the mean precipitation patterns for the two sites, they do not reveal the extreme variability inherent with winter rainfall. Indeed, some winters can be quite dry, whereas others bring an onslaught of sequential storms that produce major flooding and hillside erosion. This climatic feature has been particularly prominent since the 1970s in southern California, which has experienced several record drought years along with extremely wet winters.

Winter temperatures in Mediterranean climates are usually mild, especially right along the coast. Inland temperatures occasionally drop below freezing and sometimes threaten fruit-growing areas with widespread crop damage. On the other hand, precipitation results mainly from the passage of midlatitude cyclones, which transport relatively warm, maritime polar air. Thus, precipitation in the lower elevations falls almost exclusively as rain and not snow. Figure 15–21 shows vegetation common in the California version of the Mediterranean climate.

OVER THE OCEANS: MEDITERRANEAN While Mediterranean climates occupy relatively small portions of Earth's land mass, they extend considerably westward into ocean basins of the lower-middle latitudes. In fact, ocean areas represent about 87 percent of this climate. Ocean temperatures are not

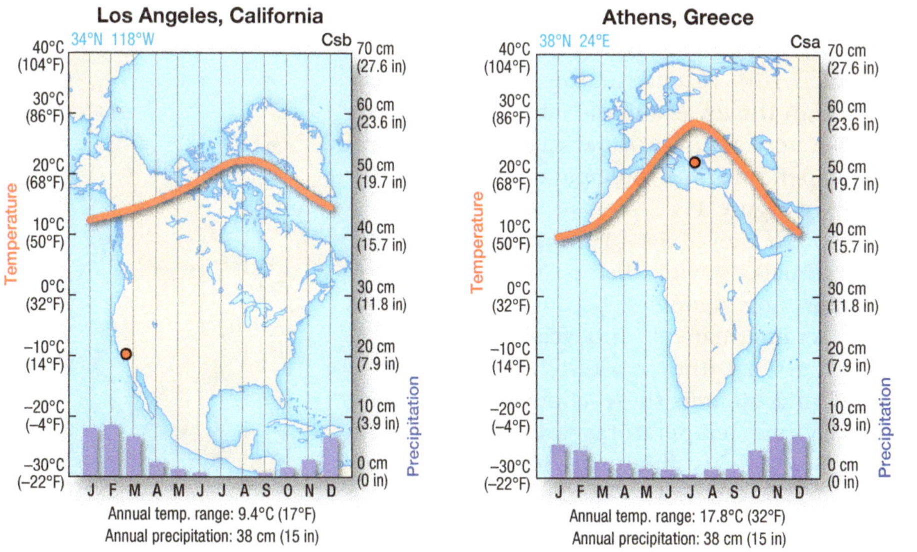

▲ **FIGURE 15–20 Mediterranean Climates.** Climographs for Los Angeles, California, and Athens, Greece, representative of Mediterranean climates.

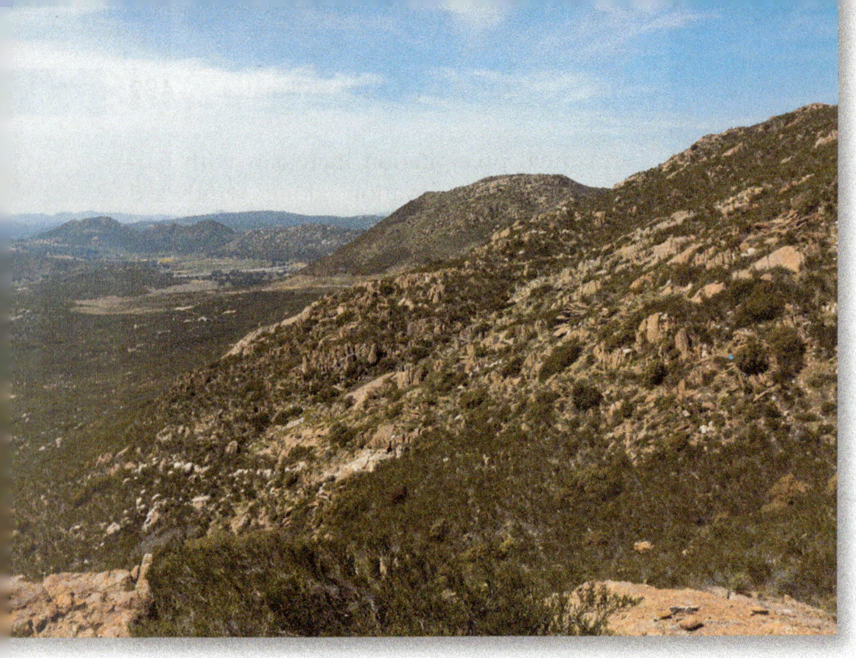

▲ **FIGURE 15–21 Mediterranean Shrubland in Southern California.**

as high in summer as they are on land and so the type is mostly Csb rather than Csa. However, in summer they experience the same effects from subtropical high pressure, and so they have the same seasonality in rainfall as their land counterparts.

DID YOU KNOW?

Mediterranean climates are far from uniform. Summer temperatures range from mild to hot, with daily highs typically decreasing toward the coast and with increasing latitudes. Central California provides a good example of how variable the summer temperatures can be in Mediterranean climates. San Francisco is noted for particularly cool summers, while about 120 km (70 mi) to the northeast, Sacramento experiences hot, dry conditions.

Humid Subtropical (Cfa, Cwa)

This is the largest of the C group by far (14 percent of Earth). Humid subtropical climates occur within the lower middle latitudes of eastern North America, South America, and Asia. Typical temperature and precipitation patterns are shown in Figure 15–22.

Though they exist in the middle latitudes, these climates have a distinct tropical feel during their long summers. Located to the west of large semipermanent anticyclones and warm ocean currents, the prevailing winds circulate hot, humid air into these climatic zones. Summer daytime temperatures are usually in the lower 30s Celsius (high 80s to low 90s Fahrenheit), and dew points in the mid-20s Celsius (mid-70s Fahrenheit) help retard nighttime cooling. As a result, hot, muggy conditions remain throughout the day and night, especially along the equatorward boundaries of the climates. Fortunately, afternoon convectional thundershowers are common in these areas and bring temporary relief from the extreme heat.

Winter temperatures are typically lower than those of Mediterranean climates farther to the west because of their greater continentality, and subfreezing temperatures are not uncommon. The occurrence of frost and snow decreases toward the lower latitudes, but even south Florida is not completely immune.

Humid subtropical areas receive abundant precipitation, ranging from about 75 to 250 cm (30 to 100 in.) per year. Over most areas the maximum precipitation is concentrated in the summer, but this generalization does not always hold. Over most of the southeastern United States, for example, summer is the wettest season. But the area extending from east Texas into Tennessee and Kentucky has a winter precipitation maximum. Regardless of which season receives most precipitation, summer is always the season of moisture deficit, because of greater potential evapotranspiration.

Precipitation during the summer is largely convectional in nature and tends to be scattered and brief. Winter precipitation, on the other hand, is usually associated with midlatitude cyclones. Over the coastal areas, tropical storms and hurricanes are capable of bringing extreme rainfall from time to time during the late summer and early fall.

OVER THE OCEANS: HUMID SUBTROPICAL Like Mediterranean climates, humid subtropical climates extend out into adjacent parts of the ocean in the lower middle latitudes, so much so the subtype is 75 percent oceanic. These oceanic parts are almost exclusively Cfa with evenly distributed rainfall. They generally exist in the western portion of the oceans, while Mediterranean climates are found further to the east.

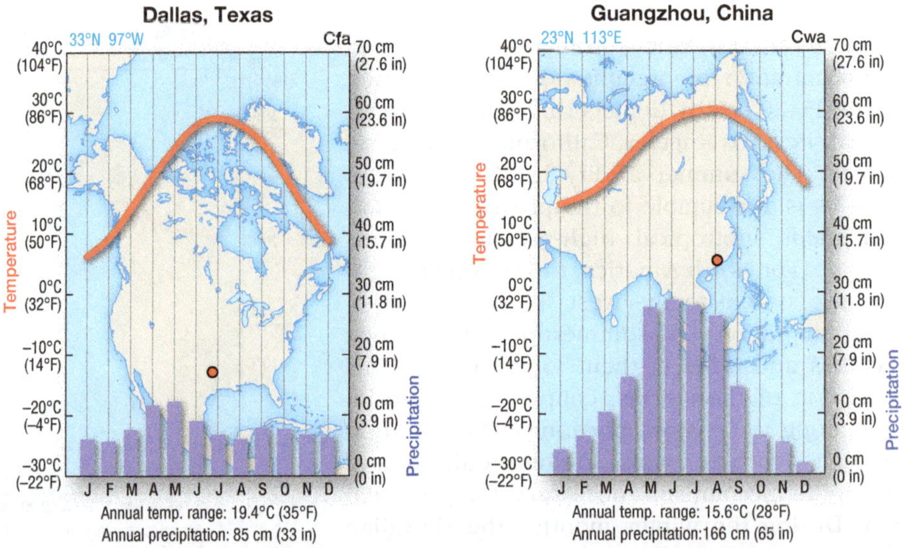

▲ **FIGURE 15–22 Humid Subtropical Climates.** Climographs for Dallas, Texas, and Guangzhou, China, representative of humid subtropical climates.

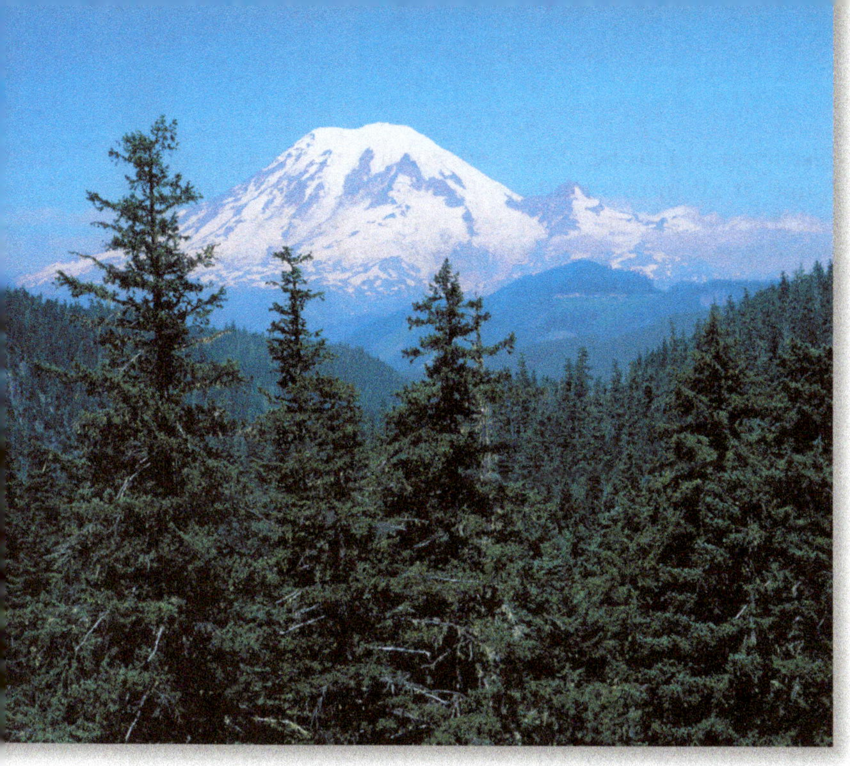

▲ **FIGURE 15–23 Coniferous Forest.** Marine west coast climates in North America are dominated by coniferous evergreen trees such as these at Mount Rainier in Washington.

Marine West Coast (Cfb, Cfc)

Marine west coast climates normally occur poleward of Mediterranean climates and, as the name would suggest, along the west coasts of continents. But inspection of Figure 15–19 shows that the latter is not always the case, as areas having the climate are found covering New Zealand and along southeastern Australia and southeast Africa. While not truly located on west coasts, these areas are located east of fairly narrow strips of land that do not greatly modify the maritime nature of the predominantly westerly flow. This exposure to air passing over cold ocean currents and their location in the path of eastward-migrating midlatitude cyclones gives these climates their characteristic features. Overall they cover less than 2 percent of the planet.

Both summers and winters are typically mild. Extremely high summer temperatures are certainly not unknown in these climates, but they are the exception rather than the norm. From northern California through Alaska, for example, low and midlevel cloud decks frequently keep daytime temperatures down and often bring drizzle or light rain to coastal regions. Along the lower elevations, the percentage of days experiencing precipitation can be high, but the total amount of rainfall is usually light. On the other hand, the Coast Ranges parallel the Pacific coastline and create a major barrier to the westerly flow. This creates a strong orographic effect that causes some areas to have extremely heavy precipitation amounts that rival or exceed those of the tropics. The Olympic Peninsula of Washington State is an excellent example of this phenomenon, where extremely heavy rainfalls are sufficient to support a lush, coniferous evergreen rain forest (Figure 15–23). The marine west coast climate of Europe, in contrast, is dominated by deciduous trees.

Just as maritime conditions moderate temperatures during the summer, they also allow for mild winter conditions at surprisingly high latitudes. Even as far north as Sitka, Alaska (57° N), the coldest month of the year has a mean temperature above the freezing point of water (Figure 15–24). At many lower-latitude locations, low-elevation snowfall is a rarity; when it does occur, it usually melts away within a short time. In Europe, where the marine west coast climate extends farther inland than it does in North America, snow is more frequent and remains on the ground for a longer period of time.

DID YOU KNOW?

Seattle, Washington, has a reputation for experiencing a lot of rain each year. But it actually receives less annual precipitation on average—93 cm (37 in.)—than does New York City, with 124 cm (50 in.). The difference is that Seattle has an average of 150 rainy days each year, but these are usually of low intensity. New York averages only 120 days with measurable precipitation annually, but that city usually experiences more intense precipitation events.

Given the fact that both summer and winter temperatures are mild, it follows that these climates have low annual temperature ranges. This contrasts sharply with the situation for climates in the eastern portion of continents at the same range of latitudes. Those locations have moderately high summer temperatures, but when combined with the extreme cold of winter, they the yield extremely high temperature ranges of the severe midlatitude group of climates.

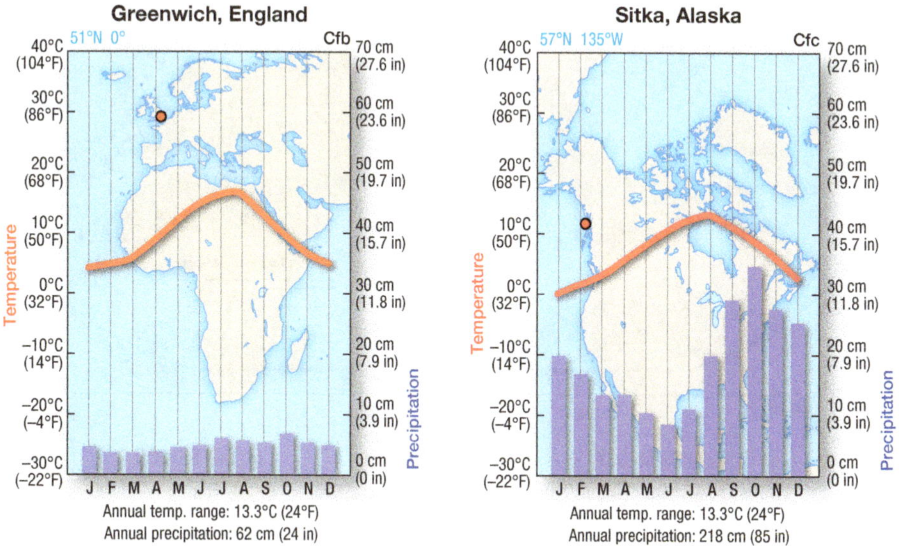

▲ **FIGURE 15–24 Marine West Coast Climates.** Climographs for Greenwich, England, and Sitka, Alaska, representative of marine west coast climates.

OVER THE OCEANS: MARINE WEST COAST Marine west coast climates have been so named because of their position on land masses (keep in mind that for many decades the Koeppen system has been applied only to land areas). But if we examine the existence of this climate type over land and water combined, we see that the name isn't really all that fitting. They extend all the way across the Atlantic and Pacific Oceans on the poleward sides of the Mediterranean and humid subtropical climates. The visual impression of the maps is accurate; these climates are 95 percent oceanic!

CHECKPOINT

15.13 In the context of the Koeppen classification system, what does it mean to say that a midlatitude region has a mild climate?

15.14 How are the Mediterranean and marine west coast climates similar? How are they different?

Severe Midlatitude Climates

15.6 Describe the characteristics of severe midlatitude climates.

The severe midlatitude climate group includes two climates, humid continental and subarctic, both of which are marked by very cold winters. These climates require large continental areas within the high-middle latitudes—between about 40° and 70°—in order to avoid the moderating effects of an ocean (Figure 15–25). Thus, they are restricted to Europe, Asia, and North America and are not found at all in the Southern Hemisphere. Neither subtype has even 10 percent of its area over ocean, and so we will not discuss the ocean component of this group. As expected from climates resulting from strong continentality, both types of D climates exhibit large annual temperature ranges.

Both of the severe midlatitude climates receive precipitation throughout the year and have no true dry season. In many locations, precipitation is greater in the summer than winter. Precipitation during the summer can result from local convection or by the passage of midlatitude cyclones; winter precipitation (mostly as snow) results almost entirely from cyclonic activity.

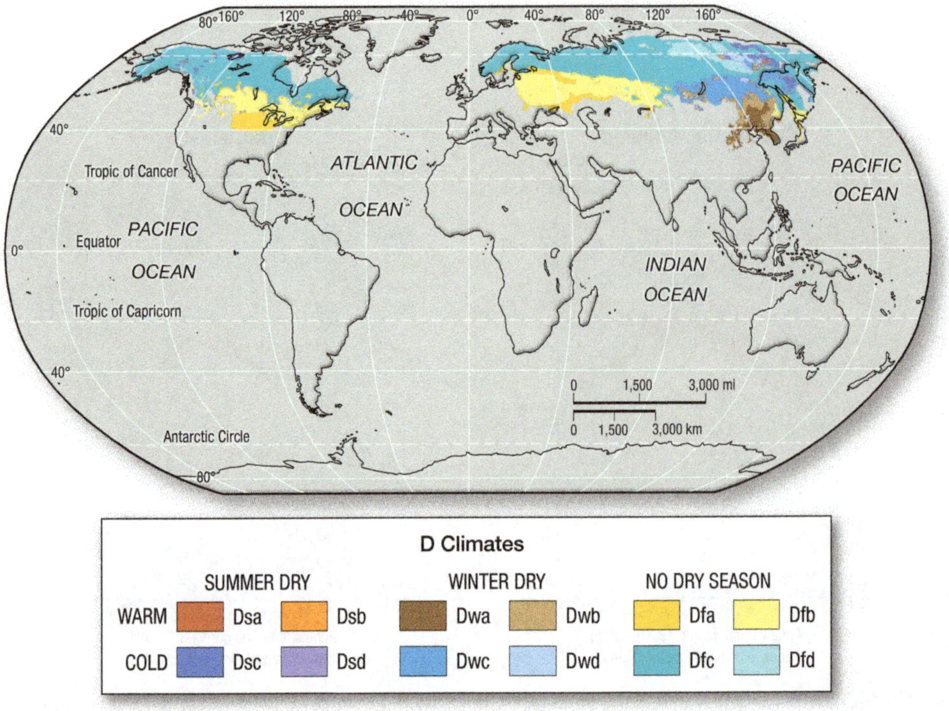

▲ **FIGURE 15–25 D Climates.** The distribution of severe midlatitude climates.

Humid Continental (Dfa, Dfb, Dwa, Dwb)

Although it has only 35 percent of the D group's area, a huge segment of the population of the United States, Canada, eastern Europe, and Asia lives in the **humid continental** climate (Figure 15–26), including the inhabitants of New York, Chicago, Toronto, Montreal, Moscow, Warsaw, and Stockholm. This is the more temperate of the two severe midlatitude climates, normally found between about 40° N and 55° N in the eastern parts of continents. Summers are warm and often hot. New York City, for example, has an average high temperature in August of 29 °C (84 °F). But winter is another matter altogether, with a mean February maximum temperature of 4 °C (40 °F). Note that New York City is located near the southern boundary of the climate and on the coast, so temperatures are milder than those of most other locations with a humid continental climate.

Mean annual precipitation in these climates usually ranges between 50 and 100 cm (20 and 40 in.). Over the United States and southern Canada, there is a conspicuous decrease in precipitation with increasing latitude and distance from the Atlantic shoreline, reflecting a reduced moisture content in the atmosphere.

Subarctic (Dfc, Dfd, Dwc, Dwd)

Subarctic climates occupy the northernmost extent of the severe midlatitude regions, with more than half of the area of Alaska and Canada having this type of climate. In North America, the coniferous forest that dominates the region is referred to as the **boreal forest** (Figure 15–27); in Asia, it goes by the name **taiga**. Summer temperatures are somewhat lower than those of the adjacent humid continental regions, but

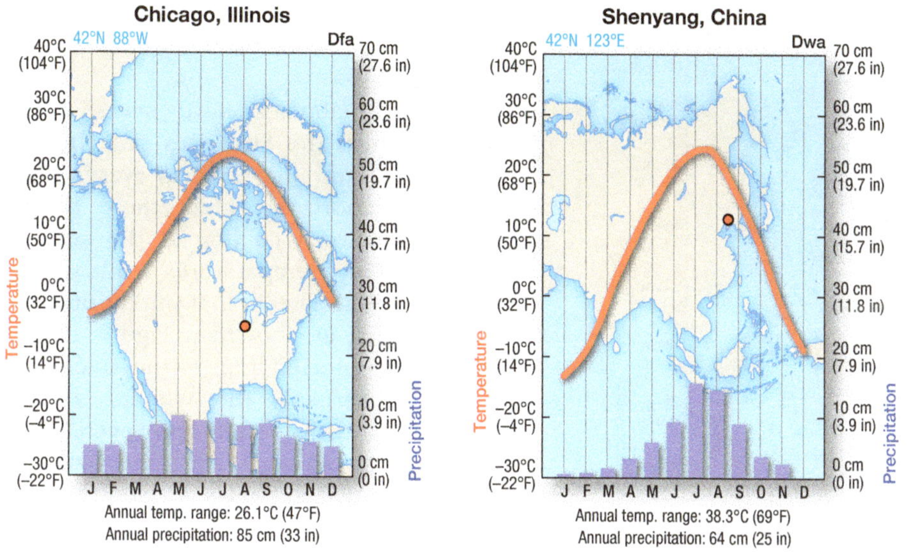

◂ **FIGURE 15–26 Humid Continental Climates.** Climographs for Chicago, Illinois, and Shenyang, China, representative of humid continental climates.

the major difference occurs in winter, when mean monthly temperatures can be extremely low. At many locations, the monthly mean temperature can remain below the freezing level for up to seven months. Thus, winter is long and separated from the brief summer only by a short-lived autumn and spring. Figure 15–28 highlights the very large annual temperature ranges that result from the mild summers and severe winters.

Typically, precipitation is greater in the summer than winter, mostly because of the more poleward displacement of midlatitude cyclone tracks in summer. Nonetheless, annual precipitation is usually low, ranging from about 12 to 50 cm (5 to 20 in.).

Ocean bodies remain relatively mild during the winter, even at high latitudes. For that reason we do not see D climates over the oceans.

▾ **FIGURE 15–27 Subarctic Vegetation.** Vast tracts of the Canadian and Siberian subarctic climates are covered by evergreen coniferous forests.

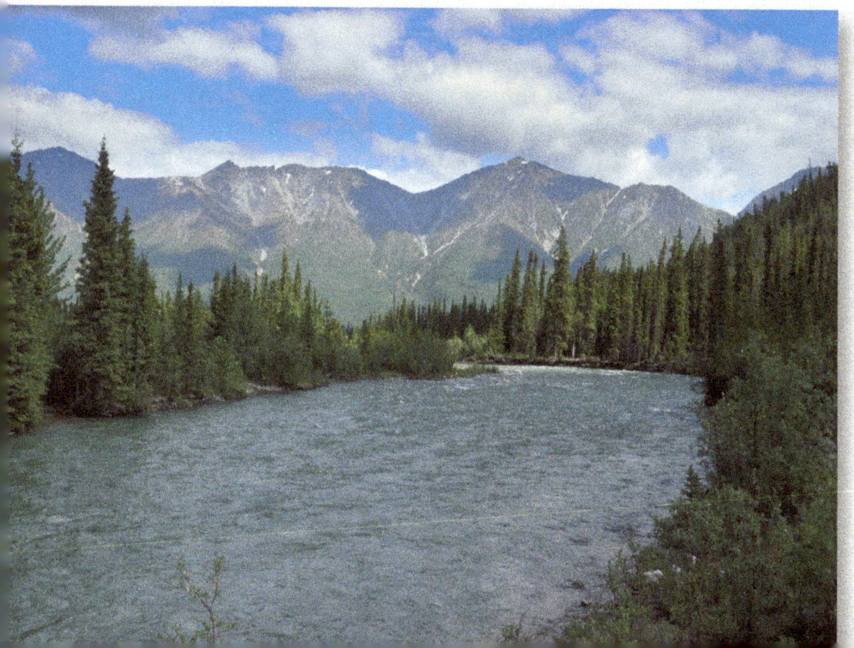

CHECKPOINT

15.15 What differentiates the two kinds of severe midlatitude climates?

15.16 Why don't severe midlatitude climates exist over the oceans?

Polar Climates

15.7 Describe the characteristics of polar and highland climates.

Polar climates exist in the highest latitudes—typically areas poleward of about 70° (Figure 15–29). Such regions occur in the Northern Hemisphere, across northern Canada, Alaska, Asia, and Greenland. In the Southern Hemisphere they are almost entirely confined to the continent of Antarctica. In total about 18 percent of Earth is a polar climate, so this is another fairly large group. As we mentioned earlier, the climate boundaries of the Koeppen system were set to coincide with boundaries of differing natural vegetation types. In this case, polar climates begin at the high-latitude boundaries of the subarctic climates, where vast expanses of coniferous forest give way to a generally treeless landscape. The polar climate group consists of two distinct types. The most equatorward and milder of the two is the **tundra**, which has at least one month with an average temperature above 0 °C (32 °F). At the most poleward regions of the globe lie the true ice cap climates.

DID YOU KNOW?

The polar front is a vitally important factor in much midlatitude weather and climate. Yet it can be found year-round in only one continent: North America. Continental polar air can be blocked from making contact with maritime polar air over Eurasia by the Himalayan Mountains. Australia and South America often do not have continental polar air because of their small amount of land mass at high latitudes.

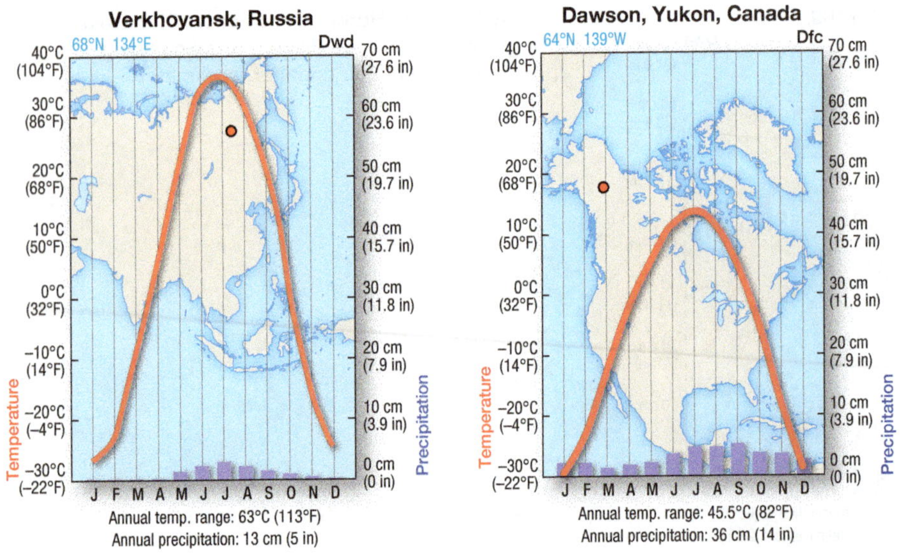

▲ **FIGURE 15–28 Subarctic Climates.** Climographs for Verkhoyansk, Russia, and Dawson, Yukon, Canada, representative of subarctic climates.

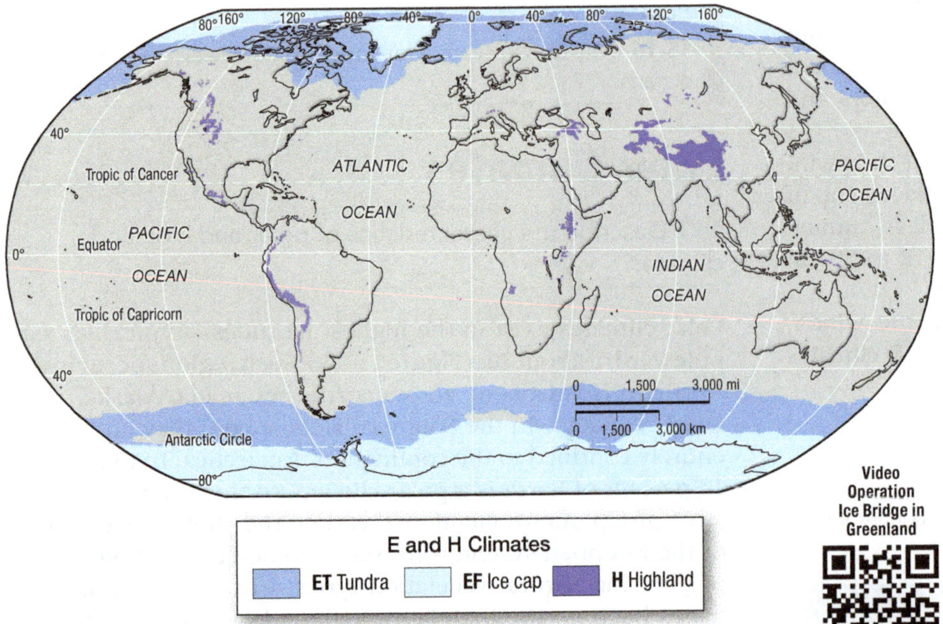

▲ **FIGURE 15–29 E and H Climates.** The distribution of polar and highland climates

Video
Operation
Ice Bridge in
Greenland
http://goo.gl/fybvT8

Tundra (ET)

Tundra climates (the larger of the two polar climates) are named for the associated vegetation type that consists primarily of low-growing mosses, lichens, and flowering plants, with few woody shrubs and trees. The boundary dividing the tundra from the subarctic climate occurs where the mean monthly temperature of the warmest month does not exceed 10 °C (50 °F). In the Northern Hemisphere, this occurs in the general vicinity of 60° N. This climate is almost entirely missing from the Southern Hemisphere, where there is only minimal land coverage at this latitude.

Tundra climates have severe winters in which the Sun rises only briefly each day and never gets very high above the horizon. Under these conditions, radiational cooling at the surface leads to low temperatures and strong stability—both factors that inhibit precipitation. Thus, as shown in Figure 15–30, low winter precipitation is the rule. Surprisingly, perhaps, the winters in the tundra regions are often less severe than those of the adjacent and lower-latitude subarctic climates. This is because the tundra regions of North America, Greenland, and Asia are located nearer to major water bodies and thus have a lesser degree of continentality. (Even when ice covered, an ocean has a somewhat moderating effect on winter temperatures.)

During the summer, tundra regions have very long periods of daylight, but again the Sun never gets very high above the horizon. As a result, temperatures are normally mild in the midafternoon, not very much greater than those of the predawn period. Thus, despite the fact that annual temperature ranges are fairly high, daily temperature ranges are low.

One very conspicuous feature of tundra regions is the existence of **permafrost**, a perennially frozen layer below the surface. During the winter, conditions are so cold that the entire soil is completely frozen, in some places to a depth of several hundred meters. When summer arrives, there is enough warming right at the surface to melt the uppermost soil, and a relatively shallow layer of thawed soil overlies the completely impermeable frozen layer several tens of centimeters below. As a result, any rain or snowmelt that occurs at the surface is unable to permeate deeply into the soil, and the upper part of the surface becomes saturated. This combination of water-logged soil near the surface and a solid ice layer below precludes the establishment of any deeply rooted plants, and only low-growing vegetation can take hold (Figure 15–31).

Ice Cap (EF)

As the name rightfully suggests, continental **polar ice cap** areas exist where ice is present throughout the entire year (Figure 15–32). On land they are confined to the

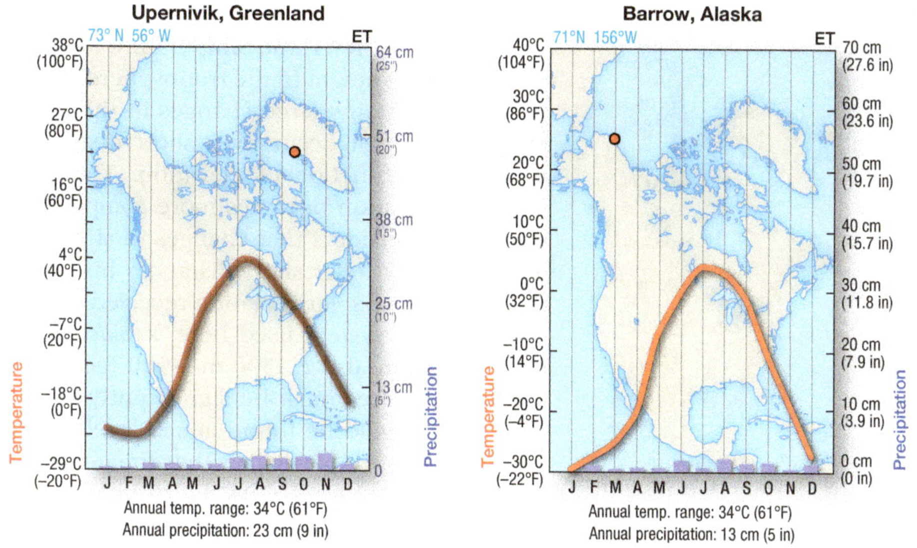

▲ **FIGURE 15–30 Tundra Climates.** Climographs for Upernivik, Greenland, and Barrow, Alaska, representative of tundra climates.

Greenland interior and most of Antarctica, and over ocean they are found in the interior of the Arctic sea. The two parts total over 6 percent of the planet. They exist where the mean temperature of the warmest month does not go above the melting temperature for ice, 0 °C (32 °F). On the continents ice accumulations over these regions can exceed several kilometers, so in addition to being found at high latitudes they also occupy high elevations—a situation perfect for maintaining extraordinarily low winter temperatures (Figure 15–33).

The cold air overlying polar ice caps (also called *continental glaciers*) becomes extremely dense and frequently flows down the margins of the ice under the force of gravity. When funneled through narrow canyons, these katabatic winds (discussed in Chapter 8) can become extremely strong.

Most areas of ice cap receive little precipitation because of the intense cold. On the other hand, all precipitation falls as snow and has the potential to accumulate onto existing ice. Except along the margins, there is no melt and only a small amount of sublimation to remove the ice. Yet some ice does get removed. Beneath the surface, the individual ice crystals merge into an almost solid block of ice. When subjected to the constant pressure of the overlying mass, the ice gradually deforms and expands outward toward the margins of the ice sheet. At the continental margins, the ice often breaks off into large icebergs that float in the Arctic and Antarctic waters.

OVER THE OCEANS: TUNDRA AND ICE CAP The descriptions of the tundra and ice cap climates given above deal exclusively with land regions. But inspection of Table 15–1 shows that tundra is about 90 percent oceanic and ice cap is not far behind at 73 percent. A look at Figure 15–29 reveals that these two climate types cover virtually all of the Arctic and Antarctic Oceans, with the ice cap climates located further poleward. Large parts of the Southern Ocean north of Antarctica are tundra, as are large parts of the North Atlantic and North Pacific. By and large these are cold, stormy, inhospitable places for those inclined to venture onto the ocean at such high latitudes. Midlatitude cyclones batter the warmer parts, and the colder sections are close to perpetual

▲ **FIGURE 15–31 Arctic Tundra in Denali National Park, Alaska.** Long days in the tundra provide a short growing season suitable for flowers, grasses, and low-growing plants, but woody vegetation is nearly absent.

▲ **FIGURE 15–32 Ice Caps in Wiencke Island, Palmer Archipelago, Antarctic.** In most of the ice cap climate, snow persists throughout the year, building large continental ice sheets.

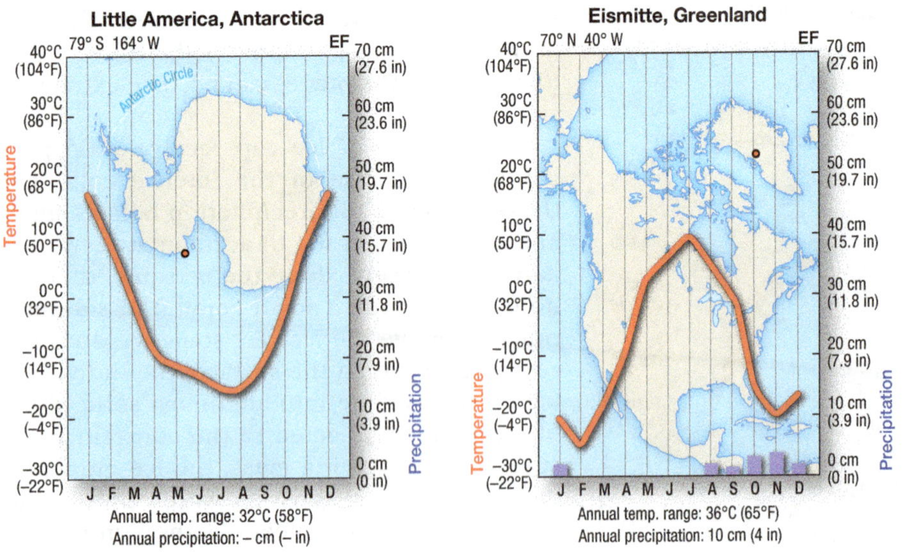

▲ **FIGURE 15–33 Ice Cap Climates.** Climographs for Little America, Antarctica, and Eismitte, Greenland, representative of ice cap climates.

ice cap. As for the ice cap climate itself, over-the-ocean ice thickness does not approach what is seen on land because the water below provides heat and resistance to temperature change that rock and soil cannot. Year-round ice insulates the surface from the warmer water below, but not perfectly. As a result temperatures are not as low in winter as on land. We see that, even when solid, water has a moderating effect on temperature extremes.

CHECKPOINT

15.17 What is permafrost and what climate conditions are responsible for its existence?

15.18 What temperature criterion delimits ice cap climates?

Highland Climates (H)

Highland climates are unique among those in the Koeppen system because their distribution is not governed by geographic location but rather by topography. These climates are found in large mountain or plateau areas. As we saw in Chapter 1, the temperature tends to decrease with height in the troposphere. Thus, in high mountains large changes in the mean temperature can occur over short distances, simply as a function of elevation. The temperature situation becomes more complicated when one considers that changes in slope angle and aspect over short distances influence the intensity of solar radiation received. In the classification system used here, a place must have elevation above 1500 m (4900 feet), and the temperatures must be significantly colder than what would be found at sea level in the same location. In particular, the climate group expected at sea level must be different than what is found. In other words, elevation must shift the climate to a colder group, or the climate will not be considered as significantly affected by topography. About 1 percent of Earth meets these criteria.

Precipitation type and intensity also vary spatially across highlands. Mountain slopes can enhance precipitation on their windward sides and simultaneously create a rain shadow downwind. Elevation difference also affects the ratio of precipitation falling as snow versus rain, with higher elevations favoring a greater amount of snow. Mount Kilimanjaro in Tanzania provides a striking example of this phenomenon. Located very near the equator at 3° S, Kilimanjaro possesses an extraordinarily wide range of climates from its base near sea level to its peak at 5895 m (19,340 ft). Near the surface, the mountain has a tropical wet and dry climate similar to the surrounding area. But the climate changes with height and the mountain eventually becomes covered with perennial ice over its higher reaches. This is an example of *vertical zonation*, the layering of climatic types with elevation in mountainous environments. Thus, although we use a single designation H, it must be understood that this category contains an extremely rich collection of climates. This is true when considering zones within a single H climate and also when comparing one H climate with another. Thus, the vertical zonation of the tropical Andes is distinctly different from that in the Canadian Rockies.

DID YOU KNOW?

Mount Washington, New Hampshire, set three world wind speed records on a single day—April 12, 1934. The peak gust of 373 km/hr (231 mph) and the 5-minute average speed of 303 km/hr (188 mph) were new records, while the mean wind speed for the entire day—206 km/hr (128 mph)—tied the record set the day before at the same location. To put these wind speeds in perspective, recall that the minimum wind speed for a Category 5 hurricane is 250 km/hr (155 mph).

CHECKPOINT

15.19 What are the characteristics of the subarctic climate such as that of Verkhoyansk, Russia?

15.20 Why is there relatively little precipitation in polar climates?

15.21 Why are highlands treated as a separate climate zone?

Summary

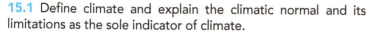

15.1 Define climate and explain the climatic normal and its limitations as the sole indicator of climate.

- Climate involves a *climatic normal*—the kind of conditions expected at a given place and time—but also involves variability and the frequency of temperature and precipitation extremes.

15.2 Identify the main climate groups in the Koeppen system.

- The Koeppen system delineates six groups: four of them (A, C, D, and E) are based on temperature characteristics, one on the lack of precipitation (B), and one on elevation (H). These groups are further subdivided into specific types.
- Climate types vary greatly in their proportion of land and ocean. Generally speaking, warmer climates are found farther poleward over ocean than on land.

15.3 Describe the characteristics of tropical climates.

- Tropical climates are found in low-latitude regions. They are marked by high temperatures throughout the year and usually exhibit high precipitation.
- Categories include tropical wet, monsoonal, and tropical wet and dry.

15.4 Describe the characteristics of dry climates.

- Dry climates exist in subtropical and some midlatitude regions. In these climates precipitation values are low relative to the potential evaporation.
- They fall into the categories of subtropical and midlatitude deserts and steppes.

15.5 Describe the characteristics of mild midlatitude climates.

- Mild midlatitude climates are those in which winters are not very severe. Their precipitation values and summer temperatures vary considerably.
- The three types of midlatitude climates are Mediterranean, humid subtropical, and marine west coast.

15.6 Describe the characteristics of severe midlatitude climates.

- As the name asserts, these climates are subject to severe winter temperatures though their summer conditions may be hot or warm.
- The two subtypes are humid continental and subarctic.

15.7 Describe the characteristics of polar and highland climates.

- Polar climates exist only in high latitudes and fall into the subcategories of tundra and ice cap.
- Highland climates are subject to great spatial variability due to the influence of elevation changes.

Key Terms

boreal forest *p. 492*
climate *p. 474*
climatic normal *p. 474*
climograph *p. 480*
highland *p. 496*
humid continental *p. 492*
humid subtropical *p. 488*
Koeppen system *p. 476*
marine west coast *p. 488*
Mediterranean *p. 488*
midlatitude desert *p. 484*
midlatitude steppe *p. 484*
monsoonal *p. 479*
permafrost *p. 494*
polar ice cap *p. 494*
savanna *p. 482*
semidesert *p. 483*
steppe *p. 483*
subarctic *p. 492*
subtropical desert *p. 484*
subtropical steppe *p. 484*
taiga *p. 492*
Thornthwaite's classification system *p. 476*
tropical wet *p. 479*
tropical wet and dry *p. 479*
tundra *p. 493*

Review Questions

1. Describe what is meant by *climate*.
2. Describe the general criteria by which the Koeppen system delineates climates.
3. The first-order grouping of climates in the Koeppen system is based mainly on temperature. Which climate type departs from that rule?
4. Describe the geographical distribution of tropical climates. What features distinguish this particular group?
5. Briefly describe the fundamental differences between Af, Am, and Aw climates.
6. Of the three types of tropical climates, which occupies the smallest portion of Earth's land surface?
7. Despite their low latitudes, tropical climates are not among the hottest on Earth. Why not?
8. Where are the various dry climates located, and what geographical characteristics cause them to occur where they do?

9. What factor other than annual precipitation is involved in a climate being defined as dry?
10. Describe the four types of dry climates and explain how they differ from each other.
11. Are the mild midlatitude climates really mild? Explain.
12. Describe the various types of mild midlatitude climates and their distribution. Why is it that two of them locate mostly along the west coast of continents, while the other tends to be on the eastern side?
13. What are the two types of severe midlatitude climates, and how do they differ?
14. Describe the three types of polar climates and their distributions.
15. Why are severe midlatitude climates missing from the Southern Hemisphere?

Critical Thinking

1. Do you anticipate that global warming, if it continues as expected, will substantially alter the location of the boundaries of the various Koeppen climates?
2. Although the Koeppen system is intended to create distinct climate classes, its boundaries are based on the boundaries between vegetation types. Is this really a problem? Can you think of any alternative methods for delineating climates?
3. Western Kansas is located near the junction of B, C, and D climates. Do you suppose that a person driving around this region would notice substantial climatic differences as she crossed from one climate zone to another? What does this tell us about the applicability of large-scale classification schemes to smaller-scale analysis?

Problems & Exercises

1. Check Figure 15–4 and determine the type of climate you live in. Then log on to http://ggweather.com/normals/index.htm and find the climate information for the location nearest to where you live. Create a climograph for that location and compare it to the example given in the book for the type of climate in which you live. How closely do they match?
2. Go to the web page for the National Weather Service office nearest to where you live. Compare the monthly temperature and precipitation values observed there during the past year to the average for that location. Were they markedly different? How much variance do you expect to encounter between monthly observed values and climatological averages?

Visual Analysis

Examine the satellite composite map of North American vegetation and answer the following questions:

15.1. How many types of Koeppen climates are likely expressed in this image?

15.2. Do the locations of the vegetation correspond well with those of the Koeppen climates (compare with Figure 15–4)? Explain.

15.3. Where on the map would you expect to find the coldest and hottest conditions? Wettest and driest?

CHAPTER

16 Climate Changes: Past and Future

Glaciers in many locales have undergone retreat, and many other types of climatic indicators also signal change during the last century or so. These include direct measurements of temperature and a host of indirect measures typified by the opening photos. For example, satellite measurements show that Arctic sea ice has decreased in recent years to the point that the Northwest Passage connecting the Atlantic and Pacific oceans has opened repeatedly since 2007. The fact that no openings could be inferred on the basis of satellites prior to 2007 fueled press articles about global warming. The Northwest Passage story illustrates some of the issues regarding climate change. First, satellite measurements go back only to 1978. Is it possible the Arctic was open in prior years? Extrapolating a short record is fraught with difficulty. Second, there were repeated crossings between 1978 and 2007, which shows the satellite record is not perfect, and illustrates the general problem of drawing conclusions from indirect measures of climate. Third, how can we be sure the imputed warming is climate change and not merely a string of somewhat warmer years? Finally, if a climate change has occurred, how can we identify the underlying cause?

This chapter explores issues such as these.

◄ *Retreating Glaciers. Across the world there has been a retreat of glaciers. The large photo shows a recent position of the Athabasca glacier, while the inset shows its extent in 1952.*

Learning Outcomes

After reading this chapter, you should be able to:

16.1 Define climate change and explain how it can be described in terms of changes to boundary conditions in Earth's climate system.

16.2 Explain how scientists understand Earth's past climates.

16.3 Summarize major climate changes that have occurred over geologic time.

16.4 Evaluate natural and anthropogenic factors involved in climate change, including changes in solar output; Earth's orbit; land configuration and surface; atmospheric turbidity; and radiation-absorbing gases.

16.5 Analyze feedback mechanisms and Earth-system interactions involved in climate change.

16.6 List the interactions between land and ocean related to climate change.

16.7 Describe the role of general circulation models in identifying causes and effects of climate change and projecting future changes in climate.

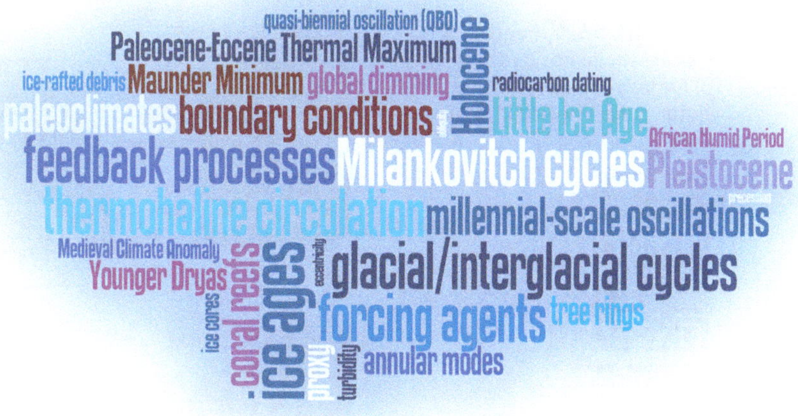

Defining Climate Change

16.1 Define climate change and explain how it can be described in terms of changes to boundary conditions in Earth's climate system.

We saw in Chapter 15 that climate is defined as the statistical properties of atmospheric variables, including temperature, precipitation, and wind. Thus, climate change can be defined as a change in any statistical property of the atmosphere, such as a change in mean temperature. We also saw in Chapter 15 that climate is more than just an average, or mean, value. Year-to-year variations, seasonal variations, and the tendency for above- or below-normal years to occur in sequence are also a major component of climate. Thus, changes in climate may occur even though the mean values of precipitation, temperature, and wind remain the same over time. For example, even with no change in annual average precipitation, changes in the timing of drought and heavy rainfall years could have profound consequences for people and would certainly be considered climatic change.

Consider how changes in climate might occur, thinking first in very general terms. In some ways a planet's climate is like a system that responds to the configuration of external factors, often called **boundary conditions**. These boundary conditions include factors like the gaseous composition of the atmosphere, its surface pressure, the average amount of incoming solar radiation, and even the length of a year. If any of these were to change, we would expect the system to adjust accordingly. Thus, for example, if the Sun's output were to increase, we would expect an increase in global average temperature. Some years might be cooler than before, but on average we would expect a warmer climate, which is, of course, a change in a statistical property. External conditions might vary too quickly for any sort of climatic equilibrium to be obtained, but we can nevertheless think of the external conditions as driving climatic change, or as acting as **forcing agents**.

Note that the response to external factors can be heavily modified by internal **feedback processes**. Think, for example, about the factors governing winter climates in northern Canada. It should be obvious that as solar radiation inputs decline following the summer solstice, temperature declines as well. Once the surface becomes snow covered, a smaller fraction of solar radiation is absorbed, contributing to further cold. The high albedo of snow amplifies the effect of decreasing incoming solar radiation (Figure 16–1). This is an example of a *positive* feedback, one that amplifies a given change. A *negative* feedback does just the opposite, namely, it damps out or at least partially offsets a change. In our Canada example the increasing temperature gradient across the middle latitudes gives rise to a negative feedback. A strong gradient favors more storminess (chiefly midlatitude cyclones), which acts to move heat northward. This makes Canadian winters warmer than they would be otherwise and clearly works contrary to decreasing sunlight. As will be seen later, a host of amplifying and damping feedbacks are involved in climate change.

An important question to consider is: How can we detect climatic change? We might, for example, average temperature values over successive 100-year periods to get an idea of century-to-century change. We wonder: Is it possible for statistics computed like this to change over time, even with no change in boundary conditions? The answer is "yes." Thus, for example, even though the true mean is unchanging, the individual 100-year means rise and fall from one century to the next. So, if we observe the 100-year mean temperature changing, it is not clear that we are seeing climatic change—it might be an artifact of working with finite-length samples. Obviously, there is no way around this problem; using longer averaging periods reduces the sampling error but increases the likelihood of blurring truly distinct climates (those arising

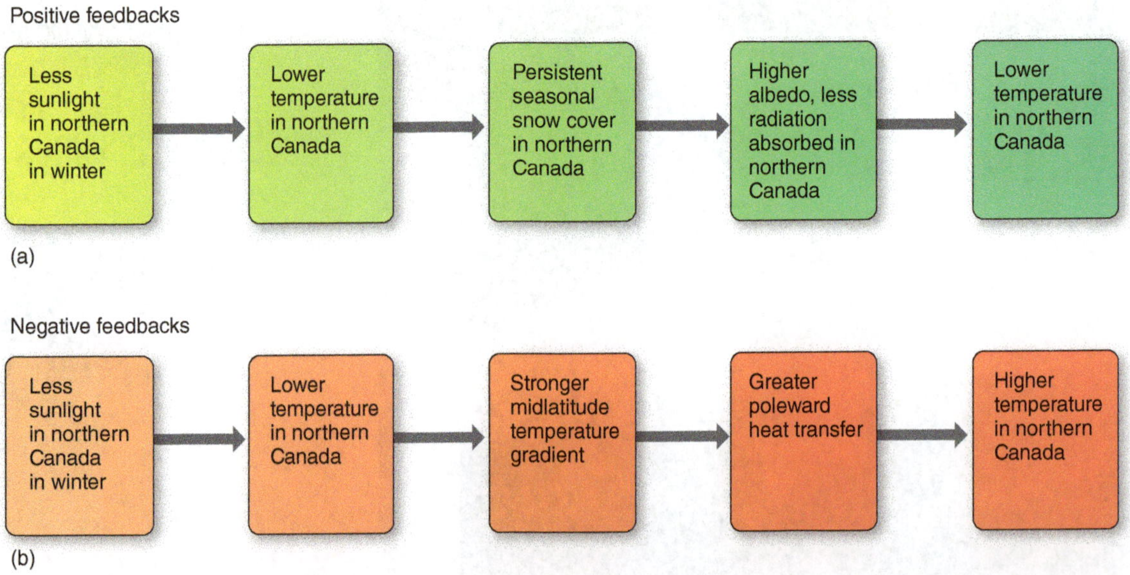

▶ **FIGURE 16–1 Feedback Processes.** (a) Positive and (b) negative feedbacks affecting winter temperatures in Canada.

Video
Climate Change through Native Alaskan Eyes
http://goo.gl/pzslqe

from changing external conditions). On the other hand, if there are physical reasons to expect that changes in the environment (like the observed increases in greenhouse gases) could lead to particular climate responses and those responses are in fact observed, we have very strong reason to expect that decadal or century-to-century changes reflect actual climate change rather than random variation.

Blaming Climate Change

When unusual weather events occur—record heat, a Category 5 hurricane, prolonged summer drought or flooding—the public naturally asks if climate change is the reason. There is no good answer to this kind of question. For example, although it might be tempting to fault global warming for a heat wave or blame ENSO for a powerful storm, there is almost always no way to connect climate change to a particular event.

We know that the intrinsic variability of the atmosphere ensures that both routine and unusual events will occur in the absence of any long-term change. For example, global warming might increase the mean temperature at a location and also the occurrence of weeklong runs of high temperature, say, greater than 38 °C (100 °F). If so, we can be sure that more frequent heat waves are part of global warming. But knowing that heat waves are more common does not allow one to say that a particular heat wave would not have occurred anyway.

That said, the overwhelming consensus among atmospheric scientists is that an increase in temperature across most of the world and an increase in the frequency of many atmospheric events (like heat waves, droughts, and particularly strong hurricanes) have in fact been observed and are due to a very large extent to human activities. So what about the doubts expressed by "climate skeptics"? As a general rule a certain degree of skepticism is not a bad thing. Science advances better when accepted ideas are open to scrutiny. But it is also important to remember that people on any side of the political spectrum are capable of mistaking (purposely or otherwise) their preconceived notions for reasoned analysis. And equally important: Warnings about climate change are neither a hoax nor a conspiracy. Concerns about climate change have arisen as a result of decades of research by thousands of scientists around the world.

CHECKPOINT

16.1 What are the boundary conditions of a system?

16.2 How do positive and negative feedback processes affect changes in climate induced by outside agents? Explain.

Methods for Determining Past Climates

16.2 Explain how scientists understand Earth's past climates.

Though huge gaps exist in our knowledge of past climates, scientists have learned much about the climatic history of Earth. It is really quite remarkable that we are able to describe past climates as well as we can. Consider the fact that although Earth formed about 4.5 billion years ago, people are believed to have inhabited the planet for only about 250,000 years, or for about 0.006 percent of the planet's existence.

You can also view the relative time spans another way: Imagine the planet formed at midnight on January 1 of some vast cosmic year, and it is now midnight one year later. In this scenario people would have first appeared on the scene about half an hour ago, at 11:31 P.M., December 31. The first thermometer would have appeared about 2 seconds ago. Thus, throughout the vast majority of Earth's history, nobody was around to observe and record the climate. The lack of firsthand information is further worsened by the fact that, throughout most of human history, no direct records of climate were preserved.

We continue to gain insights into past climates based on information left in the geological and biological records through what are called **proxy** indicators—evidence derived from sources other than human measurements. Proxy measures have been obtained in different places, using different techniques, and brought together by scientists in various disciplines. Taken together, proxy data gives us some idea of what **paleoclimates**, the climates of the past, must have been like. Not surprisingly, the further back we go in time, the less detailed is the information provided by the various indicators.

We now briefly describe some methods for studying past climates, beginning with techniques useful for uncovering change on the longest time scales.

Oceanic Deposits

Scientists have been drilling into the ocean floor since the 1970s. They extract deep cores of material that has been deposited over very long time periods, with more recent material constantly burying older material previously laid down (Figure 16–2). Included in the deposited material are the bones and shells of plankton and other animal life, made largely of calcium carbonate ($CaCO_3$). The information contained in the oxygen in the calcium carbonate is most important for determining past climates.

Most oxygen atoms have an atomic weight of 16 (^{16}O), but a small percentage of oxygen atoms contain two additional neutrons per atom, making their atomic weight 18 (^{18}O). Both *isotopes* exist in ocean water, and both are incorporated into the shells and bones of marine animals. If the ratio of ^{18}O to ^{16}O in the water is relatively high, it will also be high in the sea life living in that water. Because ^{16}O is lighter than ^{18}O, water containing ^{16}O evaporates more readily than does water containing ^{18}O. Thus, if glaciers are expanding, the oceans (and the oxygen-containing calcium carbonate in shells and bones) will have relatively high $^{18}O/^{16}O$ ratios, as more of the ^{16}O water is removed from the ocean and deposited as snow onto the growing ice sheets. When the organisms die, they sink to the ocean bottom, where their calcium carbonate is deposited. The ocean bottom thereby maintains a record of climate through the varying $^{18}O/^{16}O$ ratios in its layers. Scientists

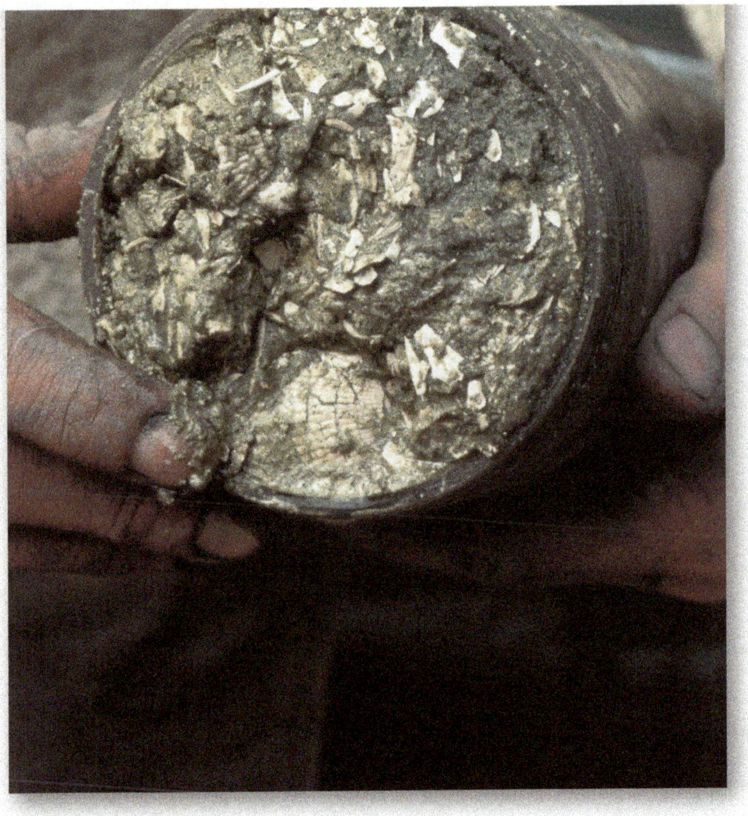

▲ **FIGURE 16–2 Evidence from Ocean Deposits.** Ocean sediment cores contain the remains of long-dead animals that reveal climate conditions prevailing during their lifetime. In addition, rocks deposited by melting icebergs give clues about continental ice sheets.

extract cores of the ocean-bottom material, note the isotope ratios, and infer past changes in global ice volume. Information obtained from sea cores was instrumental in overturning a long-held, but mistaken, idea that there were four glaciation episodes in the last 2 million years.

Other ocean indicators of past climates come in the form of material removed from land surfaces by glaciers. Icebergs shed by continental glaciers carry rocks, pebbles, and other debris equatorward, and this material is deposited on the ocean floor when the iceberg melts. Called **ice-rafted debris**, this material was a key in discovering a number of very rapid changes in Northern Hemisphere ice sheets.

Ice Cores

Scientists have also determined $^{18}O/^{16}O$ ratios for deep **ice cores** obtained from the Greenland and Antarctic ice sheets and from alpine glaciers at lower latitudes. When snow is deposited onto existing glaciers, the oxygen in the H_2O may be either ^{18}O or ^{16}O. Snow that falls under relatively warm conditions contains a greater amount of the heavier isotope. On the ice caps, scientists from Europe and the United States have been drilling through the ice and extracting cores nearly 3 km (1.8 mi) deep to infer past temperature patterns.

In addition to the temperature data obtained from isotope ratios, ice cores provide information on the past chemistry of the atmosphere and on the incidence of past volcanic eruptions. As new snow falls onto a glacier, bubbles of the ambient air become permanently trapped in the ice. The concentration of carbon dioxide and other trace gases in these bubbles yields a long-term record of their varying levels in the air. Among the more interesting results from chemical analyses at the cores is the strong correlation between past temperatures and concentrations of carbon dioxide and methane (Figure 16–3). Past periods of high temperature coincide with high concentrations, whereas glacial periods coincide with reduced concentrations of these greenhouse gases. Generally the changes in gas concentration lag behind the temperature changes by 800 to 1000 years, which implies that changes in composition are not the root cause of the temperature changes. Instead, this indicates a positive feedback process in which changes in the greenhouse gases amplify the climate change. Thus, for example, warming triggers release of the gases to the atmosphere, and the elevated greenhouse concentration produces more warming.

DID YOU KNOW?

Dripping water in caves leaves behind mineral deposits such as stalagmites growing upward from the floor. Known collectively as speleothems, these rocks can be dated and analyzed for changes in oxygen isotope ratio and other indicators of climate. They have recently become an important source of information about past climates on continents, where (obviously) ocean cores are unavailable.

Ice sheets also provide valuable information about major volcanic eruptions. When such eruptions occur, some of the dust ejected into the atmosphere settles on the tops of glaciers. Researchers can determine when heavy volcanic activity occurred by noting the depth of the dust layers within the cores. The chemical composition of aerosols deposited in glacial ice also provides information on past events, with high acidities implying increased volcanic activity. This information may prove to be valuable in determining the importance of volcanic activity as a causal factor for climatic change.

Remnant Landforms

All of Earth's landforms are the end result of processes that build up and wear down features at the surface. The largest features, such as mountains and valleys, are produced by *tectonic* forces, those that produce a deformation of Earth's crusts. Once formed, all such features become subject to erosional processes that remove material at the surface and transport it to other locations. When the forces transporting the material are no longer capable of moving the material, *deposition* occurs. There are several mechanisms for eroding and depositing material, including the movement of water, the slow-moving ice sheets expanding across the surface, wave action along coastlines, wind, and floating icebergs carrying land debris. Each of these mechanisms leaves certain telltale characteristics, and trained field scientists can use the evidence to infer climatic conditions at the time of erosion or deposition.

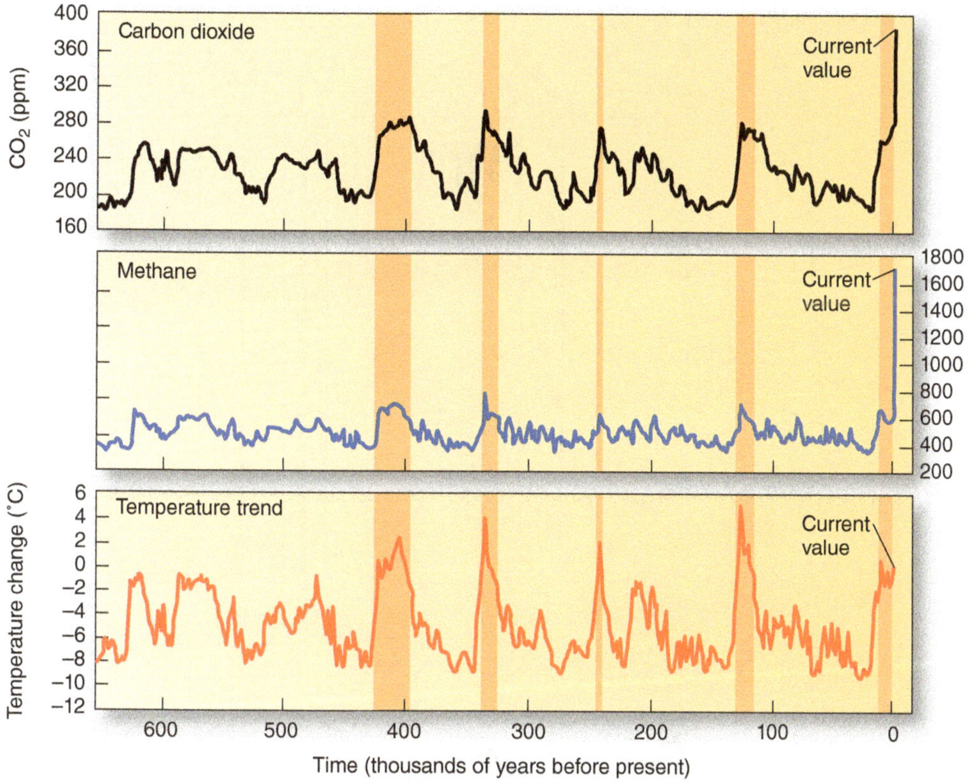

◄ **FIGURE 16–3 Evidence from Ice Cores.** Temperature, carbon dioxide, and methane variations for the past 650,000 years based on ice cores. Shaded periods are times of elevated greenhouse gases and warmth.

FEATURES ASSOCIATED WITH ICE AND WATER Obviously, glaciers can exist only under cold climatic conditions. Fortunately for the earth scientists, erosional and depositional features caused by the movement of glaciers often leave very distinct signatures on the landforms that last long after the glaciers have fully retreated. Some of the features associated with glaciation are related to erosion. For example, as alpine glaciers expand down preexisting valleys, they widen out the lower reaches of the valleys. This can transform typical V-shaped valleys (looking up or down the valleys), associated with those cut by running water, to U-shaped valleys. Glaciers also often leave scratch marks on solid rock walls and valley floors, or polish exposed rocks to a very smooth finish.

Unlike running water, flowing ice sheets are capable of moving very large-sized sediment as effectively as small particles of sand, silt, and clay. When the ice sheet melts away, its deposits (called *till*) can contain a very wide assortment of sediment sizes, from microscopically small clays to very large boulders. Thus, poorly sorted deposits often exist along the past margin of an ice sheet. Where glaciers terminate in an ocean, land materials are rafted away and eventually deposited on the seafloor as icebergs melt.

Streams and rivers also provide useful evidence for reconstructing past climates. Depositional features can be particularly useful in this regard, because the size of material that can be transported by running water depends on the speed of flow. Rapidly moving streams can carry large rocks. If the flow of water in a stream begins to slow down, the maximum size of sediment it can transport decreases, and the largest material it contains will be deposited. This allows us to infer that layers made up of very large sediment must have been deposited by large stream flows. Thus, by examining the layering, or *stratigraphy*, of stream banks or road cuts, one can get some idea of the sequencing of high and low precipitation episodes.

Waves along coastlines also leave distinct features that can be left behind after sea level rises or lowers. Thus, by examining the elevation of terraces or submerged coastal platforms, we can infer how much glacial ice has accumulated or melted.

CORAL REEFS **Coral reefs** are hard ridges extending from the ocean floor to just below the water surface, along the shallow margins of warm, tropical oceans. They form by the growth of small marine organisms (coral) having hard shells, composed largely of calcium. Colonies of coral live together atop the hard material left behind by past coral. When these coral die, their shells remain at the top of the reef and provide the foundation on which subsequent coral grow.

Coral reefs have been useful in several ways for the inference of past conditions. Because they exist along shallow waters, relic coral reefs can provide information on the location of past sea levels. Moreover, because the chemical composition of growing coral is affected by the water temperature, analysis of the changing chemistry of coral reefs with depth provides useful information on past conditions. They have been particularly useful in providing information on past El Niño events.

Past Vegetation

Climate is by no means the only important factor affecting the distribution of vegetation, but it does exert a strong influence on the distribution of vegetation communities. When

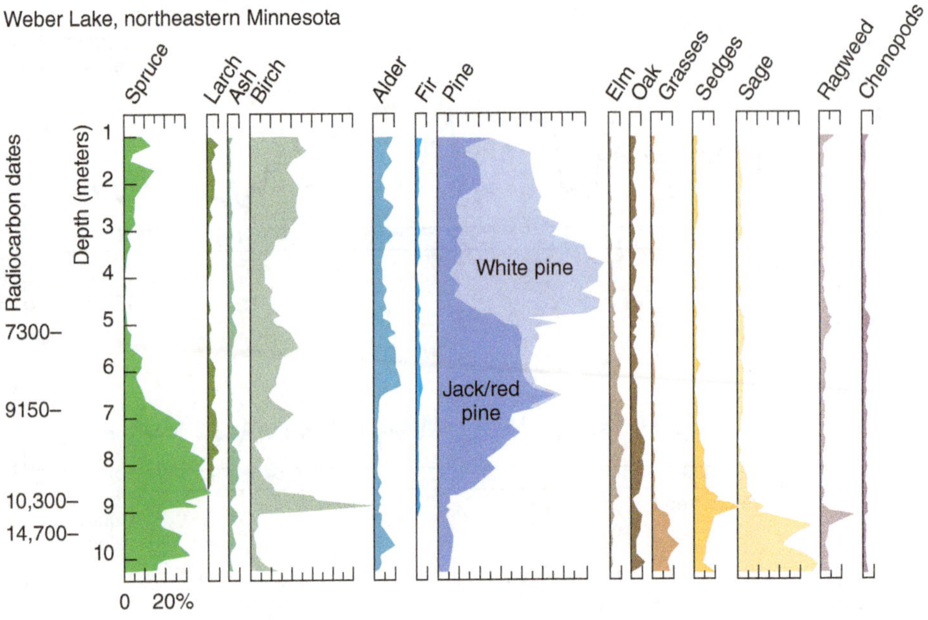

▸ **FIGURE 16–4 Evidence from Pollen.** Diagrams of pollen provide information on past vegetation at a site, which is useful for determining past climates.

Video
20,000 years of Pine Pollen
http://goo.gl/iCAe1

a vegetation community occupies a region, some of its pollen and spores can be deposited and preserved indefinitely in lake beds or bogs. This preserved pollen can be extracted and identified by *palynologists*. Organic material deposited with the pollen is often subjected to a technique called **radiocarbon dating**. Radiocarbon dating provides a good estimate of the age of material younger than about 50,000 years and, along with the pollen data, allows a determination of the distribution of vegetation species that existed at various times during the past. This information can then be displayed in pollen diagrams (Figure 16–4) that depict the sequence of past vegetation assemblages. Because many types of vegetation stands are identified with particular climate types, these diagrams provide useful information for deciphering the climatic history of an area.

Much information about past climates extending back for several thousand years can also be obtained from **tree rings**. Each year many trees increase the width of their trunks by the growth of concentric rings, each distinct from the previous rings (Figure 16–5). The width of each ring depends on how favorable temperature and/or moisture conditions were during a given year for the particular tree species. Under climatic stress conditions resulting from a lack of moisture or excessive warmth, the growth of these rings will be retarded. When conditions are favorable for growth, the rings will be relatively thick.

Some tree species are sensitive to temperature variations, whereas other species are affected primarily by changing moisture conditions. In either case, the extraction of cores from the trunks of very old trees (the oldest ones date back more than 5,000 years) yields a continuous record of annual tree growth, which correlates with precipitation and/or temperature. The correlation depends on the species and on the environmental setting. At high elevations, for example, low temperatures might slow annual growth, while for the same species at lower elevation, warm temperatures might create slower growth through increased moisture stress. Although obviously a complicating factor, differential climatic response can be used to advantage by permitting researchers to isolate various types of climatic change from one another. Indeed, such records have been obtained from old stands of trees around the world and provide climatologists with a wealth of information about past conditions. In recent years isotopic analysis of tree cores has added to what can be learned from ring widths alone, and thereby greatly expanded the usefulness of this general method.

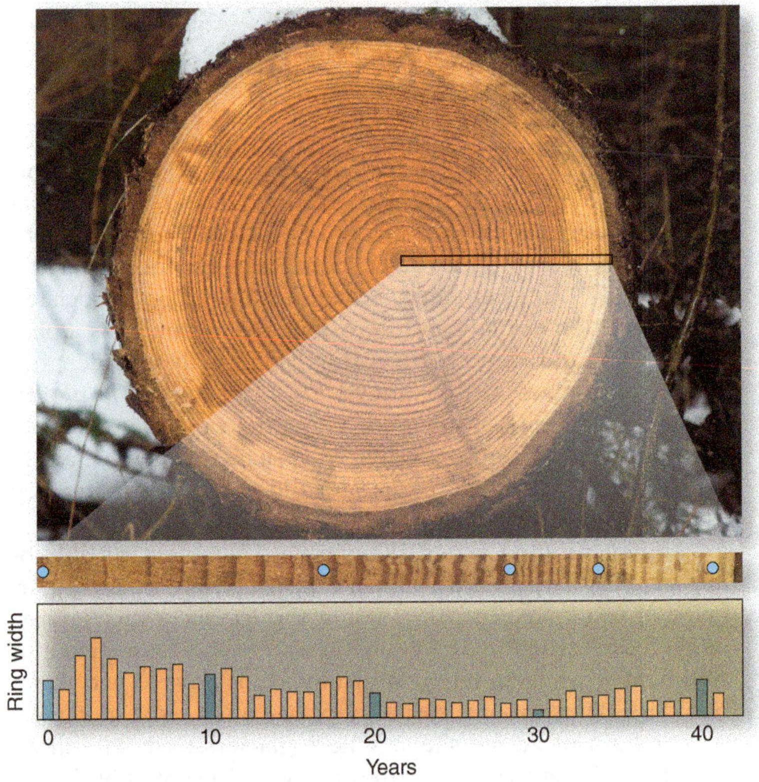

▴ **FIGURE 16–5 Evidence from Tree Rings.** The blue dots on the middle panel correspond to each tenth-year.

CHECKPOINT

16.3 How is it possible to infer past climate conditions for times before climate records were kept?

16.4 Explain how the following can provide evidence for past climates: ocean-floor rocks, ice cores, relict landforms, tree rings.

Past Climates

16.3 Describe major climate changes that have occurred over geologic time.

Earth scientists have devised a widely used scheme to divide the planet's natural history into distinct time frames. The geologic column shown Figure 16–6 uses a hierarchical system dividing time into *eras, periods,* and *epochs*. These time segments are not based on climatic characteristics but rather on geologic and fossil evidence indicative of past environmental conditions and events. Thus, while climatic episodes are sometimes associated with particular eras, periods, or epochs, the time segments should not be considered to have uniform climatic conditions. In addition, significant climate events sometimes cross boundaries in the column, so the divisions do not mark starting or ending times. Finally, the geologic column is not absolute—the names, numbers, and durations of the time intervals are revised periodically to reflect improved knowledge. Figure 16–6 shows the most recent geologic scale, which was adopted in 2009. For our purposes, the terminology is best used as a kind of calendar, to identify points in time, rather than to identify particular climates or events. With this in mind, we can discuss some significant episodes of Earth's climatic history.

DID YOU KNOW?

When sand grains are thrown into the air, solar radiation resets an internal radiometric clock that can be used to date the time of exposure. Relict dunes provide evidence of dry episodes, and their age tells scientists when those episodes occurred. The dating technique, optically stimulated luminescence, was developed in the mid-1990s. It can date sediments with ages ranging from 300 to 100,000 years ago.

Warm Intervals and Ice Ages

We tend to think of our own time as "normal," but this is really not the case at all. Looking at a broad span of Earth's history, we would have to describe the present climate as highly unusual. Most of the time our planet has been considerably warmer than it is today and largely free of permanent (year-round) ice. Think of Earth as a warm planet whose history has been punctuated by multiple, relatively brief **ice ages**. Though varying, the warm times persist for hundreds of millions of years to billions of years, whereas the ice ages last on the order of tens of millions of years to perhaps a hundred million years. The earliest known ice age dates to about 2.8 billion years ago, followed by several others beginning at 2.4 billion years ago (Figure 16–7).

Some periods of warmth have been particularly prominent, such as the warmth of the mid-Cretaceous period, extending from about 120 to 90 million years ago (MYA). In those times, dinosaurs roamed beyond the Arctic Circle. With very little water held in land ice, sea level was perhaps 150 to 200 m (490 to 650 ft) higher, flooding about 20 percent

Era	Period	Epoch	Duration in Millions of Years	Millions of Years Ago
CENOZOIC	Neogene	Holocene	0.0117	0.0117
		Pleistocene	2.6	2.588
		Pliocene	2.7	5.332
		Miocene	17.7	23.03
	Paleogene	Oligocene	10.9	33.9 ± 0.1
		Eocene	21.9	55.8 ± 0.2
		Paleocene	9.7	65.5 ± 0.3
MESOZOIC	Cretaceous		80	145.5 ± 4.0
	Jurassic		54.1	199.6 ± 0.6
	Triassic		51.4	251.0 ± 0.4
PALEOZOIC	Permian		48	299.0 ± 0.8
	Carboniferous	Pennsylvanian	19.1	318.1 ± 1.3
		Mississippian	41.1	359.2 ± 2.5
	Devonian		56.8	416.0 ± 2.8
	Silurian		27.7	443.7 ± 1.5
	Ordovician		44.6	488.3 ± 1.7
	Cambrian		53.7	542.0 ± 1.0
PRECAMBRIAN				

▲ **FIGURE 16–6** The Geologic Column.

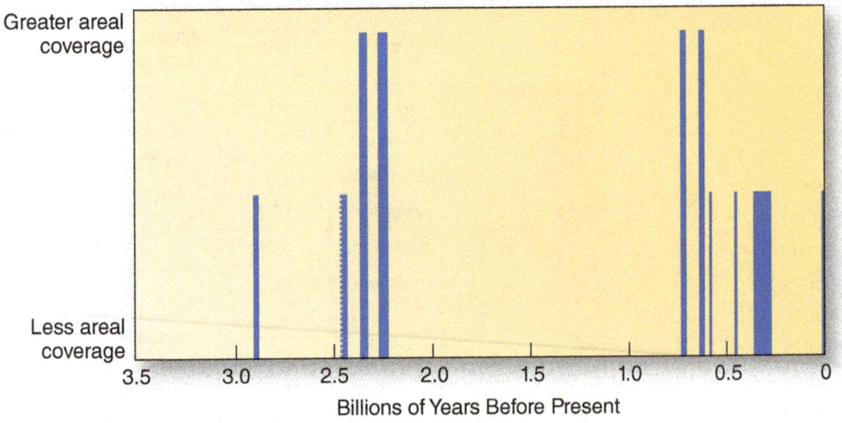

▲ **FIGURE 16–7 Warm Intervals and Major Ice Ages Throughout Earth History.** Heights of bars indicate the spatial extent of glaciations.

of current continental areas. Global average temperature is thought to have been anywhere from 5 °C to 15 °C (9 °F to 27 °F) warmer than at present. Another significant warm period existed from about 45 to 55 MYA (Figure 16–8).

All ice ages stand in stark contrast to "warm ages" such as the mid-Cretaceous and Eocene. At one extreme, nearly the entire planet might have been ice covered during the ice age of about 700 MYA, in a condition aptly called "snowball Earth." Evidence for total ice cover at that time comes from a variety of diverse indicators, but the snowball hypothesis remains controversial insofar as the amount of ice is concerned.

Regardless of whether the entire planet has ever been completely frozen, it is clear all ice ages have abundant year-round ice. The enormous ice sheets draw their ice from what had previously been sea water. As a result, sea levels are inevitably much lower during ice ages.

Temperature differences between ice ages and warm ages vary greatly by latitude, with high latitudes showing greater changes than the tropics. For example, during an ice age, polar sea-surface temperatures (SSTs) might be 10 °C (18 °F) colder but tropical sea-surface temperatures only 1 °C to 5 °C (2 °F to 9 °F) colder. Beyond these generalities, little can be said with certainty about the six earliest ice ages. On the other hand, we have substantial information regarding conditions during the most recent ice age.

The Pleistocene

What people often refer to when they speak of "the ice age" is an epoch that occupies most of the last 2.5 million years, called the **Pleistocene** (see the top of Figure 16–6 for reference). As shown in Figure 16–9a, the world's climate has undergone many changes during the last few million years. Beginning about 2.5 MYA, these oscillations began increasing in amplitude, and about 800,000 years ago the amplitude increased dramatically, becoming about twice as large as in the preceding million years (Figure 16–9b). These oscillations in temperature and ice cover are called **glacial/interglacial cycles.** As can be seen in the diagram, they are quite irregular. For most cycles ice volume increases slowly and then terminates rapidly in a warming event. The cycles generally last about 100,000 years, with a relatively small part of each cycle, typically only 20 percent, spent in the warm interglacial phase. The last 150,000 years are shown in Figure 16–9c. Neither ice growth nor decay is uniform; "quivering change" is a better descriptor, with short-term oscillations superimposed on the longer cycles. Taking the last 2 million years as a whole, there have been about 30 cycles altogether, with associated global temperature changes of perhaps 5 °C (9 °F).

Ice volume changes have been largest in the Northern Hemisphere, with the size of ice sheets growing and shrinking by a factor of three or so with each glacial cycle. Although the Antarctic sheet has changed far less, Antarctic air temperature changes are believed to match those of the north polar latitudes, with glacial period temperatures about 10 °C (18 °F) lower than an interglacial. As is clear from Figure 16–9c, the planet is now in a warm interglacial, rivaled only a few times in the last 2 million years. Interestingly, one of those times was the last interglacial, which reached its peak about 130,000 years ago and might have set the record for Pleistocene warmth. Global sea level was about 6 m (20 ft) higher than now, and there is evidence that midlatitude continental

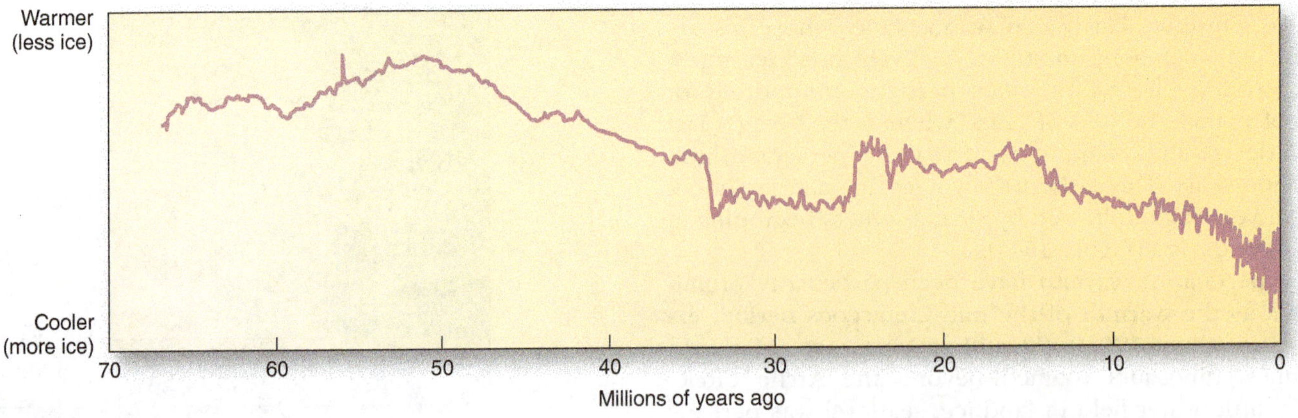

▲ **FIGURE 16–8 Global Temperatures.** Global temperature indicator based on Earth's surface ice content for the past 70 million years.

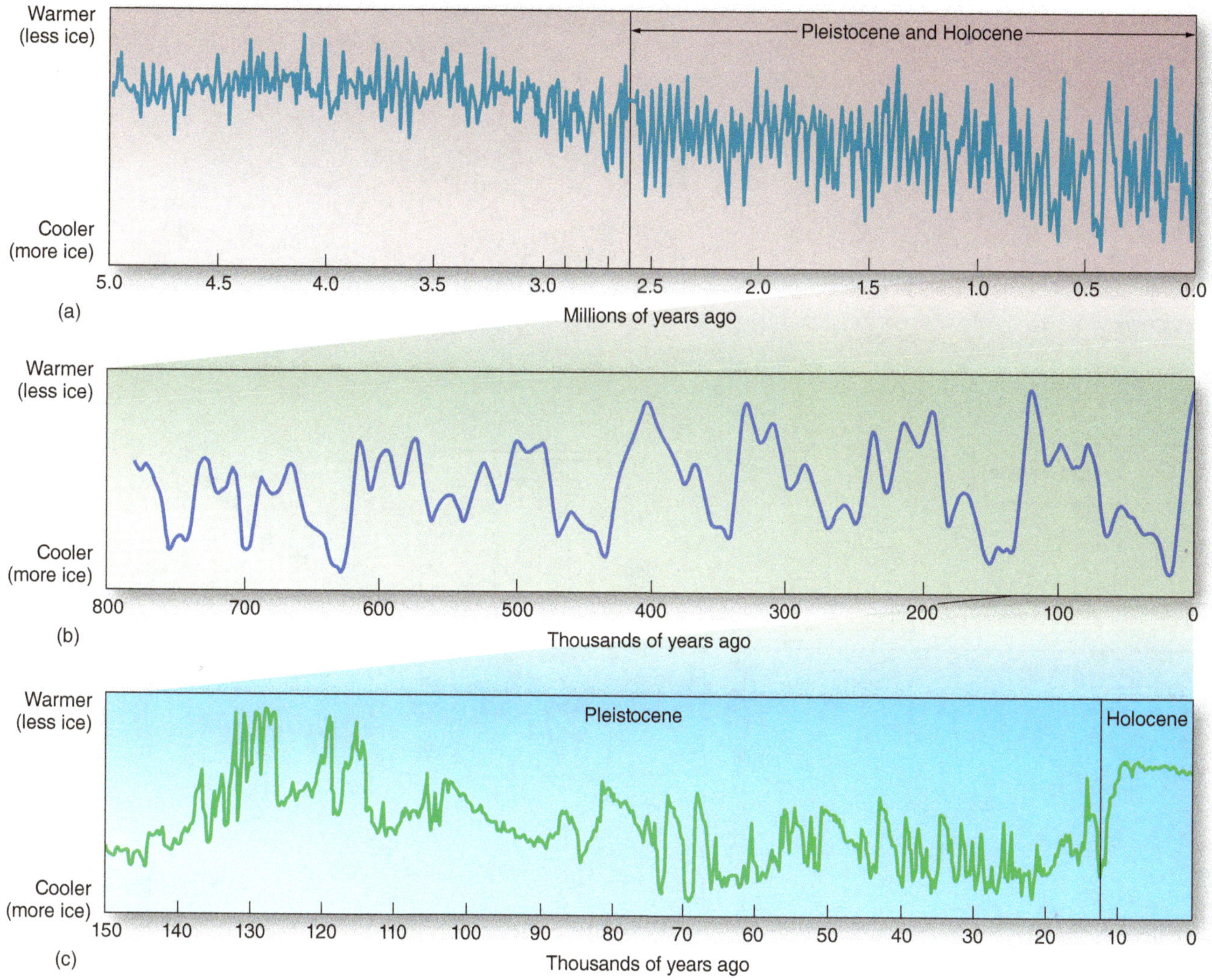

▲ **FIGURE 16–9 Climate Indicators for the Last 50 Million Years.** Curves (a) and (b) indicate mainly global changes in ice volume. Curve (c) reflects climatic conditions over the North Atlantic but is broadly consistent with global changes, as can be seen by comparison with later portions of (b).

areas were 1 °C to 3 °C (2 °F to 5 °F) warmer. In contrast, SSTs were not too different from what they are now. Between these two warm periods sits the most recent glaciation, an event that reached its maximum about 20,000 years ago.

CHECKPOINT

16.5 What does the term "snowball Earth" refer to?

16.6 What is one major trend in Earth's climate during the last 50 million years? Explain.

16.7 Should Earth be considered a "warm" planet or a "cold" planet? Explain.

The Last Glacial Maximum

In the depths of the last glaciation, around 20,000 years ago, many aspects of the Earth–atmosphere system were different. Most dramatically, of course, land ice covered much more area, as seen in Figure 16–10.

Ice sheets advanced to within 40° of the equator, covering the ground to a depth of several kilometers. In North America the ice reached about as far south as present-day St. Louis, but only to the latitude of New York and Seattle on the East and West Coasts. Tremendous quantities of water were transferred from ocean to land, building sheets 3500 to 4000 m (2.1 to 2.5 mi) thick. Given enough time, this would be enough to depress the continental crust by more than 800 m (0.5 mi). When the ice melted, the land surface gradually expanded upward toward its original level. (Even now, continents have yet to fully rebound from glacial depression.) The Laurentide ice sheet in North America was in some ways equivalent to a huge mountain range, running from the Rocky Mountains to the Atlantic. Sea level was about 120 m (394 ft) lower than it is now, so that a land bridge existed between Siberia and Alaska. (Given enough time, movement of water from the ocean would cause oceanic floors to rise by about 35 m—115 ft.) There were also significant changes in sea ice, especially in the Antarctic Ocean, where winter sea ice covered about twice the area it now does.

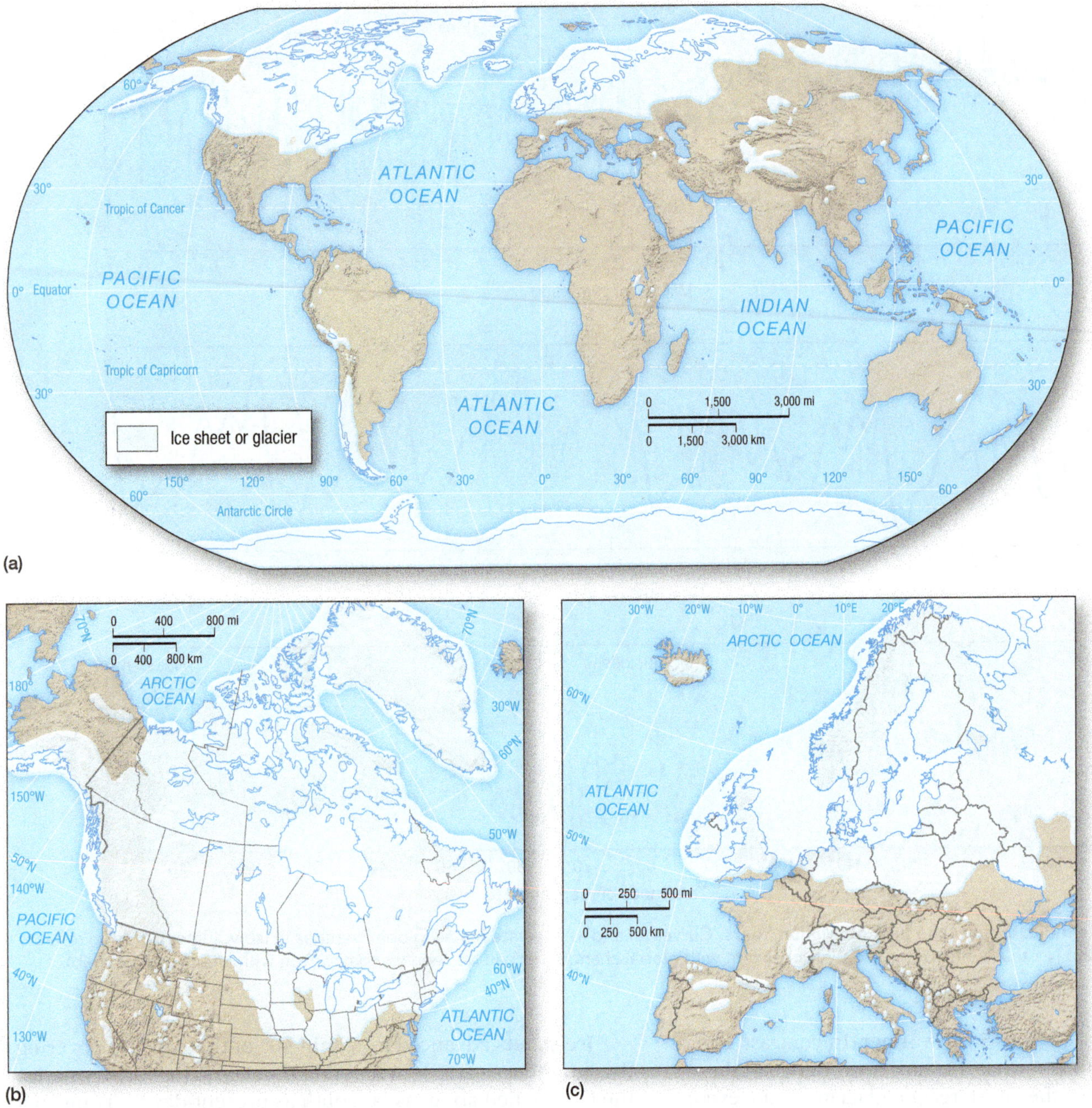

▲ **FIGURE 16–10 Earth During the Ice Age.** The map shows the extent of ice and coastlines at the time of the last glacial maximum.

Video
Retreat of Continental Ice Sheets
http://goo.gl/BqLnkX

On land, temperature changes varied greatly by proximity to the ice sheets and to the ocean. For example, in western North America, maritime air masses kept temperatures to within 4 °C to 5 °C (7 °F to 9 °F) of modern values. This contrasts sharply with the area that is now Tennessee and South Carolina, where temperatures were 15 °C to 20 °C (27 °F to 36 °F) colder than they are now. Temperature changes in the tropics were smaller, perhaps 4 °C to 5 °C (7 °F to 9 °F). Snowline was about 1000 m (3300 ft) lower, which translates into a temperature decrease of 5 °C to 6 °C (9 °F to 11 °F) for elevations above 2000 m (6600 ft).

Most places were not only colder but they seem also to have been drier. This is especially true for the high latitudes, where precipitation amounts were about 50 percent below today's. Some desert areas of both South America and Africa were larger, and lake levels were lower in tropical Africa and Central America. A cold desert covered much of the section of western Europe that was not under ice. Precipitation changes were undoubtedly determined to a large degree by circulation changes. Where prevailing winds shifted such that they blew from high latitudes (as for eastern North America), conditions were drier.

It must be emphasized that this glacial period was hardly uniform. In fact, abrupt climate changes were very common throughout this period, with polar temperatures changing by 8 °C to 16 °C (14 °F to 29 °F) over the course of just decades to centuries (see Figure 16–9c). Because changes such as these are not confined to the glacial period, they are covered in more detail in a later section.

The Holocene

Like other glacial-to-interglacial transitions, warming following the last glacial maximum was fast, at least compared to times of cooling. Warming began about 15,000 years ago (prior to the start of the **Holocene**), only to be interrupted about 2000 years later when colder conditions returned. Called the **Younger Dryas**, this colder period lasted for about 1200 years. Following the Younger Dryas, starting about 11,800 years ago, came another period of abrupt warming, with temperatures in Greenland increasing by about 1 °C per decade, bringing climate into the interglacial we enjoy today.

The Holocene has not been uniformly warm, as reflected in the so-called **Little Ice Age**. Spanning perhaps 1400 to 1850, this was a cold period for western Europe. During these years, alpine glaciers advanced as temperatures fell by about 0.5 °C to 1 °C (0.9 °F to 1.8 °F). Historical records indicate that this seemingly small decrease in mean temperature had a considerable effect on living conditions throughout Europe. Shortened growing seasons led to reductions in agricultural productivity, especially in northern Europe, though its impact was felt in mountain records from around the world. Although in no sense is it a true "ice age," it does represent the largest temperature change during historical times and is considered a global event.

Within the Holocene, considerable attention has been devoted to understanding the last 1000 years, in large part because it sets the context for presumed global warming occurring in the last hundred or so years. There have been a number of climate reconstructions for this, as seen in Figure 16–11. The latest of those shown used what are believed to be more reliable indicators and more advanced data analysis methods. Compared to the others, it suggests a more intense Little Ice Age than most other curves imply.

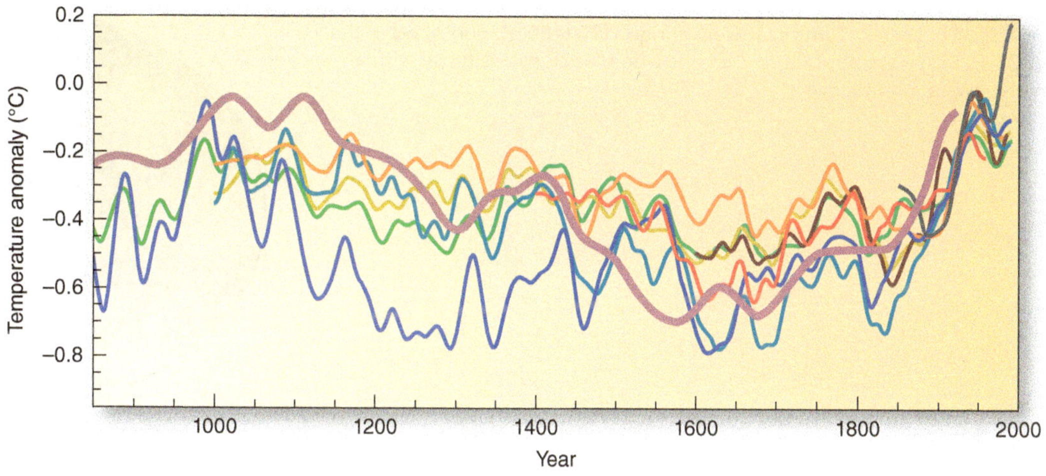

▲ **FIGURE 16–11 Temperature Reconstructions for the Last 1000 Years.** Temperatures are expressed as departures from the 1961–1990 average. The heavy pink line is believed to be the most reliable.

The Last Century

During the 20th century, a growing network of meteorological stations was established around the world. Although there are many problems with the resulting measurements, such as the movement of meteorological instruments at a given site, the data give firsthand accounts of temperature and precipitation patterns. Augmenting the land-based measurements are observations from ships, and much more recently, satellite estimates of temperature, precipitation, wind, and other parameters. As discussed in Chapter 3, these measurements give an unambiguous picture of rapid warming during the last few decades. Figure 16–12 shows the geographic distribution of the warming for the 1901–2012 period, using one of several data sets. The map depicts the change in annual temperature in degrees Celsius, assuming a linear change over the entire period. Figure 16–12 (and other diagrams in this chapter) was published in a 2013 report by a group of climate experts known as the Intergovernmental Panel on Climate Change. The IPCC panel is described in *Box 16–1, Focus on the Environment and Societal Impacts: Intergovernmental Panel on Climate Change.*

Notice that some of the North Atlantic shows slight cooling, whereas the interiors of Asia and northern North America show warming much above global average values. Clearly, there is considerable spatial variation in warming on an annual basis. Similar conclusions hold for individual seasons. That is, different places have warmed at different rates in different seasons (again, with some places cooling in some seasons). All this is to say that the term "global warming" must be used carefully as a reference to change in the global average value.

DID YOU KNOW?

Current levels of carbon dioxide and methane are higher than at any time in the last 650,000 years. A nearly 3 km (2 mi) long ice core from the interior of Antarctica shows that today's concentrations of CO_2 and CH_4 are, respectively, 27 and 130 percent higher than what occurred naturally over that time span. Other evidence suggests that current levels of CO_2 have not been matched in the past 10 million or more years. Thus, our time has seen the shattering of a truly long-standing record for this important greenhouse gas.

DID YOU KNOW?

Snowpacks in western North America have declined during the last few decades and are projected to fall more with global warming. Growth rates for some tree species are sensitive to the snowpack that accumulates in the preceding winter. Using rings from trees collected at various elevations, scientists have used this relationship to estimate snowpacks in the Rocky Mountain headwater areas of the Columbia and Missouri Rivers. Their analysis shows that recent 20th-century droughts have been rivaled only twice in the last 800 years.

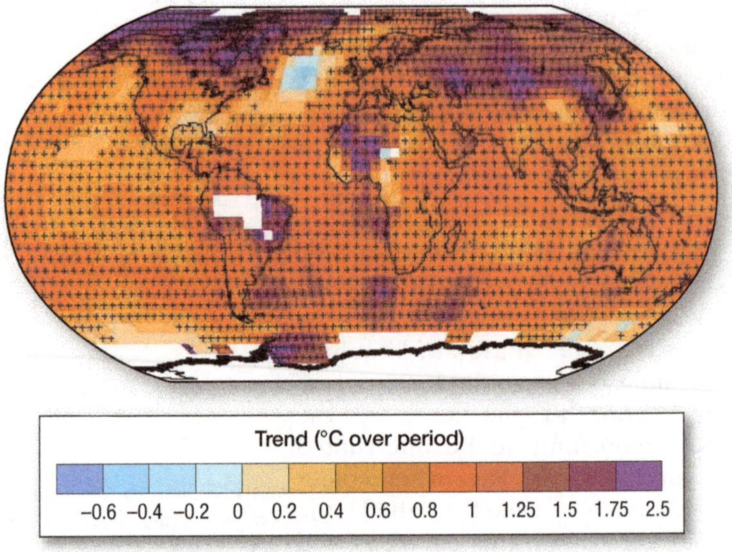

▸ **FIGURE 16–12 Temperature Change, 1901–2012.** Published in the IPCC Fifth Assessment Report. Zones with plus signs are those with a 90% level of confidence.

Animation
Global Warming
http://goo.gl/E1JEYu

16–1 FOCUS ON THE ENVIRONMENT AND SOCIETAL IMPACTS

Intergovernmental Panel on Climate Change

You have probably heard about the current debates around climate change and global warming. You may think of climate change as a problem in remote polar regions and mountain ranges, where ice sheets and glaciers are melting, and wonder if it's true that climate change is already affecting the regions where most people live. The fact is that it has indeed occurred and will continue to do so. And these changes are not restricted to increases in temperature and the melting of ice sheets and glaciers; very strong evidence leads us to the conclusion that there have been and will continue to be changes in the occurrence of extremely heavy precipitation events, the frequency and intensity of droughts, and the concomitant changes in the cost of food and energy. These climate changes are discussed later in the chapter. The best single source for documenting these changes is the current report from the Intergovernmental Panel on Climate Change (IPCC).

The IPCC was established in 1988 under the auspices of the World Meteorological Organization (WMO) and the United Nations Environment Programme (UNEP). Its mandates included the assessment of the current knowledge about climate changes (including an analysis of what is not known), discussion of how these changes impact the global environment and socioeconomic activity, and the development of policy recommendations.

The panel is composed of three working groups: Working Group I (WGI) analyzes the state of scientific knowledge of climate change, Working Group II (WGII) reports on its human and environmental impacts, and Working Group III (WGIII) offers mitigation strategies. None of the working groups is charged with undertaking new research; their responsibility is to present an overview of the state of knowledge in each of their respective areas. This is particularly significant because the IPCC reports do not reflect the views of just a select group of people, but rather a consensus of the scientific community at large about what it knows and does not know about climate change.

The assessment report released in 2013 was the fifth since the inception of the panel. Like the earlier reports, it incorporated the most recent findings into its analysis and made use of improvements in computer modeling, enhanced satellite data acquisition, and further observational evidence. The report was a massive collective effort; it was authored by more than 150 expert climate scientists from 30 countries and was subject to review by another 600 experts. Extensive opportunity was provided for comment by governments, organizations, and individuals. Indeed, more than 30,000 written comments were submitted, and editors were required to ensure an adequate response to all substantive submissions. Draft documents were given two rounds of review. Each assessment report also described the level of uncertainty in its analyses, with explicit definitions for qualitative characterizations. (The "extremely likely" role of humans in the observed warming mentioned previously is one such example.)

In summary, the IPCC reports are not position papers prepared by a group of scientists intent on espousing a particular viewpoint. Rather, they are the scientific community's response to a request for informed analysis of the current state of the physical science surrounding climate change. The IPCC reports represent the most comprehensive and authoritative summary available on a topic of intense interest and great importance for both humans and the natural realm. In 2007 the group was collectively awarded the Nobel Peace Prize, along with former United States Vice President Al Gore, who has campaigned to spread information on the seriousness of global warming.

16–1–1 Summarize the purpose and makeup of the IPCC.

16–1–2 Describe the process used to assemble the 2013 IPCC report. What aspects of the process were designed to ensure its objectivity and accuracy?

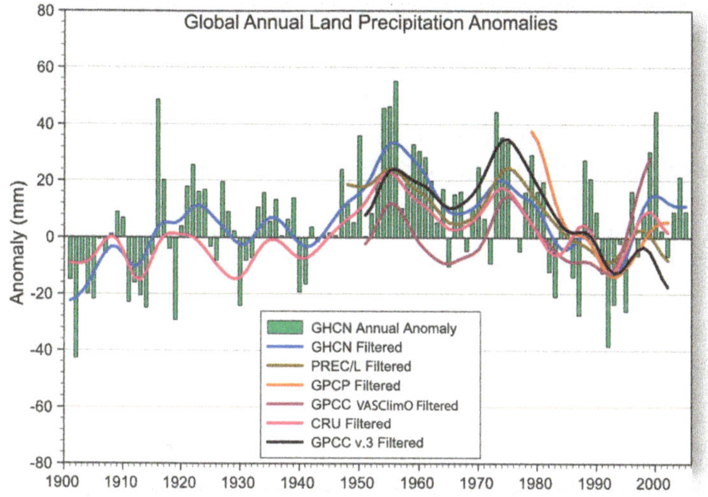

▲ **FIGURE 16–13 Global Average Precipitation Changes.** Values are plotted in millimeters as departures from the long-term mean. The curves correspond to various data processing methods and meteorological organizations.

EFFECTS OF WARMING ON PRECIPITATION Global warming is undoubtedly not the only climate trend in the last 100 years, but it is probably the least ambiguous. For example, precipitation—the other major climate variable—exhibited nothing like the progressive change seen for temperature (Figure 16–13). Although there are episodes of apparently wetter and drier than average global conditions, there is no strong trend for the record as a whole. Spatial patterns of precipitation trends are even less clear (Figure 16–14). Looking at the map, notice how much of the world lacks sufficient data to calculate a trend (gray areas). Notice also that of those places where data are sufficient, the calculated trend value is statistically significant in only a few locations (places marked with a black mark). The difficulty is that precipitation exhibits extreme variability from year to year and from place to place. If long-term changes are to be detected statistically, they must be very large to stand out against this background of variation. For most of the world that simply did not happen over the period of record.

EFFECTS OF WARMING ON TEMPERATURE-RELATED VARIABLES Narrowly construed, the term "warming" refers to temperature. However, during the last century many other variables have participated in changes consistent with a warmer planet. A few of those mentioned in the most recent and earlier IPCC reports are as follows:

- The number of days with frost has decreased over many parts of the midlatitude regions.
- There has been a decrease in the number of extreme cold events across much of the world, and extreme warm events have become more frequent.
- Snow cover has decreased in most areas (especially in the Northern Hemisphere), and those decreases have mostly been driven by increasing temperature. In places where snow cover increased, increasing precipitation (rather than cooling) was the cause.
- The breakup date for river and lake ice occurred earlier by an average of 6.5 days per century, and the freeze-up date occurred later by an average of 5.8 days per century.
- From 1901 to 2002 the maximum extent of seasonally frozen ground declined by about 7 percent in the Northern Hemisphere.

CHECKPOINT

16.8 How have climates of the Holocene generally differed from those in the preceding epoch?

16.9 How do you think the climate of the Little Ice Age may have affected human populations in Europe?

16.10 How does global climate in the last century compare with average conditions during the last 1000 years?

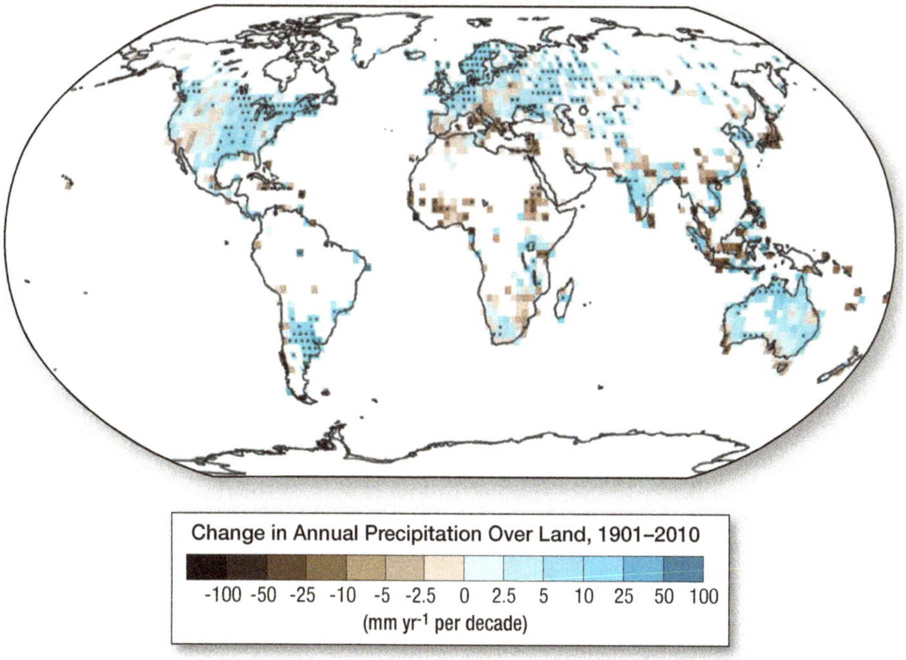

▶ **FIGURE 16–14 Rate of Precipitation Change (Millimeters per year per Decade), 1901–2010.** Published in the fifth IPCC report. Areas with insufficient data are shown in gray.

Factors Involved in Climatic Change

16.4 Explain natural and anthropogenic factors involved in climate change, including changes in solar output; Earth's orbit; land configuration and surface; atmospheric turbidity; and radiation-absorbing gases.

As we have seen, Earth's climate has undergone significant changes of varying magnitudes and time scales over the course of its existence. The big question, of course, is *why?* Several possible causes are easy to identify. These include variations in the intensity of radiation emitted by the Sun, changes in Earth's orbit, land surface changes, and differences in the gaseous and aerosol composition of the atmosphere. Each operates on a different time scale, as we will now see.

While examining the list of factors affecting climate change, it is important to understand that many of them do not operate independently. This means, first of all, that they operate simultaneously. Thus, for example, while one agent might be leading to warming, another might counteract or enhance that warming. Second, it means that agents of change might interact with one another, so that the effects are not merely additive. For example, the effect of tropospheric aerosols produced by humans might differ depending on whether or not they are overlaid by a plume of stratospheric volcanic aerosols. The issues of simultaneous and interacting factors will become especially evident in the discussion about changes in Earth's orbital characteristics.

Variations in Solar Output

As we implied earlier, Earth's climate is quite sensitive to the Sun's output. Considering that the amount of energy emitted by the Sun is not truly constant, this mechanism of climate change has considerable theoretical appeal. For example, some changes in the solar output, on the order of 0.1 to 0.2 percent, appear to be related to the occurrence of sunspots. As mentioned in Chapter 2, sunspots are relatively cold regions of the photosphere, about the size of Earth's diameter. The abundance of sunspots rises and falls on several time scales, including the very striking 10.7-year cycle. (As is customary, we will refer to this as the 11-year cycle.)

Satellite measurements show that when measured over the course of a few weeks, solar radiation decreases as sunspots increase. This is consistent with sunspots being cold—when more of the Sun is covered by cold regions, less radiation is emitted. But at longer time scales, such as those of a complete 11-year cycle, increasing sunspots are correlated with more radiation. Clearly, at these longer time scales, there must be solar changes that compensate for the increased area of sunspots. (The most likely explanation is an increase in surrounding bright areas.) The link between sunspot activity and solar output has led to considerable speculation that such phenomena could account for some of the climatic changes that have occurred on Earth. For example, droughts in the Great Plains of the United States have shown some tendency to recur at an interval that roughly corresponds to a double sunspot cycle, and earlier research has noted similar periodicities in Nile River flows. In addition, air temperatures over eastern North America rise and fall by about 0.2 °C (0.4 °F) in apparent synchronicity with the 11-year cycle. Some supporting evidence for the connection between climate and solar activity is also given by the fact that the **Maunder Minimum**, the period of minimal sunspot activity between about 1645 to 1715, coincided with one of the coldest periods of the Little Ice Age (see *Box 2–2, Physical Principles: The Sun,* on page 68). However, there have been other episodes in which variations in sunspot activity did not coincide with changes in climate, and alternative explanations have been offered for the apparent 22-year climate changes.

The equivocal evidence for a Sun–climate connection became stronger in the late 1980s, when several scientists observed that the relationship between tropospheric conditions and sunspot activity was much stronger when the direction of stratospheric winds over the tropics was taken into account. These winds tend to reverse their direction in approximately two-year cycles in a pattern known as the **quasi-biennial oscillation (QBO)**. When the QBO is in its west-to-east mode, for example, there appears to be a relationship between the number of sunspots and winter conditions over northern Canada. Surface pressure rises and falls with sunspot number, and the mean storm track shifts north and south. When the QBO is in its east-to-west phase, however, no such connection is evident. Although these associations are intriguing and statistically very strong, no causal mechanisms have been proven to explain these relationships.

Somewhat stronger evidence for an Earth–Sun connection comes from consideration of millennial-scale variations in solar output and climate. Data for the last 12,000 years show a definite 1500-year cycle in solar radiation that matches well with debris deposited on the floor of the North Atlantic by icebergs. This has led some researchers to conclude that solar forcing underlies at least part of the temperature changes observed in the Holocene.

Changes in Earth's Orbit

In Chapter 2 we saw that the seasons occur primarily because of the tilt of Earth's axis relative to the Sun. If we imagine a plane on which Earth makes its revolution around the Sun, we can see that the axis of rotation is oriented 23.5° from the perpendicular to the plane (that is, it has an obliquity of 23.5°). The orientation of the axis is constant through the course of the year, so no matter where Earth is relative to the Sun, its axis points toward the North Star, Polaris. For the six months following the March equinox, the Northern Hemisphere is inclined toward the Sun, and during the rest of the year the Southern Hemisphere has a greater exposure to the Sun. This is the primary factor in causing the seasons. Obviously, if the axis of rotation were greater than 23.5°, this effect would be stronger and lead to a greater seasonality. Likewise, the effect would disappear if the axis were exactly perpendicular to the plane of the orbit.

We also saw that Earth's orbit is elliptical, rather than circular, so that on about January 4 Earth is about 3 percent closer to the Sun than on July 4 (see Figure 2–11). This change in Earth–Sun distance causes the planet as a whole to receive about 7 percent more solar radiation at the top of the atmosphere in early January (perihelion) than during early July (aphelion). It should be readily apparent that a greater eccentricity would result in greater differences in incoming radiation available at the top of the atmosphere during the course of a year.

As far as the Northern Hemisphere is concerned, the Earth–Sun distance is largest during the summer and smallest during the winter, causing winters to be somewhat warmer and summers to be cooler than they otherwise would be. Thus, not only is the amount of eccentricity important with regard to seasonality, but so is the timing of the minimum and maximum Earth–Sun distances with respect to the equinoxes and solstices (as will soon be explained).

Three astronomical factors influence the timing and intensity of the seasons: eccentricity in the orbit, the tilt of Earth's axis off the perpendicular to the plane of the orbit, and the timing of aphelion and perihelion relative to the timing of the equinoxes. As it happens, these three factors all change cyclically over time on a variety of time scales. These three cycles are collectively called the **Milankovitch cycles**, in honor of the early 20th-century astronomer who explained their potential influence on Earth's climate.

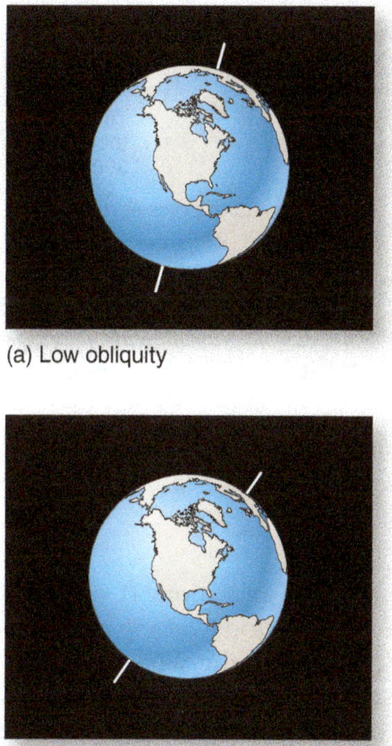

(a) Low obliquity

(b) High obliquity

Animation
Orbital Variations and Climate Change
http://goo.gl/Qi1VvR

▲ **FIGURE 16–16 Obliquity.** The degree of tilt of a planet's axis is its obliquity, which can be (a) relatively low or (b) high.

ECCENTRICITY The **eccentricity** of Earth's orbit changes cyclically on several time scales, with a cycle of about 100,000 years being especially prominent. Figure 16–15 shows the difference between a circular orbit with no eccentricity at all, and an orbit with considerable eccentricity. Though the Earth–Sun distance at aphelion is currently about 3 percent greater than at perihelion, the relative distance has varied between about 1 and 11 percent during the last 600,000 years. Over about the last 15,000 years, there has been a steady decrease in eccentricity, which will continue for about another 35,000 years.

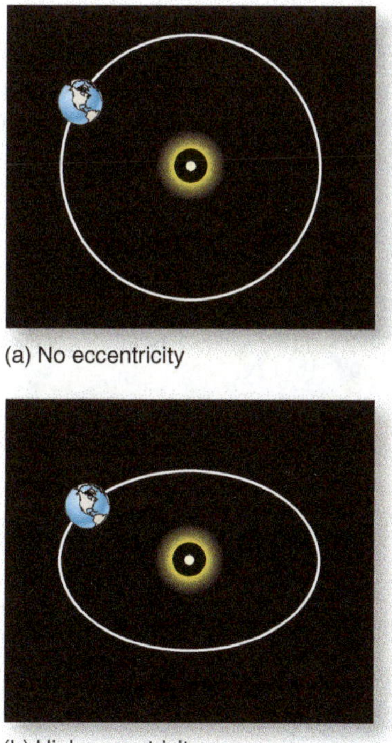

(a) No eccentricity

(b) High eccentricity

▲ **FIGURE 16–15 Eccentricity.** Planetary orbits need not be (a) perfectly circular, but (b) may have a certain degree of eccentricity.

OBLIQUITY The tilt of the Earth's axis, **obliquity** (Figure 16–16), also varies cyclically, but with a dominant period of about 41,000 years, during which it varies between 22.1° and 24.5° off the perpendicular. While the range of the axis tilt may seem small, it is capable of producing substantial differences in summer and winter insolation. In particular, high-latitude regions can undergo changes in available solar radiation at the top of the atmosphere of about 15 percent due to variations in obliquity. The most recent peak in obliquity occurred roughly 10,000 years ago. Thus, we are about midway in the half cycle from maximum to minimum obliquity.

PRECESSION Though the summer solstice for the Northern Hemisphere currently occurs near the time of aphelion, this changes through time because the axis wobbles on a 27,000-year cycle. In other words, the axis of rotation gyrates so that in about 13,500 years it will point to a different star, Vega, instead of Polaris (Figure 16–17). This change in orientation of the Earth axis, called **precession**, directly alters the timing and intensity of the seasons. Combined with changes in the orientation of the elliptical orbit, the result

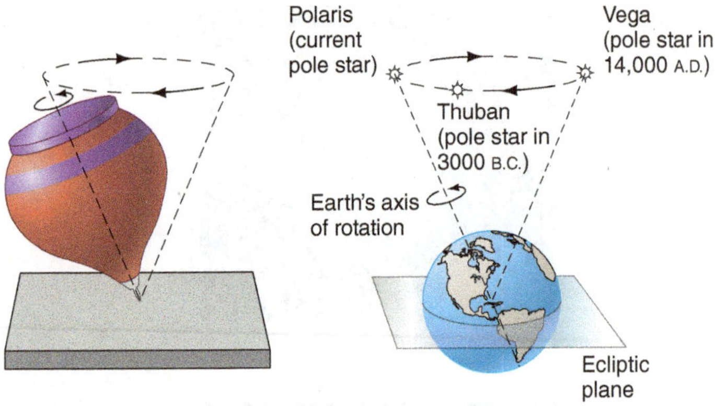

▲ **FIGURE 16–17 Precession.** Over long periods of time Earth precesses on its axis like a spinning top, and thus the north pole points toward different stars.

is a 23,000-year cycle in radiation. If orientation of the axis toward Vega were to exist today along with the current timing of aphelion and perihelion, the winter solstice for the Northern Hemisphere would nearly coincide with aphelion. The resultant increase in seasonality would cause warmer summers and cooler winters in the Northern Hemisphere. At the same time, the Southern Hemisphere would experience less seasonality because its summer solstice would occur near aphelion.

It is important to note that the importance of precession in influencing radiation receipts depends on the magnitude of the eccentricity of the orbit. Low values of eccentricity (nearly circular orbits) mitigate the importance of precession; greater eccentricity amplifies it. Given the current trend toward decreasing eccentricities, we can expect that this effect will be relatively small over the next 50,000 or so years.

Most scientists believe the Milankovitch cycles have played an important role in the expansion and retreat of glaciers during the Pleistocene because of the way they work together to influence seasonality. Large glaciers, such as those currently occupying most of Greenland and Antarctica, are most likely to expand when seasonality is low. With less seasonality, warmer winter temperatures foster a greater amount of snowfall over much of the ice sheets due to a greater availability of water vapor. Cooler summers also promote glaciation because the rate of melt along the margins of the ice sheets is slowed.

Considerable observational evidence for the role of Milankovitch cycles is seen in the climate record of the Pleistocene. Using radiation in the northern middle latitudes as a surrogate for orbital forcing, there is good agreement between the timing of ice advances and retreats. In addition, when the variability in the climate record is broken down according to various time scales, the three main Milankovitch periods emerge as containing most of the climate variation. There is also some understanding of the pathways by which orbital forcing influences climate. For example, computer modeling suggests that increased summer radiation during the early Holocene led to enhanced monsoon circulations, especially in North Africa. These would have brought increased moisture to the continent, as is observed in the climate record. For these and other reasons, changing orbital parameters are widely accepted as driving glacial/interglacial cycles.

Changes in Land Configuration and Surface Characteristics

Many climatologists believe that climatic changes occurring over the longest time spans were at least partly in response to changes in the size and location of Earth's continents (Figure 16–18). Support for an older view, that continental drift was the primary forcing variable, has eroded in the face of quantitative estimates of change obtained from computer simulations. The breakup of the early supercontinent Pangaea and the slow movement of the resultant continents undoubtedly caused major climatic changes, even if not as large as observed in the geologic record. Such would have to be the case, of course, because all the factors that affect temperature and other climate variables (such as latitude and continentality) were themselves greatly affected by the movement of the continents. Though dramatic, the climatic changes resulting

▶ **FIGURE 16–18 Distribution of Land and Ocean 150 MYA.**

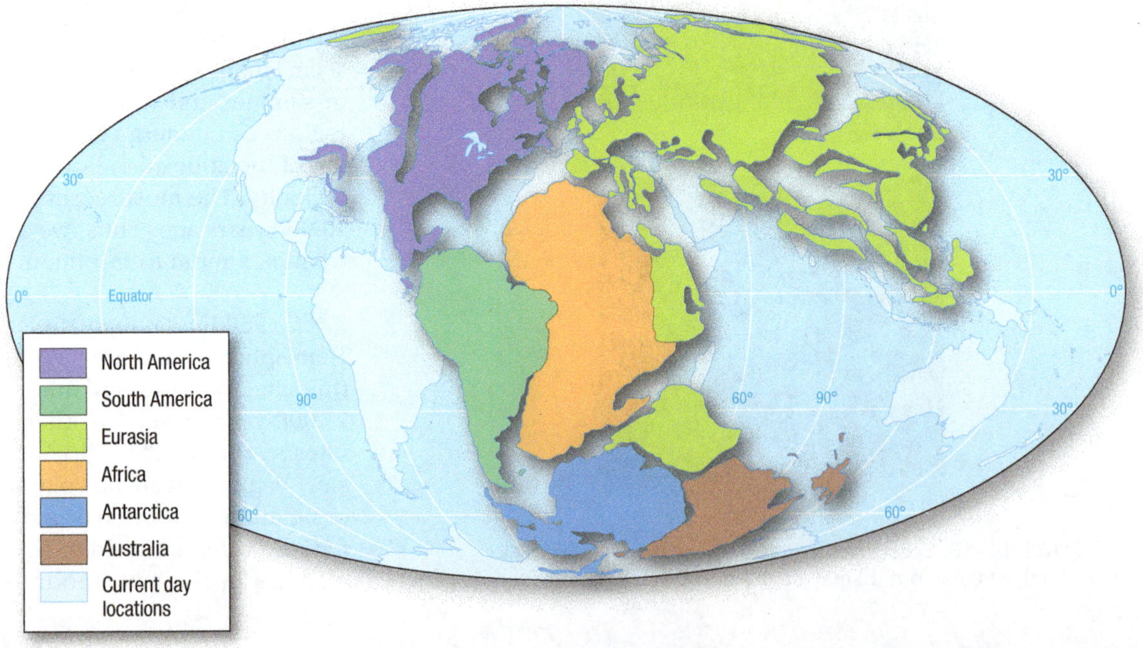

from continental displacement would be extremely slow. Like changes in position, episodes of continental mountain building have almost certainly produced significant climate change. For example, computer modeling suggests that the presence of large mountain regions (the Rockies, Himalayas, and Andes) would amplify Rossby waves during the winter season and promote enhanced monsoon circulations in the summer, both of which are consistent with observational data.

MODIFICATION OF EARTH'S SURFACE BY HUMAN ACTIVITY At shorter time scales, modification of Earth's surface, especially by human activity, can greatly influence the disposition of solar radiation. One such activity is deforestation, in which large tracts of land are cleared of trees. The loss of vegetation reduces evapotranspiration from the surface. This in turn leads to higher temperatures near the surface, as the amount of energy channeled into the latent heat of evaporation is reduced and also decreases precipitation. In addition, decomposition of cleared vegetation directly increases atmospheric CO_2, an important greenhouse gas. Compounding this is the loss of vegetative surfaces formerly providing photosynthesis, a process that removes CO_2 from the atmosphere. Even if reforestation occurs, the conversion of mature to immature trees still reduces the rate of CO_2 removal from the atmosphere.

Alteration of arid and semiarid land surfaces by the overgrazing of cattle may also lead to changes in the regional climate. Soil compaction associated with grazing can increase the amount of runoff, thus making less water available for evaporation into the atmosphere. Also, one hypothesis suggests that the removal of vegetation leads to an increase in the surface albedo. As the albedo increases, according to this line of reasoning, the reduction in the energy absorbed by the ground causes cooling of the surface and a reduction in the environmental lapse rate. Because a lowered lapse rate would make the air more stable and less susceptible to convectional precipitation, overgrazing could enhance the vulnerability of these regions to drought. On the other hand, the reduction in vegetation due to overgrazing could have the opposite effect, in which a reduction in evaporation leads to a general warming of the surface and an increase in the environmental lapse rate. Of course, the instability associated with the increased lapse rate will have little effect if there is not enough water available to yield precipitation. Yet another effect arises from the change in surface roughness, which alters the momentum transfer between atmosphere and ground.

CHECKPOINT

16.11 Define the following terms: eccentricity, obliquity, precession.

16.12 How do cyclical variations in Earth's orbit, including Milankovitch cycles, affect climate? Explain.

Changes in Atmospheric Turbidity

Atmospheric **turbidity** refers to the amount of suspended solid and liquid material (aerosols) contained in the air. Some aerosols are released into the atmosphere through natural processes, such as large-scale volcanic eruptions; others are released by human activity, such as smokestack emissions. Aerosols can enter the atmosphere directly as in the examples just named. They can also enter the atmosphere indirectly as a result of chemical processes in which certain gases—usually sulfates, but also nitrates and hydrocarbons—react in the presence of sunlight to form solid and liquid aerosols. Some aerosols are no more than tiny clumps of molecules, whereas others are millimeters across. This is a huge range in size, roughly comparable to the difference between tennis balls and planets. Regardless of their source, composition, or size, however, prolonged variations in aerosol contents can have important ramifications for climate.

Existing both in the stratosphere and the troposphere, aerosols directly affect the transmission and absorption of both solar and infrared radiation. By absorbing incoming sunlight they can cause a heating of the atmosphere around the aerosols, but they can also increase the amount of backscattering and thereby reduce the amount of radiation reaching the surface. The relative effects of aerosol absorption vs. backscatter are hard to assess and depend on numerous factors, including the albedo of the aerosols and the underlying surface. Note, however, that both absorption and backscatter reduce solar radiation reaching the surface. For example, during the period from about 1970 to 1990, the amount of solar radiation reaching Earth's surface *decreased* by several percent—a phenomenon referred to as **global dimming**. This dimming is attributed to heavy aerosol releases during that era. More recently, as increasingly stringent pollution controls took hold, dimming was replaced by modest "brightening."

To a lesser extent, aerosols can increase the absorption of outgoing longwave radiation that would otherwise escape to space and thereby increase the longwave radiation radiated from the atmosphere to the surface. In this way, aerosols can have the direct effect of increasing nighttime temperatures.

Aerosols can also affect climate indirectly, through their ability to serve as cloud condensation nuclei. A cloud with a greater number of condensation nuclei might contain as much liquid water as one with few nuclei, but it will have more droplets of mostly smaller sizes. Clouds containing a large number of small droplets are less likely to yield precipitation and, therefore, are likely to persist for longer periods. In addition, the total droplet surface area of such clouds is greater, and this greater surface area increases the reflectivity of the cloud (assuming the same amount of liquid water). Thus, tropospheric aerosols might give rise to more extensive, longer-lived, brighter clouds. Aerosols can also *reduce* cloud amount through their warming effect in the troposphere. As we saw in Chapter 7, the presence of warmer air aloft inhibits the vertical motions that lead to cloud development. Returning to the case of global dimming, the current scientific understanding is that indirect effects of aerosols were more important than direct effects in the observed dimming.

Observational support for the indirect effect of aerosols comes from satellite images of ship trails, which are tracks of low-level clouds produced by oceangoing vessels (Figure 16–19). In the relatively clean ocean atmosphere, smokestack exhaust aerosols enhance condensation, creating long cloud trails that stand out against the surrounding clear sky. With their high albedo, these clouds reduce the amount

▲ **FIGURE 16–19 "Global Dimming" Effect of Aerosols.** Ship tracks appear as white streaks embedded in a low-level cloud deck of speckled light gray clouds. The image shows an area just offshore of the northwestern United States (colored green), with small amounts of cloud-free ocean (blue).

of solar radiation reaching the surface. Additional support for indirect aerosol effects comes from a recent study of pollution tracks on land, which appear on satellite images as plumes of brighter clouds downwind of cities. By analyzing data from a number of sensors, it was discovered that aerosols substantially reduce precipitation in downwind areas by suppressing both ice formation and droplet coalescence. (Recall that these are important precipitation growth mechanisms.)

Moving beyond these generalities, let us now look in more detail at the climatological effects of tropospheric and stratospheric aerosols as agents of climatic change.

TROPOSPHERIC AEROSOLS Natural sources of tropospheric aerosols include the spraying of salt particles by ocean waves and bubbles, soot and gases from fires, the erosion of soil by wind, the dispersal of spores and pollen, the emission of sulfides by marine plankton, and other biospheric processes. Of course, volcanic eruptions can also discharge a huge amount of material into the troposphere, but it settles and precipitates out soon after the eruption and thus exerts no long-term effects on climate (unless many volcanoes erupt sequentially). Humans produce particulate matter in a number of ways, such as combustion of fossil fuels, burning wood and other biofuels, burning forests for conversion to agriculture, and plowing fields that become sources of dust. For example, we have all seen soot—a mixture of dark carbon-bearing particles—billowing out the exhaust of diesel trucks. Many tropospheric aerosols originate from combustion gases that are subsequently converted to particles by processes within the atmosphere. For example, coal burning releases sulfur dioxide gas (SO_2) that is converted to sulfuric acid (H_2SO_4), which ultimately condenses to form sulfate aerosols. Some aerosols of natural origin also begin as gases.

Although we have no direct long-term records of tropospheric aerosols, human activities have unquestionably increased their concentrations. This is especially true over industrialized land areas. Unlike some gases, however, which have very long residence times (Chapter 1), individual aerosols do not have long life spans in the troposphere. Aerosols produced over urban or industrial centers settle out before they can be widely dispersed across the globe, thereby leading to concentrations that vary widely both temporally and spatially. As a result, their climatic effects tend to be isolated to areas near the pollution sources.

We have seen that aerosols both absorb and scatter incoming solar radiation. The combination of aerosol backscatter and absorption results in the brown haze we recognize as evidence of a polluted atmosphere and gives rise to the term *atmospheric brown cloud* (*ABC*), which simply refers to a local accumulation of aerosols. Thus, one immediate impact of aerosols is to reduce solar radiation at the surface, which promotes surface cooling. Another direct impact of ABCs is to warm the lower atmosphere through increased absorption of solar radiation, and for larger particles, increased absorption of longwave radiation. Another direct effect of tropospheric aerosols is to modify absorption of solar radiation by the surface. Where black carbon from aerosols is deposited on snow or ice, the surface albedo falls, which leads to warming. Indeed, recent studies have shown that for some Arctic locations, the warming effect of aerosols is several times greater than the warming effect of increasing CO_2.

Until recently it was not known whether the direct effects of tropospheric aerosol increases would lead to an overall warming or cooling near the surface. Numerical models now tell us that increased aerosol contents have the net effect of reducing surface temperatures globally. In fact, the radiative cooling due to anthropogenic aerosols appears to have been half as large as the radiative warming that would theoretically accompany observed increases in greenhouse gases over the industrial period. Thus, the warming effect of additional greenhouse gases has been masked somewhat by the cooling effects of aerosols. But we cannot expect the aerosol effect to offset the greenhouse gas effect indefinitely. If there were to be a stabilization in the release of anthropogenic emissions, atmospheric CO_2 contents could take up to 200 years to reach a new equilibrium level, whereas aerosol levels would respond almost immediately. Thus, the greenhouse effect would continue to increase while the aerosol effect would stabilize and cease to offset the opposing effect. Despite the possibly beneficial effects of aerosols on climate change, remember that aerosols exert their own negative effects on humans and the environment (Chapter 14).

DID YOU KNOW?

Huge fires triggered by even a small nuclear exchange could lead to climate changes larger than any experienced in recent human history. Research published in 2007 and 2008 shows that smoke (lifted in part by heating through enhanced absorption of solar radiation) would reach the upper stratosphere and darken the skies for years. Cooling would be immediate and would lower planetary temperatures below those of the Little Ice Age.

STRATOSPHERIC AEROSOLS Unlike tropospheric aerosols, those in the stratosphere result primarily from natural processes; human activities are not as important. The stratosphere maintains a background level of aerosols introduced by the upward diffusion of sulfur gases from the troposphere, which then undergo gas-to-particle conversion. Though background levels remain fairly constant through time, significant increases can occur in the months following volcanic eruptions.

Stratospheric aerosols can remain in the stratosphere longer than their tropospheric counterparts for two reasons: First, they tend to be smaller and therefore have lower terminal velocities. More importantly, the most effective agent of aerosol removal, precipitation, is not important in the stratosphere.

As previously stated, aerosols affect the energy balance by absorbing and backscattering solar radiation and by absorbing and radiating longwave radiation. The reduction in solar radiation reaching the surface would favor a lowering of tropospheric air temperatures, while the absorption of outgoing longwave radiation and the increase in downward emission would promote surface warming. So which actually dominates in the case of stratospheric aerosols? It turns out that the relative importance of shortwave and longwave radiation changes depending on the size of the aerosols. If the stratospheric aerosols are small, the reduction in solar radiation reaching the surface exceeds the gain in longwave radiation. In fact, the aerosols *are* small and promote lower temperatures for up to several years after volcanic eruptions.

Several recent volcanic eruptions have provided scientists with useful observational data on the effects of stratospheric aerosols. Mount St. Helens in Washington underwent a major eruption on May 18, 1980. Despite the massive ejection of solid material (much of one side of the mountain was blown away), it caused little if any major impact on hemispheric weather because it released relatively small amounts of sulfuric gases, which can ultimately transform to aerosols. Though the daytime sunlight was blocked out downwind of the eruption for a few days, no effects were noted much beyond that time frame.

The April 4, 1982, eruption of El Chichón in Mexico was far more violent than that of Mount St. Helens; and, more significantly, it released a particularly large amount of sulfuric gases. The result was an increase in atmospheric albedo, which decreased global temperatures by about 0.2 °C (0.4 °F) for several months. The Mt. Pinatubo (Philippines) eruption of June 12, 1991 (Figure 16–20), is believed to have released about twice as much sulfuric gas as did El Chichón, along with the predictable effect of an even greater reduction in global temperatures. The eruption increased the atmospheric albedo both directly (by enhanced aerosol backscatter) and indirectly (by an increase in the reflectivity of clouds). The cooling effect was greatest by about August 1992, when the global mean tropospheric temperature decreased by about 0.73 °C (1.6 °F) from that of June 1991 (despite the fact that August is climatologically warmer than June in the Northern Hemisphere). Figure 16–21 shows the change in aerosol content over the globe prior to (a) and following (b–d) the eruption. Aerosol contents increased 100-fold during the peak concentration.

▲ **FIGURE 16–20 Mt. Pinatubo.** When Mt. Pinatubo erupted in 1991, it injected huge amounts of sulfur into the atmosphere, thereby lowering global temperatures for months.

AEROSOLS VS. GREENHOUSE GASES—COMPARING THEIR IMPACTS We must emphasize that the role of anthropogenic (and natural) aerosols in climate is subject to great uncertainty. This comes partly from the fact that the sources are highly variable over space and time and are not known with great precision. As we have seen, aerosols have both cooling and warming effects, and these vary by location, season, and altitude. Thus, while it is believed that globally ABCs have had a cooling effect during industrial times, in some places warming effects have dominated (particularly in parts of Asia). Understanding the climatic response to aerosol loading is the subject of intense research, as it is so closely related to the issue of greenhouse warming. There is, however, one very important difference. The short lifetime of aerosols means that changes in release rates have an almost immediate impact. This stands in strong contrast to greenhouse gas emissions, whose residence times are on the order of decades to centuries. In the case of greenhouse gas warming, our actions of the last few decades have committed the planet to future climate changes that can only slowly be controlled by reductions in emission.

Changes in Radiation-Absorbing Gases

The issue of climatic warming due to increases in carbon dioxide and other greenhouse gases has been the subject of

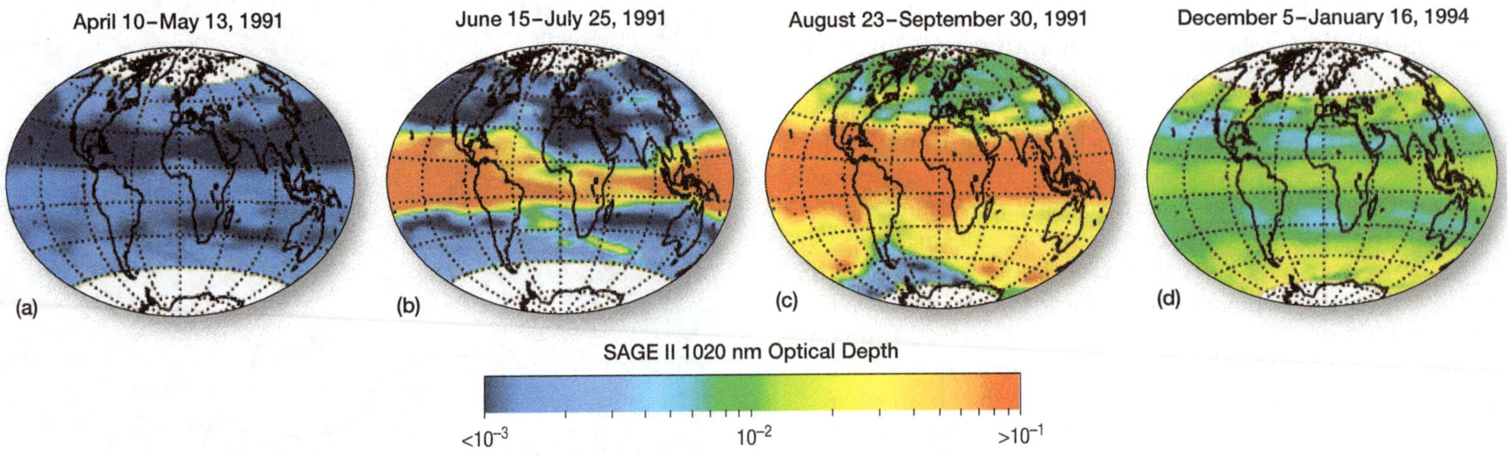

▲ **FIGURE 16–21 Volcanic Aerosols.** The map shows average aerosol content over the globe prior to and after the June 12, 1991, Mt. Pinatubo eruption. Note that part (b) represents a two-month period almost immediately following the eruption. Redder areas indicate greater aerosol contents.

intense scientific scrutiny and political controversy. On both human and geologic time scales, there is little doubt that large changes in CO_2 have occurred. For example, ice cores show that CO_2 rises and falls in concert with glacial/interglacial cycles. (The CO_2 changes lag behind the ice volume changes, and thus they are believed to amplify the glacial cycles, but they are not thought to be the fundamental cause of those cycles.) On human time scales the burning of fossil fuels and the clearing of forested areas have led to a steady increase in the carbon dioxide content of the atmosphere (see Figure 1–6).

THE MECHANISM OF GREENHOUSE WARMING What is the physical mechanism behind greenhouse warming? The simple answer is that because greenhouse gases absorb longwave radiation, less surface emission escapes to space and more is retained by the atmosphere. This is true, but a more complete explanation is somewhat more involved. Recall that temperature in the troposphere decreases with altitude, and that as temperature decreases, emitted radiation also decreases. With this in mind, suppose your eyes were sensitive only to longwave radiation and that you looked down from space on a planet like ours. Because the atmosphere absorbs so strongly at longwave wavelengths, you would see little of the surface. Instead, because most of the upwelling radiation originates high in the atmosphere, you would see a colder radiation source, as depicted in Figure 16–22a. In other words, emission to space is less than would be expected from knowing the warm surface temperatures enjoyed by us surface dwellers. The presence of a greenhouse gas raises the effective radiating (emission) altitude, and because temperature decreases with altitude, there is less emission to space than would otherwise occur. This is consistent with the simple explanation of greenhouse warming. Of course, assuming thermal equilibrium, the amount of emitted radiation does balance absorbed solar radiation, as we saw in Chapter 3.

Now suppose greenhouse gases increase. The immediate effect is to darken the atmosphere even more at longwave wavelengths, which raises the effective radiating altitude further, which in turn reduces longwave losses to space (Figure 16–22b). Absorbed solar radiation now exceeds emission, which means the planet warms and emits more longwave radiation. Increases in both temperature and emission of longwave continue to increase until a new radiation balance is achieved (Figure 16–22c). Obviously, that new equilibrium will be a warmer climate.

DID YOU KNOW?

A significant greenhouse gas not regulated by the Kyoto Protocol is released in manufacturing flat-panel displays. Because nitrogen trifluoride (NF_3) is far more absorbing of longwave radiation than CO_2, the release of just 4000 tons each year is equivalent to 67 million tons of CO_2 (about 4 percent of the CO_2 release).

This basic mechanism is modified by feedback mechanisms described in a later section. Before discussing those, we provide an overview of recent changes in greenhouse gas concentration.

RECENT CHANGES IN GREENHOUSE GASES Since the middle of the 19th century, there has been an exponential increase in the input of carbon dioxide to the atmosphere by industrial activities (chiefly fossil fuel combustion and cement manufacture). Not surprisingly, most emissions have come from the developed nations, with 90 percent originating in the Northern Hemisphere. The growth rate has varied with business cycles but has averaged a few percent per year. Over the period from 1999 to 2005, the growth rate was 3 percent per year. These industrial releases have been augmented by smaller but still significant releases resulting from deforestation. Deforestation adds CO_2 to the atmosphere through direct burning of the logged trees, decomposition of biomass, and other processes. Moreover, the removal of trees reduces the ability to remove subsequent inputs of carbon dioxide by photosynthesis. CO_2 releases by deforestation (and land-use changes generally) were relatively more important early in the

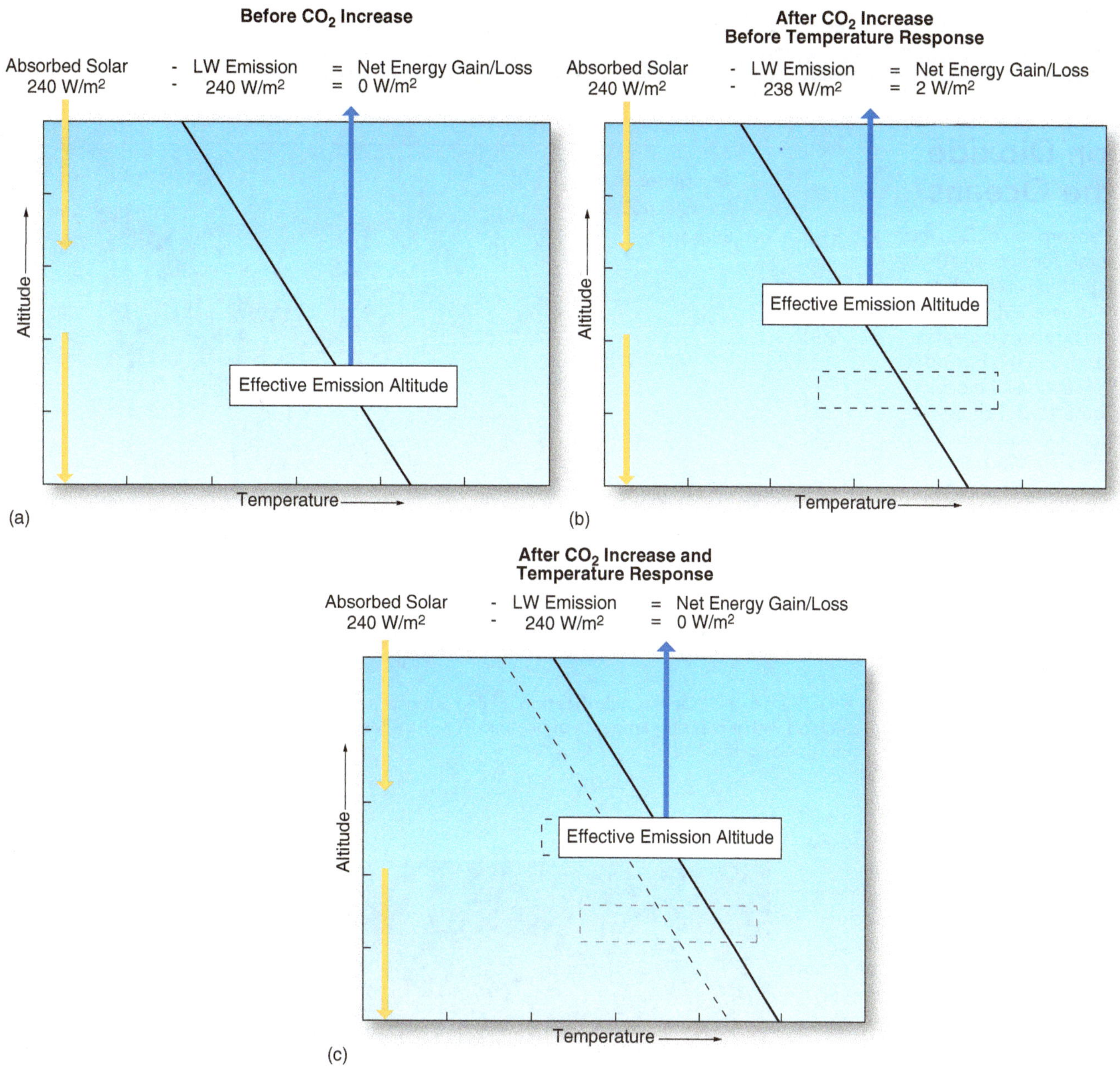

▲ **FIGURE 16–22 Simplified Depiction of How Increasing Greenhouse Gas Levels Lead to Warming.** In (a), before a greenhouse gas increase, the planet has a radiation balance and the effective altitude of emission is low. In (b), an increase in some greenhouse gas raises the effective emission altitude. With emission from a higher, colder layer, the planet emits less radiation. In (c), the temperature warms sufficiently to restore a radiation balance.

period and do not show the explosive exponential growth of industrial CO_2. Thus, fossil fuel use has been the main driver of CO_2 emission for the last few decades, and it is expected to be so for the foreseeable future.

Although much of the media discussion about the current increase in greenhouse gases centers on carbon dioxide, CO_2 is only one of several anthropogenic greenhouse gases that absorb outgoing longwave radiation. Methane (CH_4), nitrous oxide (N_2O), and chlorofluorocarbons (CFCs) are also effective absorbers whose contents have increased over historical times. In fact, during the industrial period these gases have been responsible for about 35 percent of the greenhouse-induced increase in net radiation.

Of course, greenhouse gas emission is not the whole story; in some cases removal processes are extremely important in governing the amount of a particular gas in the atmosphere. For example, much of the anthropogenic CO_2 released to the atmosphere has been taken up by the oceans and lost from the atmosphere in other ways (See *Box 16–2, Focus on the Environment and Societal Impacts: Carbon Dioxide and the Oceans*). The best current estimate is that only about 40 percent of the CO_2 emission for any year remains in the atmosphere. Thus, changes in atmospheric concentration do not necessarily mirror the emission patterns.

Still, as seen in Figure 16–23, measurements of CO_2, CH_4, and N_2O show the unmistakable imprint of human activity,

16–2 FOCUS ON THE ENVIRONMENT AND SOCIETAL IMPACTS

Carbon Dioxide and the Oceans

While the absorption of CO_2 may sound entirely fortuitous, it brings about the problem of acidification of the oceans. When CO_2 is absorbed by water, it creates carbonic acid, and this has led to some acidification of the ocean waters. Figure 16–2–1 shows the changes in atmospheric carbon dioxide since 1960, along with amount of oceanic carbon dioxide and pH since the 1980s. There is a very close correspondence between oceanic and atmospheric levels. At the same time, the dissolved CO_2 causes the water to have a greater acidity (lower pH).

This can have deleterious effects on oceanic organisms with calcareous shells and exoskeletons, such as coral (Figure 16–2–2). As described earlier in this chapter, coral are small marine organisms with hard shells composed largely of calcium carbonate. But as oceans become more acidic, coral have a more difficult time accumulating calcium carbonate. This increased acidity has already caused a substantial reduction in the growth rate of corals in the Great Barrier Reef of Australia, one of the world's great natural wonders.

We know with certainty that atmospheric carbon dioxide contents will continue to increase during this century and that some CO_2 emissions will continue to diffuse into the oceans. But there is less certainty about whether the oceans will continue to absorb atmospheric contents at their current rate. Several factors are involved in this issue. The first is that greater atmosphere–ocean differences in the concentration of any particular gas increase the rate of transfer. In other words, an increase in atmospheric CO_2 by itself would lead to greater diffusion into the oceans. But counter to this is the fact that warmer waters are less capable of maintaining carbon dioxide, and oceanic surface temperatures will also be increasing. The situation is further muddled by the complexity and uncertainty of future movements of oceanic currents—in particular, patterns of sinking and upwelling—as well as the ability of plants and plankton to obtain CO_2 through photosynthesis.

Despite these uncertainties, we do know that carbon dioxide will continue to diffuse into the oceans and lead to increasing acidity and associated ecological impacts.

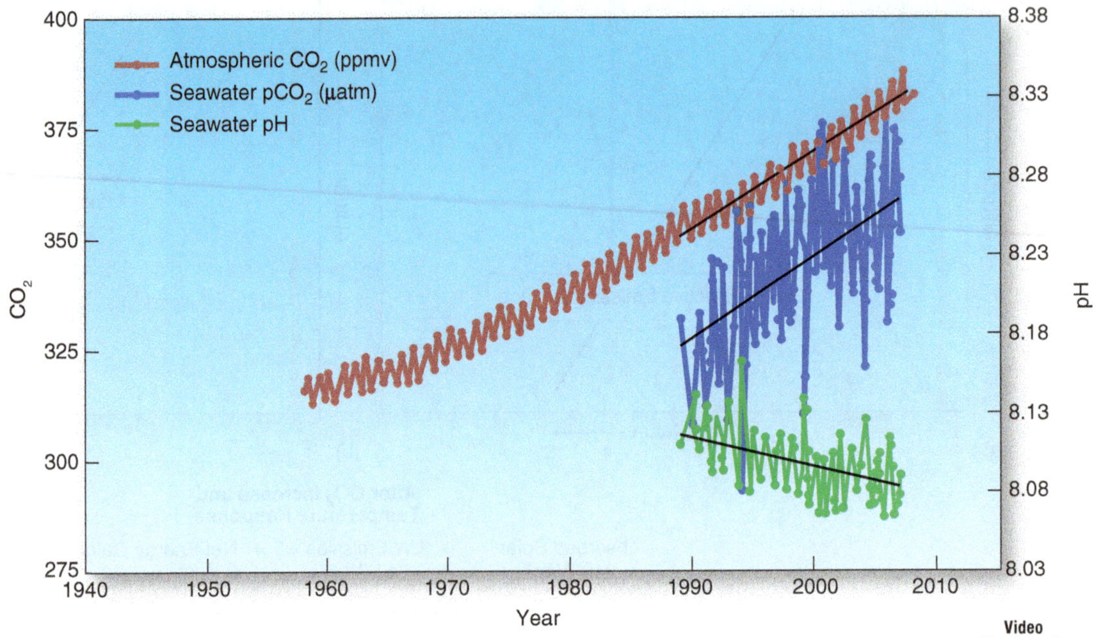

▲ **FIGURE 16–2–1 Ocean Acidification.** Rising atmospheric carbon dioxide contents have coincided with increasing amounts of oceanic CO_2. As a result, oceans have become more acidic (lower pH).

Video
The Ocean's Green Machines
http://goo.gl/z1MwpP

▲ **FIGURE 16–2–2 A Coral Reef.**

16–2–1 **How have changes in the CO_2 content of the atmosphere affected the chemistry of ocean water?**

16–2–2 **Why are changes in the CO_2 content of ocean water a problem for some ocean organisms?**

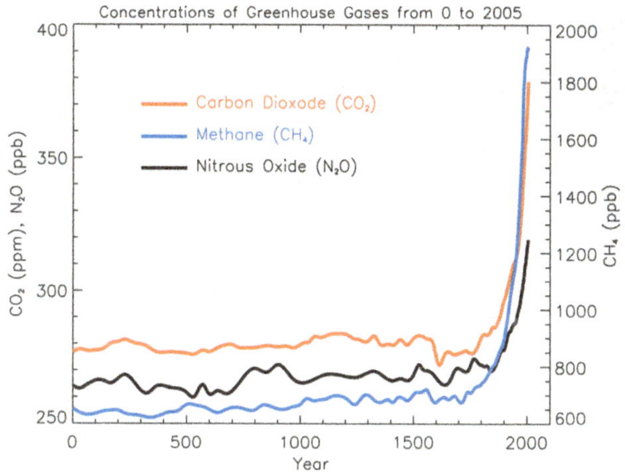

▲ **FIGURE 16–23 Time Series of Important Greenhouse Gases from A.D. 0 to the Present.** Published by the IPCC.

with concentration for all gases far above background levels. Although difficult to see in the graph, methane concentrations slowed beginning about 1990, leveled off by the end of the century, but have been rising since 2006. A number of hypotheses have been advanced, but the reasons for this are not understood, and it is not clear what the future holds for methane. CFCs (not shown on the graph) represent a bright spot in this picture; their abundance is actually decreasing as a result of regulations intended to protect the ozone layer.

The changes in greenhouse gas concentrations seen in Figure 16–23 are just part of the puzzle of greenhouse-induced climatic change. Our ability to understand their role in past changes and predict future changes is complicated by the interconnections of different components of the atmospheric systems through feedbacks, as discussed in the next section.

CHECKPOINT

16.13 Explain how can tropospheric aerosols can promote both cooling and warming.

16.14 How does an increase in greenhouse gases change the effective emission altitude? What role does this effect play in global warming?

16.15 List three other anthropogenic gases in addition to CO_2. What is their contribution to global warming?

Feedback Mechanisms

16.5 Explain feedback mechanisms and Earth-system interactions involved in climate change.

As mentioned earlier in the chapter, the Earth system involves many feedbacks, some positive, some negative. The situation is further confounded by the fact that the various feedbacks operate at vastly different rates: Some are instantaneous, whereas others operate over thousands of years. In some cases relationships seem relatively straightforward, but for others it is difficult to determine if a single connection between two variables is positive or negative. With these points in mind, we now examine some of the important feedback mechanisms related to climatic change.

Ice-Albedo Feedback

Much of Earth's surface is occupied by large continental ice sheets and floating sea ice (Figures 16–24 and 16–25). If the atmosphere were to cool along the margins of these ice masses, there could be an expansion of the ice-covered area. Similarly, any warming would likely lead to melting and a retreat of the ice margin. Because ice has a higher albedo than most other natural surfaces, an expansion of the ice would lead to a reduction in the amount of insolation absorbed by the surface, which in turn would lead to further cooling (a positive feedback). Likewise, a retreat of the ice associated with a warming climate would increase the amount of surface no longer covered by the reflective ice and, therefore, lead to enhanced warming. Thus, a positive feedback mechanism is at work that favors continued warming or cooling once such a temperature trend is initiated. It is largely for this reason that middle and high latitudes show larger changes than low latitudes. Where there is little prospect for ice (the tropics), this feedback does not operate. Of course, ice advances and retreats do not continue unchecked, because other feedback mechanisms prevent the ice from totally covering the globe or from completely disappearing.

▼ **FIGURE 16–24 Arctic Sea Ice in Norway.**

▲ **FIGURE 16–25 Distribution of Northern Hemisphere Sea Ice.** The extent of Arctic sea ice on August 26, 2012. This date marked the smallest extent of ice cover ever determined in the three decades of satellite coverage. The solid line represents the average extent of ice coverage between 1979 and 2010.

Water Vapor and Cloud Feedbacks

There are two important feedback mechanisms involving water vapor in the atmosphere. First, observations and calculations show that relative humidity tends to be surprisingly constant when temperature changes by even large amounts. For example, the cold winter of a middle latitude might have a relative humidity not too different from that of the summer. This means, of course, that the summer has a higher specific humidity—there is more moisture in the summer atmosphere. The same thing plays out when temperature increases from one year to the next. Again, the warmer atmosphere contains more water vapor. As already mentioned, water vapor is a very strong greenhouse gas, so this contributes to further warming. This direct water vapor effect is a positive feedback, because the initial change (warming) is amplified by the process.

The second water-related feedback concerns changes in the lapse rate of the troposphere. If the planet warms and its atmosphere contains more water, the lapse rate at low latitudes will decrease. Air aloft will remain colder than air below, but the difference will not be as great as in the prior cooler climate. The reason for this lapse rate change is related to the value of the saturated lapse rate. Recall that the saturated rate is variable and is smaller for warmer air. That is, a warm saturated parcel cools more slowly during ascent than a cold parcel. Therefore, in a warmer climate, rising saturated parcels are expected to cool more slowly. In moist tropical climates the lapse rate of the troposphere is largely governed by the saturated rate; thus, the lower latitudes will show the largest decrease in lapse rate if the planet warms. If the lapse rate is smaller, then everything else being equal, the temperature at the effective radiating altitude is higher. A higher temperature means more emission, thus the surface does not need to warm as much in order to achieve a radiation balance. We see that this is a negative feedback because it counteracts the initial change.

Although there is near unanimous agreement on the direction in which the two water vapor feedbacks act and agreement that the lapse rate feedback is less important, there are considerable differences among various calculations regarding their magnitude. However, because they are related physically to each other, a calculation indicating a large positive direct water vapor feedback will necessarily suggest a large negative lapse rate feedback. This greatly narrows the differences between estimates of the overall feedback effect. According to the calculations referenced in the IPCC report, the water vapor feedback (direct plus lapse rate) is positive, and it amplifies expected global warming by about 50 percent.

Animation
Arctic Sea Ice Decline
http://goo.gl/mrmgAl

A final water-related feedback that needs mentioning is clouds. We have seen that clouds are very active from a radiative standpoint. Thick clouds strongly reflect solar radiation, which obviously promotes cooling. On the other hand, a cloud of even modest depth absorbs all upwelling longwave radiation and itself radiates as a nearly perfect blackbody down to the surface. Whether cloud feedback is positive or negative depends on which of these effects is larger, and whether the amount of cloud increases or decreases. Although most calculations suggest a positive cloud feedback, the range is large—from 10 to 50 percent, according to the IPCC report.

Atmosphere–Biota Interactions

Changes in climate are linked with land–vegetation patterns. The influence of climate on vegetation is easy to comprehend—banana trees do not grow in Greenland and there is no tundra vegetation in the Amazon Basin. Less obvious, perhaps, is the fact that vegetation likewise influences the climate in a number of ways. By transpiring moisture to the air, the presence of plants affects the moisture content of the atmosphere and the likelihood of precipitation. Plant assemblages also affect the surface albedo and the transfer of heat at the surface. It is easy to see why this is important in arid and semiarid locations, where the abundance and type of vegetation are very sensitive to precipitation. Less obviously, vegetation–albedo feedback effects are also important at high latitudes, because forest cover greatly reduces the high albedo that would otherwise occur during winter in a snow climate. For example, it is believed that poleward migration of forests following the last glacial maximum had a significant amplifying effect on Holocene warming.

One of the most important feedback mechanisms involved in the possible warming of the atmosphere involves the influence of CO_2 concentration on photosynthetic rates. It has been known for some time that many plant species undergo accelerated growth and enhanced photosynthesis in an environment rich in carbon dioxide, and some species also do better under warmer conditions. This could create a negative feedback (with potentially positive results as far as people are concerned) in which increasing CO_2 contents and temperatures allow plants to suppress further increases in the greenhouse gas.

Scientists believe that this "fertilization" process may already be appearing, especially in the latitudes between about 45° N and 70° N. Satellite observations indicate that the region's supply of green vegetation is increasing and that its growing season is beginning about a week earlier and ending a week later than it did prior to the 1980s. These effects would be consistent with those expected from increased warmth and CO_2 levels observed in the area. At the same time, the seasonal oscillations in Northern Hemisphere carbon dioxide contents described in Chapter 1 have undergone changes in the intensity and timing of their cycles. The spring decrease in carbon dioxide associated with the leafing of deciduous plants is also occurring about a week earlier, and the difference between springtime maxima and late-summer CO_2 minima is increasing. Changes in the seasonal cycle of CO_2 are most conspicuous near the Arctic, where the magnitude of the oscillations has increased by 40 percent during the past few decades.

On the other hand, a lack of nutrients can become a limiting factor in plants' ability to respond to an enriched CO_2 atmosphere. This has recently been demonstrated in the Alaskan Arctic, where tundra vegetation was artificially subjected to twice the carbon dioxide level of the normal atmosphere. At first, the vegetation responded to the fertilization with increased growth rates. But by the third year, elevated CO_2 ceased to have any effect.

In a separate experiment, it appeared that high CO_2 contents might even *harm* tropical rain forest plants by causing a loss of nutrients in the soil. And another recent study has shown that trees in tropical rain forests have been growing, maturing, and dying more rapidly than in the past. This leads to a more open environment that favors replacement of trees with vines requiring greater amounts of sunlight. Ironically, the vines are less effective at photosynthesis than the trees they replace, and they do not even consume enough CO_2 to offset that released by the decaying trees.

Increases in atmospheric CO_2 could lead to other unwelcome results. It is possible, for example, that certain weeds would benefit more from higher carbon dioxide concentrations than would agricultural plants. Insects and other pests might also enjoy a warmer environment associated with increased CO_2, becoming more troublesome to agriculturalists than they are today.

A warming climate is also likely to have a large impact on fire regimes. Higher temperatures wick additional moisture out of both live and dead woody vegetation during dry seasons, making them more flammable. Coupled with a shortened winter, naturally occurring fire would therefore become more common. Calculations show that for parts of the western United States and Canada, a 1 °C (1.8 °F) increase in global temperature would increase the median area burned by more than a factor of six (Figure 16–26). (For more information on climate change impacts on vegetation, see *Box 16–3, Focus on the Environment and Societal Impacts: Plant Migrations and Global Change.*)

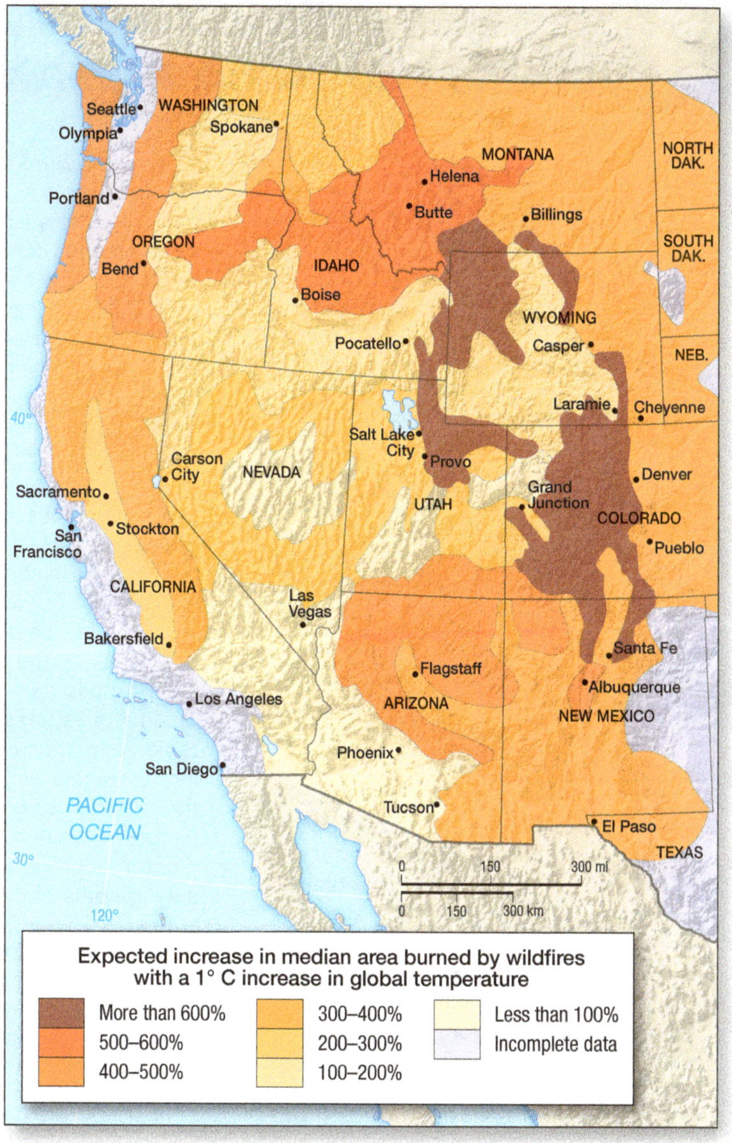

▲ **FIGURE 16–26 Projected Changes in Wildfires.** The map shows the expected percentage increase in median area burned by wildfires with a 1 °C increase in global temperature.

CHECKPOINT

16.16 Describe the ice-albedo feedback. As a positive feedback, does it necessarily lead to global warming? Explain.

16.17 How do changes in water vapor and clouds affect atmospheric temperatures? Use the concepts of positive and negative feedback in your answer.

16.18 Explain the relationship between CO_2 levels in the atmosphere and plant growth.

16–3 FOCUS ON THE ENVIRONMENT AND SOCIETAL IMPACTS

Plant Migrations and Global Change

As global temperatures have warmed and cooled, plant communities have responded by migrating poleward during periods of glacial retreat and equatorward as glaciers advanced. For example, large tracts of species such as spruce that once inhabited much of what is now the northern and northeastern United States have shifted northward into Canadian regions formerly covered by glacial ice.

The rate at which plant communities can migrate in response to changing climates is important to their survival. If plant communities are too slow to adapt to changing conditions, the vegetation type faces possible extinction. Because of the possibility of future rapid increases in global temperatures, scientists are concerned about the potential for major changes in Earth's plant communities. It is clear that some species in a plant community might be better able to expand their boundaries than others. Such changes in the rate of migration for individual species would lead to a change in the plant community, in which the overall composition of the plants in the new environment is different from what it previously was. For example, trees have longer life spans and reproductive cycles than brushy plants, and therefore take longer to adapt to changing environmental conditions.

Moreover, the expansion of suburbs and agriculture into previously undisturbed areas has caused some plant communities to become fragmented. In other words, instead of having extensive areas of a particular plant community, those vegetation associations now occur in isolated patches. The fragmentation of these communities hinders their ability to migrate in response to climate change and could make some future migrations impossible.

An interesting example of rapid vegetation change due to increasing temperatures has been observed in the grasslands of northeastern Colorado (Figure 16–3–1). A trend toward increasing nighttime temperatures has caused the average date of the last killing frost to occur earlier in the spring. This has put the normally dominant grass species—the blue grama, which historically has accounted for 90 percent of the ground cover—at a competitive disadvantage compared to various weeds. The weeds that are taking over the grasslands landscape are far more susceptible to drought. This is highly significant to ranchers who rely on the grass cover to supply most of their livestock's food needs.

▲ **FIGURE 16–3–1** Prairie at Pawnee National Grassland, northeastern Colorado.

16–3–1 How does climate change pose challenges to plant communities?

16–3–2 Describe one way in which human activities make it difficult for plant species to migrate in response to climate change.

Atmosphere–Ocean Interactions

16.6 Describe the interactions between land and ocean related to climate change.

Temperature increases and other changes in the oceans associated with natural and anthropogenic climate change affect the atmosphere–ocean system on a global scale. These effects involve changes in sea levels and in ocean circulation.

Warming Temperatures and Sea Levels

One obvious relationship between the oceans and the atmosphere relates to the effect warming temperatures have on sea levels. Increases in global temperatures can raise the mean sea level, primarily by the expansion of warmer waters. Increases in atmospheric temperature can also contribute to rising sea levels by causing glaciers to melt and release water back to the oceans. A recent study estimates that increasing temperatures associated with a projected doubling of carbon dioxide within the next half century could cause the average sea level to increase by 19 cm (7 in.), but the rise in sea level would not be uniform throughout the oceans, and the rise could be as much as 35 cm (14 in.) off the coast of Europe.

Melting of the West Antarctic ice sheet is now occurring, which could raise sea levels by 5 m (16 ft) or more in upcoming centuries. Unlike any other ice sheet, the West Antarctic sheet is a concern because its base is below sea level. It is therefore susceptible to melting by ocean waters eroding its periphery. Until lately this idea was not widely accepted, but information gleaned from sediments beneath the ice sheet strongly suggest that part or all of the sheet collapsed sometime in the last 2 million years, most likely in the interglacial period 400,000 years ago. More recently, a pair of papers by research teams were published in highly influential journals on the same day in May 2014 that documented the "irreversible" retreat of the ice sheets due to the upwelling of warm water along the ice margins. Sea-level changes, whether large or small, feed back on climate patterns through a number of mechanisms, including effects on surface water currents and the movement of deep water below the surface, the proportion of land and ocean, and marine and terrestrial productivity, to name a few.

Changes in Ocean Circulation

A very important feedback concerns global-scale ocean circulations. As we discussed in Chapter 8, there is a large flow of water northward in Atlantic surface waters that delivers tremendous heat to the overlying atmosphere. After cooling and sinking in the North Atlantic, water flows southward in a much wider stream at great depth to the rim of Antarctica, where it joins deep water forming at the edge of the ice cap. Other branches of deep water flow northward into the Indian and Pacific Oceans, eventually rising and returning to the southern ocean at the surface. This huge, beltlike circulation system is partly driven by temperature differences but also by salinity variations; hence, it is called **thermohaline circulation** (see Figure 16–27).

Much attention has focused on the Atlantic portion of this system, known as the Atlantic Meridional Overturning Circulation (MOC). There is now convincing evidence to suggest that the MOC is not steady but instead makes rapid jumps from one mode of operation to another. Rates and locations of sinking water change abruptly, which in turn affect other aspects of the pattern. The new configuration lasts for some time before changing abruptly again. Numerous reorganizations of such ocean circulation have been found in a marine record running from about 60,000 years ago until about 10,000 years ago. In this way, changes in ocean circulation are implicated in abrupt climate jumps, at least during the last glacial period. For example, particularly cold episodes are believed to be triggered by pulses of freshwater released during melting episodes of the Laurentide ice sheet. A layer of less dense freshwater at the surface shuts off downwelling, which in turn interrupts the flow of warm surface waters northward across the Atlantic.

DID YOU KNOW?

Reservoirs around the world have partially counteracted sea-level rise from melting ice caps. By storing roughly twice the water volume of Lake Michigan, the roughly 30,000 artificial lakes dotting Earth have prevented about 30 mm of sea-level rise in the last 50 years.

Feedback from atmosphere to ocean arises as the changing climate alters patterns of runoff, precipitation, and evaporation, which in turn influence ocean temperature and salinity. In addition, changing winds and other atmospheric processes affect the distribution of sea ice, which has a huge impact on the ocean heat balance. If these changes are large enough, ocean circulation shifts to a new mode, which in turn influences global climate. If true, this feedback mechanism provokes a concern that relatively modest global warming might cause another rapid reorganization perhaps 100 years from now, when world population will demand about three times today's food supply. If the resulting climate change turns out to be as large as the one 8200 years ago (which occurred when the climate was even warmer than now), the results could be disastrous.

THERMAL INERTIA OF THE OCEANS Note here that the huge mass and high specific heat of the oceans, along with their vertical and horizontal mixing, create a strong thermal inertia. This means that warming of the ocean in response to increasing atmospheric temperatures would be very slow. Thus, oceanic temperature changes would be retarded in

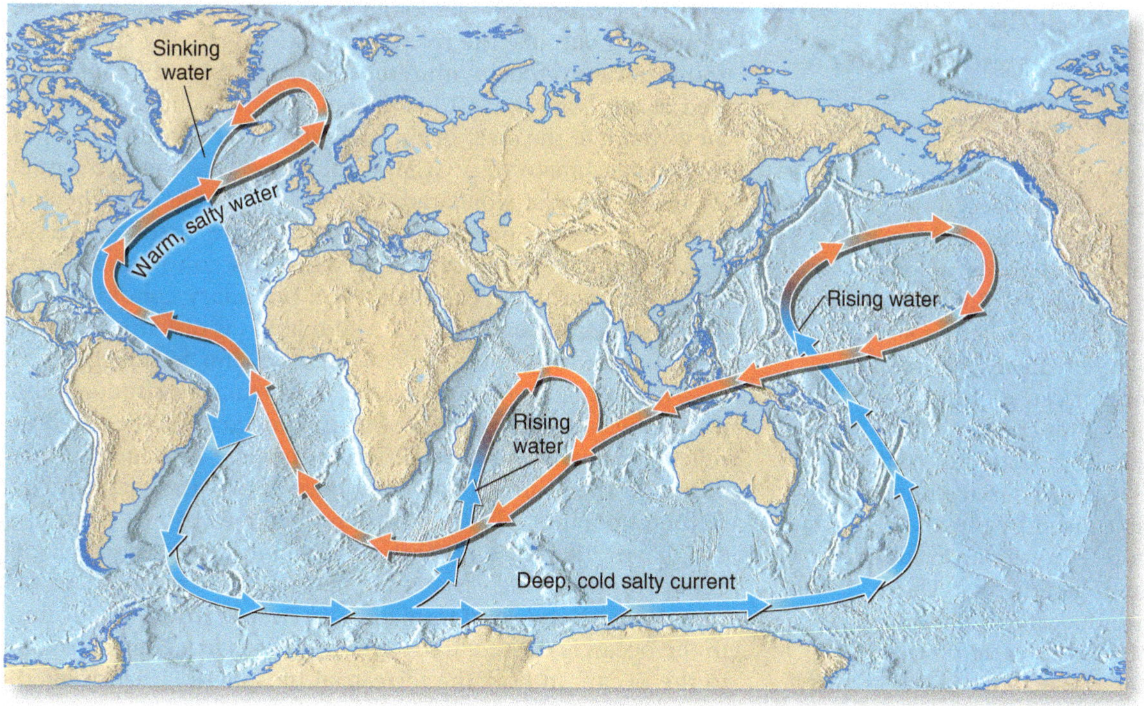

◄ **FIGURE 16–27** **The Ocean's Thermohaline Circulation.**

Video
The Thermohaline Circulation
http://goo.gl/50mrj0

Animation
North Atlantic Deep Water Circulation
http://goo.gl/mgRLOZ

response to global warming, and there could be a time lag of several decades before any increase in atmospheric temperature would be observed in the oceans. Once the warming is realized, it would modify atmospheric pressure and wind distributions.

CHECKPOINT

16.19 Why have scientists predicted that the West Antarctic ice sheet is melting?

16.20 What is the thermohaline circulation?

16.21 Climatologists have hypothesized that, in the past, pulses of freshwater from melting glaciers in what is now Canada may have disrupted the flow of warm surface waters across the North Atlantic Ocean toward western Europe. How might this have affected that region's climate?

Projecting Climate Changes

16.7 Describe the role of general circulation models in identifying causes and effects of climate change and projecting future changes in climate.

Given the extraordinary complexity of atmospheric systems, you may wonder how climate scientists have been able to reach conclusions about the causes and effects of climate change. However, researchers have used their understanding of the many variables that influence climate change, along with data on present and past climates, to model the natural processes involved and how human activities affect those processes. Using climate models, scientists can make projections of possible future changes in climate.

Identifying the Causes of Climate Change

The Fifth Assessment Report of the IPCC presents the most recent state of knowledge about observed and predicted climate changes and their impacts. It states quite succinctly:

> Human influence has been detected in warming of the atmosphere and the ocean, in changes in the global water cycle, in reductions in snow and ice, in global mean sea level rise, and in changes in some climate extremes (see Figure SPM.6 and Table SPM.1). This evidence for human influence has grown since AR4 [its last report]. It is extremely likely that human influence has been the dominant cause of the observed warming since the mid-20th century.
>
> It is extremely likely that more than half of the observed increase in global average surface temperature from 1951 to 2010 was caused by the anthropogenic increase in greenhouse gas concentrations and other anthropogenic forcings together. The best estimate of the human-induced contribution to warming is similar to the observed warming over this period.

So the bottom line is that the sum of all scientific evidence has continued to clarify the role of human activities in accounting for observed climate changes. People are responsible for climate changes and the important environmental impacts they could have.

General Circulation Models

In both attributing observed changes to particular causes and in making predictions about possible futures, climate science relies heavily on **atmosphere-ocean general circulation models (GCMs)**. GCMs are mathematical representations of the Earth atmosphere–ocean–land system that run on supercomputers. They use well-established physical theory (such as conservation of energy, mass, and momentum and radiative transfer laws) to calculate the three-dimensional motion of the atmosphere and ocean for the entire globe. GCMs are somewhat similar to the models used in numerical weather prediction, but they differ greatly in data needs, internal structure, and purpose. In particular, GCMs estimate the response of Earth's climate to a set of given external conditions known as boundary conditions. Examples of such boundary conditions include Earth topography, tilt of Earth's axis, solar output, volcanic aerosol loading, and greenhouse gas emissions. The programs compute the state of the atmosphere, oceans, and land surface conditions at discrete points in time (for example, every 30 minutes), using the current state as a starting point for the next time step.

Because of the enormous computing resources required, GCMs have rather limited spatial detail. The atmospheric component of a GCM, for instance, might provide values only every degree or two in latitude and longitude. The limited spatial detail means that small-scale processes (e.g., storm development) can be represented only approximately as *parameterizations*. Similarly, the need to minimize computation means that many other processes are highly parameterized or even ignored. For example, although GCMs compute soil moisture inputs and outputs, the role of vegetation is highly simplified, and water movement within the soil column is often not modeled at all. Note that GCMs do not compute climate directly, they compute the state of the atmosphere at various points in time. The climate of a GCM is found by running the model for many years of simulated time and averaging over the end of the simulation period. Multiple years are needed because just as with the real atmosphere, there is considerable variability from year to year even when boundary conditions are unchanging. The early part of a simulation is discarded so that arbitrary starting values do not taint the climate statistics.

Numerous GCMs are in use at research institutions and universities throughout the world. Indeed, the IPCC report relied on results from 40 different models. Although all are based on the same fundamental laws of physics, they employ different parameterizations, different computational methods, and have different temporal and spatial resolutions. For example, there is wide variation in how sea ice is modeled, in assumptions about cloud formation, in how vertical overturning in the ocean is treated, in freshwater inputs from land surfaces to the ocean, in how the terrestrial biosphere is represented, and in how the atmosphere and oceans are coupled (to name just a few of the many differences!).

In light of all these differences, and considering the obvious deficiencies of GCMs, what reason is there for confidence in their output? First, GCMs have advanced to the point that they do a good job of reproducing many aspects of today's large-scale climate. They successfully simulate observed seasonal patterns of air temperature, major rainfall and desert regions, ocean currents, the seasonal migration of storm tracks and tropical monsoon circulations, and the extent of sea ice. In addition, GCMs exhibit various forms of year-to-year variability that are typical of the observed Earth system (including ENSO events). A third source of confidence comes from GCM success at simulating past climates formed under different boundary conditions. For example, models have simulated the response of tropical monsoon circulations to the greater Northern Hemisphere seasonality in solar radiation that prevailed 6000 to 10,000 years ago. Fourth, GCMs have had success at reproducing the observed global temperature curve seen in Figure 3–32. Although the greenhouse gas and other boundary forcings were relatively modest during this period, their effects are captured by GCMs. Finally, some confidence in GCM output comes from realizing that they rest, as much as possible, on physical theory. Their development has been a story of increasing progression of realism in processes represented, and that has resulted in a steady progression in the quality of their output. It is important to realize that GCM projections are not mere extrapolations of past behavior, nor do they depend on purely statistical associations, which may or may not reflect causal relations.

GCMs provide what many see as the strongest evidence for the role of humans in the warming of the last century. As seen in Figure 16–28a, the models successfully reproduce the observed warming. The simulated curve is the average from 58 simulations using 14 different models, all driven by a combination of natural and human forcings. Agreement between the curves implies that observed changes have resulted from both natural and anthropogenic factors. Natural factors included in the models are changes in volcanic activity and solar output. Human agents included greenhouse gases, aerosols, and some runs also considered changes in land use. There is considerable uncertainty in some of these forcings—aerosols are a good example—so the inputs should be considered as merely the best available estimates for potential drivers of recent climate. The models differed significantly in what boundary condition changes were included (e.g., some included tropospheric ozone while others did not) and in the details of their representation (e.g., there were differences in the treatment of aerosol effect on radiation). Figure 16–28b shows a similar curve from 19 simulations using five models but with only natural forcings included. We see that warming in the last few decades is absent, and in fact the models show modest cooling in that period. In other words, without the anthropogenic inputs the planet would have cooled, not warmed. Of the two human inputs, greenhouse gases warmed the planet, whereas aerosols were an overall cooling agent.

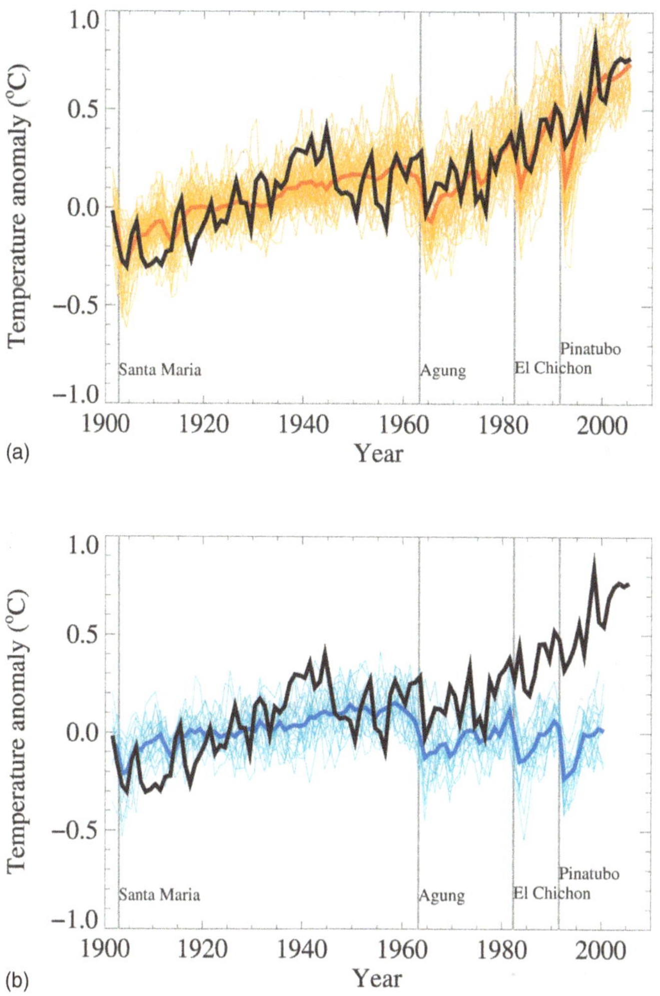

▲ **FIGURE 16–28 Projected Changes in Temperature.** The graphs show global mean temperature observed (black curve) and modeled using a suite of GCMs as published by the IPCC. In (a), natural and anthropogenic forcings were used, whereas in (b) only natural forcings were included. Light yellow and blue curves show individual models. Heavy lines are averages from groups of models.

Predicted Temperature Trends Through the 21st Century

General circulation models have been used in a very large number of scientific studies to examine how the atmosphere might respond to increasing greenhouse gas concentrations in the future. But one of the inherent problems in doing such studies is that we do not know what the rate of input of greenhouse gas emissions will be in the future. Although most of the world's countries have agreed to reduce their emissions, the rate of future CO_2 and other greenhouse gas emissions may fall anywhere within a fairly wide range of values. So the IPCC examined climate forecasts put out by multiple GCMs, each considering a variety of greenhouse emission scenarios during the next century.

In the Fifth Assessment Report the IPCC used model outputs assuming varying amounts of **radiative forcing**. To understand radiative forcing, consider the balance of incoming net radiation at the top of the troposphere as it was at a point

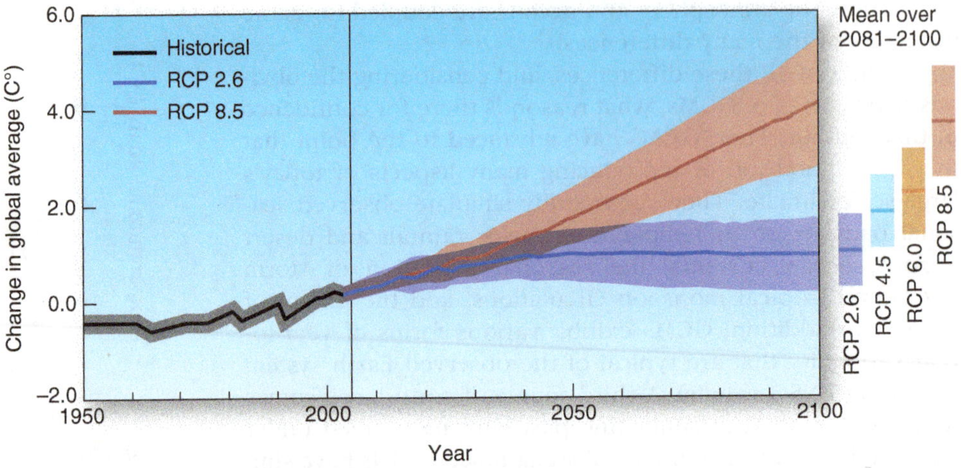

▸ **FIGURE 16–29 IPCC Global Mean Temperature Projections for Various Emission Scenarios.** Observed and predicted global temperatures for the RCP2.6 (blue) and 8.5 (red) scenario. Solid lines show the mean of the model projections and the shaded areas describe the range of possible values. The bars on the right show the mean and range of predictions for the period 2081–2100 for the four scenarios.

Video
Taking Earth's Temperature
http://goo.gl/2zgsrL

in time (the IPCC used the year 1750 as its reference point). If one or more boundary conditions were to change, like an increase in solar emission or an increase in greenhouse gasses, there would be a shift from the balance between incoming and outgoing net radiation, expressed in watts per square meter (W/m^2). An increase in the net radiation would create a disequilibrium that would lead to warming (with a positive radiative forcing) or cooling (with a negative radiative forcing). The IPCC used modeled projections with representative concentration pathways (RCPs) designated as RCP2.6 (for a radiative forcing of 2.6 W/m^2), RCP4.5, RCP6.0, and RCP8.5. These forcings each correspond to increasing amounts of presumed greenhouse gas concentrations.

CHANGE IN MEAN TEMPERATURE Figure 16–29 from the IPCC Fifth Assessment Report plots the range of global temperature predictions from a number of GCMs to the year 2100. The curve to the left of the vertical line represents temperatures up to 2005, with a narrow range of decreasing uncertainty represented by the thickness of the shaded area. To the right, solid curves represent the average of all the predictions for the global temperatures based on RCP8.5 (red) and RCP2.6 (blue). The range of mean temperatures predicted by the GCMs is shown by the width of the shaded areas. The temperature values are measured in degrees Celsius relative to the average temperature for the period 1986–2005. The figure tells us that under the low radiative forcing assumption there would likely be an increase in global mean temperature of about 1 °C (1.8 °F) that flattens out at about midcentury. The scenario depicted under the RCP8.5 scenario is much more drastic, with temperatures increasing by about 4 °C (7.2 °F) by the end of the century, with no apparent cessation in warming by that time.

Looking at the projected changes more broadly, the IPCC has concluded that by the year 2100 the global average temperature will "likely" have increased by at least 1.5 °C (2.7 °F) relative to that of the 1850–1900 period, under every RCP scenario except RCP2.6. The increase in global temperature will "likely" be greater than 2 °C (3.6 °F) under the RCP6.0 and RCP8.5 scenarios:

> Global surface temperature change for the end of the 21st century is *likely* to exceed 1.5°C relative to 1850 to 1900 for all RCP scenarios except RCP2.6. It is *likely* to exceed 2°C for RCP6.0 and RCP8.5, and *more likely than not* to exceed 2°C for RCP4.5. Warming will continue beyond 2100 under all RCP scenarios except RCP2.6. Warming will continue to exhibit interannual-to-decadal variability and will not be regionally uniform (see Figures SPM.7 and SPM.8).

TABLE 16–1

IPCC Temperature Increase (°C) Predictions Relative to 1986–2005

	2046–2065		2081–2100	
	Mean	Likely Range	Mean	Likely Range
RCP2.6	1.0	0.4 to 1.6	1.0	0.3 to 1.7
RCP4.5	1.4	0.9 to 2.0	1.8	1.1 to 2.6
RCP6.0	1.3	0.8 to 1.8	2.2	1.4 to 3.1
RCP8.5	2.0	1.4 to 2.6	3.7	2.6 to 4.8

The predicted and likely range of temperature increases relative to 1986–2005 is given in Table 16–1.

GEOGRAPHIC VARIATIONS IN WARMING The amount of warming experienced over the 20th century has not been uniform across the globe, nor will it be in the 21st century. Figure 16–30 shows projections under the four RCP scenarios for temperature increases for the periods 2016–2035 and 2081–2100. Notice that land areas generally increase by greater amounts than do oceanic regions. Warming is also greatest in the Arctic region, though large areas of interior North America, Asia, Africa, and South America also show high amounts of warming, as does virtually all of Australia.

Other Effects of Climate Change

As we've said earlier, climate is defined as more than just the long-term averages of atmospheric variables. It also includes ranges of variability and the frequency of extreme events. Table 16–2 summarizes IPCC analyses of how the incidence of various atmospheric phenomena have changed since the mid-20th century, the likelihood that human activities have contributed to those changes, and the likelihood of further

◂ **FIGURE 16–30 IPCC Air Temperature Projections for Selected Time Periods and Emission Scenarios.**

MapMaster World Physical Environment Global Surface Warming Worst Case Projections

changes (discussed below) by the end of the 21st century. (See *Box 16–4, Focus on the Environment and Societal Impacts:* Challenges, Mitigation, and Adaptation, for a discussion of climate change issues as related specifically to the United States.)

EVAPORATION AND ATMOSPHERIC MOISTURE The rate of evaporation from the surface is strongly influenced by the air temperature, with higher temperatures promoting enhanced evaporation. At the same time, the amount of water vapor that can be maintained in the air is greater at higher temperatures (Chapter 5). Thus the average global specific humidity is expected (with medium confidence, according to the IPCC) to increase with atmospheric warming, as has already been observed in recent decades. While the specific humidity is likely to increase, the atmosphere's increased saturation

TABLE 16–2

Extreme Weather and Climate Events Projected for 21st Century

Phenomenon and Direction of Trend	Assessment That Changes Occurred (Typically Since 1950 Unless Otherwise Indicated)	Assessment of a Human Contribution to Observed Changes	Likelihood of Further Changes	
			Early 21st Century	Late 21st Century
Warmer and/or fewer cold days and nights over most land areas	*Very likely*	*Very likely*	*Likely*	*Virtually certain*
	Very likely	*Likely*		*Virtually certain*
	Very likely	*Likely*		*Virtually certain*
Warmer and/or more frequent hot days and nights over most land areas	*Very likely*	*Very likely*	*Likely*	*Virtually certain*
	Very likely	*Likely*		*Virtually certain*
	Very likely	*Likely* (nights only)		*Virtually certain*
Warm spells/heat waves. Frequency and/or duration increases over most land areas	*Medium confidence* on a global scale *Likely* in large parts of Europe, Asia and Australia	*Likely*	Not formally assessed	*Very likely*
	Medium confidence in many (but not all) regions	Not formally assessed		*Very likely*
	Likely	*More likely than not*		*Very likely*
Heavy precipitation events. Increase in the frequency, intensity, and/or amount of heavy precipitation	*Likely* more land areas with increases than decreases	*Medium confidence*	*Likely* over many land areas	*Very likely* over most of the mid-latitude land masses and over wet tropical regions
	Likely more land areas with increases than decreases	*Medium confidence*		*Likely* over many areas
	Likely over most land areas	*More likely than not*		*Very likely* over most land areas
Increases in intensity and/or duration of drought	*Low confidence* on a global scale *Likely* changes in some regions	*Low confidence*	*Low confidence*	*Likely (medium confidence)* on a regional to global scale
	Medium confidence in some regions	*Medium confidence*		*Medium confidence* in some regions
	Likely in many regions, since 1970	*More likely than not*		*Likely*
Increases in intense tropical cyclone activity	*Low confidence* in long term (centennial) changes *Virtually certain* in North Atlantic since 1970	*Low confidence*	*Low confidence*	*More likely than not* in the Western North Pacific and North Atlantic
	Low confidence	*Low confidence*		*More likely than not* in some basins
	Likely in some regions, since 1970	*More likely than not*		*Likely*
Increased incidence and/or magnitude of extreme high sea level	*Likely* (since 1970)	*Likely*	*Likely*	*Very likely*
	Likely (late 20th century)	*Likely*		*Very likely*
	Likely	*More likely than not*		*Likely*

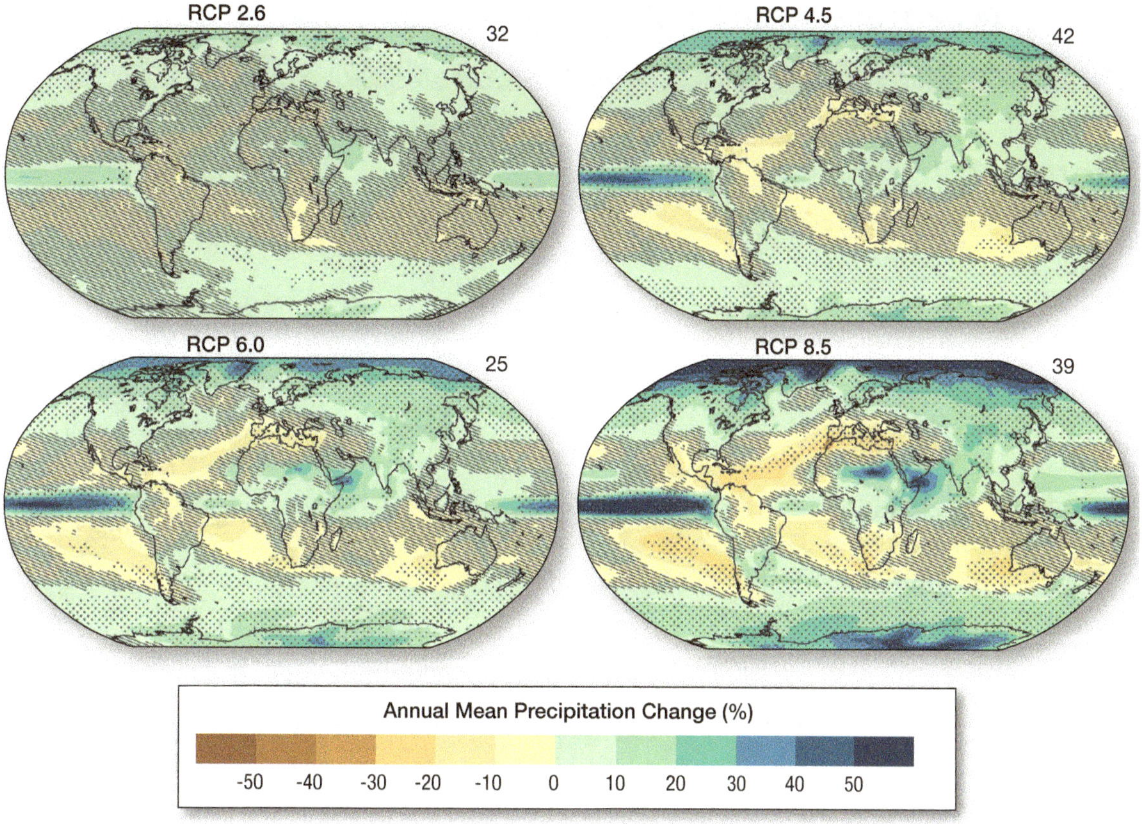

◄ **FIGURE 16–31 Projected Changes in Precipitation.** The mean of climate model outputs for precipitation during the last two decades of the century. Zones marked by stippling represent the highest level of confidence, while hatched areas are of lesser confidence.

specific humidity will increase such that the average relative humidity is likely to remain about constant.

PRECIPITATION The IPCC has determined that it is virtually certain that global precipitation will increase into the latter part of the 21st century. For all scenarios other than the RCP2.6 scenario, the IPCC anticipates a likely increase in global precipitation of 1 to 3 percent for every 1 °C temperature increase. Regional changes in precipitation for the four scenarios are shown in Figure 16–31, and typically show the same overall patterns but with greatest changes occurring with increasing radiative forcing (higher RCP values). In viewing the maps it is important to note that the degree of confidence in the predictions is indicated by the presence of stippling (dots) or hatching (slashed lines), with the former representing higher confidence and the latter denoting less confidence. There is a high level of confidence that precipitation will increase substantially over the already wet equatorial Pacific and over much of the high latitudes. Many dry regions, such as Mexico and the Mediterranean region, are likely to become even dryer.

In addition to regional and global changes in the amount of precipitation, there will likely be changes in the nature of precipitation. The IPCC predicts that there will likely be a shift in the nature of short-term precipitation events, with a higher occurrence of intense storms and a decrease in the number of weak storms. Moreover, most midlatitude land areas are very likely to experience more intense and more frequent extreme precipitation events.

CRYOSPHERE The amount of sea ice, snow cover, and permafrost are all sensitive to changing temperatures, as warmer conditions make it more difficult for the maintenance of water in its solid phase. Floating sea ice will likely continue to decrease substantially in the Arctic. Permafrost layers, in which the soil at some depth below the surface is perennially frozen, happen to exist in the part of the world subject to the greatest amount of predicted warming. Thus, the area covered by permafrost can be expected to decrease substantially. This has strong implications locally and globally: First, the weakened subsoil may become unable to support overlying buildings (Figure 16–32). Second, the thawed soil is likely to release

▼ **FIGURE 16–32** Melting permafrost causes these two buildings in Dawson City, Yukon, to lean toward each other.

16–4 FOCUS ON THE ENVIRONMENTAL AND SOCIETAL IMPACTS

Challenges, Mitigation, and Adaptation

Although the IPCC assessment reports are authoritative and comprehensive, they are not the sole compendium on climate change. For example, a team of over 300 distinguished experts assembled the 2014 Third National Climate Assessment (NCA) to summarize the impacts of climate change on the United States (www.globalchange.gov). The report discussed much of what had been released in the fifth IPCC report, but emphasized effects of climate change in the United States, both in terms of geographical regions and individual sectors, such as energy, transportation, agriculture, ecosystems, and human health. It also discussed measures that can be taken to mitigate climate change, and ways society can adapt to a new climatic regime.

Temperature

The NCA reported that the global increase in temperature has occurred in the United States, just as it has over the most of the world. The group estimates that since 1895 U.S. temperatures have increased by 0.7 to 1.1 °C (1.3 to 1.9 °F), with the majority of the warming having occurred after 1970. As has been the case worldwide, the most dramatic of the North American warming has occurred in the higher latitudes—Alaska and the Canadian north. Specific projections of future U.S. temperatures are subject to even greater uncertainty than those of global values, but the range of temperature increases varies from 1.7 to 2.8 °C (3 to 5 °F) under low emission scenarios to as much as 2.8 to 5.6 °C (5 to 10 °F) under high greenhouse gas conditions by the end of the century.

Precipitation

The percentage of annual precipitation that occurs from extremely heavy events (those in the top 1 percent for each location) has increased in recent decades over all of the United States, particularly in the extreme Northeast (Figure 16–4–1). If this trend continues these regions will become susceptible to increased urban and flash flooding. In the dry Southwest no major change has been noted in the percentage of precipitation occurring during extreme events, but there has been a high incidence of extreme, extensive, and prolonged droughts. This has led to direct water resource issues as the demand increases while supplies dwindle. There is also the indirect effect in which drought leads to the drying of vegetation and the susceptibility to devastating wildfires.

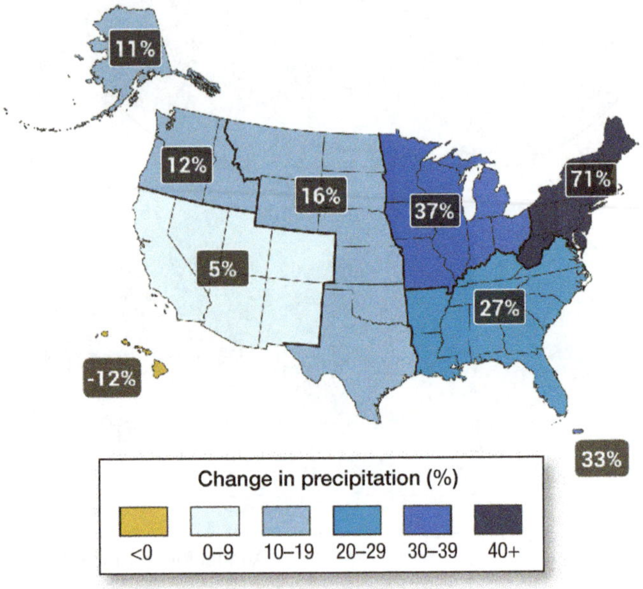

▲ **FIGURE 16–4–1 Annual Precipitation Changes, 1958–2012.** Precipitation has increased over most of the continental United States in the last half century, particularly over the Northeast.

Impacts on Economic Sectors

The effects of climate change show up in virtually every sector of modern society. The NCA elaborates on many of these issues in depth, but here we provide just a brief overview of a few impacts.

Energy

While we often think about the consumption of energy as a source of greenhouse gas emissions, climate changes impact the demand for energy as well.

carbon dioxide and methane into the air, thus promoting further warming. Projected declines in snow cover and permafrost areas are shown in Figure 16–33.

Much observational evidence illustrates the retreat in alpine and continental glaciers over both hemispheres, and such a decrease in glacial ice is expected to continue through the century. Forecasts of the amount of glacial ice that will be lost are hard to make, but it is believed that under the RCP8.5 scenario the decrease in glacial volume could be anywhere from 35 to 85 percent by the end of the century.

CHANGE IN SEA LEVEL In addition to warming of the atmosphere, oceanic temperatures have increased. The IPCC estimates that the upper 75 m (246 ft) has warmed by about 0.44 °C (0.8 °F) between 1971 and 2010. Warming of the oceans is expected to continue through the century, with estimated values for the upper 100 m (328 ft) ranging between 0.6 °C (1.1 °F) for the RCP2.6 scenario to 2.0 °C (3.6 °F) for the RCP8.5. Because water expands with increasing temperatures, the mean sea level will inevitably increase. The increase in sea level by thermal expansion will be augmented by the melting of glaciers. Figure 16–34

As temperatures increase the amount of energy needed for heating is expected to decrease, but the demand for indoor air conditioning will increase. As the U.S. population continues to move to places like Arizona and Texas, we can expect to see the demand for energy to meet that need increase through time.

Agriculture

Video
Climate, Crops, and Bees
http://goo.gl/jNcba

Some effects of climate change on agriculture are fairly direct and obvious. A decrease in the frequency of freezing temperatures might be beneficial to Florida citrus growers, and some crops could benefit by the longer growing season associated with warmer climates. But at the same time, higher temperatures can lead to greater evaporation rates and an increased demand for irrigation water, which might not always be met. Indirect effects include the uncertain ways that weeds and pests might thrive and impact crop production. It has also been shown that higher nighttime temperatures during the pollinating season have led to lower productivity and quality of many crops. With regard to precipitation, the projected increases in the incidence of extremely heavy rainfall events might reasonably be expected to accelerate soil erosion.

Health

Between 1978 and 2007 extreme heat has killed nearly as many people in the United States as tornadoes, hurricanes, and lightning put together, which makes the projected increase in temperatures a major direct threat to human health. People without access to air conditioning—often those who are poor and elderly—face an enhanced risk of becoming heat--related fatalities. But higher temperatures can also be associated with increased air pollution, and many insects that spread disease proliferate under hot, humid conditions.

Mitigation and Adaptation

Being aware of the threat of climate change gives us the opportunity to undertake measures that might slow down the rate of change (mitigation) and reduce its negative impacts (adaptation).

Mitigation

By far the most important measure that can be taken to slow the rate of anthropogenic climate change is to reduce greenhouse gas emissions. This can be promoted or even mandated by governments, but such actions face resistance due to economic costs that might be incurred. Some reductions can occur independent of mandates, for example, by a switch from higher polluting sources to "cleaner" fuels. This has happened to some extent through the increased use of natural gas in the United States. Economic slowdowns can naturally reduce emissions by reducing manufacturing and transportation activities. It is important to realize that carbon dioxide has a long residence time (Chapter 1) in the atmosphere, so even drastic cutbacks would not lower its concentration in the atmosphere for decades. On the other hand, other pollutants like methane and soot have much shorter residence times. Cutting back their emissions could have more immediate impacts on the rate of climate change.

Adaptation

The first step in adapting to a changing world is recognizing that such change is happening. Once people accept the fact that climate change is happening—even if many questions remain regarding its specific manifestations—we can begin to take earlier actions rather than having to react later in crisis mode. Given the many different ways that climate changes will impact different regions, there is no single adaptive strategy for universal use. Coastal regions vulnerable to inundation need to plan for such occurrences, while some heartland areas must prepare for more damaging heat waves and droughts. In much of the West, an ongoing threat will be drought and wildfires. Regardless, understanding of the problem and a willingness to tackle it are prerequisites.

16–4–1 In what ways did the National Climate Assessment augment the findings of the IPCC?

16–4–2 Describe some of the impacts of climate change in the United States.

16–4–3 What are the different approaches involved in mitigation and adaptation to climate change?

shows the range of projected sea-level increases under each of the RCP scenarios. Though there is considerable uncertainty in the amount of sea-level increase through the century, it is virtually certain that substantial increases will occur.

An increase in sea level of about half a meter may not sound all that threatening, but that increase makes coastal cities far more vulnerable to inundation during the passage of systems such as "Superstorm" Sandy. Water from this hurricane's storm surge completely filled up traffic tunnels and subways in the New York City area while also taking out the underground electrical transformers. Given the inevitability of such events in the future (Figure 16–35), coastal cities worldwide need to consider mitigation measures.

TROPICAL CYCLONES Tropical cyclones begin as disturbances over warm, tropical ocean waters. Although it might seem reasonable to expect that warmer sea-surface temperatures would lead to more tropical cyclones, there is currently no evidence to suggest that the frequency of tropical cyclones will increase, with the IPCC concluding that there will likely be a decrease or no change in their incidence. However, it is also likely that an increase in the intensity of winds and precipitation will occur, as higher ocean surface temperatures will provide increased latent heat to those storms that do occur.

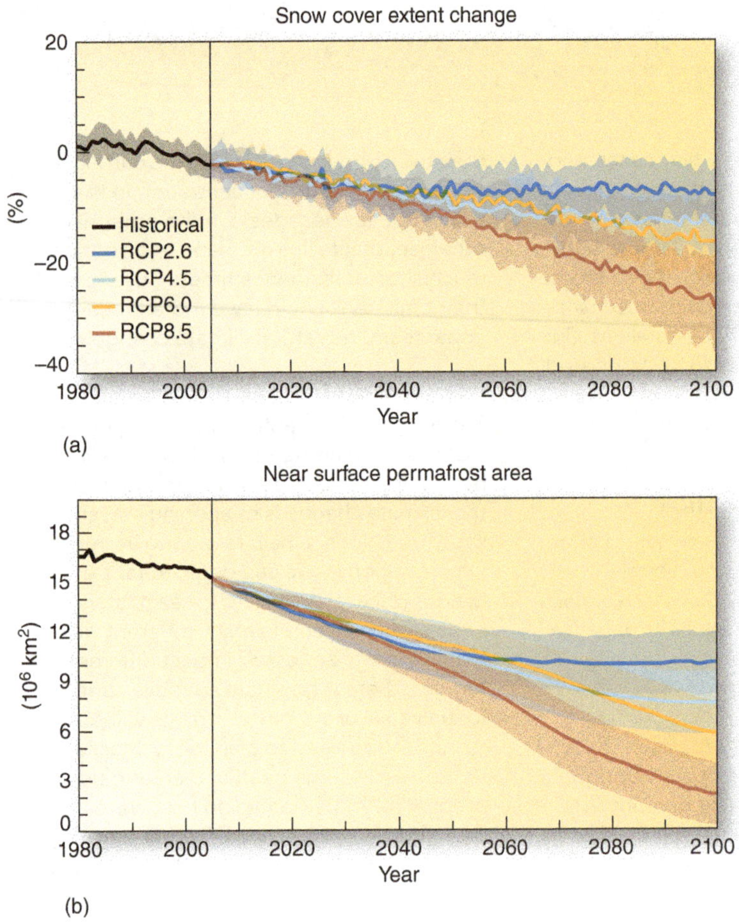

▲ **FIGURE 16–33 Projected Changes in (a) Areal Snow Cover and (b) Permafrost.**

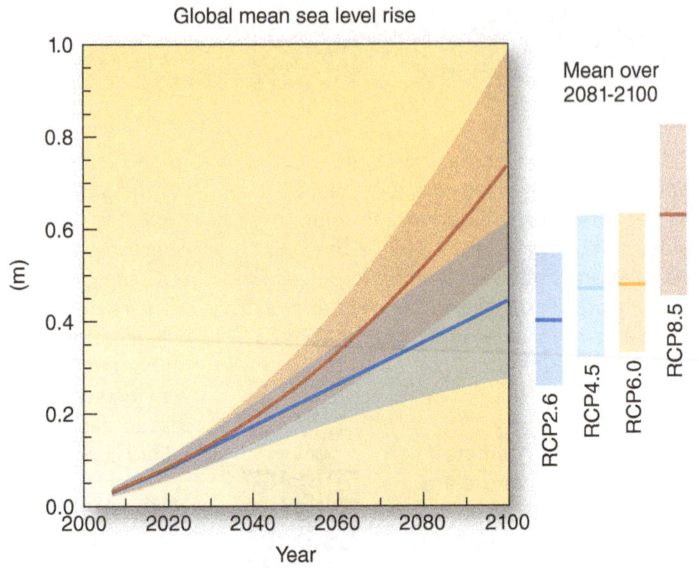

▲ **FIGURE 16–34 Projected Changes in Global Sea Level.** The curves depict the mean and range of values projected for the RCP2.6 (blue) and 8.5 (red) scenarios. The bars on the right show the mean and range of predictions for the period 2081–2100 for four scenarios.

▲ **FIGURE 16–35** Ipanema Beach, Rio de Janeiro, Brazil.

Video
Sea Level Rise
http://goo.gl/zRrhgp

CHECKPOINT

16.22 What physical principles and boundary conditions are used in building general circulation models of the atmosphere?

16.23 Look at the two graphs in Figure 16–28. How do the differences between the two graphs help in deciding whether the global warming is attributable to natural or human causes?

16.24 Which event(s) in Table 16–2 might be most likely to cause problems in the area where you live during this century? Explain.

Summary

16.1 Define climate change and explain how it can be described in terms of changes to boundary conditions in Earth's climate system.

- The climate of Earth has undergone numerous changes that have occurred over differing time scales, and there is every reason to expect that the climate will continue to change long into the future.
- Climate changes in response to external factors, called boundary conditions, and as a result of positive and negative feedback processes.

16.2 Explain how scientists understand Earth's past climates.

- Most of what we know about past climates has been inferred through indirect evidence of past conditions. Such evidence includes remnant landforms, botanical information, the presence of old soils, oceanic deposits, and cores obtained from the Greenland and Antarctic ice sheets.
- Climatic changes have occurred with differing magnitudes over varying time frames.

16.3 Describe major climate changes that have occurred over geologic time.

- At the longest time scales, Earth has been mostly warm, with little permanent ice.
- The planet has experienced an ice age at least seven times, the most recent of which continues to the present. Within an ice age, glaciers grow and shrink in what are called glacial/interglacial cycles.
- Present times correspond to an interglacial, which followed the last glacial maximum of 20,000 years ago. Within this period (the Holocene), the most notable changes were a large abrupt cooling period 8200 years ago and another much smaller one from 1400 to 1850 (the Little Ice Age).
- Over the period in which data have been collected, temperature records suggest global warming by a few tenths of a degree Celsius.

16.4 Explain natural and anthropogenic factors involved in climate change, including changes in solar output; Earth's orbit; land configuration and surface; atmospheric turbidity; and radiation-absorbing gases.

- Several processes can cause the climate to change.
- Changes in solar radiation receipts over hundreds of thousands of years are caused by variations in Earth's orbit around the Sun—the Milankovitch cycles.
- Radiation receipts vary in response to changes in radiation emitted by the Sun.
- Changes in the land surface of the planet may have been responsible for shifts in climate over both very long and relatively short time scales.
- The absorption of solar and thermal radiation by aerosols and radiation-absorbing gases, including anthropogenic greenhouse gases such as carbon dioxide (CO_2), methane (CH_4), nitrous oxide (NO_2), and chlorofluorocarbons (CFCs), may be important determinants in recent and future climatic conditions.
- Since the middle of the 19th century, Earth has experienced an exponential increase in the input of carbon dioxide to the atmosphere by industrial activities (chiefly fossil fuel combustion and cement manufacture).

16.5 Explain feedback mechanisms and Earth-system interactions involved in climate change.

- All of the factors that influence climatic conditions do not act separately, but as part of complex feedback mechanisms connecting many parts of the Earth system.

16.6 Describe the interactions between land and ocean related to climate change.

- The melting of glaciers along with increasing ocean temperatures has led to rises in sea level, which are expected to continue through this century and beyond.
- The thermohaline circulation is a three-dimensional movement of water that is affected by and in turn influences atmospheric climatic conditions.

16.7 Describe the role of general circulation models in identifying causes and effects of climate change and projecting future changes in climate.

- The effects of these feedbacks can be examined by the use of general circulation models (GCMs).
- The use of GCMs has advanced to the point that they can now simulate major features of our present climate as well as global temperature change during the last 100 years. Taken with theory and observations, they leave little doubt that most of the warming over the industrial period has been the result of increasing anthropogenic greenhouse gases.
- GCMs are also key in projecting future climates resulting from various assumptions about greenhouse emissions.
- Projected possible effects of climate change during the 21st century include increases in temperatures over land, heavy precipitation events, droughts, sea levels, and tropical cyclone activity.

Key Terms

Review Questions

1. Define *climatic change* and discuss the practical difficulties of identifying climatic change from observations.
2. Describe two types of remnant landforms that can provide information on past climates in a region.
3. Explain how cores taken from ocean deposits and ice sheets can be used to infer past climate conditions.
4. How are tree cores used as indicators of past climates?
5. Describe how pollen samples obtained from old soils provide information on past climates.
6. Explain how it is possible that the global climate can be both cooling and warming at the same time.
7. How does the present climate compare to past climates over the course of geologic history? Is it correct to say the ice age is over?
8. Describe the frequency at which glacial/interglacial cycles have occurred during the Pleistocene.
9. What time frames constitute the Pleistocene and Holocene epochs?
10. What magnitude of mean temperature differences coincided with the various glacial/interglacial episodes of the Pleistocene?
11. Which regions of North America experienced the greatest temperature differences from current values during the last glacial episode of the Pleistocene?
12. Describe the factors that can lead to variations in the amount of solar radiation available at the top of Earth's atmosphere.
13. List the factors that can lead to climatic change. At what time scales do each of these occur?
14. What evidence is there that variations in sunspot activity do or do not lead to climate changes on Earth?
15. How do changes in eccentricity and obliquity and precession interact to influence Earth's climate? What time scales apply to each?
16. What types of occurrences on Earth can affect atmospheric turbidity? How do turbidity differences affect global temperatures?
17. Describe positive and negative feedbacks and provide examples of each.
18. What reasons are there to think projections from GCMs can be trusted?
19. What is a general circulation model?

Critical Thinking

1. There is little doubt most of the warming of recent decades is due to the activities of humans, but some are not convinced that the effect of this warming will be problematic for society. How would you argue for and against this viewpoint?
2. In addition to changes in average temperature and precipitation, it is believed that the frequency of unusual events (such as droughts and floods) might be more common as a result of global warming. What regions of North America are more vulnerable to economic losses from these events than from increased temperatures?
3. Explain how climate scientists used multiple simulations based on several different GCMs to determine the relative contributions of natural and human forcings to climate change.
4. As recently as 20,000 years ago a continental glacier extended as far south as the central United States. Describe what impacts the ice might have had on the average position and magnitude of the polar jet stream.
5. What are two ways in which positive feedback mechanisms involving the cryosphere could enhance temperature increases in the Arctic in coming decades?

Problems & Exercises

1. Check your favorite newspaper or news magazine to see if there are any articles describing new findings about climate change or political issues related to the topic. How would you rank the importance of climate change relative to other major political issues?
2. Go to the website www.pewclimate.org and check to see if it includes any new press releases. If so, what are the major findings contained in them? Are these findings actually new or do they expand on known information?

Visual Analysis

The attached maps show the current distribution of quaking aspen trees and their projected distribution for 2090, assuming a substantial increase in greenhouse gas concentrations.

16.1. What general change in distribution is expected to take place?

16.2. Given the fact that temperatures will increase substantially by 2090, what does that tell you about the type of climate conditions favored by quaking aspen?

16.3. This projected distribution change involves the movement of a type of tree to different elevation ranges. What kind of changes might happen for trees that exist outside of mountainous regions?

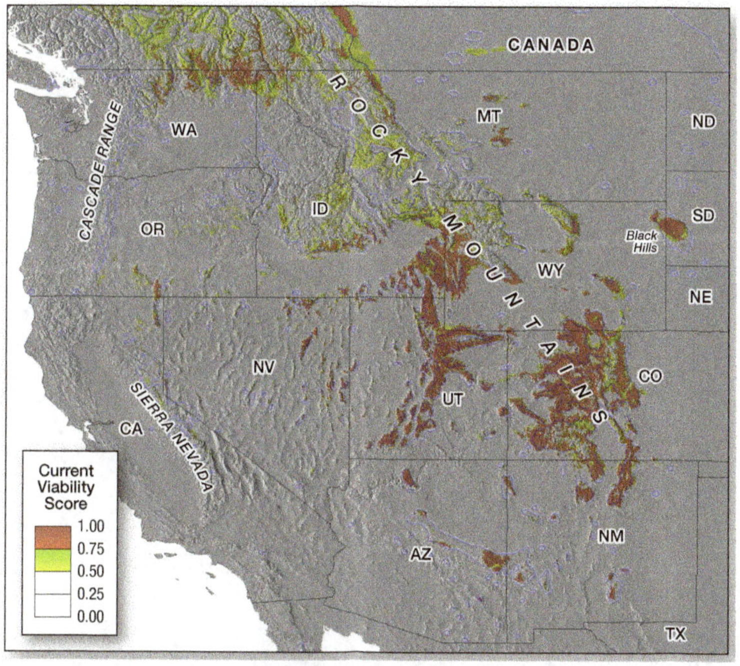

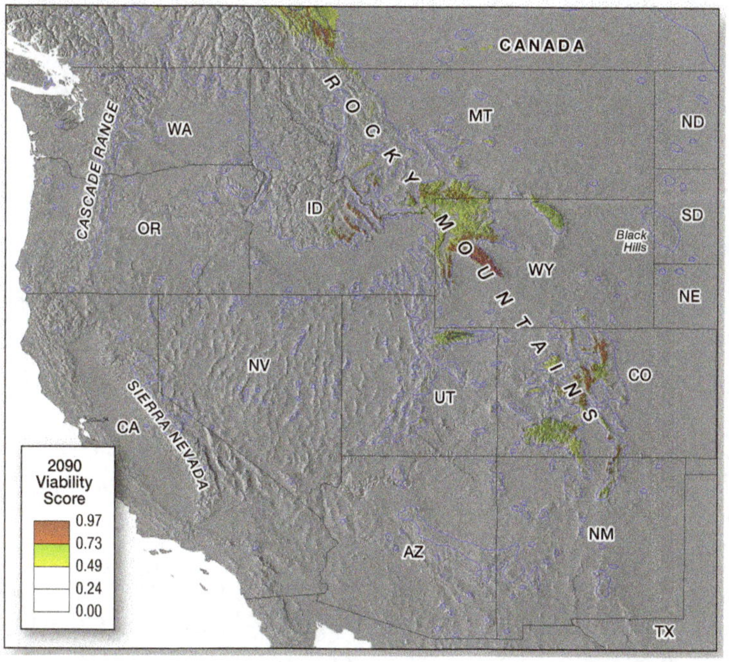

PART

7 Atmospheric Optics and Appendices

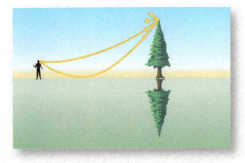

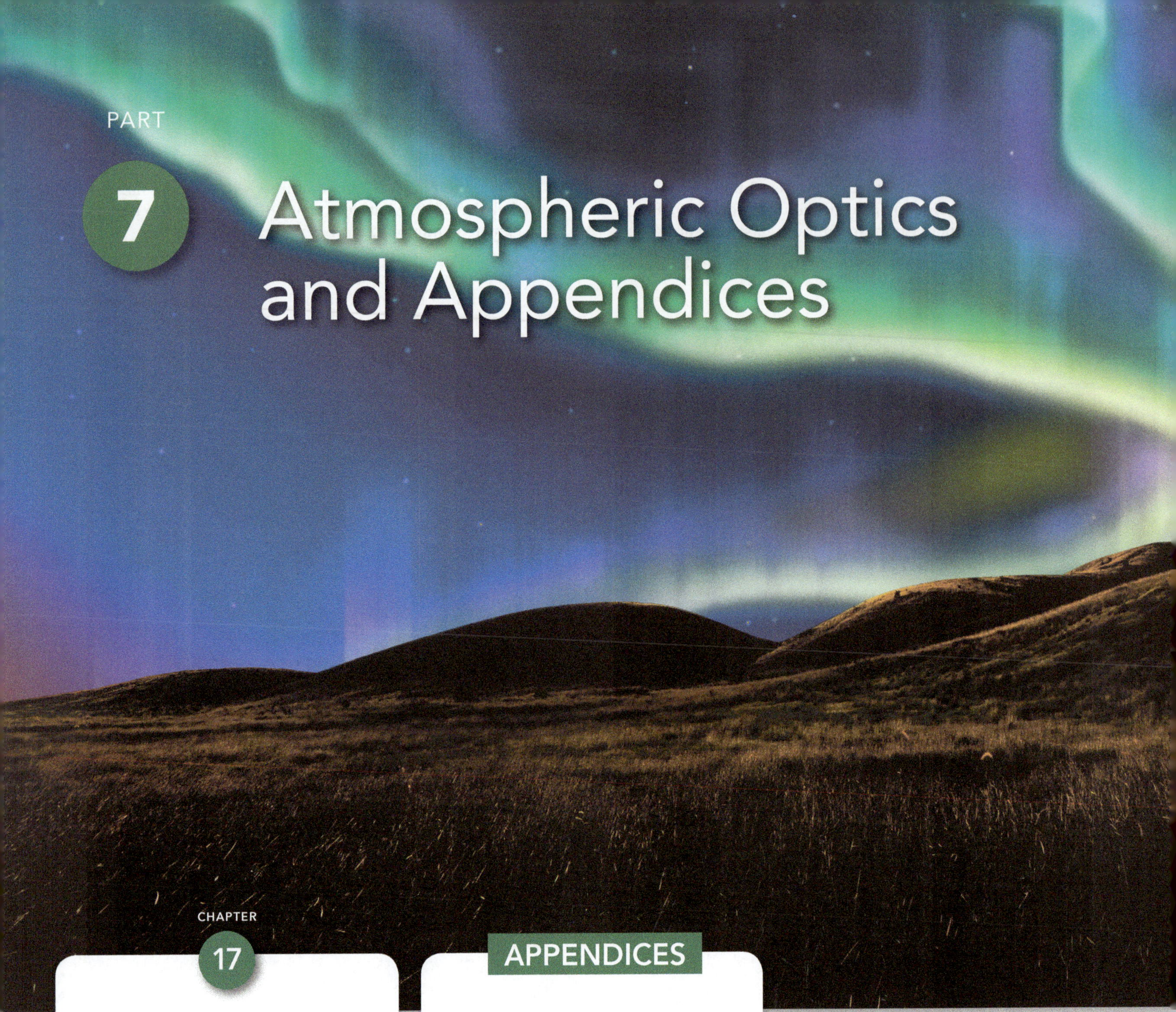

◂ *The Northern Lights. The optical phenomena considered in this part of the book depend fundamentally on the position of the observer, and therefore cannot be said to exist in a particular place like the aurora pictured here.*

The previous parts of this book have covered the issues of the mass and energy of the atmosphere, water in its various states, the distribution and movement of air, disturbances, and human activities. We now conclude the discussion of atmospheric phenomena with a chapter on atmospheric optics and several appendices that provide additional detail for some of the topics covered earlier.

CHAPTER

17

Atmospheric Optics

Imagine yourself driving along a straight, two-lane highway on a hot, sunny, summer afternoon. The car ahead of you is moving just a little too slowly, so you take a look into the oncoming traffic lane and you decide it is safe to pass. But as you cross over into the lane of oncoming traffic, another car appears, seemingly out of nowhere. Though you are surprised by the sudden appearance of the other vehicle, you have enough time to get back into your own lane—perhaps a bit unnerved but otherwise no worse off. Yet you can't help but wonder why you didn't see that car before you started to pass. Were your eyes playing tricks on you? Or perhaps the visibility was not as great as you thought it was. The answer could very well be that the atmosphere altered the path of the visible radiation reflected off the oncoming car so the rays were deflected away from your eyes, and that it was not until

◂ *A mirage gives the false impression of a wet road on an Arizona highway.*

you got sufficiently close to the vehicle that the light was able to meet with your eyes. Sometimes the atmosphere can indeed make objects appear to be in a different position from where they really are or even alter their appearance entirely. This final chapter describes the processes by which the atmosphere affects the path of visible radiation passing through it and the resultant images we see. We refer to these topics collectively as **atmospheric optics**.

Learning Outcomes

After reading this chapter, you should be able to:

17.1 Explain atmospheric refraction and the resulting optical effects that can be seen in clear air.

17.2 Explain the processes that produce optical phenomena involving clouds and precipitation.

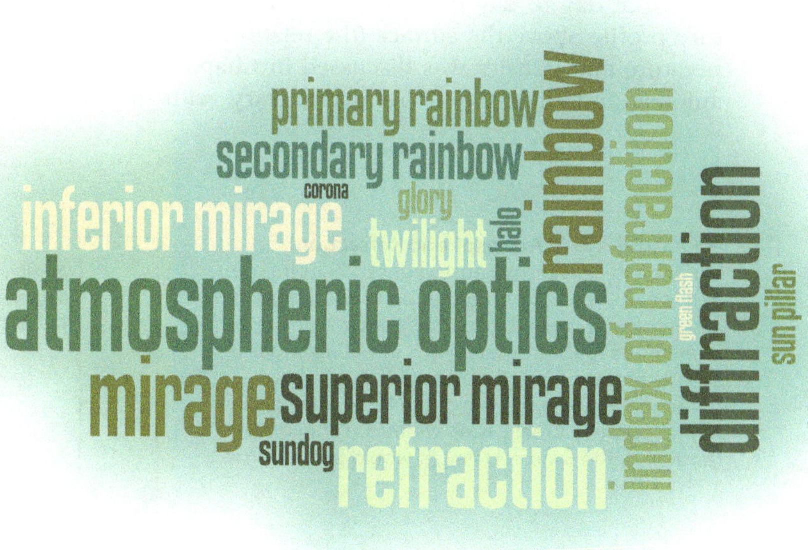

Refraction

17.1 Explain atmospheric refraction and the resulting optical effects that can be seen in clear air.

In Chapter 3, we saw that the atmosphere scatters and absorbs incoming solar radiation. Both processes have an important effect on the energy obtained by the surface and the atmosphere. In addition to scattering and absorbing, the atmosphere also *refracts* solar radiation, where **refraction** is defined as the bending of rays as they pass through a medium.

Refraction occurs whenever radiation travels through a medium whose density varies or whenever it passes from one medium to another having a different density. Refraction occurs because radiation speed varies with density—the denser the medium, the slower the radiation. The degree of slowing determines the **index of refraction**, which is simply the speed of light in a vacuum divided by the speed of light in the medium. For example, the speed of light in water is about 75 percent of that in a vacuum, and so the index of refraction for water is $1/0.75$ or about 1.33. The index of refraction for air is much smaller, about 1.0003. To visualize how differential speeds cause the radiation to bend, imagine two people in a canoe, with one of them paddling on the right side of the canoe and the other on the left. If the person on the right paddles more vigorously, the canoe steers to the left. The same thing happens to a ray of light passing through the atmosphere when the density of air varies across the direction of travel. The ray is slowed more on the denser side than on the other side, and the ray turns toward the denser part of the air.

Under most circumstances the density of the atmosphere decreases with height above the surface. This causes the radiation to refract slightly, forming an arc with the concave side oriented downward (Figure 17–1). The rate at which density changes with height varies considerably from place to place and from day to day because of differences in the temperature profile above the surface (the relationship between temperature and density was discussed in Chapter 4). Thus, the amount and direction of refraction vary with atmospheric conditions. The index of refraction also varies slightly with wavelength, and thus refractive bending is slightly different for every color of light. In particular, shorter wavelengths are refracted more than longer wavelengths.

DID YOU KNOW?

People who are very nearsighted or farsighted often pay a premium for eyeglasses with a high index of refraction. A higher index of refraction means the lenses bend light more, and so the high-index lenses can be much thinner than standard lenses and yet provide the same corrective benefit.

When light moves from one medium to another such as from air into water, there is refraction at the point of transition providing the light does not enter at a 90-degree angle. The prism in Figure 17–2 is a good example. Notice the direction of bending. If you imagine the light entering the glass and finding higher density on the right of its path, you can see why the light is bent to the right. At the exit point from the glass, the higher density and lower speed are to the right of the travel path—in the glass—so the ray again turns to the right. Notice also that there is more bending for the shorter blue and purple wavelengths than for yellow and red. This is expected, given the higher index of refraction for shorter wavelengths.

Suppose light moves directly toward higher density within a medium or into a medium of higher density at a right angle. In other words, suppose density does not vary from left to right across the path. The light will slow down, but uniformly with no side-to-side differences. This is somewhat like both canoe paddlers relaxing by the same amount. The canoe slows down but doesn't turn. Similarly, when light moves along a density gradient rather than across a gradient, it slows down but is not refracted.

From this it stands to reason that if the Sun is directly overhead, there will be little refraction during its downward path. What refraction does occur will be more-or-less random due to small regions of varying density. Of course, such variation occurs at all angles and so it adds and subtracts to whatever

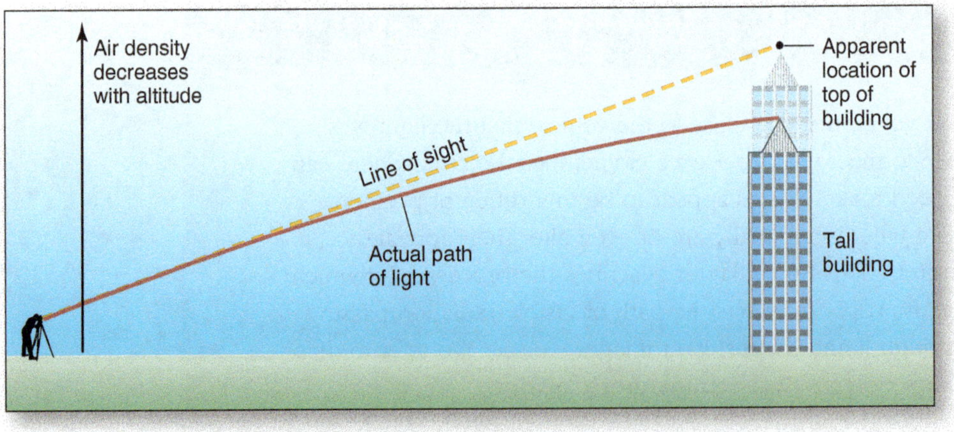

▸ **FIGURE 17–1 Atmospheric Refraction.** Variation in air density causes light to be refracted. In this case, the light from the top of the tall building is bent downward, so its path is concave downward. The light reaches the surveyor's eye at an angle slightly greater than what it would be without refraction, making the top of the building appear higher up than it really is. Without correcting for refraction, the reported height would be too large.

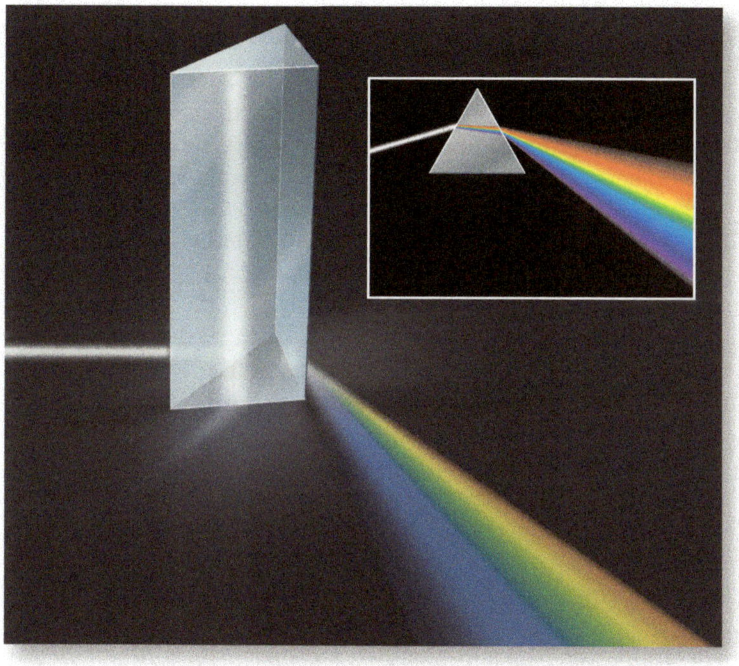

▲ **FIGURE 17–2 Refraction by a Prism.** As light passing through air enters a glass prism, the greater density of glass causes refraction. The same thing occurs as light traveling through the prism exits it. But shorter wavelengths of light undergo more refraction than longer wavelengths, resulting in a separation of colors.

refraction occurs over the full optical path. We notice such random variation in refraction at night in the twinkling of stars. Layers of air with varying density come and go along the light path and change the apparent position of the stars.

Let's look next at several more dramatic results of refraction in the clear atmosphere.

CHECKPOINT

17.1 What happens to light if it enters a medium of higher density?

17.2 What direction does light bend if it moves through a medium of varying density?

Refraction and the Setting or Rising Sun

Refraction of incoming solar radiation is greatest when the Sun is low over the horizon, because the low solar angle causes the rays to pass through a greater amount of atmosphere (as described in Chapter 3). At sunset, refraction is sufficient to cause direct rays to be visible even after the Sun has dropped below the horizon (this is also true just prior to sunrise). In Figure 17–3, the Sun is positioned below the horizon. Without an atmosphere this would bring nightfall, but refraction causes the Sun to appear to be just above the horizon. When the Sun is positioned slightly farther below the horizon, its direct rays cannot be seen at the surface, but diffuse radiation can illuminate the sky to create **twilight** conditions. A further lowering of the Sun puts it far enough below the horizon to bring total nighttime. The period of twilight varies; it is longest during the high Sun season and increases with latitude.

In addition to slightly shifting the apparent position of the Sun near the horizon, refraction can also affect its apparent shape and color. You may have noticed that near dawn or sunset the Sun seems to have horizontal bands of different colors, with redder coloration near the bottom. This occurs because longer wavelength colors (such as reds and oranges) undergo less refraction than do shorter wavelength colors (such as blue and green), as described above. The shorter wavelengths, undergoing a greater amount of refraction, concentrate near the top of the apparent sun, while the longer wavelengths locate near the bottom. Under some atmospheric conditions the Sun appears momentarily to be capped by a bright green spot, known as the **green flash** (Figure 17–4).

CHECKPOINT

17.3 At sunset is the Sun above or below the horizon? Explain why.

17.4 What accounts for the reddish color of light near sunrise or sunset?

DID YOU KNOW?

The apparently large size of "harvest moons" is a psychological illusion. Because of refraction the Moon actually occupies a little less of the sky when it is near the horizon than overhead. Thus, if photographed with a time-lapse camera, the orb would shrink a tiny bit as it approaches the horizon. However, we invariably see it as larger when low in the sky! Several explanations have been suggested, but the reason behind this common illusion has yet to be definitively proven.

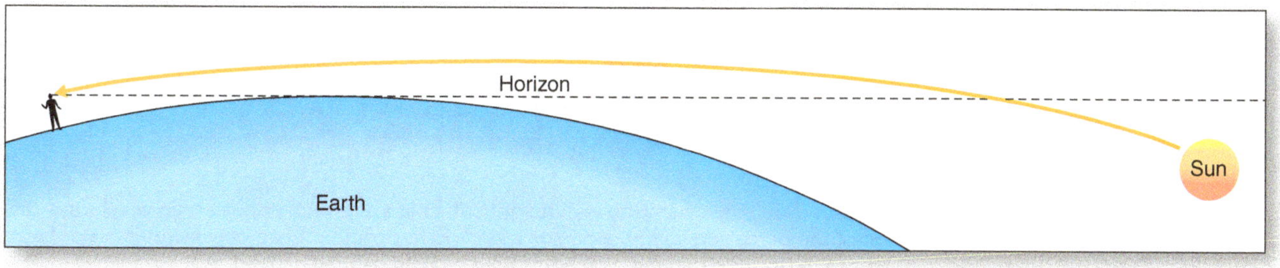

▲ **FIGURE 17–3 Refraction at Sunrise and Sunset.** When the Sun is slightly below the horizon, it can still be seen at Earth's surface because of refraction. The various wavelengths are refracted differentially so that the bottom of the Sun appears redder than the top.

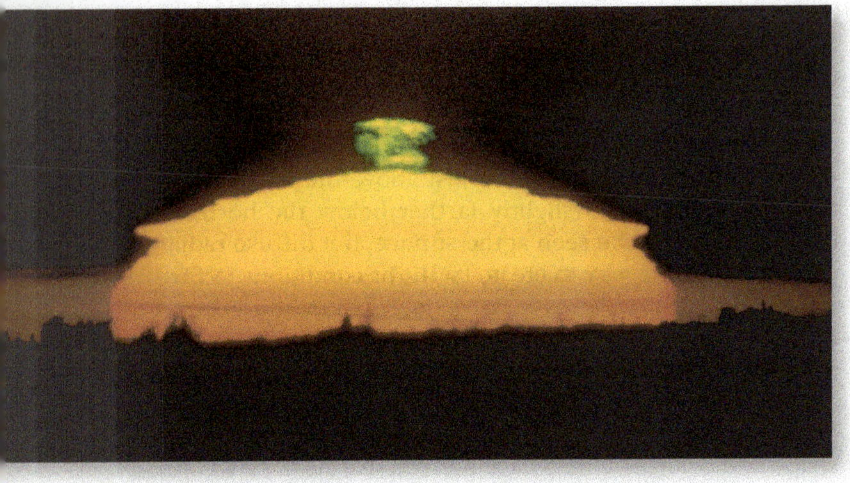

▲ **FIGURE 17–4 Green Flash.** Produced by differential refraction, these last only a few seconds.

Mirages

We are all familiar with movie and cartoon images of a parched, emaciated man crawling through a hot desert who perceives an oasis on the horizon, only to be disappointed by the reality of it being a mere **mirage**. Such mirages are caused not by excessive optimism but rather by the refraction of visible light when the temperature decreases rapidly with increasing height. And contrary to what some people think, the false oasis is not the only type of mirage; that term applies to any apparent upward or downward displacement of an object due to refraction.

To understand how mirages form, let's first consider a midday situation in which the air temperature near the surface decreases rapidly with increasing height much faster than the dry adiabatic lapse rate, as shown in Figure 17–5a. Near the surface, vertical changes in air density with height are controlled primarily by vertical temperature gradients. With a very high temperature near thc surface, density is low. The steep temperature gradient in this example results in increasing density with height, so the paths of the rays moving through the atmosphere are curved upward, as indicated by the yellow lines on Figure 17–5b.

In Figure 17–5b the viewer at the left (let's call her Lauren) perceives distant objects to be slightly lower than they actually are. This is because the light rays reflected off distant objects approach Lauren's eyes at an angle slightly below horizontal. Thus, a person standing at position A (we will call him Al) appears slightly shorter than he really is. Despite the minor distortion in the perceived height of Al, Lauren has no trouble seeing him in his entirety.

Now look at how one type of mirage develops as Al moves over to position B. The lower portion of his body appears to have disappeared because the light reflected off his legs is bent all the way to the ground, where it is absorbed before it can reach Lauren. As Al walks farther away, more and more of his body disappears from the bottom upward until he completely vanishes from sight at position C.

A different type of mirage appears from still more intense heating near the surface, such as that over an asphalt road on a hot afternoon. The heated air in the shallow layer just above the surface has an extremely steep temperature profile, while the air immediately above the shallow layer is somewhat cooler and has a less steep vertical temperature profile (Figure 17–6a). The steeper temperature gradient of the lower layer causes it to refract air more strongly than does the air above it. When this happens, a two-image **inferior mirage** can be seen, in which the viewer perceives not only a true image of an object but also an inverted image directly below. This can be seen in Figure 17–6b, where light is reflected off the top of the tree in all directions. Some of the light is directed toward the viewer's eyes after undergoing only a small amount of refraction. This produces the regular image of the tree. But some of the reflected light is also directed toward the ground, where the steep temperature gradient causes very strong refraction. This light is refracted upward and reaches the viewer's eye from below, making the top of the tree appear below the ground and upside-down as if the viewer wcre looking at the tree's reflection on the calm surface of a pond. Light reflected off all parts of the

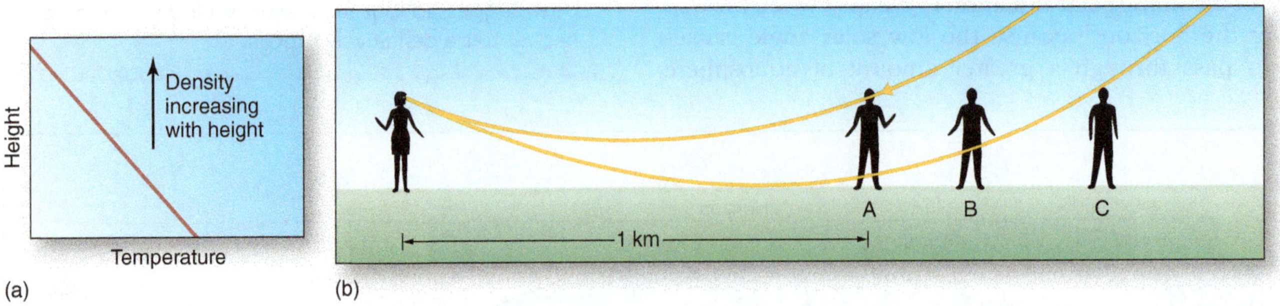

▲ **FIGURE 17–5 Upward Refraction.** (a) A steady, steep drop in temperature with height can cause a refraction pattern in which rays are bent concave upward. (b) Though there is some distortion of the perceived image, a person standing at position A can be viewed in his or her entirety by the person on the left. As the person on the right moves to position B, the visible light reflected off the lower legs does not reach the viewer on the left because it is absorbed at the ground. At position C, the person's image disappears entirely to the viewer on the left. Note the extreme vertical exaggeration of the diagram.

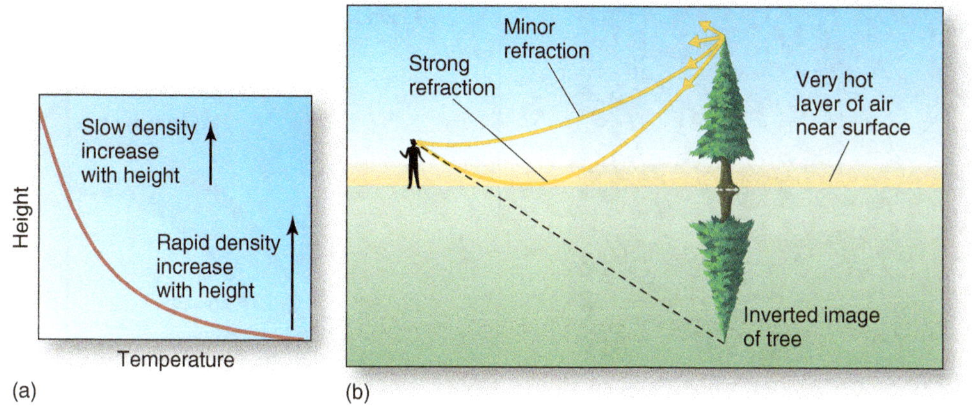

FIGURE 17–6 Inferior Mirage. (a) If there is a very steep decrease in temperature immediately near the surface along with a lesser temperature gradient just above, a two-image inferior mirage can occur. (b) Diffuse, visible radiation off the top of the tree goes in many directions. Some of the light passes directly to the viewer with minimal refraction (the upper arc from the treetop to the viewer). Some of the light approaches the surface, where intense refraction (the lower arc) gives a second image of the tree. This creates an inverted image beneath the true image.

tree undergoes this type of refraction to produce a mirror image in the "pond."

So what actually happens when one sees a mirage that resembles a puddle of water? In that case a viewer perceives an inferior mirage of the sky, which looks very similar to a water surface. This chapter's opening photo of the hot desert road provides a good example.

Note that the amount of refraction need not be the same for light leaving the top and bottom of the object. If light reflected by the bottom is refracted more strongly, the image will appear vertically stretched, or taller than reality. But if refraction increases vertically, the object will appear compressed. Obviously, the question of whether an object is stretched or compressed is independent of displacement, which is always downward for an inferior mirage.

A **superior mirage** forms when images are displaced upward. Light rays are bent concave downward as a result of decreasing density with increasing height. This is the normal situation described earlier that caused the Sun to appear higher above the horizon than it really is. But for a mirage to be noticeable, the normal density gradient must be enhanced by a temperature profile in which warm (less dense) air lies above cold air. This may happen, for example, over cool water surfaces with warmer air just above. As with inferior mirages, there may be stretching or compression of the image, depending on how refraction varies with altitude. Here, however, there is compression when refraction decreases with altitude and stretching when refraction increases vertically. In extreme situations, small surface features can be lifted and stretched, giving the appearance of floating cities or mountains. In Figure 17–7 small whitecaps are stretched to the point where they appear as small waterfalls.

CHECKPOINT

17.5 What are inferior mirages and superior mirages?

17.6 Can mirages be seen only in the desert? Explain why or why not.

Cloud and Precipitation Optics

17.2 Explain the processes that produce optical phenomena involving clouds and precipitation.

Refraction is not limited to clear air, nor is it the only process that can create interesting optical effects. In the remainder of this chapter, we will see how refraction and two other processes produce some very familiar (and not so familiar) optical effects.

Rainbows

One of the most striking features to appear in the atmosphere is the familiar **rainbow** (Figure 17–8). Rainbows are sweeping arcs of light that exhibit changes in color from the inner part of the ring to the outer part. Rainbows only appear when rain is falling some distance away and with a clear sky above and behind the viewer that allows sunlight to reach the surface unobstructed. You might have observed that rainbows are always located in exactly the opposite direction of the Sun. In other words, if the Sun is to your back in the southwest, your shadow will point toward the center of the rainbow to the northeast.

FIGURE 17–7 Superior Mirage. Whitecaps and intervening darker parts of the water surface have been stretched vertically, giving the appearance of a striped wall.

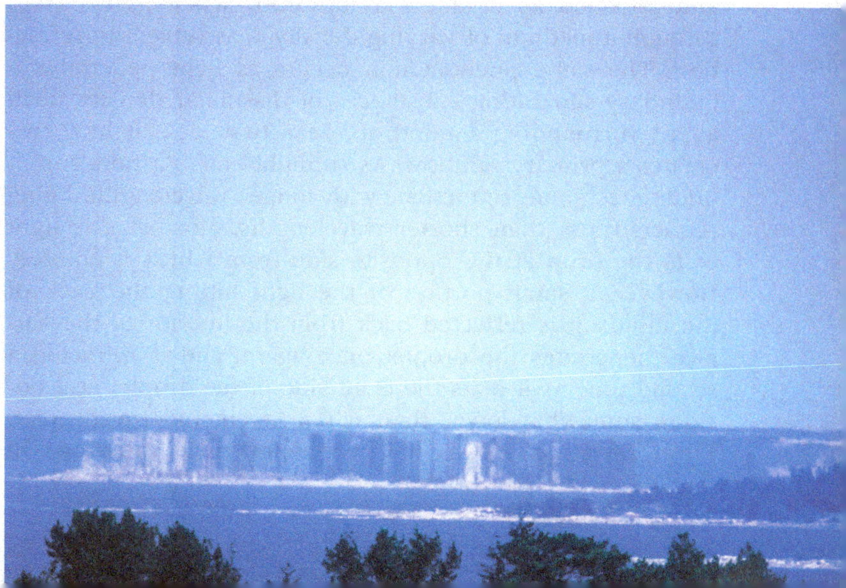

(a)

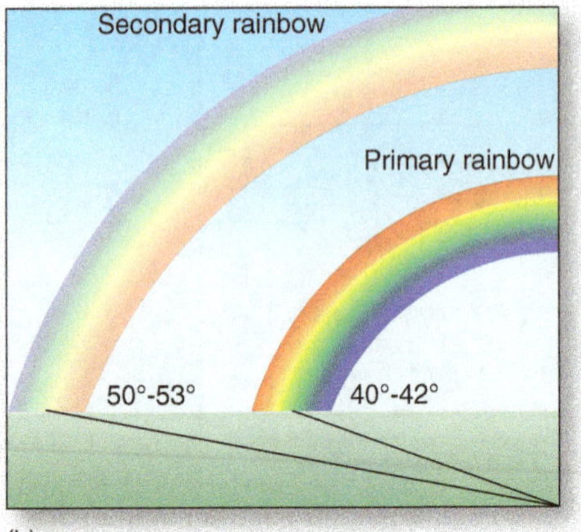

(b)

◂ **FIGURE 17–8 Rainbows in Rocky Mountain National Park.** (a) Note that the brighter primary rainbow is surrounded by a dimmer secondary rainbow with a reversed color sequence. (b) Measured from the center of their arcs, primary and secondary rainbows have an angular size of 40-42 degrees and 50-53 degrees respectively.

PRIMARY AND SECONDARY RAINBOWS The brightest and most common rainbows are **primary rainbows**. These rainbows are always the same angular size, so at the horizon the angular distance from one end to the other extends about 82 degrees of angle (for visualization purposes, think of this as an angle that extends almost all the way from due north to due east). In a primary rainbow, the shortest wavelengths of visible light (violet and blue) appear at the innermost portion of the ring, and the longer wavelengths (orange and red) frame the outermost portion. A primary rainbow is often surrounded by a less distinct **secondary rainbow** that covers about 102 degrees of arc at the horizon and has the reverse color scheme of the primary rainbow (that is, the reds appear on the inner portion of the ring and the blues on the outer portion). Of course, if the precipitation shaft is not large enough or is too far in the distance, only a partial rainbow will appear.

The big question, of course, is how do these form? The answer lies in the way in which sunlight is refracted (bent) and reflected as it enters a raindrop. When light passes through a medium of varying density, it is subject to refraction. The same phenomenon occurs as light penetrates a boundary separating substances of dissimilar density, such as air surrounding a raindrop. Let's first look at how this creates a primary rainbow. As sunlight enters a raindrop, it undergoes some refraction, with longer wavelengths being refracted less than shorter wavelengths. Most of this light exits the drop at the opposite side from which it entered. However, a small portion of the light hitting the back of the raindrop is reflected back from the interior of the surface, penetrates the droplet once again, and is refracted a second time as it exits the front side of the droplet at a position somewhat lower than where it entered the droplet. This process is shown for two hypothetical raindrops in Figure 17–9a. Because each wavelength is refracted differentially, only one particular wavelength of light exiting a raindrop is directed toward a particular viewer at a particular location. Thus the upper raindrop directs red light toward the viewer, while the lower raindrop directs violet light to that person. Because the lower raindrop appears at a lower angle above the horizon, its violet light forms the lower (inner) side of the ring, and the red light from the upper raindrop appears at the upper (outer) portion. The red light is refracted 42.3° and the violet light is refracted 40.6°. As a result, the ring is only 1.7° wide. This assumes the Sun's rays are parallel. That assumption is not strictly true because the Sun is not an infinitely small point in space. Because solar radiation originates from the entire solar disk, rainbow rings are about 2.4° wide.

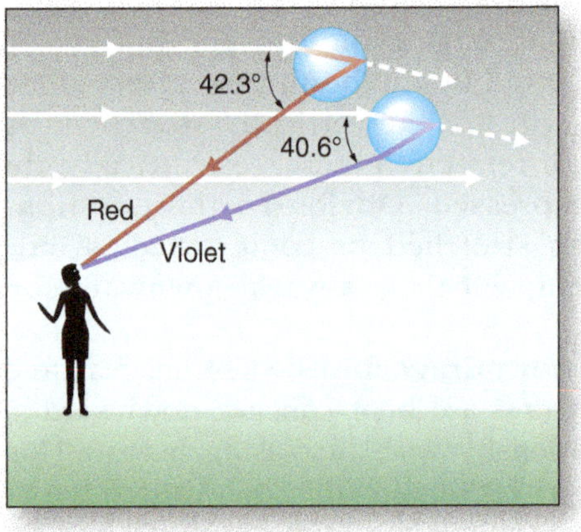

▴ **FIGURE 17–9 Primary Rainbow Reflection and Refraction.** Sunlight from behind the viewer undergoes reflection and refraction to produce a primary rainbow. The amount of total refraction is different for each wavelength, causing the familiar color separation of a rainbow.

▲ **FIGURE 17–10 Reflections for a Secondary Rainbow.** A secondary rainbow requires two reflections within raindrops. The second reflection inverts the color sequence seen in the primary rainbow.

Secondary rainbows are formed in much the same manner as are primary rainbows, except that two reflections occur at the back of the raindrop, as shown in Figure 17–10. This results in a reverse color scheme than that of the primary rainbow, with longer wavelengths of light situated on the inner portion of the band and with the shorter wavelengths on the outer portion. Sunlight exits the same side of the raindrop as it enters but upon leaving forms an angle between 50° and 53° with the original ray. Thus, secondary rainbows have an angular size of about 100° (Figure 17–9b).

THE ARC OF A RAINBOW Rainbows are obviously circular arcs of concentric colors, but why? Why, for example, can't they be horizontal bands, or elongated vertically or horizontally? To understand this, consider a single color of light reflected toward an observer as in Figure 17–11a. Every ray of that light forms the same angle with the original ray of light, perhaps 41° as for the blue light of Figure 17–11a. Finding the origin of blue light, therefore, amounts to finding all drops that can reflect light at that angle. All of the drops shown in Figure 17–11a result in the same angle of reflected light. If the angle seems different, it is because the viewing angle is different. Rays near the top of the arc are seen in vertical cross section, whereas those lower in the sky are seen more from the side.

A little more thought shows why light of a given color has to appear circular to the observer. Only some drops can reflect light with a particular angle back to the observer. For example, a raindrop directly above or below the highest drop in Figure 17–11a would reflect blue light with the same angle, but those rays would hit above or below the observer's eye and thus would not be seen. As Figure 17–11b shows, blue rays will originate from drops lying on a cone. The cone's exterior forms a 41° angle with the incoming sunlight. All of the drops on that cone send light to the observer. Looking along the cone, the observer sees it as a circular arc. Other colors will originate on other "cones" of raindrops, but with differing angles.

For any given color, the angle between the Sun's rays and the reflected light is constant. It follows that the reflected light has to change orientation with respect to an observer as the Sun moves. If the Sun moves higher in the sky, reflected rays get closer to horizontal. Eventually none of the reflected rays reaches an observer on the ground, which is why rainbows don't occur near noontime. As you probably realize, they are most common late in the afternoon and early evening. This is partly because the Sun is lower, and partly because thunderstorms are more common then.

CHECKPOINT

17.7 Describe the spatial relationship among the Sun, raindrops, and an observer for a primary rainbow.

17.8 What accounts for the fact that a primary rainbow has an angular size of about 85°?

17.9 Why are bands in a rainbow concentric circles rather than some other shape?

DID YOU KNOW?

You are never seeing *the* rainbow, you are seeing one of trillions. When the Sun shines on rain, reflections with angles needed to form a rainbow occur from every drop. This produces a vast number of potentially visible rainbows, only one of which is visible from a single position. Naturally, the one visible to you is what you call "the rainbow." A friend standing next to you sees a different rainbow, and most likely she also calls it "the rainbow."

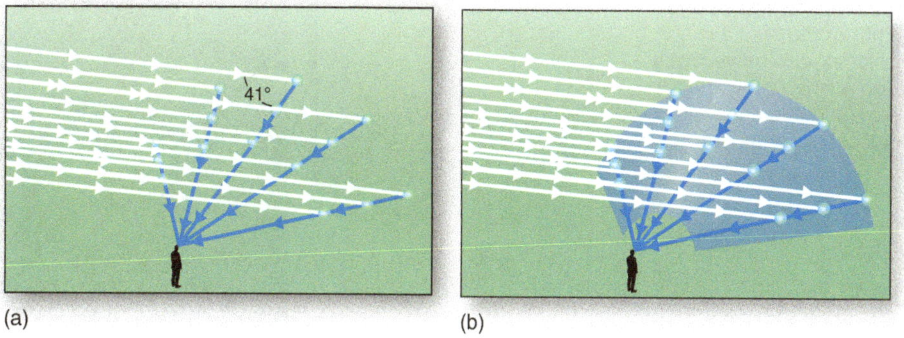

▶ **FIGURE 17–11 Arc of a Rainbow.** Blue light is reflected by all raindrops, but only rays forming a 41° angle with incoming sunlight reach the observer. (a) Reflection from a few of those drops. (b) Every drop sending blue light to the observer lies on a cone whose exterior angle is 41°. Looking along the cone, the person sees a partial circle of blue light. Other colors behave similarly, resulting in concentric circular arcs of varying colors.

Halos, Sundogs, and Sun Pillars

Cirrostratus clouds produce circular bands of light that surround the Sun or Moon, called **halos** (shown in Figure 6–19), with radii of 22° or 46° (46° halos are less common and not as bright as 22° halos). Unlike rainbows, whose appearance requires that the Sun be directly behind the viewer, halos occur when ice crystals are between the viewer and the Sun or Moon. Figure 17–12a illustrates the refraction within ice crystals that produces a 22° halo. Sunlight (or moonlight) passes through the sides of column-shaped and platelike ice crystals, in which each of the six edges form a 60° angle. The ice crystal acts as a prism that refracts the sunlight 22°. Crystals that are 22° away from the sight path of the Sun or Moon and have the necessary orientation will refract light toward the viewer. Because ice crystals are so numerous and randomly aligned within the cloud, a sufficient number will direct the light toward the observer to make the halo bright enough to be visible from the ground. Figure 17–12b shows how column-shaped ice crystals refract light at 46° when the crystal is oriented lengthwise toward the incoming light. The 46° halo can only result from column-shaped crystals, not from platelike crystals.

Platelike ice crystals larger than about 30 μm across tend to align themselves horizontally. If the Sun is slightly above the horizon and behind these crystals, bright spots appear 22° to the right and left of the Sun (Figure 17–13). These **sundogs** (or *parhelia*) often appear as whitish spots in the sky, but sometimes they exhibit color differentiation, with redder colors located on the side of the sundog nearest the sun and the blues and violets located on the outer side. Platelike crystals between a low Sun and an observer can also *reflect* (as opposed to refract) sunlight off their tops and bottoms to produce **sun pillars** (Figures 17–13 and 17–14). The many ice crystals are aligned almost, but not exactly, horizontally, with each reflecting a portion of the incoming light differently to produce the apparent columns stretching upward and downward from the Sun.

▸ **FIGURE 17–12 Refraction and Crystal Shape.** (a) Column-shaped and platelike crystals refract light to produce a 22° halo. (b) Refraction where column-shaped ice crystals have 90° angles produces a 46° halo.

Video
Weather
Satellites in Orbit

http://goo.gl/YNnyxU

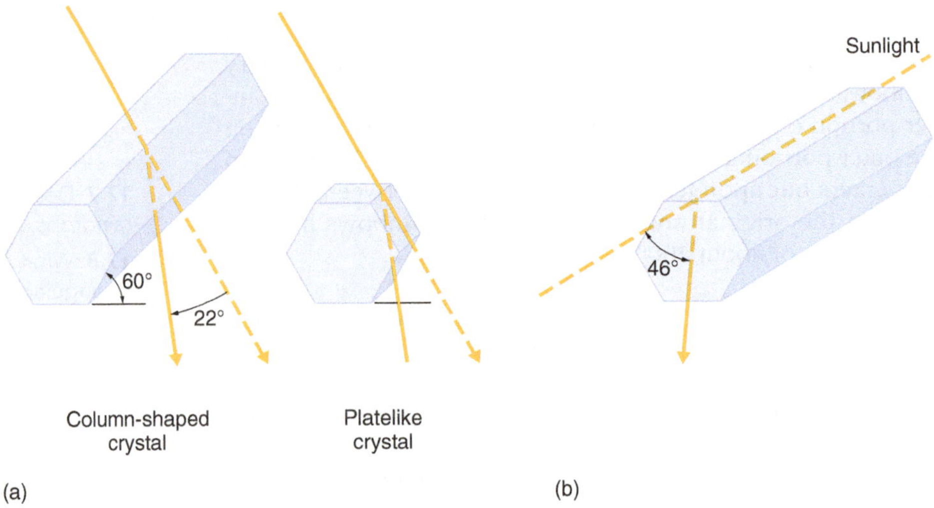

▸ **FIGURE 17–13 Sun Dogs.** (a) Nearly horizontally oriented ice crystals refract light from a setting or rising Sun to produce sundogs so named because of the way they accompany the Sun. (b) This photo taken in Manitoba, Canada has sundogs to the left and right of the Sun, a sun pillar extending vertically above and below the Sun, and a faint halo surrounding the Sun.

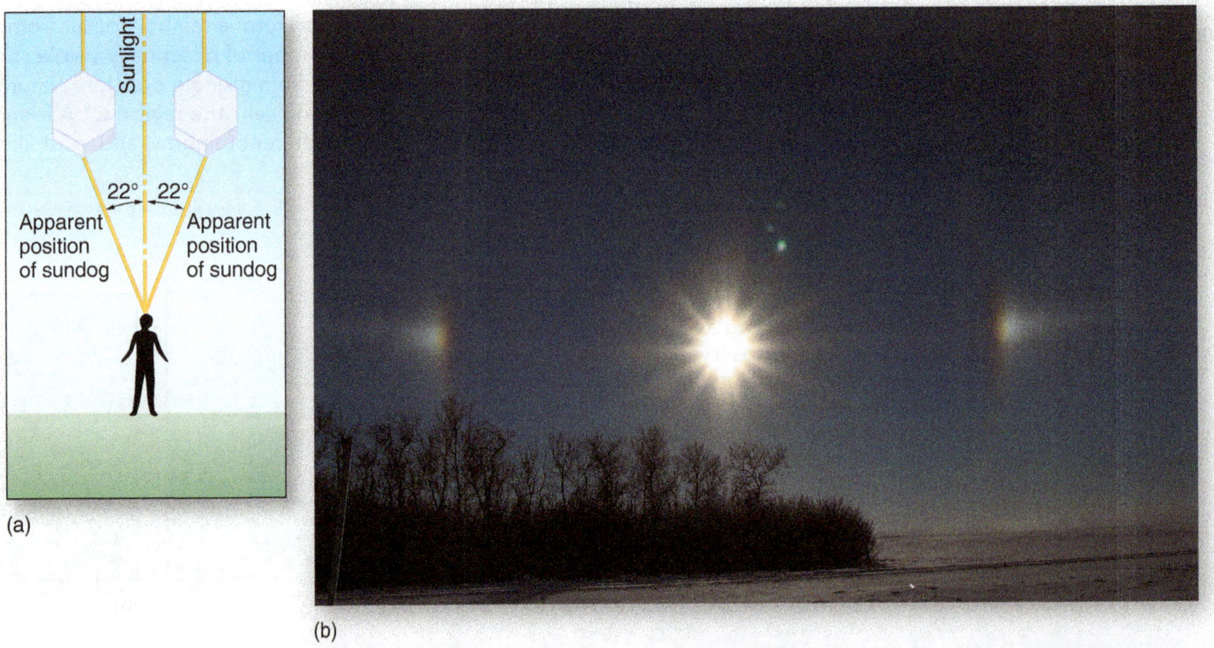

▲ **FIGURE 17–14 Sun Pillar at Voyageurs National Park, Minnesota.** These columns of bright light can occur when looking in the direction of the Sun near the horizon through clouds composed primarily of platelike crystals arranged horizontally.

▲ **FIGURE 17–15 Corona.** Clouds consisting of fairly uniform size water droplets can cause circular zones of illumination around the moon or Sun.

CHECKPOINT

17.10 What is the relationship between the observer, a cloud, and the Sun when a halo is seen?

17.11 Explain the two sizes of halos.

Coronas and Glories

Coronas and glories are optical phenomena resulting from the bending of light as it passes around water droplets (**diffraction**). The **corona** (Figure 17–15) is a circular illumination of the sky immediately surrounding the Moon—or in rarer instances, the Sun. Clouds having uniform droplet sizes cause highly circular coronas that concentrate shorter wavelength (bluish) colors on their innermost portions and longer (redder) wavelengths on their outer margins. When the cloud contains a wide assortment of droplet sizes, the illumination appears white and irregularly shaped. The size of the corona is also related to droplet size, with larger droplets producing smaller coronas.

If you are ever in an aircraft flying above a cloud deck, look for the plane's shadow on the clouds. You may see a series of rings called a **glory** (Figure 17–16). Glories have an angular size of about 5° to 20° and so are much smaller than the arcs of a rainbow (which span 85° of sky). Exactly how glories form is unknown, but it is certain they are caused by sunlight interacting with water droplets. One theory says that they are formed by interaction of light rays within and on the surface of the drop. A newer theory, supported by a detailed mathematical analysis, depends on effects from light waves that pass by but do not actually enter the drops. This seemingly impossible behavior, known as *tunneling*, is well known in physics. A complete explanation is complicated, and we won't go into the details here. Instead we will close by pointing out that glories can form around the shadow of anything, including people on high peaks above the cloud. The shadow actually has nothing to do with the glory, but rather is a nearly inescapable consequence of facing the reflecting clouds with the Sun behind.

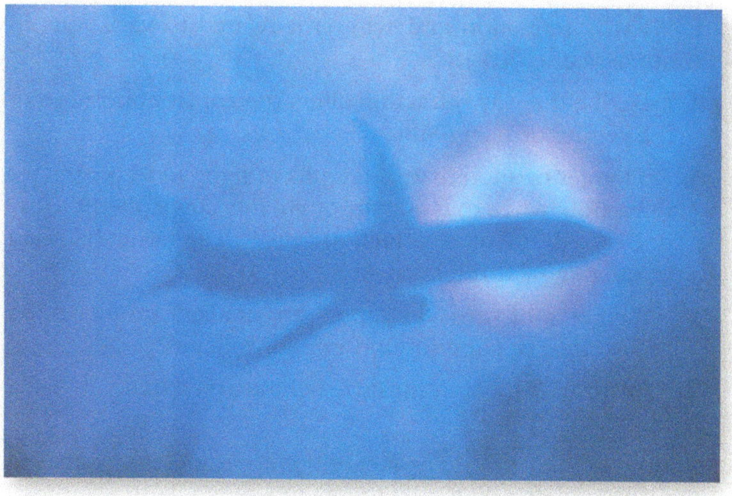

▲ **FIGURE 17–16 Glories.** Glories can appear as rings of light surrounding an aircraft's shadow on the top of a cloud.

CHECKPOINT

17.12 What factors account for the differences among halos, sun dogs, and sun pillars?

17.13 What process produces coronas and glories? Explain.

Summary

17.1 Explain atmospheric refraction and the resulting optical effects that can be seen in clear air.

- Refraction is the bending of light caused by variations in density along the light path.
- Light is bent either left or right toward the region of higher density.
- Mirages, the apparent shortening of tall objects, and the visibility of objects below the horizon are examples of phenomena caused by refraction within the atmosphere.

17.2 Explain the processes that produce optical phenomena involving clouds and precipitation.

- Light is refracted when it enters or leaves a water droplet or ice crystal.
- Rainbows form when sunlight passes over the viewer and is reflected back by raindrops. Because refraction varies with wavelength, each color originates from droplets at slightly different angles from the viewer's eye.
- Ice crystals are responsible for halos, sundogs and sun pillars. In all cases ice is between the light source (Sun or Moon) and the viewer.
- Coronas and glories involve diffraction, when light is bent as it passes around a water droplet. Coronas can form around the Sun or the Moon, and require ice between the light source and the observer. Glories rely on sunlight reflected by a cloud back toward the observer.

Key Terms

atmospheric optics *p. 543*
corona *p. 551*
diffraction *p. 551*
glory *p. 551*
green flash *p. 545*
halo *p. 550*
index of refraction *p. 544*
inferior mirage *p. 546*
mirage *p. 546*
primary rainbow *p. 548*
rainbow *p. 547*
refraction *p. 544*
secondary rainbow *p. 548*
sun pillar *p. 550*
sundog *p. 550*
superior mirage *p. 547*
twilight *p. 545*

Review Questions

1. What is refraction and why is it related to variations in atmospheric density?
2. Describe the way refraction alters the apparent position of the setting or rising Sun.
3. Do longer or shorter wavelengths of light undergo greater refraction when passing through the atmosphere? How does the differential refraction cause an apparent banding of the Sun near the horizon?
4. Which type of vertical temperature gradient promotes the appearance of superior and inferior mirages?
5. Explain why the Sun must be behind you when you see a rainbow.
6. How do some mirages create the appearance of standing water on hot days?
7. Describe the difference in the way primary and secondary rainbows form.
8. Why are rainbow bands concentric circles?
9. How does the color pattern of a secondary rainbow differ from that of a primary rainbow?
10. In addition to refraction, what process must occur within raindrops to produce a rainbow?
11. Describe the formation of sun pillars. Does refraction play a role in their formation?
12. How are sundogs formed? Describe the color patterns associated with them.
13. Explain how coronas are formed around the Sun or Moon. What factor or factors determine their size?
14. What are glories, and how are they formed? Are they the result of refraction alone or is another process also involved?

Critical Thinking

1. Consider the way the apparent position of the Sun sweeps across the sky over the course of the day (see Chapter 2). How will the period of twilight vary between summer and winter where you live? Will twilight conditions generally last longer in the tropics or in the high latitudes? After answering this question, go to Question 1 in the Problems and Exercises section and check to see if your answer was correct.
2. Can altostratus clouds produce halos? Explain why or why not.
3. Can falling ice crystals produce rainbows? Explain why or why not.
4. Which of the optical phenomena described in this chapter are most likely to occur where you live? Are they equally likely to appear at all times of the year?
5. In Chapter 3 we discussed Rayleigh, Mie, and nonselective scattering. What similarities and dissimilarities exist among those scattering processes and the optical effects caused by refraction, reflection, and diffraction that were discussed in this chapter?
6. Explain why superior mirages do not occur over land on hot, sunny days.

Problems & Exercises

1. Refer to the website http://aa.usno.navy.mil/faq/docs/RST_defs.php#top, and look up the definitions of civil, nautical, and astronomical twilight. Then use the available tables to determine the length of day where you live for March 21, June 21, September 21, and December 21 of the current year. Does the length of day show significant differences using each of the three definitions of twilight? How do these differences vary throughout the year?
2. On hot, sunny days, look for the presence of mirages. Are they equally apparent in all directions? If not, why do you think that might be the case? Also, check to see how long they remain visible. Do they still persist at sunset?

Visual Analysis

17.1. Observe the optical phenomenon shown in this image and identify the type of cloud that must be present to create that feature.

17.2. What do you know about the composition of the cloud, based on the presence of this optical feature?

17.3. What type of change in the cloud would be needed for it to produce a corona instead of the existing feature?

17.4. What conditions would have to exist with regard to the cloud composition and its orientation with regard to a viewer for it to produce a sun pillar?

APPENDIX A

Units of Measurement and Conversions

SI UNITS

I. Basic

Length	meter (m)
Mass	kilogram (kg)
Time	second (s)
Electrical Current	ampere (A)
Temperature	kelvin (K)

II. Derived

Force	newton (N = kg m/s^2)
Pressure	pascal (Pa = N/m^2)
Energy	joule (J = N m)
Power	watt (W = J/s)
Electrical Potential Difference	volt (V = J/C)
Electrical Charge	coulomb (C)

III. Some Useful Conversions

Length

1 centimeter = 0.39 inches

1 meter = 3.281 feet
= 39.37 inches

1 kilometer = 0.62 miles

1 inch = 2.54 centimeters

1 foot = 30.48 centimeters
= 0.305 meters

1 mile = 1.61 kilometers

Mass/Weight

1 gram = 0.035 ounces

1 kilogram = 2.2 pounds

1 ounce = 28.35 grams

1 pound = 0.454 kilograms

Speed

1 meter/second = 2.24 miles/hour
= 3.60 km/hour

1 mile/hour = 0.45 meters/second
= 1.61 km/hour

Temperature

Celsius Temperature = (°F − 32)/1.8
= K− 273.15

Fahrenheit Temperature = 1.8 × °C + 32

Kelvin Temperature = °C + 273.15

Energy

1 joule = 0.239 calories

1 calorie = 4.186 joules

APPENDIX B

The Standard Atmosphere

Altitude (km)	Temperature (°C)	Pressure (mb)	p/p_0*	Density (kg/m^3)	ρ/ρ_0*
30.00	−46.60	11.97	0.01	0.02	0.02
25.00	−51.60	25.49	0.03	0.04	0.03
20.00	−56.50	55.29	0.05	0.09	0.07
19.00	−56.50	64.67	0.06	0.10	0.08
18.00	−56.50	75.65	0.07	0.12	0.09
17.00	−56.50	88.49	0.09	0.14	0.12
16.00	−56.60	103.52	0.10	0.17	0.14
15.00	−56.50	121.11	0.12	0.20	0.16
14.00	−56.50	141.70	0.14	0.23	0.19
13.00	−56.50	165.79	0.16	0.27	0.22
12.00	−56.50	193.99	0.19	0.31	0.25
11.00	−56.40	226.99	0.22	0.37	0.30
10.00	−49.90	264.99	0.26	0.41	0.34
9.50	−46.70	285.84	0.28	0.44	0.36
9.00	−43.40	308.00	0.30	0.47	0.38
8.50	−40.20	331.54	0.33	0.50	0.40
8.00	−36.90	356.51	0.35	0.53	0.43
7.50	−33.70	382.99	0.38	0.56	0.45
7.00	−30.50	411.05	0.41	0.59	0.48
6.50	−27.20	440.75	0.43	0.62	0.50
6.00	−23.90	472.17	0.47	0.66	0.54
5.50	−20.70	505.39	0.50	0.70	0.57
5.00	−17.50	540.48	0.53	0.74	0.60
4.50	−14.20	577.52	0.57	0.78	0.63
4.00	−11.00	616.60	0.61	0.82	0.67
3.50	−7.70	657.80	0.65	0.86	0.70
3.00	−4.50	701.21	0.69	0.91	0.74
2.50	−1.20	746.91	0.74	0.96	0.78
2.00	2.00	795.01	0.78	1.01	0.82
1.50	5.30	845.59	0.83	1.06	0.86
1.00	8.50	898.76	0.89	1.11	0.91
0.50	11.80	954.61	0.94	1.17	0.95
0.00	15.00	1013.25	1.00	1.23	1.00

*p/p_0 = ratio of air pressure to sea level value; ρ/ρ_0 = ratio of air density to sea level value

APPENDIX C

Simplified Weather Map Symbols

Example of Simplified Station Model

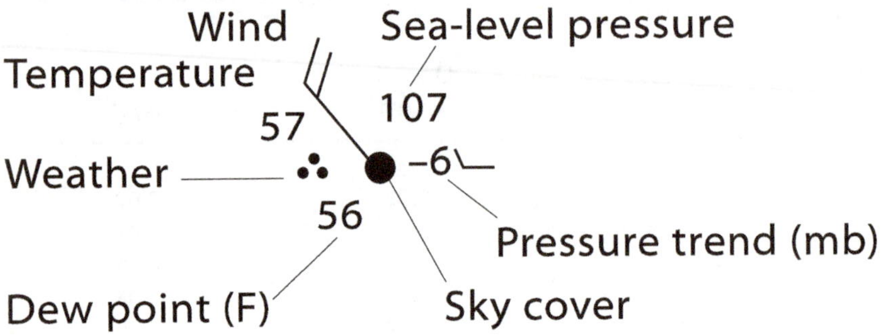

Temperature = 57° F
Wind: 15–20 mph from Northwest
Sea level pressure = 1010.7 mb
Pressure steady, then decreasing by 0.6 mb over last 3 hours
No cloud cover
Dew point = 56° F
Weather condition: moderate continuous rain

Fronts

Fronts are shown on surface weather maps by the symbols below. (Arrows—not shown on maps—indicate direction of motion of front.)

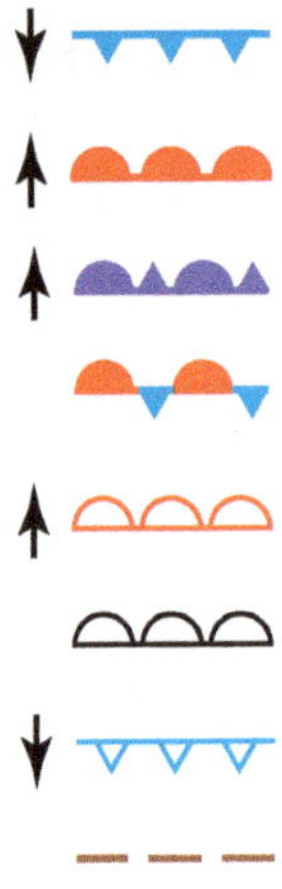

- Cold front (surface)
- Warm front (surface)
- Occluded front (surface)
- Stationary front (surface)
- Warm front (aloft)
- Dryline symbol
- Cold front (aloft)
- Upper-level trough

Air Pressure Tendency

- Rising, then falling; same as or higher than 3 hr ago
- Rising, then steady; or rising, then rising more slowly
- Rising steadily, or unsteadily
- Falling or steady, then rising; or rising, then rising more rapidly
- Steady; same as 3 hr ago

- Falling, then rising; same as or lower than 3 hr ago
- Falling, then steady; or falling, then falling more slowly
- Falling steadily, or unsteadily
- Steady or rising, then falling; or falling, then falling more rapidly

Cloud Abbreviations

St	stratus
Fra	fractus
Sc	stratocumulus
Ns	nimbostratus
As	altostratus
Ac	altocumulus
Ci	cirrus
Cs	cirrostratus
Cc	cirrocumulus
Cu	cumulus
Cb	cumulonimbus

Height of Base of Lowest Cloud

Code	*Feet*	*Meters*
0	0–149	0–49
1	150–299	50–99
2	300–599	100–199
3	600–999	200–299
4	1000–1999	300–599
5	2000–3499	600–999
6	3500–4999	1000–1499
7	5000–6499	1500–1999
8	6500–7999	2000–2499
9	8000 or above or no clouds	2500 or above or no clouds

Cloud Cover

- No clouds
- 1/8
- Scattered
- 3/8
- 4/8
- 5/8
- Broken
- 7/8
- Overcast
- Sky obscured

Cloud Types

- Cu of fair weather, little vertical development and seemingly flattened
- Cu of considerable development, generally towering, with or without other Cu or Sc, bases all at same level
- Cb with tops lacking clear-cut outlines, but distinctly not cirriform or anvil-shaped; with or without Cu, Sc, or St
- Sc formed by spreading out of Cu; Cu often present also
- Sc not formed by spreading out of Cu
- St or StFra, but no StFra of bad weather
- StFra and/or CuFra of bad weather (scud)
- Cu and Sc (not formed by spreading out of Cu) with bases at different levels
- Cb having a clearly fibrous (cirriform) top, often anvil-shaped, with or without Cu, Sc, St, or scud
- Thin As (most of cloud layer semitransparent)

- Thick As, greater part sufficiently dense to hide sun (or moon), or Ns
- Thin Ac, mostly semitransparent; cloud elements not changing much and at a single level
- Thin Ac in patches; cloud elements continually changing and/or occurring at more than one level
- Thin Ac in bands or in a layer gradually spreading over sky and usually thickening as a whole
- Ac formed by the spreading out of Cu or Cb
- Double-layered Ac, or a thick layer of Ac, not increasing; or Ac with As and/or Ns
- Ac in the form of Cu-shaped tufts or Ac with turrets
- Ac of a chaotic sky, usually at different levels; patches of dense Ci usually present
- Filaments of Ci, or "mares' tails," scattered and not increasing

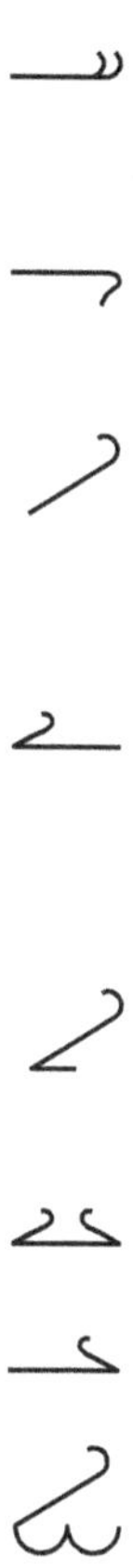

- Dense Ci in patches or twisted sheaves, usually not increasing, sometimes like remains of Cb; or towers or tufts
- Dense Ci, often anvil-shaped, derived from or associated with Cb
- Ci, often hook-shaped, gradually spreading over the sky and usually thickening as a whole
- Ci and Cs, often in converging bands, or Cs alone; generally overspreading and growing denser; the continuous layer not reaching 45° altitude
- Ci and Cs, often in converging bands, or Cs alone; generally overspreading and growing denser; the continuous layer exceeding 45° altitude
- Veil of Cs covering the entire sky
- Cs not increasing and not covering entire sky
- Cc alone or Cc with some Ci or Cs, but the Cc being the main cirriform cloud

Wind Speed

Knots (kt) per hour	*Kilometers per hour*
Calm	Calm
1–2	1–3
3–8	4–13
9–14	14–19
15–20	20–32
21–25	33–40

Wind Speed

Knots (kt) per hour	*Kilometers per hour*
26–31	41–50
32–37	51–60
38–43	61–69
44–49	70–79
50–54	80–87
55–60	88–96

Wind Speed

Knots (kt) per hour	*Kilometers per hour*
61–66	97–106
67–71	107–114
72–77	115–124
78–83	125–134
84–89	135–143
119–123	192–198

1 kt = 1.15 mph

Weather Conditions

Cloud development NOT observed or NOT observable during past hour	Clouds generally dissolving or becoming less developed during past hour	State of sky on the whole unchanged during past hour	Clouds generally forming or developing during past hour	Visibility reduced by smoke
Light fog (mist)	Patches of shallow fog at station, NOT deeper than 6 feet on land	More or less continuous shallow fog at station, NOT deeper than 6 feet on land	Lightning visible, no thunder heard	Precipitation within sight, but NOT reaching the ground
Drizzle (NOT freezing) or snow grains (NOT falling as showers) during past hour, but NOT at time of observation	Rain (NOT freezing and NOT falling as showers) during past hour, but NOT at time of observation	Snow (NOT falling as showers) during past hour, but NOT at time of observation	Rain and snow or ice pellets (NOT falling as showers) during past hour, but NOT at time of observation	Freezing drizzle or freezing rain (NOT falling as showers) during past hour, but NOT at time of observation
Slight or moderate dust storm or sandstorm, has decreased during past hour	Slight or moderate dust storm or sandstorm, no appreciable change during past hour	Slight or moderate dust storm or sandstorm has begun or increased during past hour	Severe dust storm or sandstorm, has decreased during past hour	Severe dust storm or sandstorm, no appreciable change during past hour
Fog or ice fog at distance at time of observation, but NOT at station during past hour	Fog or ice fog in patches	Fog or ice fog, sky discernible, has become thinner during past hour	Fog or ice fog, sky NOT discernible, has become thinner during past hour	Fog or ice fog, sky discernible, no appreciable change during past hour
Intermittent fall of snowflakes, slight at time of observation	Continuous fall of snowflakes, slight at time of observation	Intermittent fall of snowflakes, moderate at time of observation	Continuous fall of snowflakes, moderate at time of observation	Intermittent fall of snowflakes, heavy at time of observation
Continuous fall of snowflakes, heavy at time of observation	Ice prisms (with or without fog)	Snow grains (with or without fog)	Isolated starlike snow crystals (with or without fog)	Ice pellets or snow pellets
Slight snow shower(s)	Moderate or heavy snow shower(s)	Slight shower(s) of snow pellets, or ice pellets with or without rain, or rain and snow mixed	Moderate or heavy shower(s) of snow pellets, or ice pellets, or ice pellets with or without rain or rain and snow mixed	Slight shower(s) of hail, with or without rain or rain and snow mixed, not associated with thunder
Slight or moderate thunderstorm without hail, but with rain, and/or snow at time of observation	Slight or moderate thunderstorm, with hail at time of observation	Heavy thunderstorm, without hail, but with rain and/or snow at time of observation	Thunderstorm combined with dust storm or sandstorm at time of observation	Heavy thunderstorm with hail at time of observation
Moderate or heavy shower(s) of hail, with or without rain, or rain and snow mixed, not associated with thunder	Slight rain at time of observation; thunderstorm during past hour, but NOT at time of observation	Moderate or heavy rain at time of observation; thunderstorm during past hour, but NOT at time of observation	Slight snow, or rain and snow mixed, or hail at time of observation; thunderstorm during past hour, but NOT at time of observation	Moderate or heavy snow, or rain and snow mixed, or hail at time of observation; thunderstorm during past hour, but NOT at time of observation

Weather Conditions

Haze	Widespread dust in suspension in the air, NOT raised by wind, at time of observation	Dust or sand raised by wind at time of observation	Well-developed dust whirl(s) within past hour	Dust storm or sandstorm within sight of or at station during past hour
Precipitation within sight, reaching the ground but distant from station	Precipitation within sight, reaching the ground, near to but NOT at station	Thunderstorm, but no precipitation at the station	Squall(s) within sight during past hour or at time of observation	Funnel cloud(s) within sight of station at time of observation
Showers of rain during past hour, but NOT at time of observation	Showers of snow, or of rain and snow, during past hour, but NOT at time of observation	Showers of hail, or of hail and rain, during past hour, but NOT at time of observation	Fog during past hour, but NOT at time of observation	Thunderstorm (with or without precipitation) during past hour, but NOT at time of observation
Severe dust storm or sandstorm has begun or increased during past hour	Slight or moderate drifting snow, generally low (less than 6 ft)	Heavy drifting snow, generally low	Slight or moderate blowing snow, generally high (more than 6 ft)	Heavy blowing snow, generally high
Fog or ice fog, sky NOT discernible, no appreciable change during past hour	Fog or ice fog, sky discernible, has begun or become thicker during past hour	Fog or ice fog, sky NOT discernible, has begun or become thicker during past hour	Fog depositing rime, sky discernible	Fog depositing rime, sky NOT discernible
Continuous drizzle (NOT freezing), heavy at time of observation	Slight freezing drizzle	Moderate or heavy freezing drizzle	Drizzle and rain, slight	Drizzle and rain, moderate or heavy
Intermittent drizzle (NOT freezing), slight at time of observation	Continuous drizzle (NOT freezing), slight at time of observation	Intermittent drizzle (NOT freezing), moderate at time of observation	Continuous drizzle (NOT freezing), moderate at time of observation	Intermittent drizzle (NOT freezing), heavy at time of observation
Continuous rain (NOT freezing), heavy at time of observation	Slight freezing rain	Moderate or heavy freezing rain	Rain or drizzle and snow, slight	Rain or drizzle and snow, moderate or heavy
Intermittent rain (NOT freezing), slight at time of observation	Continuous rain (NOT freezing), slight at time of observation	Intermittent rain (NOT freezing), moderate at time of observation	Continuous rain (NOT freezing), moderate at time of observation	Intermittent rain (NOT freezing), heavy at time of observation
Slight rain shower(s)	Moderate or heavy rain shower(s)	Violent rain shower(s)	Slight shower(s) of rain and snow mixed	Moderate or heavy shower(s) of rain and snow mixed
Slight or moderate thunderstorm without hail, but with rain, and/or snow at time of observation	Slight or moderate thunderstorm, with hail at time of observation	Heavy thunderstorm, without hail, but with rain and/or snow at time of observation	Thunderstorm combined with dust storm or sandstorm at time of observation	Heavy thunderstorm with hail at time of observation

APPENDIX D

Weather Extremes

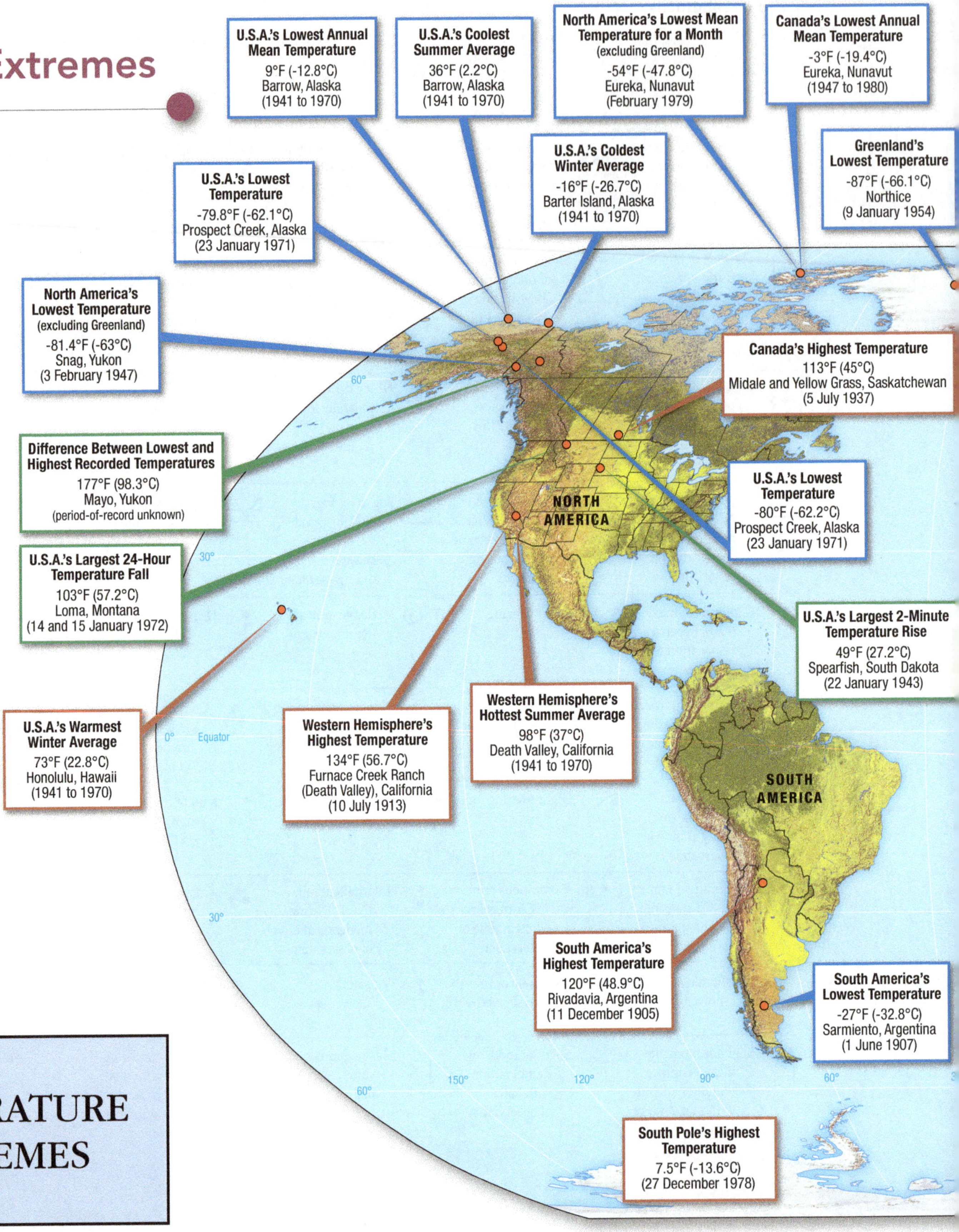
U.S.A.'s Lowest Annual Mean Temperature
9°F (-12.8°C)
Barrow, Alaska
(1941 to 1970)
U.S.A.'s Coolest Summer Average
36°F (2.2°C)
Barrow, Alaska
(1941 to 1970)
North America's Lowest Mean Temperature for a Month
(excluding Greenland)
-54°F (-47.8°C)
Eureka, Nunavut
(February 1979)
Canada's Lowest Annual Mean Temperature
-3°F (-19.4°C)
Eureka, Nunavut
(1947 to 1980)
U.S.A.'s Coldest Winter Average
-16°F (-26.7°C)
Barter Island, Alaska
(1941 to 1970)
Greenland's Lowest Temperature
-87°F (-66.1°C)
Northice
(9 January 1954)
U.S.A.'s Lowest Temperature
-79.8°F (-62.1°C)
Prospect Creek, Alaska
(23 January 1971)
North America's Lowest Temperature
(excluding Greenland)
-81.4°F (-63°C)
Snag, Yukon
(3 February 1947)
Canada's Highest Temperature
113°F (45°C)
Midale and Yellow Grass, Saskatchewan
(5 July 1937)
Difference Between Lowest and Highest Recorded Temperatures
177°F (98.3°C)
Mayo, Yukon
(period-of-record unknown)
U.S.A.'s Lowest Temperature
-80°F (-62.2°C)
Prospect Creek, Alaska
(23 January 1971)
NORTH AMERICA
U.S.A.'s Largest 24-Hour Temperature Fall
103°F (57.2°C)
Loma, Montana
(14 and 15 January 1972)
U.S.A.'s Largest 2-Minute Temperature Rise
49°F (27.2°C)
Spearfish, South Dakota
(22 January 1943)
U.S.A.'s Warmest Winter Average
73°F (22.8°C)
Honolulu, Hawaii
(1941 to 1970)
Western Hemisphere's Highest Temperature
134°F (56.7°C)
Furnace Creek Ranch
(Death Valley), California
(10 July 1913)
Western Hemisphere's Hottest Summer Average
98°F (37°C)
Death Valley, California
(1941 to 1970)
SOUTH AMERICA
South America's Highest Temperature
120°F (48.9°C)
Rivadavia, Argentina
(11 December 1905)
South America's Lowest Temperature
-27°F (-32.8°C)
Sarmiento, Argentina
(1 June 1907)
South Pole's Highest Temperature
7.5°F (-13.6°C)
(27 December 1978)
60°
30°
0°
Equator
30°
60°
150°
120°
90°
60°

TEMPERATURE EXTREMES

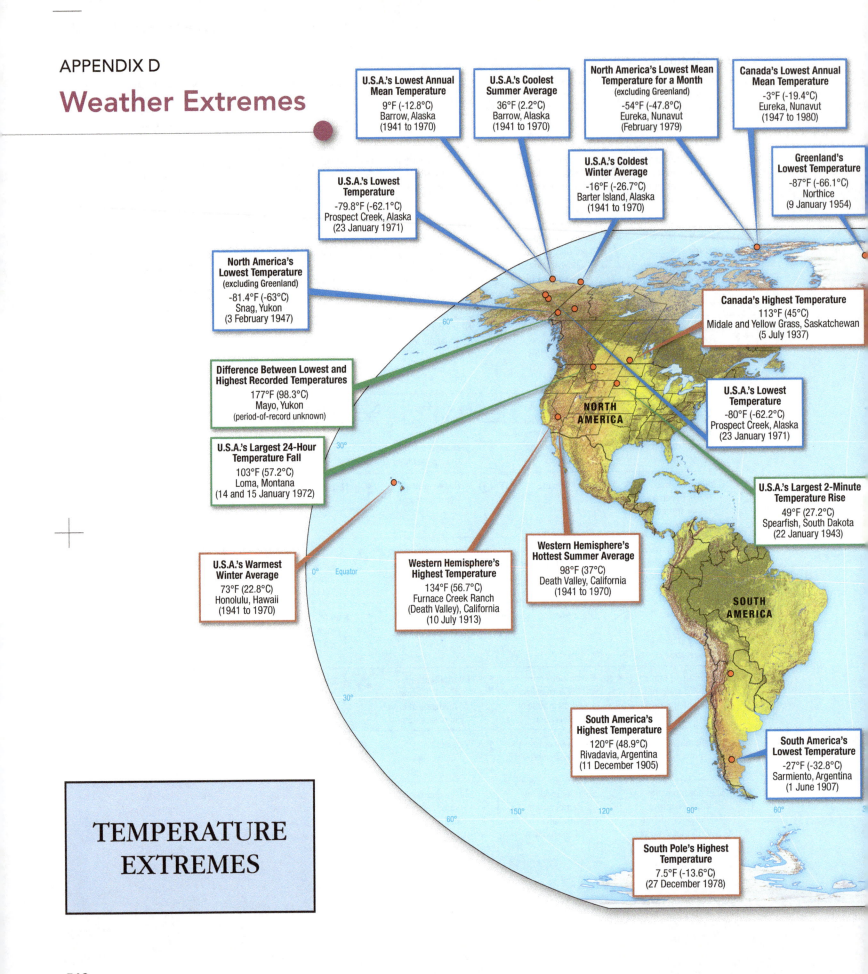

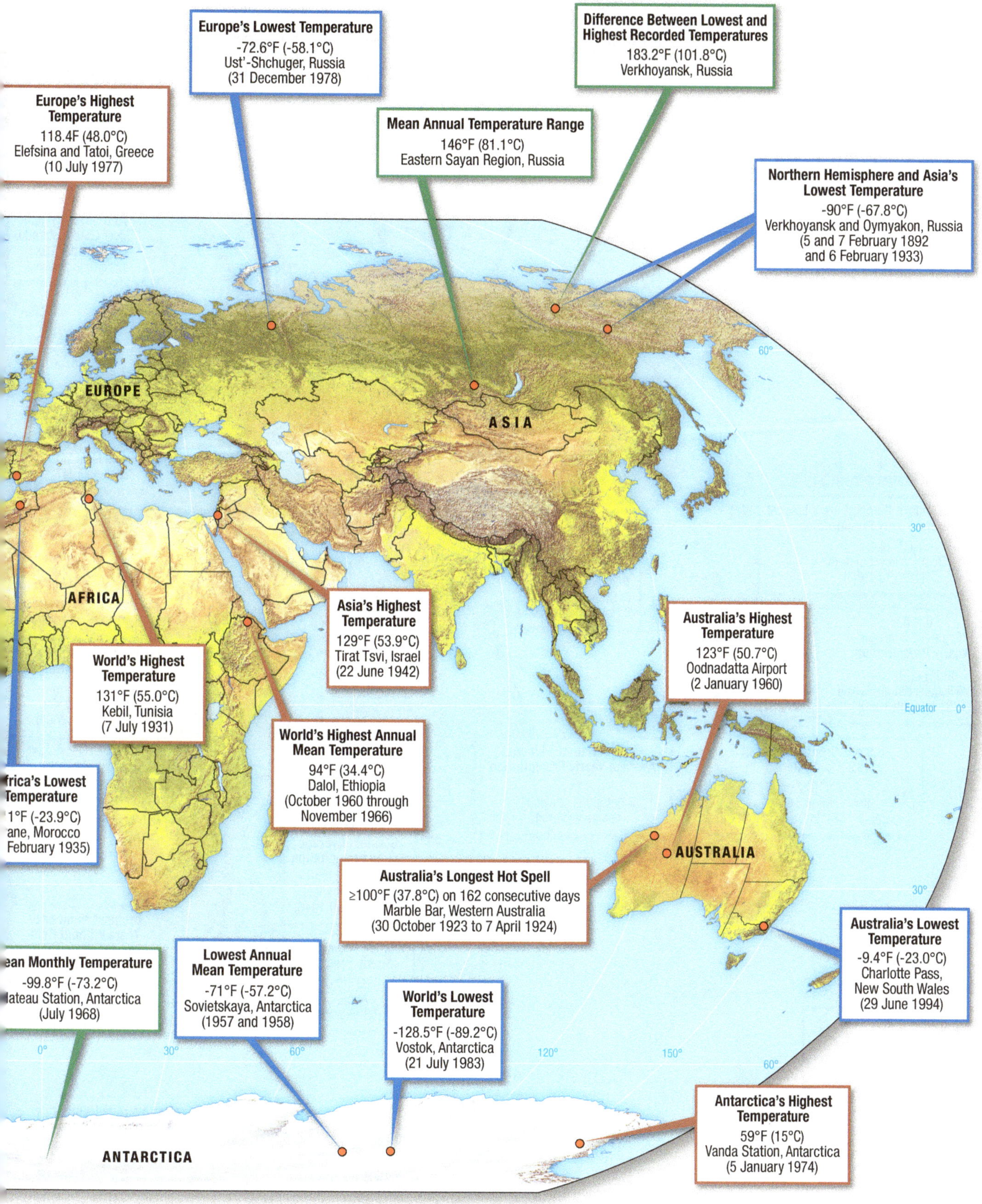
Europe's Lowest Temperature
-72.6°F (-58.1°C)
Ust'-Shchuger, Russia
(31 December 1978)
Difference Between Lowest and Highest Recorded Temperatures
183.2°F (101.8°C)
Verkhoyansk, Russia
Europe's Highest Temperature
118.4F (48.0°C)
Elefsina and Tatoi, Greece
(10 July 1977)
Mean Annual Temperature Range
146°F (81.1°C)
Eastern Sayan Region, Russia
Northern Hemisphere and Asia's Lowest Temperature
-90°F (-67.8°C)
Verkhoyansk and Oymyakon, Russia
(5 and 7 February 1892
and 6 February 1933)
EUROPE
ASIA
AFRICA
Asia's Highest Temperature
129°F (53.9°C)
Tirat Tsvi, Israel
(22 June 1942)
Australia's Highest Temperature
123°F (50.7°C)
Oodnadatta Airport
(2 January 1960)
World's Highest Temperature
131°F (55.0°C)
Kebil, Tunisia
(7 July 1931)
World's Highest Annual Mean Temperature
94°F (34.4°C)
Dalol, Ethiopia
(October 1960 through
November 1966)
frica's Lowest Temperature
1°F (-23.9°C)
ane, Morocco
February 1935)
AUSTRALIA
Australia's Longest Hot Spell
≥100°F (37.8°C) on 162 consecutive days
Marble Bar, Western Australia
(30 October 1923 to 7 April 1924)
Australia's Lowest Temperature
-9.4°F (-23.0°C)
Charlotte Pass,
New South Wales
(29 June 1994)
ean Monthly Temperature
-99.8°F (-73.2°C)
lateau Station, Antarctica
(July 1968)
Lowest Annual Mean Temperature
-71°F (-57.2°C)
Sovietskaya, Antarctica
(1957 and 1958)
World's Lowest Temperature
-128.5°F (-89.2°C)
Vostok, Antarctica
(21 July 1983)
Antarctica's Highest Temperature
59°F (15°C)
Vanda Station, Antarctica
(5 January 1974)
ANTARCTICA
60°
30°
Equator 0°
30°
60°
0°
30°
60°
120°
150°

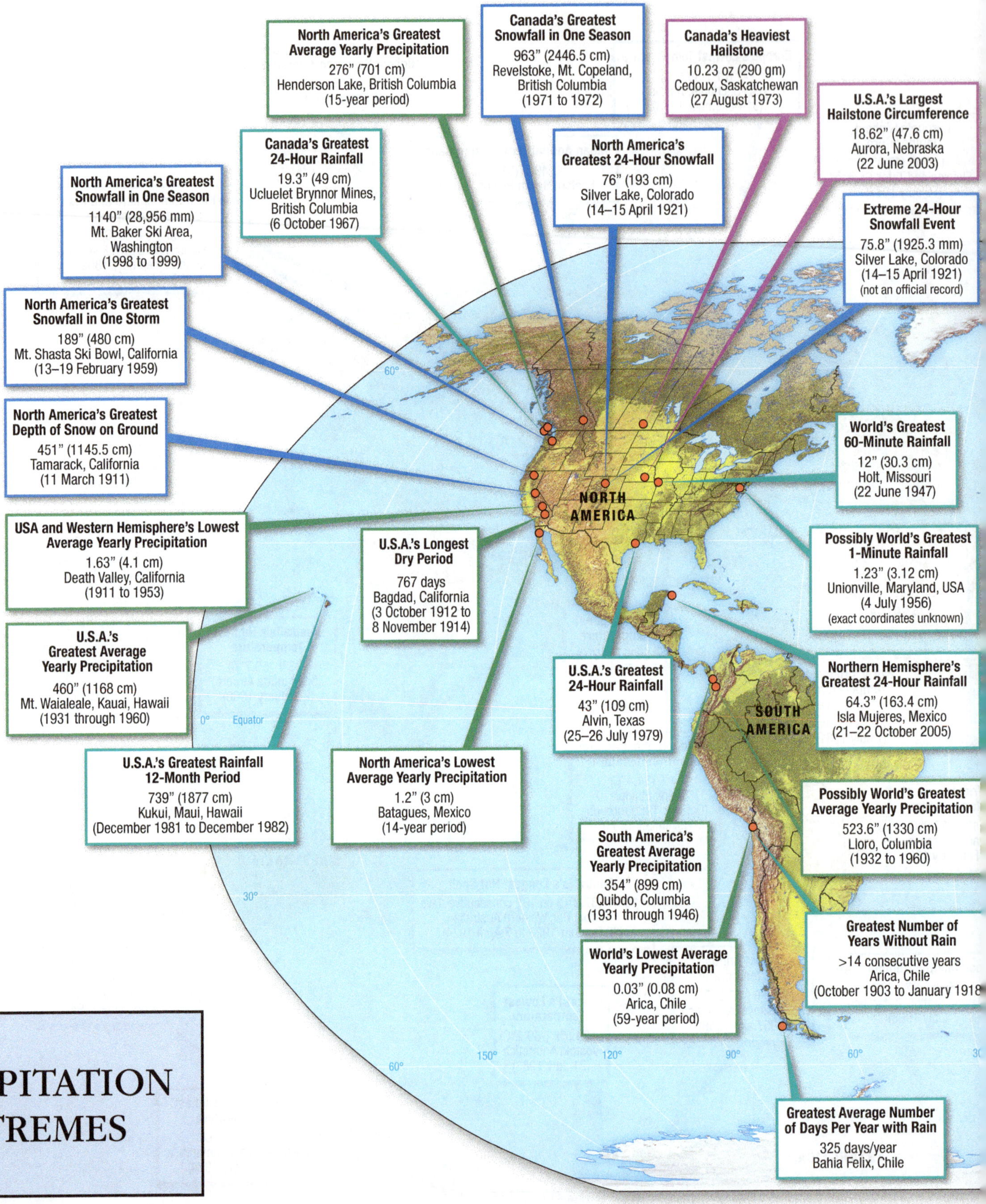
North America's Greatest Average Yearly Precipitation
276" (701 cm)
Henderson Lake, British Columbia
(15-year period)
Canada's Greatest Snowfall in One Season
963" (2446.5 cm)
Revelstoke, Mt. Copeland, British Columbia
(1971 to 1972)
Canada's Heaviest Hailstone
10.23 oz (290 gm)
Cedoux, Saskatchewan
(27 August 1973)
U.S.A.'s Largest Hailstone Circumference
18.62" (47.6 cm)
Aurora, Nebraska
(22 June 2003)
Canada's Greatest 24-Hour Rainfall
19.3" (49 cm)
Ucluelet Brynnor Mines, British Columbia
(6 October 1967)
North America's Greatest 24-Hour Snowfall
76" (193 cm)
Silver Lake, Colorado
(14–15 April 1921)
North America's Greatest Snowfall in One Season
1140" (28,956 mm)
Mt. Baker Ski Area, Washington
(1998 to 1999)
Extreme 24-Hour Snowfall Event
75.8" (1925.3 mm)
Silver Lake, Colorado
(14–15 April 1921)
(not an official record)
North America's Greatest Snowfall in One Storm
189" (480 cm)
Mt. Shasta Ski Bowl, California
(13–19 February 1959)
60°
North America's Greatest Depth of Snow on Ground
451" (1145.5 cm)
Tamarack, California
(11 March 1911)
World's Greatest 60-Minute Rainfall
12" (30.3 cm)
Holt, Missouri
(22 June 1947)
NORTH AMERICA
USA and Western Hemisphere's Lowest Average Yearly Precipitation
1.63" (4.1 cm)
Death Valley, California
(1911 to 1953)
U.S.A.'s Longest Dry Period
767 days
Bagdad, California
(3 October 1912 to 8 November 1914)
Possibly World's Greatest 1-Minute Rainfall
1.23" (3.12 cm)
Unionville, Maryland, USA
(4 July 1956)
(exact coordinates unknown)
U.S.A.'s Greatest Average Yearly Precipitation
460" (1168 cm)
Mt. Waialeale, Kauai, Hawaii
(1931 through 1960)
0° Equator
U.S.A.'s Greatest 24-Hour Rainfall
43" (109 cm)
Alvin, Texas
(25–26 July 1979)
SOUTH AMERICA
Northern Hemisphere's Greatest 24-Hour Rainfall
64.3" (163.4 cm)
Isla Mujeres, Mexico
(21–22 October 2005)
U.S.A.'s Greatest Rainfall 12-Month Period
739" (1877 cm)
Kukui, Maui, Hawaii
(December 1981 to December 1982)
North America's Lowest Average Yearly Precipitation
1.2" (3 cm)
Batagues, Mexico
(14-year period)
Possibly World's Greatest Average Yearly Precipitation
523.6" (1330 cm)
Lloro, Columbia
(1932 to 1960)
South America's Greatest Average Yearly Precipitation
354" (899 cm)
Quibdo, Columbia
(1931 through 1946)
30°
Greatest Number of Years Without Rain
>14 consecutive years
Arica, Chile
(October 1903 to January 1918
World's Lowest Average Yearly Precipitation
0.03" (0.08 cm)
Arica, Chile
(59-year period)
60°
150°
120°
90°
60°
Greatest Average Number of Days Per Year with Rain
325 days/year
Bahia Felix, Chile

PRECIPITATION EXTREMES

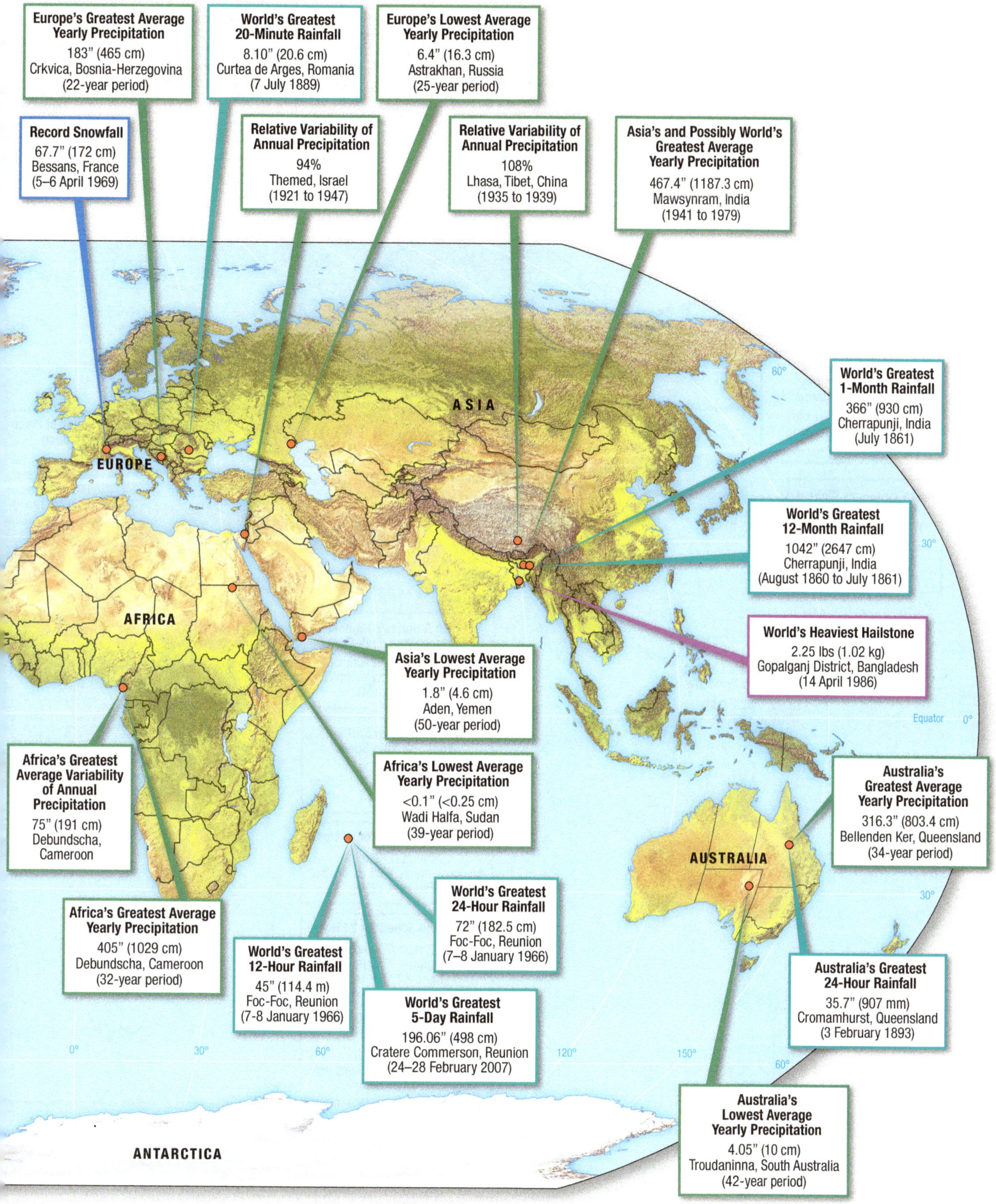
Europe's Greatest Average Yearly Precipitation
183" (465 cm)
Crkvica, Bosnia-Herzegovina
(22-year period)
World's Greatest 20-Minute Rainfall
8.10" (20.6 cm)
Curtea de Arges, Romania
(7 July 1889)
Europe's Lowest Average Yearly Precipitation
6.4" (16.3 cm)
Astrakhan, Russia
(25-year period)
Record Snowfall
67.7" (172 cm)
Bessans, France
(5–6 April 1969)
Relative Variability of Annual Precipitation
94%
Themed, Israel
(1921 to 1947)
Relative Variability of Annual Precipitation
108%
Lhasa, Tibet, China
(1935 to 1939)
Asia's and Possibly World's Greatest Average Yearly Precipitation
467.4" (1187.3 cm)
Mawsynram, India
(1941 to 1979)
World's Greatest 1-Month Rainfall
366" (930 cm)
Cherrapunji, India
(July 1861)
World's Greatest 12-Month Rainfall
1042" (2647 cm)
Cherrapunji, India
(August 1860 to July 1861)
World's Heaviest Hailstone
2.25 lbs (1.02 kg)
Gopalganj District, Bangladesh
(14 April 1986)
Asia's Lowest Average Yearly Precipitation
1.8" (4.6 cm)
Aden, Yemen
(50-year period)
Africa's Lowest Average Yearly Precipitation
<0.1" (<0.25 cm)
Wadi Halfa, Sudan
(39-year period)
Africa's Greatest Average Variability of Annual Precipitation
75" (191 cm)
Debundscha, Cameroon
Africa's Greatest Average Yearly Precipitation
405" (1029 cm)
Debundscha, Cameroon
(32-year period)
World's Greatest 12-Hour Rainfall
45" (114.4 m)
Foc-Foc, Reunion
(7-8 January 1966)
World's Greatest 24-Hour Rainfall
72" (182.5 cm)
Foc-Foc, Reunion
(7–8 January 1966)
World's Greatest 5-Day Rainfall
196.06" (498 cm)
Cratere Commerson, Reunion
(24–28 February 2007)
Australia's Greatest Average Yearly Precipitation
316.3" (803.4 cm)
Bellenden Ker, Queensland
(34-year period)
Australia's Greatest 24-Hour Rainfall
35.7" (907 mm)
Cromamhurst, Queensland
(3 February 1893)
Australia's Lowest Average Yearly Precipitation
4.05" (10 cm)
Troudaninna, South Australia
(42-year period)
EUROPE
ASIA
AFRICA
AUSTRALIA
ANTARCTICA
Equator 0°
60°
30°
30°
60°
0°
30°
60°
120°
150°

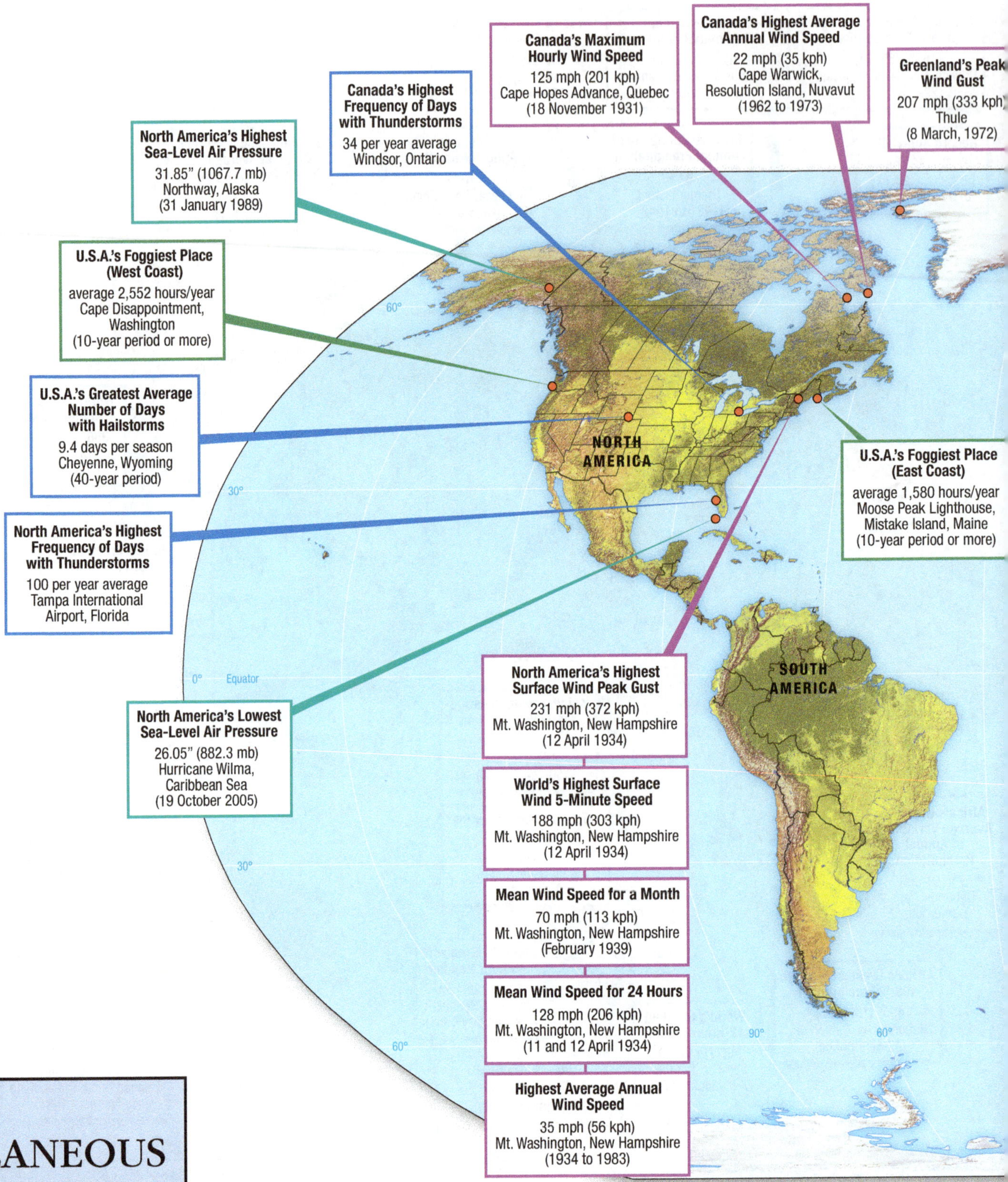
Canada's Maximum Hourly Wind Speed
125 mph (201 kph) Cape Hopes Advance, Quebec (18 November 1931)
Canada's Highest Average Annual Wind Speed
22 mph (35 kph) Cape Warwick, Resolution Island, Nuvavut (1962 to 1973)
Greenland's Peak Wind Gust
207 mph (333 kph) Thule (8 March, 1972)
Canada's Highest Frequency of Days with Thunderstorms
34 per year average Windsor, Ontario
North America's Highest Sea-Level Air Pressure
31.85" (1067.7 mb) Northway, Alaska (31 January 1989)
U.S.A.'s Foggiest Place (West Coast)
average 2,552 hours/year Cape Disappointment, Washington (10-year period or more)
U.S.A.'s Greatest Average Number of Days with Hailstorms
9.4 days per season Cheyenne, Wyoming (40-year period)
North America's Highest Frequency of Days with Thunderstorms
100 per year average Tampa International Airport, Florida
U.S.A.'s Foggiest Place (East Coast)
average 1,580 hours/year Moose Peak Lighthouse, Mistake Island, Maine (10-year period or more)
NORTH AMERICA
SOUTH AMERICA
60°
30°
0° Equator
30°
60°
90°
60°
North America's Lowest Sea-Level Air Pressure
26.05" (882.3 mb) Hurricane Wilma, Caribbean Sea (19 October 2005)
North America's Highest Surface Wind Peak Gust
231 mph (372 kph) Mt. Washington, New Hampshire (12 April 1934)
World's Highest Surface Wind 5-Minute Speed
188 mph (303 kph) Mt. Washington, New Hampshire (12 April 1934)
Mean Wind Speed for a Month
70 mph (113 kph) Mt. Washington, New Hampshire (February 1939)
Mean Wind Speed for 24 Hours
128 mph (206 kph) Mt. Washington, New Hampshire (11 and 12 April 1934)
Highest Average Annual Wind Speed
35 mph (56 kph) Mt. Washington, New Hampshire (1934 to 1983)

MISCELLANEOUS EXTREMES

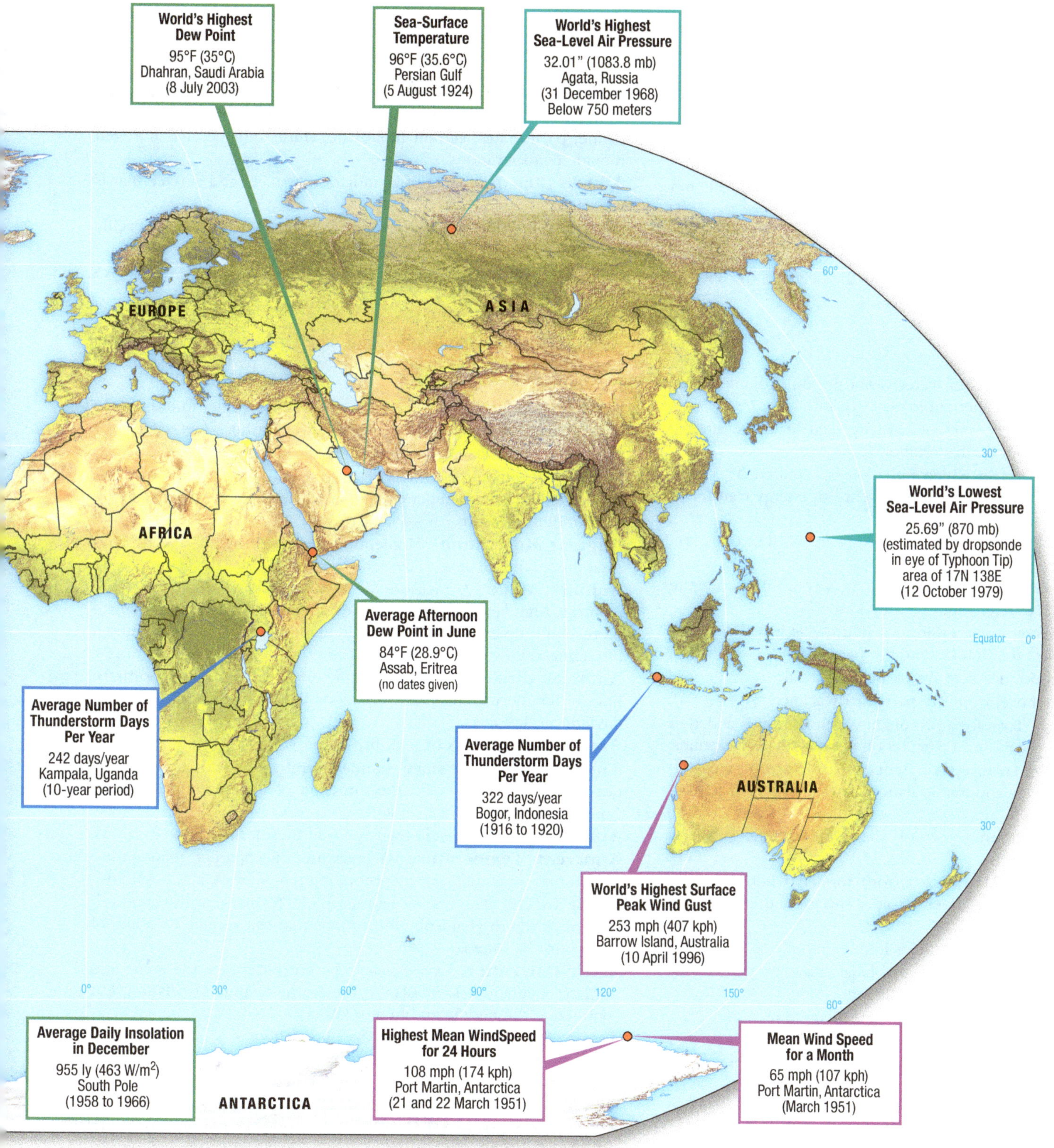
World's Highest Dew Point
95°F (35°C)
Dhahran, Saudi Arabia
(8 July 2003)
Sea-Surface Temperature
96°F (35.6°C)
Persian Gulf
(5 August 1924)
World's Highest Sea-Level Air Pressure
32.01" (1083.8 mb)
Agata, Russia
(31 December 1968)
Below 750 meters
EUROPE
ASIA
AFRICA
AUSTRALIA
ANTARCTICA
World's Lowest Sea-Level Air Pressure
25.69" (870 mb)
(estimated by dropsonde in eye of Typhoon Tip)
area of 17N 138E
(12 October 1979)
Average Afternoon Dew Point in June
84°F (28.9°C)
Assab, Eritrea
(no dates given)
Average Number of Thunderstorm Days Per Year
242 days/year
Kampala, Uganda
(10-year period)
Average Number of Thunderstorm Days Per Year
322 days/year
Bogor, Indonesia
(1916 to 1920)
World's Highest Surface Peak Wind Gust
253 mph (407 kph)
Barrow Island, Australia
(10 April 1996)
Average Daily Insolation in December
955 ly (463 W/m²)
South Pole
(1958 to 1966)
Highest Mean WindSpeed for 24 Hours
108 mph (174 kph)
Port Martin, Antarctica
(21 and 22 March 1951)
Mean Wind Speed for a Month
65 mph (107 kph)
Port Martin, Antarctica
(March 1951)
Equator
0°
30°
60°
90°
120°
150°

Glossary

200 mb Map A weather map that depicts the height of the 200 mb level.
300 mb Map A weather map that depicts the height of the 300 mb level.
500 mb Map A weather map that depicts the height of the 500 mb level.
700 mb Map A weather map that depicts the height of the 700 mb level.
850 mb Map A weather map that depicts the height of the 850 mb level.

A

Absolute Humidity Mass of water vapor per unit volume of air, usually expressed in grams per cubic meter (g/m^3).
Absolute Vorticity The sum of vorticity relative to the surface and vorticity arising from Earth's rotation.
Absorption A process in which radiation is captured by a molecule. Unlike reflection, absorption represents an energy transfer to the absorbing molecule.
Acceleration A change in velocity: a change in speed or direction, or both.
Acceleration of Gravity Acting alone, gravity would accelerate all objects by the same rate, about 9.8 m/sec/sec.
Accretion The growth of a falling ice particle as it collides with nearby water droplets that freeze onto the particle.
Acid Fog Fog whose droplets are particularly acidic.
Acid Rain Rain whose drops are particularly acidic.
Adiabatic Process Term for a process in which no heat is added or removed. For example, a rising air parcel cools adiabatically as it expands.
Advanced Weather Interactive Processing System (AWIPS) A system for the display and manipulation of weather information at Weather Service Offices.
Advection Horizontal transport of some atmospheric property (heat, moisture, etc.).
Advection Fogs Very low cloud (fog) formed when relatively warm air moves (is advected) over a colder surface and is cooled to the dew point.
Aerosols Small, suspended particles in the atmosphere.
Aerosol Brown Cloud Tropospheric haze caused by particulates absorbing and back-scattering solar radiation.
Aerovane Device for measuring wind speed and direction.
Aggregation The process in which ice crystals join together to form snowflakes. If these snowflakes melt as they fall, they arrive as rain.
Air Mass A large body of air having little horizontal variation in temperature and moisture.
Air Mass Thunderstorms Relatively small, short-lived thunderstorms that do not produce very strong winds, large hail, or tornadoes.
Air Parcels Boxes or bundles of air conceived as not able to exchange energy or matter with their surroundings.
Air Pollution Gaseous and particulate contamination of the atmosphere.
Air Quality Index (AQI) An index used for air pollution monitoring across the United States. It is based on a formula incorporating ozone, particulates, carbon monoxide, sulfur dioxide, and nitrogen dioxide concentrations.
Albedo The fraction of solar radiation arriving at a surface that is reflected.
Aleutian Low A semipermanent cell found in the North Pacific in winter.
Altocumulus A midlevel layered cloud with some rolls or patches of vertical development.
Altostratus A midlevel layered cloud.
Analog Approach A forecasting approach based on comparing current conditions to similar conditions in the past and the weather that followed.
Analysis Phase The first step in numerical weather forecasting in which data are assembled and corrected for further processing.
Anemometer Device for measuring wind speed.
Aneroid Barometer A device used to measure air pressure. The barometer's elastic chamber expands and contracts in response to the surrounding pressure.
Angular Momentum Momentum possessed by an object that is rotating around an axis.
Annular Mode A measure of hemispheric wind and pressure patterns. Annular modes exist in both hemispheres and encapsulate more extratropical variability in circulation than any other known phenomenon.
Antarctic Circle The line of latitude 65.5° S, marking the northern limit of Southern Hemisphere locations that can receive 24 hours of daylight or darkness.
Anticyclones Areas of high pressure.
Anvil Upper part of some thunderstorm clouds with a wedge-like (anvil) shape extending in the direction of the upper-level winds. Comprised entirely of ice crystals.
Aphelion Earth's position when it is farthest from the Sun (~July 3).
Apparent Temperature A measure of apparent temperature used for warm conditions, incorporating temperature and humidity.
Arctic Circle The line of latitude 65.5° N, marking the southern limit of Northern Hemisphere locations that can receive 24 hours of daylight or darkness.
Arctic Oscillation A see-saw in pressure between the arctic and northern midlatitudes. Similar to the North Atlantic Oscillation, but confined to the middle and higher latitudes.
Argon An inert gas found in the atmosphere at a nearly constant proportion of about 0.9%.
Aspirated Psychrometer Device for measuring humidity containing a fan that circulates air across a moist wick (see sling psychrometer)
Atlantic Multidecadal Oscillation A periodic reversal of North Atlantic sea surface temperature anomalies on the order of several decades. Associated with the frequency of Atlantic hurricanes.
Atmosphere The gases, droplets, and particles surrounding Earth's surface.
Atmosphere–Ocean General Circulation Models (GCMs) Computer models that simulate Earth's climate and ocean characteristics.

Atmospheric Optics The effect that the atmosphere's gases, aerosols, cloud droplets, and precipitation have on the light in the atmosphere.

Atmospheric Rivers Large air streams of high moisture content.

Atmospheric Window The range of wavelengths (about 8 to 12 μm) that are not readily absorbed by the gases of the atmosphere.

Aurora Borealis or Aurora Australis An illumination of the sky found in the high northern (borealis) or southern (australis) latitudes, which is produced as charged particles arriving from the Sun react with the upper atmosphere.

Automated Surface Observing System (ASOS) A type of automated weather-sensing platform used at major airports and National Weather Service offices.

Automated Weather Observation Systems (AWOS) A type of automated weather-sensing platform used at Canadian airports.

AWIPS Acronym for Advanced Weather Interactive Processing System, the computer system used by NWS forecasters for the display of weather maps, satellite and radar imagery, and other types of data.

Azimuth Direction of wind flow, measured as a clockwise angle from due north.

B

Ball Lightning A rare and short-lived type of electrification that resembles a glowing mass of air about the size of a basketball.

Banner Clouds A cloud formed near the top of a topographic barrier by orographic uplift.

Baroclinic A situation that results in upper-level weather maps having height contours and isotherms that intersect each other. This situation may enhance or reduce upper-level divergence or convergence.

Baroclinic Instability A situation in which warm air advection enhances vertical motions similar to those associated with statically unstable air.

Barograph Device that produces a continuous record of air pressure in the form of a chart.

Barometer An instrument for measuring air pressure.

Barometric Pressure Pressure exerted by the atmosphere.

Barotropic A situation that results in upper-level weather maps having height contours and isotherms parallel to each other.

Beam Spreading The process whereby a beam of radiation is distributed over a larger horizontal area as the angle of incidence departs from vertical. Reduces the intensity of radiation absorption by the surface.

Bergeron Process The primary mechanism for precipitation formation outside the tropics; this process involves the coexistence of ice crystals and supercooled water droplets.

Bermuda-Azores High A semipermanent cell found in the Atlantic in summer.

Bimetallic Strip Two strips of metal bonded together. Used in thermographs.

Blackbody An object or substance that is perfectly efficient at absorbing and radiating radiation. Blackbodies do not exist in nature, but represent an ideal.

Blizzard Continuous episode of heavy snow and winds exceeding 56 km/hr (35 mph).

Blue Jets Upward-moving electrical ejections from the tops of thunderstorms.

Boreal Forest An evergreen forest that typically dominates in subarctic climates.

Boundary Conditions A configuration of external factors that influence a system such as Earth's atmosphere.

Bounded Weak Echo Region An apparently empty area on a radar display, where moist air enters a supercell thunderstorm. Water droplets are abundant but are too small to provide a strong radar echo. Also called a *bounded weak echo region.*

Buoyancy Tendency of an air parcel to rise because it is less dense than surrounding air.

C

Calorie Amount of heat required to raise the temperature of 1 gram of water 1 °C (about 4.2 J).

Canadian Meteorological Centre A branch of the Meteorological Service of Canada that provides forecast guidance to national and regional prediction centers.

Canary Current Cold southward-moving ocean current in the Northern Hemisphere Eastern Pacific Ocean.

Carbon Dioxide An important variable gas in the atmosphere, made up of one atom of carbon bound to two atoms of oxygen. An important greenhouse gas.

Carbon Monoxide (CO) An air pollutant with molecules that have one carbon and one oxygen atom.

Carbon Oxides A general class of gases consisting of carbon and oxygen atoms.

Celsius Scale The temperature scale that designates 0° as the freezing point and 100° as the sea level boiling point of water.

Central Pacific (CP) Type El Niño Warming of the Pacific Ocean centered near the International Date Line.

Chaos A condition that makes it impossible to precisely predict how a system (such as the state of the atmosphere) will appear at some time in the future. This can occur when small errors in the input value of a variable become magnified over time.

Charge Separation The separation of positive and negative ions into different parts of a cloud. A necessary precursor for lightning.

Chinook Winds Downslope winds that warm because of adiabatic compression.

Chromosphere The layer of the Sun immediately surrounding the photosphere.

Cirrocumulus A high cloud composed of ice that is generally layered but with some rolls or pockets of vertical development.

Cirrostratus A high, layered cloud consisting of ice crystals.

Cirrus A high cloud made up entirely of ice crystals.

Climate The statistical properties of the atmosphere, including measures of average conditions, variability, etc.

Climate Prediction Center (CPC) The National Weather Service agency involved in making long-term forecasts.

Climatic Normal The average value of some weather variable, such as daily high temperature, for a recent 30-year period.

Climatological Forecasts Weather forecasts based on the use of long-term climatic averages.

Climatology The study of long-term atmospheric conditions.

Climograph A graph depicting average precipitation and temperature values for each month for a given location.

Cloud An area of the atmosphere containing sufficient concentration of water droplets and/or ice crystals to be visible.

Cloud Seeding An attempt to stimulate precipitation by introducing certain materials into existing clouds.

Cloud-to-Cloud Lightning Lightning that flows from one part of a cloud to another or from one cloud to another. Distinct from cloud-to-ground lightning.

Cloud-to-Ground Lightning Lightning that moves from a cloud to the ground.

Clouds with Extensive Vertical Development Category of cloud forms that includes cumuliform, cumulus, and cumulonimbus.

Coalescence Merging of two droplets in a cloud.

Cold Air Advection Air flow in which cold air is flowing into a region of warmer air.

Cold Cloud A cloud with a temperature below 0 °C from top to bottom.

Cold Front Transition zone between cold and warm air masses; forms when a cold air mass advances on a warm air mass.

Collector Drop A relatively large falling raindrop that collides and coalesces with smaller, slower-moving droplets beneath it.

Collision When one droplet strikes another in a cloud.

Collision–Coalescence Process Precipitation growth process where droplets collide and merge into a larger droplet.

Community Collaborative Rain, Hail and Snow Network Program for precipitation data collection relying on volunteers for primary measurements.

Condensation Nuclei Small, airborne particles that enhance condensation. Without condensation nuclei, condensation would occur only at very high relative humidity (at about 200 percent or more), while condensation nuclei allow condensation to occur at or slightly below 100 percent relative humidity.

Condensation Change from vapor to liquid phase. Condensation releases the energy required for evaporation.

Conduction Heat transfer from molecule to molecule, without significant movement of the molecules.

Confluence A type of horizontal convergence that occurs when streamlines come closer together in the downstream direction.

Continental Arctic (cA) Air Mass Extremely cold and dry air mass that forms in polar regions of the Northern Hemisphere, usually north of the Arctic Circle. *See also* Air Mass.

Continental Polar (cP) Air Mass Cold, dry air mass that forms in northern Canada or Siberia. Not as extreme as a continental arctic air mass. *See also* Air Mass.

Continental Tropical (cT) Air Mass Hot, dry air mass that forms over a subtropical continent, particularly in summer.

Contrails Clouds that form behind aircraft (contraction of condensation trail).

Convection Heat transfer by fluid flow (movement of a gas or liquid).

Convection Zone An internal layer of the Sun where upwelling gases carry energy from the core toward the surface.

Convective Outlooks Information on the probability of different types of severe storms issued by the Storm Prediction Center.

Convergence Horizontal motions of air resulting in a net inflow of air (with more air imported than exported). Causes rising or sinking motions.

Conveyor Belt Model The modern description of air flow through midlatitude cyclones.

Cool Cloud A cloud in which the lower reaches have temperatures above 0 °C and in which the temperatures in the upper portions are below 0 °C.

Cooling Degree-Day An index of the amount of seasonal air conditioning required for a location. Cooling degree-days are calculated by subtracting a base temperature (usually 65 °F) from the daily mean temperature and summing the differences. Days with mean temperature below the base are ignored in the calculation.

Coral Reefs Ridges extending from the ocean floor to just below the water surface, along the shallow margins of warm, tropical oceans, formed by the growth of small, hard-shelled marine organisms.

Core The interior of the Sun, where nuclear fusion produces energy that is ultimately radiated to Earth.

Coriolis Force An imaginary deflective force arising from Earth's rotation that is necessary to account for motions measured relative to the surface.

Corona A circular illumination of the sky immediately surrounding the Moon, or in some instances the Sun, caused by diffraction.

Coverage Fraction of sky obscured by clouds.

Cumuliform A type of cloud with extensive vertical development

Cumulonimbus A cumulus cloud with very deep vertical development extending into the lower stratosphere, distinguished by an anvil at its top consisting of ice crystals.

Cumulus Any cloud having substantial vertical development.

Cumulus Stage The earliest stage of air mass thunderstorms in which clouds begin to grow but have not yet begun to produce lightning or precipitation.

Cutoff Low An upper-level area of low pressure that takes on a circular flow distinct from the general flow around it.

Cyclogenesis The beginning of cyclone formation.

Cyclone A region of low pressure relative to the surrounding area.

D

Dalton's Law This law of physical science states that the pressure of a combination of gases is equal to the sum of the partial pressures of each of the gases.

Dart Leader A zone of ionized air that serves as a conduit for a lightning stroke subsequent to the initial stepped leader in a lightning flash.

December Solstice Day of the year when the Sun is overhead at 23.5° south latitude, on or about December 21. Also called the winter solstice in the Northern Hemisphere.

Density The mass of a substance per unit volume, expressed as kilograms per cubic meter (kg/m^3) in the International System of Units (SI).

Deposition Change from the vapor phase to the solid phase (frost is an example). Deposition releases the energy of vaporization and fusion.

Derechos Powerful, large-scale winds that flow in a straight line.

Dew Condensation deposited on a surface (grass, windshields, etc.).

Dew Point The temperature to which the air must be cooled to become saturated.

Dew Point Lapse Rate Rate of change of dew point with altitude.

Dew Point Temperature Also called *dew point.* The temperature at which saturation will occur, given sufficient cooling.

Diabatic Process Process that involves the addition or removal of heat. For example, air in contact with a cold surface loses heat diabatically by conduction.

Diffluence A type of horizontal divergence that occurs when streamlines spread apart in the downstream direction.

Diffraction The bending of light rays.

Diffuse Radiation Sunlight that is scattered downward to the surface. *See* scattering.

Diffuse Reflection or Scattering Redirection (reflection) of radiation in all directions, in contrast to specular reflection.

Diffuse Solar Radiation *See* Diffuse Radiation.

Direct Radiation Sunlight that passes through the atmosphere without absorption or scattering.

Direct Solar Radiation *See* Direct Radiation.

Dissipative Stage The final stage of an air mass thunderstorm in which there is insufficient inflow of water vapor to offset the loss of moisture by precipitation.

Divergence Horizontal motions of air resulting in a net outflow of air (with more air exported than imported). Causes rising or sinking motions.

Domain In numerical forecasting the area represented by a model.

Doppler Effect The shortening or elongation of sound waves caused as an object approaches or moves away from the listener. A similar effect is used to detect the movement of storms as electromagnetic radiation is shortened or lengthened after hitting the target precipitation.

Doppler Radar A type of radar that can measure horizontal motions as well as the internal characteristics of clouds.

Double Eye Wall A feature on a hurricane in which there are two eye walls—an inner one and an outer one. The appearance of a double eye wall is often a precursor to hurricane intensification.

Downbursts Strong downward movements of air due to the drag caused by heavy precipitation and the cooling of air as falling precipitation evaporates.

Downwelling Downward movement of water in the ocean.

Drag Frictional force experienced by a falling raindrop or other precipitation particle.

Dry Adiabatic Lapse Rate (DALR) Temperature decrease experienced by a rising unsaturated parcel (about 1 °C/100 m). Sinking parcels warm at the same rate. The DALR is a constant.

Dry Bulb Thermometer Temperature of dry thermometer in a psychrometer (same as air temperature).

Dryline A boundary between humid air and denser dry air. A favored location for thunderstorm development.

Dual-Polarization Radar Precipitation radar that emits radiation with both vertical and horizontal orientations.

Dynamic Low A low-pressure system created by divergence in the middle or upper troposphere.

E

Earth Vorticity The spin of the atmosphere due to the rotation of the planet. It is zero at the equator and increases to a maximum at either pole.

East Greenland Drift An ocean current that flows southward in the North Atlantic.

Easterly Waves Disruptions in the normal trade wind pattern that often move tropical disturbances westward.

Eastern Pacific (EP) Type El Niño Warming of the Pacific Ocean centered near South America.

Eccentricity The degree to which an elliptically orbiting planet's orbit departs from a true circle.

Ecliptic Plane The imaginary surface swept by Earth as it orbits the Sun.

Ekman Spiral Progressive turning of an ocean current at increasing depth caused by the Coriolis force. Turning is to the right in the Northern Hemisphere and to the left in the Southern Hemisphere.

El Niño A recurrent event in the tropical eastern Pacific in which sea surface temperatures are significantly above normal. The inverse event (cold sea surface temperatures) is called a *La Niña*.

Electromagnetic Radiation Energy emitted by virtue of an object's temperature. Radiation is unique in that it does not require a transfer medium and can travel through a vacuum. The energy transfer is accomplished by oscillations in an electric field and a magnetic field.

Emissivity The property of a substance or object that expresses, as a fraction or percentage, how efficient it is at emitting radiation.

Energy The ability to do work. Can be in many forms such as kinetic, potential, chemical.

Energy Balance The condition in which energy gains equal energy losses and temperature is unchanging.

Enhanced Fujita Scale The scale currently used to infer maximum tornadic winds based on observed damage.

Ensemble Forecasting A forecasting technique in which a numerical model is run multiple times using slightly varying input values. Useful for assessing the reliability of a forecast.

ENSO An acronym for the El Niño Southern Oscillation phenomenon. Involves the interaction of Tropical Pacific Sea Surface Temperatures and atmospheric pressures.

Entrainment The incorporation of surrounding, unsaturated air into a cloud.

Environmental Lapse Rate (ELR) The rate of vertical temperature decrease in the air column. The value is highly variable, depending on local conditions. For the troposphere, the global average is about 0.65 °C/100 m.

Equation of State The equation relating air pressure to temperature and density.

Equatorial Countercurrent Eastward-flowing surface current found along the equator.

Equatorial Low Region of low air pressure at the surface found along the equator.

Equinoxes The two days of the year on which Earth's axis is not tilted toward or away from the Sun. On the equinoxes every latitude receives 12 hours of sunlight, and the Sun is overhead at the equator. The equinoxes occur March 21–22 and September 22–23.

Escape Velocity The rate of movement required of an air molecule to escape Earth's gravity.

European Center for Medium-Range Weather Forecasting Weather forecasting agency for the European Union.

Evaporation The change in phase of liquid water to water vapor.

Evapotranspiration The combined processes of evaporation and transpiration; the delivery of water to the atmosphere by vegetation and by direct evaporation from wet surfaces.

Extraterrestrial Radiation The solar radiation incident at the top of Earth's atmosphere.

Eye The center of a hurricane, marked by generally clear skies and light winds.

Eye Wall Very cloudy portion of a hurricane immediately adjacent to the eye; usually the region of highest wind speed and most intense precipitation.

Eye Wall Replacement The dissipation and replacement of a hurricane's eye wall.

F

Fahrenheit Scale A temperature scale that assigns values of 32° to the freezing point of water and 212° to the sea level boiling point of water.

Fair-Weather Electric Field The ever-present voltage difference between the surface and the ionosphere, which serves as the background situation for events such as lightning. Also called *mean electric field*.

Feedback Processes In natural systems the manner in which a change in one variable is muted or enhanced by its impact on a another variable.

Ferrel Cell The midlatitude circulation cell of the three-cell model.

Fetch Distance traveled by wind over a uniform surface, such as a water body.

First Law of Thermodynamics Most generally, the law that states that energy is a conserved property. In a meteorological context it states that heat added to a gas results in some combination of expansion of the gas and an increase in its internal energy.

Flares Intensely hot eruptions on the solar surface.

Flash The rapid combination of lightning strokes. The rapidity and nearness of the strokes often cause the individual strokes to appear as a single flickering stroke.

Flash Floods Floods that occur over very brief time periods, often as a heavy thunderstorm remains in place and delivers heavy precipitation.

Floods The inundation of an area by excessive water. This may occur from any combination of a variety of factors, such as excessive snow melt, heavy precipitation, and waterlogged soils.

Flurries Snowfall of short duration leading to little accumulation of snow on the ground.

Foehn Wind A synoptic-scale wind that flows downslope and warms by compression.

Fog Air that is adjacent to the surface and contains suspended water droplets, usually formed by diabatic cooling.

Force The product of mass and acceleration, as expressed in Newton's law ($F = ma$).

Forced Convection or Mechanical Turbulence Mixing of the air caused by horizontal movements (wind).

Forcing Agents External conditions that can drive climate changes.

Forecast Accuracy A measure of how closely the forecast resembles the actual weather outcome.

Forecast Bias The propensity for a forecast method to systematically overpredict or underpredict some weather variable such as precipitation.

Forecast Quality The degree to which a forecast represents conditions that eventually occur.

Forecast Skill The improvement provided by a forecast over and above some reference accuracy.

Forecast Value The degree to which a forecast assists in a decision about a particular problem.

Free Atmosphere Region of the atmosphere far enough above the surface that friction can be neglected, generally 1.5 km and higher.

Free Convection A mixing process due to buoyancy, when parcels of air rise because they are less dense than surrounding air.

Freezing Rain A form of precipitation in which rain droplets freeze as they fall below an inversion and pass into air having a temperature below 0 °C.

Friction Force that acts to slow wind but does not change its direction. Friction develops between the atmosphere and surface and between layers of air moving at different velocities.

Frictional Dissipation Conversion of kinetic energy to heat by friction.

Front A transition zone between two dissimilar air masses (that is, air masses with differing temperature, moisture, or density).

Frost A coating of ice crystals on a surface when the air adjacent to the surface becomes saturated at temperatures below 0 °C.

Frost Point The temperature at which saturation occurs, provided that temperature is less than 0 °C.

Frozen Dew A coating of ice on a surface that occurs when a layer of dew freezes as temperatures drop below 0 °C.

Fujita Scale The scale for categorizing tornado intensity.

Funnel Cloud A column of rapidly rotating air similar to a tornado, except that the column has not extended to the ground.

G

Gamma Rays Electromagnetic radiation at wavelengths far shorter than those of visible light (from about 0.0000001 μm to 0.000001 μm). Gamma rays, which make up only a tiny proportion of the Sun's energy, are absorbed hundreds of kilometers above the surface.

General Circulation A term that refers to planetary-scale winds and pressure, features that appear in the time-averaged state.

Geopotential Height Loosely defined, the altitude at which atmospheric pressure takes on a particular value, as in "500 mb height." Because pressure reflects the mass of overlying atmosphere, geopotential heights reflect the potential energy of the atmosphere above that height.

Geostrophic Flow An idealized condition in which the upper-level air flows at constant speed and direction, parallel to straight isobars. There is no acceleration in geostrophic flow, and frictional forces are negligible.

Geostrophic Wind *See* Geostrophic Flow.

Glacial/Interglacial Cycles Episodic climate changes associated with the advance and retreat of glaciers.

Global Dimming A measured decrease in solar radiation reaching the surface, which lasted from about 1970 until 1990. It was probably not global in extent, and it did not persist beyond the early 1990s.

Global Scale Horizontal dimensions of the size of continents and larger (thousands of kilometers).

Glory A series of rings formed around the shadow of an aircraft on the top of a cloud, formed by refraction, reflection, and diffraction.

Gradient A change in some quantity (temperature, moisture, pressure) over space. For example, the temperature gradient is the rate of change of temperature per unit distance and can be expressed in degrees Celsius per kilometer (°C/km).

Gradient Flow *See* Geostrophic wind.

Gradient Wind Wind flowing parallel to curved isobars. Frictional forces are negligible. With gradient flow there is a constant adjustment between the pressure gradient force and Coriolis force, causing the wind to change speed and direction as it flows along the isobars.

Granules Tops of convection cells seen on the surface of the Sun, analogous to bubbles on the surface of boiling water.

Graupel Ice crystals that have grown by riming to produce a spongy, somewhat translucent particle.

Gravity The force that attracts objects to Earth's surface. Although the acceleration of gravity is constant, the force of gravity varies from object to object. The force of gravity per unit volume of air is directly proportional to density. *See* Acceleration of Gravity.

Graybodies Bodies or substances that are not 100 percent efficient at absorbing or radiating energy. In reality, all bodies are graybodies.

Green Flash The brief appearance of green light near the top of the Sun sometimes observed at sunrise or sunset.

Greenhouse Effect The result of clouds and greenhouse gases (mainly water vapor and carbon dioxide), which absorb longwave radiation and cause near-surface temperatures to be much higher than they would otherwise be.

Greenwich Mean Time Also called *universal time* (*UT*). An international reference for time-keeping used for weather observations, satellite imaging, etc. It corresponds to local time at 0° longitude, a meridian passing through Greenwich, England.

Grid Representation The assignment of grid points to a large area to be used in numerical predictions as opposed to the use of spectral wave techniques.

Growing Degree-Day An index for estimating when crops will have undergone enough growth to send them to market. Growing degree-days are calculated by subtracting a base temperature for a particular crop from the daily mean temperature and summing the differences.

Gulf Stream Warm ocean current in the Northern Hemisphere Atlantic Ocean found off the East Coast of the United States.

Gust Front The boundary of rapidly advancing cold air originating as strong downdrafts that spread out after hitting the surface.

Gyres Large basin-wide, nearly circular ocean circulation cells that form around subtropical high-pressure zones.

H

Haboobs Desert dust storms triggered by intense downdrafts that cause strong horizontal winds.

Hadley Cell A somewhat idealized, large-scale wind and pressure pattern found in tropical latitudes of both hemispheres. Air rises above the equator, flows poleward to about 25° latitude, subsides, and flows back to the equator at low levels.

Hail Precipitation in the form of ice crystals, almost always associated with thunderstorms. Hail falls rapidly to the surface and thus does not melt during its descent.

Hail Cascade A localized area of falling hail.

Hair Hygrometer Device for measuring humidity relying on contraction and expansion of human hair.

Halo A circular band of light surrounding the Sun or Moon, caused by ice crystal refraction.

Hawaiian High A semipermanent cell found in the Pacific, most notably during the summer months.

Haze Particles suspended in the air that scatter light and reduce visibility.

Heat The kinetic energy of atoms or molecules that make up a substance.

Heat Index *See* Apparent Temperature.

Heat Lightning Lightning that is far enough away that thunder sound waves are not heard. Not a particular kind of lightning.

Heating Degree-Days An index of the amount of seasonal heating required for a location. Heating degree-days are calculated by subtracting the daily mean temperature from a base temperature (usually 65 °F) and summing the differences. Days with mean temperature above the base are ignored in the calculation.

Heterogeneous Nucleation The condensation of liquid droplets or the deposition of ice crystals onto condensation or ice nuclei.

Heterosphere The high atmosphere (above 80 km or so), where gases are not well mixed but rather are stratified according to molecular weight. Vertical motions are too weak to overcome gravitational settling, so heavier gases are found beneath lighter gases.

High Another term for *anticyclone*.

High, Middle, and Low Clouds Cloud categories based on altitude.

Highland An area at high elevation.

Holocene The most recent geologic epoch. Includes the most recent 11,800 years.

Homogeneous Nucleation The condensation of water droplets or deposition of ice crystals without condensation or ice nuclei.

Homosphere The lowest 80 km (50 mi) of the atmosphere in which the relative abundance of the permanent gases is constant.

Hook or Hook Echo An image on a weather radar display that has a hook shape. Important because the presence of a hook echo indicates an elevated likelihood of tornadic activity.

Horse Latitudes Areas associated with the oceanic subtropical highs, generally characterized by clear skies and light winds.

Hot Tower An area of intense convection within a hurricane. Often associated with hurricane intensification.

Humid Continental A type of severe midlatitude climate found across much of North America and Asia.

Humid Subtropical A type of mild midlatitude climate marked by hot summers but generally mild winters. Usually found on the eastern portions of continents.

Humidity An expression of the amount of water vapor in the air.

Hurricane An intense tropical cyclone (warm-core low), with sustained winds of at least 120 km/hr.

Hurricane Warning A notice issued by the National Hurricane Center advising about the probability of hurricane force winds over particular areas. Issued 36 hours in advance of the anticipated onset of tropical-storm-force winds.

Hurricane Watch Similar to a hurricane warning, but issued 48 hours in advance of the expected onset of tropical-storm-force winds.

Hydrocarbons Gases consisting only of oxygen and hydrogen atoms.

Hydrologic Cycle The perennial movement of water in its three phases between the atmosphere, Earth's surface, and groundwater.

Hydrostatic Equation Equation that applies when the vertical pressure gradient force equals the force of gravity, and there is no acceleration of air upward or downward.

Hydrostatic Equilibrium When the vertical pressure gradient force is balanced by the force of gravity. Because the forces balance, there is no acceleration upward or downward.

Hygrometer An instrument that measures humidity.

Hygroscopic The ability to permit condensation at relative humidity below 100%.

Hygroscopic Nuclei Airborne particles having an affinity for water, serving as condensation nuclei.

Hygrothermograph An instrument that records humidity and temperature.

I

Ice Ages Cold episodes in Earth's history.

Ice Cores Extractions from glaciers used to obtain information about past climates.

Ice Nuclei Particles onto which ice crystals can form when the air becomes saturated. In the absence of freezing nuclei, water droplets freeze only at very low temperatures (near 240 °C). Ice nuclei allow ice to form at relatively "high" temperatures (around 210 °C).

Ice Storm A weather event that results in freezing rain.

Ice-Rafted Debris Rock material transported by icebergs that is eventually deposited on the ocean floor when the iceberg melts. Useful for helping to infer past climates.

Icelandic Low A semipermanent cell found in the North Atlantic in winter.

Ideal Gas Law Also known as the *equation of state.* Important law describing the relationship between pressure, temperature, and density.

Index of Refraction A measure of the degree of slowing of radiation as it passes through a medium. Obtained by dividing the speed of light in a vacuum by its speed in the medium.

Inferior Mirage A mirage that is an inverted image of an object or objects at the surface.

Infrared Images Satellite images based on the observation of infrared radiation emitted by the surface and layers of the atmosphere.

Infrared Radiation Electromagnetic radiation at wavelengths longer than visible radiation, from about 0.7 μm to 1000 μm.

Insolation Incident, or incoming, solar radiation.

Interglacial A warmer segment of time between episodes of glaciation.

Intergovernmental Panel on Climate Change (IPCC) The agency of the World Meteorological Organization charged with assessing and disseminating current knowledge about climate change.

Intertropical Convergence Zone (ITCZ) Another name for the Equatorial Low, with winds that converge from the north and south.

Inverse Square Law Equation showing that radiation intensity decreases with the square of the distance from the emitter.

Inversion Condition in which temperature increases with increasing altitude.

Ionosphere Region in the upper atmosphere from about 80 to 500 km (50 to 300 mi) where charged particles (ions) are relatively abundant.

Ions Electrically charged atom or group of atoms.

Isobar A line on a weather map connecting points of equal pressure. Moving along an isobar, there is no change in pressure. The pressure gradient force acts perpendicular to isobars.

Isotachs Lines on a map representing the distribution of wind speeds.

Isotherm A line on a weather map connecting points of equal temperature. Moving along an isotherm, there is no change in temperature. Temperature gradients are perpendicular to isotherms.

J

Joule Basic unit of energy in the International System of Units (SI). A joule (J) is the energy needed to accelerate 1 kg at a rate of 1 m/sec/sec across a distance of 1 m. A joule is equivalent to about 0.25 calories. See *calorie.*

June Solstice Day of the year when the Sun is overhead at 23.5° north latitude, on or about June 21. Also called the summer solstice in the Northern Hemisphere.

K

K-Index A numerical index used to help assess the likelihood of air mass thunderstorms and heavy rain.

Katabatic Winds Airflow down a slope under the influence of gravity.

Kelvin Scale An absolute temperature scale, where a value of 0 K implies an absence of thermal energy. The Kelvin scale assigns 100 units between the melting and sea level boiling points of water.

Kilopascal A unit of pressure equal to 1000 pascals or 0.1 millibars.

Kinetic Energy Energy of motion.

Koeppen System A widely used system for classifying climates into six major groups and numerous subgroups, based primarily on temperature and precipitation characteristics.

L

La Niña The opposite pattern to an El Niño, in which below-normal sea surface temperatures exist in the tropical eastern Pacific.

Labrador Current Cold, southward-flowing ocean current moving from the Arctic Ocean along the coast of Labrador.

Lake Breeze Onshore flow of air, commonly caused by strong solar heating of the land.

Lake-Effect or Lake-Enhanced Snow Anomalously heavy snowfall caused by a combination of factors related to lakes. Confined to a relatively narrow strip of land adjacent to the water body.

Laminar Boundary Layer Thin layer of air adjacent to the surface with very little vertical motion, a few millimeters thick.

Land Breeze A wind that blows from the land toward the water along the coastal zone during the night and early morning. The result of differential cooling between land and sea.

Latent Heat (1) Energy present in water vapor, used in converting water from liquid to gas. Latent heat is released upon condensation. (2) Energy associated with the change of phase of a substance. *See* Deposition, latent heat of fusion *and* latent heat of vaporization.

Latent Heat Flux Heat transfer that occurs whenever water vapor moves from one place to another. Energy used to evaporate water travels with the water vapor and is released upon condensation.

Latent Heat of Fusion Also called *latent heat of melting.* Energy released when a substance freezes, consumed when a substance melts. For water, the latent heat of fusion is about 334,000 J/kg.

Latent Heat of Vaporization Also called *latent heat of condensation.* Energy consumed when a substance evaporates, released when a substance condenses. For water, the latent heat of vaporization is about 2,500,000 J/kg. *See* Condensation.

Leader A column of ionized air that approaches the surface and precedes cloud–ground lightning.

Lenticular Cloud A lens-shaped cloud that usually forms downwind of topographic barriers.

Level of Free Convection The level to which conditionally unstable air must be lifted so that it can continue to rise due to its own buoyancy.

Lifted Index A numerical index used to help assess the potential for thunderstorm development.

Lifting Condensation Level Altitude to which an air parcel would need to be lifted for condensation to occur.

Lightning An electrical discharge from a cloud to the ground, within a cloud, between clouds, or from a cloud to the surrounding air.

Limb-Darkening The phenomenon in which the edge of the Sun appears darker than its center.

Little Ice Age A period in Earth's history from about 1400 to 1850 characterized by low temperatures.

Local Vertical Direction The direction that is straight up for any location, in the opposite direction as the force of gravity. Also known as the zenith direction.

London-Type Smog A type of air pollution in which smoke combines with fog.

Long-Range Forecast A weather prediction extending beyond 7 days.

Long Waves Also known as *Rossby waves*. Waves in the midlatitude westerlies having wavelengths on the order of thousands of kilometers. Often a series of Rossby waves circle the planet, forming a pattern of ridges and troughs.

Longwave Radiation Another term for *infrared radiation*.

Los Angeles–Type Smog Synonymous with photochemical smog, a type of air pollution that forms from chemical reactions in the presence of sunlight, most commonly in dry air.

Low Another term for *cyclone*.

M

Mammatus A feature on parts of some cumulonimbus clouds consisting of round, downward-extending protrusions.

March Equinox One of two days every year when the Sun is overhead at the equator, on or about March 21. Also called the spring equinox in the Northern Hemisphere.

Marine Layer A layer of moist, cool air beneath an inversion layer.

Marine West Coast A type of mild midlatitude climate typically found in higher-midlatitude regions on the western portions of continents over land. They also extend seaward for considerable distances. They have mild summer and winter temperatures and no dry season.

Maritime Polar (mP) Air Mass Air mass that forms over an ocean in the high latitudes. *See also* Air Mass.

Maritime Tropical (mT) Air Mass Air mass that forms over a low-latitude ocean. *See also* Air Mass.

Mature Cyclone The stage of a midlatitude cyclone in which cold and warm fronts are fully developed and air rotates around a low-pressure system.

Mature Stage The stage in an air mass thunderstorm marked by heavy storm activity, with strong updrafts, lightning, and heavy precipitation.

Maunder Minimum A period of Earth's history between about 1645 and 1715 characterized by minimal sunspot activity.

Maximum Thermometer Thermometer for measuring highest temperature during the day.

Mean Electric Field *See* Fair-Weather Electric Field.

Mean Free Path The average distance traveled by molecules before colliding with adjacent molecules; increases with altitude.

Mechanical Turbulence *See* Forced Convection.

Mediterranean A type of mild midlatitude climate marked by mild winters and dry summers. Found generally on the west coast of continents at lower-middle latitudes, extending considerable distances seaward.

Medium-Range Forecast Weather forecasts for predictions 3 to 7 days in advance.

Mercury Barometer The standard instrument for the measurement of atmospheric pressure.

Meridional Wind Wind flowing north–south parallel to a line of longitude. Actual winds are seldom completely meridional, but usually have both a meridional and *zonal* component.

Mesocyclone A rotating region within a cumulonimbus cloud where tornadoes often form.

Mesoscale A scale of meteorological phenomena typically having horizontal extents of several tens of kilometers.

Mesoscale Convective Complex (MCC) A type of mesoscale convective system having an oval or nearly circular shape.

Mesoscale Convective System (MCS) A general clustering of thunderstorms on the order of a few hundred kilometers across.

Mesosphere Region of the atmosphere from about 50 km to 80 km (30 to 50 mi), characterized by decreasing temperature with increasing altitude.

Meteorological Service of Canada The official meteorological agency for Canada.

Meteorology The science that studies the atmosphere.

Methane An important greenhouse gas comprised of one carbon atom and four hydrogen atoms, with present-day concentration of about 1.8 parts per million.

Microburst A small but severe downburst, whose wind shear is capable of causing air crashes.

Micrometers or Microns One millionth of a meter.

Microscale The smallest scale of meteorological phenomena, such as that which might surround a leaf.

Microwave Radiation Electromagnetic radiation with wavelengths between about 1000 and 1,000,000 μm. Weather radars use microwave radiation for imaging.

Midlatitude Cyclone A low-pressure system characterized by the presence of frontal boundaries.

Midlatitude Desert A very dry region found outside the tropics, usually due to extreme continentality.

Midlatitude Steppe A dry region outside the tropics but of lesser aridity than a midlatitude desert.

Mie Scattering Scattering of visible radiation caused by particulates.

Milankovitch Cycles Variations in Earth's orbital characteristics having periodicities of tens of thousands of years and longer.

Millibar A unit of atmospheric pressure, abbreviated as mb. Sea level pressure is about 1013 mb.

Minimum Thermometer Thermometer for measuring lowest temperature during the day.

Mirage The apparent displacement of an object's true position due to refraction.

Mixed Layer That part of the lower atmosphere in which vertical motion (convection) is strong enough for even dispersal of pollutants.

Mixing Ratio A measure of atmospheric moisture: the mass of water vapor per unit mass of dry air, usually expressed in grams per kilogram (g/kg).

Moist Adiabatic Lapse Rate Another term for the *Saturated Adiabatic Lapse Rate* (*SALR*).

Monsoon A regional circulation pattern in which there is a seasonal reversal of wind and pressure, generally characterized by onshore flow during the summer and offshore flow during the winter.

Monsoon Depressions Areas of low pressure superimposed in the southeasterly airflow out of the Bay of Bengal.

Mountain Breeze A breeze that flows down a hill at night.

Multicell Thunderstorms Thunderstorms that consist of multiple cells or zones of updrafts.

N

Nacreous Clouds Multicolored, pearlescent clouds found in the stratosphere. Also called *mother-of-pearl clouds*, these consist of ice crystals or supercooled water.

National Centers for Environmental Prediction (NCEP) The unit within the National Weather Service that develops and applies numerical models for use by National Weather Service offices.

National Oceanic and Atmospheric Administration (NOAA) *The U.S. office that oversees the National Weather Service and other environmental agencies.*

National Weather Service The official meteorological agency for the United States.

National Weather Service (NWS) *The U.S. organization tasked with weather forecasting and other activities related to weather.*

NCDC Stands for National Climate Data Center.

NCEP The initials of the National Centers for Environmental Prediction.

Nested Grid Model A particular numerical weather prediction model.

Net All-Wave Radiation The difference between absorbed radiation and emitted radiation. Includes both solar and terrestrial radiation. Also called *net radiation*.

Net Longwave Radiation Difference between absorbed and emitted longwave (or terrestrial) radiation.

Net Radiation *See* Net All-Wave Radiation.

Neutral Stability A condition in which a lifted parcel of air does not return to its original position nor continue to rise. Neutral stability occurs when the environmental lapse rate is equal to the appropriate adiabatic rate, so that temperatures inside the parcel match those of the surroundings.

Newton's Second Law An expression of the conservation of momentum that is stated as: net force equals mass times acceleration ($F = ma$).

NEXRAD Stands for Next Generation Weather Radar, a network of Doppler radar units established by the U.S. National Weather Service.

NHC Stands for National Hurricane Center.

Nimbostratus A low, layered cloud that yields light precipitation.

Nimbus Rain-producing clouds.

Nitric Oxide (NO) A compound consisting of one nitrogen and one oxygen atom.

Nitrogen (N_2) An inert gas comprising about 78% of the atmosphere.

Nitrogen Dioxide (NO_2) An air pollutant consisting of one nitrogen atom and two oxygen atoms that gives air a yellow to reddish-brown tinge.

Nitrogen Oxides A general class of gases consisting only of nitrogen and oxygen atoms.

NOAA Stands for National Oceanic and Atmospheric Administration.

Noctilucent Cloud A type of cloud that exists in the mesosphere, visible just after sunset (or before sunrise), when the surface and the lower atmosphere are in Earth's shadow.

Nonselective Scattering Scattering of radiation in which all wavelengths are scattered about equally. This type of scattering causes clouds to appear white.

North American Mesoscale Forecast System (NAM) One of the numerical models used by the National Weather Service.

North Atlantic Drift Continuation of the Gulf Stream into the North Atlantic Ocean. Carries heat eastward toward Europe.

North Atlantic Oscillation A see-saw in pressure between the *Icelandic Low* and the *Bermuda-Azores High*. Similar to the *Arctic Oscillation*, but confined to middle and sub-tropical latitudes.

North Equatorial Current Eastward-flowing ocean current in the Northern Atlantic and Pacific.

Northeast Trade Winds Northern Hemisphere trade winds.

Northeaster or Nor'easter A winter weather condition of the Atlantic Coast of the United States and Canada associated with the passage of midlatitude cyclones. The strong northeasterly winds are usually coupled with blizzard conditions.

NSSFC Stands for National Severe Storms Forecast Center.

Nuclear Fusion The thermonuclear process in which extreme heat and pressure cause atoms to combine, forming a different (heavier) element. A small part of the original mass is converted to tremendous quantities of energy and released to the environment.

Numerical Weather Forecasting Weather forecasting using the output from computer models.

Numerical Weather Prediction Weather prediction based on equations representing physical processes (as opposed to statistical relations).

NWS The initials of the National Weather Service.

O

Obliquity The degree of tilt of Earth's axis relative to the ecliptic plane, currently about 23.5°.

Occluded Front A front found in the late stages of a midlatitude cyclone.

Occlusion An occurrence in the latter stage of a midlatitude cyclone characterized by a merging of cold and warm fronts, with warmer air aloft.

Ocean Current The horizontal movement of surface waters caused by prevailing winds.

Oceanic Niño Index (ONI) A measure of El Niño/La Niña strength based on sea-surface temperatures in the Equatorial Pacific.

Oceanography Science of the oceans.

Omega High A pressure pattern depicted on upper-level weather maps by a pattern resembling the Greek letter Ω.

Orographic Lifting Rising motions caused by airflow over a mountain range or other topographic barrier.

Orographic Uplift or Orographic Effect *See* Orographic Lifting.

Outflow Boundary The boundary separating cooler surface air flowing outward from a downdraft and warmer ambient air.

Outgassing The emission of gases that accompanies volcanic eruptions.

Overrunning Warm air sliding over a dense cold air mass; the characteristic flow associated with a warm front.

Oxides of Carbon A general class of air pollutants consisting of oxygen and carbon.

Oxygen (O_2) A reactive gas essential for life comprising about 21% of the atmosphere. Has limited role in atmospheric processes.

Ozone Molecules consisting of three oxygen atoms, most abundant in the middle and upper stratosphere.

Ozone Hole Ozone depletions found at high latitudes (especially over Antarctica) in the spring of each year.

Ozone Layer The portion of the stratosphere where ozone is relatively abundant, reaching a few parts per million.

P

Pacific Decadal Oscillation An alternating pattern of sea surface temperature in the Pacific that reverses itself over periods of several decades.

Paleoclimates Former Earth climates.

Particulates *See* aerosols.

Pascal The standard unit of pressure in most scientific applications, equal to 1 N/m^2.

Perihelion Earth's closest approach to the Sun (~January 3).

Permafrost A condition found in particular polar climates in which the subsoil is permanently frozen with an overlying soil layer that undergoes thaw during the summer.

Permanent Gases Those gases whose relative abundance is constant within the homosphere.

Persistence Forecast A weather forecast made by assuming some existing trend continues into the future.

Photochemical Smog Secondary air pollutants formed by chemical reactions in the presence of sunlight.

Photodissociation Splitting of molecules into atoms or submolecules by radiation. For example, in the thermosphere, ultraviolet radiation dissociates molecular oxygen (O_2) into atomic oxygen (O).

Photon Elementary particle responsible for electromagnetic radiation.

Photosphere That part of the Sun that emits most of the energy reaching Earth. It is the "visible" part of the Sun, a layer representing about 0.05 percent of the solar radius.

Photosynthesis The growth process of green plants, whereby water and carbon dioxide are converted to carbohydrate, releasing oxygen.

Planetary Albedo Proportion of incoming solar radiation reflected by Earth, with approximate value of 30%.

Planetary Boundary Layer or Boundary Layer The lowest 1.5 km or so of the atmosphere, where friction is an important force in affecting wind.

Pleistocene The geologic epoch prior to the Holocene, extending from about 2.6 million years before the present to about 11,800 years before the present.

PM_{10} Designation given to particulates with diameters smaller than 10 μm, which are believed to have major health consequences for humans.

$PM_{2.5}$ Designation given to particulates smaller than 2.5 μm in diameter. Major attention has recently been given to this class of particulates as possibly the most damaging to human health.

Polar Cell High-latitude circulation cell in the three-cell model having high pressure at the surface and low-level winds flowing equatorward.

Polar Easterlies Low-level winds originating in the polar highs, a feature of the general circulation of the atmosphere, often very weak or absent.

Polar Front Transition zone between cold polar air and warmer air of the midlatitudes.

Polar Front Theory The theory postulated in the early part of the twentieth century describing the formation, development, and dissipation of midlatitude cyclones. Many of the features of the theory are still considered valid.

Polar Highs Low-level anticyclones of the Arctic and Antarctic. A feature of the general circulation of the atmosphere, often absent or weakly developed.

Polar Ice Cap A climate type that when occurring over a land body is covered in permanent ice.

Polar Jet Stream A jet stream found in the upper troposphere above the polar front, a result of the strong temperature contrast across the front.

Polaris The North Star.

Postprocessing Phase The step in numerical prediction that involves the production of forecast maps and other graphical outputs.

Potential Energy Energy possessed by virtue of an object's position above some reference level. Potential energy is available for conversion to kinetic energy.

Potential (Convective) Instability The condition in which a layer of air can become statically unstable if lifted sufficiently. Also called convective instability.

Power The rate at which work is done or energy expended. The standard unit is the watt, equal to 1 J/sec.

Precession The wobble of Earth's axis that has a periodicity of about 27,000 years. Combined with changes in the orientation of Earth's orbit, it yields radiation cycles of about 23,000 years.

Precipitable Water Vapor A measure of the total water vapor content of the atmosphere. The depth of water that would result if all the water in the column were to condense. Global average precipitable water vapor is about 2.5 cm (1 in.).

Precipitation Liquid water or ice that falls to Earth's surface. Rain is considered precipitation, but dew is not.

Precipitation Fog A type of fog that develops when falling raindrops evaporate enough water vapor into the air to saturate it.

Prediction Phase The step in numerical prediction that involves making a huge number of calculations based on input data and the application of physical laws.

Pressure Force exerted per unit area. In most sciences the standard unit of measurement is the pascal (Pa), equal to 1 N/m^2. In daily meteorological applications, however, the *millibar* (mb) is frequently used in the United States and the kilopascal in Canada.

Pressure Gradient The rate of change of pressure over distance. Gives rise to the pressure gradient force that sets air in motion.

Pressure Gradient Force A force that arises from spatial variation in pressure. Acting alone, the pressure gradient force would cause air to blow from an area of high pressure toward an area of low pressure. The vertical pressure gradient force is always present but is nearly balanced by gravity most of the time. Much weaker horizontal pressure gradients are the ultimate cause of wind.

Primary Pollutants Substances that pollute the atmosphere upon release.

Primary Rainbow The most common type of rainbow in which the shortest wavelength colors appear in the innermost portion of the arch and the longer wavelengths on the outer part of the arch.

Probability Forecasts A type of weather forecast that states the likelihood of a particular event occurring.

Proxy In climate change science some indicator of previous conditions used in the absence of direct measurements.

Psychrometer An instrument for measuring atmospheric moisture.

Pyranometer An instrument for measuring solar radiation.

Q

Qualitative Forecasts A type of weather forecast that simply places the occurrence or nonoccurrence of an event into a particular category, such as rain or no rain.

Quantitative Forecasts A weather forecast that assigns a numerical value to an expected occurrence, such as the amount of rain or snow fall.

Quasi-Biennial Oscillation (QBO) A periodic reversal in the direction of lower stratospheric winds over the tropics occurring on an approximately 2-year cycle.

R

Radar A device that uses microwave radiation for imaging the atmosphere.

Radar Images Graphical representations of the data obtained from weather radar.

Radiation Another term for *electromagnetic radiation*.

Radiation Fog A low-level cloud formed diabatically when the atmosphere loses heat by radiation upward.

Radiation Zone Layer of the Sun where energy moves outward by radiation from the core to the convection zone.

Radiative Forcing The change in the net radiation for the troposphere that would occur as a result of a change in some external factor, such as the amount of energy radiated by the Sun or by an increase in atmospheric carbon dioxide.

Radiocarbon Dating A method to infer the dates at which some remnant vegetation was alive. Knowledge of past vegetation can provide useful information regarding past climates.

Radiosonde An instrument package carried by balloon, used to measure vertical profiles of temperature, moisture, and pressure. Measurements are radioed to the ground from the instrument cluster.

Rain Precipitation arriving at the surface in the form of liquid drops, usually between 0.5 and 5 mm (0.02 and 0.2 in.). Outside of the tropics, rain usually begins in the ice stage and melts before reaching the surface. Rain that freezes on contact with the surface, forming a layer of ice, is called *freezing rain*.

Rain Gauge Device for measuring rainfall.

Rain Shadow An area on the lee (downwind) side of a mountain barrier having relatively low precipitation.

Rainbow A wide, sweeping band of light caused by refraction of sunlight by raindrops.

Rapid Refresh (RAP) Model One of the numerical models used for short-term forecasting.

Rawinsonde A radiosonde tracked by radar to provide wind information.

Rayleigh Scattering The scattering of radiation by agents substantially smaller than the radiation's wavelength. In the case of the atmosphere, this applies to the scattering of visible radiation by air molecules.

Reflection A process in which radiation arriving at a surface bounces back, without being absorbed or transmitted. Reflection does not heat the reflector, because there is no net energy transfer to the surface.

Refraction The bending of light within a medium or as it passes from one medium to another. Refraction results from density differences within/between the transfer media.

Regional Weather Centre A Meteorological Service of Canada office that combines forecasts and other services for regions of Canada.

Relative Humidity The measure of the amount of water vapor in the air as a fraction of saturation, often expressed as a percentage. Because the saturation point is temperature-dependent, relative humidity depends on both the moisture content and the temperature of the air.

Relative Vorticity Atmospheric rotation relative to Earth's surface and independent of Earth's rotation around its axis.

Resistance Thermometer Device that uses variations in electrical resistance to measure temperature.

Respiration Biological process that combines oxygen with carbohydrate to produce energy, releasing water vapor and carbon dioxide as by-products.

Return Stroke Synonymous with *lightning stroke*.

Revolution Movement of Earth in its annual orbit around the Sun.

Ridge An elongated axis of high pressure.

Riming *See* Accretion.

Roll Cloud A rotating cloud often found within the leading edge of a gust front.

Rossby Waves *See* Long Waves.

S

Saffir-Simpson Scale A scheme for classifying the intensity of hurricanes.

Santa Ana Wind A local name for a foehn wind in California.

Saturated Adiabatic Lapse Rate (SALR) Rate of temperature decrease for a rising saturated parcel of air. The value ranges from about 4 to 10 °C/km, depending mainly on temperature.

Saturation The maximum amount of water that can exist in the atmosphere as a vapor. More precisely, saturation occurs when a flat surface of pure water is in equilibrium with the overlying atmosphere. The evaporation rate equals the condensation rate, so the vapor content of the air is unchanging. The saturation point increases with increasing temperature.

Saturation Mixing Ratio The mixing ratio of the atmosphere when it is saturated.

Saturation Specific Humidity The specific humidity of the atmosphere when it is saturated.

Saturation Vapor Pressure The vapor pressure of the atmosphere when it is saturated.

Savanna A vegetation type marked by wide areas of grasses interspersed with widely separated trees.

Scalar A quantity or property that possesses magnitude but has no direction. Examples are temperature, pressure, and density.

Scattering The dispersion or redirection of radiation by gases, dust, water drops, ice, and other particulates. Scattering does not heat the atmosphere because there is no energy transfer to the scattering agent. *See* Diffuse Radiation.

Scientific Method A framework used to make scientific discoveries about the physical world.

Scientific Theory Scientific hypotheses that have withstood repeated tests and have not been contradicted and are generally accepted as scientific fact.

Sea Breeze A flow of air from the water toward land along a coastal region.

Sea Breeze Front A boundary separating warm air inland from approaching cooler marine air associated with a sea breeze.

Sea Level Pressure The pressure that would presumably exist at a point if it were at sea level. This involves a conversion of observed surface air pressure.

Seasonal Outlook A forecast of the distribution of temperatures and precipitation for an upcoming multiple-month period.

Second Law of Thermodynamics Fundamental physical principal stating that heat moves from higher temperature toward lower temperature.

Secondary Pollutants Pollutants that form in the atmosphere from reactions involving anthropogenic or natural substances. See *Primary Pollutants*.

Secondary Rainbow A less distinct type of rainbow that sometimes accompanies a primary rainbow. These are always more extensive than primary rainbows.

Semidesert A type of dry climate that annually receives enough precipitation to distinguish it from a true desert.

Semipermanent Cell Large area of high or low pressure present year after year, usually with size and location changing seasonally.

Sensible Heat The energy contained in air that can be sensed via its temperature.

September Equinox One of two days every year when the Sun is overhead at the equator, on or about September 21. Also called the fall equinox in the Northern Hemisphere.

Severe Storm and Tornado Watches Public notices issued by the Storm Prediction Center advising of the possibility of tornadoes or other severe weather for particular regions.

Severe Thunderstorm A thunderstorm that produces either very strong winds, large hail, or tornadoes.

Severe Thunderstorm Warning An advisory issued by a local office of the National Weather Service indicating that severe thunderstorms are occurring or imminent.

Severe Weather Warning Advisory issued when severe weather is observed. In the case of a hurricane warning, the potential for landfall exists within a 24-hour period.

Severe Weather Watch Advisory issued when atmospheric conditions are favorable for severe weather. In the case of a hurricane watch, the potential for landfall exists outside a 24-hour period.

Sheet Lightning Lightning that is seen as an illumination of clouds by in-cloud or cloud-to-cloud discharges.

Shelf Cloud A portion of a severe thunderstorm cloud that protrudes ahead of the main portion of the cloud and above a gust front.

Short Wave A small wave in the midlatitude westerlies. Often superimposed on Rossby waves, they move more quickly, and thus travel through the large-scale pattern.

Shortwave Radiation Electromagnetic energy having wavelengths shorter than about 4 μm.

Shower Precipitation of short duration, especially rainfall.

Siberian High A semipermanent cell found in North Asia during winter.

Single-Cell Model A simple conception of global atmospheric circulation calling for one circulation cell in each hemisphere.

Sleet Precipitation in the form of ice pellets, resulting when raindrops freeze before reaching the surface.

Sling Psychrometer Device for measure humidity using a thermometer that is wetted and spun through the air.

Snow Frozen, crystalline precipitation that forms and remains in the ice stage throughout its descent.

Snow Pillow Mattress filled with antifreeze for measuring snow depth.

Snow Squalls Brief period of moderate to heavy snow.

Solar Altitude The angle between the horizon and the Sun. When the Sun is overhead, the solar altitude is 90°.

Solar Constant The amount of radiation reaching the top of the atmosphere when Earth is at its average distance from the Sun. This is not a pure constant, but rises and falls with changes in solar emission. Its value is about 1376 W/m^2.

Solar Declination The latitude of overhead Sun; the place where one would go to find the Sun directly overhead at noon.

Solar Disk The visible part of the Sun appearing as a circle in the sky. Responsible for most radiation emitted to Earth.

Solar Wind A continuous stream of particles (mostly protons and electrons) emitted by the Sun, traveling about one-third to one-half the speed of light.

Solstices The two times each year that mark the northern and southern limits of the latitude of overhead Sun. On the June solstice (approximately June 21) Northern Hemisphere latitudes have their longest day of the year. On the December solstice (approximately December 22), Southern Hemisphere latitudes have their longest day of the year.

Source Region A large area of land or ocean of more or less uniform characteristics, above which an air mass can form.

South Equatorial Current Eastward-flowing ocean current found in the Southern Atlantic, Pacific, and Indian Oceans.

Southeast Trade Winds Southern Hemisphere trade winds.

Southern Oscillation The reversal of surface pressure patterns over the tropical Pacific associated with El Niño events.

Specific Heat The amount of energy required to raise the temperature of a given mass of a substance by a given amount.

Specific Humidity A measure of atmospheric moisture. The mass of water vapor per unit mass of air, usually expressed in grams per kilogram (g/kg).

Specular Reflection Redirection of radiation (reflection) in a preferred direction, in contrast to diffuse reflection.

Speed A scalar property representing the rate of motion.

Speed Convergence The compaction of air due to decreasing wind speed in the downwind direction.

Speed Divergence The spreading of air due to increasing wind speed in the downwind direction.

Sprites Large, brief bursts of light emanating from cloud tops as lightning occurs.

Squall Line A linear band of thunderstorms, often found several hundred kilometers ahead of a cold front.

St. Elmo's Fire An electrical occurrence in which tall objects like church steeples or ships' masts glow and emit a continuous barrage of sparks accompanied by a hissing sound.

Stable Air Air that, when displaced vertically, returns to its initial position. Stable air resists uplift.

Standard Atmosphere The mean structure of the atmosphere with regard to temperature and pressure.

Static Stability The condition of the atmosphere that inhibits or favors vertical displacement of air parcels due to the effects of buoyancy.

Station Model A plotting on weather maps for individual locations depicting current temperature, dew point, pressure, and other meteorological information.

Stationary Front A transition zone between dissimilar air masses (a front) showing little or no tendency to move.

Steam Fog Fog that forms when cold air moves over a warmer water surface.

Stefan-Boltzmann Law The law for blackbody emission stating that the total energy emitted over all wavelengths is proportional to the fourth power of absolute temperature.

Steppe Another term for *semidesert*.

Stepped Leader A narrow zone of ionized air that serves as a conduit for an initial lightning stroke.

Storm Surge A potentially damaging influx of coastal waters brought about by high winds and low pressures associated with hurricanes.

Stratocumulus A low, layered cloud having superimposed rows or cells of vertical development evidenced by areas of differing whiteness.

Stratopause Upper limit of the stratosphere; the transition between the stratosphere and mesosphere.

Stratosphere A layer of the atmosphere between about 16 and 50 km (10 and 30 mi), characterized by generally increasing temperature with increasing altitude.

Stratus A cloud with a layered structure.

Streamlines Lines that depict the direction of airflow. Air parcels are envisioned to flow along streamlines. *See* confluence *and* diffluence.

Stroke or Return Stroke The flow of electricity associated with an individual bolt of lightning.

Structure The layering of the atmosphere combined with the reduction in density with altitude.

Stuve Diagram A particular type of thermodynamic diagram used for plotting temperature and moisture profiles.

Subarctic One of the two types of severe midlatitude climates present over North America and Asia.

Sublimation Change from a solid into a vapor without passing through the liquid phase. Also used to describe the reverse (vapor to solid).

Subpolar Low A belt of low pressure in the three-cell model, between the polar easterlies and midlatitude westerlies.

Subtropical Desert Low-latitude arid regions formed primarily by subsiding air associated with the Hadley cell.

Subtropical High A semipermanent cell that occupies large areas of the midlatitude oceans, especially in the warm season.

Subtropical Jet Stream A jet stream common in the upper troposphere on the poleward side of the Hadley cells, produced by the conservation of angular momentum.

Subtropical Steppe A semiarid, low-latitude climate.

Suction Vortex A zone of intense rotation within a large tornado that often causes the most devastation.

Sulfur Dioxide (SO_2) An air pollutant consisting of a sulfur atom and two oxygen atoms.

Sulfur Oxides A general class of pollutants consisting of sulfur and oxygen.

Sulfur Trioxide (SO_3) A compound consisting of a sulfur atom and three oxygen atoms. Not harmful by itself but it can combine with water droplets to form acid fog or acid rain.

Sun Pillars Bands of light stretching vertically from the Sun caused by reflection off almost horizontally aligned ice crystals.

Sundogs Paired bright spots found 22° to the right or left of the Sun caused by ice crystal refraction.

Sunspots Magnetic storms of the Sun, appearing as dark (Earth-sized) spots on the photosphere.

Supercells A very large thunderstorm formed from an extremely powerful updraft.

Supercell Thunderstorm *See* Supercells.

Supercooled Water Water existing in the liquid phase with a temperature less than 0 °C.

Superior Mirage An atmospheric effect on visible light that causes objects at the surface to appear taller or higher in the horizon than they really are.

Supersaturated A relative humidity greater than 100 percent, when the atmosphere is more than saturated with water vapor. Requires a very clean atmosphere, where condensation nuclei are lacking.

Supersaturation *See* Supersaturated.

Surface Maps Weather maps that describe the weather at the surface.

Surface Pressure Atmospheric pressure near the surface, usually measured 2 m above the surface along with temperature and other variables in an instrument shelter.

Sweep One rotation of a radar transmitter/receiver unit to gather data.

Synoptic Scale The scale of meteorological phenomena having areas on the order of hundreds or thousands of square kilometers.

T

Taiga Another name for the boreal forest.

Teleconnection Relationship between weather or climate patterns at two widely separated locations.

Temperature Advection The importation of warm or cold air as air flows.

Temperature Gradient Temperature change per unit distance. A strong temperature gradient implies that temperature changes rapidly over a short distance.

Temperature Inversion *See* Inversion.

Temperature An index of the average kinetic energy of the molecules comprising a substance.

Terminal Velocity The final speed obtained by an object falling through the atmosphere, when friction with the surrounding air balances the force of gravity.

Thermal Low Low-pressure cell produced by heating of the surface.

Thermistor An object whose electrical resistance changes with temperature, thus allowing temperature to be determined by measuring changes in electrical current.

Thermodynamic Diagram A diagram showing the relationship between pressure, temperature, density, and water vapor content, such that characteristics of air parcels can be determined as they ascend and descend.

Thermoelectric Effect A theory of lightning formation in which separation of charge is produced by positive ions migrating from warmer particles to colder ice crystals.

Thermograph Device that produces a continuous temperature record in the form of a chart.

Thermohaline Circulation A movement of surface waters in the oceans due to variations in temperature and salt content.

Thermometer Instrument used to measure temperature.

Thermosphere Outermost reaches of the atmosphere, beginning at about 80 km (50 mi), characterized by increasing temperature with increasing altitude and by extremely low density.

Thornthwaite's Classification System The most widely used system for classifying general climate zones.

Threat Score A measure of precipitation forecast skill that considers the area correctly forecast relative to that under threat of precipitation.

Three-Cell Model A generalized description of global-scale circulation that calls for three large cells in each hemisphere. The cells rotate on a vertical plane with axes parallel to latitude lines, thereby moving heat and moisture in a north–south direction.

Thunder Sound produced when lightning discharges heat into the surrounding air, causing pressure waves to emanate outward.

Tibetan Low A semipermanent cell found in southern Asia in summer.

Tipping-Bucket Gage A type of automated rain gage.

Tornado Outbreak A severe storm that has spawned at least six tornadoes.

Tornado A rotating column of air with extreme horizontal winds.

Tornado Warning An advisory issued by a local office of the National Weather Service indicating that a tornado is occurring or imminent.

Trade Wind Inversion A temperature inversion (layer of air having increasing temperature with altitude) commonly found at subtropical and tropical latitudes.

Trade Winds Prevailing lower troposphere winds of the tropics, associated with Hadley circulation. Strongest in the respective winter season, the trades blow from the northeast in the Northern Hemisphere and from the southeast in the Southern Hemisphere. *See also* Northeast Trade Winds *and* Southeast Trade Winds.

Transpiration Transfer of water to the atmosphere by vegetation, mostly by water evaporating and escaping plant tissues through leaf pores.

Tree Rings Patterns of growth inside tree trunks, with each ring representing one year's worth of growth. Examination of tree ring widths can provide proxy information about past climates.

Tropic of Cancer A line of latitude at 23.5° N, the northern limit of solar declination.

Tropic of Capricorn A line of latitude at 23.5° S, the southern limit of solar declination.

Tropical Depression A closed zone of low pressure with wind speeds less than 60 km/hr (35 mph).

Tropical Disturbance A disorganized group of thunderstorms with weak pressure gradients and little or no rotation.

Tropical Storm Warning A notice that sustained winds of 63 to 118 km/hr (39 to 73 mph) are expected somewhere within the specified area within 36 hours.

Tropical Storm Watch Similar to a tropical storm warning, but issued 48 hours in advance of the expected onset of tropical-storm-force winds.

Tropical Storm A storm that originates in tropical regions and has wind speeds between 60 and 120 km/hr (35 and 70 mph).

Tropical Wet and Dry Low-latitude climates that frequently border arid regions. They are wet most of the year but have a distinct dry season.

Tropical Wet Near-equatorial climates that are rainy year-round.

Tropopause Upper limit of the tropopause; the transition between the troposphere and stratosphere.

Troposphere The lowest temperature layer of the atmosphere, from the surface to about 16 km, characterized by generally decreasing temperatures with increasing altitude.

Trough An elongated axis of low pressure.

Tundra A low-lying vegetation type associated with permafrost regions.

Turbidity Loosely speaking, the "dustiness" of the atmosphere, including the effect of all particulates that reduce visibility.

Twilight The diffuse radiation seen near the horizon after the Sun has set.

Typhoon A strong tropical system similar to a hurricane but located in the western Pacific Ocean.

U

Ultraviolet Radiation Electromagnetic radiation at wavelengths too short to be visible, from about 0.001 to 0.4 μm.

Unstable Air Air that experiences a buoyant force following a vertical displacement, causing it to rise. Uplift is promoted by instability.

Upslope Fog Very low cloud (fog) formed when air flows upward along sloping ground and cools adiabatically to the dew point.

Upwelling Movement of ocean or lake water from lower levels toward the surface.

Urban Heat Island Increased local temperatures that result from urbanization.

V

Valley Breeze A low-level movement of air in an upslope direction, developing during the daylight hours as result of solar heating.

Vapor Pressure A measure of atmospheric moisture, the partial pressure exerted by water vapor.

Variable Gases Gases present in amounts that vary greatly in abundance, either vertically, horizontally, or seasonally. Water vapor is the most important variable gas.

Vault *See* Bounded Weak Echo Region.

Vector A quantity or property possessing both magnitude and direction. Examples are wind velocity, the Coriolis force, and the pressure gradient force.

Veering Wind Wind that changes direction in a clockwise sense.

Velocity A vector property that includes speed and motion of an object or substance.

Vertical Pressure Gradient Force The upward-directed force that arises because pressure always decreases with increasing altitude.

Virga Rain that falls into a layer of warm dry air and evaporates before reaching the surface.

Visible Images Satellite images based on observations of reflected visible radiation from the surface and clouds.

Visible Radiation Electromagnetic radiation between about 0.4 and 0.7 μm, which is detectable by the human eye.

Volatile Organic Compounds Carbon-hydrogen molecules, both anthropogenic and naturally produced, which can be gaseous or particulate. A precursor to photochemical smog.

Volume The result of multiple sweeps of a radar transmitter/receiver unit, each directed at different angles relative to the horizon, to provide a three-dimensional view of storm activity.

Vorticity The turning of an object (such as an air parcel), usually with respect to the vertical direction. This is important to meteorology because of its association with areas of divergence and convergence.

W

Walker Circulation An east–west circulation pattern of the tropics, characterized by several cells of rising and sinking air connected by horizontal motions along more or less parallel lines of latitude.

Wall Cloud Thick cloud beneath a rotating thunderstorm, a place where severe weather often develops.

Warm Advection Heat carried by airflow across isotherms from warm to cold.

Warm Cloud Clouds with temperature above 0°C throughout.

Warm Core High High-pressure cells with higher temperature than surrounding air. Also called *warm core anticyclones*, they originate from atmospheric motion, not from differences in heating.

Warm Core Low Low-pressure cells that are warmer than surrounding air, produced by heating. Also called *warm core cyclones*.

Warm Front Transition zone between two air masses of different temperature, produced when warm air advances on and overruns cold air.

Water Vapor Images Images depicting the distribution of water vapor and cloud cover.

Water Vapor Water in its gaseous phase, not to be mistaken for small water droplets. Colorless and odorless, it seldom amounts to more than a few percent of the total atmospheric mass.

Waterspout A rather weak whirlwind (narrow rotating column of air) forming over a water surface. Rising and condensing air makes the waterspout visible.

Watt The SI unit of power, abbreviated as W, with dimensions of energy per unit time; 1 W 5 1 J/sec.

Wave Cyclone Low-pressure storm common in midlatitudes, typically formed in a favored position along a Rossby wave.

Wavelength Distance between successive peaks of a wave, or successive troughs, or between any two corresponding points along a wavetrain.

Weather Forecast Office A United States Weather Service facility that issues local and area forecasts.

Weather Day-to-day conditions of the atmosphere.

Weighing-Bucket Rain Gage A type of automated gage for the measurement of precipitation.

West Greenland Drift A branch of a current flowing southward in the North Atlantic.

Westerlies Winds belts found in the middle latitudes of both hemispheres that have a strong west-to-east component.

Wet Bulb Depression The difference between air temperature and wet bulb temperature. Large values (large differences between dry and wet bulb temperatures) are indicative of low humidity.

Wet Bulb Temperature The minimum temperature achieved as water evaporates from a wick surrounding a thermometer's bulb.

Wien's Law Law for blackbody emission that states that the wavelength of maximum emission is inversely proportional to absolute temperature.

Wind The horizontal movement of air.

Wind Chill Temperature Index An index of apparent temperature used for cold conditions that incorporates air temperature and wind speed.

Wind Profiler A type of Doppler radar unit that provides information about vertical changes in wind speed and direction.

Wind Shear Large change in wind direction or speed over a short distance.

Wind Vane An instrument for measuring or indicating wind direction, often a pivoting arrow that continually points into the wind.

WMO Stands for the World Meteorological Organization, the agency of the United Nations charged with collecting and disseminating meteorological data.

World Meteorological Centers Offices in Moscow, Russia; Washington, D.C.; and Melbourne, Australia; that disseminate weather information to national weather agencies around the world.

World Meteorological Organization (WMO) The agency of the United Nations charged with collecting and disseminating meteorological data.

X

X-Ray Radiation Electromagnetic radiation at wavelengths between about 0.00001 μm and about 0.01 μm.

Y

Younger Dryas A 1200-year cold period that interrupted a period of warming that began about 15,000 years ago and eventually gave way to the current Holocene epoch.

Z

Zenith Angle The angle between the Sun and the vertical direction. When the Sun is overhead, the zenith angle is zero. *See* solar altitude. *See* Solar Altitude.

Zonal Wind (Flow) Wind flowing east–west, parallel to a line of latitude. Actual winds are seldom completely zonal but usually have both a meridional and zonal component.

Zone Forecast A weather forecast issued at regular intervals for a particular region.

Credits

Chapter 1

Photographs

Part 1 Opener, Henglein and Steets/Age Fotostock;- Chapter 1 Opener, Castle Light Images/Alamy Live News; 1–1a, Terry Donnelly/Alamy; 1–1b, Pwollinga/Fotolia; 1–1c, Simanovskiy/Fotolia; 1–1d, Jason Friend/Alamy; 1–1–1, Jessica Rinaldi/Reuters; 1–3, NASA; 1–3–1, NASA; 1–5a, Space Science and Engineering Center/SSEC; 1–5b, Space Science and Engineering Center/SSEC; 1–7, Guentermanaus/Shutterstock; 1–9, Andrew Pielage/ZUMA Press/Newscom; 1–11, Andrew Lambert Photography/Science Source; 1–14, Bob Gurr/All Canada Photos/SuperStock; 1–15, Pi-Lens/Shutterstock; 1–16a, National Oceanic and Atmospheric Administration; 1–16b, National Oceanic and Atmospheric Administration; 1–18a, Oxford Science Archive/Heritage Images/Glow Images; 1–18b, James MacPherson/AP Images

Illustrations

1–2b, Data from Travis, David J., Andrew M. Carleton, Ryan G. Lauritsen, "Contralils reduce daily temperature range", Nature, 418 (2002), p. 601.; 1–6, Courtesy of National Oceanographic and Atmospheric Agency; 1–8, Pearson Education; 1–16a, Courtesy of National Oceanic and Atmospheric Administration (NOAA)

Tables

1–1, National Oceanic and Atmospheric Administration (NOAA) http://www.ncdc.noaa.gov/billions/events

Chapter 2

Photographs

Chapter 2 Opener, Naoki Mutai/Getty Images; 2–5a, National Optical Astronomy Observatories; 2–5b, National Optical Astronomy Observatories; 2–9, National Oceanic and Atmospheric Administration; 2–16, T. Walker/Photri Images/Alamy; 2–Visual Analysis, Stasique/Shutterstock

Illustrations

2–1, Pearson Education; 2–B1–1, Pearson Education; 2–B2–1, Pearson Education; 2–B2–2, Pearson Education; 2–7, Pearson Education; 2–8, Based on Ahrens, Donald. 2012. Meteorology Today 10e. Cengage: pg. 41.

Chapter 3

Photographs

Chapter 3 Opener, Justin Sullivan/Getty Images; 3–1, TerraGraphic/Alamy; 3–4, NASA; 3–6, Lawrence Wee/Shutterstock; 3–11b, Stephen Mcsweeny/Shutterstock; 3–21, Edward Aguado; 3–26, All Weather, Inc.; 3–27, Edward Aguado

Illustrations

3–7, National Renewable Energy Laboratory; 3–10b, J. N. Howard, Proc. I. R. E., v.47, 1959 and R. M. Goody and G. D. Robinson, Quart. J. Roy. Meterol. Soc. v. 77, 1951 as modified by R. G Fleagle and J. A. Businger, An Introduction to Atmospheric Physics, 2nd Editon, ©1980 Acedemic Press, New York, NY page 352, Academic Press, Inc.; 3–16, Based on Christopherson, Geosystems, An Introduction to Physical Geography, 5e, pg. 131, Figure 5.14 Prentice Hall, Upper Saddle River, NJ, Prentice-Hall (Pearson Prentice-Hall) 2005; 3–18, a, b, c, Based on Christopherson, Geosystems,An Introduction to Physical Geography, 5e, pg. 131, Figure 5.14 Prentice Hall, Upper Saddle River, NJ; 3–29b, From The Atlas of Canada, Natural Resources Canada: http://geogratis.gc.ca/api/en/nrcan-rncan/ess-sst/fd8efb83-b73d-5442-ab60-7987c824f5fd.html?pk_campaign=recentItem; 3–32, National Oceanic and Atmospheric Administration (NOAA). National Climatic Data Center (NCDC)/National Environmental Satellite, Data, and Information Service (NESDIS)/NOAH; 3–33, IPCC, 2013: Summary for Policymakers. In: Climate Change 2013: The Physical Science Basis. Working Group I Contribution to the Fifth Assessment Report of the Intergovernmental Panel on Climate Change, Figure SPM.5. Cambridge University Press. Source: Courtesy of the University of Wyoming, Department of Atmospheric Science. www.climatesource.com/map_gallery.html

Tables

3–1, Based on Peixoto and Oort, Physics of Climate, American Institute of Physics, 1992.

Chapter 4

Photographs

Chapter 4 Opener, Nie Jianjiang/Xinhua/Photoshot/Newscom; Chapter 4 Opener-Additional, Courtesy Jeff Schmaltz/EOSDIS MODIS Rapid Response Team/NASA; 4–2, Momesso/Fotolia; 4–3–2, Phillip Roullard/ZUMA Press/Newscom; 4–4b, Charles D. Winters/Science Source; 4–5a, Edward Aguado; 4–5c, Edward Aguado; 4–7a, Phil Hossack/Winnipeg Free Press; 4–7b, Cade Malone; 4–16, NOAA/NASA; 4–26, Edward Aguado; 4–Visual Analysis, NOAA

Illustrations

4–5b, Based on Morgan and Moran, Meteorology, The Atmosphere and the the Science of Weather, 5th Edition, ©1997 Prentice Hall, Upper Saddle River; 4–24, Courtesy of the National Oceanic and Atmospheric Administration (NOAA); 4–6–1, Data from Pilot Outlook: http://www.pilotoutlook.com/aviation_weather/atmospheric_pressure_and_altimetry.; 4–6–2, Data from Pilot Outlook: http://www.pilotoutlook.com/aviation_weather/atmospheric_pressure_and_altimetry.

Chapter 5

Photographs

Part 2 Opener, SDC/Shutterstock; Chapter 5 Opener, Horns Rev 1 owned by Vattenfall. Photographer Christian Steiness; 5–3, Arttdi/Shutterstock; 5–7–1, Chuck Berman/Chicago Tribune/MCT/Newscom; 5–9a, Bobbe Christopherson; 5–9b, Yen-yu Shih/Fotolia; 5–12, Oksana/Perkins/Shutterstock; 5–15, Philippe Lopez/AFP/Getty Images; 5–19a, Gnatuk/Shutterstock; 5–19b, Anette Linnea Rasmussen/Shutterstock; 5–19c, Age Fotostock/SuperStock; 5–20, Michael Hare/Shutterstock; 5–21, NASA; 5–22, Kavram/Shutterstock; 5–26, Raminder Pal Singh/EPA/Alamy

Illustrations

5–4–2, Courtesy of the National Oceanic and Atmospheric Administration (NOAA); 5–10, National Climatic Data Center/National Oceanic and Atmospheric Administration (NOAA); Source: Understanding Humidity. USA Today. www.usatoday.com/weather/whumdef.htm

Tables

5–5, Courtesy of the National Oceanic and Atmospheric Administration (NOAA)

Chapter 6

Photographs

Chapter 6 Opener, JR Hott/Splash News/Corbis; 6–1b, Schimmelpfennig/Premium Stock Photography GmbH/Alamy; 6–1c, Jose Antonio Santiso Fernandez/Getty Images; 6–2a, Edward Aguado; 6–2b, Edward Aguado; 6–2c, Edward Aguado; 6–3–1a, Mark Crosse/The Fresno Bee/AP Images; 6–3–1b, Bill Ingram/ZUMA Press/Corbis; 6–16, Joyce Photographics/Science Source; 6–17, Jerome Wilson/Alamy; 6–18, Maurice Nimmo/Frank Lane Picture Agency Limited; 6–19, Gianni Muratore/Alamy; 6–20, National Oceanic and Atmospheric Administration; 6–21, Brian Cosgrove/Dorling Kindersley; 6–22, Pekka Parviainen/Science Source; 6–23, Edward Aguado; 6–24, Brian Cosgrove/Dorling Kindersley; 6–25, James Steinberg/Science Source; 6–26, Brian Cosgrove/Dorling Kindersley; 6–27, Science Source; 6–28, BanksPhotos/Getty Images; 6–29, Dean Pennala/Shutterstock; 6–30, TRVScenics/Alamy; 6–31, LorenRyePhoto/Alamy; 6–32, Chad Carpenter/ZUMA Press/Newscom; 6–33, Pekka Parviainen/Science Source; 6–34, University Corporation for Atmospheric Research; 6–35a, b, c, National Oceanic and Atmospheric Administration; 6–Visual Analysis, Stian Olsen/Alamy

Chapter 7

Photographs

Chapter 7 Opener, Rogelio V. Solis/AP Images; 7–3–1, Michael Gallagher/AP Images; 7–4–1, Gerald Herbert/AP Images; 7–4–1, Bruce Crummy/ZUMA Press, Inc./Alamy; 7–5, Doug Millar/Science Source; 7–9a, Anest/Shutterstock; 7–9b, Science Source; 7–9c, Ted Kinsman/Science Source; 7–11, Rick Olivo/The Daily Press/AP Images; 7–16, EdgeOfReason/Shutterstock; 7–18a, Travis Heying/The Wichita Eagle/AP Images; 7–18b, University Corporation for Atmospheric Research/Science Source; 7–23, Kent D. Johnson/AP Images; 7–25a, Edward Aguado; 7–Visual Analysis, National Oceanic and Atmospheric Administration

Illustrations

7–1, Adapted from McDonald, The Physics of Cloud Formation, Advances in Geophysics, v.5, Academic Press, NY, NY, as modified by Rogers, A Short Course in Cloud Physics, © 1979, Pergamon Press, NY, NY; 7–13, Based on Severe and Hazardous Weather: An Introduction to High Impact Meteorology, Rauber et al, Kendall Hunt Publishing; 3rd edition (April 7, 2009), Kendall Hunt Publishing Company; 7–14, National Oceanic and Atmospheric Administration (NOAA) Observed snowfall, 01/27/2014 Buffalo, NY; 7–24, Based on Severe and Hazardous Weather: An Introduction to High Impact Meteorology, Rauber et al, Kendall Hunt Publishing; 3rd edition (April 7, 2009), Kendall Hunt Publishing Company; Visual Analysis, National Oceanic and Atmospheric Administration (NOAA)

Chapter 8

Photographs

Part 3 Opener, Bill Gozansky/Alamy; Chapter 8 Opener, California Department of Water Resources; Chapter 8 Opener-1, California Department of Water Resources; 8–2a, NASA; 8–2b, NASA; 8–12, NASA; 8–17a, National Oceanic and Atmospheric Administration; 8–20, Rosenstiel School of Marine and Atmospheric Science; 8–26, Rob Sheppard/Danita Delimont Photography/Newscom; 8–27, Chris Carlson/AP Images; 8–28, NASA; 8–31, NASA

Illustrations

8–1, National Drought Mitigation Center at the University of Nebraska-Lincoln, the United States Department of Agriculture, and the National Oceanic and Atmospheric Administration http://droughtmonitor.unl.edu; 8–6, a, b, Based on Christopherson, Geosystems, An Introduction to Physical Geography, 3rd Ed. ©1997 Prentice Hall, Inc., Upper Saddle River, NJ; 8–8 a, b., Based on Miller et al., Elements of Meteorology, ©1983, Merrill Publishing Co., Columbus, OH; 8–17b, Copyright © 2014 Gimeno, Nieto, Vázquez and Lavers, Gimeno L, Nieto R, Vázquez M and Lavers DA (2014) Atmospheric rivers: a mini-review. Front. Earth Sci. 2:2. doi: 10.3389/feart.2014.00002; 8–18, Based on Trujillo and Thurman, Essentials of Oceanography, 6th Ed., ©1999 Prentice Hall, Inc., Upper Saddle River, NJ; 8–19, Based on Trujillo and Thurman, Essentials of Oceanography, 6th Ed., ©1999 Prentice Hall, Inc., Upper Saddle River, NJ; 8–20b, Based on Trujillo and Thurman, Essentials of Oceanography, 6th Ed., ©1999 Prentice Hall, Inc., Upper Saddle River, NJ; 8–22 a (map), Based on Trujillo and Thurman, Essentials of Oceanography, 6th Ed., ©1999 Prentice Hall, Inc., Upper Saddle River, NJ; 8–22 b (map), Based on Trujillo and Thurman, Essentials of Oceanography, 6th Ed., ©1999 Prentice Hall, Inc., Upper Saddle River, NJ; 8–33, Based on Peixoto and Oort, Physics of Climate, New York: American

Institute of Physics ©1992, p. 419; 8–35, Courtesy of the Climatic Diagnostics Center/ National Oceanic and Atmospheric Administration (NOAA); 8–36, Courtesy of the Climatic Diagnostics Center/ National Oceanic and Atmospheric Administration (NOAA); 8–37, Courtesy of the Climatic Diagnostics Center/ National Oceanic and Atmospheric Administration (NOAA); 8–38, Based on the National Oceanic and Atmospheric Administration. http://www.cpc.ncep.noaa.gov/products/analysis_monitoring/ensostuff/ensoyears.shtml; 8–39 a, b, c, d, Courtesy of the Climatic Prediction Center/National Oceanic and Atmospheric Administration (NOAA); 8–40 a, b, c, d., Courtesy of the Climatic Diagnostics Center/National Oceanic and Atmospheric Administration (NOAA); 8–41, Based on Yu et al, The changing impact of El Nino on US winter temperatures, Geophysical Research Letters, Vol 39, 2012; 8–42, TRMM/NASA/Goddard Space Flight Center Australian Rainfall Contributed by Tropical Cyclones http://trmm.gsfc.nasa.gov/publications_dir/australia_tc_rain_apr_2011.html; 8–43, Courtesy of National Aeronautics and Space Administration (NASA)/Jet Propulsion Laboratory(JPL)/California Institute of Technology; 8–46, Climate Prediction Center/National Oceanic and Atmospheric Administration (NOAA). http://www.cpc.ncep.noaa.gov/data/teledoc/nao_ts.shtml; Visual Analysis, Courtesy of the National Oceanic and Atmospheric Administration (NOAA)/National Weather Service (NWS)/National Centers for Environmental Prediction (NCEP)/Environmental Modeling Center (EMC) Marine Modeling and Analysis Branch

Tables

8–1, Source: California Air Resources Board.

Chapter 9

Photographs

Chapter 9 Opener, David McNew/Getty Images; 9–2–1, Dennis Chesters/GOES/NASA; 9–4–3, Craig Ruttle/AP Images; 9–4–4b, NASA; 9–8a, Plymouth State Archive; 9–8b, Plymouth State Archive

Illustrations

9–3–1, 9, Courtesy of National Oceanic and Atmospheric Administration (NOAA); 9–4–1, 9, Courtesy of the National Oceanic and Atmospheric Administration (NOAA); 9–8a and b, Courtesy of the National Weather Service; 9–5–1, Density Altitude, Federal Aviation Administration; 9–5–2, Density Altitude, Federal Aviation Administration

Tables

9–1, Modifed from Moran and Morgan, Meteorology, The Atmosphere and the Science of Weather, Fifth Edition, © 1997, Prentice Hall, Inc., Upper Saddle River, NJ

Chapter 10

Photographs

Part 4 Opener, Glenn Bartley/All Canada Photos/Alamy; Chapter 10 Opener, US Army Photo/Alamy; 10–1–2, Kristina Barker/Rapid City Journal/AP Images; 10–11c, NOAA; 10–12c, NOAA; 10–13c, NOAA; 10–14c, NOAA; 10–18B, NOAA

Illustrations

10–1–1, Courtesy of the National Oceanic and Atmospheric Administration (NOAA); 10–7, Adapted from Lutgens and Tarbuck, The Atmosphere, An Introduction to Meteorology, Ninth Edition, © 2004, Prentice Hall, Inc., Upper Saddle River, NJ.; Source: Courtesy of the University of Wyoming, Department of Atmospheric Science. weather.uwyo.edu/upperair/uamap.html; Source: Courtesy of the University of Wyoming, Department of Atmospheric Science. weather.uwyo.edu/upperair/uamap.html.

Chapter 11

Photographs

Chapter 11 Opener, Corey Hochachka/Age Fotostock; 11–1, National Oceanic and Atmospheric Administration; 11–3, Alexey Stiop/Fotolia; 11–3–1, Westend61 GmbH/Alamy; 11–4–3, Carsten Peter/National Geographic/Getty Images; 11–5, North Wind Picture Archives/Alamy; 11–6, U.S. Air Force; 11–6–1, Jason Clark/Southcreek Global/ZUMA/Alamy; 11–6–2, J.B. Forbes/MCT/Newscom; 11–6–3, Simon Brewer/Corbis; 11–7, UAF/NASA; 11–7–1, Vincent Deligny/AFP/Getty Images/Newscom; 11–8, Victor Habbick Visions/Science Source; 11–10, Larry Miller/Science Source; 11–11, NOAA; 11–16, Jason Patrick Ross/Shutterstock; 11–18, Howard Bluestein/Science Source; 11–24, DOD Photo/Alamy; 11–28a, Erik Nguyen/Eureka Premium/Corbis; 11–28b, Mike Theiss/National Geographic/Getty Images; 11–30a, Minerva Studio/Fotolia; 11–30b, Shawn Yorks/The Guymon Daily Herald/AP Images; 11–37b, Shalyn Phillips/Demotix/Corbis; 11–38, Bardell Andreas/ZUMA/Newscom; 11–45, Karen Anderson Photography/Flickr/Getty Images

Illustrations

11–2, 11–14, Based on Wallace and Hobbs, Atmospheric Science, An Introductory Survey, © 1977, Academic Press, New York, NY; 11–9, Based on Gedzelman, The Science and Wonders of the Atmosphere, © 1980, Wiley, Inc., New York, NY; 11–12, Image from Roger Edwards/ National Oceanic and Atmospheric Administration (NOAA) Storm Prediction Center; 11–19, Based on Djuric, Weather Analysis, © 1994, Prentice Hall, Inc., Upper Saddle River, NJ; 11–23, Courtesy of the National Oceanic and Atmospheric Administration (NOAA); 11–27, "The North American Lightning Detection Network (NALDN)—First Results: 1998–2000" by R. Orville in Monthly Weather Review 130(8): 2098–2109, © 2002 American Meteorological Society.; 11–32, 11–33, Based on Brandes, in Church et al., eds., The Tornado: Its Structure, Dynamics, Prediction, and Hazards, © 1993, American Geophysical Union, Washington, DC; 11–34, Based on McKnight, Physical Geography, A Landscape Appreciation, © 1996, Prentice Hall, Inc., Upper Saddle River, NJ; 11–35, Based on Christopherson, Geosystems: An Introduction to Physical Geography, Sixth Edition, © 2006, Prentice Hall, Inc., Upper Saddle River, NJ.; 11–36, Courtesy of the National Oceanic and Atmospheric Administration (NOAA); 11–37a, Based on Fujita, Journal of Atmospheric Sciences, © 1981, American Meteorological Society, Boston, MA, as modified by Battan, Fundamentals of Meteorology, © 1984, Prentice Hall, Inc., Upper Saddle River, NJ; 11–40, Courtesy of the National Oceanic and Atmospheric Administration (NOAA); 11–41, Courtesy of the National Oceanic and Atmospheric Administration (NOAA)

Chapter 12

Photographs

Chapter 12 Opener, NASA; 12–1, NASA; 12–2b, Anam Collection/Alamy; 12–3–1, Michael Reynolds/EPA/Newscom; 12–4–2, NOAA; 12–4–3, NOAA; 12–4–4, Rick Wilking/Reuters; 12–5–1, Corbis; 12–5–2, Eric Kayne/Houston Chronicle/Rapport Press/Newscom; 12–6–1, Neil Cooper/Alamy; 12–6–2, NASA; 12–7–2, U.S. Coast Guard; 12–9a, National Oceanic and Atmospheric Administration; 12–9b, National Oceanic and Atmospheric Administration; 12–9c, National Oceanic and Atmospheric Administration; 12–19, Jennifer Weisbord/Polaris/Newscom; 12–20a, USGS Information Services; 12–20b, National Oceanic and Atmospheric Administration; 12–21, NASA; 12–22a, Wilfredo Lee/AP Images; 12–24, Bob Pearson/EPA/Newscom

Illustrations

12–2–1, National Hurricane Center/National Oceanic and Atmospheric Administration (NOAA); 12–3–1 a-f, National Weather Service/National Oceanic and Atmospheric Administration (NOAA); 12–3–3, National Weather Service/National Oceanic and Atmospheric Administration (NOAA); 12–4, From Christopherson, Geosystems, 9e; 12–6, Adapted from Miller, Science, v. 157, 1967, American Association for the Advancement of Science, as modified by Eagleman, Meteorology: The Atmosphere in Action, Second Edition, ©1985, Wadsworth, Inc., Belmont, CA, Eagleman, Thomas R.; 12–6–2 a, b, c, d, National Oceanic and Atmospheric Administration (NOAA); 12–6–3, National Oceanic and Atmospheric Administration (NOAA); 12–7, Anthes et al., Three-Dimensional Particle Trajectories in a Model Hurrican." Weatherwise, Vol. 24, 1971; 12–8, National Aeronautics and Space Administration (NASA), Earth Observatory; 12–10, National Aeronautics and Space Administration (NASA)Tropical Rainfall Measuring Mission (TRMM); 12–12, National Oceanic and Atmospheric Administration; 12–13, National Oceanic and Atmospheric Administration's National Climatic Data Center 2010; 12–14, Based on Moran and Morgan, Meteorology: The Atmosphere and the Science of Weather, 5th Edition, copyrights © 1997, Prentice Hall, Inc., Upper Saddle River; 12–7–1, Data from the Bureau of Transportation Statistics, Airline On-Time Data. http://www.rita.dot.gov/bts/sites/rita.dot.gov.bts/files/subject_areas/airline_information/special_features/hurricane_sandy/index.html.; 12–7–3, Data from IATA Economic Brief: The Impact of Hurricane Sandy. Novemeber 2012. International Air Transport Association.; 12–15, a, b, c, National Hurricane Center/National Oceanic and Atmospheric Administration (NOAA); 12–15, National Hurricane Center/National Oceanic and Atmospheric Administration (NOAA); 12–17, Adapted from Novian and Gray, Monthly Weather Review, © 1974 American Meteorological Society, Boston, MA, as modified by unpublished National Weather Service Training Center pamphlet, American Meteorological Society; 12–23, National Oceanic and Atmospheric Administration (NOAA); 12–VA12–1, National Hurricane Center/ National Oceanic and Atmospheric Administration (NOAA); 12–VA–12–2, National Hurricane Center/ National Oceanic and Atmospheric Administration (NOAA)

Tables

12–1, Source: Atlantic Oceanographic and Meteorological Laboratory, National Oceanic and Atmospheric Administration (NOAA)

Chapter 13

Photographs

Part 5 Opener, Daniel Leal-Olivas/Corbis; Chapter 13 Opener, Robert Galbraith/Reuters/Corbis; 13–1–1, Marvin Gentry/Reuters /Landov; 13–2, Edward Aguado; 13–4, Edward Aguado; 13–7, Edward Aguado; 13–20a, Space Science and Engineering Center, University of Wisconsin, Madison; 13–20b, Space Science and Engineering Center, University of Wisconsin, Madison; 13–20c, Space Science and Engineering Center, University of Wisconsin, Madison

Illustrations

13–1, © Pearson Education, Inc.; 13–2–1, Aviation Weather Center/National Oceanic and Atmospheric Administration (NOAA); 13–3, Department of Commerce/National Oceanic and Atmospheric Administration/National Centers for Environmental Prediction/DOC/NOAA/NECP/Weather Prediction CenterDOC/NOAA/NECP/Weather Prediction Center; 13–3, National Oceanic and Atmospheric Administration (NOAA); 13–6, National Oceanic and Atmospheric Administration (NOAA); 13–7, Adapted from NOAA and Thomas DeFelice, Introduction to Meteorological Instrumentation and Measurement, ©1998, Prentice Hall, Inc. Upper Saddle River, NJ. By permission of Thomas DeFelice; 13–9a. & b., National Oceanic and Atmospheric Administration (NOAA); 13–10, National Centers for Environmental Prediction/NCEP/National Oceanic and Atmospheric Administration (NOAA); 13–11 a & b, Climate Prediction Center/National Oceanic and Atmospheric Administration (NOAA); 13–12a & b, Climate Prediction Center/National Oceanic and Atmospheric Administration (NOAA); 13–13, Weather Prediction Center/National Oceanic and Atmospheric Administration (NOAA); 13–15, National Weather Service Storm Prediction Center/National Oceanic and Atmospheric Administration (NOAA; 13–16, National Weather Service Storm Prediction Center/National Oceanic and Atmospheric Administration (NOAA; 13–17, National Oceanic and Atmospheric Administration (NOAA); 13–18, National Oceanic and Atmospheric Administration (NOAA); 13–21, National Oceanic and Atmospheric Administration (NOAA); 13–22, 72501 OKX Upton, 12Z 22 Jan 2014, National Oceanic and Atmospheric Administration (NOAA)

Chapter 14

Photographs

Chapter 14 Opener, Liu Chang/Legal Evening/Reuters; 14–1–1a, Carnegie Library of Pittsburgh; 14–1–1b, Arthur Smith; 14–2–1, Stocktrek Images/Alamy; 14–4a, Science Source; 14–4b, Mary Terriberry/Shutterstock; 14–6, Horst Mahr/imagebroker/Alamy; 14–7, Paul Rushton/Alamy; 14–8a, Jose Gil/Fotolia; 14–8b, Trekandshoot/Fotolia; 14–12, Julio Etchart/Alamy; 14–18b, Josef

Hanus/Shutterstock; 14–Visual Analysis, Zhou Changguo/Imaginechina/AP Images

Illustrations

14–1, US Environmental Protection Agency (EPA), National Emissions Inventory, 2012 http://www.epa.gov; 14–2, US Environmental Protection Agency (EPA), https://www.nh.gov/epht/topics/part_matter.htm; 14–2–3, US Environmental Protection Agency (EPA): http://www.epa.gov/airtrends/; 14–3, US Environmental Protection Agency (EPA): http://www.epa.gov/acidrain/education/site_students/phscale.htm; 14–13, US Environmental Protection Agency (EPA): http://www.epa.gov/airtrends/aqtrends.html; 14–14a, b, c & d, US Environmental Protection Agency (EPA): http://www.epa.gov/airtrends/; 14–15, US Environmental Protection Agency (EPA): http://www.epa.gov/airquality/greenbook/mapnpoll.html; 14–16, National Aeronautics and Space Administration (NASA): http://www.nasa.gov/topics/earth/features/health-sapping.htm; 14–17, © Pearson Education, Inc.; 14–18a, © Pearson Education, Inc.; 14–19, © Pearson Education, Inc.

Chapter 15

Photographs

Part 6 Opener, Katja Kircher/Getty Images; Chapter 15 Opener, Peter Schickert/Alamy; 15–1, Joel Sartore/Getty Images; 15–2, Evenfh/Fotolia; 15–2–1, Turtix/Shutterstock; 15–7, Stephane Bidouze/Shutterstock; 15–10, Adrian Assalve/Getty Images; 15–12, McPhoto/Age Fotostock; 15–18, Claudio Del Luongo/Fotolia; 15–21, Patrick Poendl/Shutterstock; 15–23, Jean Carter/Age Fotostock/Alamy; 15–27, PiLensPhoto/Fotolia; 15–31, Ron Niebrugge/Alamy; 15–32, Thomas Haltner/Colouria Media/Alamy

Illustrations

15–4, Data from K. Fennig, A. Andersson, S. Bakan, C. Klepp, M. Schroeder, 2012: Hamburg Ocean Atmosphere Parameters and Fluxes from Satellite Data - HOAPS 3.2 - Monthly Means/6-Hourly Composites. Satellite Application Facility on Climate Monitoring. doi:10.5676/EUM_SAF_CM/HOAPS/V001. Matsuura, K. and C.J. Willmott, Terrestrial Air Temperature and Precipitation: Monthly Climatologies, Version 4.01, Center for Climatic Research, Department of Geography, University of Delaware; 15–5, Data from K. Fennig, A. Andersson, S. Bakan, C. Klepp, M. Schroeder, 2012: Hamburg Ocean Atmosphere Parameters and Fluxes from Satellite Data - HOAPS 3.2 - Monthly Means/6-Hourly Composites. Satellite Application Facility on Climate Monitoring. doi:10.5676/EUM_SAF_CM/HOAPS/V001. Matsuura, K. and C.J. Willmott, Terrestrial Air Temperature and Precipitation: Monthly Climatologies, Version 4.01, Center for Climatic Research, Department of Geography, University of Delaware; 15–6, Based on McKnight, Physical Geography, A Landscape Appreciation, 5th Ed. ©1996, Prentice Hall, Inc., Upper Saddle River, NJ; 15–8, Based on McKnight, Physical Geography, A Landscape Appreciation, 5th Ed. ©1996, Prentice Hall, Inc., Upper Saddle River, NJ; 15–9, Based on McKnight, Physical Geography, A Landscape Appreciation, 5th Ed. ©1996, Prentice Hall, Inc., Upper Saddle River, NJ; 15–11, Data from K. Fennig, A. Andersson, S. Bakan, C. Klepp, M. Schroeder, 2012: Hamburg Ocean Atmosphere Parameters and Fluxes from Satellite Data - HOAPS 3.2 - Monthly Means/6-Hourly Composites. Satellite Application Facility on Climate Monitoring. doi:10.5676/EUM_SAF_CM/HOAPS/V001. Matsuura, K. and C.J. Willmott, Terrestrial Air Temperature and Precipitation: Monthly Climatologies, Version 4.01, Center for Climatic Research, Department of Geography, University of Delaware; 15–13, Based on McKnight, Physical Geography, A Landscape Appreciation, 5th Ed. ©1996, Prentice Hall, Inc., Upper Saddle River, NJ; 15–14, Courtesy of the National Oceanic and Atmospheric Administration (NOAA); 15–15, Based on McKnight, Physical Geography, A Landscape Appreciation, 5th Ed. ©1996, Prentice Hall, Inc., Upper Saddle River, NJ; 15–16, Based on McKnight, Physical Geography, A Landscape Appreciation, 5th Ed. ©1996, Prentice Hall, Inc., Upper Saddle River, NJ; 15–17, Based on McKnight, Physical Geography, A Landscape Appreciation, 5th Ed. ©1996, Prentice Hall, Inc., Upper Saddle River, NJ; 15–19, Data from K. Fennig, A. Andersson, S. Bakan, C. Klepp, M. Schroeder, 2012: Hamburg Ocean Atmosphere Parameters and Fluxes from Satellite Data - HOAPS 3.2 - Monthly Means/ 6-Hourly Composites. Satellite Application Facility on Climate Monitoring. doi:10.5676/EUM_SAF_CM/HOAPS/V001. Matsuura, K. and C.J. Willmott, Terrestrial Air Temperature and Precipitation: Monthly Climatologies, Version 4.01, Center for Climatic Research, Department of Geography, University of Delaware; 15–20, Based on McKnight, Physical Geography, A Landscape Appreciation, 5th Ed. ©1996, Prentice Hall, Inc., Upper Saddle River, NJ; 15–22, Based on McKnight, Physical Geography, A Landscape Appreciation, 5th Ed. ©1996, Prentice Hall, Inc., Upper Saddle River, NJ; 15–24, Based on McKnight, Physical Geography, A Landscape Appreciation, 5th Ed. ©1996, Prentice Hall, Inc., Upper Saddle River, NJ; 15–25, Data from K. Fennig, A. Andersson, S. Bakan, C. Klepp, M. Schroeder, 2012: Hamburg Ocean Atmosphere Parameters and Fluxes from Satellite Data - HOAPS 3.2 - Monthly Means/6-Hourly Composites. Satellite Application Facility on Climate Monitoring. doi:10.5676/EUM_SAF_CM/HOAPS/V001. Matsuura, K. and C.J. Willmott, Terrestrial Air Temperature and Precipitation: Monthly Climatologies, Version 4.01, Center for Climatic Research, Department of Geography, University of Delaware; 15–26, Based on McKnight, Physical Geography, A Landscape Appreciation, 5th Ed. ©1996, Prentice Hall, Inc., Upper Saddle River, NJ; 15–28, Based on McKnight, Physical Geography, A Landscape Appreciation, 5th Ed. ©1996, Prentice Hall, Inc., Upper Saddle River, NJ; 15–29, Data from K. Fennig, A. Andersson, S. Bakan, C. Klepp, M. Schroeder, 2012: Hamburg Ocean Atmosphere Parameters and Fluxes from Satellite Data - HOAPS 3.2 - Monthly Means/6-Hourly Composites. Satellite Application Facility on Climate Monitoring. doi:10.5676/EUM_SAF_CM/HOAPS/V001. Matsuura, K. and C.J. Willmott, Terrestrial Air Temperature and Precipitation: Monthly Climatologies, Version 4.01, Center for Climatic Research, Department of Geography, University of Delaware; 15–30 right, Based on McKnight, Physical Geography, A Landscape Appreciation, 5th Ed. ©1996, Prentice Hall, Inc., Upper Saddle River, NJ; 15–30 left, Based on Darrel Hess, Physical Geography, California Edition (Custom) - 3rd edition, page 233; 15–33, Based on McKnight, Physical Geography, A Landscape Appreciation, 5th Ed. ©1996, Prentice Hall, Inc., Upper Saddle River, NJ

Chapter 16

Photographs

Chapter 16 Opener a, Videowokart/Shutterstock; Chapter 16 Opener b, Andreas Feininger/The LIFE Picture Collection/Getty Images; 16–2, Lowell Georgia/Corbis; 16–2–2, Vlad61/Shutterstock; 16–3–1, Arina P Habich/Shutterstock; 16–5, Matthijs Wetterauw/Shutterstock; 16–20, Arlan Naeg/AFP/Getty Images/Newscom; 16–24, Johner Images/Alamy; 16–25, NASA; 16–32, Timothy Epp/Shutterstock; 16–35, Karl Thomas/Robert Harding World Imagery/Corbis

Illustrations

IPCC Fifth Assessment Report: Climate Change 2013–2014; 16–2–1, Modified after R. A. Feely, Bulletin of the American Meteorological Socieyt, July 2008 source http: http://pmel.noaa.gov/co2/file/Hawaii+Carbon+Dioxide+Time-Series; 16–4–1, From Climate Chang Impacts in the United States, National Science and Technology Council and the U.S. Global Change Research Program. May 2014, p.6. http://www.globalchange.gov/sites/globalchange/files/NCA3_Highlights_LowRes-small-FINAL_posting.pdf figure source: updated from Karl et all. 2009; 16–6, Based on Hamblin, Earths' Dynamic Systems, 8th Ed. ©1998, Prentice Hall, Inc., Upper Saddle River, NJ; 16–9 b, c, d, Based on "Past Environmental Changes: Characteristic Features of Quaternary Climate Variations" by Patrice J. Bartlein from Past and Future Rapid Environmental Changes, edited by B. Huntley et al., @ 1997; 16–10, Based on McKnight, Physical Geology, A Landscape Appreciation, Fight Edition, ©1996 Prentice Hall, Inc., Upper Saddle River, NJ; 16–11, Based on Millennium's Hottest Decade Retains its Title, for Now" by Richard A. Kerr in SCIENCE 307(5711):828–829, February 11, 2005/©2005 AAAS; 16–12, Fifth IPCC Assessment Report, echnical Summary Fig. TS.2, Intergovernmental Panel of Climate Change, IPCC; 16–13, Climate Change 2007: Report of the Intergovernmental Panel of Climate Change, IPCC; 16–14, 2013–2014 IPCC Climate Change Report Fig. SPM.2, Intergovernmental Panel of Climate Change, IPCC; 16–17, Based on Chaisson and McMillan, Astronomy Today, 2nd Ed., ©1997 Prentice Hall, Upper Saddle River, NJ; 16–21 a, b, c, d., National Aeronautics and Space Administration (NASA); 16–23, Climate Change 2007: Report of the Intergovernmental Panel of Climate Change, IPCC; 16–25, National Aeronautics and Space Administration (NASA http://www.nasa.gov/topics/earth/features/arctic-seaice-2012.html; 16–28, 2013–2014 IPCC Cliamte Change Report Fig. SPM.2. Climate Change 2007: Report of the Intergovernmental Panel of Climate Change, IPCC; 16–29, 2013–2014 IPCC Climate Change Report Fig. SPM.7 fig (a). Intergovernmental Panel of Climate Change, IPCC; 16–31, Climate Change 2013–2014, The Physical Science Basis, Summary for Policymakers, ICPP http://www.ipcc.ch/report/ar5/wg1/docs/WGIAR5_SPM_brochure_en.pdf, page 20 © 2013 Intergovernmental Panel on Climate Change; 16–33, Climate Change 2013–2014 IPCC Technical Summary/IPCC/Stocker, T.F., D. Qin, G.-K. Plattner, L.V. Alexander, S.K. Allen, N.L. Bindoff, F.-M. Bréon, J.A. Church, U. Cubasch, S. Emori, P. Forster, P. Friedlingstein, N. Gillett, J.M. Gregory, D.L. Hartmann, E. Jansen, B. Kirtman, R. Knutti, K. Krishna Kumar, P. Lemke, J. Marotzke, V. Masson-Delmotte, G.A. Meehl, I.I. Mokhov, S. Piao, V. Ramaswamy, D. Randall, M. Rhein, M. Rojas, C. Sabine, D. Shindell, L.D. Talley, D.G. Vaughan and S.-P. Xie, 2013: Technical Summary. In: Climate Change 2013: The Physical Science Basis. Contribution of Working Group I to the Fifth Assessment Report of the Intergovernmental Panel on Climate Change [Stocker, T.F., D. Qin, G.-K. Plattner, M. Tignor, S.K. Allen, J. Boschung, A. Nauels, Y. Xia, V. Bex and P.M. Midgley (eds.)]. Cambridge University Press, Cambridge, United Kingdom and New York, NY, USA.; 16–34, Climate Change 2013–2014, The Physical Science Basis Summary for Policy Makers, Intergovernmental Panel on Climate Change http://www.ipcc.ch/report/ar5/wg1/docs/WGIAR5_SPM_brochure_en.pdf; 16–Visual Analysis, Courtesy of the National Oceanic and Atmospheric Administration

Tables

16–2, Climate Change 2007: Report of the Intergovernmental Panel of Climate Change, IPCC; 16–3, Climate Change 2007: Report of the Intergovernmental Panel of Climate Change, IPCC

Chapter 17

Photographs

Chapter 17 Opener, Jeremy Woodhouse/Getty Images; 17–2, Grafonom/Fotolia; 17–4, Pekka Parvlainen/Science Photo Library/Science source; 17–7, Jim Steinberg/Science Source; 17–8, Science Source; 17–13b, Mark Duffy/Alamy; 17–14, Clint Farlinger/Alamy; 17–15, Pekka Parvlainen/Science Photo Library/Science source; 17–16, Peter Barritt/Alamy; 17–VA, Juhku/Fotolia

Photographs

CC01, Gianni Muratore/Alamy; CC02, Brian Cosgrove/Dorling Kindersley; CC03, Brian Cosgrove/Dorling Kindersley; CC04, Brian Cosgrove/Dorling Kindersley; CC05, Richard Newstead/Corbis; 1–FM, NASA Earth Observatory; 2–FM, Suomi NPP/VIIRS/NASA; 3–FM, NASA; 4–FM, NASA; WT01, Anthony Harvie/Stone/Getty Images; Focus on Aviation icon, Iurii/Shutterstock; Focus on severe weather icon, Minerva Studio/Shutterstock; Focus on the environment and societal impacts icon, Vaclav Volrab/Shutterstock; Forecasting icon, Pixfly/Shutterstock; Physical principles icon, Terric Delayn/Shutterstock; Special Interest icon, Natapong Paopijit/Shutterstock; Part 7 Opener, Atiketta Sangasaeng/Shutterstock

Illustrations

UN APP01, Courtesy of the National Oceanic and Atmospheric Administration (NOAA); Back End Map, IPCC Fifth Assessment Report: Climate Change 2014: Impacts, Adaptation, Vulnerability Summary for Policymakers. Intergovernmental Panel on Climate Change. Figure SPM.4.

Index

Note: Page numbers followed by b indicate boxed material; those followed by f indicate figures; those followed by n indicate footnotes; those followed by t indicate tables.

A

B

C

I

N

O

P

Q

R

S

T

U

V

Cloud Guide

High Clouds: Cloud Bases Above 6 km (20,000 ft)

SHUTTERSTOCK

Cirrus These clouds are made exclusively of ice crystals. They are not as horizontally extensive as cirrostratus clouds.

SHUTTERSTOCK

Cirrocumulus These high clouds can produce striking skies. Composed of ice crystals, they often contain linear bands, numerous patches of greater vertical development, or both.

GIANNI MURATORE/ALAMY

Cirrostratus These are thin layered clouds composed of ice crystals. They are relatively indistinct and give the sky a whitish appearance.

RICHARD NEWSTEAD/CORBIS

Contrails A contrail is a long, narrow cloud that is formed as exhaust from a jet aircraft condenses in cold air at high altitude. Upper level winds may gradually cause contrails to spread out.

Middle Clouds: Cloud Bases 2–6 km (6,500–20,000 ft)

SHUTTERSTOCK

Altocumulus These midlevel clouds are horizontally layered but exhibit varying thicknesses across their bases. Thicker areas can be arranged as parallel linear bands or as a series of individual puffs.

SHUTTERSTOCK

Altocumulus (Lenticular) These clouds are marked by their lens-shaped appearance. They usually form downwind of mountain barriers as horizontal airflow is disrupted into a sequence of waves.

BRIAN COSGROVE/DORLING KINDERSLEY

Altostratus These are midlevel, layered clouds that produce gray skies and obscure the Sun or Moon enough to make them appear as poorly defined bright spots. In this example, the setting sun brightens the clouds near the horizon but the gray appearance remains elsewhere.

BRIAN COSGROVE/DORLING KINDERSLEY

Altostratus (Multilayer) These are midlevel layered clouds that are dense enough to completely hide the Sun or Moon.

© 2015 by Pearson Education, Inc.

Cloud Guide

Low Clouds and Clouds of Vertical Development: Cloud Bases 0–2 km (0–6,500 ft)

JIM LEE/NOAA

Cumulus These clouds often have flat bottoms, rounded tops, and a "cellular" structure made up of individual clouds. (The word "cumulus" comes from the Latin word for "heap.") Cumulus clouds tend to grow vertically.

JIM LEE/NOAA

Cumulus Humilis Often called fair-weather cumulus, these small white individual masses lack conspicuous vertical development and rarely produce precipitation.

BRIAN COSGROVE/DORLING KINDERSLEY

Nimbostratus These low clouds are thick gray layers that contain sufficient water to yield light-to-moderate precipitation.

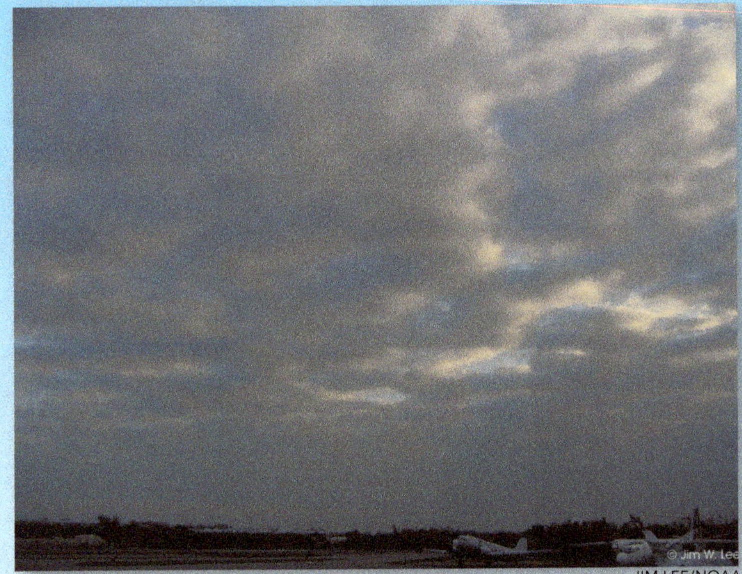

JIM LEE/NOAA

Stratocumulus These are low, layered clouds that have regions of some vertical development. Differences in thickness create varying degrees of darkness when seen from below.

JIM LEE/NOAA

Cumulus Congestus These clouds have considerably more vertical development than cumulus humilis. They may produce heavy precipitation, but not the severe weather associated with some cumulonimbus clouds.

SHUTTERSTOCK

Cumulonimbus These clouds result from very strong updrafts that may push the cloud tops up to several kilometers into the stratosphere. Their characteristic feature is the anvil, a zone of ice crystals extending outward from the main portion of the cloud.

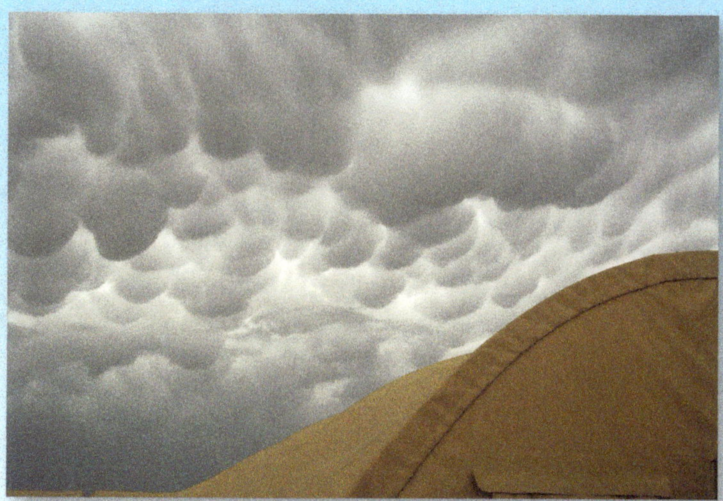

PUBLIC DOMAIN

Cumulonimbus with Mammatus These are dramatic features associated with some cumulonimbus, resulting from strong downdrafts and turbulence along the bases or margins of the clouds.

NOAA

Cumulonimbus with Wall Cloud A feature associated with some cumulonimbus clouds. When wall clouds are present, heavy rain, hail, and sometimes tornadoes can be expected.

© 2015 by Pearson Education, Inc.

Temperature Change

The impact of rising concentrations of atmospheric carbon dioxide and other greenhouse gases will depend heavily on what efforts are made to limit their release by human activities. This map shows a projected outcome of continued heavy reliance on coal and other fossil fuels. Occurring over less than a century, mapped values are fifty to one hundred times faster than anything seen on Earth in the last 65 million years, and will be highly disruptive for both natural ecosystems and human society.

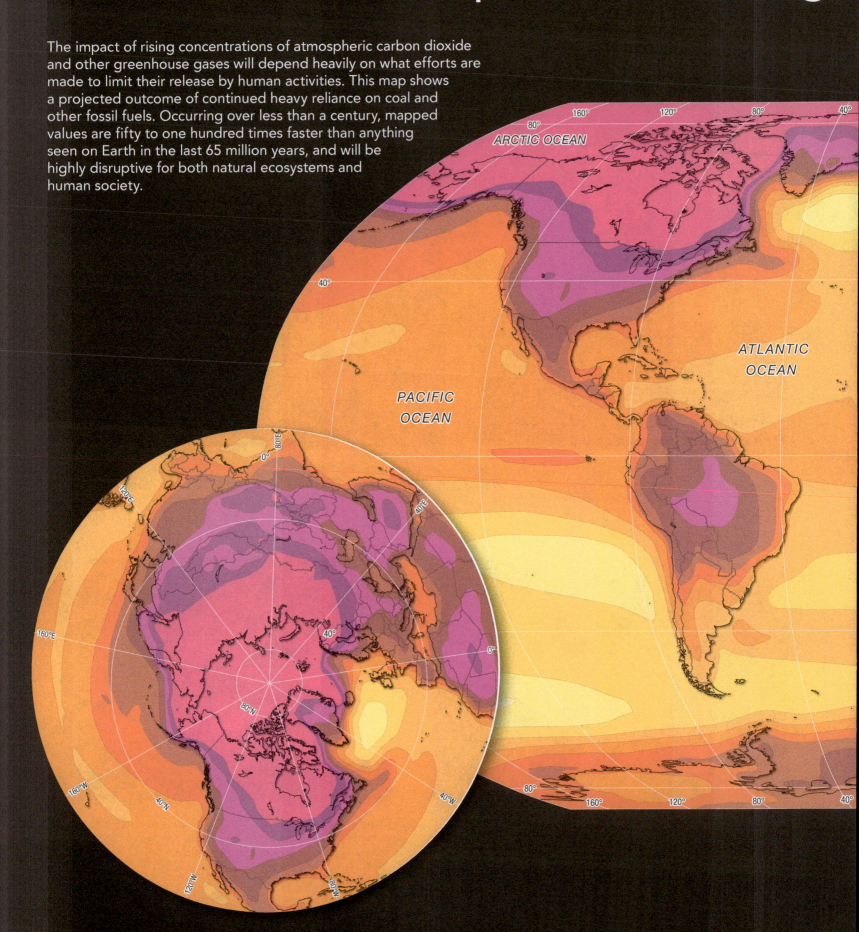

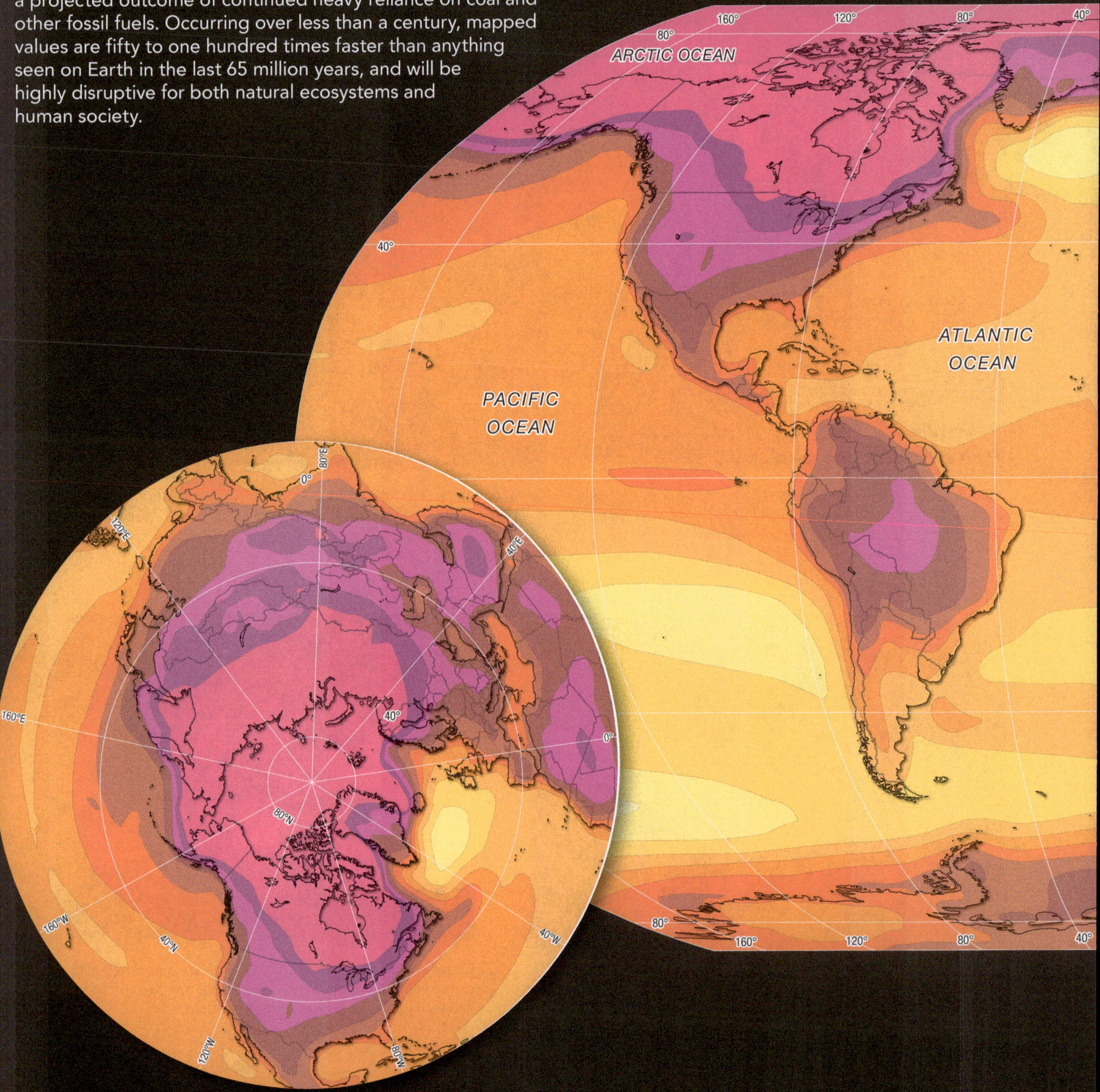

2081–2100 (Projected)

This map represents an average based on more than two dozen models of the Earth System. Notice how projected changes are smaller over oceans than continents, and are very large in the higher northern latitudes. The atmospheric, oceanic, and land surface processes represented in such models are discussed in *Understanding Weather and Climate,* with explanations for how these processes—along with feedbacks among the various Earth spheres—work to produce the highly uneven pattern of heating seen in this map.

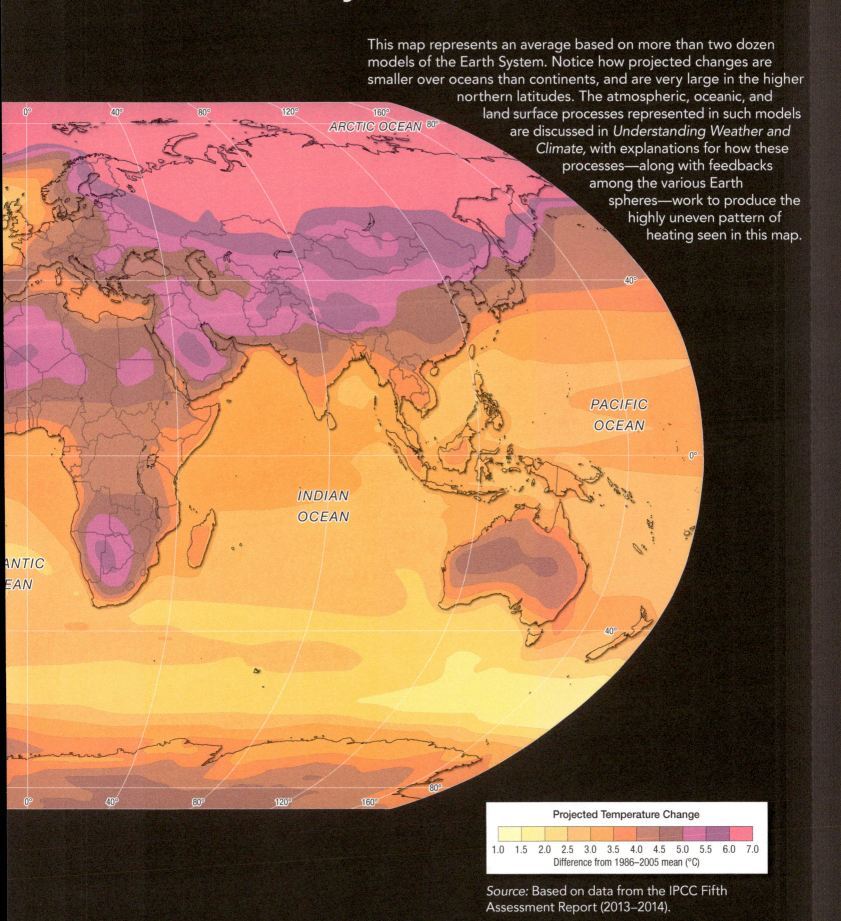

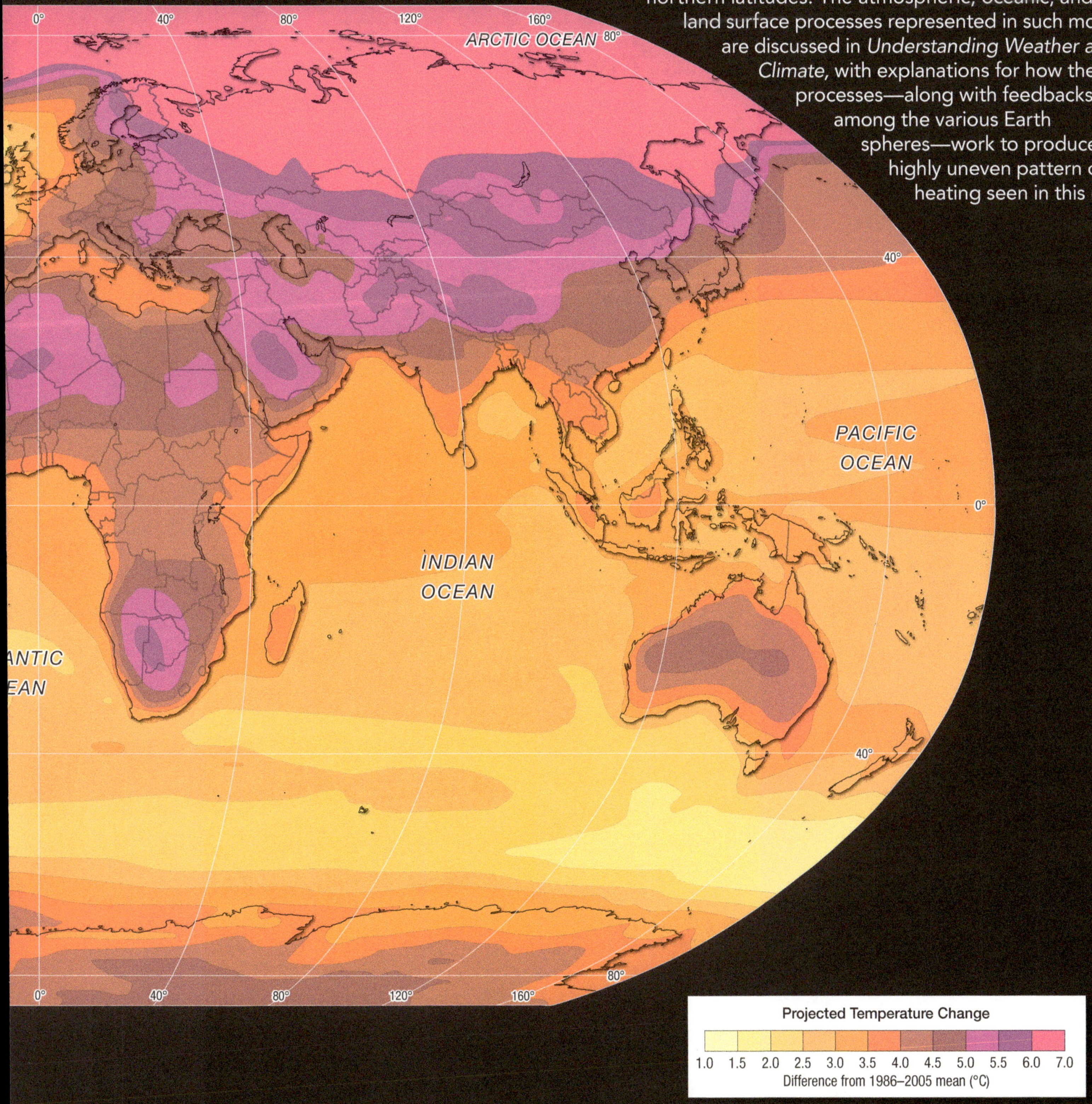

Source: Based on data from the IPCC Fifth Assessment Report (2013–2014).